Standard Atomic Weights of the Elements 2005, IUPAC

Based on Relative Atomic Mass of ^{12}C = 12, where ^{12}C is a neutral atom in its nuclear and electronic ground state.†

Name	Symbol	Atomic Number	Atomic Weight
Actinium*	Ac	89	(227)
Aluminum	Al	13	26.9815386(8)
Americium*	Am	95	(243)
Antimony	Sb	51	121.760(1)
Argon	Ar	18	39.948(1)
Arsenic	As	33	74.92160(2)
Astatine*	At	85	(210)
Barium	Ba	56	137.327(7)
Berkelium*	Bk	97	(247)
Beryllium	Be	4	9.012182(3)
Bismuth	Bi	83	208.98040(1)
Bohrium*	Bh	107	(264)
Boron	B	5	10.811(7)
Bromine	Br	35	79.904(1)
Cadmium	Cd	48	112.411(8)
Calcium	Ca	20	40.078(4)
Californium*	Cf	98	(251)
Carbon	C	6	12.0107(8)
Cerium	Ce	58	140.116(1)
Cesium	Cs	55	132.9054519(2)
Chlorine	Cl	17	35.453(2)
Chromium	Cr	24	51.9961(6)
Cobalt	Co	27	58.933195(5)
Copper	Cu	29	63.546(3)
Curium*	Cm	96	(247)
Darmstadtium*	Ds	110	(281)
Dubnium*	Db	105	(262)
Dysprosium	Dy	66	162.500(1)
Einsteinium*	Es	99	(252)
Erbium	Er	68	167.259(3)
Europium	Eu	63	151.964(1)
Fermium*	Fm	100	(257)
Fluorine	F	9	18.9984032(5)
Francium*	Fr	87	(223)
Gadolinium	Gd	64	157.25(3)
Gallium	Ga	31	69.723(1)
Germanium	Ge	32	72.64(1)
Gold	Au	79	196.966569(4)
Hafnium	Hf	72	178.49(2)
Hassium*	Hs	108	(277)
Helium	He	2	4.002602(2)
Holmium	Ho	67	164.93032(2)
Hydrogen	H	1	1.00794(7)
Indium	In	49	114.818(3)
Iodine	I	53	126.90447(3)
Iridium	Ir	77	192.217(3)
Iron	Fe	26	55.845(2)
Krypton	Kr	36	83.798(2)
Lanthanum	La	57	138.90547(7)
Lawrencium*	Lr	103	(262)
Lead	Pb	82	207.2(1)
Lithium	Li	3	[6.941(2)]†
Lutetium	Lu	71	174.967(1)
Magnesium	Mg	12	24.3050(6)
Manganese	Mn	25	54.938045(5)
Meitnerium*	Mt	109	(268)
Mendelevium*	Md	101	(258)
Mercury	Hg	80	200.59(2)
Molybdenum	Mo	42	95.94(2)
Neodymium	Nd	60	144.242(3)
Neon	Ne	10	20.1797(6)
Neptunium*	Np	93	(237)
Nickel	Ni	28	58.6934(2)
Niobium	Nb	41	92.90638(2)
Nitrogen	N	7	14.0067(2)
Nobelium*	No	102	(259)
Osmium	Os	76	190.23(3)
Oxygen	O	8	15.9994(3)
Palladium	Pd	46	106.42(1)
Phosphorus	P	15	30.973762(2)
Platinum	Pt	78	195.084(9)
Plutonium*	Pu	94	(244)
Polonium*	Po	84	(209)
Potassium	K	19	39.0983(1)
Praseodymium	Pr	59	140.90765(2)
Promethium*	Pm	61	(145)
Protactinium*	Pa	91	231.03588(2)
Radium*	Ra	88	(226)
Radon*	Rn	86	(222)
Rhenium	Re	75	186.207(1)
Rhodium	Rh	45	102.90550(2)
Roentgenium*	Rg	111	(272)
Rubidium	Rb	37	85.4678(3)
Ruthenium	Ru	44	101.07(2)
Rutherfordium*	Rf	104	(261)
Samarium	Sm	62	150.36(2)
Scandium	Sc	21	44.955912(6)
Seaborgium*	Sg	106	(266)
Selenium	Se	34	78.96(3)
Silicon	Si	14	28.0855(3)
Silver	Ag	47	107.8682(2)
Sodium	Na	11	22.98976928(2)
Strontium	Sr	38	87.62(1)
Sulfur	S	16	32.065(5)
Tantalum	Ta	73	180.94788(2)
Technetium*	Tc	43	(98)
Tellurium	Te	52	127.60(3)
Terbium	Tb	65	158.92535(2)
Thallium	Tl	81	204.3833(2)
Thorium*	Th	90	232.03806(2)
Thulium	Tm	69	168.93421(2)
Tin	Sn	50	118.710(7)
Titanium	Ti	22	47.867(1)
Tungsten	W	74	183.84(1)
Uranium*	U	92	238.02891(3)
Vanadium	V	23	50.9415(1)
Xenon	Xe	54	131.293(6)
Ytterbium	Yb	70	173.04(3)
Yttrium	Y	39	88.90585(2)
Zinc	Zn	30	65.409(4)
Zirconium	Zr	40	91.224(2)
—‡*		112	(285)
—‡*		113	(284)
—‡*		114	(289)
—‡*		115	(288)

† The atomic weights of many elements vary depending on the origin and treatment of the sample. This is particularly true for Li; commercially available lithium-containing materials have Li atomic weights in the range of 6.939 and 6.996. Uncertainties are given in parentheses following the last significant figure to which they are attributed.

* Elements with no stable nuclide; the value given in parentheses is the atomic mass number of the isotope of longest known half-life. However, three such elements (Th, Pa, and U) have a characteristic terrestrial isotopic composition, and the atomic weight is tabulated for these.

‡ Not yet named.

Study smarter—and make every minute count!

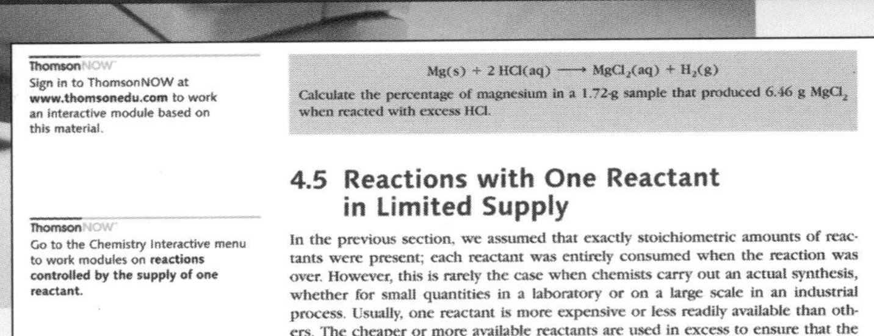

ThomsonNOW
Sign in to ThomsonNOW at **www.thomsonedu.com** to work an interactive module based on this material.

ThomsonNOW
Go to the Chemistry Interactive menu to work modules on **reactions controlled by the supply of one reactant.**

$$Mg(s) + 2\,HCl(aq) \longrightarrow MgCl_2(aq) + H_2(g)$$

Calculate the percentage of magnesium in a 1.72-g sample that produced 6.46 g $MgCl_2$ when reacted with excess HCl.

4.5 Reactions with One Reactant in Limited Supply

In the previous section, we assumed that exactly stoichiometric amounts of reactants were present; each reactant was entirely consumed when the reaction was over. However, this is rarely the case when chemists carry out an actual synthesis, whether for small quantities in a laboratory or on a large scale in an industrial process. Usually, one reactant is more expensive or less readily available than others. The cheaper or more available reactants are used in excess to ensure that the more expensive material is completely converted to product.

The industrial production of methanol, CH_3OH, is such a case. Methanol, an important industrial product, is manufactured by the reaction of carbon monoxide and hydrogen.

$$CO(g) + 2\,H_2(g) \longrightarrow CH_3OH(\ell)$$

Succeed in your course with the help of this best-selling text and its powerful diagnostic self-study tool. The Third Edition of ***Chemistry: The Molecular Science*** carefully and thoroughly integrates powerful multimedia learning components with the text, providing a seamless learning system that will improve your understanding of course concepts.

ThomsonNOW™ for *Chemistry: The Molecular Science*

Link to interactive tutorials, simulations, and animations based on your unique level of understanding! **ThomsonNOW** is an online assessment-centered learning tool that gives you a *Personalized Study* plan based on a diagnostic pre-test. Your unique study plan can help you prepare for exams by targeting your study needs and helping you to visualize, practice, and master the material in the text.

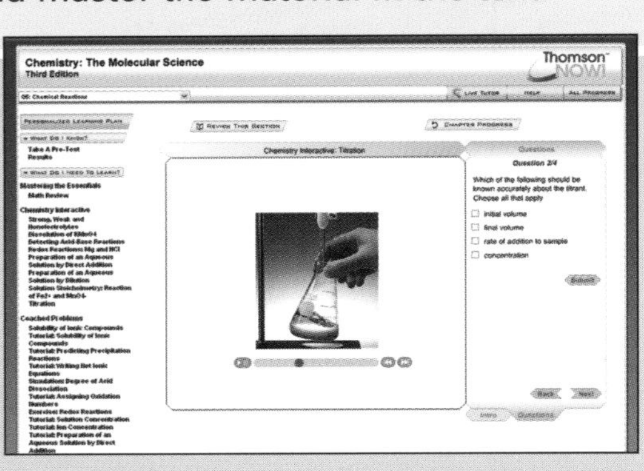

How does it work?

Just sign in to **ThomsonNOW** by using the access code packaged with your text.* With a click of the mouse, you can build a complete *Personalized Study* plan for yourself with the program's three powerful components:

What I Know

After you've completed the *Pre-Test,* a detailed *Personalized Study* plan (based on your test results) outlines the concepts you need to review and presents links to specific pages in your textbook that discuss that concept. You find out exactly what you need to study!

What I Need to Know

Working from your *Personalized Study* plan, you are guided to media-enhanced activities including *Interactive Modules, Coached Problems, Active Figures,* tutorials, and exercises. These activities can also be accessed at any time without taking the *Pre-Test.*

What I've Learned

An optional *Post-Test* ensures that you've mastered the concepts in each chapter. As with the *Pre-Test,* your individual results may be e-mailed to your instructor to help you both assess your progress. If you need to improve your score, **ThomsonNOW** will work with you as you continue to build your knowledge.

Sign in at **www.thomsonedu.com** to view chemistry tutorials and simulations, develop problem-solving skills, and test your knowledge with these interactive self-study resources.

* If you purchased a new book, you may already have access to **ThomsonNOW**. Register by following the directions on the card that came with this text. If you did not purchase a new book, you can obtain an access code at **www.thomsonedu.com/thomsonnow/buy**.

Chemistry

THE MOLECULAR SCIENCE

Third Edition

John W. Moore
University of Wisconsin—Madison

Conrad L. Stanitski
Franklin and Marshall College

Peter C. Jurs
Pennsylvania State University

Australia · Brazil · Canada · Mexico · Singapore · Spain · United Kingdom · United States

Chemistry: The Molecular Science, Third Edition
John W. Moore, Conrad L. Stanitski, Peter C. Jurs

Publisher: David Harris
Acquisitions Editor: Lisa Lockwood
Development Editor: Peter McGahey
Assistant Editor: Brandi Kirksey
Editorial Assistant: Toriana Holmes
Technology Project Manager: Lisa Weber
Marketing Manager: Amee Mosley
Marketing Assistant: Melissa Wong
Marketing Communications Manager: Bryan Vann
Project Manager, Editorial Production: Jennifer Risden, Lisa Weber
Creative Director: Rob Hugel
Art Directors: Lee Friedman, John Walker
Print Buyer: Becky Cross
Permissions Editor: Sarah D'Stair
Production Service: Lachina Publishing Services
Text Designer: Mark Ong
Photo Researcher: Dena Digilio Betz, Robin Samper
Copy Editor: Amy Mayfield, Lachina Publishing Services
Illustrators: Greg Gambino; Matt Fornadel, Lachina Publishing Services
OWL Producers: Stephen Battisti, Cindy Stein, David Hart (Center for Educational Software Development, University of Massachusetts, Amherst)
Cover Designer: Michele Wetherbee
Cover Images: Background photo: Sean McKenzie/WWI/Peter Arnold, Inc.; inset photos (left to right): Clive Freeman/Biosym Technologies/Photo Researchers, Inc.; Courtesy of IBM Research, Almaden Research Center. Unauthorized uses prohibited.; Kenneth Eward/Photo Researchers, Inc.; Clive Freeman/Biosym Technologies/Photo Researchers, Inc.
Cover Printer: Transcontinental Printing/Interglobe
Compositor: Lachina Publishing Services
Printer: Transcontinental Printing/Interglobe

Printed in Canada
1 2 3 4 5 6 7 11 10 09 08 07

Thomson Higher Education
10 Davis Drive
Belmont, CA 94002-3098
USA

Library of Congress Control Number: 2006936871

Student Edition:
ISBN-13: 978-0-495-10521-3
ISBN-10: 0-495-10521-X

Instructor Edition:
ISBN-13: 978-0-495-11255-6
ISBN-10: 0-495-11255-0

International Instructor Edition:
ISBN-13: 978-0-495-11256-3
ISBN-10: 0-495-11256-9

Volume 1:
ISBN-13: 978-0-495-11598-4
ISBN-10: 0-495-11598-3

Volume 2:
ISBN-13: 978-0-495-11601-1
ISBN-10: 0-495-11601-7

To All Students of Chemistry

We intend that this book will help you
to discover that chemistry is
relevant to your lives and careers,
full of beautiful ideas and phenomena,
and of great benefit to society.
May your study of this fascinating subject
be exciting, successful, and fun!

About the Authors

JOHN W. MOORE received an A.B. magna cum laude from Franklin and Marshall College and a Ph.D. from Northwestern University. He held a National Science Foundation (NSF) postdoctoral fellowship at the University of Copenhagen and taught at Indiana University and Eastern Michigan University, before joining the faculty of the University of Wisconsin–Madison in 1989. At the University of Wisconsin, Dr. Moore is W. T. Lippincott Professor of Chemistry, Director of the Institute for Chemical Education, and Chair of the General Chemistry Division. He has been Editor of the *Journal of Chemical Education (JCE)* since 1996. He has won the ACS George C. Pimentel Award in Chemical Education and the [illegible]mes Flack Norris Award for Excellence in Teaching Chemistry. [illegible]003, he won the Benjamin Smith Reynolds Award at the Uni-[illegible]y of Wisconsin–Madison in recognition of his excellence in [illegible]g chemistry to engineering students. Dr. Moore has [illegible]received the second of two major grants from the NSF to [illegible]evelopment of a chemistry pathway for the NSF-[illegible]ational Science Digital Library.

[illegible]NITSKI is Distinguished Emeritus Professor of [illegible]niversity of Central Arkansas and is currently [illegible]ranklin and Marshall College. He received his [illegible]ion from Bloomsburg State College, M.A. in [illegible]m the University of Northern Iowa, and [illegible]try from the University of Connecticut. [illegible]y textbooks for science majors, allied-[illegible]ience majors, and high school chem-[illegible]as won many teaching awards, including the CMA CATALYST National Award for Excellence in Chemistry Teaching; the Gustav Ohaus–National Science Teachers Association Award for Creative Innovations in College Science Teaching; the Thomas R. Branch Award for Teaching Excellence and the Samuel Nelson Gray Distinguished Professor Award from Randolph-Macon College; and the 2002 Western Connecticut ACS Section Visiting Scientist Award. He was Chair of the American Chemical Society Division of Chemical Education during 2001 and has been an elected Councilor for that division. An instrumental and vocal performer, he also enjoys jogging, tennis, rowing, and reading.

PETER C. JURS is Professor of Chemistry at the Pennsylvania State University. Dr. Jurs earned his B.S. in Chemistry from Stanford University and his Ph.D. in Chemistry from the University of Washington. He then joined the faculty of Pennsylvania State University, where he has been Professor of Chemistry since 1978. Jurs's research interests have focused on the application of computational methods to chemical and biological problems, including the development of models linking molecular structure to chemical or biological properties (drug design). For this work he was awarded the A.C.S. Award for Computers in Chemistry in 1990. Dr. Jurs has been Assistant Head for Undergraduate Education at Penn State, and he works with the Chemical Education Interest Group to enhance and improve the undergraduate program. In 1995, he was awarded the C. I. Noll Award for Outstanding Undergraduate Teaching. Dr. Jurs serves as an elected Councilor for the American Chemical Society Computer Division.

Contents Overview

Detailed Contents

© Scott Camazine/Photo Researchers, Inc.

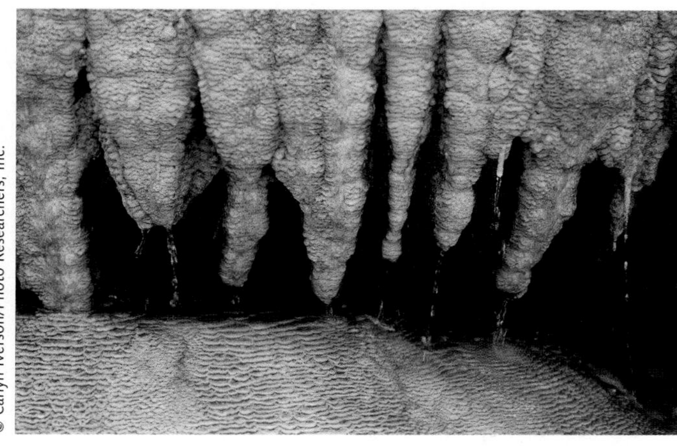

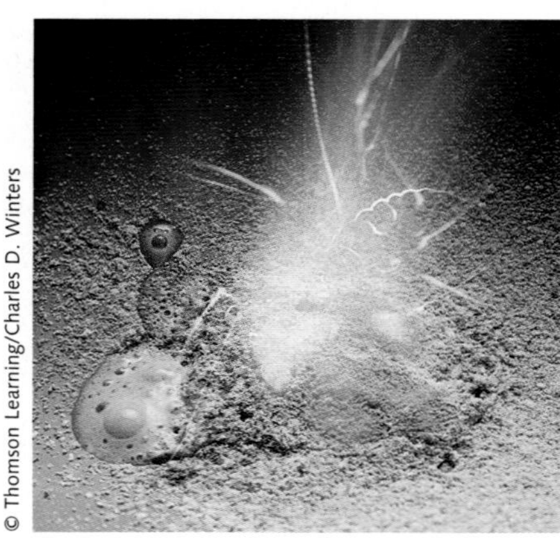

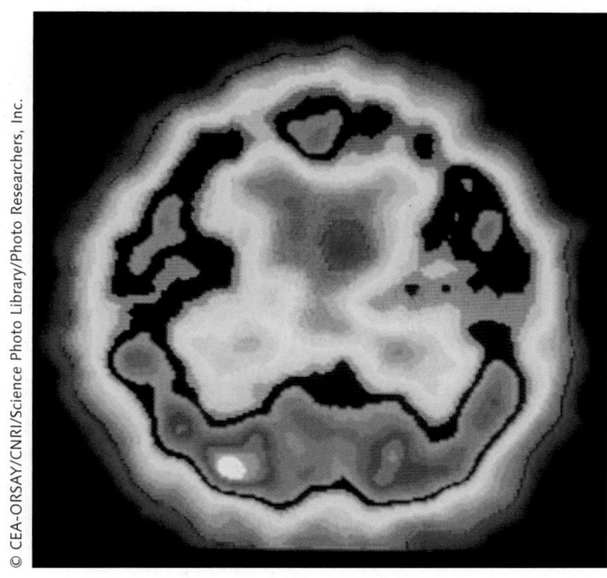

Preface

Chemistry is a mature science, yet new chemistry and new ways to apply chemical principles are reported every day. Chemical research is helping to solve problems as diverse as how to make electronic circuits on the molecular scale; how to design and synthesize new, more effective drugs; and how to create metals, plastics, and other materials that have exactly the properties we want. All of these problems require the chemist's unique, molecular-scale viewpoint—a perspective whose value has been proved many times over during the past century.

Because it is so broadly applicable, much of today's cutting-edge chemical research involves collaborations with biochemists, biologists, pharmacologists, physicians, geologists, atmospheric scientists, physicists, materials scientists, engineers, and others. It is crucial that students in first-year chemistry courses recognize our discipline's ability to solve important problems and its important contributions to other disciplines. We acted on that premise when writing this textbook, and we have now updated it to include many recent chemical innovations.

Goals

Our overarching goal is to involve science and engineering students in active study of what modern chemistry is, how it applies to a broad range of disciplines, and what effects it has on their own lives. We maintain a high level of rigor so that students in mainstream general chemistry courses for science majors and engineers will learn the concepts and develop the problem-solving skills essential to their future ability to use chemical ideas effectively. We have selected and carefully refined the book's many unique features in support of this goal.

More specifically, we intend that this textbook will help students develop:

- A broad overview of chemistry and chemical reactions,
- An understanding of the most important concepts and models used by chemists and those in chemistry-related fields,
- The ability to apply the facts, concepts, and models of chemistry appropriately to new situations in chemistry, to other sciences and engineering, and to other disciplines,
- Knowledge of the many practical applications of chemistry in other sciences, in engineering, and in other fields,
- An appreciation of the many ways that chemistry affects the daily lives of all people, students included, and
- Motivation to study in ways that help all students achieve real learning that results in long-term retention of facts and concepts and how to apply them.

Because modern chemistry is inextricably entwined with so many other disciplines, we have integrated organic chemistry, biochemistry, environmental chemistry, industrial chemistry, and materials chemistry into the discussions of chemical principles and facts. Applications in these areas are discussed together with the principles on which they are based. This approach serves to motivate students whose interests lie in related disciplines and also gives a more accurate picture of the multidisciplinary collaborations so prevalent in contemporary chemical research and modern industrial chemistry.

Audience

Chemistry: The Molecular Science is intended for mainstream general chemistry courses for students who expect to pursue further study in science, engineering, or science-related disciplines. Those planning to major in chemistry, biochemistry, biological sciences, engineering, geological sciences, agricultural sciences, materials science, physics, and many related areas will benefit from this book and its approach. We assume that the students who use this book have a basic foundation in mathematics (algebra and geometry) and in general science. Almost all will also have had a chemistry course before coming to college. The book is suitable for the typical two-semester sequence of general chemistry, and it has also been used quite successfully in a one-semester accelerated course that presumes students have a strong background in chemistry and mathematics.

New in This Edition

Users of the first two editions of this book have been most enthusiastic about many of its features and as a result have provided superb feedback that we have taken into account to enhance its usefulness to their students. Reviewers have also been helpful in pointing out things we should improve. Like the second edition, this third edition is a complete revision. Although the art program in the first edition won the coveted Talbot award for visual excellence, we continue to incorporate into all chapters new art to further enhance the student's ability to visualize molecular-scale processes and to connect these processes with real-world phenomena. We have also enhanced popular, pedagogically sound features, such as *Chemistry in the News, Chemistry You Can Do, Estimation, Portrait of a Scientist,* and *Tools of Chemistry.* Most of these features have been updated, and many are entirely new.

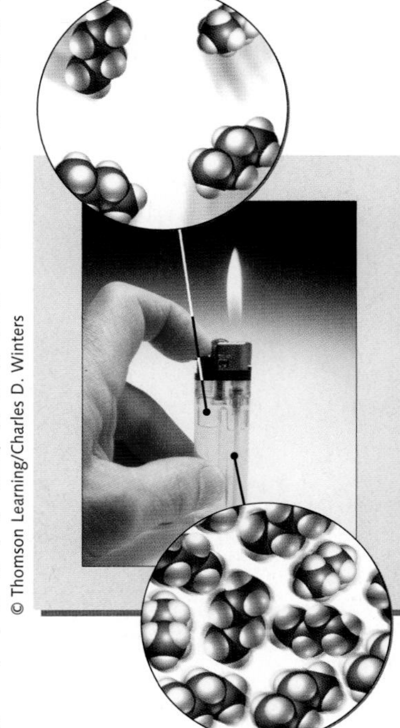
© Thomson Learning/Charles D. Winters

Our emphasis on conceptual understanding continues, and we have revised the entire text to help students to gain a thorough mastery of the important chemical principles. We have moved some sections from one chapter to another and reorganized content to present the material in the most logical way possible. In addition, we continue to use recently published pedagogical research that points the way toward teaching methods and writing characteristics that are most effective in helping students learn chemistry and retain their knowledge over the long term. For example, we have consolidated in Chapter 12 material on biomolecules (carbohydrates and fats) from Chapter 3, material on triglycerides and cis-trans isomerism in fats and oils from Chapter 8, and material on chirality from Chapter 9 to provide a more cohesive presentation of biomolecular and organic chemistry. Atmospheric chemistry has been consolidated into Chapter 10, also juxtaposing related material and providing better organization. Based on recent articles in the *Journal of Chemical Education,* Chapter 14 indicates that activities are required for a true equilibrium constant and Chapter 16 notes that the accepted definition of pH does not involve a logarithm of concentration.

In response to comments from users and reviewers, we have greatly revised the introduction to the methods of science in Chapter 1, the material in Chapter 7 on the quantum mechanical model of the atom (including a new section on the shapes of atomic orbitals), and the material in Chapter 8 on covalent bonding. Carbon nanotubes are discussed along with other network solids in Chapter 11, and the discussions of thermochemistry and entropy have been further refined. The new material enhances our unique program of integrating organic chemistry, biochemistry, materials chemistry, environmental chemistry, and other applications of chemistry in related disciplines. These integrative efforts have proved invaluable in helping students recognize the importance of chemistry to them personally, to society, and to the other disciplines in which many students seek careers.

Specifically, we have made these changes from the second edition:

- Replaced the majority of Problem-Solving Examples with new problems; there are 120 new Problem-Solving Examples
- Revised or added more than a dozen new Exercises, some of them conceptual (a unique feature of this text)
- Revised or expanded many chapter-end summary problems
- Revised the end-of-chapter questions to provide better organization and increased the number of questions by 152
- Greatly increased the number of paired, closely related end-of-chapter questions where only one of the pair is answered in the book
- Included 20 new *Chemistry in the News* features, each of which provides the latest information about a chemistry topic that is important to society
- Added one *Chemistry You Can Do* and one *Estimation* feature, enhancing these two unique features of the book

- Expanded the already significant number of descriptions of women and minority chemists
- Revised the example of chemistry in action and the description of the methods of science in Chapter 1
- Rewrote the section on alkyl groups in Chapter 3 to enhance students' understanding of the material
- Revised the section on energy and enthalpy in Chapter 6 to improve clarity and organization
- Updated the material on quantum theory of atomic structure and added a section on shapes of atomic orbitals in Chapter 7
- Revised the treatment of covalent bonding in Chapter 8
- Consolidated material on atmospheric chemistry along with gas laws in Chapter 10
- Added material on carbon nanotubes to Chapter 11
- Enhanced and updated the already excellent treatment of automobile fuels in Chapter 12 to improve clarity and include E-85
- Added four new sections to Chapter 12, consolidating material on biomolecular structure of a more cohesive presentation
- Described E-85 fuel and updated material on oxygenated and reformulated gasoline
- Updated material on entropy in Chapters 14 and 18 to reflect the latest pedagogical approaches
- Updated material on equilibrium constants and pH, bringing in the concept of activity
- Updated and reorganized material on buffers and pH change upon addition of strong acid or base in Chapter 17
- Updated material on hybrid cars in Chapter 19
- Introduced material on radioactivity in common foods in Chapter 20
- Added an appendix on ground state electron configurations of atoms
- Updated the definitions in the extensive glossary

Features

We strongly encourage students to understand concepts and to learn to apply those concepts to problem solving. We believe that such understanding is essential if students are to be able to use what they learn in subsequent courses and in their future careers. All too often we hear professors in courses for which general chemistry is a prerequisite complain that students have not retained what we have taught them. This book is unique in its thoughtful choice of features that address this issue and help students achieve long-term retention of the material.

Problem Solving

Problem solving is introduced in Chapter 1, and a framework is built there that is followed throughout the book. Each chapter contains many worked-out **Problem-Solving Examples**—a total of 257 in the book as a whole. Most consist of five parts: a Question (problem); an Answer, stated briefly; a Strategy and Explanation section that outlines one approach to solving the problem and provides significant help for students whose answer did not agree with ours; a Reasonable Answer Check section marked with a ✔ that indicates how a student could check whether a result is reasonable; and a **Problem-Solving Practice** that provides a similar question or questions, with answers appearing only in an appendix. We explicitly encourage students first to define the problem, develop a plan, and work out an answer without looking at either the Answer or the Explanation, and only then to compare their answer with ours. If their answer did not agree with ours, students are asked to repeat their work. Only then do we suggest that they look at the Strategy and Explanation, which is couched in conceptual as well as numeric terms so that it will improve students' understanding, not just their ability to answer an identical question on an exam. The Reasonable Answer Check section helps students learn how to use estimated results and other criteria to decide whether an answer is reasonable, an ability that will serve them well in the future. By providing similar practice problems that are answered in the back of the book, we encourage students to immediately consolidate their thinking and improve their ability to apply their new understanding to related problems.

Enhancing students' abilities to estimate results is the goal of the ***Estimation*** boxes found in many chapters. These are a unique feature of this book. Each *Estimation* poses a problem that relates to the content of the chapter in which it appears and for which a rough calculation suffices. Students gain knowledge of various means of approximation, such as back-of-the-envelope calculations and graphing, and are encouraged to use diverse sources of information, such as encyclopedias, handbooks, and the Internet.

To further ensure that students do not merely memorize algorithmic solutions to specific problems, we provide 340 **Exercises,** which immediately follow introduction of new concepts within each chapter. Often the results that students obtain from a numeric Exercise provide insights into the concepts. Most Exercises are thought provoking and require that students apply conceptual thinking. Exercises that are conceptual rather than mathematical are clearly designated.

Examples, Practice problems, *Estimation* boxes, and Exercises are all intended to stimulate active thinking and participation by students as they read the text and to help them hone their understanding of concepts. The grand total of more than 600 of these **active-learning items** exceeds the number found in any similar textbook.

Conceptual Understanding

We believe that a sound conceptual foundation is the best means by which students can approach and solve a wide variety of real-world problems. This approach is supported by considerable evidence in the literature: Students learn better and retain what they learn longer when they have mastered fundamental concepts. Chemistry requires familiarity with at least three conceptual levels:

- **Macroscale** (laboratory and real-world phenomena)
- **Nanoscale** (models involving particles: atoms, molecules, and ions)
- **Symbolic** (chemical formulas and equations)

These three conceptual levels are explicitly defined in Chapter 1. This chapter emphasizes the value of the chemist's unique nanoscale perspective on science and the world with a specific example of how chemical thinking can help solve a real-world problem—how the anticancer agent paclitaxel (Taxol®) was discovered and synthesized in large quantities for use as a drug. This theme of conceptual understanding and its application to problems continues throughout the book. Many of the problem-solving features already mentioned have been specifically designed to support conceptual understanding.

Units are introduced on a need-to-know basis at the first point in the book where they contribute to the discussion. Units for length and mass are defined in Chapter 2, in conjunction with the discussion of the sizes and masses of atoms and subatomic particles. Energy units are defined in Chapter 6, where they are first needed to deal with kinetic and potential energy, work, and heat. In each case, defining units at the time when the need for them can be made clear allows definitions that would otherwise appear pointless and arbitrary to support the development of closely related concepts.

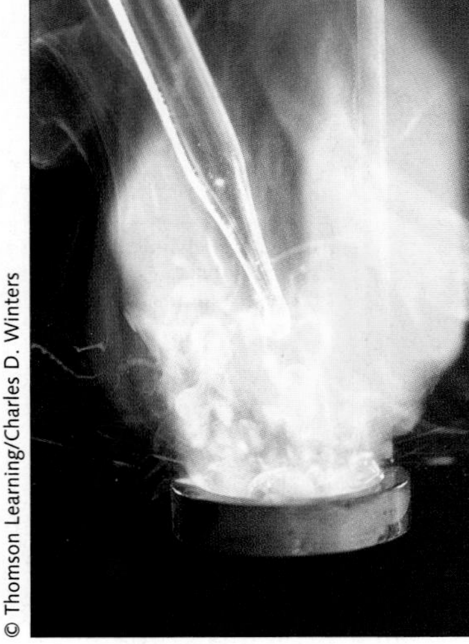
© Thomson Learning/Charles D. Winters

Whenever possible, both in the text and in the end-of-chapter questions, **we use real chemical systems in examples and problems.** In the kinetics chapter, for example, the text and problems utilize real reactions and real data from which to determine reaction rates or reaction orders. Instead of $A + B \longrightarrow C + D$, students will find $I^- + CH_3Br \longrightarrow CH_3I + Br^-$. Some data have been taken from the recent research literature. The same approach is employed in many other chapters, where real chemical systems are used as examples.

Most important, we provide **clear, direct, thorough, and understandable explanations** of all topics, including those such as kinetics, thermodynamics, and electrochemistry that many students find daunting. The methods of science and concepts such as chemical and physical properties; purification and separation; the relation of macroscale, nanoscale, and symbolic representations; elements and compounds; and kinetic-molecular theory are introduced in Chapter 1 so that they can be used throughout the later discussion. Rather than being bogged down with discussions of units and nomenclature, students begin this book with an overview of what real chemistry is about—together with fundamental ideas that they will need to understand it.

Visualization for Understanding

The **illustrations** in *Chemistry: The Molecular Science* have been designed to engage today's visually oriented students. The success of the illustration program is exemplified by the fact that

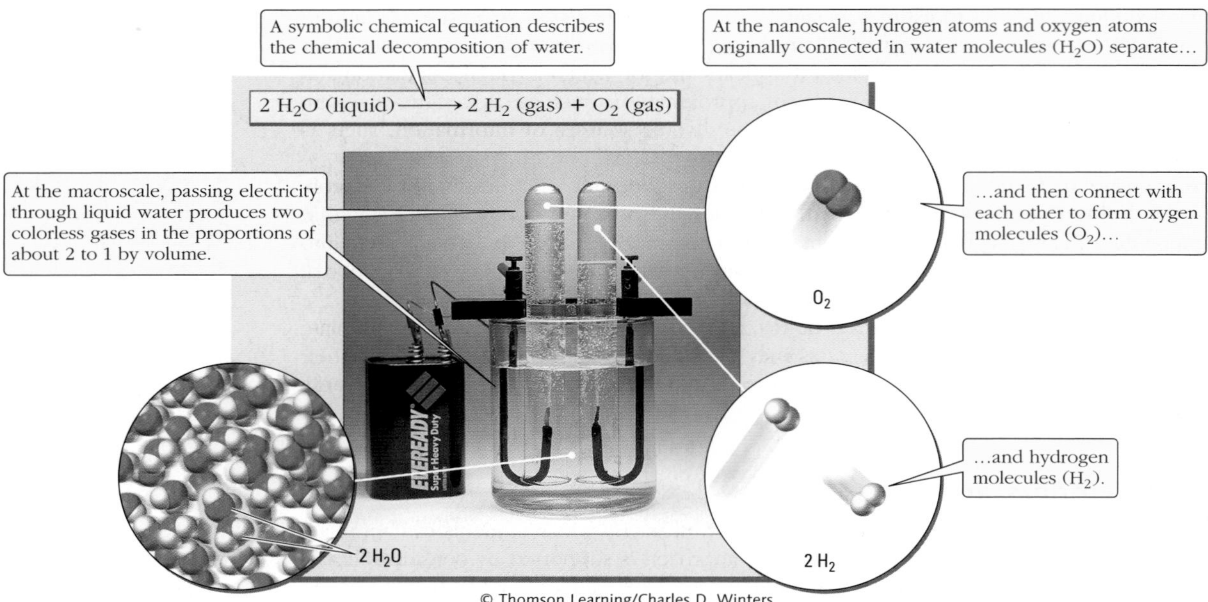

A symbolic chemical equation describes the chemical decomposition of water.
2 H2O (liquid) ⟶ 2 H2 (gas) + O2 (gas)
At the nanoscale, hydrogen atoms and oxygen atoms originally connected in water molecules (H2O) separate...
At the macroscale, passing electricity through liquid water produces two colorless gases in the proportions of about 2 to 1 by volume.
...and then connect with each other to form oxygen molecules (O2)...
O2
...and hydrogen molecules (H2).
2 H2
2 H2O

© Thomson Learning/Charles D. Winters

the first edition was awarded the coveted Talbot prize for visual excellence by the Society of Academic Authors. Illustrations help students to visualize atoms and molecules and to make connections among macroscale observations, nanoscale models, and symbolic representations of chemistry. Excellent color photographs of substances and reactions, many by Charles D. Winters, are presented together with greatly magnified illustrations of the atoms, molecules, and/or ions involved that have been created by J/B Woolsey Associates LLC. New drawings for this edition have been created by Greg Gambino. Often these are accompanied by the symbolic formula for a substance or equation for a reaction, as in the example shown above. These **nanoscale views of atoms, molecules, and ions** have been generated with molecular modeling software and then combined by a skilled artist with the photographs and formulas or equations. Similar illustrations appear in exercises, examples, and end-of-chapter problems, thereby ensuring that students are tested on the ideas they represent. The result provides an exceptionally effective way for students to learn how chemists think about the nanoscale world of atoms, molecules, and ions.

Often the story is carried solely by an illustration and accompanying text that points out the most important parts of the figure. An example is the visual story of molecular structure shown below. In other cases, text in balloons explains the operation of instruments, apparatus, and experiments; clarifies the development of a mathematical derivation; or points out salient features of graphs or nanoscale pictures. Throughout the book visual interest is high, and visualizations of many kinds are used to support conceptual development.

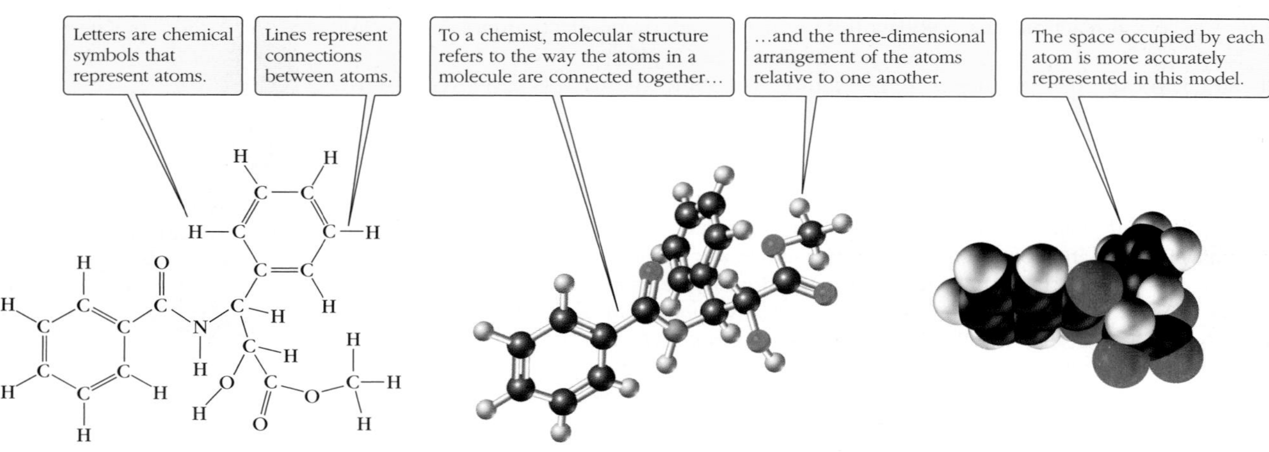

Letters are chemical symbols that represent atoms.
Lines represent connections between atoms.

The space occupied by each atom is more accurately represented in this model.

Ball-and-stick model

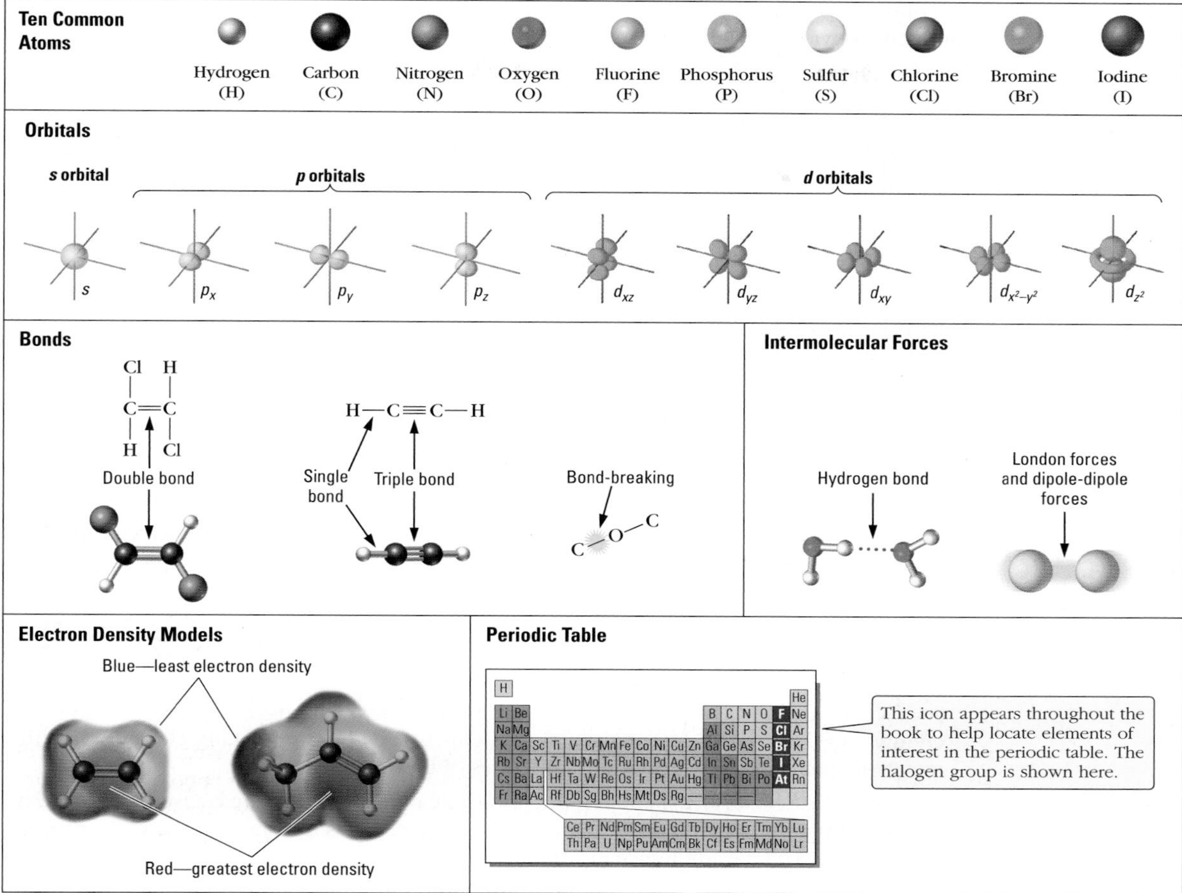

Integrated Media

The third edition again integrates the ThomsonNOW™ Web-based review and study tools included with new copies of the text.

- Margin notes throughout each chapter indicate tutorials and exercises relevant to the discussion.
- Active Figures in each chapter provide an animated version of a text figure illustrating an important concept accompanied by an exercise.
- A new set of modules based on the Estimation boxes allows for continued work developing approximation skills.
- Select end-of-chapter Questions for Review and Thought, indicated by the icon ■, are now available as tutors in ThomsonNOW and are assignable in the OWL homework system for those who use it. Many of these tutors are parameterized.
- The *In Closing* review at the end of each chapter has new references to the end-of-chapter questions available in ThomsonNOW and OWL.

Interdisciplinary Applications

Whenever possible we include practical applications, especially those applications that students will revisit when they study other natural science and engineering disciplines. Applications have been integrated where they are relevant, rather than being relegated to isolated chapters and separated from the principles and facts on which they are based. We intend that students should see that chemistry is a lively, relevant subject that is fundamental to a broad range of disciplines and that can help solve important, real-world problems.

We have especially emphasized the **integration of organic chemistry and biochemistry** throughout the book. In many areas, such as stoichiometry and molecular formulas, organic compounds provide excellent examples. To take advantage of this synergy, we have incorporated basic organic topics into the text beginning with Chapter 3 and used them wherever they are appropriate. In the discussion of molecules and the properties of molecular compounds, for example, the concepts of structural formulas, functional groups, and isomers are developed naturally and effectively. Many of the principles that students encounter in general chemistry are directly applicable to biochemistry, and a large percentage of the students in most general chemistry courses are planning careers in biological or medical areas that make constant use of biochemistry. For this reason, we have chosen to deal with fundamental biochemical topics in juxtaposition with the general chemistry principles that underlie them.

Here are some examples of integration of organic and biochemistry; the book contains many more:

- Section 3.3, *Hydrocarbons,* and Section 3.4, *Alkanes and Their Isomers,* introduce simple hydrocarbons and the concept of isomerism as a natural part of the discussion of molecular compounds.
- Section 6.12, *Foods: Fuels for Our Bodies,* applies thermochemical and calorimetric principles learned earlier in the chapter to the caloric values of proteins, fats, and carbohydrates in food.
- Chapter 12, *Fuels, Organic Chemicals, and Polymers,* builds on principles and facts introduced earlier, applying them to organic molecules and functional groups selected for their relevance to synthetic and natural polymers. Proteins and polysaccharides illustrate the importance of biopolymers.
- Section 13.9, *Enzymes: Biological Catalysts,* applies kinetic principles developed earlier in the chapter and ideas about molecular structure from earlier chapters to enzyme catalysis and the way in which it is influenced by protein structure.
- Section 18.9, *Gibbs Free Energy and Biological Systems,* discusses the role of Gibbs free energy and coupling of thermodynamic systems in metabolism, making clear the fact that metabolic pathways are governed by the rules of thermodynamics.
- Section 19.8, *Neuron Cells,* applies electrochemical principles to the transmission of nerve impulses from one neuron to another, showing that changes in concentrations of ions result in changes in voltage and hence electrical signals.

Environmental and industrial chemistry are also integrated. In Chapter 6, *Energy and Chemical Reactions,* thermochemical principles are used to evaluate the energy densities of fuels. In Chapter 10, *Gases and the Atmosphere,* a discussion of gas phase chemical reactions leads into the stories of stratospheric ozone depletion and air pollution. Chapter 10 also deals with the consequences of combustion in a section on global warming. In Chapter 13, *Chemical Kinetics: Rates and Reactions,* the importance of catalysts is illustrated by several industrial processes and exhaust-emission control on automobiles. In Chapter 16, *Acids and Bases,* practical acid-base chemistry illustrates many of the principles developed in the same chapter. In Chapter 21, *The Chemistry of the Main Group Elements,* and Chapter 22, *Chemistry of Selected Transition Metals and Coordination Compounds,* principles developed in earlier chapters are applied to uses of the elements and to extraction of elements from their ores. Students in a variety of disciplines will discover that chemistry is fundamental to their other studies.

Special Features

Another of our fundamental beliefs is that students should be involved in doing chemistry, and they ought to learn that common household materials are also chemicals. Most chapters include a ***Chemistry You Can Do*** experiment that can be performed in a kitchen or dorm room and that illustrates a topic included in the chapter. *Chemistry You Can Do* experiments require only simple equipment and familiar chemicals available at home or on a college campus.

Chemistry in the News boxes bring the latest discoveries in chemistry and applications of chemistry to the attention of students, making clear that chemistry is continually changing and developing—it is not merely a static compendium of items to memorize. These boxes have been updated, and 20 are new to this edition. ***Tools of Chemistry*** boxes provide examples of how chemists use modern instrumentation to solve challenging problems. Like any other human pursuit, chemistry depends on people, so we include in nearly every chapter **biographical sketches** of men and women who have advanced our understanding or applied chemistry imaginatively to important problems.

End-of-Chapter Study Aids

At the end of each chapter, students will find many ways to test and consolidate their learning. Every chapter except Chapter 1 ends with a **Summary Problem** that brings together concepts and problem-solving skills from throughout the chapter. Students are challenged to answer a multi-faceted question that builds on and is relevant to the chapter's content. The **In Closing** section highlights the learning goals for the chapter, provides references to the sections in the chapter that address each goal, and includes references to end-of-chapter questions available in ThomsonNOW and OWL. **Key Terms** are listed, with references to the sections where they are defined.

A broad range of **Questions for Review and Thought** are provided to serve as a basis for homework or in-class problem solving. **Review Questions,** which are not answered in the back of the book, test vocabulary and simple concepts. Next come **Topical Questions,** which are keyed to the major topics in the chapter and listed under headings that correspond with each section in the chapter. Questions are often accompanied by photographs, graphs, and diagrams that make the situations described more concrete and realistic. Usually a question that is answered at the end of the book is paired with a similar one that is not. There are also many **General Questions** that are not explicitly keyed to chapter topics. Often these require students to integrate several concepts. **Applying Concepts** includes questions specifically designed to test conceptual learning. Many of these questions include diagrams of atoms, molecules, or ions and require students to relate macroscopic observations, atomic-scale models, and symbolic formulas and equations. **More Challenging Questions** require students to apply more thought and to better integrate multiple concepts than do typical end-of-chapter questions. **Conceptual Challenge Problems,** most of which were written by H. Graden Kirksey of the University of Memphis, are especially important in helping students assess and improve their conceptual thinking ability. Designed for group work, the Conceptual Challenge Problems are rigorous and thought provoking. Much effective learning can be induced by dividing a class into groups of three or four students and then assigning these groups to work collaboratively on these problems.

Organization

The order of chapters reflects the most common division of content between the first and second semesters of a typical general chemistry course. The first few chapters briefly review basic material that most students should have encountered in high school. Next, the book develops the ideas of chemical reactions, stoichiometry, and energy transfers during reactions. Throughout these early chapters, organic chemistry, biochemistry, and applications of chemistry are integrated. We then deal with the electronic structure of atoms, bonding and molecular structures, and the way in which structure affects properties. To finish up a first-semester course, there are adjacent chapters on gases and on liquids and solids.

Next, we extend our integration of organic chemistry in a chapter that describes the role of organic chemicals in fuels, polymers, and biopolymers. Chapters on kinetics and equilibrium establish fundamental understanding of how fast reactions will go and what concentrations of reactants and products will remain when equilibrium is reached. These ideas are then applied to solutions, as well as to acid-base and solubility equilibria in aqueous solutions. A chapter on thermodynamics and Gibbs free energy is followed by one on electrochemistry, which makes use of thermodynamic ideas. Finally, the book focuses on nuclear chemistry and the descriptive chemistry of main group and transition elements.

To help students connect chemical ideas that are closely related but are presented in different chapters, we have included **numerous cross references (indicated by the ⇐ symbol).** These cross references will help students link a concept being developed in the chapter they are currently reading with an earlier, related principle or fact. They also provide many opportunities for students to review material encountered earlier.

A number of variations in the order of presentation are possible. For example, in the classes of one of the authors, the first six sections of Chapter 18 on thermodynamics follow Chapter 13 on chemical kinetics and precede Chapter 14 on equilibrium. Section 14.7 is omitted, and the last five sections of Chapter 18 follow Chapter 14. The material on thermochemistry in Chapter 6 could be postponed and combined with Chapter 18 on thermodynamics with only minor adjustments in the teaching of other chapters, so long as the treatment of thermochemistry precedes the material in Chapter 12, which focuses on energy and fuels, and Chapter 13, which uses thermochemical concepts in the discussion of activation energy. Many other reorderings of chapters or sections within chapters are possible. The numerous cross references will aid students in picking up concepts that they would be assumed to know, had the chapters been taught consecutively.

At the University of Wisconsin–Madison, this textbook is used in a one-semester accelerated course that is required for most engineering students. We assume substantial high school background in both chemistry and mathematics, and the syllabus includes Chapters 1, 7, 8, 9, 12, 13, 18, 14, 16, 17, and 19. This presentation strategy works quite well, and some engineering students have commented favorably on the inclusion of practical applications of chemistry, such as octane rating and catalysis, in which they were interested.

Chemistry: The Molecular Science can be divided into a number of sections, each of which treats an important aspect of chemistry:

Fundamental Ideas of Chemistry

Chapter 1, The Nature of Chemistry, is designed to capture students' interest from the start by concentrating on chemistry (not on math, units, and significant figures, which are treated comprehensively in an appendix). It asks Why Care About Chemistry? and then tells a story of modern drug discovery and development that illustrates interdisciplinary chemical research. It also introduces major concepts that bear on all of chemistry, emphasizing the three conceptual levels with which students must be familiar—macroscale, nanoscale, and symbolic.

Chapter 2, Atoms and Elements, introduces units and dimensional analysis on a need-to-know basis in the context of the sizes of atoms. It concentrates on thorough, understandable treatment of the concepts of atomic structure, atomic weight, and moles of elements, making the connections among them clear. It concludes by introducing the periodic table and highlighting the periodicity of properties of elements.

Chapter 3, Chemical Compounds, distinguishes ionic compounds from molecular compounds and illustrates molecular compounds with the simplest alkanes. The important theme of structure is reinforced by showing several ways that organic structures can be written. Charges of monatomic ions are related to the periodic table, which is also used to show elements that are important in living systems. Molar masses of compounds and determining formulas fit logically into the chapter's structure.

Chemical Reactions

Chapter 4, Quantities of Reactants and Products, begins a three-chapter sequence that treats chemical reactions qualitatively and quantitatively. Students learn how to balance equations and to use typical inorganic reaction patterns to predict products. A single stepwise method is provided for solving all stoichiometry problems, and 11 examples demonstrate a broad range of stoichiometry calculations.

Chapter 5, Chemical Reactions, has a strong descriptive chemistry focus, dealing with exchange reactions, acid-base reactions, and oxidation-reduction reactions in aqueous solutions. It includes real-world occurrences of each type of reaction. Students learn how to recognize a redox reaction from the chemical nature of the reactants (not just by using oxidation numbers) and how to do titration calculations.

Chapter 6, Energy and Chemical Reactions, begins with a thorough and straightforward introduction to forms of energy, conservation of energy, heat and work, system and surroundings, and exothermic and endothermic processes. Carefully designed figures help students to understand thermodynamic principles. Heat capacity, heats of changes of state, and heats of reactions are clearly explained, as are calorimetry and standard enthalpy changes. These ideas are then applied to fossil fuel combustion and to metabolism of biochemical fuels (proteins, carbohydrates, and fats).

Electrons, Bonding, and Structure

Chapter 7, Electron Configurations and the Periodic Table, introduces spectra, quantum theory, and quantum numbers, using color-coded illustrations to visualize the different energy levels of *s, p, d,* and *f* orbitals. The *s*-, *p*-, *d*-, and *f*-block locations in the periodic table are used to predict electron configurations.

Chapter 8, Covalent Bonding, provides simple stepwise guidelines for writing Lewis structures, with many examples of how to use them. The role of single and multiple bonds in hydrocarbons is smoothly integrated with the introduction to covalent bonding. The discussion of polar bonds is enhanced by molecular models that show variations in electron density. Molecular orbital theory is introduced as well.

Chapter 9, Molecular Structures, provides a thorough presentation of valence-shell electron-pair repulsion (VSEPR) theory and orbital hybridization. Molecular geometry and polarity are extensively illustrated with computer-generated models, and the relation of structure, polarity, and hydrogen bonding to attractions among molecules is clearly developed and illustrated in solved problems. The importance of noncovalent interactions is emphasized early and then reinforced by describing how noncovalent attractions determine the structure of DNA.

States of Matter

Chapter 10, Gases and the Atmosphere, uses kinetic-molecular theory to interpret the behavior of gases and then describes each of the individual gas laws. Mathematical problem solving focuses on the ideal gas law or the combined gas law, and many conceptual Exercises throughout the chapter emphasize qualitative understanding of gas properties. Gas stoichiometry is presented in a uniquely concise and clear manner. Then the properties of gases are applied to chemical reactions in the atmosphere, the role of ozone in both the troposphere and the stratosphere, industrial and photochemical smog, and global warming.

Chapter 11, Liquids, Solids, and Materials, begins by discussing the properties of liquids and the nature of phase changes. The unique and vitally important properties of water are covered thoroughly. The principles of crystal structure are introduced using cubic unit cells only. The fact that much current chemical research involves materials is illustrated by the discussions of metals, *n*- and *p*-type semiconductors, insulators, superconductors, network solids, carbon nanotubes, cement, ceramics and ceramic composites, and glasses, including optical fibers.

Important Industrial, Environmental, and Biological Molecules

Chapter 12, Fuels, Organic Chemicals, and Polymers, offers a distinctive combination of topics of major relevance to industrial, energy, and environmental concerns. Petroleum, natural gas, and coal are discussed as resources for energy and chemical materials. Enough organic functional groups are introduced so that students can understand polymer formation, and the idea of condensation polymerization is extended to carbohydrates and proteins, which are compared with synthetic polymers.

Reactions: How Fast and How Far?

Chapter 13, Chemical Kinetics: Rates of Reactions, presents one of the most difficult topics in the course with extraordinary clarity. Defining reaction rate, finding rate laws from initial rates and integrated rate laws, and using the Arrhenius equation are thoroughly developed. How molecular changes during unimolecular and bimolecular elementary reactions relate to activation energy initiates the treatment of reaction mechanisms (including those with an initial fast equilibrium). Catalysis is shown to involve changing a reaction mechanism. Both enzymes and industrial catalysts are described using concepts developed earlier in the chapter.

Chapter 14, Chemical Equilibrium, emphasizes equally a qualitative understanding of the nature of equilibrium and the solving of mathematical problems. That equilibrium results from equal but opposite reaction rates is fully explained. Both Le Chatelier's principle and the reaction quotient, *Q*, are used to predict shifts in equilibria. A unique section on equilibrium at the nanoscale introduces briefly and qualitatively how enthalpy changes and entropy changes affect equilibria. Optimizing the yield of the Haber-Bosch ammonia synthesis elegantly illustrates how kinetics, equilibrium, and enthalpy and entropy changes control the outcome of a chemical reaction.

Reactions in Aqueous Solution

Chapter 15, The Chemistry of Solutes and Solutions, builds on principles previously introduced, showing the influence of enthalpy and entropy on solution properties. Understanding of solubility, Henry's law, concentration units (including ppm and ppb), and colligative properties (including osmosis) is reinforced by applying these ideas to water as a resource, hard water, and municipal water treatment.

Chapter 16, Acids and Bases, concentrates initially on the Brønsted-Lowry acid-base concept, clearly delineating proton transfers using color coding and molecular models. In addition to a full exploration of pH and the meaning and use of K_a and K_b, acid strength is related to molecular structure, and the acid-base properties of carboxylic acids, amines, and amino acids are introduced. Lewis acids and bases are defined and illustrated using examples. Student interest is enhanced by a discussion of everyday uses of acids and bases.

Chapter 17, Additional Aqueous Equilibria, extends the treatment of acid-base and solubility equilibria to buffers, titration, and precipitation. The Henderson-Hasselbalch equation, which is widely used in biochemistry, is applied to buffer pH. Calculations of points on titration curves are shown, and the interpretation of several types of titration curves provides conceptual understanding. Acid-base concepts are applied to the formation of acid rain. The final section deals with the various factors that affect solubility (pH, common ions, complex ions, and amphoterism) and with selective precipitation.

Thermodynamics and Electrochemistry

Chapter 18, Thermodynamics: Directionality of Chemical Reactions, explores the nature and significance of entropy, both qualitatively and quantitatively. The signs of Gibbs free energy changes are related to the easily understood classification of reactions as reactant- or product-favored, with the discussion deliberately avoiding the often-misinterpreted term "spontaneous." The thermodynamic significance of coupling one reaction with another is illustrated using industrial, metabolic, and photosynthetic examples. Energy conservation is defined thermodynamically. A closing section reinforces the important distinction between thermodynamic and kinetic stability.

Chapter 19, Electrochemistry and Its Applications, defines redox reactions and uses half-reactions to balance redox equations. Electrochemical cells, cell voltage, standard cell potentials, the relation of cell potential to Gibbs free energy, and the effect of concentrations on cell potential are all explored. These ideas are then applied to the transmission of nerve impulses. Practical applications include batteries, fuel cells, electrolysis, and corrosion.

Nuclear Chemistry

Chapter 20, Nuclear Chemistry, deals with radioactivity, nuclear reactions, nuclear stability, and rates of disintegration reactions. Also provided are a thorough description of nuclear fission and nuclear fusion and a thorough discussion of nuclear radiation, background radiation, and applications of radioisotopes.

More Descriptive Chemistry

Chapter 21, The Chemistry of the Main Group Elements, consists of two main parts. The first part tells the interesting story of how the elements were formed and which are most important on earth. The physical separation of nitrogen, oxygen, and sulfur from natural sources, and the extraction of sodium, chlorine, magnesium, and aluminum by electrolysis, provide important industrial examples as well as an opportunity for students to apply principles learned earlier in the book. The second part (Section 21.6) discusses the properties, chemistry, and uses of the elements of Groups 1A–7A and their compounds in a systematic way, based on groups of the periodic table. Trends in atomic and ionic radii, melting points and boiling points, and densities of each group's elements are summarized. Group 8A is covered briefly.

Chapter 22, Chemistry of Selected Transition Elements and Coordination Compounds, treats a few important elements in depth and integrates the review of principles learned earlier. Iron, copper, chromium, silver, and gold provide an interesting, motivating collection of elements from which students can learn the principles of transition metal chemistry. In addition to the treatment of complex ions and coordination compounds, this chapter includes an extensive section on crystal-field theory, electron configurations, color, and magnetism in coordination complexes.

Supporting Materials

For Students and Instructors

ThomsonNOW™

ThomsonNOW at www.thomsonedu.com is an online assessment-centered system that helps students master material by directing them to interactive tutorials, Active Figures, exercises, and simulations that enhance students' personal conceptual understanding and problem-solving skills.

Students can access the material for each chapter using the margin annotations in the text or a diagnostic pre-test that has been carefully crafted to assess students' understanding of the chapter material. Upon completing a pre-test, students receive feedback and personalized study with links to interactive media content based on their unique needs. Other Web-based tools include hundreds of interactive molecular models, a plotting tool, molecular mass and molarity calculators, and an extensive database of compounds with their thermodynamic properties.

NEW! For the third edition, a selection of end-of-chapter questions (marked with ■ in the text) are available as tutors in ThomsonNOW; many of these tutors are parameterized. An access code is required for ThomsonNOW and may be packaged with a new copy of the text or purchased separately. Register at **www.thomsonedu.com/login** or purchase an access code at **www.thomsonedu.com/buy**.

vMentor is an online live tutoring service from Thomson Brooks/Cole in partnership with Elluminate that is included in ThomsonNOW. Whether it's one-to-one tutoring help with daily homework or exam review tutorials, vMentor lets students interact with experienced tutors right from the students' own computers at school or at home. All tutors have specialized degrees in the particular subject area (biology, chemistry, mathematics, physics, or statistics) as well as extensive teaching experience. Each tutor also has a copy of the textbook the student is using in class. Students can ask as many questions as they want when they access vMentor—and they don't need to set up appointments in advance! Access is provided with vClass, an Internet-based virtual classroom featuring two-way voice, a shared whiteboard, chat, and more. vMentor is available only to proprietary, college, and university adopters.

OWL: Online Web-based Learning
Authored by Roberta Day and Beatrice Botch of the University of Massachusetts, Amherst, and William Vining of the State University of New York at Oneonta. Used by more than 300 institutions and proven reliable for tens of thousands of students, **OWL** offers unsurpassed ease of use, reliability, and dedicated training and service. **OWL** makes homework management a breeze and helps students improve their problem-solving skills and visualize concepts, providing instant analysis and feedback on a variety of homework problems, including tutors, simulations, and chemically and/or numerically parameterized short-answer questions. **OWL** is the only system specifically designed to support mastery learning, where students work as long as they need to master each chemical concept and skill. New to this edition, approximately 15 end-of-chapter questions (marked in the text with ■) and 25 new tutorials based on *Estimation* boxes in the text can be assigned in OWL. A fee-based access code is required for **OWL. OWL** is only available to North American adopters.

NEW! **A complete e-Book!**
The **Moore e-Book in OWL** includes the complete textbook as an assignable resource that is fully linked to **OWL** homework content. This new **e-Book in OWL** is an exclusive option that will be available to all your students if you choose it. Access codes can be bundled with the text and/or ordered as a text replacement. Please consult your Thomson Brooks/Cole representative for pricing details.

To learn more about **OWL,** visit **http://owl.thomsonlearning.com** or contact your Thomson Brooks/Cole representative.

For the Student

Visit the *Chemistry: The Molecular Science* Web site at **www.thomsonedu.com/chemistry/moore3** to purchase items online and see sample materials.

Student Solutions Manual by Judy L. Ozment, Pennsylvania State University.
ISBN-13 978-0-495-11253-2
Contains fully worked-out solutions to end-of-chapter questions that have blue, boldfaced numbers. Solutions match the problem-solving strategies used in the main text. A sample is available on the student companion Web site at **www.thomsonedu.com/chemistry/moore3**.

Study Guide by Michael J. Sanger, Middle Tennessee State University. ISBN-13 978-0-495-11254-9
Contains learning tools such as brief notes on chapter sections with examples, reviews of key

terms, and practice tests with answers provided. A sample is available on the student companion Web site at www.thomsonedu.com/chemistry/moore3.

NEW! **General Chemistry: Guided Explorations** by David Hanson, State University of New York at Stony Brook. ISBN-13 978-0-495-11599-1
This student workbook, new for the third edition, is designed to support Process Oriented Guided Inquiry Learning (POGIL) with activities that promote a student-focused active classroom. It is an excellent accompaniment to *Chemistry: The Molecular Science* or any other general chemistry text.

Essential Math for Chemistry Students, Second Edition by David W. Ball, Cleveland State University. ISBN-13 978-0-495-01327-3
This book focuses on the algebra skills needed to survive in general chemistry, with worked examples showing how these skills translate into successful chemical problem solving. It's an ideal tool for students who lack the confidence or competency in the essential algebra skills required for general chemistry. The second edition includes references to OWL, our Web-based tutorial program, offering students access to online algebra skills exercises.

The Survival Guide for General Chemistry with Math Review and Proficiency Questions, 2nd Edition by Charles H. Atwood, University of Georgia. ISBN-13 978-0-495-38751-0
Designed to help students gain a better understanding of the basic problem-solving skills and concepts of general chemistry, this guide assists students who lack confidence and/or competency in the essential skills necessary to survive general chemistry. The second edition includes new proficiency questions that will help students assess their level of understanding prior to an exam.

ChemPages Laboratory CD-ROM (available separately from *JCE Software;* see **http://jce.divched.org/JCESoft/Programs/CPL/index.html**). A collection of videos with voiceover and text showing how to perform the most common laboratory techniques used by students in first-year chemistry courses.

General Chemistry Collection CD-ROM (available separately from *JCE Software;* see **http://jce.divched.org/JCESoft/Programs/GCC/index.html**). Contains many software programs, animations, and videos that correlate with the content of this book. Arrangements can be made to make this item available to students at very low cost. Call (800) 991-5534 for more information about *JCE* products.

For the Instructor

Supporting materials for instructors are available to qualified adopters. Please consult your local Thomson Brooks/Cole sales representative for details.

Visit the *Chemistry: The Molecular Science* Web site at www.thomsonedu.com/chemistry/moore3 to:

- See sample materials
- Request a desk copy
- Locate your sales representative
- Download electronic files of select materials

Instructor's Solutions Manual by Judy L. Ozment, Pennsylvania State University. ISBN-13 978-0-495-11246-4
Contains fully worked-out solutions to all end-of-chapter questions, Summary Problems, and Conceptual Challenge Problems. Solutions match the problem-solving strategies used in the text. Available in electronic format on the instructor's Multimedia Manager CD-ROM.

Test Bank by Paul Deroo, Drexel University and James Rudd, California State University, Los Angeles. IBSN-13 978-0-495-11247-1
Contains more than 1100 questions, all carefully matched to the corresponding text sections. The ExamView® computerized version of the Test Bank is also available (see next page). Electronic files of the Test Bank are available on the Instructor's PowerLecture CD-ROM.

Overhead Transparencies Set. ISBN-13 978-0-495-11249-5
Contains 150 acetates of key illustrations from the text.

Instructor's PowerLecture CD-ROM for *Chemistry: The Molecular Science,* third edition. ISBN-13 978-0-495-11250-1
This one-stop digital library and presentation tool—a cross-platform CD-ROM—includes text, art, photos, and tables in a variety of electronic formats that are easily exported into other software packages. This enhanced CD-ROM also contains simulations, molecular models, and QuickTime™ movies to supplement your lectures. You can customize your presentations by importing your personal lecture slides or other material you choose. PowerLecture also includes electronic files of select print ancillaries such as the Instructor's Solutions Manual and the Test Bank. PowerLecture also includes:

- **ExamView® Computerized Test Bank.** Using the contents of the print Test Bank, ExamView® allows instructors to create, deliver, and customize tests and study guides (both print and online) in minutes with this assessment and tutorial system. ExamView offers both a Quick Test Wizard and an Online Test Wizard that guide you step by step through the process of creating tests.
- Our book-specific **JoinIn**™ content for student classroom response systems allows you to transform your classroom and assess your students' progress with instant in-class quizzes and polls. This software lets you pose book-specific questions and display students' answers seamlessly within the Microsoft® PowerPoint® slides of your own lecture in conjunction with the "clicker" hardware of your choice. Enhance how your students interact with you, your lecture, and each other. Please consult your Thomson Brooks/Cole representative for further details.

Thomson Custom Solutions develops personalized solutions to meet your course needs. Match your learning materials to your syllabus and create the perfect learning solution—your customized text will contain the same thought-provoking, scientifically sound content, superior authorship, and stunning art that you've come to expect from Thomson Brooks/Cole texts, yet in a more flexible format. Visit www.thomsoncustom.com to start building your book today.

WebCT/Blackboard ThomsonNOW can be fully integrated with WebCT and Blackboard providing instructors using either platform access to assessments and content powered by iLrn without an extra login. Please contact your local Thomson Brooks/Cole representative for more information.

Chemistry Comes Alive! CD-ROM Series. This series of eight CDs (available separately from *JCE Software;* see **http://jce.divched.org/JCESoft/CCA/index.html**) includes HTML-format access to a broad range of videos and animations suitable for use in lecture presentations, for independent study, or for incorporation into the instructor's own tutorials.

***JCE* QBank.** (Available separately from the *Journal of Chemical Education;* see **http://www.jce.divched.org/JCEDLib/QBank/index.html**). Contains more than 3500 homework and quiz questions suitable for delivery via WebCT, Desire2Learn, or Moodle course management systems, hundreds of ConcepTest questions that can be used with "clickers" to make lectures more interactive, and a collection of conceptual questions together with a discussion of how to write conceptual questions. Available to all *JCE* subscribers.

For the Laboratory

Thomson Brooks/Cole Lab Manuals. We offer a variety of printed manuals to meet all your general chemistry laboratory needs. Instructors can visit the chemistry site at www.thomsonedu.com/chemistry for a full listing and description of these laboratory manuals and laboratory notebooks.

Thomson Custom Labs . . . for the customized laboratory (www.thomsoncustom.com/labs)
Thomson Custom Labs combines the resources of Thomson Brooks/Cole, CER, and Outernet Publishing to provide you unparalleled service in creating your ideal customized lab program. Select the experiments and artwork you need from our collection of content and imagery to find the perfect labs to match your course.

ChemPages Laboratory CD-ROM (available separately from *JCE Software;* see **http://jce.divched.org/JCESoft/Programs/CPL/index.html**). A collection of videos with voiceover and text showing how to perform the most common laboratory techniques used by students in first-year chemistry courses.

Note: Unless otherwise noted, the Web site domain names (URLS) provided here are not published by Thomson Brooks/Cole and the Publisher can accept no responsibility or liability for these sites' content. Because of the dynamic nature of the Internet, Thomson Brooks/Cole cannot in any case guarantee the continued availability of third-party Web sites.

Reviewers

Reviewers have played a critical role in the preparation of this textbook. The individuals listed below helped to shape this text into one that is not merely accurate and up to date, but a valuable practical resource for teaching and testing students.

Editorial Advisory Board

David Grainger, *University of Utah*
Benjamin R. Martin, *Texas State University, San Marcos*
David Miller, *California State University, Northridge*
Michael J. Sanger, *Middle Tennessee State University*
Sherril Soman, *Grand Valley State University*
Richard T. Toomey, *Northwest Missouri State University*

Reviewers of the Third Edition

Patricia Amateis, *Virginia Tech*
Debra Boehmler, *University of Maryland*
Norman C. Craig, *Oberlin College*
Michael G. Finnegan, *Washington State University*
Milton D. Johnson, *University of South Florida*
Katherine R. Miller, *Salisbury University*
Robert Milofsky, *Fort Lewis College*
Mark E. Ott, *Jackson Community College*
Philip J. Reid, *University of Washington*
Joel Tellinghuisen, *Vanderbilt University*
Richard T. Toomey, *Northwest Missouri State University*
Peter A. Wade, *Drexel University*
Keith A. Walters, *Northern Kentucky University*

Reviewers of the Second Edition

Ruth Ann Armitage, *Eastern Michigan University*
Margaret Asirvatham, *University of Colorado*
David Ball, *Cleveland State University*
Debbie J. Beard, *Mississippi State University*
Mary Jo Bojan, *Pennsylvania State University*
Simon Bott, *University of Houston*
Judith N. Burstyn, *University of Wisconsin, Madison*
Kathy Carrigan, *Portland Community College*
James A. Collier, *Truckee Meadows Community College*
Susan Collins, *California State University, Northridge*
Roberta Day, *University of Massachusetts–Amherst*
Norman Dean, *California State University, Northridge*
Barbara L. Edgar, *University of Minnesota, Twin Cities*
Paul Edwards, *Edinboro University of Pennsylvania*
Amina K. El-Ashmawy, *Collin County Community College*
Thomas P. Fehlner, *University of Notre Dame*
Daniel Fraser, *University of Toledo*
Mark B. Freilich, *The University of Memphis*
Noel George, *Ryerson University*
Stephen Z. Goldberg, *Adelphi University*
Gregory V. Hartland, *University of Notre Dame*
Ronald C. Johnson, *Emory University*
Jeffrey Kovac, *University of Tennessee*
John Z. Larese, *University of Tennessee*
Joe March, *University of Alabama at Birmingham*
Lyle V. McAfee, *The Citadel*
David Miller, *California State University, Northridge*
Wyatt R. Murphy, Jr., *Seton Hall University*
Mary-Ann Pearsall, *Drew University*
Vicente Talanquer, *University of Arizona*
Wayne Tikkanen, *California State University, Los Angeles*
Patricia Metthe Todebush, *Northwestern University*

Andrew V. Wells, *Chabot Community College*
Steven M. Wietstock, *Indiana University*
Martel Zeldin, *Hobart & William Smith Colleges*
William H. Zoller, *University of Washington*

Reviewers of the First Edition
Margaret Asirvatham, *University of Colorado-Boulder*
Donald Berry, *University of Pennsylvania*
Barbara Burke, *California State Polytechnic University, Pomona*
Dana Chatellier, *University of Delaware*
Mapi Cuevas, *Santa Fe Community College*
Cheryl Dammann, *University of North Carolina-Charlotte*
John DeKorte, *Glendale Community College*
Russ Geanangel, *University of Houston*
Peter Gold, *Pennsylvania State University*
Albert Martin, *Moravian College*
Marcy McDonald, *University of Alabama-Tuscaloosa*
Charles W. McLaughlin, *University of Nebraska*
David Metcalf, *University of Virginia*
David Miller, *California State University, Northridge*
Kathleen Murphy, *Daemen College*
William Reinhardt, *University of Washington*
Eugene Rochow, *Fort Myers, Florida*
Steven Socol, *McHenry County College*
Richard Thompson, *University of Missouri-Columbia*
Sheryl Tucker, *University of Missouri-Columbia*
Jose Vites, *Eastern Michigan University*
Sarah West, *University of Notre Dame*
Rick White, *Sam Houston State University*

We also thank the following people who were dedicated to checking the accuracy of the text and art.

Accuracy Reviewers of the Third Edition
Julie B. Ealy, *Pennsylvania State University*
Stephen Z. Goldberg, *Adelphi University*
David Shinn, *University of Hawaii at Manoa*
Barbara Mowery, *York College of Pennsylvania*

Accuracy Reviewers of the Second Edition
Larry Fishel, *East Lansing, Michigan*
Stephen Z. Goldberg, *Adelphi University*
Robert Milofsky, *Fort Lewis College*
Barbara D. Mowery, *Thomas Nelson Community College*

Accuracy Reviewers of the First Edition
John DeKorte, *Glendale Community College*
Larry Fishel, *East Lansing, Michigan*
Leslie Kinsland, *Cornell University*
Judy L. Ozment, *Pennsylvania State University-Abington*
Gary Riley, *St. Louis School of Pharmacy*

Acknowledgments

No project on the scale of a textbook revision is accomplished solely by the authors. We have had assistance of the very highest quality in all aspects of production of this book, and we extend hearty thanks to everyone who contributed to the project.

David Harris, publisher, and Lisa Lockwood, chemistry editor, have overseen the entire project and have collaborated effectively with the author team on decisions and initiatives that have greatly improved what was already an excellent, rigorous, mainstream general chemistry textbook. They are also responsible for assembling the excellent editorial team that provided strong support for the authors.

Peter McGahey, development editor, provided advice and active support throughout the revision and was always available when things needed to be done or authors needed to be prompted to provide copy. He assembled an excellent group of expert reviewers, obtained reviews from them in timely fashion, and provided feedback based on their comments that was invaluable. He has also served as a calm, conscientious, and caring interface between the authors and the many other members of the production staff. Thanks, Peter!

Jennifer Risden, content project manager, and Lisa Weber, who held that position for the first part of the project, both helped keep the authors on track and provided timely queries and suggestions regarding editing, layout, and appearance of the book. We thank them for their invaluable contribution. In the latter part of the project, Lisa served as technology project manager, and her ability to organize all the multimedia elements and the references to them in the printed book is much appreciated. Wyatt Murphy, Sacred Heart University, reviewed the ThomsonNOW media annotations and wrote ThomsonNOW quizzes. Brandi Kirksey, assistant editor, has ably handled all of the ancillary print materials.

The success of a book such as this one depends also on its being adopted and read. Amee Mosley, marketing manager, directs the marketing and sales programs, and many local representatives throughout the country have helped and will help get this book to students who can benefit from it.

This book is beautiful to look at, and its beauty is more than skin deep. The illustration program has been carefully designed to support student learning in every possible way. The many photographs of Charles D. Winters of Oneonta, New York, provide students with close-up views of chemistry in action. We thank Charlie for doing many new shoots (one involving bromine—somewhat of an adventure) for this new edition. Dena Digilio Betz and Robin Samper carried out photo research in a most effective and friendly fashion, and we thank them for helping to improve the illustration program that won a Talbot award in a previous edition.

Mandy Hetrick and the staff at Lachina Publishing Services have handled copy editing, layout, and production of the book. Mandy worked calmly and effectively with the authors to make certain that this book will be of the highest possible quality. Special thanks go to copy editor Amy Mayfield, who removed infelicities, made the entire book consistent, and even discovered typos that had made it through two previous editions. We thank all of the staff at Lachina who contributed to this edition.

Many of the take-home *Chemistry You Can Do* experiments in this book were adapted from activities published by the Institute for Chemical Education as *Fun with Chemistry: Volumes I and II,* by Mickey and Jerry Sarquis of Miami University (Ohio). Some were adapted from Classroom Activities published in the *Journal of Chemical Education.* Conceptual Challenge Problems at the end of most chapters were written by H. Graden Kirksey, University of Memphis, and we very much appreciate his contribution. The active-learning, conceptual approach of this book has been greatly influenced by the systemic curriculum enhancement project, *Establishing New Traditions: Revitalizing the Curriculum,* funded by the National Science Foundation, Directorate for Education and Human Resources, Division of Undergraduate Education, grant DUE-9455928.

We also thank the many teachers, colleagues, students, and others who have contributed to our knowledge of chemistry and helped us devise better ways to help others learn it. Collectively, the authors of this book have many years of experience teaching and learning, and we have tried to incorporate as much of that as possible into our presentation of chemistry.

Finally, we thank our families and friends who have supported all of our efforts—and who can reasonably expect more of our time and attention now that this new edition is complete.

We hope that using this book results in a lively and productive experience for both faculty and students.

John W. Moore
Madison, Wisconsin

Conrad L. Stanitski
Lancaster, Pennsylvania

Peter C. Jurs
State College, Pennsylvania

Special Features

CHEMISTRY IN THE NEWS

CHEMISTRY YOU CAN DO

ESTIMATION

TOOLS OF CHEMISTRY

PORTRAIT OF A SCIENTIST

12 Fuels, Organic Chemicals, and Polymers

The snowboard in this picture is made from several synthetic polymers, as are the clothing and protective gear worn by its rider. Polymers such as Kevlar and laminated polyethylene used in the snowboard make it lighter and more durable than wood.

ThomsonNOW
Throughout the text, this icon indicates an opportunity to test yourself on key concepts and to explore interactive modules by signing in to ThomsonNOW at **www.thomsonedu.com**.

The combustion of fossil fuels—coal, natural gas, and petroleum—provides nearly 85% of all the energy used in the world. When these substances burn, the carbon they contain is released into the atmosphere as CO_2. Photosynthesis converts CO_2 back into other carbon-containing compounds. Many of the carbon compounds produced by photosynthesis are directly or indirectly very useful as energy sources for humans and animals. But such carbon compounds are not as convenient to use in power plants or automobiles as are fossil fuels, which are burned in prodigious quantities daily. As fossil fuel becomes more scarce, and if stringent conservation measures are not taken, conventional petroleum reserves could last for no more than into the last half of the 21st century. If this occurs, fuel costs will inevitably rise due to the increasing scarcity of petroleum.

The law of supply and demand dictates cost.

The great Russian chemist Dimitri Mendeleev recognized the importance of petroleum as a source from which to make valuable carbon compounds and not merely to be used as a fuel. On visiting the oil fields of Pennsylvania and Azerbaijan, he supposedly remarked that burning petroleum as a fuel "would be akin to firing up a kitchen stove with bank notes."

Fossil fuels are also, by far, the largest source of hydrocarbons and their derivatives. Only about 3% of the petroleum refined today is the source of most of the organic chemicals used to make consumer products such as plastics, pharmaceuticals, synthetic rubber, synthetic fibers, and hundreds of other products we rely on. For this reason, the organic chemical industry is often referred to as the petrochemical industry. In this chapter we will discuss a few of the major classes of organic compounds and some of their reactions, especially those used to furnish energy and to make polymers, synthetic as well as natural ones.

12.1 Petroleum

Petroleum is a complex mixture. There are at least 20,000 different compounds in petroleum, about the same number as genes in the human body.

Petroleum is a complex mixture of alkanes, cycloalkanes, alkenes, and aromatic hydrocarbons formed from the remains of plants and animals from millions of years ago. Thousands of compounds, almost all of them hydrocarbons, are present in *crude oil,* the form of petroleum that is pumped from the ground. Crude oil's composition and color vary with the location in which it is found. Pennsylvania crude oils are primarily straight-chain hydrocarbons, whereas California crude oil contains a larger portion of aromatic hydrocarbons.

The different classes of hydrocarbons (compounds of hydrogen and carbon) were introduced earlier ***(⇐ p. 339).*** Alkanes have C—C bonds; alkenes have one or more C=C bonds; aromatics include benzene-like rings.

The early uses for petroleum components were mainly for lubrication and as kerosene burned in lamps. The development of automobiles with internal combustion engines created the need for liquid fuels that would burn efficiently in these engines. To meet our need for gasoline and other petroleum products, it is necessary to refine crude oil—that is, to separate the various useful components from the complex mixture and to modify their properties.

Petroleum Refining

Distillation can separate substances that have different boiling points from liquid mixtures and solutions. For example, simple distillation can separate a mixture of the liquids cyclohexane (b.p. 80.7 °C) and toluene (b.p. 110.6 °C). The mixture is heated to slightly above 80.7 °C, at which point cyclohexane vaporizes from the mixture. The cyclohexane vapor is condensed as a pure liquid and collected in a separate container (Figure 12.1).

The difference between simple distillation and fractional distillation is the degree of separation achieved.

Petroleum fractions are each mixtures of hundreds of hydrocarbons with boiling points within a certain range.

Petroleum is a much more complex mixture. Petroleum contains thousands of different hydrocarbons and their separation as individual pure compounds is neither economically feasible nor necessary. Instead, petroleum refining uses *fractional distillation* to separate petroleum into **petroleum fractions.** Each petroleum fraction is a mixture of hundreds of hydrocarbons that have boiling points within certain ranges. Such separation is possible because the boiling point increases as the number of carbon atoms increases: Larger hydrocarbon molecules (larger number of electrons and greater polarizability) have greater noncovalent intermolecular forces and higher boiling points than smaller ones ***(⇐ p. 409).***

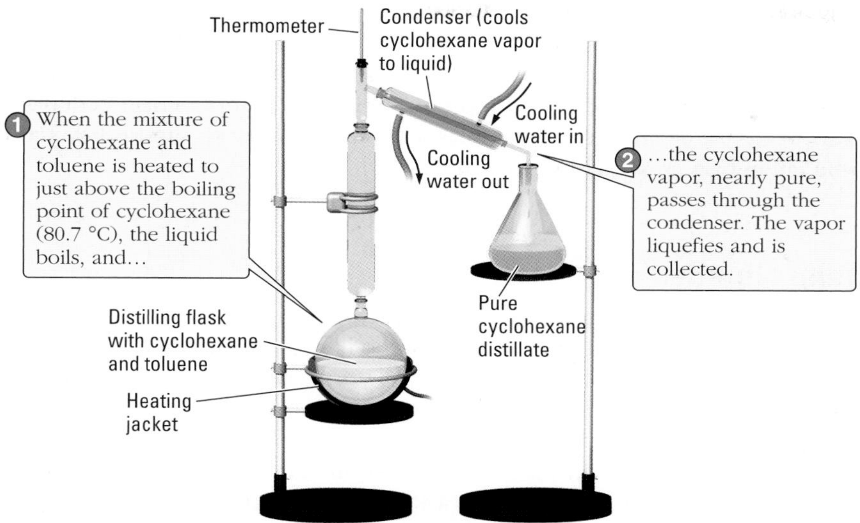
Thermometer
Condenser (cools cyclohexane vapor to liquid)
Cooling water in
Cooling water out
1 When the mixture of cyclohexane and toluene is heated to just above the boiling point of cyclohexane (80.7 °C), the liquid boils, and...
2 ...the cyclohexane vapor, nearly pure, passes through the condenser. The vapor liquefies and is collected.
Distilling flask with cyclohexane and toluene
Heating jacket
Pure cyclohexane distillate

Figure 12.1 **Distillation apparatus.**

Marathon Ashland Petroleum, LLC

A petroleum fractional distillation tower.

In fractional distillation, the crude oil is first heated to about 400 °C to produce a hot vapor mixture that enters the bottom of the fractionating tower (Figure 12.2). The temperature decreases as the vapor rises up the tower. Consequently, different gaseous components condense at various points in the tower. The more volatile, lower-boiling petroleum fractions remain in the vapor state longer than the less volatile, higher-boiling fractions. The smallest hydrocarbon molecules do not condense and are drawn off the top of the tower as gases. The largest hydrocarbon molecules do not vaporize even at 400 °C and are collected at the bottom of the tower as liquids and dissolved solids.

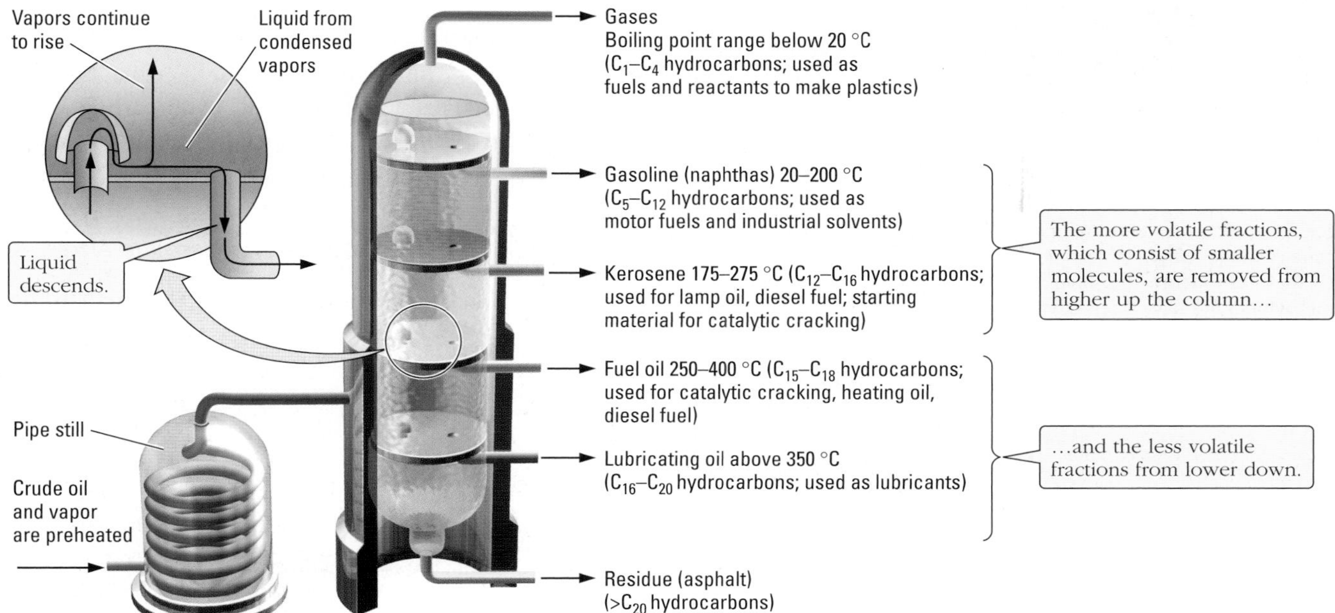
Vapors continue to rise
Liquid from condensed vapors
Liquid descends.
Pipe still
Crude oil and vapor are preheated
Gases
Boiling point range below 20 °C (C_1–C_4 hydrocarbons; used as fuels and reactants to make plastics)
Gasoline (naphthas) 20–200 °C (C_5–C_{12} hydrocarbons; used as motor fuels and industrial solvents)
Kerosene 175–275 °C (C_{12}–C_{16} hydrocarbons; used for lamp oil, diesel fuel; starting material for catalytic cracking)
Fuel oil 250–400 °C (C_{15}–C_{18} hydrocarbons; used for catalytic cracking, heating oil, diesel fuel)
Lubricating oil above 350 °C (C_{16}–C_{20} hydrocarbons; used as lubricants)
Residue (asphalt) (>C_{20} hydrocarbons)
The more volatile fractions, which consist of smaller molecules, are removed from higher up the column...
...and the less volatile fractions from lower down.

Figure 12.2 **Petroleum fractional distillation.** Crude oil is first heated to 400 °C in the pipe still. The vapors then enter the fractionation tower. As they rise, the vapors cool and condense so that different fractions can be removed at different heights in the tower.

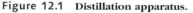

The properties and consequently the uses of various fractions differ, as shown in Figure 12.2. About 83% (35 gallons) of a refined barrel of crude oil (42 gallons) is simply burned for transportation and heating. The remaining 17% is used for nonfuel purposes, including the very important petroleum components needed as reactants to make the synthetic fabrics, plastics, pharmaceuticals, and other synthetic organic chemicals on which we depend.

Many years ago, petroleum was shipped in barrels. Today it seldom is seen in barrels, but rather is shipped in pipelines and ocean-going tankers. Nevertheless, the barrel remains as the common unit of volume measure for petroleum. A petroleum barrel contains 42 U.S. gallons.

Octane Number

The **octane number** of a gasoline is a measure of its ability to burn efficiently in an internal combustion engine. A typical automobile engine uses the gasoline fraction of refined petroleum—a mixture of C_5 to C_{12} hydrocarbons used primarily as motor fuels. This hydrocarbon mixture contains relatively small molecules and has a fairly high autoignition temperature, the temperature at which the liquid hydrocarbon will ignite and burn without a source of ignition. Thus, because of its fairly high ignition temperature, gasoline requires a source of ignition—a spark plug—to burn efficiently in an engine. Diesel engines do not have spark plugs, but use autoignition to ignite the fuel, so diesel fuel consists of larger hydrocarbon molecules.

The octane-number rating of a gasoline is determined by comparing the burning characteristics when the gasoline burns in a one-cylinder test engine with those obtained for mixtures of heptane and 2,2,4-trimethylpentane (often called isooctane). Heptane does not burn smoothly and is arbitrarily assigned an octane number of 0, whereas 2,2,4-trimethylpentane burns smoothly and is assigned an octane number of 100 (Table 12.1). A gasoline assigned an octane number of 87 has the same premature autoignition characteristics as a mixture of 13% heptane and 87% 2,2,4-trimethylpentane. This octane rating corresponds to the octane number of regular unleaded gasoline currently available in the United States. Other, higher grades of gasoline available at gas stations have octane numbers of 89 (regular plus) and 92 (premium).

Typical octane ratings of commercially available gasoline.

Straight-chain alkanes are less thermally stable and burn less smoothly than branched-chain alkanes. For example, the "straight-run" gasoline fraction obtained directly from the fractional distillation of petroleum is a poor motor fuel. It needs additional refining because it consists of primarily straight-chain hydrocarbons that autoignite too readily. The octane number of a gasoline can be increased either by increasing the percentage of branched-chain and aromatic hydrocarbon components or by adding octane enhancers. Table 12.1 lists octane numbers for some hydrocarbons and octane enhancers.

Since the method for determining octane numbers was established, fuels that are superior to 2,2,4-trimethylpentane have been developed and thus have octane numbers greater than 100. Note in Table 12.1 that octane burns even less smoothly than heptane.

Table 12.1 Octane Numbers of Some Hydrocarbons and Gasoline Additives

Name	Class of Compound	Octane Number
Octane	Alkane	−20
Heptane	Alkane	0
Hexane	Alkane	25
Pentane	Alkane	62
1-Pentene	Alkene	91
2,2,4-Trimethylpentane (isooctane)	Alkane	100
Benzene	Aromatic hydrocarbon	106
Methanol	Alcohol	107
Ethanol	Alcohol	108
Tertiary-butyl alcohol	Alcohol	113
Toluene	Aromatic hydrocarbon	118

In addition to fractional distillation, petroleum refining also includes converting the components of various fractions into more economically important products through catalytic cracking and catalytic reforming. Among these products are compounds that can be added to gasoline to increase its octane rating.

Catalytic Cracking

During petroleum refining, the percentage of each fraction collected is adjusted to match the market demand. For example, there is greater demand for gasoline than for kerosene and diesel fuel. Demands also vary seasonally. In winter, the need for home heating oil is high; in summer, when more people take vacations, demand for gasoline is higher. In summer, refiners use chemical reactions in a process called "cracking" to convert some of the larger, kerosene-fraction molecules (C_{12} to C_{16}) into smaller molecules in the gasoline range (C_5 to C_{12}). **Catalytic cracking** uses a catalyst, high temperatures, and pressure to break long-chain hydrocarbons into shorter-chain hydrocarbons that include alkanes and alkenes, many in the gasoline range. A **catalyst** is a substance that increases the rate of a chemical reaction without being consumed as a reactant would be. (The role of catalysts in chemical reactions is further discussed in Section 13.8.)

$$\underset{\text{An alkane}}{C_{16}H_{34}} \xrightarrow{\text{catalyst, pressure, and heat}} \underset{\text{An alkene}}{C_8H_{16}} + \underset{\text{An alkane}}{C_8H_{18}}$$

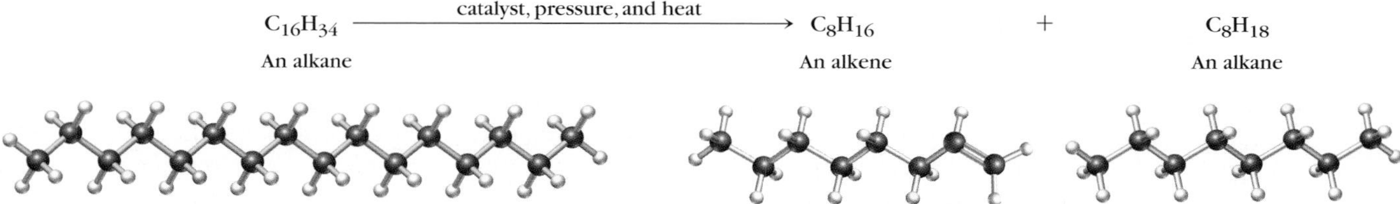

Since alkenes have higher octane numbers than alkanes, catalytic cracking also increases the octane number of the mixture. Catalytic cracking is beneficial in another way. Unlike alkanes, alkenes have C=C bonds, which makes them much more reactive than alkanes. Thus, as seen in Figure 12.3, alkenes such as ethylene and propylene can be used as starting materials to make other organic compounds, many of which are the raw materials for making plastics (Section 12.8).

Catalytic Reforming

After its separation from crude oil by fractional distillation, the gasoline fraction (straight-run gasoline) has an octane number of only 50, unsatisfactory for a motor vehicle fuel. The octane rating of gasoline can be increased by using **catalytic reforming** to convert straight-chain hydrocarbons to branched-chain hydrocarbons and aromatics. In catalytic reforming, certain catalysts, such as finely divided platinum on a support of Al_2O_3, reform straight-chain hydrocarbons with low octane numbers into their branched-chain isomers, which have higher octane numbers.

$$\underset{\substack{\text{pentane}\\ \text{62 octane}\\ (C_5H_{12})}}{CH_3CH_2CH_2CH_2CH_3} \xrightarrow{\text{catalyst}} \underset{\substack{\text{2-methylbutane}\\ \text{94 octane}\\ (C_5H_{12})}}{CH_3\overset{\displaystyle CH_3}{\overset{|}{C}}HCH_2CH_3}$$

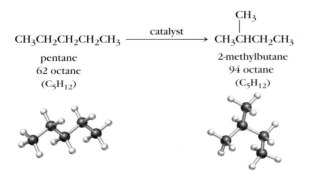

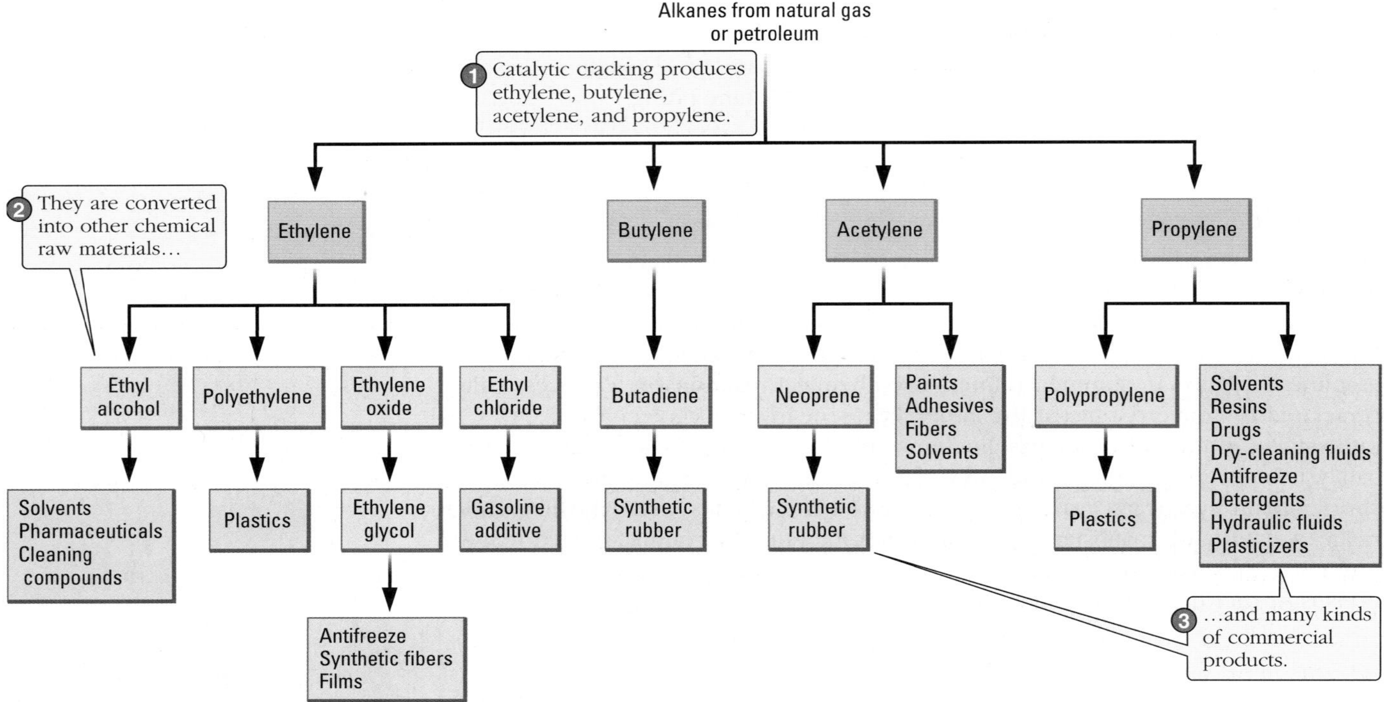

Figure 12.3 **Some chemical raw materials and commercial products made from petroleum or natural gas.**

The additional cost of higher-octane gasoline is due to the extra processing required to make higher-octane compounds.

By using different catalysts and petroleum mixtures, catalytic reforming also produces aromatic hydrocarbons, which have high octane numbers. For example, catalytic reforming converts a high percentage of straight-run gasoline, kerosene, and light oil fractions into a mixture of aromatic hydrocarbons including benzene, toluene, and xylenes. This process can be represented by the equation for converting hexane into benzene.

$$\underset{\substack{\text{hexane} \\ \text{25 octane} \\ (C_6H_{14})}}{CH_3CH_2CH_2CH_2CH_2CH_3} \xrightarrow{\text{catalyst}} \underset{\substack{\text{benzene} \\ \text{106 octane} \\ (C_6H_6)}}{C_6H_6} + 4\,H_2$$

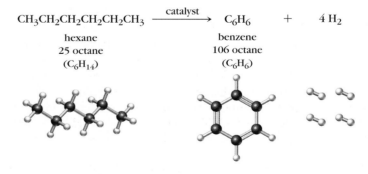

It is no exaggeration to state that our economy relies on the hydrocarbons present in petroleum or derived from it. In addition to being the source of our fuels for transportation, petroleum provides raw materials to make an impressive array of consumer products, as shown in Figure 12.3.

CONCEPTUAL EXERCISE 12.1 Rearranging Hydrocarbons

Heptane (C_7H_{16}) can be catalytically reformed to make toluene ($C_6H_5CH_3$), another seven-carbon molecule.

(a) How many hydrogen molecules are produced for every toluene molecule derived from heptane?
(b) Write a balanced chemical equation for this reaction.
(c) Why would it be profitable to convert heptane into toluene?

A catalytic converter. Semiprecious metals, like platinum and rhodium, coat the surface of ceramic beads in the catalytic converter. The metals catalyze the complete combustion of CO to carbon dioxide in the exhaust. Other metal catalysts in the converter speed up the rate at which nitrogen oxides in the exhaust are decomposed to nitrogen and oxygen.

Octane Enhancers

The octane number of a given blend of gasoline is increased by adding octane enhancers to prevent premature autoignition. In the United States, prior to 1975, the most widely used anti-knock agent was tetraethyllead, $(C_2H_5)_4Pb$.

The exhaust emissions of internal combustion engines contain carbon monoxide, nitrogen oxides, and unburned hydrocarbons, all of which contribute to air pollution. As urban air pollution worsened, Congress passed the Clean Air Act of 1970, which required that 1975-model-year cars emit no more than 10% of the carbon monoxide and hydrocarbons emitted by 1970 models. The solution to lowering these emissions was a platinum-based catalytic converter, which accelerates the conversion of carbon monoxide to carbon dioxide and the more complete burning of hydrocarbons. The only problem was that the catalyic converter required lead-free gasolines, since lead deactivates the platinum catalyst by coating its surface. As a result, automobiles manufactured since 1975 have been required to use lead-free gasoline to protect the catalytic converter. Because tetraethyllead can no longer be used in the United States and a few other countries, other octane enhancers are now added to gasoline. These include toluene, 2-methyl-2-propanol (also called *tertiary*-butyl alcohol), methanol, and ethanol.

Premature autoignition is also known as "knocking."

As little as two tanks of leaded gasoline can destroy the activity of a catalytic converter.

Regrettably, most countries of the world still allow the use of leaded gasoline, putting millions of children at risk for lead poisoning from automobile exhaust fumes.

Oxygenated and Reformulated Gasolines

The 1990 amendments to the Clean Air Act of 1970 require cities with excessive carbon monoxide pollution to use oxygenated gasoline during the winter. *Oxygenated gasoline* is a blend of gasoline to which organic compounds that contain oxygen, such as methanol, ethanol, and *tertiary*-butyl alcohol, have been added. Tests conducted by the Environmental Protection Agency (EPA) indicate that in cold weather oxygenated gasoline burns more completely than nonoxygenated gasoline, thus potentially reducing carbon monoxide emissions in urban areas by up to 17%.

Oxygenated gasoline is produced by adding oxygen-containing organic compounds to refined gasoline. Reformulated gasoline, which contains such oxygenates, also requires changes in the refining process to alter its percentages of various hydrocarbons, particularly alkenes and aromatics.

Reformulated gasoline (RFG) is oxygenated gasoline that contains a lower percentage of aromatic hydrocarbons and has a lower volatility than ordinary (nonoxygenated) gasoline. Reformulated gasoline is now used in 17 states and the District of Columbia. The 1990 regulations require RFG to contain a minimum of 2.7% oxygen by mass. Ethanol is the only gasoline oxygenate additive now used in RFG by U.S. refineries.

All gasolines are highly volatile and have vapors that can be ignited, allowing you to start your car even in the coldest of weather. However, this volatility means that some hydrocarbons get into the atmosphere as a result of accidental spills and evaporation during normal filling operations at the gas station, especially so during warm weather. Recent federal regulations require that the vapor pressure of RFG meet a limit of 9.0 psi from June 1 to September 15, the summer ozone season. More stringent requirements (7.8 psi) may apply in areas not meeting federal guidelines for air quality. Because atmospheric hydrocarbons play an important role in a series of reactions that contribute to urban air pollution (← p. 470) including tropospheric ozone formation (especially in heavy-traffic metropolitan areas), reduction of hydrocarbon emissions improves air quality. The EPA estimates that RPG use has reduced smog-forming emissions by 105 million tons, the equivalent of such emissions from 16 million vehicles.

EXERCISE 12.2 Percent Oxygen in Ethanol

Calculate the percent oxygen by mass in ethanol, CH_3CH_2OH. Is it less than 3%? How can ethanol help to meet the 2.7% oxygen requirement for RFG?

Engineering improvements such as emission control systems have succeeded in decreasing the emissions of ozone-forming nitrogen oxides and hydrocarbons from automobiles per mile traveled. Today's cars typically emit 70% less nitrogen oxides and 80–90% less hydrocarbons over their lifetimes compared with automobiles produced 30 to 40 years ago. Nevertheless, tropospheric ozone levels remain high for two reasons: (1) Since 1970 the number of vehicles in the United States has doubled, as has the number of miles they travel (Figure 12.4), and (2) The EPA attributes a major portion (approximately 25%) of ozone-forming hydrocarbons to a relatively small percentage of "super-dirty" vehicles with faulty emission control systems.

PROBLEM-SOLVING EXAMPLE 12.1 The Enthalpy of Combustion of Ethanol Compared to That of Octane

Calculate the enthalpy of combustion of ethanol in kilojoules per mole (kJ/mol) and compare the value with that of octane. Then, using the densities of the liquids, calculate the thermal energy liberated on burning a liter of each liquid fuel. The densities of octane and ethanol are 0.703 g/mL and 0.789 g/mL, respectively. The enthalpy of formation of octane(g) is −208.0 kJ/mol.

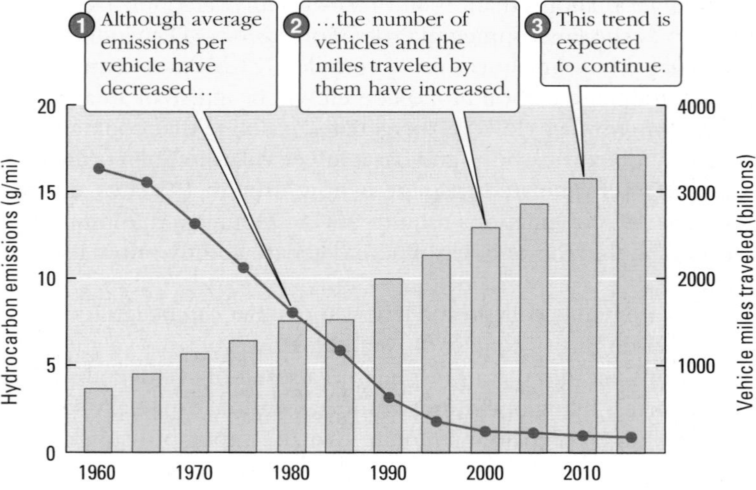

Figure 12.4 Average per-vehicle emissions (grams of hydrocarbons/mile) and mileage increases since 1960.

Answer

$$\text{Ethanol:} \quad \Delta H_{comb} = -1277.37 \text{ kJ/mol} = -2.19 \times 10^4 \text{ kJ/L}$$
$$\text{Octane:} \quad \Delta H_{comb} = -5116.4 \text{ kJ/mol} = -3.15 \times 10^4 \text{ kJ/L}$$

Strategy and Explanation All internal combustion engines burn fuel vapors. The balanced equation for the combustion of ethanol vapor is

$$C_2H_5OH(g) + 3\ O_2(g) \longrightarrow 2\ CO_2(g) + 3\ H_2O(g)$$

Using the standard molar enthalpies of formation from Appendix J and Hess's law (← *p. 246*) for both the products and the reactants, we find that the enthalpy of combustion, ΔH_{comb}, for ethanol is

$$\Delta H_{comb} = 2[\Delta H_f^\circ\ CO_2(g)] + 3[\Delta H_f^\circ H_2O(g)] - 1[\Delta H_f^\circ\ C_2H_5OH(g)]$$
$$= 2(-393.509 \text{ kJ/mol}) + 3(-241.818 \text{ kJ/mol}) - (-235.10 \text{ kJ/mol})$$
$$= -1277.37 \text{ kJ/mol}$$

The balanced combustion reaction for octane vapor is

$$C_8H_{18}(g) + \tfrac{25}{2}\ O_2(g) \longrightarrow 8\ CO_2(g) + 9\ H_2O(g)$$

Using Hess's law, the heat of combustion of octane is

$$\Delta H_{comb} = 8[\Delta H_f^\circ\ CO_2(g)] + 9[\Delta H_f^\circ\ H_2O(g)] - 1[\Delta H_f^\circ\ C_8H_{18}(g)]$$
$$= 8(-393.509 \text{ kJ/mol}) + 9(-241.818 \text{ kJ/mol}) - (-208.0 \text{ kJ/mol})$$
$$= -5116.4 \text{ kJ/mol}$$

The molar masses of ethanol and octane are 46.069 g/mol and 114.23 g/mol, respectively. The thermal energy liberated per liter for each is calculated as follows:

$$\text{Ethanol:} \quad -1277.37 \text{ kJ/mol} \times \frac{1 \text{ mol}}{46.069 \text{ g}} \times \frac{0.789 \text{ g}}{\text{mL}} \times \frac{1000 \text{ mL}}{\text{L}} = -2.19 \times 10^4 \text{ kJ/L}$$

$$\text{Octane:} \quad -5116.4 \text{ kJ/mol} \times \frac{1 \text{ mol}}{114.23 \text{ g}} \times \frac{0.703 \text{ g}}{\text{mL}} \times \frac{1000 \text{ mL}}{\text{L}} = -3.15 \times 10^4 \text{ kJ/L}$$

✓ **Reasonable Answer Check** There are about 17 mol ethanol in a liter of ethanol [(789 g ethanol/L ethanol)(1 mol ethanol/46 g ethanol) = 17 mol] and about 6 mol octane in a liter of octane [(703 g octane/L octane)(1 mol octane/114 g octane) = 6 mol]. But nearly four times as much energy is liberated by burning a mole of octane. Applying this factor on a liter-to-liter basis:

$$\frac{5000 \text{ kJ/mol octane}}{1200 \text{ kJ/mol ethanol}} \times \frac{6 \text{ mol octane/L}}{17 \text{ mol ethanol/L}} = \frac{1.5 \text{ kJ/L octane}}{1.0 \text{ kJ/L ethanol}} = 1.5$$

The 1.5 indicates that about 50% more energy is released by combustion of 1 L octane than from 1 L ethanol. This is close to the 44% difference in the energy per liter calculated for the two fuels, $(3.15 \times 10^4/2.19 \times 10^4) \times 100\% = 1.44\%$.

PROBLEM-SOLVING PRACTICE 12.1

Calculate the enthalpy of combustion of methanol and the energy released per liter using data from Appendix J. The density of methanol is 0.791 g/mL.

EXERCISE 12.3 Carbon Monoxide from Ethanol and from Toluene

(a) The combustion of C_2H_5OH can produce carbon monoxide and water. Use the balanced chemical equation for this reaction as the basis to calculate the mass in grams of CO produced per gram of ethanol burned.

(b) Repeat balancing the equation and the calculation this time using toluene, $C_6H_5CH_3$, instead of ethanol.

(c) What conclusions can you draw from your answers about using ethanol to reduce carbon monoxide emissions?

© AP Images

© David R. Frazier Photolibrary, Inc./Alamy

Using ethanol as an automotive fuel. The ethanol is mixed with gasoline.

© Peter McGahey

A natural gas–powered bus. Natural gas powers this bus, which can refuel at a natural gas pump.

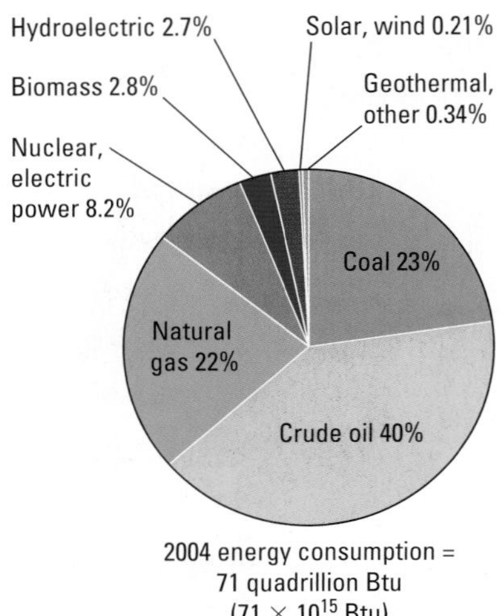

Figure 12.5 Fossil fuels predominate as sources of U.S. energy. *Source:* **EIA.DOE.GOV** Energy Information Administration

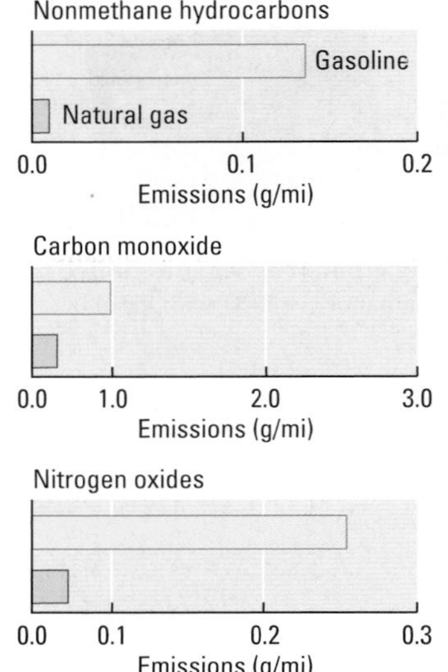

Figure 12.6 Comparisons of tailpipe emissions (grams per mile) from natural gas—and from gasoline-powered vehicles.

The mass percent of carbon in coal varies from about 40% to more than 80%, depending on the type of coal.

The usefulness of coal tar to make other materials is summed up in this bit of doggerel from a 1905 organic chemistry textbook:

You can make anything
From a salve to a star
If you only know how
From black coal tar*

*Cornish, *Organic Chemistry,* 1905.

12.2 Natural Gas and Coal

Natural Gas

Natural gas is a mixture of low-molar-mass hydrocarbons and other gases trapped with petroleum in the earth's crust. It can be recovered from oil wells or from gas wells to which the gases have migrated through the surrounding rock. The natural gas found in North America is a mixture of C_1 to C_4 alkanes [methane, CH_4 (60–90%); ethane, C_2H_6 (5–9%); propane, C_3H_8 (3–18%); and butane, C_4H_{10} (1–2%)] with small and varying amounts of other gases, such as CO_2, N_2, H_2S, and the noble gases, mainly helium. In Europe and Japan, natural gas is essentially only methane. The widespread use of natural gas as a fuel did not occur until a vast underground network of pipelines was built to distribute it from natural gas wells to consumers. Before then, most of it was simply burned at the wellheads. Oil, natural gas, and coal predominate as U.S. energy sources (Figure 12.5).

For more than three decades, the U.S. production of natural gas has exceeded that of U.S.-produced oil. Of all the natural gas produced, nearly half is used in industrial applications, 22% for home heating and cooling, 14% in commercial applications, and about 11% as a fuel to generate electricity. About half of the homes in the United States are heated by natural gas.

Natural gas is now also used as a vehicle fuel. Worldwide, nearly one million vehicles are powered by natural gas, including 30,000 in the United States. The U.S. Postal Service is the largest user of natural gas vehicles, with a fleet of more than 3000 such vehicles. California and several other states are encouraging the use of natural gas vehicles to help meet new air quality regulations. It has been shown that vehicles powered by natural gas emit significantly less carbon monoxide, hydrocarbons, and nitrogen oxides per mile compared to gasoline-powered cars and trucks (Figure 12.6). The main disadvantages of natural gas vehicles include the need for a pressurized gas tank and the lack of service stations that sell the gas in liquefied form.

CONCEPTUAL EXERCISE 12.4 Natural Gas as a CO Source

It has been asserted that burning natural gas produces 30% less carbon monoxide than the burning of gasoline. Assume that natural gas is only methane and that gasoline is only octane, C_8H_{18}. Based on the assumption that the same fraction of the carbon in each fuel is converted to CO, do calculations that either confirm or refute the assertion.

Coal

Known world coal reserves are far greater than petroleum reserves. About 90% of our annual coal production is burned to produce electricity. The use of coal as the fuel for commercial power plants is on the rise. But the burning of coal for home heating has declined because it is a relatively dirty fuel and bulky to handle. Burning coal is a major cause of air pollution because of coal's 3–4% sulfur content.

Coal consists of a complex and irregular array of partially hydrogenated six-membered carbon rings and other structures, some of which contain oxygen, nitrogen, and sulfur atoms. Like petroleum, coal supplies raw materials to the chemical industry. Most of the useful compounds obtained from coal are aromatic hydrocarbons. Heating coal at high temperatures in the absence of air, a process called *pyrolysis,* produces a mixture of coke (mostly carbon), coal tar, and coal gas. Fractional distillation of coal tar produces aromatic hydrocarbons such as benzene, toluene, and xylene. They serve as the starting materials for a large variety of important commercial products such as paints, solvents, pharmaceuticals, plastics, dyes, and synthetic detergents.

TOOLS OF CHEMISTRY

Gas Chromatography

Gas chromatography (GC) is an important method to separate, identify, and quantify chemical compounds in complex mixtures. Widely used in chemistry, biochemistry, environmental analysis, and forensic science, GC is applicable to compounds that have appreciable volatility and are thermally stable at temperatures up to several hundred degrees Celsius. We focus here on capillary column gas chromatography.

An analysis begins by injecting the sample into an inlet at the head of the GC column, where the sample is vaporized. A flow of inert carrier gas, generally helium, sweeps the vaporized sample along the column where the components are separated. The carrier gas stream containing the separated components passes through a detector, which allows identification of individual compounds in the gas stream. The resulting signal produces a *chromatogram*—a plot of concentration of each component versus time.

For capillary GC, the carrier gas usually flows at a rate in the range of 0.5 to 15 mL/min. A measured volume of sample (0.1 to 10 μL) is injected using a microsyringe. Capillary GC columns are made from specially purified silica tubes, typically 10 to 100 m long, that are drawn to have an inside diameter of 0.1 to 0.5 mm. The column is flexible, coiled, and put inside an oven for temperature control. The inside of the column tube is coated with an immobilized liquid, the *stationary phase.* Many factors determine the coating used to separate a particular mixture, but ultimately the choice is dictated by the molecular structure of the stationary liquid phase material in relation to that of the components being separated.

The detector must respond to very low concentrations of separated compounds in the flowing carrier gas. Two widely used detectors are (a) flame-ionization detectors, which atomize and ionize the sample and measure the flow of current that results, and (b) thermal conductivity detectors, which measure the thermal conductivity of the gas stream. Mass spectrometers ***(← p. 56)*** are also used as detectors. The resulting chromatogram, a series of peaks, is a plot of the concentration of each component as a function of time. The position of each peak on the time (x) axis qualitatively identifies the components in the sample; the area under each peak provides a quantitative measure of the amount of each compound.

Compounds in a mixture are separated by GC due to their interactions with the stationary liquid phase on the column. Molecular structure dictates how quickly each component moves along the column to the detector. This depends on the magnitude of intermolecular interactions—polar interactions and London forces ***(← p. 408)***—between components of the mixture and the stationary phase. Components having strong attractions for the stationary phase are retained longer on the column; those with weaker interactions reach the detector sooner. Thus, as the sample moves along the column, the components are progressively separated before they reach the detector.

If polar compounds are to be separated from a mixture, then a polar stationary phase works best, one that contains polar functional groups such as —CN, > C=O, and —OH. A nonpolar stationary phase contains hydrocarbon groups only. When the polarity of the stationary phase is matched well with that of the sample, the components exit from the column in order of their boiling points.

Capillary column GC can readily identify and measure components in gasoline. In the chromatogram below, thirty-one compounds ranging in size and complexity from butane and pentane to larger, ring compounds such as xylenes and toluene are separated from a gasoline sample.

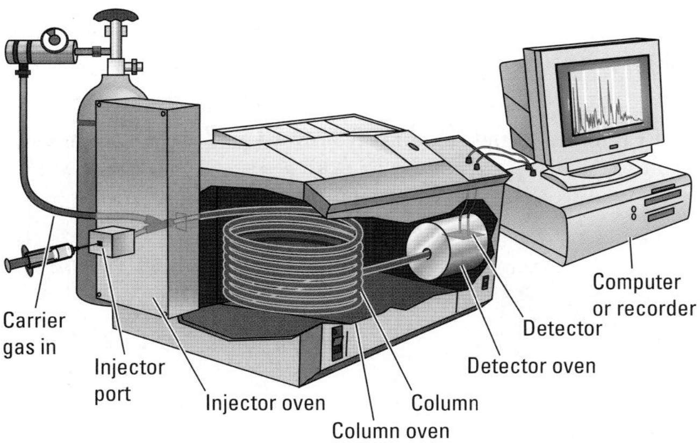

An illustration of a gas chromatograph.

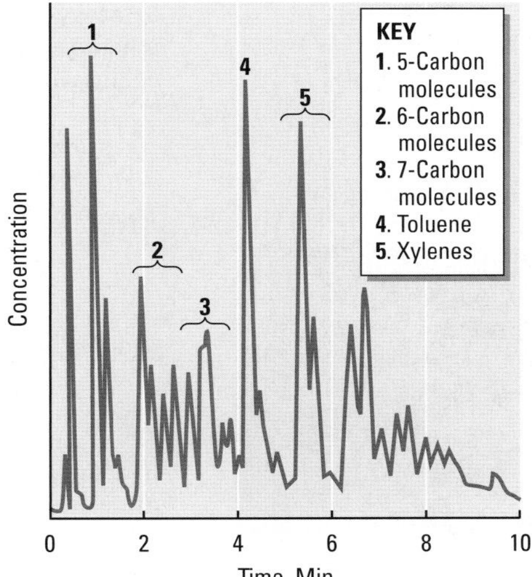

The chromatogram for a gasoline analysis. (Column: Petrocol A; Carrier: helium, 2.5 mL/min; Detector: flame ionization.) Peaks 1 are for components with five-carbon molecules, peaks 2 for six-carbon components, and peaks 3 for seven-carbon components. The last two peaks are for aromatic compounds—peak 4 indicates the presence of toluene, $C_6H_5CH_3$, and peak 5 the various xylenes, $C_6H_4(CH_3)_2$. These more complex aromatic hydrocarbons are held on the column longer than the nonaromatic hydrocarbons.

12.3 Energy Conversions

By now it should be apparent that our energy comes from a variety of sources. When you drive your car, the gasoline that moves it may have come from the United States or a foreign country. Some people are already driving cars powered by natural gas, electricity, or a combination of electricity and gasoline, as in hybrid vehicles. The natural gas may be imported and the electricity may have come from burning coal, natural gas, fuel oil, or even from a nuclear reactor (Section 20.6).

The convenience of use and the quantity of energy derived from different fuels vary depending on their chemical composition and properties. Two important measures of a fuel are its *fuel value* (quantity of energy released per gram of fuel burned) and its *energy density* (quantity of energy released per unit volume of fuel burned) ***(*** ***⇐ p. 255)******.*** Problem-Solving Example 12.2 gives you the opportunity to investigate the relationship between these two criteria.

PROBLEM-SOLVING EXAMPLE 12.2 Comparing Fuels

Evaluate each of the fuels listed below on the basis of fuel value and energy density. Use data from Table 6.2 or Appendix J. Which fuel provides the largest fuel value? Which provides the greatest energy density? Assume that when the fuels burn, carbon is converted to gaseous CO_2 and hydrogen to water vapor. The densities are given with the substances.

(a) Methane, $CH_4(g)$ (0.656 g/L)
(b) Ethanol, $C_8H_{18}(\ell)$ (0.789 g/mL)
(c) Hydrogen, $H_2(g)$ (0.082 g/L)

Answer

(a) 50.013 kJ/g CH_4, 3.28×10^1 kJ/L CH_4
(b) 26.807 kJ/g C_2H_5OH, 2.19×10^4 kJ/L C_2H_5OH
(c) 119.96 kJ/g H_2, 9.84 kJ/L H_2

Hydrogen has the highest fuel value; ethanol has the highest energy density.

Strategy and Explanation In each case, the balanced chemical equation must be written and then used in conjunction with the thermodynamic data to calculate the enthalpy change for the reaction. For methane:

$$CH_4(g) + 2\,O_2(g) \longrightarrow CO_2(g) + 2\,H_2O(g)$$

$$\Delta H^\circ = [(-393.509\text{ kJ}) + 2(-241.818\text{ kJ}) - (-74.81\text{ kJ})] = -802.34\text{ kJ}$$

Therefore, the fuel value of methane is

$$\frac{802.34\text{ kJ}}{1\text{ mol } CH_4} \times \frac{1\text{ mol } CH_4}{16.0426\text{ g } CH_4} = 50.013\text{ kJ/g } CH_4$$

and its energy density is

$$\frac{50.013\text{ kJ}}{1\text{ g } CH_4} \times \frac{0.656\text{ g } CH_4}{1\text{ L } CH_4} = 3.28 \times 10^1\text{ kJ/L } CH_4$$

In a similar manner the fuel values and energy densities for ethanol and hydrogen were calculated.

Energy efficiency comparison guide. Cost per kilowatt-hour of a refrigerator-freezer, fully self-defrosting.

PROBLEM-SOLVING PRACTICE 12.2

Evaluate each of the fuels listed below on the basis of fuel value and energy density. Use data from Appendix J. Which fuel provides the largest fuel value? Which provides the greatest energy density? Assume that when the fuels burn, carbon is converted to gaseous CO_2, hydrogen to water vapor, and nitrogen to N_2 gas. The densities are given with the substances.

(a) Octane, $C_8H_{18}(\ell)$ (0.703 g/mL)
(b) Hydrazine, $N_2H_4(\ell)$ (1.004 g/mL)
(c) Glucose, $C_6H_{12}O_6(s)$ (1.56 g/mL)

Table 12.2 A Chart of Energy Units*

Cubic Feet of Natural Gas	Barrels of Oil	Tons of Bituminous Coal	Kilowatt Hours of Electricity	Joules	Btu
1	0.00018	0.00004	0.293	1.055×10^6	1.00×10^3
1000	0.18	0.04	293	1.055×10^9	1.00×10^6
5556	1	0.22	1628	5.9×10^9	5.6×10^6
25,000	4.50	1	7326	26.4×10^9	25.0×10^7
1×10^6	180	40	293,000	1.055×10^{12}	1.00×10^9
3.41×10^6	614	137	1×10^6	3.6×10^{12}	3.4×10^9
1×10^9	180,000	40,000	293×10^6	1.055×10^{15}	1.00×10^{12}
1×10^{12}	180×10^6	40×10^6	293×10^9	1.055×10^{18}	1.00×10^{15}

*Based on normal fuel-heating values. 10^6 = 1 million; 10^9 = 1 billion; 10^{12} = 1 trillion; 10^{15} = quadrillion (quad).

1 Btu (British thermal unit) is the quantity of energy required to raise the temperature of 1 lb water from 63 °F to 64 °F; 1 Btu = 252 cal = 1.05×10^3 J.

All forms of energy are, in principle, interchangeable. The work a given quantity of energy can do is the same, no matter what the energy source. The energy contents of natural gas, fuel oil, and coal in terms of their equivalents in various units are compared in Table 12.2. As a consumer, it doesn't matter to you whether you travel using energy from a fossil fuel in the form of gasoline or whether you use electricity created by burning coal, although the energy is measured in different units. A trip might require 1 gal gasoline for the regular automobile. An electric automobile would require about 39 kilowatt-hours (kWh) of electricity to travel the same distance.

All fuel-burning engines, including automobile engines and electrical power plants, are less than 100% efficient; there is always wasted thermal energy. For example, the generation and distribution of electricity is only about 33% efficient overall. This means that for every 100 J energy produced by burning a fossil fuel such as coal, only 33 J electrical energy reaches the consumer.

One watt (W) of power is 1 J/s, so a 100-W electric lamp uses energy at the rate of 100 J every second. Using data from Table 12.2 we find that one cubic foot of natural gas supplies sufficient energy (1.055×10^6 J) to light a 100-W bulb for about 3 hours; a barrel of oil has enough energy (5.9×10^9 J) to light 1000 100-W light bulbs for almost 16 hours.

PROBLEM-SOLVING EXAMPLE 12.3 Energy Conversions

A large ocean-going tanker holds 1.5 million barrels (bbl) of crude oil.
(a) What is the energy equivalent of this oil in joules?
(b) How many tons of coal is this oil equivalent to?

Answer

(a) 8.9×10^{15} J
(b) 3.3×10^5 ton coal

Strategy and Explanation

(a) Table 12.2 gives the thermal energy equivalent to 1 barrel of oil as 5.9×10^9 J. Use this information as a conversion factor to calculate the first answer.

$$1.5 \times 10^6 \text{ bbl} \times \frac{5.9 \times 10^9 \text{ J}}{1 \text{ bbl}} = 8.9 \times 10^{15} \text{ J}$$

(b) Look in Table 12.2 and find the conversion between barrels of oil and tons of coal. One barrel of oil is equivalent to 0.22 ton coal. Use this conversion factor to calculate the answer.

$$1.5 \times 10^6 \text{ bbl} \times \frac{0.22 \text{ ton coal}}{1 \text{ bbl}} = 3.3 \times 10^5 \text{ ton coal}$$

✓ **Reasonable Answer Check** It takes only 0.22 ton coal (approximately one-fourth ton) to furnish the energy equivalent of a barrel of oil. Therefore, the coal equivalent of 1.5×10^6 barrels of oil should be about one fourth that number of barrels of oil, or about $(0.25)(1.5 \times 10^6) = 4 \times 10^5$, which is close to the calculated value.

PROBLEM-SOLVING PRACTICE 12.3

How much energy, in joules, can be obtained by burning 4.2×10^9 ton coal? This is equivalent to how many cubic feet of natural gas?

EXERCISE 12.5 Energy from Burning Oil

A large ocean-going oil tanker holds 1.5×10^6 barrels of crude oil. How many BTUs of energy are released by burning this oil? How many kilowatt-hours of electricity would be delivered to the consumer by burning this much oil, assuming a 33% overall efficiency?

CONCEPTUAL EXERCISE 12.6 The Energy Value of CO_2

Explain why CO_2 has no fuel energy value.

12.4 Organic Chemicals

Organic chemicals were once obtained only from plants, animals, and fossil fuels. A few living organisms are still direct sources of useful hydrocarbons. Rubber trees produce latex that contains rubber, a familiar hydrocarbon, and other plants produce an oil that burns almost as well in a diesel engine as diesel fuel. As important

ESTIMATION

Burning Coal

A coal-fired electric power plant burns about 1.5 million tons of coal a year. Coal has an approximate composition of $C_{135}H_{96}O_9NS$. When coal burns, it releases about 30 kJ/g. We can approximate how much energy (kJ) is released by this plant in a year and the mass of CO_2 released in that year by burning 1.5 million tons of coal.

The mass of coal is

$$1.5 \times 10^6 \text{ ton} \times \frac{2000 \text{ lb}}{1 \text{ ton}} \times \frac{1 \text{ kg}}{2.2 \text{ lb}} \approx 1 \times 10^9 \text{ kg}$$

This is approximately 1×10^9 kg, or 1×10^{12} g. Burning the coal at 30 kJ/g yields about 3×10^{13} kJ per year.

$$1 \times 10^{12} \text{ g} \times \frac{30 \text{ kJ}}{1 \text{ g}} = 3 \times 10^{13} \text{ kJ}$$

There are 135 mol carbon (1620 g C) in 1 mol coal (1906 g coal/mol). Thus, the ratio of grams of carbon to grams of coal is roughly 0.9 and so in 1×10^{12} g coal there is approximately $0.9 \times 1 \times 10^{12}$ g carbon, which is roughly 1×10^{12} g of carbon. When burned, each gram of carbon produces about 4 g CO_2 because there are 44 g CO_2 formed per 12 g C.

$$\frac{44 \text{ g } CO_2}{12 \text{ g C}} \approx 4 \text{ g } CO_2/\text{g C}$$

Thus, in one year the plant generates

$$(1 \times 10^{12} \text{ g C})\left(\frac{4 \text{ g } CO_2}{1 \text{ g C}}\right) = 4 \times 10^{12} \text{ g } CO_2$$

This is approximately the mass of 2 million SUVs.

ThomsonNOW™

Sign in to ThomsonNOW at **www.thomsonedu.com** to work an interactive module based on this material.

as these examples are, the development of synthetic organic chemistry has led to cheaper methods of making copies of naturally occurring substances and to many substances that have no counterpart in nature.

Prior to 1828, it was widely believed that chemical compounds found in living matter could not be made without living matter—a "vital force" was thought to be necessary for their synthesis. In 1828 a young German chemist, Friedrich Wöhler, dispelled the vital force theory when he prepared the organic compound urea, a major product in urine, by heating an aqueous solution of ammonium cyanate, a compound obtained from mineral sources.

$$\underset{\text{ammonium cyanate}}{[NH_4]^+[NCO]^-(aq)} \xrightarrow{\text{heat}} \underset{\text{urea}}{H_2N{-}C({=}O){-}NH_2}$$

Soon thereafter, other chemists began to prepare more and more organic chemicals without using a living system. Catalytic cracking of petroleum is very important for the production of small, unsaturated hydrocarbons such as ethylene ($CH_2{=}CH_2$), propylene ($CH_3CH{=}CH_2$), butylene ($CH_3CH_2CH{=}CH_2$), and acetylene ($HC{\equiv}CH$) (*p. 344*). These reactive molecules are starting materials in the organic chemical industry for the production of a substantial variety of substances we use daily. Aromatic compounds—such as benzene, toluene, and xylene—derived from coal tar and catalytic reforming are also important compounds from which a vast array of commercial products are produced.

Because about 85% of all known compounds are organic compounds, it is natural to ask: Why are there so many organic compounds, almost all of which contain carbon and hydrogen atoms (and commonly other kinds of atoms as well)? As discussed in earlier sections on bonding and isomerism (*p. 339, p. 345*), two reasons are (1) the ability of as many as thousands of carbon atoms to be linked to each other in a single molecule by stable C—C bonds, and (2) the occurrence of structural isomers. A third reason—the variety of functional groups that bond to carbon atoms—is further illustrated in this chapter. A **functional group** is a distinctive grouping of atoms that, as part of an organic molecule, imparts specific properties to the molecule. In addition to these three reasons, many molecules also display a very subtle type of molecular isomerism related to "handedness."

A table of functional groups and further information on how organic compounds are named is given in Appendix E.

The alcohol, aldehyde, ketone, carboxylic acid, ester, amine, and amide functional groups are discussed in Sections 12.5–12.8.

Chiral Molecules

Are you right-handed or left-handed? Regardless of the preference, we learn at a very early age that a right-handed glove doesn't fit the left hand, and vice versa. Our hands are mirror images of one another and are not superimposable (Figure 12.7). *An object that cannot be superimposed on its mirror image is called* **chiral.** Objects that are superimposable on their mirror images are **achiral.** Stop and think about the extent to which chirality is a part of our everyday life. We've already discussed the chirality of hands (and feet). Helical seashells are chiral (most spiral to the right like a right-handed screw), and many creeping vines show a chirality when they wind around a tree or post.

What is not as well known is that a large number of the molecules in plants and animals are chiral, and usually only one form (left-handed or right-handed) of the chiral molecule is found in nature. For example, all but one of the 20 amino acids from which proteins are made are chiral, and the left-handed amino acids predominate in nature! Most natural sugars are right-handed, including glucose and sucrose and deoxyribose, the sugar found in DNA.

A chiral molecule and its *nonsuperimposable* mirror image are called **enantiomers;** they are two different molecules, just as your left and right hands are

Figure 12.7 Nonsuperimposable mirror images. Right- and left-handed seashells are mirror images, as are the hands holding them.

In *A Midsummer Night's Dream*, Shakespeare wrote of the chirality of honeysuckle and woodbine plants. Queen Titania says to Bottom, ". . . Sleep thou, and I will wind thee in my arms . . . So doth the woodbine the sweet honeysuckle gently entwist. . . ." Woodbine spirals clockwise and honeysuckle twists in a counterclockwise direction.

In *Through the Looking Glass* (the companion volume to *Alice in Wonderland*), Alice speculates to her cat that ". . . perhaps looking glass milk is not good to drink."

different. The simplest case of chirality is a tetrahedral carbon atom that is bonded to four *different* atoms or groups of atoms. Such a carbon atom is said to be a *chiral center*, and a molecule with just one chiral center is always a chiral molecule.

Lactic acid is one of the compounds found in nature in both enantiomeric forms under different circumstances. During the contraction of muscles, the body produces only one enantiomer of lactic acid. The other enantiomer is produced when milk sours. Lactic acid has its central C atom bonded to four different groups: —CH_3, —OH, —H, and —COOH (Figure 12.8a). As a result of the tetrahedral geometry around the central carbon atom, it is possible to have two different arrangements of the four groups. If a lactic acid molecule is placed so that the C—H bond is vertical, as illustrated in Figure 12.8a, one possible arrangement of the remaining groups would be that in which —OH, —CH_3, and —COOH are attached in a clockwise sequence (Isomer I). Alternatively, these groups can be attached in a counterclockwise sequence (Isomer II), as in Figure 12.8b.

To illustrate further that the arrangements are different, we place Isomer I in front of a mirror (Figure 12.8c). Now you see that Isomer II is the mirror image of Isomer I. What is important, however, is that these mirror image molecules *cannot be superimposed* on one another. *These two nonsuperimposable, mirror image chiral molecules are enantiomers.*

The "handedness" of enantiomers is sometimes represented by D for right-handed (D stands for "dextro" from the Latin *dexter*, meaning "right") and L for left-handed (L stands for "levo" from the Latin *laevus*, meaning "left"). Nature's preference for one enantiomer of amino acids (the L form) has provoked much discussion and speculation among scientists since Louis Pasteur's discovery of molecular chirality in 1848. However, there is still no widely accepted explanation of this preference.

Large organic molecules may have a number of chiral carbon atoms within the same molecule. At each such carbon atom (a chiral center), two arrangements of the molecule are possible. The maximum number of possible isomers increases exponentially with the number of different chiral carbon atoms; with n different chiral carbon atoms there are a maximum of 2^n possible isomers. (The word "maximum" is used because in some cases a mirror image may not be different from the mirrored structure.) The widely used artificial sweetener aspartame (NutraSweet) has two chiral centers and two pairs of enantiomers. One enantiomer has a sweet taste, while another enantiomer is bitter, indicating that the receptor sites on our taste buds must be chiral, because they respond differently to the "handedness" of aspartame enantiomers! In another example, D-glucose is sweet and nutritious, while L-glucose is tasteless and cannot be metabolized by the body.

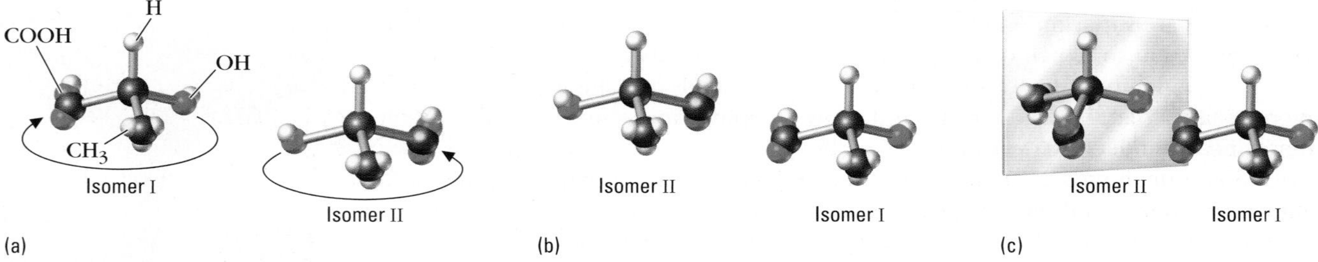

Figure 12.8 The enantiomers of lactic acid. (a) Isomer I: —OH, —CH_3, and —COOH, are attached in a clockwise manner. Isomer II: —OH, —CH_3, and —COOH are attached in a counterclockwise manner. (b) The isomers are nonsuperimposable. (c) Isomer I is placed in front of a mirror, and its mirror image is Isomer II. You could superimpose C, H, and CH_3, but then OH and COOH would not superimpose.

PROBLEM-SOLVING EXAMPLE 12.4 Handedness in Asparagine

Asparagine is a naturally occurring amino acid first isolated as a bitter-tasting white powder from asparagus juice. Later, a sweet-tasting second form of asparagine was isolated from a sprouting vetch plant. D-Asparagine is bitter; the L isomer is sweet. The Lewis structure of asparagine is

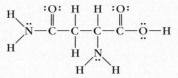

```
        :O:  H   H  :O:
H        ‖   |   |   ‖
  \ ..   |   |   |   |   ..
   N — C — C — C — C — O — H
  /          |   |       ..
H            H   N
                /..\
               H    H
```

Identify which carbon atom in asparagine is the chiral center (chiral carbon).

Answer

```
        :O:  H   H  :O:
H        ‖   |   |   ‖
  \ ..   |   |   |*  |   ..
   N — C — C — C — C — O — H      *Chiral
  /          |   |       ..        center
H            H   N
                /..\
               H    H
```

Strategy and Explanation Carefully examine the Lewis structure to identify a carbon atom with four different atoms or groups attached to it; that carbon is the chiral center. The carbon next to the —COOH group is the chiral center. It has four different atoms or groups of atoms attached to it: (1) a hydrogen atom, (2) a —COOH group, (3) an $—NH_2$ group, and (4) the rest of the molecule, starting with the $—CH_2—$ group.

PROBLEM-SOLVING PRACTICE 12.4

Aspartame, a widely used sugar substitute, is a chiral molecule with the structural formula

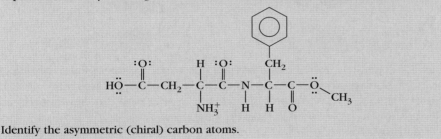

```
                                     (C6H5 ring)
                                          |
      :O:               H  :O:           CH2
       ‖                |   ‖             |
..     |                |   |             |        ..
HO — C — CH2 — C — C — N — C — C — O
..                      |        |    |    ‖    ..  \
                       NH3+      H    H    O         CH3
```

Identify the asymmetric (chiral) carbon atoms.

We turn next to compounds containing particular functional groups. Some of these very useful compounds are alcohols, acids, esters, and the natural and synthetic polymers made from them. These are discussed in Sections 12.5 through 12.8.

12.5 Alcohols and Their Oxidation Products

Alcohols are a major class of organic compounds. All alcohols, both natural and synthetic, contain the characteristic —OH functional group bonded directly to a carbon atom. Some alcohols contain more than one —OH group. Examples of commercially important alcohols and their uses are listed in Table 12.3.

ThomsonNOW™
Sign in to ThomsonNOW at **www.thomsonedu.com** to access a set of interactive molecular models including over 200 models of organic compounds organized by **functional group.**

Methanol and Ethanol

Methanol, CH_3OH, the simplest of all alcohols, has just one carbon atom. Because methanol is so useful in making other products, more than 8 billion pounds of it are

Table 12.3 Some Important Alcohols

Condensed Formula	b.p. (°C)	Systematic Name	Common Name	Use
CH_3OH	65.0	Methanol	Methyl alcohol	Fuel, gasoline additive, making formaldehyde
CH_3CH_2OH	78.5	Ethanol	Ethyl alcohol	Beverages, gasoline additive, solvent
$CH_3CH_2CH_2OH$	97.4	1-propanol	Propyl alcohol	Industrial solvent
$CH_3CH(OH)CH_3$	82.4	2-propanol	Isopropyl alcohol	Rubbing alcohol
$CH_2(OH)CH_2(OH)$	198	1,2-ethanediol	Ethylene glycol	Antifreeze
$CH_2(OH){-}CH(OH){-}CH_2(OH)$	290	1,2,3-propanetriol	Glycerol (glycerin)	Moisturizer in foods

produced annually in the United States by the reaction of carbon monoxide with hydrogen in the presence of a catalyst at 300 °C.

$$CO(g) + 2\,H_2(g) \xrightarrow{\text{catalyst, 300 °C}} CH_3OH(g)$$

 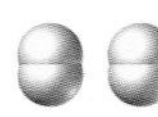 

Methanol is sometimes called *wood alcohol* because the old method of producing it was by heating a hardwood such as beech, hickory, maple, or birch in the absence of air. Methanol is highly toxic. Drinking as little as 30 mL can cause death, and smaller quantities (10 to 15 mL) can cause blindness.

About 50% of methanol is used in the production of formaldehyde (HCHO), which is used to make plastics, embalming fluid, germicides, and fungicides; 30% is used in the production of other chemicals; and the remaining 20% is used for jet fuels, antifreeze solvent mixtures, and gasoline additives.

ThomsonNOW
Go to the Coached Problems menu for simulations and tutorials on **naming alcohols.**

Because it can be made from coal, methanol will likely increase in importance as petroleum and natural gas become too expensive as sources of both energy and chemicals. Since methanol is relatively cheap, its potential as a fuel and as a starting material for the synthesis of other chemicals is receiving more attention. The technology for methanol-powered vehicles has existed for many years, particularly for racing cars that burn methanol because of its high octane rating (107). Methanol has both advantages and disadvantages as a replacement for gasoline. As a motor fuel, methanol burns more cleanly than gasoline, thus reducing levels of troublesome pollutants. Also, burning methanol emits no unburned hydrocarbons, which contribute significantly to air pollution. However, methanol has only about half the energy density of the same volume of gasoline. Therefore, twice as much methanol must be burned to give the same distance per tankful as gasoline. This disadvantage is partially compensated for by the fact that methanol costs about half as much to produce as gasoline. Another disadvantage is the tendency for methanol to corrode ordinary steel. Therefore, it is necessary to use stainless steel or a methanol-resistant coating for the fuel system.

Some racing cars burn methanol.

PROBLEM-SOLVING EXAMPLE 12.5 Methanol and MTBE

Methanol reacts with 2-methylpropene to yield methyl *tert*-butyl ether (MTBE), which was used as an octane enhancer and in reformulated gasolines.

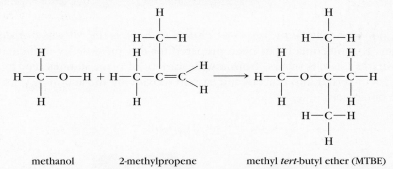

```
                          H                                  H
                          |                                  |
                        H—C—H                              H—C—H
       H                  H  |                        H      |   H
       |                  |  |     /H                 |      |   |
    H—C—O—H  +  H—C—C=C          ———→    H—C—O—C—C—H
       |                  |        \H                 |      |   |
       H                  H                           H      |   H
                                                           H—C—H
                                                             |
                                                             H

    methanol        2-methylpropene          methyl tert-butyl ether (MTBE)
```

Given these bond enthalpy data, estimate the enthalpy change, ΔH, for the synthesis of MTBE. Is the reaction endothermic or exothermic?

Bond	Bond Enthalpy (kJ/mol)
C—H	416
C—O	336
O—H	467
C—C	356
C=C	598

Answer -43.0 kJ; the reaction is exothermic.

Strategy and Explanation We can think of the reaction as being one in which bonds in the reactants are broken (endothermic process) and then bonds are formed in the products (exothermic). The balanced chemical equation for the reaction is given using Lewis structures, so we can directly count the number of moles of each kind of bond to be broken in the reactants and to be formed in the product. We then multiply these amounts by their respective bond enthalpies, D, to estimate the enthalpy change for the reaction (← ***Section 8.6***).

$$\begin{aligned}
\Delta H^\circ &= \textstyle\sum[(\text{moles of bonds}) \times D(\text{bonds broken})] \\
&\quad - \textstyle\sum[(\text{moles of bonds}) \times D(\text{bonds formed})] \\
&= \{[(3\text{ mol C—H})(416\text{ kJ/mol}) + (1\text{ mol C—O})(336\text{ kJ/mol}) \\
&\quad + (1\text{ mol O—H})(467\text{ kJ/mol})] + [(8\text{ mol C—H})(416\text{ kJ/mol}) \\
&\quad + (2\text{ mol C—C})(356\text{ kJ/mol}) + (1\text{ mol C=C})(598\text{ kJ/mol})]\} \\
&\quad - [(12\text{ C—H})(416\text{ kJ/mol}) + (2\text{ mol C—O})(336\text{ kJ/mol}) \\
&\quad + (3\text{ mol C—C})(356\text{ kJ/mol})] \\
&= (6689\text{ kJ}) - (6732) = -43.0\text{ kJ}
\end{aligned}$$

The negative sign indicates that the reaction is exothermic.

✓ **Reasonable Answer Check** The net change in the reaction is the conversion of 1 mol C=C bonds and 1 mol O—H bonds into 1 mol C—H bonds, 1 mol C—C bonds, and 1 mol C—O bonds. The other bonds are unchanged. The enthalpy change for the conversions is

$$\begin{aligned}
\Delta H^\circ &= [(1\text{ mol C=C})(598\text{ kJ/mol}) + (1\text{ mol O—H})(467\text{ kJ/mol})] \\
&\quad - [(1\text{ mol C—H})(416\text{ kJ/mol}) + (1\text{ mol C—C})\ (356\text{ kJ/mol}) \\
&\quad + (1\text{ mol (C—O})(336\text{ kJ/mol})]
\end{aligned}$$

which equals -43.0 kJ, the correct answer.

PROBLEM-SOLVING PRACTICE 12.5

Using Table 8.2 of Average Bond Enthalpies, estimate the enthalpy change for the industrial synthesis of methanol by the catalyzed reaction of carbon monoxide with hydrogen.

Ethanol, also called *ethyl alcohol* or *grain alcohol,* is the "alcohol" of alcoholic beverages. For millennia it has been prepared for this purpose by fermentation of carbohydrates (starch, sugars) from a wide variety of plant sources. For example, glucose is converted into ethanol and carbon dioxide by the action of yeast in the absence of oxygen.

$$C_6H_{12}O_6 \xrightarrow{\text{Yeast}} 2\ C_2H_5OH + 2\ CO_2$$

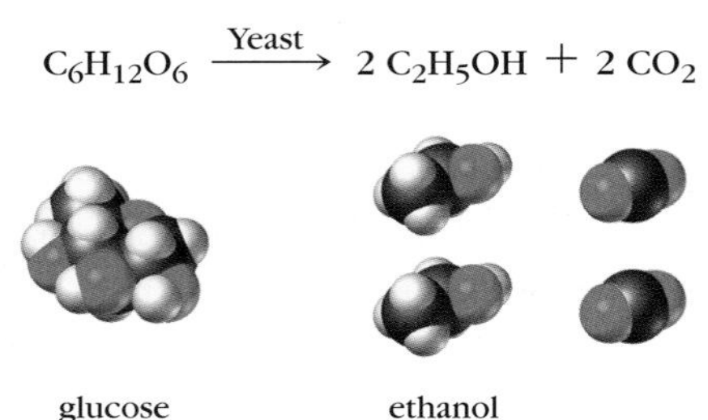

glucose ethanol

Alcohol concentration in wine.

Fermentation eventually stops when the alcohol concentration reaches a level sufficient to inhibit the yeast cells. The "proof" of an alcoholic beverage is twice the volume percent of ethanol; 80-proof vodka, for example, contains 40% ethanol by volume. Although ethanol is not as toxic as methanol, one pint of pure ethanol, rapidly ingested, will kill most people. Ethanol is a depressant; the effects of various blood levels of alcohol are shown in Table 12.4. Rapid consumption of two 1-oz "shots" of 90-proof whiskey, two 12-oz beers, or two 4-oz glasses of wine can cause one's blood alcohol level to reach 0.05%.

In the United States, the federal tax on alcoholic beverages is about \$13 per gallon, which is about 20 cents per ounce of ethanol in the beverage. Since the cost of producing ethanol is only about \$1 per gallon, ethanol intended for industrial use must be *denatured* to avoid the beverage tax and to prevent people from drinking the alcohol. Ethanol is denatured by adding to it small amounts of a toxic substance, such as methanol or gasoline, that cannot be removed easily by chemical or physical means.

Ethanol, like methanol, is receiving increased attention as an alternative fuel. At present, most of it is used in a blend of 90% gasoline and 10% ethanol (known as *gasohol* when introduced in the 1970s). Ethanol is also used as an oxygenated fuel to add to gasoline. The fermentation of corn, an abundant (but not limitless) source of carbohydrate, is used to produce ethanol for fuel use.

Ninety-five percent ethanol is the maximum ethanol concentration that can be obtained by distillation of alcohol/water mixtures because ethanol and water form a mixture that boils without changing composition.

Ethanol as a vehicle fuel is also available as Ethanol85, a mixture of 85% ethanol and 15% gasoline. There are now more than four million FlexFuel vehicles in the United States using gasohol and Ethanol85 in conjunction with regular gasoline.

Table 12.4 Blood Alcohol Levels and Their Effects

% by Volume	Effect
0.05–0.15	Lack of coordination
0.08	Commonly defined point for "driving while intoxicated"
0.15–0.20	Obvious intoxication
0.30–0.40	Unconsciousness
0.50	Possible death

Hydrogen Bonding in Alcohols

The physical properties of liquid alcohols are a direct result of the effects of hydrogen bonding among alcohol molecules. Alcohols can be considered as compounds in which alkyl groups such as CH_3— and CH_3CH_2— are substituted for one of the hydrogens of water. The change from the hydrogen in water to the alkyl group in an alcohol reduces the influence of hydrogen bonding in the progression from water (HOH) to methanol (CH_3OH) to ethanol (CH_3CH_2OH) and higher-molar-mass alcohols. The boiling points of methanol (32 g/mol; 65 °C) and ethanol (46 g/mol; 78.5 °C) are lower than that of water (18 g/mol; 100 °C) because methanol and ethanol have only one —OH hydrogen atom available for hydrogen bonding; water has two. The higher boiling point of ethylene glycol (198 °C) can be attributed to the presence of two —OH groups per molecule. Glycerol, with three —OH groups, has an even higher boiling point (290 °C). Hydrogen bonding among ethanol molecules explains why it is a liquid, while propane, which has a similar number of electrons (and molar mass) but no hydrogen bonding, is a gas at the same temperature. The alcohols listed in Table 12.3 are very water-soluble because of hydrogen bonding between water molecules and the —OH alcohol group.

See Table 12.3 (p. 562) for the structural formulas of these compounds.

CONCEPTUAL EXERCISE 12.7 Water Solubility of Alcohols

What would you expect concerning the water solubility of an alcohol containing ten carbon atoms and one —OH group? Recall that hydrocarbons such as octane are not water-soluble.

CONCEPTUAL EXERCISE 12.8 One Alcohol in Another

Methanol dissolves in glycerol. Use Lewis structures to illustrate the hydrogen bonding between methanol and glycerol molecules.

Oxidation of Alcohols

Alcohols are classified according to the number of carbon atoms *directly* bonded to the —C—OH carbon. If no carbon atom or one carbon atom is bonded directly, the compound is a *primary* alcohol. If two are bonded directly, it is a *secondary* alcohol; with three it is a *tertiary* alcohol, as illustrated below. The use of R, R′, and R″ to represent alkyl groups indicates that the alkyl groups can be different.

ThomsonNOW Go to the Coached Problems menu for a simulation on **reactivity of alcohols.**

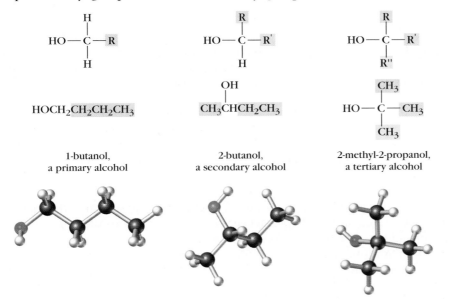

HO—C(H)(H)—R	HO—C(R)(H)—R′	HO—C(R)(R″)—R′
$HOCH_2CH_2CH_2CH_3$	$CH_3CH(OH)CH_2CH_3$	HO—C(CH_3)(CH_3)—CH_3
1-butanol, a primary alcohol	2-butanol, a secondary alcohol	2-methyl-2-propanol, a tertiary alcohol

In naming alcohols, the longest chain of carbon atoms is numbered so that the carbon with the —OH attached has the lowest possible number. A table of functional groups and further information on how organic compounds are named is given in Appendix E.

Stepwise oxidation of primary alcohols produces compounds called aldehydes, which are then oxidized to carboxylic acids.

$$\text{Primary alcohol} \longrightarrow \text{aldehyde} \longrightarrow \text{carboxylic acid}$$

Oxidation of organic compounds generally results from either the loss of two hydrogen atoms or the gain of one oxygen atom. For example, the stepwise oxidation of

Vinegar is a 5% solution of acetic acid in water.

Oxidation of organic compounds is usually the removal of two hydrogens or the addition of one oxygen; reduction is usually the addition of two hydrogens or the removal of one oxygen.

Functional Group Class	Functional Group	Example
Aldehyde	$-\overset{O}{\overset{\|}{C}}H$	Formaldehyde $H-\overset{O}{\overset{\|}{C}}H$
Carboxylic acid	$-\overset{O}{\overset{\|}{C}}-OH$	Acetic acid $CH_3-\overset{O}{\overset{\|}{C}}-OH$
Ketone	$-\overset{O}{\overset{\|}{C}}-$	Acetone $CH_3-\overset{O}{\overset{\|}{C}}-CH_3$

ethanol with aqueous potassium permanganate as the oxidizing agent first produces acetaldehyde, a member of the **aldehyde** functional group class, which contains a —CHO group.

$$CH_3-\underset{H}{\overset{H}{C}}-OH \xrightarrow{KMnO_4(aq)} CH_3-C\overset{O}{\underset{H}{\lessgtr}} \xrightarrow{KMnO_4(aq)} CH_3-C\overset{O}{\underset{OH}{\lessgtr}}$$

Aldehyde group

Carboxylic acid group

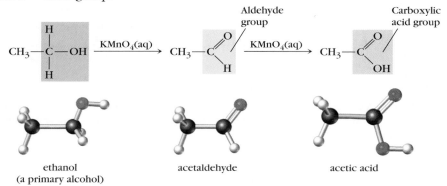

ethanol (a primary alcohol) acetaldehyde acetic acid

The acetaldehyde is then oxidized to acetic acid, a member of the **carboxylic acid** functional group class, which contains the —COOH group (also represented as $-CO_2H$). When ethanol is ingested, enzymes in the liver produce the same products. Acetaldehyde, the intermediate product, contributes to the toxic effects of alcoholism. In the presence of oxygen in air, ethanol in wine is oxidized naturally to acetic acid, converting the wine from a beverage into something more suitable for a salad dressing.

> **CONCEPTUAL EXERCISE 12.9 Looking at the Oxidation of Primary Alcohols**
>
> Look carefully at the formulas for ethanol and acetaldehyde. Explain the change in structure when ethanol is oxidized to acetaldehyde. Do the same for the oxidation of acetaldehyde to acetic acid. How do your explanations differ for the different products?

Alcohols, aldehydes, carboxylic acids, and ketones are important biologically. Oxidation of *secondary* alcohols by the loss of two hydrogens produces **ketones,** which contain the C═O group. In the laboratory an acidic solution of potassium dichromate, $K_2Cr_2O_7$, is a common oxidizing agent for this reaction.

$$H-\underset{H}{\overset{H}{C}}-\underset{H}{\overset{OH}{C}}-\underset{H}{\overset{H}{C}}-H \xrightarrow{H^+/Cr_2O_7^{2-}} H-\underset{H}{\overset{H}{C}}-\overset{O}{\overset{\|}{C}}-\underset{H}{\overset{H}{C}}-H$$

Secondary alcohol ⟶ ketone

This type of reaction is important during strenuous exercise, when the lungs and circulatory system are unable to deliver sufficient oxygen to the muscles. Under these conditions, lactic acid builds up in muscle tissue, causing soreness and exhaustion. After the exercising is finished, enzymes help to oxidize the secondary alcohol group in lactic acid to a ketone group, thereby converting the lactic acid to pyruvic acid. Pyruvic acid is an important biological compound that is further metabolized to provide energy.

$$CH_3-\underset{H}{\overset{OH}{C}}-\overset{O}{\overset{\|}{C}}-OH \xrightarrow{-2\,H} CH_3-\overset{O}{\overset{\|}{C}}-\overset{O}{\overset{\|}{C}}-OH$$

lactic acid pyruvic acid

Tertiary alcohols, having no hydrogen atoms directly bonded to the carbon bearing the —OH group, are not oxidized easily.

PROBLEM-SOLVING EXAMPLE 12.6 Oxidation of Alcohols

Write the condensed structural formulas of the alcohols that can be oxidized to make these compounds:

(a) $CH_3CH_2CH_2C(=O)—H$ (b) $CH_3CH_2C(=O)—CH_3$ (c) $CH_3CH_2CH_2CH_2C(=O)—OH$

Answer

(a) $CH_3CH_2CH_2CH_2OH$ (b) $CH_3CH_2CH(OH)CH_3$ (c) $CH_3CH_2CH_2CH_2CH_2OH$

Strategy and Explanation An aldehyde or carboxylic acid group is formed by the oxidation of a primary alcohol in which the —OH group is at the end of the molecule. A ketone group is made by the oxidation of a secondary alcohol, a compound in which the —OH group is not on a terminal (end) carbon atom.

(a) The oxidation of the four-carbon primary alcohol, 1-butanol, will produce this aldehyde.
(b) To produce this ketone, choose the secondary alcohol, 2-butanol, which has two carbons to the left and one carbon to the right of the C—OH group.
(c) The oxidation of the five-carbon primary alcohol, 1-pentanol, will produce this carboxylic acid.

PROBLEM-SOLVING PRACTICE 12.6

Draw the structural formulas of the expected oxidation products of these compounds.

(a) $CH_3CH_2CH_2OH$ (b) $CH_3CH(OH)CH_2CH_3$

CONCEPTUAL EXERCISE 12.10 Aldehydes and Combustion Products

Write the equation for the formation of an aldehyde by the oxidation of methanol. Critics of the use of methanol as a fuel or fuel oxygenate have cited the formation of this aldehyde as a major health threat. Use the Internet to find some of the toxic properties of the aldehyde.

CONCEPTUAL EXERCISE 12.11 Working Backward

When oxidized, a certain alcohol yields an aldehyde containing three carbon atoms. Further oxidation forms a three-carbon acid with the acid functional group located on the number one carbon of the chain. What are the structural formula and the name of the alcohol?

Biologically Important Compounds Containing Alcohol Groups

Many natural organic compounds are cyclic. They can be considered to be derivatives of hydrocarbons that consist of aromatic rings (⇐ ***p. 363***), cycloalkane (⇐ ***p. 340***), or fused cycloalkene rings. In many of these compounds, two or more rings are fused, that is, the rings share carbon atoms. Steroids, all of which have the four-ring structure shown next, are an example of these fused-ring structures.

Cycloalkenes are nonaromatic ring compounds that contain at least one C=C double bond.

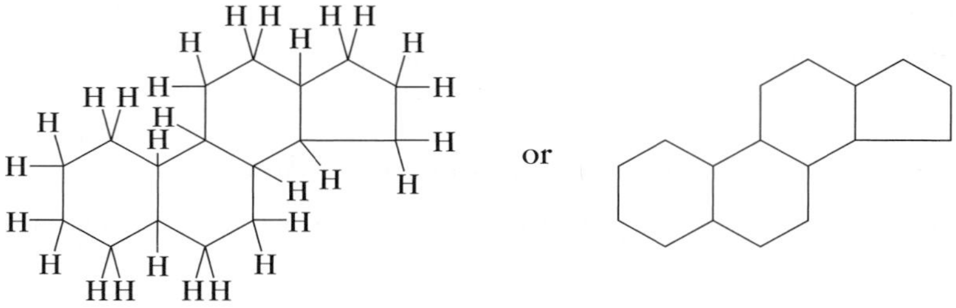

or

Many steroids—including cholesterol, the female sex hormones estradiol and estrone, and the male sex hormone testosterone—contain alcohol functional groups. Cholesterol is the most abundant steroid in the human body, where it serves as a major component of cell membranes and as the starting point for the production of the steroid hormones. The elongated, flat, rigid structure of cholesterol helps to make cell membranes sturdier. However, elevated levels of cholesterol in the blood are associated with *atherosclerosis,* a thickening of arterial walls that can lead to medical problems such as strokes and heart attacks.

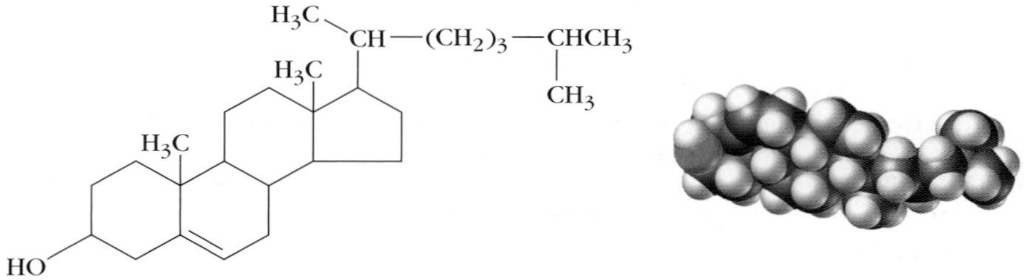

cholesterol

The presence and position of the C=C bond and the presence of the aromatic ring differ among steroids, as seen from the structural formulas of cholesterol, estradiol, and testosterone. Estradiol and estrone are female sex hormones responsible for the development of secondary sexual characteristics during puberty; testosterone is the male counterpart. From their structural formulas you can see that steroids are large molecules made up predominantly of carbon and hydrogen. Thus, steroids are not soluble in water.

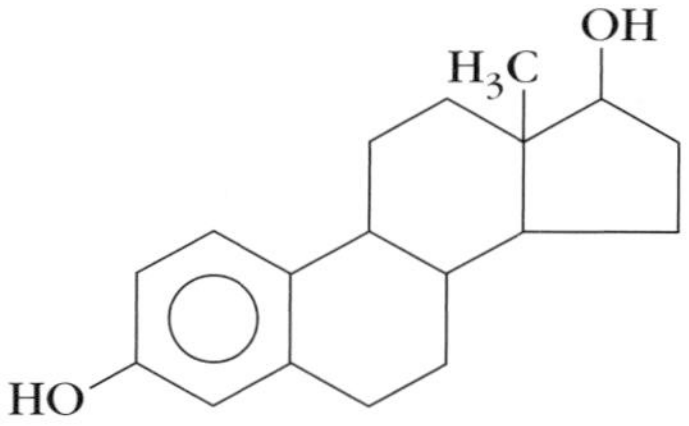

estradiol (female hormone, an estrogen)

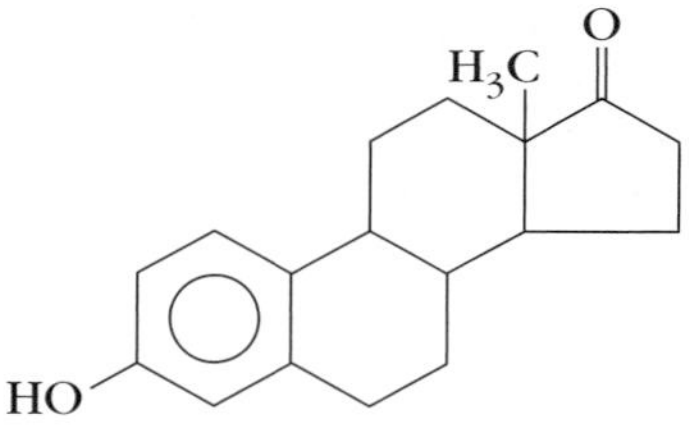

estrone (female hormone, an estrogen)

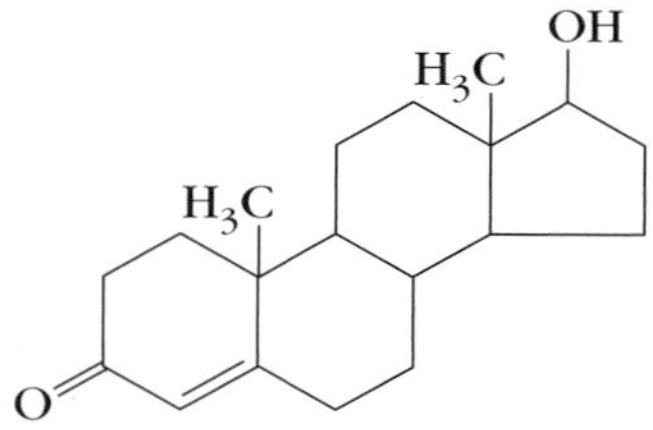

testosterone (male hormone)

CONCEPTUAL EXERCISE 12.12 A Closer Look

Look carefully at the structures for estradiol and estrone.

(a) What does the *-diol* suffix indicate in estradiol?

(b) What kind of alcohol group is on the five-membered ring in estradiol?

(c) What process converts the alcohol group of the five-membered ring in estradiol to the functional group present in that same ring in estrone?

(d) Write a description of the differences in the molecular structures between the male and female sex hormones, testosterone and estradiol, respectively.

EXERCISE 12.13 Take Me Out to the Ball Game

Home run slugger Barry Bonds and other athletes have been accused of using synthetic anabolic steroids to enhance their athletic performance. Androstenedione is such a steroid; its molecular structure is

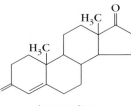

androstenedione

Compare it with the molecular structure of testosterone. In the body, what simple chemical process converts androstenedione to testosterone?

DePauw University Archives

Percy Lavon Julian
1899–1975

After receiving a doctorate in chemistry in Vienna in 1931, Percy Julian spent 18 years as a research director in the chemical industry, and then directed his own research institute. He was granted more than 100 patents and was the first to synthesize hydrocortisone (a steroid) and to isolate from soybean oil the compounds used to make the first synthetic sex hormone (progesterone). These amazing accomplishments came from a person who had to leave his home in Montgomery, Alabama, after eighth grade for further studies—no more public education was available there for an African American man. Julian enrolled as a "subfreshman" at DePauw University in Indiana. On his first day, a white student welcomed him with a handshake. Julian later related his reaction: "In the shake of a hand my life was changed. I soon learned to smile and act like I believed they all liked me, whether they wanted to or not."

12.6 Carboxylic Acids and Esters

Carboxylic Acids

Carboxylic acids contain the $-\overset{\overset{\large O}{\|}}{C}-OH$ ($-COOH$ or $-CO_2H$) functional group and are prepared by the oxidation of aldehydes or the complete oxidation of primary alcohols (p. 565). All carboxylic acids react with bases to form salts; for example, sodium lactate is formed by the neutralization reaction of lactic acid and sodium hydroxide.

$$\underset{\text{lactic acid}}{CH_3CH(OH)-\overset{\overset{O}{\|}}{C}-OH(aq)} + NaOH(aq) \longrightarrow \underset{\text{sodium lactate}}{Na^+\left[CH_3-CH(OH)-\overset{\overset{O}{\|}}{C}-O\right]^-(aq)} + H_2O(\ell)$$

Carboxylic acid molecules are polar and form hydrogen bonds with each other. This hydrogen bonding results in relatively high boiling points for the acids—even higher than those of alcohols of comparable molecular size. For example, formic acid (46 g/mol) has a boiling point of 101 °C, while ethanol (46 g/mol) has a boiling point of only 78.5 °C. Both formic acid and ethanol form hydrogen bonds, but those in formic acid are stronger.

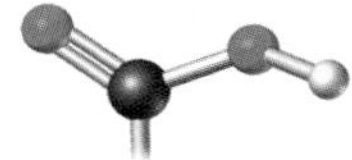

carboxylic acid group

The carboxylic acid group is present in all organic acids. Acids of all kinds are an important class of compounds, as discussed in Chapter 16.

EXERCISE 12.14 Hydrogen Bonding in Formic Acid

Use the structural formula of formic acid given in Table 12.5 to illustrate hydrogen bonding in formic acid.

A large number of carboxylic acids are found in nature and have been known for many years. As a result, some of the familiar carboxylic acids are almost always referred to by their common names (Table 12.5).

Acids in foods such as citrus fruits, strawberries, and vinegar cause them to have a sour taste.

Three commercially important carboxylic acids—adipic acid, terephthalic acid, and phthalic acid—have two acid groups per molecule and are known as *dicarboxylic acids.* These three acids are used as starting materials to make vast quantities of synthetic polymers (Section 12.8).

In *Les Miserables,* Victor Hugo comments about naturally occurring acids when he writes, "Comrades, we will overthrow the government, as sure as there are fifteen acids intermediate between margaric acid and formic acid." Formic acid, the simplest organic acid, has just one carbon atom. Margaric acid, from which margarine is derived, has seventeen carbon atoms.

Table 12.5 Some Naturally Occurring Carboxylic Acids

Structure	Common Name	b.p. (°C)	Natural Source
$HCOOH$ ($\mathrm{H{-}C(=O)OH}$)	Formic acid	101	Ants
CH_3COOH ($\mathrm{CH_3{-}C(=O)OH}$)	Acetic acid	118	Fermented fruit
CH_3CH_2COOH ($\mathrm{CH_3CH_2{-}C(=O)OH}$)	Propionic acid	141	Dairy products
$C_6H_5{-}C(=O)OH$	Benzoic acid	250	Berries
		m.p. (°C)	
$\mathrm{HOOC{-}CH_2{-}C(OH)(COOH){-}CH_2{-}COOH}$	Citric acid	153	Citrus fruits
$\mathrm{HOOC{-}CH_2{-}CH(OH){-}COOH}$	Malic acid	131	Apples
$\mathrm{HOOC{-}CH(OH){-}CH(OH){-}COOH}$	Tartaric acid	168–170	Grape juice, wine

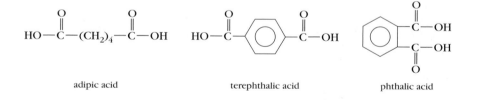

$$\mathrm{HO{-}C(=O){-}(CH_2)_4{-}C(=O){-}OH}$$
adipic acid

$$\mathrm{HO{-}C(=O){-}C_6H_4{-}C(=O){-}OH}$$
terephthalic acid

$$\mathrm{C_6H_4(C(=O){-}OH)_2}$$
phthalic acid

Esters

Carboxylic acids react with alcohols in the presence of strong acids (such as sulfuric acid) to produce **esters,** which contain the

$$\mathrm{{-}C(=O){-}O{-}R}$$

functional group. In an ester, the —OH of the carboxylic acid is replaced by the —OR group from the alcohol. The general equation for ester formation (esterification) is

$$\underset{\text{alcohol}}{\mathrm{R{-}OH}} + \underset{\text{organic acid}}{\mathrm{H{-}O{-}C(=O){-}R'}} \xrightarrow{H^+} \underset{\text{ester}}{\mathrm{R{-}O{-}C(=O){-}R'}} + \underset{\text{water}}{\mathrm{H_2O}}$$

Carboxylic acids in foods. These citrus fruits contain citric acid as well as other naturally occurring acids.

Esterification is an example of a condensation reaction. In a **condensation reaction,** two molecules combine to form a larger molecule (in esterification it is the ester) while simultaneously splitting out a small molecule (such as water). For example, just over a century ago, Felix Hoffmann, a chemist at the Bayer Chemical Company in Germany, synthesized aspirin, the world's most common pain reliever, by esterifying an alcohol group on salicylic acid with acetic acid in strong acid solution to form acetylsalicylic acid, which is aspirin. The ester group makes aspirin less irritating to the stomach lining than salicylic acid. The carboxylic acid group originally in salicylic acid remains intact in aspirin.

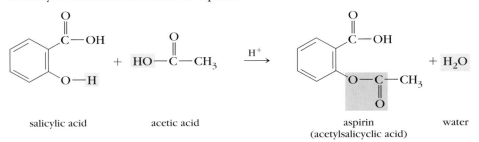

$$C_6H_4(COOH)(O{-}H) + HO{-}\overset{O}{\overset{\|}{C}}{-}CH_3 \xrightarrow{H^+} C_6H_4(COOH)(O{-}\underset{\underset{O}{\|}}{C}{-}CH_3) + H_2O$$

salicylic acid — acetic acid — aspirin (acetylsalicyclic acid) — water

The systematic name for asprin (acetylsalicylic acid) is 2-(acetyloxy)-benzoic acid.

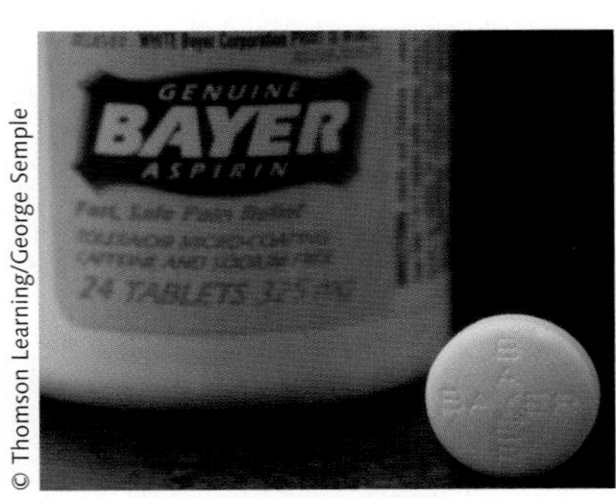

Aspirin (acetylsalicylic acid).

Triglycerides: Biologically Important Esters

One benefit of understanding organic molecules is in applying this understanding to the molecules involved in life—organic molecules that we eat, that provide the structure of our bodies, and that allow our bodies to function. Fats and oils are such molecules. *Fats* are solids at room temperature; *oils* are liquids at this temperature. Edible fats and oils are all **triglycerides** because they share the common structural feature of a three-carbon backbone from glycerol to which three long-chain fatty acids are bonded. Fatty acids generally have an even number of carbon atoms, ranging from 4 to 20. Triglycerides are formed by the esterification reaction of three moles of fatty acids with one mole of glycerol.

$$3 \text{ Fatty acid} + \text{glycerol} \longrightarrow \text{triglyceride} + 3\ H_2O$$

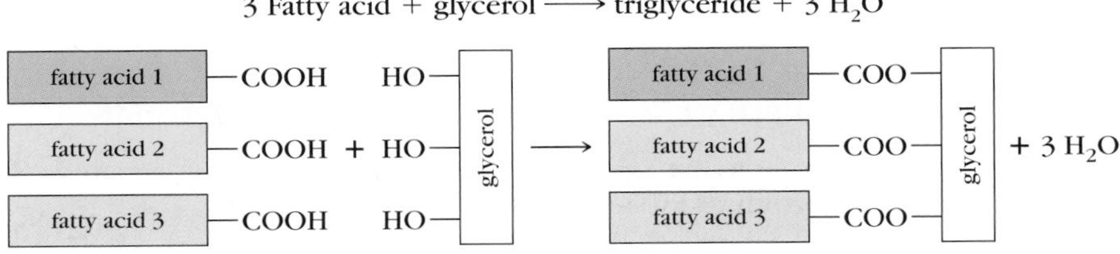

To illustrate this process, consider the reaction of a glycerol molecule with three molecules of stearic acid to produce tristearin, a very common animal fat. Stearic acid is representative of fatty acids, all of which are molecules with long chains of carbon atoms with their attached hydrogens and, at the end of the carbon chain, a —COOH *carboxylic acid* group characteristic of all organic acids. In tristearin formation, the carboxylic acid group in each of three molecules of stearic acid reacts with an —OH group (hydroxyl group) of a glycerol molecule. A water molecule is formed and released as each stearic acid molecule is joined to the glycerol molecule:

$$\text{Glycerol} + 3 \text{ stearic acid} \longrightarrow \text{tristearin} + 3 \text{ water}$$

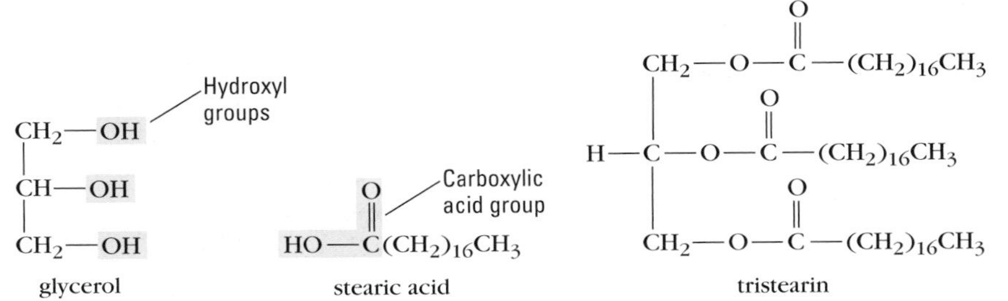

Cooking oils are liquids at room temperature. Liquid oils and solid fats differ in the connections among the carbon atoms within their molecular structures.

Linoleic and linolenic acids are essential fatty acids; we must have them in our diet. The body cannot manufacture them.

Animal and vegetable fats and oils vary considerably because their fatty acids differ in the length of the hydrocarbon chain and the number of double bonds in the chain. **Saturated fats,** such as stearic acid, contain only C—C single bonds in their hydrocarbon chains along with C—H bonds; **unsaturated fats** contain one or more C=C double bonds in their hydrocarbon chains. Oleic acid, with one C=C bond per molecule, is classified as a *monounsaturated* fatty acid, whereas fatty acids such as linoleic acid, with two or more C=C double bonds per molecule, are termed *polyunsaturated.*

The three fatty acids in a triglyceride can be the same (stearic acid in tristearin) or can be different such as stearic acid, oleic acid, and linoleic acid.

Fatty acid 1

$$HOOC{-}CH_2CH_2CH_2CH_2CH_2CH_2CH_2CH_2CH_2CH_2CH_2CH_2CH_2CH_2CH_2CH_2CH_3$$

stearic acid, a saturated fatty acid

Fatty acid 2

$$HOOC{-}(CH_2)_7{-}CH{=}CH(CH_2)_7{-}CH_3$$

oleic acid, a monounsaturated fatty acid

Fatty acid 3

$$HOOC{-}(CH_2)_7{-}CH{=}CH{-}CH_2{-}CH{=}CH{-}(CH_2)_4{-}CH_3$$

linoleic acid, a polyunsaturated fatty acid

The resulting triglyceride is

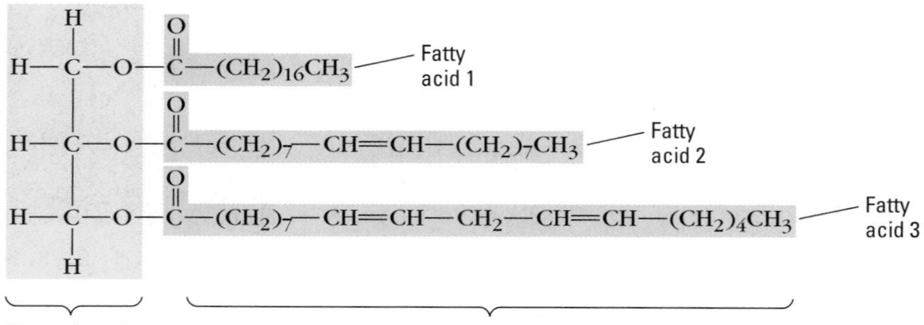

$$\begin{array}{l} H \\ H{-}C{-}O{-}\overset{O}{\overset{\|}{C}}{-}(CH_2)_{16}CH_3 \quad \text{Fatty acid 1} \\ H{-}C{-}O{-}\overset{O}{\overset{\|}{C}}{-}(CH_2)_7{-}CH{=}CH{-}(CH_2)_7CH_3 \quad \text{Fatty acid 2} \\ H{-}C{-}O{-}\overset{O}{\overset{\|}{C}}{-}(CH_2)_7{-}CH{=}CH{-}CH_2{-}CH{=}CH{-}(CH_2)_4CH_3 \quad \text{Fatty acid 3} \\ H \end{array}$$

Glycerol portion

Fatty acid portion

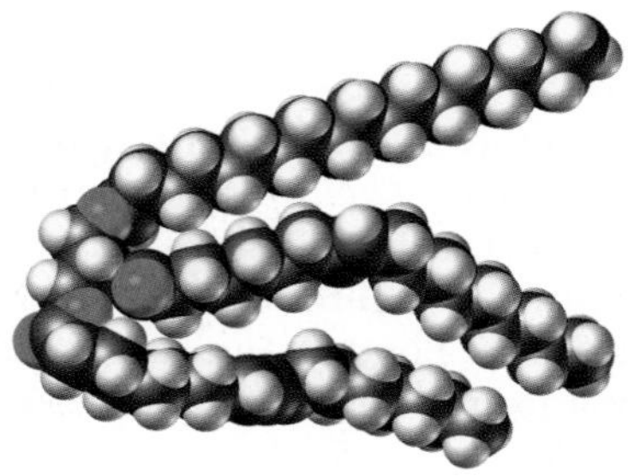

Triglyceride

PROBLEM-SOLVING EXAMPLE 12.7 Putting Together a Triglyceride

Using structural formulas, write the equation for the formation of a triglyceride by the reaction of 1 mol glycerol with 1 mol stearic acid and 2 mol oleic acid.

Answer

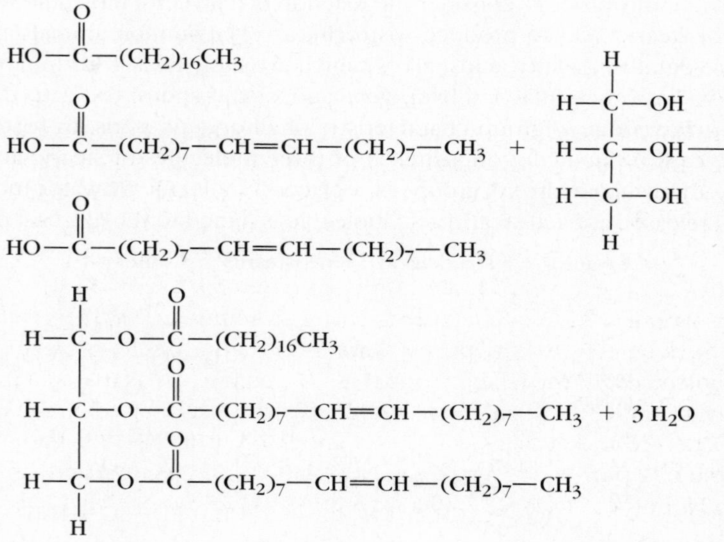

$$\begin{array}{l} HO{-}\overset{O}{\overset{\|}{C}}{-}(CH_2)_{16}CH_3 \\ HO{-}\overset{O}{\overset{\|}{C}}{-}(CH_2)_7{-}CH{=}CH{-}(CH_2)_7{-}CH_3 \\ HO{-}\overset{O}{\overset{\|}{C}}{-}(CH_2)_7{-}CH{=}CH{-}(CH_2)_7{-}CH_3 \end{array} + \begin{array}{l} \quad H \\ H{-}C{-}OH \\ H{-}C{-}OH \\ H{-}C{-}OH \\ \quad H \end{array} \longrightarrow$$

$$\begin{array}{l} \quad H \\ H{-}C{-}O{-}\overset{O}{\overset{\|}{C}}{-}(CH_2)_{16}CH_3 \\ H{-}C{-}O{-}\overset{O}{\overset{\|}{C}}{-}(CH_2)_7{-}CH{=}CH{-}(CH_2)_7{-}CH_3 \\ H{-}C{-}O{-}\overset{O}{\overset{\|}{C}}{-}(CH_2)_7{-}CH{=}CH{-}(CH_2)_7{-}CH_3 \\ \quad H \end{array} + 3\,H_2O$$

Strategy and Explanation Use the structural formulas given previously for the acids and glycerol and combine them through an esterification reaction. A water molecule is eliminated for each ester bond formed. Several different triglycerides are possible in this case depending on the position of the component fatty acids in the glycerol carbon chain. In the example shown in the answer, stearic acid is attached to the "first" (top) carbon in glycerol, with oleic acid attached to carbons 2 and 3. See Answer above.

PROBLEM-SOLVING PRACTICE 12.7

Use structural formulas to write the equation for the formation of a triglyceride by the reaction of 1 mol glycerol with 1 mol stearic acid and 2 mol oleic acid other than the triglyceride given in Problem-Solving Example 12.7.

Diets consisting of moderate amounts of fats and oils containing mono- and polyunsaturated fatty acids are considered better for good health than diets containing only saturated fats. In spite of this fact, there is a demand for solid or semisolid fats because of their texture and spreadability. The figure below indicates the relative proportions of saturated, monounsaturated, and polyunsaturated fats in some familiar cooking oils and solids.

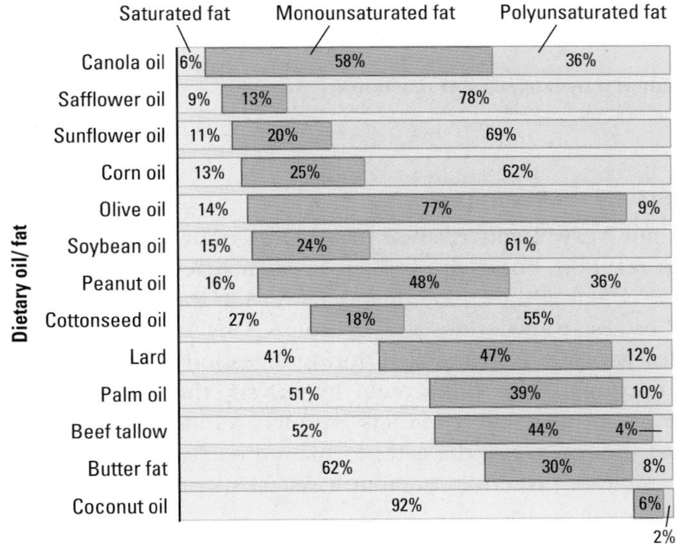

Saturated fats: No C=C

Monounsaturated fats: One C=C bond per molecule

Polyunsaturated fats: Several C=C bonds per molecule

The American Heart Association recommends that no more than 30% of dietary calories come from fat. Fats account for almost 40% of the calories in the average American's diet.

As seen from the diagram above, most dietary vegetable oils are high in unsaturated fatty acids, while most dietary animal fats contain mainly saturated fatty acids. In general, the greater the percentage of unsaturated fatty acids in a triglyceride, the lower its melting point. Thus, highly unsaturated fats are liquids at room temperature (oils); animal fats high in saturated fats are solids at that temperature. The melting points of fats reflect the shape of the molecules. The C—C single bonds in the hydrocarbon portion of saturated fats are all the same, and they allow the molecule to adopt a rather linear shape, which fits nicely into a solid packing arrangement. The C=C double bonds in unsaturated fats create a different shape. Most natural unsaturated fats are in the *cis* configuration at the C=C double bond ***(← Section 8.5)***, which puts a "kink" into the molecule, preventing unsaturated fat molecules from packing regularly into a solid. *Trans* fatty acid molecules are more linear, similar to those of saturated fats, which has raised health concerns about the widespread dietary use of foods containing *trans* fatty acids.

Vegetable oils can be converted to semisolids by *hydrogenation*—reaction of the oil with H_2 in the presence of a metal such as palladium, which accelerates the

Hydrogenation converts

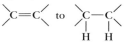

$$\rangle C{=}C\langle \text{ to } \rangle \underset{H}{C}{-}\underset{H}{C}\langle$$

reaction. The hydrogen reacts by addition to some of the C=C double bonds converting them to C—C single bonds as shown to the left. The degree of saturation in the partially hydrogenated product is limited by carefully controlling reaction conditions so that the product has the proper softness and spreadability. Margarine, shortening (for example, Crisco), and many snack foods contain partially hydrogenated fats, noted on the label as "partially hydrogenated soybean oil" (or cottonseed or other vegetable oils).

Hydrogenation: H_2 added to C=C bond

A portion of an unsaturated vegetable oil

$$\left\{ -O-\overset{O}{\overset{\|}{C}}-(CH_2)_7-\overset{H}{C}=\overset{H}{C}-CH_2-\overset{H}{C}=\overset{H}{C}-(CH_2)_4CH_3 \right\}$$

(H_2 added at each C=C)

$$\downarrow$$

A portion of a hydrogenated vegetable oil

$$\left\{ -O-\overset{O}{\overset{\|}{C}}-(CH_2)_7-\underset{H}{\overset{H}{C}}-\underset{H}{\overset{H}{C}}-CH_2-\underset{H}{\overset{H}{C}}-\underset{H}{\overset{H}{C}}-(CH_2)_4CH_3 \right\}$$

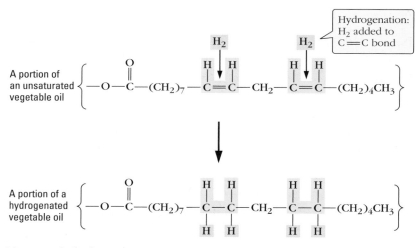

Diagram of a hydrogenation reaction.

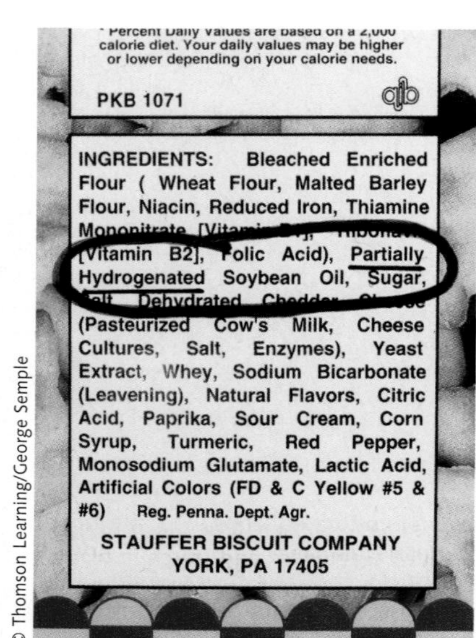

Partially hydrogenated fats. Most snack foods contain partially hydrogenated fats.

Hydrolysis of Esters

Esters are not very reactive; their most important reaction is hydrolysis. In a **hydrolysis** reaction, bonds are broken by their reaction with water, and the H— and —OH of water add to the atoms that were in the bond broken by hydrolysis. Like other esters, triglycerides are hydrolyzed by strong acid or aqueous base. Triglycerides we eat are hydrolyzed during digestion in the same way by enzyme-catalyzed reactions. With strong acid hydrolysis, the three ester linkages in the triglyceride are broken; the products are glycerol and three moles of fatty acids, which is just the reverse of the ester formation reaction. For example, acid hydrolysis of one mole of glycerol tristearate (tristearin) produces one mole of glycerol and three moles of stearic acid.

$$\begin{array}{l} H_2C-O-\overset{O}{\overset{\|}{C}}-(CH_2)_{16}CH_3 \\ H-C-O-\overset{O}{\overset{\|}{C}}-(CH_2)_{16}CH_3 \\ H_2C-O-\overset{O}{\overset{\|}{C}}-(CH_2)_{16}CH_3 \end{array} + 3\,H_2O \xrightarrow{H^+} \begin{array}{l} H_2C-OH \\ H-C-OH \\ H_2C-OH \end{array} + 3\,HO-\overset{O}{\overset{\|}{C}}-(CH_2)_{16}CH_3$$

glycerol tristearate (tristearin)   glycerol   stearic acid

During hydrolysis of a triglyceride with an aqueous base such as NaOH, glycerol is formed and the fatty acids released then react with the base, converting the fatty acids to their salts. In aqueous base, the hydrolysis process is called **saponification,** and the salts formed are called soaps. In biological systems, enzymes assist with saponification reactions that digest fats and oils. The saponification of one mole of glycerol tristearate forms one mole of glycerol and three moles of sodium stearate, a soap.

$$\begin{array}{l} H_2C-O-\overset{\overset{\large O}{\|}}{C}-(CH_2)_{16}CH_3 \\ \;\,| \\ HC-O-\overset{\overset{\large O}{\|}}{C}-(CH_2)_{16}CH_3 \\ \;\,| \\ H_2C-O-\overset{\overset{\large O}{\|}}{C}-(CH_2)_{16}CH_3 \end{array} + 3\ NaOH \xrightarrow{\text{heat}} \begin{array}{l} \quad H_2C-OH \\ \qquad | \\ H-C-OH \\ \qquad | \\ \quad H_2C-OH \end{array} + 3\ Na^{+\,-}O-\overset{\overset{\large O}{\|}}{C}-(CH_2)_{16}CH_3$$

glycerol tristearate, a fat — NaOH — glycerol — sodium stearate, a soap

Note that the saponification reaction requires three moles of base per mole of triglyceride.

EXERCISE 12.15 Triglyceride Structures

(a) Draw the structural formula for the triglyceride formed when glycerol reacts with linoleic acid. Circle the ester linkages in this triester molecule.

(b) Using structural formulas, write the equation for the hydrolysis in aqueous NaOH of the triglyceride formed in part (a).

12.7 Synthetic Organic Polymers

As important as simple organic molecules are, they are also used as starting materials to make an important class of compounds called polymers. **Polymers** (*poly,* "many"; *mer,* "part") are large molecules composed of smaller repeating units, usually arranged in a chain-like structure. For example, polyethylene consists of long chains of $-CH_2-CH_2-$ groups made from repeating ethylene units, $CH_2{=}CH_2$. Polymers occur in nature and are synthesized by chemists as well. It is virtually impossible for us to get through a day without using a dozen or more synthetic organic polymers. Today, synthetic polymers are used to make our clothes, package our foods, and build our computers, phones, and cars. You, a friend, or a family member may be alive because of a medical application of synthetic polymers. Synthetic polymers are so important that approximately 80% of the organic chemical industry is devoted to their production.

Nature makes many different polymers, including cellulose and starch in plants and proteins in both plants and animals.

The discussion in this section answers the question posed in Chapter 1 (← p. 3), "How are plastics made and why are there so many different kinds?"

There are two broad categories of the synthetic polymers commonly known as "plastics." One type, called **thermoplastics,** softens and flows when heated; when it is cooled, it hardens again. Common plastic materials that undergo such reversible changes when heated and cooled are polyethylene (milk jugs), polystyrene (inexpensive sunglasses and toys), and polycarbonates (CD audio discs). The other general type is **thermosetting plastics.** When first heated, a thermosetting plastic flows like a thermoplastic. When heated further, however, it forms a rigid structure that will not remelt. Bowling balls, football helmets, and some kitchen countertops are examples of thermosetting plastics.

Common household items made from plastic.

Some of the most useful synthetic polymers have resulted from copying natural polymers. Synthetic rubber, used in almost every automobile tire, is a copy of the molecule found in natural rubber. Many other useful synthetic polymers, such as polystyrene, nylon, Teflon, and Dacron, have no natural analogs.

Both synthetic and natural polymers are made by chemically joining many small molecules, called **monomers,** into giant polymer molecules known as **macromolecules,** which have molar masses ranging from thousands to millions. In nature, polymerization reactions usually are controlled by enzymes, and in animal cells biopolymer synthesis takes place rapidly at body temperature. Making synthetic polymers often requires high temperatures and pressures and lengthy reaction times. Synthetic and natural polymers are formed by addition or condensation

A macromolecule is a molecule with a very high molar mass.

reactions. **Addition polymers** are made by adding monomer units directly together; **condensation polymers** are produced when monomer units combine to form the polymer along with a small molecule, usually water, one molecule of which is formed and released per condensation.

ThomsonNOW™
Go to the Chemistry Interactive menu for a module on **addition polymerization.**

Addition Polymers

In addition polymerization, the monomer units are added *directly* to each other, hooking together like boxcars on a train to form the polymer; no other products are formed. The monomers for making addition polymers usually contain one or more C=C double bonds.

Polyethylene

The simplest monomer for addition polymerization is ethylene, $CH_2{=}CH_2$, which polymerizes to form *polyethylene*. When heated to 100 to 250 °C at a pressure of 1000 to 3000 atm in the presence of a catalyst, ethylene forms polyethylene chains with molar masses of up to many thousands. Polymerization involves three steps: initiation, propagation, and chain termination.

An organic peroxide, RO—OR, produces two free radicals, RO·, each with an unpaired electron.

Initiation. The first step, *initiation* of the polymerization, involves chemicals such as organic peroxides (R—O—O—R) that are unstable and easily break apart into free radicals, ·OR, each with an unpaired electron (⇐ p. 361). The free radicals react readily with molecules containing carbon-carbon double bonds to produce new free radicals.

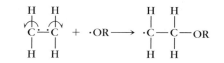

$$\mathrm{H_2C{=}CH_2} + \cdot\mathrm{OR} \longrightarrow \cdot\mathrm{CH_2{-}CH_2{-}OR}$$

Propagation. The polyethylene chain begins to grow as the unpaired electron radical formed in the initiation step bonds to a double-bond electron in another ethylene molecule.

$$\mathrm{H_2C{=}CH_2} + \cdot\mathrm{CH_2{-}CH_2{-}OR} \longrightarrow \cdot\mathrm{CH_2{-}CH_2{-}CH_2{-}CH_2{-}OR}$$

This forms another radical containing an unpaired electron that can bond with yet another ethylene molecule. The process continues, with the chain growing to form a huge polymer molecule.

$$n\,\mathrm{CH_2{=}CH_2} \longrightarrow \left(\mathrm{CH_2{-}CH_2}\right)_n$$

polyethylene

Parentheses enclose the repeating unit of the polymer.

n ranges from 1000 to 50,000.

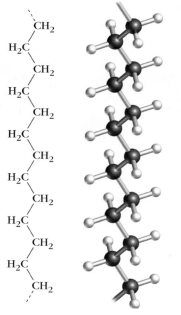

A portion of a polyethylene molecule.

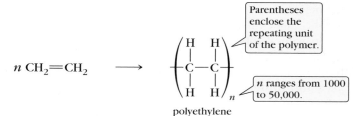

The C=C double bonds in the ethylene monomer have been changed to C—C single bonds in the polyethylene chain.

Chain Termination. Eventually, production of the polymer chain stops. This occurs when the supply of monomer is consumed or when no more free radicals are produced, such as when two chains containing free radicals combine. The value of n is called the **degree of polymerization,** the number of repeating units in the polymer chain.

$$R\text{-}(CH_2)_n CH_2\cdot + \cdot CH_2\text{-}(CH_2)_{n^*}R' \longrightarrow R\text{-}(CH_2)_n CH_2CH_2\text{-}(CH_2)_{n^*}R'$$

n and n^* represent different numbers of CH_2 groups.

CONCEPTUAL EXERCISE 12.16 What Is at the Ends of the Polymer Chains?

What do you think is attached at the ends of the polymer chains when all of the ethylene monomer molecules have been polymerized to form polyethylene?

Branched polyethylene (low density)

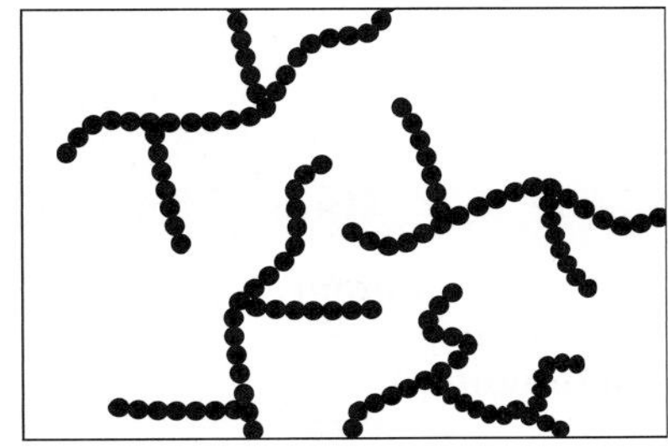

Linear polyethylene (high density)

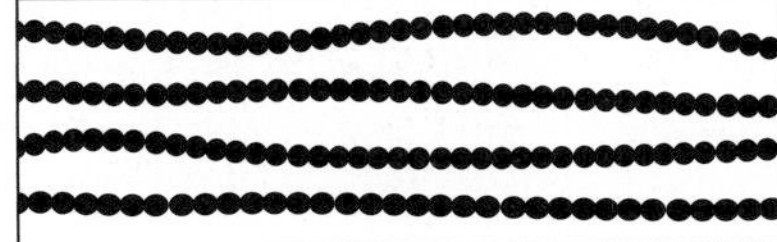

Linear and branched polyethylene.

Changing pressures and catalytic conditions produces polyethylenes of different molecular structures and hence different physical properties. For example, chromium oxide as a catalyst yields almost exclusively the linear polyethylene shown in the margin—a polymer with no branches on the carbon chain. Because the bonds around each carbon in the saturated polyethylene chain are in a tetrahedral arrangement, the zigzag structure (shown on page 576) more closely represents the shape of the chain than does a linear drawing. When ethylene is heated to 230 °C at a pressure of 200 atm without the chromium oxide catalyst, free radicals attack the chain at random positions, causing irregular branching.

Polyethylene is the world's most widely used polymer (Figure 12.9). Long linear chains of polyethylene can pack closely together to give a material with high density (0.97 g/mL) and high molar mass, referred to as high-density polyethylene (HDPE). This material is hard, tough, and semirigid; it is used in plastic milk jugs. Branched chains of polyethylene cannot pack as closely together as the linear chains in HDPE,

Chris Springmann/The Stock Market/Corbis

(a)

© Thomson Learning/Charles D. Winters

(b)

Figure 12.9 Polyethylene. (a) Production of polyethylene film. The film is manufactured by forcing the polymer through a thin space and inflating it with air. (b) The wide range of properties of different structural types of polyethylene leads to a wide variety of applications.

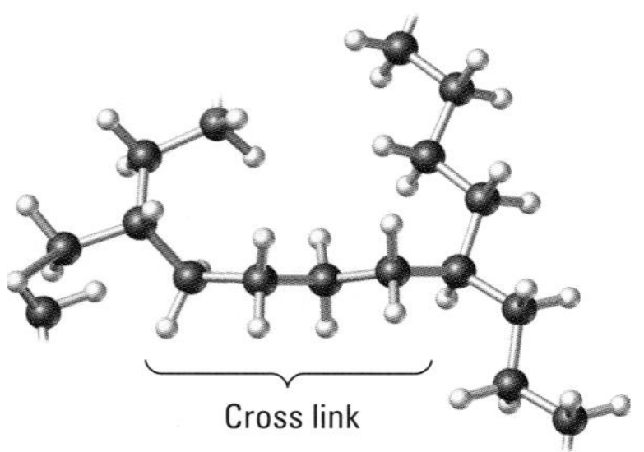

Figure 12.10 **Model of cross-linked polyethylene.**

so the resulting material has a lower density (0.92 g/mL) and is called low-density polyethylene (LDPE). This material is soft and flexible because of its weaker intermolecular forces. It is used to make sandwich bags, for example. If the linear chains of polyethylene are treated in a way that causes short chains of $—CH_2—$ groups to connect adjacent chains, the result is cross-linked polyethylene (CLPE), a very tough material, used for synthetic ice rinks and soft-drink bottle caps (Figure 12.10).

Other Addition Polymers

Many different kinds of addition polymers are made from monomers in which one or more of the hydrogen atoms in ethylene have been replaced with a halogen atom or an organic group. Table 12.6 gives information on some of these monomers and their

Table 12.6 Ethylene Derivatives That Undergo Addition Polymerization

Formula (*X* atom or group is highlighted)	Monomer Common Name	Polymer Name (trade names)	Uses
$H_2C=CH(H)$	Ethylene	Polyethylene (Polythene)	Squeeze bottles, bags, films, toys and molded objects, electrical insulation
$H_2C=CH(CH_3)$	Propylene	Polypropylene (Vectra, Herculon)	Bottles, films, indoor-outdoor carpets
$H_2C=CH(CN)$	Vinyl chloride	Poly(vinyl chloride) (PVC)	Floor tile, raincoats, pipe
$H_2C=CH(CN)$	Acrylonitrile	Polyacrylonitrile (Orlon, Acrilan)	Rugs, fabrics
$H_2C=CH(C_6H_5)$	Styrene	Polystyrene (Styrene, Styrofoam, Styron)	Food and drink coolers, building material insulation
$H_2C=CH(O-C(=O)-CH_3)$	Vinyl acetate	Poly(vinyl acetate) (PVA)	Latex paint, adhesives, textile coatings
$H_2C=C(CH_3)(C(=O)-O-CH_3)$	Methyl methacrylate	Poly(methyl methacrylate) (Plexiglas, Lucite)	High-quality transparent objects, latex paints, contact lenses
$F_2C=CF(F)$	Tetrafluoroethylene	Polytetrafluoroethylene (Teflon)	Gaskets, insulation, bearings, cooking pan coatings

addition polymers. In each case, the replacement atom or group is represented by *X* in the equation below.

$$n\ \mathrm{H_2C{=}CHX} \longrightarrow \left(\mathrm{-CH_2-CHX-CH_2-CHX-CH_2-CHX-}\right)_n$$

The growing chain

X is an aromatic ring in styrene, the monomer for making polystyrene.

X can be an atom, such as Cl in vinyl chloride, or a group of atoms, such as $-CH_3$ in propylene or $-CN$ in acrylonitrile. For example, vinyl chloride units combine to form poly(vinyl chloride). Each monomer is incorporated directly into the growing polymer chain.

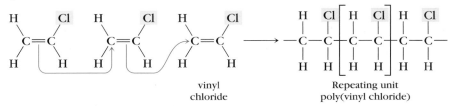

$$\mathrm{H_2C{=}CHCl} + \mathrm{H_2C{=}CHCl} + \mathrm{H_2C{=}CHCl} \longrightarrow \mathrm{-CH_2-CHCl-[CH_2-CHCl]-CH_2-CHCl-}$$

vinyl chloride; Repeating unit poly(vinyl chloride)

Figure 12.11 **Polystyrene.** Expanded polystyrene coffee cup (*left*) is soft. Clear polystyrene cup (*right*) is brittle.

In polystyrene, *n* is typically about 5700. Polystyrene is a clear, hard, colorless, solid thermoplastic that can be molded easily at 250 °C. Nearly seven billion pounds of polystyrene are used annually in the United States alone to make food containers, toys, electrical parts, and many other items. The variation in properties shown by polystyrene products is typical of synthetic polymers. For example, a clear polystyrene drinking glass that is brittle and breaks into sharp pieces somewhat like glass is quite different from an expanded polystyrene coffee cup that is soft and pliable (Figure 12.11).

$$n\ \mathrm{H_2C{=}CH(C_6H_5)} \longrightarrow \left(\mathrm{-CH_2-CH(C_6H_5)-CH_2-CH(C_6H_5)-CH_2-CH(C_6H_5)-}\right)_n$$

Styrene; The growing chain

A major use of polystyrene is in the production of Styrofoam by "expansion molding." In this process, polystyrene beads are placed in a mold and heated with steam or hot air. The tiny beads contain 4–7% by weight of a low-boiling liquid such as pentane. The steam causes the low-boiling liquid to vaporize and expand the beads. As the foamed particles expand, they are molded into the shape of the mold cavity. Styrofoam is used for meat trays, coffee cups, and many kinds of packing material.

The numerous variations in chain length, branching, and crosslinking make it possible to produce a variety of properties for each type of addition polymer. Chemists and chemical engineers can fine-tune the properties of the polymer to match the desired properties by appropriate selection of monomer and reaction conditions, thus accounting for the widespread and growing use of synthetic polymers.

PROBLEM-SOLVING EXAMPLE 12.8 Adding Up the Monomers

Kel-F is an addition polymer made from the monomer $FClC{=}CF_2$.

(a) Write the Lewis structure of this monomer.

(b) Write the Lewis structure of a portion of the Kel-F polymer containing three monomer units.

Answer

(a) 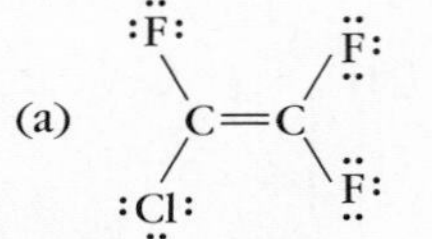

(b) $-\mathrm{CFCl}-\mathrm{CF_2}-\mathrm{CFCl}-\mathrm{CF_2}-\mathrm{CFCl}-\mathrm{CF_2}-$

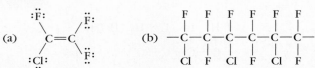

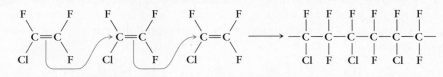

Strategy and Explanation This monomer, which is derived from ethylene, forms an addition polymer. Three monomer units combine to form the beginning of a long polymer chain with each monomer unit coupled directly to the others. No other products form.

$$3\ \mathrm{ClFC{=}CF_2} \longrightarrow -\mathrm{CFCl}-\mathrm{CF_2}-\mathrm{CFCl}-\mathrm{CF_2}-\mathrm{CFCl}-\mathrm{CF_2}-$$

PROBLEM-SOLVING PRACTICE 12.8

Draw the structural formula of a polymer chain formed by the combination of three vinyl acetate monomers. The structural formula of vinyl acetate is in Table 12.6.

PROBLEM-SOLVING EXAMPLE 12.9 Identify the Monomer

Use Table 12.6 to identify the monomer used to make each of the addition polymers, a portion of whose molecule is represented by the structures below.

(a) PVC $\left(-\mathrm{CH_2}-\mathrm{CHCl}-\right)_n$ (b) Acrilan $\left(-\mathrm{CH_2}-\mathrm{CH(CN)}-\right)_n$ (c) Polypropylene $\left(-\mathrm{CH_2}-\mathrm{CH(CH_3)}-\right)_n$

Answer

(a) Vinyl chloride, $CH_2{=}CHCl$ (b) Acrylonitrile, $CH_2{=}CHCN$
(c) Propylene, $CH_2{=}CHCH_3$

Strategy and Explanation Each polymer has a repeating unit, which is derived from its monomer. Each of the monomers has a C=C double bond, but they differ in the nonhydrogen atom or group attached to the carbon atom.

(a) $CH_2{=}CHCl$ (b) $CH_2{=}CHCN$
(c) $CH_2{=}CHCH_3$

PROBLEM-SOLVING PRACTICE 12.9

Draw the structural formula of the repeating unit for each of these addition polymers.
(a) Polypropylene (b) Poly(vinyl acetate) (c) Poly(vinyl alcohol)

Natural latex coming from a rubber tree.

Natural and Synthetic Rubbers

Natural rubber is a hydrocarbon whose monomer unit has the empirical formula C_5H_8. When rubber is decomposed in the absence of oxygen, the monomer 2-methyl-1,3-butadiene (isoprene) is obtained.

$$\mathrm{H_2C{=}C(CH_3)-CH{=}CH_2}$$

2-methyl-1,3-butadiene (isoprene)

CHEMISTRY YOU CAN DO

Making "Gluep"

White school glue, such as Elmer's glue, contains poly(vinyl acetate) and other ingredients. A "gluep" similar to Silly Putty can be made by mixing $\frac{1}{2}$ cup of glue with $\frac{1}{2}$ cup of water and then adding $\frac{1}{2}$ cup of liquid starch and stirring the mixture. After stirring it, work the mixture in your hands until it has a putty consistency. Roll it into a ball and let it sit on a flat surface. Shape a piece into a ball and drop it on a hard surface.

The "gluep" can be stored in a sealed plastic bag for several weeks. Although "gluep" does not readily stick to clothes, it leaves a water mark on wooden furniture, so be careful where you set it. Mold will form on the "gluep" after a few weeks, but adding a few drops of Lysol to it will retard mold formation.

1. What did you observe about "gluep" when it was left sitting on a flat surface?
2. Did the ball of "gluep" bounce when it was dropped?

Natural rubber occurs as *latex* (an emulsion of rubber particles in water) that oozes from rubber trees when the bark is cut. Precipitation of the rubber particles yields a gummy mass that is not only elastic and water-repellent but also very sticky, especially when warm. In 1839, after five years' work on natural rubber, Charles Goodyear (1800–1860) discovered that heating gum rubber with sulfur produces a material that is no longer sticky but is still elastic, water-repellent, and resilient.

Vulcanized rubber, as the type of rubber Goodyear discovered is now known, contains short chains of sulfur atoms that bond together the polymer chains of the natural rubber and reduce its unsaturation. The sulfur chains help to align the polymer chains, so the material does not undergo a permanent change when stretched, but springs back to its original shape and size when the stress is removed. Substances that behave this way are called *elastomers.*

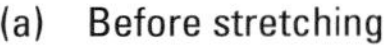

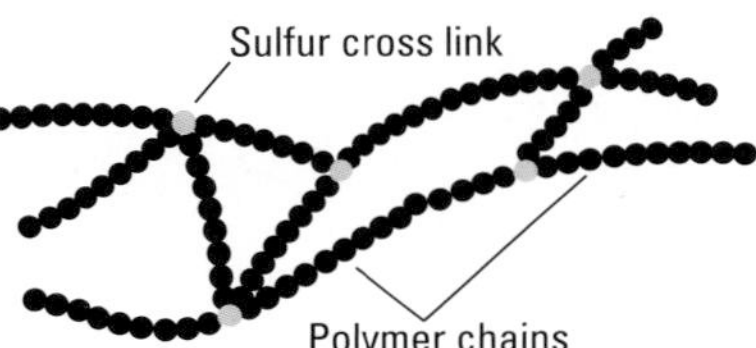

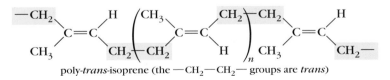

(b) After stretching

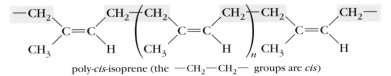

The behavior of natural rubber (polyisoprene) is due to the specific molecular geometry within the polymer chain. We can write the formula for polyisoprene with the $-CH_2CH_2-$ groups on opposite sides of the double bond (the *trans* arrangement)

$$-CH_2(CH_3)C{=}C(H)CH_2-\left[CH_2(CH_3)C{=}C(H)CH_2\right]_n-CH_2(CH_3)C{=}C(H)CH_2-$$

poly-*trans*-isoprene (the $-CH_2-CH_2-$ groups are *trans*)

or with the $-CH_2CH_2-$ groups on the same side of the double bond in a *cis* arrangement (⇐ p. 344),

$$-CH_2(CH_3)C{=}C(H)CH_2-\left[CH_2(CH_3)C{=}C(H)CH_2\right]_n-CH_2(CH_3)C{=}C(H)CH_2-$$

poly-*cis*-isoprene (the $-CH_2-CH_2-$ groups are *cis*)

Natural rubber is poly-*cis*-isoprene. However, the *trans* material also occurs in nature in the leaves and bark of the sapotacea tree and is known as *gutta-percha.* It is brittle and hard and is used for golf ball cores and electrical insulation.

In 1955, chemists at the Goodyear and Firestone companies almost simultaneously discovered how to prepare synthetic poly-*cis*-isoprene. This material is structurally identical to natural rubber. Today, synthetic poly-*cis*-isoprene can be manufactured cheaply and is used when natural rubber is in short supply.

Copolymers

Saran Wrap is an example of a copolymer of vinyl chloride with 1,1-dichloroethylene.

Many commercially important addition polymers are **copolymers,** polymers obtained by polymerizing a mixture of two or more different monomers. A copolymer of styrene with butadiene is the most important synthetic rubber produced in the United States. More than 1.5 million tons of styrene-butadiene rubber (SBR) are produced each year in the United States for making tires. A 3:1 mole ratio of butadiene to styrene is used to make SBR.

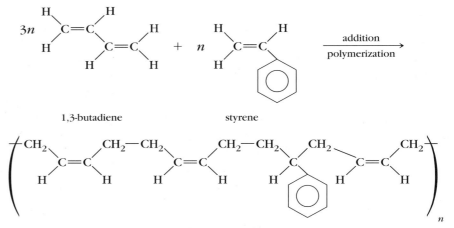

$$3n\ H_2C{=}CH{-}CH{=}CH_2 + n\ H_2C{=}CH(C_6H_5) \xrightarrow[\text{polymerization}]{\text{addition}}$$

1,3-butadiene styrene

$$\left(\!-CH_2{-}CH{=}CH{-}CH_2{-}CH_2{-}CH{=}CH{-}CH_2{-}CH_2{-}CH(C_6H_5){-}CH_2{-}CH{=}CH{-}CH_2-\!\right)_n$$

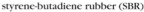

styrene-butadiene rubber (SBR)

Another important copolymer is made by polymerizing mixtures of acrylonitrile, butadiene, and styrene (ABS) to produce a sturdy material used in car bumpers and computer cases.

Ethylene can also be copolymerized with vinyl acetate

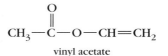

$$CH_3{-}\overset{\overset{\displaystyle O}{\|}}{C}{-}O{-}CH{=}CH_2$$

vinyl acetate

to form poly(ethylene vinylacetate) (EVA), used for athletic shoe innersoles because of its resilience and durability.

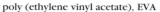

$$\left(\!-CH_2{-}\underset{\underset{\displaystyle C=O}{\overset{|}{\underset{|}{O}}}}{CH}{-}CH_2{-}CH{-}CH_2-\!\right)_n \quad \text{(each CH bears } {-}O{-}C(=O){-}CH_3\text{)}$$

poly (ethylene vinyl acetate), EVA

EVA innersole.

PROBLEM-SOLVING EXAMPLE 12.10 Macromolecular Masses

Polytetrafluoroethylene (Teflon) is an addition polymer made of chains of $CF_2{=}CF_2$ monomer units. This polymer has a molar mass of 1.0×10^6 g/mol. Calculate the number of such units in the polymer—that is, the degree of polymerization.

Answer 10,000 monomer units

Strategy and Explanation Each $CF_2{=}CF_2$ monomer unit has a molar mass of 100.0 g/mol. Therefore, the number of monomer units can be calculated using the molar mass of the polymer.

$$1 \times 10^6 \text{ g} \times \frac{1 \text{ monomer u}}{}\text{(aq)}$$

PROBLEM-SOLVING PRACTICE 12.10

Calculate the degree of polymerization of poly(vinyl chloride) that has a molar mass of 150,000 g/mol.

Condensation Polymers

The reactions of alcohols with carboxylic acids to give esters (p. 570) are examples of condensation reactions. Condensation polymers are formed by condensation polymerization reactions between monomers. Unlike addition polymerization reactions, condensation polymerization reactions do not depend on the presence of a double bond in the reacting molecules. Rather, condensation polymerization reactions generally require two different functional groups present in the same monomer or in two different monomers. Each of the two monomers has a functional group at each end of the monomer. As the functional groups at each end of one monomer react with the groups of the other monomer, long-chain condensation polymers are produced.

For example, molecules with two carboxylic acid groups (—COOH), such as terephthalic acid, and other molecules with two alcohol groups, such as ethylene glycol, can react with each other at both ends to form ester linkages (shaded part of the structure). Water is the other product of this condensation polymerization reaction; it is produced during the formation of each ester linkage in the polymer.

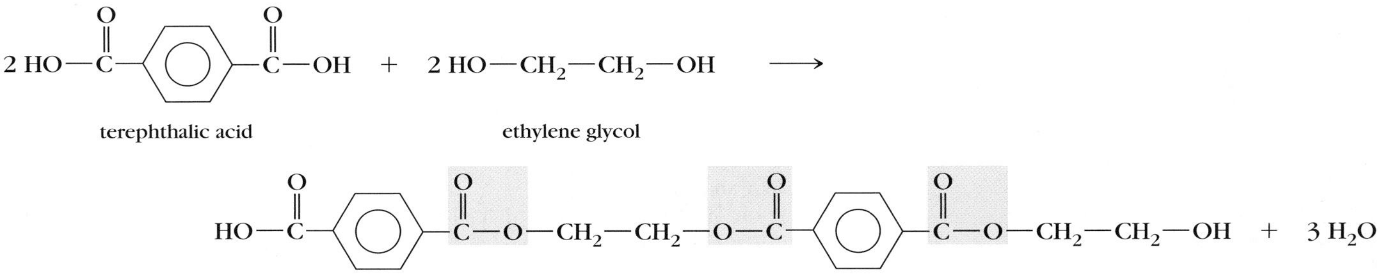

$$2\,HO{-}\overset{O}{\overset{\|}{C}}{-}C_6H_4{-}\overset{O}{\overset{\|}{C}}{-}OH \;+\; 2\,HO{-}CH_2{-}CH_2{-}OH \longrightarrow$$

terephthalic acid ethylene glycol

$$HO{-}\overset{O}{\overset{\|}{C}}{-}C_6H_4{-}\overset{O}{\overset{\|}{C}}{-}O{-}CH_2{-}CH_2{-}O{-}\overset{O}{\overset{\|}{C}}{-}C_6H_4{-}\overset{O}{\overset{\|}{C}}{-}O{-}CH_2{-}CH_2{-}OH \;+\; 3\,H_2O$$

terephthalic acid-ethylene glycol ester

Because this ester has a —COOH group on one end and an —OH group on the other end, the —COOH group can react with an —OH group of another ethylene glycol molecule. Similarly, the remaining alcohol group on the ester can react with another terephthalic acid molecule. This process continues, forming long chains of poly(ethylene terephthalate), a **polyester** commonly known as PET.

ThomsonNOW™
Go to the Chemistry Interactive menu for a module on **polyester formation.**

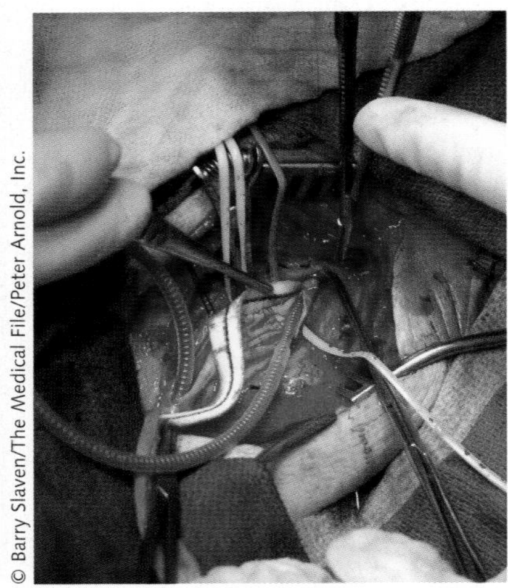

Medical uses of Dacron. A Dacron patch is used to close an atrial septal defect in a heart patient.

$$\left[CH_2-CH_2-O-\overset{O}{\overset{\|}{C}}-C_6H_4-\overset{O}{\overset{\|}{C}}-O \right]_n$$

the repeating unit poly(ethylene terephthalate), PET

Each year more than one billion pounds of PET are produced in the United States for making beverage bottles, apparel, tire cord, film for photography and magnetic recording, food packaging, coatings for microwave and conventional ovens, and home furnishings. A variety of trade names is associated with these applications. PET textile fibers are marketed under such names as Dacron and Terylene. Films of the same polyester, when magnetically coated, are used to make audio and TV tapes. This film, Mylar, has unusual strength and can be rolled into sheets one thirtieth the thickness of a human hair. The inert, nontoxic, noninflammatory, and non-blood-clotting characteristics of PET polymers make Dacron tubing an excellent substitute for human blood vessels in heart bypass operations. Dacron sheets are also used as a skin substitute for burn victims.

PROBLEM-SOLVING EXAMPLE 12.11 Condensation Polymerization

Poly(ethylene naphthalate) (PEN) is used for bar code labels. This condensation polymer is made by the reaction between naphthalic acid and ethylene glycol,

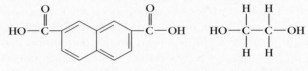

$$HO-\overset{O}{\overset{\|}{C}}-C_{10}H_6-\overset{O}{\overset{\|}{C}}-OH \qquad HO-CH_2-CH_2-OH$$

naphthalic acid — ethylene glycol

(a) Write the structural formula of the molecule formed after two ethylene glycol molecules have polymerized with two naphthalic acid molecules.

(b) Write the structural formula of the repeating unit of the polymer.

Answer

(a)

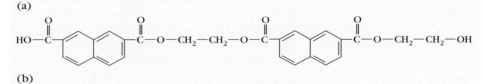

$$HO-\overset{O}{\overset{\|}{C}}-C_{10}H_6-\overset{O}{\overset{\|}{C}}-O-CH_2-CH_2-O-\overset{O}{\overset{\|}{C}}-C_{10}H_6-\overset{O}{\overset{\|}{C}}-O-CH_2-CH_2-OH$$

(b)

$$\left[O-\overset{O}{\overset{\|}{C}}-C_{10}H_6-\overset{O}{\overset{\|}{C}}-O-CH_2-CH_2-O \right]_n$$

Strategy and Explanation

(a) The formation of PEN is similar to that of PET; both polymers are polyesters. Ethylene glycol is one of the monomers in both of them. The other monomer in PET is terephthalic acid; it is naphthalic acid in PEN. The formation of PEN occurs when the carboxylic acid groups of naphthalic acid react with the alcohol groups of ethylene glycol to form ester linkages.

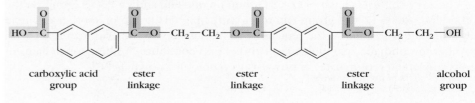

$$HO-\overset{O}{\overset{\|}{C}}-C_{10}H_6-\overset{O}{\overset{\|}{C}}-O-CH_2-CH_2-O-\overset{O}{\overset{\|}{C}}-C_{10}H_6-\overset{O}{\overset{\|}{C}}-O-CH_2-CH_2-OH$$

carboxylic acid group — ester linkage — ester linkage — ester linkage — alcohol group

Notice that there are unreacted carboxylic acid and alcohol groups at each end of the molecule. These react with other ethylene glycol and naphthalic acid groups, respectively. The growing polymer chain continues reacting at each end until it becomes very long.

(b) The repeating unit is the smallest unit that contains both monomers. In this case, the repeating unit is formed by combining one naphthalic acid molecule with one ethylene glycol molecule.

PROBLEM-SOLVING PRACTICE 12.11

A condensation polymer can be made from the single monomer glycolic acid. A portion of the polymer is given below.

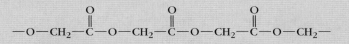

$$-O-CH_2-\overset{\overset{\displaystyle O}{\|}}{C}-O-CH_2-\overset{\overset{\displaystyle O}{\|}}{C}-O-CH_2-\overset{\overset{\displaystyle O}{\|}}{C}-O-CH_2-$$

Write the structural formula of glycolic acid.

CHEMISTRY IN THE NEWS

Superabsorbent Polymers

Disposable diapers, a common commodity, have been greatly improved because of the development of superabsorbent polymers (SAPs), which absorb up to 1000 times their weight in liquid, making them an ideal absorbent for urine in diapers. Disposable diapers using SAPs came on the market in the mid-1980s and were an immediate success. Today approximately 95% of all diapers sold in the United States are SAP-based.

Although the exact formulation of the SAPs used in disposable diapers is proprietary, some general details are known about their composition and production. Modern SAPs are made from partially neutralized, cross-linked poly(acrylic acid). Their ionic nature and interconnected structures allow them to absorb large quantities of liquids without dissolving. Most SAPs are manufactured by a solution polymerization process that involves joining acrylic acid with a cross-linking material. Manufacturing is a four-step process. In the first step, monomers and an initiator are dissolved in water. The monomers undergo condensation polymerization to form the polymer, a rubbery, colorless gel. In subsequent steps, the wet gel is dried and then ground into its final form, a granular, amorphous powder. The fine powder particles are further cross-linked by the addition of small amounts of a cross-linking agent to form the final product. The amount of cross linker used determines the extent of absorbency.

When wetted, the SAPs form a solid, rubbery gel that prevents urine leaking from it. Superabsorbent diapers generally contain 10 to 15 g of SAPs, enough to retain 30 times their weight in urine while remaining soft and pliable. In 2003, more than 22 billion SAP-based diapers were sold in the United States and Canada. Quite a remarkable development for SAPs, developed nearly 70 years ago, that remained a curiosity until applied to disposable diapers.

SAPs can also be used by firefighters to form a watery gel that prevents or retards the spread of flames. Underground cables and wires coated with SAPs prevent the penetration of water into areas where the wires or cables are spliced or cut.

SOURCE: American Chemical Society, *Chemistry*, Spring 2005; pp. 19–22.

Courtesy of BASF

Superabsorbent polymers (SAPs) can absorb prodigious amounts of liquid per gram of SAP. *Source*: American Chemical Society, *Chemistry*, Spring 2005; p. 20.

Amines, Amides, and Polyamides

Amines are organic compounds containing an $—NH_2$ functional group and are classified as primary, secondary, or tertiary amines according to how many of the H atoms in the $—NH_2$ group are replaced by alkyl groups. There can be one, two, or three alkyl groups covalently bonded to the nitrogen atom, for example as in methylamine, dimethylamine, and trimethylamine:

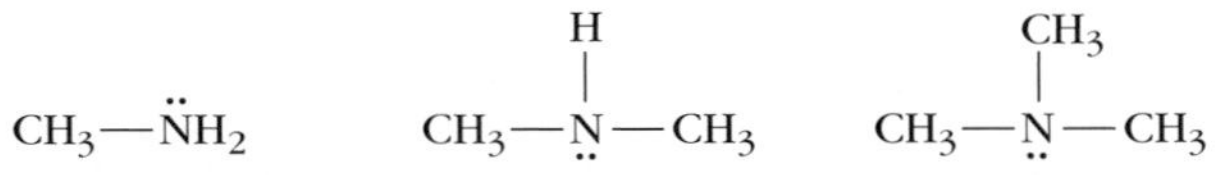

RNH_2 is primary; R_2NH is secondary; and R_3N is tertiary.

$$CH_3—\ddot{N}H_2 \qquad CH_3—\underset{\cdot\cdot}{\overset{\overset{H}{|}}{N}}—CH_3 \qquad CH_3—\underset{\cdot\cdot}{\overset{\overset{CH_3}{|}}{N}}—CH_3$$

methylamine — dimethylamine — trimethylamine

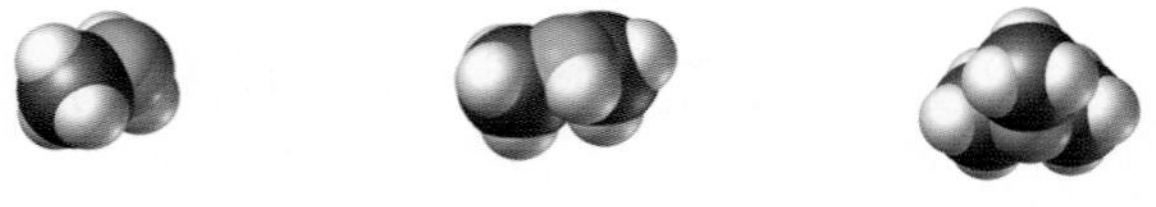

An important condensation reaction occurs when an amine reacts with a carboxylic acid at high temperature to split out a water molecule and form an **amide:**

Amides contain the $—\overset{\overset{O}{\|}}{C}—\overset{\overset{H}{|}}{N}—$ functional group.

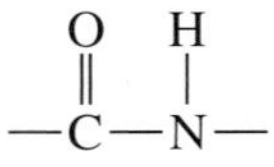

$$R—\overset{\overset{O}{\|}}{C}—OH + H—\underset{\underset{H}{|}}{N}—R' \xrightarrow{\text{heat}} R—\overset{\overset{O}{\|}}{C}—\underset{\underset{H}{|}}{N}—R' + H_2O$$

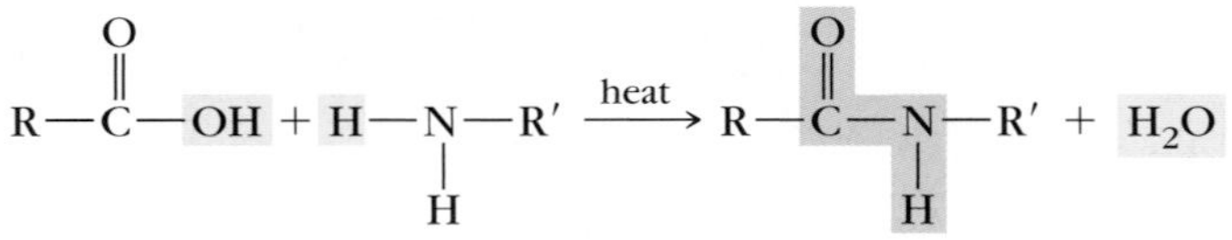

carboxylic acid — amine — amide — water

Dr. Wallace Carothers of the DuPont Company discovered a useful and important type of condensation polymerization reaction that occurs when diamines (compounds containing two $—NH_2$ groups) react with dicarboxylic acids (compounds containing two —COOH groups) to form polymers called **polyamides** or nylons. In February 1935 his research yielded a product known as nylon-66 prepared from adipic acid (a dicarboxylic acid) and hexamethylenediamine (a diamine). Adipoyl chloride, a derivative of adipic acid, is also used in place of adipic acid (Figure 12.12).

The name of nylon-66 is based on the number of carbon atoms in the diamine and diacid, respectively, that are used to make the polymer. Since hexamethylenediamine and adipic acid each have six carbon atoms, the product is called nylon-66.

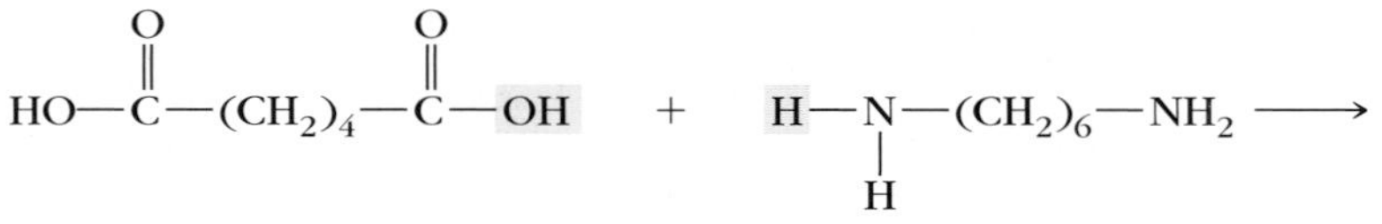

$$HO—\overset{\overset{O}{\|}}{C}—(CH_2)_4—\overset{\overset{O}{\|}}{C}—OH \quad + \quad H—\underset{\underset{H}{|}}{N}—(CH_2)_6—NH_2 \longrightarrow$$

adipic acid — hexamethylenediamine

$$HO—\overset{\overset{O}{\|}}{C}—(CH_2)_4—\overset{\overset{O}{\|}}{C}—\underset{\underset{H}{|}}{N}—(CH_2)_6—NH_2 \quad + \quad H_2O$$

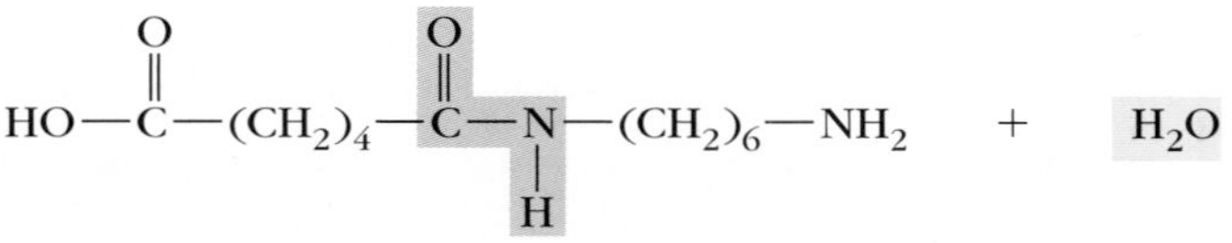

The reactions continue, extending the polymer chain, which consists of alternating adipic acid and hexamethylenediamine units. The overall equation is

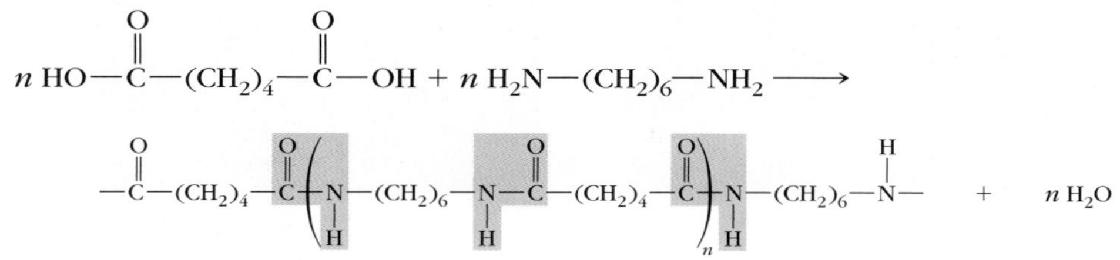

$$n\ HO—\overset{\overset{O}{\|}}{C}—(CH_2)_4—\overset{\overset{O}{\|}}{C}—OH + n\ H_2N—(CH_2)_6—NH_2 \longrightarrow$$

$$—\overset{\overset{O}{\|}}{C}—(CH_2)_4—\overset{\overset{O}{\|}}{C}\left(—\underset{\underset{H}{|}}{N}—(CH_2)_6—\underset{\underset{H}{|}}{N}—\overset{\overset{O}{\|}}{C}—(CH_2)_4—\overset{\overset{O}{\|}}{C}\right)_n—\underset{\underset{H}{|}}{N}—(CH_2)_6—\overset{\overset{H}{|}}{N}— \quad + \quad n\ H_2O$$

nylon

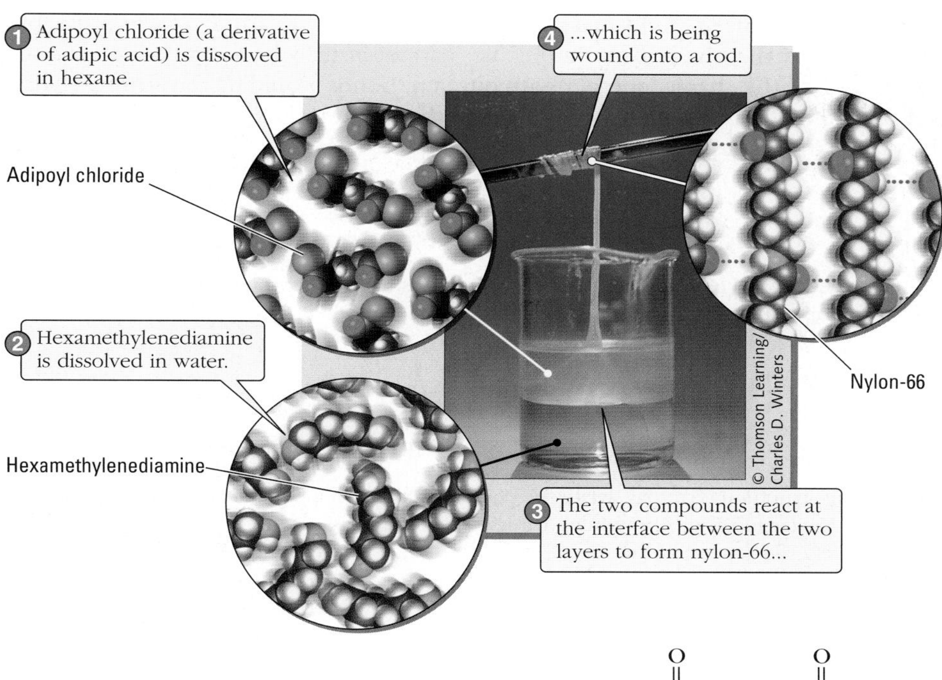
1 Adipoyl chloride (a derivative of adipic acid) is dissolved in hexane.
Adipoyl chloride
2 Hexamethylenediamine is dissolved in water.
Hexamethylenediamine
4 ...which is being wound onto a rod.
Nylon-66
3 The two compounds react at the interface between the two layers to form nylon-66...
© Thomson Learning/Charles D. Winters

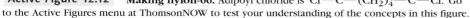

Active Figure 12.12 **Making nylon-66.** Adipoyl chloride is $Cl{-}\overset{\overset{\displaystyle O}{\|}}{C}{-}(CH_2)_4{-}\overset{\overset{\displaystyle O}{\|}}{C}{-}Cl$. Go to the Active Figures menu at ThomsonNOW to test your understanding of the concepts in this figure.

The functional group $-\overset{\overset{\displaystyle O}{\|}}{C}-\overset{\overset{\displaystyle H}{|}}{N}-$ between the monomers is the **amide linkage;** hence, the polymers are known as polyamides.

Figure 12.13 illustrates another facet of the structure of nylon—hydrogen bonding—that explains why nylon makes such good fibers. To have good tensile

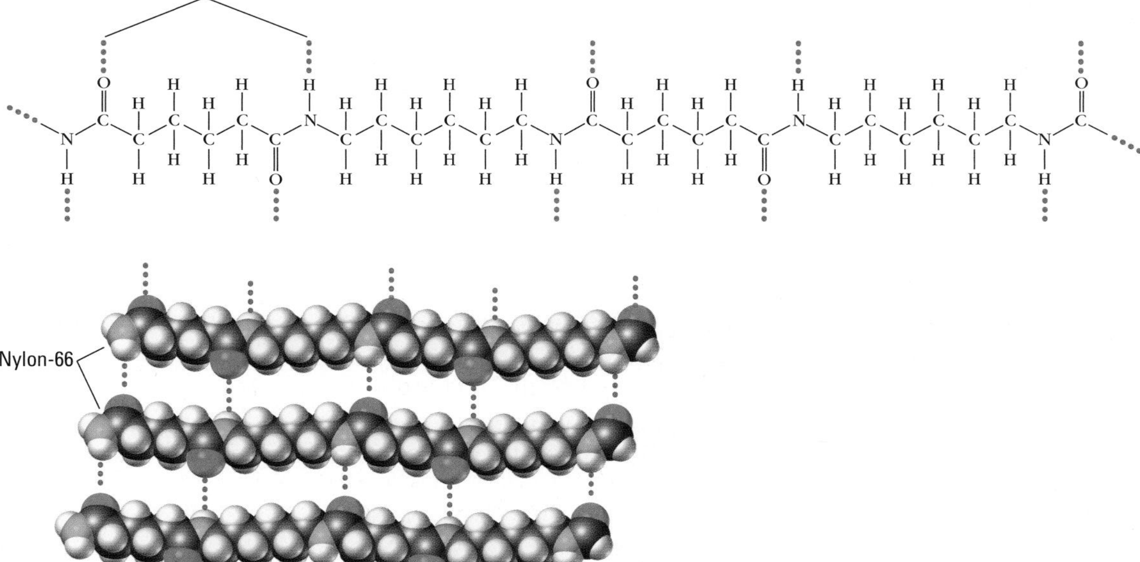
Hydrogen bonds to adjacent nylon-66 molecules.
Nylon-66

Figure 12.13 **Hydrogen bonding in nylon-66.**

A Kevlar helmet. A bullet creased, but did not penetrate, the helmet.

strength, the chains of atoms in a polymer should be able to attract one another, but not so strongly that the polymer molecules cannot initially be extended to form the fibers. Linking the chains together with covalent bonds would be too strong. Hydrogen bonds, with a strength about one tenth that of an ordinary covalent bond, join the chains in the desired manner.

PROBLEM-SOLVING EXAMPLE 12.12 Kevlar, a Condensation Polyamide

Kevlar, used to make bulletproof vests and helmets, canoes, and baseball batting gloves, is made from *p*-phenylenediamine and terephthalic acid.

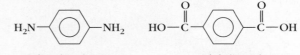

$$H_2N{-}C_6H_4{-}NH_2 \qquad HO{-}\overset{\displaystyle O}{\overset{\|}{C}}{-}C_6H_4{-}\overset{\displaystyle O}{\overset{\|}{C}}{-}OH$$

p-phenylenediamine terephthalic acid

(a) Using structural formulas, write the chemical equation for the formation of a segment of a Kevlar molecule containing three amide linkages.

(b) Identify the repeating unit for Kevlar.

Answer

(a) $$2\,HO{-}\overset{\displaystyle O}{\overset{\|}{C}}{-}C_6H_4{-}\overset{\displaystyle O}{\overset{\|}{C}}{-}OH + 2\,H_2N{-}C_6H_4{-}NH_2 \longrightarrow$$

$$HO{-}\overset{\displaystyle O}{\overset{\|}{C}}{-}C_6H_4{-}\overset{\displaystyle O}{\overset{\|}{C}}{-}\underset{\displaystyle H}{\underset{|}{N}}{-}C_6H_4{-}\underset{\displaystyle H}{\underset{|}{N}}{-}\overset{\displaystyle O}{\overset{\|}{C}}{-}C_6H_4{-}\overset{\displaystyle O}{\overset{\|}{C}}{-}\underset{\displaystyle H}{\underset{|}{N}}{-}C_6H_4{-}NH_2 + 3\,H_2O$$

(b) $$\left(\overset{\displaystyle O}{\overset{\|}{C}}{-}C_6H_4{-}\overset{\displaystyle O}{\overset{\|}{C}}{-}\underset{\displaystyle H}{\underset{|}{N}}{-}C_6H_4{-}\underset{\displaystyle H}{\underset{|}{N}}\right)_n$$

Strategy and Explanation

(a) The amide linkage joining Kevlar units together is formed by the condensation reaction between the amine functional group of *p*-phenylenediamine and the carboxylic acid group of terephthalic acid. The reaction continues between additional amine and acid groups to form three amide linkages as shown in Answer (a).

(b) The repeating unit in Kevlar contains an amide linkage and the remaining portions of the reacting acid and amine molecules (Answer (b)).

PROBLEM-SOLVING PRACTICE 12.12

How many moles of water are formed from the condensation reaction between 6 mol *p*-phenylenediamine and 10 mol terephthalic acid?

Stephanie Louise Kwolek

1923–

In 1946, Stephanie Kwolek received a Bachelor of Science degree from Carnegie Tech (now Carnegie Mellon University). Although wanting to study medicine, she couldn't afford it and decided to take a temporary job at DuPont. She liked her work so well that she stayed for 40 years, retiring in 1986. Kwolek is best known for the development of Kevlar fiber, which is five times stronger than steel on an equal weight basis. She has received many awards, including the Perkin Medal in 1997, considered one of the most prestigious awards a chemist can receive in the United States.

CONCEPTUAL EXERCISE 12.17 Nylon from 3-Aminopropanoic Acid

Polyamides can also be formed from a single monomer that contains both an amine and a carboxylic acid group. For example, the compound 3-aminopropanoic acid can polymerize to form a nylon. Write the general formula for this polymer. Write a formula for the other product that is formed.

$$H_2N{-}CH_2{-}CH_2{-}\overset{\displaystyle O}{\overset{\|}{C}}{-}OH$$

3-aminopropanoic acid

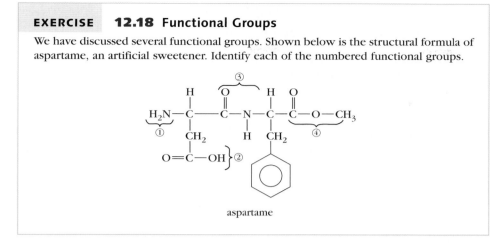

EXERCISE 12.18 Functional Groups

We have discussed several functional groups. Shown below is the structural formula of aspartame, an artificial sweetener. Identify each of the numbered functional groups.

```
                      ③
          H       O        H   O
          |       ‖        |   ‖
    H2N — C ————— C — N — C — C — O — CH3
    ①     |           |   |   ④
          CH2         H   CH2
          |               |
    O ═ C — OH } ②      C6H5 (benzene ring)
```

aspartame

Recycling Plastics

Polyethylene terephthalate (PET), widely used for soft-drink bottles, and high-density polyethylene (HDPE) are the most commonly recycled plastics. Major end uses for recycled PET include fiberfill for ski jackets and sleeping bags, carpet fibers, and tennis balls. It takes just five recycled 2-L PET soft-drink bottles to make a T-shirt and only three for the insulation in a ski jacket. High-density polyethylene (HDPE) is the second most widely recycled plastic; principal sources are milk, juice, and water jugs. In the United States, more than 200 million recycled HDPE milk and water jugs have been converted into a fiber to make Tyvek, which is used for sportswear, insulating wrap for new buildings, and very durable shipping envelopes.

Through recycling, more than 850,000 tons of PET did not end up in landfills in 2005.

Codes are stamped on plastic containers to help consumers identify and sort their recyclable plastics (Table 12.7). There has been a dramatic increase in the recycling of plastics in recent years. Still further increases require a sufficient demand for products made partially or completely from recycled materials. The use of some recycled, postconsumer plastics has been mandated by law in some places. Since 1995 in California, for example, all HDPE packaging must contain 25% recycled material. A real challenge lies ahead in finding economically viable methods to recycle the plastics from the growing, massive numbers of obsolete personal computers, CD players, and cellular phones.

These jackets are made from recycled PET soda bottles.

Table 12.7 Plastic Container Codes

Code	Material	Code	Material
1 PETE	Polyethylene terephthalate (PET)*	5 PP	Polypropylene
2 HDPE	High-density polyethylene	6 PS	Polystyrene
3 V	Poly(vinyl chloride) (PVC)*	7 OTHER	All other resins and layered multimaterial
4 LDPE	Low-density polyethylene		

*Bottle codes are different from standard industrial identification to avoid confusion with registered trademarks.

A Tyvek-wrapped house under construction.

ThomsonNOW
Sign in to ThomsonNOW at **www.thomsonedu.com** to access a set of interactive molecular models of **biologically important molecules.**

12.8 Biopolymers: Proteins and Polysaccharides

Biopolymers—naturally occurring polymers—are an integral part of living things. Many advances in creating and understanding synthetic polymers came from studying biopolymers.

Cellulose and starch, made by plants, resemble a synthetic polymer in that the monomer molecules—glucose—are all alike. On the other hand, proteins, which are made by both plants and animals, are very different from synthetic polymers because they include many different monomers. Also, the occurrences of the different monomers along the protein polymer chain are anything but regular. As a result, proteins are extremely complex copolymers.

Amino Acids to Proteins

Amino acids, the monomer units in proteins, each contain a carboxylic acid group and an amine group. The 20 amino acids from which proteins are made have the general formula

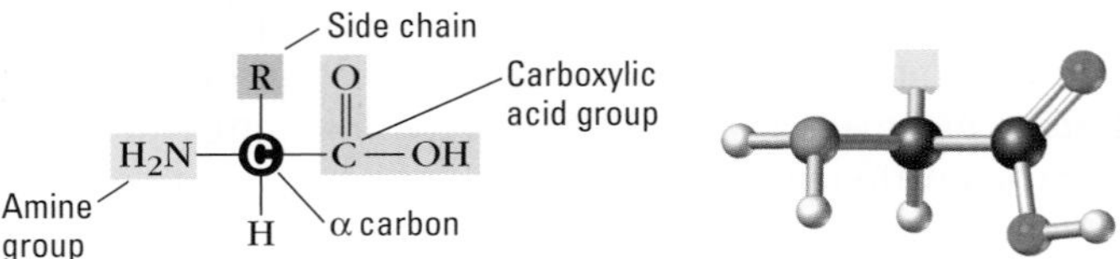

and are described as α-amino acids because the amine ($—NH_2$) group is attached to the **alpha carbon,** the first carbon next to the carboxylic acid (—COOH) group. Each amino acid has a different R group, called a side chain (Table 12.8, p. 591). Glycine, the simplest amino acid, has just hydrogen as its R group. Note from the table that some amino acid R groups contain only carbon and hydrogen, while others contain carboxylic acid, amine, or other functional groups. The amino acids are grouped in Table 12.8 according to whether the R group is nonpolar, polar, acidic, or basic. Each amino acid has a three-letter abbreviation for its name.

CONCEPTUAL EXERCISE **12.19 Hydrogen Bonding Between Amino Acids in Proteins**

Pick two amino acids from Table 12.8 whose R groups could hydrogen-bond with one another if they were close together in a protein chain or in two adjacent protein chains. Then pick two whose R groups would not hydrogen-bond under similar circumstances.

Like nylon, proteins are polyamides. The amide bond in a protein is formed by the condensation polymerization reaction between the amine group of one amino acid and the carboxylic acid group of another. In proteins, the amide linkage is called a **peptide linkage.** Relatively small amino acid polymers (up to about 50 amino acids) are known as **polypeptides.** Proteins are polypeptides containing hundreds to thousands of amino acids bonded together.

Peptide linkages are also referred to as peptide bonds.

$$\underset{\text{amino acid 1}}{H_2N{-}\overset{\overset{R}{|}}{\underset{\underset{H}{|}}{C}}{-}\overset{\overset{O}{\|}}{C}{-}OH} + \underset{\text{amino acid 2}}{H{-}\underset{\underset{H}{|}}{N}{-}\overset{\overset{R}{|}}{\underset{\underset{H}{|}}{C}}{-}\overset{\overset{O}{\|}}{C}{-}OH} \longrightarrow \underset{\text{dipeptide}}{H_2N{-}\overset{\overset{R}{|}}{\underset{\underset{H}{|}}{C}}{-}\overset{\overset{O}{\|}}{C}{-}\underset{\underset{H}{|}}{N}{-}\overset{\overset{R}{|}}{\underset{\underset{H}{|}}{C}}{-}\overset{\overset{O}{\|}}{C}{-}OH} + H_2O$$

Peptide linkage

Table 12.8 Common L-Amino Acids Found in Proteins†

Amino Acid	Abbreviation	Structure	Amino Acid	Abbreviation	Structure
Nonpolar R groups					
Glycine	Gly	$H{-}CH(NH_2){-}COOH$	*Isoleucine	Ile	$CH_3{-}CH_2{-}CH(CH_3){-}CH(NH_2){-}COOH$
Alanine	Ala	$CH_3{-}CH(NH_2){-}COOH$	Proline	Pro	$H_2C{-}CH_2$, $H_2C{-}N(H){-}CH{-}COOH$ (ring: H_2C—CH_2—CH—N(H)—H_2C)
*Valine	Val	$CH_3{-}CH(CH_3){-}CH(NH_2){-}COOH$	*Phenylalanine	Phe	$C_6H_5{-}CH_2{-}CH(NH_2){-}COOH$
*Leucine	Leu	$CH_3{-}CH(CH_3){-}CH_2{-}CH(NH_2){-}COOH$	*Methionine	Met	$CH_3{-}S{-}CH_2CH_2{-}CH(NH_2){-}COOH$
			*Tryptophan	Trp	(indole ring, N—H)$-CH_2{-}CH(NH_2){-}COOH$
Polar but neutral R groups					
Serine	Ser	$HO{-}CH_2{-}CH(NH_2){-}COOH$	Asparagine	Asn	$H_2N{-}C(=O){-}CH_2{-}CH(NH_2){-}COOH$
*Threonine	Thr	$CH_3{-}CH(OH){-}CH(NH_2){-}COOH$	Glutamine	Gln	$H_2N{-}C(=O){-}CH_2CH_2{-}CH(NH_2){-}COOH$
Cysteine	Cys	$HS{-}CH_2{-}CH(NH_2){-}COOH$	Tyrosine	Tyr	$HO{-}C_6H_4{-}CH_2{-}CH(NH_2){-}COOH$
Acidic R groups			**Basic R groups**		
Glutamic acid	Glu	$HO{-}C(=O){-}CH_2CH_2{-}CH(NH_2){-}COOH$	*Lysine	Lys	$H_2N{-}CH_2CH_2CH_2CH_2{-}CH(NH_2){-}COOH$
Aspartic acid	Asp	$HO{-}C(=O){-}CH_2{-}CH(NH_2){-}COOH$	‡Arginine	Arg	$H_2N{-}C(=NH){-}NH{-}CH_2CH_2CH_2{-}CH(NH_2){-}COOH$
			Histidine	His	(imidazole ring, N, N—H)$-CH_2{-}CH(NH_2){-}COOH$

*Essential amino acids that must be part of the human diet. The other amino acids can be synthesized by the body.

†The R group in each amino acid is highlighted.

‡Growing children also require arginine in their diet.

Note that in glycylalanine the amine group of glycine is unreacted as is the carboxylic acid group of alanine. This differs from alanylglycine, which is formed by the condensation reaction of the amine group of glycine (its carboxylic acid group is unreacted) and the carboxylic acid group of alanine (its amine group is unreacted).

Any two amino acids can react to form two different dipeptides, depending on which amine and acid groups react from each amino acid. For example, glycine and alanine can react in either of the following two ways:

$$\underset{\text{glycine}}{H_2N-CH(H)-C(=O)-OH} + \underset{\text{alanine}}{H-N(H)-CH(CH_3)-C(=O)-OH} \longrightarrow \underset{\text{glycylalanine}}{H_2N-CH(H)-C(=O)-N(H)-CH(CH_3)-C(=O)-OH} + H_2O$$

$$\underset{\text{alanine}}{H_2N-CH(CH_3)-C(=O)-OH} + \underset{\text{glycine}}{H-N(H)-CH(H)-C(=O)-OH} \longrightarrow \underset{\text{alanylglycine}}{H_2N-CH(CH_3)-C(=O)-N(H)-CH(H)-C(=O)-OH} + H_2O$$

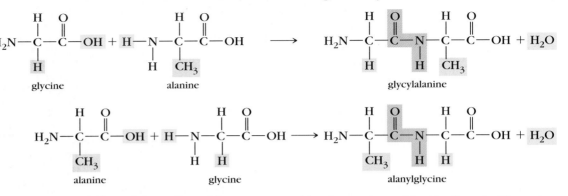

Either of these dipeptides can react with other amino acids at both ends. The extensive chains of amino acid units in proteins are built up by such condensation polymerization reactions.

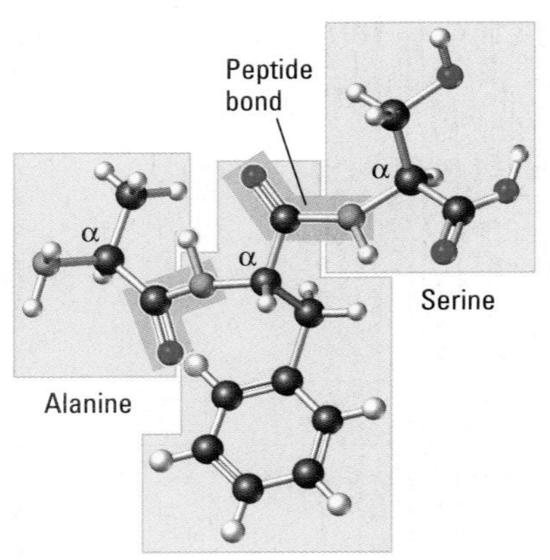

The Ala-Phe-Ser tripeptide

As the number of amino acids in the chain increases and the polypeptide chain lengthens, the number of variations quickly increases to a degree of complexity not generally found in synthetic polymers. As we have just seen, two different dipeptides can form from two different amino acids. Six *tri*peptides are possible if three different amino acids (for example, phenylalanine, Phe; alanine, Ala; and serine, Ser) are linked in combinations that contain all three amino acids, if each is used only once:

Phe-Ala-Ser	Ser-Ala-Phe	Ala-Ser-Phe
Phe-Ser-Ala	Ser-Phe-Ala	Ala-Phe-Ser

If n different amino acids are present, the number of arrangements is $n!$ (n factorial). For four different amino acids, the number of different tetrapeptides is 4!, or $4 \times 3 \times 2 \times 1 = 24$. With five different amino acids, the number of different arrangements is 5!, or 120. If all 20 different naturally occurring amino acids were bonded in one polypeptide, the sequences would make $20! = 2.43 \times 10^{18}$ (2.43 quintillion) unique 20-monomer molecules. *Because proteins can also include more than one molecule of a given amino acid, the number of possible combinations is astronomical.* It is truly remarkable that of the many different proteins that could be made from a set of amino acids, a living cell makes only the relatively small number of proteins it needs.

PROBLEM-SOLVING EXAMPLE 12.13 Peptides

Using information from Table 12.8, draw the structural formula of the tripeptide represented by Ala-Ser-Gly. Explain why it is a different compound from the tripeptide with the amino acids joined in the order Gly-Ala-Ser.

Answer

$$H_2N-CH(CH_3)-C(=O)-N(H)-CH(CH_2OH)-C(=O)-N(H)-CH(H)-C(=O)-OH$$

Ala-Ser-Gly

The structure Gly-Ala-Ser differs because the free $-NH_2$ group is on the glycine part of the molecule and the free $-COOH$ group is on the serine part of the molecule.

Strategy and Explanation The amino acid sequence in the abbreviated name shows that alanine should be written at the left with a free H_2N— group, glycine should be written at the right with a free —COOH group, and both should be connected to serine by peptide bonds. Writing the structure of Gly-Ala-Ser shows how the two tripeptides are different.

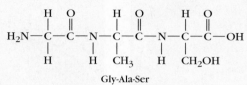

```
      H   O      H   O      H   O
      |   ||     |   ||     |   ||
H2N—C—C—N—C—C—N—C—C—OH
      |      |   |      |   |
      H      H   CH3    H   CH2OH
```

Gly-Ala-Ser

PROBLEM-SOLVING PRACTICE 12.13

Draw the structural formula of the tetrapeptide Cys-Phe-Ser-Ala.

EXERCISE 12.20 Peptide Sequences

Draw the structural formula of the tetrapeptide Ala-Ser-Phe-Cys.

Primary, Secondary, and Tertiary Structure of a Protein

The shape of a protein, and consequently its function, is determined by the sequence of amino acids along the chain, known as the **primary structure** of the protein. All proteins, no matter how large, have a peptide backbone of amino acid units covalently bonded to each other through peptide linkages.

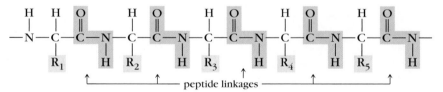

```
 H  H  O     H  O     H  O     H  O     H  O
 |  |  ||    |  ||    |  ||    |  ||    |  ||
—N—C—C—N—C—C—N—C—C—N—C—C—N—C—C—N—
    |     |  |     |  |     |  |     |  |     |
    R1    H  R2    H  R3    H  R4    H  R5    H
```

peptide linkages

The distinctions that determine the shape and function of a protein lie in the sequence of amino acids and the order of the R groups along the backbone. Even one amino acid out of place can create dramatic changes in the shape of a protein, which can lead to serious medical conditions. For example, sickle cell anemia results from an alteration in the primary structure of hemoglobin, the molecule that carries oxygen in red blood cells. Glutamic acid, the seventh amino acid in a 146-amino acid chain in hemoglobin, is replaced by valine. Replacement of the carboxylic acid side chain of glutamic acid with the nonpolar side chain of valine (Table 12.8) disturbs the intramolecular attractions in hemoglobin and changes its shape. As a result, hemoglobin molecules gather into fibrous chains that distort the red blood cells into a sickle shape.

Protein chains are not long and floppy, as you might imagine. Instead, regular patterns of **secondary structure** are created by hydrogen bonding. There are two major types of secondary protein structure: (1) the alpha (α)-helix, which occurs within a protein molecule, and (2) the beta (β)-pleated sheet, which occurs between adjacent protein chains. In the alpha-helix, hydrogen bonding occurs between the N—H hydrogen of an amine group in a peptide bond with a lone electron pair on a C=O oxygen of a peptide bond four amino acid units farther down the peptide backbone (Figure 12.14). In effect, the hydrogen in the hydrogen bond "looks over its shoulder," causing the protein to curl into a helix, much like the coiling of a

The change in structure of hemoglobin that causes sickle-cell anemia answers the question, "How can a disease be caused or cured by a tiny change in a molecule?" that was posed in Chapter 1 (⇐ *p. 2*).

Wool is primarily the protein keratin, which has an alpha-helix secondary structure. Fibroin, the principal protein in silk, is largely beta-pleated sheets.

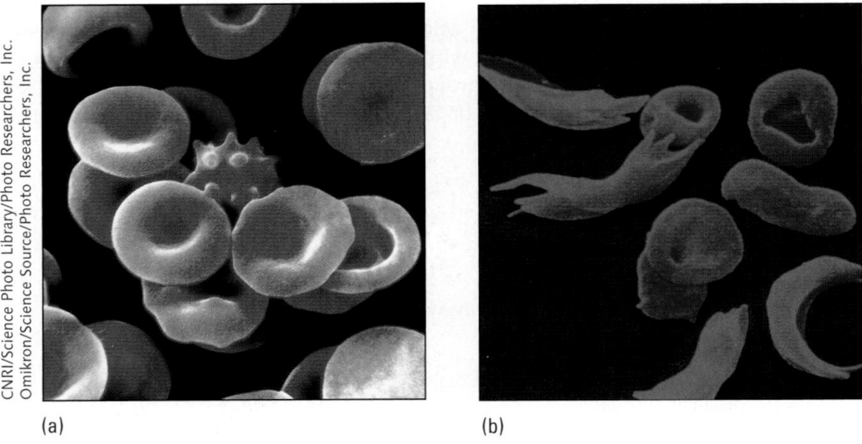

CNRI/Science Photo Library/Photo Researchers, Inc.
Omikron/Science Source/Photo Researchers, Inc.
(a)
(b)

Red blood cells. (a) Normal cells; (b) sickled cells.

telephone cord. In the alpha-helix, the R groups on the amino acid units are on the outside of the helix. In contrast, hydrogen bonding in the beta-pleated sheet occurs between the peptide backbones of neighboring protein chains, creating a zigzag pattern resembling a pleated sheet with the R groups above and below the sheet (Figure 12.14b).

The overall three-dimensional arrangement that accounts for all the twists and turns and folding of a protein is called its **tertiary structure.** The twists and turns of the tertiary structure result in a protein molecule of maximum stability. Tertiary

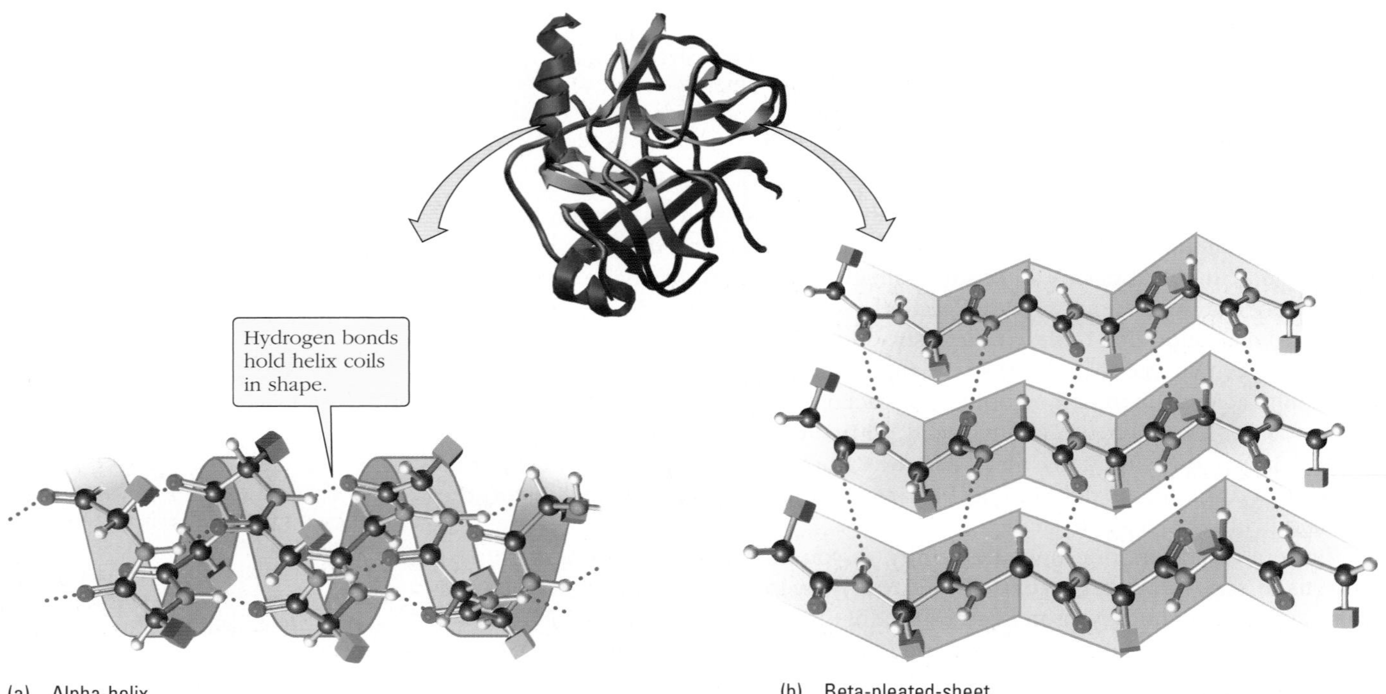

Hydrogen bonds hold helix coils in shape.

Figure 12.14 Alpha-helix (a) and beta-pleated sheet (b) structures of proteins. The enzyme molecule has regions of both alpha-helix and beta-pleated sheet protein structures. R groups are represented by green blocks.

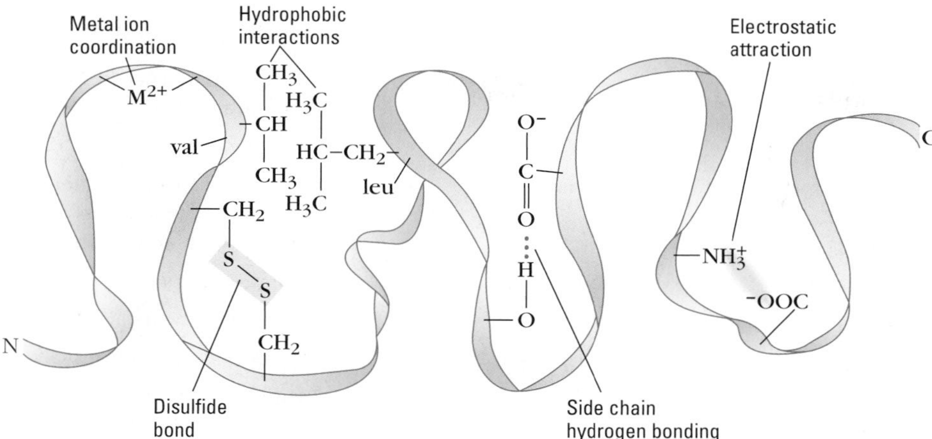

Figure 12.15 Noncovalent forces and bonds that stabilize protein tertiary structure. The tertiary structure of a protein is stabilized by several different forces.

structure is determined by interactions among the side chains strategically placed along the backbone of the chain. These interactions include noncovalent forces of attraction and covalent bonding, as illustrated in Figure 12.15.

Noncovalent Attractions	Covalent Bonds
Hydrogen bonding between side-chain groups	Disulfide bonds (—S—S—)
Hydrophobic (water-hating) interactions between nonpolar side-chain hydrocarbon groups	Coordinate covalent bonding between metal ions and electron pairs of side-chain N and O atoms
Electrostatic attractions between $—NH_3^+$ and $—COO^-$ side-chain groups	

The hydrogen bonding, hydrophobic interactions, and electrostatic attractions bring the groups closer together. Metal ions, such as Fe^{2+} in hemoglobin, are incorporated into a protein by the donation of lone pair electrons from oxygen or nitrogen atoms in side-chain groups to form coordinate covalent bonds to the metal ions. By loss of H, the —SH groups in proximate cysteine units form disulfide (—S—S—) covalent bonds that cross-link regions of the peptide backbone. The number and proximity of disulfide bonds help to limit the flexibility of a protein.

Proteins can be divided into two broad categories—fibrous proteins and globular proteins—that reflect differences in their tertiary structures. *Fibrous proteins,* such as hair, muscle fibers, and fingernails, are rod-like, with the coils or sheets of protein aligned into parallel bundles, making for tough, water-insoluble materials. In contrast, *globular proteins* are highly folded, with hydrophilic (water-loving) side chains on the outside, making these proteins water-soluble. Hemoglobin and chymotrypsin (Figure 12.16) are globular proteins, as are most enzymes.

Because of the complexity and the variety of properties provided by the different R groups and associated molecules or ions, proteins are able to perform widely diverse functions in the body. Consider some of them.

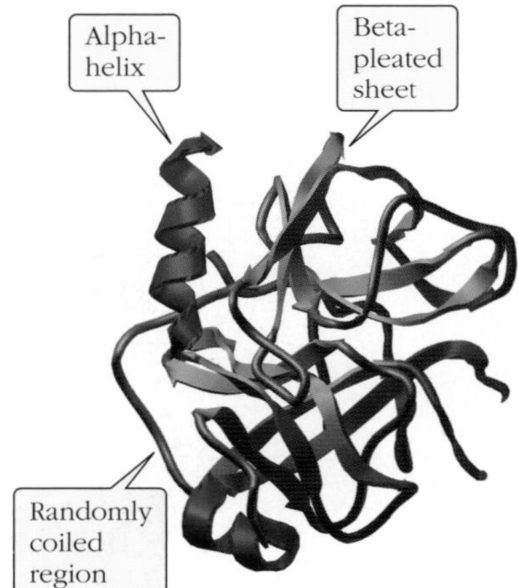

Figure 12.16 Protein folding. The protein chymotrypsin is shown in a ribbon structure that illustrates only the polypeptide backbone. Alpha-helix regions are shown in blue, beta-pleated sheet regions are shown in green, and randomly coiled regions are shown in copper.

Class of Protein	Example	Description of Function
Hormones	Insulin, growth hormone	Regulate vital processes according to needs
Muscle tissue	Myosin	Does mechanical work
Transport proteins	Albumin	Carry fatty acids and other hydrophobic molecules through the bloodstream
Clotting proteins	Fibrin	Form blood clots
Enzymes	Chymotrypsin	Catalyze biochemical reactions
Messengers	Endorphins	Transmit nerve impulses
Immune system proteins	Immunoglobulins	Fight off disease

Monosaccharides to Polysaccharides

Nature makes an abundance of compounds with the general formula $C_x(H_2O)_y$, in which x and y are whole numbers, such as in glucose, $C_6H_{12}O_6$, where $x = 6$, $y = 6$. These compounds are variously known as sugars, carbohydrates, and mono-, di-, or polysaccharides (from the Latin *saccharum,* "sugar"—because they taste sweet). As noted in Figure 12.17, the simplest carbohydrates, such as glucose, are **monosaccharides,** also called simple sugars, because they contain only one type of saccharide molecule. **Disaccharides** consist of two monosaccharide units joined together, such as glucose and fructose linked to form sucrose (table sugar) or two glucose units to form maltose. **Polysaccharides** are polymers containing many monosaccharide units, up to several thousand.

1,6-glycosidic linkages are also possible.

The monomer units of disaccharides and polysaccharides are joined by a C—O—C arrangement called a **glycosidic linkage** between carbons 1 and 4 of adjacent monosaccharide units. Polysaccharides contain many monosaccharide monomers joined together via glycosidic linkages into a very large polymer. The glycosidic linkages in disaccharides and polysaccharides are formed by a condensation reaction like the condensation reactions by which synthetic polymers are formed. A water molecule is released during the formation of each glycosidic bond from the reaction between the —OH groups of the monosaccharide monomers.

glucose + glucose → maltose (Glycosidic linkage) + H_2O water

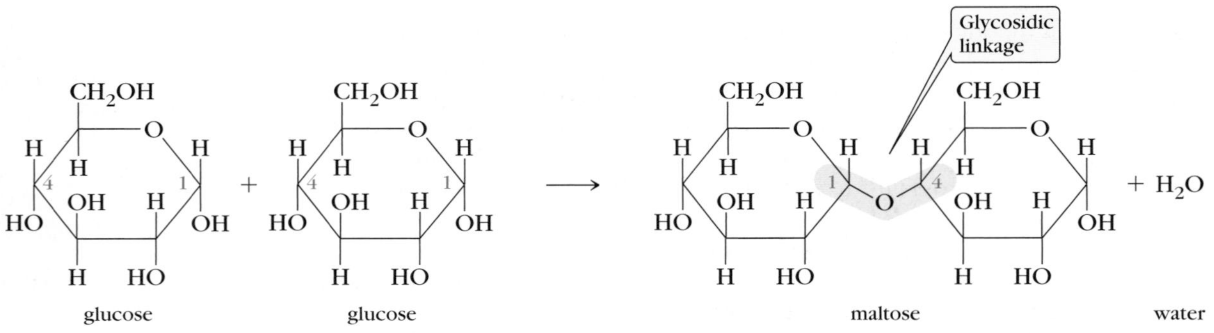

Notice that maltose has two fewer hydrogen atoms and one less oxygen atom than the combined formulas of two glucose molecules. This difference arises because a water molecule is eliminated as a product of the condensation reaction.

Starches and cellulose are the most abundant natural polysaccharides. D-glucose is the monomer in each of these polymers, which can contain as many as 5000 glucose units.

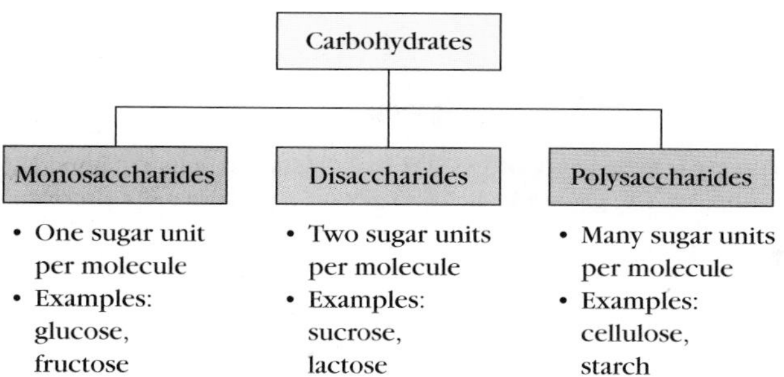

Figure 12.17 **The classes of carbohydrates.**

CONCEPTUAL EXERCISE **12.21 Sucrose Solubility**

On a chemical basis, explain why table sugar (sucrose) is soluble in water.

Polysaccharides: Starches and Glycogen

Plant starch is stored in protein-covered granules until glucose is needed for synthesis of new molecules or for energy production. If these granules are ruptured by heat, they yield two starches—*amylose* and *amylopectin.* Natural starches contain about 75% amylopectin and 25% amylose, both of which are polymers of glucose units joined by glycosidic linkages. Structurally, amylose is a condensation polymer with an average of about 200 glucose monomers per molecule arranged in a straight chain, like pearls on a necklace. Amylopectin is a branched-chain polymer analogous to the branched-chain synthetic polymers discussed earlier. A typical amylopectin molecule has about 1000 glucose monomers arranged into branched chains. The branched chains of glucose units give amylopectin properties different from those of the unbranched amylose. The main difference is their water solubilities. A family of enzymes called amylases helps to break down starches sequentially into a mixture of small branched-chain polysaccharides called dextrins and ultimately into glucose. Dextrins are used as food additives and in mucilage, paste, and finishes for paper and fabrics.

Amylose turns blue-black when tested with iodine solution, whereas amylopectin turns red.

In animals, *glycogen* serves the same storage function as starch does in plants. Glycogen, the reserve carbohydrate in the body, is stored in the liver and muscle tissues and provides glucose for "instant" energy until the process of fat metabolism can take over and serve as the energy source **(⇐ *p. 256*)**. The chains of D-glucose units in glycogen are more highly branched than those in amylopectin.

Animals store energy as fats rather than carbohydrates because fats have more energy per gram.

Cellulose, a Polysaccharide

Cellulose is the most abundant organic compound on earth, found as the woody part of trees and the supporting material in plants and leaves. Cotton is the purest natural form of cellulose. Like amylose, cellulose is composed of D-glucose units. The difference between the structures of cellulose and amylose lies in the orientation of the glycosidic linkages between the glucose units. In cellulose, the —OH groups at carbons 1 and 4 are in the *trans* position, so the glycosidic linkages between glucose units alternate in direction; thus, every other glucose unit is turned over (Figure 12.18). In amylose, the —OH groups at carbons 1 and 4 are in the *cis* position, so all the glycosidic linkages are in the same direction. This subtle

Paper and cotton are essentially cellulose.

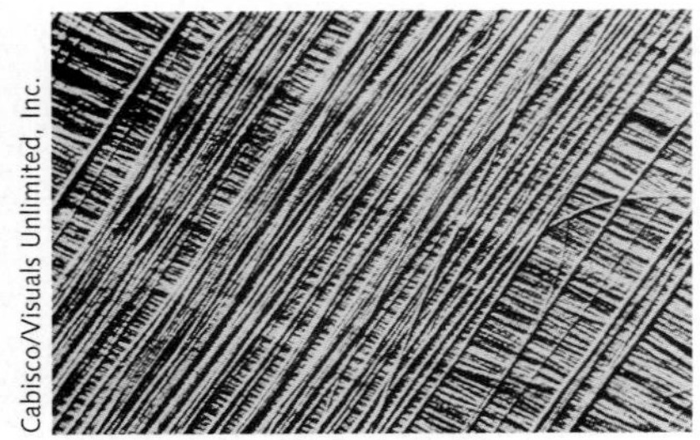

Cabisco/Visuals Unlimited, Inc.

Parallel strands of cellulose in a plant fiber.

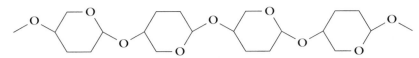

Figure 12.18 The structure of cellulose. About 280 glucose units are bonded together to form a chain structure in cellulose. The glycosidic linkages between glucose units alternate in their orientation, unlike those in amylose.

structural difference allows humans to digest starch, but not cellulose; we lack the enzyme necessary to break the *trans* glycosidic linkages in cellulose. However, termites, a few species of cockroaches, and ruminant mammals such as cows, sheep, goats, and camels do have the proper internal chemistry for this purpose. Because cellulose is so abundant, it would be advantageous if humans could use it, as well as starch, for food.

CONCEPTUAL EXERCISE 12.22 Digestion of Cellulose

Explain why humans cannot digest cellulose. Consult a reference on the Internet and explain why ruminant animals can digest cellulose.

CONCEPTUAL EXERCISE 12.23 What If Humans Could Digest Cellulose?

Think of some of the implications if humans could digest cellulose. What would be some desirable consequences? What would be some undesirable ones?

SUMMARY PROBLEM

Part 1

This chapter has described many kinds of carbon-containing compounds from a variety of sources. Several classes of carbon compounds have molecules with a carbon-to-oxygen bond, which gives the compound certain chemical properties. Use the information in the chapter to complete the sequence of chemical changes for the compound

$$CH_3(CH_2)_8CH_2OH$$

(a) Suppose this compound reacts with aqueous $KMnO_4$. Write the structural formulas of the initial and final oxidation products of this compound.

(b) Use structural formulas to write the equation for the reaction of the initial reactant in Question (a) with the final oxidation product in Question (a).

(c) Into what class of compounds does the product of the reaction in Question (b) fall?

(d) The final product in Question (b) reacts with aqueous sodium hydroxide. Write the equation for this reaction. What types of compounds are formed by this reaction? What would the products be if the reaction took place under acidic conditions instead of with aqueous base?

(e) The final oxidation product in Question (a) reacts with dimethylamine, $(CH_3)_2NH$. Write the structural formula of the product of this reaction. What class of compound is formed?

(f) Can the two reactants in Question (b) form a polymer? Explain.

Part 2

The citric acid cycle is a series of steps involving the degradation and reformation of oxaloacetic acid. Two steps in the cycle involve a reaction studied in this chapter. One step converts isocitric acid into oxalosuccinic acid.

$$\text{HO{-}C(=O){-}CH_2{-}CH(C(=O){-}OH){-}CH(OH){-}C(=O){-}OH} \longrightarrow \text{HO{-}C(=O){-}CH_2{-}CH(C(=O){-}OH){-}C(=O){-}C(=O){-}OH}$$

isocitric acid — oxalosuccinic acid

In the last step of the cycle, L-malic acid is converted into oxaloacetic acid.

$$\text{HO{-}C(=O){-}CH(OH){-}CH_2{-}C(=O){-}OH} \longrightarrow \text{HO{-}C(=O){-}C(=O){-}CH_2{-}C(=O){-}OH}$$

L-malic acid — oxaloacetic acid

(a) Identify the general type of reaction in each step.

(b) What do the two steps have in common?

(c) Identify any chiral carbon atoms in the structures above in Part 2.

Part 3

Select any four amino acids from Table 12.8. Write the structural formulas of four different tetrapeptides that could be formed from these four amino acids.

IN CLOSING

Having studied this chapter, you should be able to . . .

- Describe petroleum refining and methods used to improve the gasoline fraction (Section 12.1).
- Identify the major components of natural gas (Section 12.2).
- Identify processes used to obtain organic chemicals from coal, and name some of their products (Section 12.2).
- Convert equivalent energy units (Section 12.3). ThomsonNOW homework: Study Question 103
- Identify major organic chemicals of industrial and economic importance (Section 12.4).
- Define and describe the nature of chiral molecules and enantiomers (Section 12.4). ThomsonNOW homework: Study Question 26
- Name and draw the structures of three functional groups produced by the oxidation of alcohols (Section 12.5). ThomsonNOW homework: Study Question 39
- Name and give examples of the uses of some important alcohols (Section 12.5). ThomsonNOW homework: Study Question 35
- List some properties of carboxylic acids, and write equations for the formation of esters from carboxylic acids and alcohols (Section 12.6). ThomsonNOW homework: Study Question 45

ThomsonNOW™
Sign in to ThomsonNOW at **www.thomsonedu.com** to check your readiness for an exam by taking the Pre-Test and exploring the modules recommended in your Personalized Learning Plan.

- Use Lewis structures to illustrate triglyceride formation (12.6).
- Differentiate among saturated, monounsaturated, and polyunsaturated fats (Section 12.6). ThomsonNOW homework: Study Question 54
- Explain the formation of polymers by addition or condensation polymerization; give examples of synthetic polymers formed by each type of reaction (Section 12.7). ThomsonNOW homework: Study Questions 56, 69, 98
- Draw the structures of the repeating units in some common types of synthetic polymers and identify the monomers that form them (Section 12.7).
- Identify or write the structures of the functional groups in alcohols, aldehydes, ketones, carboxylic acids, esters, and amines (Sections 12.5–12.7).
- Identify the types of plastics being recycled most successfully (Section 12.7). ThomsonNOW homework: Study Questions 74, 75
- Illustrate the basics of protein structures and how peptide linkages hold amino acids together in proteins (Section 12.8). ThomsonNOW homework: Study Question 65
- Differentiate among the primary, secondary, and tertiary structures of proteins (Section 12.8). ThomsonNOW homework: Study Question 78
- Identify polysaccharides, their sources, the different ways they are linked, and the different uses resulting from these linkages (Section 12.8).

KEY TERMS

achiral *(12.4)*

addition polymer *(12.7)*

aldehyde *(12.5)*

alpha carbon *(12.8)*

amide *(12.7)*

amide linkage *(12.7)*

amine *(12.7)*

amino acid *(12.8)*

carboxylic acid *(12.6)*

catalyst *(12.1)*

catalytic cracking *(12.1)*

catalytic reforming *(12.1)*

chiral *(12.4)*

condensation polymer *(12.7)*

condensation reaction *(12.6)*

copolymer *(12.7)*

degree of polymerization, *n* *(12.7)*

disaccharide *(12.8)*

enantiomers *(12.4)*

ester *(12.6)*

functional group *(12.4)*

glycosidic linkage *(12.8)*

hydrolysis *(12.6)*

ketone *(12.5)*

macromolecule *(12.7)*

monomer *(12.7)*

monosaccharide *(12.8)*

octane number *(12.1)*

peptide linkage *(12.8)*

petroleum fractions *(12.1)*

polyamides *(12.7)*

polyester *(12.7)*

polymer *(12.7)*

polypeptide *(12.8)*

polysaccharide *(12.8)*

primary structure (proteins) *(12.8)*

saponification *(12.6)*

saturated fats *(12.6)*

secondary structure (proteins) *(12.8)*

tertiary structure (proteins) *(12.8)*

thermoplastics *(12.7)*

thermosetting plastics *(12.7)*

triglycerides *(12.6)*

unsaturated fats *(12.6)*

QUESTIONS FOR REVIEW AND THOUGHT

■ denotes questions available in ThomsonNOW and assignable in OWL.

Blue-numbered questions have short answers at the back of this book and fully worked solutions in the *Student Solutions Manual.*

ThomsonNOW™
Assess your understanding of this chapter's topics with sample tests and other resources found by signing in to ThomsonNOW at **www.thomsonedu.com**.

Review Questions

1. Why is the organic chemical industry referred to as the petrochemical industry?
2. What products are produced by the petrochemical industry?
3. What is the difference between catalytic cracking and catalytic reforming?
4. Explain how the octane number of a gasoline is determined.
5. Why is coal receiving increased attention as a source of organic compounds?
6. What is the difference between oxygenated gasoline and reformulated gasoline? Why are they being produced?
7. Explain why world use of natural gas is predicted to double between 1990 and 2010.
8. Table 12.1 lists several compounds with octane numbers above 100 and one compound with an octane number below zero. Explain why such values are possible.
9. Explain why methanol has a lower boiling point (65.0 °C) than water (100 °C).
10. Explain why ethylene glycol has a higher boiling point than ethanol.
11. Outline the steps necessary to obtain 89-octane gasoline, starting with a barrel of crude oil.
12. What is the major difference between crude oil and coal as a source of hydrocarbons?
13. Write the structural formula of a representative compound for each of these classes of organic compounds: alcohols, aldehydes, ketones, carboxylic acids, esters, and amines.
14. Explain why esters have lower boiling points than carboxylic acids with the same number of electrons.
15. What structural feature must a molecule have to undergo addition polymerization?
16. What feature do all condensation polymerization reactions have in common?
17. Give examples of (a) a synthetic addition polymer, (b) a synthetic condensation polymer, and (c) a natural addition polymer.
18. How does *cis-trans* isomerism affect the properties of rubber?
19. Discuss which two plastics are currently being recycled the most successfully, and give examples of some products being made from these recycled plastics.
20. What is the difference between the formation of an addition polymer and a condensation polymer?

Topical Questions

Petroleum and Fuels

21. What are petroleum fractions? What process is used to produce them?
22. (a) What is the boiling point range for the petroleum fraction containing the hydrocarbons that will provide fuel for your car?
    (b) Would you expect the octane rating from the "straight-run" gasoline obtained by fractional distillation of petroleum to be greater than 87 octane? Explain your answer.
    (c) Would you use this fraction to fuel your car? Why or why not?
23. (a) Draw the Lewis structure for the hydrocarbon that is assigned an octane rating of 0.
    (b) Draw the Lewis structure for the hydrocarbon that is assigned an octane rating of 100.
    (c) What is the boiling point for each of these hydrocarbons?
24. Explain what is meant by this statement: "All gasolines are highly volatile."
25. What would be the advantage of removing the higher-octane components such as aromatics and alkenes from oxygenated gasolines?

Chirality in Organic Compounds

26. ■ Circle the chiral centers, if any, in these molecules.

```
         O  HO HO  O
         ‖  |  |   ‖
(a) HO—C—C—C—C—H
            |  |
            H  H

            O   O
            ‖   ‖
(b) CH3—C—C—OH

                  H   O
                  |   ‖
(c) CH3—CH2—C—C—OH
                  |
                 NH2
```

27. Circle the chiral centers, if any, in these molecules.

```
         HO  OH
          |   |
(a) CH3—C—C—H        (b) H—C=C—CH2—OH
          |   |                |  |
          H   H                H  H

          Cl  F
          |   |
(c) CH3—C—C—Cl
          |   |
          Cl  H
```

28. ■ Circle the chiral centers, if any, in these compounds.

```
       H  Cl   H
       |  |    |
(a) H—C—C—C—H
       |  |    |
       H  CH3  H

        H  H    H
        |  |    |
(b) Cl—C—C—C—H
        |  |    |
        H  CH3  H

       H  Br  H  H  H
       |  |   |  |  |
(c) H—C—C—C—C—C—H
       |  |   |  |  |
       H  H   H  H  H

       H  H  Br  H  H
       |  |  |   |  |
(d) H—C—C—C—C—C—H
       |  |  |   |  |
       H  H  H   H  H
```

29. Circle the chiral centers, if any, in these compounds.

(a)
```
     H  Br  H  H
     |  |   |  |
  H—C——C——C——C—H
     |  |   |  |
     H  Cl  H  H
```

(b)
```
     H  CH3  H
     |  |    |
  H—C——C———C—H
     |  |    |
     H  H    H
```

(c)
```
     H  H  H  H  H
     |  |  |  |  |
  H—C——C=C——C——C—H
     |        |  |
     H        Br H
```

30. How can you tell from its structural formula whether a compound can exist as a pair of enantiomers?

31. What conditions must be met for a molecule to be chiral?

32. Which of these is not superimposable on its mirror image?

(a) Nail	(b) Screw
(c) Shoe	(d) Sock
(e) Golf club	(f) Football
(g) Your ear	(h) Helix
(i) Baseball bat	(j) Sweater

Alcohols

33. Give an example of (a) a primary alcohol, (b) a secondary alcohol, and (c) a tertiary alcohol. Draw Lewis structures for each example.

34. Classify each of these alcohols as primary, secondary, or tertiary.

(a) $CH_3CH_2CH_2CH_2OH$

(b)
```
CH3CHCH2CH2OH
   |
   OH
```

(c)
```
   CH3
   |
CH3CCH3
   |
   OH
```

(d)
```
CH3CHCH2CH3
   |
   OH
```

(e)
```
   CH3
   |
CH3CCH2CH3
   |
   OH
```

35. ■ Classify each of these alcohols as primary, secondary, or tertiary.

(a)
```
     H  H  H  OH   H
     |  |  |  |    |
  H—C——C——C——C———C—H
     |  |  |  |    |
     H  H  H  CH3  H
```

(b)
```
     H  H    H    H
     |  |    |    |
  H—C——C———C———C—OH
     |  |    |    |
     H  CH3  CH3  H
```

(c)
```
     H  H    H  OH  H
     |  |    |  |   |
  H—C——C———C——C——C—H
     |  |    |  |   |
     H  CH3  H  H   H
```

(d)
```
     H  H  OH  H    H
     |  |  |   |    |
  H—C——C——C——C———C—H
     |  |  |   |    |
     H  H  H   CH3  H
```

(e)
```
          H
          |
       H—C—H
     H    |
     |    |
  H—C——C—OH
     |    |
     H    |
       H—C—H
          |
          H
```

(f)
```
     H  H
     |  |
  H—C——C—OH
     |  |
     H  |
     H—C—H
        |
        H
```

36. Explain what *oxidation* of organic compounds usually involves. What is meant by *reduction* of organic compounds?

37. ■ Draw the structures of the first two oxidation products of each of these alcohols.

(a) CH_3CH_2OH (b) $CH_3CH_2CH_2CH_2OH$

38. Draw the structures of the oxidation products of each of these alcohols.

(a)
```
     H  H  OH  H
     |  |  |   |
  H—C——C——C——C—H
     |  |  |   |
     H  H  H   H
```

(b)
```
     H  H    H  OH  H
     |  |    |  |   |
  H—C——C———C——C——C—H
     |  |    |  |   |
     H  CH3  H  H   H
```

39. ■ Write the condensed structural formula of the alcohols that can be oxidized to make these compounds.

(a)
```
                   O
                   ‖
CH3CH—CH2—C—H
   |
   CH3
```

(b)
```
               O
               ‖
CH3—CH2—C—CH2—CH3
```

(c)
```
                   O
                   ‖
CH3—CH2—CH—C—OH
           |
           CH3
```

40. What is the percentage of ethanol in 90-proof vodka?
41. Explain how the common names *wood alcohol* for methanol and *grain alcohol* for ethanol came about.
42. What is denatured alcohol? Why is it made?
43. Many biological molecules, including steroids and carbohydrates, contain many —OH groups. What need might biological systems have for this particular functional group?

Carboxylic Acids and Esters

44. Explain why the boiling points for carboxylic acids are higher than those for alcohols with comparable molar masses.
45. ■ Write the structural formula of the esters that can be formed from these reactions.
    (a) $CH_3COOH + CH_3CH_2OH$
    (b) $CH_3CH_2COOH + CH_3CH_2CH_2OH$
    (c) $CH_3CH_2COOH + CH_3OH$
46. Write the structural formula for the esters that can be produced by these reactions.
    (a) Formic acid + methanol
    (b) Butyric acid + ethanol
    (c) Acetic acid + 1-butanol
    (d) Propanoic acid + 2-propanol
47. Write the condensed formula of the alcohol and acid that will react to form each of these esters.
    (a) $CH_3CH_2C(=O)—OCH_3$
    (b) $HC(=O)—OCH_2CH_3$
    (c) $CH_3C(=O)—OCH_2CH_3$
48. Explain why carboxylic acids are more soluble in water than are esters with the same molar mass.

Organic Polymers

49. What are some examples of thermoplastics? What are the properties of thermoplastics when heated and cooled?
50. What are some examples of thermosetting plastics? What are the properties of thermosetting plastics when heated and cooled?
51. Draw the structure of the repeating unit in a polymer in which the monomer is
    (a) 1-butene
    (b) 1,1-dichloroethylene
    (c) Vinyl acetate
52. What is the principal structural difference between low-density and high-density polyethylene? Is polyethylene an addition or a condensation polymer?
53. Methyl methacrylate has the structural formula shown in Table 12.6. When polymerized, it is very transparent, and it is sold in the United States under the trade names Lucite and Plexiglas.
    (a) Use structural formulas to write the chemical equation for the formation of a segment of poly(methyl methacrylate) having four monomer units.
    (b) Identify the repeating unit in the polymer.
54. ■ Use structural formulas and data from Table 12.6 to write the chemical equation for the formation of a polymer of five monomer units from these monomers.
    (a) Acrylonitrile
    (b) Propylene
55. Use structural formulas and data from Table 12.6 to write the chemical equation for the formation of a polymer of five monomer units from these monomers.
    (a) Vinylidene chloride, $H_2C{=}CCl_2$
    (b) Tetrafluoroethylene
56. ■ What monomers are used to prepare these polymers?
    (a) $—CH_2CH_2CH_2CH_2CH_2CH_2CH_2CH_2CH_2—$
    (b) $—CH(CH_3)CH_2CH(CH_3)CH_2CH(CH_3)CH_2—$
    (c) $—CH_2—CH(C_6H_5)—CH_2—CH(C_6H_5)—CH_2—CH(C_6H_5)—CH_2—CH(C_6H_5)—$
    (d) $—CH_2—C(CH_3)(Cl)—CH_2—C(CH_3)(Cl)—CH_2—C(CH_3)(Cl)—$
    (e) $—CH_2CH(C{=}O\,OC_2H_5)CH_2CH(C{=}O\,OC_2H_5)CH_2CH(C{=}O\,OC_2H_5)CH_2CH(C{=}O\,OC_2H_5)—$
57. What is the monomer in *natural rubber*? Which isomer is present in natural rubber, *cis* or *trans*?
58. Write the structural formula of four units of the polymer made from the reaction of $H_2N—(CH_2)_4—NH_2$ and HOOCCOOH. Indicate the repeating unit of the polymer.
59. ■ 4-hydroxybenzoic acid can form a condensation polymer. Use structural formulas to write the chemical equation for the formation of four units of the polymer. Indicate the repeating unit of the polymer.

$HO—C_6H_4—C(=O)—OH$

4-hydroxybenzoic acid

60. What are the two monomers used to make SBR?
61. The formation of polyesters involves which two functional groups?
62. Name one important polyester polymer and its uses.
63. Polyamides are made by condensing which functional groups? Name the most common example of this class of synthetic polymers.
64. How are amide linkages and peptide linkages similar? How are they different?
65. ■ State one major difference between proteins and unstructured polyamides.

66. Draw structures of monomers that could form each of these condensation polymers.

(a) 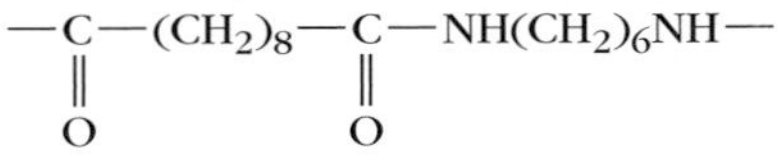 $—C(=O)—(CH_2)_8—C(=O)—NH(CH_2)_6NH—$

(b) 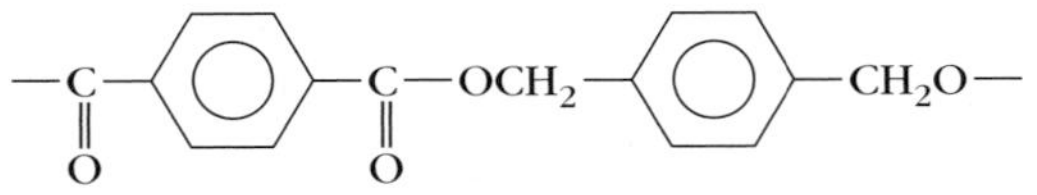 $—C(=O)—C_6H_4—C(=O)—OCH_2—C_6H_4—CH_2O—$

67. Orlon has this polymeric chain structure:

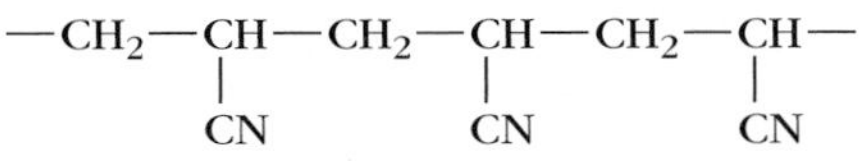

$$—CH_2—CH(CN)—CH_2—CH(CN)—CH_2—CH(CN)—$$

What is the monomer from which this structure can be made?

68. How many ethylene units are in a polyethylene molecule that has a molecular weight of approximately 42,000?

69. ■ A sample of high-molecular-weight (HMW) polyethylene has a molecular weight of approximately 450,000. Determine its degree of polymerization.

70. Ultrahigh-molecular-weight polyethyelene has high wear resistance and is used in conveyor belts. A sample of the polymer has approximately 150,000 monomer units per chain. Calculate the approximate molecular weight of this polymer.

71. Write the structural formulas for the repeating units of these compounds.

(a) Natural rubber

(b) Neoprene $—CH_2—C(Cl)=CH—CH_2—CH_2—C(Cl)=CH—CH_2—$

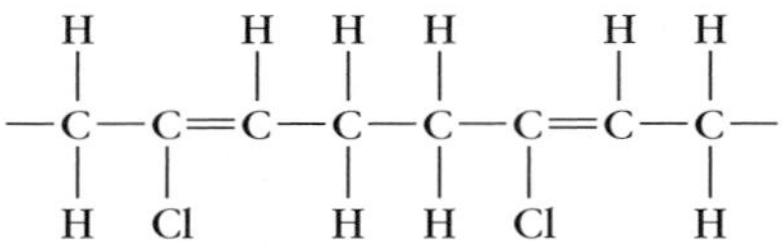

(c) Polybutadiene

72. What are some major end uses for recycled PET and HDPE polymers?

Proteins and Polysaccharides

73. Which biological molecules have monomer units that are all alike, as in synthetic polymers?

74. ■ Which biological molecules have monomer units that are not all alike, as in synthetic copolymers?

75. ■ Identify and name all the functional groups in this tripeptide.

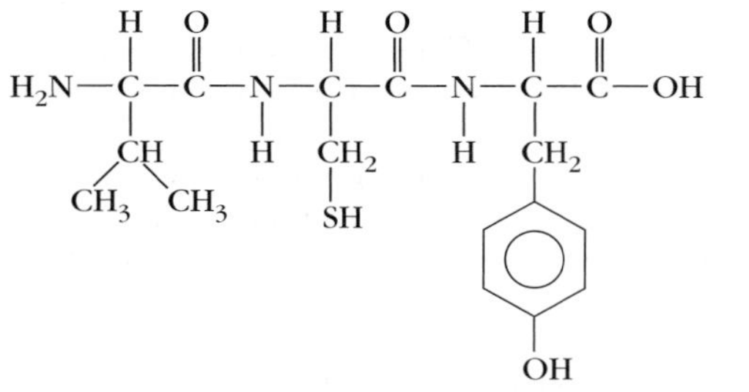

$$H_2N—CH(CH(CH_3)_2)—C(=O)—NH—CH(CH_2SH)—C(=O)—NH—CH(CH_2—C_6H_4—OH)—C(=O)—OH$$

76. Draw the structural formula of alanylglycylphenylalanine.

77. Draw the structural formula of leucylmethionylalanylserine.

78. ■ Explain the difference between (a) monosaccharides and disaccharides, and (b) disaccharides and polysaccharides.

79. What is the chief function of glycogen in animal tissue?

80. What polysaccharides yield only D-glucose upon complete hydrolysis?

81. (a) How do amylose and amylopectin differ?
(b) How are they similar?
(c) Are amylose and glycogen similar?

82. (a) Explain why humans can use glycogen but not cellulose for energy.
(b) Why can cows digest cellulose?

Tools of Chemistry: Gas Chromatography

83. If you were to analyze an oxygenated gasoline using GC, would you use a less polar or more polar stationary phase than the one used for unoxygenated gasoline?

84. Would polar compounds appear earlier or later on a chromatogram if a nonpolar stationary phase were used? Explain your answer.

General Questions

85. Compounds A and B both have the molecular formula C_2H_6O. The boiling points of compounds A and B are 78.5 °C and −23.7 °C, respectively. Use the table of functional groups in Appendix E and write the structural formulas and names of the two compounds.

86. Explain why ethanol, CH_3CH_2OH, is soluble in water in all proportions, but decanol, $CH_3(CH_2)_9OH$, is almost insoluble in water.

87. Nitrile rubber (Buna N) is a copolymer of two parts 1,3-butadiene to one part acrylonitrile. Draw the repeating unit of this polymer.

88. How are rubber molecules modified by vulcanization?

89. Write the condensed structural formula for 3-ethyl-5-methyl-3-hexanol. Is this a primary, secondary, or tertiary alcohol?

90. Using structural formulas, write a reaction for the hydrolysis of a triglyceride that contains fatty acid chains, each consisting of 16 total carbon atoms.

91. Is the plastic wrap used in covering food a thermoplastic or thermosetting plastic? Explain.

92. ■ Assume that a car burns pure octane,

$$C_8H_{18}\ (d = 0.703\ \text{g/cm}^3)$$

(a) Write the balanced equation for burning octane in air, forming CO_2 and H_2O.
(b) If the car has a fuel efficiency of 32 miles per gallon of octane, what volume of CO_2 at 25 °C and 1.0 atm is generated when the car goes on a 10.0-mile trip? (The volume of 1 mol CO_2(g) at 25 °C and 1 atm is 24.5 L.)

93. Perform the same calculations as in Question 92, but use methanol, CH_3OH (d = 0.791 g/cm^3) as the fuel. Assume the fuel efficiency is 20.0 miles per gallon.

94. Show structurally why glycogen forms granules when stored in the liver, but cellulose is found in cell walls as sheets.

95. Polytetrafluoroethylene (Teflon) is made by first treating HF with chloroform, then cracking the resultant difluorochloromethane.

$$CHCl_3 + 2\ HF \longrightarrow CHClF_2 + 2\ HCl$$

$$2\ CHClF_2 + \text{heat} \longrightarrow F_2C{=}CF_2 + 2\ HCl$$

$$F_2C{=}CF_2 \xrightarrow{\text{peroxide catalyst}} \text{Teflon}$$

If you wish to make 1.0 kg Teflon, what mass of chloroform and HF must you use to make the starting material, $CHClF_2$? (Although it is not realistic, assume that each reaction step proceeds to a 100% yield.)

Applying Concepts

96. Hydrogen bonds can form between propanoic acid molecules and between 1-butanol molecules. Draw all the propanoic acid molecules that can hydrogen-bond to the one shown below. Draw all the 1-butanol molecules that can hydrogen-bond to the one shown below. Use dotted lines to represent the hydrogen bonds.

$CH_3CH_2C(=O)—OH$ (propanoic acid) $CH_3CH_2CH_2CH_2—OH$ (1-butanol)

Based on your drawings, which should have the higher boiling point, propanoic acid or 1-butanol? Explain your reasoning.

97. Both propanoic acid and ethyl methanoate form hydrogen bonds with water. Draw all the water molecules that can hydrogen-bond to these molecules. Use dotted lines to represent the hydrogen bonds.

$CH_3CH_2C(=O)—OH$ (propanoic acid) $H—C(=O)—O—CH_2CH_3$ (ethyl methanoate)

Based on your drawings, which should be more soluble in water: propanoic acid or ethyl methanoate? Explain your reasoning.

98. ■ What monomer formed this polymer?

$—C(CH_3){=}C(H)—C(CH_3){=}C(H)—C(CH_3){=}C(H)—C(CH_3){=}C(H)—C(CH_3){=}C(H)—$

99. The illustrations below represent two different samples of polyethylene, each with the same number of monomer units. Based on the concept of density and not structure, which one is high-density polyethylene and which is low-density polyethylene? Write a brief explanation.

(a)

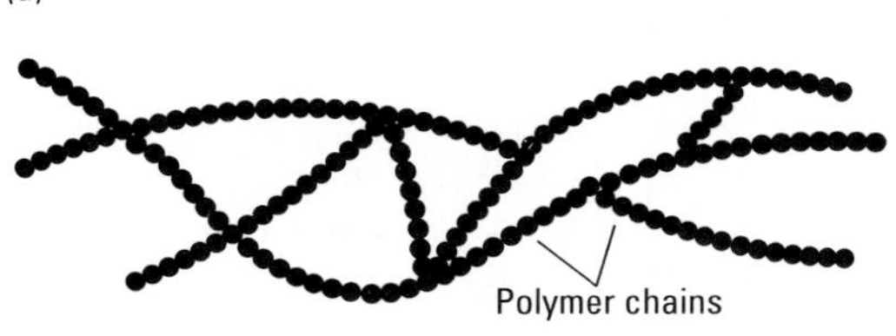

(b)

100. The backbone of a DNA molecule is a polymer of alternating sugar (deoxyribose) and phosphoric acid units held together by a phosphate ester bond. Draw a segment of the polymer consisting of at least two sugar and two phosphate units. Circle the phosphate ester bonds.

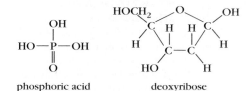

phosphoric acid: $HO—P(=O)(OH)—OH$ deoxyribose: $HOCH_2$ ring structure with O, OH, H, HO substituents

101. Draw the structure of a molecule that could undergo a condensation reaction with itself to form a polyester. Draw a segment of the polymer consisting of at least five monomer units.

102. It has been asserted that the photosynthesis of the trees in a forest the size of Australia would be needed to compensate for the additional CO_2 put into the atmosphere each year from burning fossil fuels. Identify the data that would be required to check whether this assertion is valid.

103. ■ A large oil refinery runs 400,000 barrels of crude oil per day through fractional distillation. Assume that all the crude oil was simply burned to produce energy. Use data from Table 12.2 to answer the following.
    (a) How many 100-W light bulbs could be lighted for one year by this amount of energy? (Assume that the energy can be converted to electrical energy with 100% efficiency.)
    (b) The average toaster uses about 39 kWh per year. How many toasters would this amount of energy operate for a year?

More Challenging Questions

104. The total mass of carbon in living systems on earth is estimated to be 7.5×10^{17} g. Given that the total mass of carbon on the earth is estimated as 7.5×10^{22} g, what is the concentration of carbon atoms on earth expressed in percent and in parts per million (ppm)?

105. In the laboratory, a student was given three bottles labeled X, Y, and Z. One of the bottles contained acetic acid, the second acetaldehyde, and the third ethanol. The student ran a series of experiments to determine the contents of each bottle. Substance X reacted with substance Y to form an ester under certain conditions and substance Y formed an acidic solution when dissolved in water. (a) Identify which compound is in which bottle, and write the Lewis structural formula for each compound. (b) Using structural formulas, write a chemical equation for the ester formation reaction and for the ionization of the acid. (c) The student confirmed the identifications by treating each compound with a strong oxidizing agent and found that compound X required roughly twice the amount of oxidizing agent as compound Z. Explain why this was the case.

106. In his 1989 essay "The End of Nature," William McKibben states that ". . . a clean burning automobile engine will emit 5.5 lb of carbon in the form of carbon dioxide for every gallon of gasoline it consumes." Assume that the gasoline is C_8H_{18} and its density is 0.703 g/mL.

(a) Write a balanced equation for the complete combustion of gasoline.
(b) Do calculations that either corroborate or refute McKibben's assertion. Is his assertion refuted or corroborated?

107. A 1.685-g sample of a hydrocarbon is burned completely to form 5.287 g carbon dioxide and 2.164 g water.
(a) Determine the empirical formula of the hydrocarbon.
(b) Identify whether it is an alkane or an alkene.
(c) Write a plausible Lewis structure for it.

108. Suppose 2.511 g of a hydrocarbon is burned completely to form 7.720 g carbon dioxide and 3.612 g water.
(a) Determine the empirical formula of the hydrocarbon.
(b) Identify whether it is an alkane or an alkene.
(c) Write a plausible Lewis structure for it.

109. Consider these data:

Hydrocarbon	Formula	Enthalpy of Combustion (kJ/mol)
Butane	C_4H_{10}	2853.9
Pentane	C_5H_{12}	3505.8
Hexane	C_6H_{14}	4159.5
Heptane	C_7H_{16}	4812.8
Octane	C_8H_{18}	5465.7

Graph these data with enthalpy of combustion (kJ/mol C) on the *y*-axis and number of carbon atoms on the *x*-axis. Use the graph to estimate the enthalpy of combustion of propane (C_3H_8), nonane (C_9H_{20}), and hexadecane ($C_{16}H_{34}$).

110. Two different compounds each have the formula $C_5H_{10}O_2$. Each contains a C=O group. Write the Lewis structure of each compound and identify the functional group in each one.

111. Using structural formulas, show how vinyl chloride can polymerize to form three types of PVC polymers.

112. Explain why an amino acid has a higher boiling point than an amine with the same number of electrons.

113. Silicones are Si-containing polymers used in waterproof caulking. They do not contain carbon-to-carbon bonds.
(a) Silly Putty is a polymer made from the monomer dimethylsilanol, $(CH_3)_2Si(OH)_2$. Use structural formulas to write the structure of four units of this polymer.
(b) Identify the type of polymerization that occurs.

114. Polyacetylene is an electrically conducting polymer of acetylene, HC≡CH. Now used in batteries, the polymer was discovered accidentally in 1970 when far too much catalyst was used in an experimental attempt to polymerize acetylene into synthetic rubber. Write a plausible Lewis structure for a portion of a polyacetylene molecule.

115. Glycolic acid, $HOCH_2COOH$, and lactic acid form a copolymer used as absorbable stitches in surgery. Use structural formulas to write a chemical equation showing the formation of four units of the copolymer.

116. A particular condensation copolymer forms a silk-like synthetic fabric. A portion of the polymer's structure is shown below.

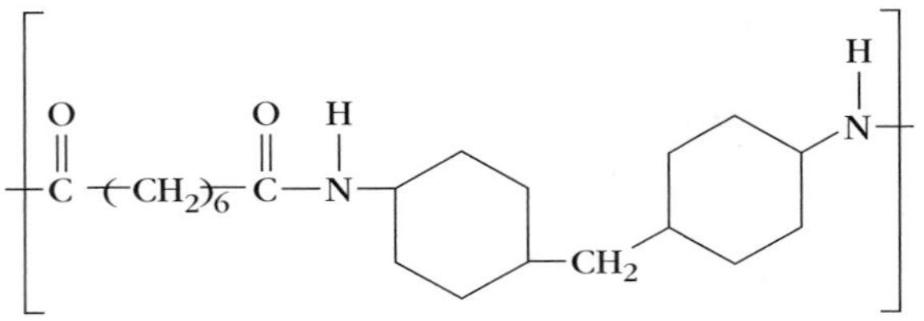

Write the structural formulas of the two monomers used to synthesize the polymer.

Conceptual Challenge Problems

CP12.A (Section 12.1) How are the boiling points of hydrocarbons during the distillation of petroleum related to their molecular size?

CP12.B (Section 12.5) Even though millions of organic compounds exist and each compound may have 10, 100, or even thousands of atoms bonded together to make one molecule, the reactions of organic compounds can be studied and even predicted for compounds yet to be discovered. What characteristic of organic compounds allows their reactions to be studied and predicted?

CP12.C (Section 12.6) What is the advantage of animals storing chemical potential energy in their bodies as triesters of glycerol and long-chain fatty acids, known as fats, instead of as carbohydrates?

13 Chemical Kinetics: Rates of Reactions

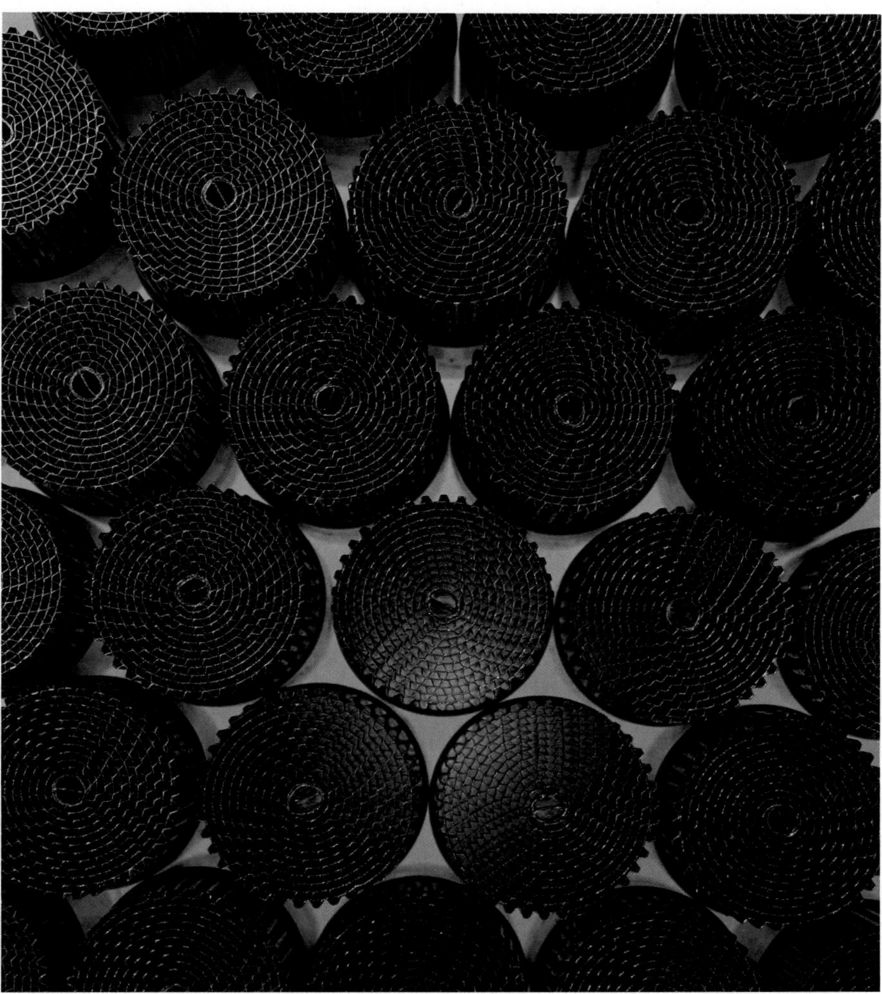

Catalysis is an exciting and profitable field of research that affects our lives every day. In the chemical industry, catalysts speed up reactions that produce new substances such as ammonia fertilizer. In the pharmaceutical industry, catalysts enable reactions that produce medicines. Catalytic converters, such as the ones for motorcycles shown here, remove noxious gases from automobile exhaust, thereby reducing air pollution. In our bodies, enzyme catalysts turn reactions on and off by changing their rates, thereby maintaining proper concentrations of many substances and transmitting nerve impulses.

Johnson Matthey Platinum Today (www.platinum.matthey.com)

ThomsonNOW™
Throughout the text, this icon indicates an opportunity to test yourself on key concepts and to explore interactive modules by signing in to ThomsonNOW at **www.thomsonedu.com**.

Turn on the valve of a Bunsen burner in your laboratory, bring up a lighted match, and a rapid combustion reaction begins with a whoosh:

$$CH_4(g) + 2\,O_2(g) \longrightarrow CO_2(g) + 2\,H_2O(g) \qquad \Delta H° = -802.34 \text{ kJ}$$

What would happen if you didn't put a lighted match in the methane-air stream? Nothing obvious. At room temperature the reaction of methane with oxygen is so slow that the two potential reactants can be mixed in a closed flask and stored unreacted for centuries. These facts about combustion of methane lie within the realm of **chemical kinetics**—*the study of the speeds of reactions and the nanoscale pathways or rearrangements by which atoms and molecules are transformed from reactants to products.*

Chemical kinetics is extremely important, because knowing about kinetics enables us to control many kinds of reactions in addition to combustion. In pharmaceutical chemistry an important problem is devising drugs that remain in their active form long enough to get to the site in the body where they are intended to act. Consequently, it is important to know whether a drug will react with other substances in the body and how long it will take to do so. In environmental chemistry, there was more than a decade of controversy over whether stratospheric ozone is being depleted by chlorofluorocarbons. Much of this hinged on verifying the sequence and rates of reactions by which stratospheric ozone is produced and consumed. Their careful studies of such reactions led to a Nobel Prize in Chemistry for Sherwood Rowland, Mario Molina, and Paul Crutzen ***(← p. 464).***

This chapter focuses on the factors that affect the speeds of reactions, the nanoscale basis for understanding those factors, and their importance in modern society, from industrial plants to cars to the cells of our bodies.

13.1 Reaction Rate

For a chemical reaction to occur, reactant molecules must come together so that their atoms can be exchanged or rearranged. Atoms and molecules are more mobile in the gas phase or in solution than in the solid phase, so reactions are often carried out in a mixture of gases or among solutes in a solution. For a **homogeneous reaction,** one in which reactants and products are all in the same phase (gas or solution, for example), four factors affect the speed of a reaction:

- The *properties* of reactants and products—in particular, their molecular structure and bonding
- The *concentrations* of the reactants and sometimes the products
- The *temperature* at which the reaction occurs
- The *presence* of a catalyst ***(← p. 549)*** and, if one is present, its concentration

A catalyst (described in Section 13.8) speeds up a reaction but undergoes no net chemical change itself. The catalytic converter in an automobile speeds up reactions that remove pollutants from the exhaust gases.

Many important reactions, including the ones in catalytic converters that remove air pollutants from automobile exhaust, are **heterogeneous reactions.** They take place at a surface—at an interface between two different phases (solid and gas, for example). The speed of a heterogeneous reaction depends on the four factors listed above as well as on the area and nature of the surface at which the reaction occurs. For example, very finely divided metal powder can burn very rapidly, whereas a pile of powder with much less surface exposed to oxygen in the air is difficult to ignite (Figure 13.1). The much more rapid reaction when greater surface is exposed has been responsible for explosions in grain elevators and coal bins where finely divided, combustible solids are blown into the air and are exposed to a spark or flame.

Figure 13.1 Combustion of iron powder. (a) The very finely divided metal powder burns slowly in a pile where only a small surface area is exposed to air. (b) Spraying the same powder through the air greatly increases the exposed surface. When the powder enters a flame, combustion is rapid—even explosive.

The speed of any process is expressed as its **rate,** which is *the change in some measurable quantity per unit of time*. A car's rate of travel, for example, is found by measuring the change in its position, Δx, during a given time interval, Δt. Suppose you are driving on an interstate highway. If you pass mile marker 43 at 2:00 PM and mile marker 173 at 4:00 PM, $\Delta x = (173 - 43)$ mi $= 130$ mi and $\Delta t = 2.00$ h. You are traveling at an average rate of $\Delta x/\Delta t = 65$ mi/h. For a chemical process, the **reaction rate** is defined as *the change in concentration of a reactant or product per unit time*. (Time can be measured in seconds, hours, days, or whatever unit is most convenient for the speed of the reaction.)

As an example of measurements made in chemical kinetics, consider Figure 13.2, which shows the violet-colored dye crystal violet reacting with aqueous sodium hydroxide to form a colorless product. The dye's violet color disappears over time, and the intensity of color can be used to determine the concentration of the dye. Crystal violet consists of polyatomic positive ions that we abbreviate as Cv^+,

Recall that the Greek letter Δ (delta) means that a change in some quantity has been measured (← ***p. 220***). As usual, Δ means to subtract the initial value of the quantity from the final value.

Change in concentration is used (rather than change in total amount of reactant) because using change in concentration makes the rate independent of the volume of the reaction mixture.

Figure 13.2 Disappearance of a dye. Violet-colored crystal violet dye in aqueous solution reacts with aqueous sodium hydroxide, which converts the dye into a colorless product. The intensity of the solution's color decreases and eventually the color disappears. The rate of the reaction can be determined by repeated, simultaneous measurements of the intensity of color and the time. From the intensity of color the concentration of dye can be calculated, so concentration can be determined as a function of time.

Table 13.1 Concentration-Time Data for Reaction of Crystal Violet with 0.10 M NaOH(aq) at 23 °C

Time, t (s)	Concentration of Crystal Violet Cation, $[Cv^+]$ (mol/L)	Average Rate (mol L^{-1} s^{-1})
0.0	5.000×10^{-5}	
		13.2×10^{-7}
10.0	3.680×10^{-5}	
		9.70×10^{-7}
20.0	2.710×10^{-5}	
		7.20×10^{-7}
30.0	1.990×10^{-5}	
		5.30×10^{-7}
40.0	1.460×10^{-5}	
		3.82×10^{-7}
50.0	1.078×10^{-5}	
		2.85×10^{-7}
60.0	0.793×10^{-5}	
		1.82×10^{-7}
80.0	0.429×10^{-5}	
		0.985×10^{-7}
100.0	0.232×10^{-5}	

and chloride ions, Cl^-. Its formula is therefore represented by CvCl. The Cv^+ ions cause the beautiful violet color. They combine with hydroxide ions, OH^-, to form a colorless, uncharged product abbreviated as CvOH. The reaction can be represented by the net ionic equation

$$Cv^+(aq) + OH^-(aq) \longrightarrow CvOH(aq) \qquad [13.1]$$

Concentrations in moles per liter are represented by square brackets surrounding the formula of the substance. For example, the concentration of crystal violet cation is represented as $[Cv^+]$.

The rate at which this reaction occurs can be calculated by dividing the change in concentration of crystal violet cation, $\Delta[Cv^+]$, by the elapsed time, Δt. For example, if the concentration of crystal violet is measured at some time t_1 to give $[Cv^+]_1$, and the measurement is repeated at a subsequent time t_2 to give $[Cv^+]_2$, then the rate of reaction is

$$\text{Rate of change of concentration of } Cv^+ = \frac{\text{change in concentration of } Cv^+}{\text{elapsed time}}$$

$$= \frac{\Delta[Cv^+]}{\Delta t} = \frac{[Cv^+]_2 - [Cv^+]_1}{t_2 - t_1}$$

The experimentally measured concentration of crystal violet as a function of time is shown in Table 13.1. Because the concentration of crystal violet decreases as time increases, $[Cv^+]_2$ is smaller than $[Cv^+]_1$. Therefore $\Delta[Cv^+]/\Delta t$ is negative. By convention, reaction rate is defined as a positive quantity. Therefore, for the crystal violet reaction the rate is defined as

The negative sign converts a negative $\Delta[Cv^+]/\Delta t$ value to a positive one.

$$\text{Reaction rate} = -\frac{\Delta[Cv^+]}{\Delta t}$$

Table 13.1 also shows calculated values of reaction rates for each time interval. Because calculating the rate involves dividing a concentration difference by a time difference, the units of reaction rate are units of concentration divided by units of time, in this case mol/L divided by s (mol L^{-1} s^{-1}).

PROBLEM-SOLVING EXAMPLE 13.1 Calculating Average Rates

Using the data in the first two columns of Table 13.1, calculate $\Delta[Cv^+]$, Δt, and the average rate of reaction for each time interval given. Use the numbers given in the third column to check your results. (A good way to do so is to use a computer spreadsheet program.)

Answer See the third column in Table 13.1.

Strategy and Explanation The time interval from 80.0 s to 100.0 s provides an example of the calculation.

$$\Delta[Cv^+] = 0.232 \times 10^{-5}\ mol/L - 0.429 \times 10^{-5}\ mol/L = -0.197 \times 10^{-5}\ mol/L$$

$$\Delta t = 100.0\ s - 80.0\ s = 20.0\ s$$

$$\text{Rate} = -\frac{\Delta[Cv^+]}{\Delta t} = -\frac{(-0.197 \times 10^{-5}\ mol/L)}{20.0\ s} = 0.985 \times 10^{-7}\ mol\ L^{-1}\ s^{-1}$$

Do the other calculations in a similar way.

✓ **Reasonable Answer Check** The rates of reaction should be positive numbers and should have units of concentration divided by time (such as mol $L^{-1}\ s^{-1}$). Both of these conditions are met.

PROBLEM-SOLVING PRACTICE 13.1

For the reaction of crystal violet with NaOH(aq), the measured rate of reaction is 1.27×10^{-6} mol $L^{-1}\ s^{-1}$ when the concentration of crystal violet cation is 4.13×10^{-5} mol/L.

(a) Estimate how long it will take for the concentration of crystal violet to drop from 4.30×10^{-5} mol/L to 3.96×10^{-5} mol/L.

(b) Could you use the same method to make an accurate estimate of how long it would take for the concentration of crystal violet to drop from 4.30×10^{-5} mol/L to 0.43×10^{-5} mol/L? Explain why or why not.

EXERCISE 13.1 Rates of Reaction

(a) From data in Table 13.1, calculate the rate of reaction for each time interval: (i) from 40.0 s to 60.0 s; (ii) from 20.0 s to 80.0 s; (iii) from 0.0 to 100.0 s.

(b) Use all of the data in the first two columns of Table 13.1 to draw a graph with time on the horizontal (x) axis and concentration on the vertical (y) axis. Draw a smooth curve through the data. On the graph, draw lines that correspond to $\Delta[Cv^+]/\Delta t$ for each interval.

(c) Why is the rate not the same for each time interval in part (a), even though the average time for each interval is 50.0 s? (That is, for interval i, the average time is (40.0 s + 60.0 s)/2 = 50.0 s.) Write an explanation of the reason for a friend who is taking this course, and ask your friend to evaluate what you have written.

Reaction Rates and Stoichiometry

From the stoichiometry of the crystal violet reaction (Reaction 13.1), it is clear that for every mole of crystal violet that reacts, a mole of CvOH product is formed, because the coefficients of Cv^+(aq), OH^-(aq), and CvOH(aq) are the same. Thus the rate of appearance of CvOH(aq) equals the rate of disappearance of Cv^+(aq). In many reactions, however, the coefficients are not all the same. For example, in the reaction

$$2\ NO_2(g) \longrightarrow 2\ NO(g) + O_2(g) \qquad [13.2]$$

for every mole of O_2 formed, two moles of NO_2 react. Thus the rate of disappearance of NO_2 is twice as great as the rate of appearance of O_2. We would like to define the rate of reaction in a way that does not depend on which substance's concentration change is measured. If we multiply $-\frac{\Delta[NO_2]}{\Delta t}$ by $\frac{1}{2}$ in the definition of the rate, then the rate is the same whether expressed in terms of $[O_2]$ or $[NO_2]$. Notice that $\frac{1}{2}$ is the reciprocal of the stoichiometric coefficient of NO_2 in Equation 13.2. For the general reaction equation

$$a\,A + b\,B \longrightarrow c\,C + d\,D$$

Because reaction rates are related to Δconcentration/Δ*t* as shown in Equation 13.3, it is important to know the exact chemical equation for which a rate is reported. If the coefficients in the equation are changed, for example, by doubling all of them, then the definition of reaction rate also changes.

where A, B, C, and D represent formulas of substances and *a*, *b*, *c*, and *d* are coefficients, the reaction rate can be defined uniformly in terms of each substance as

$$\text{Rate} = -\frac{1}{a}\frac{\Delta[A]}{\Delta t} = -\frac{1}{b}\frac{\Delta[B]}{\Delta t} = \frac{1}{c}\frac{\Delta[C]}{\Delta t} = \frac{1}{d}\frac{\Delta[D]}{\Delta t} \qquad [13.3]$$

That is, *the rate of change in concentration of any of the reactants or products is multiplied by the reciprocal of the stoichiometric coefficient to find the rate of reaction*. Because the concentrations of reactants decrease with time, their rates of change are given negative signs.

PROBLEM-SOLVING EXAMPLE 13.2 Rates and Stoichiometry

Equation 13.2 shows the decomposition of $NO_2(g)$ to form NO(g) and $O_2(g)$.

(a) Define the rate of reaction in terms of the rate of change in concentration of each reactant and product.

(b) If the rate of appearance of $O_2(g)$ is 0.023 mol L^{-1} s^{-1}, what is the rate of disappearance of $NO_2(g)$?

Answer

(a) $\text{Rate} = \frac{1}{2}\frac{\Delta[NO]}{\Delta t} = \frac{\Delta[O_2]}{\Delta t} = -\frac{1}{2}\frac{\Delta[NO_2]}{\Delta t}$

(b) 0.046 mol L^{-1} s^{-1}

Strategy and Explanation

(a) Define the rate of reaction as change in concentration divided by time elapsed (Δt). Multiply the rate for each substance by the reciprocal of the stoichiometric coefficient, and place a negative sign in front of the rate for each reactant.

(b) From the coefficients, 2 mol NO forms for every 1 mol O_2. Therefore, the rate of formation of NO is twice the rate of formation of O_2, which equals the rate of reaction. Algebraically,

$$\frac{1}{2}\frac{\Delta[NO]}{\Delta t} = \frac{1}{1}\frac{\Delta[O_2]}{\Delta t}$$

and

$$\frac{\Delta[NO]}{\Delta t} = 2 \times \frac{\Delta[O_2]}{\Delta t} = 2 \times 0.023 \text{ mol L}^{-1}\text{ s}^{-1} = 0.046 \text{ mol L}^{-1}\text{ s}^{-1}$$

PROBLEM-SOLVING PRACTICE 13.2

For the reaction

$$4\,NO_2(g) + O_2(g) \longrightarrow 2\,N_2O_5(g)$$

(a) Express the rate of formation of N_2O_5 in terms of the rate of disappearance of O_2.

(b) If the rate of disappearance of O_2 is 0.0037 mol L^{-1} s^{-1}, what is the rate of disappearance of NO_2?

Average Rate and Instantaneous Rate

A reaction rate calculated from a change in concentration divided by a change in time is called the **average reaction rate** over the time interval from which it was calculated. For example, the average reaction rate at 23 °C for the crystal violet reaction over the interval from 0.0 to 10.0 s is 13.2×10^{-7} mol L^{-1} s^{-1}. The data in Table 13.1 indicate that as the concentration of Cv^+ decreases, the average rate also decreases. Most reactions are like this: The average rate becomes smaller as the concentration of one or more reactants decreases. Your results in Exercise 13.1 should have been different for each range of time over which you calculated. Because the average reaction rate changes over time, the rate you calculate depends on when, and for what range of time, you calculate. If you want to know the rate that corresponds to a particular concentration of Cv^+ (and therefore to a particular time after the reaction began), the average rate is not appropriate, because it depends on the size of the time interval.

The **instantaneous reaction rate** is *the rate at a particular time after a reaction has begun*. To obtain it, the rate must be calculated over a very small interval around the time or concentration for which the rate is desired. For example, to calculate the rate at which Cv^+ is disappearing when its concentration is 4.29×10^{-6} mol/L, you would need to calculate $\Delta[Cv^+]/\Delta t$ at exactly 80 s from the start of the reaction. A good way to do so is shown in Figure 13.3. ***The instantaneous rate is the slope of a line tangent to the concentration-time curve at the point corresponding to the specified concentration and time.*** For a particular concentration of the same reactant at the same temperature and the same concentrations of other species, the instantaneous rate has a specific value. As you saw in Exercise 13.1c, the value of the average rate depends on the size of the Δt used to calculate the average rate.

If you are familiar with calculus, then you may recognize that in the limit of very small time intervals, $\Delta[A]/\Delta t$, where A represents a substance, becomes the same as the derivative of concentration with respect to time. That is,

$$\lim_{\Delta t \to 0} \frac{\Delta[A]}{\Delta t} = \frac{d[A]}{dt}$$

This also means that the rate of reaction at any time can be found from the *slope* (at that time) of the tangent to a curve of concentration versus time, such as the curve in Figure 13.3. Appendix A.8 discusses how to determine the slope and intercept of a graph. The slope of a tangent to the graph at a given point is referred to as the derivative of the graph at that point and can be obtained using scientific graphing programs.

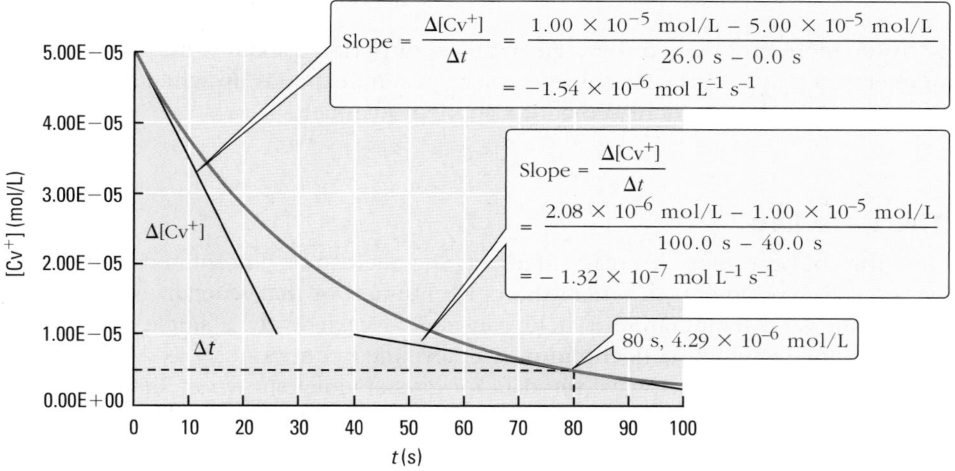

Figure 13.3 **Instantaneous reaction rates.** The experimentally measured concentration of Cv^+ is plotted as a function of time during the reaction in which Cv^+ reacts with OH^- in aqueous solution. The slopes at time 0 s and at 80 s are indicated on the graph. From these slopes the instantaneous rates 1.54×10^{-6} mol L^{-1} s^{-1} and 1.32×10^{-7} mol L^{-1} s^{-1} can be obtained.

CONCEPTUAL EXERCISE 13.2 Instantaneous Rates

Instantaneous rates for the reaction of hydroxide ion with Cv^+ can be determined from the slope of the curve in Figure 13.3 at various concentrations. They are

(1) At 4.0×10^{-5} mol/L, rate $= 12.3 \times 10^{-7}$ mol L^{-1} s^{-1}
(2) At 3.0×10^{-5} mol/L, rate $= 9.25 \times 10^{-7}$ mol L^{-1} s^{-1}
(3) At 2.0×10^{-5} mol/L, rate $= 6.16 \times 10^{-7}$ mol L^{-1} s^{-1}
(4) At 1.5×10^{-5} mol/L, rate $= 4.60 \times 10^{-7}$ mol L^{-1} s^{-1}
(5) At 1.0×10^{-5} mol/L, rate $= 3.09 \times 10^{-7}$ mol L^{-1} s^{-1}

(a) What is the relationship between the rates in (1) and (3)? Between (2) and (4)? Between (3) and (5)?
(b) What is the relationship between the concentrations in each of these cases?
(c) Is the rate of the reaction proportional to the concentration of Cv^+?

ThomsonNOW
Go to the Chemistry Interactive menu for a module on **concentration dependence of reaction rate.**

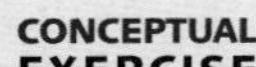

CONCEPTUAL EXERCISE 13.3 Graphing Concentrations versus Time

Consider the decomposition of $N_2O_5(g)$,

$$2\ N_2O_5(g) \longrightarrow 4\ NO_2(g) + O_2(g)$$

Assume that the initial concentration of $N_2O_5(g)$ is 0.02 mol/L and that none of the products are present. Make a graph that shows concentrations of $N_2O_5(g)$, $NO_2(g)$, and $O_2(g)$ as a function of time, all on the same set of axes and roughly to scale.

(a)

(b)

Figure 13.4 Reaction of aqueous potassium permanganate with aqueous hydrogen peroxide. The rate of the reaction of potassium permanganate with hydrogen peroxide depends on the permanganate concentration. With dilute $KMnO_4$ the reaction is slow (a); it is more rapid in more concentrated $KMnO_4$ (b). (In both cases the temperature is the same, so the difference in rate must be due to concentration of permanganate.)

13.2 Effect of Concentration on Reaction Rate

The rates of most reactions change when reactant concentrations change, just as we found for the crystal violet reaction. Figure 13.4 shows another example. The oxidation of hydrogen peroxide by permanganate ion in acidic aqueous solution

$$2\ MnO_4^-(aq) + 5\ H_2O_2(aq) + 6\ H_3O^+(aq) \longrightarrow 2\ Mn^{2+}(aq) + 5\ O_2(g) + 14\ H_2O(\ell)$$

is visibly more rapid when the concentration of permanganate is higher. One goal of chemical kinetics is to find out whether a reaction speeds up when the concentration of a reactant is increased and, if so, by how much.

The Rate Law

How the concentration of a reactant affects the rate can be determined by performing several experiments in which the concentration of that reactant is varied systematically (and temperature is held constant). Alternatively, a single experiment can be done in which concentration is determined continuously as a function of time. The latter approach gave the data for crystal violet shown in Table 13.1 and Figure 13.3, which you analyzed in Problem-Solving Example 13.1 and Exercise 13.2. You should have discovered that if the concentration of crystal violet cation is halved, the reaction rate is also halved. If the concentration of crystal violet cation is doubled, then the reaction rate is doubled. This leads to the expression

$$\text{Rate} \propto [Cv^+]$$

It says that the rate is directly proportional to (symbol $\propto$) the concentration of one of the reactants, crystal violet cation.

This proportionality can be changed to a mathematical equation by including a proportionality constant, *k*. *A mathematical equation that summarizes the rela-*

tionship between reactant concentration and reaction rate is called a **rate law** (or *rate equation*). For the crystal violet reaction the rate law is

$$\text{Rate} = k \times [\text{Cv}^+]$$

The proportionality constant, k, is called the **rate constant.** The rate constant is independent of concentration, but it has different values at different temperatures, usually becoming larger the higher the temperature. The rate constant applies only to the specific reaction being studied and it applies at a specific temperature. Thus the chemical equation and the temperature for the reaction should be given along with the rate constant. In this case we write

$$\text{Cv}^+(\text{aq}) + \text{OH}^-(\text{aq}) \longrightarrow \text{CvOH}(\text{aq}) \qquad k = 3.07 \times 10^{-2}\ \text{s}^{-1}\ (\text{at } 25\ ^\circ\text{C})$$

Determining Rate Laws from Initial Rates

The relation between rate and concentration (the rate law) must be determined experimentally. One way to do so was illustrated in Exercise 13.3, but it is difficult to determine rates from tangents to a curve such as that in Figure 13.3. Another way is to measure initial rates. The **initial rate** of a reaction is *the instantaneous rate determined at the very beginning of the reaction*. A good approximation to the initial rate is to calculate $-\Delta[\text{reactant}]/\Delta t$ after no more than 2% of the limiting reactant has been consumed.

An advantage of measuring initial rates is that the concentrations of products are low early in the process. As a reaction proceeds, more and more products are formed. In some cases products can alter the rate; comparing initial rates with rates when products are present can reveal such a complication.

Many reactions can be started by mixing two different solutions or two different gas samples. Usually the concentrations of the reactants are known before they are mixed, so the initial rate corresponds to a known set of reactant concentrations. Several experiments can then be done in which initial concentrations are varied, and the change in the reaction rate can be correlated with changes in the concentration of each reactant. As an example, consider the reaction of a base with methyl acetate, CH_3COOCH_3, an ester (***p. 570***). This reaction produces acetate ion and methanol.

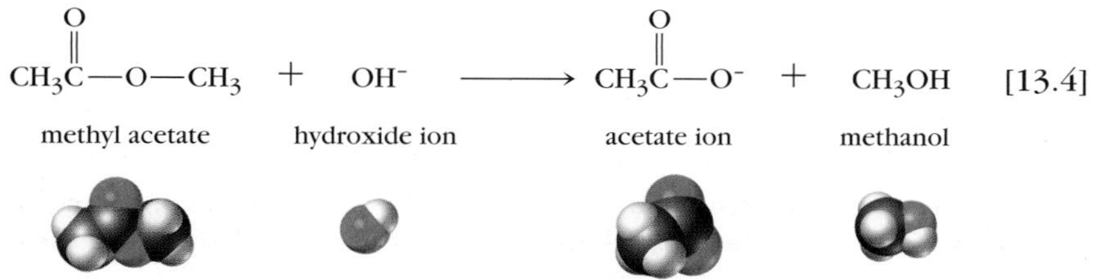

To control for the effect of temperature on rate, several experiments were done at the same temperature:

Experiment	Initial Concentration (mol/L) $[CH_3COOCH_3]$	$[OH^-]$	Initial Rate (mol L^{-1} s^{-1})
1	0.040	0.040	0.00022
	↓ no change	↓ × 2	↓ × 2
2	0.040	0.080	0.00045
	↓ × 2	↓ no change	↓ × 2
3	0.080	0.080	0.00090

Notice that in Experiments 1 and 2 the initial concentration of methyl acetate is the same. In Experiments 2 and 3 the initial concentration of hydroxide is the same.

To determine the rate law, compare two experiments in which only a single initial concentration changed. In Experiments 1 and 2, the $[CH_3COOCH_3]$ remained constant and the $[OH^-]$ doubled. The rate also doubled, which means that the rate is *directly proportional* to the $[OH^-]$. In Experiments 2 and 3, the $[OH^-]$ remained the same, the $[CH_3COOCH_3]$ doubled, and the rate doubled, indicating that the rate is also proportional to the $[CH_3COOCH_3]$. Therefore, the experimental data show that the rate is proportional to the *product* of the two concentrations, and the rate law is

$$\text{Rate} = k\,[CH_3COOCH_3][OH^-]$$

This equation also tells us that doubling both initial concentrations at the same time would cause the rate to go up by a factor of 4, which it does from Experiment 1 to Experiment 3.

ThomsonNOW™
Go to the Coached Problems menu for tutorials on:

- **concentration dependence of reaction rates**
- **determining the rate law using initial rates**

Another way to approach this problem involves proportions. As before, choose two experiments in which one concentration did not change. Then calculate the ratio of the other concentrations and the ratio of rates. For the methyl acetate reaction, using Experiments 1 and 2 where the $[CH_3COOCH_3]$ was constant,

$$\frac{[OH^-]_2}{[OH^-]_1} = \frac{0.080\text{ M}}{0.040\text{ M}} = 2.0 \quad \text{and} \quad \frac{\text{rate}_2}{\text{rate}_1} = \frac{0.00045\text{ mol L}^{-1}\text{ s}^{-1}}{0.00022\text{ mol L}^{-1}\text{ s}^{-1}} = 2.0$$

it is clear that both the concentrations and the rates change in the same proportion. This same method could be applied to analyze results of Experiments 2 and 3, where the initial $[OH^-]$ was constant and the initial $[CH_3COOCH_3]$ changed.

Once the rate law is known, a value for k, the rate constant, can be found by substituting rate and initial concentration data for any one experiment into the rate law. For example, a value of k for the methyl acetate-hydroxide ion reaction could be obtained from data for the first experiment,

rate — $[CH_3COOCH_3]$ — $[OH^-]$

$$0.00022\text{ mol L}^{-1}\text{ s}^{-1} = k(0.040\text{ mol/L})(0.040\text{ mol/L})$$

$$k = \frac{0.00022\text{ mol L}^{-1}\text{ s}^{-1}}{(0.040\text{ mol/L})(0.040\text{ mol/L})} = 0.14\text{ L mol}^{-1}\text{ s}^{-1}$$

A more precise value for k can be obtained by using all available experimental data—that is, by calculating a k for each experiment and then averaging the k values to obtain an overall result.

EXERCISE 13.4 Rates and Concentrations

Use the rate law for the reaction of methyl acetate with base to predict the effect on the rate of reaction if the concentration of methyl acetate is doubled and the concentration of hydroxide ions is halved.

For the reaction as written in Problem-Solving Example 13.3,

$$\text{Rate} = -\frac{1}{2}\frac{\Delta[NO]}{\Delta t} = -\frac{\Delta[Cl_2]}{\Delta t} = +\frac{1}{2}\frac{\Delta[NOCl]}{\Delta t}$$

PROBLEM-SOLVING EXAMPLE 13.3 Rate Law from Initial Rates

Initial rates $\left(-\frac{\Delta[Cl_2]}{\Delta t}\right)$ for the reaction of nitrogen monoxide and chlorine

$$2\,NO(g) + Cl_2(g) \longrightarrow 2\,NOCl(g)$$

were measured at 27 °C starting with various concentrations of NO and Cl_2. These data were collected.

Experiment	Initial Concentrations (mol/L) [NO]	$[Cl_2]$	Initial Rate (mol L^{-1} s^{-1})
1	0.020	0.010	8.27×10^{-5}
2	0.020	0.020	1.65×10^{-4}
3	0.020	0.040	3.31×10^{-4}
4	0.040	0.020	6.60×10^{-4}
5	0.010	0.020	4.10×10^{-5}

(a) What is the rate law?
(b) What is the value of the rate constant k?

Answer
(a) Rate $= k[Cl_2][NO]^2$
(b) $k = 2.1 \times 10^1$ L^2 mol^{-2} s^{-1}

Strategy and Explanation (a) Analyze data from experiments in which one concentration remains the same. In Experiments 1, 2, and 3, the concentration of NO is constant, while the Cl_2 concentration increases from 0.010 to 0.020 to 0.040 mol/L. Each time $[Cl_2]$ is doubled, the initial rate also doubles. For example, when $[Cl_2]$ is doubled from 0.020 to 0.040 mol/L in Experiments 2 and 3, the initial rate doubles from 1.65×10^{-4} to 3.31×10^{-4} mol L^{-1} s^{-1}. The initial rate is directly proportional to $[Cl_2]$.

In Experiments 2, 4, and 5, $[Cl_2]$ is constant, while [NO] varies. In Experiments 2 and 4, [NO] is doubled, but the initial rate increases by a factor of 4, or 2^2.

$$\frac{\text{Experiment 4 rate}}{\text{Experiment 2 rate}} = \frac{6.60 \times 10^{-4}\ \text{mol L}^{-1}\ \text{s}^{-1}}{1.65 \times 10^{-4}\ \text{mol L}^{-1}\ \text{s}^{-1}} = \frac{4}{1} = \frac{2^2}{1}$$

When comparing one experiment with another, as is done in the Strategy and Explanation of Problem-Solving Example 13.3, it is usually convenient to put the larger rate in the numerator. Otherwise fractions are obtained instead of whole numbers.

This same result is found in Experiments 2 and 5. Thus the initial rate is proportional to the *square* of [NO]. Therefore, the rate law is

$$\text{Rate} = k[Cl_2][NO]^2$$

(b) Once the rate law is known, the rate constant k can be calculated. For Experiment 1, for example,

$$8.27 \times 10^{-5}\ \text{mol L}^{-1}\ \text{s}^{-1} = k(0.010\ \text{mol/L})(0.020\ \text{mol/L})^2$$

$$k = \frac{8.27 \times 10^{-5}\ \text{mol L}^{-1}\ \text{s}^{-1}}{(0.010\ \text{mol/L})(0.020\ \text{mol/L})^2} = 2.1 \times 10^1\ \text{mol}^{-2}\ \text{L}^2\ \text{s}^{-1}$$

For this reaction the rate is proportional to one concentration and to the square of another. The rate constant equals the rate (units of mol L^{-1} s^{-1}) divided by three concentration terms multiplied together (units of mol^3 L^{-3}). Thus, the rate constant has units of

$$\frac{\text{mol L}^{-1}\ \text{s}^{-1}}{\text{mol}^3\ \text{L}^{-3}} = \text{mol}^{-2}\ \text{L}^2\ \text{s}^{-1}.$$

For Experiments 2, 3, 4, and 5, the rate constants are 2.1×10^3, 2.1×10^3, 2.1×10^3, and 2.0×10^3 L^2 mol^{-2} s^{-1}, respectively. The average of these values is 2.1×10^3 L^2 mol^{-2} s^{-1}, which can be used to calculate the rate for any set of NO and Cl_2 concentrations at 27 °C.

✓ **Reasonable Answer Check** The five calculated k values are nearly equal. If the rate law were incorrect, or if an error were made in one or more calculations, some k values would be quite different from the others.

PROBLEM-SOLVING PRACTICE 13.3

At 23 °C, these data were collected for the crystal violet reaction

$$Cv^+(aq) + OH^-(aq) \longrightarrow CvOH(aq)$$

Experiment	Initial Concentrations (mol/L)		Initial Rate
	$[Cv^+]$	$[OH^-]$	(mol L^{-1} s^{-1})
1	4.3×10^{-5}	0.10	1.3×10^{-6}
2	2.2×10^{-5}	0.10	6.7×10^{-7}
3	1.1×10^{-5}	0.10	3.3×10^{-7}

(a) Is it possible to determine the complete rate law from the data given? Why or why not?
(b) Assume that the rate does not depend on the concentration of hydroxide ion. What is the rate law?
(c) Calculate the rate constant, again assuming that the rate does not depend on the concentration of hydroxide ion.
(d) Calculate the initial rate of reaction when the concentration of crystal violet cation is 0.00045 M and the concentration of hydroxide ion is 0.10 M. Report your results in mol L^{-1} s^{-1}.
(e) Calculate the rate when the concentration of Cv^+ is half the initial value of 0.00045 M.

EXERCISE 13.5 Determining the Rate Law Using Logarithms

For the reaction in Problem-Solving Example 13.3, assume that the rate law is of the form Rate = $k[NO]^x[Cl_2]^y$. Show mathematically that by taking logarithms of both sides of the rate law and comparing Experiments 2 and 4, where the concentration of chlorine is the same, x, is given by

$$x = \frac{\log\left(\frac{\text{Rate}_4}{\text{Rate}_2}\right)}{\log\left(\frac{[NO]_4}{[NO]_2}\right)}$$

ThomsonNOW

Go to the Coached Problems menu for tutorials on:

- **concentration time relationships**
- **determining the rate law using the graphical method**

13.3 Rate Law and Order of Reaction

For many (but not all) homogeneous reactions, the rate law has the general form

$$\text{Rate} = k[A]^m[B]^n \ldots$$

where concentrations of substances, [A], [B], . . . are raised to powers, m, n, ... The substances A, B, . . . might be reactants, products, or catalysts. The exponents m, n, . . . are usually positive whole numbers but might be negative numbers or fractions. These exponents define the **order of the reaction** with respect to each reactant. If n is 1, for example, the reaction is first-order with respect to B; if m is 2, then the reaction is second-order with respect to A. The sum of m and n (plus the exponents on any other concentration terms in the rate equation) gives the **overall reaction order.** (The reaction in Problem-Solving Example 13.3 is first-order in Cl_2, second-order in NO, and third-order overall.) A very important point to remember is that ***the rate law and reaction orders must be determined experimentally; they cannot be predicted from stoichiometric coefficients in the balanced overall chemical equation.***

None of the reactions in Problem-Solving Example 13.4 has a rate law that can be derived correctly from the stoichiometric equation. For example, H_2 has a coefficient of 2 in Reaction (a), but the rate law involves $[H_2]$ to the first power.

PROBLEM-SOLVING EXAMPLE 13.4 Reaction Order and Rate Law

For each reaction and experimentally determined rate law listed below, determine the order with respect to each reactant and the overall order.

(a) $2\,NO(g) + 2\,H_2(g) \longrightarrow N_2(g) + 2\,H_2O(g)$ Rate = $k[NO]^2[H_2]$

(b) $14\,H_3O^+(aq) + 2\,HCrO_4^-(aq) + 6\,I^-(aq) \longrightarrow 2\,Cr^{3+}(aq) + 3\,I_2(aq) + 22\,H_2O(\ell)$
Rate = $k[HCrO_4^-][I^-]^2[H_3O^+]^2$

(c) *cis*-2-butene(g) $\longrightarrow$ *trans*-2-butene(g) (catalytic concentration of I_2 present)
Rate = k[*cis*-2-butene]$[I_2]^{1/2}$

Answer

(a) First-order in H_2, second-order in NO, third-order overall
(b) First-order in $HCrO_4^-$, second-order in I^-, second-order in H_3O^+, fifth-order overall
(c) First-order in *cis*-2-butene, 0.5-order in I_2, 1.5-order overall

Strategy and Explanation Use the exponents in the rate law—not the stoichiometric coefficients—to determine the order.

(a) The rate law has a single term that is raised to the first power and another term that is squared, so the reaction is first-order in H_2, second-order in NO, and third-order overall.
(b) The rate law contains three terms. Since the $HCrO_4^-$ term is raised to the first power, the reaction is first-order in $HCrO_4^-$. The other two terms are squared, so the reaction is second-order in I^- and second-order in H_3O^+. The exponents sum to five, so the reaction is fifth-order overall.

(c) In this case the rate of reaction depends on the concentration of the reactant and also on the square root ($\frac{1}{2}$ power) of the concentration of a catalyst, I_2. The reaction is therefore first-order in *cis*-2-butene, 0.5-order in I_2, and 1.5-order overall.

PROBLEM-SOLVING PRACTICE 13.4

In Problem-Solving Example 13.3 the rate law for reaction of NO with Cl_2 was found to be

$$2\ NO(g) + Cl_2(g) \longrightarrow 2\ NOCl(g) \qquad \text{Rate} = k[NO]^2[Cl_2]$$

(a) What is the order of the reaction with respect to NO? With respect to Cl_2?
(b) Suppose that you triple the concentration of NO and simultaneously decrease the concentration of Cl_2 by a factor of 8. Will the reaction be faster or slower under the new conditions? How much faster or slower? (Assume that the temperature is the same in both sets of conditions.)

The Integrated Rate Law

Another approach to experimental determination of the rate law and rate constant for a reaction uses calculus to derive what is called the integrated rate law. As an example of the integrated rate law method, suppose that we have a reaction in which a single substance A reacts to form products.

$$A \longrightarrow \text{products}$$

First-Order Reaction. If the rate law is first-order, then

$$\text{Rate} = -\frac{\Delta[A]}{\Delta t} = k[A]$$

This expression can be transformed, using calculus, to the integrated first-order rate law,

$$\ln[A]_t = -kt + \ln[A]_0 \qquad [13.5]$$

where $[A]_t$ represents the concentration of A at time t, $[A]_0$ represents the initial concentration of A (when $t = 0$), and ln represents the natural logarithm function. (Logarithms are discussed in Appendix A.6.)

Equation 13.5 has the same form as the general equation for a straight line, $y = mx + b$, in which m is the slope and b is the y-intercept.

y-axis variable | slope | *x*-axis variable | *y*-intercept

$$\ln[A]_t = -kt + \ln[A]_0$$

$$y = mx + b \qquad [13.5]$$

If the reaction is actually first-order, then a graph of ln[A] on the vertical (y) axis versus t on the horizontal (x) axis should be a straight line. A linear graph, such as the one in Figure 13.5a, is evidence that the reaction is first-order.

Second-Order Reaction. For the same reaction

$$A \longrightarrow \text{products}$$

suppose that the rate depends on the square of the concentration of the reactant; that is, suppose the rate law is second-order.

$$\text{Rate} = k[A]^2$$

The integrated rate law derived using calculus is

$$\frac{1}{[A]_t} = kt + \frac{1}{[A]_0}$$

We have already seen that the rate law allows us to calculate the rate of reaction from the concentration of the reactants (and perhaps other substances). The integrated rate law allows us to calculate the concentration of a reactant (or perhaps another substance) as a function of time.

You do not have to know calculus to use the results that constitute the integrated rate law method. If you do know calculus, however, you will be able to derive the results for yourself.

Using calculus,

$$-\frac{d[A]}{dt} = k[A] \quad \text{and} \quad \frac{d[A]}{[A]} = -k\,dt$$

$$\int_{[A]_0}^{[A]_t} \frac{d[A]}{[A]} = -k\int_0^t dt$$

$$\ln[A]_t - \ln[A]_0 = -k(t_t - t_0)$$

If the reaction starts at time t_0, then $t_t - t_0 = t$, the elapsed time, and $\ln[A]_t - \ln[A]_0 = -kt$ or $\ln[A]_t = -kt + \ln[A]_0$.

Using calculus,

$$-\frac{d[A]}{dt} = k[A]^2 \quad \text{and} \quad \frac{d[A]}{[A]^2} = -k\,dt$$

$$\int_{[A]_0}^{[A]_t} \frac{d[A]}{[A]^2} = -k\int_0^t dt$$

$$-\frac{1}{[A]_t} - \left(-\frac{1}{[A]_0}\right) = -k(t_t - t_0)$$

If the reaction starts at time t_0, then $t_t - t_0 = t$, the elapsed time, and

$$\frac{1}{[A]_t} = kt + \frac{1}{[A]_0}$$

First-order reaction

Intercept = $\ln[A]_0$

Slope = $-k$

$\ln[A]$

Time, t

(a)

Second-order reaction

Slope = k

Intercept = $\frac{1}{[A]_0}$

$1/[A]$

Time, t

(b)

Zeroth-order reaction

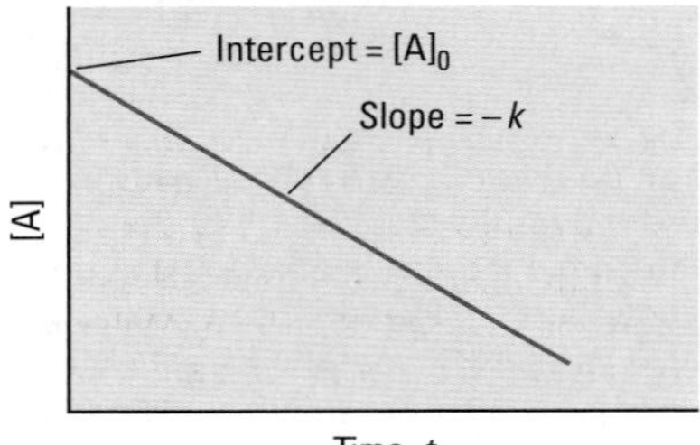

(c)

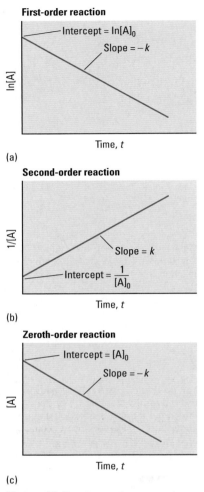

Figure 13.5 First-order, second-order, and zeroth-order plots. (a) If a reaction is first-order in reactant A, plotting $\ln[A]_t$ versus t gives a straight line. (b) If a reaction is second-order in reactant A, plotting $1/[A]_t$ versus t gives a straight line. (c) If a reaction is zeroth-order in reactant A, plotting $[A]_t$ versus t gives a straight line.

This equation is also of the form $y = mx + b$. If a reaction is second-order, a graph of $1/[A]_t$ versus t will be linear with slope $= k$ and y-intercept $= 1/[A]_0$. Such a straight-line graph is evidence that a reaction is second-order. An example graph is shown in Figure 13.5b.

Zeroth-Order Reaction. There are a few reactions for which the rate does not depend on the concentration of a reactant at all. These are called zeroth-order reactions, because the rate law can be written as a rate constant k times a concentration to the zeroth power. (Recall that anything raised to the zeroth power equals 1.)

$$\text{Rate} = -\frac{\Delta[A]}{\Delta t} = k[A]^0 = k$$

This rate law says that for a zeroth-order reaction the rate is always the same no matter what the concentration of reactant. The rate is equal to the rate constant k. For a zeroth-order reaction you can derive the integrated form without calculus. Simply use algebra to rearrange the rate law just given.

$$\Delta[A] = -k\Delta t$$
$$[A]_t - [A]_0 = -k(t_t - t_0) = -kt$$
$$[A]_t = -kt + [A]_0$$

Again, the equation is of the form $y = mx + b$. If a reaction is zeroth-order, graphing A_t versus t gives a straight line with slope $= -k$ and y-intercept $= [A]_0$, as shown in Figure 13.5c.

To summarize these three situations, a rate law that involves powers of the reactant concentration can be written as

$$\text{Rate} = -\frac{\Delta[A]}{\Delta t} = k[A]^m$$

where m is the order of the reaction as defined earlier. The integrated rate law depends on the value of m, and the results for integer values of m from 0 up to 2 are given in Table 13.2.

To determine the order of reaction, then, we collect concentration-time data and make the three plots listed in Table 13.2 and shown in Figure 13.5. Only one (or perhaps none) of the plots will be a straight line. If one is straight, then it indicates the order, and the rate constant can be calculated from its slope. The units of the rate constant are those of the slope of the line; they are given in the last column of Table 13.2. The units of the rate constant depend on the order of the reaction.

Table 13.2 Integrated Rate Laws

Order	Rate Equals	Integrated Rate Law*	Straight-line Plot	Slope of Plot	Units of k
0	$k[A]^0 = k$	$[A]_t = -kt + [A]_0$	$[A]_t$ vs t	$-k$	conc. time^{-1}
1	$k[A]$	$\ln[A]_t = -kt + \ln[A]_0$	$\ln[A]_t$ vs t	$-k$	time^{-1}
2	$k[A]^2$	$\frac{1}{[A]_t} = kt + \frac{1}{[A]_0}$	$\frac{1}{[A]_t}$ vs t	k	conc.$^{-1}$ time^{-1}

*In the table, $[A]_0$ indicates the initial concentration of substance A, that is, the concentration of A at $t = 0$, the time when the reaction was started.

PROBLEM-SOLVING EXAMPLE 13.5 Reaction Order and Rate Constant from Integrated Rate Law

These data were obtained for decomposition of cyclopentene at 825 K.

$$C_5H_8(g) \longrightarrow C_5H_6(g) + H_2(g)$$

Time (s)	$[C_5H_8]$ (mol/L)	Time (s)	$[C_5H_8]$ (mol/L)
0	0.0200	300	0.0084
20	0.0189	400	0.0063
50	0.0173	500	0.0047
100	0.0149	700	0.0027
200	0.0112	1000	0.0011

Cyclopentene has the structure

Obtain the order of the reaction and the rate constant.

Answer The reaction is first-order in C_5H_8; $k = 2.88 \times 10^{-3}\ s^{-1}$.

Strategy and Explanation Use the integrated rate law method, making graphs to test for zeroth-, first-, and second-order. Such plots are shown in Figure 13.6. The zeroth-order and second-order plots are curved, while the first-order plot is a straight line. Thus the reaction must be first-order. From the first-order plot, calculate the slope using the points marked on the graph as open circles.

$$\text{Slope} = \frac{\{-6.56 - (-4.15)\}}{\{910 - 74\}\ s} = -2.88 \times 10^{-3}\ s^{-1}$$

It is important to use two points on the straight line through the experimental data (open circles in Figure 13.6b), not two of the data points themselves, when you calculate the slope. Making a graph is similar to averaging, because the straight line and its slope are based on all ten points in the data table, not just two of them.

The slope of $-2.88 \times 10^{-3}\ s^{-1}$ is the negative of the rate constant (see Table 13.2), which means that $k = 2.88 \times 10^{-3}\ s^{-1}$.

✓ **Reasonable Answer Check** The units are s^{-1}, which corresponds to the reciprocal time units indicated in Table 13.2 for a first-order rate constant.

PROBLEM-SOLVING PRACTICE 13.5

Use the concentration-time data in Table 13.1 for the reaction of Cv^+ with OH^- (p. 610) to deduce the order of the reaction with respect to Cv^+. Determine the rate constant. (Assume that $[OH^-]$ is constant during the reaction.)

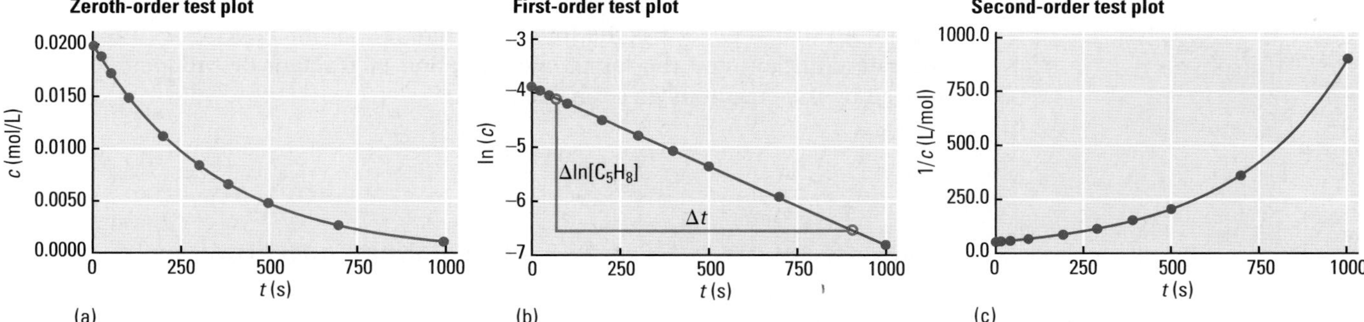

Figure 13.6 Integrated rate law plots for cyclopentene decomposition reaction. (a) Zeroth-order, (b) first-order, and (c) second-order plots for the decomposition reaction of cyclopentene in aqueous solution at 825 K.

CHEMISTRY YOU CAN DO

Simulating First-Order and Zeroth-Order Reactions

Obtain 100 pennies, several dripless birthday candles, a rigid piece of cardboard or plastic at least six inches square, matches, a watch that displays seconds, and a ruler. You will do two separate experiments.

In the first experiment, shake the pennies in a container and toss them onto a flat surface. Count how many pennies land heads-up. Record the number of tosses (n_t = 1 so far) and the number of heads (n_h). Pick up the heads-up pennies and set them aside. Repeat the toss with the remaining pennies, again counting the number of heads, setting them aside, and recording the results. Continue this process until the number of pennies remaining is three or less. Graph the number of heads as a function of the number of tosses. Also graph the natural logarithm of the number of heads as a function of number of tosses.

In the second experiment, melt the bottom of a birthday candle a little and stick it to the rigid piece of cardboard so that it stands vertically. Fasten the ruler vertically and close enough to the candle so that you can measure the length of the candle (the wax part, not counting the wick). Record the length of the candle. Make certain that no combustible material (including the ruler) is close enough to the candle that a flame might ignite it. Light the candle and record the time to the nearest second. Read the length of the candle every 30 s and record your measurement. After five minutes, extinguish the candle. Plot the length of the candle wax as a function of time. Plot the natural logarithm of the length of the candle wax versus time.

1. In the experiment with the pennies, which variable corresponds with time in a real kinetics experiment? Which variable corresponds with concentration?

© Thomson Learning/Charles D. Winters

2. What is the order of the penny-tossing process?
3. For each data point in the penny-tossing process, calculate $\Delta n_h/n_h$, where Δn_h is the difference between the number of heads for the current toss and the previous one. (Use 100 for the number of heads in the zeroth toss.) What do you observe about the values of $\Delta n_h/n_h$? Why do you think these values are the way they are?
4. What is the order of the candle-burning process?
5. What can you observe about the change in the length of the candle for each 30-s interval? Is this consistent with your answer to Question 4? If so, why?

SOURCES: Based on Sanger, M. J., Wiley, R. A. Jr., Richter, E. W., and Phelps, A. J. *Journal of Chemical Education,* Vol. 79, 2002; pp. 989–991; and Sanger, M. J. *Journal of Chemical Education,* Vol. 80, 2003; pp. 304A–304B.

Calculating Concentration or Time from Rate Law

Once the rate law has been determined experimentally, it provides a way to calculate the concentration of a reactant or product at any time after the reaction has begun. All that is needed is the integrated rate law (from Table 13.2), the value of the rate constant, and the initial concentration of reactant or product. These are related by the equations given in Table 13.2.

PROBLEM-SOLVING EXAMPLE 13.6 Calculating Concentrations

The first-order rate constant is $1.87 \times 10^{-3}\ \text{min}^{-1}$ at 37 °C (body temperature) for reaction of cisplatin, a cancer chemotherapy agent, with water. The reaction is

$$\text{cisplatin} + H_2O \longrightarrow \text{cisplatinOH}_2^+ + Cl^-$$

Suppose that the concentration of cisplatin in the bloodstream of a cancer patient is 4.73×10^{-4} mol/L. What will the concentration be exactly 24 hours later?

Answer 3.20×10^{-5} mol/L

Strategy and Explanation Assume that the rate of reaction in the blood is the same as in water. Let [cisplatin] represent the concentration of cisplatin at any time and $[\text{cisplatin}]_0$ represent the initial concentration. The reaction is first-order, so from Table 13.2,

$$\ln[\text{cisplatin}] = -kt + \ln[\text{cisplatin}]_0$$

Rearrange the equation algebraically to

$$\ln[\text{cisplatin}] - \ln[\text{cisplatin}]_0 = -kt$$

$$\ln\left\{\frac{[\text{cisplatin}]}{[\text{cisplatin}]_0}\right\} = -kt$$

$$\frac{[\text{cisplatin}]}{[\text{cisplatin}]_0} = \text{anti}\ \ln(-kt) = e^{-kt}$$

$$[\text{cisplatin}] = [\text{cisplatin}]_0\, e^{-kt}$$

$$[\text{cisplatin}] = (4.73 \times 10^{-4}\ \text{mol/L})\, e^{-(1.87\times 10^{-3}\ \text{min}^{-1}\times 24\ \text{h})}$$

$$[\text{cisplatin}] = (4.73 \times 10^{-4}\ \text{mol/L})\, e^{\left(\frac{1.87\times 10^{-3}}{\text{min}}\times 24\ \text{h}\times\frac{60\ \text{min}}{1\ \text{h}}\right)}$$

$$[\text{cisplatin}] = (4.73 \times 10^{-4}\ \text{mol/L})(6.77 \times 10^{-2})$$

$$= 3.20 \times 10^{-5}\ \text{mol/L}$$

Mathematical operations involving logarithms and exponentials (antilogarithms) are discussed in Appendix A.6.

Because the solution to this example involves a ratio of concentrations in which the units divide out, the same approach can be taken in problems that involve the number of moles, the mass, or the number of atoms or molecules at two different times.

✓ **Reasonable Answer Check** If the drug is to be effective, it almost certainly needs to be in the body for some time—several minutes to hours or more. The concentration has dropped to a little less than 10% of its initial value in 24 hours, which is reasonable.

PROBLEM-SOLVING PRACTICE 13.6

The first-order rate constant for decomposition of an insecticide in the environment is $3.43 \times 10^{-2}\ \text{d}^{-1}$. How long does it take for the concentration of insecticide to drop to $\frac{1}{10}$ of its initial value?

Half-Life

The **half-life** of a reaction, $t_{1/2}$, is *the time required for the concentration of a reactant A to fall to one half of its initial value.* That is, $[\text{A}]_{t_{1/2}} = \frac{1}{2}[\text{A}]_0$. For a first-order reaction the half-life has the same value, no matter what the initial concentration is. For other reaction orders this is not true.

The half-life is related to the first-order rate constant. To see how, use algebra to rearrange Equation 13.5 (which was associated with a reaction A ⟶ products):

$$\ln[\text{A}]_t = -kt + \ln[\text{A}]_0 \qquad [13.5]$$

$$\ln[\text{A}]_t - \ln[\text{A}]_0 = -kt$$

$$\ln[\text{A}]_{t_{1/2}} - \ln[\text{A}]_0 = -kt_{1/2} \qquad [13.6]$$

Remember that Equation 13.5 was derived for a reaction of the form A ⟶ products. Had the equation been 2 A ⟶ products, that is, had the coefficient of A been 2 in the chemical equation, the result would be slightly different. (See Question 131 at the end of this chapter.)

Because $[\text{A}]_{t_{1/2}} = \frac{1}{2}[\text{A}]_0$, Equation 13.5 can be rewritten as

$$\ln([\text{A}]_0/2) - \ln[\text{A}]_0 = -kt_{1/2}$$

and so

$$-kt_{1/2} = \ln[\text{A}]_0 - \ln(2) - \ln[\text{A}]_0 = -\ln 2$$

$$t_{1/2} = \frac{-\ln 2}{-k} = \frac{0.693}{k} \qquad [13.7]$$

This means that measuring the half-life of a first-order reaction determines the rate constant, and vice versa. Radioactive decay (Section 20.4) is a first-order process, and half-life is typically used to report the rate of decay of radioactive nuclei.

PROBLEM-SOLVING EXAMPLE 13.7 Half-Life and Rate Constant

In Problem-Solving Example 13.5, the rate constant for decomposition of cyclopentene at 825 K was found to be $2.88 \times 10^{-3}\ s^{-1}$. The reaction gave a linear first-order plot. What is the half-life in seconds for this reaction?

Answer 241 s

Strategy and Explanation The reaction is first-order and is of the form A $\longrightarrow$ products, so Equation 13.7 can be used.

$$t_{1/2} = \frac{0.693}{k} = \frac{0.693}{2.88 \times 10^{-3}\ s^{-1}} = 2.41 \times 10^2\ s = 241\ s$$

✓ **Reasonable Answer Check** The rate constant is about $3 \times 10^{-3}\ s^{-1}$. If the concentration of cyclohexene were 1.0 mol L^{-1}, then the reaction rate would be 0.003 $s^{-1} \times$ 1.0 mol L^{-1} = 0.003 mol $L^{-1}\ s^{-1}$. This means that 0.003 mol L^{-1} would react every second. For the concentration to drop to half of 1.0 mol L^{-1}, the change in concentration would be 0.500 mol L^{-1}. Therefore, it would take at least $\frac{0.500\ mol\ L^{-1}}{0.003\ mol\ L^{-1}\ s^{-1}} = 167$ s for the concentration to drop to half its initial value, and the half-life should be at least 167 s. (Because the rate decreases as concentration decreases, it will take longer than 167 s.) A half-life of 241 s is reasonable.

PROBLEM-SOLVING PRACTICE 13.7

From Figure 13.3 determine the time required for the concentration of Cv^+ to fall to one half the initial value. Verify that the same period is required for the concentration to fall from one half to one fourth the initial value. From this half-life, calculate the rate constant.

13.4 A Nanoscale View: Elementary Reactions

Macroscale experimental observations reveal that reactant concentrations, temperature, and catalysts can affect reaction rates. But how can we interpret such observations in terms of nanoscale models? We will use the ***kinetic-molecular theory of matter,*** which was first introduced in Section 1.8 (***p. 19***) and developed further in Section 10.3 (***p. 436***), together with the ideas about molecular structure developed in Chapters 8 and 9. These concepts provide a good basis for understanding how atoms and molecules move and chemical bonds are made or broken during the very short time it takes for reactant molecules to be converted into product molecules.

According to kinetic-molecular theory, molecules are in constant motion. In a gas or liquid they bump into one another; in a solid they vibrate about specific locations. Molecules also rotate, flex, or vibrate around or along the bonds that hold the atoms together. These motions produce the transformations of molecules that occur during chemical reactions. It turns out that ***there are only two important types of molecular transformations:*** unimolecular and bimolecular. In a **unimolecular reaction** *the structure of a single particle (molecule or ion) rearranges to produce a different particle or particles.* A unimolecular reaction might involve breaking a bond and forming two new molecules, or it might involve rearrangement of one isomeric structure into another. In a **bimolecular reaction** *two particles (atoms, molecules, or ions) collide and rearrange into products.* In a bimolecular reaction new bonds may be formed between the reactant particles, and existing bonds may be broken. Sometimes the two particles combine to form a new, larger one. Sometimes two or more new molecules are formed from the original two.

ESTIMATION

Pesticide Decay

There are usually several different ways that a pesticide can decompose in an ecosystem. For this reason, it is difficult to define an accurate rate of decomposition and even more difficult to define a rate law. Often it is assumed that decomposition is first-order, and an approximate half-life is reported.

Organochlorine pesticides such as DDT, lindane, and dieldrin may have half-lives as long as 10 years in the environment. The maximum contaminant level (MCL) for lindane is 0.2 ppb (parts per billion). Suppose that an ecosystem has been contaminated with lindane at a concentration of 200 ppb. How long would you have to wait before it would be safe to enter the ecosystem without protection from the pesticide?

The level of contamination is 200 ppb/0.2 ppb = 1000 times the MCL. Presumably it would be safe to wait until the level had dropped to 1/1000 of its initial value. You can estimate how long this would be by using powers of 2, because the number of half-lives required is n, where $(1/2)^n = 1/1000$. That is, as soon as 2^n exceeds 1000, you have waited long enough. Computer scientists, who deal with binary arithmetic, can easily tell you that $2^{10} = 1024$, so $n = 10$. You can verify this using your calculator's y^x-key, or you could simply raise 2 to a power until a value greater than 1000 was calculated. If you did not have a calculator, you could multiply $2 \times 2 \times 2 \ldots$ in your head until you had enough factors to multiply out to a number bigger than 1000. Ten half-lives means 10×10 years, or 100 years, so after 100 years the ecosystem would be free of significant lindane contamination.

ThomsonNOW™
Sign in to ThomsonNOW at **www.thomsonedu.com** to work an interactive module based on this material.

All chemical reactions can be understood in terms of simple reactions such as those just described. Very complicated reactions can be built up from combinations of unimolecular and bimolecular reactions, just as complicated compounds can be built from chemical elements. For example, hundreds of such reactions are needed to understand how smog is produced in a city such as Los Angeles or to understand why chlorofluorocarbons can deplete stratospheric ozone (***p. 470***). Like the chemical elements, the simplest nanoscale reactions are building blocks, so they are referred to as **elementary reactions.** ***The equation for an elementary reaction shows exactly which molecules, atoms, or ions take part in the elementary reaction.*** The next two sections describe the two important types of elementary reactions in more detail.

EXERCISE 13.6 Unimolecular and Bimolecular Reactions

For each of the nanoscale molecular diagrams below, write a balanced equation using chemical formulas. Which of the reactions are unimolecular reactions? Which are bimolecular?

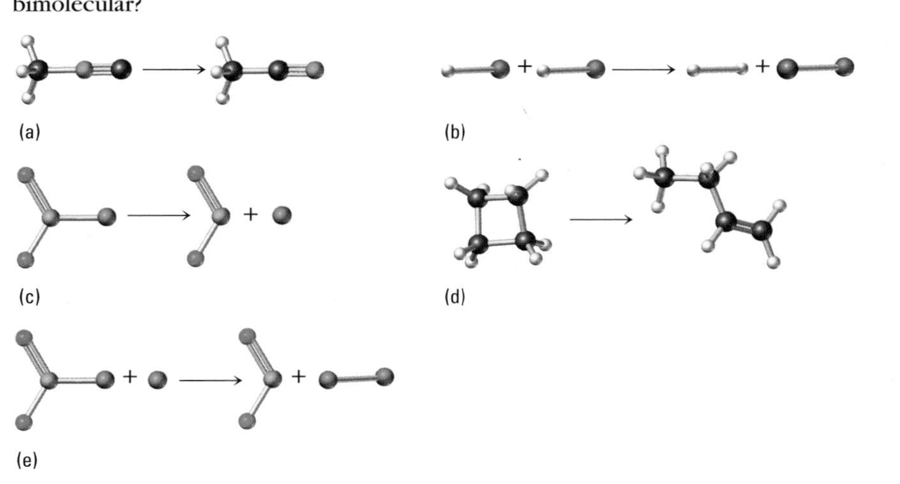

Remember the color codes for atoms:

carbon,	black	nitrogen,	blue
hydrogen,	white	chlorine,	green
oxygen,	red	iodine,	violet

Unimolecular Reactions

An example of a unimolecular reaction is the conversion of *cis*-2-butene to *trans*-2-butene.

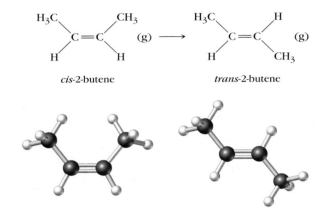

$$\mathrm{H_3C(H)C{=}C(H)CH_3}\ (g) \longrightarrow \mathrm{H_3C(H)C{=}C(CH_3)H}\ (g)$$

cis-2-butene *trans*-2-butene

Cis-2-butene and *trans*-2-butene are *cis-trans* stereoisomers (***p. 345***). The difference between the two molecules is the orientation of the methyl groups, which are on the same side of the double bond in the *cis* structure and on opposite sides in the *trans* structure. If we could grab one end of the molecule and twist it 180° around the axis of the double bond, we would get the other molecule. Thus, it is a reasonable hypothesis that the molecular pathway by which *cis*-2-butene changes to *trans*-2-butene involves twisting the molecule around the double bond. The angle of twist around the double-bond axis measures the progress of the reaction on the nanoscale. The greater the angle, the less the molecule is like *cis*-2-butene and the more it is like *trans*-2-butene, until an angle of 180° is reached and it has become *trans*-2-butene.

Figure 13.7 is similar to Figure 6.15 (***p. 242***), which showed the energy change when bonds were broken and formed as H_2 and Cl_2 changed into HCl.

Such a twist requires that the reactant molecule have sufficient energy. Chemical bonds are like springs. They can be stretched, twisted, and bent, but these changes raise the potential energy. Consequently, some kinetic energy must be converted to potential energy when one end of the *cis*-2-butene molecule twists relative to the other, just as it would if a spring were twisted. At room temperature most of the molecules do not have enough energy to twist far enough to change *cis*-2-butene into *trans*-2-butene. Therefore, *cis*-2-butene can be kept in a sealed flask at room temperature for a long time without any appreciable quantity of *trans*-2-butene being formed. However, as the temperature is raised, more and more molecules have sufficient energy to react, and the reaction gets faster and faster.

Since molecules are very small, the energy required to twist one *cis*-2-butene molecule is very small. However, if we wanted to twist a mole of molecules all at once, it would take a lot of energy. The energy required to reach the top of the "hill" is often reported per mole of molecules—that is, as (435×10^{-21} J/molecule) $\times$ (6.022×10^{23} molecules/mol) = 262 kJ/mol.

Figure 13.7 shows a plot of potential energy versus the angle of twist in *cis*- and *trans*-2-butene. The potential energy is 435×10^{-21} J higher when one end of a *cis*-2-butene molecule is twisted by 90° from the initial flat molecule. This is similar to the increased potential energy that an object such as a car has at the top of a hill compared with its energy at the bottom. Just as a car cannot reach the top of a hill unless it has enough energy, a molecule cannot reach the top of the "hill" for a reaction unless it has enough energy. Notice that the top of the hill can be approached from either side, and from the top a twisted molecule can go downhill energetically to either the *cis* or the *trans* form. *The structure at the top of an energy diagram like this one* is called the **transition state** or **activated complex.** In this case it is a molecule that has been twisted so that the methyl groups are at a 90° angle.

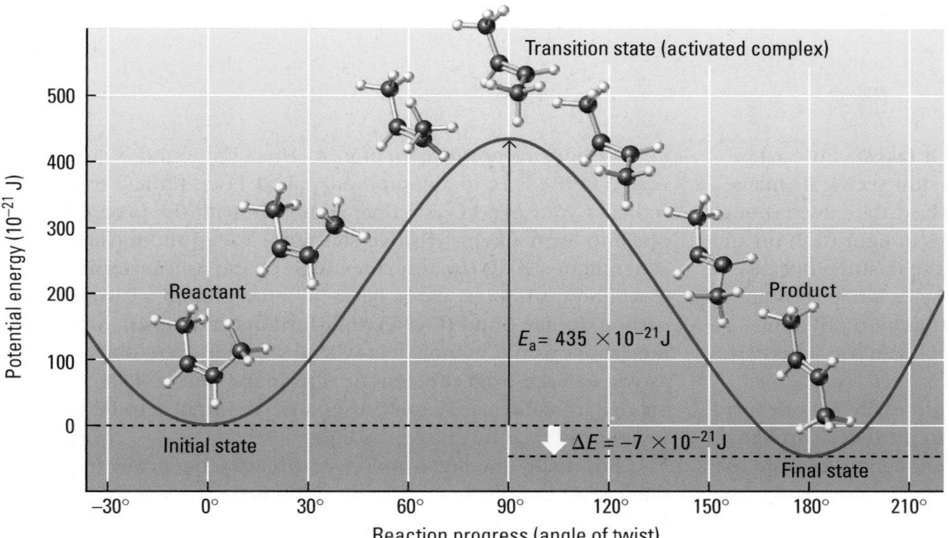

Active Figure 13.7 **Energy diagram for the conversion of *cis*-2-butene to *trans*-2-butene.** Twisting one end of the molecule with respect to the other requires an increase in potential energy because the double bond between the two central C atoms is like a spring and resists twisting. Twisting in either direction requires the same increase in potential energy, so the energy for +30° equals that for −30°. When the angle between the ends of the molecule is 90°, the potential energy has risen by 435×10^{-21} J. A molecule of *cis*-2-butene must have at least this quantity of energy before it can twist past 90° to 180°, which converts it to *trans*-2-butene. The 90° twisted structure at the top of the diagram is called the *transition state* or *activated complex*. The progress of the reaction (the change in the structure of this single molecule) is measured by the angle of twist. **Go to the Active Figures menu at ThomsonNOW to test your understanding of the concepts in this figure.**

CONCEPTUAL EXERCISE 13.7 Transition State

Methyl isonitrile reacts to form acetonitrile in a single-step elementary reaction.

$$CH_3NC(g) \longrightarrow CH_3CN(g)$$

During the reaction the nitrogen atom and one of the carbon atoms exchange places, but the rest of the molecule is unchanged. Suggest a structure for the transition state for this reaction. Draw this structure as a Lewis structure and as a ball-and-stick molecular model.

ThomsonNOW™
Go to the Coached Problems menu for a module on **reaction coordinate diagrams.**

Almost every chemical reaction has an energy barrier that must be surmounted as reactant molecules change into product molecules. The heights of such barriers vary greatly—from almost zero to hundreds of kilojoules per mole. *At a given temperature, the higher the energy barrier, the slower the reaction.* The minimum energy required to surmount the barrier is called the **activation energy, E_a,** for the reaction. For the *cis*-2-butene ⟶ *trans*-2-butene reaction the activation energy is 435×10^{-21} J/molecule, or 262 kJ/mol (see Figure 13.7).

Another interesting relationship shown in Figure 13.7 connects kinetics and thermodynamics. The energy of the product, one molecule of *trans*-2-butene, is 7×10^{-21} J *lower* than that of the reactant, one molecule of *cis*-2-butene. This means

The generalization that higher activation energy results in slower reaction applies best if the reactions are similar. For example, it applies to a group of reactions that all involve twisting around a double bond. It also applies to reactions that involve collisions of one molecule with each of a group of similar molecules. It would be less applicable if we were comparing one reaction that involved collision of two molecules with another reaction that involved twisting around a bond.

CHEMISTRY YOU CAN DO

Kinetics and Vision

Vision is not an instantaneous process. It takes a little while after a bright light goes out before you stop seeing its image. The flash of a camera blinds you for a short time even though it is on for only an instant. Some sources of light flash on and off very rapidly, but you do not notice the flashing because your eyes continue to perceive their images while they are off. However, if you can focus such a source on different parts of your retina at different times, you can see whether it is flashing. Here's how.

Find a small mirror that you can hold easily in your hand. Use the mirror to reflect the image of an incandescent light bulb onto your eye. You should be far enough away from the light so that its image is small. Now move the mirror quickly back and forth so that the image of the light bulb moves quickly across your eyeball. Does the light smear or do you see individual dots? Try the same experiment with the screen of a TV set. (Get really far away from it so the image is small.) Do you see separate images or just a smear of light? If you see individual images, it means that the light is flashing. Each time it flashes on, the moving mirror has caused it to hit a different part of your retina, and you see a separate image.

Repeat this experiment with as many different light sources as you can, and classify them as flashing or continuous. Try street lights of various kinds, car headlights, neon signs, fluorescent lights, and anything else you can think of, but *don't* try a very bright light, such as a laser pointer, that might damage your eyes. Record your observations.

Rotation around a double bond, as in the interconversion of *cis*- and *trans*-2-butene, occurs in the reactions that allow you to see ***(⟵ p. 406)***. A yellow-orange compound called β-carotene, the natural coloring agent in carrots, breaks down in your body to produce vitamin A. This compound is converted in the liver to a compound called 11-*cis*-retinal. In the retina of your eye 11-*cis*-retinal combines with the protein opsin to form a light-sensitive substance called rhodopsin. When light strikes the retina, enough energy is transferred to a rhodopsin molecule to allow rotation around a carbon-carbon double bond (C═C), transforming rhodopsin into metarhodopsin II, a molecule whose shape is quite different, as you can see from the structural formulas below. This change in molecular shape causes a nerve impulse to be sent to your brain, and you see the light.

Eventually the metarhodopsin II reacts chemically to produce a different form of retinal, which is then converted back to vitamin A, and the cycle of chemical changes can begin again. However, decomposition of metarhodopsin II is not as rapid as its formation, and an image formed on the retina persists for a tenth of a second or so. This persistence of vision allows you to perceive videos as continuously moving images, even though they actually consist of separate pictures, each painted on a screen for a thirtieth of a second.

1. For each light source that you observed, what would happen if you focused a camera on the light source, opened the shutter, and moved the camera quickly while the shutter was open?
2. What would happen if you moved the camera more slowly or very quickly?

Note: This *Chemistry You Can Do* relates to the question, "What are the molecules in my eyes doing when I watch a movie?" that was posed in Chapter 1 *(⟵ p. 2)*.

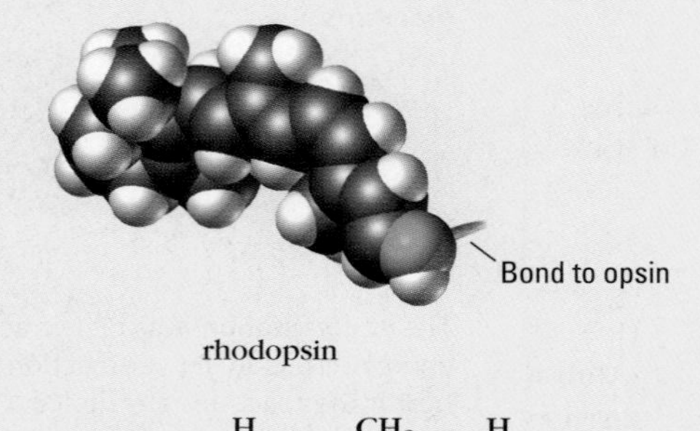

rhodopsin

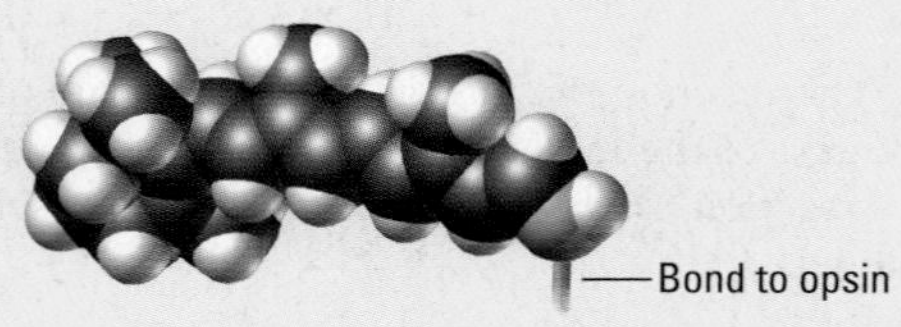

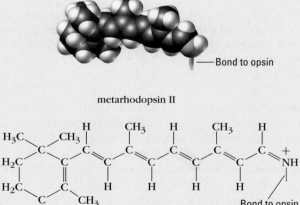

metarhodopsin II

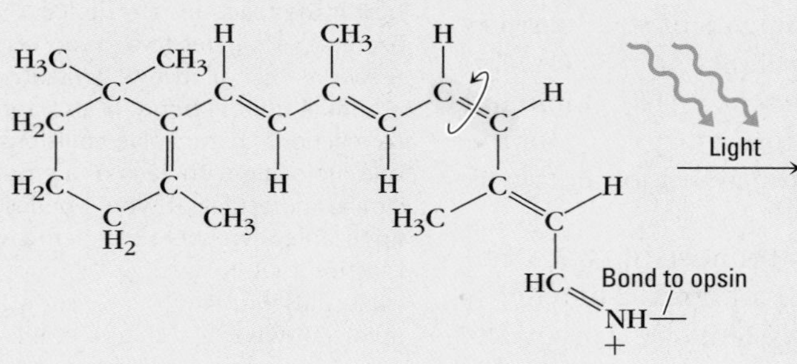

that the *cis* ⟶ *trans* reaction is *exothermic* by 7×10^{-21} J/molecule, which translates to 4 kJ/mol. Also, *cis*-2-butene is higher in energy by 7×10^{-21} J/molecule, so the reverse reaction requires that 4 kJ/mol be absorbed from the surroundings; it is *endothermic.* The height of the energy hill that must be climbed when the reverse reaction occurs is $(435 + 7) \times 10^{-21}$ J/molecule or 442×10^{-21} J/molecule (266 kJ/mol). Thus, the activation energy for the forward reaction is 4 kJ/mol less than that for the reverse reaction. For almost all reactions the activation energy for a forward reaction will differ from the activation energy of the reverse reaction, and the difference is $\Delta E°$ for the reaction.

The actual relation is $\Delta E° = E_a(\text{forward}) - E_a(\text{reverse})$. Since $\Delta E°$ differs from $\Delta H°$ only when there is a change in volume of the reaction system (under constant pressure), the difference in activation energies is often equated with the enthalpy change (⇐ *p. 233*).

Bimolecular Reactions

An example of a bimolecular process is the reaction of iodide ion, I^-, with methyl bromide, CH_3Br, in aqueous solution.

$$I^-(aq) + CH_3Br(aq) \longrightarrow ICH_3(aq) + Br^-(aq)$$

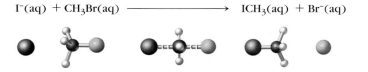

Here the equation for the elementary reaction shows that an iodide ion must collide with a methyl bromide molecule for the reaction to occur. The carbon-bromine bond does not break until after the iodine-carbon bond has begun to form. This makes sense, because just breaking a carbon-bromine bond would require a large increase in potential energy. Partially forming a carbon-iodine bond while the other bond is breaking lowers the potential energy. This helps keep the activation energy hill low. Figure 13.8 shows the energy-versus-reaction progress diagram for this reaction.

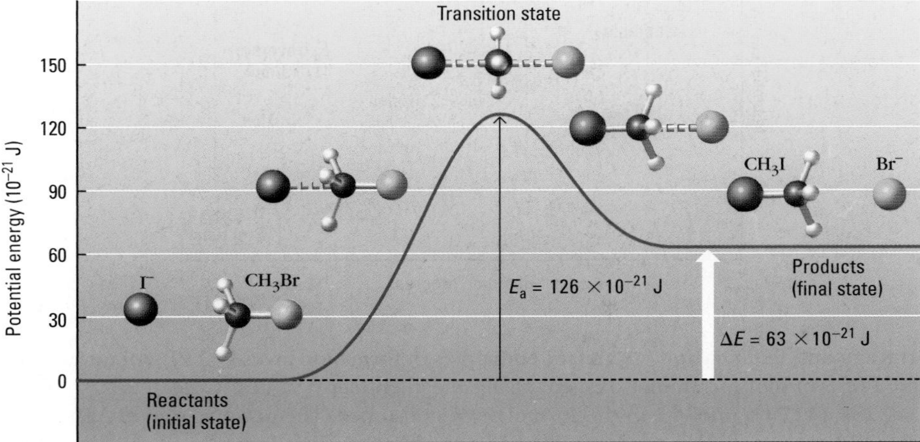

Figure 13.8 Energy diagram for iodide–methyl bromide reaction. During collision of an iodide ion with a methyl bromide molecule, a new iodine-carbon bond forms at the same time that the carbon-bromine bond is breaking. Forming the new bond lowers the potential energy, which otherwise would be raised a lot by breaking the carbon-bromine bond. This results in a lower activation energy and a faster reaction than would otherwise occur. In this case reaction progress is measured in terms of stretching of the carbon-bromine bond and formation of the iodine-carbon bond.

The methyl bromide molecule has a tetrahedral shape that is distorted because the Br atom is much larger than the H atoms. Numerous experiments suggest that the reaction occurs most rapidly in solution when the I^- ion approaches the methyl bromide from the side of the tetrahedron opposite the bromine atom. That is, approach to only one of the four sides of CH_3Br can be effective, which limits reaction to only one fourth of all the collisions at most. This factor of one fourth is called a **steric factor** because it depends on the three-dimensional shapes of the reacting molecules. For molecules much more complicated than methyl iodide, such geometry restraints mean that only a very small fraction of the total collisions can lead to reaction. No wonder some chemical reactions are slow.

The word *steric* comes from the same root as the prefix *stereo-*, which means "three-dimensional."

PROBLEM-SOLVING EXAMPLE 13.8 Reaction Energy Diagrams

A reaction by which ozone is destroyed in the stratosphere (*p. 463*) is

$$O_3(g) + O(g) \longrightarrow 2\ O_2(g)$$

(O represents atomic oxygen, which is formed in the stratosphere when photons of ultraviolet light from the sun split oxygen molecules in two.) The activation energy for ozone destruction is 19 kJ/mol of O_3 consumed. Use standard enthalpies of formation from Appendix J to calculate the enthalpy change for this reaction. Then construct an energy diagram for the reaction. Draw vertical arrows to indicate the sizes of $\Delta H°$, E_a(forward), and E_a(reverse) for the reaction. [Remember that because there are equal numbers of moles of gas phase reactants and products, there is no volume change at constant temperature, and $\Delta E° = \Delta H°$ for this reaction (*p. 230*).]

Answer $\Delta H° = \Delta E° = -392$ kJ/mol of O_3 consumed, E_a(forward) = 19 kJ/mol O_3 consumed, E_a(reverse) = 411 kJ/mol O_3 formed; the energy diagram is shown below.

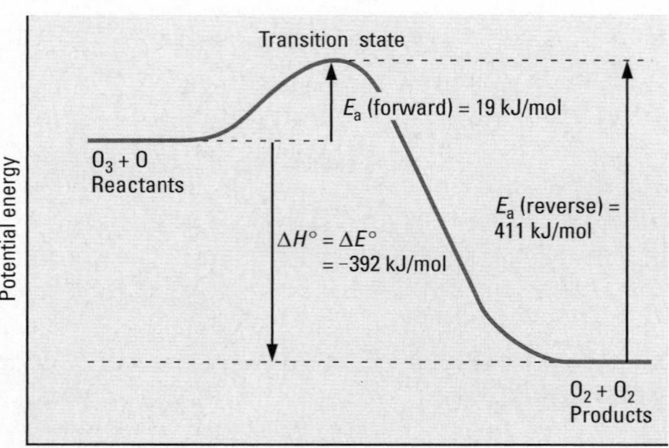

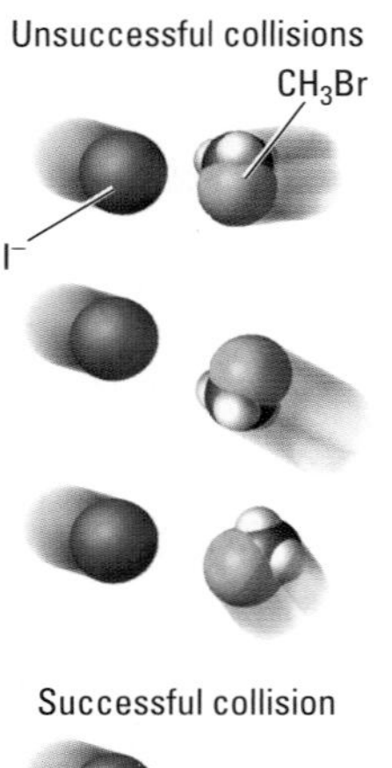

Unsuccessful collisions. In the first three collisions shown, the iodide ion does not approach the methyl bromide molecule from the side opposite the bromine atom. None of these collisions is as likely to result in a reaction as is the collision shown at the bottom.

Strategy and Explanation Standard enthalpies of formation are 249.2 kJ/mol for ozone and 142.7 kJ/mol for atomic oxygen. Using these enthalpy values, we get $\Delta H° = 0 - (249.2 + 142.7)$ kJ/mol $= -391.9$ kJ/mol of O_3 consumed. The negative sign of $\Delta H°$ indicates that the reaction is exothermic, so the products must be lower in energy than the reactants by 391.9 kJ/mol. Since E_a(forward) = 19 kJ/mol, the transition state must be this much higher in energy than the reactants. Thus, the first two arrows on the left side of the diagram can be drawn. Then the third arrow (from products to the transition state) can be drawn. It indicates that

$$E_a(\text{reverse}) = -\Delta H° + E_a(\text{forward}) = (391.9 + 19)\ \text{kJ/mol} = 411\ \text{kJ/mol}$$

PROBLEM-SOLVING PRACTICE 13.8

For the reaction

$$Cl_2(g) + 2\ NO(g) \longrightarrow 2\ NOCl(g)$$

the activation energy is 18.9 kJ/mol. For the reaction

$$2\ NOCl(g) \longrightarrow 2\ NO(g) + Cl_2(g)$$

the activation energy is 98.1 kJ/mol. Draw a diagram similar to Figure 13.7 for the reaction of NO with Cl_2 to form NOCl. Is this reaction exothermic or endothermic? Explain.

CONCEPTUAL EXERCISE 13.8 Successful and Unsuccessful Collisions

The reaction

$$2\ NOCl(g) \longrightarrow 2\ NO(g) + Cl_2(g)$$

occurs in a single bimolecular step. Draw at least four possible ways that two NOCl molecules could collide, and rank them in order of greatest likelihood that a collision will be successful in producing products.

AP Wide World Photos

Ahmed H. Zewail
1946–

For his studies of the transition states of chemical reactions using femtosecond spectroscopy, Ahmed H. Zewail of the California Institute of Technology received the 1999 Nobel Prize in chemistry. Zewail, who holds joint Egyptian and U.S. citizenship, pioneered the use of extremely short laser pulses—on the order of femtoseconds (10^{-15} s)—to study chemical kinetics. His technique has been called the world's fastest camera, and his research has enhanced understanding of many reactions, among them those involving rhodopsin and vision.

13.5 Temperature and Reaction Rate: The Arrhenius Equation

The most common way to speed up a reaction is to increase the temperature. A mixture of methane and air can be ignited by a lighted match, which raises the temperature of the mixture of reactants. This increases the reaction rate, the thermal energy evolved maintains the high temperature, and the reaction continues at a rapid rate. Reactions that speed up when the temperature is raised must slow down when the temperature is lowered. Foods are stored in refrigerators or freezers because the reactions in cells of microorganisms that produce spoilage occur more slowly at the lower temperature.

As a rough rule of thumb, the reaction rate increases by a factor of 2 to 4 for each 10-K rise in temperature.

Reaction rates increase with temperature because at a higher temperature a greater fraction of reactant molecules has enough energy to surmount the activation energy barrier. Consider again the conversion of *cis*- to *trans*-2-butene (Figure 13.7, p. 627). You learned in Section 10.3 (**← *p. 436***) that gas phase molecules are constantly in motion and have a wide distribution of speeds and energies. At room temperature relatively few *cis*-2-butene molecules have sufficient energy to surmount the energy barrier. However, as the temperature goes up, the number of molecules that have enough energy goes up rapidly, so the reaction rate increases rapidly.

The number of *cis*-2-butene molecules that have a given energy is shown by the curves in Figure 13.9. One curve is for 25 °C; the other is for 75 °C. The higher a point is on either curve, the greater is the number of molecules that have the energy corresponding to that point. The areas under the two curves are the same and represent the same total number of molecules. With a 50 °C rise in temperature, the number of molecules whose energy exceeds the activation energy is much higher, and so is the reaction rate.

A reaction is faster at a higher temperature because its rate constant is larger. That is, ***a rate constant is constant only for a given reaction at a given temperature.*** For example, for the reaction of iodide ion with methyl bromide, the data shown in Table 13.3 are found for the rate constant at different temperatures.

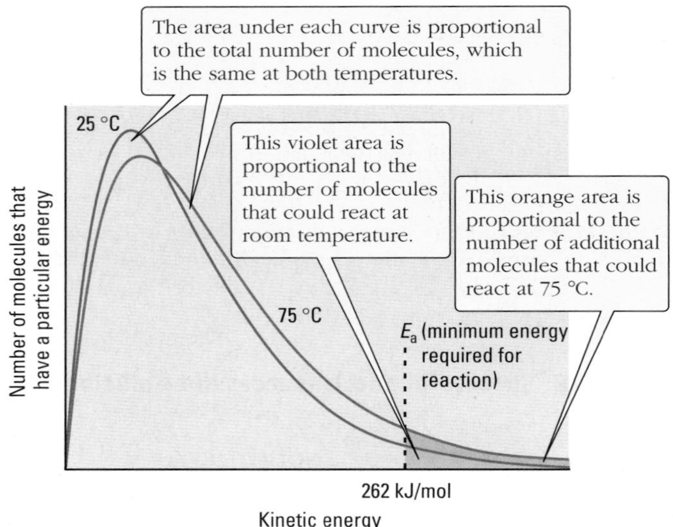

Figure 13.9 Energy distribution curves. The vertical axis gives the number of molecules that have the energy shown on the horizontal axis. Assume that a molecule reacts if it has more energy than the activation energy. The number of reactive molecules is given by the area under each curve to the right of the activation energy (262 kJ/mol). The two curves show that at 75 °C many more molecules have energies of 262 kJ/mol or higher than at 25 °C. (This graph is not to scale. Actually, 262 kJ/mol would be much farther to the right on the graph.)

Exponential curves also represent the growth of populations (such as human population) over time and are important in many other scientific fields. Logarithms and exponentials are discussed in Appendix A.6.

When the data from Table 13.3 are graphed (Figure 13.10), it is obvious that the rate constant increases very rapidly as temperature increases. A graph that is shaped like this is called an exponential curve. It can be represented by the equation

$$k = Ae^{-E_a/RT} \qquad \text{[Arrhenius equation]}$$

where A is the **frequency factor,** e is the base of the natural logarithm system (2.718 . . .), E_a is the activation energy, R is the gas law constant and has the value 8.314 J mol^{-1} K^{-1}, and T is the absolute (Kelvin) temperature (← ***p. 442***). This equation is called the **Arrhenius equation** after its discoverer, Svante Arrhenius, a Swedish chemist.

R is the constant found in the ideal gas law, $PV = nRT$ (← ***p. 445***). Here it is expressed in units of J mol^{-1} K^{-1} instead of L atm mol^{-1} K^{-1}, which is the reason that the numerical value is not 0.0821 L atm mol^{-1} K^{-1}.

The Arrhenius equation can be interpreted as follows. The frequency factor, A, depends on how often molecules collide when all concentrations are 1 mol/L and on

Table 13.3 Temperature Dependence of Rate Constant for Iodide Plus Methyl Bromide Reaction

T(K)	k(L mol^{-1} s^{-1})	T(K)	k(L mol^{-1} s^{-1})	T(K)	k(L mol^{-1} s^{-1})
273	4.18×10^{-5}	310	2.31×10^{-3}	350	6.80×10^{-2}
280	9.68×10^{-5}	320	5.82×10^{-3}	360	1.41×10^{-1}
290	2.00×10^{-4}	330	1.39×10^{-2}	370	2.81×10^{-1}
300	8.60×10^{-4}	340	3.14×10^{-2}		

whether the molecules are properly oriented when they collide. For example, in the case of the iodide–methyl bromide reaction, A includes the steric factor of $\frac{1}{4}$ that resulted because only one of the four sides of the CH_3Br molecule was appropriate for iodide to approach. The rest of the equation, $e^{-E_a/RT}$, gives the fraction of all the reactant molecules that have sufficient energy to surmount the activation energy barrier.

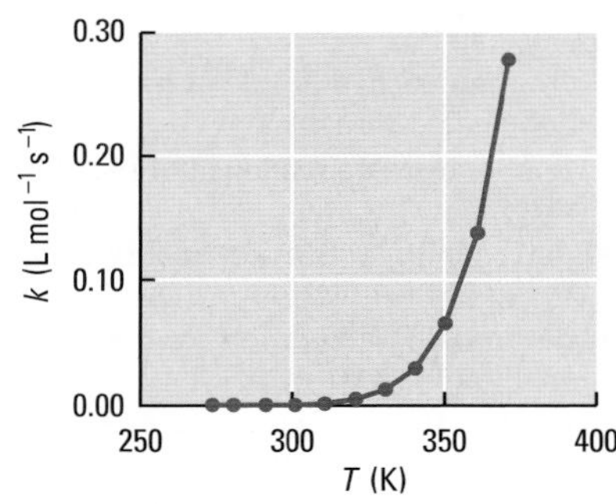

Figure 13.10 Effect of temperature on rate constant. The rate constant for the reaction of iodide ion with methyl bromide in aqueous solution is plotted as a function of temperature. The rate constant increases very rapidly with temperature, and the shape of the curve is characteristic of exponential increase.

Determining Activation Energy

The activation energy and frequency factor can be obtained from experimental measurements of rate constants as a function of temperature (such as those in Table 13.3). When a large number of experimental data pairs are given, a graph is usually a good way of obtaining information from the data. This is easier to do if the graph is linear. The activation energy equation can be modified by taking natural logarithms of both sides so that its graph is linear.

$$k = Ae^{-E_a/RT}$$

$$\ln(k) = \ln(A) + \ln(e^{-E_a/RT}) = \ln(A) + (-E_a/RT)$$

Rearranging this equation gives the equation of a straight line.

$$\ln(k) = -\left(\frac{E_a}{R} \times \frac{1}{T}\right) + \ln(A)$$

$$y = (m \times x) + b$$

That is, if we graph $\ln(k)$ on the vertical (y) axis and $1/T$ on the horizontal (x) axis, the result should be a straight line whose slope is $-E_a/R$ and whose y-intercept is $\ln(A)$. For the data in Table 13.3, such a graph is shown in Figure 13.11. It is linear, its slope is -9.29×10^3 K, and the y-intercept is 23.85. Since the slope $= -E_a/R$, the activation energy can be calculated as

$$E_a = -(\text{slope}) \times R = -(-9.29 \times 10^3\ \text{K})\left(\frac{8.314\ \text{J}}{\text{K mol}}\right)\left(\frac{1\ \text{kJ}}{1000\ \text{J}}\right) = 77.2\ \text{kJ/mol}$$

The vertical axis plots $\ln(k)$, and k has units of L mol^{-1} s^{-1}. At the y-intercept, $\ln(A)$ equals $\ln(k)$, and therefore A must have the same units as k. Since $\ln(A) = 23.85$, the frequency factor, A, is

$$e^{23.85}\ \text{L mol}^{-1}\ \text{s}^{-1} = 2.28 \times 10^{10}\ \text{L mol}^{-1}\ \text{s}^{-1}$$

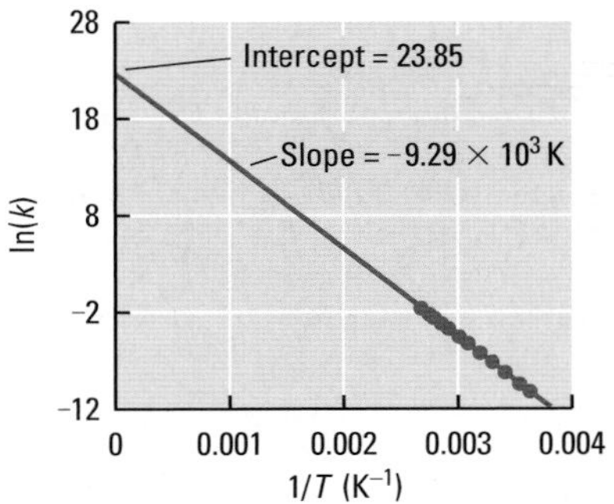

Figure 13.11 Determining activation energy graphically. A graph of $\ln(k)$ versus $1/T$ gives a straight line for the iodide–methyl bromide reaction. The activation energy can be obtained from the slope of the line, and the frequency factor from its y-intercept. Because the line through the experimental data has to be extrapolated a long way to reach the y-intercept, determining A accurately requires measuring k over a wide range of temperatures.

The Arrhenius equation can be used to calculate the rate constant at any temperature. For example, the rate constant for the reaction of iodide ion with methyl bromide can be calculated at 50. °C by substituting the temperature (in kelvins), the frequency factor, the activation energy, and the constant R into the equation.

$$k = Ae^{-E_a/RT} = (2.28 \times 10^{10}\ \text{L mol}^{-1}\ \text{s}^{-1})e^{(-77{,}200\ \text{J/mol})/(8.314\ \text{J K}^{-1}\ \text{mol}^{-1})(273.15+50.)\ \text{K}}$$

$$= (2.28 \times 10^{10}\ \text{L mol}^{-1}\ \text{s}^{-1})e^{-28.7} = 7.56 \times 10^{-3}\ \text{L mol}^{-1}\ \text{s}^{-1}$$

As a means of calculating rate constants, the Arrhenius equation works best within the range of temperatures over which the activation energy and frequency factor were determined. (For the reaction of iodide with methyl bromide, that range was 273 to 370 K.)

ThomsonNOW™
Go to the Coached Problems menu for a module and tutorials on:
- **temperature dependence of reaction rates**
- **determining activation energy using the graphical method**
- **determining activation energy using two data points**
- **using activation energy to predict k at various temperatures**

PROBLEM-SOLVING EXAMPLE 13.9 Temperature Dependence of Rate Constant

The experimental rate constant for the reaction of iodide ion with methyl bromide is 7.70×10^{-3} L mol^{-1} s^{-1} at 50 °C and 4.25×10^{-5} L mol^{-1} s^{-1} at 0 °C. Calculate the frequency factor and activation energy.

Answer $A = 1.66 \times 10^{10}$ L mol^{-1} s^{-1}; $E_a = 7.63 \times 10^4$ J mol^{-1}

Strategy and Explanation This problem could be solved by making a graph of ln(k) versus $1/T$, but with only two data pairs, a graph is not the best way. Instead, write two equations, one for each data pair.

$$k_1 = Ae^{-E_a/RT_1}$$
$$k_2 = Ae^{-E_a/RT_2}$$

Now divide the first equation by the second to eliminate A.

$$\frac{k_1}{k_2} = \frac{Ae^{-E_a/RT_1}}{Ae^{-E_a/RT_2}} = e^{-E_a/RT_1} \times e^{+E_a/RT_2}$$

$$\frac{k_1}{k_2} = e^{\frac{E_a}{R}\left(\frac{1}{T_2} - \frac{1}{T_1}\right)}$$

Next take the natural logarithm of both sides.

$$\ln\left(\frac{k_1}{k_2}\right) = \frac{E_a}{R}\left(\frac{1}{T_2} - \frac{1}{T_1}\right)$$

This equation can be solved for E_a.

$$E_a = \frac{R\ln\left(\frac{k_1}{k_2}\right)}{\frac{1}{T_2} - \frac{1}{T_1}} = \frac{(8.314\text{ J mol}^{-1}\text{ K}^{-1})\ln\left(\frac{7.70 \times 10^{-3}}{4.25 \times 10^{-5}}\right)}{\frac{1}{273.15\text{ K}} - \frac{1}{323.15\text{ K}}}$$

$$= \frac{43.23\text{ J mol}^{-1}\text{ K}^{-1}}{5.66 \times 10^{-4}\text{ K}^{-1}} = 7.63 \times 10^4\text{ J mol}^{-1}$$

Finally, using the calculated value of E_a, solve one of the rate constant expressions for A.

$$k_1 = Ae^{-E_a/RT_1}$$

$$A = \frac{k_1}{e^{-E_a/RT_1}} = k_1 e^{E_a/RT_1}$$

$$= (7.70 \times 10^{-3}\text{ L mol}^{-1}\text{ s}^{-1})e^{\left(\frac{7.63 \times 10^4\text{ J mol}^{-1}}{(8.314\text{ J mol}^{-1}\text{ K}^{-1})(323.15\text{ K})}\right)}$$

$$= (7.70 \times 10^{-3}\text{ L mol}^{-1}\text{ s}^{-1})(2.156 \times 10^{12}) = 1.66 \times 10^{10}\text{ L mol}^{-1}\text{ s}^{-1}$$

PROBLEM-SOLVING PRACTICE 13.9

Calculate the rate constant for the reaction of iodide ion with methyl bromide at a temperature of 75 °C.

Recall that the collision frequency contribution to the frequency factor is for 1 M concentrations. This equation summarizes the effects of both temperature and concentration on rate of a reaction. The temperature effect depends primarily on the large increase in the number of sufficiently energetic collisions as the temperature increases, which shows up as larger values of k at higher temperatures. The effect of concentration is clearly indicated by the concentration terms in the rate law. If the rate law is known for a reaction, and if both the A and E_a values are known, then the rate can be calculated over a wide range of conditions.

The Arrhenius equation enables us to calculate the rate constant as a function of temperature, and the rate law shows how the rate depends on concentration. To obtain a single equation that summarizes the effects of both temperature and concentration, substitute k from the Arrhenius equation into the rate law for the iodide–methyl bromide reaction, giving

$$\text{Rate} = k \times [I^-] \times [CH_3Br]$$
$$\text{Rate} = A \times e^{-E_a/RT} \times [I^-] \times [CH_3Br] \qquad [13.8]$$

Collision frequency × steric factor | Fraction of sufficiently energetic molecules | Concentrations of colliding molecules

EXERCISE 13.9 Activation Energy and Experimental Data

The frequency factor A is 6.31×10^8 L mol^{-1} s^{-1} and the activation energy is 10. kJ/mol for the gas phase reaction

$$NO(g) + O_3(g) \longrightarrow NO_2(g) + O_2(g)$$

which is important in the chemistry of stratospheric ozone depletion.

(a) Calculate the rate constant for this reaction at 370. K.

(b) Assuming that this is an elementary reaction, calculate the rate of the reaction at 370. K if [NO] = 0.0010 M and $[O_3]$ = 0.00050 M.

13.6 Rate Laws for Elementary Reactions

An elementary reaction is a one-step process whose equation describes which nanoscale particles break apart, rearrange their positions, or collide to make a reaction occur. Therefore it is possible to figure out what the rate law and reaction order are for an elementary reaction, without doing an experiment. By contrast, when an equation represents a reaction that we do not understand at the nanoscale, rate laws and reaction orders must be determined experimentally (p. 615). The macroscale rate law can then be used to help develop a hypothesis about how a particular reaction takes place at the nanoscale.

Rate Law for a Unimolecular Reaction

In Section 13.4 we used the isomerization of *cis*-2-butene as an example of a reaction in which a single reactant molecule was converted to a product molecule or molecules—a unimolecular reaction (p. 626).

$$cis\text{-2-butene} \longrightarrow trans\text{-2-butene}$$

Suppose a flask contains 0.005 mol/L of *cis*-2-butene vapor at room temperature. The molecules have a wide range of energies, but only a few of them have enough energy at this temperature to get over the activation energy barrier. Thus, during a given period only a few molecules twist sufficiently to become *trans*-2-butene. Now suppose that we double the concentration of *cis*-2-butene in the flask to 0.010 mol/L, while keeping the temperature the same. The fraction of molecules with enough energy to cross over the barrier remains the same. However, as there are now twice as many molecules, twice as many must be crossing the barrier in any given time. Therefore, the rate of the *cis* ⟶ *trans* reaction is twice as great. That is, the reaction rate is proportional to the concentration of *cis*-2-butene, and the rate law must be

Suppose that the fraction of molecules that have enough energy to react is 0.1%, or 0.001. If there are 10,000 molecules in a given volume, then $0.001 \times 10{,}000$ gives only 10 that have enough energy to react. If there are twice as many molecules in the same volume—that is, 20,000 molecules—then $0.001 \times 20{,}000$ gives 20 with enough energy to react, and the number reacting per unit volume (the rate) is twice as great.

$$\text{Rate} = k[cis\text{-2-butene}]$$

In the general case of any unimolecular elementary reaction,

$$A \longrightarrow \text{products} \qquad \text{the rate law is} \qquad \text{Rate} = k[A]$$

For any unimolecular reaction the nanoscale mechanism predicts that a first-order rate law will be observed in a macroscale laboratory experiment.

Rate Law for a Bimolecular Reaction

A good example of a reaction in which two molecules collide [a bimolecular reaction, (p. 629)] is the gas phase reaction of nitrogen monoxide and ozone that is involved in stratospheric ozone depletion and was mentioned in Exercise 13.9.

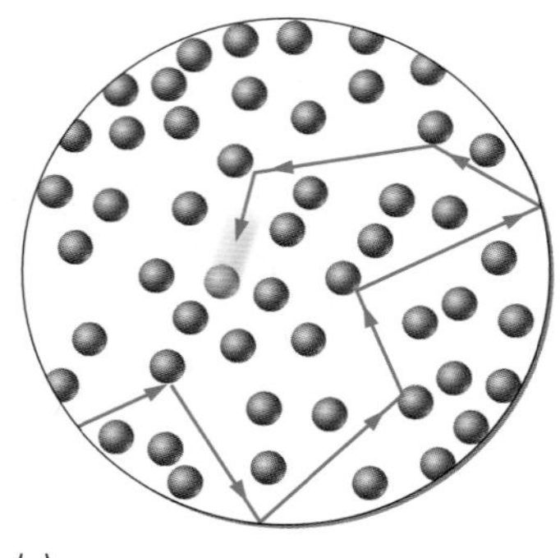

(a)

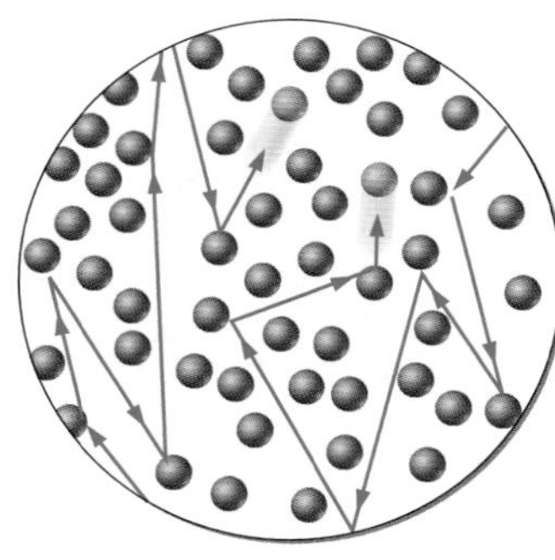

(b)

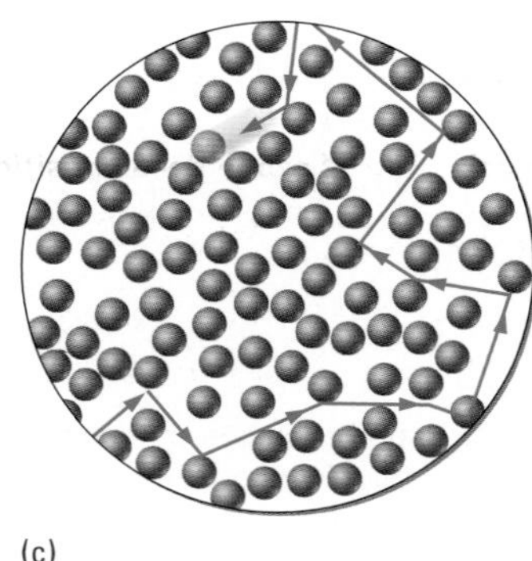

(c)

Figure 13.12 Effect of concentration on frequency of bimolecular collisions. (a) A single green molecule moves among 50 purple molecules and collides with five of them per second. (b) Two green molecules now move among 50 purple molecules, and there are 10 green-purple collisions per second. (c) If the number of purple molecules is doubled to 100, the frequency of green-purple collisions is also doubled, to 10 per second. The number of collisions is proportional to *both* the concentration of green molecules *and* the concentration of purple molecules. (In a real sample both the green and the purple molecules would be moving, but in this diagram the motion of the purple molecules is not shown.)

$$NO(g) + O_3(g) \longrightarrow NO_2(g) + O_2(g) \qquad [13.9]$$

Here the equation shows that the elementary reaction involves the collision of one NO molecule and one O_3 molecule. Since the molecules must collide to exchange atoms, the rate depends on the number of collisions per unit time.

Figure 13.12a represents one NO molecule (the green ball) and many O_3 molecules (the purple balls) in a tiny region within a flask where Reaction 13.9 is taking place. In a given time, the NO molecule collides with five O_3 molecules. If the concentration of NO molecules is doubled to two NO molecules in the same portion of the flask (Figure 13.12b), *each* NO molecule collides with five different O_3 molecules. Doubling the concentration of NO has doubled the number of collisions. This also doubles the rate, because the rate is proportional to the number of collisions (Equation 13.8, p. 634). The number of collisions also doubles when the O_3 concentration is doubled (Figure 13.12c). Thus, the rate law for this reaction must be

$$\text{Rate} = k[NO][O_3]$$

This description of the NO + O_3 reaction applies in general to bimolecular elementary reactions, even if the two molecules that must collide are of the same kind. That is, for the elementary reaction

$$A + B \longrightarrow \text{products} \qquad \text{the rate law is} \qquad \text{Rate} = k[A][B]$$

and for the elementary reaction

$$A + A \longrightarrow \text{products} \qquad \text{the rate law is} \qquad \text{Rate} = k[A]^2$$

For the NO + O_3 reaction the experimentally determined rate law is the same as the one we just derived by assuming the reaction occurs in one step. The equation for the reaction is

$$NO(g) + O_3(g) \longrightarrow NO_2(g) + O_2(g)$$

and the experimental rate law is

$$\text{Rate} = k[NO][O_3]$$

This experimental observation suggests, but does not prove, that the reaction does take place in a single step. Other evidence also suggests that this reaction is bimolecular.

In contrast, for the decomposition of hydrogen peroxide the equation is

$$2\,H_2O_2(aq) \longrightarrow 2\,H_2O(\ell) + O_2(g)$$

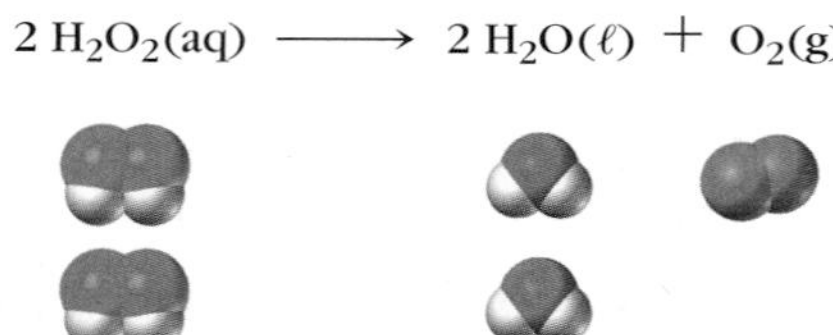

and the experimental rate law is Rate = $k[H_2O_2]$. This rate law proves that this reaction *cannot* occur in a single step that involves collision of two H_2O_2 molecules. A single-step bimolecular reaction would have a second-order rate law, but the observed rate law is first-order. This means that more than a single elementary step is needed when hydrogen peroxide decomposes.

CONCEPTUAL EXERCISE 13.10 Rate Law and Elementary Reactions

For the reaction

$$NO_2(g) + CO(g) \longrightarrow NO(g) + CO_2(g)$$

the experimentally determined rate law is

$$\text{Rate} = k[NO_2]^2$$

Does this reaction occur in a single step? Explain why or why not.

13.7 Reaction Mechanisms

Most chemical reactions do not take place in a single step. Instead, they involve a sequence of unimolecular or bimolecular elementary reactions. For each elementary reaction in the sequence we can write an equation. A set of such equations is called a **reaction mechanism** (or just a *mechanism*). For example, iodide ion can be oxidized by hydrogen peroxide in acidic solution to form iodine and water according to this overall equation:

$$2\,I^-(aq) + H_2O_2(aq) + 2\,H_3O^+(aq) \longrightarrow I_2(aq) + 4\,H_2O(\ell) \quad [13.10]$$

When the acid concentration is between 10^{-3} M and 10^{-5} M, experiments show that the rate law is

$$\text{Rate} = k[I^-][H_2O_2]$$

The reaction is first-order in the concentrations of I^- and H_2O_2, and second-order overall.

Looking at Equation 13.10 for the oxidation of iodide ion by hydrogen peroxide, you might think that two iodide ions, one hydrogen peroxide molecule, and two hydronium ions would all have to come together at once. However, the rate law corresponds with a bimolecular collision of I^- and H_2O_2. It is highly unlikely that at the same time five ions or molecules would all be at the same place, be properly oriented, and have enough energy to react. Instead, chemists who have studied this reaction propose that initially one H_2O_2 (HOOH) molecule and one I^- ion come together:

ThomsonNOW
Go to the Chemistry Interactive menu for modules on:

- **mechanism of ozone decomposition**
- **mechanism of reaction between NO_2 and CO**

Step 1: $HOOH + I^- \xrightarrow{\text{slow}} OH^- + HOI$

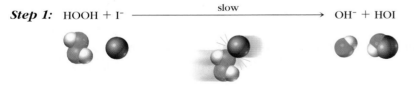

This first step forms hypoiodous acid, HOI, and hydroxide ion, both known substances. The HOI then reacts with another I^- to form the product I_2:

Step 2: $HOI + I^- \xrightarrow{\text{fast}} OH^- + I_2$

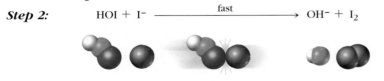

In each of Steps 1 and 2, a hydroxide ion was produced. Since the solution is acidic, these OH^- ions react immediately with H_3O^+ ions to form water.

Step 3: $2\ OH^- + 2\ H_3O^+ \xrightarrow{\text{fast}} 4\ H_2O$

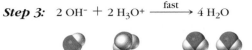

Each of the three steps in this mechanism is an elementary reaction. Each has its own activation energy, E_a, and its own rate constant, k. When the three steps are summed (by putting all the reactants on the left, putting all the products on the right, and eliminating formulas that appear as both reactants and products), the overall stoichiometric equation (Equation 13.10) is obtained. ***Any valid mechanism must consist of a series of unimolecular or bimolecular elementary reaction steps that sum to the overall reaction.***

Step 1: $HOOH + I^- \xrightarrow{\text{slow}} \cancel{HOI} + \cancel{OH^-}$

Step 2: $\cancel{HOI} + I^- \xrightarrow{\text{fast}} I_2 + \cancel{OH^-}$

Step 3: $\cancel{2\ OH^-} + 2\ H_3O^+ \xrightarrow{\text{fast}} 4\ H_2O$

Overall: $2\ I^- + HOOH + 2\ H_3O^+ \longrightarrow I_2 + 4\ H_2O$ [13.10]

Step 1 of the mechanism is slow, while Steps 2 and 3 are fast. Step 1 is called the **rate-limiting step;** because it is the slowest in the sequence, it limits the rate at which I_2 and H_2O can be produced. Steps 2 and 3 are rapid and therefore not rate-limiting. As soon as some HOI and OH^- are produced by Step 1, they are transformed into I_2 and H_2O by Steps 2 and 3. ***The rate of the overall reaction is limited by, and equal to, the rate of the slowest step in the mechanism.***

An analogy to the rate-limiting or rate-determining step is that no matter how quickly you shop in the supermarket, it seems that the time it takes to get out of the store depends on the rate at which you move through the checkout line.

Step 1 is a bimolecular elementary reaction. Therefore its rate must be first-order in HOOH and first-order in I^-. The mechanism predicts that the rate law should be

$$\text{Reaction rate} = k[\text{HOOH}][\text{I}^-]$$

which agrees with the experimentally observed rate law. ***A valid mechanism must correctly predict the experimentally observed rate law.***

EXERCISE 13.11 Rate Law for an Elementary Reaction

What is the rate law for Step 2 of the mechanism for the reaction of hydrogen peroxide and iodide ion?

If significant concentrations of an intermediate build up while a reaction is occurring, then reactants may disappear faster than products are formed (because buildup of the intermediate stores some of the used reactant before it is converted to final product). In such a case, the definition of reaction rate given in Equation 13.3 will not be correct. One way to detect formation of an intermediate is to notice that products are not formed as fast as reactants are used up.

The species HOI and OH^- are produced in Step 1 and used up in Step 2 or 3. In a mechanism, atoms, molecules, or ions that are produced in one step and consumed in a later step (or later steps) are called **reaction intermediates** (or just **intermediates**). Very small concentrations of HOI and OH^- are produced while the reaction is going on. Once the HOOH, the I^-, or both are used up, the intermediates HOI and OH^- are consumed by Steps 2 and 3 and disappear. HOI and OH^- are crucial to the reaction mechanism, but neither of them appears in the overall stoichiometric equation. If an experimenter is proficient enough to demonstrate that a particular intermediate was present, this provides additional evidence that a mechanism involving that intermediate is the correct one.

In summary, a valid reaction mechanism should

- Consist of a series of unimolecular or bimolecular elementary reactions;
- Consist of reaction steps that sum to the overall reaction equation; and
- Correctly predict the experimentally observed rate law.

ThomsonNOW™
Go to the Coached Problems menu for exercises on:
- **reaction mechanisms**
- **heterogeneous catalysis**

Mechanisms with a Fast Initial Step

An experimental rate law should include only concentrations that can be measured experimentally. Usually this means concentrations of reactants (and perhaps products), but not the concentrations of intermediates. (The concentrations of intermediates usually are very small and therefore hard to measure.) The first step in the mechanism in the previous section was rate-limiting and involved two of the reactants, so it was easy to relate the overall rate to the concentrations of reactants. But what happens if the rate-limiting step is the second or a subsequent step?

Consider the reaction

$$2\,NO(g) + Br_2(g) \longrightarrow 2\,NOBr(g) \qquad \text{Rate (experimental)} = k[NO]^2[Br_2]$$

for which the currently accepted mechanism is

Step 1: (fast) $NO(g) + Br_2(g) \rightleftharpoons NOBr_2(g)$

Step 2: (slow) $NOBr_2(g) + NO(g) \longrightarrow 2\,NOBr(g)$

Because the second step is slow, $NOBr_2$ can often break apart and reform $NO + Br_2$ before it reacts to form products. We say that Step 1 is reversible, which is indicated by the double arrow ($\rightleftharpoons$). Once $NOBr_2$ forms, it can react in either of two ways: back to the reactants, $NO + Br_2$, or forward (in Step 2) to form the product, NOBr.

The overall rate of the reaction is the rate of the rate-limiting step (Step 2):

$$\text{Rate} = k[NOBr_2][NO]$$

However, this is not a valid rate law to compare with experimental results, because an experimental rate law should include only concentrations that can be measured experimentally. (Usually, the concentration of an intermediate, such as $NOBr_2$, cannot be measured.) Therefore, we need a way to relate the concentration of $NOBr_2$ to the concentrations of the reactants, NO and Br_2. This can be done by recognizing that in the mechanism there are three reactions and three rate constants:

$$\text{Step } +1 \text{ (forward), rate constant } k_1$$

$$\text{Step } -1 \text{ (backward), rate constant } k_{-1}$$

$$\text{Step 2, rate constant } k_2$$

The concentration of $NOBr_2$ depends on all three reactions. It is increased by Step +1 and decreased by Step −1 and Step 2. Initially, the concentration of $NOBr_2$ builds up because Step +1 produces $NOBr_2$. However, when $[NOBr_2]$ gets big enough, the rates of Step −1 and Step 2 get bigger, and eventually the $[NOBr_2]$ reaches what is called a steady state—it neither increases nor decreases until the reaction is nearly over.

In the steady state, the rate of the reaction by which $NOBr_2$ is formed must equal the sum of the two rates by which it is destroyed; that is,

$$\text{Rate of Step } +1 = \text{rate of Step } -1 + \text{rate of Step } 2$$

$$k_1[NO][Br_2] = k_{-1}[NOBr_2] + k_2[NOBr_2][NO] \qquad [13.11]$$

Suppose that $k_{-1}[NOBr_2] = 0.20 \text{ mol L}^{-1} \text{ s}^{-1}$ and $k_2[NOBr_2][NO] = 0.00020 \text{ mol L}^{-1} \text{ s}^{-1}$, which is much smaller. Then $k_{-1}[NOBr_2] - k_2[NOBr_2][NO]$ $= (0.20 - 0.00020) \text{ mol L}^{-1} \text{ s}^{-1}$ $= 0.20 \text{ mol L}^{-1} \text{ s}^{-1}$. That is, the rate for Step 2 is negligible and can be ignored in the calculation.

Equation 13.11 can usually be simplified because Step 2 is much slower than either Step 1 or Step −1. This means that $k_2[NOBr_2][NO] \ll k_{-1}[NOBr_2]$ and $k_2[NOBr_2][NO]$ can be neglected in the calculation.

$$k_1[NO][Br_2] \cong k_{-1}[NOBr_2] \qquad [13.12]$$

Equation 13.12 can be rearranged algebraically to show that $[NOBr_2]$ is proportional to the concentration of each reactant.

$$[NOBr_2] = \frac{k_1}{k_{-1}}[NO][Br_2]$$

This allows the rate to be expressed in terms of the concentrations of reactants.

$$\text{Rate} = k_2[NOBr_2][NO] = k_2\left(\frac{k_1}{k_{-1}}[NO][Br_2]\right)[NO]$$

$$= \frac{k_1 k_2}{k_{-1}}[NO]^2[Br_2] = k'[NO]^2[Br_2] \qquad [13.13]$$

Equation 13.13 shows that the rate constant k' is actually a quotient of rate constants for three elementary reactions, but the rate is proportional to the concentration of Br_2 and to the square of the concentration of NO. That is, the mechanism predicts that the reaction is second-order in NO and first-order in Br_2, which agrees with the experimental rate law. For mechanisms in which the rate-limiting step is the second or a subsequent step, a mathematical relationship such as Equation 13.13 can usually be found that relates an overall rate constant, k', to the concentrations of the reactants and the rate constants for the steps up to and including the rate-limiting step.

Kinetics and Mechanism

Studying the kinetics of a chemical reaction involves collecting data on the concentrations of reactants as a function of time. From such data the rate law for the reaction and a rate constant can usually be obtained. The reaction can also be studied at several different temperatures to determine its activation energy. This allows us to predict how fast the macroscale reaction will be under a variety of experimental conditions, but it does not provide definitive information about the nanoscale mechanism by which the reaction takes place. A reaction mechanism is an educated guess—a hypothesis—about the way the reaction occurs. If the mechanism predicts correctly the overall stoichiometry of the reaction and the experimentally determined rate law, then it is a reasonable hypothesis. However, it is impossible to prove for certain that a mechanism is correct. Sometimes several mechanisms can agree with the same set of experiments. This is what makes kinetic studies one of the most interesting and rewarding areas of chemistry, but it also can provoke disputes among scientists who favor different possible mechanisms for the same reaction.

PROBLEM-SOLVING EXAMPLE 13.10 Rate Law and Reaction Mechanism

The gas phase reaction between nitrogen monoxide and oxygen,

$$2\ NO(g) + O_2(g) \longrightarrow 2\ NO_2(g)$$

is found experimentally to obey the rate law

$$\text{Rate} = k[NO]^2[O_2]$$

Decide which of these mechanisms is (are) compatible with this rate law.

(a) $NO + NO \rightleftharpoons N_2O_2$ fast
$N_2O_2 + O_2 \longrightarrow 2\ NO_2$ slow

(b) $NO + NO \longrightarrow NO_2 + N$ slow
$N + O_2 \longrightarrow NO_2$ fast

(c) $NO + O \rightleftharpoons NO_2$ fast
$NO_2 + NO \longrightarrow N_2O_3$ fast
$N_2O_3 + O \longrightarrow 2\ NO_2$ slow

(d) $NO + O_2 \rightleftharpoons NO_3$ fast
$NO_3 + NO \longrightarrow 2\ NO_2$ slow

(e) $NO + O_2 \longrightarrow NO_2 + O$ slow
$NO + O \longrightarrow NO_2$ fast

(f) $2\ NO + O_2 \longrightarrow 2\ NO_2$

Answer Mechanisms (a) and (d) are compatible with the rate law and stoichiometry.

Strategy and Explanation Examine each mechanism to see whether it (1) consists only of unimolecular and bimolecular steps, (2) agrees with the overall stoichiometry, and (3) predicts the experimental rate law. Eliminate those that do not. The remaining mechanism(s) may be correct.

Mechanism (f) involves collision of three molecules: two NO and one O_2. It can be eliminated because it does not consist of unimolecular or bimolecular steps. All of the other mechanisms consist of bimolecular steps.

Mechanism (c) does not have O_2 as a reactant in the overall stoichiometry, so it can be eliminated. All other mechanisms predict the observed overall stoichiometry.

In mechanism (a) the first step is fast and reversible. Applying the idea that the rates are approximately equal for the forward and reverse reactions in that first step gives

$$k_1[NO]^2 = k_{-1}[N_2O_2] \quad \text{and} \quad [N_2O_2] = \frac{k_1}{k_{-1}}[NO]^2$$

Since the overall rate equals the rate of the rate-limiting step,

$$\text{Rate} = k_2[N_2O_2][O_2] = k_2\frac{k_1}{k_{-1}}[NO]^2[O_2] = k'[NO]^2[O_2]$$

Consequently, mechanism (a) could be the actual mechanism.

The slow first step in mechanism (b) implies an overall rate $= k[NO]^2$, which eliminates it from consideration.

Continuing this kind of reasoning, mechanism (d) is seen to be a possibility, but mechanism (e) predicts rate $= k[NO][O_2]$ and therefore can be eliminated. Because there are still two possible mechanisms, (a) and (d), additional experiments need to be done to try to distinguish between them.

PROBLEM-SOLVING PRACTICE 13.10

The Raschig reaction produces the industrially important reducing agent hydrazine (N_2H_4) from ammonia (NH_3) and hypochlorite ion (OCl^-) in basic aqueous solution. A proposed mechanism is

Step 1: $NH_3(aq) + OCl^-(aq) \xrightarrow{\text{slow}} NH_2Cl(aq) + OH^-(aq)$

Step 2: $NH_2Cl(aq) + NH_3(aq) \xrightarrow{\text{fast}} N_2H_5^+(aq) + Cl^-(aq)$

Step 3: $N_2H_5^+(aq) + OH^-(aq) \xrightarrow{\text{fast}} N_2H_4(aq) + H_2O(\ell)$

(a) What is the overall stoichiometric equation?
(b) Which step is rate-limiting?
(c) What reaction intermediates are involved?
(d) What rate law is predicted by this mechanism?

EXERCISE 13.12 Rate Law and Mechanism

Consider the reaction mechanism

$$ICl(g) + H_2(g) \rightleftharpoons HI(g) + HCl(g) \qquad \text{fast}$$

$$HI(g) + ICl(g) \longrightarrow HCl(g) + I_2(g) \qquad \text{slow}$$

(a) What is the overall reaction equation?
(b) Derive the rate law predicted by this mechanism.
(c) Does the rate law depend on the concentration of one of the products of the reaction?
(d) Would the rate constant determined from the initial rate of this reaction equal the rate constant determined at a time when 80% of the reactants had been consumed? Explain why or why not.

(a)

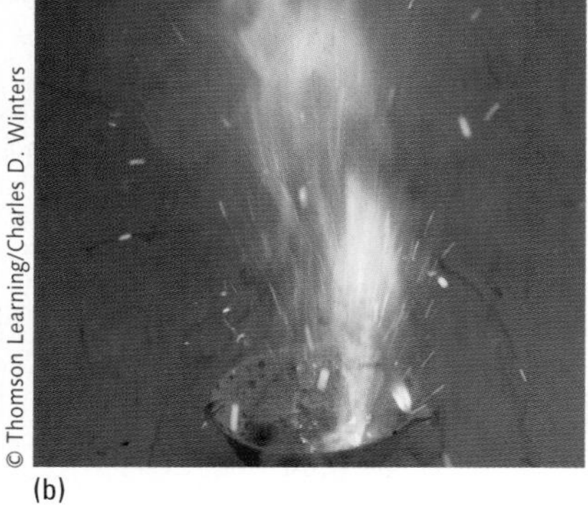

(b)

(c)

Figure 13.13 Catalysis in action. (a) A 30% aqueous solution of H_2O_2 is dropped onto a piece of liver. The liquid foams as H_2O_2 rapidly decomposes to O_2 and H_2O. Liver contains an enzyme that catalyzes the decomposition of H_2O_2. (b) Laboratory-scale explosion of a sample of ammonium nitrate, NH_4NO_3, catalyzed by chloride ion. (c) Scene following explosion of the ship *Grandcamp* in Texas City, Texas, April 16, 1947.

13.8 Catalysts and Reaction Rate

Raising the temperature increases a reaction rate because it increases the fraction of molecules that are energetic enough to surmount the activation energy barrier. Increasing reactant concentrations can also increase the rate because it increases the number of molecules per unit volume. A third way to increase reaction rates is to add a catalyst (p. 608).

For example, an aqueous solution of hydrogen peroxide can decompose to water and oxygen.

$$2\,H_2O_2(aq) \longrightarrow O_2(g) + 2\,H_2O(\ell)$$

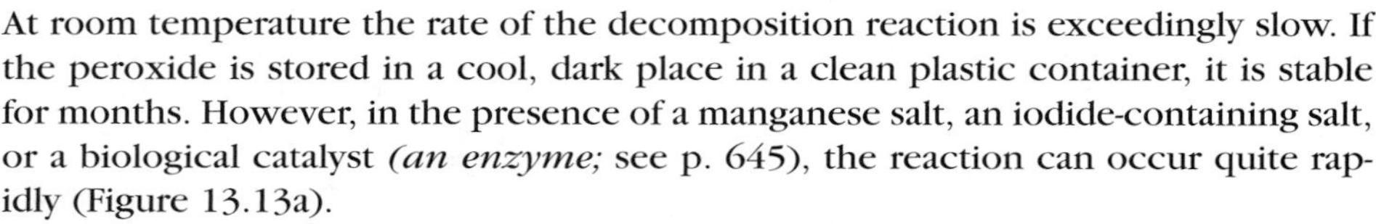

At room temperature the rate of the decomposition reaction is exceedingly slow. If the peroxide is stored in a cool, dark place in a clean plastic container, it is stable for months. However, in the presence of a manganese salt, an iodide-containing salt, or a biological catalyst *(an enzyme;* see p. 645), the reaction can occur quite rapidly (Figure 13.13a).

Ammonium nitrate is used as fertilizer and is stable at room temperature. At higher temperatures and in the presence of chloride ion as a catalyst, however, ammonium nitrate can explode with tremendous force (Figure 13.13b). Approximately 600 people were killed in Texas City, Texas, in 1947 when workers tried to extinguish a fire in the hold of the ship *Grandcamp* with salt water (which contains ~0.5 M Cl^-) and the ammonium nitrate cargo exploded (Figure 13.13c).

How does a catalyst or an enzyme help a reaction to go faster? It does so by participating in the reaction mechanism. That is, ***the mechanism for a catalyzed reaction is different from the mechanism of the same reaction without the catalyst.*** The rate-limiting step in the catalyzed mechanism has a lower activation energy and therefore is faster than the slow step for the uncatalyzed reaction. To see how this works, let us again consider conversion of *cis*- to *trans*-2-butene in the gas phase.

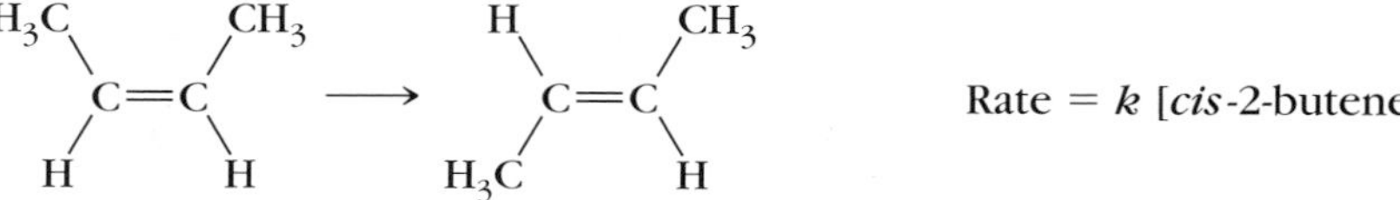

$$\text{Rate} = k\,[cis\text{-2-butene}]$$

If a trace of gaseous molecular iodine, I_2, is added to a sample of *cis*-2-butene, the iodine accelerates the change to *trans*-2-butene. The iodine is neither consumed nor produced in the overall reaction, so it does not appear in the overall balanced equation. However, because the reaction rate depends on the concentration of I_2, there is a term involving concentration of I_2 in the rate law for the catalyzed reaction.

$$\text{Rate} = k[cis\text{-2-butene}][I_2]^{1/2}$$

The exponent of $\frac{1}{2}$ for the concentration of I_2 in the rate law indicates the square root of the concentration. A square root dependence usually means that only half a molecule—in this case a single I atom—is involved in the mechanism.

The rate of the conversion of *cis*- to *trans*-2-butene changes because the presence of I_2 somehow changes the reaction mechanism. The best hypothesis is that iodine molecules first dissociate to form iodine atoms.

ThomsonNOW
Go to the Chemistry Interactive menu for a module on **catalytic decomposition of H_2O_2.**

***Step 1:* I_2 dissociation**

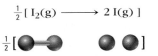

$$\tfrac{1}{2}[\,I_2(g) \longrightarrow 2\,I(g)\,]$$

(This equation is multiplied by $\frac{1}{2}$ because only one of the two I atoms from the I_2 molecule is needed in subsequent steps of the mechanism.) An iodine atom then attaches to the *cis*-2-butene molecule, breaking half of the double bond between the two central carbon atoms and allowing the ends of the molecule to twist freely relative to each other.

***Step 2:* Attachment of I atom to *cis*-2-butene**

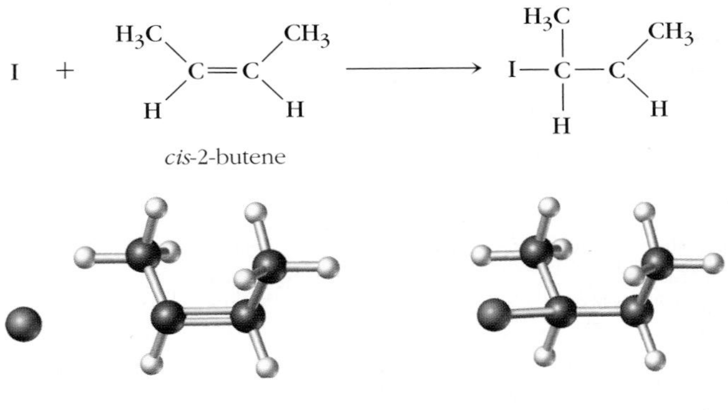

I + $H_3C(H)C{=}C(H)CH_3$ ⟶ $I{-}C(H_3C)(H){-}C(H)CH_3$

cis-2-butene

***Step 3:* Rotation around the C—C bond**

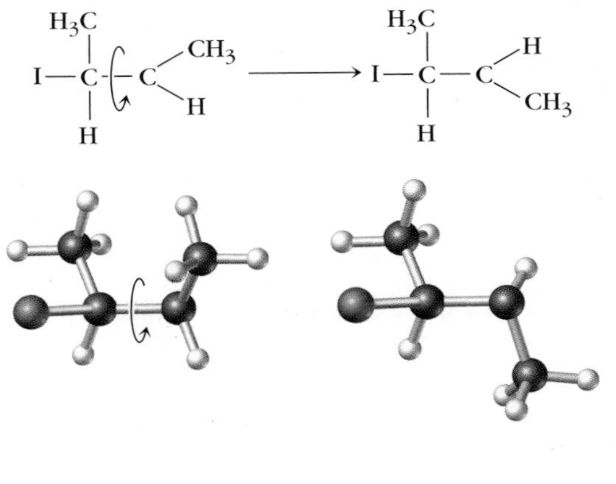

$I{-}C(H_3C)(H){-}C(CH_3)(H)$ ⟶ $I{-}C(H_3C)(H){-}C(H)(CH_3)$

Step 4: **Loss of an I atom and reformation of the carbon–carbon double bond**

$$I{-}C(H_3C)(H){-}C(H)(CH_3) \longrightarrow H_3C(H)C{=}C(H)(CH_3) + I$$

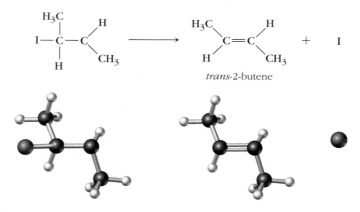

trans-2-butene

After the new double bond forms to give *trans*-2-butene and the iodine atom falls away, two iodine atoms come together to regenerate molecular iodine.

Step 5: **I_2 regeneration**

$$\tfrac{1}{2}[\,2\,I(g) \longrightarrow I_2(g)\,]$$

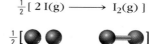

There are five important points concerning this mechanism.

- The I_2 dissociates to atoms and then reforms. To an "outside" observer the concentration of I_2 is unchanged; I_2 is not involved in the balanced stoichiometric equation even though it has appeared in the mechanism. *This is generally true of catalysts.*
- Figure 13.14 shows that the activation energy barrier is significantly lower for the catalyzed reaction (because the mechanism is different). Consequently the reaction rate is much faster. Dropping the activation energy from 262 kJ/mol for

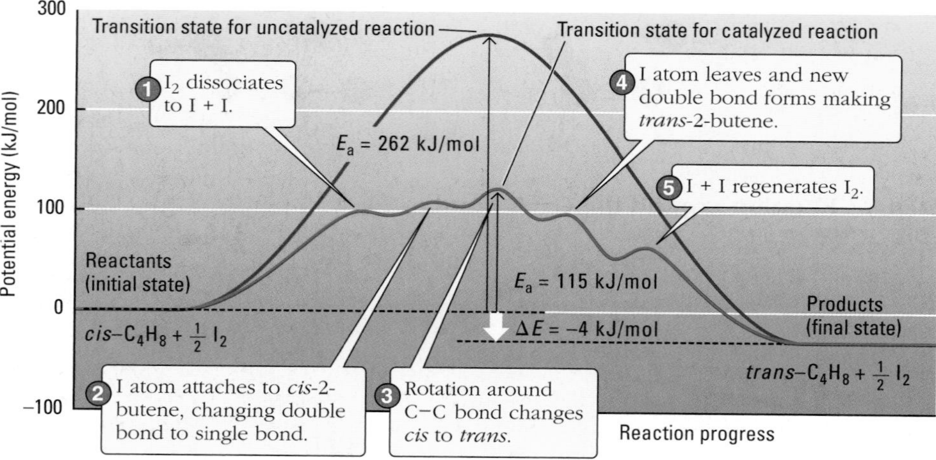

Figure 13.14 Energy diagrams for catalyzed and uncatalyzed reactions. A catalyst accelerates a reaction by altering the mechanism so that the activation energy is reduced. With a smaller barrier to overcome, more reactant molecules have enough energy to cross the barrier, and reaction occurs more readily. (The steps involved are described in the text.) Notice that the shape of the barrier has changed because the mechanism has changed. This changes the activation energy, but not ΔE for the reaction.

the uncatalyzed reaction to 115 kJ/mol for the catalyzed process makes the catalyzed reaction 10^{15} times faster at a temperature of 500. K.

- The catalyzed mechanism has five reaction steps, and its energy-versus-reaction progress diagram (Figure 13.14) has five energy barriers (five humps appear in the curve).
- The catalyst I_2 and the reactant *cis*-2-butene are both in the gas phase during the reaction. When a catalyst is present in the same phase as the reacting substance or substances, it is called a **homogeneous catalyst.**
- Although the mechanism is different, the initial and final energies for the catalyzed reaction are the same as for the uncatalyzed reaction. This means that ΔE and ΔH are the same for the catalyzed as for the uncatalyzed reaction.

Because $k = Ae^{-E_a/RT}$,

$$\frac{k_2}{k_1} = \frac{Ae^{-E_{a_2}/RT}}{Ae^{-E_{a_1}/RT}} = e^{-(E_{a_2}-E_{a_1})/RT}$$

$$= e^{\left(\frac{(262{,}000-115{,}000)\text{ J mol}^{-1}}{8.314\text{ J K}^{-1}\text{ mol}^{-1}\times 500.\text{ K}}\right)}$$

$$= e^{3.536\times 10^{1}} = 2.28 \times 10^{15}$$

EXERCISE 13.13 Catalysis

The oxidation of thallium(I) ion by cerium(IV) ion in aqueous solution has the equation

$$2\,Ce^{4+}(aq) + Tl^{+}(aq) \longrightarrow 2\,Ce^{3+}(aq) + Tl^{3+}(aq)$$

The accepted mechanism for this reaction is

Step 1: $Ce^{4+}(aq) + Mn^{2+}(aq) \longrightarrow Ce^{3+}(aq) + Mn^{3+}(aq)$

Step 2: $Ce^{4+}(aq) + Mn^{3+}(aq) \longrightarrow Ce^{3+}(aq) + Mn^{4+}(aq)$

Step 3: $Mn^{4+}(aq) + Tl^{+}(aq) \longrightarrow Mn^{2+}(aq) + Tl^{3+}(aq)$

(a) Verify that this mechanism predicts the overall reaction.
(b) Identify all intermediates in this mechanism.
(c) Identify the catalyst in this mechanism.
(d) Suppose that the first step in this mechanism is rate-limiting. What would the rate law be?
(e) Suppose that the second step in this mechanism is rate-limiting. What would the rate law be?

ThomsonNOW™
Go to the Coached Problems menu for an exercise on **identifying catalytic species.**

13.9 Enzymes: Biological Catalysts

© Royalty-Free/Corbis

Your body is a chemical factory of cells that can manufacture a broad range of compounds that are needed so that you can move, breathe, digest food, see, hear, smell, and even think. But did you ever consider how the reactions that make those compounds are controlled? And how they can all occur reasonably quickly at the relatively low body temperature of 37 °C? Oxidation of glucose powers all the systems of your body, but you would not want it to take place at the temperature it does when cellulose ***(a polymer of glucose,*** ***p. 597)*** in wood burns in a fireplace. The chemical reactions of your body are catalyzed by enzymes. An **enzyme** is *a highly efficient catalyst for one or more chemical reactions in a living system.* The presence or absence of appropriate enzymes turns these reactions on or off by speeding them up or slowing them down. This allows your body to maintain nearly constant temperature and nearly constant concentrations of a variety of molecules and ions, an absolute necessity if you are to continue functioning.

The 1989 Nobel Prize was given to Thomas Cech and Sidney Altman for the discovery that RNA molecules as well as proteins can be biological catalysts. In fact, the ribosome, the large protein/RNA complex that is responsible for all polymerization of amino acids into intact proteins, uses RNA to accomplish its reactions.

Enzymes are usually proteins, but other biological macromolecules, such as RNA, can also increase the rates of reactions in living systems. Most enzymes are globular proteins, polymers of amino acids ***(p. 590)*** in which one or more long chains of amino acids fold into a nearly spherical shape. The shape of a globular protein is determined by noncovalent interactions ***(p. 595)*** among the amino acid components (hydrogen bonds, attractions of opposite ionic charges, dipole-dipole and ion-dipole forces), a few weak covalent bonds, and the fact that nonpolar (hydrophobic) amino acid side groups congregate in the middle of the molecule, avoiding the surrounding aqueous solution.

The weak covalent bonds in some globular proteins are disulfide bonds. They occur between sulfur atoms in side chains of the amino acid cysteine. A cysteine side chain at one point in the protein can become bonded to a cysteine side chain much farther along the protein backbone ***(p. 593).***

CHEMISTRY YOU CAN DO

Enzymes: Biological Catalysts

Raw potatoes contain an enzyme called *catalase,* which converts hydrogen peroxide to water and oxygen. You can demonstrate this by performing the following experiment.

Purchase a small bottle of hydrogen peroxide at a pharmacy or find one in your medicine chest. The peroxide is usually sold as a 3% solution in water. Pour about 50 mL of the peroxide solution into a clear glass or plastic cup. Add a small slice of a fresh potato to the cup. (Since potato is less dense than water, the potato will float.)

Almost immediately you will see bubbles of oxygen gas on the potato slice. (To make the bubbles more obvious you can add some dishwashing soap.)

1. Does the rate of evolution of oxygen change with time? If so, how does it change?
2. If you cool the hydrogen peroxide solution in a refrigerator and then do the experiment, is there a perceptible change in the initial rate of O_2 evolution?
3. Is there a difference between the time at which O_2 evolution begins for warm and for cold hydrogen peroxide?
4. What happens to the rate of oxygen evolution if you heat the slice of potato on a stove or in an oven before adding it to the peroxide solution?

Enzymes are among the most effective catalysts known. They can increase reaction rates by factors of 10^9 to 10^{19}. For example, essentially every collision of the enzyme carbonic anhydrase with a carbonic acid molecule results in decomposition, and the enzyme can decompose about 36 million H_2CO_3 molecules every minute.

$$H_2CO_3(aq) \xrightarrow{\text{carbonic anhydrase}} CO_2(g) + H_2O(\ell)$$

Most enzymes are highly specific catalysts. Some act on only one or two of the hundreds of different substances found in living cells. For example, carbonic anhydrase catalyzes only the decomposition of carbonic acid. Other enzymes can speed up several reactions, but usually these reactions are all of the same type.

Some enzymes can act as catalysts entirely on their own. Others require one or more inorganic or organic molecules or ions called **cofactors** to be present before their catalytic activity becomes fully available. For example, many enzymes require nicotinamide adenine dinucleotide ion, NAD^+ (niacinamide ion). Molecules or ions that are cofactors are often derived from small quantities of minerals and vitamins in our diets. If the cofactor needed for an enzyme to catalyze a reaction is not available because of dietary deficiency, that reaction cannot occur when it is needed, and a bodily function will be impaired.

Enzyme Activity and Specificity

A molecule whose reaction is catalyzed by an enzyme is referred to as a **substrate.** In some cases there may be more than one substrate, as when an enzyme catalyzes transfer of a group of atoms from one molecule to another. Enzyme catalysis is extremely effective and specific because the structure of the enzyme is finely tuned to minimize the activation energy barrier. Usually one part of the enzyme molecule, called the **active site,** interacts with the substrate via the same kinds of noncovalent attractions that hold the enzyme in its globular structure. The nanoscale structure of an enzyme's active site is specifically suited to attract and bind a substrate molecule and to help the substrate react.

When a substrate binds to an enzyme, both molecular structures can change. Each structure adjusts to fit closely with the other, and the structures become com-

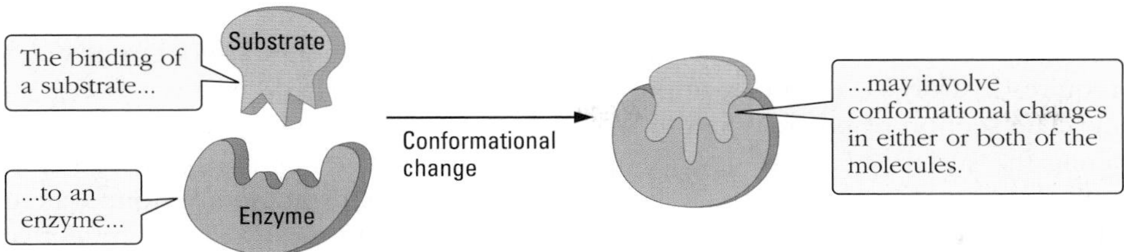

Figure 13.15 Induced fit of substrate to enzyme. Binding of a substrate to an enzyme may involve changing the shape of either or both molecules, thereby inducing them to fit together. In some cases a substrate molecule may be stretched or strained, helping bonds to break and reaction to occur.

plementary. The change in shape of either the enzyme, the substrate, or both molecules when they bind is called **induced fit.** Enzymes catalyze reactions of only certain molecules because the structures of most molecules are not close enough to the structure of the active site for an induced fit to occur. The induced fit of a substrate to an enzyme also can lower the activation energy for a reaction. For example, it may distort the substrate and stretch a bond that will be broken in the desired reaction. A schematic example of how this can work is shown in Figure 13.15. To see how it works in a specific case, consider the enzyme lysozyme, whose structure is shown in Figure 13.16 as a space-filling model with substrate in the active site. Lysozyme catalyzes hydrolysis reactions of polysaccharides ***(⇐ p. 596)*** found in bacterial cell walls. The reaction involved is

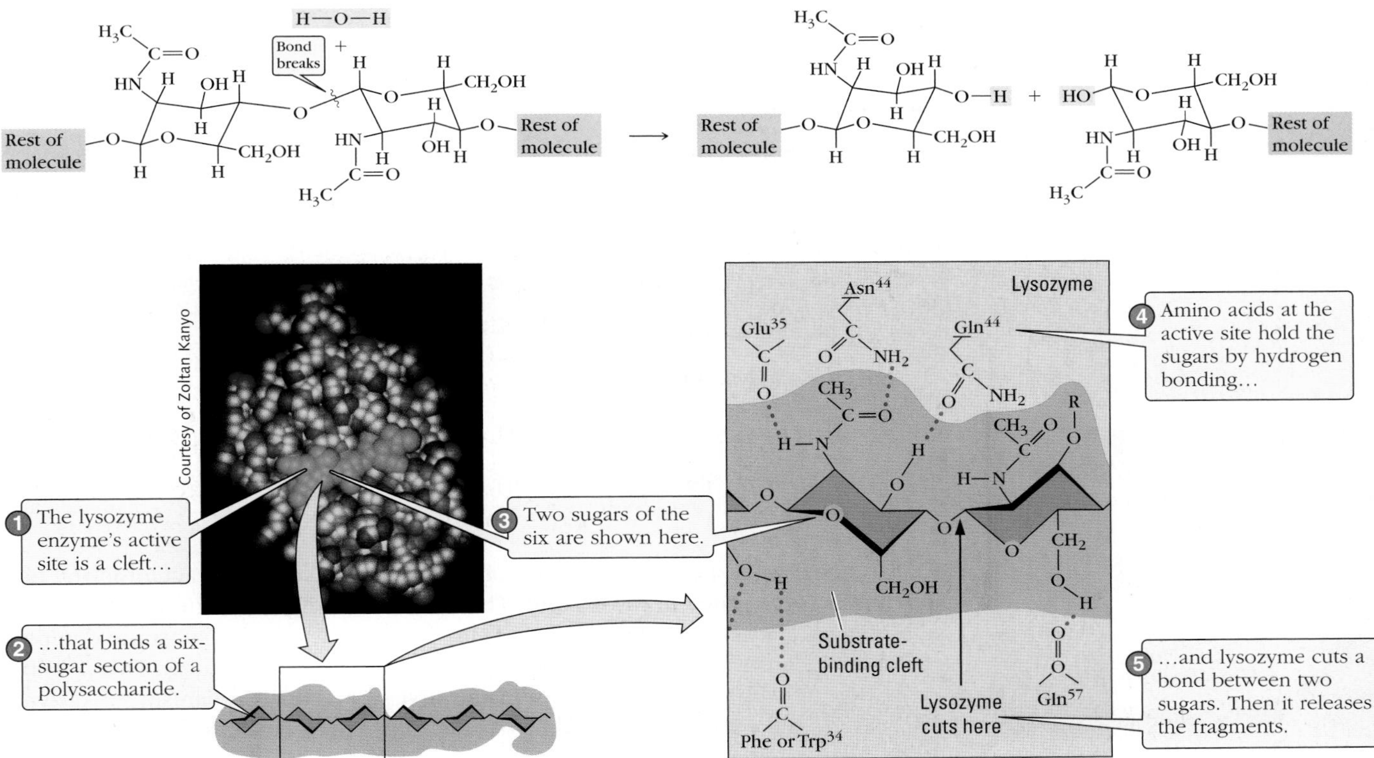

Figure 13.16 Lysozyme with substrate in the active site. The structure of lysozyme is shown at the left, as a space-filling model. The active site is a cleft in the surface of the lysozyme that stretches horizontally across the middle of the enzyme. The active site is occupied by a portion of a polysaccharide molecule, the substrate (*green atoms*). (The part of the polysaccharide not bound to the active site has been omitted so that you can see the enzyme better.) The diagram at the right shows noncovalent interactions (*red dotted lines*) that hold the substrate to the enzyme. The bond that will be broken when the substrate is hydrolyzed is marked by an arrow.

In Section 12.6 (p. 574) hydrolysis was described as a reaction in which a water molecule and some other molecule react, with both molecules splitting in two. In the lysozyme-catalyzed reaction, the H from the water ends up with one part of the substrate molecule, and the OH ends up with the other part.

A hydrolysis reaction is the opposite of a condensation reaction. Most biopolymers are formed by condensation. Breaking them into their building block molecules requires hydrolysis, and many important enzymes catalyze hydrolysis reactions.

The section of polysaccharide shown in Figure 13.16 fits nicely into the cleft along the surface of the lysozyme, but many other long-chain molecules, such as polypeptides, might fit there as well. Shape is important, but so are noncovalent attractions and their positioning so that the substrate can make the most effective use of them. The enlarged portion of Figure 13.16 shows many hydrogen bonds between enzyme and substrate. It should be clear that the specificity of the enzyme depends not only on the shape of the active site, but also on the positions of hydrogen-bonding groups and groups that participate in other noncovalent interactions so that they can adjust to complementary sites on the substrate.

To summarize, enzymes are extremely effective as catalysts for several reasons:

- Enzymes bring substrates into close proximity and hold them there while a reaction occurs.
- Enzymes hold substrates in the shape that is most effective for reaction.
- Enzymes can act as acids and bases during reaction, donating or accepting hydrogen ions from the substrate quickly and easily.
- The potential energy of a bond distorted by the induced fit of the substrate to the enzyme is already partway up the activation energy hill that must be surmounted for reaction to occur.
- Enzymes sometimes contain metal ions that are needed to help catalyze oxidation-reduction reactions.

Enzyme Kinetics

An enzyme changes the mechanism of a reaction, as does any catalyst. The first step in the mechanism for any enzyme-catalyzed reaction is binding of the substrate and the enzyme, which is referred to as formation of an **enzyme-substrate complex.** Representing enzyme by E, substrate by S, and products by P, we can write a single-step uncatalyzed mechanism and a two-step enzyme-catalyzed mechanism as follows.

Uncatalyzed mechanism: $S \longrightarrow P$

Enzyme-catalyzed mechanism:

Step 1 (fast): $S + E \rightleftharpoons ES$ (formation of enzyme-substrate complex)

Step 2 (slow): $ES \longrightarrow P + E$ (formation of products and regeneration of enzyme)

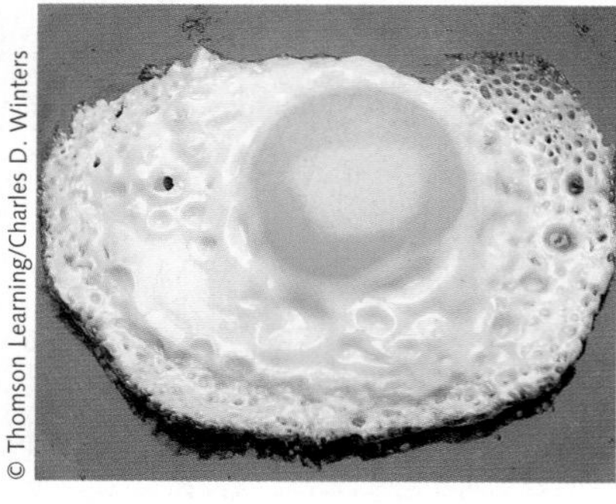

Frying an egg causes protein in the white to denature and precipitate from the nearly clear original solution to form a white flexible solid.

Denaturation of enzymes answers the question, "What happens when I hard-boil an egg?" that was posed in Chapter 1 (p. 3).

That the enzyme is a catalyst is evident from the fact that it is a reactant in the first step and is regenerated in the second. This mechanism applies to nearly all enzyme-catalyzed reactions. Because the second step is slow, the enzyme-substrate complex can often separate and reform S + E before it reacts to form products. This possibility is indicated by the double arrow in the first step. This mechanism for enzyme catalysis is similar to the mechanisms of reactions with a rapid, reversible first step that were discussed in Section 13.7 (p. 637), and it can be analyzed mathematically in the same way as those mechanisms were.

Because of the noncovalent interactions between enzyme and substrate, the activation energy is significantly lower for the enzyme-catalyzed reaction than it would be for the uncatalyzed process. This situation is shown in Figure 13.17. Even at temperatures only a little above room temperature, significant numbers of molecules have enough energy to surmount this lower barrier. Thus, enzyme-catalyzed reactions can occur reasonably quickly at body temperature.

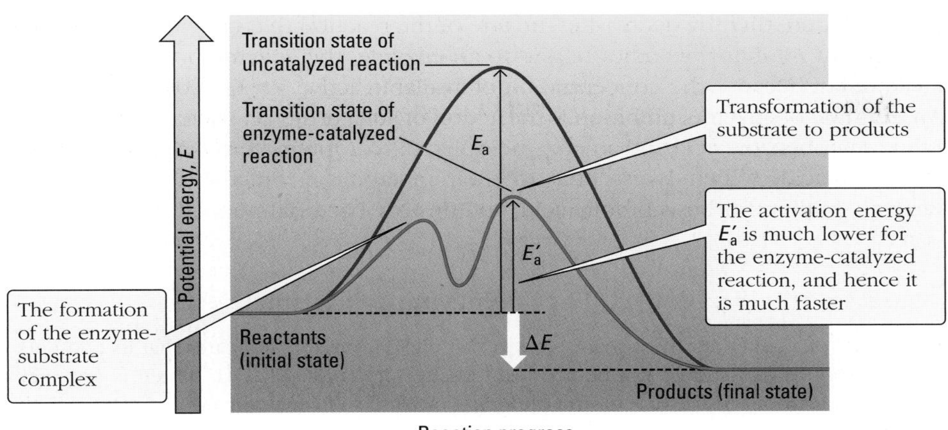

Figure 13.17 **Energy diagram for enzyme-catalyzed reaction.** The red curve is the energy profile for a typical reaction in a living system with no enzyme present. The green energy profile is drawn to the same scale for the same reaction with enzyme catalysis.

Special Features of Enzyme Catalysis

Enzyme-catalyzed reactions obey the same principles of chemical kinetics that we discussed earlier in this chapter. Nevertheless, both the enzyme itself and the mechanism of enzyme catalysis have some special features that you should be aware of. First, because of the form of the mechanism, either the enzyme or the substrate may be the limiting reactant in the first step. If the substrate is limiting, increasing the concentration of substrate produces more enzyme-substrate complex and makes the reaction go faster. This is the expected behavior: Increasing the concentration of a reactant should increase the rate proportionately. If the enzyme becomes the limiting reactant, however, it can become completely converted to enzyme-substrate complex, leaving no enzyme available for additional substrate. If this happens, a further increase in the concentration of substrate will not increase the rate of reaction. As a result there is a *maximum rate* (those who study enzyme kinetics call this the maximum velocity) for an enzyme-catalyzed reaction. The behavior of rate with increasing substrate concentration is shown in Figure 13.18.

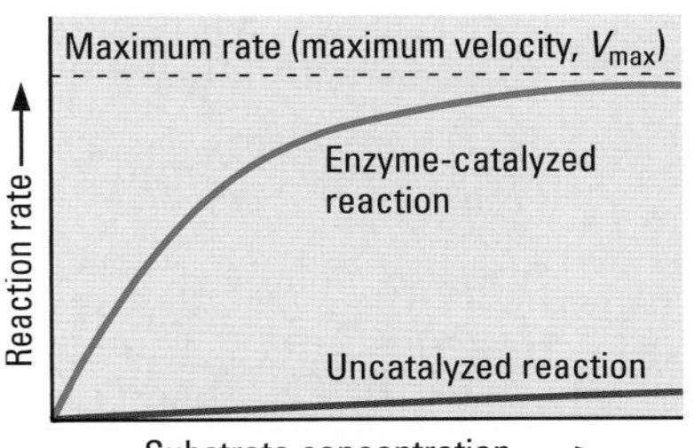

Figure 13.18 **Maximum velocity for an enzyme-catalyzed reaction.** Because there is only a limited quantity of enzyme available, increasing substrate concentration beyond the point at which the enzyme becomes the limiting reactant does not increase the rate further. There is a maximum rate (maximum velocity) for any enzyme-catalyzed reaction.

Enzyme-catalyzed reactions also behave unusually with respect to temperature. The rate does increase with increasing temperature, but if the temperature becomes high enough, there is a sudden decrease in rate, as shown in Figure 13.19. This happens because there is increased molecular and atomic motion as the temperature increases, and that motion can disrupt the structures of enzymes and other proteins. This change in protein structure is called **denaturation.** It occurs, for example, when an egg is boiled or fried. When a globular protein is denatured it loses its coiled globular structure, and its solubility and other properties change. Once an enzyme's structure has changed, the active site is no longer available, enzyme catalysis is seriously impaired, and the reaction rate falls to its uncatalyzed value. Notice that this happens only a little above 37 °C, which is body temperature for humans. Enzymes have evolved to produce maximum rates at body temperature, and slightly higher temperatures cause most of them to denature.

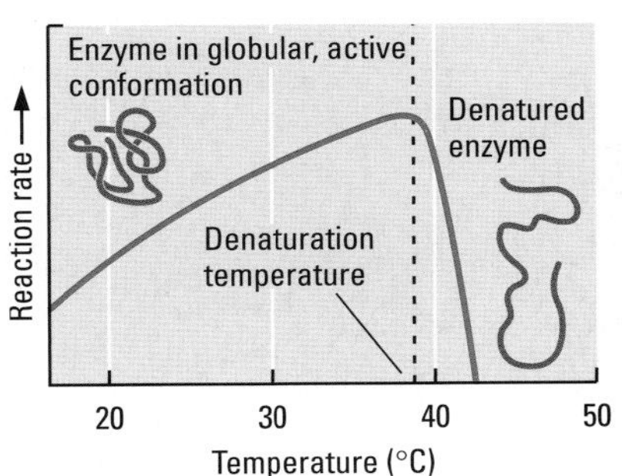

Figure 13.19 **Enzyme activity destroyed by high temperature.** At a temperature somewhat above normal body temperature, there is sufficient molecular motion to overcome the noncovalent interactions that maintain protein structure. This disrupts the structure of an enzyme, thereby destroying its catalytic activity. The process by which the enzyme structure becomes disrupted is called denaturation.

Inhibition of Enzymes

There is another way that the activity of an enzyme can be destroyed. Some molecules or ions can fit an enzyme's active site, but remain there unreacted. Such a molecule is called an **inhibitor.** An inhibitor bound to an enzyme decreases its effective concentration, thereby decreasing the rate of the reaction that the enzyme catalyzes. If sufficient inhibitor becomes bound to an enzyme, the enzyme provides little catalytic effect because the concentration of available active sites becomes very small. An example of enzyme inhibition is the action of sulfa drugs on bacteria. Bacteria use *para*-aminobenzoic acid and an enzyme called dihydropteroate synthetase to synthesize folic acid, which is essential to their metabolism. Sulfa drugs bind to this enzyme, inhibit synthesis of folic acid, and destroy bacterial populations.

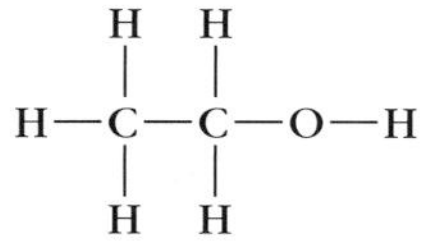

PROBLEM-SOLVING EXAMPLE 13.11 Enzyme Inhibition

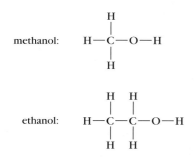

The label of a container of methanol (methyl alcohol) invariably indicates that its contents are poisonous and should not be taken internally. Methanol, CH_3OH, which is not very toxic, is metabolized by the enzyme methanol oxidase to formaldehyde, $H_2C{=}O$, which is very toxic. Methanol poisoning is sometimes treated by giving the patient ethanol, CH_3CH_2OH, which inhibits the enzyme. Identify similarities and differences in the structures of methanol and ethanol that could account for ethanol's acting as an inhibitor.

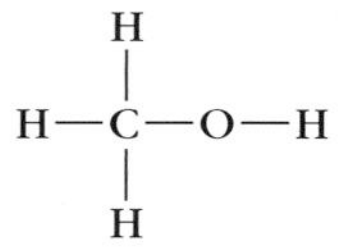

methanol: H—C(H)(H)—O—H

ethanol: H—C(H)(H)—C(H)(H)—O—H

Answer Both molecules are alcohols and can hydrogen-bond. Methanol has three hydrogens on the carbon next to the —OH group. Ethanol has only two hydrogens on the carbon adjacent to the —OH, and it has one more carbon and two more hydrogens.

Strategy and Explanation Because its shape is similar to that of methanol and because it can form hydrogen bonds of similar strength, ethanol binds to the active sites of some of the methanol oxidase catalyst molecules. Because of the extra carbon atom or the difference in number of hydrogens adjacent to the —OH group, the catalyst is unable to oxidize the ethanol, and ethanol remains bound. The smaller concentration of catalyst molecules decreases the rate of conversion of methanol to formaldehyde in the body, and the harmful effect is less.

PROBLEM-SOLVING PRACTICE 13.11

Bacteria need to use *p*-aminobenzoic acid to help synthesize folic acid in order to survive. Sulfa drugs interfere with this process. The structures of *p*-aminobenzoic acid and folic acid are

p-aminobenzoic acid

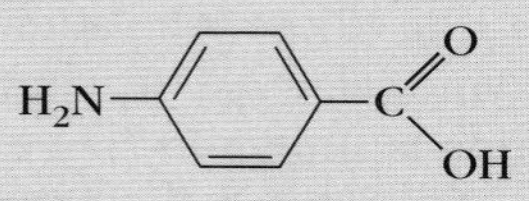

H_2N—C_6H_4—C(=O)OH

folic acid

H_2N–pterin–CH_2—NH—C_6H_4—C(=O)—(N(H)—CH(COO$^-$)—CH_2—CH_2—C(=O))$_{1-7}$—O$^-$

pterin (2-amino-4-oxopteridine) | *p*-aminobenzoic acid | glutamates

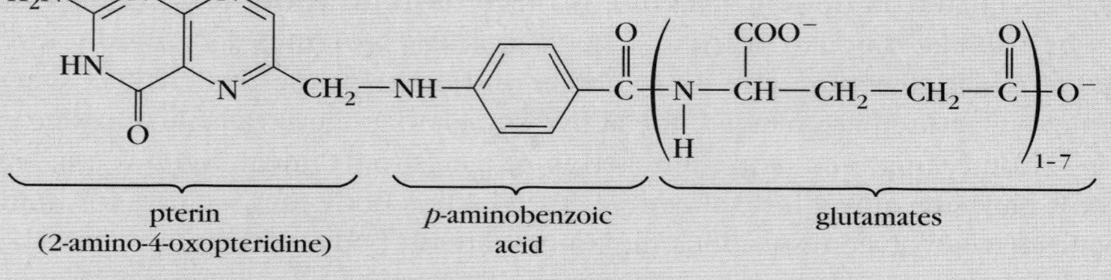

Which of these structures is most likely a sulfa drug? Explain your choice.

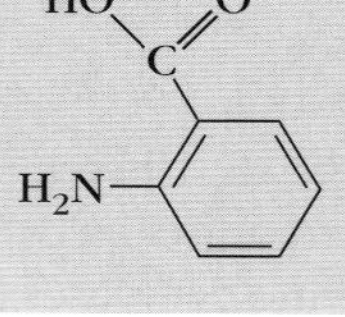

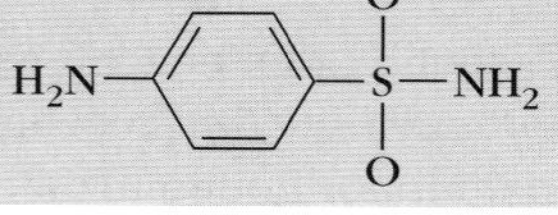

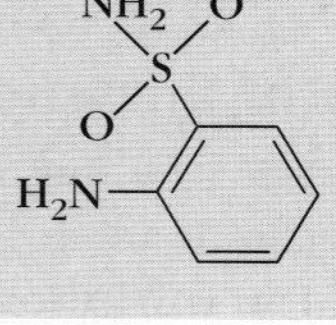

(1) H_2N—C_6H_4 with C(=O)HO ortho   (2) H_2N—C_6H_4—S(=O)(=O)—NH_2 (3) H_2N—C_6H_4 with S(=O)(=O)NH_2 ortho

CHEMISTRY IN THE NEWS

Protease Inhibitors and AIDS

A new class of drugs for treatment of AIDS has been available for several years and has been highly publicized. These drugs, called protease inhibitors, have slowed the spread of this disease considerably by inhibiting growth of human immunodeficiency virus (HIV).

As the name implies, protease inhibitors inhibit an enzyme, HIV-1 protease. This enzyme is essential for maturation of the virus because it catalyzes a reaction in which a long polypeptide chain is cut into shorter pieces. The cuts occur at specific locations along the chain, and the smaller pieces created by HIV-1 protease are proteins that are essential to the survival of HIV. Like plastic trash bags that have to be separated from a long roll before they become useful, these proteins must be cut from the long polypeptide before they can carry out their functions in HIV. Several different proteins are produced this way by HIV-1 protease. Therefore, if this enzyme could be inhibited, reproduction of the virus would be interfered with in several different ways.

The action of HIV-1 protease, and its importance to HIV, is typical of how reactions are controlled in living organisms. It is much quicker to cut a long polypeptide chain into shorter, active protein molecules than it is to synthesize lots of protein molecules on short notice. Therefore, other enzymes work in advance together with the virus's DNA to synthesize the long polypeptide, called a pre-protein. HIV-1 protease then chops the pre-protein into appropriate pieces whenever they are needed by HIV. If a lot of a particular protein is needed, it can be formed quickly by making a few cuts in the pre-protein, instead of having to put together a large number of amino acids in the proper sequence.

HIV-1 protease actually consists of two polypeptide chains held together as a dimer by noncovalent attractive forces. Because the very large number of atoms in the enzyme would obscure your view of its overall structure, the picture of HIV-1 protease shown here represents the polypeptide strands of the two monomers using ribbons and tubes. From the picture you can see that there is an open space in the middle—between the two halves of the enzyme. This is the active site. The enzyme works by having the two monomers come together to form an active site around the long pre-protein. This happens at a specific place along the pre-protein chain, and the active site cuts the polypeptide at that point by helping to break a peptide bond. Then the two monomers and the two pieces of polypeptide separate. The HIV-1 protease monomers can later cut another piece from the same or another polypeptide.

AIDS drugs that are protease inhibitors consist of molecules that can occupy the active site of HIV-1 protease, but their structures differ enough from the pre-protein structure that the protease cannot cut them. They remain in the active site, as shown in the second figure, holding the dimer together and preventing HIV-1 protease from cutting any more pre-protein molecules.

Eight protease inhibitor molecules have now been tested and are available for treatment of AIDS patients. Their efficacy is somewhat reduced because the viruses that survive treatment with protease inhibitors can adapt rapidly, forming variants that are not susceptible to treatment with the standard drugs. On the basis of kinetic analyses, one group in the Czech Republic has found a pseudopeptide HIV-protease inhibitor that effectively inhibits a wide range of variant proteases. Although this and many other similar studies have not yet produced thoroughly tested drugs that can be used in humans, they illustrate the kinds of continuing research being applied to the tragic problem of AIDS.

SOURCES: Based on information from *Science*, June 28, 1996; *FDA Consumer*, July–August 1999; *Journal of Molecular Biology*, December 6, 2002; pp. 739–754; and two Internet sites: **http://aids.org/immunet/atn.nsf/homepage** **http://www.aidsinfo.nih.gov/drugs/**

HIV protease is a dimer. It consists of two polypeptide chains. They enclose an active site in the center.
Active site

Structure of the HIV-1 protease dimer. HIV-1 protease consists of two similar parts that are held together by noncovalent attractions.

HIV-1 protease dimer with inhibitor. An inhibitor molecule, drawn as a space-filling structure, occupies the active site of HIV-1 protease.

13.10 Catalysis in Industry

An expert in the field of industrial chemistry has said that every year more than one trillion dollars' worth of goods is manufactured with the aid of manmade catalysts. Without them, fertilizers, pharmaceuticals, fuels, synthetic fibers, solvents, and detergents would be in short supply. Indeed, 90% of all manufactured items use catalysts at some stage of production. The major areas of catalyst use are in petroleum refining, industrial production of chemicals, and environmental controls. In this section we provide a few examples of the many important industrial reactions that depend on catalysis.

Many industrial reactions use **heterogeneous catalysts.** Such catalysts are present in a different phase from that of the reactants being catalyzed. Usually the catalyst is a solid and the reactants are in the gaseous or liquid phase. Heterogeneous catalysts are used in industry because they are more easily separated from the products and leftover reactants than are homogeneous catalysts. Catalysts for chemical processing are generally metal-based and often contain precious metals such as platinum and palladium. In the United States more than $600 million worth of such catalysts are used annually by the chemical processing industry, almost half of them in the preparation of polymers.

Manufacture of Acetic Acid

The importance of acetic acid, CH_3COOH, in the organic chemicals industry is comparable to that of sulfuric acid in the inorganic chemicals industry; annual production of acetic acid in the United States approaches 5 billion pounds. Acetic acid is used widely in industry to make plastics and synthetic fibers, as a fungicide, and as the starting material for preparing many dietary supplements. One way of synthesizing the acid is an excellent example of homogeneous catalysis: Rhodium(III) iodide is used to speed up the combination of carbon monoxide and methyl alcohol, both inexpensive chemicals, to form acetic acid.

$$\underset{\text{methyl alcohol}}{CH_3OH} + \underset{\text{carbon monoxide}}{CO} \xrightarrow{RhI_3\ \text{catalyst}} \underset{\text{acetic acid}}{CH_3\overset{\overset{\displaystyle O}{\|}}{C}{-}OH}$$

The role of the rhodium(III) iodide catalyst in this reaction is to bring the reactants together and allow them to rearrange to form the products. Carbon monoxide and the methyl group from the alcohol become attached to the rhodium atom, which helps transfer the methyl group to the CO. After this rearrangement, the intermediate reacts with solvent water to form acetic acid.

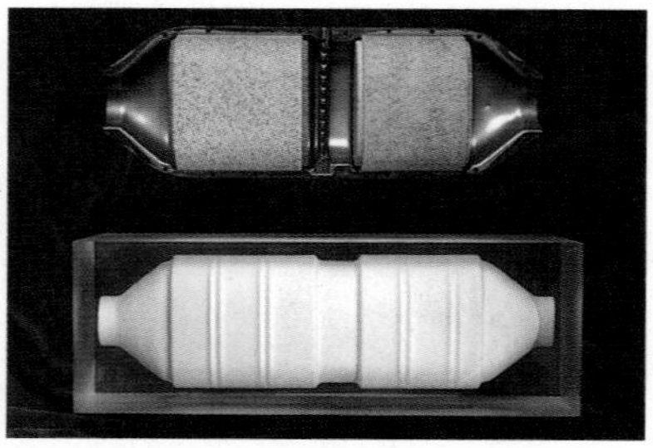

AC/GM/Peter Arnold, Inc.

Figure 13.20 Automobile catalytic converter. Catalytic converters are standard equipment on the exhaust systems of all new automobiles. This one contains two catalysts: One converts nitrogen monoxide to nitrogen and the other converts carbon monoxide and hydrocarbons to carbon dioxide and water.

Controlling Automobile Emissions

The largest growth in catalyst use is in *emissions control* for both automobiles and power plants. This market uses very large quantities of platinum group metals: platinum, palladium, rhodium, and iridium. In 2004, nearly 100,000 kg platinum and more than 100,000 kg palladium were sold worldwide for automotive uses. Demand was also high for rhodium for this same purpose. All three metals are also used in chemical processing as catalysts, and the petroleum industry uses platinum and rhodium to catalyze refining processes.

The purpose of the catalysts in the exhaust system of an automobile is to ensure that the combustion of carbon monoxide and hydrocarbons is complete (Figure 13.20).

$$2\ CO(g) + O_2(g) \xrightarrow{\text{Pt-NiO catalyst}} 2\ CO_2(g)$$

$$2\ C_8H_{18}(g) + 25\ O_2(g) \xrightarrow{\text{Pt-NiO catalyst}} 16\ CO_2(g) + 18\ H_2O(g)$$

2,2,4-trimethylpentane,
a component of gasoline

and to convert nitrogen oxides to molecules that are less harmful to the environment. At the high temperature of combustion, some N_2 from air reacts with O_2 to give NO, a serious air pollutant. Thermodynamic data show that nitrogen monoxide is unstable and should revert to N_2 and O_2. But remember that thermodynamics says nothing about rate. Unfortunately, the rate of reversion of NO to N_2 and O_2 is slow. Fortunately, catalysts have been developed that greatly speed this reaction.

$$2\ NO(g) \xrightarrow{\text{catalyst}} N_2(g) + O_2(g)$$

The role of the heterogeneous catalyst in the preceding reactions is probably to weaken the bonds of the reactants and to assist in product formation. For example, Figure 13.21 shows how NO molecules can dissociate into N and O atoms on the surface of a platinum metal catalyst.

Converting Methane to Liquid Fuel

Your home may be heated with natural gas, which consists largely of methane, CH_4. Although widely used, it is also widely wasted because much of it is found in geographical areas far removed from where fuels are consumed, and because transporting the flammable gas is expensive and dangerous. One solution to making methane

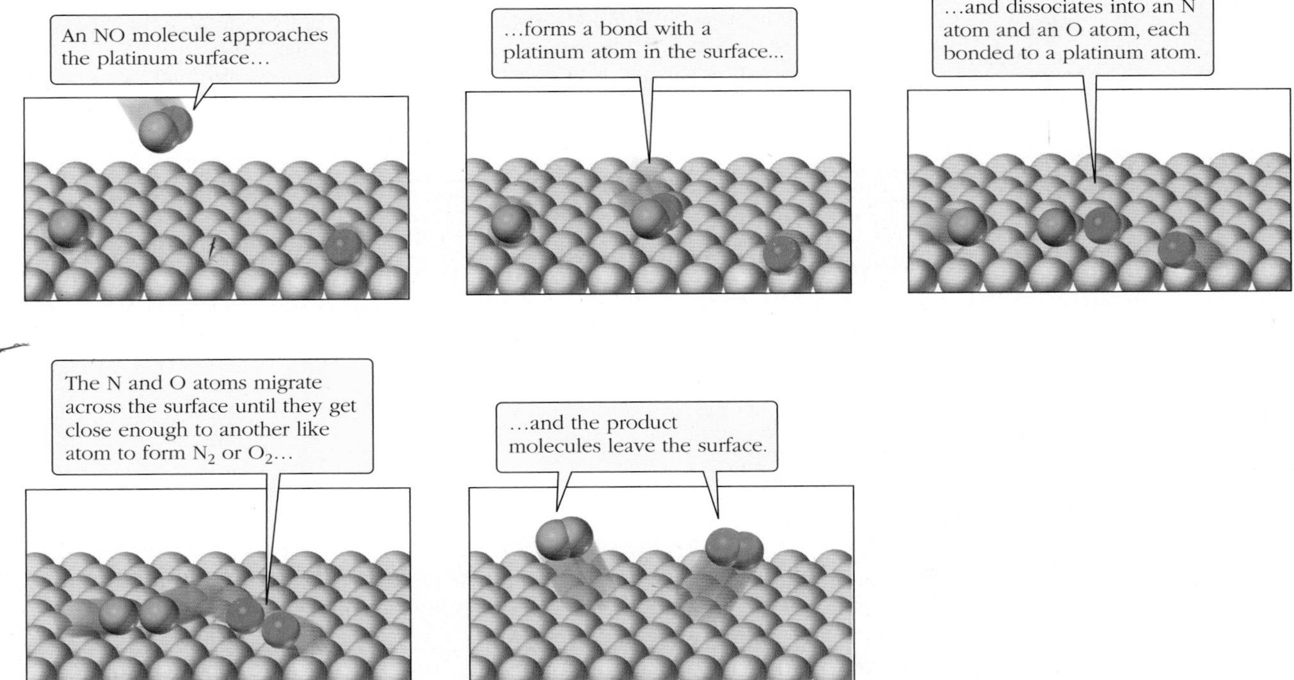

Figure 13.21 **Catalytic conversion of NO to N_2 and O_2.** A platinum surface can speed conversion of NO to N_2 and O_2 by helping to dissociate NO into N atoms and O atoms, which then travel across the surface and combine to form N_2 and O_2 molecules. The entire process of interaction with the surface and dissociation takes about 1.7×10^{-12} s.

CHEMISTRY IN THE NEWS

Catalysis and Coal

With prices for natural gas at $7 per million cubic feet and crude oil at more than $60 per barrel (up from $2 per million cubic feet and $10 per barrel not long ago), chemical companies and energy companies are looking for new sources of raw materials and fuel. One new source may actually be the fuel that fired the Industrial Revolution 150 years ago: coal.

The United States has the largest reserves of coal in the world—more than 250 billion tons—so coal is available and not subject to control by other countries. Coal has been used less during the past half-century, though, because it causes more pollution. Coal contains more impurities such as sulfur and mercury, is harder to purify prior to combustion, and has a larger fraction of carbon than other fossil fuels (which leads to more CO_2 emissions and a greater contribution to global warming [***Section 10.13,*** ***p. 473***]). But with properly designed catalysts coal might make a comeback.

Coal is mostly carbon and can be converted to gaseous hydrocarbons (gasified) by reacting it with steam; an approximate equation is

$$\text{Coal (C)} + H_2O(g) \longrightarrow CO(g) + H_2(g)$$

Impurities in the coal such as sulfur and mercury can be separated from the gaseous carbon monoxide and hydrogen. Using appropriate catalysts, the carbon monoxide and hydrogen can then be converted to either gaseous fuels, such as methane, CH_4, or liquid fuels such as alkanes or alkenes.

The Rentech Corporation uses an iron-based catalyst to convert coal into low-sulfur liquid hydrocarbon fuels that the company claims burn very cleanly. General Electric Corporation purchased coal-gasification technology in 2004 and expects to use it to create gaseous fuel for electric power plants. GreatPoint Corporation, using a proprietary catalyst, has developed a coal gasifier that the company claims can turn coal into highly pure methane that could substitute for natural gas. Because of the high price of petroleum, Eastman Chemical Corporation is considering increasing its use of coal as a raw material from which to manufacture chemicals from 20 to 40% of its production. And Dow Chemical Corporation is in discussions with Shenhua, China's largest state-run coal producer, regarding a possible plant to make ethylene and propylene from coal in China.

The Rentech Corporation process is shown schematically in the figure. In step 1, carbon-bearing materials such as coal or natural gas are converted to synthesis gas (a mixture of carbon monoxide and hydrogen). In step 2 the synthesis gas undergoes a catalytic reaction (called Fischer-Tropsch synthesis) that produces a mixture of nearly pure liquid hydrocarbons. In step 3 the hydrocarbons are refined to diesel fuel or jet fuel. A byproduct of the process is production of power.

Unless the prices of petroleum and natural gas decline, coal may very well regain its former importance as a fuel and a source of manufactured chemicals.

SOURCES: Based on stories in *The New York Times*, Tuesday, April 18, 2006, p C1; NPR's Morning Edition, Tuesday, April 25, 2006; *Science*, Vol. 312, April 14, 2006; p. 175; and two Web sites: **http://www.rentechinc.com/** **http://www.greatpointenergy.com/**

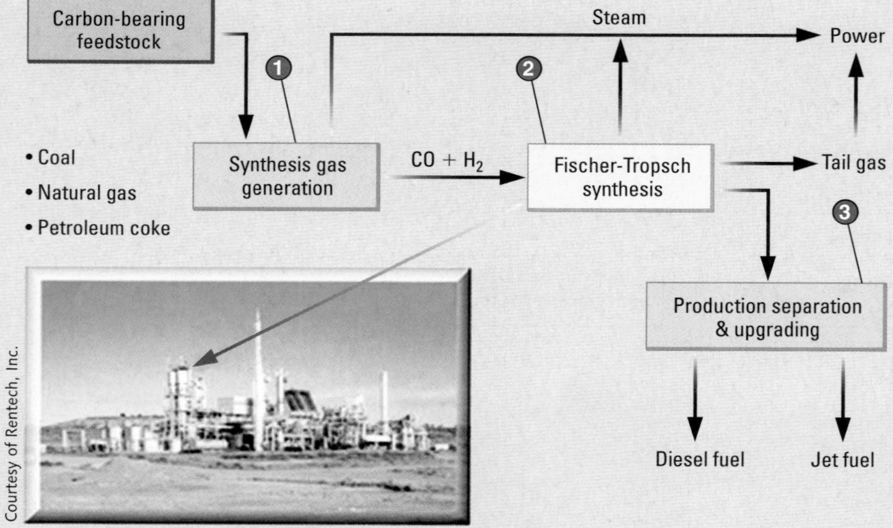

Synhytech–Pueblo, Colorado, 1992

useful is to convert it, where it is found, to a more readily transportable substance such as liquid methanol, CH_3OH. The methanol can then be used directly as a fuel, added to gasoline [as is currently done in some areas of the United States (⇐ p. 564)], or used to make other chemicals.

It has been known for some time that methane can be converted to carbon monoxide and hydrogen,

$$CH_4(g) + \frac{1}{2}O_2(g) \longrightarrow CO(g) + 2\,H_2(g)$$

and this mixture of gases can readily be turned into methanol in another step.

$$CO(g) + 2\,H_2(g) \longrightarrow CH_3OH(\ell)$$

Chemical engineers at the University of Minnesota found that methane can be converted to CO and H_2 under very mild conditions of temperature. They simply found the right catalyst. Figure 13.22 shows what happens when a room-temperature mixture of methane and oxygen flows through a heated, sponge-like ceramic disk coated with platinum or rhodium. Rather than oxidizing the methane all the way to water and carbon dioxide, the process produces a hot mixture of CO and H_2, which can be converted in good yield to methanol. It is also possible to produce other partially oxidized hydrocarbons by a similar catalytic process.

Schmidt/University of Minnesota

Figure 13.22 Methane flowing through a catalyst. The square grid of the catalyst glows yellow because of energy transferred to its surface by catalyzed combustion of the methane.

CONCEPTUAL EXERCISE 13.14 Catalysis

Which of these statements is (are) true? If any are false, change the wording to make them true.

(a) The concentration of a homogeneous catalyst may appear in the rate law.
(b) A catalyst is always consumed in the overall reaction.
(c) A catalyst must always be in the same phase as the reactants.

SUMMARY PROBLEM

An excellent way to make highly pure nickel metal for use in specialized steel alloys is to decompose $Ni(CO)_4$ by heating it in a vacuum to slightly above room temperature.

$$Ni(CO)_4(g) \longrightarrow Ni(s) + 4\,CO(g)$$

The reaction is proposed to occur in four steps, the first of which is

$$Ni(CO)_4(g) \longrightarrow Ni(CO)_3(g) + CO(g)$$

Kinetic studies of this first-order decomposition reaction have been carried out between 47.3 °C and 66.0 °C to give the results in this table.*

Temperature (°C)	Rate Constant (s^{-1})
47.3	0.233
50.9	0.354
55.0	0.606
60.0	1.022
66.0	1.873

*See Day, J. P., Basolo, F., and Pearson, R. G. *Journal of the American Chemical Society,* Vol. 90, 1968; p. 6933.

(a) What is the activation energy for this reaction?

(b) $Ni(CO)_4$ is formed by the reaction of nickel metal with carbon monoxide. If you have 2.05 g CO and you combine it with 0.125 g nickel metal, what is the maximum quantity of $Ni(CO)_4$ (in grams) that can be formed?

The replacement of CO by another molecule in $Ni(CO)_4$ (in the nonaqueous solvents toluene and hexane) was also studied to understand the general principles that govern the chemistry of such compounds.*

$$Ni(CO)_4 + P(CH_3)_3 \longrightarrow Ni(CO)_3P(CH_3)_3 + CO$$

A detailed study of the kinetics of the reaction led to the mechanism

$$Ni(CO)_4 \longrightarrow Ni(CO)_3 + CO \qquad \text{slow}$$

$$Ni(CO)_3 + P(CH_3)_3 \longrightarrow Ni(CO)_3P(CH_3)_3 \qquad \text{fast}$$

(c) Which step in the mechanism is unimolecular? Which is bimolecular?

(d) Add the steps of the mechanism to show that the result is the balanced equation for the observed reaction.

(e) Is there an intermediate in this reaction? If so, what is it?

(f) It was found that doubling the concentration of $Ni(CO)_4$ increased the reaction rate by a factor of 2. Doubling the concentration of $P(CH_3)_3$ had no effect on the reaction rate. Based on this information, write the rate equation for the reaction.

(g) Does the experimental rate equation support the proposed mechanism? Why or why not?

ThomsonNOW™
Sign in to ThomsonNOW at **www.thomsonedu.com** to check your readiness for an exam by taking the Pre-Test and exploring the modules recommended in your Personalized Learning Plan.

IN CLOSING

Having studied this chapter, you should be able to . . .

- Define reaction rate and calculate average rates (Section 13.1). ThomsonNOW homework: Study Question 11
- Describe the effect that reactant concentrations have on reaction rate, and determine rate laws and rate constants from initial rates (Section 13.2). ThomsonNOW homework: Study Questions 21, 27
- Determine reaction orders from a rate law, and use the integrated rate law method to obtain orders and rate constants (Section 13.3). ThomsonNOW homework: Study Question 29
- Calculate concentration at a given time, time to reach a certain concentration, and half-life for a first-order reaction (Section 13.3). ThomsonNOW homework: Study Questions 41, 43
- Define and give examples of unimolecular and bimolecular elementary reactions (Section 13.4). ThomsonNOW homework: Study Question 45
- Show by using an energy profile what happens as two reactant molecules interact to form product molecules (Section 13.4). ThomsonNOW homework: Study Question 65
- Define activation energy and frequency factor, and use them to calculate rate constants and rates under different conditions of temperature and concentration (Section 13.5). ThomsonNOW homework: Study Questions 51, 53, 57, 62
- Derive rate laws for unimolecular and bimolecular elementary reactions (Section 13.6). ThomsonNOW homework: Study Question 71

- Define reaction mechanism and identify rate-limiting steps and intermediates (Section 13.7). ThomsonNOW homework: Study Questions 73, 77
- Given several reaction mechanisms, decide which is (are) in agreement with experimentally determined stoichiometry and rate law (Section 13.7). ThomsonNOW homework: Study Question 79
- Explain how a catalyst can speed up a reaction; and draw energy profiles for catalyzed and uncatalyzed reaction mechanisms (Section 13.8). ThomsonNOW homework: Study Question 83
- Define the terms *enzyme, substrate,* and *inhibitor,* and identify similarities and differences between enzyme-catalyzed reactions and uncatalyzed reactions (Section 13.9). ThomsonNOW homework: Study Question 87
- Describe several important industrial processes that depend on catalysts (Section 13.10).

KEY TERMS

activated complex *(13.4)*
activation energy (E_a) *(13.4)*
active site *(13.9)*
Arrhenius equation *(13.5)*
average reaction rate *(13.1)*
bimolecular reaction *(13.4)*
chemical kinetics *(Introduction)*
cofactor *(13.9)*
denaturation *(13.9)*
elementary reaction *(13.4)*
enzyme *(13.9)*
enzyme-substrate complex *(13.9)*
frequency factor *(13.5)*
half-life *(13.3)*
heterogeneous catalyst *(13.10)*
heterogeneous reaction *(13.1)*
homogeneous catalyst *(13.8)*
homogeneous reaction *(13.1)*
induced fit *(13.9)*
inhibitor *(13.9)*
initial rate *(13.2)*
instantaneous reaction rate *(13.1)*
intermediate *(13.7)*
order of reaction *(13.3)*
overall reaction order *(13.3)*
rate *(13.1)*
rate constant *(13.2)*
rate law *(13.2)*
rate-limiting step *(13.7)*
reaction intermediate *(13.7)*
reaction mechanism *(13.7)*
reaction rate *(13.1)*
steric factor *(13.4)*
substrate *(13.9)*
transition state *(13.4)*
unimolecular reaction *(13.4)*

QUESTIONS FOR REVIEW AND THOUGHT

■ denotes questions available in ThomsonNOW and assignable in OWL.

Blue-numbered questions have short answers at the back of this book and fully worked solutions in the *Student Solutions Manual.*

ThomsonNOW™
Assess your understanding of this chapter's topics with sample tests and other resources found by signing in to ThomsonNOW at **www.thomsonedu.com**.

Review Questions

1. Which of these is appropriate for determining the rate law for a chemical reaction?
   (a) Theoretical calculations based on balanced equations
   (b) Measuring the rate of the reaction as a function of the concentrations of the reacting species
   (c) Measuring the rate of the reaction as a function of temperature
2. Name at least three factors that affect the rate of a chemical reaction.
3. Using the rate law rate = $k[A]^2[B]$, define the order of the reaction with respect to A and B and the overall reaction order.
4. Define the terms "enzyme," "substrate," and "inhibitor," and give an example of each kind of molecule.
5. Explain the difference between a homogeneous and a heterogeneous catalyst. Give an example of each.
6. Define the terms "unimolecular elementary reaction" and "bimolecular elementary reaction," and give an example of each.
7. Define the terms "activation energy" and "frequency factor." Write an equation that relates activation energy and frequency factor to reaction rate.
8. Why is reaction kinetics important to our understanding of depletion of stratospheric ozone (**← p. 462**)?

Topical Questions

Reaction Rate

9. Consider the dissolving of sugar as a simple process in which kinetics is important. Suppose that you dissolve an equal mass of each kind of sugar listed. Which dissolves the fastest? Which dissolves the slowest? Explain why in terms of rates of heterogeneous reactions. (If you are not sure which is fastest or slowest, try them all.)
   (a) Rock candy sugar (large sugar crystals)
   (b) Sugar cubes
   (c) Granular sugar
   (d) Powdered sugar
10. A cube of aluminum 1.0 cm on each edge is placed into 9 M NaOH(aq), and the rate at which H_2 gas is given off is measured.
   (a) By what factor will this reaction rate change if the aluminum cube is cut exactly in half and the two halves are placed in the solution? Assume that the reaction rate is proportional to the surface area, and that all of the surface of the aluminum is in contact with the NaOH(aq).
   (b) If you had to speed up this reaction as much as you could without raising the temperature, what would you do to the aluminum?
11. ■ Experimental data in the table are for the hypothetical reaction

$$A \longrightarrow 2\,B$$

Time (s)	[A] (mol/L)
0.00	1.000
10.0	0.833
20.0	0.714
30.0	0.625
40.0	0.555

   (a) Make a graph of concentration as a function of time, draw a smooth curve through the points, and calculate the rate of change of [A] for each 10-s interval from 0.00 to 40.0 s. Why might the rate of change decrease from one time interval to the next?
   (b) How is the rate of change of [B] related to the rate of change of [A] in the same time interval?
   (c) Calculate the rate of change of [B] for the time interval from 10.0 to 20.0 s.
12. A compound called phenyl acetate reacts with water according to the equation

$$\underset{\text{phenyl acetate}}{CH_3\overset{\overset{\displaystyle O}{\|}}{C}{-}O{-}C_6H_5} + H_2O \longrightarrow \underset{\text{acetic acid}}{CH_3\overset{\overset{\displaystyle O}{\|}}{C}{-}O{-}H} + \underset{\text{phenol}}{C_6H_5{-}O{-}H}$$

These data were collected at 5 °C.

Time (min)	[Phenyl acetate] (mol/L)
0	0.55
0.25	0.42
0.50	0.31
0.75	0.23
1.00	0.17
1.25	0.12
1.50	0.082

   (a) Make a graph of concentration as a function of time, describe the shape of the curve, and compare it with Figure 13.3.
   (b) Calculate the rate of change of the concentration of phenylacetate during the period from 0.20 to 0.40 min, and then during the period from 1.2 to 1.4 min. Compare the values and tell why one is smaller than the other.
   (c) What is the rate of change of the phenol concentration during the period from 1.00 to 1.25 min?
13. Using data given in the table for the reaction

$$N_2O_5 \longrightarrow 2\,NO_2 + \tfrac{1}{2}\,O_2$$

calculate the average rate of reaction during each of these intervals:
   (a) 0.00 to 0.50 h
   (b) 0.50 to 1.0 h
   (c) 1.0 to 2.0 h
   (d) 2.0 to 3.0 h
   (e) 3.0 to 4.0 h
   (f) 4.0 to 5.0 h

Time (h)	$[N_2O_5]$ (mol/L)	Time (h)	$[N_2O_5]$ (mol/L)
0.00	0.849	3.00	0.352
0.50	0.733	4.00	0.262
1.00	0.633	5.00	0.196
2.00	0.472		

14. Using data from Question 13, calculate the average rate over the interval 0 to 5.0 h. Compare your result with the average rates over the intervals 1.0 to 4.0 h and 2.0 to 3.0 h, all of which have the same midpoint (2.5 h from the start).
15. Using all your calculated rates from Question 13,
   (a) Show that the reaction obeys the rate law

$$\text{Rate} = -\frac{\Delta\,[N_2O_2]}{\Delta t} = k[N_2O_5]$$

   (b) Evaluate the rate constant k as an average of the values obtained for the six intervals.
16. Using the rate law and the rate constant you calculated in Question 15, calculate the reaction rate exactly 2.5 h from the start. Do your results from this and Question 14 agree with the statement in the text that the smaller the time interval, the more accurate the average rate?

17. For the reaction

$$2\ NO_2(g) \longrightarrow 2\ NO(g) + O_2(g)$$

make qualitatively correct plots of the concentrations of $NO_2(g)$, NO(g), and $O_2(g)$ versus time. Draw all three graphs on the same axes, assume that you start with $NO_2(g)$ at a concentration of 1.0 mol/L, and assume that the reaction is first-order. Explain how you would determine, from these plots,
(a) The initial rate of the reaction
(b) The final rate (i.e., the rate as time approaches infinity)

18. For the reaction

$$O_3(g) + O(g) \longrightarrow 2\ O_2(g)$$

make qualitatively correct plots of the concentrations of $O_3(g)$, O(g), and $O_2(g)$ versus time. Draw all three graphs on the same axes, assume that you start with $O_3(g)$ and O(g), each at a concentration of 1.0 μmol/L, and assume that the reaction is second-order. Explain how you would determine, from these plots,
(a) The initial rate of the reaction
(b) The final rate (i.e., the rate as time approaches infinity)

Effect of Concentration on Reaction Rates

19. If a reaction has the experimental rate law rate $= k[A]^2$, explain what happens to the rate when
(a) The concentration of A is tripled
(b) The concentration of A is halved

20. A reaction has the experimental rate law rate $= k[A]^2[B]$. If the concentration of A is doubled and the concentration of B is halved, what happens to the reaction rate?

21. ■ The reaction of $CO(g) + NO_2(g)$ is second-order in NO_2 and zeroth-order in CO at temperatures less than 500 K.
(a) Write the rate law for the reaction.
(b) How will the reaction rate change if the NO_2 concentration is halved?
(c) How will the reaction rate change if the concentration of CO is doubled?

22. Nitrosyl bromide, NOBr, is formed from NO and Br_2.

$$2\ NO(g) + Br_2(g) \longrightarrow 2\ NOBr(g)$$

Experiment shows that the reaction is first-order in Br_2 and second-order in NO.
(a) Write the rate law for the reaction.
(b) If the concentration of Br_2 is tripled, how will the reaction rate change?
(c) What happens to the reaction rate when the concentration of NO is doubled?

23. For the reaction of $Pt(NH_3)_2Cl_2$ with water,

$$Pt(NH_3)_2Cl_2 + H_2O \longrightarrow Pt(NH_3)_2(H_2O)Cl^+ + Cl^-$$

the rate law is rate $= k[Pt(NH_3)_2Cl_2]$ with $k = 0.090\ h^{-1}$.
(a) Calculate the initial rate of reaction when the concentration of $Pt(NH_3)_2Cl_2$ is
(i) 0.010 M (ii) 0.020 M (iii) 0.040 M
(b) How does the rate of disappearance of $Pt(NH_3)_2Cl_2$ change with its initial concentration?
(c) How is this related to the rate law?
(d) How does the initial concentration of $Pt(NH_3)_2Cl_2$ affect the rate of appearance of Cl^- in the solution?

24. Methyl acetate, CH_3COOCH_3, reacts with base to break one of the C—O bonds.

$$CH_3\overset{\overset{\displaystyle O}{\|}}{C}-O-CH_3(aq) + OH^-(aq) \longrightarrow CH_3\overset{\overset{\displaystyle O}{\|}}{C}-O^-(aq) + HO-CH_3(aq)$$

The rate law is rate $= k[CH_3COOCH_3][OH^-]$ where $k = 0.14\ L\ mol^{-1}\ s^{-1}$ at 25 °C.
(a) What is the initial rate at which the methyl acetate disappears when both reactants, CH_3COOCH_3 and OH^-, have a concentration of 0.025 M?
(b) How rapidly (i.e., at what rate) does the methyl alcohol, CH_3OH, initially appear in the solution?

25. Measurements of the initial rate of reaction between two compounds, triphenylmethyl hexachloroantimonate, **I**, and bis-(9-ethyl-3-carbazolyl)methane, **II**, in 1,2-dichloroethane at 40 °C yielded these data:

Initial Concentration × 10^5 (mol/L)		**Initial Rate × 10^9 ($mol\ L^{-1}\ s^{-1}$)**
[I]	**[II]**	
1.65	10.6	1.50
14.9	10.6	17.7
14.9	7.10	11.2
14.9	3.52	6.30
14.9	1.76	3.10
4.97	10.6	4.52
2.48	10.6	2.70

(a) Derive the rate law for this reaction.
(b) Calculate the rate constant k and express it in appropriate units.

26. A study of the hypothetical reaction $2\ A + B \longrightarrow C + D$ gave these results:

Experiment	**Initial Concentration (mol/L)**		**Initial Rate ($mol\ L^{-1}\ s^{-1}$)**
	[A]	**[B]**	
1	0.10	0.050	6.0×10^{-3}
2	0.20	0.050	1.2×10^{-2}
3	0.30	0.050	1.8×10^{-2}
4	0.20	0.150	1.1×10^{-1}

(a) What is the rate law for this reaction?
(b) Calculate the rate constant k and express it in appropriate units.

27. ■ For the reaction

$$2\ NO(g) + 2\ H_2(g) \longrightarrow N_2(g) + 2\ H_2O(g)$$

these data were obtained at 1100 K:

[NO] (mol/L)	$[H_2]$ (mol/L)	Initial Rate (mol L^{-1} s^{-1})
5.00×10^{-3}	2.50×10^{-3}	3.0×10^{-3}
15.0×10^{-3}	2.50×10^{-3}	9.0×10^{-3}
15.0×10^{-3}	10.0×10^{-3}	3.6×10^{-2}

(a) What is the order with respect to NO? With respect to H_2?
(b) What is the overall order?
(c) Write the rate law.
(d) Calculate the rate constant.
(e) Calculate the initial rate of this reaction at 1100 K when $[NO] = [H_2] = 8.0 \times 10^{-3}$ mol L^{-1}.

28. The hypothetical reaction

$$2\,A + 2\,B \longrightarrow C + 3\,D$$

was studied by measuring the initial rate of appearance of C. These data were obtained:

[A] (mol/L)	[B] (mol/L)	Initial Rate (mol L^{-1} s^{-1})
6.0×10^{-3}	1.0×10^{-3}	0.012
6.0×10^{-3}	2.0×10^{-3}	0.024
2.0×10^{-3}	1.5×10^{-3}	0.0020
4.0×10^{-3}	1.5×10^{-3}	0.0080

(a) What is the order of the reaction with respect to substance A?
(b) What is the order with respect to B?
(c) What is the overall order?
(d) What is the rate law?
(e) Calculate the rate constant.
(f) If at a given instant A is disappearing at a rate of 0.034 mol L^{-1} s^{-1}, what is the rate of appearance of C? What is the rate of appearance of D?

Rate Law and Order of Reaction

29. ■ For each of these rate laws, state the reaction order with respect to the hypothetical substances A and B, and give the overall order.
(a) Rate = $k[A][B]^3$ (b) Rate = $k[A][B]$
(c) Rate = $k[A]$ (d) Rate = $k[A]^3[B]$

30. For each of the rate laws below, what is the order of the reaction with respect to the hypothetical substances X, Y, and Z? What is the overall order?
(a) Rate = $k[X][Y][Z]$ (b) Rate = $k[X]^2[Y]^{1/2}[Z]$
(c) Rate = $k[X]^{1.5}[Y]^{-1}$ (d) Rate = $k[X]/[Y]^2$

31. A reaction A + B ⟶ products is found to be second-order in B. Which rate equation cannot be correct?
(a) Rate = $k[A][B]$
(b) Rate = $k[A][B]^2$
(c) Rate = $k[B]^2$

32. The reaction

$$2\,NO(g) + 2\,H_2(g) \longrightarrow N_2(g) + 2\,H_2O(g)$$

is found to be first-order in $H_2(g)$. Which rate equation cannot be correct?
(a) Rate = $k[NO]^2[H_2]$
(b) Rate = $k[H_2]$
(c) Rate = $k[NO]^2[H_2]^2$

33. For the reaction of phenyl acetate with water the concentration as a function of time was given in Question 12. Assume that the concentration of water does not change during the reaction. Analyze the data from Question 12 to determine
(a) The rate law
(b) The order of the reaction with respect to phenyl acetate
(c) The rate constant
(d) The rate of reaction when the concentration of phenyl acetate is 0.10 mol/L (assuming that the concentration of water is the same as in the experiments in the table in Question 12)

34. When phenacyl bromide and pyridine are both dissolved in methanol, they react to form phenacylpyridinium bromide.

$$C_6H_5{-}\overset{\overset{\displaystyle O}{\|}}{C}{-}CH_2Br + C_5H_5N \longrightarrow C_6H_5{-}\overset{\overset{\displaystyle O}{\|}}{C}{-}CH_2NC_5H_5^+ + Br^-$$

When equal concentrations of reactants were mixed in methanol at 35 °C, these data were obtained:

Time (min)	[Reactant] (mol/L)	Time (min)	[Reactant] (mol/L)
0	0.0385	500	0.0208
100	0.0330	600	0.0191
200	0.0288	700	0.0176
300	0.0255	800	0.0163
400	0.0220	1000	0.0143

(a) What is the rate law for this reaction?
(b) What is the overall order?
(c) What is the rate constant?
(d) What is the rate of this reaction when the concentration of each reactant is 0.030 mol/L?

35. The transfer of an oxygen atom from NO_2 to CO has been studied at 540 K:

$$CO(g) + NO_2(g) \longrightarrow CO_2(g) + NO(g)$$

These data were collected:

Initial Rate (mol L^{-1} h^{-1})	Initial Concentration (mol/L)	
	[CO]	$[NO_2]$
5.1×10^{-4}	0.35×10^{-4}	3.4×10^{-8}
5.1×10^{-4}	0.70×10^{-4}	1.7×10^{-8}
5.1×10^{-4}	0.18×10^{-4}	6.8×10^{-8}
1.0×10^{-3}	0.35×10^{-4}	6.8×10^{-8}
1.5×10^{-3}	0.35×10^{-4}	10.2×10^{-8}

Use the data in the table to
(a) Write the rate law.
(b) Determine the reaction order with respect to each reactant.
(c) Calculate the rate constant, and express it in appropriate units.

36. The compound *p*-methoxybenzonitrile *N*-oxide, which has the formula $CH_3OC_6H_4CNO$, reacts with itself to form a dimer—a molecule that consists of two *p*-methoxybenzonitrile *N*-oxide units connected together $(CH_3OC_6H_4CNO)_2$. The reaction can be represented as

$$A + A \longrightarrow B \quad \text{or} \quad 2\,A \longrightarrow B$$

where A represents *p*-methoxybenzonitrile *N*-oxide and B represents the dimer $(CH_3OC_6H_4CNO)_2$. For the reaction in carbon tetrachloride at 40 °C with an initial concentration of 0.011 M, these data were obtained:

Time (min)	Percent Reaction	Time (min)	Percent Reaction
0	0	942	60.9
60	9.1	1080	64.7
120	16.7	1212	66.6
215	26.5	1358	68.5
325	32.7	1518	70.3
565	47.3		

(a) Determine the rate law for the reaction.
(b) What is the rate constant?
(c) What is the order of the reaction with respect to A?

37. The bromination of acetone is catalyzed by acid.

$$CH_3COCH_3(aq) + Br_2(aq) + H_2O(\ell) \xrightarrow{\text{acid catalyst}}$$
$$CH_3COCH_2Br(aq) + H_3O^+(aq) + Br^-(aq)$$

The rate of disappearance of bromine was measured for several different initial concentrations of acetone, bromine, and hydronium ion.

Initial Concentration (mol/L)			Initial Rate of Change of $[Br_2]$ (mol L^{-1} s^{-1})
$[CH_3COCH_3]$	$[Br_2]$	$[H_3O^+]$	
0.30	0.05	0.05	5.7×10^{-5}
0.30	0.10	0.05	5.7×10^{-5}
0.30	0.05	0.10	12.0×10^{-5}
0.40	0.05	0.20	31.0×10^{-5}
0.40	0.05	0.05	7.6×10^{-5}

(a) Deduce the rate law for the reaction and give the order with respect to each reactant.
(b) What is the numerical value of *k*, the rate constant?
(c) If $[H_3O^+]$ is maintained at 0.050 M, whereas both $[CH_3COCH_3]$ and $[Br_2]$ are 0.10 M, what is the rate of the reaction?

38. One of the major eye irritants in smog is formaldehyde, CH_2O, formed by reaction of ozone with ethylene.

$$C_2H_4(g) + O_3(g) \longrightarrow 2\,CH_2O(g) + \tfrac{1}{2}\,O_2(g)$$

These data were collected:

Initial Concentration (mol/L)		Initial Rate of Formation of CH_2O (mol L^{-1} s^{-1})
$[O_3]$	$[C_2H_4]$	
0.50×10^{-7}	1.0×10^{-8}	1.0×10^{-12}
1.5×10^{-7}	1.0×10^{-8}	3.0×10^{-12}
1.0×10^{-7}	2.0×10^{-8}	4.0×10^{-12}

(a) Determine the rate law for the reaction using the data in the table.
(b) What is the reaction order with respect to O_3? What is the order with respect to C_2H_4?
(c) Calculate the rate constant, *k*.
(d) What is the rate of reaction when $[C_2H_4]$ and $[O_3]$ are both 2.0×10^{-7} M?

39. This question requires working with the equations of Table 13.2. Using an initial concentration $[A]_0$ of 1.0 mol/L and a rate constant *k* with a numerical value of 1.0 in appropriate units, make plots of [A] versus time over the time interval 0 to 5 s for each type of integrated rate law. Compare your results with Figure 13.5.

40. Studies of radioactive decay of nuclei show that the *decay rate* of a radioactive sample is proportional to the *amount* of the radioactive species present. Once half the radioactivity has disappeared, the radioactive decay *rate* is only half of its original value. Is radioactive decay a zeroth-order, first-order, or second-order process?

41. ■ If the initial concentration of the reactant in a first-order reaction A ⟶ products is 0.64 mol/L and the half-life is 30. s,
(a) Calculate the concentration of the reactant 60 s after initiation of the reaction.
(b) How long would it take for the concentration of the reactant to drop to one-eighth its initial value?
(c) How long would it take for the concentration of the reactant to drop to 0.040 mol L^{-1}?

42. If the initial concentration of the reactant in a first-order reaction A ⟶ products is 0.50 mol/L and the half-life is 400 s,
(a) Calculate the concentration of the reactant 1600 s after initiation of the reaction.
(b) How long would it take for the concentration of the reactant to drop to one-sixteenth its initial value?
(c) How long would it take for the concentration of the reactant to drop to 0.062 mol L^{-1}?

43. ■ The compound SO_2Cl_2 decomposes in a first-order reaction

$$SO_2Cl_2(g) \longrightarrow SO_2(g) + Cl_2(g)$$

that has a half-life of 1.47×10^4 s at 600 K. If you begin with 1.6×10^{-3} mol of pure SO_2Cl_2 in a 2.0-L flask, at what time will the amount of SO_2Cl_2 be 1.2×10^{-4} mol?

44. The hypothetical compound A decomposes in a first-order reaction that has a half-life of 2.3×10^2 s at 450. °C. If the initial concentration of A is 4.32×10^{-2} M, how long will it take for the concentration of A to drop to 3.75×10^{-3} M?

A Nanoscale View: Elementary Reactions

45. ■ Which of these reactions are unimolecular and elementary, which are bimolecular and elementary, and which are not elementary?
(a) $CH_4(g) + 2\ O_2(g) \longrightarrow CO_2(g) + 2\ H_2O(g)$
(b) $O_3(g) + O(g) \longrightarrow 2\ O_2(g)$
(c) $Mg(s) + 2\ H_2O(\ell) \longrightarrow H_2(g) + Mg(OH)_2(s)$
(d) $O_3(g) \longrightarrow O_2(g) + O(g)$

46. Which of these reactions are unimolecular and elementary, which are bimolecular and elementary, and which are not elementary?
(a) $HCl(g) + H_2O(g) \longrightarrow H_3O^+(g) + Cl^-(g)$
(b) $I^-(g) + CH_3Cl(g) \longrightarrow ICH_3(g) + Cl^-(g)$
(c) $C_2H_6(g) \longrightarrow C_2H_4(g) + H_2(g)$
(d) $N_2(g) + 3\ H_2(g) \longrightarrow 2\ NH_3(g)$
(e) $O_2(g) + O(g) \longrightarrow O_3(g)$

47. Assume that each gas phase reaction occurs via a single bimolecular step. For which reaction would you expect the steric factor to be more important? Why?

$$Cl + O_3 \longrightarrow ClO + O_2 \quad \text{or} \quad NO + O_3 \longrightarrow NO_2 + O_2$$

48. Assume that each gas phase reaction occurs via a single bimolecular step. For which reaction would you expect the steric factor to be more important? Why?

$$H_2C{=}CH_2 + H_2 \longrightarrow H_3C{-}CH_3 \quad \text{or}$$
$$(CH_3)_2C{=}CH_2 + HBr \longrightarrow (CH_3)_2CBr{-}CH_3$$

Temperature and Reaction Rate

49. From Problem-Solving Example 13.8 (p. 630), where the energy profile of the ozone plus atomic oxygen reaction was derived, obtain the activation energy. Then determine the ratio of the reaction rate for this reaction at 50. °C to the reaction rate at room temperature (25 °C). Assume that the initial concentrations are the same at both temperatures.

50. Suppose a reaction rate constant has been measured at two different temperatures, T_1 and T_2, and its value is k_1 and k_2, respectively.
(a) Write the Arrhenius equation at each temperature.
(b) By combining these two equations, derive an expression for the ratio of the two rate constants, k_1/k_2. Use this formula to answer the next two questions.

51. ■ Suppose a chemical reaction rate constant has an activation energy of 76 kJ/mol, as in the example in Figure 13.11. By what factor is the rate of the reaction at 50. °C increased over its rate at 25 °C?

52. A chemical reaction has an activation energy of 30. kJ/mol. If you had to slow down this reaction a thousandfold by cooling it from room temperature (25 °C), what would the temperature be?

53. ■ These data were obtained for the rate constant for reaction of an unknown compound with water:

T (°C)	k (s^{-1})	T (°C)	k (s^{-1})
25.0	7.95×10^{-8}	56.2	1.04×10^{-5}
30.0	2.37×10^{-7}	78.2	1.45×10^{-4}

(a) Calculate the activation energy and frequency factor for this reaction.
(b) Estimate the rate constant of the reaction at a temperature of 100.0 °C.

54. *p*-Methylphenyl acetate reacts with imidazole to produce *p*-methylphenol and acetyl imidazole. The rate constants for this second-order reaction at a series of temperatures are given in the table.

T (°C)	k (L mol^{-1} s^{-1})
10	2.34×10^{-2}
18	3.25×10^{-2}
25	4.5×10^{-2}
35	5.83×10^{-2}
42	7.5×10^{-2}
60	1.52×10^{-1}

(a) Calculate the activation energy and frequency factor for this reaction.
(b) Estimate the rate constant for this reaction at a temperature of 100.0 °C.

55. For the reaction of iodine atoms with hydrogen molecules in the gas phase, these rate constants were obtained experimentally.

$$2\ I(g) + H_2(g) \longrightarrow 2\ HI(g)$$

T (K)	10^{-5} k (L^2 mol^{-2} s^{-1})
417.9	1.12
480.7	2.60
520.1	3.96
633.2	9.38
666.8	11.50
710.3	16.10
737.9	18.54

(a) Calculate the activation energy and frequency factor for this reaction.
(b) Estimate the rate constant of the reaction at 400.0 K.

56. Make an Arrhenius plot and calculate the activation energy for the gas phase reaction

$$2\ NOCl(g) \longrightarrow 2\ NO(g) + Cl_2(g)$$

T (K)	Rate Constant (L mol^{-1} s^{-1})
400.	6.95×10^{-4}
450.	1.98×10^{-2}
500.	2.92×10^{-1}
550.	2.60
600.	16.3

57. ■ The activation energy E_a is 139.7 kJ mol^{-1} for the gas phase reaction

$$HI + CH_3I \longrightarrow CH_4 + I_2$$

Calculate the fraction of the molecules whose collisions would be energetic enough to react at
(a) 100. °C (b) 200. °C
(c) 500. °C (d) 1000. °C

58. The activation energy E_a is 10. kJ mol^{-1} for the gas phase reaction

$$NO + O_3 \longrightarrow NO_2 + O_2$$

Calculate the fraction of the molecules whose collisions would be energetic enough to react at
(a) 400. °C (b) 600. °C
(c) 800. °C (d) 1000. °C

59. For the gas phase reaction

$$CH_3CH_2I(g) \longrightarrow CH_2CH_2(g) + HI(g)$$

the activation energy E_a is 221 kJ/mol and the frequency factor A is 1.2×10^{14} s^{-1}. If the concentration of CH_3CH_2I is 0.012 mol/L, what is the rate of reaction at
(a) 400. °C? (b) 800. °C?

60. For the gas phase reaction

$$cis\text{-CHClCHCl}(g) \longrightarrow trans\text{-CHClCHCl}(g)$$

the activation energy E_a is 234 kJ/mol and the frequency factor A is 6.3×10^{12} s^{-1}. If the concentration of *cis*-CHClCHCl is 0.0043 mol/L, what is the rate of reaction at
(a) 400. °C? (b) 800. °C?

61. For the reaction

$$N_2O_5(g) \longrightarrow 2\ NO_2(g) + \tfrac{1}{2}\ O_2(g)$$

the rate constant k at 25 °C is 3.46×10^{-5} s^{-1} and at 55 °C it is 1.5×10^{-3} s^{-1}. Calculate the activation energy, E_a.

62. ■ For a hypothetical reaction the rate constant is 4.63×10^{-3} s^{-1} at 25 °C and 2.37×10^{-2} s^{-1} at 43 °C. Calculate the activation energy.

Rate Laws for Elementary Reactions

63. For the hypothetical reaction A + B ⟶ C + D, the activation energy is 32 kJ/mol. For the reverse reaction (C + D ⟶ A + B), the activation energy is 58 kJ/mol. Is the reaction A + B ⟶ C + D exothermic or endothermic?

64. Use the diagram to answer these questions.

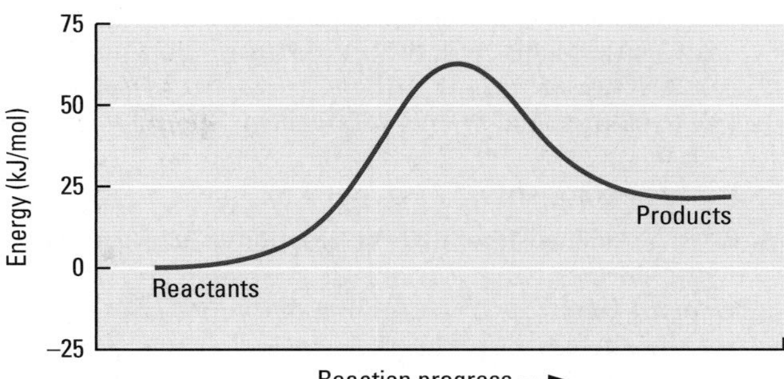

(a) Is the reaction exothermic or endothermic?
(b) What is the approximate value of ΔE for the forward reaction?
(c) What is the activation energy in each direction?
(d) A catalyst is found that lowers the activation energy of the reaction by about 10 kJ/mol. How will this catalyst affect the rate of the reverse reaction?

65. ■ Draw an energy versus reaction progress diagram (similar to the one in Question 64) for each of the reactions whose activation energy and enthalpy change are given below.
(a) $\Delta H° = -145$ kJ mol^{-1}; $E_a = 75$ kJ mol^{-1}
(b) $\Delta H° = -70$ kJ mol^{-1}; $E_a = 65$ kJ mol^{-1}
(c) $\Delta H° = 70$ kJ mol^{-1}; $E_a = 85$ kJ mol^{-1}

66. Draw an energy versus reaction progress diagram (similar to the one in Question 64) for each of the reactions whose activation energy and enthalpy change are given below.
(a) $\Delta H° = 105$ kJ mol^{-1}; $E_a = 175$ kJ mol^{-1}
(b) $\Delta H° = -43$ kJ mol^{-1}; $E_a = 95$ kJ mol^{-1}
(c) $\Delta H° = 15$ kJ mol^{-1}; $E_a = 55$ kJ mol^{-1}

67. Which of the reactions in Question 65 would be expected to (a) Occur fastest? (b) Occur slowest? (Assume equal temperatures, equal concentrations, equal frequency factors, and the same rate law for all reactions.)

68. Which of the reactions in Question 66 would be expected to (a) Occur fastest? (b) Occur slowest? (Assume equal temperatures, equal concentrations, equal frequency factors, and the same rate law for all reactions.)

69. For which of the reactions in Question 65 would the *reverse* reaction (a) Be fastest? (b) Be slowest? (Assume equal temperatures, equal concentrations, equal frequency factors, and the same rate law for all reactions.)

70. For which of the reactions in Question 66 would the *reverse* reaction (a) Be fastest? (b) Be slowest? (Assume equal temperatures, equal concentrations, equal frequency factors, and the same rate law for all reactions.)

71. ■ Assuming that each reaction is elementary, predict the rate law.
(a) $NO(g) + NO_3(g) \longrightarrow 2\ NO_2(g)$
(b) $O(g) + O_3(g) \longrightarrow 2\ O_2(g)$
(c) $(CH_3)_3CBr(aq) \longrightarrow (CH_3)_3C^+(aq) + Br^-(aq)$
(d) $2\ HI(g) \longrightarrow H_2(g) + I_2(g)$

72. Assuming that each reaction is elementary, predict the rate law.

(a) $Cl(g) + ICl(g) \longrightarrow I(g) + Cl_2(g)$
(b) $Cl(g) + H_2(g) \longrightarrow HCl(g) + H(g)$
(c) $2\ NO_2(g) \longrightarrow N_2O_4(g)$
(d) Cyclopropane(g) $\longrightarrow$ propene(g)

Reaction Mechanisms

73. ■ Experiments show that the reaction of nitrogen dioxide with fluorine

Overall reaction: $2\ NO_2(g) + F_2(g) \longrightarrow 2\ FNO_2(g)$

has the rate law

$$\text{Initial reaction rate} = k[NO_2][F_2]$$

and the reaction is thought to occur in two steps.

Step 1: $NO_2(g) + F_2(g) \longrightarrow FNO_2(g) + F(g)$

Step 2: $NO_2(g) + F(g) \longrightarrow FNO_2(g)$

(a) Show that the sum of this sequence of reactions gives the balanced equation for the overall reaction.
(b) Which step is rate-determining?

74. Nitrogen monoxide is reduced by hydrogen to give water and nitrogen

$$2\ H_2(g) + 2\ NO(g) \longrightarrow N_2(g) + 2\ H_2O(g)$$

and one possible mechanism for this reaction is a sequence of three elementary steps.

Step 1 (fast): $2\ NO(g) \rightleftharpoons N_2O_2(g)$

Step 2 (slow): $N_2O_2(g) + H_2(g) \longrightarrow N_2O_2(g) + H_2O(g)$

Step 3 (fast): $N_2O(g) + H_2(g) \longrightarrow N_2(g) + H_2O(g)$

(a) Show that the sum of these steps gives the net reaction.
(b) What is the rate law for this reaction?

75. ■ For the reaction

$$2\ NO(g) + Cl_2(g) \longrightarrow 2\ NOCl(g)$$

the currently accepted mechanism is

$NO + Cl_2 \rightleftharpoons NOCl_2$	fast
$NOCl_2 + NO \longrightarrow 2\ NOCl$	slow

(a) What is the rate law for this mechanism? (Be sure to express it in terms of concentrations of reactants or products of the overall reaction, not in terms of intermediates.)
(b) Suggest another mechanism that agrees with the same rate law.
(c) Suggest another mechanism that does not agree with the same rate law.

76. For the reaction

$$2\ N_2O_5(g) \longrightarrow 4\ NO_2(g) + O_2(g)$$

the currently accepted mechanism is

$N_2O_5 \rightleftharpoons NO_2 + NO_3$	fast
$NO_2 + NO_3 \longrightarrow NO_2 + O_2 + NO$	slow
$NO + NO_3 \longrightarrow 2\ NO_2$	fast

What is the rate law for this reaction?

77. ■ For the reaction mechanism

$H_3C{-}C({=}O){-}O{-}CH_3 + H_3O^+ \rightleftharpoons H_3C{-}C({=}O^+H){-}O{-}CH_3 + H_2O$	fast
$H_3C{-}C({=}O^+H){-}O{-}CH_3 + H_2O \rightleftharpoons H_3C{-}C(OH)(O^+H_2){-}O{-}CH_3$	slow
$H_3C{-}C(OH)(O^+H_2){-}O{-}CH_3 \rightleftharpoons H_3C{-}C(OH)(OH){-}O^+(H){-}CH_3$	fast
$H_3C{-}C(OH)(OH){-}O^+(H){-}CH_3 + H_2O \rightleftharpoons H_3C{-}C({=}O){-}OH + H{-}O{-}CH_3 + H_3O^+$	fast

(a) Write the chemical equation for the overall reaction.
(b) Write the rate law for the reaction.
(c) Is there a catalyst involved in this reaction? If so, what is it?
(d) Identify all intermediates in the reaction.

78. For the reaction mechanism

$A + B \rightleftharpoons C$	fast
$C + A \longrightarrow 2\ D$	slow

(a) Write the chemical equation for the overall reaction.
(b) Write the rate law for the reaction.
(c) Is there a catalyst involved in this reaction? If so, what is it?
(d) Identify all intermediates in the reaction.

79. ■ For the reaction

$$CH_3{-}C(CH_3)_2{-}Br + OH^- \longrightarrow CH_3{-}C(CH_3)_2{-}OH + Br^-$$

the rate law is

$$\text{Rate} = k[(CH_3)_3CBr]$$

Identify each mechanism that is compatible with the rate law.

(a) $(CH_3)_3CBr \longrightarrow (CH_3)_3C^+ + Br^-$	slow
$(CH_3)_3C^+ + OH^- \longrightarrow (CH_3)_3COH$	fast
(b) $(CH_3)_3CBr + OH^- \longrightarrow (CH_3)_3COH + Br^-$	
(c) $(CH_3)_3CBr + OH^- \longrightarrow (CH_3)_2(CH_2)CBr^- + H_2O$	fast
$(CH_3)_2(CH_2)CBr^- \longrightarrow (CH_3)_2(CH_2)C + Br^-$	slow
$(CH_3)_2(CH_2)C + H_2O \longrightarrow (CH_3)_3COH$	fast

80. Which of these mechanisms is compatible with the rate law (more than one may be chosen)

$$\text{Rate} = k[Cl_2]^{3/2}[CO]$$

(a) $\frac{1}{2}Cl_2 \rightleftharpoons Cl$	fast
$Cl + Cl_2 \longrightarrow Cl_3$	fast
$Cl_3 + CO \longrightarrow COCl_2 + Cl$	slow
$Cl \rightleftharpoons \frac{1}{2}Cl_2$	fast

(b) $Cl_2 + CO \longrightarrow CCl_2 + O$ slow
$O + Cl_2 \longrightarrow Cl_2O$ fast
$Cl_2O + CCl_2 \longrightarrow COCl_2 + Cl_2$ fast

(c) $\frac{1}{2}Cl_2 \rightleftharpoons Cl$ fast
$Cl + CO \rightleftharpoons COCl$ fast
$COCl + Cl_2 \longrightarrow COCl_2 + Cl$ slow
$Cl \rightleftharpoons \frac{1}{2}Cl_2$ fast

(d) $Cl_2 + CO \rightleftharpoons COCl + Cl$ fast
$COCl + Cl_2 \longrightarrow COCl_2 + Cl$ slow
$Cl + Cl \longrightarrow Cl_2$ fast

Catalysts and Reaction Rate

81. Which of these statements is (are) true?
   (a) The concentration of a homogeneous catalyst may appear in the rate law.
   (b) A catalyst is always consumed in the reaction.
   (c) A catalyst must always be in the same phase as the reactants.
   (d) A catalyst can change the course of a reaction and allow different products to be produced.
82. Hydrogenation reactions—processes in which H_2 is added to a molecule—are usually catalyzed. An excellent catalyst is a very finely divided metal suspended in the reaction solvent. Tell why finely divided rhodium, for example, is a much more efficient catalyst than a small block of the metal that has the same mass.
83. ■ Which of these reactions appear to involve a catalyst? In those cases where a catalyst is present, tell whether it is homogeneous or heterogeneous.
   (a) $CH_3CO_2CH_3(aq) + H_2O(\ell) \longrightarrow CH_3COOH(aq) + CH_3OH(aq)$
   Rate $= k[CH_3CO_2CH_3][H_3O^+]$
   (b) $H_2(g) + I_2(g) \longrightarrow 2\ HI(g)$
   Rate $= k[H_2][I_2]$
   (c) $2\ H_2(g) + O_2(g) \longrightarrow 2\ H_2O(g)$
   Rate $= k[H_2][O_2]$ (area of Pt surface)
   (d) $H_2(g) + CO(g) \longrightarrow H_2CO(g)$
   Rate $= k[H_2]^{1/2}[CO]$
84. In acid solution, methyl formate forms methyl alcohol and formic acid.

$$\underset{\text{methyl formate}}{HCO_2CH_3(aq)} + H_2O(\ell) \longrightarrow \underset{\text{formic acid}}{HCOOH(aq)} + \underset{\text{methyl alcohol}}{CH_3OH(aq)}$$

   The rate law is as follows: rate $= k[HCO_2CH_3][H_3O^+]$. Why does H_3O^+ appear in the rate law but not in the overall equation for the reaction?

Enzymes: Biological Catalysts

85. Write a one- or two-sentence definition in your own words for each term:

enzyme	cofactor
polypeptide	monomer
polysaccharide	lysozyme
substrate	active site
protein	inhibition

86. Write a one- or two-sentence definition in your own words for each term:

dimer	hydrolysis
HIV protease	enzyme-substrate complex
induced fit	globular protein
denaturation	maximum velocity

87. ■ When enzymes are present at very low concentration, their effect on reaction rate can be described by first-order kinetics. By what factor does the rate of an enzyme-catalyzed reaction change when the enzyme concentration is changed from 1.5×10^{-7} to 4.5×10^{-6} M?
88. When substrates are present at relatively high concentration and are catalyzed by enzymes, the effect on reaction rate of changing substrate concentration can be described by zeroth-order kinetics. By what factor does the rate of an enzyme-catalyzed reaction change when the substrate concentration is changed from 1.5×10^{-3} to 4.5×10^{-2} M?
89. The reaction

$$^{-}O(O{=})CCH_2CH_2C({=}O)O^{-} \xrightarrow{-2\,H} {}^{-}O(O{=})CCH{=}CHC({=}O)O^{-}$$

   is catalyzed by the enzyme succinate dehydrogenase. When malonate ions or oxalate ions are added to the reaction mixture, the rate decreases significantly. Try to account for this observation in terms of the description of enzyme catalysis given in the text. The structures of malonate and oxalate ions are

$$^{-}O(O{=})CCH_2C({=}O)O^{-} \qquad {}^{-}O(O{=})C{-}C({=}O)O^{-}$$

   malonate ion   oxalate ion

90. Some enzymes can be inhibited by heavy metal ions such as Hg^{2+} and Pb^{2+}. These metal ions have a large affinity for sulfur-containing groups and can react with molecules such as CH_3CH_2SH to form compounds such as $Pb(CH_3CH_2S)_2$. Based on the structures of the amino acids given in Table 12.8 (p. 591), suggest at least one amino acid that is likely to be present in enzymes that are inhibited by heavy metals.

Catalysis in Industry

91. In the first paragraph of Section 13.10 of the text, an expert in the field of industrial chemistry is quoted. Explain the expert's statement, in view of your understanding of the nature of catalysts. Why are catalysts so important?
92. Why are homogeneous catalysts harder to separate from products and leftover reactants than are heterogeneous reactants?
93. In an automobile catalytic converter the catalysis is accomplished on a surface consisting of platinum and other precious metals. The metals are deposited as a thin layer on a honeycomb-like ceramic support (see the photo).

   (a) Why is the ceramic support arranged in the honeycomb geometry?
   (b) Why are the metals deposited on the ceramic surface instead of being used as strips or rods?

94. Find all examples of reactions described in this chapter that are catalyzed by metals. Are these metals main group metals or transition metals? What type of chemical reactions are they: acid-base or oxidation-reduction? What conclusions can be drawn about metal-catalyzed chemical reactions from these examples?

General Questions

95. Draw a reaction energy diagram for an exothermic process. Mark the positions of reactants, products, and activated complex. Indicate the activation energies of the forward and reverse processes and explain how ΔE for the reaction can be calculated from the diagram.
96. Draw a reaction energy diagram for an endothermic process. Mark the positions of reactants, products, and activated complex. Indicate the activation energies of the forward and reverse processes and explain how ΔE for the reaction can be calculated from the diagram.
97. Indicate whether each of these statements is true or false. Change the wording of each false statement to make it true.
    (a) It is possible to change the rate constant for a reaction by changing the temperature.
    (b) The reaction rate remains constant as a first-order reaction proceeds at a constant temperature.
    (c) The rate constant for a reaction is independent of reactant concentrations.
    (d) As a second-order reaction proceeds at a constant temperature, the rate constant changes.
98. Consider the class of substances known as catalysts.
    (a) Define "catalyst."
    (b) What effect does a catalyst have on the energy barrier for a reaction?
    (c) What special characteristics do enzymes have that distinguish them from other catalysts?
99. Nitrogen monoxide can be reduced with hydrogen.

$$2\ H_2(g) + 2\ NO(g) \longrightarrow 2\ H_2O(g) + N_2(g)$$

Experiment shows that when the concentration of H_2 is halved, the reaction rate is halved. Furthermore, raising the concentration of NO by a factor of 3 raises the rate by a factor of 9. Write the rate equation for this reaction.

100. One reaction that may occur in air polluted with nitrogen monoxide is

$$2\ NO(g) + O_2(g) \longrightarrow 2\ NO_2(g)$$

Using the data in the table, answer the questions that follow. Assume that all experiments are done at the same temperature.

Experiment	Initial Concentration (mol/L)		Initial Rate of Formation of $[NO_2]$ (mol L^{-1} s^{-1})
	[NO]	$[O_2]$	
1	0.0010	0.0010	7.0×10^{-6}
2	0.0010	0.0020	1.4×10^{-5}
3	0.0010	0.0030	2.1×10^{-5}
4	0.0020	0.0030	8.4×10^{-5}
5	0.0030	0.0030	1.9×10^{-4}

(a) What is the order of reaction with respect to each reactant?
(b) Write the rate law for the reaction.
(c) Calculate the rate of formation of NO_2 when $[NO] = [O_2] = 0.005$ mol/L.

101. For the reaction of NO and O_2 at 660 K,

$$2\ NO(g) + O_2(g) \longrightarrow 2\ NO_2(g)$$

Concentration (mol/L)		Rate of Disappearance of NO (mol L^{-1} s^{-1})
[NO]	$[O_2]$	
0.010	0.010	2.5×10^{-5}
0.020	0.010	1.0×10^{-4}
0.010	0.020	5.0×10^{-5}

(a) Determine the order of the reaction for each reactant.
(b) Write the rate equation for the reaction.
(c) Calculate the rate constant.
(d) Calculate the rate when [NO] = 0.025 mol/L and $[O_2]$ = 0.050 mol/L.
(e) If O_2 disappears at a rate of 1.0×10^{-4} mol L^{-1} s^{-1}, what is the rate at which NO disappears? What is the rate at which NO_2 is forming?

102. Nitryl fluoride is an explosive compound that can be made by oxidizing nitrogen dioxide with fluorine:

$$2\ NO_2(g) + F_2(g) \longrightarrow 2\ NO_2F(g)$$

Several kinetics experiments, all done at the same temperature and involving formation of nitryl fluoride are summarized in this table:

Experiment	Initial Concentration (mol/L)			Initial Rate (mol L^{-1} s^{-1})
	$[NO_2]$	$[F_2]$	$[NO_2F]$	
1	0.0010	0.0050	0.0020	2.0×10^{-4}
2	0.0020	0.0050	0.0020	4.0×10^{-4}
3	0.0020	0.0020	0.0020	1.6×10^{-4}
4	0.0020	0.0020	0.0010	1.6×10^{-4}

(a) Write the rate law for the reaction.
(b) What is the order of the reaction with respect to each reactant and each product?
(c) Calculate the rate constant k and express it in appropriate units.

103. The deep blue compound $CrO(O_2)_2$ can be made from the chromate ion by using hydrogen peroxide in an acidic solution.

$$HCrO_4^-(aq) + 2\ H_2O_2(aq) + H_3O^+(aq) \longrightarrow CrO(O_2)_2(aq) + 4\ H_2O(\ell)$$

The kinetics of this reaction have been studied, and the rate equation is found to be

$$\text{Rate of disappearance of } HCrO_4^- = k[HCrO_4^-]\,[H_2O_2][H_3O^+]$$

One of the mechanisms suggested for the reaction is

$$HCrO_4^- + H_3O^+ \rightleftharpoons H_2CrO_4 + H_2O$$

$$H_2CrO_4 + H_2O_2 \longrightarrow H_2CrO(O_2)_2 + H_2O$$

$$H_2CrO(O_2)_2 + H_2O_2 \longrightarrow CrO(O_2)_2 + 2\,H_2O$$

(a) Give the order of the reaction with respect to each reactant.
(b) Show that the steps of the mechanism agree with the overall equation for the reaction.
(c) Which step in the mechanism is rate-limiting?

104. How does a chemical reaction mechanism differ from other types of mechanisms—for example, the gear-changing mechanism of a bicycle, or the mechanism of an elevator? How is it similar to these mechanisms?

105. Why is chemical kinetics important in understanding the environmental problem posed by chlorofluorocarbons in the ozone layer (⇐ *p. 464*)?

106. A reaction between molecules A and B (A + B ⟶ products) is found to be first-order in A. Which rate equation cannot be correct?
(a) Rate = k[A][B] (b) Rate = $k[A][B]^2$
(c) Rate = $k[B]^2$

Applying Concepts

107. The graph shows the change in concentration as a function of time for the reaction

$$2\,H_2O_2(g) \longrightarrow 2\,H_2O(g) + O_2(g)$$

What do each of the curves A, B, and C represent?

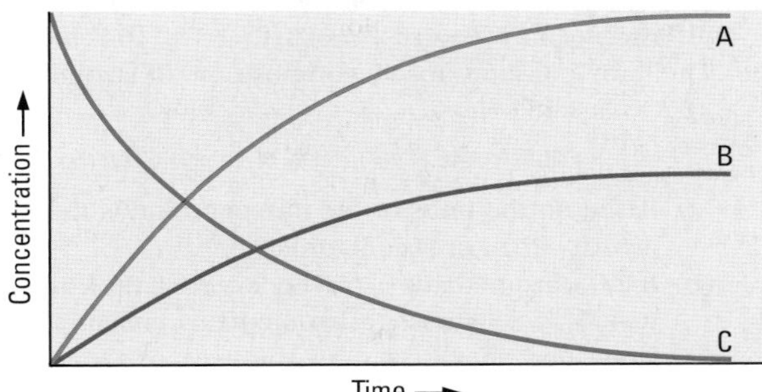

108. Draw a graph similar to the one in Question 107 for the reaction

$$2\,N_2O_5(g) \longrightarrow 4\,NO_2(g) + O_2(g)$$

109. The picture below is a "snapshot" of the reactants at time = 0 for the reaction

$$H_2(g) + I_2(g) \longrightarrow 2\,HI(g)$$

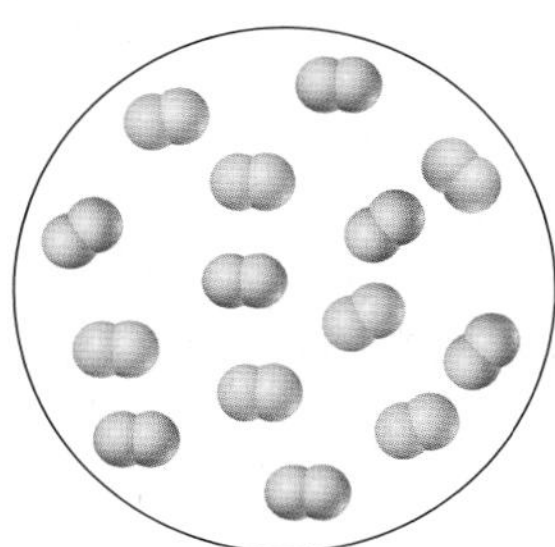

Suppose the reaction is carried out at two different temperatures and that another snapshot is taken after a constant time has elapsed.

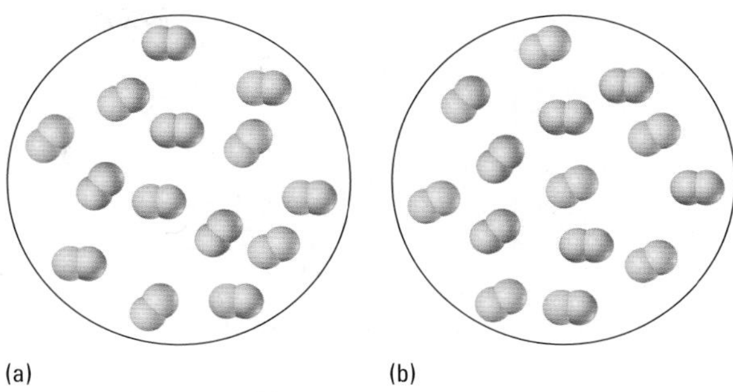

Which of these two snapshots corresponds to the lower temperature reaction condition?

110. Consider Question 109 again, only this time a catalyst is used instead of a lower temperature. Which of the two snapshots corresponds to the presence of a catalyst?

111. Initial rates for the reaction A + B + C ⟶ D + E were measured with various concentrations of A, B, and C as represented in the pictures below. Based on these data, what is the rate law?

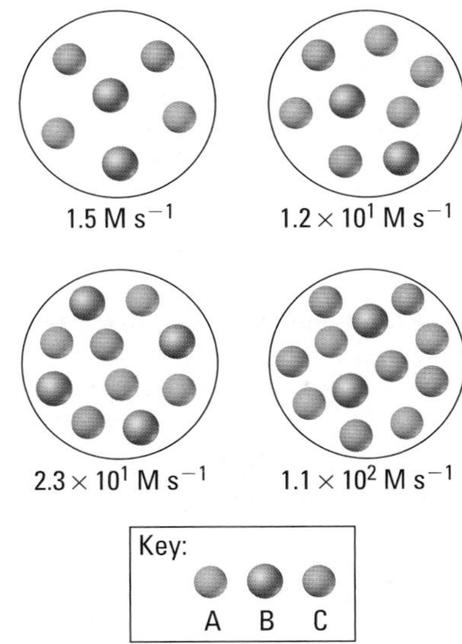

112. Using a molecular model kit, build models of *cis*-2-butene, *trans*-2-butene, and the transition state, or activated complex. How much force do you need to apply to the models to change the reactant into the product by passing through the activated complex? The answer will, of course, depend on the kind of model kit that you use.

113. The rate of decay of a radioactive solid is independent of the temperature of that solid—at least for temperatures easily obtained in the laboratory. What does this observation imply about the activation energy for this process?

114. Platinum metal is used as a catalyst in the decomposition of NO(g) into N_2(g) and O_2(g). A graph of the rate of the reaction

as a function of NO concentration is shown below. Explain why the rate stops increasing and levels out.

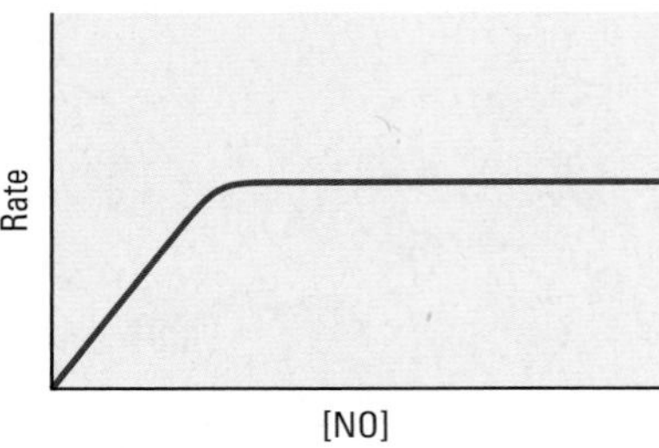

115. During the 1970s, at the time when catalytic converters began to be installed on automobiles, tetraethyllead, which was used to improve the octane rating of gasoline, was gradually removed from gasoline at the pump. In the combustion chamber of the engine the four ethyl groups in tetraethyllead are converted to carbon dioxide and water, leaving either elemental lead or lead(II) oxide. In a modern automobile, burning a single tank of leaded gasoline can ruin the catalytic converter, destroying nearly all of the catalytic activity. Based on what you learned about the mechanism of reactions in a catalytic converter, suggest a reason that lead in gasoline might destroy the catalyst's ability to speed up reactions.

More Challenging Questions

116. Let x represent the number of half-lives that have elapsed during the course of a first-order reaction. That is, the elapsed time t is x times the half-life $t_{1/2}$: $t = xt_{1/2}$.
(a) Show that at time t the concentration $[A]_t$ of reactant is related to the initial concentration $[A]_0$ by

$$[A]_t = [A]_0(\tfrac{1}{2})^x$$

(b) Use your result in part (a) to show that

$$\log\left(\frac{[A]_t}{[A]_0}\right) = \frac{t}{t_{1/2}}(-\log 2) = \frac{t}{t_{1/2}}(-0.301)$$

(c) Use your result in part (b) to show that for any first-order reaction a plot of $\log [A]_t$ versus t will be a straight line and that half-life can be obtained from the slope of the line.

117. Measurements of the initial rate of hydrolysis of benzenesulfonyl chloride in aqueous solution at 15 °C in the presence of fluoride ion yielded the results in the table for a fixed concentration of benzenesulfonyl chloride of 2×10^{-4} M. The reaction rate is known to be proportional to the concentration of benzenesulfonyl chloride.

$[F^-] \times 10^2$ (mol/L)	Initial Rate $\times 10^7$ (mol L^{-1} s^{-1})
0	2.4
0.5	5.4
1.0	7.9
2.0	13.9
3.0	20.2
4.0	25.2
5.0	32.0

Note that some reaction must be occurring in the absence of any fluoride ion, because at zero concentration of fluoride the rate is not zero. This residual rate should be subtracted from each observed rate to give the rate of the reaction being studied.
(a) Derive the complete rate law for the reaction.
(b) Calculate the rate constant k and express it in appropriate units.

118. Given these data and your result from Question 116, determine the half-life and the initial concentration of the reactant for the reaction

$$trans\text{-CHClCHCl(aq)} \longrightarrow cis\text{-CHClCHCl(g)}$$

[*trans*-CHClCHCl] (mol/L)	Time (s)
9.23×10^{-4}	30
8.51×10^{-4}	60
7.86×10^{-4}	90
7.25×10^{-4}	120
6.19×10^{-4}	180
3.82×10^{-4}	360

119. The decomposition of N_2O_5 is first-order with a rate constant of 2.5×10^{-4} s^{-1}.
(a) What is the half-life for decomposition of N_2O_5?
(b) How long does it take for the concentration of N_2O_5 to drop to $\frac{1}{32}$ of its original value?
(c) How long does it take for the concentration of N_2O_5 to drop from 3.4×10^{-3} mol/L to 2.3×10^{-5} mol/L?

120. The rate constant for decomposition of azomethane at 425 °C is 0.68 s^{-1}.

$$CH_3N{=}NCH_3(g) \longrightarrow N_2(g) + C_2H_6(g)$$

(a) Based on the units of the rate constant, is the reaction zeroth-, first-, or second-order?
(b) If 2.0 g azomethane is placed in a 2.0-L flask and heated to 425 °C, what mass of azomethane remains after 5.0 s?
(c) How long does it take for the mass of azomethane to drop to 0.24 g?
(d) What mass of nitrogen would be found in the flask after 0.5 s of reaction?

121. Cyclopropane isomerizes to propene when heated. Rate constants for the reaction

$$\text{cyclopropane} \longrightarrow \text{propene}$$

are 1.10×10^{-4} s^{-1} at 470. °C and 1.02×10^{-3} s^{-1} at 510. °C.
(a) What is the activation energy, E_a, for this reaction?
(b) At 500. °C and [cyclopropane] = 0.10 M, how long does it take for the concentration to drop to 0.023 M?

122. When heated, cyclobutane, C_4H_8, decomposes to ethylene, C_2H_4.

$$C_4H_8(g) \longrightarrow 2\ C_2H_4(g)$$

The activation energy, E_a, for this reaction is 262 kJ/mol.
(a) If the rate constant $k = 0.032$ s^{-1} at 800. K, what is the value of k at 900. K?

(b) At 850. K and an initial concentration of cyclobutane of 0.0427 M, what is the concentration of cyclobutane after 2 h?

123. For each reaction listed with its rate law, propose a reasonable mechanism.

(a) $CH_3CO_2CH_3(aq) + H_2O(\ell) \longrightarrow CH_3COOH(aq) + CH_3OH(aq)$
Rate $= k[CH_3CO_2CH_3][H_3O^+]$

(b) $H_2(g) + I_2(g) \longrightarrow 2\ HI(g)$
Rate $= k[H_2][I_2]$

(c) $2\ H_2(g) + O_2(g) \longrightarrow 2\ H_2O(g)$
Rate $= k[H_2][I_2]$ (area of Pt surface)

(d) $H_2(g) + CO(g) \longrightarrow H_2CO(g)$
Rate $= k[H_2]^{1/2}[CO]$

124. Suppose a rate constant for an uncatalyzed reaction is described by the Arrhenius equation with an activation energy E_a. The introduction of a catalyst lowers the activation energy to the value E_a' and thus increases the rate constant to the value k'. Write the Arrhenius equation for both k and k', and then derive an expression for the ratio k'/k in terms of E_a' and E_a. Assume that the frequency factor A and the temperature T are constant. Use this equation to answer the next two questions.

125. Suppose a catalyst was found for the reaction in Exercise 13.9 that reduced the activation energy to zero. By what factor would this reaction rate be increased at 370. K?

126. In the discussion of Figure 13.14, this statement is made: "Dropping the activation energy from 262 kJ/mol for the uncatalyzed reaction to 115 kJ/mole for the catalyzed process makes the catalyzed reaction 10^{15} times faster." Prove this statement.

127. When a molecule of *cis*-2-butene has rotated 90° around the axis through the two central carbon atoms, the π bond between these atoms is essentially broken. Use bond energies to estimate the activation energy for rotation around the π bond in ethylene, C_2H_4. Compare your result with the activation energy value in Figure 13.7.

128. Consider the reaction mechanism for iodine-catalyzed isomerization of *cis*-2-butene presented in Section 13.8. Use bond enthalpies (Table 8.2) to estimate the energy change for each step in the mechanism. (In Step 3 assume that rotation around the single bond requires about 5 kJ/mol.) Do your estimates agree with the energy values for the intermediates in Figure 13.14?

129. The iodine-catalyzed isomerization of *cis*-2-butene can be speeded up still further if a bright light with wavelength less than about 800 nm shines on the reaction mixture. Consider the mechanism presented in Section 13.8 and suggest a reason that the reaction is even faster in the presence of light than in the dark. Also explain why wavelengths longer than 800 nm are not effective in speeding up the reaction.

130. In a time-resolved picosecond spectroscopy experiment, Sheps, Crowther, Carrier, and Crim (*J. Phys. Chem. A*, Vol. 110, 2006; pp. 3087–3092) generated chlorine atoms in the presence of pentane. The pentane was dissolved in dichloromethane, CH_2Cl_2. The chlorine atoms are free radicals and are very reactive. After a nanosecond the chlorine atoms have reacted with pentane molecules, removing a hydrogen atom to form HCl and leaving behind a pentane radical with a single unpaired electron. The equation is

$$Cl(dcm) + C_5H_{12}(dcm) \longrightarrow HCl(dcm) + C_5H_{11}\cdot(dcm)$$

where (dcm) indicates that a substance is dissolved in dichloromethane. Measurements of the concentration of chlorine atoms were made as a function of time at three different concentrations of pentane in the dichloromethane. These results are shown in the table.

	Concentration of Chlorine Atoms (M) for Different C_5H_{12} Concentrations		
Time (ps)	**0.15 M C_5H_{12}**	**0.30 M C_5H_{12}**	**0.60 M C_5H_{12}**
100.0	4.42×10^{-5}	3.11×10^{-5}	1.77×10^{-5}
140.0	3.823×10^{-5}	2.39×10^{-5}	1.15×10^{-5}
180.0	3.38×10^{-5}	1.94×10^{-5}	8.48×10^{-6}
220.0	3.03×10^{-5}	1.49×10^{-5}	6.04×10^{-6}
260.0	2.68×10^{-5}	1.19×10^{-5}	4.12×10^{-6}
300.0	2.42×10^{-5}	9.45×10^{-6}	3.14×10^{-6}
340.0	2.08×10^{-5}	7.75×10^{-6}	2.38×10^{-6}
380.0	1.91×10^{-5}	6.35×10^{-6}	1.75×10^{-6}
420.0	1.71×10^{-5}	4.58×10^{-6}	1.61×10^{-6}
460.0	1.53×10^{-5}	3.77×10^{-6}	9.98×10^{-7}

(a) What is the order of the reaction with respect to chlorine atoms?

(b) Does the reaction rate depend on the concentration of pentane in the dichloromethane? If so, what is the order of the reaction with respect to pentane?

(c) Why does the concentration of pentane not affect the analysis of data that you performed in part a?

(d) Write the rate law for the reaction and calculate the rate of reaction for a concentration of chlorine atoms equal to 1 μM and a pentane concentration of 0.23 M.

(e) Sheps, Crowther, Carrier, and Crim found that the rate of formation of HCl matched the rate of disappearance of Cl. From this they concluded that there were no intermediates and side reactions were not important. Explain the basis for this conclusion.

131. If you know some calculus, derive the integrated first-order rate law for the reaction

$$2\ A \longrightarrow \text{products}$$

by following these steps.

(i) Define the reaction rate in terms of [A].
(ii) Write the rate law in terms of [A], k, and t.
(iii) Separate variables in the rate law.
(iv) Integrate the rate law.
(v) Write the integrated equation in the form $y = mx + b$.
(vi) Derive the half-life as was done on p. 624.

132. If you know some calculus, derive the integrated second-order rate law for the reaction

$$2\ A \longrightarrow \text{products}$$

Follow the same steps listed in Question 131.

Conceptual Challenge Problems

CP13.A (Section 13.5) A rule of thumb is that for a typical reaction, if concentrations are unchanged, a 10-K rise in temperature increases the reaction rate by two to four times. Use an average increase of three times to answer the questions below.

(a) What is the approximate activation energy of a "typical" chemical reaction at 298 K?

(b) If a catalyst increases a chemical reaction's rate by providing a mechanism that has a lower activation energy, then what change do you expect a 10-K increase in temperature to make in the rate of a reaction whose uncatalyzed activation energy of 75 kJ/mol has been lowered to one-half this value (at 298 K) by addition of a catalyst?

CP13.B (Section 13.7) A sentence in an introductory chemistry textbook reads, "Dioxygen reacts with itself to form trioxygen, ozone, according to the equation, $3\ O_2 \longrightarrow 2\ O_3$." As a student of chemistry, what would you write to criticize this sentence?

CP13.C (Section 13.7) A classmate consults you about a problem concerning the reaction of nitrogen monoxide and dioxygen in the gas phase. She has been told that the reaction is second-order in nitrogen monoxide and first-order in dioxygen; hence, the rate law may be written as rate $= k[NO]^2[O_2]$. She has been asked to propose a mechanism for this reaction. She proposes that the mechanism is this single equation:

$$NO + NO + O_2 \longrightarrow NO_2 + NO_2$$

She asks your opinion about whether this is correct. What should you tell her to explain why the answer is correct or incorrect?

CP13.D (Section 13.8) Polypropylene (*p. 578, Table 12.6*) is an important type of plastic, and over 3 million tons of it is produced every year in the United States. Properties of polypropylene can be changed by the way the polymer is made. For example, melting points between 130 °C and 160 °C can be obtained by using an appropriate catalyst to polymerize propylene (propene, C_3H_6). Two important types of polypropylene are isotactic, in which the methyl groups are all on the same side of the polymer chain, and syndiotactic, in which the methyl groups alternate between one side and the other of the chain.

Suppose that you are part of a team designing a new catalyst to polymerize propylene. Your catalyst will have a zirconium atom at the center of a structure consisting of carbon and hydrogen atoms. The zirconium atom will hold the growing polypropylene chain by bonding to one end of it. The metal atom will also attract a propylene molecule and bond to it before transferring the growing polypropylene chain to the other end of the new propylene molecule. The process is shown here.

What would be a reasonable shape for the rest of the catalyst molecule surrounding the metal atom so that isotactic polypropylene would be produced? It may help to build molecular models to see how each new propylene molecule needs to be added to the growing polymer chain to get all the methyl groups on the same side.

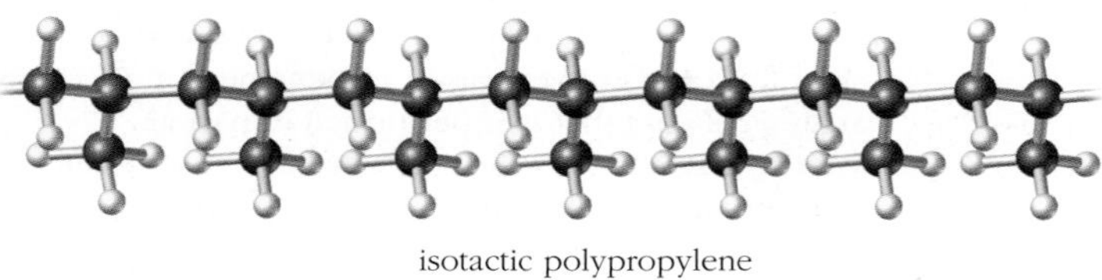

isotactic polypropylene

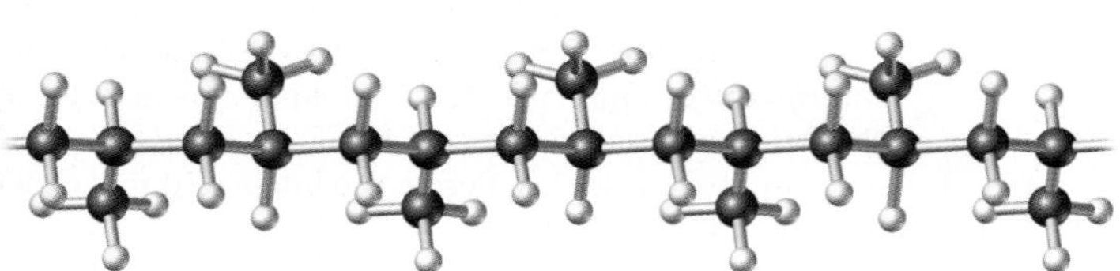

syndiotactic polypropylene

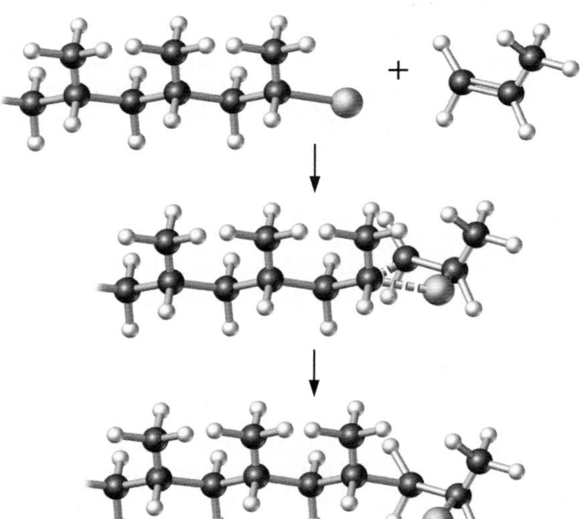

14 Chemical Equilibrium

This pinecone fish, *Monocentris gloriamaris,* has a bioluminescent area on its lower jaw. The bioluminescent area glows orange in the daytime and blue-green at night. It apparently helps to attract zooplankton upon which the pinecone fish feeds. In this daytime photograph, the orange area is clearly visible below and between the eye and the mouth. The bioluminescent area consists of a colony of rod-shaped bacteria, *Vibrio fischeri,* that are common in marine environments around the world. Colonies of *Vibrio fischeri* become luminescent only when they reach a minimum population and the bacteria in the colony sense that a sufficiently large population has been reached. This biological sensing depends on a chemical equilibrium involving acylhomoserine lactone, a substance that signals to one bacterium that others are nearby (see p. 698).

ThomsonNOW
Throughout the text, this icon indicates an opportunity to test yourself on key concepts and to explore interactive modules by signing in to ThomsonNOW at **www.thomsonedu.com**.

Chemical reactions that involve only pure solids or pure liquids are simpler than those that occur in the gas phase or in a solution. Either no reaction occurs between the solid and liquid reactants, or a reaction occurs in which at least one reactant is completely converted into products. [If one or more of the reactants are present in excess, those reactants will be left over, but the limiting reactant *(← p. 140)* will be completely reacted away.] This happens because the concentration of a pure solid or a pure liquid does not change during a reaction, provided the temperature remains constant. If the initial concentrations of reactants are large enough to cause the reaction to occur, those same concentrations will be present throughout the reaction, and the reaction will not stop until the limiting reactant is used up.

When a reaction occurs in the gas phase or in a solution, concentrations of reactants decrease as the reaction takes place *(← p. 613)*. Eventually the concentrations decrease to the point at which conversion of reactants to products is no longer favored. Then the concentrations of reactants and products stop changing, but none of the concentrations has become zero. At least a tiny bit (and often a lot) of each reactant and each product is present in the reaction mixture. Because the concentrations have stopped changing, it is often relatively easy to measure them, thus providing quantitative information about *how much* product can be obtained from the reaction. It is also possible to predict how changes in temperature, pressure, and concentrations will affect the quantity of product produced. This kind of information, combined with what you learned in Chapter 13 about factors that affect the rates of chemical reactions, enables us to predict which reactions will be useful for manufacturing a broad range of substances that enhance our quality of life.

As an example of the importance of such information, consider ammonia. The United States uses approximately 30 billion pounds of liquefied ammonia per year, mostly as fertilizer to provide nitrogen needed to support growth of a broad range of crops. Therefore, ammonia is a very important factor in providing people with food. Ammonia is synthesized directly from nitrogen and hydrogen by the *Haber-Bosch process* (Section 14.8).

$$N_2(g) + 3\,H_2(g) \rightleftharpoons 2\,NH_3(g) \qquad \Delta H = -92.2 \text{ kJ}$$

The chemists and chemical engineers who operate ammonia manufacturing plants do their best to obtain the maximum quantity of ammonia with the minimum input of reactants and the minimum consumption of energy resources. The German chemist Fritz Haber won the 1918 Nobel Prize for research that showed how to determine the best conditions for carrying out this reaction. The German engineer Carl Bosch received the Nobel Prize in 1931 (together with Friedrich Bergius) for his pioneering chemical engineering work that enabled large-scale ammonia synthesis to be successful.

In this chapter you will learn the same principles that Haber used. With them you will be able to make both qualitative and quantitative predictions about how much product will be formed under a given set of reaction conditions.

14.1 Characteristics of Chemical Equilibrium

When the concentrations of reactants stop decreasing and the concentrations of products stop increasing, we say that a chemical reaction has reached equilibrium. In a **chemical equilibrium,** *there are finite concentrations of reactants and products, and these concentrations remain constant.* An equilibrium reaction always results in smaller amounts of products than the theoretical yield predicts *(← p. 145)*, and sometimes hardly any products are produced. The concentrations of reactants and products at equilibrium provide a quantitative way of determining how successful a reaction has been. *When products predominate over reactants*, the reaction is **product-favored.** *When the equilibrium mixture consists mostly of reactants* with very little product, the reaction is **reactant-favored.**

EXERCISE 14.1 Concentrations of Pure Solids and Liquids

The introduction to this chapter states that at a given temperature the concentration of a pure solid or liquid does not depend on the quantity of substance present. Verify this assertion by calculating the concentration (in mol/L) of these solids and liquids at 20 °C. Obtain densities from Table 1.1 (← *p. 11*).

(a) Aluminum (b) Benzene (c) Water (d) Gold

Equilibrium Is Dynamic

When equilibrium is reached and concentrations of reactants and products remain constant, it appears that a chemical reaction has stopped. This is true only of the net, macroscopic reaction, however. On the nanoscale both forward and reverse reactions continue, but *the rate of the forward reaction exactly equals the rate of the reverse reaction*. To emphasize that *chemical equilibrium involves a balance between opposite reactions*, it is often referred to as a **dynamic equilibrium,** and an equilibrium reaction is usually written with a double arrow ($\rightleftharpoons$) between reactants and products.

A good example of chemical equilibrium is provided by weak acids (← *p. 172*), which ionize only partially in water.

$$\underset{\text{acetic acid}}{CH_3COOH(aq)} + \underset{\text{water}}{H_2O(\ell)} \rightleftharpoons \underset{\text{acetate ion}}{CH_3COO^-(aq)} + \underset{\text{hydronium ion}}{H_3O^+(aq)}$$

After equilibrium has been reached at room temperature, more than 90% of the acetic acid remains in molecular form (CH_3COOH) and the equilibrium concentrations of acetate ions and hydronium ions are each less than one tenth the concentration of acetic acid molecules. Nevertheless, both the forward and reverse reactions continue. Evidence to support this idea can be obtained by adding a tiny quantity of sodium acetate in which radioactive carbon-14 has been substituted into the CH_3COO^- ion to give $^{14}CH_3COO^-$. Almost immediately the radioactivity can be found in acetic acid molecules as well. This would not happen if the reaction had come to a halt, but it does happen because the reverse reaction,

$$^{14}CH_3COO^-(aq) + H_3O^+(aq) \longrightarrow {}^{14}CH_3COOH(aq) + H_2O(\ell)$$

is still taking place. To a macroscopic observer nothing seems to be happening because the reverse reaction and the forward reaction are occurring at equal rates and there is no net change in each concentration.

The *equi* in the word *equilibrium* means "equal." It refers to equal rates of forward and reverse reactions, not to equal quantities or concentrations of the substances involved. The *librium* part of the word comes from *libra*, meaning "balance." Chemical equilibrium is an equal balance between two reaction rates.

A set of double arrows, $\rightleftharpoons$, in an equation indicates a dynamic equilibrium in which forward and reverse reactions are occurring at equal rates; it also indicates that the reaction should be thought of in terms of the concepts of chemical equilibrium.

Equilibrium Is Independent of Direction of Approach

Another important characteristic of chemical equilibrium is that, *for a specific reaction at a specific temperature, the equilibrium state will be the same, no matter what the direction of approach to equilibrium*. As an example, consider again the synthesis of ammonia from N_2 and H_2.

$$N_2(g) + 3\ H_2(g) \rightleftharpoons 2\ NH_3(g)$$

Suppose that you introduce 1.0 mol $N_2(g)$ and 3.0 mol $H_2(g)$ into an empty (evacuated) 1.00-L container at 472 °C. Some (but not all) of the N_2 reacts with the H_2 to form NH_3. After equilibrium is established, you would find that the concentration of H_2 has fallen from its initial value of 3.0 mol/L to an equilibrium value of 0.89 mol/L. You would also find equilibrium concentrations of 0.30 mol/L for N_2 and 1.4 mol/L for NH_3.

Now consider a second experiment at the same temperature in which you introduce 2.0 mol NH_3(g) into an empty 1.00-L container at 472 °C. The 2.0 mol NH_3 consists of 2.0 mol N atoms and 6.0 mol H atoms—the same number of N atoms and H atoms contained in the 1.0 mol N_2 and 3.0 mol H_2 used in the first experiment. Because there is only NH_3 present initially, the reverse reaction occurs, producing some N_2 and H_2. Measuring the concentrations at equilibrium reveals that the concentration of NH_3 dropped from the initial 2.0 mol/L to 1.4 mol/L, and the concentrations of N_2 and H_2 built up to 0.30 mol/L and 0.89 mol/L. These equilibrium concentrations are the same as those achieved in the first experiment. Thus, *whether you start with reactants or products, the same equilibrium state is achieved*—as long as the number of atoms of each type, the volume of the container, and the temperature remain the same. This is shown schematically in Figure 14.1.

Catalysts Do Not Affect Equilibrium Concentrations

Another important characteristic of chemical equilibrium is that *if a catalyst is present, the same equilibrium state will be achieved, but more quickly*. A catalyst speeds up the forward reaction, but it also speeds up the reverse reaction. The overall effect is to produce exactly the same concentrations at equilibrium, whether or not a catalyst is in the reaction mixture. A catalyst can be used to speed up production of products in an industrial process, but it will not result in greater equilibrium concentrations of products, nor will it reduce the concentration of product present when the system reaches equilibrium.

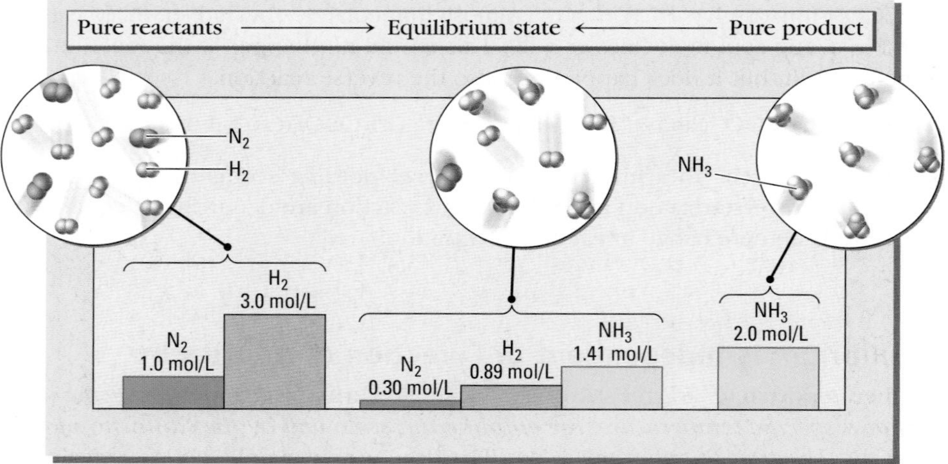

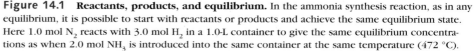

Figure 14.1 Reactants, products, and equilibrium. In the ammonia synthesis reaction, as in any equilibrium, it is possible to start with reactants or products and achieve the same equilibrium state. Here 1.0 mol N_2 reacts with 3.0 mol H_2 in a 1.0-L container to give the same equilibrium concentrations as when 2.0 mol NH_3 is introduced into the same container at the same temperature (472 °C).

CONCEPTUAL EXERCISE 14.2 Recognizing an Equilibrium State

A mixture of hydrogen gas and oxygen gas is maintained at 25 °C for 1 year. On the first day of each month the mixture is sampled and the concentrations of hydrogen and oxygen measured. In every case they are found to be 0.50 mol/L H_2 and 0.50 mol/L O_2; that is, the concentrations do not change over a long period. Is this mixture at equilibrium? If you think not, how could you do an experiment to prove it?

14.2 The Equilibrium Constant

Consider again the isomerization reaction of *cis*-2-butene to *trans*-2-butene, whose rate we discussed previously (⇐ *p. 626*).

$$\underset{cis\text{-2-butene}}{\text{H(H}_3\text{C)C=C(H)CH}_3} \rightleftharpoons \underset{trans\text{-2-butene}}{\text{H(H}_3\text{C)C=C(CH}_3\text{)H}}$$

At 500 K the reaction reaches an equilibrium in which the concentration of *trans*-2-butene is 1.65 times the concentration of *cis*-2-butene. This is a single-step, elementary process (⇐ *p. 626*). Both the forward and reverse reactions in this system involve only a single molecule and, therefore, are unimolecular and first-order (⇐ *p. 635*). Thus, the rate equations for forward and reverse reactions can be derived from the reaction equations.

$$\text{Rate}_{\text{forward}} = k_{\text{forward}}[cis\text{-2-butene}] \qquad \text{Rate}_{\text{reverse}} = k_{\text{reverse}}[trans\text{-2-butene}]$$

Suppose that we start with 0.100 mol *cis*-2-butene in a 5-L closed flask at 500 K. The *cis*-2-butene begins to react at a rate given by the forward rate equation. Initially no *trans*-2-butene is present, so the initial rate of the reverse reaction is zero. As the forward reaction proceeds, *cis*-2-butene is converted to *trans*-2-butene. The concentration of *cis*-2-butene decreases, so the forward rate decreases. As soon as some *trans*-2-butene has formed, the reverse reaction begins. As the concentration of *trans*-2-butene builds up, the reverse rate gets faster. The forward rate slows down (and the reverse rate speeds up) until the two rates are equal. At this time equilibrium has been achieved, and in the macroscopic system no further change in concentrations will be observed (Figure 14.2).

On the nanoscale, when equilibrium has been achieved, both reactions are still occurring, but the forward and reverse rates are equal. Therefore we can equate the two rates to give

$$\text{Rate}_{\text{forward}} = \text{Rate}_{\text{reverse}}$$

and, by substituting from the two previous rate equations,

$$k_{\text{forward}}[cis\text{-2-butene}] = k_{\text{reverse}}[trans\text{-2-butene}]$$

In this and subsequent equations in this chapter, square brackets designate *equilibrium* concentrations. The equation can be rearranged so that both rate constants are on one side and both equilibrium concentrations are on the other side.

$$\frac{k_{\text{forward}}}{k_{\text{reverse}}} = \frac{[trans\text{-2-butene}]}{[cis\text{-2-butene}]}$$

This expression shows that the ratio of equilibrium concentrations is equal to a ratio of rate constants. Because a ratio of two constants is also a constant, the ratio of

ThomsonNOW

Go to the Coached Problems menu for modules on:

- **equilibrium state**
- **equilibrium constant**

It is important to distinguish equilibrium concentrations, which do not change over time, from the changing concentrations of reactants and products before equilibrium is achieved. Later in this chapter we will use the notation "conc. X" to indicate the concentration of X in a system that is not at equilibrium. This distinction is important because only equilibrium concentrations can be calculated from equilibrium constants.

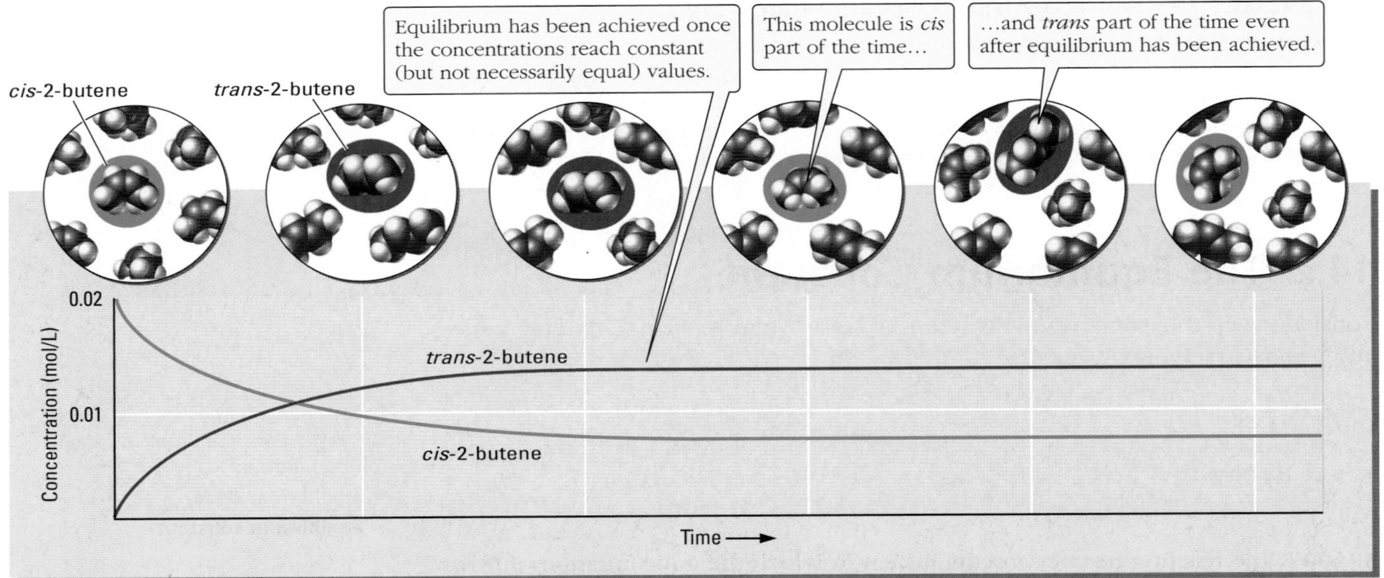
Equilibrium has been achieved once the concentrations reach constant (but not necessarily equal) values.
This molecule is cis part of the time...
...and trans part of the time even after equilibrium has been achieved.
cis-2-butene
trans-2-butene
0.02
0.01
Concentration (mol/L)
trans-2-butene
cis-2-butene
Time

Active Figure 14.2 **Approach to equilibrium.** The graph shows the concentrations of *cis*-2-butene and *trans*-2-butene as the *cis* compound reacts to form the *trans* compound at 500 K. The nanoscale diagrams above the graph show "snapshots" of the composition of a tiny portion of the reaction mixture. The same molecule has been circled in each diagram. **Go to the Active Figures menu at ThomsonNOW to test your understanding of the concepts in this figure.**

In Chapter 13 we noted that the rate law for an overall chemical reaction cannot be derived from the balanced equation, but must be determined experimentally by kinetic studies. This is not true of the equilibrium constant expression. When a reaction takes place by a mechanism that consists of a sequence of steps, the equilibrium constant expression can be obtained by multiplying together the rate constants for the forward reactions in all steps and then dividing by the rate constants for the reverse reactions in all steps. This process gives the same equilibrium constant expression that can be obtained from the coefficients of the balanced overall equilibrium equation.

It turns out that when concentrations are large enough, the value of the equilibrium "constant" expressed in terms of concentrations does not remain constant, even at the same temperature. This happens because of noncovalent interactions among molecules, and especially ions, that cause them to behave differently as their concentrations become larger. To deal with this behavior, true equilibrium constants must be expressed in terms of corrected concentrations that are called *activities*.

equilibrium concentrations must also be constant. We call this ratio K_c, where the capital letter K is used to distinguish it from the rate constants, k_{forward} and k_{reverse}, and the subscript c indicates that it is a ratio of equilibrium *concentrations*. At 500 K the experimental value of K_c is 1.65 for the *cis* $\rightleftharpoons$ *trans* butene reaction.

$$K_c = \frac{k_{\text{forward}}}{k_{\text{reverse}}} = \frac{[\textit{trans}\text{-2-butene}]}{[\textit{cis}\text{-2-butene}]} = 1.65 \qquad \text{(at 500 K)}$$

Because the values of the rate constants vary with temperature *(⊏ p. 632)*, the value of K_c also varies with temperature. For the butene isomerization reaction, K_c is 1.47 at 600 K and 1.36 at 700 K.

A quotient of equilibrium concentrations of product and reactant substances that has a constant value for a given reaction at a given temperature is called an **equilibrium constant** and is given the symbol K_c. On the next page we show how to derive a mathematical expression for the equilibrium constant directly from the chemical equation for any equilibrium process. The mathematical expression is called an **equilibrium constant expression.**

Equilibrium constants can be used to answer three important questions about a reaction:

- When equilibrium has been achieved, do products predominate over reactants?
- Given initial concentrations of reactants and products, in which direction will the reaction go to achieve equilibrium?
- What concentrations of reactants and products are present at equilibrium?

If a reaction moves quickly to equilibrium, you can use equilibrium constants to determine the composition soon after reactants are mixed. Equilibrium constants are less valuable for slow reactions. Until equilibrium has been reached, only kinetics is capable of predicting the composition of a reaction mixture.

CONCEPTUAL EXERCISE 14.3 Properties of Equilibrium

After a mixture of *cis*-2-butene and *trans*-2-butene has reached equilibrium at 600 K, where $K_c = 1.47$, half of the *cis*-2-butene is suddenly removed. Answer these questions:

(a) Is the new mixture at equilibrium? Explain why or why not.

(b) In the new mixture, which rate is faster, *cis* → *trans* or *trans* → *cis*? Or are both rates the same?

(c) In an equilibrium mixture, which concentration is larger, *cis*-2-butene or *trans*-2-butene?

(d) If the concentration of *cis*-2-butene at equilibrium is 0.10 mol/L, what will be the concentration of *trans*-2-butene?

Writing Equilibrium Constant Expressions

The equilibrium constant expression for any reaction has concentrations of products in the numerator and concentrations of reactants in the denominator. Each concentration is raised to the power of its stoichiometric coefficient in the balanced equation. *The only concentrations that appear in an equilibrium constant expression are those of gases and of solutes in dilute solutions*, because these are the only concentrations that can change as a reaction occurs. Concentrations of pure solids, pure liquids, and solvents in dilute solutions ***do not*** appear in equilibrium constant expressions.

To illustrate how this works, consider the general equilibrium reaction

$$a\,\text{A} + b\,\text{B} \rightleftharpoons c\,\text{C} + d\,\text{D} \qquad [14.1]$$

By convention we write the equilibrium constant expression for this reaction as

Equilibrium constant

Product concentrations raised to powers of coefficients

$$K_c = \frac{[\text{C}]^c[\text{D}]^d}{[\text{A}]^a[\text{B}]^b} \qquad [14.2]$$

Reactant concentrations raised to powers of coefficients

ThomsonNOW™
Go to the Coached Problems menu for a tutorial on **writing equilibrium expressions.**

Let's apply these ideas to the combination of nitrogen and oxygen gases to form nitrogen monoxide.

$$N_2(g) + O_2(g) \rightleftharpoons 2\,NO(g)$$

The reaction of nitrogen with oxygen occurs in automobile engines and other high-temperature combustion processes where air is present.

Because all substances in the reaction are gaseous, all concentrations appear in the equilibrium constant expression. The concentration of the product NO(g) is in the numerator and is squared because of the coefficient 2 in the balanced chemical equation. The concentrations of the reactants N_2(g) and O_2(g) are in the denominator. Each is raised to the first power because each has a coefficient of 1.

$$\text{Equilibrium constant} = K_c = \frac{[\text{NO}]^2}{[\text{N}_2][\text{O}_2]}$$

Equilibria Involving Pure Liquids and Solids

As another example, consider the combustion of solid yellow sulfur, which consists of S_8 molecules. The combustion reaction produces sulfur dioxide gas.

$$\tfrac{1}{8}\,S_8(s) + O_2(g) \rightleftharpoons SO_2(g)$$

This reaction occurs whenever a material that contains sulfur burns in air, and it is responsible for a good deal of sulfur dioxide air pollution. It is also the first reaction in a sequence by which sulfur is converted to sulfuric acid, the number-one industrial chemical in the world.

Placing product concentration in the numerator, reactant concentrations in the denominator, and stoichiometric coefficients as exponents gives

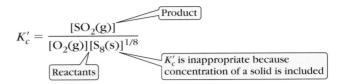

$$K_c' = \frac{[SO_2(g)]}{[O_2(g)][S_8(s)]^{1/8}}$$

Product

Reactants

K_c' is inappropriate because concentration of a solid is included

Because sulfur is a solid, the number of molecules per unit volume is fixed by the density of sulfur at any given temperature (see Exercise 14.1, p. 673). Therefore, the sulfur concentration is not changed either by reaction or by addition or removal of solid sulfur. It is an experimental fact that, as long as some solid sulfur is present, the equilibrium concentrations of O_2 and SO_2 are not affected by changes in the amount of sulfur. Therefore, the equilibrium constant expression for this reaction is properly written as

It is very important to remember that concentrations of pure solids, pure liquids, and solvents for dilute solutions do not appear in the equilibrium constant expression. Any substance designated as (s) or (ℓ) in the equilibrium equation does not appear in the equilibrium constant expression.

$$K_c = \frac{[SO_2(g)]}{[O_2(g)]} \qquad \text{At 25 °C, } K_c = 4.2 \times 10^{52}$$

Equilibria in Dilute Solutions

Consider an aqueous solution of the weak base ammonia, which contains a small concentration of ammonium ions and hydroxide ions because ammonia reacts with water.

$$NH_3(aq) + H_2O(\ell) \rightleftharpoons NH_4^+(aq) + OH^-(aq)$$

If the concentration of ammonia molecules (and consequently of ammonium ions and hydroxide ions) is small, the number of water molecules per unit volume remains essentially the same as in pure water. Because the molar concentration of water is effectively constant for reactions involving dilute solutions, the concentration of water should not be included in the equilibrium constant expression. Thus, we write

$$K_c = \frac{[NH_4^+][OH^-]}{[NH_3]}$$

and the concentration of water is not included in the denominator. At 25 °C, $K_c = 1.8 \times 10^{-5}$ for reaction of ammonia with water.

Notice that no units were included in the equilibrium constant for the ammonia ionization reaction. There are two concentrations in the numerator and only one in the denominator, which ought to give units of mol/L. However, if we always express concentrations in mol/L, the units of the equilibrium constant can be figured out from the equilibrium constant expression. Therefore it is customary to omit the units from the equilibrium constant value, even if the equilibrium constant should have units. We follow that custom in this book.

PROBLEM-SOLVING EXAMPLE 14.1 Writing Equilibrium Constant Expressions

Write an equilibrium constant expression for each chemical equation.

(a) $4\,NO_2(g) + O_2(g) + 2\,H_2O(g) \rightleftharpoons 4\,HNO_3(g)$
(b) $BaSO_4(s) \rightleftharpoons Ba^{2+}(aq) + SO_4^{2-}(aq)$
(c) $6\,NH_3(aq) + Ni^{2+}(aq) \rightleftharpoons Ni(NH_3)_6^{2+}(aq)$
(d) $2\,BaO_2(s) \rightleftharpoons 2\,BaO(s) + O_2(g)$
(e) $NH_3(g) \rightleftharpoons NH_3(\ell)$

Answer

(a) $K_c = \dfrac{[HNO_3]^4}{[NO_2]^4[O_2][H_2O]^2}$ (b) $K_c = [Ba^{2+}][SO_4^{2-}]$

(c) $K_c = \dfrac{[Ni(NH_3)_6^{2+}]}{[NH_3]^6[Ni^{2+}]}$ (d) $K_c = [O_2]$

(e) $K_c = \dfrac{1}{[NH_3]}$

Strategy and Explanation Concentrations of products go in the numerator of the fraction, and concentrations of reactants go in the denominator. Each concentration is raised to the power of the stoichiometric coefficient of the species. In part (b), one species, $BaSO_4(s)$, is a pure solid and does not appear in the expression. In part (d), two species, $BaO_2(s)$ and $BaO(s)$, are solids and do not appear. In part (e), one species, $NH_3(\ell)$, is a pure liquid and does not appear.

PROBLEM-SOLVING PRACTICE 14.1

Write an equilibrium constant expression for each equation.

(a) $CaCO_3(s) \rightleftharpoons CaO(s) + CO_2(g)$

(b) $HCl(g) + LiH(s) \rightleftharpoons H_2(g) + LiCl(s)$

(c) $CH_4(g) + H_2O(g) \rightleftharpoons CO(g) + 3\,H_2(g)$

(d) $CN^-(aq) + H_2O(\ell) \rightleftharpoons HCN(aq) + OH^-(aq)$

Equilibrium Constant Expressions for Related Reactions

Consider the equilibrium involving nitrogen, hydrogen, and ammonia.

$$N_2(g) + 3\,H_2(g) \rightleftharpoons 2\,NH_3(g) \qquad K_{c_1} = 3.5 \times 10^8 \quad (\text{at } 25\ ^\circ\text{C})$$

We could also write the equation so that 1 mol NH_3 is produced.

$$\tfrac{1}{2}\,N_2(g) + \tfrac{3}{2}\,H_2(g) \rightleftharpoons NH_3(g) \qquad K_{c_2} = ?$$

Coefficients half as big

Is the value of the equilibrium constant, K_{c_2}, for the second equation the same as the value of the equilibrium constant, K_{c_1}, for the first equation? To see the relation between K_{c_1} and K_{c_2}, write the equilibrium constant expression for each balanced equation.

$$K_{c_1} = \frac{[NH_3]^2}{[N_2][H_2]^3} = 3.5 \times 10^8 \qquad \text{and} \qquad K_{c_2} = \frac{[NH_3]}{[N_2]^{1/2}[H_2]^{3/2}} = ?$$

Concentrations raised to powers half as big

This makes it clear that K_{c_1} is the square of K_{c_2}; that is, $K_{c_1} = (K_{c_2})^2$. Therefore, the answer to our question is

$$K_{c_2} = (K_{c_1})^{1/2} = (3.5 \times 10^8)^{1/2} = 1.9 \times 10^4$$

Whenever the stoichiometric coefficients of a balanced equation are multiplied by some factor, the equilibrium constant for the new equation (K_{c_2} in this case) is the old equilibrium constant (K_{c_1}) raised to the power of the multiplication factor.

What is the value of K_{c_3}, the equilibrium constant for the decomposition of ammonia to the elements, which is the reverse of the first equation?

$$2\,NH_3(g) \rightleftharpoons N_2(g) + 3\,H_2(g) \qquad K_{c_3} = ? = \frac{[N_2][H_2]^3}{[NH_3]^2}$$

Concentration of NH_3 is in denominator

It is clear that K_{c_3} is the reciprocal of K_{c_1}. That is, $K_{c_3} = 1/K_{c_1} = 1/(3.5 \times 10^8) = 2.9 \times 10^{-9}$. ***The equilibrium constant for a reaction and that for its reverse are the reciprocals of one another.*** If a reaction has a very large equilibrium constant, the reverse reaction will have a very small one. That is, if a reaction is strongly product-favored, then its reverse is strongly reactant-favored. In the case of the production of ammonia from its elements at room temperature, the forward reaction has a large equilibrium constant (3.5×10^8). As expected, the reverse reaction, decomposition of ammonia to its elements, has a small equilibrium constant (2.9×10^{-9}).

EXERCISE 14.4 Manipulating Equilibrium Constants

The balanced equation for conversion of oxygen to ozone has a very small value of K_c.

$$3\,O_2(g) \rightleftharpoons 2\,O_3(g) \qquad K_c = 6.25 \times 10^{-58}$$

(a) What is the value of K_c if the equation is written as

$$\tfrac{3}{2}\,O_2(g) \rightleftharpoons O_3(g)$$

(b) What is the value of K_c for the conversion of ozone to oxygen?

$$2\,O_3(g) \rightleftharpoons 3\,O_2(g)$$

Equilibrium Constant for a Reaction That Combines Two or More Other Reactions

If two chemical equations can be combined to give a third, the equilibrium constant for the combined reaction can be obtained from the equilibrium constants for the two original reactions. For example, air pollution is produced when nitrogen monoxide forms from nitrogen and oxygen and then combines with additional oxygen to form nitrogen dioxide (p. 469).

(1) $N_2(g) + O_2(g) \rightleftharpoons 2\,NO(g)$ $\qquad K_{c_1} = \dfrac{[NO]^2}{[N_2][O_2]}$

(2) $2\,NO(g) + O_2(g) \rightleftharpoons 2\,NO_2(g)$ $\qquad K_{c_2} = \dfrac{[NO_2]^2}{[NO]^2[O_2]}$

The sum of these two equations is

Sum of Equations 1 and 2

(3) $N_2(g) + 2\,O_2(g) \rightleftharpoons 2\,NO_2(g)$

Product of equilibrium constants K_{c_1} and K_{c_2}

$$K_{c_3} = \frac{[NO_2]^2}{[N_2][O_2]^2} = \frac{\cancel{[NO]^2}}{[N_2][O_2]} \times \frac{[NO_2]^2}{\cancel{[NO]^2}[O_2]} = K_{c_1} \times K_{c_2}$$

Because K_{c_1} and K_{c_2} were known experimentally and could be used to calculate K_{c_3}, there is no need to measure K_{c_3} experimentally.

That is, ***if two chemical equations can be summed to give a third, the equilibrium constant for the overall equation equals the product of the two equilibrium constants for the equations that were summed.*** This is a powerful tool for obtaining equilibrium constants without having to measure them experimentally for each individual reaction.

PROBLEM-SOLVING EXAMPLE 14.2 Manipulating Equilibrium Constants

Given these equilibrium reactions and constants,

(1) $\frac{1}{4}\,S_8(s) + 2\,O_2(g) \rightleftharpoons 2\,SO_2(g)$ $\qquad K_{c_1} = 1.86 \times 10^{105}$

(2) $\frac{1}{8}\,S_8(s) + \frac{3}{2}\,O_2(g) \rightleftharpoons SO_3(g)$ $\qquad K_{c_2} = 1.77 \times 10^{53}$

calculate the equilibrium constant for this reaction, which is important in the formation of acid rain air pollution.

$$2\,SO_2(g) + O_2(g) \rightleftharpoons 2\,SO_3(g) \qquad K_{c_3} = \frac{[SO_3]^2}{[SO_2]^2[O_2]}$$

Answer 16.8

Strategy and Explanation Compare the given equations with the target equation to see how each given equation needs to be manipulated. The target equation has SO_2 on the left, so Equation 1 needs to be reversed and we need to take the reciprocal of the equilibrium constant. The target equation has SO_3 on the right. Equation 2 need not be reversed, but each coefficient must be multiplied by 2, which means squaring K_{c_2}. Once this has been done, the two new equations sum to the target equation, and we can multiply their equilibrium constants.

(1)′ $2\,SO_2(g) \rightleftharpoons \frac{1}{4}\,S_8(s) + 2\,O_2(g)$ $\qquad K'_{c_1} = \dfrac{1}{1.86 \times 10^{105}} = 5.38 \times 10^{-106}$

(2)′ $\frac{1}{4}\,S_8(s) + 3\,O_2(g) \rightleftharpoons 2\,SO_3(g)$ $\qquad K'_{c_2} = (1.77 \times 10^{53})^2 = 3.13 \times 10^{106}$

$2\,SO_2(g) + O_2(g) \rightleftharpoons 2\,SO_3(g)$ $\qquad K_{c_3} = K'_{c_1} \times K'_{c_2} = 16.8$

✓ **Reasonable Answer Check** For Equation 1 the equilibrium constant was quite large, so for the reverse reaction it should be quite small, which it is. Also check the order of magnitude of the answer by checking the powers of 10. In the square the power of 10 should be doubled, which it is. And the sum of the powers of 10 for the two equilibrium constants that were multiplied should be close to the power of 10 in the answer, which it is.

PROBLEM-SOLVING PRACTICE 14.2

When carbon dioxide dissolves in water it reacts to produce carbonic acid, $H_2CO_3(aq)$, which can ionize in two steps.

$H_2CO_3(aq) + H_2O(aq) \rightleftharpoons HCO_3^-(aq) + H_3O^+(aq)$ $\qquad K_{c_1} = 4.2 \times 10^{-7}$

$HCO_3^-(aq) + H_2O(aq) \rightleftharpoons CO_3^{2-}(aq) + H_3O^+(aq)$ $\qquad K_{c_2} = 4.8 \times 10^{-11}$

Calculate the equilibrium constant for the reaction

$$H_2CO_3(aq) + 2\,H_2O(aq) \rightleftharpoons CO_3^{2-}(aq) + 2\,H_3O^+(aq)$$

Equilibrium Constants in Terms of Pressure

In a constant-volume system, when the concentration of a gas changes, the partial pressure (⇐ *p. 454*) of the gas also changes. This follows from the ideal gas equation

$$PV = nRT$$

Solving for the partial pressure P_A of a gaseous substance, A,

[A] = n_A/V, the number of moles of A per unit volume

$$P_A = \frac{n_A}{V} RT = [A]RT \qquad [14.3]$$

Equation 14.3 allows us to express the equilibrium constant for the general reaction in Equation 14.1 (p. 677) in a form similar to Equation 14.2, but in terms of partial pressures as

$$K_P = \frac{P_C^c \times P_D^d}{P_A^a \times P_B^b} \qquad [14.4]$$

Product pressures raised to powers of coefficients

Reactant pressures raised to powers of coefficients

Because K_c is related to K_P for the same gas phase reaction, either can be used to calculate the composition of an equilibrium mixture. Most examples in this chapter involve K_c, but the same rules apply to solving problems with K_P.

The subscript on K_P indicates that the equilibrium constant has been expressed in terms of partial pressures. For some gas phase equilibria $K_c = K_P$; for many others it does not. Therefore it is useful to be able to relate one type of equilibrium constant to the other. This can be done by combining Equations 14.2, 14.3, and 14.4 to give

$\Delta n = c + d - a - b$ is the number of moles of gaseous products minus the number of moles of gaseous reactants

$$K_P = \frac{P_C^c \times P_D^d}{P_A^a \times P_B^b} = \frac{\{[C]RT\}^c\{[D]RT\}^d}{\{[A]RT\}^a\{[B]RT\}^b} = \frac{[C]^c[D]^d}{[A]^a[B]^b}(RT)^{c+d-a-b} = K_c(RT)^{\Delta n} \qquad [14.5]$$

As an example of this relation, consider the equilibrium

$$2\ NOCl(g) \rightleftharpoons 2\ NO(g) + Cl_2(g) \qquad K_c = 4.02 \times 10^{-2}\ mol/L \text{ at } 298\ K$$

For this reaction $\Delta n = 2 + 1 - 2 = 3 - 2 = 1$, because there are three moles of gas phase product molecules and only two moles of gas phase reactants. Therefore, for this reaction

$$K_P = K_c \times (RT)^1$$

$$= (4.02 \times 10^{-2}\ mol/L) \times (0.0821\ L\ atm\ K^{-1}\ mol^{-1}) \times (298\ K) = 0.984\ atm$$

EXERCISE 14.5 Relating K_c and K_P

For each of these reactions, calculate K_P from K_c.

(a) $N_2(g) + 3\ H_2(g) \rightleftharpoons 2\ NH_3(g)$ $K_c = 3.5 \times 10^8$ at 25 °C

(b) $2\ H_2(g) + O_2(g) \rightleftharpoons 2\ H_2O(g)$ $K_c = 3.2 \times 10^{81}$ at 25 °C

(c) $N_2(g) + O_2(g) \rightleftharpoons 2\ NO(g)$ $K_c = 1.7 \times 10^{-3}$ at 2300 K

(d) $2\ NO_2(g) \rightleftharpoons N_2O_4(g)$ $K_c = 1.7 \times 10^2$ at 25 °C

14.3 Determining Equilibrium Constants

To determine the value of an equilibrium constant it is necessary to know all of the equilibrium concentrations that appear in the equilibrium constant expression. This is most commonly done by allowing a system to reach equilibrium and then measuring the **equilibrium concentration** of one or more of the reactants or products. Algebra and stoichiometry are then used to obtain K_c.

Reaction Tables, Stoichiometry, and Equilibrium Concentrations

A systematic approach to calculations involving equilibrium constants involves making a table that shows initial conditions, changes that take place when a reaction occurs, and final (equilibrium) conditions. As an example, consider the colorless gas dinitrogen tetraoxide, $N_2O_4(g)$. When heated it dissociates to form red-brown $NO_2(g)$ according to the equation

$$N_2O_4(g) \rightleftharpoons 2\ NO_2(g)$$

Suppose that 2.00 mol $N_2O_4(g)$ is placed into an empty 5.00-L flask and heated to 407 K. Almost immediately a dark red-brown color appears, indicating that much of the colorless gas has been transformed into NO_2 (Figure 14.3). By measuring the intensity of color, it can be determined that the concentration of NO_2 at equilibrium is 0.525 mol/L. To use this information to calculate the equilibrium constant, follow these steps:

1. ***Write the balanced equation for the equilibrium reaction. From it derive the equilibrium constant expression.*** The balanced equation and equilibrium constant expression are

$$N_2O_4(g) \rightleftharpoons 2\ NO_2(g) \qquad K_c = \frac{[NO_2]^2}{[N_2O_4]}$$

2. ***Set up a table containing initial concentration, change in concentration, and equilibrium concentration for each substance included in the equilibrium constant expression. Enter all known information into this reaction table.*** In this case the number of moles of $N_2O_4(g)$ and the volume of the flask were given, so we first calculate the initial concentration of $N_2O_4(g)$ as (conc. N_2O_4) $= \frac{2.00\text{ mol}}{5.00\text{ L}} = 0.400$ mol/L. Because the flask initially contained no NO_2, the initial concentration of NO_2 is zero. The equilibrium concentration of NO_2 was measured as 0.525 mol/L. The reaction table looks like this:

	$N_2O_4(g)$	$\rightleftharpoons$	$2\ NO_2(g)$
Initial concentration (mol/L)	0.400		0
Change as reaction occurs (mol/L)	______		______
Equilibrium concentration (mol/L)	______		0.525

3. ***Use*** **x** ***to represent the change in concentration of one substance. Use the stoichiometric coefficients in the balanced equilibrium equation to calculate the other changes in terms of*** **x.** When the reaction proceeds from left to right, the concentrations of reactants decrease. Therefore the change in concentration of a reactant is negative. The concentrations of products increase, so the change in concentration of a product is positive. Usually it is best to begin with the reactant or product column that contains the most information. In this case that is the NO_2 column, where both initial and equilibrium concentrations are known. Therefore, we let x represent the unknown change in concentration of NO_2.

	$N_2O_4(g)$	$\rightleftharpoons$	$2\ NO_2(g)$
Initial concentration (mol/L)	0.400		0
Change as reaction occurs (mol/L)	______		x
Equilibrium concentration (mol/L)	______		0.525

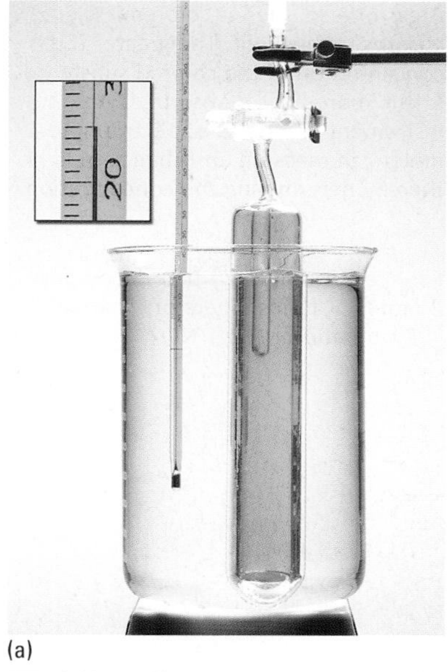

(a)

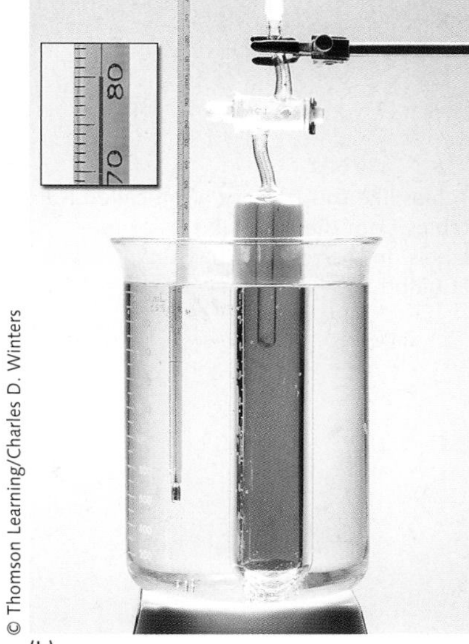

(b)

Figure 14.3 Formation of NO_2 from N_2O_4. Sealed glass tubes containing an equilibrium mixture of dinitrogen tetraoxide (colorless) and nitrogen dioxide (red-brown) are shown (a) at room temperature and (b) immersed in water at 80 °C. Notice the much darker color at the higher temperature, indicating that some dinitrogen tetraoxide has reacted to form nitrogen dioxide. The intensity of color can be used to measure the concentration of nitrogen dioxide.

ThomsonNOW™
Go to the Coached Problems menu for a tutorial on **determining the equilibrium constant.**

A gas or a solution is colored if it absorbs visible light. The greater the concentration of the colored substance is, the more light is absorbed. An instrument known as a spectrophotometer can measure absorbance of light, thereby determining the concentration.

For every 1 mol N_2O_4 that decomposes, 2 mol NO_2 forms. Therefore there is a $\frac{1}{2}$:1 mol ratio of N_2O_4:NO_2.

Next, we use the mole ratio from the balanced equation to find the change in concentration of N_2O_4 in terms of x.

$$\Delta(\text{conc. } N_2O_4) = \frac{x \text{ mol } NO_2 \text{ formed}}{\text{L}} \times \frac{1 \text{ mol } N_2O_4 \text{ reacted}}{2 \text{ mol } NO_2 \text{ formed}}$$

$$= \tfrac{1}{2}x \text{ mol } N_2O_4 \text{ reacted/L}$$

The sign of the change in concentration of N_2O_4 is *negative*, because the concentration of N_2O_4 *decreases*. The table becomes

	$N_2O_4(g)$	$\rightleftharpoons$	$2\ NO_2(g)$
Initial concentration (mol/L)	0.400		0
Change as reaction occurs (mol/L)	$-\frac{1}{2}x$		x
Equilibrium concentration (mol/L)	______		0.525

4. ***From initial concentrations and the changes in concentrations, calculate the equilibrium concentrations in terms of* x *and enter them in the table.*** The concentration of N_2O_4 at equilibrium, $[N_2O_4]$, is the sum of the initial 0.40 mol/L of N_2O_4 and the change due to reaction $-\frac{1}{2}x$ mol/L; that is, $[N_2O_4] = (0.40 - \frac{1}{2}x)$ mol/L. Similarly, the equilibrium concentration of NO_2 (which is already known to be 0.525 mol/L) is $0 + x$, and the table becomes

Tables like this one are often called ICE tables from the initial letters of the labels on the rows: Initial, Change, and Equilibrium.

	$N_2O_4(g)$	$\rightleftharpoons$	$2\ NO_2(g)$
Initial concentration (mol/L)	0.400		0
Change as reaction occurs (mol/L)	$-\frac{1}{2}x$		x
Equilibrium concentration (mol/L)	$0.400 - \frac{1}{2}x$		$0.525 = 0 + x$

5. ***Use the simplest possible equation to solve for* x. *Then use* x *to calculate the unknown you were asked to find.*** (Usually the unknown is K_c or a concentration.) In this case the simplest equation to solve for x is the last entry in the table, $0.525 = 0 + x$, and it is easy to see that $x = 0.525$. Calculate $[N_2O_4] = (0.400 - \frac{1}{2}x)$ mol/L $= (0.400 - \frac{1}{2} \times 0.525)$ mol/L $= 0.138$ mol/L. The problem stated that $[NO_2] = 0.525$ mol/L, so K_c is given by

$$K_c = \frac{[NO_2]^2}{[N_2O_4]} = \frac{(0.525 \text{ mol/L})^2}{(0.138 \text{ mol/L})} = 2.00 \qquad \text{(at 407 K)}$$

6. ***Check your answer to make certain it is reasonable.*** In this case the equilibrium concentration of product is larger than the concentration of reactant. Since the products are in the numerator of the equilibrium constant expression, we expect a value greater than 1, which is what we calculated.

PROBLEM-SOLVING EXAMPLE 14.3 Determining an Equilibrium Constant Value

Consider the gas phase reaction

$$H_2(g) + I_2(g) \rightleftharpoons 2\ HI(g)$$

Suppose that a flask containing H_2 and I_2 has been heated to 425 °C and the initial concentrations of $H_2$and I_2 were each 0.0175 mol/L. With time, the concentrations of H_2 and

I_2 decline and the concentration of HI increases. At equilibrium [HI] = 0.0276 mol/L. Use this experimental information to calculate the equilibrium constant.

Answer $K_c = 56$

Strategy and Explanation Use the information given and follow the six steps. In the table below, each entry has been color-coded to match the numbered steps.

1. ***Write the balanced equation and equilibrium constant expression.***

$$H_2(g) + I_2(g) \rightleftharpoons 2\,HI(g) \qquad K_c = \frac{[HI]^2}{[H_2][I_2]}$$

2. ***Construct a reaction (ICE) table (see below) and enter known information.***
3. ***Represent changes in concentration in terms of x.***

The best choice is to enter x in the third column, because both initial and equilibrium concentrations of HI are known, and this provides a simple way to calculate x. Next, derive the rest of the concentration changes in terms of x. If the concentration of HI increases by a given quantity, the mole ratios say that the concentrations of H_2 and I_2 must decrease only half as much:

$$\frac{x \text{ mol } \cancel{\text{HI produced}}}{\text{L}} \times \frac{1 \text{ mol } H_2 \text{ consumed}}{2 \text{ } \cancel{\text{mol HI produced}}} = \frac{1}{2}x \text{ mol/L } H_2 \text{ consumed}$$

Because the coefficients of H_2 and I_2 are equal, each of their concentrations decreases by $\frac{1}{2}x$ mol/L. The entries in the table for change in concentration of H_2 and I_2 are negative, because their concentrations decrease.

4. ***Calculate equilibrium concentrations and enter them in the table.***

	$H_2(g)$	+	$I_2(g)$	$\rightleftharpoons$	$2\,HI(g)$
Initial concentration (mol/L)	0.0175		0.0175		0
Change as reaction occurs (mol/L)	$-\frac{1}{2}x$		$-\frac{1}{2}x$		x
Equilibrium concentration (mol/L)	$0.0175 - \frac{1}{2}x$		$0.0175 - \frac{1}{2}x$		$0.0276 = 0 + x$

Color-coded entries in the table in this example correspond with color-coded steps in the explanation section.

5. ***Solve the simplest equation for x.*** The last row and column in the table contains $0.0276 = 0 + x$, which gives $x = 0.0276$. Substitute this value into the other two equations in the last row of the table to get the equilibrium concentrations, and substitute them into the equilibrium constant expression:

$$K_c = \frac{[HI]^2}{[H_2][I_2]} = \frac{(0.0276)^2}{(0.0175 - [\frac{1}{2} \times 0.0276])(0.0175 - [\frac{1}{2} \times 0.0276])}$$

$$= \frac{(0.0276)^2}{(0.0037)(0.0037)} = 56 \qquad \text{(at 424 °C)}$$

✓ **Reasonable Answer Check** The equilibrium constant is larger than 1, so there should be more products than reactants when equilibrium is reached. The equilibrium concentration of the product (0.0276 mol/L) is larger than those of the reactants (0.0037 mol/L each).

PROBLEM-SOLVING PRACTICE 14.3

Measuring the conductivity of an aqueous solution in which 0.0200 mol CH_3COOH has been dissolved in 1.00 L of solution shows that 2.96% of the acetic acid molecules have ionized to CH_3COO^- ions and H_3O^+ ions. Calculate the equilibrium constant for ionization of acetic acid and compare your result with the value given in Table 14.1.

Saying that 2.96% of the acetic acid molecules have ionized means that at equilibrium the concentration of acetate ions is 2.96/100 = 0.0296 times the initial concentration of acetic acid molecules.

Experimentally determined equilibrium constants for a few reactions are given in Table 14.1. These reactions occur to widely differing extents, as shown by the wide range of K_c values.

Table 14.1 Selected Equilibrium Constants at 25 °C

Reaction	K_c	K_P
Gas phase reactions		
$\frac{1}{8}S_8(s) + O_2(g) \rightleftharpoons SO_2(g)$	4.2×10^{52}	4.2×10^{52}
$2\ H_2(g) + O_2(g) \rightleftharpoons 2\ H_2O(g)$	3.2×10^{81}	1.3×10^{80}
$N_2(g) + 3\ H_2(g) \rightleftharpoons 2\ NH_3(g)$	3.5×10^{8}	5.8×10^{5}
$N_2(g) + O_2(g) \rightleftharpoons 2\ NO(g)$	4.5×10^{-31}	4.5×10^{-31}
	1.7×10^{-3} (at 2300 K)	
$H_2(g) + I_2(g) \rightleftharpoons 2\ HI(g)$	2.5×10^{1}	2.5×10^{1}
$2\ NO_2(g) \rightleftharpoons N_2O_4(g)$	1.7×10^{2}	7.0
$CH_4(g) + H_2O(g) \rightleftharpoons CO(g) + 3\ H_2(g)$	2.0×10^{-28}	1.25×10^{-25}
cis-2-butene(g) $\rightleftharpoons$ *trans*-2-butene(g)	3.2	3.2
Weak acids and bases		
Formic acid		
$HCOOH(aq) + H_2O(\ell) \rightleftharpoons H_3O^+(aq) + HCOO^-(aq)$	1.8×10^{-4}	—
Acetic acid		
$CH_3COOH(aq) + H_2O(\ell) \rightleftharpoons H_3O^+(aq) + CH_3COO^-(aq)$	1.8×10^{-5}	—
Carbonic acid		
$H_2CO_3(aq) + H_2O(\ell) \rightleftharpoons H_3O^+(aq) + HCO_3^-(aq)$	4.2×10^{-7}	—
Ammonia (weak base)		
$NH_3(aq) + H_2O(\ell) \rightleftharpoons NH_4^+(aq) + OH^-(aq)$	1.8×10^{-5}	—
Very slightly soluble solids		
$CaCO_3(s) \rightleftharpoons Ca^{2+}(aq) + CO_3^{2-}(aq)$	3.8×10^{-9}	—
$AgCl(s) \rightleftharpoons Ag^+(aq) + Cl^-(aq)$	1.8×10^{-10}	—
$AgI(s) \rightleftharpoons Ag^+(aq) + I^-(aq)$	1.5×10^{-16}	—

ThomsonNOW™
Go to the Coached Problems menu for a tutorial on the **magnitude of the equilibrium constant: reactions of NO with O_2 and of $Pb(NO_3)_2$ + KI.**

The symbol >> means "much greater than."

14.4 The Meaning of the Equilibrium Constant

The size of the equilibrium constant tells how far a reaction has proceeded by the time equilibrium has been achieved. In addition, it can be used to calculate how much product will be present at equilibrium. There are three important cases to consider.

Case 1. K_c >> 1: *Reaction is strongly product-favored; equilibrium concentrations of products are much greater than equilibrium concentrations of reactants.*

A large value of K_c *means that reactants have been converted almost entirely to products when equilibrium has been achieved.* That is, the products are strongly favored over the reactants. An example is the reaction of NO(g) with $O_3(g)$, which is one way that ozone is destroyed in the stratosphere (⇐ p. 462).

$$NO(g) + O_3(g) \rightleftharpoons NO_2(g) + O_2(g) \quad K_c = \frac{[NO_2][O_2]}{[NO][O_3]} = 6 \times 10^{34} \quad \text{(at 25 °C)}$$

The very large value of K_c tells us that if 1 mol each of NO and O_3 are mixed in a flask at 25 °C and allowed to come to equilibrium, $[NO_2][O_2] >> [NO][O_3]$. Virtually none of the reactants will remain, and essentially only NO_2 and O_2 will be found in the flask. For practical purposes, this reaction goes to completion, and it would not be necessary to use the equilibrium constant to calculate the quantities of products that would be obtained. The simpler methods developed in Chapter 4 (⇐ *p. 135*) would work just fine in this case.

The reaction goes to completion when equilibrium has been reached. This could take a long time if the reaction is slow.

Case 2. K_c << 1: *Reaction is strongly reactant-favored; equilibrium concentrations of reactants are greater than equilibrium concentrations of products.*

The symbol << means "much less than."

Conversely, *an extremely small K_c means that when equilibrium has been achieved, very little of the reactants have been transformed into products.* The reactants are favored over the products at equilibrium.

$$3\ O_2(g) \rightleftharpoons 2\ O_3(g) \qquad K_c = \frac{[O_3]^2}{[O_2]^3} = 6.25 \times 10^{-58} \quad \text{(at 25 °C)}$$

This means that $[O_3]^2 << [O_2]^3$ and if O_2 is placed in a flask at 25 °C, *very* little O_3 will be found when equilibrium is achieved. The concentration of O_2 would remain essentially unchanged. In the terminology of Chapter 5, we would write "N.R." and say that no reaction occurs.

Case 3. $K_c \cong 1$: *Equilibrium mixture contains significant concentrations of reactants and products; calculations are needed to determine equilibrium concentrations.*

The symbol ≅ means "approximately equal to."

If K_c is neither extremely large nor extremely small, the equilibrium constant must be used to calculate how far a reaction proceeds toward products. In contrast with the reactions in Case 1 and Case 2, dissociation of dinitrogen tetraoxide has neither a very large nor a very small equilibrium constant. At 391 K the value is 1.00, which means that significant concentrations of both N_2O_4 and NO_2 are present at equilibrium.

$$K_c = 1.00 = \frac{[NO_2]^2}{[N_2O_4]}, \text{ so } [NO_2]^2 = [N_2O_4] \quad \text{(at 391 K)} \qquad [14.6]$$

Recall that the equation for dissociation of dinitrogen tetraoxide is

$$N_2O_4(g) \rightleftharpoons 2\ NO_2(g)$$

What range of equilibrium constants represents this middle ground depends on how small a concentration is significant and on the form of the equilibrium constant. If the concentrations of N_2O_4 and NO_2 at equilibrium are both 1.0 mol/L, then the ratio of $[NO_2]^2/[N_2O_4]$ does equal the K_c value of 1.00 at 391 K. But what if the concentrations were much smaller? Would they still be equal? You can verify this by using Equation 14.6. If the equilibrium concentration of NO_2 is 0.01, then the concentration of N_2O_4 must be $(0.01)^2$, which equals 0.0001. Thus, even though K_c = 1.00, the concentration of one substance can be much larger than the concentration of the other. This happens because there is a squared term in the numerator of the equilibrium constant expression and a term to a different power (the first power) in the denominator. Whenever the total of the exponents in the numerator differs from the total in the denominator, it becomes very difficult to say whether the concentrations of the products will exceed those of the reactants without doing a calculation.

By contrast, if the total of the exponents is the same, then if $K_c > 1$, products predominate over reactants, and if $K_c < 1$, reactants predominate over products. Examples in which this is true include

ThomsonNOW™
Go to the Coached Problems menu for a module on the **meaning of the equilibrium constant.**

$$cis\text{-2-butene}(g) \rightleftharpoons trans\text{-2-butene}(g) \qquad K_c = \frac{[trans]}{[cis]} = 3.2 \quad \text{(at 25 °C)}$$

and

$$2\ HI(g) \rightleftharpoons H_2(g) + I_2(g) \qquad K_c = \frac{[H_2][I_2]}{[HI]^2} = 0.040 \quad \text{(at 25 °C)}$$

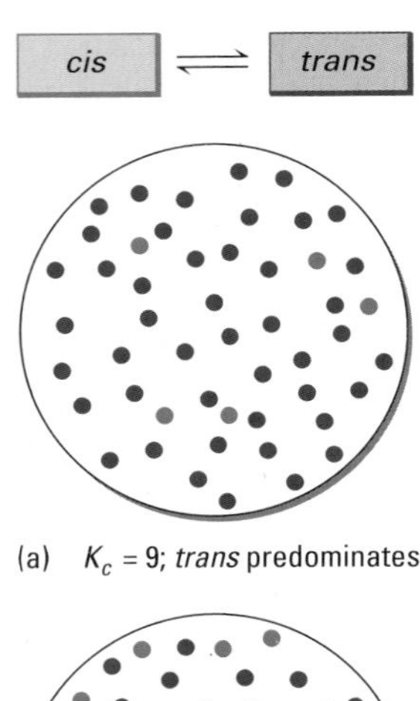

(a) $K_c = 9$; *trans* predominates

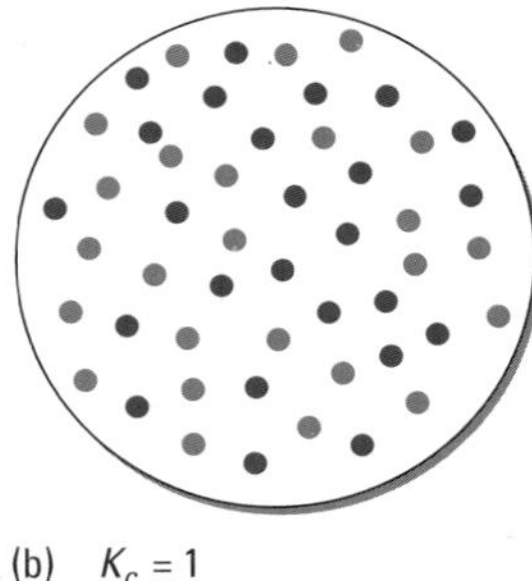

(b) $K_c = 1$

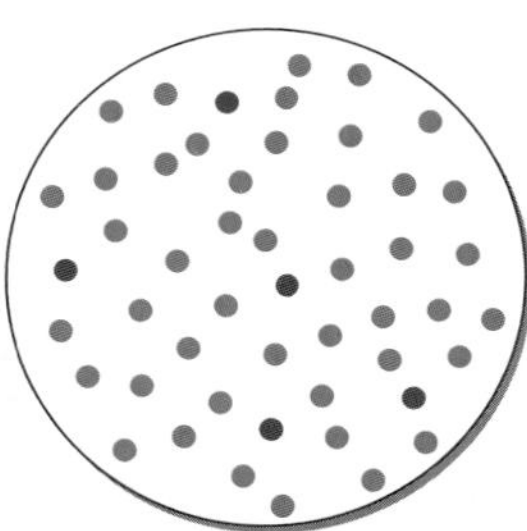

(c) $K_c = 1/9$; *cis* predominates

Figure 14.4 Equilibrium constants and concentrations of reactants and products. In the system *cis*-2-butene(g) $\rightleftharpoons$ *trans*-2-butene(g), the equilibrium constant decreases as temperature increases. (a) At a low temperature, when $K_c = 9$, the ratio [*trans*]/[*cis*] is 9/1 or 45/5. (b) At 1000 K, when $K_c = 1$, the ratio is 1/1 or 25/25. (c) At a much higher temperature, when $K_c = 0.111 = 1/9$, the ratio is 1/9 or 5/45.

Figure 14.4 diagrams reactant and product concentrations as a function of K_c for the isomerization of *cis*-2-butene in the gas phase.

You might wonder whether reactant-favored systems in which small quantities of products form are important. Many are. Examples include the acids and bases listed in Table 14.1. For acetic acid, the acidic ingredient in vinegar, the reaction is

$$CH_3COOH(aq) + H_2O(\ell) \rightleftharpoons H_3O^+(aq) + CH_3COO^-(aq)$$

$$K_c = \frac{[H_3O^+][CH_3COO^-]}{[CH_3COOH]} = 1.8 \times 10^{-5} \quad \text{(at 25 °C)}$$

The value of K_c for acetic acid is small, and at equilibrium the concentrations of the products (acetate ions and hydronium ions) are small relative to the concentration of the reactant (acetic acid molecules). This confirms that acetic acid is a weak acid. Nevertheless, vinegar tastes sour because a small percentage of the acetic acid molecules react with water to produce $H_3O^+(aq)$.

If the form of the equilibrium constant is the same for two or more different reactions, then the degree to which each of those reactions is product-favored can be compared quantitatively in a very straightforward way. For example, the equilibrium reactions are similar and the equilibrium constant expressions all have the same form for formic acid, acetic acid, and carbonic acid:

$$K_c = \frac{[H_3O^+][\text{anion}^-]}{[\text{acid}]}$$

where anion⁻ is $HCOO^-$, CH_3COO^-, or HCO_3^-, and acid is HCOOH, CH_3COOH, or H_2CO_3. From data in Table 14.1, we can say that formic acid is stronger than acetic acid (has a larger K_c value) and carbonic acid is the weakest of the three acids.

If a reaction has a large tendency to occur in one direction, then the reverse reaction has little tendency to occur. This means that the equilibrium constant for the reverse of a strongly product-favored reaction will be extremely small. Table 14.1 shows that combustion of hydrogen to form water vapor has an enormous equilibrium constant (3.2×10^{81}). This reaction is strongly product-favored. We say that it goes to completion.

The reverse reaction, decomposition of water to its elements,

$$2\,H_2O(g) \rightleftharpoons 2\,H_2(g) + O_2(g) \qquad K_c = \frac{[H_2]^2[O_2]}{[H_2O]^2} = 3.1 \times 10^{-82} \quad \text{(at 25 °C)}$$

is strongly reactant-favored, as indicated by the *very* small value of K_c.

PROBLEM-SOLVING EXAMPLE 14.4 Using Equilibrium Constants

Use equilibrium constants (Table 14.1) to predict which of the reactions below will be product-favored at 25 °C. Place all of the reactions in order from most reactant-favored to least reactant-favored.

(a) $NH_3(aq) + H_2O(\ell) \rightleftharpoons NH_4^+(aq) + OH^-(aq)$
(b) $HCOOH(aq) + H_2O(\ell) \rightleftharpoons H_3O^+ + HCOO^-(aq)$
(c) $N_2O_4(g) \rightleftharpoons 2\,NO_2(g)$

Answer All reactions are reactant-favored. The order from most reactant-favored to least reactant-favored is (a), (b), (c).

Strategy and Explanation Check whether the reactions all have equilibrium constant expressions of the same form. If they do, then the smaller the equilibrium constant is, the less product-favored (more reactant-favored) the reaction is. The equilibrium constant expressions are all of the form

$$K_c = \frac{[\text{product 1}][\text{product 2}]}{[\text{reactant}]}$$

because $H_2O(\ell)$ does not appear in the expressions for (a) and (b). The equilibrium constants for reactions (a) and (b) are 1.8×10^{-5} and 1.8×10^{-4}, respectively. The equilibrium constant for reaction (c) is not given in Table 14.1, but K_c for the reverse reaction is given as 1.7×10^2. Because the reaction is reversed it is necessary to take the reciprocal (p. 679), which gives an equilibrium constant for reaction (c) of 5.8×10^{-3}. Therefore the most reactant-favored reaction (smallest K_c) is (a), the next smallest K_c is for reaction (b), and the largest K_c is for reaction (c), which is the least reactant-favored.

PROBLEM-SOLVING PRACTICE 14.4

Suppose that solid AgCl and AgI are placed in 1.0 L water in separate beakers.

(a) In which beaker would the silver ion concentration, $[Ag^+]$, be larger?

(b) Does the volume of water in which each compound dissolves affect the equilibrium concentration?

CONCEPTUAL EXERCISE 14.6 Manipulating Equilibrium Constants

The equilibrium constant is 1.8×10^{-5} for reaction of ammonia with water.

$$NH_3(aq) + H_2O(\ell) \rightleftharpoons NH_4^+(aq) + OH^-(aq)$$

(a) Is the equilibrium constant large or small for the reverse reaction, the reaction of ammonium ion with hydroxide ion to give ammonia and water?

(b) What is the value of K_c for the reaction of ammonium ions with hydroxide ions?

(c) What does the value of this equilibrium constant tell you about the extent to which a reaction can occur between ammonium ions and hydroxide ions?

(d) Predict what would happen if you added a 1.0 M solution of ammonium chloride to a 1.0 M solution of sodium hydroxide. What observations might allow you to test your prediction in the laboratory?

14.5 Using Equilibrium Constants

Because equilibrium constants have numeric values, they can be used to predict quantitatively in which direction a reaction will proceed and how far it will go.

ThomsonNOW™
Go to the Coached Problems menu for tutorials on:

- **systems at equilibrium**
- **estimating equilibrium concentrations**

Predicting the Direction of a Reaction

Suppose that you have a mixture of 50. mmol $NO_2(g)$ and 100. mmol $N_2O_4(g)$ at 25 °C in a container with a volume of 10. L. Is the system at equilibrium? If not, in which direction will it react to achieve equilibrium? A useful way to approach such questions is to use the **reaction quotient,** Q, which *has the same mathematical form as the equilibrium constant expression but is a ratio of actual concentrations in the mixture,* instead of equilibrium concentrations. For the reaction

Recall that the metric prefix "m" means $\frac{1}{1000} = 10^{-3}$. Therefore, 1 mmol $= 1 \times 10^{-3}$ mol.

$$2\,NO_2(g) \rightleftharpoons N_2O_4(g) \qquad K_c = \frac{[N_2O_4]}{[NO_2]^2} = 1.7 \times 10^2 \quad \text{(at 25 °C)}$$

$$Q = \frac{(\text{conc. } N_2O_4)}{(\text{conc. } NO_2)^2} = \frac{(100. \times 10^{-3}\ \text{mol}/10.\ \text{L})}{(50. \times 10^{-3}\ \text{mol}/10.\ \text{L})^2} = \frac{1.0 \times 10^{-2}}{(5.0 \times 10^{-3})^2} = 4.0 \times 10^2$$

- *If Q is equal to K_c, then the reaction is at equilibrium.* The concentrations will not change.
- *If Q is less than K_c, then the concentrations of products are not as large as they would be at equilibrium.* The reaction will proceed from left to right to increase the product concentrations until they reach their equilibrium values.

In the expression for Q, we have used (conc. N_2O_4) to represent the *actual* concentration of N_2O_4 at a given time. We use $[N_2O_4]$ to represent the equilibrium concentration of N_2O_4. When the reaction is at equilibrium (conc. N_2O_4) = $[N_2O_4]$.

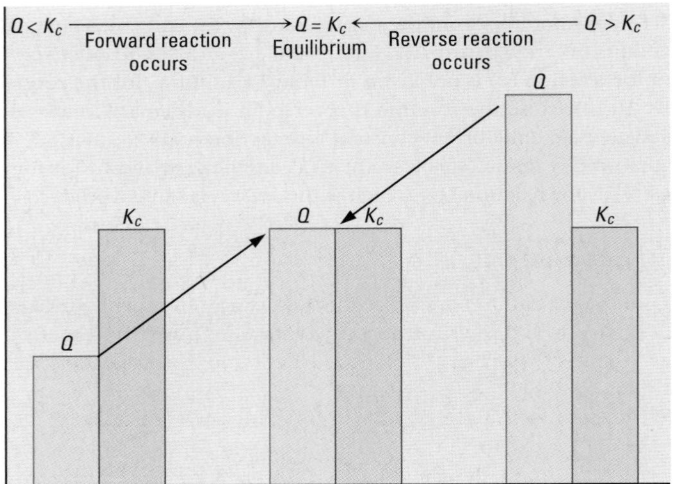

Figure 14.5 Predicting direction of a reaction. The relative sizes of the reaction quotient, Q, and the equilibrium constant, K_c, determine in which direction a mixture of substances will react to achieve equilibrium. The rule is

$Q < K_c$, reaction goes in forward direction (→)
$Q = K_c$, reaction at equilibrium
$Q > K_c$, reaction goes in reverse direction (←)

The reaction will proceed until $Q = K_c$.

- *If Q is greater than* K_c, *then the product concentrations are higher than they would be at equilibrium.* The reaction will proceed from right to left, increasing reactant concentrations, until equilibrium is achieved.

These relationships are shown schematically in Figure 14.5. In the case of the NO_2/N_2O_4 mixture described above, Q is greater than K_c, so, to establish equilibrium, some N_2O_4 will react to form NO_2. As the reverse reaction takes place, Q becomes smaller and eventually becomes equal to K_c.

PROBLEM-SOLVING EXAMPLE 14.5 Predicting Direction of Reaction

Consider this equilibrium, which is used industrially to generate hydrogen gas.

$$CH_4(g) + H_2O(g) \rightleftharpoons CO(g) + 3\,H_2(g) \qquad K_c = 0.94 \text{ at } 25\ °C$$

If 1.0 mol CH_4, 1.0 mol H_2O, 2.0 mol H_2, and 0.5 mol CO are mixed in a 10.0-L container at 25 °C, will the concentration of H_2O be greater or less than 0.10 mol/L when equilibrium is reached?

Answer Less than 0.10 mol/L

Strategy and Explanation Calculate the initial concentration of each gas and thus evaluate Q. Then compare Q with K_c.

$$(\text{conc. } CH_4) = \frac{1.0 \text{ mol}}{10.0 \text{ L}} = 0.10 \text{ mol/L} \qquad (\text{conc. } H_2) = \frac{2.0 \text{ mol}}{10.0 \text{ L}} = 0.20 \text{ mol/L}$$

$$(\text{conc. } H_2O) = \frac{1.0 \text{ mol}}{10.0 \text{ L}} = 0.10 \text{ mol/L} \qquad (\text{conc. CO}) = \frac{0.5 \text{ mol}}{10.0 \text{ L}} = 0.050 \text{ mol/L}$$

Because the units are defined by the equilibrium constant expression, leave them out.

$$Q = \frac{(\text{conc. CO})(\text{conc. } H_2)^3}{(\text{conc. } CH_4)(\text{conc. } H_2O)} = \frac{(0.050)(0.20)^3}{(0.10)(0.10)} = 0.040$$

which is much smaller than 0.94, the value of K_c. Because $Q < K_c$, the forward reaction—reaction of CH_4 with H_2O to form CO and H_2—will occur until the equilibrium concentrations are reached. The initial concentration of H_2O was 0.10 mol/L; when H_2O reacts with CH_4, the H_2O concentration decreases.

✓ **Reasonable Answer Check** Because all concentrations are fractions, multiplying concentrations will give smaller fractions. The concentrations are all approximately 10^{-1}, but the numerator has concentration to the fourth power and the denominator to the second power overall. This means that Q should be much less than 1, and it is.

PROBLEM-SOLVING PRACTICE 14.5

For the equilibrium

$$2\ SO_2(g) + O_2(g) \rightleftharpoons 2\ SO_3(g) \qquad K_c = 245 \quad \text{(at 1000 K)}$$

the equilibrium concentrations are $[SO_2] = 0.102$, $[O_2] = 0.0132$, and $[SO_3] = 0.184$. Suppose that the concentration of SO_2 is suddenly doubled. Calculate Q and use it to show that the forward reaction would take place to reach a new equilibrium.

CONCEPTUAL EXERCISE 14.7 Reaction Quotient and Pressure Equilibrium Constant

Is it possible to apply the idea of the reaction quotient to gas phase reactions where the equilibrium constant is given in terms of pressure (as in Equation 14.4, p. 682)? Define Q for such a reaction and give an appropriate set of rules by which you can predict in which direction a gas phase reaction will go to achieve equilibrium.

Calculating Equilibrium Concentrations

Equilibrium constants from Table 14.1 (or one of the appendixes, or a reference compilation) can be used to calculate how much product is formed and how much of the reactants remain once a system has reached equilibrium. To verify our earlier statement that if an equilibrium constant is very large, essentially all of the reactants are converted to products (p. 686), consider the reaction

$$\tfrac{1}{8}\ S_8(s) + O_2(g) \rightleftharpoons SO_2(g) \qquad K_c = 4.2 \times 10^{52} \quad \text{(at 25 °C)}$$

Suppose we place 4.0 mol O_2 and a large excess of sulfur in an empty 1.00-L flask and allow the system to reach equilibrium. We can calculate the quantity of O_2 left and the quantity of SO_2 formed at equilibrium by summarizing information in a table. (Because S_8 is a solid and does not appear in the equilibrium constant expression, we do not need any entries under S_8 in the table.)

	$\frac{1}{8}S_8(s)$	+	$O_2(g)$	$\rightleftharpoons$	$SO_2(g)$
Initial concentration (mol/L)			4.0		0.0
Change as reaction occurs (mol/L)			$-x$		$+x$
Equilibrium concentration (mol/L)			$4.0 - x$		$0.0 + x$

We know the concentrations of reactant and product before the reaction, but we do not know how many moles per liter of O_2 are consumed during the reaction, and so we designate this as x mol/L. (There is a minus sign in the table because O_2 is consumed.) Since the mole ratio is (1 mol SO_2)/(1 mol O_2), we know that x mol/L SO_2 is formed when x mol/L O_2 is consumed. To calculate the concentration of O_2 we take what was present initially (4.0 mol/L) minus what was consumed in the reaction (x mol/L). The equilibrium concentration of SO_2 must be the initial concentration (0 mol/L) plus what was formed by the reaction (x mol/L). Putting these values into the equilibrium constant expression, we have

$$K_c = \frac{[SO_2]}{[O_2]} = \frac{x}{4.0 - x} = 4.2 \times 10^{52}$$

Solving algebraically for x (and following the usual rules for significant figures), we find

$$x = (4.2 \times 10^{52})(4.0 - x)$$
$$x = (16.8 \times 10^{52}) - (4.2 \times 10^{52})x$$
$$x + (4.2 \times 10^{52})x = 16.8 \times 10^{52}$$

Because 4.2×10^{52} is so much larger than 1, adding 1 to it makes no appreciable change in the very large number.

Notice that $x + (4.2 \times 10^{52})x = (1 + 4.2 \times 10^{52})x$, which to a very good approximation is equal to $(4.2 \times 10^{52})x$. Thus,

$$x = \frac{16.8 \times 10^{52}}{4.2 \times 10^{52}} = 4.0$$

The fact that samples of sulfur at 25 °C can be exposed to oxygen in the air for long periods without being converted to sulfur dioxide shows the importance of chemical kinetics. This reaction is very slow at room temperature, so only a faint odor of sulfur dioxide is noticeable in the vicinity of solid sulfur.

The equilibrium concentration of SO_2 is $x = 4.0$ mol/L and that of O_2 is $(4.0 - x)$ mol/L, or 0 mol/L. That is, within the precision of our calculation, all the O_2 has been converted to SO_2. As stated earlier, a very large K_c value (4.2×10^{52} in this case) implies that essentially all of the reactants have been converted to products. The reaction is strongly product-favored and goes to completion, so the calculation could have been done using the methods in Section 4.4 (⇐ p. 134).

PROBLEM-SOLVING EXAMPLE 14.6 Calculating Equilibrium Concentrations

One way to generate hydrogen gas is to react carbon monoxide with steam.

$$CO(g) + H_2O(g) \rightleftharpoons CO_2(g) + H_2(g) \qquad K_c = 10.\ \text{at } 420\ °C$$

Suppose you place 2.5×10^{-3} mol/L CO and 2.5×10^{-3} mol/L H_2O into a container and heat the mixture to 420 °C. What is the concentration of each of the four substances when equilibrium is reached?

Answer $[CO] = [H_2O] = 6.0 \times 10^{-4}$ mol/L; $[CO_2] = [H_2] = 1.9 \times 10^{-3}$ mol/L

Strategy and Explanation Follow the procedure on p. 683. The equation is written in the statement of the problem. In this equation all of the mole ratios are 1:1. This tells us that equal numbers of moles of CO and H_2O are consumed as the reaction proceeds to equilibrium. Since both substances are in the same flask, equal numbers of moles *per liter* (equal concentrations) of reactants must also be consumed, and we designate each of these as $-x$. Mole ratios also tell us that if the concentration of CO decreases by x mol/L, then the concentration of H_2 (and the concentration of CO_2) must increase by the same quantity, x mol/L. Because their initial concentrations were the same and their mole ratio is 1:1, the equilibrium concentrations of H_2O and CO must be the same. Each is equal to the initial concentration plus the change (the change for reactants is $-x$, so this means subtracting x) as the reactants are consumed. The reaction table is

	CO	+	H_2O	⇌	CO_2	+	H_2
Initial concentration (mol/L)	2.5×10^{-3}		2.5×10^{-3}		0.000		0.000
Change as reaction occurs (mol/L)	$-x$		$-x$		$+x$		$+x$
Equilibrium concentration (mol/L)	$0.0025 - x$		$0.0025 - x$		x		x

Substituting these values into the expression for K_c, we have

$$K_c = 10.0 = \frac{[H_2][CO_2]}{[H_2O][CO]} = \frac{(x)(x)}{(0.0025 - x)(0.0025 - x)} = \frac{x^2}{(0.0025 - x)^2}$$

Solving this equation for x is not as difficult as it might seem at first glance. Because the right-hand side is a perfect square, we can take the square root of both sides, giving

$$\sqrt{K_c} = \sqrt{10.0} = 3.16 = \frac{x}{(0.0025 - x)}$$

and then solve for x.

$$x = (3.16)(0.0025 - x)$$

$$x = 0.0079 - 3.16x$$

$$4.16x = 0.0079$$

$$x = 1.9 \times 10^{-3}$$

The concentrations of the products are both equal to x mol/L—that is, to 1.9×10^{-3} mol/L—while the concentrations of the reactants that remain are both $(0.0025 - x)$ mol/L $= 6.0 \times 10^{-4}$ mol/L. This result demonstrates quantitatively that when $K_c > 1$ a reaction is product-favored because at equilibrium the product concentrations are larger than the reactant concentrations.

✓ **Reasonable Answer Check** To check the calculation, substitute these concentrations into the equilibrium constant expression and compare the calculated K_c with the value given in the statement of the problem.

$$K_c = \frac{[H_2][CO_2]}{[H_2O][CO]} = \frac{(1.9 \times 10^{-3})^2}{(6.0 \times 10^{-4})^2} = 10.0$$

which equals the value given.

PROBLEM-SOLVING PRACTICE 14.6

The equilibrium constant for dissolving the insoluble substance gold(I) iodide, AuI(s), in aqueous solution is 1.6×10^{-23} at 25 °C. Write the equilibrium constant expression and calculate the concentration of Au^+(aq) and I^-(aq) ions in a solution in which 0.345 g AuI(s) is in equilibrium with the aqueous ions.

PROBLEM-SOLVING EXAMPLE 14.7 Calculating Equilibrium Concentrations

When colorless hydrogen iodide gas is heated to 745 K, a beautiful purple color appears. This shows that some iodine gas has been formed, which means that the compound has been decomposed partially to its elements.

$$2\,HI(g) \rightleftharpoons H_2(g) + I_2(g) \qquad K_c = 0.0200 \quad (\text{at } 745\text{ K})$$

Suppose that a mixture of 1.00 mol HI(g) and 1.00 mol H_2(g) is sealed into a 10.0-L flask and heated to 745 K. What will be the concentrations of all three substances when equilibrium has been achieved?

Answer [HI] = 0.096 M, $[I_2]$ = 0.0018 M, $[H_2]$ = 0.102 M

Strategy and Explanation Follow the usual steps (p. 683). The balanced equation was given, and the equilibrium constant expression follows the table (next page). Because amounts are given instead of concentrations, divide each number of moles by the volume of the flask to get (conc. HI) = (conc. H_2) = 1.00 mol/10.0 L = 0.100 mol/L. Because no I_2 is present at the beginning, $Q = 0$. This is much less than K_c, even though K_c is small. Therefore the reaction must go from left to right, so it makes sense to let x mol/L be the concentration of I_2 when equilibrium is reached. Since the coefficients of H_2 and I_2 are the same, if x mol/L of I_2 is produced, then x mol/L of H_2 must also be produced. Therefore the change in concentration of both H_2 and I_2 is $+x$ mol/L. Because the coefficient

Iodine vapor has a beautiful purple color.

of HI is twice the coefficient of I_2, twice as many moles of HI must react as moles of I_2 formed; the change in concentration of HI is $-2x$ mol/L. The reaction table is then

In solving this problem we might have chosen $-x$ to be the change in concentration of HI, in which case the changes in concentrations of H_2 and I_2 would each have been $+\frac{1}{2}x$.

	2 HI(g)	$\rightleftharpoons$	H_2(g)	+	I_2(g)
Initial concentration (mol/L)	0.100		0.100		0.000
Change as reaction occurs (mol/L)	$-2x$		$+x$		$+x$
Equilibrium concentration (mol/L)	$(0.100 - 2x)$		$(0.100 + x)$		x

Now write the equilibrium constant expression in terms of the equilibrium concentrations calculated in the third row of the table.

$$K_c = \frac{[H_2][I_2]}{[HI]^2} = \frac{(0.100 + x)x}{(0.100 - 2x)^2} = 0.0200$$

The ratio of terms involving x is not a perfect square, so you cannot take a square root as was done in Problem-Solving Example 14.6. To solve the equation directly, multiply out the numerator and denominator to obtain

$$\frac{0.100x + x^2}{0.0100 - 0.400x + 4x^2} = 0.0200$$

Multiply both sides by the denominator and then multiply out the terms. This gives

$$0.100x + x^2 = 0.0200 \times (0.0100 - 0.400x + 4x^2)$$

$$x = 0.000200 - 0.00800x + 0.0800x^2$$

Collecting terms in x^2 and x, we have

$$0.9200x^2 + 0.10800x - 0.000200 = 0$$

This is a quadratic equation of the form $ax^2 + bx + c = 0$, where $a = 0.9200$, $b = 0.10800$, and $c = -0.000200$. The equation can be solved using the quadratic formula (Appendix A.7).

$$x = \frac{-b \pm \sqrt{b^2 - 4ac}}{2a} = \frac{-0.10800 \pm \sqrt{0.10800^2 - 4 \times 0.9200 \times (-0.000200)}}{2 \times 0.9200}$$

$$x = \frac{-0.10800 + 0.11135}{1.840} = 1.83 \times 10^{-3}$$

The other root, $x = -0.119$, can be eliminated because it would result in a negative concentration of I_2 at equilibrium, which is clearly impossible. From the equilibrium concentration row of the table,

$$[HI] = (0.100 - 2x)\text{mol/L} = 0.100 - (2 \times 1.83 \times 10^{-3}) = 0.0963 \text{ mol/L}$$

$$[H_2] = (0.100 + x)\text{mol/L} = 0.100 + (1.83 \times 10^{-3}) = 0.102 \text{ mol/L}$$

$$[I_2] = x \text{ mol/L} = 0.00183 \text{ mol/L}$$

✓ **Reasonable Answer Check** Solving this type of problem involves a lot of algebra, and it would be easy to make a mistake. It is very important to check the result by substituting the equilibrium concentrations into the equilibrium constant expression and verifying that the correct value of K_c (0.0200) results.

$$K_c = \frac{[H_2][I_2]}{[HI]^2} = \frac{(0.102)(0.00183)}{(0.0963)^2} = 0.0201$$

which is acceptable agreement because it differs from K_c by 1 in the last significant figure.

PROBLEM-SOLVING PRACTICE 14.7

Obtain the equilibrium constant for dissociation of dinitrogen tetraoxide to form nitrogen dioxide from Table 14.1. If 1.00 mol N_2O_4 and 0.500 mol NO_2 are initially placed in a container with a volume of 4.00 L, calculate the concentrations of N_2O_4(g) and NO_2(g) present when equilibrium is achieved at 25 °C.

14.6 Shifting a Chemical Equilibrium: Le Chatelier's Principle

Le Chatelier is pronounced "luh SHOT lee ay."

Suppose you are an environmental engineer, biologist, or geologist, and you have just measured the concentration of hydronium ion, H_3O^+, in a lake. You know that the H_3O^+ ions are involved in many different equilibrium reactions in the lake. How can you predict the influence of changing conditions? For example, what happens if there is a large increase in acid rainfall that has a hydronium ion concentration different from that of the lake? Or what happens if lime (calcium oxide), a strong base, is added to the lake? These questions and many others like them can be answered qualitatively by applying a useful guideline known as **Le Chatelier's principle:** ***If a system is at equilibrium and the conditions are changed so that it is no longer at equilibrium, the system will react to reach a new equilibrium in a way that partially counteracts the change.*** To adjust to a change, a system reacts in either the forward direction (producing more products) or the reverse direction (producing more reactants) until a new equilibrium state is achieved. Le Chatelier's principle applies to changes in the concentrations of reactants or products that appear in the equilibrium constant expression, the pressure or volume of a gas phase equilibrium, and the temperature. Changing conditions, thereby changing the equilibrium concentrations of reactants and products, is called **shifting an equilibrium.** If the reaction occurs in the *forward direction,* we say that the equilibrium reaction has *shifted to the right.* If the system reacts in the *reverse direction,* the reaction has *shifted to the left.*

Henri Le Chatelier (1850–1936) was a French chemist who, as a result of his studies of the chemistry of cement, developed his ideas about how altering conditions affects an equilibrium system.

Changing Concentrations of Reactants or Product

If the concentration of a reactant or a product that appears in the equilibrium constant expression is changed, a system can no longer be at equilibrium because Q must have a different value from K_c. The equilibrium will shift to use up a substance that was added, or to replenish a substance that was removed.

- If the concentration of a reactant is increased, the system will react in the forward direction.
- If the concentration of a reactant is decreased, the system will react in the reverse direction.
- If the concentration of a product is increased, the system will react in the reverse direction.
- If the concentration of a product is decreased, the system will react in the forward direction.

To see why this happens, consider the simple equilibrium discussed in Section 14.2.

$$\textit{cis}\text{-2-butene(g)} \rightleftharpoons \textit{trans}\text{-2-butene(g)} \qquad K_c = \frac{[\textit{trans}]}{[\textit{cis}]} = 1.5 \quad \text{(at 600 K)}$$

Suppose that 2 mmol *cis*-2-butene is placed into a 1.0-L container at 1000 K. The forward reaction will be faster than the reverse reaction until the concentration of *trans*-2-butene builds up to 1.5 times the concentration of *cis*-2-butene. Then equilibrium is achieved. Now suppose that half of the *cis*-2-butene is instantaneously removed from the container (see Exercise 14.3, p. 677). Because the concentration of *cis*-2-butene suddenly drops to half its former value, the forward reaction rate will also drop to half its former value. The reverse rate will not be affected, because the concentration of *trans*-2-butene has not changed. This means that the reverse reaction is twice as fast as the forward reaction, and *cis*-2-butene molecules are

As long as the temperature remains the same, the value of the equilibrium constant also remains the same. Adding or removing a reactant or a product does not change the equilibrium constant value. If the concentration of the substance added or removed appears in the equilibrium constant expression, however, the value of Q changes. Because Q no longer equals K_c, the system is no longer at equilibrium and must react to achieve a new equilibrium.

being formed twice as fast as they are being used up. Therefore the concentration of *cis*-2-butene will increase and the concentration of *trans*-2-butene will decrease until the forward and reverse rates are again equal. The graph in Figure 14.6 illustrates this situation.

The effect of changing concentration has many important consequences. For example, when the concentration of a substance needed by your body falls slightly, several enzyme-catalyzed chemical equilibria shift so as to increase the concentration of the essential substance. In industrial processes, reaction products are often continuously removed. This shifts one or more equilibria to produce more products and thereby maximize the yield of the reaction.

Hard water contains dissolved CO_2 and metal ions such as Ca^{2+} and Mg^{2+}.

In nature, slight changes in conditions are responsible for effects such as the formation of limestone stalactites and stalagmites in caves and the crust of limestone that slowly develops in a tea kettle if you boil hard water in it. Both of these examples involve calcium carbonate. Limestone, a form of calcium carbonate, $CaCO_3$, is present in underground deposits, a leftover of the ancient oceans from which it precipitated long ago or of ancient organisms, such as coral, that induced it to precipitate. Limestone reacts with an aqueous solution of CO_2 and dissolves.

$$CaCO_3(s) + CO_2(aq) + H_2O(\ell) \rightleftharpoons Ca^{2+}(aq) + 2\,HCO_3^-(aq) \qquad [14.7]$$

If groundwater that is saturated with CO_2 encounters a bed of limestone below the surface of the earth, the forward reaction can occur until equilibrium is reached, and the water subsequently contains significant concentrations of aqueous Ca^{2+} and HCO_3^- ions in addition to dissolved CO_2. These ions, often accompanied by $Mg^{2+}(aq)$, constitute hard water (Section 15.11).

ThomsonNOW
Go to the Chemistry Interactive menu for modules on:

- **Le Chatelier's principle: a water tank analogy**
- **effect of changing volume on the NO_2/N_2O_4 equilibrium**
- **effect of concentration changes on the butane/isobutane equilibrium**

As with all equilibria, a reverse reaction is occurring in addition to the forward reaction. This reverse reaction can be demonstrated by mixing aqueous solutions of $CaCl_2$ and $NaHCO_3$ (salts containing the Ca^{2+} and HCO_3^- ions) in an open beaker (Figure 14.8). You will eventually see bubbles of CO_2 gas and a precipitate of solid $CaCO_3$. Because the beaker is open to the air, any gaseous CO_2 that escapes the solution is swept away. Its removal reduces the concentration of $CO_2(aq)$ and causes the equilibrium to shift to the left to produce more CO_2. Eventually all of the dissolved Ca^{2+} and HCO_3^- ions disappear from the solution, having been converted to gaseous CO_2, solid $CaCO_3$, and water.

Suppose that water containing dissolved CO_2, Ca^{2+}, and HCO_3^- contacts the air in a cave (or hard water contacts air in your tea kettle). Carbon dioxide bubbles out

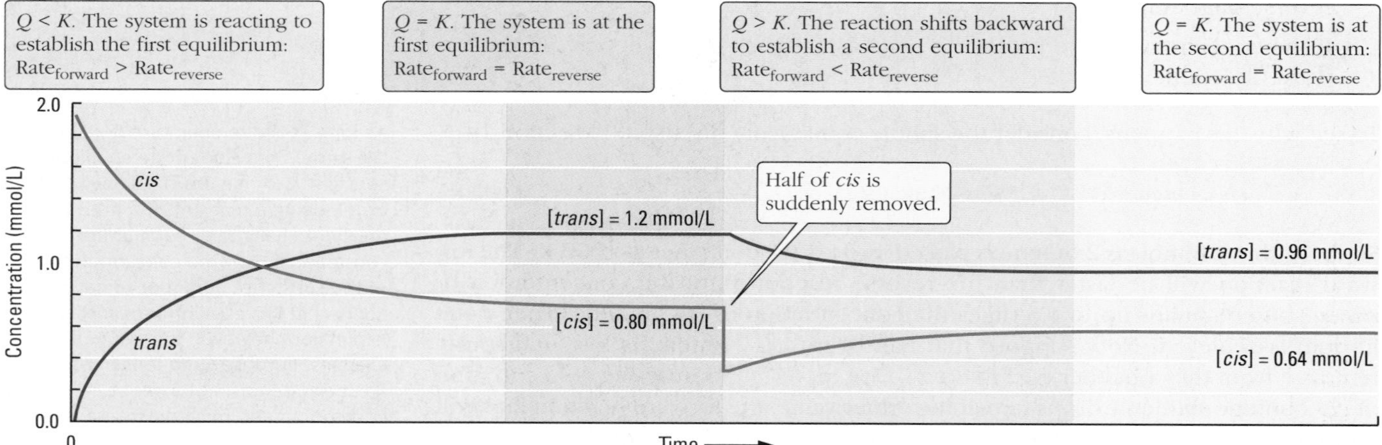

Figure 14.6 Approach to new equilibrium after a change in conditions.

of the solution, the concentration of CO_2 decreases on the reactant side of Equation 14.7, and the equilibrium shifts toward the reactants. The reverse reaction forms $CO_2(aq)$, compensating partially for the reduced concentration of $CO_2(aq)$. Some of the calcium ions and hydrogen carbonate ions combine, and some $CaCO_3(s)$ precipitates (Figure 14.7) as a beautiful formation in the cave (or as scale in your kettle).

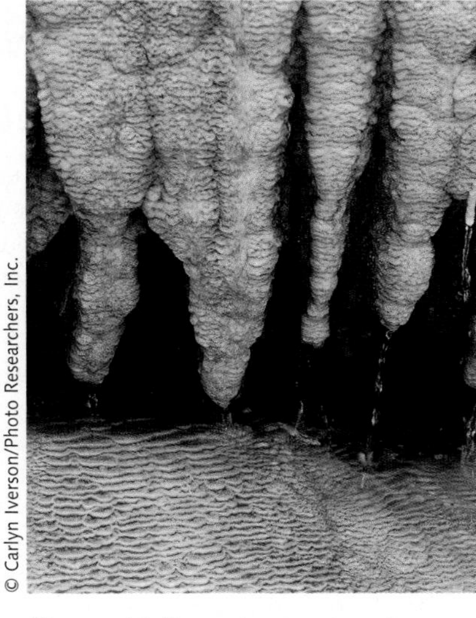

Figure 14.7 Stalactites in a limestone cave. Stalactites hang from the ceilings of caves. Stalagmites grow from the floors of caves up toward the stalactites. Both consist of limestone, $CaCO_3$. The process that produces these lovely formations is an excellent example of chemical equilibrium.

CONCEPTUAL EXERCISE 14.8 Effect of Adding a Substance

Solid phosphorus pentachloride decomposes when heated to form gaseous phosphorus trichloride and gaseous chlorine. Write the equation for the equilibrium that is set up when solid phosphorus pentachloride is introduced into a container, the container is evacuated and sealed, and the solid is heated. Once the system has reached equilibrium at a given temperature, what will be the effect on the equilibrium of the following?

(a) Adding chlorine to the container
(b) Adding phosphorus trichloride to the container
(c) Adding a small quantity of phosphorus pentachloride to the container

Changing Volume or Pressure in Gaseous Equilibria

One way to change the pressure of a gaseous equilibrium mixture is to keep the volume constant and to add or remove one or more of the substances whose concentrations appear in the equilibrium constant expression. The effect of adding or removing a substance has just been discussed. We consider here other ways of changing pressure or volume.

The pressures of all substances in a gaseous equilibrium can be changed by changing the volume of the container. Consider the effect of tripling the pressure on the equilibrium

$$N_2O_4(g) \rightleftharpoons 2\,NO_2(g) \qquad K_c = \frac{[NO_2]^2}{[N_2O_4]}$$

ThomsonNOW
Go to the Coached Problems menu for a tutorial on **disturbing a chemical equilibrium.**

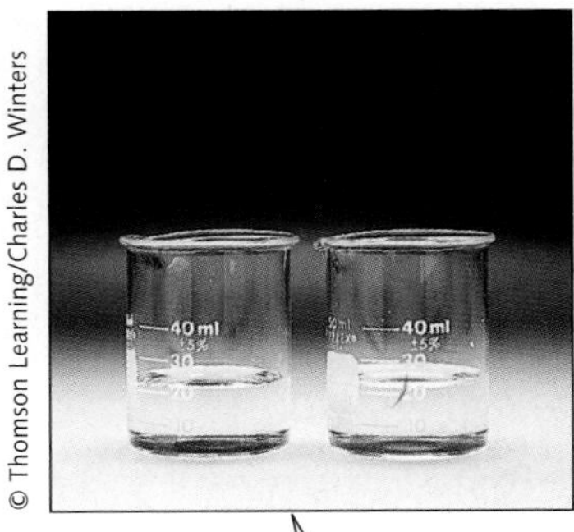

① These two salt solutions provide Ca^{2+} and HCO_3^- ions, which are products in the net ionic equation for the reaction of calcium carbonate with carbon dioxide.

② Bringing the reaction products together by mixing the two solutions shifts the equilibrium toward the reactants.

③ The CO_2 produced by the reverse reaction bubbles out of the solution into the air, thereby reducing a reactant concentration in solution and shifting the equilibrium to the left.

④ Therefore the reverse reaction continues until almost all of the Ca^{2+} and HCO_3^- ions have reacted to form CO_2, $CaCO_3$, and water.

Figure 14.8 Reaction of $CaCl_2(aq)$ with $NaHCO_3(aq)$.

CHEMISTRY IN THE NEWS

Bacteria Communicate Chemically

Stories about fish fingers that glowed in the dark after being stored too long in a dorm-room fridge sound like urban legends, but they may well be true. Certain bacteria, such as *Vibrio fischeri*, are bioluminescent when a colony gets big enough, and these bacteria are present in many fish. As a fish decomposes, the bacteria grow rapidly and may produce a green glow.

Rapid growth of many forms of bacteria involves communication among individual bacterial cells. Only after a colony reaches a certain size and the bacteria in the colony realize it has reached that size is there really rapid growth. In this group state, pathogenic bacteria initiate the majority of infections in plants and animals. How do bacteria sense when they have reached an adequate group size? And could bacterial infections be treated by preventing bacteria from realizing they actually are in a group? Answers to these questions could be very beneficial to humankind, and scientists are working to find them.

Gram negative bacteria communicate by secreting so-called signal molecules that can diffuse out of one bacterial cell and into another. In the case of *V. fischeri*, the signal molecule is an acyl-homoserine lactone (AHL) and it is produced by a protein designated LuxI. The AHL molecules diffuse out of the cell in which they were produced and can enter other *V. fischeri* cells, if any are nearby. When the signal molecule enters a cell, it can bind with another protein, LuxR, in an equilibrium process:

$$\text{LuxR} + \text{AHL} \rightleftharpoons \text{LuxR—AHL}$$

If only a few *V. fischeri* cells are close together, most of the AHL diffuses away, and its concentration remains small. But once a bacterial colony reaches a certain size, most of the cells are surrounded by other cells, all of which are secreting AHL. The increased concentration of AHL shifts the equilibrium to the right. This forms sizable concentrations of the LuxR—AHL combination, which binds to DNA and switches on genes that encode the enzyme luciferase. In the presence of O_2, luciferase produces luminescence like that of a firefly. The LuxR—AHL combination also switches on genes that code for formation of LuxI, which produces more AHL. Therefore, AHL is said to *autocatalyze* its own synthesis, and this explains why the colony of bacteria suddenly grows very rapidly (and produces the green glow) after an initial period of slow growth.

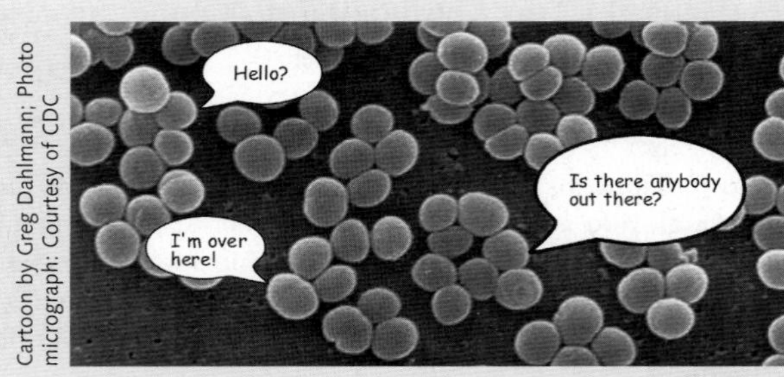

Cartoon by Greg Dahlmann; Photo micrograph: Courtesy of CDC

Bacterial communication.

This process is referred to as *quorum sensing,* because it requires a significant number of bacteria (a quorum) to initiate the autocatalysis of formation of AHL. Many bacteria display quorum-sensing behavior, and often their rapid growth has much worse consequences than a green-glowing fish stick. Quorum-sensing behavior is thought to account for more than 50% of all crop diseases and as much as 80% of all human bacterial infections. Therefore scientists are

by reducing the volume of the container to one third of its original value (at constant temperature). This situation is shown in Figure 14.9 (p. 699). Decreasing the volume increases the pressures of N_2O_4 and NO_2 to three times their equilibrium values. Decreasing the volume also increases the concentrations of N_2O_4 and NO_2 to three times their equilibrium values. Because $[NO_2]$ is squared in the equilibrium constant expression but $[N_2O_4]$ is not, tripling both concentrations increases the numerator of Q by $3^2 = 9$ but increases the denominator by only 3.

$$Q = \frac{(\text{conc. } NO_2)^2}{(\text{conc. } N_2O_4)} = \frac{(3 \times [NO_2])^2}{(3 \times [N_2O_4])} = \frac{9}{3} \times \frac{[NO_2]^2}{[N_2O_4]} = 3 \times K_c$$

Because Q is larger than K_c under the new conditions, the reaction should produce more reactant; that is, the equilibrium should shift to the left.

The same prediction is made using Le Chatelier's principle: The reaction should shift to partially compensate for the increase in pressure. That means decreasing the pressure, which can happen if the total number of gas phase molecules decreases. In the case of the N_2O_4/NO_2 equilibrium, the reverse reaction should occur, because one N_2O_4 molecule is produced for every two NO_2 molecules that react. A shift to the left reduces the number of gas phase molecules and hence the pressure.

Remember that the pressure of an ideal gas is proportional to the number of moles of gas, and therefore to the number of molecules of gas (*p. 445*).

Photo by Jason Varney Photography, Courtesy of Helen Blackwell

Helen Blackwell.

looking for a way to prevent quorum-sensing equilibria from shifting right and forming the enzyme–signal molecule combination. Substances that could do this are called quorum-sensing antagonists. They have great potential to alleviate problems of bacterial resistance to antibiotics and could bring great benefit to humankind.

One way to find quorum-sensing antagonists is being explored by the research group of Professor Helen Blackwell of the University of Wisconsin-Madison. Blackwell and her co-workers make many different compounds whose molecular structures are closely related and test those compounds for their ability to inhibit quorum-sensing behavior in bacteria. To help make the compounds Blackwell uses a laboratory version of a microwave oven such as those found in kitchens to heat food. Microwave irradiation heats reactant chemicals quickly and results in greater yields of higher-purity products. In a single day Blackwell's research group can prepare as many as 100–500 related compounds. These can subsequently be tested for activity, and the rate at which new biologically active compounds can be discovered is greatly accelerated. Recently Blackwell and her students have identified quorum-sensing antagonists that are among the most potent discovered to date.

Knowledge of quorum sensing also allows chemists to enhance the growth of bacteria, and in some cases that may be a good thing. For example, changing the concentrations of autocatalyst substances might be a way of controlling the differentiation of stem cells into different type of tissues. Or, a system can be imagined in which bacterial populations could be maintained at a desired level. A quorum-sensing signal molecule could turn on a suicide gene if the colony got too large, but when the population decreased the signal concentration would become smaller, thus turning off the gene and not completely wiping out the colony. Exploring the reasons for that green glow in the refrigerator might have incredibly beneficial consequences.

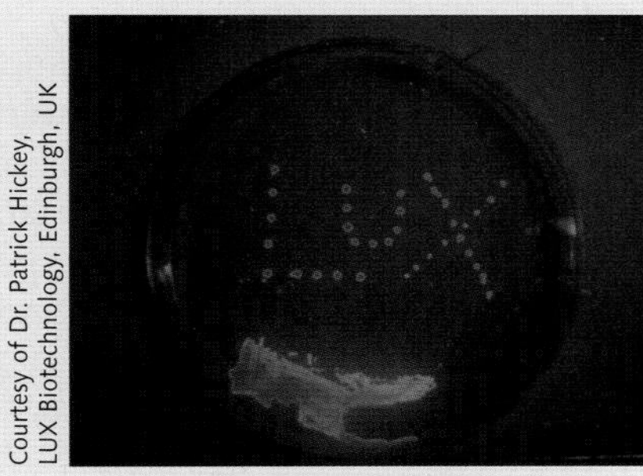
Courtesy of Dr. Patrick Hickey, LUX Biotechnology, Edinburgh, UK

Glowing *Vibrio fischeri*.

SOURCES: Based on information from *New York Times*, January 17, 2006; *The Health Show*, April 21, 2005, **http://www.healthshow.org/archive/week_2005_04_17.shtml;** Blackwell laboratory, **http://www.chem.wisc.edu/blackwell/index.htm;** and *New York Times*, February 27, 2001.

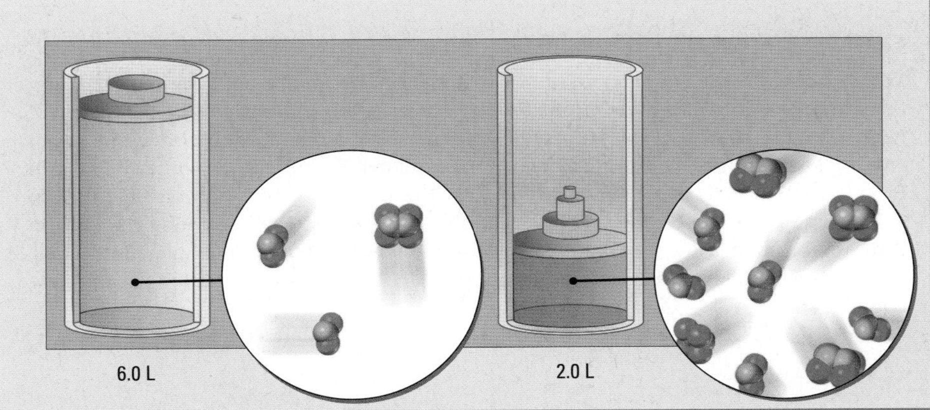

Figure 14.9 Shifting an equilibrium by changing pressure and volume. If the volume of an equilibrium mixture of NO_2 and N_2O_4 is decreased from 6.0 to 2.0 L, the equilibrium shifts toward the smaller number of molecules in the gas phase (the N_2O_4 side) to partially compensate for the increased pressure. When the new equilibrium is achieved, the concentrations of both NO_2 and N_2O_4 have increased, but the NO_2 concentration is less than three times as great, while the N_2O_4 concentration is more than three times as great. The total pressure goes from 1.00 to 2.62 atm (rather than to 3.00 atm, which was the new pressure before the equilibrium shifted).

However, consider the situation with respect to another equilibrium we have already mentioned in Problem-Solving Example 14.3 (p. 684). At 425 °C all substances are in the gas phase.

$$2\ HI(g) \rightleftharpoons H_2(g) + I_2(g)$$

Suppose that the pressure of this system were tripled by reducing its volume to one third of the original volume. What would happen to the equilibrium? In this case, all the concentrations triple, but because equal numbers of moles of gaseous substances appear on both sides of the equation, Q still has the same numeric value as K_c. That is, the system is still at equilibrium, and no shift occurs. Thus, *changing pressure by changing the volume shifts an equilibrium only if the sum of the coefficients for gas phase reactants is different from the sum of the coefficients for gas phase products.*

Finally, consider what happens if the pressure of the N_2O_4/NO_2 equilibrium system is increased by adding an inert gas such as nitrogen while retaining exactly the same volume. The total pressure of the system would increase, but since neither the amounts of N_2O_4 and NO_2 nor the volume would change, the concentrations of N_2O_4 and NO_2 would remain the same. Q still equals K_c, the system is still at equilibrium, and no shift occurs (or needs to). Thus, changing the pressure of an equilibrium system must change the concentration of at least one substance in the equilibrium constant expression if a shift in the equilibrium is to occur.

CONCEPTUAL EXERCISE 14.9 Changing Volume Does Not Always Shift an Equilibrium

In Problem-Solving Example 14.7 you found that for the reaction

$$2\ HI(g) \rightleftharpoons H_2(g) + I_2(g) \qquad K_c = 0.0200 \quad (\text{at } 745\ K)$$

the equilibrium concentrations were [HI] = 0.0963 mol/L, $[H_2]$ = 0.102 mol/L, and $[I_2]$ = 0.00183 mol/L. Use algebra to show that if each of these concentrations is tripled by reducing the volume of this equilibrium system to one third of its initial value, the system is still at equilibrium, and therefore the pressure change causes no shift in the equilibrium.

EXERCISE 14.10 Effect of Changing Volume

Verify the statement in the text that, for the N_2O_4/NO_2 equilibrium system, decreasing the volume to one third of its original value increases the equilibrium concentration of N_2O_4 by more than a factor of 3 while it increases the equilibrium concentration of NO_2 by less than a factor of 3. Start with the equilibrium conditions you calculated in Problem-Solving Practice 14.7. Then decrease the volume of the system from 4.00 L to 1.33 L, and calculate the new concentrations of N_2O_4 and NO_2 assuming the shift in the equilibrium has not yet taken place. Set up the usual table of initial concentrations, change, and equilibrium concentrations, and calculate the concentrations at the new equilibrium.

ThomsonNOW™
Go to the Coached Problems menu for tutorials on:

- **effect of temperature on equilibrium**
- **effect of addition or removal of a reagent**
- **equilibria and volume**
- **temperature dependence of the equilibrium of cobalt chloride**

Changing Temperature

When temperature changes, the values of most equilibrium constants also change, and systems will react to achieve new equilibria consistent with new values of K_c. You can make a qualitative prediction about the effect of temperature on an equilibrium if you know whether the reaction is exothermic or endothermic. As an

ESTIMATION

Generating Gaseous Fuel

The reaction of coke (mainly carbon) with steam is called the water-gas reaction. It was used for many years to generate gaseous fuel from coal. The thermochemical expression is

$$C(s) + H_2O(g) \rightleftharpoons CO(g) + H_2(g) \qquad \Delta H° = 131.293 \text{ kJ}$$

The equilibrium constant K_p has the value 9.5×10^{-17} at 298 K, 1.9×10^{-7} at 500 K, 2.60×10^{-2} at 800 K, 18.8 at 1200 K, and 500. at 1600 K. Suppose that you have an equilibrium mixture in which the partial pressure of steam is 1.00 atm. Estimate the temperature at which the partial pressures of CO and H_2 would also equal 1 atm. (The reaction would need to be carried out at temperatures roughly this high to produce appreciable quantities of products.)

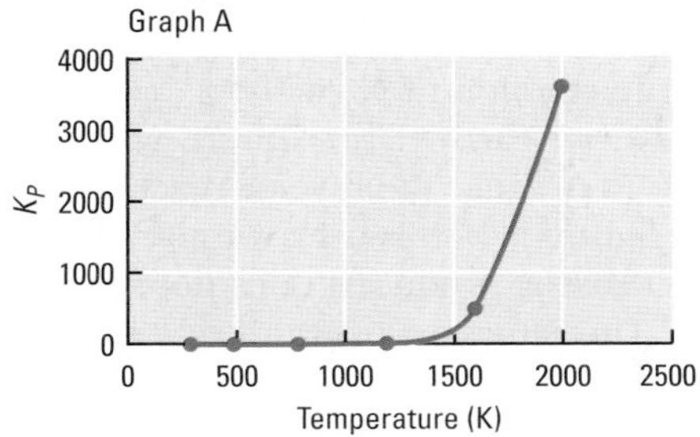

A good way to make this estimation is to plot the data and see which temperature corresponds to $K_p = 1$. This has been done in Graph A. It is clear that the range of values for the equilibrium constant is so wide that an accurate temperature cannot be read corresponding to $K_p = 1$. An estimate would be easier to make if the graph were a straight line. Because the values of K_p rise very rapidly, it is possible that taking the logarithm of each K_p value would generate a linear graph. This has been done in Graph B, where the natural logarithm of the equilibrium constant, $\ln(K_p)$, has been plotted on the vertical axis. Since $\ln(K_p) = 0$ when $K_p = 1$, we are looking for the temperature at which the graph crosses zero on the vertical axis. Although the graph is not linear, the appropriate temperature can be read from the graph and is slightly less than 1000 K.

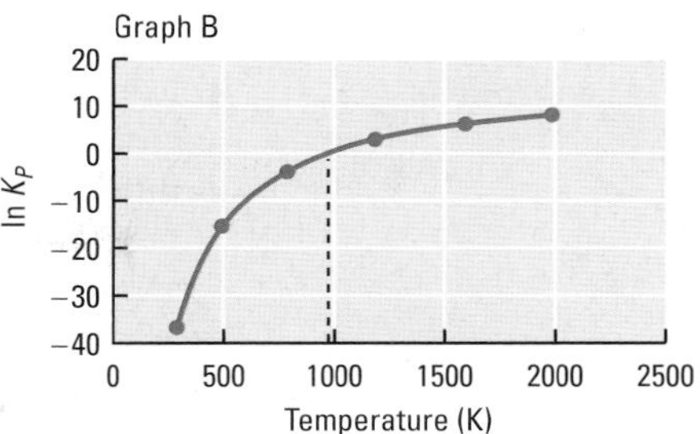

Experimenting with different functions of K_p on the vertical axis and different functions of T on the horizontal axis reveals that a linear graph is obtained when $\ln(K_p)$ is plotted against $1/T$, as in Graph C. From this graph the equation of the straight line is found to be

$$\ln(K_P) = (-1.58 \times 10^4 \text{ K})\left(\frac{1}{T}\right) + 16.1$$

Substituting $K_p = 1$, that is, $\ln(K_p) = 0$, we find that

$$T = \frac{1.58 \times 10^4 \text{ K}}{16.1} = 980 \text{ K}$$

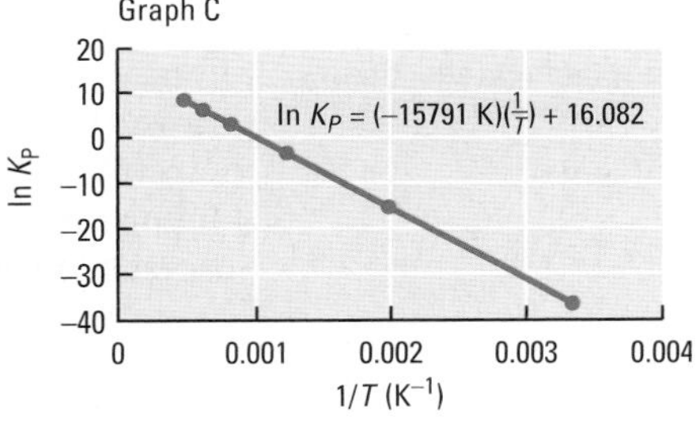

example, consider the endothermic, gas phase reaction of N_2 with O_2 to give nitrogen monoxide, NO.

$$N_2(g) + O_2(g) \rightleftharpoons 2\,NO(g) \qquad \Delta H° = 180.5 \text{ kJ}$$

$$K_c = 4.5 \times 10^{-31} \quad \text{at 298 K}$$

$$K_c = \frac{[NO]^2}{[N_2][O_2]} \qquad K_c = 6.7 \times 10^{-10} \quad \text{at 900 K}$$

$$K_c = 1.7 \times 10^{-3} \quad \text{at 2300 K}$$

In this case the equilibrium constant increases very significantly as the temperature increases. At 298 K the equilibrium constant is so small that essentially no reaction occurs. Suppose that a room-temperature equilibrium mixture were suddenly heated to 2300 K. What would happen? The equilibrium should shift to partially compensate for the temperature increase. This happens if the reaction shifts in the endothermic direction, since that change would involve a transfer of energy into the reaction system, cooling the surroundings. For the $N_2 + O_2$ reaction the forward process is endothermic ($\Delta H° > 0$), so N_2 and O_2 should react to produce more NO. At the new equilibrium, the concentration of NO should be higher and the concentrations of N_2 and O_2 should be lower. Since this makes the numerator in the K_c expression bigger and the denominator smaller, K_c should be larger at the higher temperature. This outcome corresponds with the experimental result.

The effect on K_c of raising the temperature of the $N_2 + O_2$ reaction leads us to a general conclusion: ***For an endothermic reaction, an increase in temperature always means an increase in K_c; the reaction will become more product-favored at higher temperatures.*** When equilibrium is achieved at the higher temperature, the concentration of products is greater and that of the reactants is smaller. Likewise, and as illustrated by Problem-Solving Example 14.8, the opposite is true for an exothermic reaction: ***For an exothermic reaction, an increase in temperature always means a decrease in K_c; the reaction will become less product-favored at higher temperatures.***

Another consequence of the shift toward products in the $N_2 + O_2$ reaction at high temperatures is that automobile engines emit small concentrations of NO. The NO is rapidly oxidized to brown NO_2 in the air above cities, and the NO_2 in turn produces many further reactions that create air pollution problems. Air pollution involving NO was discussed in Section 10.12 (⇐ ***p. 466***). One of the functions of catalytic converters in automobiles is to speed up reduction of these nitrogen oxides back to elemental nitrogen.

The effect of temperature on the reaction of N_2 with O_2 has important consequences. This reaction produces NO in earth's atmosphere when lightning suddenly raises the temperature of the air. Because the reverse reaction is slow at room temperature, and because after the lightning bolt is over, the air rapidly cools to normal temperatures, much of the NO that is produced does not react to re-form N_2 and O_2 as it would at equilibrium. This situation provides one natural mechanism by which nitrogen in the air can be converted into a form that can be used by plants. (Converting nitrogen into a useful form is called nitrogen fixation; see Section 21.3.) Humans have tried to use the same kind of process to produce NO and from it HNO_3 for use in manufacturing fertilizer. At the end of the nineteenth century, a chemical plant at Niagara Falls, New York (where there was plentiful electric power), operated an electric arc process for fixing nitrogen for several years. This electric arc plant was important because it was the first attempt to deal with the limitations on plant growth caused by lack of sufficient nitrogen in soils that had been heavily farmed. About a century ago scientists and many in the general public were worried that the earth's farmland could not grow enough food to support a growing population. Consequently, strenuous efforts were made to adjust the conditions of the $N_2 + O_2$ reaction so that significant yields of NO could be obtained. The very high temperature of a lightning bolt or an electric arc was one way to do this.

PROBLEM-SOLVING EXAMPLE 14.8 Le Chatelier's Principle

Consider an equilibrium mixture of nitrogen, hydrogen, and ammonia in which the reaction is

$$N_2(g) + 3\,H_2(g) \rightleftharpoons 2\,NH_3(g) \qquad \Delta H° = -92.2 \text{ kJ} \quad \text{at } 25\ °C$$

For each of the changes listed below, tell whether the value of K_c increases or decreases, and tell whether more NH_3 or less NH_3 is present at the new equilibrium established after the change.

(a) More H_2 is added (at a constant temperature of 25 °C and constant volume).
(b) The temperature is increased.
(c) The volume of the container is doubled (at constant temperature).

Answer

(a) K_c stays the same; more NH_3 is present.
(b) K_c decreases; less NH_3 is present.
(c) K_c stays the same; less NH_3 is present.

Strategy and Explanation

(a) Since the temperature does not change, the value of K_c does not change. Adding a reactant to the equilibrium mixture will shift the equilibrium toward the product, producing more NH_3. This can be seen in another way by considering the reaction quotient.

$$Q = \frac{(\text{conc. } NH_3)^2}{(\text{conc. } N_2)(\text{conc. } H_2)^3}$$

When more H_2 is added, the denominator becomes larger. This makes Q smaller than K_c, which predicts that the reaction should produce more product; that is, some of the added H_2 reacts with N_2 to make more NH_3. Notice also that the concentration of N_2 decreases, because some reacts with the H_2, but the concentration of H_2 in the new equilibrium is still greater than it was before.

(b) The reaction is exothermic. Increasing the temperature shifts the equilibrium in the endothermic direction—that is, to the left (toward the reactants). This shift leads to a decrease in the NH_3 concentration, an increase in the concentrations of H_2 and N_2, and a decrease in the value of K_c.

(c) Since the temperature is constant, the value of K_c must be constant. Doubling the volume should cause the reaction to shift toward a greater number of moles of gaseous substance—that is, toward the left. Doubling the volume would normally halve the pressure, but the shift of the equilibrium partially compensates for this effect, and the final equilibrium will be at a pressure somewhat more than half the pressure of the initial equilibrium.

PROBLEM-SOLVING PRACTICE 14.8

Consider the equilibrium between N_2O_4 and NO_2 in a closed system.

$$N_2O_4(g) \rightleftharpoons 2\ NO_2(g)$$

Draw Lewis structures for the molecules involved in this equilibrium. Based on the bonding in the molecules, predict whether the reaction is exothermic or endothermic; hence, predict whether the concentration of N_2O_4 is larger in an equilibrium system at 25 °C or at 80 °C. Verify your prediction by looking at Figure 14.3.

CONCEPTUAL EXERCISE 14.11 Summarizing Le Chatelier's Principle

Construct a table to summarize your understanding of Le Chatelier's principle. Consider these changes in conditions:

(a) Addition of a reactant
(b) Removal of a reactant
(c) Addition of a product
(d) Removal of a product
(e) Increasing pressure by decreasing volume
(f) Decreasing pressure by increasing volume
(g) Increasing temperature
(h) Decreasing temperature

For each of these changes in conditions, indicate (1) how the reaction system changes to achieve a new equilibrium, (2) in which direction the equilibrium reaction shifts, and (3) whether the value of K_c changes and, if so, in which direction. For some of these changes there are qualifications. For example, increasing pressure by decreasing volume does not always shift an equilibrium. List as many of these qualifications as you can.

14.7 Equilibrium at the Nanoscale

In Section 14.2 (p. 675), we used the isomerization of *cis*-2-butene to show that both forward and reverse reactions occur simultaneously in an equilibrium system.

$$cis\text{-2-butene(g)} \rightleftharpoons trans\text{-2-butene(g)} \qquad K_c = \frac{[trans]}{[cis]} = 2.0 \qquad \text{(at 415 K)}$$

Because the equilibrium constant is 2.0, there are twice as many *trans* molecules as *cis* molecules in an equilibrium mixture at 415 K. In other words, two thirds of the molecules have the *trans* structure and one third have the *cis* structure. Because the molecules are continually reacting in both the forward and the reverse directions, another way to think about this situation is that each molecule is *trans* two thirds of the time and *cis* one third of the time.

Based on its molecular structure, you might think that a 2-butene molecule ought to be just as likely to be *cis* as *trans,* so why is the *trans* isomer favored at 415 K? In Figure 13.7 ***(← p. 627)*** we noted that one molecule of the *trans* isomer is 7×10^{-21} J lower in energy than one molecule of the *cis* isomer, and this difference is all important. For a mole of molecules, the difference in energy is $(-7 \times 10^{-21}\ \text{J})(6.022 \times 10^{23}\ \text{mol}^{-1}) = -4 \times 10^3\ \text{J/mol} = -4\ \text{kJ/mol}$, giving the thermochemical expression

$$cis\text{-2-butene(g)} \rightleftharpoons trans\text{-2-butene(g)} \qquad \Delta H^\circ = -4\ \text{kJ}$$

Consider the rate constants for the forward and reverse reactions in the isomerization of 2-butene. Based on Figure 13.7, E_a (forward) = 262 kJ/mol and E_a (reverse) = 266 kJ/mol. This means that the rate constant for the reverse reaction will be smaller than for the forward reaction. At equilibrium the forward and reverse rates are equal, so

Larger k value and smaller concentration | Smaller k value and larger concentration

$$k_{\text{forward}} \times [cis] = k_{\text{reverse}} \times [trans]$$

If k_{reverse} is smaller than k_{forward}, then the concentration of *trans* must be larger than the concentration of *cis,* or the rates would not be equal. Because *cis*-2-butene is 4 kJ/mol higher in energy than *trans*-2-butene, it occurs only half as often as *trans*-2-butene at 415 K. We can generalize that ***in an equilibrium system, molecules that are higher in energy occur less often.*** We shall refer to this tendency as the energy effect on the position of equilibrium.

A second energy factor also affects how large an equilibrium constant is. It depends on how spread out in space is the total energy of products compared to reactants and can be illustrated by the dissociation of dinitrogen tetraoxide.

$$N_2O_4(g) \rightleftharpoons 2\ NO_2(g) \qquad \Delta H^\circ = 57.2\ \text{kJ}$$

In a constant-pressure system (such as a reaction at atmospheric pressure), the two moles of product molecules occupy twice the volume of the one mole of reactant molecules. This means that the energy of the NO_2 molecules is spread over twice the volume that would have been occupied by the N_2O_4 molecules. A thermodynamic quantity called **entropy** provides a quantitative measure of how much energy is spread out when something happens. Entropy will be defined more completely in Section 18.3. For now it suffices to say that ***if there are more product molecules than reactant molecules, entropy favors the products in an equilibrium system.*** Although it depends on spreading out of energy, we shall refer to this tendency as the entropy effect on the position of equilibrium.

Despite the fact that the entropy factor favors NO_2 in the dissociation of N_2O_4, when data from Table 14.1 are used to calculate K_c at 25 °C (298 K) the result is $1/(1.7 \times 10^2) = 5.9 \times 10^{-3}$. This is because the dissociation of N_2O_4 is endothermic ($\Delta H° = 57.2$ kJ). The enthalpy of 2 mol NO_2 is 57.2 kJ higher than the enthalpy of 1 mol N_2O_4. From our first generalization that molecules with more energy occur less often, we expect that 2 NO_2 is less likely than N_2O_4 because of the energy effect. Therefore the equilibrium constant should be smaller than the entropy argument predicts. ***Both the energy effect and the entropy effect must be taken into account to predict an equilibrium constant value.***

These ideas can also help us to understand the temperature dependence of the equilibrium constant. K_c values for the dissociation of N_2O_4 are given in the table in the margin. As the temperature rises, K_c becomes much larger. At high temperatures the molecules have lots of energy, and when the volume doubles, much more energy is dispersed. The energy difference between reactants and products (ΔH) is smaller relative to the average energy per molecule. ***The higher the temperature is, the less important the energy effect (ΔH) becomes, and the more the entropy effect (dispersal of energy) determines the position of equilibrium.***

We have shown that if the number of gas phase molecules increases when a reaction occurs, then the products will be favored by entropy. This is not the only way for the products of a reaction to have greater entropy than the reactants, but it is one of the most important. In Chapter 18 we will show how entropy can be measured and tabulated. That discussion will enable us to use enthalpy changes and entropy changes for a reaction to calculate its equilibrium constant at a variety of temperatures.

$N_2O_4(g) \rightleftharpoons 2\,NO_2(g)$

T (K)	K_c
298	5.9×10^{-3}
350	1.3×10^{-1}
400	1.5×10^{0}
500	4.6×10^{1}
600	4.6×10^{2}

PROBLEM-SOLVING EXAMPLE 14.9 Energy and Entropy Effects on Equilibria

For the equilibrium

$$CH_4(g) + H_2O(g) \rightleftharpoons CO(g) + 3\,H_2(g)$$

(a) Estimate whether the entropy increases, decreases, or remains the same when products form.
(b) Does the entropy effect favor reactants or products?
(c) Use data from Appendix J to calculate $\Delta H°$.
(d) Does the energy effect favor reactants or products?
(e) Is the reaction likely to be product-favored at high temperatures? Why or why not?

Answer

(a) Entropy increases
(b) Entropy favors products
(c) $\Delta H° = 206.10$ kJ
(d) Reactants
(e) Yes. Entropy favors products and the energy effect is less important at high temperatures.

Strategy and Explanation Consider how many gas phase reactants and products there are and apply the ideas about entropy in the previous paragraphs.

(a) Because there is 4 mol gas phase products and only 2 mol gas phase reactants, the products will have higher entropy.
(b) The products have higher entropy, and the products are favored.
(c)
$$\begin{aligned}\Delta H° &= \sum\{(\text{moles product})\Delta H_f°(\text{product})\} - \sum\{(\text{moles reactant})\Delta H_f°(\text{reactant})\}\\ &= \{(1\text{ mol})(-110.525\text{ kJ/mol})\}\\ &\quad - \{(1\text{ mol})(-74.81\text{ kJ/mol}) + (1\text{ mol})(-241.818\text{ kJ/mol})\}\\ &= -110.525\text{ kJ} - (-316.628\text{ kJ})\\ &= 206.10\text{ kJ}\end{aligned}$$

(d) Because the reactants are lower in energy, they are favored by the energy effect.

(e) The reaction is product-favored at high temperatures because the energy effect favoring reactants becomes less important, and there is a relatively large entropy effect (four product molecules for every two reactant molecules in the gas phase).

✓ **Reasonable Answer Check** In part (c) it is reasonable that the reaction is endothermic, because six bonds are broken and only four are formed when reactant molecules are changed into product molecules.

PROBLEM-SOLVING PRACTICE 14.9

For the reaction

$$2\ SOCl_2(g) + O_2(g) \rightleftharpoons 2\ SO_2(g) + 2\ Cl_2(g)$$

(a) Does the entropy effect favor products?

(b) Does the energy effect favor products?

(c) Will there be a greater concentration of $SO_2(g)$ at high temperature or at low temperature? Explain.

14.8 Controlling Chemical Reactions: The Haber-Bosch Process

The principles that allow us to control a reaction are based on our understanding of both equilibrium systems and the rates of chemical reactions. Some generalizations about equilibrium systems are

- ***A product-favored reaction has an equilibrium constant larger than 1.***
- ***If a reaction is exothermic, this energy factor favors the products.***
- ***If there is an increase in entropy when a reaction occurs, this entropy factor favors the products.***
- ***Product-favored reactions at low temperatures are usually exothermic.***
- ***Product-favored reactions at high temperatures are usually ones in which the entropy increases (energy becomes more dispersed).***

Using these general rules about equilibria, we can often predict whether a reaction is capable of yielding products. But it is also important that those products be produced rapidly. Recall these useful generalizations about reaction rates from Chapter 13.

- ***Reactions in the gas phase or in solution, where molecules of one reactant are completely mixed with molecules of another, occur more rapidly than do reactions between pure liquids or solids that do not dissolve in one another (← p. 608).***
- ***Reactions occur more rapidly at high temperatures than at low temperatures (← p. 631).***
- ***Reactions are faster when the reactant concentrations are high than when they are low (← p. 614).***
- ***Reactions between a solid and a gas, or between a solid and something dissolved in solution, are usually much faster when the solid particles are as small as possible (← p. 608).***
- ***Reactions are faster in the presence of a catalyst. Often the right catalyst makes the difference between success and failure in industrial chemistry (← p. 642).***

For gases, higher concentration corresponds to higher partial pressure.

Because finely divided solids react more rapidly, coal is ground to a powder before it is burned to generate electricity.

The current interest in biotechnology is largely driven by the fact that naturally occurring enzymes are among the most effective catalysts known (← p. 645).

One of the best examples of the application of the principles of chemical reactivity is the chemical reaction we now use for the synthesis of ammonia from its elements. Even though the earth is bathed in an atmosphere that is about 80% N_2 gas, nitrogen cannot be used by most plants until it has been fixed—that is, converted into biologically useful forms. Although nitrogen fixation is done naturally by organisms such as cyanobacteria and some field crops such as alfalfa and soybeans, most plants cannot fix N_2. They must instead obtain nitrogen from cyanobacteria, some other organism, or fertilizer. Proper fertilization is especially important for recently developed varieties of wheat, corn, and rice that have resulted in much improved food production.

It is estimated that 40–60% of the nitrogen in the average human body has come from ammonia produced by the Haber-Bosch process and that increased agricultural productivity resulting from this process supports about 40% of the world's population.

Direct combination of nitrogen and oxygen was used at the beginning of the twentieth century to provide fertilizer, but this process was not very efficient (p. 702). A much better way of manufacturing ammonia was devised by Fritz Haber and Carl Bosch, who chose the *direct synthesis of ammonia from its elements* as the basis for an industrial process.

$$N_2(g) + 3\,H_2(g) \rightleftharpoons 2\,NH_3(g)$$

EXERCISE 14.12 Ammonia Synthesis

For the ammonia synthesis reaction, predict

(a) Whether the reaction is exothermic or endothermic.
(b) Whether the reaction product is favored by entropy.
(c) Whether the reaction produces more products at low or high temperatures.
(d) What would happen if you tried to increase the rate of the reaction by increasing the temperature.

Oesper Collection in the History of Chemistry, University of Cincinnati

Fritz Haber
1868–1934

The industrial chemical process by which ammonia is manufactured was developed by Fritz Haber, a chemist, and Carl Bosch, an engineer. Haber's studies in the early 1900s revealed that direct ammonia synthesis should be possible. In 1914 the engineering problems were solved by Bosch. Haber's contract with the manufacturer of ammonia called for him to receive 1 pfennig (one hundredth of a German mark—similar to a penny) per kilogram of ammonia, and he soon became not only famous but also rich. In 1918 he was awarded a Nobel Prize for the ammonia synthesis, but the choice was criticized because of his role in developing the use of poison gases for Germany during World War I.

At first glance this reaction might seem to be a poor choice. Hydrogen is available naturally only in combined form—for example, in water or hydrocarbons—meaning that hydrogen must be extracted from these compounds at considerable expense in energy resources and money. As you discovered in Problem-Solving Practice 14.9, the ammonia synthesis reaction becomes less product-favored at higher temperatures. But higher temperatures are needed for ammonia to be produced fast enough for the process to be efficient and economical. Nonetheless, the **Haber-Bosch process** (shown schematically in Figure 14.10) has been so well developed that ammonia is very inexpensive (less than \$300 per ton). For this reason it is widely used as a fertilizer and is often among the "top five" chemicals produced in the United States. Annual U.S. production of NH_3 by the Haber-Bosch process is approximately 30 billion pounds.

Both the thermodynamics and the kinetics of the direct synthesis of ammonia have been carefully studied and fine-tuned by industry so that the maximum yield of product is obtained in a reasonable time and at a reasonable cost of both money and energy resources.

- The reaction is exothermic, and there is a decrease in entropy when it takes place. Therefore this reaction is predicted to be product-favored at low temperatures, but reactant-favored at high temperatures. (You should have made this prediction in Problem-Solving Practice 14.9, and it is in accord with Le Chatelier's principle for an exothermic process.)
- To increase the equilibrium concentration of NH_3, the reaction is carried out at high pressure (200 atm). This does not change the value of K_c, or K_P, but an increase in pressure can be compensated for by converting N_2 and H_2 to NH_3; 2 mol $NH_3(g)$ exerts less pressure than a total of 4 mol gaseous reactants [$N_2(g)$ + 3 $H_2(g)$] in the same-sized container.

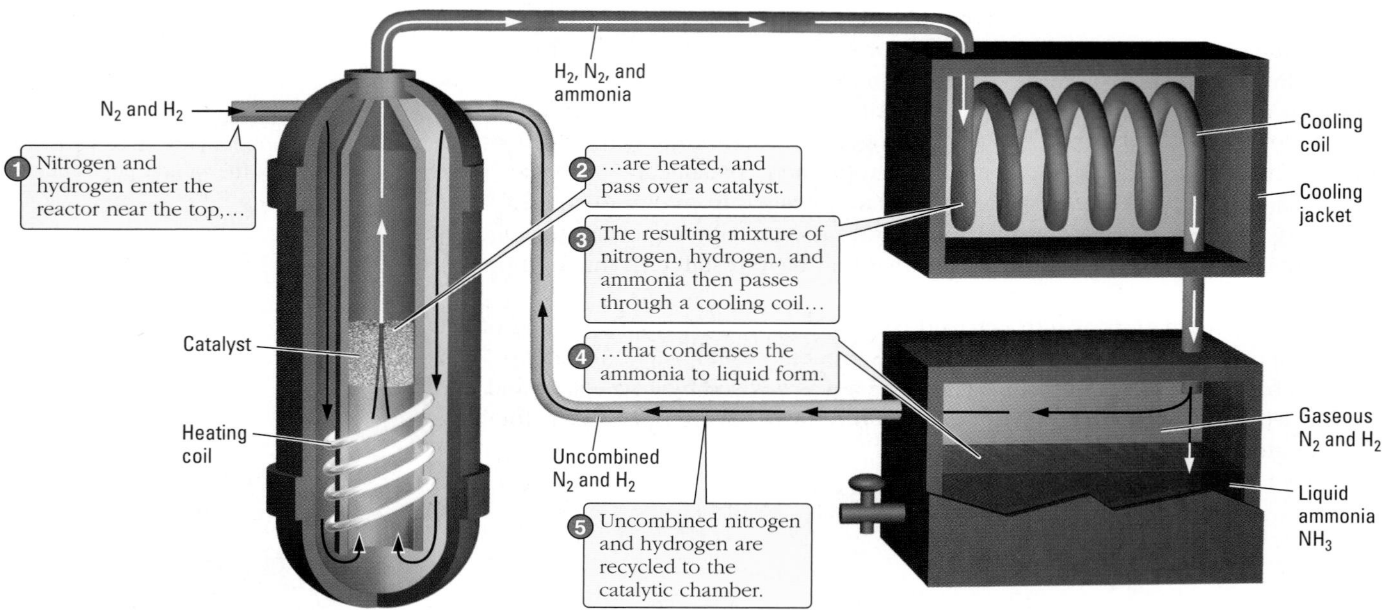

Figure 14.10 **Haber-Bosch process for synthesis of ammonia (schematic).**

- Ammonia is continually liquefied and removed from the reaction vessel, which reduces the concentration of the product of the reaction and shifts the equilibrium toward the right.
- The reaction is quite slow at room temperature, so the temperature must be raised to increase the rate. Although the rate increases with increasing temperature, the equilibrium constant declines. Thus, the faster the reaction, the smaller the yield.
- The temperature cannot be raised too much in an attempt to increase the rate, but a rate increase can be achieved with a catalyst. An effective catalyst for the Haber-Bosch process is Fe_3O_4 mixed with KOH, SiO_2, and Al_2O_3. Since the catalyst is not effective below about 400 °C, the optimum temperature, considering all the factors controlling the reaction, is about 450 °C.

Making predictions about chemical reactivity is part of the challenge, the adventure, and the art of chemistry. Many chemists enjoy the challenge of making useful new materials, which usually means choosing to make them by reactions that we believe will be product-favored and reasonably rapid. Such predictions are based on the ideas outlined in this chapter and Chapter 13.

SUMMARY PROBLEM

One approach to achieving cleaner-burning fuels and more-efficient automobiles is to extract hydrogen from gasoline and other liquid fossil fuels. The extracted hydrogen could be combined with oxygen in fuel cells (Section 19.9) like those currently used in spacecraft to generate electricity. The electricity could be used in electric motors to power automobiles and to provide air conditioning and other amenities expected by auto buyers. Because electric motors are far more efficient than current automobile engines, such a car might get 80 miles per gallon of fuel. By answering the following questions, you can explore how the ideas of chemical equilibrium and chemical kinetics can be applied to motive power for cars.

1. The hydrogen extracted from hydrocarbon fuels must be free from soot (solid carbon) and carbon monoxide, which would interfere with the operation of a fuel cell. Consider possible reactions by which hydrogen could be obtained from a hydrocarbon such as octane (C_8H_{18}). Write an equation for a reaction that you think would not be suitable and write an equation for one that you think would be suitable. Explain your choice in each case.

2. Use data from Appendix J to calculate the change in enthalpy for each of the two reactions you wrote in Question 1. Predict whether entropy increases or decreases when each reaction occurs. Is either of them ruled out because it is not product-favored? If not, continue. If either of the reactions is not product-favored, think about whether conditions could be altered to make it product-favored.

3. The chemical process by which hydrogen is obtained for use in synthesizing ammonia (Haber-Bosch process) involves treating methane (from natural gas) with steam. The first step in this process is

$$CH_4(g) + H_2O(g) \rightleftharpoons CO(g) + 3\ H_2(g)$$

   (a) Write the equilibrium constant expression for this reaction.

   (b) What is the relation between K_c and K_P for this reaction?

   (c) Calculate the enthalpy change for this reaction.

   (d) Based on the equation, predict the sign of the entropy change for this reaction.

   (e) Is the reaction product-favored at high temperatures but not at lower temperatures? Or the other way around? Explain.

4. To remove carbon monoxide from the hydrogen destined for the Haber-Bosch process, this reaction is used:

$$CO(g) + H_2O(g) \rightleftharpoons CO_2(g) + H_2(g)$$

   (a) Use bond enthalpies to estimate the enthalpy change for this reaction.

   (b) At 450 °C, K_P for this reaction is 6.48. Calculate K_c.

   (c) Suppose that 0.100 mol CO and 0.100 mol H_2O were introduced into an empty 10.0-L flask at 450 °C. Determine the concentration of $H_2(g)$ in the flask once equilibrium has been achieved.

   (d) What is the concentration of CO(g) remaining in the flask in part (c)? Is it low enough that we can say that the hydrogen is free of carbon monoxide?

5. If you were in charge of designing a system for generating hydrogen gas for use in the Haber-Bosch process, how might you obtain pure hydrogen? Assume that the process is based on the two reactions given in Questions 3 and 4. Suggest a chemical reagent that could be used to react with $CO_2(g)$ and thereby remove it from the hydrogen generated as a product in each of the two reactions. Would this same reagent work if you needed to remove $SO_2(g)$?

6. To get the highest-purity $H_2(g)$ and the maximum yield from the hydrogen-generating process, what reactant concentration(s) would you increase or decrease? How would you adjust product concentrations?

7. In the hypothetical fuel cell system for an electric automobile, hydrocarbon fuel is vaporized and partially oxidized in a limited quantity of air. In a second step the products of the first reaction are treated with steam over copper

oxide and zinc oxide catalysts. In a final purification step, more air is introduced and a platinum catalyst helps convert carbon monoxide to carbon dioxide. Write a balanced chemical equation for each of these three steps. Assume that the hydrocarbon fuel is octane.

8. What are the advantages of generating hydrogen gas from hydrocarbon fuel in an automobile rather than storing hydrogen in a fuel tank? What are the disadvantages of storing hydrogen in a car? What advantages does combining hydrogen with oxygen in a fuel cell have, as opposed to burning a hydrocarbon fuel in an internal combustion engine?

ThomsonNOW™
Sign in to ThomsonNOW at **www.thomsonedu.com** to check your readiness for an exam by taking the Pre-Test and exploring the modules recommended in your Personalized Learning Plan.

IN CLOSING

Having studied this chapter, you should be able to . . .

- Recognize a system at equilibrium and describe the properties of equilibrium systems (Section 14.1).
- Describe the dynamic nature of equilibrium and the changes in concentrations of reactants and products that occur as a system approaches equilibrium (Sections 14.1 and 14.2). ThomsonNOW homework: Study Question 7
- Write equilibrium constant expressions, given balanced chemical equations (Section 14.2). ThomsonNOW homework: Study Questions 13, 17
- Obtain equilibrium constant expressions for related reactions from the expression for one or more known reactions (Section 14.2). ThomsonNOW homework: Study Question 19
- Calculate K_P from K_c or K_c from K_P for the same equilibrium (Section 14.2). ThomsonNOW homework: Study Question 25
- Calculate a value of K_c for an equilibrium system, given information about initial concentrations and equilibrium concentrations (Section 14.3). ThomsonNOW homework: Study Questions 34, 35, 109
- Make qualitative predictions about the extent of reaction based on equilibrium constant values—that is, predict whether a reaction is product-favored or reactant-favored based on the size of the equilibrium constant (Section 14.4). ThomsonNOW homework: Study Question 37
- Calculate concentrations of reactants and products in an equilibrium system if K_c and initial concentrations are known (Section 14.5). ThomsonNOW homework: Study Questions 43, 45, 50, 55, 101
- Use the reaction quotient Q to predict in which direction a reaction will go to reach equilibrium (Section 14.5). ThomsonNOW homework: Study Questions 53, 54
- Use Le Chatelier's principle to show how changes in concentrations, pressure or volume, and temperature shift chemical equilibria (Section 14.6). ThomsonNOW homework: Study Questions 61, 63
- Use the change in enthalpy and the change in entropy qualitatively to predict whether products are favored over reactants (Section 14.7). ThomsonNOW homework: Study Questions 73, 87
- List the factors affecting chemical reactivity, and apply them to predicting optimal conditions for producing products (Section 14.8).

KEY TERMS

chemical equilibrium *(14.1)*

dynamic equilibrium *(14.1)*

entropy *(14.7)*

equilibrium concentration *(14.3)*

equilibrium constant *(14.2)*

equilibrium constant expression *(14.2)*

Haber-Bosch process *(14.8)*

Le Chatelier's principle *(14.6)*

product-favored *(14.1)*

reactant-favored *(14.1)*

reaction quotient *(14.5)*

shifting an equilibrium *(14.6)*

QUESTIONS FOR REVIEW AND THOUGHT

■ denotes questions available in ThomsonNOW and assignable in OWL.

Blue-numbered questions have short answers at the back of this book and fully worked solutions in the *Student Solutions Manual.*

ThomsonNOW™
Assess your understanding of this chapter's topics with sample tests and other resources found by signing in to ThomsonNOW at **www.thomsonedu.com**.

Review Questions

1. Define the terms *chemical equilibrium* and *dynamic equilibrium.*
2. If an equilibrium is product-favored, is its equilibrium constant large or small with respect to 1? Explain.
3. List three characteristics that you would need to verify in order to determine that a chemical system is at equilibrium.
4. The decomposition of ammonium dichromate,

$$(NH_4)_2Cr_2O_7(s),$$

yields nitrogen gas, water vapor, and solid chromium(III) oxide. The reaction is exothermic. In a closed container this process reaches an equilibrium state. Write a balanced equation for the equilibrium reaction. How is the equilibrium affected if
   (a) More ammonium dichromate is added to the equilibrium system?
   (b) More water vapor is added?
   (c) More chromium(III) oxide is added?

© Thomson Learning/Charles D. Winters

Decomposition of $(NH_4)_2Cr_2O_7$.

5. For the equilibrium reaction in Question 4, write the expression for the equilibrium constant.
   (a) How would this equilibrium constant change if the total pressure on the system were doubled?
   (b) How would the equilibrium constant change if the temperature were increased?
6. Indicate whether each statement below is true or false. If a statement is false, rewrite it to produce a closely related statement that is true.
   (a) For a given reaction, the magnitude of the equilibrium constant is independent of temperature.
   (b) If there is an increase in entropy and a decrease in enthalpy when reactants in their standard states are converted to products in their standard states, the equilibrium constant for the reaction will be negative.
   (c) The equilibrium constant for the reverse of a reaction is the reciprocal of the equilibrium constant for the reaction itself.
   (d) For the reaction

$$H_2O_2(\ell) \rightleftharpoons H_2O(\ell) + \tfrac{1}{2}O_2(g)$$

the equilibrium constant is one-half the magnitude of the equilibrium constant for the reaction

$$2\,H_2O_2(\ell) \rightleftharpoons 2\,H_2O(\ell) + O_2(g)$$

Topical Questions

Characteristics of Chemical Equilibrium

7. ■ Think of an experiment you could do to demonstrate that the equilibrium

$$2\,NO_2(g) \rightleftharpoons N_2O_4(g)$$

is a dynamic process in which the forward and reverse reactions continue to occur after equilibrium has been achieved. Describe how such an experiment might be carried out.

8. Discuss this statement: "No true chemical equilibrium can exist unless reactant molecules are constantly changing into product molecules, and vice versa."
9. Suppose you place a large piece of ice into a well-insulated thermos with some water in it, and it comes to equilibrium with part of the ice melted.
   (a) What is the temperature of the equilibrium system?
   (b) Is this a static or a dynamic equilibrium? Explain.

10. The atmosphere consists of about 80% N_2 and 20% O_2, yet there are many oxides of nitrogen that are stable and can be isolated in the laboratory.
   (a) Is the atmosphere at chemical equilibrium with respect to forming NO?
   (b) If not, why doesn't NO form? If so, how is it that NO can be made and kept in the laboratory for long periods?

The Equilibrium Constant

11. Consider the gas phase reaction of $N_2 + O_2$ to give 2 NO and the reverse reaction of 2 NO to give $N_2 + O_2$, discussed in Section 14.2. An equilibrium mixture of NO, N_2, and O_2 at 5000. K that contains equal concentrations of N_2 and O_2 has a concentration of NO about half as great. Make qualitatively correct plots of the concentrations of reactants and products versus time for these two processes, showing the initial state and the final dynamic equilibrium state. Assume a temperature of 5000. K. Don't do any calculations—just sketch how you think the plots will look.
12. After 0.1 mol of pure *cis*-2-butene is allowed to come to equilibrium with *trans*-2-butene in a closed flask at 25 °C, another 0.1 mol *cis*-2-butene is suddenly added to the flask.
   (a) Is the new mixture at equilibrium? Explain why or why not.
   (b) In the new mixture, immediately after addition of the *cis*-2-butene, which rate is faster: *cis* → *trans* or *trans* → *cis*? Or are both rates the same?
   (c) After the second 0.1 mol *cis*-2-butene has been added and the system is at equilibrium, if the concentration of *trans*-2-butene is 0.01 mol/L, what is the concentration of *cis*-2-butene?
13. ■ Write the equilibrium constant expression for each reaction.
   (a) $2\ H_2O_2(g) \rightleftharpoons 2\ H_2O(g) + O_2(g)$
   (b) $PCl_3(g) + Cl_2(g) \rightleftharpoons PCl_5(g)$
   (c) $SiO_2(s) + 3\ C(s) \rightleftharpoons SiC(s) + 2\ CO(g)$
   (d) $H_2(g) + \frac{1}{8}\ S_8(s) \rightleftharpoons H_2S(g)$
14. Write the equilibrium constant expression for each reaction.
   (a) $3\ O_2 \rightleftharpoons 2\ O_3(g)$
   (b) $SiH_4(g) + 2\ O_2(g) \rightleftharpoons SiO_2(s) + 2\ H_2O(g)$
   (c) $MgO(s) + SO_2(g) + \frac{1}{2}\ O_2(g) \rightleftharpoons MgSO_4(s)$
   (d) $2\ PbS(s) + 3\ O_2(g) \rightleftharpoons 2\ PbO(s) + 2\ SO_2(g)$
15. Write the equilibrium constant expression for each reaction.
   (a) $TlCl_3(s) \rightleftharpoons TlCl(s) + Cl_2(g)$
   (b) $CuCl_4^{2-}(aq) \rightleftharpoons Cu^{2+}(aq) + 4\ Cl^-(aq)$
   (c) $CO(g) + H_2O(g) \rightleftharpoons CO_2(g) + H_2(g)$
   (d) $4\ H_3O^+(aq) + 2\ Cl^-(aq) + MnO_2(s) \rightleftharpoons Mn^{2+}(aq) + 6\ H_2O(\ell) + Cl_2(aq)$
16. Write the equilibrium constant expression for each reaction.
   (a) The oxidation of ammonia with ClF_3 in a rocket motor
   $$NH_3(g) + ClF_3(g) \rightleftharpoons 3\ HF(g) + \tfrac{1}{2}\ N_2(g) + \tfrac{1}{2}\ Cl_2(g)$$
   (b) The simultaneous oxidation and reduction of a chlorite ion
   $$3\ ClO_2^-(aq) \rightleftharpoons 2\ ClO_3^-(aq) + Cl^-(aq)$$
   (c) $IO_3^-(aq) + 6\ OH^-(aq) + Cl_2(aq) \rightleftharpoons IO_6^{5-}(aq) + 2\ Cl^-(aq) + 3\ H_2O(\ell)$
17. ■ Write the equilibrium constant expression for each of these heterogeneous systems.
   (a) $CaSO_4 \cdot 5\ H_2O(s) \rightleftharpoons CaSO_4 \cdot 3\ H_2O(s) + 2\ H_2O(g)$
   (b) $SiF_4(g) + 2\ H_2O(g) \rightleftharpoons SiO_2(s) + 4\ HF(g)$
   (c) $LaCl_3(s) + H_2O(g) \rightleftharpoons LaClO(s) + 2\ HCl(g)$
18. Write the equilibrium constant expression for each of these heterogeneous systems.
   (a) $N_2O_4(g) + O_3(g) \rightleftharpoons N_2O_5(s) + O_2(g)$
   (b) $C(s) + 2\ N_2O(g) \rightleftharpoons CO_2(g) + 2\ N_2(g)$
   (c) $H_2O(\ell) \rightleftharpoons H_2O(g)$
19. ■ Consider these two equilibria involving $SO_2(g)$ and their corresponding equilibrium constants.
   $$SO_2(g) + \tfrac{1}{2}\ O_2(g) \rightleftharpoons SO_3(g) \qquad K_{c_1}$$
   $$2\ SO_3(g) \rightleftharpoons 2\ SO_2(g) + O_2(g) \qquad K_{c_2}$$
   Which of these expressions correctly relates K_{c_1} to K_{c_2}?
   (a) $K_{c_2} = K_{c_1}^2$ (b) $K_{c_2}^2 = K_{c_1}$
   (c) $K_{c_2} = 1/K_{c_1}$ (d) $K_{c_2} = K_{c_1}$
   (e) $K_{c_2} = 1/K_{c_1}^2$
20. The reaction of hydrazine (N_2H_4) with chlorine trifluoride (ClF_3) has been used in experimental rocket motors.
   $$N_2H_4(g) + \tfrac{4}{3}\ ClF_3(g) \rightleftharpoons 4\ HF(g) + N_2(g) + \tfrac{2}{3}\ Cl_2(g)$$
   How is the equilibrium constant, K_P, for this reaction related to K_P' for the reaction written this way?
   $$3\ N_2H_4(g) + 4\ ClF_3(g) \rightleftharpoons 12\ HF(g) + 3\ N_2(g) + 2\ Cl_2(g)$$
   (a) $K_P = K_P'$ (b) $K_P = 1/K_P'$
   (c) $K_P^3 = K_P'$ (d) $K_P = (K_P')^3$
   (e) $3K_P = K_P'$
21. Hydrogen can react with elemental sulfur to give the smelly, toxic gas H_2S according to the reaction
   $$H_2(g) + \tfrac{1}{8}\ S_8(s) \rightleftharpoons H_2S(g)$$
   If the equilibrium constant K_c for this reaction is 7.6×10^5 at 25 °C, determine the value of the equilibrium constant for the reaction written as
   $$8\ H_2(g) + S_8(s) \rightleftharpoons 8\ H_2S(g)$$
22. At 450 °C, the equilibrium constant K_c for the Haber-Bosch synthesis of ammonia is 0.16 for the reaction written as
   $$3\ H_2(g) + N_2(g) \rightleftharpoons 2\ NH_3(g)$$
   Calculate the value of K_c for the same reaction written as
   $$\tfrac{3}{2}\ H_2(g) + \tfrac{1}{2}\ N_2(g) \rightleftharpoons NH_3(g)$$
23. For each reaction in Question 13, write the equilibrium constant expression for K_P.
24. For each reaction in Question 14, write the equilibrium constant expression for K_P.
25. ■ The vapor pressure of water at 80. °C is 0.467 atm. Find the value of K_c for the process
   $$H_2O(\ell) \rightleftharpoons H_2O(g)$$
   at this temperature.
26. The value of K_c for the reaction
   $$N_2(g) + 3\ H_2(g) \rightleftharpoons 2\ NH_3(g)$$
   is 2.00 at 400. °C. Find the value of K_P for this reaction at this temperature using atmospheres as units.

Determining Equilibrium Constants

27. Isomer A is in equilibrium with isomer B, as in the reaction

$$A(g) \rightleftharpoons B(g)$$

Three experiments are done, each at the same temperature, and equilibrium concentrations are measured. For each experiment, calculate the equilibrium constant, K_c.
(a) [A] = 0.74 mol/L, [B] = 0.74 mol/L
(b) [A] = 2.0 mol/L, [B] = 2.0 mol/L
(c) [A] = 0.01 mol/L, [B] = 0.01 mol/L

28. Two molecules of A react to form one molecule of B, as in the reaction

$$2\ A(g) \rightleftharpoons B(g)$$

Three experiments are done, each at the same temperature, and equilibrium concentrations are measured. For each experiment, calculate the equilibrium constant, K_c.

(a) [A] = 0.74 mol/L, [B] = 0.74 mol/L
(b) [A] = 2.0 mol/L, [B] = 2.0 mol/L
(c) [A] = 0.01 mol/L, [B] = 0.01 mol/L

By comparing the results of Questions 27 and 28, what can you conclude about this statement: "If the concentrations of reactants and products are equal, then the equilibrium constant is always 1.0."

29. Consider the equilibrium

$$2\ A(aq) \rightleftharpoons B(aq)$$

At equilibrium, [A] = 0.056 M and [B] = 0.21 M. Calculate the equilibrium constant for the reaction as written.

30. The following reaction was examined at 250 °C.

$$PCl_5(g) \rightleftharpoons PCl_3(g) + Cl_2(g)$$

At equilibrium, $[PCl_5] = 4.2 \times 10^{-5}$ M, $[PCl_3] = 1.3 \times 10^{-2}$ M, and $[Cl_2] = 3.9 \times 10^{-3}$ M. Calculate the equilibrium constant K_c for the reaction.

31. At high temperature, hydrogen and carbon dioxide react to give water and carbon monoxide.

$$H_2(g) + CO_2(g) \rightleftharpoons H_2O(g) + CO(g)$$

Laboratory measurements at 986 °C show that there is 0.11 mol each of CO and water vapor and 0.087 mol each of H_2 and CO_2 at equilibrium in a 1.0-L container. Calculate the equilibrium constant K_P for the reaction at 986 °C.

32. Carbon dioxide reacts with carbon to give carbon monoxide according to the equation

$$C(s) + CO_2(g) \rightleftharpoons 2\ CO(g)$$

At 700. °C, a 2.0-L flask is found to contain at equilibrium 0.10 mol CO, 0.20 mol CO_2, and 0.40 mol C. Calculate the equilibrium constant K_P for this reaction at the specified temperature.

33. Assume you place 0.010 mol N_2O_4(g) in a 2.0-L flask at 50. °C. After the system reaches equilibrium, $[N_2O_4]$ = 0.00090 M. What is the value of K_c for this reaction?

$$N_2O_4(g) \rightleftharpoons 2\ NO_2(g)$$

34. ■ Nitrosyl chloride, NOCl, decomposes to NO and Cl_2 at high temperatures.

$$2\ NOCl(g) \rightleftharpoons 2\ NO(g) + Cl_2(g)$$

Suppose you place 2.00 mol NOCl in a 1.00-L flask and raise the temperature to 462 °C. When equilibrium has been established, 0.66 mol NO is present. Calculate the equilibrium constant K_c for the decomposition reaction from these data.

35. ■ An equilibrium mixture contains 3.00 mol CO, 2.00 mol Cl_2, and 9.00 mol $COCl_2$ in a 50.-L reaction flask at 800. K. Calculate the value of the equilibrium constant K_c for the reaction

$$CO(g) + Cl_2(g) \rightleftharpoons COCl_2(g)$$

at this temperature.

36. At 667 K, HI is found to be 11.4% dissociated into its elements.

$$2\ HI(g) \rightleftharpoons H_2(g) + I_2(g)$$

If 1.00 mol HI is placed in a 1.00-L container and heated to 667 K, calculate (a) the equilibrium concentration of all three substances and (b) the value of K_c for this equilibrium at this temperature.

The Meaning of the Equilibrium Constant

37. ■ Using the data of Table 14.1, predict which of these reactions will be product-favored at 25 °C. Then place all the reactions in order from most reactant-favored to most product-favored.
(a) $2\ NH_3(g) \rightleftharpoons N_2(g) + 3\ H_2(g)$
(b) $NH_4^+(aq) + OH^-(aq) \rightleftharpoons NH_3(aq) + H_2O(\ell)$
(c) $2\ NO(g) \rightleftharpoons N_2(g) + O_2(g)$

38. Using the data of Table 14.1, predict which of these reactions will be product-favored at 25 °C. Then place all the reactions in order from most reactant-favored to most product-favored.
(a) $2\ NO_2(g) \rightleftharpoons N_2O_4(g)$
(b) $H_2CO_3(aq) + H_2O(\ell) \rightleftharpoons HCO_3^-(aq) + H_3O^+(aq)$
(c) $AgI(s) \rightleftharpoons Ag^+(aq) + I^-(aq)$

39. The equilibrium constants for dissolving silver sulfate and silver sulfide in water are 1.7×10^{-5} and 6×10^{-30}, respectively.
(a) Write the balanced dissociation reaction equation and the associated equilibrium constant expression for each process.
(b) Which compound is more soluble?
(c) Which compound is less soluble?

40. The equilibrium constants for dissolving calcium carbonate, silver nitrate, and silver chloride in water are 3.8×10^{-9}, 2.0×10^2, and 1.8×10^{-10}, respectively.
(a) Write the balanced dissociation reaction equation and the associated equilibrium constant expression for each process.
(b) Which compound is most soluble?
(c) Which compound is least soluble?

Using Equilibrium Constants

41. The hydrocarbon C_4H_{10} can exist in two forms: butane and 2-methylpropane. The value of K_c for conversion of butane to 2-methylpropane is 2.5 at 25 °C.

$$CH_3{-}CH_2{-}CH_2{-}CH_3 \rightleftharpoons H{-}C(CH_3)_2{-}CH_3$$

butane — 2-methylpropane

(a) Suppose that the initial concentrations of both butane and 2-methylpropane are 0.100 mol/L. Make up a table of initial concentrations, change in concentrations, and equilibrium concentrations for this reaction.
(b) Write the equilibrium constant expression in terms of x, the change in the concentration of butane, and then solve for x.
(c) If you place 0.017 mol butane in a 0.50-L flask at 25 °C, what will be the equilibrium concentrations of the two isomers?

42. A mixture of butane and 2-methylpropane at 25 °C has [butane] = 0.025 mol/L and [2-methylpropane] = 0.035 mol/L. Is this mixture at equilibrium? If the *equilibrium* concentration of butane is 0.025 mol/L, what must [2-methylpropane] be at equilibrium? (See the reaction and K_c value in Question 41.)

43. ■ The hydrocarbon cyclohexane, C_6H_{12}, can isomerize, changing into methylcyclopentane, a compound with the same formula but with a different molecular structure.

$$\underset{\text{cyclohexane}}{C_6H_{12}(g)} \rightleftharpoons \underset{\text{methylcyclopentane}}{C_5H_9CH_3(g)}$$

The equilibrium constant K_c has been estimated to be 0.12 at 25 °C. If you had originally placed 3.79 g cyclohexane in an empty 2.80-L flask, how much cyclohexane (in grams) would be present when equilibrium is established?

44. At room temperature, the equilibrium constant K_c for the reaction

$$2\ NO(g) \rightleftharpoons N_2(g) + O_2(g)$$

is 1.4×10^{30}.
(a) Is this reaction product-favored or reactant-favored?
(b) In the atmosphere at room temperature the concentration of N_2 is 0.33 mol/L, and the concentration of O_2 is about 25% of that value. Calculate the equilibrium concentration of NO in the atmosphere produced by the reaction of N_2 and O_2.
(c) How does this affect your answer to Question 10?

45. ■ Hydrogen gas and iodine gas react via the equation

$$H_2(g) + I_2(g) \rightleftharpoons 2\ HI(g) \qquad K_c = 76 \quad (\text{at } 600.\ K)$$

If 0.050 mol HI is placed in an empty 1.0-L flask at 600. K, what are the equilibrium concentrations of HI, I_2, and H_2?

46. Consider the equilibrium

$$N_2(g) + O_2(g) \rightleftharpoons 2\ NO(g)$$

At 2300 K the equilibrium constant $K_c = 1.7 \times 10^{-3}$. If 0.15 mol NO(g) is placed into a 10.0-L flask and heated to 2300 K, what are the equilibrium concentrations of all three substances?

47. The equilibrium constant K_c for the reaction

$$H_2(g) + I_2(g) \rightleftharpoons 2\ HI(g)$$

has the value 50.0 at 745 K.
(a) When 1.00 mol I_2 and 3.00 mol H_2 are allowed to come to equilibrium at 745 K in a flask of volume 10.00 L, what amount (in moles) of HI will be produced?
(b) What amount of HI is produced in a 5.00-L flask?
(c) What total amount of HI is present at equilibrium if an additional 3.00 mol H_2 is added to the 10.00-L flask?

48. The equilibrium constant K_c for the *cis-trans* isomerization of gaseous 2-butene has the value 1.50 at 580. K.

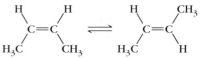

$$(H)(H_3C)C{=}C(H)(CH_3) \rightleftharpoons (H)(H_3C)C{=}C(CH_3)(H)$$

(a) Is the reaction product-favored at 580. K?
(b) Calculate the amount (in moles) of *trans* isomer produced when 1 mol *cis*-2-butene is heated to 580. K in the presence of a catalyst in a flask of volume 1.00 L and reaches equilibrium.
(c) What would the answer be if the flask had a volume of 10.0 L?

49. The equilibrium constant K_c for the reaction

$$CO(g) + H_2O(g) \rightleftharpoons CO_2(g) + H_2(g)$$

has the value 4.00 at 500. K. If a mixture of 1.00 mol CO and 1.00 mol H_2O is allowed to come to equilibrium in a flask of volume 1.00 L at 500. K,
(a) Calculate the final concentrations of all four species: CO, H_2O, CO_2, and H_2.
(b) What would be the equilibrium concentrations if an additional 1.00 mol each of CO and H_2O were added to the flask?

50. ■ At 503 K the equilibrium constant K_c for the dissociation of N_2O_4,

$$N_2O_4(g) \rightleftharpoons 2\ NO_2(g)$$

has the value 40.0.
(a) Calculate the fraction of N_2O_4 left undissociated when 1.00 mol of this gas is heated to 503 K in a 10.0-L container.
(b) If the volume is now reduced to 2.0 L, what will be the new fraction of N_2O_4 that is undissociated?

51. The equilibrium constant K_c for the reaction

$$N_2(g) + 3\ H_2(g) \rightleftharpoons 2\ NH_3(g)$$

has the value 5.97×10^{-2} at 500. °C. If 1.00 mol N_2 gas and 1.00 mol H_2 gas are heated to 500. °C in a 10.00-L flask together with a catalyst, calculate the percentage of N_2 converted to NH_3. (*Hint:* Assume that only a very small fraction of the reactants is converted to products. Obtain an approximate answer and use it to obtain a more accurate result.)

52. What percentage of N_2 would be converted to NH_3 in Question 51 if the volume of the flask were 5.00 L?

53. ■ Consider the equilibrium at 25 °C

$$2\ SO_3(g) \rightleftharpoons 2\ SO_2(g) + O_2(g) \qquad K_c = 3.58 \times 10^{-3}$$

Suppose that 0.15 mol $SO_3(g)$, 0.015 mol $SO_2(g)$, and 0.0075 mol $O_2(g)$ are placed into a 10.0-L flask at 25 °C.
(a) Is the system at equilibrium?
(b) If the system is not at equilibrium, in which direction must the reaction proceed to reach equilibrium? Explain your answer.

54. Consider the equilibrium at 25 °C

$$CH_4(g) + H_2O(g) \rightleftharpoons CO(g) + 3\ H_2(g) \qquad K_c = 9.4 \times 10^{-1}$$

Suppose that 0.25 mol $CH_4(g)$, 0.15 mol $H_2O(g)$, 0.25 mol CO(g), and 0.15 mol $H_2(g)$ are placed into a 10.0-L flask at 25 °C.
(a) Is the system at equilibrium?

(b) If the system is not at equilibrium, in which direction must the reaction proceed to reach equilibrium? Explain your answer.

55. ■ Consider the equilibrium

$$N_2(g) + O_2(g) \rightleftharpoons 2\ NO(g)$$

At 2300 K the equilibrium constant $K_c = 1.7 \times 10^{-3}$. Suppose that 0.015 mol NO(g), 0.25 mol N_2(g), and 0.25 mol O_2(g) are placed into a 10.0-L flask and heated to 2300 K.
(a) Is the system at equilibrium?
(b) If not, in which direction must the reaction proceed to reach equilibrium?
(c) Calculate the equilibrium concentrations of all three substances.

56. Consider the equilibrium

$$H_2(g) + I_2(g) \rightleftharpoons 2\ HI(g)$$

At 745 K the equilibrium constant $K_c = 50.0$. Suppose that 0.75 mol HI(g), 0.025 mol H_2(g), and 0.025 mol I_2(g) are placed into a 20.0-L flask and heated to 745 K.
(a) Is the system at equilibrium?
(b) If not, in which direction must the reaction proceed to reach equilibrium?
(c) Calculate the equilibrium concentrations of all three substances.

Shifting a Chemical Equilibrium: Le Chatelier's Principle

57. Consider this equilibrium, established in a 2.0-L flask at 25 °C:

$$N_2O_4(g) \rightleftharpoons 2\ NO_2(g) \qquad \Delta H° = +57.2\ \text{kJ}$$

What will happen to the concentration of N_2O_4 if the temperature is increased? Explain your choice.
(a) It will increase. (b) It will decrease.
(c) It will not change.
(d) It is not possible to tell from the information provided.

58. Solid barium sulfate is in equilibrium with barium ions and sulfate ions in solution.

$$BaSO_4(s) \rightleftharpoons Ba^{2+}(aq) + SO_4^{2-}(aq)$$

What will happen to the barium ion concentration if more solid $BaSO_4$ is added to the flask? Explain your choice.
(a) It will increase. (b) It will decrease.
(c) It will not change.
(d) It is not possible to tell from the information provided.

59. Hydrogen, bromine, and HBr in the gas phase are in equilibrium in a container of fixed volume.

$$H_2(g) + Br_2(g) \rightleftharpoons 2\ HBr(g) \qquad \Delta H° = -103.7\ \text{kJ}$$

How will each of these changes affect the indicated quantities? Write "increase," "decrease," or "no change."

Change	$[Br_2]$	[HBr]	K_c	K_P
Some H_2 is added to the container.	____	____	____	____
The temperature of the gases in the container is increased.	____	____	____	____
The pressure of HBr is increased.	____	____	____	____

60. Nitrogen, oxygen, and nitrogen monoxide are in equilibrium in a container of fixed volume.

$$N_2(g) + O_2(g) \rightleftharpoons 2\ NO(g) \qquad \Delta H° = 180.5\ \text{kJ}$$

How will each of these changes affect the indicated quantities? Write "increase," "decrease," or "no change."

Change	$[N_2]$	[NO]	K_c	K_P
Some NO is added to the container.	____	____	____	____
The temperature of the gases in the container is decreased.	____	____	____	____
The pressure of N_2 is decreased.	____	____	____	____

61. ■ The equilibrium constant K_c for this reaction is 0.16 at 25 °C, and the standard enthalpy change is 16.1 kJ.

$$2\ NOBr(g) \rightleftharpoons 2\ NO(g) + Br_2(\ell)$$

Predict the effect of each of these changes on the position of the equilibrium; that is, state which way the equilibrium will shift (left, right, or no change) when each of the following changes is made.
(a) Adding more Br_2
(b) Removing some NOBr
(c) Lowering the temperature

62. The formation of hydrogen sulfide from the elements is exothermic.

$$H_2(g) + \tfrac{1}{8}\ S_8(s) \rightleftharpoons H_2S(g) \qquad \Delta H° = -20.6\ \text{kJ}$$

Predict the effect of each of these changes on the position of the equilibrium; that is, state which way the equilibrium will shift (left, right, or no change) when each of the following changes is made.
(a) Adding more sulfur
(b) Adding more H_2
(c) Raising the temperature

63. ■ The oxidation of NO to NO_2,

$$2\ NO(g) + O_2(g) \rightleftharpoons 2\ NO_2(g)$$

is exothermic. Predict the effect of each of these changes on the position of the equilibrium; that is, state which way the equilibrium will shift (left, right, or no change) when each of the following changes is made.
(a) Adding more O_2
(b) Adding more NO_2
(c) Lowering the temperature

64. Phosphorus pentachloride is in equilibrium with phosphorus trichloride and chlorine in a flask.

$$PCl_5(s) \rightleftharpoons PCl_3(g) + Cl_2(g)$$

What will happen to the concentration of Cl_2 if additional PCl_5(s) is added to the flask?
(a) It will increase. (b) It will decrease.
(c) It will not change.
(d) It is impossible to tell from the information provided.

65. Consider the equilibrium

$$PbCl_2(s) \rightleftharpoons Pb^{2+}(aq) + 2\ Cl^-(aq)$$

(a) Will the equilibrium concentration of aqueous lead(II) ion increase, decrease, or remain the same if some solid NaCl is added to the flask?
(b) Make a graph like the one in Figure 14.6 to illustrate what happens to each of the concentrations after the NaCl is added.

66. Consider the transformation of butane into 2-methylpropane (see Question 41). The system is originally at equilibrium at 25 °C in a 1.0-L flask with [butane] = 0.010 M and [2-methylpropane] = 0.025 M. Suppose that 0.0050 mol of 2-methylpropane is suddenly added to the flask, and the system shifts to a new equilibrium.
(a) What is the new equilibrium concentration of each gas?
(b) Make a graph like the one in Figure 14.6 to show how the concentrations of the isomers change when the 2-methylpropane is added.

Equilibrium at the Nanoscale

67. For each of these reactions at 25 °C, indicate whether the entropy effect, the energy effect, both, or neither favors the reaction.
(a) $N_2(g) + 3\,F_2(g) \rightleftharpoons 2\,NF_3(g)$ $\Delta H° = -249$ kJ
(b) $N_2F_4(g) \rightleftharpoons 2\,NF_2(g)$ $\Delta H° = 93.3$ kJ
(c) $N_2(g) + 3\,Cl_2(g) \rightleftharpoons 2\,NCl_3(g)$ $\Delta H° = 460$ kJ

68. For each of these processes at 25 °C, indicate whether the entropy effect, the energy effect, both, or neither favors the process.
(a) $C_3H_8(g) + 5\,O_2(g) \rightleftharpoons 3\,CO_2(g) + 4\,H_2O(g)$ $\Delta H° = -2045$ kJ
(b) $Br_2(g) \rightleftharpoons Br_2(\ell)$ $\Delta H° = -31$ kJ
(c) $2\,Ag(s) + 3\,N_2(g) \rightleftharpoons 2\,AgN_3(s)$ $\Delta H° = 621$ kJ

69. For each of these chemical reactions, predict whether the equilibrium constant at 25 °C is greater than 1 or less than 1, or state that insufficient information is available. Also indicate whether each reaction is product-favored or reactant-favored.
(a) $2\,NO(g) + O_2(g) \rightleftharpoons 2\,NO_2(g)$ $\Delta H° = -115$ kJ
(b) $2\,O_3(g) \rightleftharpoons 3\,O_2(g)$ $\Delta H° = -285$ kJ
(c) $N_2(g) + 3\,Cl_2(g) \rightleftharpoons 2\,NCl_3(g)$ $\Delta H° = 460$ kJ

70. For each of these chemical reactions, predict whether the equilibrium constant at 25 °C is greater than 1 or less than 1, or state that insufficient information is available. Also indicate whether each reaction is product-favored or reactant-favored.
(a) $2\,NaCl(s) \rightleftharpoons 2\,Na(s) + Cl_2(g)$ $\Delta H° = -823$ kJ
(b) $2\,CO(g) + O_2(g) \rightleftharpoons 2\,CO_2(g)$ $\Delta H° = -566$ kJ
(c) $3\,CO_2(g) + 4\,H_2O(g) \rightleftharpoons C_3H_8(g) + 5\,O_2(g)$ $\Delta H° = 2045$ kJ

Controlling Chemical Reactions: The Haber-Bosch Process

71. Considering both the enthalpy effect and the entropy effect for the Haber-Bosch process, explain why choosing the temperature at which to run this reaction is very important.

72. Explain in your own words why it was so important to find a highly effective catalyst for the ammonia synthesis reaction before the Haber-Bosch process could become successful.

73. ■ Although ammonia is made in enormous quantities by the Haber-Bosch process, sulfuric acid is made in even greater quantities by the *contact process.* A simplified version of this process can be represented by these three reactions.

$$S(s) + O_2(g) \rightleftharpoons SO_2(g)$$
$$2\,SO_2(g) + O_2(g) \rightleftharpoons 2\,SO_3(g)$$
$$SO_3(g) + H_2O(\ell) \rightleftharpoons H_2SO_4(\ell)$$

(a) Use data from Appendix J to calculate $\Delta H°$ for each reaction.
(b) Which reactions are exothermic? Which are endothermic?
(c) In which of the reactions does entropy increase? In which does it decrease? In which does it stay about the same?
(d) For which reaction(s) do low temperatures favor formation of products?

74. Lime, CaO(s), can be produced by heating limestone, $CaCO_3(s)$, to cause a decomposition reaction.
(a) Write a balanced equation for the reaction.
(b) Predict the sign of the enthalpy change for the reaction.
(c) From the data in Appendix J, calculate $\Delta H°$ for this reaction at 25 °C to verify or contradict your prediction in part (b).
(d) Predict the entropy effect for this reaction.
(e) Is the reaction favored by entropy, energy, both, or neither?
(f) Explain in terms of Le Chatelier's principle why limestone must be heated to make lime.

General Questions

75. Write equilibrium constant expressions, in terms of reactant and product concentrations, for each of these reactions.

$H_2O(\ell) \rightleftharpoons H^+(aq) + OH^-(aq)$ $K_c = 1.0 \times 10^{-14}$

$CH_3COOH(aq) \rightleftharpoons CH_3COO^-(aq) + H^+(aq)$ $K_c = 1.8 \times 10^{-5}$

$N_2(g) + 3\,H_2(g) \rightleftharpoons 2\,NH_3(g)$ $K_c = 3.5 \times 10^8$

$CO(g) + H_2O(g) \rightleftharpoons CO_2(g) + H_2(g)$ $K_c = 4.00$ (at 500 K)

Assume that all gases and solutes have initial concentrations of 1.0 mol/L. Then let the *first* reactant in each reaction change its concentration by $-x$.
(a) Using the reaction table (ICE table) approach, write equilibrium constant expressions in terms of the unknown variable x for each reaction.
(b) Which of these expressions yield quadratic equations?
(c) How would you go about solving the others for x?

76. Write equilibrium constant expressions, in terms of reactant and product concentrations, for each of these reactions.

$2\,O_3(g) \rightleftharpoons 3\,O_2(g)$ $K_c = 7 \times 10^{56}$

$2\,NO_2(g) \rightleftharpoons N_2O_4(g)$ $K_c = 1.7 \times 10^2$

$HCOO^-(aq) + H^+(aq) \rightleftharpoons HCOOH(aq)$ $K_c = 5.6 \times 10^3$

$Ag^+(aq) + I^-(aq) \rightleftharpoons AgI(s)$ $K_c = 6.7 \times 10^{15}$

Assume that all gases and solutes have initial concentrations of 1.0 mol/L. Then let the *first* reactant in each reaction change its concentration by $-x$.

(a) Using the reaction table (ICE table) approach, write equilibrium constant expressions in terms of the unknown variable x for each reaction.

(b) Which of these expressions yield quadratic equations?

(c) How would you go about solving the others for x?

77. Many common nonmetallic elements exist as diatomic molecules at room temperature. When these elements are heated to 1500. K, the molecules break apart into atoms. A general equation for this type of reaction is

$$E_2(g) \rightleftharpoons 2\ E(g)$$

where E stands for each element. Equilibrium constants for dissociation of these molecules at 1500. K are

Species	K_c	Species	K_c
Br_2	8.9×10^{-2}	H_2	3.1×10^{-10}
Cl_2	3.4×10^{-3}	N_2	1×10^{-27}
F_2	7.4	O_2	1.6×10^{-11}

(a) If 1.00 mol of each diatomic molecule is placed in a separate 1.0-L container and heated to 1500. K, what is the equilibrium concentration of the atomic form of each element at 1500. K?

(b) From these results, predict which of the diatomic elements has the lowest bond dissociation energy, and compare your results with thermochemical calculations and with Lewis structures.

78. The chemistry of compounds composed of a transition metal and carbon monoxide has been an interesting area of research for the past 70 years. $Ni(CO)_4$ is formed by the reaction of nickel metal with carbon monoxide.

(a) If you have 2.05 g CO, and you combine it with 0.125 g nickel metal, how many grams of $Ni(CO)_4$ can be formed?

(b) An excellent way to make pure nickel metal is to decompose $Ni(CO)_4$ in a vacuum at a temperature slightly higher than room temperature. If the molar enthalpy of formation of $Ni(CO)_4$ gas is −602.9 kJ/mol, what is the enthalpy change for this decomposition reaction?

$$Ni(CO)_4(g) \longrightarrow Ni(s) + 4\ CO(g)$$

In an experiment at 100. °C it is determined that with 0.010 mol $Ni(CO)_4(g)$ initially present in a 1.0-L flask, only 0.000010 mol remains after decomposition.

(c) What is the equilibrium concentration of CO in the flask?

(d) What is the value of the equilibrium constant K_c for this reaction at 100. °C?

(e) Predict whether there is an increase or a decrease in entropy when this reaction occurs.

(f) Calculate the equilibrium constant K_p for this reaction at 100. °C.

79. A small sample of *cis*-dichloroethene in which one carbon atom is the radioactive isotope ^{14}C is added to an equilibrium mixture of the *cis* and *trans* isomers at a certain temperature. Eventually, 40% of the radioactive molecules are found to be in the *trans* configuration at any given time.

(a) What is the value of K_c for the *cis* $\rightleftharpoons$ *trans* equilibrium?

(b) What would have happened if a small sample of radioactive *trans* isomer had been added instead of the *cis* isomer?

80. In a 0.0020 M solution of acetic acid, any acetate species spends 9% of its time as an acetate ion, CH_3COO^-, and the remaining 91% of its time as acetic acid, CH_3COOH. What is the value of K_c for the reaction?

$$CH_3COOH + H_2O \rightleftharpoons CH_3COO^- + H_3O^+$$

81. The following amounts of HI, H_2, and I_2 are introduced into a 10.00-L reaction flask and heated to 745 K.

	n_{HI} (mol)	n_{H_2} (mol)	n_{I_2} (mol)
Case a	1.0	0.10	0.10
Case b	10.	1.0	1.0
Case c	10.	10.	1.0
Case d	5.62	0.381	1.75

The equilibrium constant for the reaction

$$2\ HI(g) \rightleftharpoons H_2(g) + I_2(g)$$

has the value 0.0200 at 745 K. In which cases will the concentration of HI increase as equilibrium is attained, and in which cases will the concentration of HI decrease?

82. The following amounts of CO(g), $H_2O(g)$, $CO_2(g)$, and $H_2(g)$ are introduced into a 10.00-L reaction flask and heated to 745 K.

	n_{CO} (mol)	n_{H_2O} (mol)	n_{CO_2} (mol)	n_{H_2O} (mol)
Case a	1.0	0.10	0.10	0.10
Case b	10.	1.0	1.0	1.0
Case c	10.	10.	1.0	1.0
Case d	5.62	0.381	1.75	1.75

The equilibrium constant for the reaction

$$CO(g) + H_2O(g) \rightleftharpoons CO_2(g) + H_2(g)$$

has the value $K_c = 4.00$ at 500 K. For which cases will the concentration of CO increase as equilibrium is attained, and in which cases will the concentration of CO decrease?

Applying Concepts

83. Suppose that you have heated a mixture of *cis*- and *trans*-2-pentene to 600. K, and after 1 h you find that the composition is 40% *cis*. After 4 h the composition is found to be 42% *cis*, and after 8 h it is 42% *cis*. Next, you heat the mixture to 800. K and find that the composition changes to 45% *cis*. When the mixture is cooled to 600. K and allowed to stand for 8 h, the composition is found to be 42% *cis*. Is this system at equilibrium at 600. K? Or would more experiments be needed before you could conclude that it was at equilibrium? If so, what experiments would you do?

84. In Section 14.2 the equilibrium constant for the reaction

$$\frac{1}{8}\,S_8(s) + O_2(g) \rightleftharpoons SO_2(g)$$

is given as 4.2×10^{52}. If this reaction is so product-favored, why can large piles of yellow sulfur exist in our environment (as they do in Louisiana and Texas)?

85. When pressure greater than atmospheric pressure is applied to ice at 0.00 °C, the ice melts. Explain how this is an example of Le Chatelier's principle.

86. For the reaction

$$cis\text{-2-butene} \rightleftharpoons trans\text{-2-butene}$$

K_c is 1.65 at 500. K, 1.47 at 600. K, and 1.36 at 700. K. Predict whether the conversion from the *cis* to the *trans* isomer of 2-butene is exothermic or endothermic.

Use this table to answer Questions 87 and 88.

Equilibrium Constants K_c for Some *Cis-Trans* Interconversions

$$\text{(H)(R)C=C(H)(R)} \rightleftharpoons \text{(H)(R)C=C(R)(H)}$$

Temperature (K)	R is F	R is Cl	R is CH_3
500.	0.420	0.608	1.65
600.	0.491	0.678	1.47
700.	0.549	0.732	1.36

87. ■ Based on the data in the table and these reaction equations,
 (i) *cis*-(R is F) $\rightleftharpoons$ *trans*-(R is F)
 (ii) *trans*-(R is Cl) $\rightleftharpoons$ *cis*-(R is Cl)
 (iii) *cis*-(R is CH_3) $\rightleftharpoons$ *trans*-(R is CH_3)
 (a) Which reaction is most exothermic?
 (b) Which reaction is most endothermic?

88. Based on the data in the table and these reaction equations,
 (i) *cis*-(R is F) $\rightleftharpoons$ *trans*-(R is F)
 (ii) *trans*-(R is Cl) $\rightleftharpoons$ *cis*-(R is Cl)
 (iii) *cis*-(R is CH_3) $\rightleftharpoons$ *trans*-(R is CH_3)
 (a) Which reaction is most product-favored at 500. K?
 (b) Which reaction is most reactant-favored at 700. K?

89. Figure 14.3 shows the equilibrium mixture of N_2O_4 and NO_2 at two different temperatures. Imagine that you can shrink yourself down to the size of the molecules in the two glass tubes and observe their behavior for a short period of time. Write a brief description of what you observe in each of the flasks.

90. Imagine yourself to be the size of ions and molecules inside a beaker containing this equilibrium mixture with a K_c greater than 1.

$$\underset{\text{pink}}{Co(H_2O)_6^{2+}(aq)} + 4\,Cl^-(aq) \rightleftharpoons \underset{\text{blue}}{CoCl_4^{2-}(aq)} + 6\,H_2O(\ell)$$

Write a brief description of what you observe around you before and after additional water is added to the mixture.

91. Which of the diagrams represent equilibrium mixtures for the reaction

$$A_2(g) + B_2(g) \rightleftharpoons 2\,AB(g)$$

at temperatures where $10^2 > K_c > 0.1$?

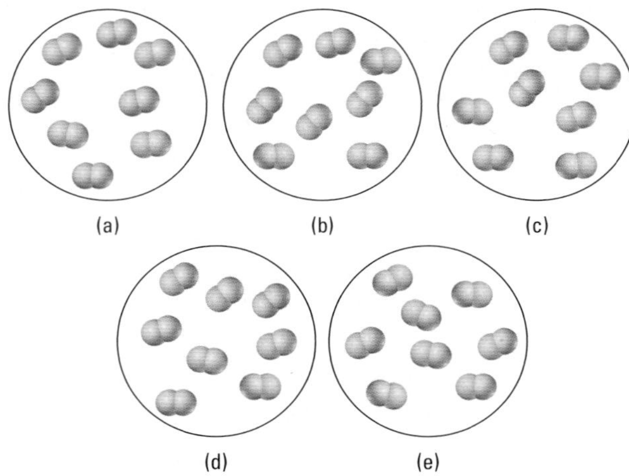

92. Which diagram in Question 91 best represents an equilibrium mixture with an equilibrium constant of
 (a) 0.44?
 (b) 4.0?
 (c) 36?

93. Draw a nanoscale (particulate) level diagram for an equilibrium mixture of

$$CO(g) + H_2O(g) \rightleftharpoons CO_2(g) + H_2(g) \qquad K_c = 4.00$$

94. The diagram below represents an equilibrium mixture for the reaction

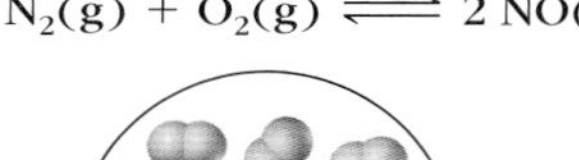

$$N_2(g) + O_2(g) \rightleftharpoons 2\,NO(g)$$

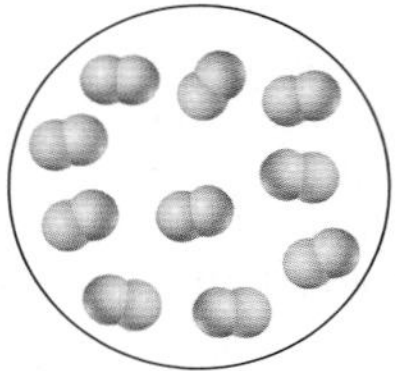

What is the equilibrium constant?

95. A sample of benzoic acid, a solid carboxylic acid, is in equilibrium with an aqueous solution of benzoic acid. A tiny quantity of D_2O, water containing the isotope ^2H, deuterium, is added to the solution. The solution is allowed to stand at constant temperature for several hours, after which some of the solid benzoic acid is removed and analyzed. The benzoic acid is found to contain a tiny quantity of deuterium, D, and the formula of the deuterium-containing molecules is C_6H_5COOD. Explain how this can happen.

96. In a second experiment with benzoic acid (see Question 95), a tiny quantity of water that contains the isotope ^{18}O is added to a saturated solution of benzoic acid in water. When some of the solid benzoic acid is analyzed, no ^{18}O is found in the benzoic acid. Compare this situation with the experiment involving deuterium, and explain how the results of the two experiments can differ as they do.

97. Samples of N_2O_4 can be prepared in which both nitrogen atoms are the heavier isotope ^{15}N. Designating this isotope as N*, we can write the formula of the molecules in such a sample as $O_2N^*\text{—}N^*O_2$ and the formula of typical N_2O_4 as $O_2N\text{—}NO_2$. When a tiny quantity of $O_2N^*\text{—}N^*O_2$ is intro-

duced into an equilibrium mixture of N_2O_4 and NO_2, the ^{15}N immediately becomes distributed among both N_2O_4 and NO_2 molecules, and in the N_2O_4 it is invariably in the form $O_2N^*—NO_2$. Explain how this observation supports the idea that equilibrium is dynamic.

98. For the equilibrium

$$\underset{\text{pink}}{Co(H_2O)_6^{2+}(aq)} + 4\ Cl^-(aq) \rightleftharpoons \underset{\text{blue}}{CoCl_4^{2-}(aq)} + 6\ H_2O(\ell)$$

K_c is somewhat greater than 1. If water is added to a blue solution of $CoCl_4^{2-}(aq)$, the color changes from blue to pink.

(a) Does water appear in the equilibrium constant expression for this reaction?

(b) How can adding water shift the equilibrium to the left?

(c) Is this shift in the equilibrium in accord with Le Chatelier's principle? Why or why not? (*Hint:* Consider what happens to the concentrations of substances that appear in the equilibrium constant expression. Calculate Q_f/Q_i, where Q_f is the final value of Q and Q_i is the initial value of Q where enough water is added to double the volume of solution, that is, to halve all of the concentrations. Why is Q affected so much by changing the volume of the solution?)

More Challenging Questions

99. A sample of nitrosyl bromide is heated to 100. °C in a 10.00-L container to decompose it partially according to the equation

$$2\ NOBr(g) \rightleftharpoons 2\ NO(g) + Br_2(g)$$

The container is found to contain 6.44 g NOBr, 3.15 g NO, and 8.38 g Br_2 at equilibrium.

(a) Find the value of K_c at 100. °C.

(b) Find the total pressure exerted by the mixture of gases.

(c) Calculate K_P for this reaction at 100. °C.

100. Exactly 5.0 mol ammonia was placed in a 2.0-L flask that was then heated to 473 K. When equilibrium was established, 0.2 mol nitrogen had been formed according to the decomposition reaction

$$2\ NH_3(g) \rightleftharpoons N_2(g) + 3\ H_2(g)$$

(a) Calculate the value of the equilibrium constant K_c for this reaction at 473 K.

(b) Calculate the total pressure exerted by the mixture of gases inside the 2.0-L flask at this temperature.

(c) Calculate K_P for this reaction at 473 K.

101. ■ The equilibrium constant K_c has a value of 3.30 at 760. K for the decomposition of phosphorus pentachloride,

$$PCl_5(g) \rightleftharpoons PCl_3(g) + Cl_2(g)$$

(a) Calculate the equilibrium concentrations of all three species arising from the decomposition of 0.75 mol PCl_5 in a 5.00-L vessel.

(b) Calculate the equilibrium concentrations of all three species resulting from an initial mixture of 0.75 mol PCl_5 and 0.75 mol PCl_3 in a 5.00-L vessel.

102. A 1.00-mol sample of CO_2 is heated to 1000. K with excess solid graphite in a container of volume 40.0 L. At 1000. K, K_c is 2.11×10^{-2} for the reaction

$$C(graphite) + CO_2(g) \rightleftharpoons 2\ CO(g)$$

(a) What is the composition of the equilibrium mixture at 1000. K?

(b) If the volume of the flask is changed and a new equilibrium established in which the amount of CO_2 is equal to the amount of CO, what is the new volume of the flask? (Assume the temperature remains 1000. K.)

103. Predict whether the equilibria listed below will be shifted to the left or the right when the following changes occur: (i) the temperature is increased; (ii) the pressure is decreased; (iii) more of the substance indicated by a colored box is added.

(a) $C(s) + H_2O(g) \rightleftharpoons \boxed{CO(g)} + H_2(g)$ $\quad \Delta H°\ (298\ K) = +131.3\ kJ\ mol^{-1}$

(b) $3\ Fe(s) + 4\ \boxed{H_2O(g)} \rightleftharpoons Fe_3O_4(s) + 4\ H_2(g)$ $\quad \Delta H°\ (298\ K) = -149.9\ kJ\ mol^{-1}$

(c) $\boxed{C(s)} + CO_2(g) \rightleftharpoons 2\ CO(g)$ $\quad \Delta H°\ (298\ K) = +172.5\ kJ\ mol^{-1}$

(d) $N_2O_4(g) \rightleftharpoons 2\ \boxed{NO_2(g)}$ $\quad \Delta H°\ (298\ K) = +54.8\ kJ\ mol^{-1}$

104. Predict whether the equilibria listed below will be shifted to the left or the right when the following changes are made: (i) the temperature is increased; (ii) the pressure is decreased; (iii) more of the substance indicated by a colored box is added.

(a) $N_2(g) + \boxed{O_2(g)} \rightleftharpoons 2\ NO(g)$ $\quad \Delta H°\ (298\ K) = +180.0\ kJ\ mol^{-1}$

(b) $CH_4(g) + 2\ O_2(g) \rightleftharpoons \boxed{CO_2(g)} + 2\ H_2O(g)$ $\quad \Delta H°\ (298\ K) = -802.3\ kJ\ mol^{-1}$

(c) $CaCO_3(s) \rightleftharpoons CaO(s) + \boxed{CO_2(g)}$ $\quad \Delta H°\ (298\ K) = +177.9\ kJ\ mol^{-1}$

105. Use the fact that the equilibrium constant K_c equals the ratio of the forward rate constant divided by the reverse rate constant, together with the Arrhenius equation $k = Ae^{-E_a/RT}$, to show that a catalyst does not affect the value of an equilibrium constant even though the catalyst increases the rates of forward and reverse reactions. Assume that the frequency factors A for forward and reverse reactions do not change, and that the catalyst lowers the activation barrier for the catalyzed reaction.

106. Consider the reaction mechanism given for formation of NOBr in Section 13.7.

Step 1: $NO(g) + Br_2(g) \rightleftharpoons NOBr_2(g)$ fast

Step 2: $NOBr_2(g) + NO(g) \longrightarrow 2\ NOBr(g)$ slow

(a) What is the overall stoichiometric equation?

(b) Assume that when the reaction reaches equilibrium, each step in the mechanism also reaches equilibrium. Write the equilibrium constant for each step in terms of the forward and reverse rate constants for that step.

(c) Use your result in part (b) to derive the equilibrium constant for the overall stoichiometric equation in terms of the rate constants for the forward and reverse reactions of each step in the mechanism.

(d) Does your result confirm the statement made on p. 676 in the margin that the equilibrium constant can be obtained by taking the product of the rate constants for all forward steps and dividing by the product of the rate constants for all reverse steps?

107. When a mixture of hydrogen and bromine is maintained at normal atmospheric pressure and heated above 200. °C, the hydrogen and bromine react to form hydrogen bromide and a gas-phase equilibrium is established.
   (a) Write a balanced chemical equation for the equilibrium reaction.
   (b) Use bond enthalpies from Table 8.2 to estimate the enthalpy change for the reaction.
   (c) Based on your answers to parts (a) and (b), which is more important in determining the position of this equilibrium, the entropy effect or the energy effect?
   (d) In which direction will the equilibrium shift as the temperature increases above 200. °C? Explain.
   (e) Suppose that the pressure were increased to triple its initial value. In which direction would the equilibrium shift?
   (f) Why is the equilibrium not established at room temperature?

108. A sample of pure SO_3 weighing 0.8312 g was placed in a 1.00-L flask and heated to 1100. K to decompose it partially.

$$2\ SO_3(g) \rightleftharpoons 2\ SO_2(g) + O_2(g)$$

If a total pressure of 1.295 atm was developed, find the value of K_c for this reaction at this temperature.

109. ■ At 25 °C the vapor pressure of water is 0.03126 atm.
   (a) Calculate K_P and K_c for

$$H_2O(\ell) \rightleftharpoons H_2O(g)$$

   (b) What is the value of K_P for this same reaction at 100. °C?
   (c) Suggest a general rule for calculating K_P for any liquid in equilibrium with its vapor at its normal boiling point.

Conceptual Challenge Problems

Conceptual Challenge Problems CP14.A, CP14.B, CP14.C, CP14.D, and CP14.E are related to the information in this paragraph. Aqueous iron(III) ions, $Fe^{3+}(aq)$, are nearly colorless. If their concentration is 0.001 M or lower, a person cannot detect their color. Thiocyanate ions, $SCN^-(aq)$, are colorless also, but monothiocyanatoiron(III) ions, $Fe(SCN)^{2+}(aq)$, can be detected at very low concentrations because of their color. These ions are light amber in very dilute solutions, but as their concentration increases, the color intensifies and appears blood-red in more concentrated solutions. Suppose you prepared a stock solution by mixing equal volumes of 1.0×10^{-3} M solutions of both iron(III) nitrate and potassium thiocyanate solutions. The equilibrium reaction is

$$\underset{\text{colorless}}{Fe^{3+}(aq)} + \underset{\text{colorlesss}}{SCN^-(aq)} \rightleftharpoons \underset{\text{amber}}{Fe(SCN)^{2+}(aq)}$$

CP14.A (Section 14.1) Describe how you would use 5-mL samples of the stock solution and additional solutions of 0.010 M $Fe^{3+}(aq)$ and 0.010 M $SCN^-(aq)$ to show experimentally that the reaction between $Fe^{3+}(aq)$ and $SCN^-(aq)$ does not go to completion but instead reaches an equilibrium state in which appreciable quantities of reactants and product are present. (Refer to the first paragraph for further information.)

CP14.B (Section 14.1) Suppose that you added 1 drop of 0.010 M $Fe^{3+}(aq)$ to a 5-mL sample of the stock solution, followed by 10 drops of 0.010 M $SCN^-(aq)$. You treated a second 5-mL sample of the stock solution by first adding 10 drops of 0.010 M $SCN^-(aq)$, followed by 1 drop of $Fe^{3+}(aq)$. How would the color intensity of these two solutions compare after the same quantities of the same solutions were added in reverse order? (Refer to the first paragraph for further information.)

CP14.C (Section 14.6) Predict what will happen if you add a small crystal of sodium acetate to a 5-mL sample of the stock solution (described in the first paragraph) so that some acetatoiron(III) ion, a coordination complex, is formed.

CP14.D (Section 14.6) Predict what will happen if you begin to add a 0.010 M solution of $Fe^{3+}(aq)$ drop by drop to a 5-mL sample of the stock solution (described in the first paragraph) until the total volume becomes 10 mL. (A 0.010 M solution of Fe^{3+} ions is pale yellow.) Predict what will happen if you do the same experiment but add 0.010 M $SCN^-(aq)$ to the stock solution. Predict what will happen if you mix 0.010 M $Fe^{3+}(aq)$ with 0.010 M $SCN^-(aq)$. Would the results of these three experiments be similar or different? Explain.

CP14.E (Section 14.6) Predict what will happen if you put a 5-mL sample of the stock solution (described in the first paragraph) in a hot water bath. Predict what will happen if it is placed in an ice bath.

CP14.F (Section 14.4) Consider the equilibrium reaction between dioxygen and trioxygen (ozone). What is the minimum volume of air (21% dioxygen by volume) at 1.00 atm and 25 °C that you would predict to have at least one molecule of trioxygen, if the only source of trioxygen were its formation from dioxygen and if the atmospheric system were at equilibrium?

$$3\ O_2(g) \rightleftharpoons 2\ O_3(g) \qquad K_c = 6.3 \times 10^{-58} \quad (\text{at } 25\ ^\circ\text{C})$$

(The volume of 1 mol air at 1 atm and 25 °C is 24.45 L.)

15

The Chemistry of Solutes and Solutions

Oceans are vast aqueous solutions containing Na^+, Mg^{2+}, and Cl^- ions plus other solutes that give ocean water its characteristic salinity, vapor pressure, and freezing point. This chapter considers the macroscale to nanoscale links that govern how solutes dissolve in solvents, how to express solute concentration in solution, and how such concentration affects the freezing point and boiling point of solutions.

ThomsonNOW™
Throughout the text, this icon indicates an opportunity to test yourself on key concepts and to explore interactive modules by signing in to ThomsonNOW at **www.thomsonedu.com**.

Every day we all encounter many solutions, such as soft drinks, juices, coffee, and gasoline. A *solution* is a homogeneous mixture of two or more substances **(***← p. 14***)**. The component present in greatest amount is the *solvent;* the other components are *solutes* **(***← p. 192***)**. In sweetened iced tea, for example, water is the solvent; sugar and soluble extracts of tea are the solutes.

Although solids dissolved in liquids or mixtures of liquids are the most common types of solutions, other kinds are possible as well, encompassing the three physical states of matter (Table 15.1). Chemistry often focuses on liquid solutions, and in particular on those in which water is the solvent (aqueous solutions). Much of this chapter is devoted to aqueous solutions because water is the most important solvent on our planet.

In this chapter we will explore in some detail the macroscale to nanoscale connections between solutes and solvents to answer questions such as:

- Why does a particular solvent readily dissolve one kind of solute, but not another? Water, for example, dissolves NaCl, but does not dissolve gasoline.
- In what ways can the concentration of a dissolved solute be expressed?
- Is thermal energy released or absorbed when a solute dissolves?
- How do factors such as temperature or pressure changes affect the solubility of a solute in a given solvent? Why, for example, does a cold, carbonated beverage become "flat" when it is opened and warmed to room temperature?

To answer these questions, we will apply what you learned from Chapter 9 about the noncovalent forces that act between molecules—London forces, dipole-dipole forces, and hydrogen bonding **(***← p. 411***)**—to the interactions among solute and solvent molecules. We will also utilize the thermodynamic and equilibrium principles presented in Chapters 6 and 14. These principles are important to understand the effect that solutes have on the vapor pressures, melting points, and boiling points of solvents. In addition, the nature of unwanted solutes in natural and polluted water is discussed.

15.1 Solubility and Intermolecular Forces

In every solution, the interplay between solute and solvent particles determines whether a solute will dissolve in a particular solvent and how much is dissolved.

Table 15.1 Types of Solutions

Type of Solution	Example
Gas in gas	Air—a mixture principally of N_2 and O_2 but containing other gases as well
Gas in liquid	Carbonated beverages (CO_2 in water)
Gas in solid	Hydrogen in palladium metal
Liquid in liquid	Motor oil—a mixture of liquid hydrocarbons; ethanol in water
Solid in liquid	The oceans (dissolved Na^+, Cl^-, and other ions)
Solid in solid	Bronze (copper and tin); pewter (tin, antimony, and lead)

Solute-Solvent Interactions

An old adage says that "oil and water don't mix." Chemists use a similar saying about solubility: "Like dissolves like," where "like" refers to solutes and solvents whose molecules attract each other by similar types of noncovalent intermolecular forces. Such forces exist between solute particles, as well as between solvent particles. Dissolving a solute in a solvent is favored when the solute-solvent intermolecular forces are stronger than the solute-solute or the solvent-solvent intermolecular forces.

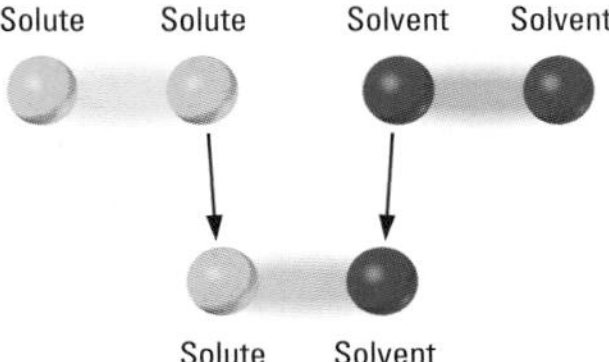

Consider, for example, dissolving the hydrocarbon octane, C_8H_{18}, in carbon tetrachloride, CCl_4 (Figure 15.1a). Both are nonpolar liquids that dissolve in each other because of the similar London forces between molecules in each compound. Liquids such as these that dissolve in each other in any proportion are said to be **miscible.** In contrast, gasoline, a mixture of nonpolar hydrocarbons, is not miscible with water, a polar substance (Figure 15.1b). The nonpolar hydrocarbons cannot hydrogen-bond to water molecules, but rather stay attracted to each other through London forces (solute-solute attractions); the water molecules remain hydrogen-bonded to each other (solvent-solvent attractions). Liquids such as gasoline and water, with noncovalent attractions so different between their molecules that they do not dissolve in each other, are described as **immiscible.**

The differing solubilities of various alcohols in water further illustrate the "like dissolves like" principle and the role of noncovalent intermolecular forces. Simple, low-molar-mass alcohols dissolve in water due to hydrogen bonding between water molecules and the —OH group of the alcohol. The hydrogen bonding forces in the solution are the same forces as those in pure water or pure ethanol. The hydrogen bonding between ethanol and water is illustrated in Figure 15.2.

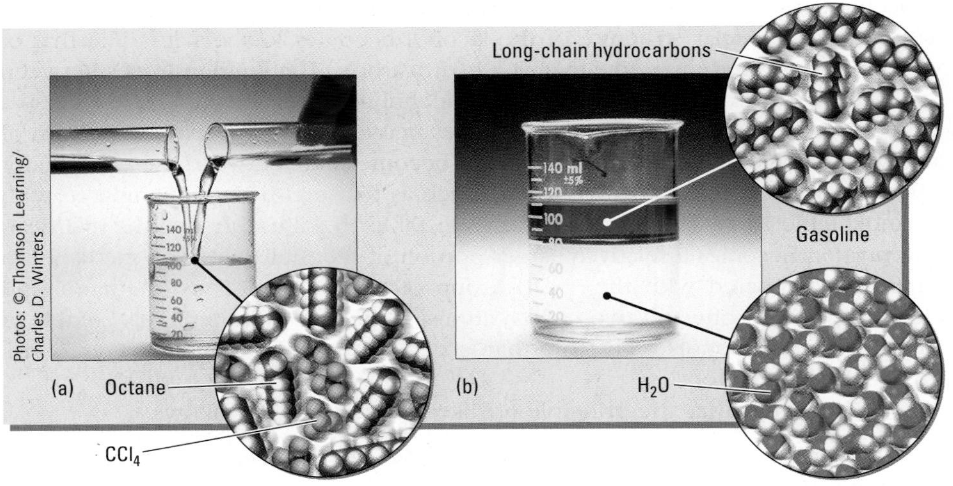

There is a dye in the gasoline to make it a different color.

Active Figure 15.1 **(a) Miscible and (b) Immiscible liquids.** When carbon tetrachloride and octane, both colorless, clear liquids, are mixed (a), each dissolves completely in the other, and there is no sign of an interface or boundary between them. (b) When gasoline and water are put together, the mixture remains as two distinct layers. **Log on to ThomsonNOW to test your understanding of the concepts in this figure.**

(a) Hydrogen bonding in ethanol

(b) Hydrogen bonding in water

(c) Hydrogen bonding between ethanol and water

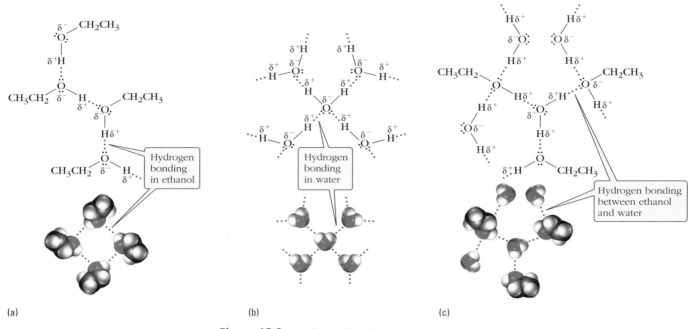

Figure 15.2 Hydrogen bonding. Hydrogen bonding (a) among ethanol molecules (solute-solute attraction); (b) among water molecules (solvent-solvent attraction); and (c) among ethanol and water molecules (solute-solvent attraction).

Alaskan oil spill. This oil spill in Prince William Sound, Alaska, is a large-scale example of the nonsolubility of crude oil, a nonpolar material, in water, a polar substance. The oil floats on top of the water, but does not dissolve in it.

Alcohol molecules contain a polar portion, the —OH group, and a nonpolar portion, the hydrocarbon part. The polar region is *hydrophilic* ("water loving"); any polar part of a molecule will be hydrophilic because of its attraction to polar water molecules. The nonpolar hydrocarbon region is *hydrophobic* ("water hating") ***(⇐ p. 416)***. In a low-molar-mass alcohol such as methanol or ethanol (Figure 15.2), hydrogen bonding of the —OH group with water molecules is stronger than the London forces between the nonpolar hydrocarbon-like parts of the alcohol molecules. As the hydrophobic hydrocarbon chain of the alcohol increases in length, the molecular structure of the alcohol becomes less and less like that of water and more and more like that of a hydrocarbon. The London forces between the nonpolar hydrocarbon portion of the alcohol molecules become stronger, so a point is reached at which the London forces between the hydrocarbon portion of alcohol molecules (solute-solute attraction) become sufficiently large that the water solubility of the alcohol becomes small, such as that of 1-hexanol compared with ethanol (Table 15.2). The Lewis structures in Table 15.2 illustrate that the hydrocarbon portion becomes a relatively larger portion of the molecule from methanol to 1-hexanol compared with the —OH group of the alcohol. Thus, methanol and ethanol, with just one and two carbon atoms, respectively, are infinitely soluble in water, whereas alcohols with more than six carbon atoms per molecule are virtually insoluble in water.

We can summarize the principle of "like dissolves like" as follows:

- ***Substances with similar noncovalent forces are likely to be soluble in each other.***
- ***Solutes do not readily dissolve in solvents whose noncovalent forces are quite different from their own.***
- ***Stronger solute-solvent attractions favor solubility. Stronger solute-solute or solvent-solvent attractions reduce solubility.***

Table 15.2 Solubilities of Some Alcohols in Water

Name	Formula	Lewis Structure	Solubility in Water (g/100 g H_2O at 20 °C)
Methanol	CH_3OH	H–C(H)(H)–O–H	Miscible
Ethanol	CH_3CH_2OH	H–C(H)(H)–C(H)(H)–O–H	Miscible
1-Propanol	$CH_3(CH_2)_2OH$	H–C(H)(H)–C(H)(H)–C(H)(H)–O–H	Miscible
1-Butanol	$CH_3(CH_2)_3OH$	H–C(H)(H)–C(H)(H)–C(H)(H)–C(H)(H)–O–H	7.9
1-Pentanol	$CH_3(CH_2)_4OH$	H–C(H)(H)–C(H)(H)–C(H)(H)–C(H)(H)–C(H)(H)–O–H	2.7
1-Hexanol	$CH_3(CH_2)_5OH$	H–C(H)(H)–C(H)(H)–C(H)(H)–C(H)(H)–C(H)(H)–C(H)(H)–O–H	0.6

CONCEPTUAL EXERCISE 15.1 Predicting Water Solubility

How could the data in Table 15.2 be used to predict the solubility in water of 1-octanol or 1-decanol?

CONCEPTUAL EXERCISE 15.2 Predicting Solubility

You have a sample of 1-octanol and a sample of methanol. Which is more water-soluble? Which is more soluble in gasoline? Explain your choices in terms of "like dissolves like."

PROBLEM-SOLVING EXAMPLE 15.1 Predicting Solubilities

A beaker initially contains three layers (Figure 15.3): In part (a) the top layer is colorless heptane, C_7H_{16} (density = 0.684 g/mL); the middle layer is a green solution of $NiCl_2$ in water (density = 1.10 g/mL); the bottom layer is colorless carbon tetrachloride, CCl_4 (density = 1.59 g/mL). After the liquids are mixed, two layers remain—a lower, colorless layer and an upper green-colored layer. Using the principle of "like dissolves like," explain what resulted when the layers were mixed.

ThomsonNOW

Log on to ThomsonNOW at **www.thomsonedu.com** and click Coached Problems for an exercise on solubility.

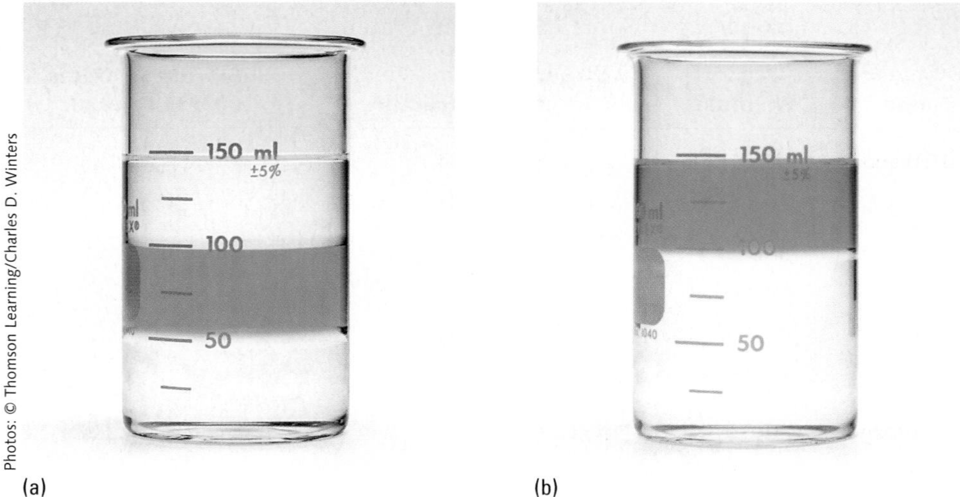

Photos: © Thomson Learning/Charles D. Winters

Figure 15.3 Miscibility of liquids. (a) This mixture was prepared by carefully layering three liquids of differing densities: a bottom layer of colorless carbon tetrachloride; a middle layer of green aqueous nickel chloride; and an upper layer of colorless heptane. (b) The mixture after the components have been mixed.

Answer The nonpolar liquids carbon tetrachloride and heptane dissolved in each other; neither the nickel chloride nor the water dissolved in the nonpolar liquids.

Strategy and Explanation To apply the "like dissolves like" principle requires identifying the polarity and intermolecular forces of the substances in the flask. Hexane is a hydrocarbon and thus nonpolar with London forces between hexane molecules. Likewise, carbon tetrachloride is a nonpolar molecule (no *net* dipole even though the C—Cl bonds are each polar) due to the symmetry of its tetrahedral shape (⇐ *p. 383*). Nickel chloride is an ionic solid that dissolves in water to form a green solution due to the attraction of polar water molecules for the Ni^{2+} cations and Cl^- anions in the compound. Thus, there is little attraction for the polar aqueous solution to mix with either of the nonpolar liquids above and below the aqueous layer. Once the mixture is stirred to mix the components, the nonpolar liquids dissolve in each other to form a bottom layer denser than the upper aqueous layer.

PROBLEM-SOLVING PRACTICE 15.1

Using the principle of "like dissolves like," predict whether
(a) Ethylene glycol, $HOCH_2CH_2OH$, dissolves in gasoline
(b) Molecular iodine dissolves in carbon tetrachloride
(c) Motor oil dissolves in carbon tetrachloride
Explain your predictions.

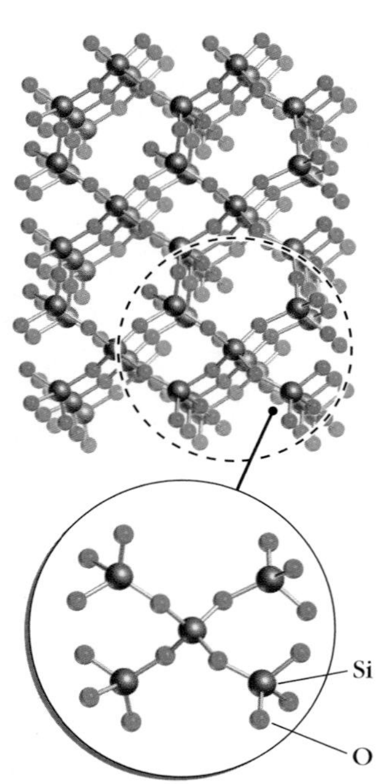

A portion of a quartz (SiO_2) structure. The structure is based on SiO_4 tetrahedra linked through shared oxygen atoms.

Solids such as quartz (SiO_2) that are held together by an extensive network of covalent bonds are generally insoluble in polar or nonpolar solvents. In quartz, the silicon and oxygen atoms form SiO_4 tetrahedra linked through covalent bonds to shared oxygen atoms. These strong covalent bonds are not broken by weaker attractions to solvent molecules. Thus, quartz (and sand derived from it) is insoluble in water or any other solvent at room temperature. Sandy beaches do not dissolve in ocean water.

The solubility of a substance in water or in the triglycerides—nonpolar fats or oily substances in our bodies (⇐ *p. 571*)—plays an important role in our body chemistry. Vitamins, for example, are either water-soluble or fat-soluble. Vitamins A, D, E, and K are known as fat-soluble vitamins because they dissolve in triglycerides. All the other vitamins are water-soluble. The major significance of this difference is the danger of overdosing on fat-soluble vitamins because they are stored in fatty tis-

ThomsonNOW™
Log on to ThomsonNOW at **www.thomsonedu.com** and click Chemistry Interactive for a module on:

- **like dissolves like: water, iodine, and carbon tetrachloride**

sues and may accumulate to harmful levels. By contrast, overdosing on water-soluble vitamins is difficult and uncommon because these vitamins are not stored in the body and any excess of them is excreted in urine.

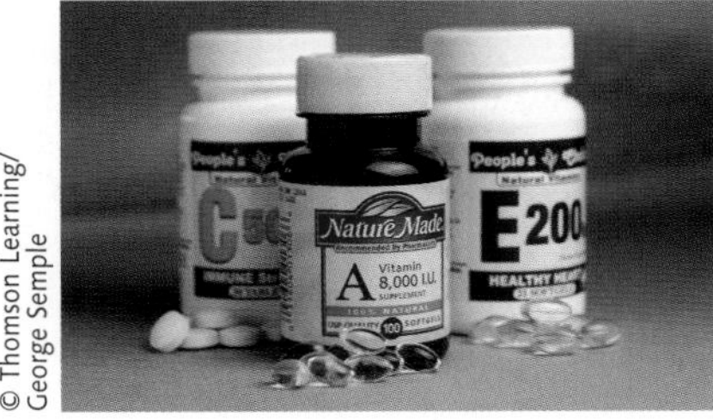

Vitamins are fat-soluble (A, D, E, K) or water-soluble (the rest).

PROBLEM-SOLVING EXAMPLE 15.2 Solubility and Noncovalent Forces

Use the structural formulas of niacin (nicotinic acid) and vitamin A to determine which vitamin is more soluble in water and which is more soluble in fat.

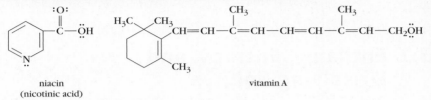

$$\text{C}_5\text{H}_4\text{N}{-}\text{C}(={:}\ddot{\text{O}}{:}){-}\ddot{\text{O}}\text{H}$$

niacin (nicotinic acid)

$$\text{(H}_3\text{C)(CH}_3\text{)(CH}_3\text{)C}_6\text{ring}{-}\text{CH}{=}\text{CH}{-}\overset{\text{CH}_3}{\text{C}}{=}\text{CH}{-}\text{CH}{=}\text{CH}{-}\overset{\text{CH}_3}{\text{C}}{=}\text{CH}{-}\text{CH}_2\ddot{\text{O}}\text{H}$$

vitamin A

Answer Niacin is water-soluble; vitamin A is fat-soluble.

Strategy and Explanation Use the structural formulas to determine whether any water-soluble or fat-soluble portions exist in the molecules.

Niacin is water-soluble because its —COOH group and nitrogen atom can hydrogen-bond with water molecules.

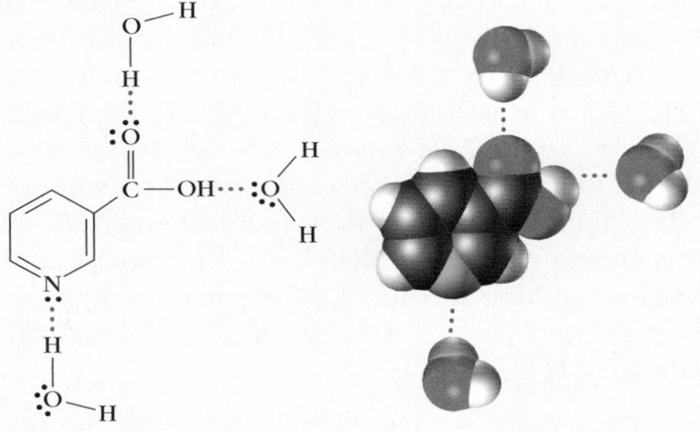

niacin
(nicotinic acid)

Niacin dissolves because the solute-solvent interaction between niacin and water is stronger than hydrogen bonding between water molecules or the dipole-dipole attractions between niacin molecules.

Vitamin A has an —OH group that can hydrogen-bond with water. However, as is the case with other long-chain alcohols, the hydrocarbon portion of the molecule is large enough to make the molecule hydrophobic. London forces between the hydrocarbon portions are more important than the hydrogen bonding, and vitamin A is insoluble in water. On the other hand, the extended hydrocarbon portion of the molecule is soluble in the long-chain hydrocarbon portion of fats.

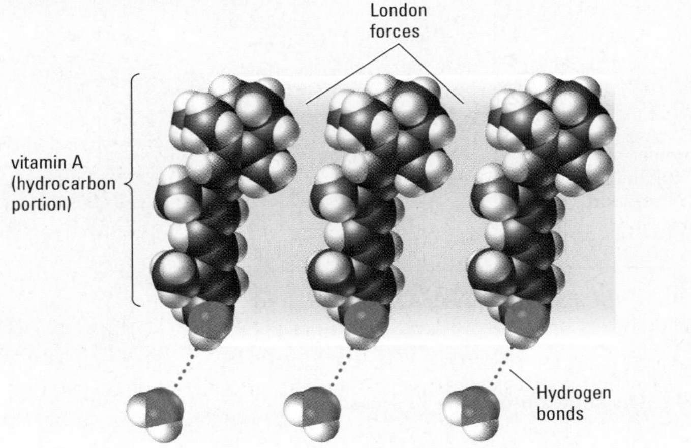

PROBLEM-SOLVING PRACTICE 15.2

Explain why vitamin C, which has the structure below, is water-soluble.

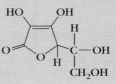

15.2 Enthalpy, Entropy, and Dissolving Solutes

ThomsonNOW
Log on to ThomsonNOW at **www.thomsonedu.com** and click Chemistry Interactive for modules on:

- **the energetics of dissolution of KF**
- **the energetics of mixing liquids**

A solution forms when atoms, molecules, or ions of one kind mix with atoms, molecules, or ions of a different kind. For the solution to form, intermolecular forces attracting solvent molecules to each other, and such forces attracting solute molecules or ions to each other must be overcome (*steps a and b* in Figure 15.4). In *step (a)* the enthalpy of the collection of solvent molecules increases due to their separation, resulting in an endothermic process with a positive ΔH ($\Delta H > 0$). Likewise, this must also occur in *step b* for the separation of solute ions or molecules. When the solute and solvent particles mix (*step c*), they attract each other and there is a decrease in enthalpy; ΔH is negative ($\Delta H < 0$) and energy is released in this exothermic step. If the enthalpy of the final solution is lower than that of the initial solute and solvent, as shown in Figure 15.4 (*left*), a net release of energy to the surroundings occurs and the solution-making process is *exothermic*. When the enthalpy of the final solution is greater than the enthalpies of the initial solute and solvent, the process is *endothermic* and the surroundings transfer energy to the system (Figure 15.4, *right*). The net energy change, either exothermic or endothermic, is called the **enthalpy of solution,** ΔH_{soln}.

The enthalpy of solution is also known as the heat of solution.

$$\text{Enthalpy of solution} = \Delta H_{soln} = \Delta H_{step\ a} + \Delta H_{step\ b} + \Delta H_{step\ c}$$

An exothermic solution-making process

Enthalpy, H
(b) Separation of solute molecules or ions (ΔH°_{solute})
$\Delta H^\circ_{solvent} + \Delta H^\circ_{solute}$ is > 0.
(c) Mixing of solvent and solute
(a) Separation of solvent molecules ($\Delta H^\circ_{solvent}$)
Solution formation (mixing of solvent and solute); ΔH° solution is exothermic ($\Delta H^\circ < 0$).
Solvent Solute
Net energy change
Solution

An endothermic solution-making process

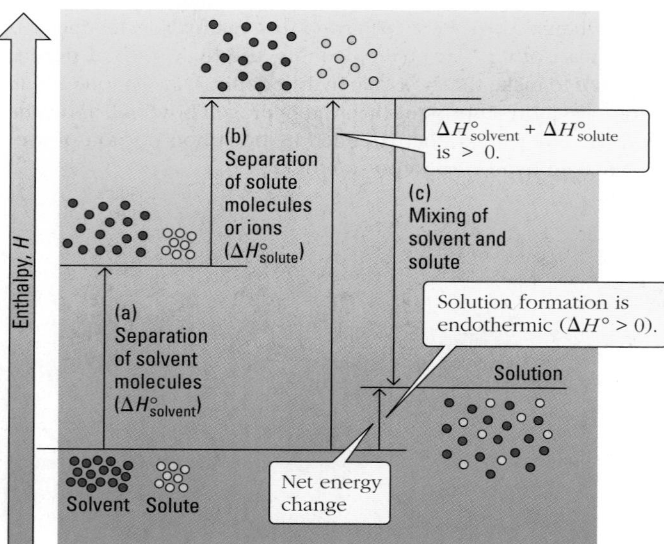

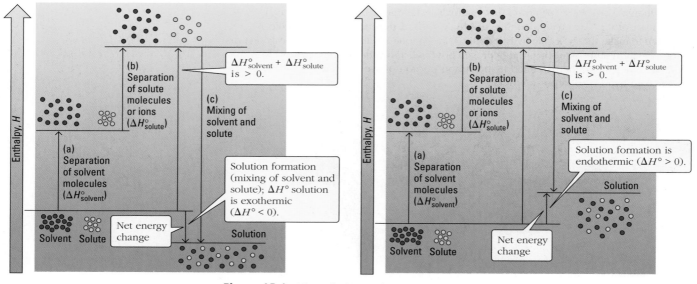

Figure 15.4 The solution-making process.

After the solute and solvent mix, the solvent-solute forces (indicated by the magnitude of $\Delta H_{\text{step c}}$) may not be strong enough to overcome the solute-solute and solvent-solvent attractions ($\Delta H_{\text{step a}} + \Delta H_{\text{step b}}$). Under these conditions, dissolving is not favored by the enthalpy effect *(⇐ p. 705).*

When solutes and solvents mix to form solutions, solute and solvent molecules usually spread out over a larger volume. This spreads their energy over a larger volume and results in an increase in entropy *(⇐ p. 704).* As described in Section 14.7, *entropy* (symbolized by S) is a measure of the energy dispersal of a system; an increase in entropy ($\Delta S > 0$) indicates a process where energy is spread out more, and dispersal of energy corresponds with dispersal of matter. *Processes in which the entropy of the system increases tend to be product-favored.* Therefore, in many cases of solution making, a large entropy increase occurs as solvent and solute molecules mix. When the enthalpy of solution is rather small, entropy is the driving force behind solution formation, such as when octane and carbon tetrachloride are mixed. In other cases, even though the enthalpy of solution is positive and significantly large ($\Delta H_{\text{soln}} > 0$), the entropy increase is large enough to cause the solution to form. This is true for some ionic solutes, such as ammonium nitrate, NH_4NO_3 (Section 15.3). Because molecules in gases are always much more dispersed than liquids, there is a significant decrease in entropy ($\Delta S < 0$) as gas solute molecules are brought closer together between solvent molecules.

Richard Megna/Fundamental Photographs

A solute (sugar) dissolving in a solvent (water). As the solute dissolves, concentration gradients are observed as wavy lines.

15.3 Solubility and Equilibrium

Some solutes dissolve to a much greater extent than others in the same mass of the same solvent. The **solubility** of a solute is the maximum quantity of solute that dissolves in a given quantity of solvent at a particular temperature (Figure 15.5). A solution can be described as saturated, unsaturated, or supersaturated depending on the quantity of solute that is dissolved in it. A **saturated solution,** as its name implies, is one whose solute concentration equals its solubility (all points *along* the curve in Figure 15.5). When a saturated solution forms, there is a *dynamic equilibrium* between undissolved and dissolved solute.

Remember that the concentration of a pure substance (solute or solvent) does not appear in the K_c expression.

$$\text{Solute} + \text{solvent} \rightleftharpoons \text{solute (in solution)} \qquad K_c = [\text{solute (in solution)}]$$

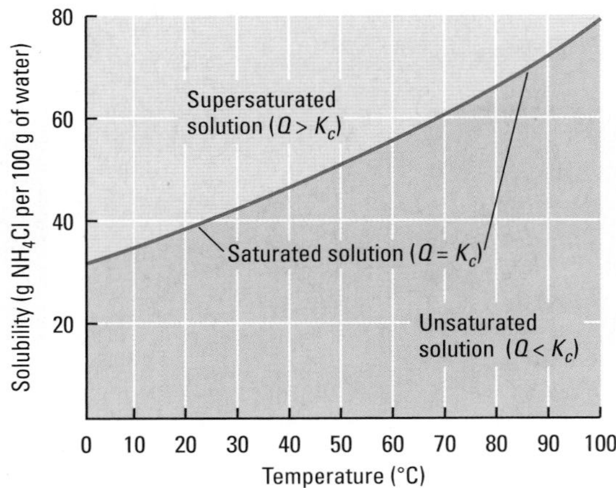

Figure 15.5 Types of solutions. The solubility of solid ammonium chloride and three types of ammonium chloride solutions—saturated (the curve), unsaturated (below the curve), and supersaturated (above the curve).

Some solute molecules or ions are mixing with solvent molecules and going into solution, while others are separating from solvent molecules and entering the pure solute phase. Both processes are going on simultaneously at identical rates. When a solid dissolves in a liquid, for example, the quantity of solid is observed to decrease until the solution becomes saturated. A saturated solution is in equilibrium with its solute and solvent, and in this case $Q = K_c$ (⇐ **p. 689**). Once equilibrium is reached the quantity of solid solute (and the quantity of dissolved solute) stops changing.

An **unsaturated solution** is one in which the solute concentration is less than its solubility; that is, an unsaturated solution can accommodate additional solute at a given temperature (region *under* the curve in Figure 15.5). Thus, for an unsaturated solution, $Q < K_c$, and more solute can dissolve. If solute continues to be added to an unsaturated solution at a given temperature, a point is reached where the solution becomes saturated, at which point $Q = K_c$.

Unsaturated solution: $Q < K_c$; saturated solution: $Q = K_c$; supersaturated solution: $Q > K_c$.

For some solutes, there is a third case. It is possible to prepare solutions that contain *more* than the equilibrium concentration of solute at a given temperature; such a solution is **supersaturated** (the region *above* the curve in Figure 15.5). For example, a supersaturated solution of ammonium chloride can be made by first making a saturated solution at 90 °C. This is done by adding 72 g of solid NH_4Cl to 100 g water and heating the solution to 90 °C. At that point the solution is saturated. To make the supersaturated solution, the temperature of the saturated solution is lowered very slowly to 25 °C. If this is done carefully, none of the NH_4Cl will crystallize out of solution, and the resulting solution will be supersaturated, holding more dissolved NH_4Cl than a saturated solution at that temperature; 72 g of the solute is dissolved at 25 °C, whereas the solution should hold a maximum of 40 g at that temperature (Figure 15.5). In equilibrium terms, $Q > K_c$, and excess solute should crystallize out of solution, but it is very slow to do so. Some supersaturated solutions, such as honey, can remain so for days or months. To form a crystal, several ions or molecules must move into an arrangement much like the appropriate crystal lattice positions, and it can take a long time for such an alignment to occur by chance. However, precipitation of a solid from a supersaturated solution occurs rapidly if a tiny crystal of the solute is added to the solution (Figure 15.6). The lattice of the added crystal provides a template onto which more ions or molecules can be added.

Fudge is made using a supersaturated sugar solution. In smooth fudge, the sugar remains uncrystallized; poor-quality fudge has a gritty texture because it contains crystallized excess sugar.

Sometimes other actions, such as stirring a supersaturated solution or scratching the inner walls of its container, will cause solute to precipitate rapidly.

(a)

(b)

(c)

(d)

(e)

Figure 15.6 Crystallization from a supersaturated solution. Sodium acetate, $Na(CH_3COO)$, easily forms supersaturated solutions in water. (a) The solution on the left looks ordinary, but it is supersaturated, holding more dissolved sodium acetate than a saturated solution at that temperature. The supersaturated solution was prepared by dissolving a quantity of sodium acetate in water at a much higher temperature and cooling the solution very slowly. (b) After a tiny seed crystal of sodium acetate is added, some of the excess dissolved sodium acetate immediately begins to crystallize. (c through e) Very soon, numerous sodium acetate crystals can be seen. If the solution remains uncovered for a long time, all of the water will evaporate, leaving behind the solid sodium acetate.

EXERCISE 15.3 Crystallizing Out of Solution

Using Figure 15.5, how many grams of excess NH_4Cl would crystallize out of the supersaturated NH_4Cl solution described previously at 25 °C?

EXERCISE 15.4 Solubility

Refer to Figure 15.5 to determine whether each of these NH_4Cl solutions in 100 g H_2O is unsaturated, saturated, or supersaturated:

(a) 30 g NH_4Cl at 70 °C (b) 60 g NH_4Cl at 60 °C

(c) 50 g NH_4Cl at 50 °C

Dissolving Ionic Solids in Liquids

Sodium chloride is an ionic compound whose crystal lattice consists of Na^+ and Cl^- ions in a cubic array *(⟵ p. 517)*. Strong electrostatic attractions between oppositely charged ions hold the ions tightly in the lattice. The enthalpy change when 1 mol of Na^+ and Cl^- ions is completely separated from a crystal lattice is referred to as the lattice energy of the ionic compound *(⟵ p. 321)*. The large lattice energy of NaCl (788 kJ/mol) accounts for sodium chloride's high melting point (800 °C). It is possible for solvent molecules to attract Na^+ ions away from Cl^- ions in a crystal, but they have to be the right kind of solvent molecules. Trying to dissolve NaCl (or any other ionic compound) with carbon tetrachloride or hexane (both nonpolar solvents) is a futile exercise because nonpolar molecules have very little attraction for ions. On the other hand, water dissolves NaCl.

ThomsonNOW

Log on to ThomsonNOW at **www.thomsonedu.com** and click Coached Problems for exercises on:

- **dissolving ionic compounds**
- **factors affecting solubility**

Water is a good solvent for an ionic compound because water molecules are small and highly polar *(⟵ p. 404)*. As shown in Figure 15.7, the partially negative

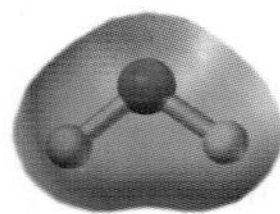

Electron density of a water molecule. The red area has high electron density and partial negative charge; the blue area has low electron density and partial positive charge.

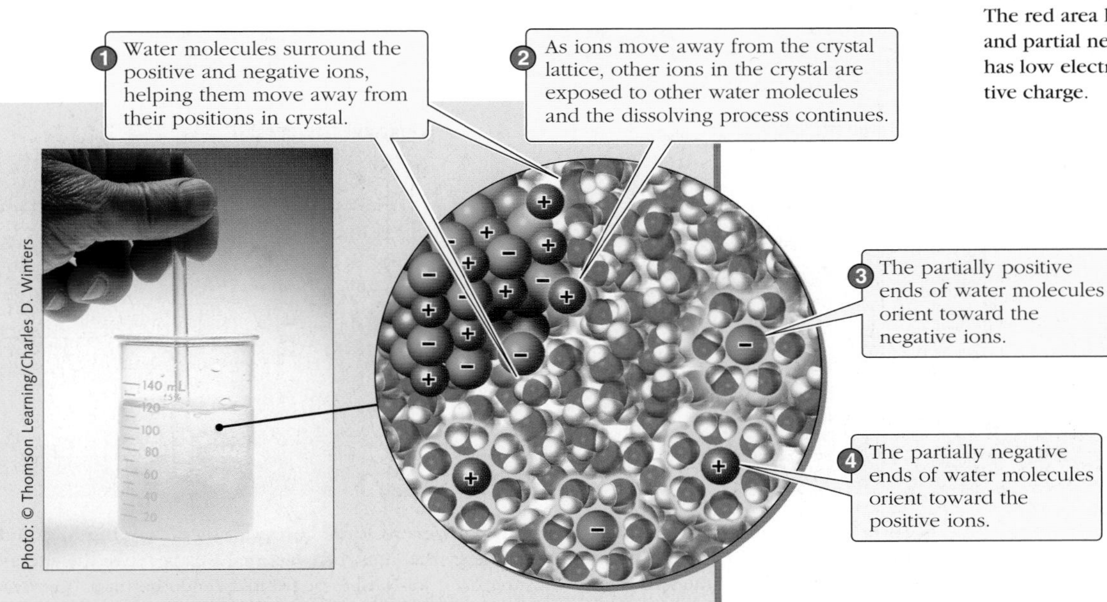

1 Water molecules surround the positive and negative ions, helping them move away from their positions in crystal.
2 As ions move away from the crystal lattice, other ions in the crystal are exposed to other water molecules and the dissolving process continues.
3 The partially positive ends of water molecules orient toward the negative ions.
4 The partially negative ends of water molecules orient toward the positive ions.
140 mL
120
100
80
60
Photo: © Thomson Learning/Charles D. Winters

Figure 15.7 Water dissolving an ionic solid. Water molecules surround the positive (gray) and negative (green) ions, helping them move away from their positions in the crystal.

Enthalpies of Hydration of Selected Ions (kJ/mol)

Cations		Anions	
H^+	−1130	F^-	−483
Li^+	−558	Cl^-	−340
Na^+	−444	Br^-	−309
Mg^{2+}	−2003		
Ca^{2+}	−1557		
Al^{3+}	−2537		

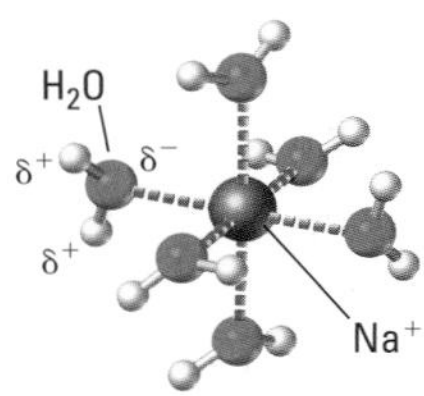

Figure 15.8 Hydration of a sodium ion. The arrangement of water molecules around this Na^+ ion is highly ordered.

Energy is always released when particles that attract each other get closer together, so the lattice energy is always negative. Energy is always required to separate particles that attract each other.

ThomsonNOW
Log on to ThomsonNOW at **www.thomsonedu.com** and click Chemistry Interactive for a module on:

- **chemical hot packs**

oxygen atoms of water molecules are attracted to positive ions and help pull them away from the crystal lattice, while the partially positive hydrogen atoms of other water molecules are attracted to the negative ions in the lattice and help pull them away from the lattice. This process, in which water molecules surround positive and negative ions, is called **hydration** (Figure 15.8). Energy known as the *enthalpy of hydration* is released when these new attractions form between ions and water molecules as they mix and get close enough to one another. Energy is always required to separate the ions to overcome their attraction, and energy is always released when ions become hydrated because noncovalent interactions are being formed. Whether dissolving a particular ionic compound is exothermic (ΔH_{soln} is negative) or endothermic (ΔH_{soln} is positive) depends on the relative sizes of the lattice energy *(⟵ p. 321)* of the ionic compound and the hydration enthalpies of its positive and negative ions. The relationship between the enthalpy of solution, the lattice energy of the ionic compound, and the enthalpies of hydration of the ions is

$$\Delta H_{soln} = -\text{lattice energy} + \Delta H_{hydration}\,(\text{cations}) + \Delta H_{hydration}\,(\text{anions})$$

Figure 15.9 shows how lattice energy and the enthalpies of hydration combine to give the enthalpy of solution.

Practical applications of endothermic and exothermic dissolution include cold packs containing NH_4NO_3 used to treat athletic injuries and hot packs containing $CaCl_2$ used to warm foods (page 733).

The solubility rules for ionic compounds in water *(⟵ p. 165)* remind us that not all ionic compounds are highly water-soluble in spite of the strong attractions between water molecules and ions. For some ionic compounds, the lattice energy is so large that water molecules cannot effectively pull ions away from the lattice. As a result, such compounds have large positive enthalpies of solution and usually are only slightly soluble.

Entropy and the Dissolving of Ionic Compounds in Water

The spreading of energy introduced when a crystal lattice breaks down, and the spreading of energy introduced by the mixing of ions with solvent molecules, both

Small endothermic $\Delta H_{solution}$

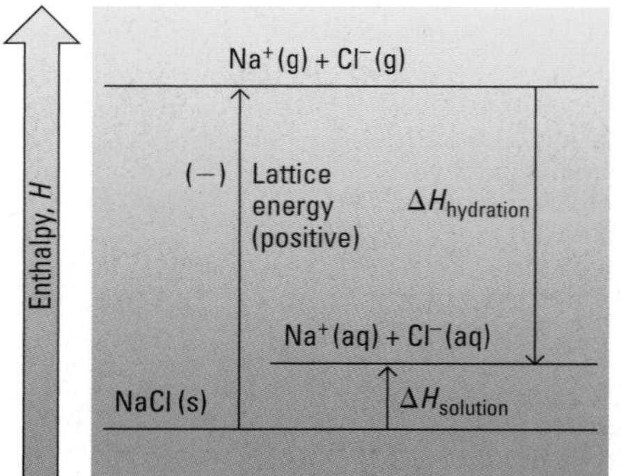

Large exothermic $\Delta H_{solution}$

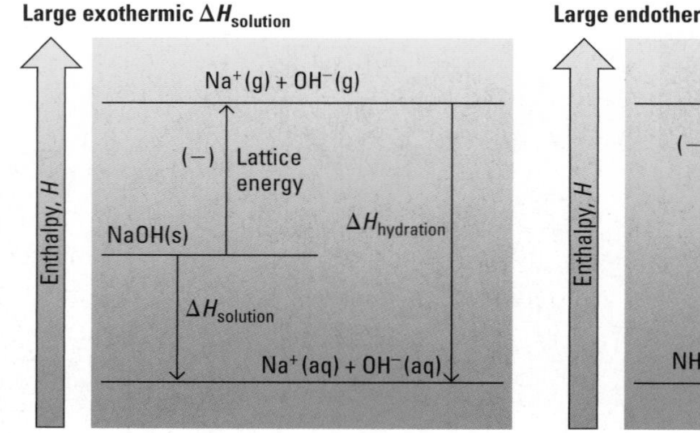

Large endothermic $\Delta H_{solution}$

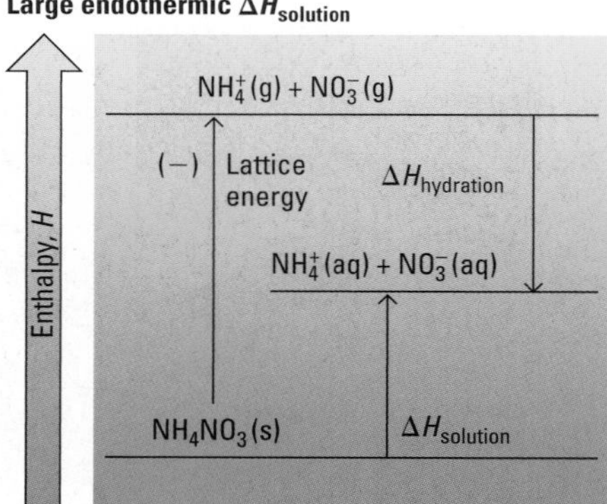

Figure 15.9 Enthalpies of solution of three different ionic compounds dissolving in equal volumes of water. *(Left)* For NaCl, the lattice energy that must be overcome is larger than the energy released on hydration of the ions. This causes the ΔH_{soln} of NaCl to be positive (endothermic). *(Center)* The lattice energy for NaOH is much smaller than the energy released when the ions become hydrated. Consequently, the ΔH_{soln} of NaOH is negative (highly exothermic). *(Right)* For NH_4NO_3, the heat of hydration of the ions is much smaller than the lattice energy, so the resulting ΔH_{soln} has a large positive value (highly endothermic).

favor the dissolving process ($\Delta S > 0$). This entropy increase is counteracted by the ordering of solvent molecules around the ions ($\Delta S < 0$) (Figure 15.8). For 1+ and 1− charged ions, the overall entropy change is positive, and dissolving is favored. For some salts that contain 2+ or 3+ ions, the charges on the ions are so large and the ions so small that water molecules are aligned in a highly organized manner around the ions. When a large number of water molecules are locked into place by this strong hydration, the entropy of solution may be negative, which does not favor solubility. Calcium oxide (only 0.131 g CaO dissolves in 100 mL of water at 10 °C) and aluminum oxide Al_2O_3 (insoluble) exemplify this effect.

Commercial cold and hot packs. *(Top)* When the desired cooling is needed, ammonium nitrate in the inner container of this cold pack is brought into contact with water in the outer container by breaking the seal of the inner container. *(Bottom)* This hot pack releases heat when the inner pouch containing either $CaCl_2$ or $MgSO_4$ is punctured and the compound dissolves in the water in the outer container.

15.4 Temperature and Solubility

Solubility of Gases

To understand how temperature affects gas solubility, we can apply Le Chatelier's principle (← *p. 695*) to the equilibrium between a pure solute gas, the solvent, and a saturated solution of the gas.

$$\text{Gas} + \text{solvent} \rightleftharpoons \text{saturated solution} \qquad \text{Usually, } \Delta H_{soln} < 0 \text{ (exothermic)}$$

When a gas dissolves to form a saturated liquid solution, the process is almost always exothermic. Gas molecules that were relatively far apart are brought much closer in the solution to other molecules that attract them, lowering the potential energy of the system and releasing some energy to the surroundings.

If the temperature of a solution of a gas in a liquid increases, the equilibrium shifts in the direction that partially counteracts the temperature rise. That is, the equilibrium shifts to the left in the preceding equation. Thus, a dissolved gas becomes less soluble with increasing temperature. Conversely, cooling a solution of a gas that is at equilibrium with undissolved gas will cause the equilibrium to shift to the right in the direction that liberates heat, so more gas dissolves. This is illustrated in Figure 15.10 with data for the solubility of oxygen in water.

Cooler water in contact with the atmosphere contains more dissolved oxygen at equilibrium than water at a higher temperature. For this reason fish seek out cooler (usually deeper) waters in the summer. Fish have an easier time obtaining

Although generally useful, Le Chatelier's principle does not always correctly predict how the solubility of ionic solutes changes with temperature.

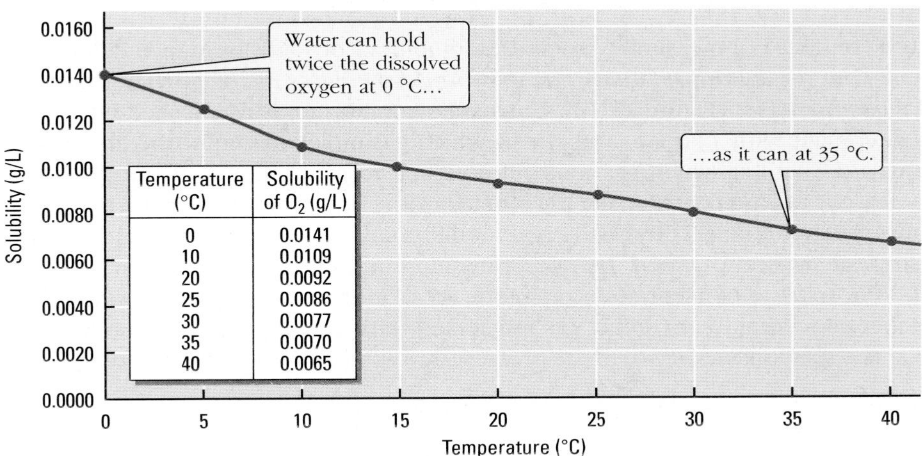

Temperature (°C)	Solubility of O_2 (g/L)
0	0.0141
10	0.0109
20	0.0092
25	0.0086
30	0.0077
35	0.0070
40	0.0065

Figure 15.10 Solubility of oxygen in water at various temperatures. The solubility of oxygen, like that of other gases, decreases with increasing temperature.

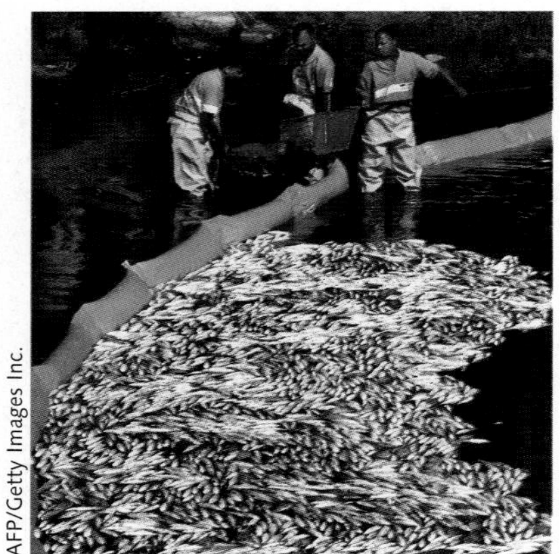

A fish kill caused by a lack of dissolved oxygen.

oxygen when its concentration in the water is higher. The decrease in gas solubility as temperature increases makes *thermal pollution* a problem for aquatic life in rivers and streams. Natural heating of water by sunlight and by warmer air can usually be accommodated. But excess heat from extended periods of very hot weather or from sources such as industrial facilities and electrical power plants can warm the water sufficiently that the concentration of dissolved oxygen is reduced to the point where some species of fish die.

CONCEPTUAL EXERCISE 15.5 Carbonated Beverages

Explain on a molecular basis why a carbonated beverage goes "flat" once it is opened and warms to room temperature.

CONCEPTUAL EXERCISE 15.6 Warming Water

Explain why water that has been used to cool a nuclear power plant, and thus is at a relatively high temperature, must be cooled before it is put back into the lake or river from which it came.

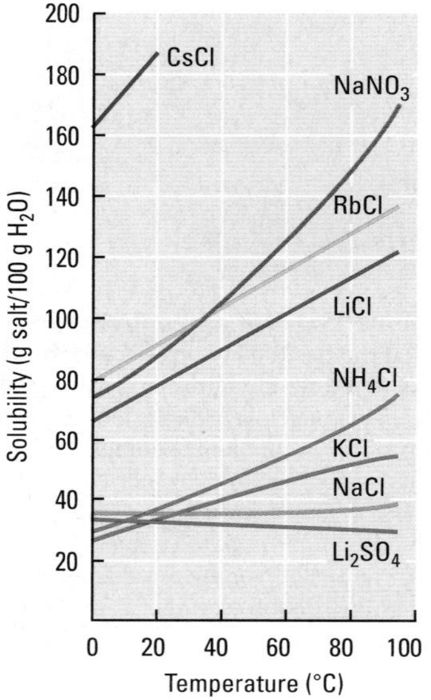

Figure 15.11 Solubility of ionic compounds and temperature. The solubility of most ionic compounds in water depends on the temperature.

Solubility of Solids

Common experience tells us that more sugar (sucrose) can dissolve in hot coffee or hot tea than in cooler coffee or tea. This is an example of the fact that the aqueous solubility of most solid solutes, including ionic compounds, increases with increasing temperature (Figure 15.11). Although predictions of solubility based on enthalpies of solution usually work, there are notable exceptions, such as Li_2SO_4.

CONCEPTUAL EXERCISE 15.7 Temperature and Solubility

If a substance has a positive enthalpy of solution, which would likely cause more of it to dissolve, hot solvent or cold solvent? Explain.

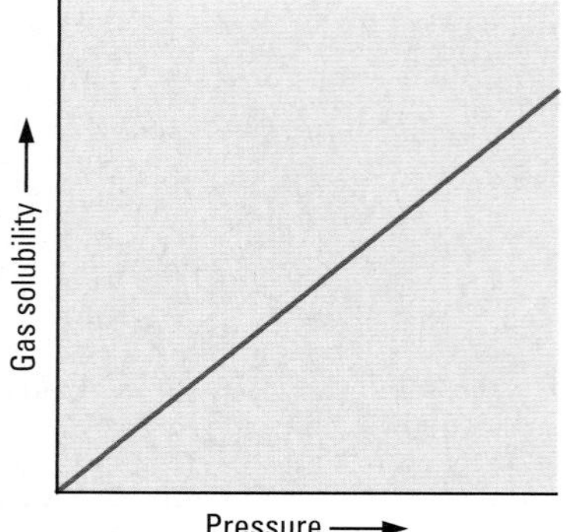

Figure 15.12 Pressure dependence of the solubility of a gas in a liquid. All gases that do not react with the solvent behave this way.

15.5 Pressure and Dissolving Gases in Liquids: Henry's Law

Pressure does not measurably affect the solubilities of solids or liquids in liquid solvents, but *the solubility of any gas in a liquid increases as the partial pressure of the gas increases* (Figure 15.12). A dynamic equilibrium is established when a gas is in contact with a liquid—the rate at which gas molecules enter the liquid phase equals the rate at which gas molecules escape from the liquid. If the pressure is increased, gas molecules strike the surface of the liquid more often, increasing the rate of dissolution of the gas. A new equilibrium is established when the rate of escape increases to match the rate of dissolution. The rate of escape is first order (⇐ ***p. 618***) in concentration of solute, so a higher rate of gas escape requires a higher concentration of solute gas molecules—that is, a higher gas solubility.

The relationship between gas pressure and solubility is known as **Henry's law:**

$$S_g = k_H P_g$$

where S_g is the solubility of the gas in the liquid, and P_g is the pressure of the gas above the solution (or the partial pressure of the gas if the solution is in contact with a mixture of gases). The value of k_H, known as the Henry's law constant, depends on the identities of both the solute and the solvent and on the temperature (Table 15.3). It has units of moles per liter per millimeter of mercury (mm Hg).

Figure 15.13 illustrates how gas solubility depends on pressure. The behavior of a carbonated drink when the cap is opened is an everyday illustration of the solubility of gases in liquids under pressure. The drink fizzes when opened because the partial pressure of CO_2 over the solution drops, the solubility of the gas decreases, and dissolved gas escapes from the solution.

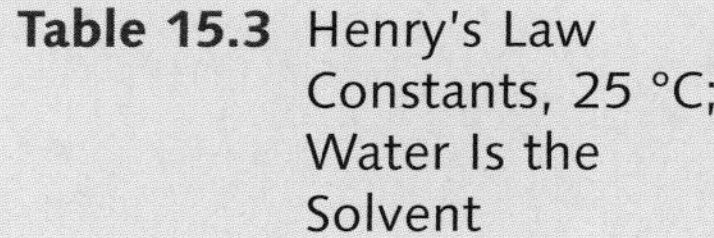

Table 15.3 Henry's Law Constants, 25 °C; Water Is the Solvent

Gas	k_H mol L^{-1} mm Hg^{-1}
N_2	8.42×10^{-7}
O_2	1.66×10^{-6}
CO_2	3.4×10^{-2}
He	3.7×10^{-4}
Air	6.5×10^{-1}

Henry's law. The greater the partial pressure of CO_2 over the soft drink in the bottle, the greater the concentration of dissolved CO_2. When the bottle is opened, the partial pressure of CO_2 drops and CO_2 bubbles out of the solution.

PROBLEM-SOLVING EXAMPLE 15.3 Using Henry's Law

The "bends" is a condition that can occur in scuba divers due to the solubility of N_2 in the blood. Calculate

(a) the number of grams of dissolved N_2 in 1 L blood at 25 °C.

(b) the volume (mL) of N_2 dissolved in 1 L blood at 25 °C. The Henry's law constant for N_2 in water at 25 °C is 8.42×10^{-7} mol L^{-1} mm Hg^{-1}. Assume that the solubility of N_2 in blood is approximately that of N_2 in water. The partial pressure of N_2 in the atmosphere is 0.78 atm.

Answer (a) 1.40×10^{-2} g N_2 (b) 16 mL of N_2

Strategy and Explanation The solubility of N_2 in blood (g/L) can be calculated, but first the solubility must be calculated in terms of molarity (mol/L) using Henry's law, and then the result is converted to grams by using the molar mass. The volume (mL) of N_2 in 1 L blood can then be calculated using the ideal gas law.

(a) We assume that the dissolved N_2 is in equilibrium with nitrogen in the air. Converting atm to mm Hg we have

$$0.78 \text{ atm} \times \frac{760 \text{ mm Hg}}{1 \text{ atm}} = 593 \text{ mm Hg}$$

Use Henry's law to calculate the solubility of N_2, $S_{N_2} = k_H P_{N_2}$, in mol/L:

$$S_{N_2} = k_H P_{N_2} = (8.42 \times 10^{-7} \text{ mol L}^{-1} \text{ mm Hg}^{-1})(593 \text{ mm Hg}) = 4.99 \times 10^{-4} \text{ mol/L}$$

$$\text{mass of } N_2 = (4.99 \times 10^{-4} \text{ mol/L})\left(\frac{28.02 \text{ g}}{1 \text{ mol}}\right) = 1.40 \times 10^{-2}\ g_{N_2}/\text{L}$$

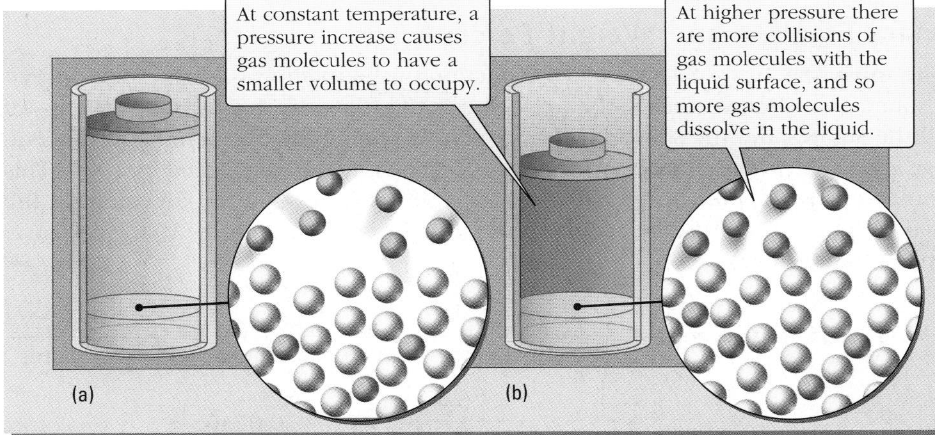

At constant temperature, a pressure increase causes gas molecules to have a smaller volume to occupy.
At higher pressure there are more collisions of gas molecules with the liquid surface, and so more gas molecules dissolve in the liquid.
(a)
(b)

Figure 15.13 **Henry's law.** (a) A gas sample in a liquid under pressure in a closed container. (b) The pressure is increased at constant temperature, causing more gas to dissolve.

The partial pressure of a gas in a mixture of gases is the pressure that a pure sample of the gas would exert if it occupied the same volume as the mixture. Partial pressure is proportional to the mole fraction of the gas (← p. 455).

(b) Apply the ideal gas law to calculate the volume (mL) of the 4.99×10^{-4} mol N_2 dissolved in 1 L blood. Remember that the temperature must be in kelvins.

$$V = \frac{nRT}{P} = \frac{(4.99 \times 10^{-4}\ \text{mol})(0.0821\ \text{L atm mol}^{-1}\ \text{K}^{-1})(298\ \text{K})}{0.78\ \text{atm}}$$

$$= 0.0157\ \text{L} = 16\ \text{mL}$$

This is a significant volume of nitrogen gas dissolved in one liter of blood, which can cause serious pain when the gaseous nitrogen is released from the blood as the scuba diver rises to the surface of the water.

ThomsonNOW
Log on to ThomsonNow at **www.thomsonedu.com** and click Coached Problems for a module on **Henry's law: the Diet Coke video.**

PROBLEM-SOLVING PRACTICE 15.3

The Henry's law constant for oxygen in water at 25 °C is 1.66×10^{-6} mol L^{-1} mm Hg^{-1}. Suppose that a trout stream at 25 °C is in equilibrium with air at normal atmospheric pressure. What is the concentration of O_2 in this stream? Express the result in milligrams per liter (mg/L). The mole percent of oxygen in air is 21%.

15.6 Solution Concentration: Keeping Track of Units

The terms *unsaturated, saturated*, and *supersaturated* describe a solution with respect to the quantity of solute in a given quantity of solvent at a certain temperature. Sometimes the terms *concentrated* or *dilute* are also used to describe a solution. Although useful, these are all qualitative descriptors of solution concentration, and it is often important to specify the concentration of a solute in a solution more precisely. Several concentration units are used to do so, including mass fraction, weight percent, molarity, and molality. Often it is useful to be able to quantitatively describe quantities of solute across a wide range of concentrations, from rather large to very small. For example, a variety of units are used to express the concentrations of dilute solutions containing unwanted—even potentially harmful—ionic solutes, such as those containing lead, mercury, selenium, and nitrate, and organic compounds in drinking water. These units are also useful clinically in discussing solute concentrations in blood or urine.

Mass Fraction and Weight Percent

The **mass fraction** of a solute is the fraction of the total mass of the solution that a solute contributes—that is, the mass of a single solute divided by the total mass of all the solutes and the solvent. Mass fraction is commonly expressed as a percentage and called **weight percent,** which is the mass fraction multiplied by 100%. This is the same as the number of grams of solute per 100 g of solution. For example, the mass fraction of sucrose in a solution consisting of 25.0 g sucrose, 10.0 g fructose, and 300. g water is

The symbol % means "per hundred," that is, divided by 100. Thus, 100% is $\frac{100}{100} = 1$.

$$\text{Mass fraction of sucrose} = \frac{\text{mass of sucrose}}{\text{mass of sucrose} + \text{mass of fructose} + \text{mass of water}}$$

$$= \frac{25.0\ \text{g}}{25.0\ \text{g} + 10.0\ \text{g} + 300.\ \text{g}} = 0.0746$$

Weight percent can be thought of as parts per hundred.

The weight percent of sucrose in the solution is $(0.0746) \times 100\% = 7.46\%$.

CHEMISTRY IN THE NEWS

Bubbling Away: Delicate and Stout

In the movie *Gigi,* Maurice Chevalier sings of "the night they invented champagne," that fizzy, celebratory beverage. As bubbles rise in a glass of champagne, their fizz and pop capture our attention. Recently, researchers in France and Australia have investigated more closely the bubbles in champagne and in beer (the body of the liquid, not the foamy "head" on beer). In both drinks the bubbles are carbon dioxide, produced as a by-product of the natural process that produces alcohol via fermentation of sugars in grapes (wine, champagne) or grains (beer).

$$\underset{\text{glucose}}{C_6H_{12}O_6} \longrightarrow 2\ \underset{\text{ethanol}}{CH_3CH_2OH} + 2\ CO_2$$

The properties of the bubbles are affected by the concentration of dissolved gas, the alcohol content of the drink, and the concentration of large molecules, including proteins, present in the drink. Champagne has three times the CO_2 concentration and twice the alcohol content of beer. Beer has as much as 30 times more large molecules than champagne.

Bubbles in beverages generally start to grow from what are called *nucleation sites*—places where the gas comes out of solution—such as on dust particles or on tiny imperfections on the inside walls of a glass. When a bubble gets sufficiently buoyant, it breaks away from the nucleation site. As CO_2 diffuses into the bubble, it gets larger, its buoyancy increases, and it moves more quickly to the surface. Gerard Liger-Belair and his colleagues at the University of Reims in France have used high-speed cameras to measure the growth and ascent rate of champagne bubbles, whose radii grow at the rate of 120 to 240 μm per second as they ascend from the sides of a champagne flute.

© Thomson Learning/George Semple

(a) Glass of Guinness stout.

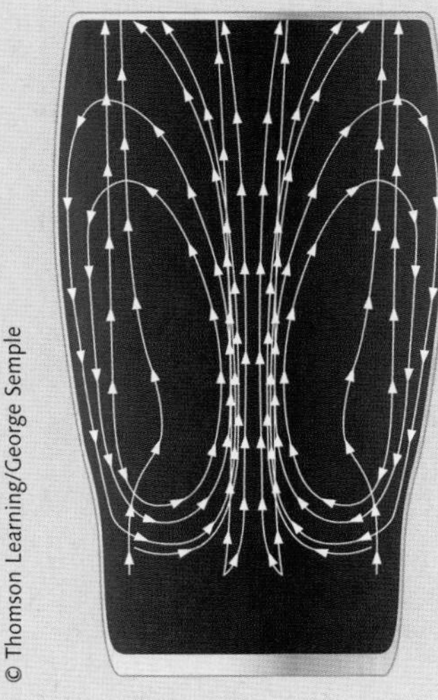

© Thomson Learning/George Semple

(b) Computer simulation of bubbles traveling in Guinness stout.

© Thomson Learning/Charles D. Winters

Rising bubbles in champagne.

Whereas bubbles in champagne always rise, those in the Irish beer Guinness stout rise and fall within the body of the beer, seemingly defying the laws of nature. For years, pub regulars argued whether the bubbles in stout actually were falling as well as rising. Was it perhaps an optical illusion or an alcohol-induced error of the observer? The higher protein concentration in beer causes more "drag" on the bubbles, slowing their rate of ascent much more than in champagne. At the higher protein concentration, the bubbles in beer act as rigid spheres because they collect a protein "skin" as they rise. The lower protein content in champagne does not create bubbles that are as rigid. Champagne bubbles would have to travel more than 10 cm (about 4 in) to pick up enough protein to become as rigid as beer bubbles. This is longer than a typical champagne flute. That explains differences for rising bubbles. But do the bubbles in stout actually descend? Using computer modeling software, Australian researchers Clive Fletcher and colleagues demonstrated that, unlike those in champagne, bubbles in Guinness stout ascend *and* descend. The critical issue is the rigidity, size, and bouyancy of the bubbles. Beer bubbles with a radius of about 30 μm sink near the sides of the glass while 500-μm ones only rise. The barrel-chested shape of the typical Guinness glass also seems to be a factor.

Richard Zare, a Stanford University chemistry professor who also has published articles about beer bubbles, points out that "Once you begin to learn about the nature of beer bubbles, you will never again look at a glass of beer in quite the same way."

SOURCE: Weiss, P. "The Physics of Fizz." *Science News*, Vol. 157, May 6, 2000; p. 300.

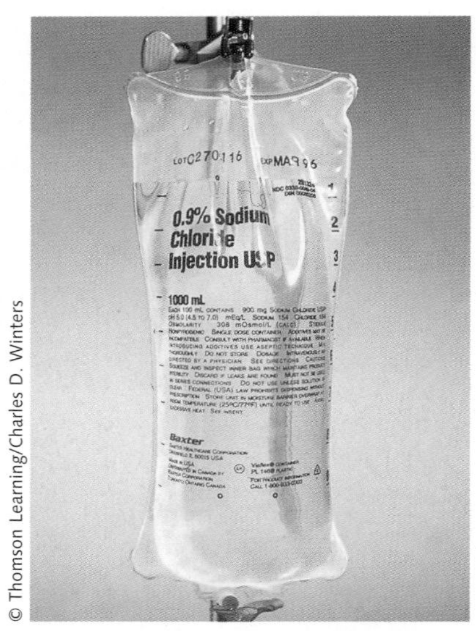

Sterile saline solution. Saline solutions like this one are routinely given to patients who have lost body fluids.

PROBLEM-SOLVING EXAMPLE 15.4 Mass Fraction and Weight Percent

Sterile saline solutions containing NaCl in water are often used in medicine. What is the weight percent of NaCl in a solution made by dissolving 4.6 g NaCl in 500. g pure water?

Answer 0.91%

Strategy and Explanation We apply the definitions of mass fraction and weight percent to the solute and solution stated in the problem.

$$\text{Mass fraction of NaCl} = \frac{4.6\text{ g NaCl}}{4.6\text{ g NaCl} + 500.\text{ g H}_2\text{O}} = 0.0091$$

$$\text{Weight percent of NaCl} = \text{mass fraction of NaCl} \times 100\%$$

$$= 0.0091 \times 100\% = 0.91\%$$

✓ **Reasonable Answer Check** The mass fraction is about 5 g NaCl in about 500 g solution, or about 0.01, which agrees with the more accurate result.

PROBLEM-SOLVING PRACTICE 15.4

What is the weight percent of glucose in a solution containing 21.5 g glucose ($C_6H_{12}O_6$) in 750. g pure water?

The density of liquid water is essentially 1 g/mL, 10^3 g/L, or 1 kg/L.

EXERCISE 15.8 Mass Fraction and Weight Percent

Ringer's solution is used in physiology experiments. One liter of the solution contains 6.5 g NaCl, 0.20 g $NaHCO_3$, 0.10 g $CaCl_2$, and 0.10 g KCl. Assume that one liter of water has been used to prepare the solution and that the density of water is 1.00 g/mL.

(a) Calculate the mass fraction and weight percent of $NaHCO_3$ in the solution.
(b) Which solute has the lowest weight percent in the solution?

Parts per Million, Billion, and Trillion

1 ppm is equivalent to one penny in $10,000; 1 ppb is one penny in $10,000,000.

1 μg (microgram) = 10^{-6} g; 1 ng (nanogram) = 10^{-9} g.

Solutes in very dilute solutions have very low mass fractions. Consequently, the mass fraction in such solutions is often expressed in **parts per million** (abbreviated ppm). One part per million is equivalent to one gram of solute per one million grams of solution, or proportionally, one milligram of solute per one thousand grams of solution (1 mg/kg). Thus, a commercial bottled water with a calcium ion concentration of 66 ppm contains 66 mg Ca^{2+} per 1000 g water, essentially one liter. For even smaller mass fractions, **parts per billion** (1 ppb = one microgram, μg, of solute per thousand grams of solution) and **parts per trillion** (1 ppt = one nanogram, ng, of solute in one thousand grams of solution) are often used. As the names imply, a mass fraction converts to parts per billion by multiplying by 10^9 ppb and to parts per trillion by multiplying by 10^{12} ppt.

PROBLEM-SOLVING EXAMPLE 15.5 ppm, ppb, and Mass Fraction

Most of the wells in the country of Bangladesh are contaminated with arsenic, a toxic material in sufficient dose, with levels well above the World Health Organization's (WHO) guideline maximum value of 0.010 mg As/L. The WHO estimates that 28–35 million Bangladesh citizens have been exposed to high arsenic levels by drinking water from wells in which the arsenic concentration in the water is at least 0.050 mg/L.

(a) Calculate the mass of arsenic per liter of solution in the 0.050 mg/L solution. Assume that the solution is almost entirely water and the density of the solution is 1.0 g/mL.

(b) Calculate the concentration of arsenic in ppb.

Answer (a) 5.0×10^{-5} g arsenic/L solution (b) 50. ppb

Strategy and Explanation

(a) Assuming that the solution is almost entirely water, 1 L of the solution has a mass of 1.0×10^3 g. The mass of arsenic in this 1-L water sample is calculated from its mass fraction, expressed as the ratio of grams of arsenic per 10^6 g of solution.

$$\left(\frac{0.050 \text{ g arsenic}}{1 \times 10^6 \text{ g solution}}\right)\left(\frac{1.0 \times 10^3 \text{ g solution}}{\text{L solution}}\right) = 5.0 \times 10^{-5} \text{ g arsenic/L solution}$$

(b) This answer can be converted to ppb by first converting it to micrograms of arsenic per liter.

$$\left(\frac{5.0 \times 10^{-5} \text{ g arsenic}}{1 \text{ L}}\right)\left(\frac{1 \text{ μg}}{1 \times 10^{-6} \text{ g}}\right) = 50. \text{ μg arsenic/L}$$

Thus, a mass fraction of 5.0×10^{-2} g arsenic per 10^6 g of solution (0.050 ppm) corresponds to a concentration of 50. μg arsenic/L, which is the equivalent of 50. ppb.

✓ **Reasonable Answer Check** Used correctly, the conversion factors change 0.050 mg/L directly to 5.0×10^{-5} g arsenic/L and 50 ppb arsenic.

To express the mass fraction as parts per billion, multiply it by 10^9 ppb. This makes a very small number bigger and easier to handle.

A $1 million prize has been offered to create an inexpensive system to reduce arsenic levels in drinking water in developing countries.

PROBLEM-SOLVING PRACTICE 15.5

Drinking water often contains small concentrations of selenium (Se). If a sample of water contains 30 ppb Se, how many micrograms of Se are present in 100. mL of this water?

CONCEPTUAL EXERCISE 15.9 Lead in Drinking Water

One drinking-water sample has a lead concentration of 20 ppb; another has a concentration of 0.003 ppm.

(a) Which sample has the higher lead concentration?

(b) The current EPA acceptable limit for lead in drinking water is 0.015 mg/L. Compare each of the water sample's lead concentration with the acceptable limit.

At the height of the Roman Empire, worldwide lead production was about 80,000 tons per year. Today it is about 3 million tons annually. Lead was first used for water pipes in ancient Rome. The Latin name for lead, *plumbum,* gave us the word "plumber."

EXERCISE 15.10 Bottled Water as a Magnesium Source

A 500-mL bottle of Evian bottled water contains 12 mg of magnesium. The recommended daily allowance of magnesium for adult women is 280 mg/day. How many 1-L bottles of Evian would a woman have to drink to obtain her total daily allowance of magnesium solely in this way?

EXERCISE 15.11 Striking It Rich in the Oceans?

The concentration of gold in seawater is about 1×10^{-3} ppm. The earth's oceans contain 3.5×10^{20} gal of seawater. Approximately how many pounds of gold are in the oceans? 1 gal = 3.785 L; 1 lb = 454 g.

Molarity

As defined in Section 5.6 (⇐ *p. 192*), the *molarity* of a solution is

$$\text{Molarity} = \frac{\text{number of moles of solute}}{\text{number of liters of solution}}$$

Multiplying the volume of a solution (L) by its molarity yields the number of moles of solute in that volume of solution. For example, the number of moles of KNO_3 in 250. mL of 0.0200 M KNO_3 is

$$0.250\ \text{L} \times \left(\frac{0.0200\ \text{mol KNO}_3}{1\ \text{L}}\right) = 5.00 \times 10^{-3}\ \text{mol KNO}_3$$

from which the number of grams can be determined using the molar mass of potassium nitrate.

$$5.00 \times 10^{-3}\ \text{mol KNO}_3 \times \frac{101.1\ \text{g KNO}_3}{1\ \text{mol KNO}_3} = 0.506\ \text{g KNO}_3$$

Thus, to make 250. mL of 0.0200 M KNO_3 solution, you would weigh 0.506 g KNO_3, put it into a 250-mL volumetric flask, and add to it sufficient water to bring the volume of the solution to 250. mL (⇐ ***p. 194***).

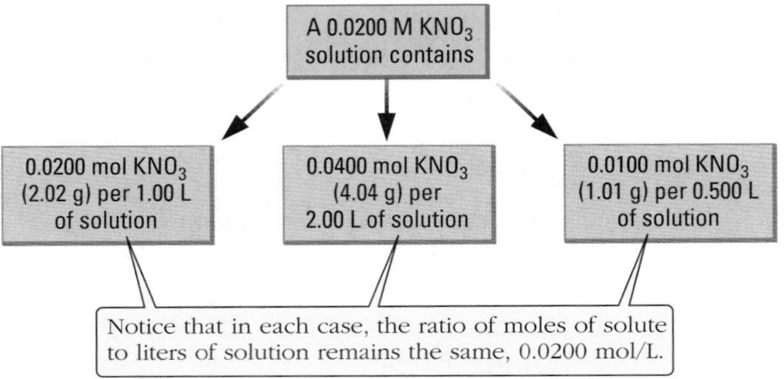

PROBLEM-SOLVING EXAMPLE 15.6 Molarity

(a) Calculate the mass in grams of $NiCl_2$ needed to prepare 500. mL of 0.125 M $NiCl_2$.
(b) How many milliliters of this solution are required to prepare 250 mL of 0.0300 M $NiCl_2$?

Answer (a) 8.10 g $NiCl_2$ (b) 60.0 mL

Strategy and Explanation To solve this problem we apply the concept of molarity in part (a) and use the dilution relationship (⇐ ***p. 195***) in part (b).

(a) To find the number of grams of $NiCl_2$, the number of moles of $NiCl_2$ must first be calculated. This can be done by multiplying the volume (L) times the molarity.

$$0.500\ \text{L} \times \frac{0.125\ \text{mol NiCl}_2}{1\ \text{L}} = 0.0625\ \text{mol NiCl}_2$$

The number of grams of $NiCl_2$ can be determined from the number of moles by using the molar mass of $NiCl_2$, 129.6 g/mol.

$$0.0625\ \text{mol NiCl}_2 \times \frac{129.6\ \text{g NiCl}_2}{1\ \text{mol NiCl}_2} = 8.10\ \text{g NiCl}_2$$

(b) A more concentrated solution is being diluted to a less concentrated one, so the number of moles of $NiCl_2$ remain the same; only the volume of solution changes. To calculate the volume of the more concentrated (undiluted) solution required we can use the relation

$$\text{Molarity}_{\text{conc.}} \times V_{\text{conc.}} = \text{Molarity}_{\text{dil}} \times V_{\text{dil}}$$

in which $\text{Molarity}_{\text{conc.}}$ and $V_{\text{conc.}}$ represent the molarity and volume of the initial (undiluted) solution and $\text{Molarity}_{\text{dil}}$ and V_{dil} are the molarity and volume of the final (diluted) solution (⇐ ***p. 195***). In this case,

$$\text{Molarity}_{\text{conc.}} = 0.125 \text{ M} \qquad V_{\text{conc.}} = \text{volume to be determined}$$

$$\text{Molarity}_{\text{dil}} = 0.0300 \text{ M} \qquad V_{\text{dil}} = 250 \text{ mL} = 0.250 \text{ L}$$

Solving for $V_{\text{conc.}}$

$$V_{\text{conc.}} = \frac{\text{Molarity}_{\text{dil}} \times V_{\text{dil}}}{\text{Molarity}_{\text{conc.}}} = \frac{(0.0300 \text{ M})(0.250 \text{ L})}{0.125 \text{ M}} = 0.0600 \text{ L} = 60.0 \text{ mL}$$

✓ **Reasonable Answer Check** The molar mass of $NiCl_2$ is approximately 130 g/mol, so 1 L of a 0.125 M $NiCl_2$ solution would contain about 0.125 mol or 16 g $NiCl_2$. One half of a liter (500 mL) of that solution would contain one half as much $NiCl_2$, or about 8 g, which is close to the calculated value of 8.10 g. The diluted solution is 0.0300 M, which is only about one fourth the undiluted concentration. Therefore, in the dilution the volume must change by a factor of about 4, from 60.0 mL to 250. mL.

PROBLEM-SOLVING PRACTICE 15.6

(a) Calculate the mass in grams of NaBr needed to prepare 250. mL of 0.0750 M NaBr.
(b) How many milliliters of this solution are required to prepare 500. mL of 0.00150 M NaBr?

PROBLEM-SOLVING EXAMPLE 15.7 Molarity and ppm

Seawater contains 19,000 ppm Cl^- making chloride the most abundant anion in the oceans. Assume that the density of seawater is 1.03 g/mL. Calculate
(a) the mass in grams of chloride ion per liter of seawater.
(b) the molarity of chloride in seawater.

Answer (a) 20. g Cl^-/L seawater (b) 0.56 mol Cl^-/L seawater

Strategy and Explanation

(a) Calculate the mass of Cl^- per liter using the definition of ppm and the density of seawater.

$$\frac{(1.9 \times 10^4 \text{ g Cl}^-)}{(1 \times 10^6 \text{ g seawater})} \times \frac{(1.03 \times 10^3 \text{ g seawater})}{(1 \text{ L seawater})} = 20. \text{ g Cl}^-/\text{L seawater}$$

(b) Calculate the molarity of chloride ion by converting grams to moles.

$$\frac{(20. \text{ g Cl}^-)}{(1 \text{ L seawater})} \times \frac{(1 \text{ mol Cl}^-)}{(35.5 \text{ g Cl}^-)} = 0.56 \text{ mol Cl}^-/\text{L seawater}$$

✓ **Reasonable Answer Check** There is about 0.02 g Cl^- per gram of seawater. A liter of seawater is about 1000 g of seawater, which contains about 20 g Cl^-; this is close to the calculated answer.

PROBLEM-SOLVING PRACTICE 15.7

The concentration of magnesium ion in seawater is 0.0556 M making magnesium the second most abundant cation in oceans. Express the Mg^{2+} concentration as (a) mass fraction and (b) ppm. Assume that the density of seawater is 1.03 g/mL.

PROBLEM-SOLVING EXAMPLE 15.8 Weight Percent and Molarity

Hydrochloric acid is sold as a concentrated aqueous solution of HCl with a density of 1.18 g/mL. The concentrated acid contains 38% HCl by mass. Calculate the molarity of hydrochloric acid in this solution.

Answer 12.4 M

Molarity and molality. The photo shows a 0.10 molal solution (0.10 mol/kg) of potassium chromate (*flask at right*) and a 0.10 molar solution (0.10 mol/L) of potassium chromate (*flask at left*). Each solution contains 0.10 mol (19.4 g) of yellow K_2CrO_4, shown in the dish at the front. The 0.10 molar (0.10 mol/L) solution on the left was made by placing the solid in the flask and adding enough water to make 1.0 L solution. The 0.10 molal (0.10 mol/kg) solution on the right was made by placing the solid in the flask and adding 1000 g (1 kg) water. Adding 1 kg water produces a solution that has a volume greater than 1 L.

Strategy and Explanation To calculate the molarity, the number of moles of HCl and the volume of the solution in liters must be determined. The 38.0% hydrochloric acid solution means that 100. g solution contains 38.0 g HCl. Therefore,

$$\text{Amount (moles) of HCl} = 38.0 \text{ g HCl} \times \frac{1 \text{ mol HCl}}{36.5 \text{ g HCl}} = 1.04 \text{ mol HCl}$$

The molarity of the solution can be calculated using the number of moles of HCl and the density of the solution to determine its volume.

$$\text{Molarity} = \frac{1.04 \text{ mol}}{100. \text{ g solution}} \times \frac{1.18 \text{ g solution}}{1 \text{ mL solution}} \times \frac{1000 \text{ mL solution}}{1 \text{ L solution}}$$

$$= 12.4 \frac{\text{mol}}{\text{L}} = 12.4 \text{ M}$$

✓ **Reasonable Answer Check** The molar mass of HCl, 36.5 g/mol, is close to the number of grams of HCl in 100 g solution. The density of the solution is about 1, so there is about 1 mol HCl in 100 mL of this solution. Consequently, there are approximately 10 mol HCl in 1 L solution—approximately a 10 M HCl solution. The calculated answer is a bit higher because we rounded the actual density (1.18 g/mL) to 1.

PROBLEM-SOLVING PRACTICE 15.8

The density of a commercial 30.0% hydrogen peroxide (H_2O_2) solution is 1.11 g/mL at 25 °C. Calculate the molarity of hydrogen peroxide in this solution.

Molality

Solute concentration can also be expressed in terms of the moles of solute in relation to the mass of the solvent. **Molality** (abbreviated m), is defined as the amount (moles) of solute per kilogram of *solvent* (not solution).

$$\text{Molality of solute A} = m_A = \frac{\text{moles of solute A}}{\text{kilograms of solvent}}$$

For example, we can calculate the molality of a solution prepared by dissolving 0.413 g methanol (CH_3OH) in 1.50×10^3 g water by first determining the number of moles of methanol solute and the number of kilograms of solvent, water.

$$0.413 \text{ g methanol} \times \frac{1 \text{ mol methanol}}{32.042 \text{ g methanol}} = 0.0129 \text{ mol methanol}$$

The mass of water is 1.50×10^3 g, which is 1.50 kg. Substituting these values into the molality equation gives

$$\text{Molality} = m = \frac{0.0129 \text{ mol methanol}}{1.50 \text{ kg water}} = 0.00860 \text{ mol/kg}$$

Molality will be used in Section 15.7 when we deal with the effect of solute concentration on decreasing the freezing point and raising the boiling point of a solution.

PROBLEM-SOLVING EXAMPLE 15.9 Molarity and Molality

Automobile lead storage batteries contain an aqueous solution of sulfuric acid. The solution has a density of 1.230 g/mL at 25 °C and contains 368. g H_2SO_4 per liter.

Calculate
(a) the molarity of sulfuric acid in the solution.
(b) the molality of sulfuric acid in the solution.

Answer (a) Molarity = 3.75 mol H_2SO_4/L (b) Molality = 4.35 mol H_2SO_4/kg solvent

Strategy and Explanation

(a) Determine the number of moles of solute in a liter of solution and calculate the molarity.

$$368.\text{ g } H_2SO_4 \times \frac{1 \text{ mol } H_2SO_4}{98.1 \text{ g } H_2SO_4} = 3.75 \text{ mol } H_2SO_4$$

This is the amount of solute in 1 L solution; the molarity is 3.75 mol/L, 3.75 M.

(b) To calculate the molality we first need to determine the mass (kg) of solvent (water) and then divide that into the moles of solute. The density of the solution is 1.230 g/mL, which is equivalent to 1.230×10^3 g solution/L; 1 L solution contains 1.230×10^3 g of the water and sulfuric acid mixture, 368. g of which is sulfuric acid. Therefore, the mass of water can be determined.

$$1230 \text{ g solution} - 368 \text{ g } H_2SO_4 = 862 \text{ g water} = 0.862 \text{ kg water}$$

We can now calculate the molality of sulfuric acid in the solution.

$$\text{Molality of } H_2SO_4 = \frac{\text{moles of } H_2SO_4}{\text{kg of solvent}} = \frac{3.75 \text{ mol } H_2SO_4}{0.862 \text{ kg solvent}}$$
$$= 4.35 \text{ mol } H_2SO_4/\text{kg solvent}$$

✓ **Reasonable Answer Check** The molar mass of sulfuric acid is about 100 g/mol and so 368 g of the acid would be about 3.7 mol. There is about 4 mol of the acid in about a kilogram of water, so the molality is about 4, close to the more accurate answer of 4.35 mol/kg. Because the density of the solution is not 1.00 g/ml, the mass of solvent and the volume of the solution are not equal, so the molarity and the molality differ.

Molality and molarity are *not* the same, although the difference becomes negligibly small for dilute solutions, those less than 0.01 mol/L.

PROBLEM-SOLVING PRACTICE 15.9

Calculate the molarity and the molality of NaCl in a 20% aqueous NaCl solution whose density is 1.148 g/mL at 25 °C.

Table 15.4 summarizes the types of concentration units we have used so far, comparing their units, advantages, and disadvantages.

Table 15.4 Comparison of Concentration Expressions

Concentration Expression	Units of Concentration	Advantages	Disadvantages
Mass fraction	None	Independent of temperature; used in special applications	Density must be known to convert mass fraction to molarity
Weight percent	Percent	Independent of temperature; useful in wide range of applications	Density must be known to convert weight percent to molarity
Parts per million, parts per billion, parts per trillion	ppm, ppb, ppt	Independent of temperature; widely used in environmental applications	Must be determined by very exacting analytical methods
Molarity	$\frac{\text{Moles solute}}{\text{liter solution}}$	Volume measurements are easy and the results readily used in stoichiometric calculations	Temperature dependent; density must be known to determine solvent mass
Molality	$\frac{\text{Moles solute}}{\text{kg solvent}}$	Molality of solution does not change with temperature	Mass of a volatile solvent changes by evaporation*

*This is also a disadvantage of the other concentration expressions.

CHEMISTRY IN THE NEWS

Buckyballs and the Environment

The discovery of buckyballs surprised scientists and won the Nobel Prize in chemistry for their discoverers (← *p. 28*). Buckyballs, C_{60}, are molecules whose carbon atoms are arranged in a cage-like structure. Scientists initially thought that buckyball molecules, being made only of carbon, would be insoluble in water. Recently, however, researchers have discovered that in water these molecules aggregate into nanocrystals, called nano-C_{60}, which are hydrophilic due to their weakly negative-charged surface. At concentrations as low as 0.4 ppm, the nano-C_{60}s inhibit the growth of two types of common soil bacteria. At higher concentrations the nanocrystals inhibit bacterial respiration. Although the antibacterial action may be beneficial in certain applications, concern has arisen that the presence of nano-C_{60}s in the environment could damage ecosystems. There are no current guidelines for handling or disposing of buckyballs or their nanocrystals; regulations only exist based on the properties of bulk carbon black. Joseph B. Hughes, professor of environmental engineering at Georgia Institute of Technology suggests that "As information becomes available, we have to be ready to modify these regulations and best practices for safety."

SOURCE: *Chemical & Engineering News*, May 16, 2005; p. 11.

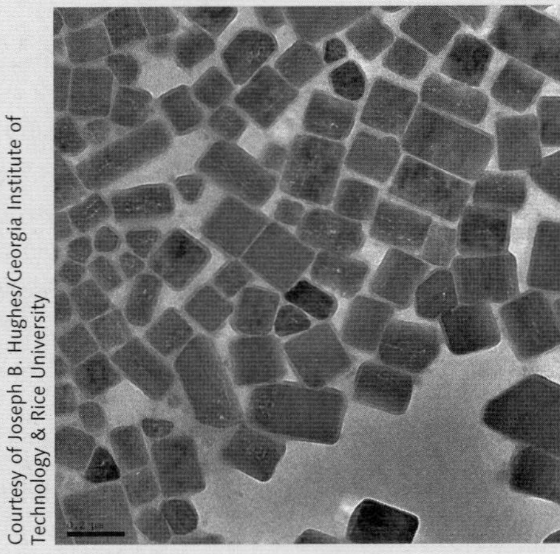

Courtesy of Joseph B. Hughes/Georgia Institute of Technology & Rice University

Nanocrystalline arrays of buckyballs in water.

CONCEPTUAL EXERCISE 15.12 Molality

The "proof" of an alcoholic beverage is defined as twice the percent by volume of alcohol in the beverage. What is the molality of ethanol, C_2H_5OH, in a quart of a 90-proof alcoholic beverage? Assume that ethanol is the only solute. The density of ethanol is 0.79 g/mL; the density of the alcoholic beverage is 0.861 g/mL; 1 L = 1.057 qt.

CONCEPTUAL EXERCISE 15.13 Molality and Molarity

(a) What information is required to calculate the molality of a solution?
(b) What information is needed to calculate the molarity of a solution if the solution's composition is given in weight percent?

ThomsonNOW™
Log on to ThomsonNOW at **www.thomsonedu.com** and click Coached Problems for tutorials on:

- **vapor pressure changes and Raoult's law**
- **boiling point elevation**
- **freezing point depression**
- **the effect of ionic compounds on boiling point**
- **the effect of ionic compounds on freezing point**
- **vapor pressure and boiling point elevation**

15.7 Vapor Pressures, Boiling Points, and Freezing Points of Solutions

Up to this point, we have discussed solutions in terms of the nature of the solute and the nature of the solvent. Some properties of solutions do not depend on the nature of the solute or solvent, but rather depend only on the *number* of dissolved solute particles—ions or molecules—per unit volume.

In liquid solutions, solute molecules or ions disrupt solvent-solvent noncovalent attractions, causing changes in solvent properties that depend on these attractions. For example, when solute is added to a solvent, the freezing point of the resulting solution is lower and its boiling point is higher than that of the pure solvent. How much the properties of the solution differ from those of the pure solvent depends only on the concentration of the solute particles. **Colligative properties** of solutions are those that *depend only on the concentration of solute particles* (ions or molecules) in the solution, regardless of what kinds of particles are present. We will consider four colligative properties: vapor pressure lowering, boiling point elevation, freezing point depression, and osmotic pressure. These are all quite common and important in the world around us.

Osmotic pressure is discussed in Section 15.8.

Vapor Pressure Lowering

In a closed container a dynamic equilibrium exists between a pure liquid and its vapor—the rate at which molecules escape the liquid phase equals the rate at which vapor phase molecules return to the liquid. This equilibrium gives rise to a vapor pressure that depends on the temperature (⇐ ***p. 492***). But the vapor pressure of a pure liquid differs from that of the liquid when a solute has been dissolved in it. Compare a small portion of the liquid/vapor boundary for pure water with that for seawater (mainly an aqueous sodium chloride solution), as shown at the molecular scale in Figure 15.14. For an aqueous solution such as seawater, in which sodium ions and chloride ions (and many other kinds of ions and molecules) are present, the vapor pressure of water is lower than for a sample of pure water.

The vapor pressure of any pure solvent will be lowered by the addition of a nonvolatile solute to the solvent. How much a dissolved solute lowers the vapor pressure of a solvent depends on the solute concentration and is expressed by **Raoult's law:**

Raoult's law works best with dilute solutions. Deviations from Raoult's law occur when solute-solvent intermolecular forces are either much weaker or much stronger than the solvent-solvent and solute-solute intermolecular forces.

$$P_1 = X_1 P_1^0$$

where P_1 is the vapor pressure of the solvent over the *solution*, P_1^0 is the vapor pressure of the *pure* solvent at the same temperature, and X_1 is the mole fraction of *solvent* in the solution. For example, suppose you want to know the vapor pressure over a sucrose solution at 25 °C in which the mole fraction of water is 0.986. The vapor pressure of pure water at 25 °C is 23.76 mm Hg. From these data, the vapor pressure of water over the solution can be calculated using Raoult's law.

$$P_{\text{water}} = (X_{\text{water}})(P^0_{\text{water}}) = (0.986)(23.76 \text{ mm Hg}) = 23.42 \text{ mm Hg}$$

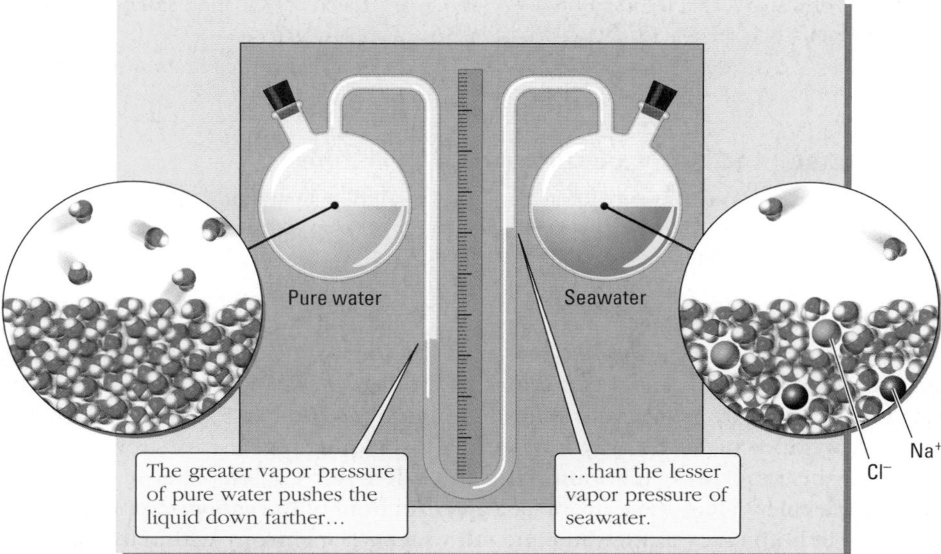
Pure water
Seawater
The greater vapor pressure of pure water pushes the liquid down farther...
...than the lesser vapor pressure of seawater.
Na⁺
Cl⁻

Figure 15.14 The vapor pressure of pure water and that of seawater. Seawater is an aqueous solution of NaCl and many other salts. The vapor pressure over an aqueous solution is not as great as that over pure water at the same temperature.

Raoult's law can also be applied to solutions in which the solvent and the solute are both volatile so that an appreciable amount of each can be in the vapor above the solution. We will not consider such cases.

The vapor pressure has been lowered by (23.76 − 23.42) mm Hg = 0.34 mm Hg. Therefore, the vapor pressure of water over the solution is only 98.6% that of pure water at 25 °C.

PROBLEM-SOLVING EXAMPLE 15.10 Raoult's Law

Ethylene glycol, $HOCH_2CH_2OH$, is used as an antifreeze. What is the vapor pressure of water above a solution of 100.0 mL ethylene glycol and 100.0 mL water at 90 °C? Densities: ethylene glycol, 1.15 g/mL; water, 1.00 g/mL. The vapor pressure of pure water at 90 °C is 525.8 mm Hg.

ethylene glycol
$HOCH_2CH_2OH$

Answer 394 mm Hg

Strategy and Explanation To determine the vapor pressure of water over the solution using Raoult's law, we must first calculate the mole fraction of water in the solution. For the mole fraction, the milliliters of ethylene glycol and water must be converted to grams and then to moles.

Mole fraction of A, X_A, $= \dfrac{\text{moles of A}}{\text{total number of moles}} = \dfrac{\text{moles A}}{\text{moles A} + \text{moles B} + \cdots}$

$$100.0 \text{ mL eth. gly.} \times \frac{1.15 \text{ g eth. gly.}}{1.00 \text{ mL eth. gly.}} \times \frac{1 \text{ mol eth. gly.}}{62.0 \text{ g eth. gly.}} = 1.86 \text{ mol eth. gly.}$$

$$100.0 \text{ mL water} \times \frac{1.00 \text{ g water}}{1.00 \text{ mL water}} \times \frac{1 \text{ mol water}}{18.0 \text{ g water}} = 5.56 \text{ mol water}$$

$$X_{\text{water}} = \frac{5.56}{5.56 + 1.86} = \frac{5.56}{7.42} = 0.749$$

Applying Raoult's law:

$$P_{\text{water}} = (X_{\text{water}})(P^0_{\text{water}}) = (0.749)(525.8 \text{ mm Hg}) = 394 \text{ mm Hg}$$

✓ **Reasonable Answer Check** Because there are nearly three times the number of moles of water as there are moles of ethylene glycol, the mole fraction of water should be greater than the mole fraction of ethylene glycol, and it is (mole fraction of ethylene glycol = 1 − 0.749 = 0.251). Therefore, the vapor pressure of water over the solution should be about 75% that of pure water (394/525.8 × 100%), which it is. The answer is reasonable.

PROBLEM-SOLVING PRACTICE 15.10

The vapor pressure of an aqueous solution of urea, CH_4N_2O, is 291.2 mm Hg. The vapor pressure of water at that temperature is 355.1 mm Hg. Calculate the mole fraction of each component.

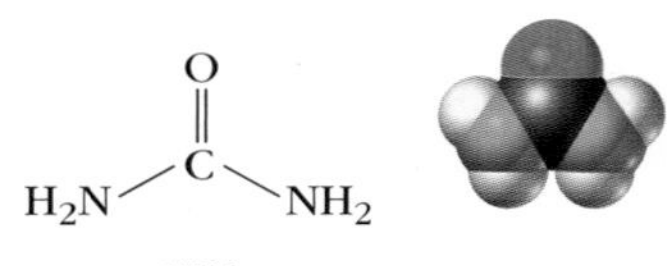

urea

EXERCISE 15.14 Vapor Pressure of a Mixture

Calculate the vapor pressure of water over a solution containing 50.0 g sucrose, $C_{12}H_{22}O_{11}$, and 100.0 g water at 45 °C. The vapor pressure of pure water at this temperature is 71.88 mm Hg.

Entropy plays a role in vapor pressure lowering. Consider the entropy change for the vaporization of pure water with that for the vaporization of a corresponding quantity of water from a sodium chloride solution (Figure 15.15). Because NaCl has a very low vapor pressure, very few sodium or chloride ions escape from the solution and the vapor in equilibrium with the salt water consists almost entirely of water molecules. Thus, the entropy of a given amount of the vapor is approximately the same in both cases (pure water and salt water). The entropy within the salt solution, however, is greater than that in pure water.

Now consider what happens when water vaporizes from pure water and from the salt solution. As with any change from a liquid to a gas, there is a significant

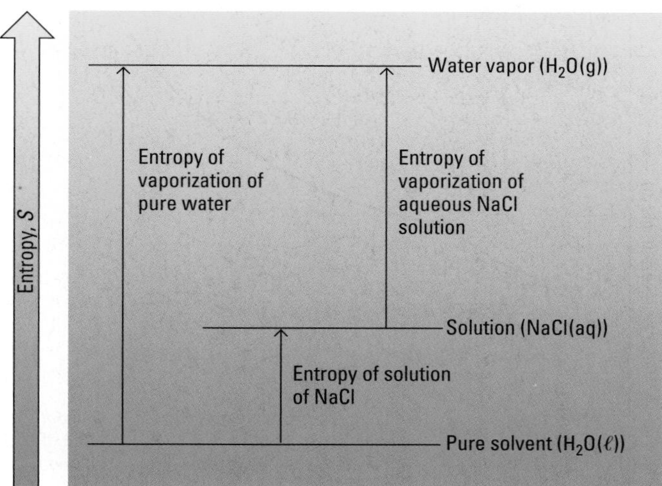

Figure 15.15 Vapor pressure lowering and entropies of solution and vaporization. The entropy of vaporization of pure water is greater than the entropy of solution.

increase in entropy of the water vaporizing from either source (pure water or salt solution) (p. 745). But the entropy of vaporization is not the same in each case. As illustrated in Figure 15.15, the entropy of vaporization is larger for the water vapor from pure water than from the salt solution because the entropy of pure water was already smaller than that of the salt solution. The main point is that a bigger entropy increase corresponds to a more product-favored process. The result is that the vaporization from pure water creates a higher pressure of water vapor (that is, more water molecules per unit volume) in equilibrium with pure water than does the vaporization of water from a salt solution.

Boiling Point Elevation

As a result of vapor pressure lowering, the vapor pressure of an aqueous solution of a nonvolatile solute at 100 °C is less than 760 mm Hg (1 atm). For the solution to boil, it must be heated *above* 100 °C. The **boiling point elevation,** ΔT_b, is the difference between the normal boiling point of water and the higher boiling point of an aqueous solution of a nonvolatile nonelectrolyte solute (Figure 15.16).

The increase in boiling point is proportional to the concentration of the solute, expressed as molality, and can be calculated from this relationship.

$$\Delta T_b = T_b\,(\text{solution}) - T_b\,(\text{solvent}) = K_b m_{\text{solute}}$$

where ΔT_b is the boiling point of solution minus the boiling point of pure solvent. The value of K_b, the *molal boiling point elevation constant* of the *solvent,* depends only on the solvent. For example, K_b for water is 0.52 °C kg mol^{-1}; that of benzene is 2.53 °C kg mol^{-1}.

The units of the molal boiling point elevation constant are abbreviated °C kg mol^{-1}.

Molality, rather than molarity, is used in boiling point elevation determinations because the molality of a solution does not change with temperature changes, but the molarity of a solution does. Molality is based on the masses of solute *and* solvent, which are unaffected by temperature changes. The volume of a solution, which is used in molarity, expands or contracts when the solution is heated or cooled.

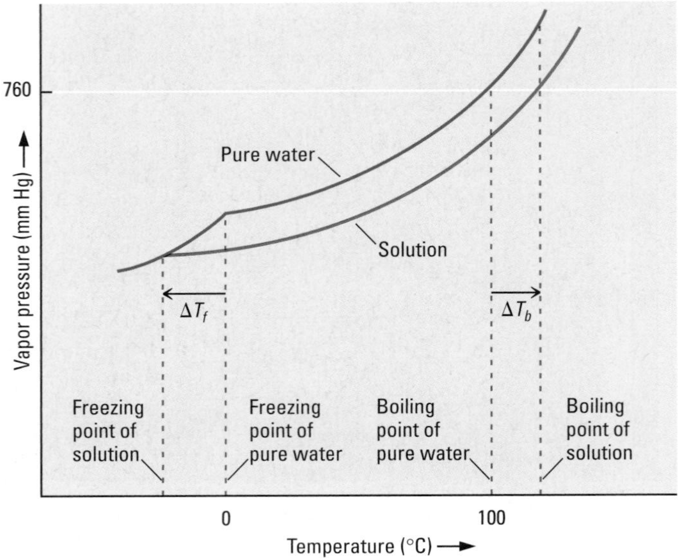

Figure 15.16 Boiling point elevation (ΔT_b) and freezing point lowering (ΔT_f) for aqueous solutions. Addition of solute to a pure solvent raises its boiling point and lowers its freezing point.

EXERCISE 15.15 Calculating the Boiling Point of a Solution

The boiling point elevation constant for benzene is 2.53 °C kg mol^{-1}. The boiling point of pure benzene is 80.10 °C. If a solute's concentration in benzene is 0.10 mol/kg, what will be the boiling point of the solution?

Freezing Point Lowering

A pure liquid begins to freeze when the temperature is lowered to the substance's freezing point and the first few molecules cluster together into a crystal lattice forming a tiny quantity of solid. As long as both solid and liquid phases are present and the temperature is at the freezing point, there is a dynamic equilibrium in which the rate of crystallization equals the rate of melting. When a *solution* freezes, a few molecules of solvent cluster together to form pure solid *solvent* (Figure 15.17), and a dynamic equilibrium is set up between solution and solid solvent.

The molecules or ions in the liquid in contact with the frozen solvent in a freezing solution are not all solvent molecules. This causes a slower rate at which particles move from solution to solid than the rate in the pure liquid solvent. To achieve dynamic equilibrium, a correspondingly slower rate of escape of molecules from the solid crystal lattice must occur. According to the kinetic-molecular theory, this slower rate occurs at a lower temperature, so the freezing point of the solution is lower than that of the pure liquid solvent (Figure 15.16).

The **freezing point lowering,** ΔT_f, is proportional to the concentration of the solute (molality) in the same way as the boiling point elevation.

$$\Delta T_f = K_f m_{\text{solute}}$$

As with K_b, the proportionality constant K_f depends only on the solvent and not the type of solute. For water, the freezing point constant is 1.86 °C kg mol^{-1}; by comparison, that of benzene is 5.10 °C kg mol^{-1} and that of cyclohexane is 20.2 °C kg mol^{-1}.

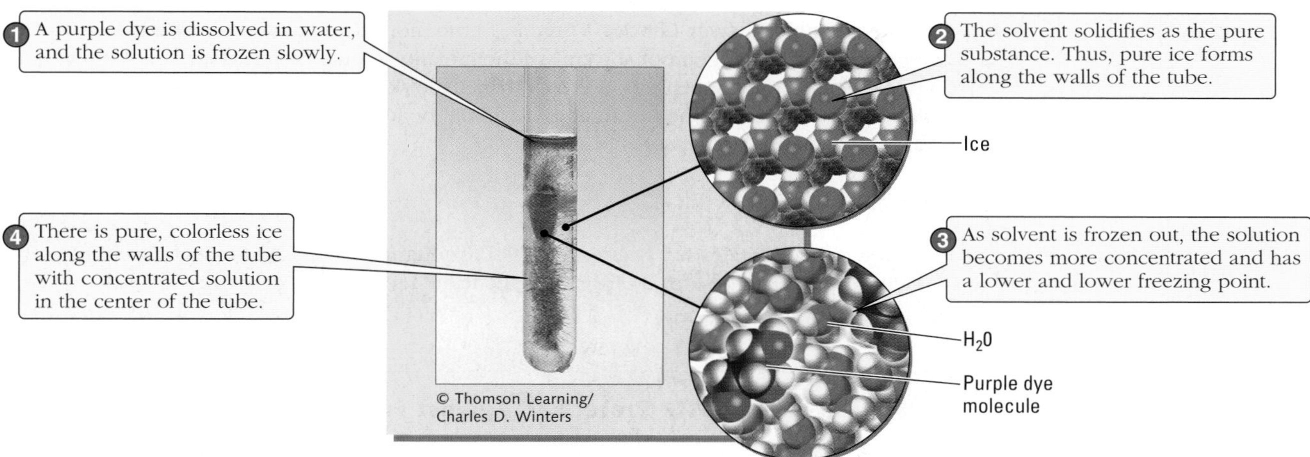

1 A purple dye is dissolved in water, and the solution is frozen slowly.
2 The solvent solidifies as the pure substance. Thus, pure ice forms along the walls of the tube.
Ice
3 As solvent is frozen out, the solution becomes more concentrated and has a lower and lower freezing point.
H2O
Purple dye molecule
4 There is pure, colorless ice along the walls of the tube with concentrated solution in the center of the tube.
© Thomson Learning/ Charles D. Winters

Figure 15.17 **Solvent freezing.**

Using ethylene glycol, $HOCH_2CH_2OH$, a relatively nonvolatile alcohol, in automobile cooling systems is a practical application of boiling point elevation and freezing point lowering. Ethylene glycol raises the boiling temperature of the coolant mixture of ethylene glycol and water to a level that prevents engine overheating in hot weather. Ethylene glycol also lowers the freezing point of the coolant, thereby keeping the solution from freezing in the winter.

Another practical application of freezing point lowering is adding salt (NaCl) to ice when making homemade ice cream. Lowering the freezing temperature of the ice-salt water mixture freezes the ice cream more quickly.

PROBLEM-SOLVING EXAMPLE 15.11 Boiling Point Elevation and Freezing Point Lowering

Calculate the boiling and freezing points of an aqueous solution containing 39.5 g ethylene glycol ($HOCH_2CH_2OH$) dissolved in 750. mL water. Assume the density of water to be 1.00 g/mL; its K_b = 0.52 °C kg mol^{-1} and its K_f = 1.86 °C kg mol^{-1}.

Answer Boiling point = 100.44 °C; freezing point = −1.58 °C

Strategy and Explanation To use the equations for freezing point and boiling point changes, the molality of the solution must be determined. The molar mass of ethylene glycol is 62.07 g/mol, and the number of moles of ethylene glycol is

$$39.5 \text{ g} \times \frac{1 \text{ mol}}{62.07 \text{ g}} = 0.636 \text{ mol}$$

The mass of solvent is

$$750. \text{ mL} \times \frac{1.00 \text{ g}}{\text{mL}} \times \frac{1 \text{ kg}}{10^3 \text{ g}} = 0.750 \text{ kg}$$

The molality of the solution is

$$\frac{0.636 \text{ mol}}{0.750 \text{ kg}} = 0.848 \text{ mol/kg}$$

The boiling point elevation is

$$\Delta T_b = (0.52 \text{ °C kg mol}^{-1})(0.848 \text{ mol/kg}) = 0.44 \text{ °C}$$

Therefore, the solution boils at 100.44 °C.

The freezing point lowering is

$$\Delta T_f = (1.86 \text{ °C kg mol}^{-1})(0.848 \text{ mol/kg}) = 1.58 \text{ °C}$$

The freezing point is lowered by 1.58 °C and so the solution freezes at −1.58 °C.

✓ **Reasonable Answer Check** There is a little more than half a mole of ethylene glycol in three-fourths kilogram of solvent, so the molality should be less than 1 mol/kg, which it is. Because the concentration is less than 1 mol/kg, the boiling point should be raised less than 0.52 °C and the freezing point should be lowered less than 1.86 °C, which they are. The answers are reasonable.

PROBLEM-SOLVING PRACTICE 15.11

A water tank contains 6.50 kg water. Will the addition of 1.20 kg ethylene glycol be sufficient to prevent the solution from freezing if the temperature drops to −25 °C?

Ethylene glycol is toxic and should not be allowed to get into drinking water supplies.

EXERCISE 15.16 Protection Against Freezing

Suppose that you are closing a cabin in the north woods for the winter and you do not want the water in the toilet tank to freeze. You know that the temperature might get as low as −30. °C, and you want to protect about 4.0 L water in the toilet tank from freezing. What volume of ethylene glycol (density = 1.113 g/mL; molar mass = 62.1 g/mol) should you add to the 4.0 L water?

PROBLEM-SOLVING EXAMPLE 15.12 Molar Mass from Freezing Point Lowering

A researcher prepares a new compound and uses freezing point depression measurements to determine the molar mass of the compound. The researcher dissolves 1.50 g compound in 75.0 g cyclohexane, which has a freezing point of 6.50 °C and a freezing point depression constant of 20.2 °C kg mol^{-1}. The freezing point of the solution is measured as 2.70 °C. Use these data to calculate the molar mass of the new compound.

Answer 106 g/mol

Strategy and Explanation To calculate the molar mass, we need to calculate the change in freezing point and the molality of the solution. The change in the freezing point, ΔT_f, is 6.50 °C − 2.70 °C = 3.80 °C. Rearranging the freezing point lowering equation, we can solve for the molality of the solution.

$$\text{Molality} = \frac{\Delta T_f}{K_f} = \frac{3.80\ ^\circ\text{C}}{20.2\ ^\circ\text{C kg mol}^{-1}} = 0.188\ \text{mol/kg}$$

From the molality we can calculate the molar mass of the new compound from the relation

$$\text{Molality, } m = \frac{\text{moles solute}}{\text{kg solvent}} = \frac{\text{grams solute/molar mass}}{\text{kg solvent}}$$

This equation can be solved for the molar mass, g/mol.

$$\text{Molar mass} = \frac{\text{grams solute}}{(\text{molality})(\text{kg solvent})} = \frac{1.50\ \text{g}}{(0.188)(0.0750)\ \text{mol}} = 106\ \text{g/mol}$$

✓ **Reasonable Answer Check** About 1.5 g compound in 0.0750 kg solvent, equivalent to 20 g compound per kilogram of solvent, forms a 0.188 mol/kg solution. Thus, a solution containing 1 mol compound per kilogram would contain 20 g/kg solvent × $\frac{1\ \text{kg solvent}}{0.188\ \text{mol}}$, which equals 106 g/mol.

PROBLEM-SOLVING PRACTICE 15.12

A student determines the freezing point to be 5.15 °C for a solution made from 0.180 g of a nonelectrolyte in 50.0 g benzene. Calculate the molar mass of the solute. The freezing point constant of benzene is 5.10 °C kg mol^{-1}. The freezing pont of benzene is 5.50 °C.

Colligative Properties of Electrolytes

Experimentally, the vapor pressures of 1 M aqueous solutions of sucrose, NaCl, and $CaCl_2$ are all less than that of water at the same temperature, which is to be expected because solutes lower the vapor pressure of the pure solvent. However,

$$\text{vp pure water} > \text{vp 1 M sucrose} > \text{vp 1 M NaCl} > \text{vp 1 M } CaCl_2$$

Because colligative properties of dilute solutions are proportional to the concentration of solute *particles,* this vapor pressure order is not surprising. Electrolytes such as NaCl and $CaCl_2$ contribute more particles per mole than do nonelectrolytes such as sucrose or ethanol. Whereas 1 mol sucrose contributes 1 mol of particles (sucrose molecules) to solution, 1 mol NaCl contributes 2 mol of particles (1 mol Na^+ and 1 mol Cl^-), and 1 mol $CaCl_2$ produces 3 mol of particles (1 mol Ca^{2+} and 2 mol Cl^-). Therefore, electrolytes have a greater effect on boiling point than nonelectrolytes do.

For solutions of electrolytes, the boiling point elevation equation can be written as

$$\Delta T_b = K_b\, m_{\text{solute}}\, i_{\text{solute}}$$

and the freezing point lowering equation becomes

$$\Delta T_f = K_f\, m_{\text{solute}}\, i_{\text{solute}}$$

The i_{solute} factor gives the number of particles per formula unit of solute. It is called the van't Hoff factor, named after Jacobus Henricus van't Hoff (1852–1911). The value of i is 1 for nonelectrolytes because these molecular solutes, such as ethanol, sucrose, benzene, and carbon tetrachloride, do *not* dissociate in solution. For soluble ionic solutes (strong electrolytes), i equals the number of ions per formula unit of the ionic compound. In extremely dilute solutions $i_{\text{solute}} = 2$ for NaCl and $i_{\text{solute}} = 3$ for calcium chloride. The actual i_{solute} value must be determined experimentally. The theoretical i value assumes that the ions act independently in solution, which is achieved only in extremely dilute solutions where the interaction between cations and anions is minimal. In more concentrated solutions, cations and anions interact and $i_{\text{expt}} < i_{\text{theor}}$. For example, in aqueous $MgSO_4$ solutions, $i_{\text{theor}} = 2$, whereas in 0.50 M $MgSO_4$, $i_{\text{expt}} = 1.07$; in 0.005 M $MgSO_4$, $i_{\text{expt}} = 1.72$.

Edgar Fahs Smith Memorial Collection/ University of Pennsylvania

1852–1911

Van't Hoff was one of the founders of physical chemistry. While still a graduate student, he proposed an explanation of optical isomerism based on the tetrahedral nature of the carbon atom (*p. 559*). Physical chemistry is the branch of chemistry that applies the laws of physics to the understanding of chemical phenomena. Van't Hoff conducted seminal experimental studies in chemical kinetics, chemical equilibrium, osmotic pressure, and chemical affinity. Van't Hoff received the first Nobel Prize in chemistry (1901) for his fundamental discoveries in physical chemistry, including his work on the colligative properties of solutions.

CHEMISTRY IN THE NEWS

Snow, Salt, and Environmental Damage

During winter, road salt, principally sodium chloride, is applied liberally to roads to lower the freezing point of ice and snow, making it easier to clear the roads. The dissolved salt, however, runs off into adjacent streams, raising their chloride ion concentration. Measurements taken in rural New Hampshire, in the Hudson River Valley, and around Baltimore, MD show increasing chloride ion concentrations in streams in those areas. During the past 25 years, chloride ion concentration has tripled in some Baltimore area streams to 30 mg/L. Streams in areas unaffected by road salt runoff have 2 to 8 mg/L chloride. In rural areas bordering an interstate highway in New Hampshire, chloride concentration has nearly quadrupled during the past two decades to 70 mg/L.

Peter M. Groffman of the Institute for Ecosystem Studies in Millbrook, NY is concerned by such increases in chloride ion concentrations. Groffman projects that at the current rate of increase, streams in many Northeastern sites will exceed 250 mg/L chloride by 2100. At such a high chloride concentration, water is not potable for humans and is toxic to some aquatic life.

SOURCE: S. Perkins, *Science News*, Vol. 168, September 24, 2005; pp. 195–196.

© Mark Joseph/Stone/Getty Images

Salt helps to lower the melting point of snow and ice, facilitating their removal from roads.

This answers the question raised in Chapter 1 (← p. 2): "Why does salt help to clear snow and ice from roads?"

A buildup of dissolved NaCl or $CaCl_2$ along roads is environmentally hazardous because the excessive Cl^- concentration is harmful to roadside plants.

Another practical application of freezing point lowering can be seen in areas where winter weather produces lots of frozen precipitation. To remove snow and particularly ice, roads and walkways are often salted. Although sodium chloride is usually used, calcium chloride is particularly good for this purpose because it has three ions per formula unit and dissolves exothermically. Not only is the freezing point of water lowered, but the heat of solution helps melt the ice.

CONCEPTUAL EXERCISE **15.17 Freezing Point Lowering**

The freezing point of a 2.0 mol/kg $CaCl_2$ solution is measured as −4.78 °C. Calculate the *i* factor and use it to approximate the degree of dissociation of $CaCl_2$ in this solution.

15.8 Osmotic Pressure of Solutions

A *membrane* is a thin layer of material that allows molecules or ions to pass through it. A **semipermeable membrane** allows only certain kinds of molecules or ions to pass through while excluding others (Figure 15.18). Examples of semipermeable membranes are animal bladders, cell membranes in plants and animals, and cellophane, a polymer derived from cellulose. When two solutions containing the same solvent are separated by a membrane permeable only to solvent molecules, osmosis will occur. **Osmosis** is *the movement of a solvent through a semipermeable membrane from a region of lower solute concentration (higher solvent concentration) to a region of higher solute concentration (lower solvent concentration).* The **osmotic pressure** of a solution is *the pressure that must be applied to the solution to stop osmosis from a sample of pure solvent.*

Consider the osmosis example shown in Figure 15.19. A 5% aqueous sugar solution is placed in a bag attached to a glass tube. The bag is made of a semipermeable membrane that allows water but not sugar molecules to pass through it. When the bag is submerged in pure water, water flows into the bag by osmosis and raises the liquid level in the tube. When the bag is first submerged, more collisions of solvent

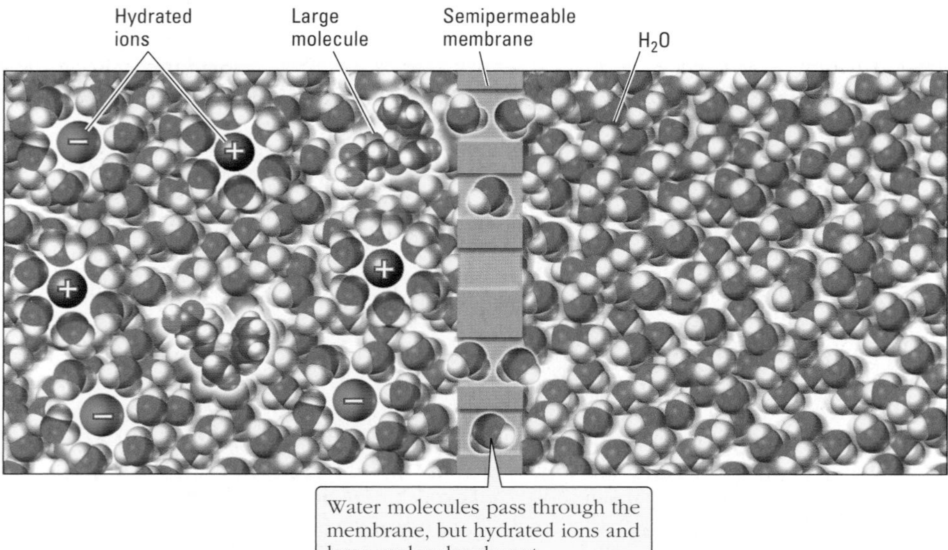

Figure 15.18 **Osmotic flow of a solvent through a semipermeable membrane to a solution.** The semipermeable membrane is shown acting as a sieve. Many membranes operate in different ways, but the ultimate effect is the same.

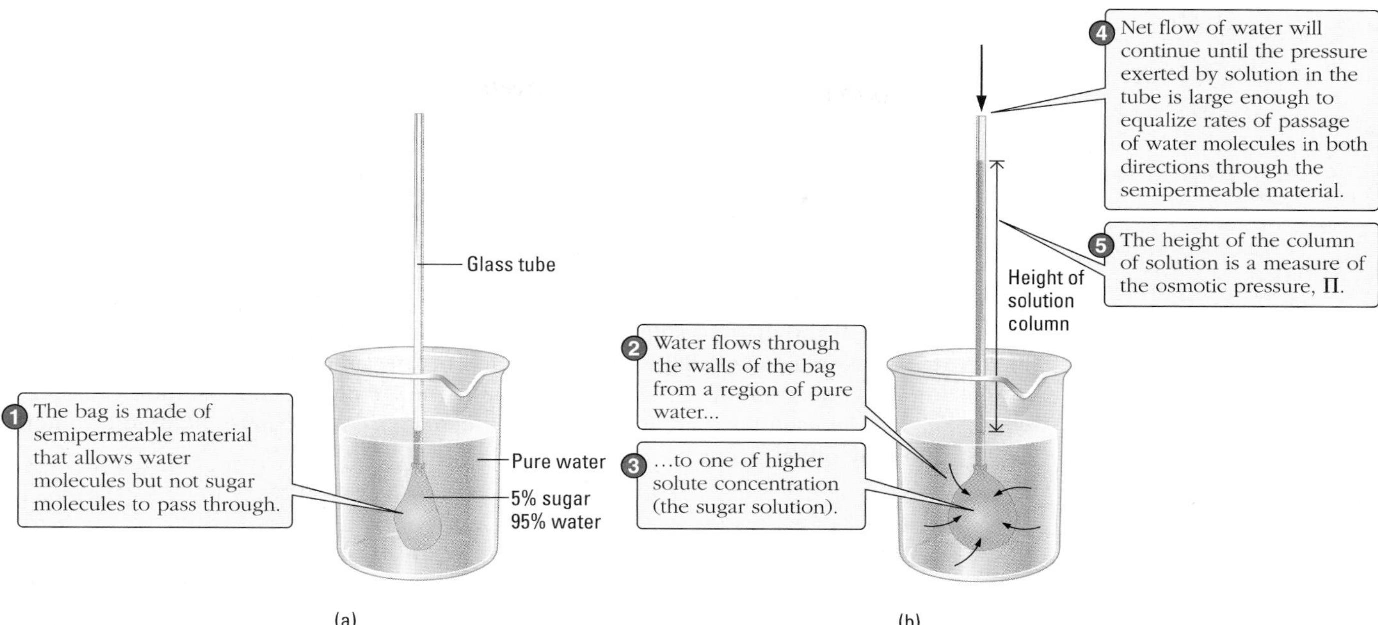

Figure 15.19 **Demonstration of osmotic pressure.**

molecules per unit area of the membrane occur on the pure solvent side than there are on the solution side (where there are fewer solvent molecules per unit volume). Hence, water moves through the membrane from the beaker where water is in greater concentration into the solution in the bag, where the water concentration is lower. As this continues, the number of water molecules in the solution increases, the number of collisions of water molecules within the solution increases, and water rises in the tube as pressure builds up in the bag. A dynamic equilibrium is achieved when the pressure in the bag equals the osmotic pressure, at which point the rate of passing water molecules is the same in both directions. The height of the water column then remains unchanged, and is a measure of the osmotic pressure.

Osmotic pressure—like vapor pressure lowering, boiling point elevation, and freezing point lowering—results from the unequal rates at which solvent molecules pass through an interface or boundary. In the case of evaporation and boiling, it is the solution/vapor interface; for freezing, it is the solution/solid interface. The semipermeable membrane is the interface for osmosis.

All colligative properties can be understood in terms of differences in entropy between a pure solvent and a solution. This is perhaps most easily seen in the case of osmosis. When solvent and solute molecules mix, entropy usually increases. If pure solvent is added to a solution, a higher entropy state will be achieved as solvent and solute molecules diffuse among one another to form a more dilute solution. Unless strong noncovalent intermolecular forces exist between the solute and solvent, there will be a negligible enthalpy change, so the increase in entropy makes mixing of solvent and solution a product-favored process. A semipermeable membrane prevents solute molecules from passing into pure solvent, so the only way mixing can occur (and entropy can increase) is for solvent to flow into the solution, and it does.

The more concentrated the solution, the more product-favored the mixing is and the greater is the pressure required to prevent it. Osmotic pressure (Π) is proportional to the molarity of the solution, c,

$$\Pi = cRTi$$

where R is the gas constant, T is the absolute temperature (in kelvins), and i is the number of particles per formula unit of solute.

The osmotic pressure equation is similar to the ideal gas law equation, $PV = nRT$, which can be rearranged to $P = (n/V)RT = cRT$, where n/V is the molar concentration of the gas.

ThomsonNOW
Log on to ThomsonNOW at **www.thomsonedu.com** and click Coached Problems for a tutorial on **osmotic pressure.**

Even though the solution concentration is small, osmotic pressure can be quite large. For example, the osmotic pressure of a 0.020 M solution of a molecular solute ($i = 1$) at 25 °C is

$$\Pi = cRTi = \left(\frac{0.020\text{ mol}}{\text{L}}\right)(0.0821\text{ L atm mol}^{-1}\text{ K}^{-1})(298\text{ K})(1) = 0.49\text{ atm}$$

This pressure would support a water column more than 15 ft high. One way to determine osmotic pressure is to measure the height of a column of solution in a tube, as shown in Figure 15.19. Heights of a few centimeters can be measured accurately, so quite small concentrations can be determined by osmotic pressure experiments. If the mass of solute dissolved in a measured volume of solution is known, it is possible to calculate the molar mass of the solute by using the definition of molar concentration, $c = n/V$ = amount (mol)/volume (L). Osmotic pressure is especially useful in studying large molecules whose molar mass is difficult to determine by other means.

Freezing point lowering and boiling point elevation measurements can also be used to find the molar mass in the same manner as shown in Problem-Solving Example 15.13 for osmotic pressure measurements.

PROBLEM-SOLVING EXAMPLE 15.13 Molar Mass from Osmotic Pressure

A solution containing 2.50 g of a nonelectrolyte polymer dissolved in 150. mL solution has an osmotic pressure of 1.25×10^{-2} atm at 25 °C. Calculate the molar mass of the polymer.

Answer 3.26×10^{4} g/mol

Strategy and Explanation We can calculate the molarity of the solution using the osmotic pressure equation. Because the polymer is a nonelectrolyte, $i = 1$.

$$c = \frac{\Pi}{RTi} = \frac{1.25 \times 10^{-2}\text{ atm}}{(0.0821\text{ L atm mol}^{-1}\text{ K}^{-1})\ (298\text{ K})(1)} = 5.11 \times 10^{-4}\text{ mol/L}$$

The volume of solution is 0.150 L, so the number of moles of polymer is

$$\text{Amount polymer} = (0.150\text{ L})(5.11 \times 10^{-4}\text{ mol/L}) = 7.67 \times 10^{-5}\text{ mol}$$

This is the number of moles in 2.50 g of the polymer, therefore the average molar mass of the polymer is

$$\text{Average molar mass of polymer} = \frac{2.50\text{ g polymer}}{7.67 \times 10^{-5}\text{ mol polymer}} = 3.26 \times 10^{4}\text{ g/mol}$$

✓ **Reasonable Answer Check** Because the molarity is low, the molar mass of the polymer must be relatively large to create an osmotic pressure of 1.25×10^{-2} atm. In Section 12.7, polymers were described as long chains of linked monomer units, so a molar mass of over 30,000 g/mol is not unreasonable for the polymer.

PROBLEM-SOLVING PRACTICE 15.13

The osmotic pressure of a solution of 5.0 g of horse hemoglobin (a protein) in 1.0 L water is 1.8×10^{-3} atm at 25 °C. What is the molar mass of the hemoglobin?

Blood and other fluids inside living cells contain many different solutes, and the osmotic pressures of these solutions play an important role in the distribution and balance of solutes within the body. Dehydrated patients are often given water and nutrients intravenously. However, pure water cannot simply be dripped into a patient's veins. The water would flow into the red blood cells by osmosis, causing them to burst (Figure 15.20c). A solution that causes this condition is called a **hypotonic** solution. To prevent cells from bursting, an **isotonic** (or iso-osmotic) intravenous solution must be used. Such a solution has the same total concentration of solutes and therefore the same osmotic pressure as the patient's

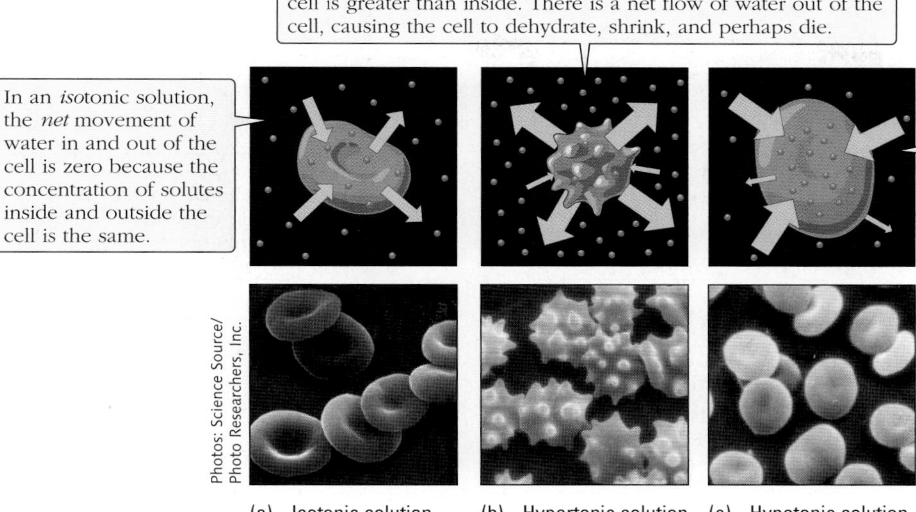

In a *hyper*tonic solution, the concentration of solutes outside the cell is greater than inside. There is a net flow of water out of the cell, causing the cell to dehydrate, shrink, and perhaps die.
In an *iso*tonic solution, the *net* movement of water in and out of the cell is zero because the concentration of solutes inside and outside the cell is the same.
In a *hypo*tonic solution, the concentration of solutes outside the cell is less than inside. There is a net flow of water into the cell, causing the cell to swell and perhaps to burst.
Photos: Science Source/ Photo Researchers, Inc.
(a) Isotonic solution
(b) Hypertonic solution
(c) Hypotonic solution

Figure 15.20 **Osmosis and the living cell.**

blood (Figure 15.20a). A solution of 0.9% sodium chloride is isotonic with fluids inside cells in the body.

If an intravenous solution more concentrated than the solution inside a red blood cell were added to blood, the cell would lose water to the solution and shrivel up. A solution that causes this condition is a **hypertonic** solution (Figure 15.20b). Cell-shriveling by osmosis happens when vegetables or meats are cured in *brine,* a concentrated solution of NaCl. If you put a fresh cucumber into brine, water will flow out of its cells and into the brine, leaving behind a shriveled vegetable (Figure 15.21). With proper spices added to the brine, a cucumber will become a tasty pickle.

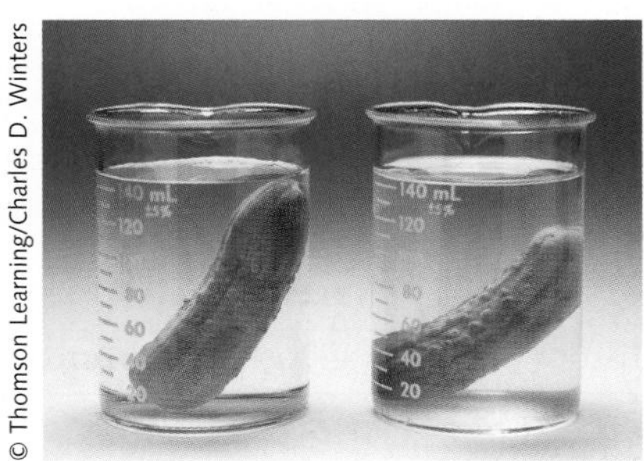

Figure 15.21 **Osmosis in vegetables.** When a cucumber is soaked in a concentrated salt solution, water flows from the plant cells into the salt solution by osmosis, converting the cucumber into a pickle. A cucumber soaked in a concentrated salt solution (*right*) has lost much water and shrivels into a pickle. A cucumber soaked in pure water (*left*) is affected very little.

Reverse Osmosis

Reverse osmosis occurs when pressure greater than the osmotic pressure is applied and solvent is made to flow through a semipermeable membrane from a concentrated solution to a dilute solution. In effect, the semipermeable membrane serves as a filter with very tiny pores through which only the solvent can pass. Reverse osmosis can be used to remove small molecules or ions to obtain highly purified water. Seawater contains a high concentration of dissolved salts; its osmotic pressure is 24.8 atm. If a pressure greater than 24.8 atm is applied to a chamber containing seawater, water molecules can be forced to flow from seawater through a semipermeable membrane to a region containing purer water (Figure 15.22). Pressures up to 100 atm are used to provide reasonable rates of seawater purification. Seawater, which contains upward of 35,000 ppm of dissolved salts (Table 15.5), can be purified by reverse osmosis to between 400 and 500 ppm of solutes, which is well within the World Health Organization's limits for drinking water. Large reverse osmosis plants in places like the Persian Gulf countries and Florida can purify more than 100 million gallons of water per day. Nearly 50% of Saudi Arabia's fresh water is provided by the world's largest desalination plant at Jubai. Reverse osmosis purifies nearly 15 million gallons of brackish underground water daily for the city of Cape Coral, Florida. That facility is one of more than 100 in the state of Florida.

This reverse osmosis plant in Israel can annually convert 100 million cubic meters of salt water to fresh water.

Small reverse osmosis units are used to make the ultrapure water for some "spotless" car washes.

Brackish water is a mixture of salt water and fresh water.

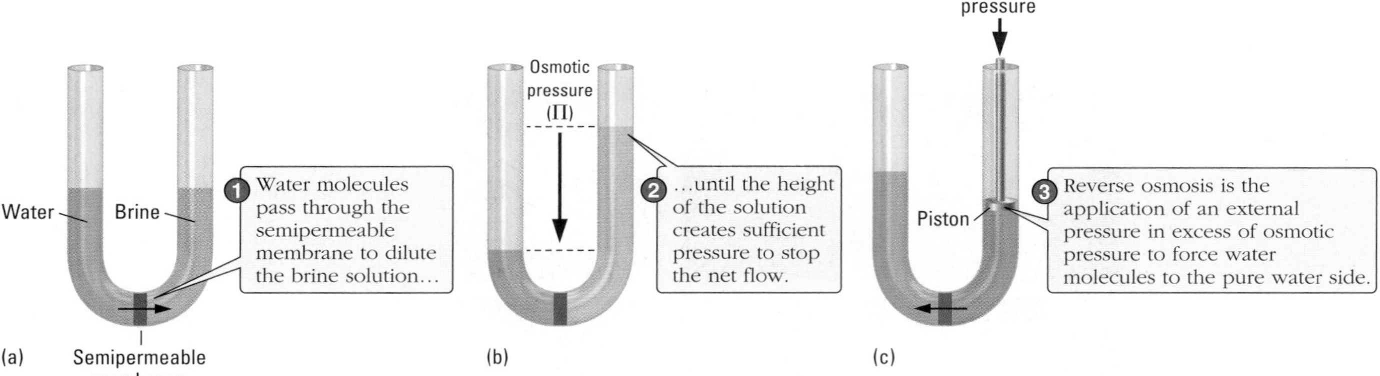

Figure 15.22 Normal and reverse osmosis. Normal osmosis is represented in (a) and (b). Reverse osmosis is represented in (c).

Table 15.5 Ions Present in Seawater at 100 ppm or More

	Mass Fraction	
Ion	**g/kg**	**ppm**
Cl^-	19.35	19,350
Na^+	10.76	10,760
SO_4^{2-}	2.710	2710
Mg^{2+}	1.290	1290
Ca^{2+}	0.410	410
K^+	0.400	400
HCO_3^-, CO_3^{2-}	0.106	106
Total	35.026	35,026

Blue sky results from light scattering by colloidal-sized particles in air. Because blue light is more strongly scattered than is red light, the sky appears to be a diffuse blue. This answers the question from Chapter 1 (p. 2): "Why is the sky blue?"

15.9 Colloids

Around 1860, Thomas Graham found that starch, gelatin, glue, and egg albumin diffuse only very slowly in water and do not diffuse through a thin membrane, but sugar or salt do. He coined the word "colloid" to describe a class of substances distinctly different from sugar and salt and similar materials. **Colloids** are now understood to be mixtures in which relatively large particles, the **dispersed phase,** are distributed uniformly throughout a solvent-like medium called the **continuous phase,** or the dispersing medium. Like true solutions, colloids are found in the gas, liquid, and solid states. Although both true solutions and colloids appear homogeneous to the naked eye, at the microscopic level colloids are not homogeneous.

In colloids, the dispersed phase particles might be as large as 10 to nearly 1000 times the size of a single small molecule. Colloidal particles are larger than those found in true solutions and smaller than the ones in suspensions.

	Smaller Particles ⟶ Larger Particles		
	True Solution	**Colloidal Dispersion**	**Suspension**
Particles	Ions and molecules	Colloids	Large-sized particles
Particle size	0.2–2.0 nm	2–2000 nm	>2000 nm
Properties	• Don't settle out on standing • Not separable using ordinary filters	• Don't settle out on standing • Not filterable	• Settle out on standing • Filterable
Example	Seawater	Fog	River silt

Colloidal particles can be so large—about 2000 nm in diameter—that they scatter visible light passing through the continuous medium, a phenomenon known as the **Tyndall effect.** Figure 15.23 shows a beam of light passing through three glass cuvets. The cuvets on the left and right contain a colloidal mixture of a gelatin in water; the center cuvet holds a solution of NaCl. The light scattering is clearly seen in the colloidal suspensions; particles in the NaCl solution are too small to scatter the light, and the Tyndall effect is not seen. Colloidal particles of dust and smoke in the

Table 15.6 Types of Colloids

Continuous Phase	Dispersed Phase	Type	Examples
Gas	Liquid	Aerosol	Fog, clouds, aerosol sprays
Gas	Solid	Aerosol	Smoke, airborne viruses, automobile exhaust
Liquid	Gas	Foam	Shaving cream, whipped cream
Liquid	Liquid	Emulsion	Mayonnaise, milk, face cream
Liquid	Solid	Sol	Gold in water, milk of magnesia, mud
Solid	Gas	Foam	Foam rubber, sponge, pumice
Solid	Liquid	Gel	Jelly, cheese, butter
Solid	Solid	Solid sol	Milk glass, many alloys such as steel, some colored gemstones

The Tyndall effect. Shafts of light are visible through the mist in this forest.

air of a room can easily be observed in a beam of sunlight because they scatter the light; you have probably seen such a well-defined sunbeam many times. A common colloid, *fog,* consists of water droplets (the dispersed phase) in air (the continuous phase, and itself a solution). The lack of visibility in fog is due to the Tyndall effect.

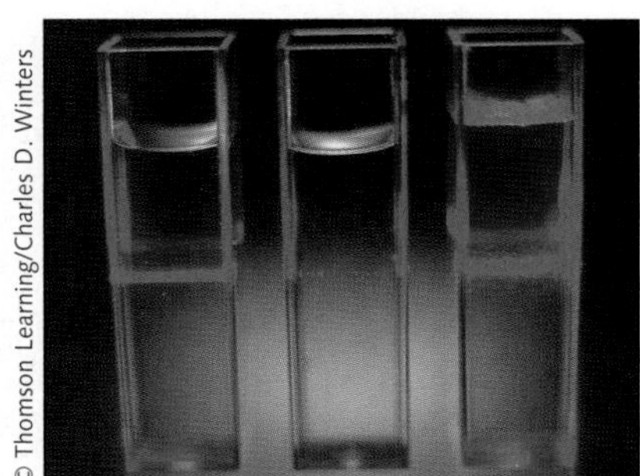

Figure 15.23 Light scattering by a colloidal dispersion. A narrow beam of light half way up from the bottom of each cuvet passes through a colloidal mixture (*left*), then through a salt solution, and finally through another colloidal mixture. The Tyndall effect is seen in the colloidal mixtures, but not in the salt solution. This illustrates the light-scattering ability of colloid-sized particles.

Types of Colloids

Colloids are classified according to the state of the dispersed phase (solid, liquid, or gas) and the state of the continuous phase. Table 15.6 lists several types of colloids and some examples of each. Liquid-liquid colloids form only in the presence of an emulsifier—a third substance that coats and stabilizes the particles of the dispersed phase. Such colloidal dispersions are called **emulsions.** In mayonnaise, for example, egg yolk contains a protein that stabilizes the tiny drops of oil that are dispersed in the aqueous continuous phase. As you can see from examples listed in Table 15.6, colloids are very common in everyday life.

Colloids with water as the continuous phase are either hydrophilic or hydrophobic. In a hydrophilic colloid there is *a strong attraction between the dispersed phase and the continuous (aqueous) phase*. Hydrophilic colloids form when the molecules of the dispersed phase have multiple sites that interact with water through hydrogen bonding and dipole-dipole attractions. Proteins in aqueous media are hydrophilic colloids.

CHEMISTRY YOU CAN DO

Curdled Colloids

Regular (whole) milk contains about 4% fat; skim milk contains considerably less. In addition, milk contains protein. Both the fat and the proteins are in the form of colloids.

Add about 2 tablespoons of vinegar or lemon juice to about 100 mL of whole milk (do the same with skim milk and 1% or 2% butterfat milk, if you have it), stir, and watch what happens. Let it stand overnight at room temperature. Observe and record the results.

An additional experiment you can try is adding salt to similar samples of milk and recording your observations.

When you finish, discard this milk down the drain. You should *not* drink it because it has been unrefrigerated and might contain harmful bacteria.

1. Write an explanation for what you observed.
2. Does the salt have the same effect on milk as the acid does?

NASA/Peter Arnold, Inc.

Figure 15.24 Aerial view of silt formation in a river delta from colloidal soil particles. Silt forms at a river delta (lower left) as colloidal particles in the river water meet the salt water as the river enters an ocean or a salt-water bay. The higher salt concentration causes the colloidal particles to coagulate.

In a hydrophobic colloid there is a *lack of attraction between the dispersed phase and the continuous phase*. Although you might assume that such colloids would tend to separate quickly, hydrophobic colloids can be quite stable once they are formed. A colloidal solution (sol) of gold particles prepared in 1857 is still preserved in the British Museum.

A stable hydrophobic colloid coagulates when ions come into contact with the dispersed phase. Soil particles carried in rivers are hydrophobic. When river water containing large amounts of colloidally suspended soil particles meets seawater with a high ion concentration, the colloidal particles coagulate to form silt. The deltas of the Mississippi and Nile Rivers have been and continue to be formed in this way (Figure 15.24).

15.10 Surfactants

Molecules that have a hydrophobic part and a hydrophilic part are called **surfactants** (surface-active agents) because they tend to act at the surface of a substance that is in contact with the solution that contains the surfactant. Soap, the classic surfactant, dates back to the Sumerians in 2500 BC. Chemically, soaps are salts of fatty acids and have always been made by the reaction of a fat with an alkali, a process known as *saponification* (⇐ p. 574).

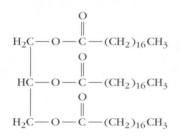

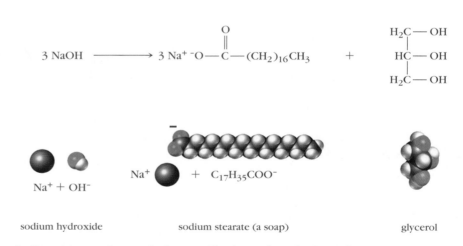

$$\begin{array}{l} H_2C-O-\overset{\overset{O}{\|}}{C}-(CH_2)_{16}CH_3 \\ HC-O-\overset{\overset{O}{\|}}{C}-(CH_2)_{16}CH_3 \\ H_2C-O-\overset{\overset{O}{\|}}{C}-(CH_2)_{16}CH_3 \end{array} + 3\,NaOH \longrightarrow 3\,Na^+\,{}^-O-\overset{\overset{O}{\|}}{C}-(CH_2)_{16}CH_3 + \begin{array}{l} H_2C-OH \\ HC-OH \\ H_2C-OH \end{array}$$

$Na^+ + OH^-$ $\qquad$ $Na^+ + C_{17}H_{35}COO^-$

tristearin (glyceryl tristearate) $\qquad$ sodium hydroxide $\qquad$ sodium stearate (a soap) $\qquad$ glycerol

Sodium stearate is a typical soap. The long-chain hydrocarbon part of the molecule is hydrophobic, while the polar carboxylate group is hydrophilic.

Hand soaps are pure soaps to which dyes and perfumes are added.

$$CH_3CH_2CH_2CH_2CH_2CH_2CH_2CH_2CH_2CH_2CH_2CH_2CH_2CH_2CH_2CH_2CH_2\;\overset{\overset{O}{\|}}{C}-O^-Na^+$$

Hydrophobic end $\qquad$ Hydrophilic end

sodium stearate

Water, oil, and a surfactant together form an emulsion. The surfactant acts as the emulsifying agent, such as in the case of bile salts emulsifying dietary fats.

Bile salts are important surfactants in the body that help to break up ingested fats. The hydrophobic portion of bile salts is incorporated into the surface of the ingested fat molecules, where it acts as an emulsifying agent, keeping fat molecules dispersed from each other. The hydrophilic portion of the bile salt keeps the emulsified fats in solution.

Synthetic surfactants called **detergents** are made from refined petroleum or coal products. Detergents have a long hydrocarbon chain that is hydrophobic and a polar end that is hydrophilic, somewhat like those of soaps. One common detergent is sodium lauryl sulfate, which is used in many shampoos.

$$CH_3CH_2CH_2CH_2CH_2CH_2CH_2CH_2CH_2CH_2CH_2CH_2OSO_3^-Na^+$$
sodium lauryl sulfate

CONCEPTUAL EXERCISE 15.18 Estimating Osmotic Pressure

Which solution would have the higher osmotic pressure, a 0.02 M sucrose solution or a 0.02 M typical soap solution? Explain your answer.

ThomsonNOW
Log on to ThomsonNOW at **www.thomsonedu.com** and click Chemistry Interactive for modules on:
- **surfactant-oil interactions**
- **fabric softener and intermolecular forces**

In water solutions, surfactants tend to aggregate to form hollow, colloid-sized spherical particles called *micelles* (Figure 15.25) that can transport various materials within them. The hydrophobic ends of the surfactant molecules point inward to the center of the micelle, while their hydrophilic heads point outward so that they interact with water molecules. Ordinary soap, a surfactant, forms micelles in water. Soaps cleanse because hydrophobic oil on clothing or skin joins to the hydrophobic centers of the soap micelles and is rinsed away (Figure 15.25). When ordinary soaps are used with hard water (p. 762), the soaps react with Ca^{2+}, Mg^{2+}, and Fe^{2+} ions in the hard water to produce insoluble salts that form an unsightly scum around the bathtub or on clothes. Detergents do not form these insoluble salts, even with hard water.

This explanation of soaps and detergents answers the question posed in Chapter 1 (*p. 3*): "How does soap help clothes get clean?"

15.11 Water: Natural, Clean, and Otherwise

More than 97% of water, the most abundant substance on the earth's surface, is found in the oceans, which cover about three quarters of the earth's surface (Figure 15.26, p. 760). Glaciers, ice caps, and snow pack account for 1.72% of water. *Surface water*—lakes, rivers, streams, and reservoirs—makes up only a very small portion (0.008%). *Groundwater*, water that is held in large underground natural *aquifers*, makes up 0.77% of water.

". . . Water, water everywhere, nor any drop to drink." *The Rime of the Ancient Mariner* by Samuel Taylor Coleridge alludes to the fact that the dissolved salt in seawater renders it useless for drinking.

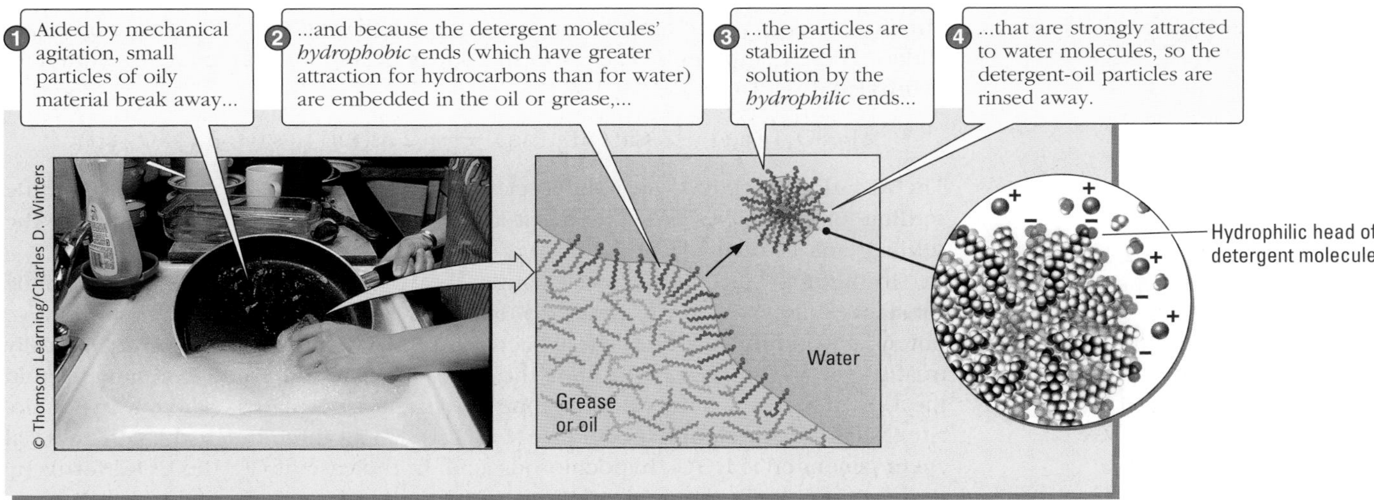
1 Aided by mechanical agitation, small particles of oily material break away...
2 ...and because the detergent molecules' hydrophobic ends (which have greater attraction for hydrocarbons than for water) are embedded in the oil or grease,...
3 ...the particles are stabilized in solution by the hydrophilic ends...
4 ...that are strongly attracted to water molecules, so the detergent-oil particles are rinsed away.
Water
Grease or oil
Hydrophilic head of detergent molecule
© Thomson Learning/Charles D. Winters

Figure 15.25 **The cleansing action of soaps and detergents.**

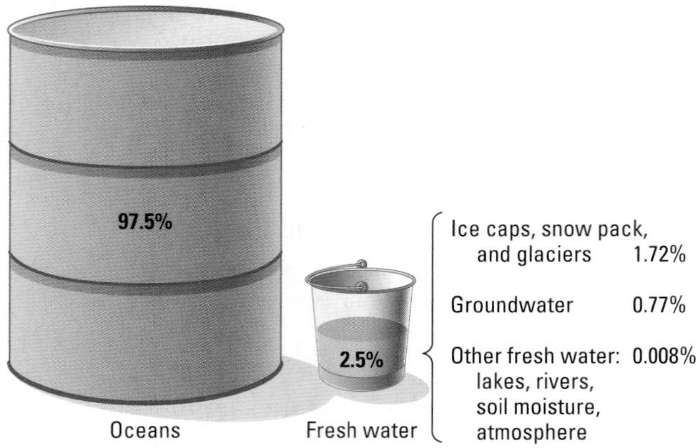

Figure 15.26 The earth's water supply. Of the 2.5% that is fresh water, less than 1% is available as groundwater or surface water.

Note from Figure 15.26 that fresh water makes up less than 3% of the earth's surface water, with most of it tied up in glaciers. Indeed, the amount of fresh water available to satisfy our demands (we can't live without it) is relatively limited. In the United States, a bit more than half (57%) of fresh water is used in industry and about a third (34%) in agriculture to irrigate crops; only 9% is available for domestic and municipal purposes. Even when a sufficient quantity of water is available, the quality of the water needs to be assured before it is safe to drink. Water must first be treated to make it potable.

Municipal Drinking Water Purification

Fresh water, even from natural sources, contains dissolved materials that must be removed or decreased to make water fit for domestic use, or for agricultural and industrial purposes. The Safe Water Drinking Act of 1974 established required standards of purity and safety for public water supplies. The EPA sets limits for contaminants that may be present in drinking water and requires continual monitoring of municipal water supplies.

Municipal water purification takes place in a series of steps (Figure 15.27). After a coarse filter (a screen) removes large objects such as tires, tree limbs, bottles, and so on, the water goes into a settling tank, where small clay and dirt particles settle out. To speed up the sedimentation, $Al_2(SO_4)_3$ (alum) and $Ca(OH)_2$ (slaked lime) are added. These compounds react to form a sticky gelatinous precipitate of aluminum hydroxide.

Calcium hydroxide (slaked lime) forms when lime (CaO) reacts with water.

$$Al_2(SO_4)_3(aq) + 3\ Ca(OH)_2(aq) \longrightarrow 2\ Al(OH)_3(s) + 3\ CaSO_4(aq)$$

The $Al(OH)_3$ gel collects suspended clay and dirt particles as it sinks slowly in the settling tank. Particles not settled out are removed by passing the water from the settling tank through a sand filter.

In the next step, the water is aerated—sprayed into the air to oxidize organic substances dissolved in the water. To this point, nothing has been done to remove potentially harmful bacteria. These bacteria are killed in the final step by chemically treating the water—*chlorination* is the most common bactericidal method used in the United States. Chlorination is done by adding chlorine gas, sodium hypochlorite, NaOCl, or calcium hypochlorite, $Ca(OCl)_2$. In all three cases, the antibacterial agent generated is HOCl, hypochlorous acid. In the case of Cl_2, the HOCl forms by the reaction of chlorine with water.

$$Cl_2(g) + H_2O(\ell) \longrightarrow HOCl(aq) + H^+(aq) + Cl^-(aq)$$

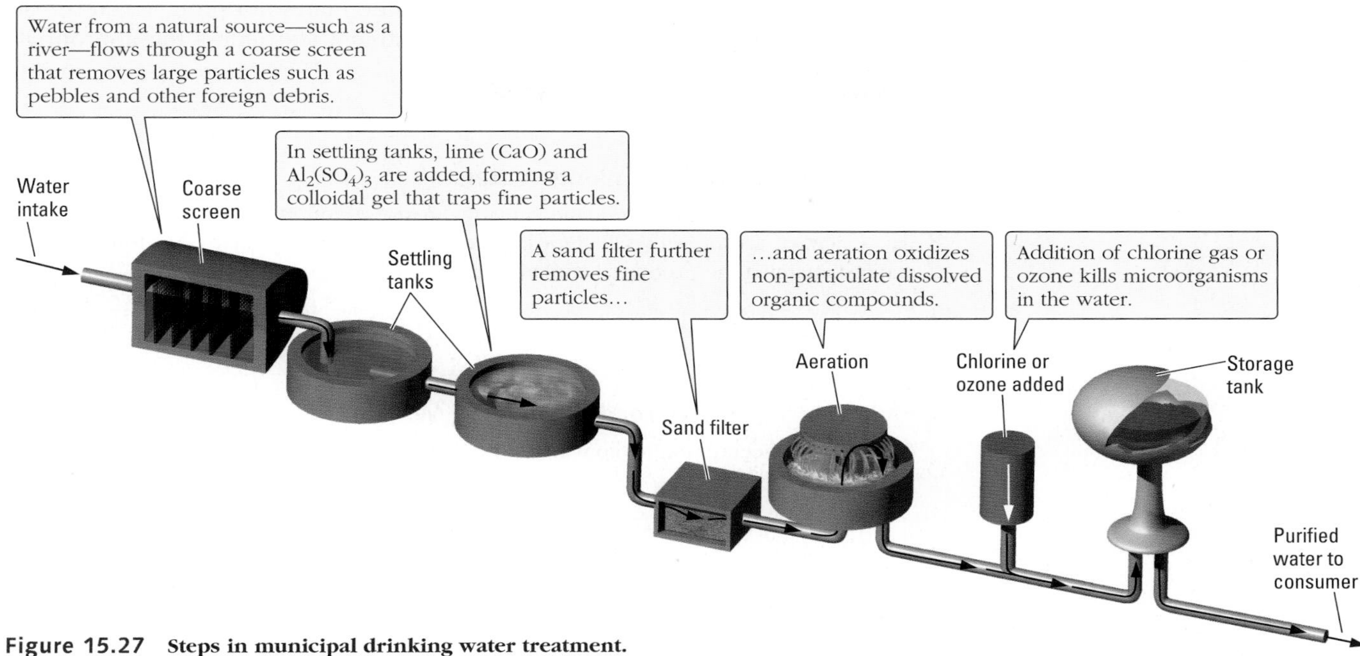

Figure 15.27 **Steps in municipal drinking water treatment.**

The extent of chlorination is adjusted so that between 0.075 and 0.600 ppm of HOCl remains in solution as the water leaves the treatment plant, which is sufficient to ensure that bacterial contamination does not occur before the water reaches the user.

Chlorination was first used for drinking water supplies in the early 1900s, with a resulting drop in the number of deaths in the United States caused by typhoid and other water-borne diseases from 35/100,000 population in 1900 to 3/100,000 population in 1930. Chlorination is the principal means of preventing water-borne diseases spread by bacteria, including cholera, typhoid, paratyphoid, and dysentery.

In spite of chlorination, most city water supplies are not bacteria-free, but only rarely do these surviving bacteria cause disease. In the United States the most common water-borne bacterial disease is giardiasis, a gastrointestinal disorder. Most often this disease is caused by bacteria in surface water that has leaked into drinking water supplies, but on occasion it can be traced to city water systems.

Chlorination is not without its own small risk because of by-products it forms. Even the best-designed purification systems allow some organic compounds to pass through, which then become chlorinated. In particular, humic acids, breakdown products of plant materials always present in surface water, react with residual HOCl. The reaction forms a class of compounds known as trihalomethanes (THMs), such as chloroform, $CHCl_3$. Most drinking water meets the current maximum contaminant level standard of 80 ppb THMs. Chloroform is the trihalomethane of chief concern because it is suspected of causing liver cancer. Information is still being evaluated about the seriousness of this potential threat, especially given the fact that the THM level in drinking water is normally less than 1 ppm.

The 80 ppb standard for THMs was established in 1998 by the U.S. Environmental Protection Agency. The previous standard had been 100 ppb. The national average for THMs in drinking water is 51 ppb.

Gaseous ozone, O_3, is used in many European cities to disinfect municipal water supplies. Ozone is an even more effective bactericide than chlorine, so less of it is needed to purify the water. The disadvantages of ozone are that it must be generated on site, and that ozone does not remain in the water as long as chlorine, allowing the possibility of recontamination.

Disinfection of water by using ultraviolet radiation is becoming more popular. It is fast, is economical for small installations such as rural homes, and leaves no residual by-products. Like ozonation, UV disinfection does not protect against bacterial contamination after water leaves the treatment site unless the appropriate amount of chlorine is added.

EXERCISE 15.19 Drinking Water

Selenium poisoning can occur in individuals who ingest more than 400 μg selenium per day. Calculate the mass of selenium ingested daily by an individual who drinks 3 qt water containing the maximum contaminant level of selenium, 0.050 ppm.

EXERCISE 15.20 Lead in Drinking Water

If the Pb^{2+} concentration in tap water is 0.025 ppm, how many liters of this water contain 100.0 μg Pb^{2+}?

Hard Water: Natural Impurities

Degrees of hardness:
Soft water: < 65 mg of metal ion/gal
Slightly hard: 65–228 mg/gal
Moderately hard: 228–455 mg/gal
Hard: 455–682 mg/gal
Very hard: > 682 mg/gal

A relatively high concentration of Ca^{2+}, Mg^{2+}, Fe^{2+}, or Mn^{2+} ions imparts "hardness" to water. Water hardness is objectionable because it (1) causes precipitates (called scale) to form in boilers and hot-water pipes, (2) causes soaps to form insoluble curds, and (3) may impart a disagreeable taste to the water.

Water hardness is produced when surface water containing carbon dioxide trickles through limestone or dolomite, releasing calcium or magnesium ions as their soluble bicarbonates.

$$\underset{\text{limestone}}{CaCO_3(s)} + CO_2(g) + H_2O(\ell) \longrightarrow Ca^{2+}(aq) + 2\ HCO_3^-(aq)$$

$$\underset{\text{dolomite}}{CaCO_3 \cdot MgCO_3(s)} + 2\ CO_2(g) + 2\ H_2O(\ell) \longrightarrow Ca^{2+}(aq) + Mg^{2+}(aq) + 4\ HCO_3^-(aq)$$

Such hard water can be softened by removing these ions by two principal methods: (1) the lime-soda process or (2) ion exchange.

In the lime-soda process, hydrated lime, $Ca(OH)_2$, and soda, Na_2CO_3, are added to the water. Several reactions take place, which can be summarized as follows:

In hard water	Added		
$HCO_3^-(aq)$	$+\ OH^-(aq)$	$\longrightarrow$	$CO_3^{2-}(aq) + H_2O(\ell)$
$Ca^{2+}(aq)$	$+\ Na_2CO_3(aq)$	$\longrightarrow$	$CaCO_3(s) + 2\ Na^+(aq)$
$Mg^{2+}(aq)$	$+\ 2\ OH^-(aq)$	$\longrightarrow$	$Mg(OH)_2(s)$

The lime-soda process works because calcium carbonate, $CaCO_3$, is much less soluble than calcium bicarbonate, $Ca(HCO_3)_2$, and because magnesium hydroxide, $Mg(OH)_2$, is much less soluble than magnesium bicarbonate, $Mg(HCO_3)_2$. The overall result of the lime-soda process is to precipitate almost all the calcium and magnesium ions and to leave sodium ions (non-hard water ions) as replacements.

Iron present as Fe^{2+} and manganese present as Mn^{2+} can be removed from water by aeration to produce higher oxidation states. If the water is neutral or slightly alkaline (either naturally or from the addition of lime), insoluble compounds $Fe(OH)_3$ and $MnO_2(H_2O)_x$ form and precipitate from solution.

Ion exchange is another way to remove ions that cause water hardness. In a cation-exchange resin containing either H^+ or Na^+ ions, these ions replace the ions that cause hard water. Home water treatment ion-exchange units usually replace hardness ions with Na^+ ions. The exchange resin is a polymer containing numerous negatively charged $—SO_3^-$ functional groups that have Na^+ ions attached. When smaller, more positive ions like Mg^{2+} and Ca^{2+} in hard water flow over the resin, they displace Na^+ ions from the resin. The process happens many times, resulting in water that contains Na^+ ions in place of the Ca^{2+} and Mg^{2+} ions, which are bound to the resin. Two $—SO_3^-$ groups are required for every 2+ ion removed from solution; two Na^+ ions are released for every 2+ ion removed.

For persons on low-sodium diets, lime-soda–treated water may provide too high a daily dose of Na^+.

$$\underset{\text{ion-exchange resin}}{2\,(\text{Polymer—SO}_3^-)\text{Na}^+} + \underset{\text{from hard water}}{\text{Ca}^{2+}} \rightleftharpoons (\text{Polymer—SO}_3^-)_2\text{Ca}^{2+} + 2\,\text{Na}^+(\text{aq})$$

The same equation applies if Mg^{2+} is substituted for Ca^{2+}.

Because Na^+ ions do not harden water, the resulting water is called "soft" water.

When all its sodium ions have been replaced, the ion-exchange resin becomes saturated with hard water ions. The resin must be regenerated by treating it with a highly concentrated NaCl solution. This reverses the process given in the previous equation and the hard water ions released from the resin are rinsed down the drain.

© Thomson Learning/Charles D. Winters

A home water softener. Hard water is softened by substituting sodium ions for calcium and magnesium ions.

SUMMARY PROBLEM

You are asked to prepare three mixtures at 25 °C.

Mixture I: 25.0 g CCl_4 and 100. mL of water

Mixture II: 15.0 g $CaCl_2$ in 125 mL of water

Mixture III: 21 g ethylene glycol ($HOCH_2CH_2OH$) in 150. mL water

Answer these questions about these mixtures. If one of the solutes fails to dissolve in water, some of the questions will not be applicable.

(a) What is the weight percent of the mixture?

(b) What is the mass fraction of the mixture?

(c) Is a solution formed? (If a solution is formed, answer the remaining questions. You may assume a density of the solution of 1.0 g/mL.)

(d) Name the dissolved species in solution and draw a diagram representing how the solvent (water) molecules interact with these species.

(e) Express the concentration of the solution in ppm.

(f) Express the concentration of the solution in molality.

(g) Calculate the vapor pressure of water in equilibrium with the solution.

(h) Calculate the boiling point of the solution.

(i) Calculate the freezing point of the solution.

(j) Calculate the osmotic pressure of the solution.

IN CLOSING

Having studied this chapter, you should be able to . . .

- Describe how liquids, solids, and gases dissolve in a solvent (Section 15.1). ThomsonNOW homework: Study Question 1

ThomsonNOW

Log on to ThomsonNOW at **www.thomsonedu.com** to check your readiness for an exam by taking the Pre-Test and exploring the modules recommended in your Personalized Learning Plan.

- Predict solubility based on properties of solute and solvent (Section 15.1). ThomsonNOW homework: Study Questions 25, 29
- Interpret the dissolving of solutes in terms of enthalpy and entropy changes (Section 15.2). ThomsonNOW homework: Study Question 28
- Differentiate among unsaturated, saturated, and supersaturated solutions (Section 15.3). ThomsonNOW homework: Study Question 104
- Describe how ionic compounds dissolve in water (Section 15.3).
- Predict how temperature affects the solubility of ionic compounds (Section 15.4).
- Predict the effects of temperature (Section 15.4) and pressure on the solubility of gases in liquids (Section 15.5). ThomsonNOW homework: Study Question 8
- Describe the compositions of solutions in terms of weight percent, mass fraction, parts per million, parts per billion, parts per trillion, molarity, and molality (Section 15.6). ThomsonNOW homework: Study Questions 36, 47, 53
- Interpret vapor pressure lowering in terms of Raoult's law (Section 15.7). ThomsonNOW homework: Study Question 65
- Use molality to calculate the colligative properties: freezing point lowering and boiling point elevation (Section 15.7). ThomsonNOW homework: Study Questions 59, 69, 73, 114
- Differentiate the colligative properties of nonelectrolytes and electrolytes (Section 15.7). ThomsonNOW homework: Study Questions 61, 95
- Explain the phenomena of osmosis and reverse osmosis, and calculate osmotic pressure (Section 15.8). ThomsonNOW homework: Study Question 77
- Describe the various kinds of colloids and their properties (Section 15.9).
- Explain how surfactants work (Section 15.10).
- Discuss the earth's water supply and the sources of fresh water (Section 15.11).
- Discuss how municipal drinking water is purified (Section 15.11).
- Describe what causes hard water and explain how water can be softened (Section 15.11). ThomsonNOW homework: Study Question 81

KEY TERMS

boiling point elevation *(15.7)*
colligative properties *(15.7)*
colloid *(15.9)*
continuous phase *(15.9)*
detergents *(15.10)*
dispersed phase *(15.9)*
emulsion *(15.9)*
enthalpy of solution *(15.2)*
freezing point lowering *(15.7)*
Henry's law *(15.5)*
hydration *(15.3)*
hypertonic *(15.8)*
hypotonic *(15.8)*
immiscible *(15.1)*
isotonic *(15.8)*
mass fraction *(15.6)*
miscible *(15.1)*
molality *(15.6)*
osmosis *(15.8)*
osmotic pressure *(15.8)*
parts per billion *(15.6)*
parts per million *(15.6)*
parts per trillion *(15.6)*
Raoult's law *(15.7)*
reverse osmosis *(15.8)*
saturated solution *(15.3)*
semipermeable membrane *(15.8)*
solubility *(15.3)*
supersaturated solution *(15.3)*
surfactants *(15.10)*
Tyndall effect *(15.9)*
unsaturated solution *(15.3)*
weight percent *(15.6)*

QUESTIONS FOR REVIEW AND THOUGHT

■ denotes questions available in ThomsonNOW and assignable in OWL.

Blue-numbered questions have short answers at the back of this book and fully worked solutions in the *Student Solutions Manual.*

ThomsonNOW™

Assess your understanding of this chapter's topics with sample tests and other resources found by signing in to ThomsonNOW at **www.thomsonedu.com**.

Review Questions

1. ■ Which of these general types of substances would you expect to dissolve readily in water?
   (a) Alcohols (b) Hydrocarbons
   (c) Metals (d) Nonpolar molecules
   (e) Polar molecules (f) Salts
2. Explain on a molecular basis why the components of blended motor oils remain dissolved and do not separate.
3. Explain why gasoline and motor oil are miscible, as in the fuel mixtures used in two-cycle lawn mower engines.
4. Describe the differences among solutions that are unsaturated, saturated, and supersaturated in terms of amount of solute.
5. Describe the differences among unsaturated, saturated, and supersaturated solutions in terms of Q and K_c.
6. State Henry's law. Name three factors that govern the solubility of a gas in a liquid.
7. In general, how does the water solubility of most ionic compounds change as the temperature is increased?
8. ■ How does the solubility of gases in liquids change with increased temperature? Explain why.
9. Which is the highest solute concentration: 50 ppm, 500 ppb, or 0.05% by weight?
10. Estimate your concentration on campus in parts per million and parts per thousand.
11. Define molality. How does it differ from molarity?
12. Explain the difference between the mass fraction and the mole fraction of solute in a solution.
13. Explain why the vapor pressure of a solvent is lowered by the presence of a nonvolatile solute.
14. Why is a higher temperature required for boiling a solution containing a nonvolatile solute than for boiling the pure solvent?
15. Which would have the lowest freezing point?
    (a) A 1.0 mol/kg NaCl solution in water
    (b) A 1.0 mol/kg $CaCl_2$ solution in water
    (c) A 1.0 mol/kg methanol solution in water
    Explain your choice.
16. Write the osmotic pressure equation and explain all terms.
17. Explain the difference between (a) a hypotonic and an isotonic solution; (b) an isotonic and a hypertonic solution.
18. Explain how reverse osmosis works.
19. How do colloids differ from suspensions?
20. Explain why the Tyndall effect is not observed with solutions.
21. How can the presence of a strong electrolyte cause a hydrophobic colloid to coagulate?
22. Sketch an illustration of a soap molecule. Based on its structure, why is it considered a surfactant?
23. Surfactant molecules have what common structural features?

Topical Questions

How Substances Dissolve

24. Explain why some liquids are miscible in each other while other liquids are immiscible. Using only three liquids, give an example of a miscible pair and an immiscible pair.
25. ■ Why would the same solid readily dissolve in one liquid and be almost insoluble in another liquid? Give an example of such behavior.
26. Knowing that the solubility of oxalic acid at 25 °C is 1 g per 7 g water, how would you prepare 1 L of a saturated oxalic acid solution?
27. ■ A saturated solution of NH_4Cl was prepared by adding solid NH_4Cl to water until no more solid NH_4Cl would dissolve. The resulting mixture felt very cold and had a layer of undissolved NH_4Cl on the bottom. When the mixture reached room temperature, no solid NH_4Cl was present. Explain what happened. Was the solution still saturated?
28. ■ The lattice energy of $CaCl_2$ is −2258 kJ/mol, and its enthalpy of hydration is +2175 kJ/mol. Is the process of dissolving $CaCl_2$ in water endothermic or exothermic?
29. ■ Simple acids such as formic acid, HCOOH, and acetic acid, CH_3COOH, are very soluble in water; however, fatty acids such as stearic acid, $CH_3(CH_2)_{16}COOH$, and palmitic acid, $CH_3(CH_2)_{14}COOH$, are water-insoluble. Based on what you know about the solubility of alcohols, explain the solubility of these organic acids.
30. If a solution of a certain salt in water is saturated at some temperature and a few crystals of the salt are added to the solution, what do you expect will happen? What happens if the same quantity of the same salt crystals is added to an unsaturated solution of the salt? What would you expect to happen if the temperature of this second salt solution is slowly lowered?
31. Describe what happens when an ionic solid dissolves in water. Sketch an illustration that includes at least three positive ions, three negative ions, and a dozen or so water molecules in the vicinity of the ions.
32. The partial pressure of O_2 in your lungs varies from 25 mm Hg to 40 mm Hg. What concentration of O_2 (in grams per liter) can dissolve in water at 37 °C when the O_2 partial pressure is 40. mm Hg? The Henry's law constant for O_2 at 37 °C is 1.5×10^{-6} mol L^{-1} mm Hg^{-1}.
33. The Henry's law constant for nitrogen in blood serum is approximately 8×10^{-7} mol L^{-1} mm Hg^{-1}. What is the N_2 concentration in a diver's blood at a depth where the total pressure is 2.5 atm? The air the diver is breathing is 78% N_2 by volume.
34. The Henry's law constant for N_2 in water at 25 °C is 8.4×10^{-7} mol L^{-1} mm Hg^{-1}. What is the solubility of N_2 in mol/L if its partial pressure is 1520 mm Hg? What is the solubility when the N_2 partial pressure is 20. mm Hg?

Concentration Units

35. Convert 2.5 ppm to weight percent.
36. ■ Convert 73.2 ppm to weight percent.
37. Show mathematically how 1 ppm is equivalent to 1 mg/1 kg.
38. Show mathematically how 1 ppb is equivalent to 1 μg/1 kg.
39. What mass (in grams) of sucrose is in 1.0 kg of a 0.25% sucrose solution?
40. How many grams of ethanol are in 750. mL of a 12% ethanol solution? (Assume its density is the same as that of water.)
41. A sample of water is found to contain 0.010 ppm lead ions (Pb^{2+}).
    (a) What is the mass of lead ions per liter of this solution? (Assume the density of the water solution is 1.0 g/mL.)
    (b) What is the lead concentration in ppb?
42. A sample of lead-based paint is found to contain 60.5 ppm lead. The density of the paint is 8.0 lb/gal. What mass of lead (in grams) would be present in 50. gal of this paint?
43. A paint contains 200. ppm lead. Approximately what mass of lead (in grams) will be in 1.0 cm^2 of this paint (density = 8.0 lb/gal) when 1 gal is uniformly applied to 500. ft^2 of a wall?
44. Calculate the mass in grams of solute needed to prepare each of these solutions.
    (a) 250. mL of 0.50 M NaCl
    (b) 0.50 L of 0.15 M sucrose, $C_{12}H_{22}O_{11}$
    (c) 200. mL of 0.20 M $NaHCO_3$
45. Calculate the mass in grams of solute required to prepare each of these solutions.
    (a) 750. mL of 4.00 M NH_4Cl
    (b) 1.50 L of 0.750 M KCl
    (c) 150. mL of 0.350 M Na_2SO_4
46. Calculate the molarity of the solute in a solution containing
    (a) 14.2 g KCl in 250. mL solution
    (b) 5.08 g K_2CrO_4 in 150. mL solution
    (c) 0.799 g $KMnO_4$ in 400. mL solution
    (d) 15.0 g $C_6H_{12}O_6$ in 500. mL solution
47. ■ Calculate the molarity of the solute in a solution containing
    (a) 6.18 g $MgNH_4PO_4$ in 250. mL solution
    (b) 16.8 g $NaCH_3COO$ in 300. mL solution
    (c) 2.50 g CaC_2O_4 in 750. mL solution
    (d) 2.20 g $(NH_4)_2SO_4$ in 400. mL solution
48. Concentrated nitric acid is a 70.0% solution of nitric acid, HNO_3, in water. The density of the solution is 1.41 g/mL at 25 °C. What is the molarity of nitric acid in this solution?
49. Concentrated sulfuric acid has a density of 1.84 g/cm^3 and is 18 M. What is the weight percent of H_2SO_4 in the solution?
50. Consider a 13.0% solution of sulfuric acid, H_2SO_4, whose density is 1.090 g/mL.
    (a) Calculate the molarity of this solution.
    (b) To what volume should 100. mL of this solution be diluted to prepare a 1.10 M solution?
51. A 0.6 mL teardrop contains 4 mg NaCl. Calculate the molarity of NaCl in the teardrop.
52. You need a 0.050 mol/kg aqueous solution of methanol (CH_3OH). What mass of methanol would you need to dissolve in 500. g water to make this solution?
53. ■ You want to prepare a 1.0 mol/kg solution of ethylene glycol, $C_2H_4(OH)_2$, in water. What mass of ethylene glycol do you need to mix with 950. g water?
54. A 23.2% by weight aqueous solution of sucrose has a density of 1.127 g/mL. Calculate the molarity of sucrose in this solution.
55. An aqueous beverage has a lead concentration of 25 ppb. Express this lead concentration as molarity.
56. Calculate the number of grams of KI required to prepare 100. mL of 0.0200 M KI. How many milliliters of this solution are required to produce 250. mL of 0.00100 M KI?
57. Calculate the molality of a solution made by dissolving 6.58 g NaCl in 250. mL water.

Colligative Properties

58. What is the boiling point of a solution containing 0.200 mol of a nonvolatile nonelectrolyte solute in 100. g benzene? The normal boiling point of benzene is 80.10 °C, and K_b = 2.53 °C kg mol^{-1}.
59. ■ What is the boiling point of a solution composed of 15.0 g urea, $(NH_2)_2CO$, in 0.500 kg water?
60. Place these aqueous solutions in order of increasing boiling point.
    (a) 0.10 mol KCl/kg
    (b) 0.10 mol glucose/kg
    (c) 0.080 mol $MgCl_2$/kg
61. ■ List these aqueous solutions in order of decreasing freezing point.
    (a) 0.10 mol methanol/kg (b) 0.10 mol KCl/kg
    (c) 0.080 mol $BaCl_2$/kg (d) 0.040 mol Na_2SO_4/kg
    (Assume that all of the salts dissociate completely into their ions in solution.)
62. Calculate the boiling point at 760 mm Hg and the freezing point of these solutions.
    (a) 20.0 g citric acid, $C_6H_8O_7$, in 100.0 g water
    (b) 3.00 g CH_3I in 20.0 g benzene (K_b benzene = 2.53 °C kg mol^{-1}; K_f benzene = 5.10 °C kg mol^{-1})
63. Calculate the freezing and boiling points (at 760 mm Hg) of a solution of 4.00 g urea, $CO(NH_2)_2$, dissolved in 75.0 g water.
64. Calculate the mass in grams of urea that must be added to 150. g water to give a solution whose vapor pressure is 2.5 mm Hg less than that of pure water at 40 °C (vp H_2O at 40 °C = 55.34 mm Hg).
65. ■ At 60 °C the vapor pressure of pure water is 149.44 mm Hg and that above an aqueous sucrose ($C_{12}H_{22}O_{12}$) solution is 119.55 mm Hg. Calculate the mole fraction of water and the mass in grams of sucrose in the solution if the mass of water is 150. g.
66. At 760 mm Hg, a solution of 5.52 g glycerol in 40.0 g water has a boiling point of 100.777 °C. Calculate the molar mass of glycerol.
67. The boiling point of benzene is increased by 0.65 °C when 5.0 g of an unknown organic compound (a nonelectrolyte) is dissolved in 100. g benzene. Calculate the approximate molar mass of the organic compound. (K_b benzene = 2.53 °C kg mol^{-1}.)
68. The freezing point of *p*-dichlorobenzene is 53.1 °C, and its K_f is 7.10 °C kg mol^{-1}. A solution of 1.52 g of the drug sulfanilamide in 10.0 g *p*-dichlorobenzene freezes at 46.7 °C. What is the molar mass of sulfanilamide?
69. ■ You add 0.255 g of an orange crystalline compound with an empirical formula of $C_{10}H_8Fe$ to 11.12 g benzene. The

boiling point of the solution is 80.26 °C. The normal boiling point of benzene is 80.10 °C and its K_b = 2.53 °C kg mol^{-1}. What are the molar mass and molecular formula of the compound?

70. If you use only water and pure ethylene glycol, $HOCH_2CH_2OH$, in your car's cooling system, what mass (in grams) of the glycol must you add to each quart of water to give freezing protection down to −31.0 °C?
71. Anthracene, a hydrocarbon obtained from coal, has an empirical formula of C_7H_5. To find its molecular formula you dissolve 0.500 g anthracene in 30.0 g benzene. The boiling point of the solution is 80.34 °C. The normal boiling point of benzene is 80.10 °C, and K_b = 2.53 °C kg mol^{-1}. What are the molar mass and molecular formula of anthracene?
72. A 1.00 mol/kg aqueous sulfuric acid solution, H_2SO_4, freezes at −4.04 °C. Calculate i, the van't Hoff factor, for sulfuric acid in this solution.
73. ■ Some ethylene glycol, $HOCH_2CH_2OH$, was added to your car's cooling system along with 5.0 kg water.
   (a) If the freezing point of the solution is −15.0 °C, what mass (in grams) of the glycol must have been added?
   (b) What is the boiling point of the coolant mixture?
74. Calculate the concentration of nonelectrolyte solute particles in human blood if the osmotic pressure is 7.53 atm at 37 °C, the temperature of the body.
75. The blood of cold-blooded animals and fish is isotonic with seawater. If seawater freezes at −2.3 °C, what is the osmotic pressure of the blood of these animals at 20.0 °C? (Assume the density is that of pure water.)
76. The molar mass of a polymer was determined by measuring the osmotic pressure, 7.6 mm Hg, of a solution containing 5.0 g of the polymer dissolved in 1.0 L benzene. What is the molar mass of the polymer? Assume a temperature of 298.15 K.
77. ■ The osmotic pressure at 25 °C is 1.79 atm for a solution prepared by dissolving 2.50 g sucrose, empirical formula $C_{12}H_{22}O_{11}$, in enough water to give a solution volume of 100 mL. Use the osmotic pressure equation to show that the empirical formula for sucrose is the same as its molecular formula.

Water: Purification and Solutions

78. Dietitians recommend drinking six 8-oz glasses of water each day. If your drinking water contains the maximum contamination level for arsenic, 0.050 ppm, how much arsenic would you consume in a week following this recommendation?
79. When lead is present in drinking water, in what form does it exist?
80. The maximum contamination level (MCL) for chlordane is 0.002 ppm. A sample of well water contained 5 ppb chlordane. Is the sample within the MCL for chlordane?
81. ■ In a home, 200. gal of hard water containing 500. mg Ca^{2+}/gal passed through the Na^+-based ion-exchange water softener. If the ion-exchange resin operates at 100% efficiency, what mass of Na^+ ions is displaced from the resin?
82. How do the lime-soda and ion-exchange processes differ in treating hard water?
83. Explain how hard water produces "ring around the bathtub."
84. During municipal drinking water treatment, water is sprayed into the air. Why is this done?
85. Discuss the risks and benefits of using ozone to treat municipal drinking water.
86. Why is it necessary to bubble air through an aquarium?

General Questions

87. What is the difference between solubility and miscibility?
88. If 5 g solvent, 0.2 g solute A, and 0.3 g solute B are mixed to form a solution, what is the weight percent concentration of A?
89. A chemistry classmate tells you that a supersaturated solution is also saturated. Is the student correct? What would you tell the student about her/his statement?
90. In *The Rime of the Ancient Mariner* the poet Samuel Taylor Coleridge wrote, ". . . Water, water, everywhere/And all the boards did shrink. . . ." Explain this effect in terms of osmosis.
91. A 10.0 M aqueous solution of NaOH has a density of 1.33 g/cm^3 at 20 °C. Calculate the weight percent of NaOH in the solution.
92. Concentrated aqueous ammonia is 14.8 M and has a density of 0.90 g/cm^3. Calculate the weight percent of NH_3 in the solution.
93. (a) What is the molality of a solution made by dissolving 115.0 g ethylene glycol, $HOCH_2CH_2OH$, in 500. mL water? The density of water at this temperature is 0.978 g/mL.
   (b) What is the molarity of the solution?
94. Dimethylglyoxime (DMG) reacts with nickel(II) ion in aqueous solution to form a bright red compound. However, DMG is insoluble in water. To get it into aqueous solution where it can encounter Ni^{2+} ions, it must first be dissolved in a suitable solvent, such as ethanol. Suppose you dissolve 45.0 g DMG ($C_4H_8N_2O_2$) in 500. mL ethanol (C_2H_5OH; density = 0.7893 g/mL). What are the molality and weight percent of DMG in this solution?
95. ■ Arrange these aqueous solutions in order of increasing boiling point.
   (a) 0.20 mol ethylene glycol/kg
   (b) 0.12 mol K_2SO_4/kg
   (c) 0.10 mol $BaCl_2$/kg
   (d) 0.12 mol KBr/kg
96. Arrange these aqueous solutions in order of decreasing freezing point.
   (a) 0.20 mol ethylene glycol/kg
   (b) 0.12 mol Na_2SO_4/kg
   (c) 0.10 mol NaBr/kg
   (d) 0.12 mol KI/kg
97. The solubility of NaCl in water at 100 °C is 39.1 g/100. g of water. Calculate the boiling point of a saturated solution of NaCl.
98. The organic salt $[(C_4H_9)_4N][ClO_4]$ consists of the ions $(C_4H_9)_4N^+$ and ClO_4^-. The salt dissolves in chloroform. What mass (in grams) of the salt must have been dissolved if the boiling point of a solution of the salt in 25.0 g chloroform is 63.20 °C? The normal boiling point of chloroform is

61.70 °C and K_b = 3.63 °C kg mol^{-1}. Assume that the salt dissociates completely into its ions in solution.

99. A solution, prepared by dissolving 9.41 g $NaHSO_3$ in 1.00 kg water, freezes at −0.33 °C. From these data, decide which of these equations is the correct expression for the dissociation of the salt.
(a) $NaHSO_3(aq) \longrightarrow Na^+(aq) + HSO_3^-(aq)$
(b) $NaHSO_3(aq) \longrightarrow Na^+(aq) + H^+(aq) + SO_3^{2-}(aq)$

100. A 0.250 M sodium sulfate solution is added to a 0.200 M barium nitrate solution and 0.700 g barium sulfate precipitates.
(a) Write the balanced equation for this reaction.
(b) Calculate the minimum volume of barium nitrate solution that was used.
(c) Calculate the minimum volume of sodium sulfate needed to precipitate 0.700 g barium sulfate. Assume 100% yield.

101. Aluminum chloride reacts with phosphoric acid to give aluminum phosphate, $AlPO_4$, which is used industrially as the basis of adhesives, binders, and cements.
(a) Write a balanced equation for the reaction of aluminum chloride and phosphoric acid.
(b) If you begin with 152. g aluminum chloride and 3.00 L of 0.750 M phosphoric acid, what mass of $AlPO_4$ can be isolated?
(c) Use the solubility table (Table 5.1) to predict the solubility of $AlPO_4$.

Applying Concepts

102. Using these symbols,

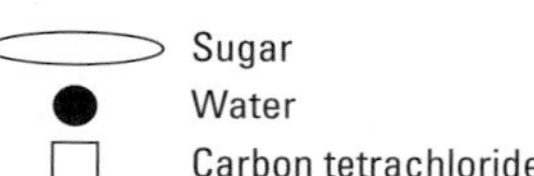

draw nanoscale diagrams for the contents of a beaker containing
(a) Water and sugar
(b) Carbon tetrachloride and sugar

103. Using these symbols,

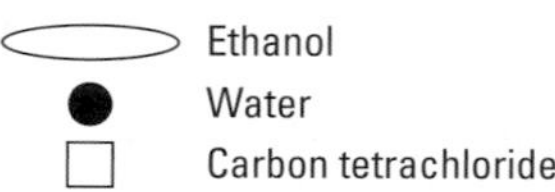

draw nanoscale diagrams for the contents of a beaker containing
(a) Water and ethanol
(b) Water and carbon tetrachloride

104. ■ Refer to Figure 15.11 to determine whether these situations would result in an unsaturated, saturated, or supersaturated solution.
(a) 40. g NH_4Cl is added to 100. g H_2O at 80 °C.
(b) 100. g LiCl is dissolved in 100. g H_2O at 30 °C.
(c) 120. g $NaNO_3$ is added to 100. g H_2O at 40 °C.
(d) 50. g Li_2SO_4 is dissolved in 200. g H_2O at 50 °C.

105. Refer to Figure 15.11 to determine whether these situations would result in an unsaturated, saturated, or supersaturated solution.
(a) 120. g RbCl is added to 100. g H_2O at 50 °C.
(b) 30. g KCl is dissolved in 100. g H_2O at 70 °C.
(c) 20. g NaCl is dissolved in 50. g H_2O at 60 °C.
(d) 150. g CsCl is added to 100. g H_2O at 10 °C.

106. Complete this table.

Compound	Mass of Compound	Mass of Water	Mass Fraction of Solute	Weight Percent of Solute	ppm of Solute
Lye	______	125 g	0.375	______	______
Glycerol	33 g	200. g	______	______	______
Acetylene	0.0015 g	______	______	0.0009%	______

107. Complete this table.

Compound	Mass of Compound	Mass of Water	Mass Fraction	Weight Percent	ppm
Table salt	52 g	175 g	______	______	______
Glucose	15 g	______	______	______	7×10^4
Methane	______	100. g	______	0.0025%	______

108. What happens on the molecular level when a liquid freezes? What effect does a nonvolatile solute have on this process? Comment on the purity of water obtained by melting an iceberg.

109. If KI is added to a saturated solution of SrI_2, will the amount of solid SrI_2 present decrease, increase, or remain unchanged? What about the concentration of Sr^{2+} ion in solution?

110. In your own words, explain why
(a) Seawater has a lower freezing point than fresh water.
(b) Salt is added to the ice in an ice cream maker to freeze the ice cream faster.

111. Criticize these statements.
(a) A saturated solution is always a concentrated one.
(b) A 0.10 mol/kg sucrose solution and a 0.10 mol KCl/kg solution have the same osmotic pressure.

More Challenging Questions

112. In chemical research, newly synthesized compounds are often sent to commercial laboratories for analysis that determines the weight percent of C and H by burning the compound and collecting the evolved CO_2 and H_2O. The molar mass is determined by measuring the osmotic pressure of a solution of the compound. Calculate the empirical and molecular formulas of a compound, C_xH_yCr, given this information:
(a) The compound contains 73.94% C and 8.27% H; the remainder is chromium.
(b) At 25 °C, the osmotic pressure of 5.00 mg of the unknown dissolved in 100. mL of chloroform solution is 3.17 mm Hg.

113. An osmotic pressure of 5.15 atm is developed by a solution containing 4.80 g dioxane (a nonelectrolyte) dissolved in 250. mL water at 15.0 °C. The empirical formula of dioxane is C_2H_4O. Use the osmotic pressure data to show that the empirical formula and the molecular formula of dioxane are not the same.

114. ■ The "proof" of an alcohol-containing beverage is twice the volume percentage of ethanol, C_2H_5OH (density = 0.789 g/mL), in the beverage. A bottle of 100-proof vodka is left outside where the temperature is −15 °C.
(a) Will the vodka freeze?
(b) What is the boiling point of the vodka?

115. A martini is a 5-oz (142-g) cocktail containing 30% by mass of alcohol. When the martini is consumed, about 15% of it passes directly into the bloodstream (blood volume = 7.0 L in an adult). Consider an adult who drinks two martinis with lunch. Estimate the blood alcohol concentration in this person after the two martinis have been consumed. An adult with a blood alcohol concentration of 3.0×10^{-3} g/mL or more is considered intoxicated. Is the person intoxicated?

116. Osmosis is responsible for sap rising in trees. Calculate the approximate height to which it could rise if the sap were 0.13 M in sugar and the dissolved solids in the water outside the tree were at a concentration of 0.020 M. *Hint:* A column of liquid exerts a pressure directly proportional to its density.

117. A 0.100 mol NaCl/kg solution has a van't Hoff factor of 1.87. Calculate the freezing point, the boiling point, and the osmotic pressure of this solution, whose density is 1.01 g/mL.

118. The osmotic pressure of human blood at 37 °C is 7.7 atm. Calculate what the molarity of a glucose solution should be if it is to be safely administered intravenously.

119. A 0.109 mol/kg aqueous formic acid solution, HCOOH, freezes at −0.210 °C. Calculate the percent dissociation of formic acid.

120. Consider these data for a solution of water and ethanol, C_2H_5OH.

Mass Percent Ethanol	Boiling Point (°C)
25.0	84.5
50.0	81.2
67.0	80.0
85.0	78.4

(a) Plot these data and use the graph to determine the boiling point of a solution that is 59.0% ethanol by mass.
(b) Calculate the molarity and the molality of the 67.0% solution.

121. A 0.050 mol/kg aqueous solution of iodic acid, HIO_3, freezes at −3.12 °C; a 0.20 mol/kg solution of the acid freezes at −2.71 °C; a 1.0 mol/kg solution freezes at −1.72 °C.
(a) Plot these data and use the graph to determine the boiling point of a solution that is 0.5 mol iodic acid/kg.
(b) Calculate the percent dissociation of iodic acid in each solution.

122. Consider these data for aqueous solutions of ammonium chloride, NH_4Cl.

Molality (mol/kg)	Freezing Point (°C)
0.0050	−3.617
0.020	−3.544
0.20	−3.392
1.0	−3.33

(a) Plot these data and from the graph determine the freezing point of a 0.50 mol ammonium chloride/kg solution.
(b) Calculate the percent dissociation of ammonium chloride in each solution.

123. Maple syrup sap is 3% sugar (sucrose) and 97% water by mass. Maple syrup is produced by heating the sap to evaporate a certain amount of the water.
(a) Describe what happens to the composition and boiling point of the solution as evaporation takes place.
(b) A rule of thumb among maple syrup producers is that the finished syrup should boil about 4 °C higher than the original sap being boiled. Explain the chemistry behind this guideline.
(c) If the finished product boils 4 °C higher than the original sap, calculate the concentration of sugar in the final product. Assume that sugar is the only solute and the operation is done at 1 atm pressure.

124. A 0.63% by weight aqueous tin(II) fluoride, SnF_2, solution is used as an oral rinse in dentistry to decrease tooth decay.
(a) Calculate the SnF_2 concentration in ppm and ppb.
(b) What is the molarity of SnF_2 in the solution?
(c) Large quantities of tin are produced commercially by the reduction of cassiterite, SnO_2, the principal tin ore, with carbon.

$$SnO_2(s) + 2\,C(s) \longrightarrow Sn(s) + 2\,CO(g)$$

Once purified, the tin is reacted with hydrogen fluoride vapor to produce SnF_2. Consider that one metric ton of cassiterite ore is reduced with sufficient carbon to tin metal in 80% yield. The tin is purified and reacted with sufficient hydrogen fluoride to produce SnF_2 in 94% yield. Calculate how many 250.-mL bottles of 0.63% SnF_2 solution could be prepared from the one metric ton of cassiterite by these steps.

Conceptual Challenge Problems

CP15.A (Section 15.6) Concentrations expressed in units of parts per million and parts per billion often have no meaning for people until they relate these small and large numbers to their own experiences.
(a) What time in seconds is 1 ppm of a year?
(b) What time in seconds is 1 ppb of a 70-year lifetime?

CP15.B (Section 15.4) Bodies of water with an abundance of nutrients that support a blooming growth of plants are said to be eutrophoric. In general, fish do not thrive for long in eutrophoric waters because little oxygen is available for them. Suppose someone asked you why this was true, given the fact that growing plants produce oxygen as a product of photosynthesis. How would you respond to this person's inquiry?

CP15.C (Section 15.7) Suppose that you want to produce the lowest temperature possible by using ice, sodium chloride, and water to chill homemade ice cream made in a 1.5-L metal cylinder surrounded by a coolant held in a wooden bucket. You have all the ice, salt, and water you want. How would you plan to do this?

16 Acids and Bases

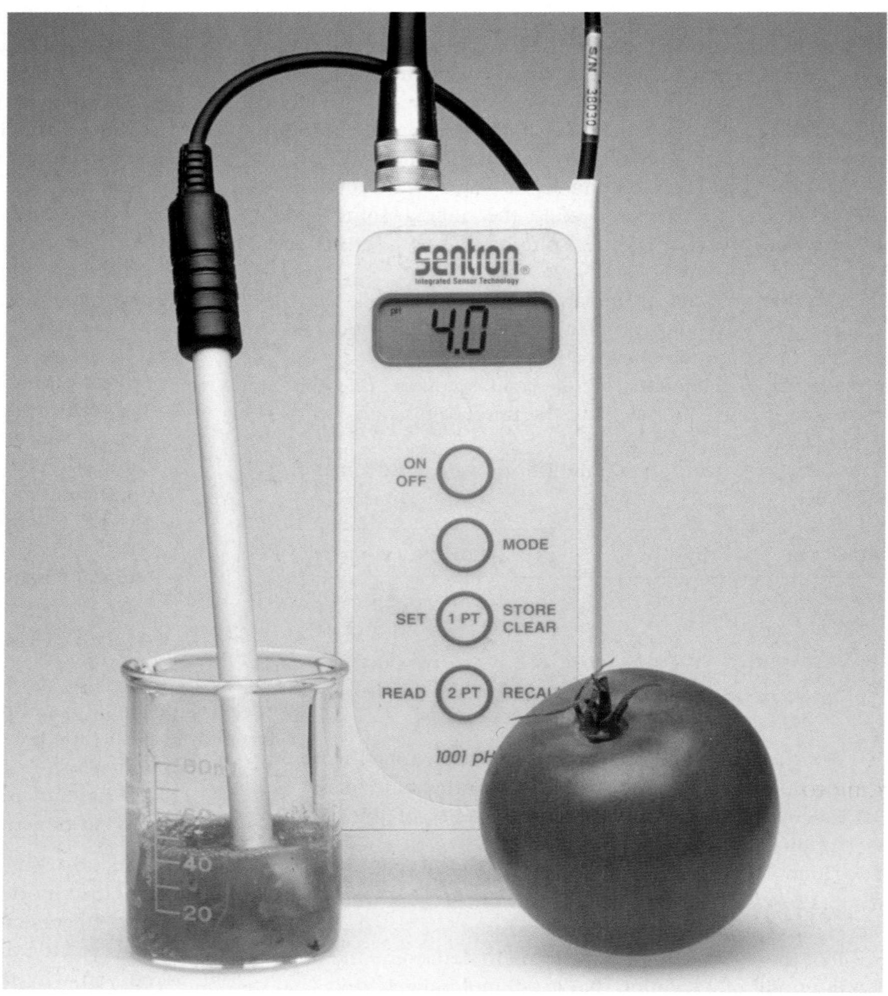

Tomato juice has a pH of 4.0 as indicated by the pH meter. Measurement of the acidity or alkalinity of substances, expressed as pH, is important in commerce, medicine, and studies of the environment. The measurement and uses of pH are discussed in this chapter.

It is difficult to overstate the importance of acids and bases. Aqueous solutions, which abound in our environment and in all living organisms, are almost always acidic or basic to some degree. Photosynthesis and respiration, the two most important biological processes on earth, depend on acid-base reactions. Carbon dioxide (CO_2) is the most important acid-producing compound in nature. Rainwater is generally slightly acidic because of dissolved CO_2, and acid rain results from further acidification of rainwater by acids formed by the gaseous pollutants SO_2 and NO_2. The oceans are slightly basic, as are many ground and surface waters. Natural waters can also be acidic; the more acidic the water, the more easily metals such as lead can be dissolved from water pipes or soldered joints.

ThomsonNOW™
Throughout the text, this icon indicates an opportunity to test yourself on key concepts and to explore interactive modules by signing in to ThomsonNOW at **www.thomsonedu.com**.

Because of their importance, the properties of acids and bases have been studied extensively. In 1677 Antoine Lavoisier proposed that oxygen made an acid acidic; he even derived the name *oxygen* from Greek words meaning "acid former." But in 1808 it was discovered that the gaseous compound HCl, which dissolves in water to give hydrochloric acid, contains only hydrogen and chlorine. It later became clear that hydrogen, not oxygen, is common to all acids in aqueous solution. It was also shown that aqueous solutions of both acids and bases conduct an electrical current because acids and bases are electrolytes—they release ions into solution. In 1887 the Swedish chemist Svante Arrhenius proposed that acids *ionize in aqueous solution to produce hydrogen ions (protons) and anions;* bases *ionize to produce hydroxide ions and cations.* We now rely on a more general acid-base concept: ***hydronium ions,*** H_3O^+***, are responsible for the properties of acidic aqueous solutions, and hydroxide ions,*** OH^-***, are responsible for the properties of basic aqueous solutions.***

Because of their small size and high charge density, hydrogen ions (protons) are always associated with water molecules in aqueous solution and are usually represented as H_3O^+, the hydronium ion.

$$H^+ + H_2O \longrightarrow H_3O^+$$

16.1 The Brønsted-Lowry Concept of Acids and Bases

A major problem with the Arrhenius acid-base concept is that certain substances, such as ammonia (NH_3), produce basic solutions and react with acids, yet contain no hydroxide ions. In 1923 J. N. Brønsted in Denmark and T. M. Lowry in England independently proposed a new way of defining acids and bases in aqueous solutions:

- **Brønsted-Lowry acids** are hydrogen ion donors.
- **Brønsted-Lowry bases** are hydrogen ion acceptors.

According to the Brønsted-Lowry concept, an acid can donate a hydrogen ion, H^+ (a proton), to another substance, while a base can accept an H^+ ion from another substance. In **acid-base reactions,** acids *donate* H^+ ions, and bases *accept* them. To accept an H^+ and serve as a Brønsted-Lowry base, a molecule or ion must have an *unshared pair of electrons.* For example, in aqueous solutions, H^+ ions from acids such as nitric acid, HNO_3, react with water molecules, which accept protons to form hydronium ions, H_3O^+.

ThomsonNOW™
Go to the Coached Problems menu for tutorials on:
- **Brønsted-Lowry acids and bases**
- **strong acids and strong bases**

Acid: H^+ donor; Base: H^+ acceptor

$$HNO_3(aq) + H_2O(\ell) \longrightarrow H_3O^+(aq) + NO_3^-(aq)$$

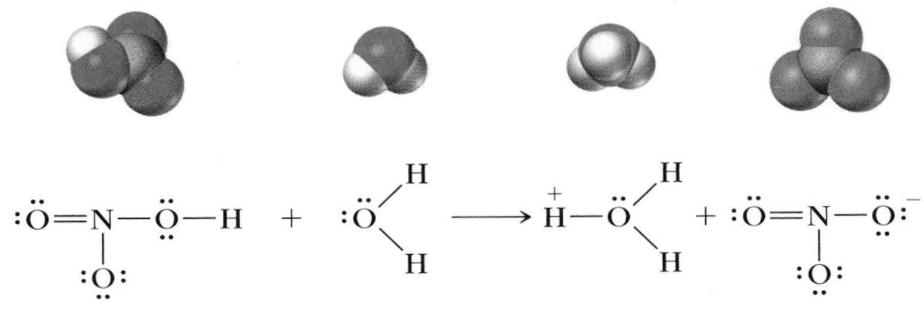

:Ö=N—Ö—H (with :Ö: on N) + :Ö(H)(H) ⟶ H—Ö⁺(H)(H) + :Ö=N—Ö:⁻ (with :Ö: on N)

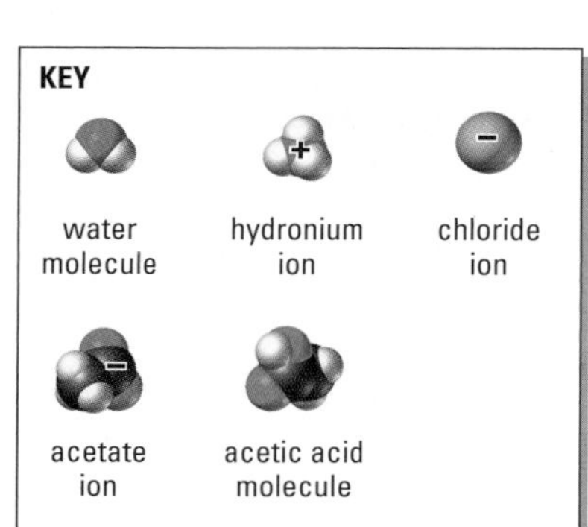

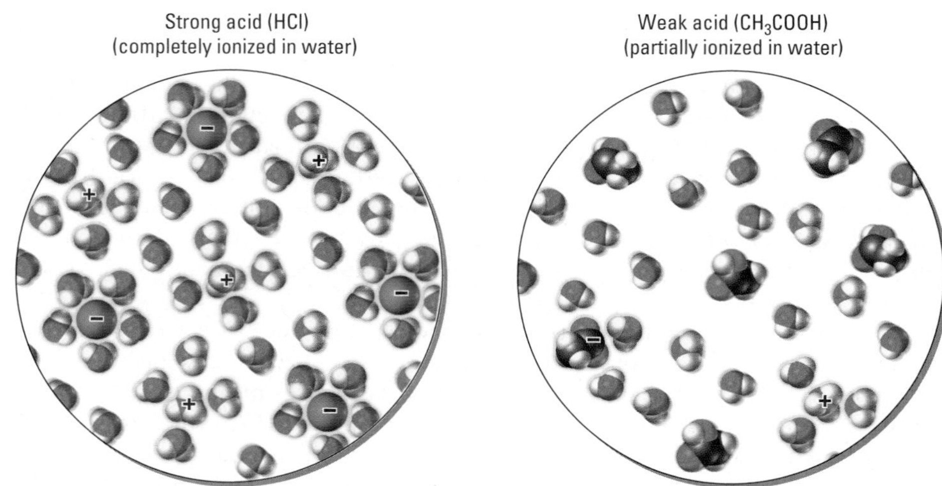

Ionization of acids in water. A strong acid such as hydrochloric acid (HCl) is completely ionized in water; a weak acid such as acetic acid (CH_3COOH) is only partially ionized in water.

An ionic substance that dissolves in water is a strong electrolyte (⟸ ***p. 100).*** A strong acid ionizes completely and is a strong electrolyte.

In reacting with an acid, water acts as a Brønsted-Lowry base by using an unshared electron pair to accept the H^+. Because nitric acid is a strong acid, it is 100% ionized in aqueous solution, and the equation on page 771 is strongly product-favored.

In contrast, a weak acid does *not* ionize completely and therefore is a weak electrolyte and at equilibrium appreciable quantities of un-ionized acid molecules are present. For example, hydrofluoric acid, HF, is a weak acid, and its ionization in water is written as

$$\underset{\text{hydrofluoric acid, a weak acid}}{HF(aq)} + H_2O(\ell) \rightleftharpoons H_3O^+(aq) + F^-(aq)$$

The double arrow indicates an equilibrium between the reactants and the products (⟸ ***p. 672).*** Because ionization of HF is much less than 100%, this means that at equilibrium most HF molecules are un-ionized and there are relatively few hydronium and fluoride ions in solution. Thus, the ionization of *weak* acids is a reactant-favored process.

Ammonia, NH_3, is a compound consisting of electrically neutral molecules; the ammonium ion, NH_4^+, is a polyatomic ion.

Ammonia, NH_3, is a base because it accepts an H^+ from water (an acid) to form an ammonium ion, NH_4^+. Water, having donated an H^+, is converted into a hydroxide ion, OH^-.

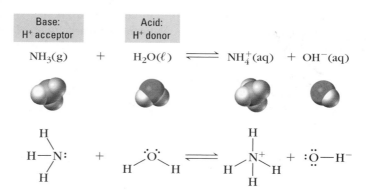

Base: H^+ acceptor
Acid: H^+ donor

$$NH_3(g) + H_2O(\ell) \rightleftharpoons NH_4^+(aq) + OH^-(aq)$$

$$H{-}\ddot{N}H_2 + H{-}\ddot{O}{-}H \rightleftharpoons [NH_4]^+ + :\ddot{O}{-}H^-$$

Ammonia establishes an equilibrium with water, ammonium ions, and hydroxide ions and is therefore a weak base (⟸ ***p. 173).*** At equilibrium the solution contains far more ammonia molecules than ammonium and hydroxide ions. Therefore, ammonium hydroxide is not an appropriate name for an aqueous solution of ammonia.

CONCEPTUAL EXERCISE 16.1 Brønsted-Lowry Acids and Bases

Identify each molecule or ion as a Brønsted-Lowry acid or base.

(a) HBr (b) Br^- (c) HNO_2 (d) CH_3NH_2

CONCEPTUAL EXERCISE 16.2 Using Le Chatelier's Principle

Use Le Chatelier's principle (← *p. 695*) to explain why a larger percentage of NH_3 will react with water in a very dilute solution than in a less dilute solution.

Water's Role as Acid or Base

In aqueous solution, all Brønsted-Lowry acids and bases react with water molecules. As we have seen, a water molecule *accepts* an H^+ from an acid such as nitric acid, while a water molecule *donates* an H^+ to a base such as an ammonia molecule. According to the Brønsted-Lowry definitions, water serves as a base (an H^+ acceptor) when an acid is present and as an acid (an H^+ donor) when a base is present. Therefore, water displays both acid and base properties—*it can donate or accept H^+ ions,* depending on the circumstances. The general reactions of water with acids (HA) and molecular bases (B) are

Water acting as a base

$$\underset{\text{acid}}{HA} + \underset{\text{base}}{H_2O} \rightleftharpoons H_3O^+ + A^-$$

Water acting as an acid

$$\underset{\text{base}}{B} + \underset{\text{acid}}{H_2O} \rightleftharpoons BH^+ + OH^-$$

If the base is an anion, B^-, it accepts H^+ from water to form BH.

A substance, like water, that can donate or accept H^+ is said to be **amphiprotic.**

EXERCISE 16.3 Acids and Bases

Complete these equations. (*Hint:* CH_3NH_2 and $(CH_3)_2NH$ are amines, so they are bases.)

(a) $HCN + H_2O \longrightarrow$ (b) $HBr + H_2O \longrightarrow$

(c) $CH_3NH_2 + H_2O \longrightarrow$ (d) $(CH_3)_2NH + H_2O \longrightarrow$

Conjugate Acid-Base Pairs

Whenever an acid donates H^+ to a base, a new acid and a new base are formed. We illustrate this using the reaction between acetic acid, CH_3COOH, and water (Active Figure p. 774). Acetic acid is an H^+ ion donor (an acid), and water is an H^+ ion acceptor (a base). The products of the reaction are a new acid, H_3O^+, and a new base, CH_3COO^-. In the reverse reaction, H_3O^+ acts as an acid (H^+ donor), and acetate ion as a base (H^+ acceptor). The structures of CH_3COOH and CH_3COO^- differ from one another by only a single H^+, just as the structures of H_2O and H_3O^+ do.

A pair of molecules or ions *related to each other by the loss or gain of a single H^+* is called a **conjugate acid-base pair.** Every Brønsted-Lowry acid has its conjugate base, and every Brønsted-Lowry base has its conjugate acid.

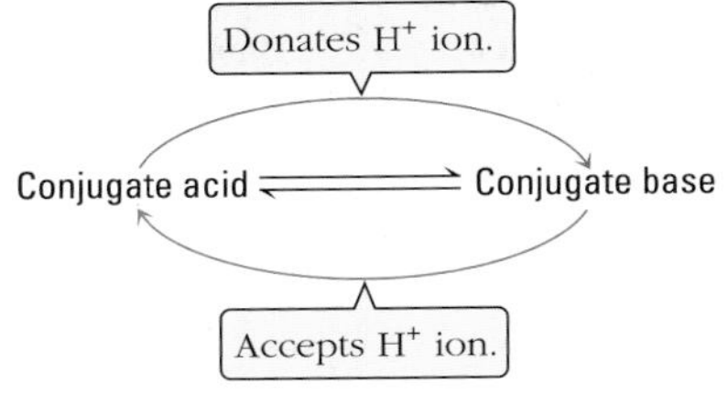

ThomsonNOW™

Go to the Chemistry Interactive menu for modules on:

- **Brønsted-Lowry acid-base reactions**
- **acid-base conjugate pairs**

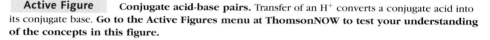

In the Active Figure, acetic acid and acetate ion are a conjugate acid-base pair.

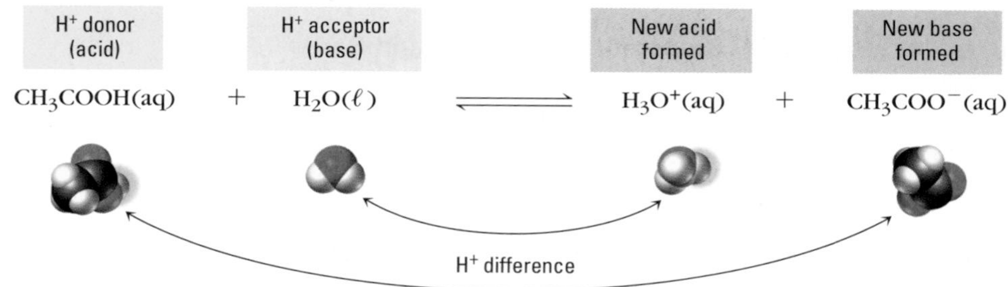

Active Figure **Conjugate acid-base pairs.** Transfer of an H^+ converts a conjugate acid into its conjugate base. **Go to the Active Figures menu at ThomsonNOW to test your understanding of the concepts in this figure.**

Removing an H^+ ion from the acid forms the conjugate base making the charge of the remaining portion of the acid one unit more negative. For example, hydrofluoric acid, HF, has the F^- (fluoride) ion as its conjugate base; HF and F^- are a conjugate acid-base *pair*.

ACID Donates $H^+ \longrightarrow$

$$HF(aq) + H_2O(\ell) \rightleftharpoons H_3O^+(aq) + F^-(aq)$$

ACID base acid BASE

BASE $\longleftarrow$ Accepts H^+

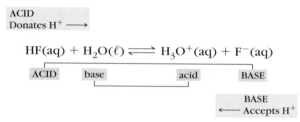

In the forward reaction, HF is a Brønsted-Lowry acid. It donates an H^+ to water, which, by accepting the H^+, acts as a Brønsted-Lowry base. In the reverse reaction, fluoride ion is the Brønsted-Lowry base, accepting an H^+ from H_3O^+, the acid. As noted in the equation, there are two conjugate acid-base pairs: (1) HF and F^- and (2) H_2O and H_3O^+. *One member of a conjugate acid-base pair is always a reactant and the other is always a product; they are never both products or both reactants.*

The conjugate acid formula can be derived from the formula of its conjugate base by adding an H^+ ion to the conjugate base and making the charge of the resulting conjugate acid one unit more positive than the conjugate base. Therefore, the conjugate acid of Cl^- ion is HCl; the conjugate acid of NH_3 is NH_4^+.

BASE Accepts $H^+ \longrightarrow$

$$NH_3(aq) + H_2O(\ell) \rightleftharpoons NH_4^+(aq) + OH^-(aq)$$

BASE acid ACID base

ACID $\longleftarrow$ Donates H^+

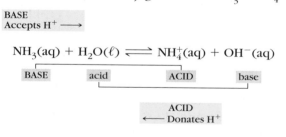

PROBLEM-SOLVING EXAMPLE 16.1 Conjugate Acid-Base Pairs

Complete this table by identifying the correct conjugate acid or conjugate base.

Acid	Its Conjugate Base	Base	Its Conjugate Acid
HCOOH	________	CN^-	________
H_2S	________	________	HSO_4^-
PH_4^+	________	________	H_2SO_3
________	ClO^-	S^{2-}	________

Answer

Acid	Its Conjugate Base	Base	Its Conjugate Acid
HCOOH	$HCOO^-$	CN^-	HCN
H_2S	HS^-	SO_4^{2-}	HSO_4^-
PH_4^+	PH_3	HSO_3^-	H_2SO_3
HClO	ClO^-	S^{2-}	HS^-

Strategy and Explanation In each of these cases, apply the relationship

ACID
Donates H^+ ⟶

Conjugate acid ⇌ Conjugate base

BASE
⟵ Accepts H^+

The conjugate acid can be identified by adding H^+ to the conjugate base; the conjugate base forms by loss of H^+ from the conjugate acid. For example, because CN^- has no H^+ to donate, it must be a base—the conjugate base of HCN, its conjugate acid. Likewise, HSO_3^- is the conjugate base of its conjugate acid, H_2SO_3. The other conjugate acid-base pairs can be worked out similarly.

PROBLEM-SOLVING PRACTICE 16.1

Complete the table.

Acid	Its Conjugate Base	Base	Its Conjugate Acid
$H_2PO_4^-$	______	______	HPO_4^{2-}
______	H^-	NH_2^-	______
HSO_3^-	______	ClO_4^-	______
HF	______	______	HBr

EXERCISE 16.4 HSO_4^- as a Base

(a) Write the equation for HSO_4^- ion acting as a base in water.
(b) Identify the conjugate acid-base pairs.
(c) Is HSO_4^- amphiprotic? Explain your answer.

Relative Strengths of Acids and Bases

Strong acids are better H^+ ion donors than weak acids. Correspondingly, strong bases are better H^+ ion acceptors than weak bases. Thus, *stronger acids have weaker conjugate bases and weaker acids have stronger conjugate bases.* For example, compare HCl, a strong acid, with HF, a weak acid.

$$HCl(aq) + H_2O(\ell) \longrightarrow H_3O^+(aq) + Cl^-(aq)$$

$$HF(aq) + H_2O(\ell) \rightleftharpoons H_3O^+(aq) + F^-(aq)$$

In a dilute solution (<1.0 M) the ionization of HCl is virtually 100%; essentially all of the HCl molecules react with water to form H_3O^+ and Cl^-. The Cl^- ion exhibits virtually no tendency to accept H^+ from H_3O^+. On the other hand, the reverse of the ionization of HF is significant; F^- ion readily accepts H^+ from H_3O^+, and hydrofluoric acid is a weak acid that is mainly un-ionized.

By measuring the extent to which various acids donate H^+ ions to water, chemists have developed an extensive tabulation of the relative strengths of acids

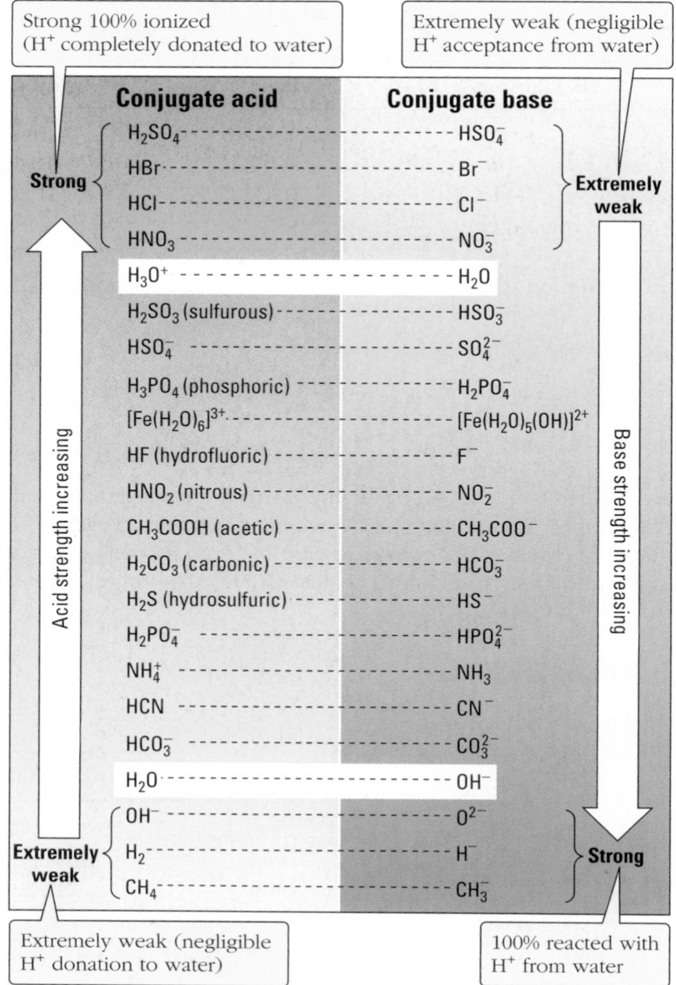

Figure 16.1 Relative strengths of conjugate acids and bases in water.

and their conjugate bases. An abbreviated table is given in Figure 16.1, where the strongest acids are at the top left and the weakest bases at the top right. The weakest acids are at the bottom left, with the strongest bases at the bottom right. From Figure 16.1 we can draw two important generalizations regarding conjugate acid-base pairs:

- ***As acid strength decreases, conjugate base strength increases; the weaker the acid, the stronger its conjugate base.***
- ***As base strength decreases, conjugate acid strength increases; the weaker the base, the stronger its conjugate acid.***

Knowing the relative acid and base strengths of the reactants, we can predict the direction of an acid-base reaction. ***The stronger acid and the stronger base will always react to form a weaker conjugate base and a weaker conjugate acid.*** Strong Brønsted-Lowry bases such as hydride ion (H^-), sulfide ion (S^{2-}), oxide ion (O^{2-}), amide ion (NH_2^-), and hydroxide ion (OH^-) readily accept H^+ ions, while weaker Brønsted-Lowry bases do so less readily. For example, the reaction of calcium hydride, CaH_2, with water is highly exothermic because of the extremely strong basic properties of the hydride ion, H^-, which avidly accepts H^+

from water to produce H_2, the extremely weak conjugate acid of hydride ion. In this reaction, hydride ion is a stronger base than OH^-, and water is a stronger acid than H_2, so the forward reaction is favored.

$$H^-(aq) + H_2O(\ell) \longrightarrow H_2(g) + OH^-(aq)$$

A great many weak bases are anions, such as CN^- (cyanide), F^- (fluoride), and CH_3COO^- (acetate).

We can apply information from Figure 16.1 to consider whether the forward or the reverse reaction is favored in an equilibrium. For example, consider the equilibrium

$$HSO_4^-(aq) + CO_3^{2-}(aq) \rightleftharpoons HCO_3^-(aq) + SO_4^{2-}(aq)$$

From Figure 16.1 we note that HSO_4^- is a stronger acid than HCO_3^-, and CO_3^{2-} is a stronger base than SO_4^{2-}. Since acid-base reactions favor going from the stronger to the weaker member of each conjugate acid-base pair, the forward reaction is favored and H^+ will be transferred from HSO_4^- to CO_3^{2-} to form bicarbonate (HCO_3^-) and sulfate (SO_4^{2-}) ions.

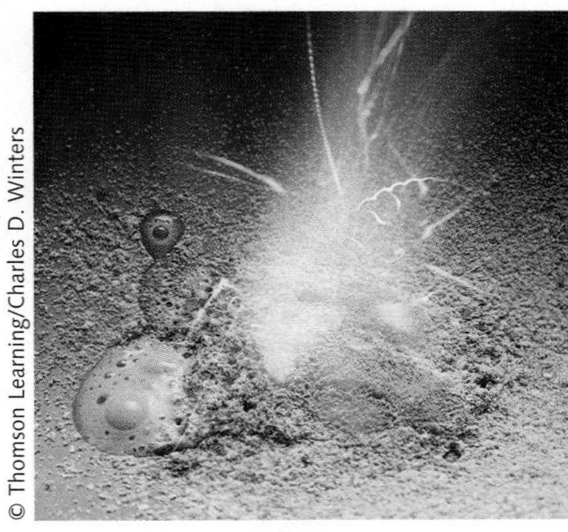

The strongly basic properties of hydride ion. The reaction of calcium hydride with water is highly exothermic due to the strongly basic hydride ion.

CONCEPTUAL EXERCISE 16.5 Conjugate Acid-Base Strength

Use Figure 16.1 to predict whether the forward or reverse reaction is favored for the equilibrium

$$CH_3COOH(aq) + SO_4^{2-}(aq) \rightleftharpoons HSO_4^-(aq) + CH_3COO^-(aq)$$

16.2 Carboxylic Acids and Amines

Many weak acids such as acetic, lactic, and pyruvic acids are organic acids.

$CH_3{-}C({=}O){-}OH$ acetic acid

$CH_3C(OH)(H){-}C({=}O){-}OH$ lactic acid

$CH_3C({=}O){-}C({=}O){-}OH$ pyruvic acid

Carboxylic acid group

These acids all contain the carboxylic acid (—COOH) functional group (⟸ *p. 569*). Although carboxylic acid molecules usually contain many other hydrogen atoms, only the hydrogen atom bound to the oxygen atom of the carboxylic acid group is sufficiently positive to be donated as an H^+ ion in aqueous solution. The two highly electronegative oxygen atoms of the carboxylic acid group pull electron density away from the hydrogen atom. As a result, the —O—H bond of the acid is even more polar, and its hydrogen atom more acidic. The C—H bonds in organic acids are relatively nonpolar and strong, and these hydrogen atoms are *not* acidic, as can be seen with butanoic acid.

$H{-}CH_2{-}CH_2{-}CH_2{-}C({=}O){-}O{-}H$

nonacidic hydrogens (the C—H hydrogens); acidic hydrogen (the O—H hydrogen)

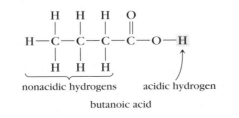

butanoic acid

Anions formed by loss of an H^+ from a —COOH group, such as acetate ion, CH_3COO^-, from acetic acid, CH_3COOH, are stabilized by resonance (⇐ *p. 357*).

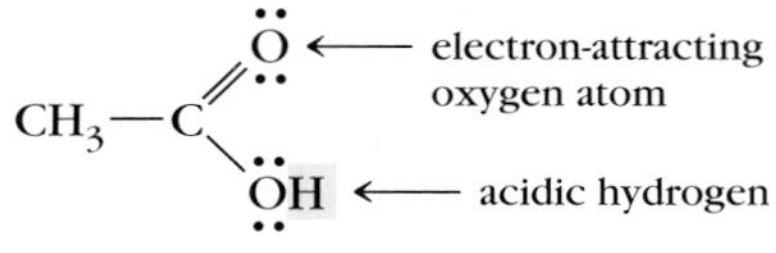

acetic acid (electron-attracting oxygen atom; acidic hydrogen)

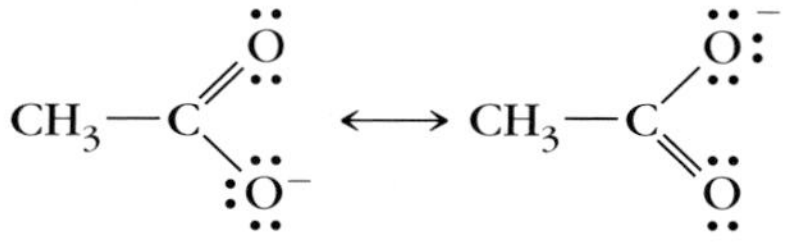

acetate ion

$CH_3—\ddot{N}H_2$

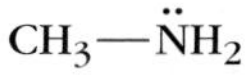

methylamine

Amines, such as methylamine, CH_3NH_2, are compounds that, like ammonia, have a nitrogen atom with three of its valence electrons in covalent bonds and an unshared electron pair on the nitrogen atom. The lone pair of electrons can accept an H^+, and so, like ammonia, amines react as weak bases with water, accepting an H^+ from water.

$$R—NH_2(aq) + H_2O(\ell) \rightleftharpoons R—NH_3^+(aq) + OH^-(aq)$$

Amines can have one, two, or three groups covalently bonded to the nitrogen atom (⇐ *p. 586*), for example, as in methylamine, dimethylamine, and trimethylamine. Many natural biochemicals and drugs are amines, such as epinephrine (adrenaline, a natural hormone) and Novocain (a local anesthetic).

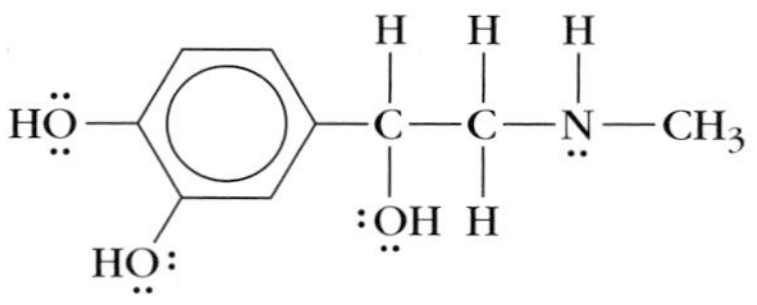

epinephrine

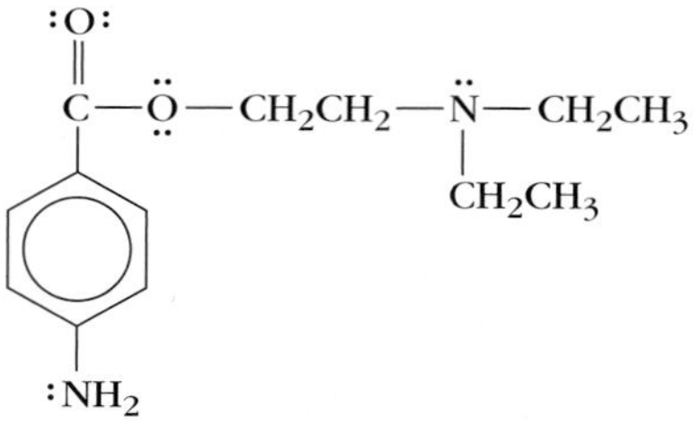

Novocain

CHEMISTRY IN THE NEWS

Acids in Hippo Sweat

Research chemists at Kyoto Pharmaceutical University (Japan) have isolated and identified the compounds that give hippopotamus sweat its characteristic rusty color. Samples were collected from the head and back of a hippo during a week-long period. The samples had to be kept cool, moist, and dilute to prevent the molecules, initially colorless, from oxidizing and polymerizing into reddish-brown pigments. After the samples were purified, they were analyzed spectroscopically. The key components contain three-ring structures. The colors of the oxidized products were due to two compounds: The red pigment is called hipposudoric acid; the orange one norhipposudoric acid. Because the compounds absorb light in the UV-visible range (200–600 nm), the researchers speculate that the pigments may act as a sunscreen, protecting the hippo's thin outer skin from UV radiation damage. Hipposudoric acid was also found to have antibacterial properties, even at low concentrations.

SOURCE: *Chemistry & Engineering News,* May 31, 2004; p. 9.

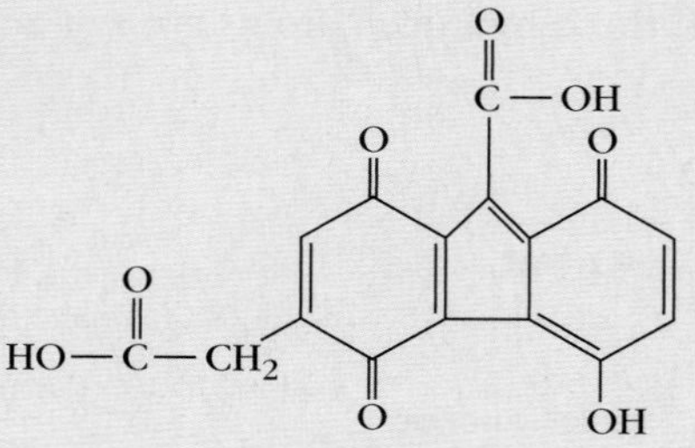

hipposudoric acid

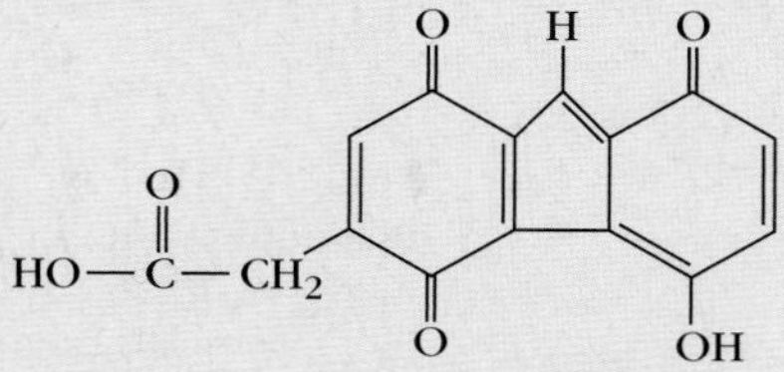

norhipposudoric acid

CONCEPTUAL EXERCISE 16.6 Piperidine, an Analog of Ammonia

Write the equation for the reaction of piperidine, a component of pepper, with (a) water and with (b) hydrochloric acid.

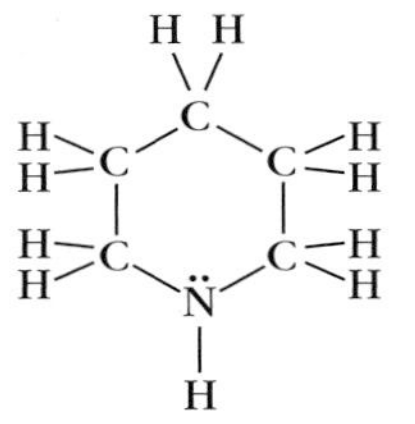

piperidine

EXERCISE 16.7 Classifying Compounds

Identify each of these molecules as a carboxylic acid, an amine, or neither.

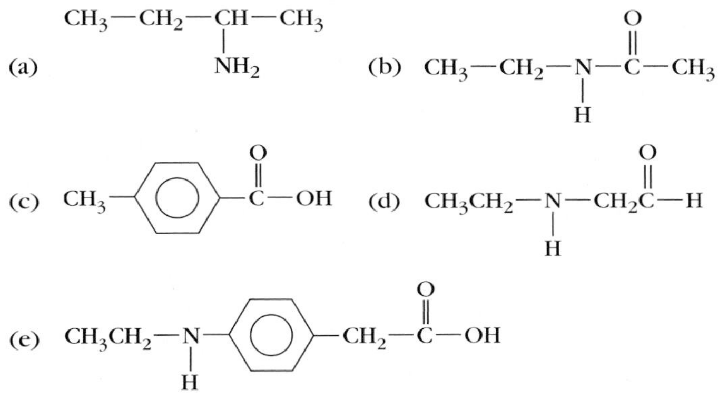

(a) $CH_3-CH_2-CH(NH_2)-CH_3$

(b) $CH_3-CH_2-N(H)-C(=O)-CH_3$

(c) $CH_3-C_6H_4-C(=O)-OH$

(d) $CH_3CH_2-N(H)-CH_2C(=O)-H$

(e) $CH_3CH_2-N(H)-C_6H_4-CH_2-C(=O)-OH$

16.3 The Autoionization of Water

Even highly purified water conducts a very small electrical current, which indicates that pure water contains a very small concentration of ions. These ions are formed when water molecules react to produce hydronium ions and hydroxide ions in a process called **autoionization.**

$$H_2O(\ell) + H_2O(\ell) \rightleftharpoons H_3O^+(aq) + OH^-(aq)$$

BASE acid ACID base

In this reaction, one water molecule serves as an H^+ acceptor (base) while the other is an H^+ donor (acid). The equilibrium between the water molecules and the hydronium and hydroxide ions is very reactant-favored. Therefore, the concentrations of these ions in pure water are *very* low. Nevertheless, autoionization of water is very important in understanding how acids and bases function in aqueous solutions. As in the case of any equilibrium reaction, an equilibrium constant expression can be written for the autoionization of water.

$$2\,H_2O(\ell) \rightleftharpoons H_3O^+(aq) + OH^-(aq) \qquad K_w = [H_3O^+][OH^-]$$

ThomsonNOW™
Go to the Chemistry Interactive menu for a module on **autoionization between water molecules.**

Like all equilibrium constant expressions, the one for K_w includes concentrations of solutes but not the concentration of the solvent, which in this case is water.

Table 16.1 Temperature Dependence of K_w for Water

T (°C)	K_w
10	0.29×10^{-14}
15	0.45×10^{-14}
20	0.68×10^{-14}
25	1.01×10^{-14}
30	1.47×10^{-14}
50	5.48×10^{-14}

In aqueous solutions, the H_3O^+ and OH^- concentrations are inversely related; as one increases, the other must decrease. Their product always equals 1.0×10^{-14} at 25 °C.

The term *alkaline* is also used to describe basic solutions.

This equilibrium constant K_w is known as the **ionization constant for water.** From electrical conductivity measurements of pure water, we know that $[H_3O^+] = [OH^-] = 1.0 \times 10^{-7}$ M at 25 °C. Hence

$$K_w = [H_3O^+][OH^-] = (1.0 \times 10^{-7})(1.0 \times 10^{-7}) = 1.0 \times 10^{-14} \quad \text{(at 25 °C)}$$

The equation $K_w = [H_3O^+][OH^-]$ applies to pure water and all aqueous solutions. Like other equilibrium constants, the value of K_w is temperature-dependent (Table 16.1).

According to the K_w expression, the product of the hydronium ion concentration times the hydroxide ion concentration will always remain the same at a given temperature. If the hydronium ion concentration increases (because an acid was added to the water, for example), then the hydroxide ion concentration must decrease, and vice versa. The equation also tells us that if we know one concentration, we can calculate the other.

The relative concentrations of H_3O^+ and OH^- also indicate the acidic, neutral, or basic nature of the aqueous solution. For all aqueous solutions there are three possibilities.

Neutral solution: $[H_3O^+] = [OH^-]$ both equal to 1.0×10^{-7} M at 25 °C

Acidic solution: $[H_3O^+] > 1.0 \times 10^{-7}$ M $[OH^-] < 1.0 \times 10^{-7}$ M

Basic solution: $[H_3O^+] < 1.0 \times 10^{-7}$ M $[OH^-] > 1.0 \times 10^{-7}$ M

When a solution has equal concentrations of $[H_3O^+]$ and $[OH^-]$, it is said to be **neutral.** If either an acid or a base is added to a neutral solution, the autoionization equilibrium between H_3O^+ and OH^- will be disturbed. Recall that according to Le Chatelier's principle ***(⇐ p. 695),*** an equilibrium shifts in such a way as to offset the effect of any disturbance. When an acid is added, the concentration of H_3O^+ ions increases. To oppose this increase, some added H_3O^+ ions react with OH^- ions in water to form H_2O, thereby reducing the $[OH^-]$. When equilibrium is re-established, $[H_3O^+] > [OH^-]$ and the solution is **acidic;** however, the mathematical product $[H_3O^+][OH^-]$ is still equal to 1.0×10^{-14} at 25 °C. Similarly, if a base is added to water, some of the added OH^- ions react with H_3O^+ ions in water to form H_2O, thereby decreasing the $[H_3O^+]$. When equilibrium is re-established, $[H_3O^+] < [OH^-]$ and the solution is **basic;** $[H_3O^+][OH^-]$ still equals 1.0×10^{-14}.

PROBLEM-SOLVING EXAMPLE 16.2 Hydronium and Hydroxide Concentrations

Calculate:

(a) The hydroxide ion concentration at 25 °C in 0.10 M HCl, a strong acid.

(b) The hydronium ion concentration at 25 °C in 0.010 M KOH, a strong base.

Based on your calculations, explain why the 0.10 M HCl solution is acidic and the 0.010 M KOH solution is basic.

Answer (a) $[OH^-] = 1.0 \times 10^{-13}$ M (b) $[H_3O^+] = 1.0 \times 10^{-12}$ M

The 0.10 M HCl solution is acidic because its hydronium concentration is greater than its hydroxide concentration: $[H_3O^+] > [OH^-]$. The 0.010 M KOH solution is basic because $[OH^-] > [H_3O^+]$.

Strategy and Explanation In each case, the relationship $[H_3O^+][OH^-] = 1.0 \times 10^{-14}$ applies.

(a) Being a strong acid, hydrochloric acid is 100% ionized and so the $[H_3O^+]$ is 1.0×10^{-1} M. The $[OH^-]$ can readily be calculated.

$$[H_3O^+][OH^-] = 1.0 \times 10^{-14} = (1.0 \times 10^{-1}\ \text{M})\ [OH^-]$$

$$[OH^-] = \frac{1.0 \times 10^{-14}}{1.0 \times 10^{-1}} = 1.0 \times 10^{-13}\ \text{M}$$

(b) The strong base KOH is completely ionized and the $[OH^-]$ is 1.0×10^{-2} M. The hydronium concentration can be calculated directly:

$$[H_3O]^+ = \frac{1.0 \times 10^{-14}}{[OH^-]} = \frac{1.0 \times 10^{-14}}{1.0 \times 10^{-2}} = 1.0 \times 10^{-12} \text{ M}$$

PROBLEM-SOLVING PRACTICE 16.2

Which is more acidic, a solution whose H_3O^+ concentration is 5.0×10^{-4} M or one that has an OH^- concentration of 3.0×10^{-8} M?

PROBLEM-SOLVING EXAMPLE 16.3 $[H_3O^+]$ and $[OH^-]$ Concentrations

Calculate the hydroxide ion concentration at 25 °C in 6.0 M HNO_3 and the hydronium ion concentration in 6.0 M NaOH.

Answer $[OH^-]$ of 6.0 M $HNO_3 = 1.7 \times 10^{-15}$ M; $[H_3O^+]$ of 6.0 M NaOH $= 1.7 \times 10^{-15}$ M

Strategy and Explanation Assume that nitric acid and sodium hydroxide, both strong electrolytes, are 100% ionized (Figure 16.1). Therefore, a 6.0 M nitric acid solution has $[H_3O^+] = 6.0$ M and its $[OH^-]$ can be calculated.

$$[H_3O^+][OH^-] = (6.0)[OH^-] = 1.0 \times 10^{-14}$$

$$[OH^-] = \frac{1.0 \times 10^{-14}}{6.0} = 1.7 \times 10^{-15} \text{ M}$$

Because NaOH is a strong base, 6.0 M sodium hydroxide has $[OH^-] = 6.0$ M.

$$[H_3O^+][OH^-] = [H_3O^+](6.0) = 1.0 \times 10^{-14}$$

$$[H_3O^+] = \frac{1.0 \times 10^{-14}}{6.0} = 1.7 \times 10^{-15} \text{ M}$$

✓ **Reasonable Answer Check** Note that at the high hydronium ion concentration of 6.0 M nitric acid, the hydroxide ion concentration is very, very low, which is to be expected of a highly acidic solution. In contrast, the hydronium ion concentration is exceedingly low in 6.0 M NaOH, a highly basic solution.

PROBLEM-SOLVING PRACTICE 16.3

Which is more basic, a solution whose H_3O^+ concentration is 2.0×10^{-5} M or one that has an OH^- concentration of 5.0×10^{-9} M?

16.4 The pH Scale

The $[H_3O^+]$ and $[OH^-]$ in an aqueous solution vary widely depending on the acid or base present and its concentration. In general, the $[H_3O^+]$ in aqueous solutions can range from about 10 mol/L down to about 10^{-15} mol/L. The $[OH^-]$ can also vary over the same range in aqueous solution.

Because these concentrations can be so small, they have very large negative exponents. It is more convenient to express these concentrations in terms of logarithms. We define the **pH** of a solution as ***the negative of the base 10 logarithm (log) of the hydronium ion concentration (mol/L).***

$$\mathbf{pH = -\log[H_3O^+]}$$

The *negative* logarithm of the small concentration values is used since it gives a positive pH value. Thus, the pH of pure water at 25 °C is given by

$$\text{pH} = -\log[1.0 \times 10^{-7}] = -[\log(1.0) + \log(10^{-7})]$$
$$= -[0 + (-7.00)] = 7.00$$

The definition $\text{pH} = -\log[H_3O^+]$ is accurate only for small concentrations of hydronium ions. A more accurate definition is $\text{pH} = -\log a_{H_3O^+}$, where $a_{H_3O^+}$ represents the *activity* of hydronium ions. Activity represents an effective concentration that has been corrected for noncovalent interactions among ions and molecules in solutions. For examples where the definition involving concentration fails, see McCarty, C. G.; Vitz, E. *Journal of Chemical Education*, Vol. 83, 2006; pp. 752–757. A complete definition of pH is quite complicated (see http://www.iupac.org/goldbook/P04524.pdf and Galster, H. *pH Measurement: Fundamentals, Methods, Applications, Instrumentation;* VCH: New York, 1991).

The p in pH is derived from French "puissance" meaning "power." Thus, pH is the "power of hydrogen."

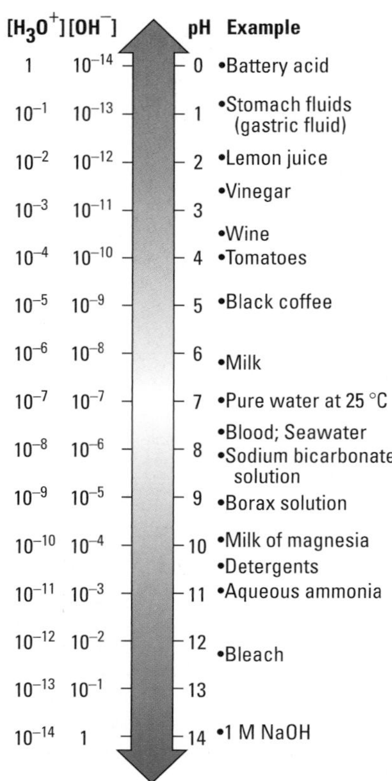

Figure 16.2 The pH of aqueous solutions. The relationship of pH to the concentrations of H_3O^+ and OH^- (in moles/liter at 25 °C) is shown. The pH values of some common substances are also included in the diagram.

In terms of pH, for aqueous solutions at 25 °C we can write

Neutral solution	pH = 7.00
Acidic solution	pH < 7.00
Basic (alkaline) solution	pH > 7.00

Figure 16.2 shows the pH values along with the corresponding H_3O^+ and OH^- concentrations of some common solutions. Notice that, for example, $-\log(1 \times 10^{-x}) = x$.

$$\text{Lemon juice:} \quad [H_3O^+] = 1 \times 10^{-2}\ M; \quad pH = -\log(1 \times 10^{-2}) = 2$$

$$\text{Black coffee:} \quad [H_3O^+] = 1 \times 10^{-5}\ M; \quad pH = -\log(1 \times 10^{-5}) = 5$$

Keep in mind that a change of *one* pH unit represents a *ten-fold* change in H_3O^+ concentration, a change of two pH units represents a 100-fold change, and so on. Thus, according to Figure 16.2, the $[H_3O^+]$ in lemon juice (pH = 2) is 1000 times *greater* than that in black coffee (pH = 5).

For solutions in which $[H_3O^+]$ or $[OH^-]$ has a value other than an exact power of 10 ($1, 1 \times 10^{-1}, 1 \times 10^{-2}, \ldots$) a calculator is convenient for finding the pH. For example, the pH is 2.35 for a solution that contains 0.0045 mol of the strong acid HNO_3 per liter.

$$pH = -\log(4.50 \times 10^{-3}) = 2.35$$

See Appendix A.6 for more about using logarithms.

See Appendix A.3 for a discussion of significant figures.

The digits to the left of the decimal point in a pH represent a power of 10. Only the digits to the right of the decimal point are significant. In Problem-Solving Example 16.4, where $pH = -\log(2.95 \times 10^{-2}) = -\log(2.95) + (-\log 10^{-2}) = -0.470 + 2.000 = 1.530$, there are three significant figures in 1.530, the result, because there are three significant figures in 2.95.

Notice that, as in the case of equilibrium constants, the concentration units of mol/L are ignored when the logarithm is taken. It is not possible to take the logarithm of a unit.

PROBLEM-SOLVING EXAMPLE 16.4 Calculating pH from $[H_3O^+]$

Calculate the pH of an aqueous HNO_3 solution that has a volume of 250. mL and contains 0.4649 g HNO_3.

Answer 1.530 (*Note:* This pH has been calculated to three significant figures, those to the right of the decimal point. In actual measurements, pH values are seldom measurable to this degree of accuracy.)

Strategy and Explanation Nitric acid is a strong acid, so every mole of HNO_3 that dissolves produces a mole of H_3O^+ and a mole of NO_3^-. First, determine the number of moles of HNO_3 and then calculate the H_3O^+ concentration.

$$0.465\ g\ HNO_3 \times \frac{1\ mol\ HNO_3}{63.012\ g\ HNO_3} = 0.007380\ mol\ HNO_3$$

$$[H_3O^+] = \frac{0.007380\ mol\ HNO_3}{0.250\ L} = 0.0295\ M$$

Then, express this concentration as pH.

$$pH = -\log(2.95 \times 10^{-2}) = 1.530$$

✓ **Reasonable Answer Check** If the $[H_3O^+]$ were 0.10 M, the pH would be 1.00; the pH would be 2.00 for an H_3O^+ concentration of 0.010 M. Therefore, a solution with an H_3O^+ concentration of 0.0295 M, which is between these two values, should have a pH between 1.00 and 2.00, which it does.

PROBLEM-SOLVING PRACTICE 16.4

Calculate the pH of a 0.040 M NaOH solution.

ThomsonNOW™
Go to the Coached Problems menu for a tutorial on the **pH scale.**

The calculation done in Problem-Solving Example 16.4 can be reversed; the hydronium ion concentration of a solution can be calculated from its pH value as shown in Problem-Solving Example 16.5.

PROBLEM-SOLVING EXAMPLE 16.5 Calculating $[H_3O^+]$ from pH

A hospital patient's blood sample has a pH of 7.40.
(a) Calculate the sample's H_3O^+ concentration.
(b) What is its OH^- concentration?
(c) Is the sample acidic, neutral, or basic?

Answer (a) 4.0×10^{-8} M (b) 2.5×10^{-7} M (c) Basic

Strategy and Explanation Use the pH value to find the hydronium ion concentration. From that concentration, calculate the OH^- concentration from the $[H_3O^+][OH^-] = 1.0 \times 10^{-14}$ relationship.

(a) Substituting into the definition of pH,

$$-\log[H_3O^+] = 7.40, \quad \text{so} \quad \log[H_3O^+] = -7.40$$

By the rules of logarithms, $10^{\log(x)} = x$, so we can write $10^{\log[H_3O^+]} = 10^{-pH} = [H_3O^+]$. Finding $[H_3O^+]$ therefore requires finding the antilogarithm of -7.40 (Appendix A.6).

$$[H_3O^+] = 10^{-7.40} = 4.0 \times 10^{-8}\ M$$

(b) Rearrange the $[H_3O^+][OH^-] = 1.0 \times 10^{-14}$ equation to solve for $[OH^-]$.

$$[OH^-] = \frac{1.0 \times 10^{-14}}{[H_3O^+]} = \frac{1.0 \times 10^{-14}}{4.0 \times 10^{-8}} = 2.5 \times 10^{-7}\ M$$

(c) Because the pH is greater than 7.0, the sample is basic.

✓ **Reasonable Answer Check** Because the pH of 7.40 is slightly above 7.00, the hydroxide ion concentration should be a bit higher than 1.0×10^{-7} M, that of a neutral solution, which it is. Therefore, the sample is slightly basic.

PROBLEM-SOLVING PRACTICE 16.5

In a hospital laboratory the pH of a bile sample is measured as 7.90.
(a) What is the H_3O^+ concentration? (b) Is the sample acidic or basic?

Courtesy of the Arnold and Mabel Beckman Foundation, Photo by Yana Bridle

Arnold Beckman
1900–2004

When he invented the first electronic pH meter in 1934, Arnold Beckman revolutionized pH measurement. Beckman, a professor at the California Institute of Technology at the time, developed the instrument in response to a request from the California Fruit Growers' Association for a quicker, more accurate way to measure the acidity of lemon juice. He went on to found the highly successful Beckman Instrument Company, a firm that invented the first widely used infrared and ultraviolet spectrophotometers and other laboratory instruments. Arnold Beckman and his wife Mabel have donated millions of dollars to advance chemical research and education nationwide.

CONCEPTUAL EXERCISE 16.8 pH of Solutions of Different Acids

Would the pH of a 0.1 M solution of the strong acid HNO_3 be the same as the pH of a 0.1 M solution of the strong acid HCl? Explain.

EXERCISE 16.9 Super Acidic

Recently, a pH sensor has been developed that operates under extremely acidic conditions, such as those found in environments like Iron Mountain, California. At this abandoned mine site, the groundwater has a pH of -3.6 (not a typo; it is a minus!!). Calculate the H_3O^+ concentration (molarity) of this groundwater. Is this answer reasonable? (Compare your result with the number of moles of pure water per liter.)

The OH^- concentration can also be expressed in exponential terms as pOH.

$$pOH = -\log[OH^-]$$

The $[OH^-]$ of pure water at 25 °C is 1.0×10^{-7} M, and therefore its pOH is

$$pOH = -\log(1 \times 10^{-7}) = -(-7.00) = 7.00$$

Because the values of $[H_3O^+]$ and $[OH^-]$ are related by the K_w expression, for all aqueous solutions at 25 °C, we can write

$$K_w = [H_3O^+][OH^-] = 1.0 \times 10^{-14}$$

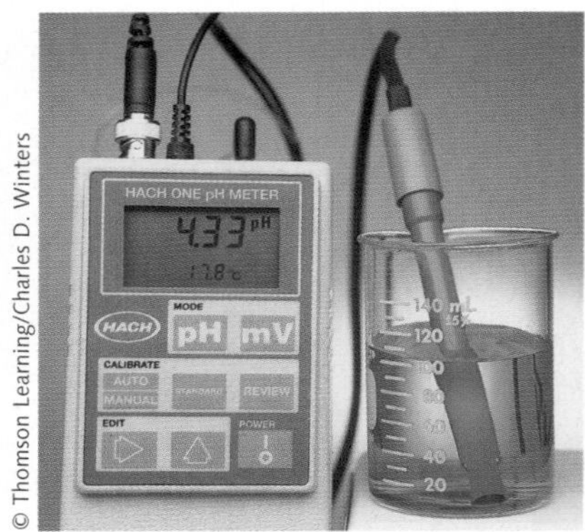

© Thomson Learning/Charles D. Winters

Figure 16.3 A pH meter. A pH meter can quickly and accurately determine the pH of a sample. How a pH meter works is described in Section 19.7.

$-\log K_w = pK_w$;
$-\log[H_3O^+] = pH$;
$-\log[OH^-] = pOH$

Knowing the pH, then the pOH is just 14.00 − pH; knowing the pOH, the pH is 14.00 − pOH.

This equation can be rewritten by taking −log of each side

$$-\log K_w = -\log[H_3O^+] + (-\log[OH^-]) = -\log(1.0 \times 10^{-14})$$

or

$$pK_w = pH + pOH = 14.00$$

The relation between pH and pOH can be used to find one value when the other is known. A 0.0010 M solution of the strong base NaOH, for example, has an OH^- concentration of 0.0010 M and a pOH given by

$$pOH = -\log(1.0 \times 10^{-3}) = 3.00$$

and therefore

$$pH = 14.00 - pOH = 14.00 - 3.00 = 11.00$$

EXERCISE 16.10 pOH and pH

Which solution is more basic, one that has a pH of 5.5 or one with a pOH of 8.5? What is the H_3O^+ concentration in each solution?

Measuring pH

The pH of a solution is readily measured using a pH meter (Figure 16.3). This device consists of a pair of electrodes (often in one probe) that detect the H_3O^+ concentration of the test solution, convert it into an electrical signal, and display it directly as the pH value. The meter is initially calibrated using standard solutions of known pH. The pH of body fluids, soil, environmental and industrial samples, and other substances can be measured easily and accurately with a pH meter.

Acid-base indicators are a much older and less precise (but more convenient) method to determine the pH of a sample. Such indicators are substances that change color within a narrow pH range, generally 1 to 2 pH units, by the loss or gain of an H^+ ion that changes the indicator's molecular structure so that it absorbs light in different regions of the visible spectrum. The indicator is one color at a lower pH (its "acid" form), and a different color at a higher pH (its "base" form). Consider the indicator bromthymol blue (Figure 16.4). At or below pH 6 it is yellow (its acid form); at pH 8 and above it is blue (its base form). Between pH 6 and 7, the indicator changes from pure yellow to a yellow-green color. At pH 7, it is a mixture of 50% yellow and 50% blue, so it appears green. As the pH changes from 7 to 8, the color becomes pure blue. Thus, the pH of a sample that turns bromthymol blue to green has a pH of about 7. If the indicator color is blue, the pH of the sample is at least 8, and it could be much higher.

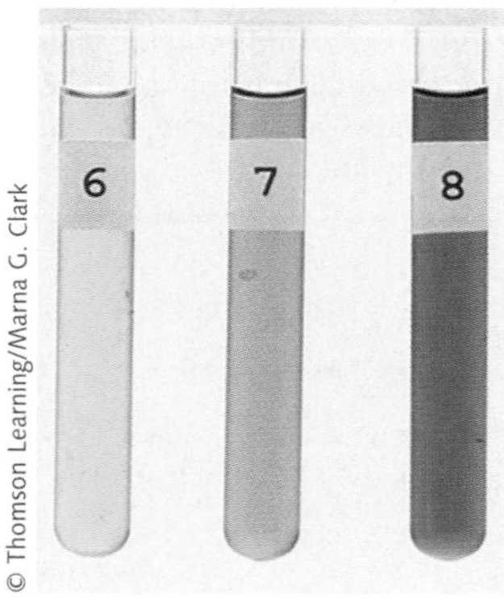

© Thomson Learning/Marna G. Clark

Figure 16.4 Bromthymol blue indicator. At or below a pH of 6 the indicator is yellow. At pH 7 it is pale green, and at pH 8 and above, the color is blue.

Strips of paper impregnated with acid-base indicators are also used to test the pH of many substances. The color of the paper after it has been dampened by the solution to be tested is compared with a set of colors at known pHs.

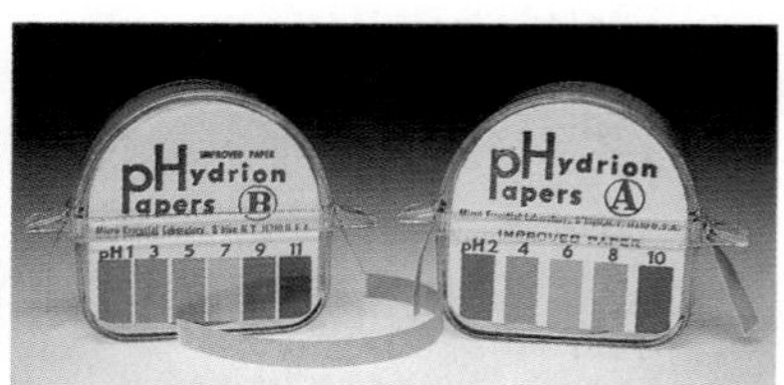

© Thomson Learning/Charles D. Winters

Indicator paper strips. Strips of paper impregnated with indicator are used to find an approximate pH.

16.5 Ionization Constants of Acids and Bases

You learned earlier that the greater the value of the equilibrium constant for a reaction, the more product-favored that reaction is. In an acid-base reaction, the stronger the reactant acid and the reactant base, the more product-favored the reaction (**⇐ p. 671**). Consequently, the magnitude of equilibrium constants can give us an idea about the relative strengths of weak acids and bases. For example, *the larger the equilibrium constant for an acid's ionization, the stronger the acid.*

Acid Ionization Constants

An ionization equation for the transfer to water of H^+ from any acid represented by the general formula HA is

$$\underset{\text{conjugate acid}}{HA(aq)} + H_2O(\ell) \rightleftharpoons H_3O^+(aq) + \underset{\text{conjugate base}}{A^-(aq)}$$

The corresponding **acid ionization constant expression** is

$$K_a = \frac{[H_3O^+][\text{conjugate base}]}{[\text{conjugate acid}]} = \frac{[H_3O^+][A^-]}{[HA]}$$

Remember that the solvent (water, in this case) does not appear in the equilibrium constant expression.

In the acid ionization constant expression, the *equilibrium* concentrations of conjugate base and hydronium ion appear in the numerator; the *equilibrium* concentration of *un-ionized* conjugate acid appears in the denominator. As with other equilibrium constant expressions, pure solids and liquids, such as water, are not included.

The equilibrium constant K_a is called the **acid ionization constant.** As more acid ionizes, the [HA] denominator term in the acid ionization constant expression gets smaller as the numerator terms increase. Consequently, the ratio gets larger. For strong acids such as hydrochloric acid, the equilibrium is so product-favored that the acid ionization constant value is much larger than 1. In contrast with strong acids, weak acids such as acetic acid ionize to a much smaller extent, establishing equilibria in which significant concentrations of un-ionized weak acid molecules are still present in the solution. All weak acids have K_a values less than 1 because the ionization of a weak acid is reactant-favored; weak acids are weak electrolytes.

Acid ionization constants are also called acid dissociation constants.

The larger its acid ionization constant, the stronger the acid.

The common strong acids are hydrochloric (HCl), nitric (HNO_3), sulfuric (H_2SO_4), perchloric ($HClO_4$), hydrobromic (HBr), and hydroiodic (HI).

PROBLEM-SOLVING EXAMPLE 16.6 Acid Ionization Constant Expressions

Write the ionization equation and the ionization constant expression for these weak acids.

(a) HF (b) HBrO (c) $H_2PO_4^-$

Answer

(a) $HF(aq) + H_2O(\ell) \rightleftharpoons H_3O^+(aq) + F^-(aq)$ $\quad K_a = \frac{[H_3O^+][F^-]}{[HF]}$

(b) $HBrO(aq) + H_2O(\ell) \rightleftharpoons H_3O^+(aq) + BrO^-(aq)$ $\quad K_a = \frac{[H_3O^+][BrO^-]}{[HBrO]}$

(c) $H_2PO_4^-(aq) + H_2O(\ell) \rightleftharpoons H_3O^+(aq) + HPO_4^{2-}(aq)$ $\quad K_a = \frac{[H_3O^+][HPO_4^{2-}]}{[H_2PO_4^-]}$

Strategy and Explanation In each case, the ionization equation represents the transfer of an H^+ ion from an acid to water, creating a hydronium ion and the conjugate base of the acid. In the K_a expression, the product concentrations at equilibrium are divided by the reactant concentrations at equilibrium; $[H_2O]$ is not included. For example, in the case of HF:

$$HF(aq) + H_2O(\ell) \rightleftharpoons H_3O^+(aq) + F^-(aq) \qquad K_a = \frac{[H_3O^+][F^-]}{[HF]}$$

PROBLEM-SOLVING PRACTICE 16.6

Write the ionization equation and ionization constant expression for each of these weak acids:

(a) Hydrazoic acid, HN_3 (b) Formic acid, HCOOH

(c) Chlorous acid, $HClO_2$

That weak acids are only slightly ionized can be shown by measuring the pH of their aqueous solutions. The pH of a 0.10 M acetic acid solution is 2.88, which corresponds to a concentration of H_3O^+ of only 1.3×10^{-3} M. Compare this value with the 0.10 M concentration of H_3O^+ ions in a 0.10 M solution of HCl, a strong acid that has pH = 1.00. In a 0.10 M acetic acid solution, only 1.3% of the initial concentration of acetic acid is ionized:

Stronger acid than CH_3COOH — Stronger base than H_2O

$$CH_3COOH(aq) + H_2O(\ell) \rightleftharpoons \underbrace{H_3O^+(aq)}_{\text{Stronger acid than } CH_3COOH} + \underbrace{CH_3COO^-(aq)}_{\text{Stronger base than } H_2O}$$

$$\%\ \text{ionization} = \frac{[H_3O^+]\ \text{at equilibrium}}{\text{initial acid conc.}} \times 100\% = \frac{1.3 \times 10^{-3}}{1.0 \times 10^{-1}} \times 100\% = 1.3\%$$

Therefore, almost 99% of the acetic acid remains in the un-ionized molecular form, CH_3COOH. This is why weak acids (and bases) are weak electrolytes.

In an acetic acid solution, or an aqueous solution of any weak acid, two different bases compete for H^+ ions that can be donated from two different acids. In the equation above, the two bases are water and acetate ion; the two acids are acetic acid and hydronium ion. Since acetic acid is a weak acid, its K_a is much less than 1 and the equilibrium favors the reactants. The acetate ion must be a stronger H^+ acceptor than the water molecule. Another way of looking at the same reaction is that the hydronium ion must be a stronger H^+ donor than the acetic acid molecule. Both of these statements are true. Recall from Section 16.1 that acid-base reactions favor going from the stronger to the weaker member of each conjugate acid-base pair. Thus, the acetic acid equilibrium is reactant-favored; a significant concentration of un-ionized acetic acid molecules is present in solution at equilibrium.

Base Ionization Constants

A general equation analogous to that for the donation of H^+ to water by acids can be written for the *acceptance* of an H^+ *from* water by a molecular base, B, to form its conjugate acid, BH^+.

$$\underset{\text{conjugate base}}{B(aq)} + H_2O(\ell) \rightleftharpoons \underset{\text{conjugate acid}}{BH^+(aq)} + OH^-(aq)$$

If the base B were NH_3, then BH^+ would be NH_4^+.

The corresponding equilibrium constant expression is

$$K_b = \frac{[\text{conjugate acid}][OH^-]}{[\text{conjugate base}]} = \frac{[BH^+][OH^-]}{[B]}$$

The equilibrium constant K_b is called the **base ionization constant,** a term that can be misleading. Notice from the chemical equation above that the base does not ionize. Rather, K_b and its equilibrium constant expression refer to *the reaction in which a base forms its conjugate acid by removing an* H^+ *ion from water.*

When the base is an anion, A^- (such as the anion of a weak acid), the general equation is

$$\underset{\text{conjugate base}}{A^-(aq)} + H_2O(\ell) \rightleftharpoons \underset{\text{conjugate acid}}{HA(aq)} + OH^-(aq)$$

If the base A^- were CH_3COO^-, then HA would be CH_3COOH. The corresponding **base ionization constant expression** is

$$K_b = \frac{[\text{conjugate acid}][OH^-]}{[\text{conjugate base}]} = \frac{[HA][OH^-]}{[A^-]}$$

The magnitude of the K_b value indicates the extent to which the base removes H^+ ions from water to produce OH^- ions. The larger the base ionization constant, K_b, the stronger the base, the more product-favored the H^+ transfer reaction from water, and the greater the OH^- concentration produced. For a strong base, the base ionization constant is greater than 1. For a weak base, the ionization constant is less than 1, sometimes considerably less than 1, because at equilibrium there is a significant concentration of unreacted weak conjugate base and a much smaller concentration of its conjugate acid and OH^- ions in solution.

PROBLEM-SOLVING EXAMPLE 16.7 Base Ionization

For each of these weak bases, write the equation for the reaction of the base with water and the companion K_b expression.

(a) C_5H_5N (b) NH_2OH (c) F^-

Answer

(a) $C_5H_5N(aq) + H_2O(\ell) \rightleftharpoons C_5H_5NH^+(aq) + OH^-(aq)$ $K_b = \frac{[C_5H_5NH^+][OH^-]}{[C_5H_5N]}$

(b) $NH_2OH(aq) + H_2O(\ell) \rightleftharpoons NH_3OH^+(aq) + OH^-(aq)$ $K_b = \frac{[NH_3OH^+][OH^-]}{[NH_2OH]}$

(c) $F^-(aq) + H_2O(\ell) \rightleftharpoons HF(aq) + OH^-(aq)$ $K_b = \frac{[HF][OH^-]}{[F^-]}$

Strategy and Explanation The general reaction is the same for each of the bases: The base removes an H^+ from water to form the corresponding conjugate acid. In the first two parts, the base is a neutral molecule to which an H^+ is added, forming a positively charged conjugate acid. In part (c), the H^+ adds to a negatively charged ion, F^-, resulting in a conjugate acid, HF, with no net charge.

PROBLEM-SOLVING PRACTICE 16.7

Write the ionization equation and the K_b expression for these weak bases.

(a) CH_3NH_2
(b) Phosphine, PH_3
(c) NO_2^-

Values of Acid and Base Ionization Constants

Table 16.2 summarizes the ionization constants for a number of acids and their conjugate bases. The ionization constants for strong acids (those above H_3O^+ in Table 16.2) and strong bases (those below OH^- in Table 16.2) are too large to be measured easily. Fortunately, because their ionization reactions are virtually complete, these K_a and K_b values are hardly ever needed. For weak acids, K_a values show relative strengths quantitatively; for weak bases, K_b values do the same.

Consider acetic acid and boric acid. Boric acid is below acetic acid in Table 16.2, so boric acid must be a weaker acid than acetic acid; the K_a values tell us how much weaker. The K_a for boric acid is 7.3×10^{-10}; that for acetic acid is 1.8×10^{-5},

Table 16.2 Ionization Constants for Some Acids and Their Conjugate Bases at 25 °C

Acid Name	Acid	$K_a = \frac{[H_3O^+]\left[\text{conj base}\right]}{[\text{conj acid}]}$	Base Name	Base	$K_b = \frac{\left[\text{conj acid}\right][OH^-]}{[\text{conj base}]}$
Perchloric acid	$HClO_4$	Large	Perchlorate ion	ClO_4^-	Very small
Sulfuric acid	H_2SO_4	Large	Hydrogen sulfate ion	HSO_4^-	Very small
Hydrochloric acid	HCl	Large	Chloride ion	Cl^-	Very small
Nitric acid	HNO_3	≈20	Nitrate ion	NO_3^-	$\approx 5 \times 10^{-16}$
Hydronium ion	H_3O^+	1.0	Water	H_2O	1.0×10^{-14}
Sulfurous acid	H_2SO_3	1.2×10^{-2}	Hydrogen sulfite ion	HSO_3^-	8.3×10^{-13}
Hydrogen sulfate ion	HSO_4^-	1.2×10^{-2}	Sulfate ion	SO_4^{2-}	8.3×10^{-13}
Phosphoric acid	H_3PO_4	7.5×10^{-3}	Dihydrogen phosphate ion	$H_2PO_4^-$	1.3×10^{-12}
Hexaaquairon(III) ion	$Fe(H_2O)_6^{3+}$	6.3×10^{-3}	Pentaaquahydroxoiron(III) ion	$Fe(H_2O)_5OH^{2+}$	1.6×10^{-12}
Hydrofluoric acid	HF	7.2×10^{-4}	Fluoride ion	F^-	1.4×10^{-11}
Nitrous acid	HNO_2	4.5×10^{-4}	Nitrite ion	NO_2^-	2.2×10^{-11}
Formic acid	$HCOOH$	1.8×10^{-4}	Formate ion	$HCOO^-$	5.6×10^{-11}
Benzoic acid	C_6H_5COOH	6.3×10^{-5}	Benzoate ion	$C_6H_5COO^-$	1.6×10^{-10}
Acetic acid	CH_3COOH	1.8×10^{-5}	Acetate ion	CH_3COO^-	5.6×10^{-10}
Propanoic acid	CH_3CH_2COOH	1.4×10^{-5}	Propanoate ion	$CH_3CH_2COO^-$	7.1×10^{-10}
Hexaaquaaluminum ion	$Al(H_2O)_6^{3+}$	7.9×10^{-6}	Pentaaquahydroxoaluminum ion	$Al(H_2O)_5OH^{2+}$	1.3×10^{-9}
Carbonic acid	H_2CO_3	4.2×10^{-7}	Hydrogen carbonate ion	HCO_3^-	2.4×10^{-8}
Hexaaquacopper(II) ion	$Cu(H_2O)_6^{2+}$	1.6×10^{-7}	Pentaaquahydroxocopper(II) ion	$Cu(H_2O)_5OH^+$	6.25×10^{-8}
Hydrogen sulfide	H_2S	1×10^{-7}	Hydrogen sulfide ion	HS^-	1×10^{-7}
Dihydrogen phosphate ion	$H_2PO_4^-$	6.2×10^{-8}	Hydrogen phosphate ion	HPO_4^{2-}	1.6×10^{-7}
Hydrogen sulfite ion	HSO_3^-	6.2×10^{-8}	Sulfite ion	SO_3^{2-}	1.6×10^{-7}
Hypochlorous acid	$HClO$	3.5×10^{-8}	Hypochlorite ion	ClO^-	2.9×10^{-7}
Hexaaqualead(II) ion	$Pb(H_2O)_6^{2+}$	1.5×10^{-8}	Pentaaquahydroxolead(II) ion	$Pb(H_2O)_5OH^+$	6.7×10^{-7}
Hexaaquacobalt(II) ion	$Co(H_2O)_6^{2+}$	1.3×10^{-9}	Pentaaquahydroxocobalt(II) ion	$Co(H_2O)_5OH^+$	7.7×10^{-6}
Boric acid	$B(OH)_3(H_2O)$	7.3×10^{-10}	Tetrahydroxoborate ion	$B(OH)_4^-$	1.4×10^{-5}
Ammonium ion	NH_4^+	5.6×10^{-10}	Ammonia	NH_3	1.8×10^{-5}
Hydrocyanic acid	HCN	4.0×10^{-10}	Cyanide ion	CN^-	2.5×10^{-5}
Hexaaquairon(II) ion	$Fe(H_2O)_6^{2+}$	3.2×10^{-10}	Pentaaquahydroxoiron(II) ion	$Fe(H_2O)_5OH^+$	3.1×10^{-5}
Hydrogen carbonate ion	HCO_3^-	4.8×10^{-11}	Carbonate ion	CO_3^{2-}	2.1×10^{-4}
Hexaaquanickel(II) ion	$Ni(H_2O)_6^{2+}$	2.5×10^{-11}	Pentaaquahydroxonickel(II) ion	$Ni(H_2O)_5OH^+$	4.0×10^{-4}
Hydrogen phosphate	HPO_4^{2-}	3.6×10^{-13}	Phosphate ion	PO_4^{3-}	2.8×10^{-2}
Water	H_2O	1.0×10^{-14}	Hydroxide ion	OH^-	1.0
Hydrogen sulfide ion	HS^-	1×10^{-19}	Sulfide ion	S^{2-}	1×10^{5}
Ethanol	C_2H_5OH	Very small	Ethoxide ion	$C_2H_5O^-$	Large
Ammonia	NH_3	Very small	Amide ion	NH_2^-	Large
Hydrogen	H_2	Very small	Hydride ion	H^-	Large
Methane	CH_4	Very small	Methide ion	CH_3^-	Large

Increasing Acid Strength (↑) — Increasing Base Strength (↓)

which shows that boric acid is somewhat more than 10^4 times weaker than acetic acid. In fact, boric acid is such a weak acid that a dilute solution of it can be used safely as an eyewash. Don't try that with acetic acid!

The smaller the K_a value, the weaker the acid; the smaller the K_b value, the weaker the base.

CONCEPTUAL EXERCISE 16.11 Acid Strengths

The K_a of lactic acid is 1.5×10^{-4}; that of pyruvic acid is 3.2×10^{-3}.
(a) Which of these acids is the stronger acid?
(b) Which acid's ionization reaction is more reactant-favored?

K_a Values for Polyprotic Acids

So far we have concentrated on **monoprotic acids** such as hydrogen fluoride (HF), hydrogen chloride (HCl), and nitric acid (HNO_3)—acids that can donate a single H^+ per molecule.

sulfuric acid

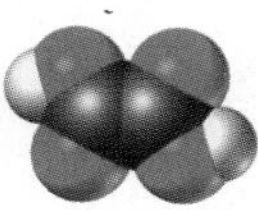
oxalic acid

phosphoric acid

Some acids, called **polyprotic acids,** can donate more than one H^+ per molecule. These include sulfuric acid (H_2SO_4), carbonic acid (H_2CO_3), and phosphoric acid (H_3PO_4). Oxalic acid ($H_2C_2O_4$ or HOOC—COOH) and other organic acids with two or more carboxylic acid (—COOH) groups are also polyprotic acids (Table 16.3).

In aqueous solution, a polyprotic acid donates its H^+ ions to water molecules in a stepwise manner. In the first step for sulfuric acid, hydrogen sulfate ion, HSO_4^-, is formed. Sulfuric acid is a strong acid, so this first ionization is essentially complete.

$$\underset{\text{ACID}}{H_2SO_4(aq)} + \underset{\text{base}}{H_2O(\ell)} \longrightarrow \underset{\text{acid}}{H_3O^+(aq)} + \underset{\text{BASE}}{HSO_4^-(aq)}$$

Hydrogen sulfate ion is the conjugate base of sulfuric acid.

In the next ionization step, hydrogen sulfate ion acting as an acid donates an H^+ ion to another water molecule. In this case, hydrogen sulfate ion is a weak acid ($K_a < 1$) and, as with other weak acids, an equilibrium is established. Sulfate ion, SO_4^{2-}, is the conjugate base of HSO_4^-, its conjugate acid.

$$\underset{\text{ACID}}{HSO_4^-(aq)} + \underset{\text{base}}{H_2O(\ell)} \rightleftharpoons \underset{\text{acid}}{H_3O^+(aq)} + \underset{\text{BASE}}{SO_4^{2-}(aq)}$$

Many chemical reactions occur in steps that can be represented by individual equations. Sometimes only the overall equation is written.

CONCEPTUAL EXERCISE 16.12 Explaining Acid Strengths

Look at the charge on the hydrogen sulfate ion. What does the charge have to do with the fact that this ion is a weaker acid than H_2SO_4?

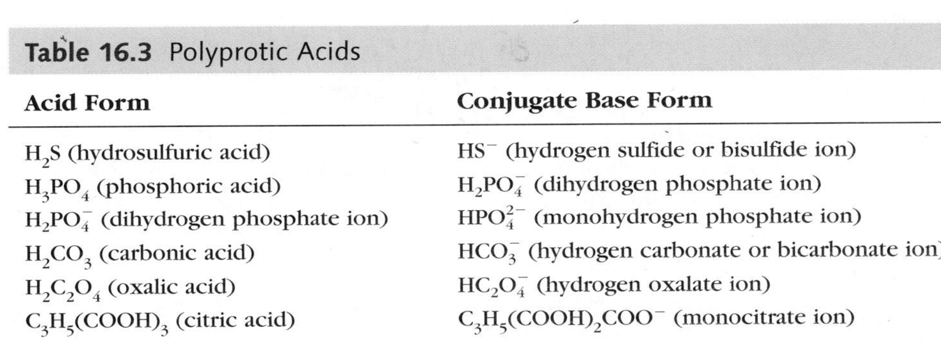

Table 16.3 Polyprotic Acids

Acid Form	Conjugate Base Form
H_2S (hydrosulfuric acid)	HS^- (hydrogen sulfide or bisulfide ion)
H_3PO_4 (phosphoric acid)	$H_2PO_4^-$ (dihydrogen phosphate ion)
$H_2PO_4^-$ (dihydrogen phosphate ion)	HPO_4^{2-} (monohydrogen phosphate ion)
H_2CO_3 (carbonic acid)	HCO_3^- (hydrogen carbonate or bicarbonate ion)
$H_2C_2O_4$ (oxalic acid)	$HC_2O_4^-$ (hydrogen oxalate ion)
$C_3H_5(COOH)_3$ (citric acid)	$C_3H_5(COOH)_2COO^-$ (monocitrate ion)

There are polyprotic bases, such as CO_3^{2-}, that can accept more than one H^+ per molecule of base. We will not discuss polyprotic bases here.

The weak acid H_3PO_4 can donate three H^+ ions per molecule through these three ionization reactions.

Both $H_2PO_4^-$ ion and HPO_4^{2-} ion are amphiprotic because they can gain or lose a proton (H^+ ion).

First ionization

$$H_3PO_4(aq) + H_2O(\ell) \rightleftharpoons H_3O^+(aq) + H_2PO_4^-(aq) \qquad K_a = 7.5 \times 10^{-3}$$

Second ionization

$$H_2PO_4^-(aq) + H_2O(\ell) \rightleftharpoons H_3O^+(aq) + HPO_4^{2-}(aq) \qquad K_a = 6.2 \times 10^{-8}$$

Third ionization

$$HPO_4^{2-}(aq) + H_2O(\ell) \rightleftharpoons H_3O^+(aq) + PO_4^{3-}(aq) \qquad K_a = 3.6 \times 10^{-13}$$

The successive K_a values for the ionization of a polyprotic acid decrease by a factor of 10^4 to 10^5, indicating that each ionization step occurs to a lesser extent than the one before it. The $H_2PO_4^-$ ion ($K_a = 6.2 \times 10^{-8}$) is a much weaker acid than phosphoric acid ($K_a = 7.5 \times 10^{-3}$), and the HPO_4^{2-} ion ($K_a = 3.6 \times 10^{-13}$) is an even weaker acid than $H_2PO_4^-$. The K_a values indicate that it is more difficult to remove H^+ from a negatively charged $H_2PO_4^-$ ion than from a neutral H_3PO_4 molecule and even more difficult to remove H^+ from a doubly negative HPO_4^{2-} ion.

EXERCISE 16.13 Polyprotic Acids

Write equations for the stepwise ionization in aqueous solution of (a) oxalic acid and (b) citric acid. (Formulas for these acids are given in Table 16.3.)

Hydrated Metal Ions as Acids

Lone pairs of electrons on water molecules form coordinate covalent bonds with the metal ion (p. 806).

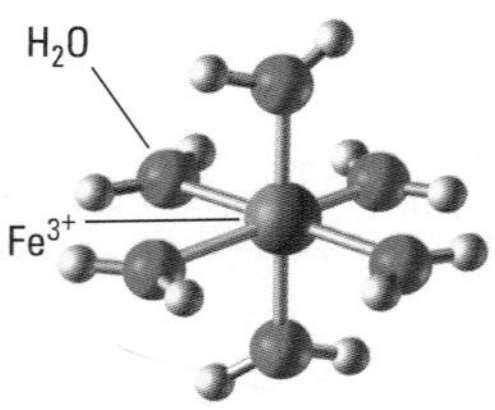

Water molecules bonded to an Fe^{3+} ion in $[Fe(H_2O)_6]^{3+}$. Square brackets in the formula indicate that ions or molecules within the brackets are bonded to the metal ion.

Some hydrated metal ions, especially those of the transition metals, are also weak acids. When a salt containing a metal ion dissolves in water, the metal ion becomes hydrated, often with six water molecules around it, $[M(H_2O)_6]^{n+}$, where M represents a metal ion whose charge is $n+$. There are M—O—H bonds in the hydrated ion, as shown by the structure in the margin. Metal ions other than those in Groups 1A and 2A have large enough charges and small enough sizes to attract the shared electron pair in the M—O bond to themselves. This weakens the O—H bond, making the hydrogen in a M—O—H bond more acidic than it would be in the O—H bond of a water molecule that is not bonded to a metal ion. Thus, the $[M(H_2O)_6]^{n+}$ ion can donate H^+, the solution becomes acidic, and the positive charge of the remaining hydrated metal ion has been decreased by one.

The ionization reaction and acid ionization constant expression for a hydrated metal ion such as Fe^{3+} can be written as

$$[Fe(H_2O)_6]^{3+}(aq) + H_2O(\ell) \rightleftharpoons [Fe(H_2O)_5(OH)]^{2+}(aq) + H_3O^+(aq)$$

$$K_a = \frac{[[Fe(H_2O)_5(OH)]^{2+}][H_3O^+]}{[[Fe(H_2O)_6]^{3+}]} = 6.3 \times 10^{-3}$$

Look carefully at the Fe-containing reactant and product in the equation. There are six water molecules in the reactant ion but only five in the product ion. The other water molecule has lost an H^+ to become an OH^- ion. As a result, the net charge on the product ion is one less than that on the reactant ion.

This K_a value shows that a solution of $FeCl_3$ will have about the same pH as a solution of phosphoric acid ($K_a = 7.5 \times 10^{-3}$) of equal concentration. Many metal ions form weakly acidic aqueous solutions, and this property is important in the chemistry of such ions in the environment. For example, Al^{3+} ions in soils react with water to produce an acidic environment that can be detrimental to tree growth.

16.6 Molecular Structure and Acid Strength

If all acids donate H^+ ions, why are some acids strong while others are weak? Why is there such a broad range of K_a values? To answer these questions we turn to the relationship of an acid's strength to its molecular structure. In doing so, we will consider a wide range of acids, from simple binary ones like HBr to structurally more complex ones containing oxygen, carbon, and other elements.

Factors Affecting Acid Strength

All acids have their acidic hydrogen bonded to some other atom, call it A, which can be bonded to other atoms as well. The H—A bond must be broken for the acid to transfer its hydrogen as an H^+ to water, and that will occur only if the H—A bond is polar.

$$\overset{\delta^+ \ \ \delta^-}{\longleftrightarrow}$$
$$\text{H—A}$$

An H—A bond with very little polarity, such as the H—C bond in methane, CH_4, makes the hydrogens nonacidic, and methane is not a significant H^+ donor to water.

The simplest case of an acid is a *binary acid,* such as HBr, one that contains just hydrogen and one other element. In this case, A is bromine.

$$\overset{\delta^+ \ \ \delta^-}{\longleftrightarrow}$$
$$\text{H—A}$$

Blue represents high partial positive charge; red indicates high partial negative charge.

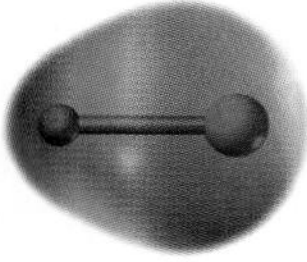

HBr

The H—Br bond is polar and HBr is a strong acid ($K_a \approx 10^8$). For acids with A atoms from the same group in the periodic table, for example HF, HCl, HBr, and HI, the H—A bond energies determine the relative acid strengths for binary acids. *As H—A bond energies decrease down a group, the H—A bond weakens, and binary acid strengths increase*.

For a series of binary acids in which the A atoms are in the same period, H—A bond energies do not vary much, and therefore the H—A bond polarity is the principal factor affecting acid strengths for binary acids. As the electronegativity of A increases across a period, the H—A bond polarity also gets larger due to greater electronegativity differences. Correspondingly, the hydrogen becomes more acidic, as seen for Period 3 nonmetals:

- The H—Si bond is relatively nonpolar and SiH_4 is nonacidic.
- The H—P bond is only slightly polar and the unshared electron pair on phosphorus makes PH_3 accept H^+ ions rather than donate them.
- The H—S bond is slightly polar and H_2S is a weak acid.
- The H—Cl bond is very polar and HCl is a strong acid.

Strengths of Oxoacids

Acids in which the acidic hydrogen is bonded directly to oxygen in an H—O— bond are called **oxoacids.** Three of the strong acids—nitric (HNO_3), perchloric ($HClO_4$), and sulfuric (H_2SO_4)—are oxoacids.

nitric acid perchloric acid sulfuric acid

Like other oxoacids, they have at least one hydrogen atom bonded to an oxygen and have the general formula

$$\mathrm{H{-}O{-}Z}\begin{smallmatrix}/\\ -\\ \backslash\end{smallmatrix}$$

The nature of Z and other atoms that may be attached to it are important in determining the strength of the H—O bond and thus the strength of an oxoacid. In general, *acid strength decreases with the decreasing electronegativity of Z.* This is reflected in the differences among the K_a values of HOCl, HOBr, and HOI, as the electronegativity of the halogen decreases from chlorine (3.0) to bromine (2.8) to iodine (2.5).

Acid:	HOCl	HOBr	HOI

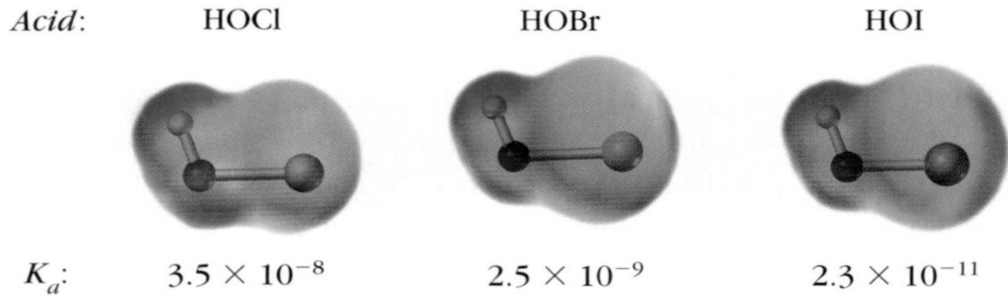

K_a:	3.5×10^{-8}	2.5×10^{-9}	2.3×10^{-11}

The number of oxygen atoms attached to Z also significantly affects the strength of the H—O bond and oxoacid strength: ***The acid strength increases as the number of oxygen atoms attached to Z increases.*** The terminal oxygen atoms (those not in an H—O bond) are sufficiently electronegative, along with Z, to withdraw electron density from the H—O bond. This weakens the bond, promoting the transfer of an H^+ ion to water. The more terminal oxygen atoms present, the greater the electron density shift and the greater the acid strength. A particularly striking example of this trend is seen with the oxoacids of chlorine from the weakest, hypochlorous acid (HOCl), to the strongest, perchloric acid ($HClO_4$).

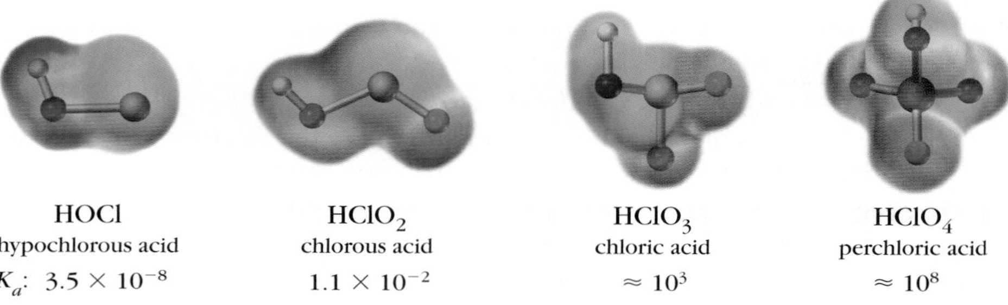

HOCl	$HClO_2$	$HClO_3$	$HClO_4$
hypochlorous acid	chlorous acid	chloric acid	perchloric acid
K_a: 3.5×10^{-8}	1.1×10^{-2}	$\approx 10^3$	$\approx 10^8$

The chemical formula for perchloric acid can be considered as $HOClO_3$ in relation to its structural formula.

To be a strong acid, an inorganic oxoacid must have at least two more oxygen atoms than acidic hydrogen atoms in the molecule. Thus, sulfuric acid is a strong acid. In contrast, the inorganic oxoacid phosphoric acid (H_3PO_4) has only four oxygen atoms for three hydrogen atoms and is a weak acid.

Strengths of Carboxylic Acids

All carboxylic acids contain the carboxylic acid group, —COOH, although all such acids do not have the same acid strength.

$$\mathrm{R{-}\overset{\overset{\displaystyle O}{\|}}{C}{-}O{-}H}$$

The differences in carboxylic acid strength are due to differences in the R group attached to the —COOH group. When R is simply a hydrocarbon group, there is little effect on acid strength; K_a values are similar, as seen by comparing acetic acid, CH_3COOH, and hexanoic acid, $CH_3(CH_2)_4COOH$.

$$\begin{array}{ccccc} & H & O & & \\ & | & \| & & \\ H - & C & - C & - OH \\ & | & & & \\ & H & & & \end{array}$$

acetic acid
$K_a = 1.8 \times 10^{-5}$

$$\begin{array}{cccccccc} & H & H & H & H & H & O & \\ & | & | & | & | & | & \| & \\ H - & C - & C - & C - & C - & C - & C & - OH \\ & | & | & | & | & | & & \\ & H & H & H & H & H & & \end{array}$$

hexanoic acid
$K_a = 1.4 \times 10^{-5}$

Acid strength, however, is affected by the addition of a highly electronegative atom to the R group. The electronegative atom causes electron density to shift from the O—H bond, weakening it and thus increasing the acid's strength. For example, replacing a hydrogen in the CH_3 group of acetic acid with chlorine, a more electronegative atom, forms chloroacetic acid, $ClCH_2COOH$. The K_a for this acid is 1.4 $\times\ 10^{-3}$, nearly 100 times larger than that of acetic acid. Replacing all three hydrogens of the CH_3 group in acetic acid with chlorines converts acetic acid to trichloroacetic acid, Cl_3CCOOH, which is about 10,000 times stronger than acetic acid and 100 times stronger than chloroacetic acid.

$$\begin{array}{ccccc} & H & O & & \\ & | & \| & & \\ H - & C & - C & - OH \\ & | & & & \\ & H & & & \end{array}$$

acetic acid
$K_a = 1.8 \times 10^{-5}$

$$\begin{array}{ccccc} & H & O & & \\ & | & \| & & \\ Cl - & C & - C & - OH \\ & | & & & \\ & H & & & \end{array}$$

chloroacetic acid
$K_a = 1.4 \times 10^{-3}$

$$\begin{array}{ccccc} & Cl & O & & \\ & | & \| & & \\ Cl - & C & - C & - OH \\ & | & & & \\ & Cl & & & \end{array}$$

trichloroacetic acid
$K_a = 3 \times 10^{-1}$

CONCEPTUAL EXERCISE 16.14 Molecular Structure and Acid Strength

Which has the larger K_a,

(a) Fluorobenzoic acid, C_6H_4FCOOH, or benzoic acid, C_6H_5COOH?
(b) Chloroacetic acid or bromoacetic acid, $BrCH_2COOH$?

Explain your answer.

CONCEPTUAL EXERCISE 16.15 Molecular Structure and Acid Strength

Consider acetic acid, CH_3COOH, and oxalic acid, HOOC—COOH. Which is the stronger acid? (Consider only the first ionization.) Explain your answer.

Amino Acids and Zwitterions

Amino acids are the monomers from which proteins are assembled (⇐ ***p. 590***). The general structural formula for amino acids is

$$\begin{array}{ccccc} & H & O & & \\ & | & \| & & \\ R - & C & - C & - OH \\ & | & & & \\ & NH_2 & & & \end{array}$$

The amino acids are called alpha (α) amino acids because the R group is on the alpha carbon—the one next to the —COOH group.

Unlike acetic acid (CH_3COOH) and ethylamine ($CH_3CH_2NH_2$), which are liquids at 25 °C, amino acids are crystalline solids that generally melt above 200 °C. What makes amino acids more like salts than simple organic compounds? The general structural formula given for amino acids does not predict such properties. Consider, however, the functional groups in an amino acid. All amino acids contain at least one acidic carboxylic acid group *and* at least one basic amine group. An intramolecular Brønsted-Lowry acid-base reaction occurs in amino acids in which H^+ is transferred from the carboxylic acid group to the amine group. This resulting dipolar structure, called a **zwitterion,** is a salt-like substance. In the case of alanine the change is

The term "zwitterion" is derived from the German word *zwitter*, meaning "a hybrid."

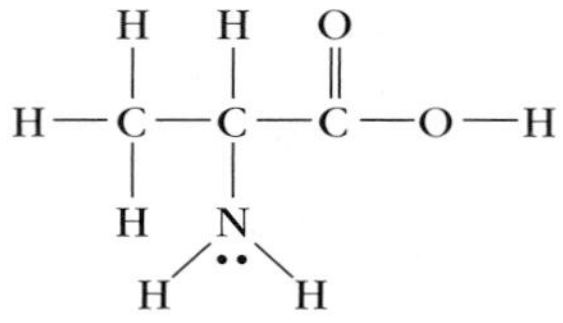

```
     H   H   O                      H   H   O
     |   |   ||                     |   |   ||
 H—C—C—C—O—H               H—C—C—C—O⁻
     |   |                          |   |
     H   N                          H   NH3+
        /··\
       H    H
```

alanine, molecular form 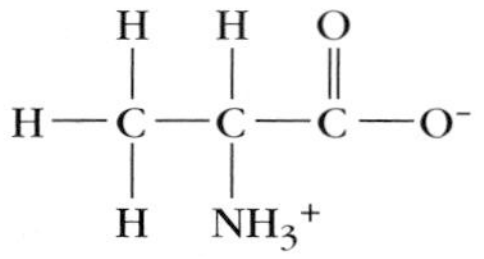 — alanine, zwitterion form

The alanine zwitterion, for example, has no net charge, although regions of positive and negative charge exist in the molecule.

Alanine and other amino acids in solution can undergo additional acid-base reactions depending on the pH of the solution. Under acidic conditions, the H^+ concentration is large enough to add an H^+ ion to the negative carboxylate group ($-COO^-$), resulting in the formation of a $-COOH$ group. Under these conditions, the amino acid has a net positive charge. In basic solution, the concentration of OH^- ions is sufficient to remove the H^+ attached to the nitrogen of the zwitterion, resulting in a net negative charge on alanine.

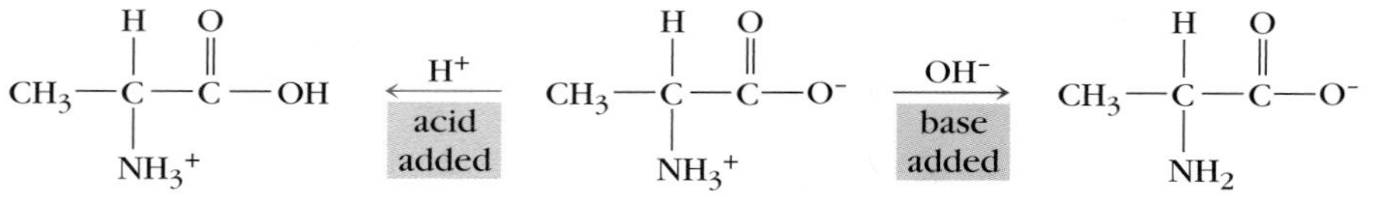

```
      H   O                       H   O                        H   O
      |   ||        H+            |   ||        OH-            |   ||
CH3—C—C—OH    <-----------  CH3—C—C—O⁻   ----------->  CH3—C—C—O⁻
      |          acid added       |          base added        |
      NH3+                        NH3+                         NH2
```

Positively charged — alanine zwitterion no net charge — Negatively charged

The relative concentrations of the three alanine forms, or the analogous forms of any amino acid, depend on the pH of the body fluids or other aqueous solutions in which they are present.

CONCEPTUAL EXERCISE 16.16 Glycine and Its Zwitterion

Glycine is the simplest amino acid.

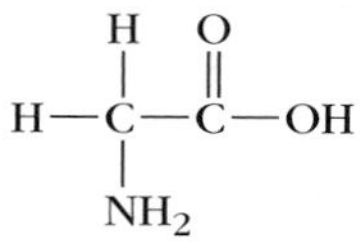

```
    H   O
    |   ||
H—C—C—OH
    |
    NH2
```

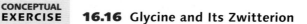

(a) Write the structural formula for its zwitterion form.

(b) Write the structural formula for glycine in solution at a pH of 2 and at a pH of 10.

CONCEPTUAL EXERCISE 16.17 H^+ Transfers

The following tripeptide, made from three amino acids, is shown in its neutral form.

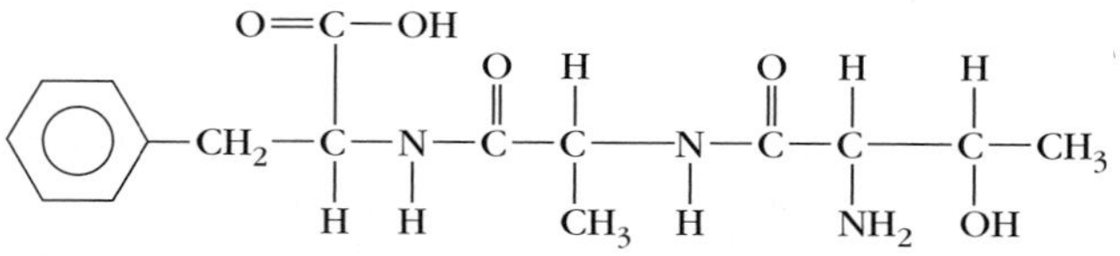

```
                 O=C—OH
                   |      O   H        O   H     H
                   |      ||  |        ||  |     |
(C6H5)—CH2—C—N—C—C——N—C—C——C—CH3
                   |  |       |   |        |     |
                   H  H       CH3 H        NH2   OH
```

Write the structural formula for the tripeptide with all the acidic or basic groups in their charged forms.

16.7 Problem Solving Using K_a and K_b

Calculations with K_a or K_b follow the same patterns as those of other equilibrium calculations illustrated earlier (**p. 683**). Similar important relationships apply to these calculations.

- *Starting with only reactants, equilibrium can be achieved only if some amount of the reactants is converted to products; that is,* ***products are formed at the expense of reactants.***
- *The chemical equilibrium equation for the ionization of the acid or the reaction of the base with water is the basis for the acid ionization or base ionization constant expression.*
- *The concentrations in the acid ionization or base ionization expression, expressed as molarity (mol/L), are those* ***at equilibrium.***
- *The magnitude of the K_a or K_b value indicates how far the forward reaction occurs at equilibrium (K_a: H^+ donation* ***to*** *water by an acid; K_b: H^+ gain by a base* ***from*** *water).*

There are several experimental methods for determining acid or base ionization constants. The simplest is based on measuring the pH of an acid solution of known concentration. If both the acid concentration and the pH are known, the K_a for the acid can be calculated, as illustrated in Problem-Solving Example 16.8.

ThomsonNOW™

Go to the Coached Problems menu for a tutorial on **determining K_a and K_b values.**

PROBLEM-SOLVING EXAMPLE 16.8 K_a from pH

What is the K_a of butanoic acid, $CH_3CH_2CH_2COOH$, a weak organic acid, if a 0.025 M butanoic acid solution has a pH of 3.21 at 25 °C?

Answer 1.6×10^{-5}

Strategy and Explanation We can determine the hydronium concentration from the pH. The hydronium concentration equals the butanoate ion ($CH_3CH_2CH_2COO^-$) concentration because the acid ionizes according to the balanced equation

$$CH_3CH_2CH_2COOH(aq) + H_2O(\ell) \rightleftharpoons H_3O^+(aq) + CH_3CH_2CH_2COO^-(aq)$$

Using the definition of pH, the equilibrium concentration of H_3O^+ is calculated to be 0.00062 M; $[H_3O^+] = 10^{-pH} = 10^{-3.21} = 0.00062$ M.

Butanoic acid dissociates to give equal concentrations of hydronium ions and butanoate ions, in this case 0.00062 M. At equilibrium the concentration of un-ionized butanoic acid equals the original concentration minus the amount that dissociated. The equilibrium concentrations of the species are represented by using a reaction (ICE) table.

	$CH_3CH_2CH_2COOH$	H_3O^+	$CH_3CH_2COO^-$
*I*nitial concentration (mol/L)	0.025	1.0×10^{-7} (from water)*	0
*C*hange as reaction occurs (mol/L)	−0.00062	+0.00062	+0.00062
*E*quilibrium concentration (mol/L)	0.025 − 0.00062	0.00062	0.00062

*This concentration can be ignored because it is so small.

From the measured pH we have calculated H_3O^+ to be 6.2×10^{-4} mol/L, which is also the $CH_3CH_2CH_2COO^-$ concentration at equilibrium because the ions are formed in equal amounts as butanoic acid ionizes. The concentration of the un-ionized acid at equilibrium is $0.025 - 0.00062 = 0.0244$ M. Using these values, we can now calculate K_a for butanoic acid.

$$K_a = \frac{[H_3O^+][CH_3CH_2CH_2COO^-]}{[CH_3CH_2CH_2COOH]}$$
$$= \frac{[0.00062][0.00062]}{0.0244} = 1.6 \times 10^{-5}$$

✓ **Reasonable Answer Check** This K_a is small, indicating that butanoic acid is a weak acid, as reflected by the fact that a 0.025 M butanoic acid solution has an $[H_3O^+]$ of just 0.00062 M. Thus, the answer makes sense; butanoic acid is only slightly ionized. Butanoic acid is similar to acetic acid in strength, as expected from its similar structure, and a 0.025 M solution has a pH identical to that of 0.025 M acetic acid (pH = 3.21).

PROBLEM-SOLVING PRACTICE 16.8

Lactic acid is a monoprotic acid that occurs naturally in sour milk and also forms by metabolism in the human body. A 0.10 M aqueous solution of lactic acid, $CH_3CH(OH)COOH$, has a pH of 2.43. What is the value of K_a for lactic acid? Is lactic acid stronger or weaker than propanoic acid?

Acid-base ionization constants such as those in Table 16.2 can be used to calculate the pH of a solution of a weak acid or a weak base from its concentration.

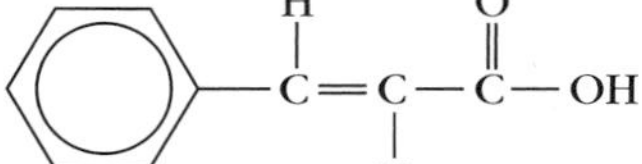

trans-cinnamic acid

© Thomson Learning/Charles D. Winters

PROBLEM-SOLVING EXAMPLE 16.9 pH from K_a

(a) Calculate the pH of a 0.020 M solution of *trans*-cinnamic acid, whose $K_a = 3.6 \times 10^{-5}$ at 25 °C. For simplicity, we will symbolize *trans*-cinnamic acid as HtCA.

(b) What percent of the acid has ionized in this solution?

Answer (a) pH = 3.07 (b) 4.3% ionized

Strategy and Explanation (a) We first must relate the information to the ionization of HtCA to release hydronium ions and *trans*-cinnamate ions, (tCA^-), into solution. Start by writing the equilibrium equation and equilibrium constant expression for *trans*-cinnamic acid.

$$HtCA(aq) + H_2O(\ell) \rightleftharpoons H_3O^+(aq) + tCA^-(aq) \qquad K_a = \frac{[H_3O^+][tCA^-]}{[HtCA]}$$

Next, define equilibrium concentrations and organize the known information in the usual ICE table. In this case, let x equal the H_3O^+ concentration at equilibrium. At equilibrium, the tCA^- ion concentration is also equal to x because the reaction produces H_3O^+ and tCA^- in equal amounts.

	HtCA	H_3O^+	tCA^-
Initial concentration (mol/L)	0.020	1.0×10^{-7} (from water)*	0
Change as reaction occurs (mol/L)	$-x$	$+x$	$+x$
Equilibrium concentration (mol/L)	$0.020 - x$	x	x

*This concentration can be ignored because it is so small.

The equilibrium constant expression can be rewritten using the values from the table where all equilibrium concentrations are defined in terms of the single unknown, x.

$$K_a = 3.6 \times 10^{-5} = \frac{[H_3O^+][tCA^-]}{[HtCA]} = \frac{[x][x]}{0.020 - x}$$

Because K_a is very small, the reaction is reactant-favored and therefore, not very much product will form. Consequently, at equilibrium the concentrations of H_3O^+ and tCA^- will be very small. Therefore, x must be quite small compared with 0.020. When x is subtracted from 0.020, the result will still be almost exactly 0.020, and so we can approximate $0.020 - x$ as 0.020 to get

$$\frac{x^2}{0.020} = 3.6 \times 10^{-5}$$

Solving for x gives

$$x = \sqrt{(3.6 \times 10^{-5})(0.020)} = \sqrt{7.2 \times 10^{-7}} = 8.5 \times 10^{-4} = [H_3O^+]$$

$$pH = -\log[H_3O^+] = -\log(8.5 \times 10^{-4}) = 3.07$$

The solution is acidic.

(b) The ionization of the acid is the major source of H_3O^+ ions (the concentration from water is insignificant). Therefore, the percent ionization is calculated by comparing the H_3O^+ concentration at equilibrium with the initial concentration of the acid.

$$\%\text{ ionization} = \frac{[H_3O^+]}{(HtCA)_{initial}} \times 100\% = \frac{8.5 \times 10^{-4}}{0.020} \times 100\% = 4.3\%$$

✓ **Reasonable Answer Check** Its K_a value of 3.6×10^{-5} indicates that *trans*-cinnamic acid is a weak acid, similar to acetic acid ($K_a = 1.8 \times 10^{-5}$) in strength and should be only slightly ionized. That 0.020 M *trans*-cinnamic acid is only 4.3% ionized and has a $[H_3O^+]$ of 8.5×10^{-4} M and a pH of 3.07 is reasonable. To check if the approximation was valid, substitute the equilibrium values into the equilibrium constant expression. The calculated result should equal the K_a.

ThomsonNOW

Go to the Coached Problems menu for tutorials on:

- **estimating the pH of weak acid solutions**
- **estimating the pH of weak base solutions**

$$H—\ddot{N}=N=\ddot{N}:$$

hydrazoic acid

PROBLEM-SOLVING PRACTICE 16.9

(a) What is the pH of a 0.015 M solution of hydrazoic acid, HN_3 ($K_a = 1.9 \times 10^{-5}$), at 25 °C?

(b) What percent of the acid has ionized in this solution?

EXERCISE 16.18 The pH of a Solution of $Ni(NO_3)_2$

When anhydrous nickel(II) nitrate dissolves in water, the Ni^{2+} ions become hydrated, forming $[Ni(H_2O)_6]^{2+}$ ions. What is the pH of a solution that is 0.15 M in nickel nitrate? K_a for $[Ni(H_2O)_6]^{2+}$ is 2.5×10^{-11}.

An analogous calculation can be done to find the pH of a solution of a weak base, such as piperidine, a component of pepper.

PROBLEM-SOLVING EXAMPLE 16.10 pH of a Weak Base from K_b

Piperidine, $C_5H_{11}N$, a nitrogen-containing base analogous to ammonia, has a $K_b = 1.3 \times 10^{-3}$. Calculate the OH^- concentration and the pH of a 0.025 M solution of piperidine.

Answer $[OH^-] = 5.1 \times 10^{-3}$ M; pH = 11.71

Strategy and Explanation We use the balanced equilibrium equation to provide the K_b expression. Using the value of K_b we can calculate $[OH^-]$ and then pOH from the OH^- concentration. Knowing the pOH, the pH can then be derived from the relation pH + pOH = 14.

Piperidine reacts with water to transfer H^+ ions from water to form hydroxide ions and piperidinium ions, $C_5H_{11}NH^+$, according to the equation

$$C_5H_{11}N(aq) + H_2O(\ell) \rightleftharpoons C_5H_{11}NH^+(aq) + OH^-(aq)$$

$$K_b = \frac{[C_5H_{11}NH^+][OH^-]}{[C_5H_{11}N]} = 1.3 \times 10^{-3}$$

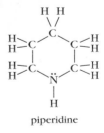

piperidine

Piperidinium ion, $C_5H_{11}NH^+$, is analogous to ammonium ion, NH_4^+.

A table can be set up like the one in Problem-Solving Example 16.9, letting x be the concentration of OH^- and of piperidinium ions at equilibrium because the forward reaction produces them in equal amounts. The equilibrium concentration of *unreacted* piperidine will be its initial concentration, 0.025 mol/L, minus x, the amount per liter that has reacted with water.

	$C_5H_{11}N$	$C_5H_{11}NH^+$	OH^-
Initial concentration (mol/L)	0.025	0	1.0×10^{-7}*
Change as reaction occurs (mol/L)	$-x$	$+x$	$+x$
Equilibrium concentration (mol/L)	$(0.025 - x)$	x	x

* The low concentration can be ignored, as it was in the K_a calculations.

Substitution into the base ionization constant expression gives

$$K_b = \frac{[C_5H_{11}NH^+][OH^-]}{[C_5H_{11}N]} = \frac{x^2}{0.025 - x} = 1.3 \times 10^{-3}$$

Generally in K_a and K_b calculations, if $\frac{x}{\text{initial concentration}} \times 100\% > 5\%$, the x term cannot be dropped from the denominator in the equilibrium constant expression and the quadratic equation is used. As seen below, in this case x is not negligible compared with 0.025 because piperidine reacts with water sufficiently and so x, the OH^- concentration, must be found by using the quadratic formula.

Multiplying out the terms gives Equation A.

$$x^2 = (0.025 - x)(1.3 \times 10^{-3}) = 3.3 \times 10^{-5} - (1.3 \times 10^{-3}x) \quad \text{(A)}$$

Rearranging Equation A into the quadratic form $ax^2 + bx + c = 0$ then gives Equation B.

Appendix A.6 reviews the use of the quadratic equation.

$$x^2 + (1.3 \times 10^{-3}x) - (3.3 \times 10^{-5}) = 0 \quad \text{(B)}$$

Solving for x using the quadratic formula gives

$$x = \frac{-(1.3 \times 10^{-3}) \pm \sqrt{(1.3 \times 10^{-3})^2 - (4 \times 1)(-3.3 \times 10^{-5})}}{2(1)}$$

$$= \frac{-(1.3 \times 10^{-3}) \pm \sqrt{1.34 \times 10^{-4}}}{2}$$

$$= \frac{-(1.3 \times 10^{-3}) \pm (1.16 \times 10^{-2})}{2}$$

$$= \frac{1.03 \times 10^{-2}}{2} = 5.1 \times 10^{-3}$$

Therefore x, the OH^- concentration, equals 5.1×10^{-3}. (The negative root in the solution of the quadratic equation is disregarded because concentration cannot be negative; you can't have less than 0 mol/L of a substance.)

Note that $\frac{5.1 \times 10^{-3}}{0.025} \times 100\% = 20\%$, which is greater than 5%, so the quadratic equation was necessary in this case; the approximation of

$$\frac{x^2}{0.025 - x} \approx \frac{x^2}{0.025}$$

would not have given the correct answer.

The pOH can be calculated from the OH^- concentration.

$$\text{pOH} = -\log(5.1 \times 10^{-3}) = 2.29$$

$$\text{pH} = 14.00 - \text{pOH} = 14.00 - 2.29 = 11.71$$

Therefore, piperidine reacts sufficiently with water to generate a fairly basic solution.

✓ **Reasonable Answer Check** Both the initial concentration and the K_b of piperidine are small. Consequently, the pH should be less than that of a 0.025 M solution of a strong base like NaOH, which would be 12.40.

$$[OH^-] = 0.025 \text{ M; pOH} = -\log(0.025) = 1.60\text{; pH} = 12.40$$

A pH of 11.71 for 0.025 M piperidine is reasonable.

PROBLEM-SOLVING PRACTICE 16.10

Calculate the OH^- concentration and the pH of a 0.015 M solution of cyclohexylamine, $C_6H_{11}NH_2$. $K_b = 4.6 \times 10^{-4}$.

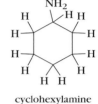

cyclohexylamine

Relationship between K_a and K_b Values

The right-hand side of Table 16.2 (p. 788) gives K_b values for the conjugate base of each acid. Try an experiment with these data: Multiply a few of the K_a values by K_b values for their conjugate bases. What do you find? Within a very small error you ought to find that $K_a \times K_b = 1.0 \times 10^{-14}$. This value is the same as K_w, the autoionization constant for water. To see why, multiply the equilibrium constant expressions for K_a and K_b.

$$K_a \times K_b = \left(\frac{[H_3O^+][A^-]}{[HA]}\right)\left(\frac{[HA][OH^-]}{[A^-]}\right)$$

Canceling like terms in the numerator and denominator of this expression gives

$$K_a \times K_b = \left(\frac{[H_3O^+]\cancel{[A^-]}}{\cancel{[HA]}}\right)\left(\frac{\cancel{[HA]}[OH^-]}{\cancel{[A^-]}}\right) = [H_3O^+][OH^-] = K_w$$

This relation shows that if you know K_a for an acid, you can find K_b for its conjugate base by using K_w. Furthermore, the larger the K_a, the smaller the K_b, and vice versa (because when multiplied they always have to give the same product, K_w). For example, K_a for HCN is 4.0×10^{-10}. The value of K_b for the conjugate base, CN^-, is

$$K_b\,(\text{for } CN^-) = \frac{K_w}{K_a\,(\text{for HCN})} = \frac{1.0 \times 10^{-14}}{4.0 \times 10^{-10}} = 2.5 \times 10^{-5}$$

HCN has a relatively small K_a and lies fairly far down in Table 16.2, which means it is a relatively weak weak acid. However, CN^- is a fairly strong weak base; its K_b of 2.5×10^{-5} is nearly the same as the K_b for ammonia (1.8×10^{-5}), making CN^- a slightly stronger base than ammonia. In general, if $K_a > K_b$, the acid is stronger than its conjugate base. Alternatively, if $K_b > K_a$, the conjugate base is stronger than its conjugate acid. For example, hypochlorite ion, OCl^- ($K_b = 2.9 \times 10^{-7}$) is a stronger base than hypochlorous acid, HOCl, is an acid ($K_a = 3.5 \times 10^{-8}$).

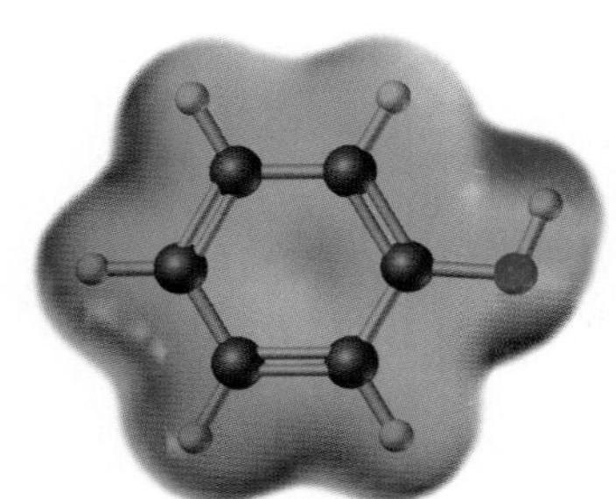

phenol

EXERCISE 16.19 K_b from K_a

Phenol, or carbolic acid, C_6H_5OH, is a weak acid, $K_a = 1.3 \times 10^{-10}$. Calculate K_b for the phenolate ion, $C_6H_5O^-$. Which base in Table 16.2 is closest in strength to the phenolate ion? How did you make your choice?

16.8 Acid-Base Reactions of Salts

An exchange reaction between an acid and a base produces a salt plus water ***(⇐ p. 174)***. The salt's positive ion comes from the base and its negative ion comes from the acid. In the case of a metal hydroxide as a base, the salt-forming general reaction is

$$\underset{\text{acid}}{HX(aq)} + \underset{\text{base}}{MOH(aq)} \longrightarrow \underset{\text{salt}}{MX(aq)} + HOH(\ell)$$

Now that you know more about the Brønsted-Lowry acid-base concept and the strengths of acids and bases, it is useful to consider acid-base reactions and salt formation in more detail.

Salts of Strong Bases and Strong Acids

Strong acids react with strong bases to form *neutral* salts. Consider the reaction of the strong acid HCl with the strong base NaOH to form the salt NaCl. If the amounts of HCl and NaOH are in the correct stoichiometric ratio (1 mol HCl per 1 mol NaOH), this reaction occurs with the complete neutralization of the acidic properties of HCl and the basic properties of NaOH. The reaction can be described first by an overall equation, then by a complete ionic equation, and finally by a net ionic equation ***(⇐ p. 168)***. Each of these equations contains useful information.

Overall equation $$HCl(aq) + NaOH(aq) \longrightarrow NaCl(aq) + H_2O(\ell)$$

Complete ionic equation $$H_3O^+(aq) + Cl^-(aq) + Na^+(aq) + OH^-(aq) \longrightarrow Na^+(aq) + Cl^-(aq) + 2\,H_2O(\ell)$$

Net ionic equation $$\underset{\text{ACID}}{H_3O^+(aq)} + \underset{\text{base}}{OH^-(aq)} \longrightarrow \underset{\text{BASE}}{H_2O(\ell)} + \underset{\text{acid}}{H_2O(\ell)}$$

The overall equation shows the substances that were dissolved or that could be recovered at the end of the reaction. The complete ionic equation indicates all of the ions that are present before and after reaction. The net ionic equation emphasizes that a Brønsted-Lowry acid (H_3O^+) is reacting with a Brønsted-Lowry base (OH^-); the spectator ions, Na^+ and Cl^-, are omitted. This reaction goes to completion because H_3O^+ is a strong acid, OH^- is a strong base, and water is a very weak acid and a very weak base.

The resulting solution contains only sodium ions and chloride ions, with a few more water molecules than before. Its properties are the same as if it had been prepared by simply dissolving some NaCl(s) in water. It has a neutral pH because it contains no significant concentrations of acids or bases. The Cl^- ion is the conjugate base of a strong acid and hence is such a weak base that it does not react with water. The Na^+ ion also does not react as either an acid or a base with water. Examples of some other salts of this type are given in Table 16.4. These salts all form neutral solutions.

Table 16.4 Some Salts Formed by Neutralization of Strong Acids with Strong Bases

	Base		
Acid	**NaOH**	**KOH**	**$Ba(OH)_2$**
	Salts	**Salts**	**Salts**
HCl	NaCl	KCl	$BaCl_2$
HNO_3	$NaNO_3$	KNO_3	$Ba(NO_3)_2$
H_2SO_4	Na_2SO_4	K_2SO_4	$BaSO_4$
$HClO_4$	$NaClO_4$	$KClO_4$	$Ba(ClO_4)_2$

Salts of Strong Bases and Weak Acids

Strong bases react with weak acids to form *basic* salts. Suppose, for example, that 0.010 mol NaOH is added to 0.010 mol of the weak acid acetic acid in 1 L of solution. The three equations are

$$CH_3COOH(aq) + NaOH(aq) \longrightarrow NaCH_3COO(aq) + H_2O(\ell) \quad \text{Overall equation}$$

$$CH_3COOH(aq) + Na^+(aq) + OH^-(aq) \longrightarrow Na^+(aq) + CH_3COO^-(aq) + H_2O(\ell) \quad \text{Complete ionic equation}$$

$$\underset{\text{weak acid}}{CH_3COOH(aq)} + \underset{\text{strong base}}{OH^-(aq)} \longrightarrow \underset{\text{base}}{CH_3COO^-(aq)} + \underset{\text{acid}}{H_2O(\ell)} \quad \text{Net ionic equation}$$

In this case, acetate ion, a weaker base than OH^-, has been formed by the reaction. Therefore, the solution is slightly basic (pH > 7), even though exactly the stoichiometric amount of acetic acid was added to the sodium hydroxide. The reaction that makes the solution basic is the reaction of water with acetate ion, a weak Brønsted-Lowry base.

$$CH_3COO^-(aq) + H_2O(\ell) \rightleftharpoons CH_3COOH(aq) + OH^-(aq)$$

This is a **hydrolysis** reaction, one in which a water molecule is split—in this case, into an H^+ ion and an OH^- ion. An H^+ ion is donated to the acetate ion to form acetic acid. The extent of hydrolysis is determined by the value of K_b for acetate ion.

All of the weak bases in Table 16.2, except for the very weak bases (NO_3^-, Cl^-, HSO_4^-, and ClO_4^-) above water in the next to last column, undergo hydrolysis reactions in aqueous solution. The larger the K_b value of a base, the more basic the solutions it produces. The pH of a solution of a salt of a strong base and a weak acid can be calculated from K_b, as shown in Problem-Solving Example 16.11.

The term "hydrolysis" is derived from *hydro*, meaning "water," and *lysis*, meaning "to break apart." The hydrolysis reaction of a molecular compound results in the addition of H— and —OH to the molecules produced by breaking a covalent bond.

As K_b of the base increases, the pH of the solution increases.

ThomsonNOW™

Go to the Coached Problems menu for a module on **hydrolysis.**

PROBLEM-SOLVING EXAMPLE 16.11 pH of a Salt Solution

Sodium benzoate, $NaC_7H_5O_2$, is used as a preservative in foods. Calculate the pH of a 0.025 M solution of $NaC_7H_5O_2$ ($K_b = 1.6 \times 10^{-10}$).

Answer 8.30

Strategy and Explanation To calculate the pH we must first calculate the hydroxide ion concentration in the solution. Sodium benzoate is a basic salt that could be synthesized from a strong base (NaOH) and a weak acid, benzoic acid. The Na^+ ion does not react with water, but benzoate ion, $C_7H_5O_2^-$, the conjugate base of a weak acid ($HC_7H_5O_2$), reacts with water to produce a basic solution.

$$C_7H_5O_2^-(aq) + H_2O(\ell) \rightleftharpoons HC_7H_5O_2(aq) + OH^-(aq)$$

$$K_b = 1.6 \times 10^{-10} = \frac{[\text{conjugate acid}][OH^-]}{[\text{conjugate base}]} = \frac{[HC_7H_5O_2][OH^-]}{[C_7H_5O_2^-]}$$

The concentrations of benzoate ion, benzoic acid, and hydroxide ion initially and at equilibrium are summarized in the following table. We let x be equal to the equilibrium concentration of OH^- as well as that of $C_7H_5O_2^-$, because they are formed in equal amounts.

	$C_7H_5O_2^-$	$C_7H_6O_2$	OH^-
Initial concentration (mol/L)	0.025	0	1.0×10^{-7} (from water)
Change as reaction occurs (mol/L)	$-x$	$+x$	$+x$
Equilibrium concentration (mol/L)	$0.025 - x$	x	x

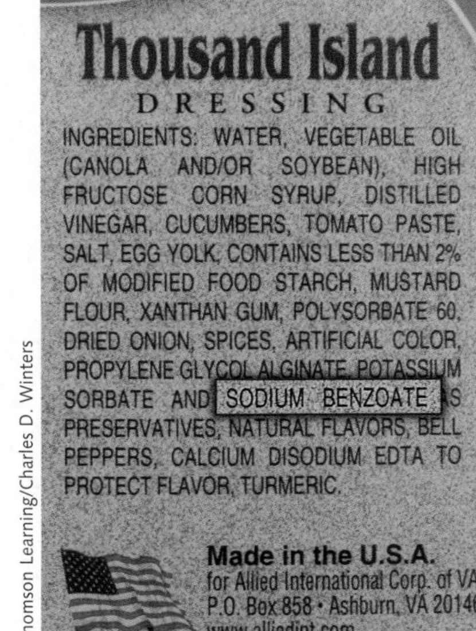

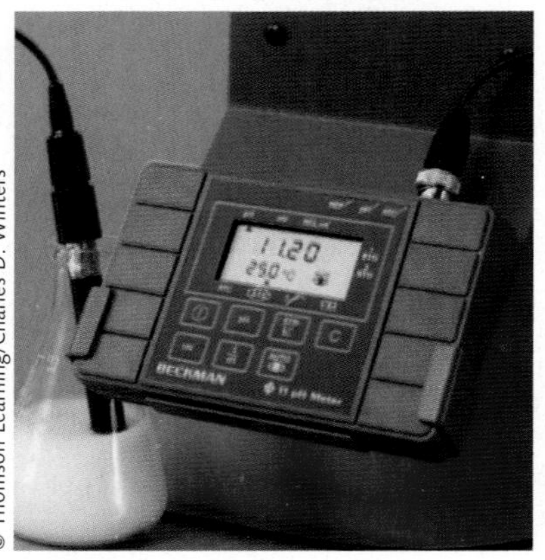

Hydrolysis of salts of strong bases and weak acids in aqueous solution. The pH meter readings indicate that aqueous solutions of sodium acetate ($NaCH_3COO$, *top*) and sodium carbonate (Na_2CO_3, *bottom*) are basic. If their concentrations are equal, the Na_2CO_3 solution is more basic (pH = 11.20) because carbonate ion, CO_3^{2-}, is a stronger base than is acetate ion, CH_3COO^- (pH = 8.88); that is, K_b of $CO_3^{2-} > K_b$ of CH_3COO^-. An inert, insoluble solid has been added to each flask to enhance the visibility of the liquid in the flask.

Benzoate ion has a very small K_b (1.6×10^{-10}) and thus is a very weak base, so it is safe to assume that x will be negligibly small compared with 0.025, and $0.025 - x \approx 0.025$.

$$K_b = 1.6 \times 10^{-10} \approx \frac{x^2}{0.025}$$

Solving for x gives $x = 2.0 \times 10^{-6}$, which is equal to the hydroxide and benzoate ion concentrations. Because $0.025 - 2.0 \times 10^{-6} = 0.025$ (using the significant figures rules), our assumption that x is negligible compared with 0.025 is justified. Therefore, at equilibrium

$$OH^- = [HC_7H_5O_2] = 2.0 \times 10^{-6}\ \text{mol/L};\ [C_7H_5O_2^-] = 0.025\ \text{mol/L}$$

Finally, the pH of the solution can be calculated.

$$K_w = [H_3O^+][OH^-] = [H_3O^+](2.0 \times 10^{-6}) = 1.0 \times 10^{-14}$$

$$[H_3O^+] = \frac{1.0 \times 10^{-14}}{2.0 \times 10^{-6}} = 5.0 \times 10^{-9}$$

$$pH = -\log(5.0 \times 10^{-9}) = 8.30$$

As expected, the solution is basic.

✓ **Reasonable Answer Check** The reaction of benzoate ion with water produces hydroxide ions in addition to those from the dissociation of water. The excess hydroxide ions cause the solution to become basic, as indicated by the pH greater than 7. This is expected because the salt is formed from a strong base and a weak acid.

PROBLEM-SOLVING PRACTICE 16.11

Sodium carbonate is an environmentally benign paint stripper. It is water-soluble, and carbonate ion is a strong enough base to loosen paint so it can be scraped off. What is the pH of a 1.0 M solution of Na_2CO_3?

CONCEPTUAL EXERCISE 16.20 pH of Soap Solutions

Ordinary soaps are often sodium salts of fatty acids, which are weak organic acids. Would you expect the pH of a soap solution to be greater than or less than 7? Explain your answer.

Salts of Weak Bases and Strong Acids

When a weak base reacts with a strong acid, the resulting salt solution is acidic. The conjugate acid of the weak base determines the pH of the solution. For example, suppose equal volumes of 0.10 M NH_3 and 0.10 M HCl are mixed. The reaction, shown in overall, complete ionic, and net ionic equations, is

$$NH_3(aq) + HCl(aq) \longrightarrow NH_4Cl(aq)$$

$$NH_3(aq) + H_3O^+(aq) + Cl^-(aq) \longrightarrow NH_4^+(aq) + Cl^-(aq) + H_2O(\ell)$$

$$\underset{\text{weak base}}{NH_3(aq)} + \underset{\text{strong acid}}{H_3O^+(aq)} \longrightarrow \underset{\text{acid}}{NH_4^+} + \underset{\text{base}}{H_2O(\ell)}$$

As soon as it is formed, the weak acid NH_4^+ reacts with water and establishes an equilibrium. The resulting solution is slightly acidic, because the reaction produces hydronium ions.

$$NH_4^+(aq) + H_2O(\ell) \rightleftharpoons NH_3(aq) + H_3O^+(aq)$$

PROBLEM-SOLVING EXAMPLE 16.12 pH of Another Salt Solution

Ammonium nitrate, NH_4NO_3, is a salt used in fertilizers and in making matches. The salt is made by the reaction of ammonia, NH_3, and nitric acid, HNO_3. Calculate the pH of a 0.15 M solution of ammonium nitrate; $K_a\,(NH_4^+) = 5.6 \times 10^{-10}$.

Answer 5.04

Strategy and Explanation You should first recognize that ammonium nitrate is the salt of a weak base and a strong acid so the solution will be acidic, with a pH less than 7.0. Because ammonium ion is a weak acid, ammonium ions react with water to form ammonia and hydronium ions, making the solution acidic.

$$NH_4^+(aq) + H_2O(\ell) \rightleftharpoons NH_3(aq) + H_3O^+(aq)$$

The following table summarizes the concentrations of ammonium ion, ammonia, and hydronium ions initially and at equilibrium. We let x be equal to the equilibrium concentration of H_3O^+ as well as that of NH_3 because they are formed in equal amounts.

	NH_4^+	NH_3	H_3O^+
Initial concentration (mol/L)	0.15	0	1.0×10^{-7} (from water)
Change as reaction occurs (mol/L)	$-x$	$+x$	$+x$
Equilibrium concentration (mol/L)	$0.15 - x$	x	x

As in prior problems, we substitute the equilibrium values into the K_a expression.

$$K_a = \frac{[NH_3][H_3O^+]}{[NH_4^+]} = \frac{(x)(x)}{(0.15 - x)} = 5.6 \times 10^{-10}$$

Because ammonium ion is such a weak acid, as indicated by its very low K_a, we can simplify the equation

$$K_a = \frac{(x)(x)}{(0.15)} \approx 5.6 \times 10^{-10}$$

and solve for x, the H_3O^+ concentration.

$$x^2 = (0.15)(5.6 \times 10^{-10})$$
$$x = 9.2 \times 10^{-6} = [H_3O^+]$$

The pH can be calculated from the hydronium ion concentration.

$$pH = -\log[H_3O^+] = -\log(9.2 \times 10^{-6}) = 5.04$$

✓ **Reasonable Answer Check** As predicted, the pH of the solution is less than 7.0, which it should be because of the release of hydronium ions by the forward reaction.

PROBLEM-SOLVING PRACTICE 16.12

Calculate the pH of a 0.10 M solution of ammonium chloride, NH_4Cl.

Strong base + strong acid yields a neutral salt (solution pH = 7.0)

Strong base + weak acid yields a basic salt (solution pH > 7.0)

Strong acid + weak base yields an acidic salt (solution pH < 7.0)

Many drugs are high-molecular-weight amines that are weak bases. Such amines are not soluble in water, which limits the ways they can be administered and means that they are not soluble in body fluids such as blood plasma and cerebrospinal fluid. By reaction with hydrochloric acid, the amines are converted to soluble hydrochloride salts that can be administered by injection or dissolved in liquid oral

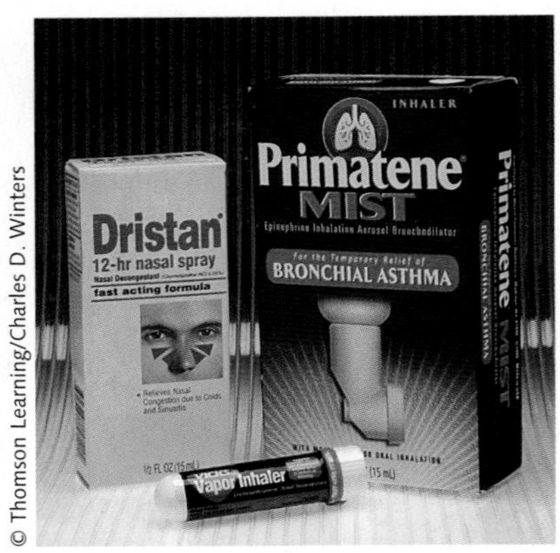

Various pharmaceutical amines.

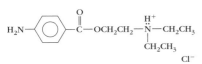

$H_2N-C_6H_4-C(=O)-OCH_2CH_2-\ddot{N}H^{+}(CH_2CH_3)-CH_2CH_3$ Cl^-

Novocain hydrochloride

medications. The resulting hydrochloride salts have the general formula BH^+Cl^-, where B represents the basic amine. This formula is like that of ammonium chloride, $NH_4^+Cl^-$. Two examples are

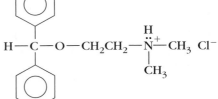

$HO-C_6H_4-CH(OH)-CH_2-\ddot{N}H_2^{+}-CH_3$ Cl^-

phenylephrine hydrochloride (Neo-synephrine), a decongestant

$(C_6H_5)_2CH-O-CH_2CH_2-\ddot{N}H^{+}(CH_3)-CH_3$ Cl^-

diphenhydramine hydrochloride (Benadryl), an antihistamine

The amine hydrochloride salt of a drug is much more water-soluble than the amine form of the drug. For example, only 0.5 g Novocain dissolves in 100 g water, whereas 100 g Novocain hydrochloride dissolves in 100 g water.

EXERCISE 16.21 Forming a Drug Hydrochloride

To stymie illicit methamphetamine synthesis, phenylephrine is substituted for pseudoephedrine, an active ingredient in over-the-counter decongestants. Use structural formulas to write the equation for the formation of phenylephrine hydrochloride from phenylephrine.

$HO-C_6H_4-CH(OH)-CH_2-\ddot{N}H-CH_3$

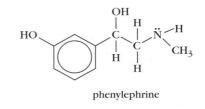

phenylephrine

Salts of Weak Bases and Weak Acids

What is the pH of a solution of a salt (such as NH_4F or $Ni(CH_3COO)_2$) containing an acidic cation and a basic anion? The salt could be formed by the reaction of a weak acid and a weak base. There are two possible reactions that can determine the pH of the NH_4F solution: formation of H_3O^+ by H^+ transfer from the cation; and formation of OH^- by hydrolysis of the anion. In the case of NH_4F,

$$NH_4^+(aq) + H_2O(\ell) \rightleftharpoons H_3O^+(aq) + NH_3(aq) \qquad K_a(NH_4^+) = 5.6 \times 10^{-10}$$

$$F^-(aq) + H_2O(\ell) \rightleftharpoons HF(aq) + OH^-(aq) \qquad K_b(F^-) = 1.4 \times 10^{-11}$$

Since $K_a(NH_4^+) > K_b(F^-)$, the reaction of ammonium ions with water to produce hydronium ions is the more favorable reaction. Therefore, the resulting solution is slightly acidic. For $Ni(CH_3COO)_2$ the possible reactions are

$$Ni(H_2O)_6^{2+}(aq) + H_2O(\ell) \rightleftharpoons Ni(H_2O)_5(OH)^+(aq) + H_3O^+(aq)$$

$$K_a(Ni(H_2O)_6^{2+}) = 2.5 \times 10^{-11}$$

$$CH_3COO^-(aq) + H_2O(\ell) \rightleftharpoons CH_3COOH(aq) + OH^-(aq)$$

$$K_b(CH_3COO^-) = 5.6 \times 10^{-10}$$

Since $K_b(CH_3COO^-) > K_a(Ni(H_2O)_6^{2+})$, the hydrolysis of CH_3COO^- ions is more favorable, and the resulting solution is slightly basic.

In general, the K_a of the weak acid and the K_b of the weak base need to be considered to determine whether the aqueous solution of a salt of a weak acid and weak base will be acidic or basic.

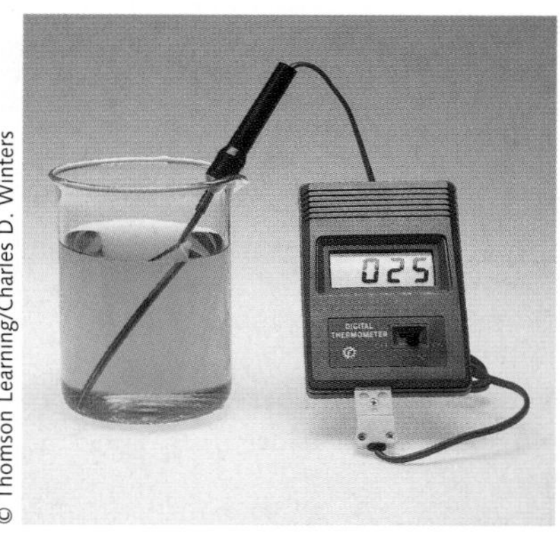

Acidic pH of an aqueous copper(II) sulfate solution. The blue solution of this copper salt is acidic (pH = 2.5) due to hydrolysis of the $Cu(H_2O)_6^{2+}$ ion.

CONCEPTUAL EXERCISE 16.22 Hydrolysis of a Salt of a Weak Acid and a Weak Base

Name a salt of a weak acid and a weak base where $K_a = K_b$. What should the pH of a solution of this salt be?

The following generalizations can be made about acid-base neutralization reactions in aqueous solution and the pH of the resulting salt solutions.

- Solution of strong acid + solution of strong base ⟶ salt solution with pH = 7 (neutral)
- Solution of strong acid + solution of weak base ⟶ salt solution with pH < 7 (acidic)
- Solution of weak acid + solution of strong base ⟶ salt solution with pH > 7 (basic)
- Solution of weak acid + solution of weak base ⟶ salt solution with pH determined by relative strengths of conjugate base and conjugate acid formed

Table 16.5 summarizes the acid-base behavior of many different ions in aqueous solution.

16.9 Lewis Acids and Bases

In 1923 when Brønsted and Lowry independently proposed their acid-base concept, Gilbert N. Lewis also was developing a new concept of acids and bases. By the early 1930s Lewis had proposed definitions of acids and bases that are more general than those of Brønsted and Lowry because they are based on sharing of electron pairs rather than on H^+ ion transfers. A **Lewis acid** is *a substance that can accept a pair of electrons to form a new bond*, and a **Lewis base** is *a substance that can donate a pair of electrons to form a new bond*. Those definitions mean that in the Lewis sense, an acid-base reaction occurs when a molecule (or ion) that has a lone pair of electrons that can be donated (a Lewis base) reacts with a molecule (or ion)

ThomsonNOW™
Go to the Chemistry Interactive menu for modules on:
- **Lewis acid-base reactions**
- **carbon dioxide as a Lewis acid**

Table 16.5 Acid-Base Properties of Typical Ions in Aqueous Solution

	Neutral		Basic			Acidic
Anions	Cl^-	NO_3^-	CH_3COO^-	CN^-	SO_4^{2-}	HSO_4^-
	Br^-	ClO_4^-	$HCOO^-$	PO_4^{3-}	HPO_4^{2-}	$H_2PO_4^-$
	I^-		CO_3^{2-}	HCO_3^-	SO_3^{2-}	HSO_3^-
			S^{2-}	HS^-	ClO^-	
			F^-	NO_2^-		
Cations	Li^+	Mg^{2+}		*None*		Al^{3+}
	Na^+	Ca^{2+}				NH_4^+
	K^+	Ba^{2+}				Transition metal ions

that can accept an electron pair (a Lewis acid). In general, Lewis acids are cations or neutral molecules with an available, empty orbital; Lewis bases are anions or neutral molecules with a lone pair of electrons. When both electrons in an electron-pair bond were originally associated with one of the bonded atoms (the Lewis base), the bond is called a **coordinate covalent bond.**

ThomsonNOW
Go to the Coached Problems menu for tutorials on:

- **Lewis acids and bases**
- **neutral Lewis acids**

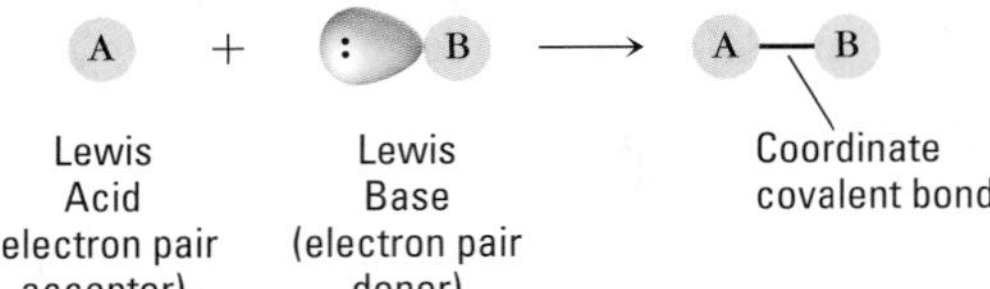

Lewis Acid (electron pair acceptor) — Lewis Base (electron pair donor) — Coordinate covalent bond

A simple example of a Lewis acid-base reaction is the formation of a hydronium ion from H^+ and water. The H^+ ion has no electrons, while the water molecule has two lone pairs of electrons on the oxygen atom. One of the lone pairs can be shared between H^+ and oxygen, thus forming an O—H bond.

A Brønsted-Lowry base (H^+ ion acceptor) must also be a Lewis base by donating an electron pair to bond with the H^+.

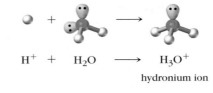

$$H^+ + H_2O \longrightarrow H_3O^+$$

hydronium ion

ThomsonNOW
Go to the Coached Problems menu for a tutorial on **cationic Lewis acids.**

Positive Metal Ions as Lewis Acids

All metal cations are potential Lewis acids. Not only do the positively charged metal cations attract electrons, but all such cations have at least one empty orbital. This empty orbital can accommodate an electron pair donated by a base, thereby forming a coordinate covalent bond. Consequently, metal ions readily form coordination complexes (Section 22.6) and also are hydrated in aqueous solution (Section 16.5). When a metal ion becomes hydrated, one of the lone pairs on the oxygen atom in each of several water molecules forms a coordinate covalent bond to the metal ion; the metal ion acts as a Lewis acid, and water acts as a Lewis base. The combination of a metal ion and a Lewis base forms a **complex ion.** For example, $Ag(NH_3)_2^+$ is a complex ion in which a silver ion (Lewis acid) is bonded to two ammonia molecules (Lewis base) through donation of an electron pair from each ammonia.

Complex ions are discussed in detail in Section 22.6.

The hydroxide ion (OH^-) is an excellent Lewis base and so it binds readily to metal cations to give metal hydroxides. An important feature of the chemistry of many metal hydroxides is that they are **amphoteric,** meaning that they can react as both a base and an acid (see Table 16.6). The amphoteric aluminum hydroxide, for example, behaves as a Lewis acid when it dissolves in a basic solution by forming a complex ion containing one additional OH^- ion, a Lewis base.

$$Al(OH)_3(s) + OH^-(aq) \rightleftharpoons [Al(OH)_4]^-(aq)$$

Table 16.6 Some Common Amphoteric Metal Hydroxides

Hydroxide	Reaction as a Base	Reaction as an Acid
$Al(OH)_3$	$Al(OH)_3(s) + 3\ H_3O^+(aq) \longrightarrow Al^{3+}(aq) + 6\ H_2O(\ell)$	$Al(OH)_3(s) + OH^-(aq) \longrightarrow [Al(OH)_4]^-(aq)$
$Zn(OH)_2$	$Zn(OH)_2(s) + 2\ H_3O^+(aq) \longrightarrow Zn^{2+}(aq) + 4\ H_2O(\ell)$	$Zn(OH)_2(s) + 2\ OH^-(aq) \longrightarrow [Zn(OH)_4]^{2-}(aq)$
$Sn(OH)_4$	$Sn(OH)_4(s) + 4\ H_3O^+(aq) \longrightarrow Sn^{4+}(aq) + 8\ H_2O(\ell)$	$Sn(OH)_4(s) + 2\ OH^-(aq) \longrightarrow [Sn(OH)_6]^{2-}(aq)$
$Cr(OH)_3$	$Cr(OH)_3(s) + 3\ H_3O^+(aq) \longrightarrow Cr^{3+}(aq) + 6\ H_2O(\ell)$	$Cr(OH)_3(s) + OH^-(aq) \longrightarrow [Cr(OH)_4]^-(aq)$

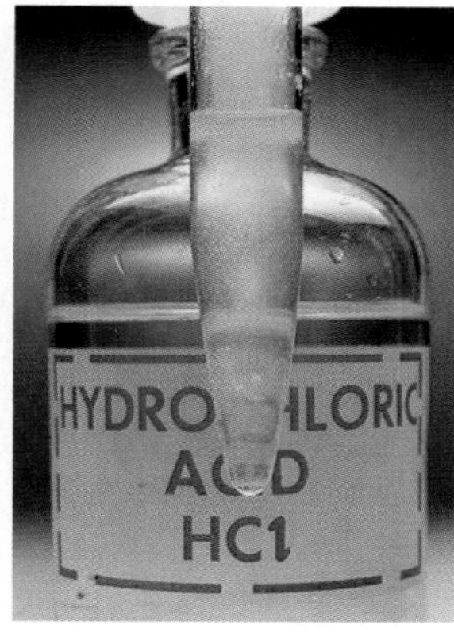

Photos: © Thomson Learning/Charles D. Winters

(a) (b) (c)

Figure 16.5 The amphoteric nature of $Al(OH)_3$. (a) Adding aqueous ammonia to a solution of Al^{3+} (left test tube) causes formation of a precipitate of $Al(OH)_3$ (right test tube). (b) Adding a strong base (NaOH) to the $Al(OH)_3$ dissolves the precipitate. Here the aluminum hydroxide acts as a Lewis acid toward the Lewis base OH^- and forms a soluble salt of the complex ion $[Al(OH)_4]^-$. (c) If we begin again with freshly precipitated $Al(OH)_3$, it dissolves as strong acid (HCl) is added. In this case $Al(OH)_3$ acts as a Brønsted-Lowry base and forms a soluble aluminum salt and water.

This reaction is shown in Figure 16.5 (b). The same compound behaves as a Brønsted-Lowry base when it reacts with a Brønsted-Lowry acid, as seen in Figure 16.5 (c).

$$Al(OH)_3(s) + 3\,H_3O^+(aq) \rightleftharpoons Al^{3+}(aq) + 6\,H_2O(\ell)$$

Metal ions also form many complex ions with the Lewis base ammonia, $:NH_3$. For example, as noted above, silver ion readily forms a water-soluble, colorless complex ion in liquid ammonia or in aqueous ammonia.

$$Ag^+(aq) + 2\,:NH_3(aq) \longrightarrow [H_3N:Ag:NH_3]^+(aq)$$

Indeed, this complex is so stable that the very water-insoluble compound AgCl can be dissolved in aqueous ammonia.

$$AgCl(s) + 2\,:NH_3(aq) \longrightarrow [H_3N:Ag:NH_3]^+(aq) + Cl^-(aq)$$

Neutral Molecules as Lewis Acids

Lewis's ideas about acids and bases account nicely for the fact that oxides of nonmetals behave as acids. Two important examples are carbon dioxide and sulfur dioxide, whose Lewis structures are

$$:\ddot{O}=C=\ddot{O}: \qquad :\ddot{O}=\ddot{S}-\ddot{\underset{..}{O}}: \longleftrightarrow :\underset{..}{\ddot{O}}-\ddot{S}=\ddot{O}:$$

carbon dioxide sulfur dioxide

In each case, there is a double bond; an "extra" pair of electrons is being shared between an oxygen atom and the central atom. Because oxygen is highly electronegative, electrons in these bonds are attracted away from the central atom, which becomes slightly positively charged. This makes the central atom a likely site to attract a pair of electrons. A Lewis base such as OH^- can bond to the carbon atom in CO_2 to give bicarbonate ion, HCO_3^-. This bonding displaces one double-bond pair of electrons back onto an oxygen atom.

© Thomson Learning/Charles D. Winters

Carbon dioxide in air reacts with spilled base such as NaOH to form Na_2CO_3. If the mouth of a glass-stoppered bottle such as the one shown here is not routinely cleaned, the sodium carbonate formed can virtually cement the top of the bottle to the neck, making it difficult to open the bottle.

$$:\ddot{O}=C=\ddot{O}: + :\ddot{\underset{\cdot\cdot}{O}}-H^- \longrightarrow :\ddot{O}=C(-\ddot{O}-H)(-\ddot{\underset{\cdot\cdot}{O}}:^-)$$

bicarbonate ion

Carbon dioxide from the air can react to form sodium carbonate around the mouth of a bottle of sodium hydroxide. Sulfur dioxide can react similarly with hydroxide ion.

The pH of some household substances. The liquids in the flasks are club soda, vinegar, and a household cleaner. The colors of the acid-base indicators in the flasks show that vinegar is more acidic than club soda, and the cleaner is much more basic than the other two liquids.

EXERCISE 16.23 Lewis Acids and Bases

Predict whether each of these is a Lewis acid or a Lewis base. (*Hint:* Drawing a Lewis structure for a molecule or ion is often helpful in making such a prediction.)

(a) PH_3 (b) BCl_3 (c) H_2S (d) NO_2 (e) Ni^{2+} (f) CO

16.10 Practical Acid-Base Chemistry

In addition to their uses in industry, various acids and bases find many applications around the home. Antacids are used to neutralize excess stomach acidity. Gardeners use acid salts such as sodium hydrogen sulfate ($NaHSO_4$) to help acidify soil and bases such as lime (CaO) to make soil more basic. In the kitchen, baking soda and baking powders are used to make biscuit dough and cake batter rise. Mild acids and bases are used to clean everything from dishes and clothes to vehicles and the family dog.

Neutralizing Stomach Acidity

Chloride ions secreted by the stomach lining come mostly from the salty foods we eat and the salt we add to our foods.

Human stomach fluids have a pH of approximately 1. This very acidic pH is caused by HCl, which is secreted by thousands of cells in the stomach lining that specialize in transporting $H_3O^+(aq)$ and $Cl^-(aq)$ from the blood. The main purpose of this acid is to suppress the growth of bacteria and to aid in the digestion of certain foods. The hydrochloric acid does not harm the stomach because its inner lining is replaced at the rate of about half a million cells per minute. However, when too much food is eaten and the stomach is stretched, or when the stomach is irritated by very spicy food, some of its acidic contents can flow back into the esophagus (gastroesophageal reflux); the burning sensation that results is called *heartburn.*

An antacid is a base that is used to neutralize excess stomach acid. The recommended dose is the amount of the base required to neutralize *some,* but not all, of the stomach acid. Several antacids and their acid-base reactions are shown in Table 16.7. People who need to restrict the quantity of sodium (Na^+) in their diets should avoid sodium-containing antacids such as sodium bicarbonate.

CONCEPTUAL EXERCISE 16.24 Strong Antacids?

Explain why strong bases such as NaOH or KOH are never used as antacids.

PROBLEM-SOLVING EXAMPLE 16.13 Neutralizing Stomach Acid

How many moles (and what mass) of HCl could be neutralized by 0.750 g of the antacid $CaCO_3$?

Answer 1.50×10^{-2} mol HCl; 0.546 g HCl

Table 16.7 The Acid-Base Chemistry of Some Antacids

Compound	Reaction in Stomach	Examples of Commercial Products
Milk of magnesia: $Mg(OH)_2$ in water	$Mg(OH)_2(s) + 2\ H_3O^+(aq) \longrightarrow Mg^{2+}(aq) + 4\ H_2O(\ell)$	Phillips' Milk of Magnesia
Calcium carbonate: $CaCO_3$	$CaCO_3(s) + 2\ H_3O^+(aq) \longrightarrow Ca^{2+}(aq) + 3\ H_2O(\ell) + CO_2(g)$	Tums, Di-Gel
Sodium bicarbonate: $NaHCO_3$	$NaHCO_3(s) + H_3O^+(aq) \longrightarrow Na^+(aq) + 2\ H_2O(\ell) + CO_2(g)$	Baking soda, Alka-Seltzer
Aluminum hydroxide: $Al(OH)_3$	$Al(OH)_3(s) + 3\ H_3O^+(aq) \longrightarrow Al^{3+}(aq) + 6\ H_2O(\ell)$	Amphojel
Dihydroxyaluminum sodium carbonate: $NaAl(OH)_2CO_3$	$NaAl(OH)_2CO_3(s) + 4\ H_3O^+(aq) \longrightarrow Na^+(aq) + Al^{3+}(aq) + 7\ H_2O(\ell) + CO_2(g)$	Rolaids

Strategy and Explanation The balanced equation for this reaction is given in Table 16.7. From the equation we see that 1 mol of the antacid reacts with 2 mol HCl, molar mass = 36.46 g/mol.

$$0.750\ \text{g}\ CaCO_3 \times \frac{1\ \text{mol}\ CaCO_3}{100.1\ \text{g}\ CaCO_3} = 7.49 \times 10^{-3}\ \text{mol}\ CaCO_3$$

$$7.49 \times 10^{-3}\ \text{mol}\ CaCO_3 \times 2\ \text{mol HCl}/1\ \text{mol}\ CaCO_3 = 1.50 \times 10^{-2}\ \text{mol HCl}$$

Using the stoichiometric mole ratio,

$$7.49 \times 10^{-3}\ \text{mol}\ CaCO_3 \times \frac{2\ \text{mol HCl}}{1\ \text{mol}\ CaCO_3} \times \frac{36.46\ \text{g HCl}}{1\ \text{mol HCl}} = 0.546\ \text{g HCl}$$

✓ **Reasonable Answer Check** Two moles of HCl are required to react with each mole of $CaCO_3$, so to neutralize 0.00749 mol $CaCO_3$ requires 0.0150 mol HCl, which is just over 0.5 g (0.546 g) HCl.

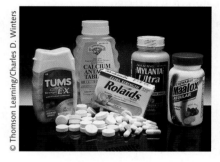

PROBLEM-SOLVING PRACTICE 16.13

Using the reactions in Table 16.7, determine which antacid, on a per gram basis, can neutralize the most stomach acid (assume 0.10 M HCl).

Commercial antacid products. These antacids neutralize excess stomach acid and relieve heartburn.

Acid-Base Chemistry in the Kitchen

Cooking and baking use chemical reactions, often involving acids and bases. Vinegar is an approximately 4 to 5% aqueous acetic acid solution present in almost all salad dressings. Lemon juice, handy for flavoring tea and cooked fish and for making salad dressings, contains citric acid, as do all citrus fruits.

One of the most useful substances in the kitchen is carbon dioxide gas, produced by chemical reactions as it is needed. Pockets of the gas are generated in bread dough and cake batter. The expanding gas makes the resulting biscuits, breads, and cakes rise, producing lighter and more palatable baked goods.

Various methods are used to generate CO_2. One is the addition of yeast, which causes bread dough to rise by catalyzing the fermentation of carbohydrates to produce ethyl alcohol and carbon dioxide.

$$C_6H_{12}O_6 \xrightarrow{\text{yeast}} 2\ CH_3CH_2OH + 2\ CO_2$$

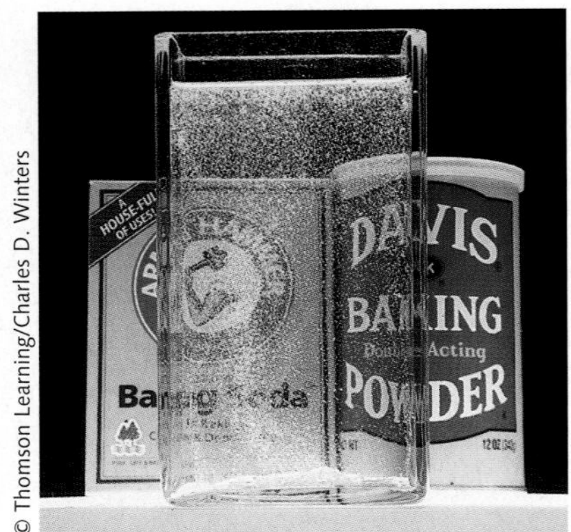

Action of baking powder. Baking powder contains the weak acid dihydrogen phosphate ion and the weak base bicarbonate ion. When mixed together in water, the acid and base react to produce carbon dioxide gas. Bubbles of the gas are seen in the picture.

Many commercial breads and homemade dinner rolls use this method to make their doughs rise.

Because CO_2 production by fermentation is slow, it is sometimes necessary to use another method, the reaction of a bicarbonate salt such as sodium bicarbonate, $NaHCO_3$ (also known as *baking soda*), with acid. But which acid should be used? A weak acid is needed; if a strong acid were used, complete neutralization of the acid would be required to make the food safe to eat. Although vinegar could be used, it would impart an undesirable taste. Long ago it was discovered that lactic acid, $CH_3CH(OH)COOH$, present in milk and formed in larger quantities when milk sours to form buttermilk, is a good source of acid for reacting with bicarbonate.

$$\underset{\text{lactic acid}}{CH_3\underset{\displaystyle\overset{|}{OH}}{C}HCOOH(aq)} + HCO_3^-(aq) \longrightarrow \underset{\text{lactate ion}}{CH_3\underset{\displaystyle\overset{|}{OH}}{C}HCOO^-(aq)} + H_2CO_3(aq)$$

$$\underset{\text{carbonic acid}}{H_2CO_3(aq)} \longrightarrow CO_2(g) + H_2O(\ell)$$

When buttermilk is not available, or when a different taste is desired, dihydrogen phosphate ion, $H_2PO_4^-$, is a convenient acid to react with the bicarbonate ion. Baking powders are a mixture of sodium or potassium dihydrogen phosphate and sodium bicarbonate. When dry, the two salts in baking powder do not react with one another. But when mixed with water in the dough, they react to produce CO_2, a reaction that occurs even more quickly in a heated oven.

$$H_2PO_4^-(aq) \rightleftharpoons H^+(aq) + HPO_4^{2-}(aq)$$

$$H^+(aq) + HCO_3^-(aq) \longrightarrow H_2O(\ell) + CO_2(g)$$

$$\text{Net reaction: } H_2PO_4^-(aq) + HCO_3^-(aq) \longrightarrow H_2O(\ell) + CO_2(g) + HPO_4^{2-}(aq)$$

Buttermilk Blueberry Muffins

$2\frac{1}{2}$ C flour
$1\frac{1}{2}$ tsp baking powder
$\frac{1}{2}$ tsp baking soda
$\frac{1}{2}$ C sugar
$\frac{1}{4}$ tsp salt
2 eggs, beaten
1 C buttermilk
3 oz butter
$1\frac{1}{2}$ C blueberries (added for flavor)

EXERCISE 16.25 Acids and Muffins

Consider the list of ingredients for buttermilk blueberry muffins in the margin. There are two sources of acid and two sources of bicarbonate, the reaction of which produces the CO_2 to make these muffins rise. What are the sources?

Household Cleaners

Most cleaning compounds such as dishwashing detergents, scouring powders, laundry detergents, and oven cleaners are basic. Synthetic detergents are derived from organic molecules designed to have even better cleaning action than soaps, but less reaction with the doubly positive ions (Mg^{2+} and Ca^{2+}) found in hard water.

The molecular structure of a synthetic detergent molecule, like that of a soap, consists of a long oil-soluble (hydrophobic) group, and a water-soluble (hydrophilic) group (⊏ ***p. 758***).

A Typical Synthetic Detergent Molecule

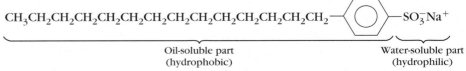

$$\underbrace{CH_3CH_2CH_2CH_2CH_2CH_2CH_2CH_2CH_2CH_2CH_2CH_2CH_2CH_2-C_6H_4}_{\text{Oil-soluble part (hydrophobic)}}\underbrace{-SO_3^-Na^+}_{\text{Water-soluble part (hydrophilic)}}$$

Typical hydrophilic groups in detergents include negatively charged sulfate ($-OSO_3^-$), sulfonate ($-SO_3^-$), and phosphate ($-OPO_3^{2-}$) groups. Compounds with these groups are called *anionic surfactants*. There are also *cationic surfac-*

ESTIMATION

Using an Antacid

Estimate how many Rolaids tablets (Table 16.7) it would take to neutralize the acidity in one glass (250 mL) of a regular cola drink. Assume the pH of the cola is 3.0. One Rolaids tablet contains 334 mg $NaAl(OH)_2CO_3$.

With a pH of 3.0, the cola has 1×10^{-3} mol acid per liter of cola, so 0.250 L cola has one fourth that much acid, or about 3×10^{-4} mol acid. To neutralize this amount of acid requires 3×10^{-4} mol base (1 mol base for every 1 mol acid). There are two bases in Rolaids—hydroxide ions and carbonate ions. Each mole of hydroxide neutralizes 1 mol acid, and each mole of carbonate neutralizes 2 mol acid.

$$H^+(aq) + OH^-(aq) \longrightarrow H_2O(\ell)$$

$$2\,H^+(aq) + CO_3^{2-}(aq) \longrightarrow H_2O(\ell) + CO_2(g)$$

Because each mole of $NaAl(OH)_2CO_3$ contains 2 mol OH^- ions and 1 mol CO_3^{2-} ions, 1 mol $NaAl(OH)_2CO_3$ neutralizes 4 mol acid. The molar mass of $NaAl(OH)_2CO_3$ is 144 g/mol, so one Rolaids tablet contains about 0.002 mol of the antacid.

$$\frac{0.334\text{ g antacid}}{1\text{ antacid tablet}} \times \frac{1\text{ mol antacid}}{144\text{ g antacid}}$$
$$= 0.00232\text{ mol antacid/tablet}$$

This tablet can neutralize four times that many moles of acid, or about 0.01 mol acid. To neutralize the 3×10^{-4} mol acid in the cola requires about 0.03 tablet.

$$3 \times 10^{-4}\text{ mol antacid} \times \frac{1\text{ tablet}}{0.01\text{ mol acid}} \approx 0.03\text{ tablet}$$

It would take only a small portion of a tablet to do the job.

ThomsonNOW™
Sign in to ThomsonNOW at **www.thomsonedu.com** to work an interactive module based on this material.

tants, almost all of which are quaternary ammonium halides in which a central nitrogen atom has four organic groups attached to it, one of which is a long hydrocarbon chain and another that frequently includes an —OH group. Some detergents are *nonionic;* they have an uncharged hydrophilic polar group attached to a large organic group of low polarity, for example,

$$CH_3(CH_2)_4\overset{\displaystyle O}{\overset{\|}{C}}—O—(CH_2)_2O(CH_2CH_2O)_2CH_2CH_2OH$$

hydrocarbon chain (hydrophobic) | ester link (hydrophilic) | ether link | ether link (hydrophilic) | alcohol group (hydrophilic)

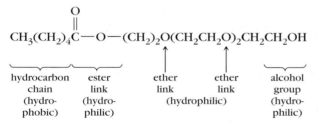

The carbon chain in this molecule is oil-soluble and hydrophobic; the rest of the molecule is hydrophilic, the combination of properties needed for a molecule to be a detergent. Nonionic detergents have several advantages over ionic detergents. With no ionic groups, nonionics cannot form salts with calcium, magnesium, and iron ions; consequently they are totally unaffected by hard water. For the same reason, nonionic detergents do not react with acids and may be used even in relatively strong acid solutions, which makes them useful in toilet bowl cleaners.

Many decades ago homemade lye soap was used to clean clothes as well as people's skins. This type of soap was made using either pure lye (NaOH, also called caustic soda) or sodium carbonate (Na_2CO_3, also called soda ash) and potassium carbonate, K_2CO_3 (also called potash) from wood ashes. Most of the soap made this way contained considerable amounts of unreacted base. This base was considered desirable because it helped raise the pH and break up the heavy soil particles common on fabrics in those days. In addition, the fabrics then were much more durable than fabrics today.

The closest thing to lye soap we see today is the typical dishwashing detergent. The first three ingredients in Table 16.8 contain anions that react with water to

Table 16.8 Formulation of a Dishwashing Detergent

Ingredient	% by Weight
Sodium carbonate, Na_2CO_3	37.5
Sodium tripolyphosphate, $Na_5P_3O_{10}$	30
Sodium metasilicate, Na_2SiO_3	30
Low-foam surfactant	0.5
Sodium dichloroisocyanurate (Cl_2 source)	1.5
Other ingredients such as colorants	0.5

produce OH^-. Together, these three salts produce solutions inside the dishwasher that have a pH near 12.5, high enough to quickly break away animal and vegetable oils from surfaces during the agitation cycle. The detergent helps to dissolve these oily particles and carry them away in the rinse water.

EXERCISE 16.26 pH of a Basic Cleaning Solution

Calculate the pH of a solution that is 5.2 M in sodium carbonate.

Corrosive Household Cleaners

Muriatic acid (hydrochloric acid) is used for cleaning bricks and concrete in new home construction or when remodeling is done. The strong acid should be handled with extreme caution.

Really tough cleaning jobs around the home require chemically aggressive cleaners. In those circumstances, either very acidic or highly basic cleaners are used to get rid of dirt, grease, or stains. Hydrochloric acid along with phosphoric acid and oxalic acid are used in toilet bowl cleaners to help get rid of stains. This combination of acids makes the pH of such cleaners very low, around 2. They should be handled cautiously because of their high acidity and reactivity. For example, labels on bottles of bleach warn against mixing bleach with other cleaners such as acidic toilet bowl cleaners. Bleach commonly contains sodium hypochlorite, NaOCl, and generally has a pH above 8; OCl^- and Cl^- are present in the bleach solution. In the presence of acidic toilet bowl cleaners, however, OCl^- and Cl^- are converted to toxic Cl_2 gas, which can erupt from the mixture. Bleach also should not be mixed with any cleaning agents containing ammonia. The chlorine in the bleach reacts with ammonia to produce fumes of chloramines such as NH_2Cl and $NHCl_2$, which can cause respiratory distress.

This answers the question posed in Chapter 1 (⇐ p. 3), "Why is there a warning on a container of household bleach that says not to mix the bleach with other cleaners, such as toilet-bowl cleaner?"

Drain and oven cleaners are at the other end of the pH spectrum, having very basic pHs of 12 or higher. Deposits of hair, grease, and fats build up inside pipes and eventually clog the drain. When this occurs, the drain has to be dismantled and cleaned (a messy job), or a drain cleaner is added to dissolve the clog.

Drain cleaners generally contain a strong base such as NaOH that reacts with fats and grease to form a soluble soap (⇐ *p. 575*).

$$\begin{matrix} CH_3(CH_2)_{16}C(=O){-}O{-}CH_2 \\ | \\ CH_3(CH_2)_{16}C(=O){-}O{-}CH \\ | \\ CH_3(CH_2)_{16}C(=O){-}O{-}CH_2 \end{matrix} + 3\ NaOH \longrightarrow 3\ CH_3(CH_2)_{16}C(=O){-}O^-Na^+ + \begin{matrix} HO{-}CH_2 \\ | \\ HO{-}CH \\ | \\ HO{-}CH_2 \end{matrix}$$

tristearin (glyceryl tristearate) — sodium stearate (a soap) — glycerol

This reaction converts the grease or fat into water-soluble products (a sodium salt and glycerol) that are washed down the now-opened drain. The reaction is exothermic, and the released heat helps soften the grease or fats, which helps their removal. The strong base also decomposes and rinses away any hair trapped in the clog.

Grocery and hardware stores have an abundance of sodium hydroxide-based drain cleaners. Some are almost pure solid NaOH (pellets or flakes). Liquid drain cleaners are often 50% or more NaOH by weight in water. These solutions, being more dense than water, sink to the bottom of the drain trap to start working quickly. Although easy to use, the liquid cleaners become diluted when running water is put into the drain, reducing their efficiency. A particularly aggressive form of drain cleaner contains small bits of aluminum metal along with solid NaOH. In water, the aluminum reacts with the sodium hydroxide to produce hydrogen gas, whose bubbles help to unseat the clogged material. The net ionic equation for the reaction is

$$2\ Al(s) + 2\ OH^-(aq) + 2\ H_2O(\ell) \longrightarrow 3\ H_2(g) + 2\ AlO_2^-(aq)$$

The hydrogen gas is flammable, so no flames or sparks should be present when this type of cleaner is used. If a pipe is weak and the clog very strong, sufficient gas pressure can build up to rupture the pipe (one way—but not a very desirable one—to unclog the drain).

Spray-on oven cleaners contain an NaOH solution mixed with a detergent and a propellant to apply the mixture to the soiled places. This mixture is sufficiently viscous to adhere to the oven surfaces long enough for the strongly basic solution to react with the baked-on food. If baked at a high temperature, the food has probably carbonized. If so, the cleaner will not be effective, and only scraping will remove the deposits.

You should always be cautious with household acids and bases. They are usually just as concentrated and harmful as industrial chemicals. All strongly acidic and basic solutions, both in the lab and in the home, can be hazardous. Acids, interestingly, are somewhat less dangerous than solutions of bases because the H_3O^+ ion tends to *denature* **(⇐ p. 649)** proteins in skin. The denatured proteins harden, forming a protective layer that prevents further attack by the acid, unless it is hot or highly concentrated. Basic solutions, by contrast, tend to dissolve proteins slowly and so produce little, if any, pain, causing considerable harm before any problem is noticed. If you should get acids or bases on your skin, wash with water for at least 15 min. If acid or base splashes into your eyes, have someone call a physician while you begin gently washing the affected area with lots of water. The international warning placard that is required for shipments of acids and bases in quantities of 1001 lb or more illustrates schematically the personal dangers of these substances.

A corrosive chemical warning placard. This type of placard is required by the U.S. Department of Transportation on loads of 1001 lb or more of acids or bases transported by highway or rail. Such a placard makes fairly clear the reactions with human skin and metals.

CHEMISTRY YOU CAN DO

Aspirin and Digestion

Aspirin is a potent drug capable of relieving pain, fever, and inflammation. Recent studies indicate that it may also decrease blood clotting and heart disease. It is made from salicylic acid, which is found naturally in a variety of plants. The effect of pure salicylic acid on your stomach, however, makes it quite unpleasant as a pain remedy. Commercial aspirin is a derivative of salicylic acid, called acetylsalicylic acid ***(⇐ p. 571)***, that has all the benefits of salicylic acid with less discomfort.

Aspirin is still somewhat acidic, however, and can sometimes cause discomfort or worse in people susceptible to stomach irritation. As a result, several different forms of aspirin are marketed today. The first and most common is plain aspirin. For people with stomach problems, there is also buffered aspirin which includes a buffer in the aspirin tablet to lessen the effect of aspirin's acidity (Section 17.1). A more recent development is enteric-coated aspirin, which is plain aspirin in a tablet with a coating that prevents the aspirin from dissolving in stomach acid, but does allow it to dissolve in the small intestine, which is alkaline.

For this experiment, obtain at least three tablets of each kind of aspirin. Examples of regular aspirin are Bayer and Anacin. Buffered aspirin can be found as Bufferin or Bayer Plus. Enteric aspirin is most commonly available as Bayer Enteric or Ecotrin. Fill three transparent cups or glasses with water. Drop one intact tablet of plain aspirin into the first glass, one tablet of buffered aspirin into the second glass, and one tablet of enteric aspirin into the third glass. Note the changes in the tablets at 1-min intervals until no further changes occur. Now repeat this experiment using vinegar instead of water to dissolve the tablets. Observe each of the tablets in the vinegar at 30-s intervals until no further change is seen. For the final experiment, fill the glasses with water and add 2 tsp of baking soda and stir until dissolved. Once you have made your baking soda solution, add one of each type of aspirin tablet and observe what happens.

1. How did each of the tablets react to each of the different solutions? The acidity of the vinegar should have mimicked the acidity of your stomach. The basic nature of the baking soda should have mimicked the environment of your intestines, which follow your stomach in the digestive system.
2. How do you suppose each type of aspirin works according to its ability to dissolve in your experiments?
3. Does this lead you to think about the kind of aspirin you take?

SUMMARY PROBLEM

Lactic acid, $CH_3CH(OH)COOH$, is a weak monoprotic acid with a melting point of 53 °C. It exists as two enantiomers *(⇐ p. 560)* that have slightly different K_a values. The L form has a K_a of 1.6×10^{-4} and the D form has a K_a of 1.5×10^{-4}. The D form is found in molasses, beer, wines, and souring milk. The L form is produced in muscle cells during anaerobic metabolism in which glucose molecules are broken down into lactic acid and molecules of adenosine triphosphate (ATP) are formed. When lactic acid builds up too rapidly in muscle tissue, severe pain results.

(a) Which form of lactic acid (D or L) is the stronger acid?

(b) What would be the measured pK_a of a 50:50 mixture of the two forms of lactic acid? $pK_a = -\log K_a$

(c) A solution of D-lactic acid is prepared. Use HL as a general formula for lactic acid, and write the equation for the ionization of lactic acid in water.

(d) If 0.1 M solutions of these two acids (D and L) were prepared, what would be the pH of each solution?

(e) Before any lactic acid dissolves in the water, what reaction determines the pH?

(f) Calculate the pH of a solution made by dissolving 4.46 g D-lactic acid in 500. mL of water.

(g) How many milliliters of a 1.15 M NaOH solution would be required to completely neutralize 4.46 g of pure lactic acid?

(h) What would be the pH of the solution made by the neutralization if the lactic acid were the D form? The L form? A 50:50 mixture of the two forms?

ThomsonNOW

Sign in to ThomsonNOW at **www.thomsonedu.com** to check your readiness for an exam by taking the Pre-Test and exploring the modules recommended in your Personalized Learning Plan.

IN CLOSING

Having studied this chapter, you should be able to . . .

- Describe water's role in aqueous acid-base chemistry (Section 16.1). ThomsonNOW homework: Study Questions 12, 16
- Identify the conjugate base of an acid, the conjugate acid of a base, and the relationship between conjugate acid and base strengths (Section 16.1). ThomsonNOW homework: Study Questions 26, 29
- Recognize how amines act as bases and how carboxylic acids ionize in aqueous solution (Section 16.2). ThomsonNOW homework: Study Question 16
- Use the autoionization of water and show how this equilibrium takes place in aqueous solutions of acids and bases (Section 16.3).
- Classify an aqueous solution as acidic, neutral, or basic based on its concentration of H_3O^+ or OH^- and its pH or pOH (Section 16.4). ThomsonNOW homework: Study Question 43
- Calculate pH (or pOH) given $[H_3O^+]$, or $[OH^-]$ given pH (or pOH) (Section 16.4). ThomsonNOW homework: Study Questions 33, 37
- Estimate acid and base strengths from K_a and K_b values (Section 16.5). ThomsonNOW homework: Study Question 51
- Write the ionization steps of polyprotic acids (Section 16.5). ThomsonNOW homework: Study Question 31
- Describe the acidic behavior of hydrated metal ions (Section 16.5). ThomsonNOW homework: Study Question 79

- Describe the relationships between acid strength and molecular structure (16.6). ThomsonNOW homework: Study Questions 20, 123
- Explain the nature of zwitterions (Section 16.6).
- Calculate pH from K_a or K_b values and solution concentration (Section 16.7). ThomsonNOW homework: Study Question 60
- Describe the hydrolysis of salts in aqueous solution (Section 16.8). ThomsonNOW homework: Study Question 78
- Recognize Lewis acids and bases and describe how they react (Section 16.9). ThomsonNOW homework: Study Questions 90, 94
- Apply acid-base principles to the chemistry of antacids, kitchen chemistry, and household cleaners (Section 16.10).

KEY TERMS

acid-base reaction *(16.1)*

acid ionization constant *(16.5)*

acid ionization constant expression *(16.5)*

acidic solution *(16.3)*

amines *(16.2)*

amphiprotic *(16.1)*

amphoteric *(16.9)*

autoionization *(16.3)*

base ionization constant *(16.5)*

base ionization constant expression *(16.5)*

basic solution *(16.3)*

Brønsted-Lowry acid *(16.1)*

Brønsted-Lowry base *(16.1)*

complex ion *(16.9)*

conjugate acid-base pair *(16.1)*

coordinate covalent bond *(16.9)*

hydrolysis *(16.8)*

ionization constant for water *(16.3)*

Lewis acid *(16.9)*

Lewis base *(16.9)*

monoprotic acids *(16.5)*

neutral solution *(16.3)*

oxoacids *(16.6)*

pH *(16.4)*

polyprotic acids *(16.5)*

zwitterion *(16.6)*

QUESTIONS FOR REVIEW AND THOUGHT

■ denotes questions available in ThomsonNOW and assignable in OWL.

Blue-numbered questions have short answers at the back of this book and fully worked solutions in the *Student Solutions Manual.*

ThomsonNOW™

Assess your understanding of this chapter's topics with sample tests and other resources found by signing in to ThomsonNOW at **www.thomsonedu.com**.

Review Questions

1. Define a Brønsted-Lowry acid and a Brønsted-Lowry base.
2. Explain in your own words what 100% ionization means.
3. Write the chemical equation for the autoionization of water. Write the equilibrium constant expression for this reaction. What is the value of the equilibrium constant at 25 °C? What is this constant called?
4. When OH^- is the base in a conjugate acid-base pair, the acid is ________; when OH^- is the acid, the base is ________.
5. Write balanced chemical equations that show phosphoric acid, H_3PO_4, ionizing stepwise as a polyprotic acid.
6. Write ionization equations for a weak acid and its conjugate base. Show that adding these two equations gives the autoionization equation for water.
7. Designate the acid and the base on the left side of these equations, and designate the conjugate partner of each on the right side.
   (a) $HNO_3(aq) + H_2O(\ell) \longrightarrow H_3O^+(aq) + NO_3^-(aq)$
   (b) $NH_4^+(aq) + CN^-(aq) \longrightarrow NH_3(aq) + HCN(aq)$
8. Dissolving ammonium bromide in water gives an acidic solution. Write a balanced equation showing how that can occur.
9. Solution A has a pH of 8 and solution B a pH of 10. Which has the greater hydronium ion concentration? How many times greater is its concentration?
10. Contrast the main ideas of the Brønsted-Lowry and Lewis acid-base concepts. Name and write the formula for a substance that behaves as a Lewis acid but not as a Brønsted-Lowry acid.

Topical Questions

The Brønsted-Lowry Concept of Acids and Bases

11. Write an equation to describe the proton transfer that occurs when each of these acids is added to water.
   (a) HBr (b) CF_3COOH
   (c) HSO_4^- (d) HNO_2

12. ■ Write an equation to describe the proton transfer that occurs when each of these acids is added to water.
(a) HCO_3^- (b) HCl
(c) CH_3COOH (d) HCN

13. Write an equation to describe the proton transfer that occurs when each of these acids is added to water.
(a) HClO (b) CH_3CH_2COOH
(c) $HSeO_3^-$ (d) HO_2^-

14. Write an equation to describe the proton transfer that occurs when each of these acids is added to water.
(a) HIO (b) $CH_3(CH_2)_4COOH$
(c) HOOCCOOH (d) $CH_3NH_3^+$

15. Write an equation to describe the proton transfer that occurs when each of these bases is added to water.
(a) H^- (b) HCO_3^- (c) NO_2^-

16. ■ Write an equation to describe the proton transfer that occurs when each of these bases is added to water.
(a) HSO_4^- (b) CH_3NH_2
(c) I^- (d) $H_2PO_4^-$

17. Write an equation to describe the proton transfer that occurs when each of these bases is added to water.
(a) PO_4^{3-} (b) SO_3^{2-} (c) HPO_4^{2-}

18. Write an equation to describe the proton transfer that occurs when each of these bases is added to water.
(a) AsO_4^{3-} (b) S^{2-} (c) N_3^-

19. Based on formulas alone, classify each of the following oxoacids as strong or weak.
(a) H_3PO_4 (b) H_2SO_4
(c) HClO (d) $HClO_4$
(e) HNO_3 (f) H_2CO_3
(g) HNO_2

20. ■ Based on formulas alone, which is the stronger acid?
(a) H_2CO_3 or H_2SO_4 (b) HNO_3 or HNO_2
(c) $HClO_4$ or H_2SO_4 (d) H_3PO_4 or $HClO_3$
(e) H_2SO_3 or H_2SO_4

21. Write the formula and name for the conjugate partner for each acid or base.
(a) CN^- (b) SO_4^{2-}
(c) HS^- (d) S^{2-}
(e) HSO_3^- (f) HCOOH (formic acid)

22. Write the formula and name for the conjugate partner for each acid or base.
(a) HI (b) NO_3^-
(c) CO_3^{2-} (d) H_2CO_3
(e) HSO_4^- (f) SO_3^{2-}

23. Which are conjugate acid-base pairs?
(a) H_2O and H_3O^+ (b) H_3O^+ and OH^-
(c) NH_2^- and NH_4^+ (d) NH_3 and NH_4^+
(e) O^{2-} and H_2O

24. Which are conjugate acid-base pairs?
(a) NH_2^- and NH_4^+ (b) NH_3 and NH_2^-
(c) H_3O^+ and H_2O (d) OH^- and O^{2-}
(e) H_3O^+ and OH^-

25. Identify the acid and the base that are reactants in each equation; identify the conjugate base and conjugate acid on the product side of each equation.
(a) $HI(aq) + H_2O(\ell) \longrightarrow H_3O^+(aq) + I^-(aq)$
(b) $OH^-(aq) + NH_4^+(aq) \longrightarrow H_2O(\ell) + NH_3(aq)$
(c) $NH_3(aq) + H_2CO_3(aq) \longrightarrow NH_4^+(aq) + HCO_3^-(aq)$

26. ■ Identify the acid and the base that are reactants in each equation; identify the conjugate base and conjugate acid on the product side of the equation.
(a) $HS^-(aq) + H_2O(\ell) \longrightarrow H_2S(aq) + OH^-(aq)$
(b) $S^{2-}(aq) + NH_4^+(aq) \longrightarrow NH_3(g) + HS^-(aq)$
(c) $HCO_3^-(aq) + HSO_4^-(aq) \longrightarrow H_2CO_3(aq) + SO_4^{2-}(aq)$
(d) $NH_3(aq) + NH_2^-(aq) \longrightarrow NH_2^-(aq) + NH_3(aq)$

27. Identify the acid and the base that are reactants in each equation; identify the conjugate base and conjugate acid on the product side of the equation.
(a) $H_2PO_4^-(aq) + HCO_3^-(aq) \longrightarrow H_2CO_3(aq) + HPO_4^{2-}(aq)$
(b) $NH_3(aq) + NH_2^-(aq) \longrightarrow NH_2^-(aq) + NH_3(aq)$
(c) $HSO_4^-(aq) + CO_3^{2-}(aq) \longrightarrow SO_4^{2-}(aq) + HCO_3^-(aq)$

28. Identify the acid and the base that are reactants in each equation; identify the conjugate base and conjugate acid on the product side of the equation.
(a) $CN^-(aq) + CH_3COOH(aq) \longrightarrow CH_3COO^-(aq) + HCN(aq)$
(b) $O^{2-}(aq) + H_2O(\ell) \longrightarrow 2\ OH^-(aq)$
(c) $HCO_2^-(aq) + H_2O(\ell) \longrightarrow HCOOH(aq) + OH^-(aq)$

29. ■ Identify the conjugate acid or base of these acids or bases.
(a) HClO (b) O^{2-}
(c) HCOOH (d) OH^-
(e) IO_3^- (f) PH_3

30. Write stepwise equations for protonation or deprotonation of each of these polyprotic acids and bases in water.
(a) H_2SO_3 (b) S^{2-}
(c) $NH_3CH_2COOH^+$ (glycinium ion, a diprotic acid)

31. ■ Write stepwise equations for protonation or deprotonation of each of these polyprotic acids and bases in water.
(a) CO_3^{2-} (b) H_3AsO_4
(c) $NH_2CH_2COO^-$ (glycinate ion, a diprotic base)

pH Calculations

32. The pH of a popular soft drink is 3.30. What is its hydronium ion concentration? Is the drink acidic or basic?

33. ■ Milk of magnesia, $Mg(OH)_2$, has a pH of 10.5. What is the hydronium ion concentration of the solution? Is this solution acidic or basic?

34. A sample of coffee has a pH of 4.3. Calculate the hydronium ion concentration in this coffee.

35. A solution of lactic acid has a pH of 2.44. What is its hydronium ion concentration?

36. What is the pH of a 0.0013 M solution of HNO_3? What is the pOH of this solution?

37. ■ What is the pH of a solution that is 0.025 M in NaOH? What is the pOH of this solution?

38. A solution of benzyl amine, $C_7H_7NH_2$, has a hydroxide ion concentration of 2.4×10^{-3} M. What is the pH of the solution? What is its pOH?

39. The hydronium ion concentration of a cyanoacetic acid solution is 0.032 M. What is its pOH?

40. The pH of a $Ba(OH)_2$ solution is 10.66 at 25 °C. What is the hydroxide ion concentration of this solution? If the solution volume is 250. mL, how many grams of $Ba(OH)_2$ must have been used to make this solution?

41. A 1000.-mL solution of hydrochloric acid has a pH of 1.3. How many grams of HCl are dissolved in the solution?

42. Make these interconversions. In each case tell whether the solution is acidic or basic.

	pH	$[H_3O^+]$ (M)	$[OH^-]$ (M)
(a)	1.00	______	______
(b)	10.5	______	______
(c)	______	1.8×10^{-4}	______
(d)	______	______	2.3×10^{-5}

43. ■ Make these interconversions. In each case tell whether the solution is acidic or basic.

	pH	$[H_3O^+]$ (M)	$[OH^-]$ (M)
(a)	______	6.1×10^{-7}	______
(b)	______	______	2.2×10^{-9}
(c)	4.67	______	______
(d)	______	2.5×10^{-2}	______
(e)	9.12	______	______

44. Figure 16.2 shows the pH of some common solutions. How many times more acidic or basic are these compared with a neutral solution?
(a) Milk (b) Seawater
(c) Blood (d) Battery acid

45. Figure 16.2 shows the pH of some common solutions. How many times more acidic or basic are these compared with a neutral solution?
(a) Black coffee (b) Household ammonia
(c) Baking soda (d) Vinegar

46. The value of K_w varies with temperature. Use Table 16.1 to calculate the pH of a neutral solution at:
(a) 30 °C. (b) 50 °C.

47. The measured pH of a sample of seawater is 8.30.
(a) What is the H_3O^+ concentration?
(b) Is the sample acidic or basic?

Acid-Base Strengths

48. Write ionization equations and ionization constant expressions for these acids and bases.
(a) CH_3COOH (b) HCN
(c) SO_3^{2-} (d) PO_4^{3-}
(e) NH_4^+ (f) H_2SO_4

49. Write ionization equations and ionization constant expressions for these acids and bases.
(a) F^- (b) NH_3
(c) H_2CO_3 (d) H_3PO_4
(e) CH_3COO^- (f) S^{2-}

50. Which solution will be more acidic?
(a) 0.10 M H_2CO_3 or 0.10 M NH_4Cl
(b) 0.10 M HF or 0.10 M $KHSO_4$
(c) 0.1 M $NaHCO_3$ or 0.1 M Na_2HPO_4
(d) 0.1 M H_2S or 0.1 M HCN

51. ■ Which solution will be more basic?
(a) 0.10 M NH_3 or 0.10 M NaF
(b) 0.10 M K_2S or 0.10 M K_3PO_4
(c) 0.10 M $NaNO_3$ or 0.10 M $NaCH_3COO$
(d) 0.10 M NH_3 or 0.10 M KCN

52. Without doing any calculations, assign each of these 0.10 M aqueous solutions to one of these pH ranges: pH 2; pH between 2 and 6; pH between 6 and 8; pH between 8 and 12; pH 12.
(a) HNO_2 (b) NH_4Cl (c) NaF
(d) $Mg(CH_3COO)_2$ (e) BaO (f) $KHSO_4$
(g) $NaHCO_3$ (h) $BaCl_2$

53. Write the chemical equation and the K_b expression for these bases.
(a) CN^- (b) $C_6H_5NH_2$ (c) HS^-

Using K_a and K_b

54. Calculate the pH of each solution in Question 52 to verify your prediction.

55. A 0.015 M solution of cyanic acid has a pH of 2.67. What is the ionization constant, K_a, of the acid?

56. What is the K_a of butyric acid if a 0.025 M butyric acid solution has a pH of 3.21?

57. What are the equilibrium concentrations of H_3O^+, acetate ion, and acetic acid in a 0.20 M aqueous solution of acetic acid (CH_3COOH)?

58. The ionization constant of a very weak acid, HA, is 4.0×10^{-9}. Calculate the equilibrium concentrations of H_3O^+, A^-, and HA in a 0.040 M solution of the acid.

59. (a) What is the pH of a 0.050 M solution of benzoic acid, C_6H_5COOH ($K_a = 6.3 \times 10^{-5}$ at 25 °C)?
(b) What percent of the acid has ionized in this solution?

60. ■ The pH of a 0.10 M solution of propanoic acid, CH_3CH_2COOH, a weak organic acid, is measured at equilibrium and found to be 2.93 at 25 °C. Calculate the K_a of propanoic acid.

61. The weak base methylamine, CH_3NH_2, has $K_b = 5.0 \times 10^{-4}$. It reacts with water according to the equation

$$CH_3NH_2(aq) + H_2O(\ell) \rightleftharpoons CH_3NH_3^+(aq) + OH^-(aq)$$

What is the pH of a 0.23 M methylamine solution?

62. Calculate the pH of a 0.12 M aqueous solution of the base aniline, $C_6H_5NH_2$; $K_b = 4.2 \times 10^{-10}$.

63. ■ Calculate the $[OH^-]$ and the pH of a 0.024 M methylamine solution; $K_b = 4.2 \times 10^{-4}$.

64. Piperidine, $C_5H_{11}N$, is an oily liquid with a peppery smell. The pH of a 0.0250 M piperidine solution is 11.70. Calculate the K_b of piperidine.

65. Pyridine, C_5H_5N, is a weak base; $K_a = 6.5 \times 10^{-6}$. A 0.2 M solution of pyridine has a pH = 8.5.
(a) Write a balanced chemical equation to represent why the solution has this pH.
(b) Calculate the concentration of unreacted pyridine in this solution.

66. Amantadine, $C_{10}H_{15}NH_2$, is used in the treatment of Parkinson's disease. For amantadine: $K_a = 7.9 \times 10^{-11}$.
   (a) Write the chemical equation for the reaction of amantadine with water.
   (b) Calculate the pH of a 0.0010 M aqueous solution of amantadine at 25 °C.

67. ■ After doing K_a calculations, you may wish you had an aspirin. Aspirin is a weak acid with $K_a = 3.27 \times 10^{-4}$ for the reaction

$$HC_9H_7O_4(aq) + H_2O(\ell) \rightleftharpoons H_3O^+(aq) + C_9H_7O_4^-(aq)$$

Two aspirin tablets, each containing 0.325 g aspirin (along with a nonreactive "binder" to hold the tablet together), are dissolved in 200.0 mL of water. What is the pH of this solution?

68. Lactic acid, $C_3H_6O_3$, occurs in sour milk as a result of the metabolism of certain bacteria. What is the pH of a solution of 56. mg lactic acid in 250. mL water? K_a for lactic acid is 1.4×10^{-4}.

69. Boric acid is a weak acid often used as an eyewash. K_a for boric acid is 7.3×10^{-10}. Find the pH of a 0.10 M solution of boric acid.

70. Calculate the pH of a 0.0050 M solution of dimethylamine, $(CH_3)_2NH$, whose K_b is 5.9×10^{-4}.

Acid-Base Reactions

71. Complete each of these reactions by filling in the blanks. Predict whether each reaction is product-favored or reactant-favored, and explain your reasoning.
   (a) ______(aq) + Br^-(aq) $\rightleftharpoons$ NH_3(aq) + HBr(aq)
   (b) CH_3COOH(aq) + CN^-(aq) $\rightleftharpoons$ ______(aq) + HCN(aq)
   (c) ______(aq) + $H_2O(\ell)$ $\rightleftharpoons$ NH_3(aq) + OH^-(aq)

72. ■ Complete each of these reactions by filling in the blanks. Predict whether each reaction is product-favored or reactant-favored, and explain your reasoning.
   (a) ______(aq) + HSO_4^-(aq) $\rightleftharpoons$ HCN(aq) + SO_4^{2-}(aq)
   (b) H_2S(aq) + $H_2O(\ell)$ $\rightleftharpoons$ H_3O^+(aq) + ______(aq)
   (c) H^-(aq) + $H_2O(\ell)$ $\rightleftharpoons$ OH^-(aq) + ______(g)

73. Predict which of these acid-base reactions are product-favored and which are reactant-favored. In each case write a balanced equation for any reaction that might occur, even if the reaction is reactant-favored. Consult Table 16.2 if necessary.
   (a) $H_2O(\ell)$ + HNO_3(aq) (b) H_3PO_4(aq) + $H_2O(\ell)$
   (c) CN^-(aq) + HCl(aq) (d) NH_4^+(aq) + F^-(aq)

74. Predict which of these acid-base reactions are product-favored and which are reactant-favored. In each case write a balanced equation for any reaction that might occur, even if the reaction is reactant-favored. Consult Table 16.2 if necessary.
   (a) NH_4^+(aq) + HPO_4^{2-}(aq)
   (b) CH_3COOH(aq) + OH^-(aq)
   (c) HSO_4^-(aq) + $H_2PO_4^-$(aq)
   (d) CH_3COOH(aq) + F^-(aq)

75. For each salt, predict whether an aqueous solution will have a pH less than, equal to, or greater than 7. Explain your answer.
   (a) $NaHSO_4$ (b) NH_4Br (c) $KClO_4$

76. For each salt, predict whether an aqueous solution will have a pH less than, equal to, or greater than 7. Explain your answer.
   (a) $AlCl_3$ (b) Na_2S (c) $NaNO_3$

77. For each salt, predict whether an aqueous solution will have a pH less than, equal to, or greater than 7. Explain your answer.
   (a) NaH_2PO_4 (b) NH_4NO_3 (c) $SrCl_2$

78. ■ For each salt, predict whether an aqueous solution will have a pH less than, equal to, or greater than 7. Explain your answer.
   (a) Na_2HPO_4 (b) $(NH_4)_2S$ (c) KCH_3COO

79. ■ Write the formula for the conjugate base of $[Ni(H_2O)_6]^{2+}$.

80. Write the formula for the conjugate acid of $[Zn(H_2O)_3(OH)]^+$.

81. Explain why $BaCO_3$ is soluble in aqueous HCl, but $BaSO_4$, which is used to make the intestines visible in X-ray photographs, remains sufficiently insoluble in the HCl in a human stomach so that poisonous barium ions do not get into the bloodstream.

82. For which of these substances would the solubility be greater at pH = 2 than at pH = 7?
   (a) $Cu(OH)_2$ (b) $CuSO_4$ (c) $CuCO_3$
   (d) CuS (e) $Cu_3(PO_4)_2$

Practical Acid-Base Chemistry

83. Double-acting baking powder contains two salts, sodium hydrogen carbonate and potassium dihydrogen phosphate, whose anions react in water to form CO_2 gas. Write a balanced chemical equation for the reaction. Which anion is the acid and which is the base?

84. Common soap is made by reacting sodium carbonate with stearic acid, $CH_3(CH_2)_{16}COOH$. Write a balanced equation for the reaction.

85. If 1 g of each antacid in Table 16.7 reacted with equal volumes of stomach acid, which would neutralize the most stomach acid?

86. Why is it not a good idea to substitute dishwashing detergent for automobile-washing detergent?

87. Why do cleaning products containing sodium hydroxide feel slippery when you get them on your skin?

88. Some cooked fish have a "fishy" odor due to amines. Explain why putting lemon juice on the fish reduces this odor.

Lewis Acids and Bases

89. Which of these is a Lewis acid? A Lewis base?
   (a) NH_3 (b) $BeCl_2$ (c) BCl_3

90. ■ Which of these is a Lewis acid? A Lewis base?
   (a) O^{2-} (b) CO_2 (c) H^-

91. Which of these is a Lewis acid? A Lewis base?
   (a) Al^{3+} (b) H_2O (c) SCN^-

92. Which of these is a Lewis acid? A Lewis base?
   (a) Cr^{3+} (b) SO_3 (c) CH_3NH_2

93. Identify the Lewis acid and the Lewis base in each reaction.
   (a) $H_2O(\ell) + SO_2(aq) \longrightarrow H_2SO_3(aq)$
   (b) $H_3BO_3(aq) + OH^-(aq) \longrightarrow B(OH)_4^-(aq)$
   (c) $Cu^{2+}(aq) + 4\,NH_3(aq) \longrightarrow [Cu(NH_3)_4]^{2+}(aq)$
   (d) $2\,Cl^-(aq) + SnCl_2(aq) \longrightarrow SnCl_4^{2-}(aq)$

94. ■ Identify the Lewis acid and the Lewis base in each reaction.
(a) $I_2(s) + I^-(aq) \longrightarrow I_3^-(aq)$
(b) $SO_2(g) + BF_3(g) \longrightarrow O_2SBF_3(s)$
(c) $Au^+(aq) + 2\ CN^-(aq) \longrightarrow [Au(CN)_2]^-(aq)$
(d) $CO_2(g) + H_2O(\ell) \longrightarrow H_2CO_3(aq)$

95. Trimethylamine, $(CH_3)_3N$:, reacts readily with diborane, B_2H_6. The diborane dissociates to two BH_3 fragments, each of which can react with trimethylamine to form a complex, $(CH_3)_3N{:}BH_3$. Write an equation for this reaction and interpret it in terms of Lewis acid-base theory.

96. Draw a Lewis structure for ICl_3. Predict the shape of this molecule. Does it function as a Lewis acid or base when it reacts with chloride ion to form ICl_4^-? What is the structure of this ion?

General Questions

97. Classify each of these as a strong acid, weak acid, strong base, weak base, amphiprotic substance, or neither acid nor base.
(a) HCl (b) NH_4^+ (c) H_2O
(d) CH_3COO^- (e) CH_4 (f) CO_3^{2-}

98. Classify each of these as a strong acid, weak acid, strong base, weak base, amphiprotic substance, or neither acid nor base.
(a) CH_3COOH (b) Na_2O (c) H_2SO_4
(d) NH_3 (e) $Ba(OH)_2$ (f) $H_2PO_4^-$

99. Several acids and their respective equilibrium constants are:

$$HF(aq) + H_2O(\ell) \rightleftharpoons H_3O^+(aq) + F^-(aq) \qquad K_a = 7.2 \times 10^{-4}$$

$$HS^-(aq) + H_2O(\ell) \rightleftharpoons H_3O^+(aq) + S^{2-}(aq) \qquad K_a = 8 \times 10^{-18}$$

$$CH_3COOH(aq) + H_2O(\ell) \rightleftharpoons H_3O^+(aq) + CH_3COO^-(aq) \qquad K_a = 1.8 \times 10^{-5}$$

(a) Which is the strongest acid? Which is the weakest acid?
(b) Which acid has the weakest conjugate base?
(c) Which acid has the strongest conjugate base?

100. State whether equal molar amounts of these would have a pH equal to 7, less than 7, or greater than 7.
(a) A weak base and a strong acid react.
(b) A strong base and a strong acid react.
(c) A strong base and a weak acid react.

101. Sulfurous acid, H_2SO_3, is a weak diprotic acid ($K_{a_1} = 1.2 \times 10^{-2}$, $K_{a_2} = 6.2 \times 10^{-8}$). What is the pH of a 0.45 M solution of H_2SO_3? (Assume that only the first ionization is important in determining pH.)

102. Ascorbic acid (vitamin C, $C_6H_8O_6$) is a diprotic acid ($K_{a_1} = 7.9 \times 10^{-5}$, $K_{a_2} = 1.6 \times 10^{-12}$). What is the pH of a solution that contains 5.0 mg of the acid per mL of water? (Assume that only the first ionization is important in determining pH.)

103. Does the pH of the solution increase, decrease, or stay the same when you
(a) Add solid ammonium chloride to 100. mL of 0.10 M NH_3?
(b) Add solid sodium acetate to 50.0 mL of 0.015 M acetic acid?
(c) Add solid NaCl to 25.0 mL of 0.10 M NaOH?

104. Does the pH of the solution increase, decrease, or stay the same when you
(a) Add solid sodium oxalate, $Na_2C_2O_4$, to 50.0 mL of 0.015 M oxalic acid?
(b) Add solid ammonium chloride to 100. mL of 0.016 M HCl?
(c) Add 20.0 g NaCl to 1.0 L of 0.012 M sodium acetate, $NaCH_3COO$?

105. (a) Solid ammonium bromide, NH_4Br, is added to water. Will the resulting solution be acidic, basic, or neutral? Use a chemical equation to explain your answer.
(b) An aqueous ammonium bromide solution has a pH of 8.50. Calculate the ammonium bromide concentration of this solution.

106. Sodium hypochlorite, NaOCl, is used as a source of chlorine in some laundry bleaches, swimming pool disinfectants, and water treatment plants. Calculate the pH of a 0.010 M solution of NaOCl ($K_b = 2.9 \times 10^{-7}$).

Applying Concepts

107. When a 0.1 M aqueous ammonia solution is tested with a conductivity apparatus (← p. 101), the bulb glows dimly. When a 0.1 M hydrochloric acid solution is tested, the bulb glows brightly. As water is added to each of the solutions, would you expect the bulb to glow brighter, stop glowing, or stay the same? Explain your reasoning.

108. Using Table 16.1, calculate what the pH of pure water is at 10 °C, 25 °C, and 50 °C. Classify the water at each temperature as either acidic, neutral, or basic.

109. If you evaporate the water in a sodium hydroxide solution, you will end up with solid sodium hydroxide. However, if you evaporate the water in an ammonium hydroxide solution, you will not end up with solid ammonium hydroxide. Explain why. What will remain after the water is evaporated?

110. For each aqueous solution, predict what ions and molecules will be present. Without doing any calculations, list the ions and molecules in order of decreasing concentration.
(a) HCl (b) $NaClO_4$ (c) HNO_2
(d) NaClO (e) NH_4Cl (f) NaOH

111. The diagrams below are nanoscale representations of different acids.

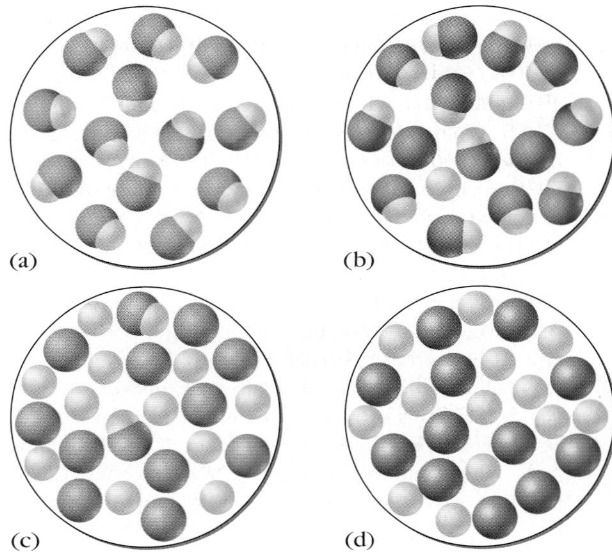

(a) Which diagram best represents hydrochloric acid? (The yellow spheres are H^+ ions and the other spheres are Cl^- ions.)
(b) Which diagram best represents acetic acid? (The yellow spheres are H^+ ions and the other spheres are CH_3COO^- ions.)

112. When asked to identify the conjugate acid-base pairs in the reaction

$$HCO_3^-(aq) + HSO_4^-(aq) \rightleftharpoons H_2CO_3(aq) + SO_4^{2-}(aq)$$

a student incorrectly wrote: "HCO_3^- is a base and HSO_4^- is its conjugate acid. H_2CO_3 is an acid and SO_4^- is its conjugate base." Write a brief explanation to the student telling why the answer is incorrect.

113. A 0.1 M aqueous solution has a pH of 5.5. Write a chemical equation that represents why the solution is acidic.

114. Explain how the Arrhenius acid-base theory and the Brønsted-Lowry theory of acids and bases are explained by the Lewis acid-base theory.

115. Calculate the K_b for the conjugate base of gallic acid, found in tea; $K_a = 3.9 \times 10^{-5}$. Identify a base from Table 16.2 with a K_b close to that of the conjugate base of gallic acid.

116. Trichloroacetic acid, CCl_3COOH, has an acid dissociation constant of 3.0×10^{-1} at 25 °C.
(a) Calculate the pH of a 0.0100 M trichloroacetic acid solution.
(b) How many times greater is the H_3O^+ concentration in this solution than in a 0.0100 M acetic acid?

117. Explain why a 0.1 M NH_4NO_3 aqueous solution has a pH of 5.4.

118. Is the dissociation of water

$$2\, H_2O(\ell) \rightleftharpoons H_3O^+(aq) + OH^-(aq)$$

exothermic or endothermic? Refer to Table 16.1. Explain your answer.

More Challenging Questions

119. If 1 g each of vinegar, lemon juice, and lactic acid react with equal masses of baking soda, which will produce the most CO_2 gas?

120. A person claimed that his stomach ruptured when he took a teaspoonful of baking soda in a glass of water to relieve heartburn after a full meal ($\frac{1}{2}$ tsp = 2.5 g $NaHCO_3$). Assume that the pH of stomach acid is 1 and that the stomach has a volume of 1 L when expanded fully. Body temperature is 37 °C. What volume of carbon dioxide gas was generated by the reaction of baking soda with stomach acids? Might his stomach have ruptured from this volume of CO_2?

121. A chilled carbonated beverage is opened and warmed to room temperature. Will the pH change and, if so, how will it change?

122. For an experiment a student needs a pH = 9.0 solution. He plans to make the solution by diluting 6.0 M HCl until the H_3O^+ concentration equals 1.0×10^{-9} M.
(a) Will this plan work? Explain your answer.
(b) Will diluting 1.00 M NaOH work? Explain your answer.

123. ■ Explain why $BrNH_2$ is a weaker base than ammonia, NH_3. Which will have the smaller K_b value? Will $ClNH_2$ be a stronger or weaker base than $BrNH_2$? Explain your answer.

124. ■ Aniline has a $pK_a = 4.63$; 3-bromoaniline has a $pK_a = 3.58$; $pK_a = -\log(K_a)$. Which is the stronger base? Explain your answer.

125. It is determined that 0.1 M solutions of the sodium salts NaM, NaQ, and NaZ have pH values of 7.0, 8.0, and 9.0, respectively. Arrange the acids HM, HQ, and HZ in order of decreasing strength. Where possible, find the K_a values of these acids.

126. Use the data in Appendix J to calculate the enthalpy change for the reaction of:
(a) 100. mL of 0.100 M NaOH with 100. mL of 1.00 M HCl.
(b) 25.0 mL of 1.00 M H_2SO_4 with 75.0 mL of 1.0 M NaOH.

127. Hydrogen peroxide, HOOH, is a powerful oxidizing agent that, in diluted form, is used as an antiseptic. For HOOH: $K_a = 2.5 \times 10^{-12}$; $K_b = 4.0 \times 10^{-3}$.
(a) Is hydrogen peroxide a strong or weak acid?
(b) Is OOH^- a strong or weak base?
(c) What is the relationship between HOOH and OOH^-?
(d) Calculate the pH of 0.100 M aqueous hydrogen peroxide.

128. *Para*-bromoaniline is used in dye production. Two different pK_b values have been reported for this compound: 10.25 and 9.98; $pK_b = -\log(K_b)$. Using these values, calculate how different the pH values are for a 0.010 M solution of *para*-bromoaniline.

Br—⌬—NH_2

129. At 25 °C, a 0.1% aqueous solution of adipic acid, $C_5H_9O_2COOH$, has a pH of 3.2. A saturated solution of the acid, which contains 1.44 g acid per 100. mL of solution, has a pH = 2.7. Calculate the percent dissociation of adipic acid in each solution.

130. Niacin, a B vitamin, can act as an acid and as a base.
(a) Write chemical equations showing the action of niacin as an acid in water. Symbolize niacin as HNc.
(b) Write chemical equations showing the action of niacin as a base in water.
(c) A 0.020 M aqueous solution of niacin has a pH = 3.26. Calculate the K_a of niacin.
(d) Calculate the K_b of niacin.
(e) Write the chemical equation that accompanies the K_b expression for niacin.

131. The loss of CO_2 from lysine forms cadaverine, an aptly named compound with an exceptionally revolting smell.

$H_2N—CH_2CH_2CH_2CH_2—C(H)(NH_2)—C(=O)—OH$

lysine

(a) Write the structural formula for cadaverine.
(b) Write the structural formula for cadaverine in its fully protonated form.
(c) Cadaverine has two pK_a values: 10.25 and 9.13; $pK_a = -\log(K_a)$. Which one is for cadaverine in its fully protonated form?

132. At normal body temperature, 37 °C, $K_w = 2.5 \times 10^{-14}$.
   (a) Calculate the pH of a neutral solution at this temperature.
   (b) The pH of blood at this temperature ranges from 7.35 to 7.45. Assuming a pH of 7.40, calculate the H_3O^+ concentration of the blood at this temperature.
   (c) By how much does the H_3O^+ concentration differ at this temperature than at 25 °C?

133. The structural formula for lysine is given in Question 131. Explain why lysine has three pK_a values: 2.18, 8.95, and 10.53. $pK_a = -\log(K_a)$.

Conceptual Challenge Problems

CP16.A (Section 16.4) Is it possible for an aqueous solution to have a pH of 0 or even less than 0? Explain your answer mathematically as well as practically based on what you know about acid solubilities.

CP16.B (Section 16.4) What is the pH of water at 200 °C? Liquid water this hot would have to be under a pressure greater than 1.0 atm and might be found in a pressurized water reactor located in a nuclear power plant.

CP16.C (Section 16.5) Develop a set of rules by which you could predict the pH for solutions of strong or weak acids and strong or weak bases without using a calculator. Your predictions need to be accurate to ±1 pH units. Assume that you know the concentration of the acid or base and that for the weak acids and bases you can look up the pK_a ($-\log K_a$) or K_a values. What rules would work to predict pH?

17 Additional Aqueous Equilibria

Precipitation reactions occur not only in the laboratory, but also on a massive scale in nature, as seen by the deposition of the mineral travertine (calcium carbonate) at Mammoth Hot Springs in Yellowstone National Park (USA). Calcium ions and hydrogen carbonate ions dissolved in the hot spring water react to form calcium carbonate (travertine), which precipitates from solution. In this chapter we will consider factors affecting the solubility of ionic compounds, as well as quantitative aspects of ionic compound solubilities, and acid-base reactions such as those in buffers and in acid-base titrations.

George D. Lepp/CORBIS

In environments as different as the interior of red blood cells, coral reefs, in ocean waters, and clouds high above the earth, important interactions occur among solutes in aqueous solution. This chapter extends the discussion of aqueous solutions begun in Chapters 15 and 16 (conjugate acid-base behavior, acid-base neutralization, the link between solubility and precipitation) to the quantitative aspects of dealing with (1) buffers, which are combinations of a weak acid and its conjugate base or a weak base and its conjugate acid; (2) acid-base titrations, which are neutralization reactions of an acid with a base; and (3) equilibria associated with solutions of slightly soluble salts.

ThomsonNOW
Throughout the text, this icon indicates an opportunity to test yourself on key concepts and to explore interactive modules by signing in to ThomsonNOW at **www.thomsonedu.com**.

17.1 Buffer Solutions

Adding a small amount of acid or base to pure water radically changes the pH. Consider what happens if 0.010 mol HCl is added to 1 L water. The pH changes from 7 to 2 because $[H_3O^+]$ changes from 10^{-7} M to 10^{-2} M. This pH change represents a *100,000-fold increase* in H_3O^+ concentration. Similarly, if 0.010 mol NaOH is added to 1 L pure water, the pH goes from 7 to 12, a *100,000-fold decrease* in $[H_3O^+]$ and a *100,000-fold increase* in $[OH^-]$. Most aquatic organisms could not survive such dramatic pH changes; the organisms can survive only within a narrow pH range. For example, if acid rain lowers the pH of a lake or stream sufficiently, fish such as trout may die.

ThomsonNOW
Go to the Chemistry Interactive menu for a module on **pH change on addition of HCl to water and a buffer.**

Unlike pure water and aqueous solutions of NaOH or HCl, there are aqueous solutions that maintain a relatively constant pH when limited amounts of base or acid are added to them. Such solutions contain a **buffer**—*a chemical system that resists change in pH*—and are referred to as **buffer solutions.** For example, a solution that contains 0.50 mol acetic acid, CH_3COOH, and 0.50 mol sodium acetate, $NaCH_3COO$, in 1.0 L of solution is a buffer with a pH of 4.74. When 0.010 mol of a strong acid is added to it, the pH changes from 4.74 to 4.72, only 0.02 pH unit; adding 0.010 mol of strong base to this buffer changes the pH from 4.74 to 4.76 (Figure 17.1). These are only slight pH changes, clearly much less than those that

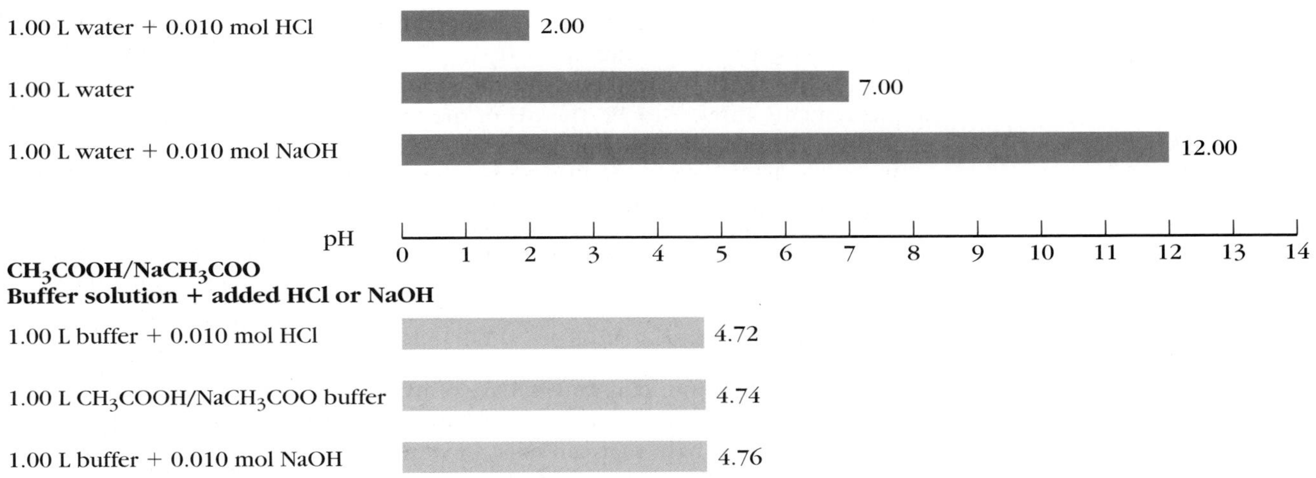

Figure 17.1 pH changes. The pH of water changes dramatically when 0.010 mol of acid (HCl) or base (NaOH) is added to it. By contrast, the pH of a sodium acetate–acetic acid buffer containing 0.50 mol acetic acid and 0.50 mol sodium acetate remains relatively constant when 0.010 mol HCl or NaOH is added to it. Only a slight change in pH occurs.

occur when the same number of moles of HCl or NaOH are added to water, as described above. How do the sodium acetate–acetic acid buffer and other buffers offset such additions of acid or base without significant change in pH?

ThomsonNOW™
Go to the Coached Problems menu for modules on:

- **buffer solutions**
- **pH of buffer solutions**
- **preparation of buffer solutions**

Buffer Action

To maintain a relatively constant pH, a buffer must contain *a weak acid that can react with added base,* and the buffer also must contain *a weak base that can react with added acid.* In addition, it is necessary that the acid and base components of a buffer solution not react with each other. A conjugate acid-base pair, such as acetic acid and acetate ion (from sodium acetate), satisfies this requirement. In a conjugate pair, if the acid and base react with each other they just produce conjugate base and conjugate acid—no observable change occurs. For example, acetic acid reacts with acetate ion to form acetate ion and acetic acid.

$$\underset{\text{CONJ ACID}}{CH_3COOH(aq)} + \underset{\text{conj base}}{CH_3COO^-(aq)} \rightleftharpoons \underset{\text{CONJ BASE}}{CH_3COO^-(aq)} + \underset{\text{conj acid}}{CH_3COOH(aq)}$$

A buffer usually consists of approximately equal quantities of a weak acid and its conjugate base, or a weak base and its conjugate acid.

The blood of mammals is an aqueous solution that maintains a constant pH. To see how a buffer works, consider human blood, whose normal pH is 7.40 ± 0.05. If the pH decreases below 7.35, a condition known as *acidosis* occurs; increasing the pH above 7.45 causes *alkalosis.* Both conditions can be life-threatening. Acidosis, for example, causes a decrease in oxygen transport by hemoglobin and depresses the central nervous system, leading in extreme cases to coma and death by creating weak and irregular cardiac contractions—symptoms of heart failure. To prevent such problems, your body must keep the pH of your blood nearly constant.

The term acidosis is used here even though the pH is not less than 7.0.

Carbon dioxide provides the most important blood buffer (but not the only one). In solution, CO_2 reacts with water to form H_2CO_3, which ionizes to produce H_3O^+ and HCO_3^- ions. The equilibria are

$$CO_2(aq) + H_2O(\ell) \rightleftharpoons H_2CO_3(aq)$$

$$H_2CO_3(aq) + H_2O(\ell) \rightleftharpoons H_3O^+(aq) + HCO_3^-(aq)$$

Since H_2CO_3 is a weak acid and HCO_3^- is its conjugate weak base, they constitute a buffer. The normal concentrations of H_2CO_3 and HCO_3^- in blood are 0.0025 M and 0.025 M, respectively—a 1:10 ratio. As long as the ratio of H_2CO_3 to HCO_3^- concentrations remains about 1 to 10, the pH of the blood remains near 7.4. (We will calculate this pH in Problem-Solving Example 17.1.)

If a strong base such as NaOH is added to this buffer, carbonic acid in the buffer will react with the added OH^-.

$$H_2CO_3(aq) + OH^-(aq) \longrightarrow HCO_3^-(aq) + H_2O(\ell)$$

$$K = \frac{1}{K_b(HCO_3^-)} = \frac{1}{2.4 \times 10^{-8}} = 4.2 \times 10^7$$

Here the equilibrium constant is $1/K_b$ of hydrogen carbonate ion, because the reaction is the reverse of the reaction of hydrogen carbonate ion with H_2O. Since OH^- is the strongest base that can exist in water solution, the reaction of hydroxide ions with carbonic acid is essentially complete, as indicated by the very large K value of 4.2×10^7.

If a strong acid such as HCl is added to this buffer, the HCO_3^- ion—the conjugate base in the buffer—will react with the hydronium ions from HCl.

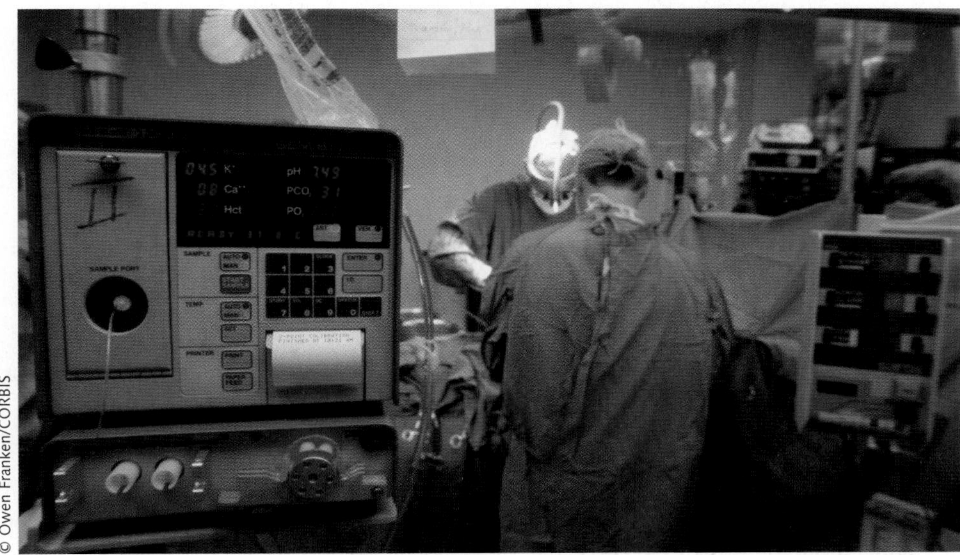

© Owen Franken/CORBIS

A blood gas analyzer. This instrument measures CO_2 level in blood, blood pH, and oxygen level. These values are related and must be within a narrow range for good health.

$$HCO_3^-(aq) + H_3O^+(aq) \longrightarrow H_2CO_3(aq) + H_2O(\ell)$$

$$K = \frac{1}{K_a(H_2CO_3)} = \frac{1}{4.2 \times 10^{-7}} = 2.4 \times 10^6$$

In this case, the equilibrium constant is $1/K_a$ of carbonic acid, because the reaction is the reverse of the ionization of carbonic acid. Since the H_3O^+ ion is such a strong acid, the reaction between HCO_3^- and H_3O^+ is essentially complete and $K = 2.4 \times 10^6$.

In many buffers, the ratio of conjugate acid:conjugate base concentrations is about 1:1. In blood, however, the $[HCO_3^-]/[H_2CO_3]$ ratio is about 10 to 1, with good reason. There are more acidic byproducts of metabolism in the blood that must be neutralized than there are basic ones.

CONCEPTUAL EXERCISE 17.1 Possible Buffers?

Could a solution of equimolar amounts of HCl and NaCl be a buffer? What about a solution of equimolar amounts of KOH and KCl? Explain each of your answers.

The pH of Buffer Solutions

The pH of a buffer solution can be calculated in two ways: (1) by using $[H_3O^+]$ in the K_a expression if the K_a and the concentrations of the conjugate acid and the conjugate base are known (Problem-Solving Example 17.1), and (2) by using the Henderson-Hasselbalch equation (Problem-Solving Example 17.2).

PROBLEM-SOLVING EXAMPLE 17.1 The pH of a Buffer from K_a

To mimic a blood buffer, a scientist prepared 1.000 L buffer containing 0.0025 mol carbonic acid and 0.025 mol hydrogen carbonate ion. Calculate the pH of the buffer. The K_a of carbonic acid is 4.2×10^{-7}.

Answer 7.38

Strategy and Explanation Carbonic acid (conjugate acid) and hydrogen carbonate ion (conjugate base) are a conjugate acid–conjugate base pair. To calculate the pH, we use the ionization constant equation for carbonic acid and its K_a value to find $[H_3O^+]$ and then pH.

$$H_2CO_3(aq) + H_2O(\ell) \rightleftharpoons H_3O^+(aq) + HCO_3^-(aq)$$

$$K_a = 4.2 \times 10^{-7} = \frac{[H_3O^+][\text{conj base}]}{[\text{conj acid}]} = \frac{[H_3O^+][HCO_3^-]}{[H_2CO_3]}$$

We can set up an "ICE" table such as we have done for other equilibrium calculations.

	H_2CO_3	+	H_2O	$\rightleftharpoons$	H_3O^+	+	HCO_3^-
Initial concentration (mol/L)	0.0025				≈0		0.025
Change as reaction occurs (mol/L)	$-x$				$+x$		$+x$
Equilibrium concentration (mol/L)	$0.0025 - x$				$+x$		$0.025 + x$

Given the small value of K_a, we will assume that $x \ll 0.0025$ and thus, $0.0025 - x \approx 0.0025$ and $0.025 + x \approx 0.025$. These values can be substituted into the K_a expression to calculate $[H_3O^+]$ and then pH.

$$K_a = 4.2 \times 10^{-7} = \frac{x(0.025 + x)}{(0.0025 - x)} \approx \frac{x(0.025)}{(0.0025)}$$

$$x = [H_3O^+] = \frac{(4.2 \times 10^{-7})(0.0025)}{0.025} = 4.2 \times 10^{-8}$$

$$\text{pH} = -\log(4.2 \times 10^{-8}) = 7.38$$

Note that $x = H_3O^+ = 4.2 \times 10^{-8}$ M, which is $\ll$0.0025. Therefore, the assumption that $x \ll 0.0025$ was valid.

✓ **Reasonable Answer Check** From the K_a expression, we see that when $[HCO_3^-] = [H_2CO_3]$, the $[H_3O^+] = K_a$. But, in this example, the concentration of HCO_3^-, the conjugate base, is ten times that of H_2CO_3, the conjugate acid. Therefore, the $[H_3O^+]$ should be ten times less than the K_a, which it is.

PROBLEM-SOLVING PRACTICE 17.1

Calculate the pH of blood containing 0.0020 M carbonic acid and 0.025 M hydrogen carbonate ion.

The *Henderson-Hasselbalch equation* can be used conveniently to calculate the pH of a buffer containing known concentrations of conjugate base and conjugate acid. The equation can also be applied to determine the ratio of conjugate base to conjugate acid concentrations needed to achieve a buffer of a given pH. This equation is derived by writing the acid ionization constant expression for a weak acid, HA, and solving for $[H_3O^+]$; A^- is the conjugate base of HA.

$$HA(aq) + H_2O(\ell) \rightleftharpoons H_3O^+(aq) + A^-(aq)$$

$$K_a = \frac{[H_3O^+][\text{conj base}]}{[\text{conj acid}]} = \frac{[H_3O^+][A^-]}{[HA]}$$

$$[H_3O^+] = \frac{K_a[\text{conj acid}]}{[\text{conj base}]} = \frac{K_a[HA]}{[A^-]}$$

The next steps convert $[H_3O^+]$ to pH. Taking the base 10 logarithm of each side of this equation gives

$$\log[H_3O^+] = \log K_a + \log \frac{[\text{conj acid}]}{[\text{conj base}]} = \log K_a + \log \frac{[HA]}{[A^-]}$$

Multiplying both sides of the equation by -1 and using the relation $-\log(x) = \log(1/x)$, we get

$$-\log[H_3O^+] = -\log K_a + \log \frac{[\text{conj base}]}{[\text{conj acid}]} = -\log K_a + \log \frac{[A^-]}{[HA]}$$

Using the definition of pH and defining $-\log K_a$ as pK_a (analogous to the definition of pH), the equation becomes the Henderson-Hasselbalch equation.

Because $-\log(x) = \log\left(\frac{1}{x}\right)$.

$$-\log \frac{[\text{conj acid}]}{[\text{conj base}]} = \log \frac{[\text{conj base}]}{[\text{conj acid}]}$$

and $-\log \frac{[HA]}{[A^-]} = \log \frac{[A^-]}{[HA]}$.

$$\text{pH} = pK_a + \log \frac{[\text{conj base}]}{[\text{conj acid}]} = pK_a + \log \frac{[A^-]}{[HA]}$$ **Henderson-Hasselbalch equation**

Using the Henderson-Hasselbalch equation to find the blood pH calculated in Problem-Solving Example 17.1 gives the same results.

$$\text{pH} = pK_a + \log \frac{[\text{conj base}]}{[\text{conj acid}]} = -\log K_a + \log \frac{[HCO_3^-]}{[H_2CO_3]}$$

$$= -\log(4.2 \times 10^{-7}) + \log\left(\frac{0.0250}{0.00250}\right)$$

$$= 6.38 + \log(10) = 6.38 + 1.00 = 7.38$$

Note that since the concentration of conjugate base is ten times that of the conjugate acid, the pH should be greater than the pK_a by 1 pH unit, and it is.

It is important to note that the Henderson-Hasselbalch equation has two important rules for when it can be applied to buffer solutions:

1. The value of the $\frac{[\text{conj base}]}{[\text{conj acid}]}$ ratio must be between 0.1 and 10.
2. The [conj base] and [conj acid] must *each* exceed the K_a by a factor of 100 or more.

EXERCISE 17.2 Blood pH

Calculate the pH of blood containing 0.0025 M $HPO_4^{2-}(aq)$ and 0.0015 M $H_2PO_4^-(aq)$. K_a of $H_2PO_4^- = 6.2 \times 10^{-8}$. (Assume that this is the only blood buffer.)

PROBLEM-SOLVING EXAMPLE 17.2 A Buffer Solution and Its pH

Calculate the pH of a buffer containing 0.050 mol/L pyruvic acid, $CH_3COCOOH$, and 0.060 mol/L sodium pyruvate, $Na^+CH_3COCOO^-$. K_a of pyruvic acid $= 3.2 \times 10^{-3}$.

Answer 2.57

Strategy and Explanation We will solve for the pH in two ways. First, we will use the K_a expression to find the H_3O^+ concentration from which the pH can be calculated. Second, we will use the Henderson-Hasselbalch equation to determine pH.

Using the K_a expression, we first write the chemical equation for the equilibrium of pyruvic acid with pyruvate ions, and its corresponding K_a expression.

$$CH_3COCOOH(aq) + H_2O(\ell) \rightleftharpoons H_3O^+(aq) + CH_3COCOO^-(aq)$$

$$K_a = 3.2 \times 10^{-3} = \frac{[H_3O^+][CH_3COCOO^-]}{[CH_3COCOOH]} = \frac{[H_3O^+](0.060)}{(0.050)}$$

Rearranging the equation to calculate $[H_3O^+]$ and then the pH gives

$$[H_3O^+] = \frac{3.2 \times 10^{-3}(0.050)}{0.060} = 2.67 \times 10^{-3}$$

$$pH = -\log(2.67 \times 10^{-3}) = 2.57$$

Applying the Henderson-Hasselbalch equation:

$$pH = pK_a + \log\frac{[\text{conj base}]}{[\text{conj acid}]} = -\log(3.2 \times 10^{-3}) + \log\frac{[CH_3COCOO^-]}{[CH_3COCOOH]}$$

$$pH = 2.49 + \log\left(\frac{0.060}{0.050}\right) = 2.49 + \log(1.2) = 2.49 + 0.079 = 2.57$$

✓ **Reasonable Answer Check** The concentration of conjugate base is a bit greater than the concentration of conjugate acid. Therefore the pH should be slightly higher than the pK_a, which it is. The answer is reasonable.

If the conjugate base and conjugate acid concentrations were equal, the concentration ratio would be one and its log would be zero. In such a case, pH = pK_a, which is not the case here.

PROBLEM-SOLVING PRACTICE 17.2

Calculate the ratio of $[HPO_4^{2-}]$ to $[H_2PO_4^-]$ in blood at a normal pH of 7.40. Assume that this is the only buffer system present.

From the Henderson-Hasselbalch equation note that when the concentrations of conjugate base and conjugate acid are equal,

$$\frac{[\text{conj base}]}{[\text{conj acid}]} = \frac{[A^-]}{[HA]} = 1 \qquad \log\frac{[\text{conj base}]}{[\text{conj acid}]} = \log\frac{[A^-]}{[HA]} = \log(1) = 0$$

and so pH = pK_a. Thus, ***a buffer's pH equals the pK_a of its weak acid when the concentrations of the acid and its conjugate base in the buffer are equal.***

A buffer for maintaining a desired pH can therefore be chosen easily by examining pK_a values, which are often tabulated along with K_a values. *Choose a conjugate acid-base pair whose conjugate acid has a pK_a near the desired pH.* Table 17.1 lists pK_a values of several common acids that could be used with their conjugate bases to prepare buffers over the pH range from 4 to 10.

To have comparable quantities of both acid and conjugate base in a buffer solution, the ratio of conjugate base to conjugate acid cannot get much smaller than 1:10 or much bigger than 10:1. ***The pH range of a buffer is limited to about one pH unit above or below the pK_a of the conjugate acid.*** In the carbonic acid/hydrogen carbonate case, for example, that would be a pH from 5.38 to 7.38 because $pK_a = -\log K_a = -\log 4.2 \times 10^{-7} = 6.38$. Other conjugate acid-base pairs, such as those in Table 17.1, can be used to prepare buffers with much different pH ranges, as determined by the pK_a value of the acid in the buffer (Table 17.1).

ThomsonNOW™

Go to the Coached Problems menu for tutorials on:

- **buffer solutions**
- **buffer solutions: preparation by acid-base reactions**
- **buffer solutions: preparation by direct addition**

PROBLEM-SOLVING EXAMPLE 17.3 Selecting an Acid-Base Pair for a Buffer Solution of Known pH

You are doing an experiment that requires a buffer solution with a pH of 4.00. Available to you are solutions needed to make buffers with these acid-base pairs: CH_3COOH/CH_3COO^-; H_2CO_3/HCO_3^-; and $CH_3CHOHCOOH/CH_3CHOHCOO^-$. Which acid-

Table 17.1 Buffer Systems That Are Useful at Various pH Values*

Desired pH	Weak Conjugate Acid	Weak Conjugate Base	K_a of Weak Conjugate Acid	pK_a
4	Lactic acid ($CH_3CHOHCOOH$)	Lactate ion ($CH_3CHOHCOO^-$)	1.4×10^{-4}	3.85
5	Acetic acid (CH_3COOH)	Acetate ion (CH_3COO^-)	1.8×10^{-5}	4.74
6	Carbonic acid (H_2CO_3)	Hydrogen carbonate ion (HCO_3^-)	4.2×10^{-7}	6.38
7	Dihydrogen phosphate ion ($H_2PO_4^-$)	Monohydrogen phosphate ion (HPO_4^{2-})	6.2×10^{-8}	7.21
8	Hypochlorous acid (HClO)	Hypochlorite ion (ClO^-)	3.5×10^{-8}	7.46
9	Ammonium ion (NH_4^+)	Ammonia (NH_3)	5.6×10^{-10}	9.25
10	Hydrogen carbonate ion (HCO_3^-)	Carbonate ion (CO_3^{2-})	4.8×10^{-11}	10.32

*Adapted from Masterton, W. L., and Hurley, C. N. *Chemistry—Principles and Reactions*, 4th ed. Philadelphia: Harcourt College Publishers, 2001; p. 416.

Notice that as K_a decreases, the pK_a increases; therefore, the weaker the acid, the larger its pK_a.

base pair should you use to make a pH 4.00 buffer solution, and what molar ratio of the compounds should you use? Use Table 17.1 for K_a and pK_a values.

Answer The $CH_3CHOHCOOH/CH_3CHOHCOO^-$ conjugate acid-base pair with a molar ratio of 0.10 mol/L lactic acid to 0.14 mol/L lactate would work.

Strategy and Explanation Because you want to have a buffer with a pH of 4.00, you need an acid-base pair whose conjugate acid has a pK_a near the desired pH of 4.00. You can use Table 17.1 to evaluate and select the conjugate acid-base pair whose pK_a is closest to pH = 4.00. Once that acid-base pair has been selected, you can use the Henderson-Hasselbalch equation to calculate the necessary conjugate acid and conjugate base concentrations to give a buffer with pH 4.00.

The acid with the pK_a (3.85) closest to the target pH is lactic acid, $CH_3CHOHCOOH$; the pK_a values of the other acids are not close enough to 4.00.

Substituting this pK_a and the lactic acid-lactate conjugate acid-base pair concentration terms into the Henderson-Hasselbalch equation gives

$$\text{pH} = 4.00 = 3.85 + \log\frac{[CH_3CHOHCOO^-]}{[CH_3CHOHCOOH]}$$

$$\log\frac{[CH_3CHOHCOO^-]}{[CH_3CHOHCOOH]} = 4.00 - 3.85 = 0.15; \quad \frac{[CH_3CHOHCOO^-]}{[CH_3CHOHCOOH]} = 10^{0.15} = 1.41$$

This means that the required concentration of lactate ion, $CH_3CHOHCOO^-$, will be roughly 1.41 times that of lactic acid, $CH_3CHOHCOOH$. The buffer could be made using lactic acid and a soluble lactate salt, such as sodium lactate, $Na^+CH_3CHOHCOO^-$. If the concentration of lactic acid is 0.10 mol/L, the required concentration of sodium lactate is 1.41(0.10 mol/L) = 0.14 mol/L.

✓ **Reasonable Answer Check** The target pH is 4.00. We can see from the Henderson-Hasselbalch equation that with equal concentrations of conjugate acid and base, the pH would equal the pK_a, 3.85. Therefore, to reach the target pH of 4.00, the concentration of lactate ion, the conjugate base, must be greater than that of lactic acid, the conjugate acid. Adding more conjugate base than conjugate acid raises the pH of the buffer pair from 3.85 to 4.00.

PROBLEM-SOLVING PRACTICE 17.3

Use the data in Table 17.1 to select a conjugate acid-base pair you could use to make buffer solutions having each of these hydrogen ion concentrations.

(a) 3.2×10^{-4} M (b) 5.0×10^{-5} M

(c) 7.0×10^{-8} M (d) 6.0×10^{-11} M

EXERCISE 17.3 Making a Buffer Solution

Use data from Table 17.1 to calculate the molar ratio of sodium acetate and acetic acid needed to make a buffer of pH 4.68.

EXERCISE 17.4 Buffers and pH

Use data from Table 17.1 to calculate the pH of these buffers.

(a) H_2CO_3 (0.10 M)/HCO_3^- (0.25 M) (b) $H_2PO_4^-$ (0.10 M)/HPO_4^{2-} (0.25 M)

When acid (H_3O^+) is added to a buffer, the acid reacts with the conjugate base of the buffer to form its conjugate acid:

Conjugate base in buffer + H_3O^+ added $\longrightarrow$ conjugate acid of the buffer base + water

When base (OH^-) is added to a buffer, the base reacts with the conjugate acid of the buffer, which is converted into its conjugate base:

Conjugate acid in buffer + OH^- added $\longrightarrow$ conjugate base of the buffer acid + water

Figure 17.2 summarizes these relationships. For a buffer made from acetic acid and sodium acetate, the changes are

$$\underset{\text{Conjugate base of the buffer}}{CH_3COO^-(aq)} + \underset{\text{added}}{H_3O^+(aq)} \longrightarrow \underset{\text{Conjugate acid of the buffer}}{CH_3COOH(aq)} + H_2O(\ell)$$

$$\underset{\text{Conjugate acid of the buffer}}{CH_3COOH(aq)} + \underset{\text{added}}{OH^-(aq)} \longrightarrow \underset{\text{Conjugate base of the buffer}}{CH_3COO^-(aq)} + H_2O(\ell)$$

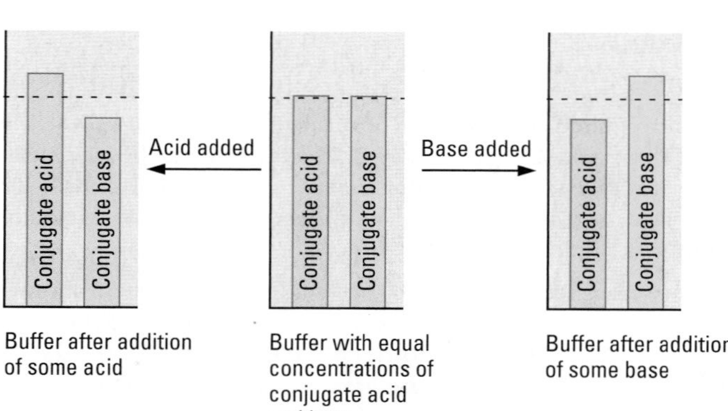

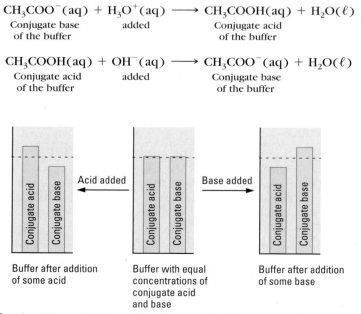

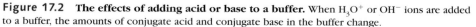

Figure 17.2 The effects of adding acid or base to a buffer. When H_3O^+ or OH^- ions are added to a buffer, the amounts of conjugate acid and conjugate base in the buffer change.

The presence of a conjugate acid together with its conjugate base can form a buffer, but not in every case; it depends on the amounts of conjugate acid and conjugate base. Consider these three examples:

(a) The combination of 0.010 mol $NaHCO_3$ with 0.010 mol Na_2CO_3 in 1.00 L solution. In this case, the conjugate acid bicarbonate ion, HCO_3^-, and the conjugate base carbonate ion, CO_3^{2-}, are in equal concentrations, so the solution is a buffer.

(b) The combination of 0.045 mol ammonia, NH_3, with 0.025 mol HCl in 1.00 L solution.

$$NH_3(aq) + HCl(g) \longrightarrow NH_4Cl(aq)$$

In this case, there are 0.045 mol of the weak base ammonia, NH_3; 0.025 mol of it reacts with 0.025 mol HCl to form 0.025 mol NH_4Cl, leaving 0.020 mol NH_3 of the original 0.045 mol unreacted. The reaction produces 0.025 mol NH_4Cl, which provides 0.025 mol ammonium ion, NH_4^+, the conjugate acid. Thus, the solution contains 0.020 mol conjugate base (NH_3) and 0.025 mol conjugate acid (NH_4^+); the solution is a buffer.

(c) The combination of 0.010 mol acetic acid, CH_3COOH, the conjugate acid, with 0.010 mol NaOH in 1.00 L solution.

$$CH_3COOH(aq) + NaOH(aq) \longrightarrow NaCH_3COO(aq) + HOH(\ell)$$

Here, 0.010 mol acetic acid reacts with 0.010 mol NaOH to form 0.010 mol sodium acetate, $NaCH_3COO$. Essentially all of the acetic acid has reacted, leaving no conjugate acid for a conjugate acid/base pair with acetate ion, CH_3COO^-. Therefore, the solution is not a buffer.

In Case (c) a small amount of acetic acid would be formed by the hydrolysis of sodium acetate, but the amount is not sufficient to act significantly as part of a buffer pair with acetate ion.

The pH Change on Addition of an Acid or a Base to a Buffer

The pH of a buffer changes when acid or base is added to it due to shifts in the concentration of conjugate acid and conjugate base in the buffer. The extent of the pH change depends on the amount of acid or base added and on the amounts of conjugate acid or conjugate base remaining in the buffer to offset additional amounts of acid or base to be added. As shown in Problem-Solving Example 17.4, the change in pH of a buffer when acid or base is added to it can be calculated in two ways: (1) by using the K_a expression in conjunction with the K_a value, or (2) by using the Henderson-Hasselbalch equation.

PROBLEM-SOLVING EXAMPLE 17.4 pH Changes in a Buffer

A buffer is prepared by adding 0.15 mol lactic acid, $CH_3CHOHCOOH$, and 0.20 mol sodium lactate, $Na^+CH_3CHOHCOO^-$, to sufficient water to make 1.00 L of buffer solution. The K_a of lactic acid is 1.4×10^{-4}.

(a) Calculate the pH of the buffer.

(b) Calculate the pH of the buffer after 0.050 mol HCl has been added (neglect volume changes).

(c) Calculate the pH of the buffer after 0.10 mol NaOH has been added to the original buffer (neglect volume changes).

Answer (a) 3.97 (b) 3.73 (c) 4.63

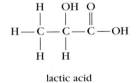

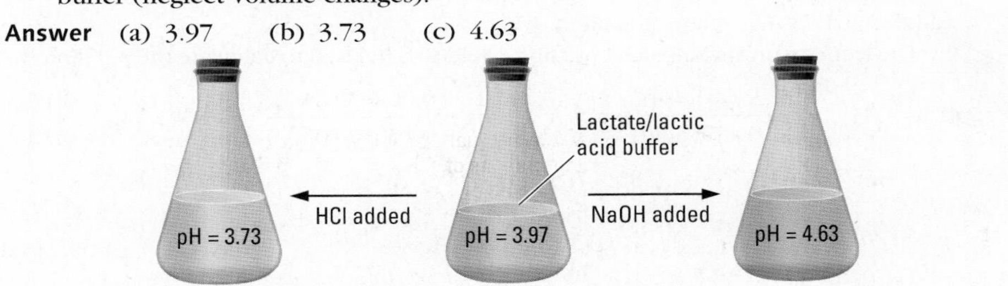

Comparison of pH of lactate/lactic acid buffer after addition of HCl or NaOH.

```
   H   OH  O
   |   |   ||
H—C—C—C—OH
   |   |
   H   H
```

lactic acid

Strategy and Explanation These calculations can be done using the K_a expression to calculate hydronium ion concentration, from which the pH can be obtained. The Henderson-Hasselbalch equation could also be used.

(a) The chemical equation for the equilibrium and the equilibrium constant expression are

$$CH_3CHOHCOOH(aq) + H_2O(\ell) \rightleftharpoons H_3O^+(aq) + CH_3CHOHCOO^-(aq)$$

$$K_a = 1.4 \times 10^{-4} = \frac{[H_3O^+][CH_3CHOHCOO^-]}{[CH_3CHOHCOOH]}$$

Solving for $[H_3O^+]$ gives the value of 1.1×10^{-4} M and a corresponding pH of 3.97.

$$[H_3O^+] = 1.4 \times 10^{-4}\frac{[CH_3CHOHCOOH]}{[CH_3CHOHCOO^-]} = 1.4 \times 10^{-4}\frac{(0.15\text{ M})}{(0.20\text{ M})} = 1.1 \times 10^{-4}\text{ M}$$

$$\text{pH} = -\log(1.1 \times 10^{-4}) = 3.97$$

Using the Henderson-Hasselbalch equation to calculate the pH begins with finding the pK_a,

$$pK_a = -\log K_a = -\log(1.4 \times 10^{-4}) = 3.85$$

and then calculating the pH,

$$\text{pH} = 3.85 + \log\frac{(0.20\text{ mol/L lactate})}{(0.15\text{ mol/L lactic acid})} = 3.85 + \log(1.33) = 3.85 + 0.12 = 3.97$$

(b) The 0.050 mol HCl added reacts with 0.050 mol lactate ion to form 0.050 mol lactic acid and water, which adds to the 0.15 mol lactic acid originally in the buffer.

$$CH_3CHOHCOO^-(aq) + H_3O^+(aq) \longrightarrow CH_3CHOHCOOH(aq) + H_2O(\ell)$$

0.050 mol from buffer	0.050 mol added	0.050 mol formed in buffer

This changes the original lactate/lactic acid ratio, so the pH changes because of the added HCl:

There is 1 L of buffer solution so that, in this case, the number of moles is the same as the number of moles per liter—the molarity.

	Moles Lactic Acid	Moles Lactate
Before reaction	0.15	0.20
After reaction	0.15 + 0.050	0.20 − 0.050

$$1.4. \times 10^{-4} = \frac{[H_3O^+][CH_3CHOHCOO^-]}{[CH_3CHOHCOOH]}$$

$$= [H_3O^+]\frac{(0.20 - 0.050)}{(0.15 + 0.050)}$$

$$= [H_3O^+](0.75)$$

$$[H_3O^+] = \frac{1.4 \times 10^{-4}}{0.75} = 1.9 \times 10^{-4}$$

$$\text{pH} = -\log(1.9 \times 10^{-4}) = 3.73$$

The pH drops from 3.97 to 3.73. Note that 0.15 mol lactate remains to react with additional acid that might be added.

The Henderson-Hasselbalch equation could also be used to calculate the pH directly.

$$\text{pH} = 3.85 + \log\frac{[\text{lactate}]}{[\text{lactic acid}]} = 3.85 + \log\frac{(0.20 - 0.050)}{(0.15 + 0.050)} = 3.85 + \log\left(\frac{0.15}{0.20}\right)$$

$$= 3.85 + \log(0.75) = 3.85 + (-0.12) = 3.73$$

(c) To offset the addition of 0.10 mol NaOH requires 0.10 mol lactic acid.

$$CH_3CHOHCOOH(aq) + OH^-(aq) \longrightarrow CH_3CHOHCOO^-(aq) + H_2O(\ell)$$

0.10 mol from buffer; 0.10 mol added; 0.10 mol formed in buffer

	Moles Lactic Acid	Moles Lactate
Before reaction	0.15	0.20
After reaction	0.15 − 0.10 = 0.005	0.20 + 0.10 = 0.30

Addition of NaOH changes the initial lactate/lactic acid ratio from $\frac{0.20}{0.15} = 1.33$ to $\frac{0.30}{0.05} = 6.0$ and a pH change occurs. We can calculate the pH by using the K_a expression to find the hydronium ion concentration, taking into account the changes in lactate and lactic acid concentrations due to the addition of 0.10 mol base.

$$K_a = 1.4 \times 10^{-4} = \frac{[H_3O^+][CH_3CHOHCOO^-]}{[CH_3CHOHCOOH]}$$

$$= \frac{[H_3O^+](0.20 + 0.10 \text{ mol/L})}{(0.15 - 0.10 \text{ mol/L})}$$

$$= [H_3O^+](6.0)$$

$$[H_3O^+] = \frac{1.4 \times 10^{-4}}{6.0} = 2.33 \times 10^{-5};\ \text{pH} = -\log(2.33 \times 10^{-5}) = 4.63$$

The pH rises from the original value of 3.97 to 4.63. This change can also be determined using the Henderson-Hasselbalch equation.

$$\text{pH} = 3.85 + \log\frac{[\text{lactate}]}{[\text{lactic acid}]} = 3.85 + \log\frac{(0.20 + 0.10)}{(0.15 - 0.10)} = 3.85 + \log\left(\frac{0.30}{0.05}\right)$$

$$= 3.85 + \log(6) = 3.85 + 0.78 = 4.63$$

✓ **Reasonable Answer Check** The pH of the initial buffer (3.97) is reasonable because the concentration of conjugate base is slightly greater than that of the conjugate acid; therefore, the ratio is greater than 1 and the log of the ratio is positive, so the pH of the buffer should be slightly greater than the pK_a of the acid (3.85). When acid is added to the buffer, we expect that if the pH changes, it should decrease, and it does—from 3.97 to 3.73. The pH change that occurs when NaOH is added is greater than that for the addition of HCl because more moles of base than acid were added. But in either case, the buffer worked because the changes were less than 1 pH unit; the answers are reasonable.

PROBLEM-SOLVING PRACTICE 17.4

For the lactate-lactic acid buffer given above, calculate the pH when these samples are added to it; assume there is no volume change.

(a) 0.075 mol HCl

(b) 0.025 mol NaOH

CONCEPTUAL EXERCISE 17.5 Blood Buffer Reaction

If an abnormally high CO_2 concentration is present in blood, which phosphorus-containing ion—$H_2PO_4^-$ or HPO_4^{2-}—can counteract the presence of excess CO_2? Explain your answer.

Buffer Capacity

When the ratio of conj base:conj acid changes to 1:10, the pH *decreases* by 1 unit. When the ratio changes to 10:1, the pH *increases* by 1 unit.

The *amounts* of conjugate acid and conjugate base in the buffer solution determine the **buffer capacity**—the quantity of added acid or base added to the buffer that the buffer can accommodate without undergoing significant pH change (more than 1 pH unit). When nearly all of the conjugate acid in a buffer has reacted with added base, adding a little more base can increase the pH significantly, because there is almost no conjugate acid left in the buffer to neutralize the added base. Similarly, if enough acid is added to a buffer to react with all of the buffer's conjugate base and excess acid remains, the pH will decrease significantly. In either case, the buffer capacity has been exceeded. For example, 1 L of a buffer solution that is 0.25 M in CH_3COOH and 0.25 M in CH_3COO^- contains 0.25 mol CH_3COOH and 0.25 mol CH_3COO^-. This buffer can accommodate the addition of up to 0.25 mol H_3O^+ or OH^-, at which point it has used up its buffer capacity. Thus, the initial buffer cannot accommodate the addition of 0.30 mol of strong acid or 0.30 mol of strong base without undergoing a major change in pH. Such additions would use up all of the buffer's conjugate base or all of its conjugate acid, respectively, and exceed the buffer's capacity. The pH would drop or rise accordingly.

PROBLEM-SOLVING EXAMPLE 17.5 Buffer Capacity

A buffer is prepared using 0.25 mol $H_2PO_4^-$ and 0.15 mol HPO_4^{2-} in 500. mL of solution.

(a) Will the buffer capacity be exceeded if 6.2 g KOH are added to it? What will be the pH of the new solution?

(b) Will the buffer capacity be exceeded if 23.0 mL of 6.0 M HCl are added to the original buffer? What is the resulting pH?

Answer (a) No, the buffer capacity will not be exceeded; pH = 7.48. (b) Yes, the buffer capacity is exceeded; pH = 5.61.

Strategy and Explanation The initial pH of the buffer can be calculated using the Henderson-Hasselbalch equation. From Table 17.1, the pK_a of $H_2PO_4^-$ is 7.21.

$$pH = 7.21 + \log\frac{[HPO_4^{2-}]}{[H_2PO_4^-]} = 7.21 + \log\frac{(0.15/0.500)}{(0.25/0.500)}$$

$$= 7.21 + \log(0.60) = 7.21 - 0.22 = 6.99$$

(a) The 6.2 g KOH is 0.11 mol KOH, which contributes 0.11 mol OH^- to be neutralized by the buffer.

$$6.2 \text{ g KOH}\left(\frac{1 \text{ mol KOH}}{56.1 \text{ g KOH}}\right) = 0.11 \text{ mol KOH}$$

The 0.11 mol OH^- added to the buffer is neutralized by reacting with 0.11 mol $H_2PO_4^-$, the conjugate acid of the buffer, to form 0.11 mol HPO_4^{2-}.

$$\underset{\substack{0.11 \text{ mol}\\ \text{from buffer}}}{H_2PO_4^-(aq)} + \underset{\substack{0.11 \text{ mol}\\ \text{from KOH}}}{OH^-} \longrightarrow \underset{\substack{0.11 \text{ mol}\\ \text{formed}}}{HPO_4^{2-}(aq)} + H_2O(\ell)$$

The reaction changes the amounts of HPO_4^{2-} and $H_2PO_4^-$ remaining in the buffer:

	Before Reaction	After Reaction
Moles HPO_4^{2-}	0.15	0.15 + 0.11 = 0.26
Moles $H_2PO_4^-$	0.25	0.25 − 0.11 = 0.14

Because there is still some $H_2PO_4^-$ remaining (0.14 mol), the buffer's capacity was not exceeded by adding 6.2 g KOH. The pH after the KOH addition is

$$pH = 7.21 + \log\frac{[HPO_4^{2-}]}{[H_2PO_4^-]} = 7.21 + \log\frac{(0.26/0.500)}{(0.14/0.500)}$$

$$= 7.21 + \log(1.9) = 7.21 + 0.27 = 7.48$$

(b) Adding 0.0200 L of 6.0 M HCl provides $(0.0230\ \text{L})\dfrac{6.0\ \text{mol H}_3\text{O}^+}{\text{L}} = 1.4 \times 10^{-1}$ mol H_3O^+, which reacts with 1.4×10^{-1} mol HPO_4^{2-} to form 1.4×10^{-1} mol $H_2PO_4^-$.

$$\underset{\text{from buffer}}{\underset{0.14\ \text{mol}}{HPO_4^{2-}(aq)}} + \underset{\text{from HCl}}{\underset{0.14\ \text{mol}}{H_3O^+(aq)}} \longrightarrow \underset{\text{formed}}{\underset{0.14\ \text{mol}}{H_2PO_4^-(aq)}} + H_2O(\ell)$$

	Before Reaction	**After Reaction**
Moles HPO_4^{2-}	0.15	$0.15 - 0.14 = 0.01$
Moles $H_2PO_4^-$	0.25	$0.25 + 0.14 = 0.39$

Thus, the composition of the buffer changes. The number of moles of HPO_4^{2-} *decreases* to 0.01 mol; the moles of $H_2PO_4^-$ *increase* from 0.25 to 0.39. The volume of the resulting solution is 0.523 L due to the addition of 0.0230 L HCl. The pH after addition of HCl is

$$\text{pH} = 7.21 + \log\frac{[HPO_4^{2-}]}{[H_2PO_4^-]} = 7.21 + \log\frac{(0.01/0.523)}{(0.39/0.523)}$$
$$= 7.21 + \log(0.026) = 7.21 - 1.6 = 5.61$$

The pH changed by 1.38 units (from 6.99 to 5.61) showing that the buffer capacity was exceeded by the addition of the HCl. The final conjugate base:conjugate acid ratio was 0.026, smaller than 1:10.

✓ Reasonable Answer Check

(a) Addition of base to a buffer should increase the pH to some extent depending on the amount of base added. The amount of KOH (0.11 mol) added was less than the amount of conjugate acid (0.25 mol) in the buffer. The conjugate base: conjugate acid ratio changed from 0.60 to 1.9; this change did not exceed ten and therefore, the buffer's capacity was not exceeded. The pH change should be less than 1 pH unit, which it is.

(b) The amount of acid added (0.14 mol) lowered the amount of conjugate base remaining in the buffer to 0.01 mol, while the amount of conjugate acid increased to 0.39 mol, thereby making the conjugate acid:conjugate base ratio greater than ten. Consequently, the pH rose by more than one pH unit and the buffer capacity was exceeded.

PROBLEM-SOLVING PRACTICE 17.5

Calculate the minimum mass (grams) of KOH that would have to be added to the initial buffer in Problem-Solving Example 17.5 to exceed its buffer capacity.

17.2 Acid-Base Titrations

In Section 5.8 (*p. 201*) acid-base titrations were described as a method by which the concentration of an acid or a base could be determined. An acid-base titration is carried out by slowly adding a measurable amount of an aqueous solution of a base or acid whose concentration is known to a known volume of an aqueous acid or base whose concentration is to be determined. For example, a standardized base could be added from a buret to a known volume of acid whose concentration is to be determined, as in Figure 17.3. The apparatus usually used for this type of titration is shown in Figure 17.3. A *standard solution* (***p. 201***), one whose concentration is known accurately, is added from the buret, a device that allows the required volume of a solution to be measured accurately. The solution in the buret is known as the **titrant.** The *equivalence point* (***p. 201***) is reached when the stoichiometric amount of titrant has been added, the amount that exactly neutralizes the acid or base being titrated.

Figure 17.3 An acid-base titration setup for titrating an acid sample with NaOH as the titrant. (a) A buret calibrated in 0.1-mL divisions contains an NaOH solution of known concentration. (b) An acid-base indicator is added to the acid solution before the titration begins. The NaOH solution is added slowly from the buret into the acid solution to be titrated. (c) The indicator changes color when the end point is reached.

Detection of the Equivalence Point

A method is needed in a titration to detect the equivalence point. This can be done by using a pH meter (Section 19.7), which electronically monitors the pH of the solution as the titration proceeds. Alternatively, the color change of an acid-base indicator can be used to detect when sufficient titrant has been added (Figure 17.4). The **end point** of a titration occurs when the indicator changes color. The goal is to use an indicator that gives an end point very close to the equivalence point.

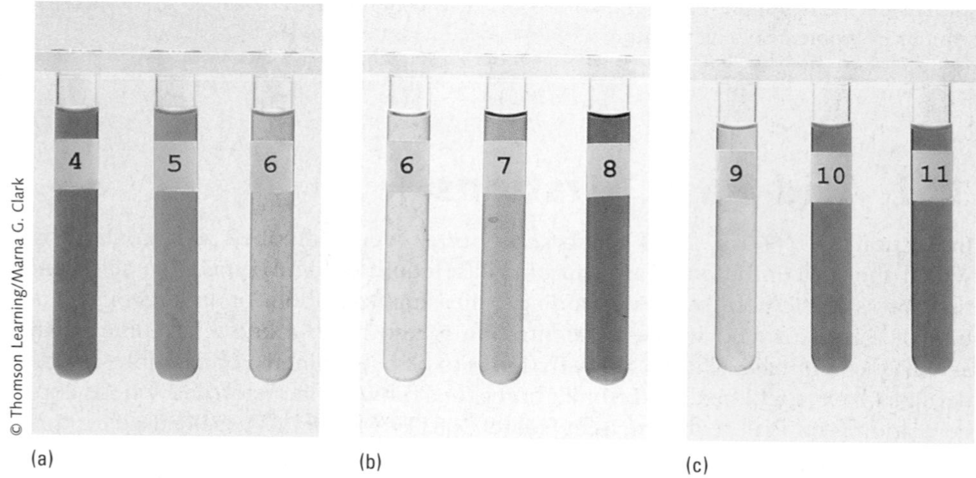

Figure 17.4 Acid-base indicators. Acid-base indicators are compounds that change color in a particular pH range. (a) Methyl red is red at pH 4 or lower, orange at pH 5, and yellow at pH 6.3 and higher. (b) Bromthymol blue changes from yellow to blue as the pH changes from 6 to 8. (c) Phenolphthalein is colorless below a pH of 8.3 and changes to red between 8.3 and 11.

Acid-base indicators, described in Section 16.4 (⇐ ***p. 784***), are typically weak organic acids (HIn) that differ in color from their conjugate bases (In^-).

$$\underset{\text{Color 1}}{HIn(aq)} + H_2O(\ell) \rightleftharpoons H_3O^+(aq) + \underset{\text{Color 2}}{In^-(aq)}$$

Removal of an H^+ from the indicator molecule changes the structure of the indicator molecule so that it absorbs light in a different region of the visible spectrum, thus changing the color of the indicator. For methyl red the reaction is

$$\underset{\text{red}}{\text{(methyl red, acid form)}}(aq) + H_2O(\ell) \rightleftharpoons \underset{\text{yellow}}{\text{(methyl red, base form)}}(aq) + H_3O^+(aq)$$

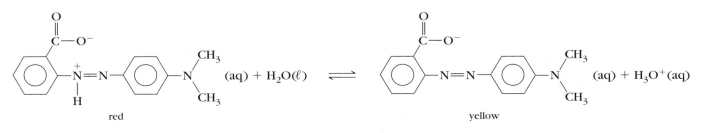

The color observed for the indicator during an acid-base titration depends on the $\frac{[HIn]}{[In^-]}$ ratio, for which three cases apply:

- When $\frac{[HIn]}{[In^-]} \geq 10$, the indicator solution is the acid color (HIn).
- When $\frac{[HIn]}{[In^-]} \leq 0.1$, the indicator solution is the conjugate base color (In^-).
- When $\frac{[HIn]}{[In^-]} \approx 1$, the indicator solution color is intermediate between the acid and the conjugate base colors.

Bromthymol blue, for example, changes from yellow to blue as the pH changes from 6 to 8 (Figure 17.4b).

We will consider in some detail three types of acid-base titrations: (1) a strong acid (HCl) titrated by a strong base (NaOH); (2) a weak acid (CH_3COOH) titrated by a strong base (NaOH); and (3) a weak base (NH_3) titrated by a strong acid (HCl). For each titration we will examine its **titration curve,** a graph of pH as a function of the volume of titrant added. In each case, we will be interested in the pH particularly at four stages of the titration:

- Prior to the addition of titrant
- After addition of titrant, but prior to the equivalence point
- At the equivalence point
- After the equivalence point

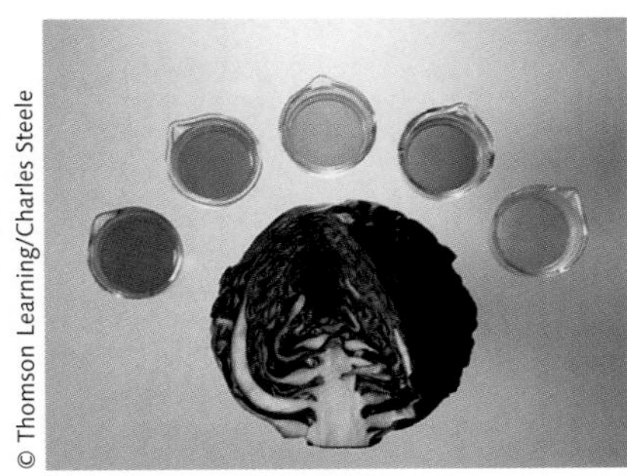

Red cabbage juice is a naturally occurring acid-base indicator. From left to right are solutions of pH 1, 4, 7, 10, and 13.

Titration of a Strong Acid with a Strong Base

The titration curve for the titration of 50.0 mL of 0.100 M HCl with 0.100 M NaOH, the titrant, is given in Figure 17.5. The titration of a strong acid with a strong base produces a neutral salt and the pH = 7.0 at the equivalence point.

In general, prior to the equivalence point in a strong acid–strong base titration, the $[H_3O^+]$ can be calculated for any volume of base added by the relation

$$[H_3O^+] = \frac{\text{original moles acid} - \text{total moles base added}}{\text{volume acid (L)} + \text{volume base added (L)}}$$

This equation assumes that the added volumes equal the exact volume of the mixed solutions.

Problem-Solving Example 17.6 illustrates calculations for the four points marked on the curve.

ThomsonNOW

Go to the Coached Problems menu for tutorials on:

- **titration curves**
- **interpreting a pH-titration curve**

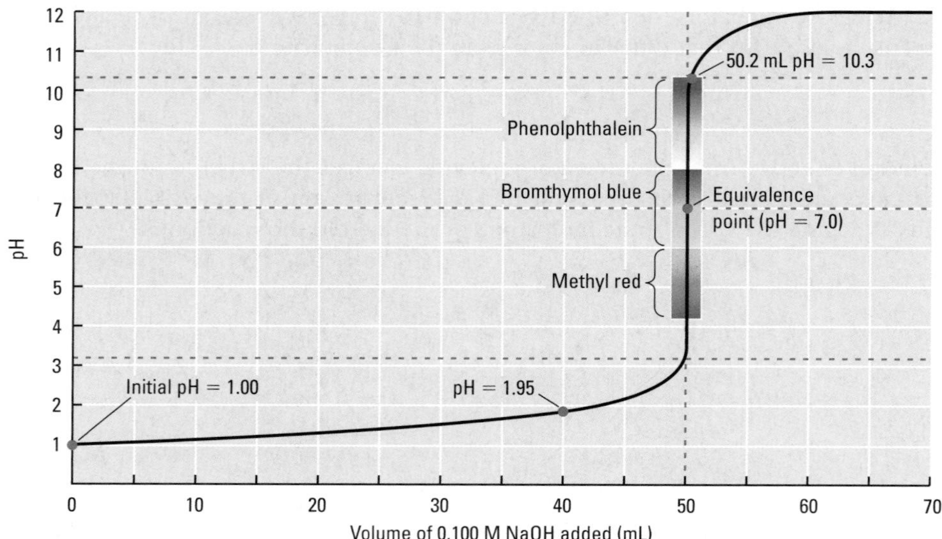

Figure 17.5 Curve for titration of 0.100 M HCl with 0.100 M NaOH. This strong acid reacts with this strong base to form a solution with a pH of 7.0 at the equivalence point.

PROBLEM-SOLVING EXAMPLE 17.6 Titration of HCl with NaOH

A 0.100 M NaOH solution is used to titrate 50.0 mL of 0.100 M HCl. Calculate the pH of the solution at these four points:

(a) Before any titrant is added

(b) After 40.0 mL of titrant has been added

(c) After 50.0 mL of NaOH has been added. What indicator—methyl red, bromthymol blue, or phenolphthalein—can be used to detect the equivalence point? See Figure 17.4.

(d) After 50.2 mL of NaOH has been added

Answer

(a) 1.00 (b) 1.95

(c) 7.00; methyl red, bromthymol blue, or phenolphthalein

(d) 10.3

Strategy and Explanation Initially only hydrochloric acid is present, so the pH is dictated by its hydronium ion concentration (part a). As the titration proceeds, the H_3O^+ concentration decreases as NaOH is added. In a strong acid–strong base titration, the $[H_3O^+]$ can be calculated prior to the equivalence point for any volume of base added by the relation

$$[H_3O^+] = \frac{\text{original moles acid} - \text{total moles base added}}{\text{volume acid (L)} + \text{volume base added (L)}}$$

We will apply that relation in part (b).

(a) Because HCl is a strong acid, the initial H_3O^+ concentration is 0.100 M and the pH is $-\log(0.100) = 1.00$. (We will round the pH values to two significant figures. See Appendix A.6 for treatment of significant figures when using logarithms.)

(b) The initial 50.0-mL solution of acid contains

$$(0.0500 \text{ L})(0.100 \text{ mol/L}) = 5.00 \times 10^{-3} \text{ mol } H_3O^+$$

As NaOH is added, the number of moles of H_3O^+ decreases due to the reaction of added OH^- ions with H_3O^+ ions in the acid.

$$\underset{\text{from HCl}}{H_3O^+(aq)} + \underset{\text{from NaOH}}{OH^-(aq)} \longrightarrow H_2O(\ell)$$

After 40.0 mL of 0.100 M NaOH is added to the original 50.0 mL of 0.100 M HCl, the $[H_3O^+]$ is

$$[H_3O^+] = \frac{(5.00 \times 10^{-3}) - (4.00 \times 10^{-3})}{0.0500\ L + 0.0400\ L} = 1.11 \times 10^{-2}\ M; \qquad pH = 1.95$$

(c) At the equivalence point, 50.0 mL of 0.100 M NaOH has been added. This amounts to (0.0500 L)(0.100 mol OH^-/L) = 5.00×10^{-3} mol OH^-, which exactly neutralizes the 5.00×10^{-3} mol H_3O^+ initially in the solution. No residual acid or excess NaOH is present. The NaCl produced is a neutral salt, so the pH is 7.00 at the equivalence point. Because the pH rises so rapidly near the equivalence point, methyl red, bromthymol blue, or phenolphthalein can be used as the indicator in a strong acid–strong base titration (Figure 17.5).

(d) Adding 50.2 mL of 0.100 M NaOH to the solution puts 5.02×10^{-3} mol OH^- into the solution: (0.0502 L)(0.100 mol OH^-/L) = 5.02×10^{-3} mol OH^-. As seen in part (c), 5.00×10^{-3} mol OH^- neutralized all of the HCl in the initial sample. Therefore, the additional 0.02×10^{-3} mol OH^-, now in 100.2 mL of solution, is not neutralized. The pH of the solution is

$$[OH^-] = \frac{0.02 \times 10^{-3}\ \text{mol OH}^-}{0.1002\ L} = 2.0 \times 10^{-4}\ M$$

$$pOH = -\log(2.0 \times 10^{-4}) = 3.70; \qquad pH = 14.00 - pOH = 14.00 - 3.70 = 10.3$$

Notice that the addition of just 0.2 mL of excess NaOH to the now unbuffered solution (part(c)) dramatically raises the pH, as seen from the titration curve (Figure 17.5).

A volume of 0.2 mL is approximately 4 drops.

PROBLEM-SOLVING PRACTICE 17.6

For the HCl-NaOH titration described above, calculate the pH when these volumes of NaOH have been added:

(a) 10.0 mL (b) 25.00 mL (c) 45.0 mL (d) 50.5 mL

EXERCISE 17.6 Titration Curve

Draw the titration curve for the titration of 50.0 mL of 0.100 M NaOH using 0.100 M HCl as the titrant.

Titration of a Weak Acid with a Strong Base

As noted in Section 16.8 (***p. 800***), the reaction of a weak acid, such as acetic acid, with a strong base, like NaOH, produces a salt—sodium acetate in this case—that has a basic anion. Like other basic anions, the acetate ions react with water to produce hydroxide ions due to the hydrolysis reaction:

$$CH_3COO^-(aq) + H_2O(\ell) \rightleftharpoons CH_3COOH(aq) + OH^-(aq)$$

As a result, when a weak acid is titrated with a strong base, the pH of the solution at the equivalence point will be greater than 7 due to hydrolysis of the basic anion formed by the titration reaction. The titration curve in Figure 17.6 for the titration of 50.0 mL of 0.100 M acetic acid with 0.100 M NaOH represents this type of titration, and Problem-Solving Example 17.7 illustrates the calculations associated with the titration curve.

Notice in Figure 17.6 that the initial pH of 0.100 M acetic acid (2.87) is higher than that of the 0.100 M HCl (1.00) in Figure 17.5. This is to be expected because acetic acid is a much weaker acid than HCl. Acetic acid is only slightly ionized (K_a of acetic acid = 1.8×10^{-5}), and the pH of 0.100 M acetic acid is 2.78, larger than $-\log(10^{-1}) = 1$ (***p. 785***). Also notice from Figure 17.6 that the rapidly rising

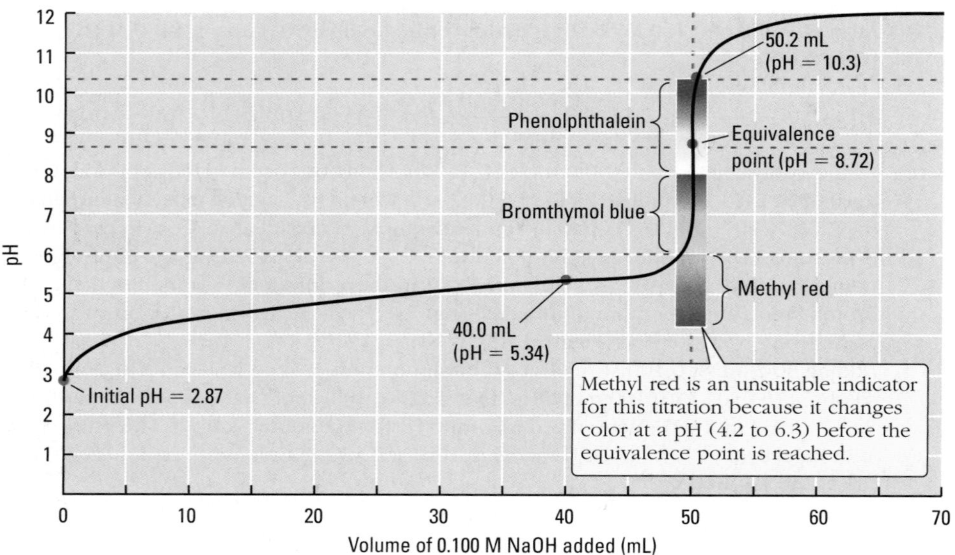

Figure 17.6 **Curve for titration of 50.0 mL of 0.100 M acetic acid with 0.100 M NaOH.**

portion of the titration curve near the equivalence point is shorter than it is for the NaOH-HCl titration (Figure 17.5). The equivalence point in Figure 17.6 is 8.72, making methyl red an unsuitable indicator because its color changes (pH 4 to 6.3) before the equivalence point is reached. Bromthymol blue or phenolphthalein can be used.

PROBLEM-SOLVING EXAMPLE 17.7 Titration of CH_3COOH with NaOH

A 0.100 M NaOH solution is used to titrate 50.0 mL of 0.100 M acetic acid. Calculate the pH of the solution at these three points:

(a) After 40.0 mL of titrant has been added
(b) After 50.0 mL of NaOH has been added
(c) After 50.2 mL of NaOH has been added

Answer (a) 5.34 (b) 8.72 (c) 10.3

Strategy and Explanation

(a) The acetic acid sample contains (0.0500 L)(0.100 mol acetic acid/L) = 5.00×10^{-3} mol acetic acid. Adding 40.0 mL of 0.100 M NaOH puts 4.00×10^{-3} mol OH^- ions into the solution, which neutralizes 4.00×10^{-3} mol acetic acid and forms 4.00×10^{-3} mol acetate ions; 1.00×10^{-3} mol acetic acid remains un-neutralized.

$$\underset{\substack{5.00 \times 10^{-3}\text{ mol}\\ \text{in acid soln}}}{CH_3COOH(aq)} + \underset{\substack{4.00 \times 10^{-3}\\ \text{mol added}}}{OH^-(aq)} \longrightarrow \underset{\substack{4.00 \times 10^{-3}\\ \text{mol formed}}}{CH_3COO^-(aq)} + H_2O(\ell)$$

The total volume of the solution is now 90.0 mL and the concentrations are

$$\text{Acetic acid} = \frac{1.00 \times 10^{-3}\text{ mol}}{0.0900\text{ L}} = 0.0111\text{ M}$$

$$\text{Acetate ion} = \frac{4.00 \times 10^{-3}\text{ mol}}{0.0900\text{ L}} = 0.0444\text{ M}$$

The pH can also be calculated using the Henderson-Hasselbalch equation.

The pH can be calculated using the K_a value and expression for acetic acid to solve for $[H_3O^+]$ and then pH.

$$K_a = \frac{[H_3O^+][CH_3COO^-]}{[CH_3COOH]} = 1.8 \times 10^{-5}$$

$$= \frac{[H_3O^+][0.00400 \text{ mol}/0.0900 \text{ L}]}{[0.00100 \text{ mol}/0.0900 \text{ L}]}$$

$$= [H_3O^+](4.00)$$

$$[H_3O^+] = \frac{1.8 \times 10^{-5}}{4.00} = 4.50 \times 10^{-6}; \qquad pH = -\log(4.50 \times 10^{-6}) = 5.34$$

(b) At this point, 5.00×10^{-3} mol OH^- has been added to 5.0×10^{-3} mol acetic acid initially present, so the stoichiometric amount of base has been added to exactly neutralize the acid in the sample. This is the equivalence point. The reaction has produced 5.00×10^{-3} mol acetate ion, whose concentration is 0.0500 M.

$$\frac{5.00 \times 10^{-3} \text{ mol acetate}}{0.100 \text{ L solution}} = 0.0500 \text{ M}$$

The pH at the equivalence point is governed by the hydrolysis of acetate ion:

$$CH_3COO^-(aq) + H_2O(\ell) \rightleftharpoons CH_3COOH(aq) + OH^-(aq)$$

We can use the K_b expression to calculate $[OH^-]$ and from it the pH. The K_b for acetate ion can be calculated from K_w and the K_a for acetic acid.

$$K_b = \frac{K_w}{K_a} = \frac{1.0 \times 10^{-14}}{1.8 \times 10^{-5}} = 5.6 \times 10^{-10}$$

Substituting into the K_b expression we let $x = [OH^-] = [CH_3COOH]$.

$$K_b = 5.6 \times 10^{-10} = \frac{[CH_3COOH][OH^-]}{[CH_3COO^-]} = \frac{x^2}{0.0500 - x}$$

Because K_b is small, we can approximate $0.0500 - x$ to be 0.0500. Solving for x,

$$K_b = 5.6 \times 10^{-10} = \frac{x^2}{0.0500}; \qquad x = 5.3 \times 10^{-6} = [OH^-]$$

which converts to a pOH of 5.28 and a pH of 8.72. This pH is in marked contrast to the equivalence point of 7.00 for the NaOH-HCl titration, where neutral NaCl was the titration product.

(c) The pH beyond the equivalence point is controlled by the OH^- concentration from excess NaOH, which is greater than the OH^- contributed by the hydrolysis of acetate ion. Therefore, the calculation for the pH beyond the equivalence point is like that for the NaOH-HCl titration with excess NaOH.

$$\text{Final } OH^- \text{ concentration} = \frac{0.02 \times 10^{-3} \text{ mol of } OH^-}{0.1002 \text{ L}} = 2.00 \times 10^{-4} \text{ M}$$

$$pOH = -\log(2.00 \times 10^{-4}) = 3.70; \qquad pH = 14.00 - pOH = 14.00 - 3.70 = 10.3$$

PROBLEM-SOLVING PRACTICE 17.7

Calculate the pH when these volumes of 0.100 M NaOH have been added when titrating 50.0 mL of 0.100 M acetic acid:

(a) 10.0 mL (b) 25.00 mL (c) 45.0 mL (d) 51.0 mL

As seen from Figures 17.5 and 17.6 and their associated Problem-Solving Examples, there are differences in the titration curves for a strong base with a strong acid

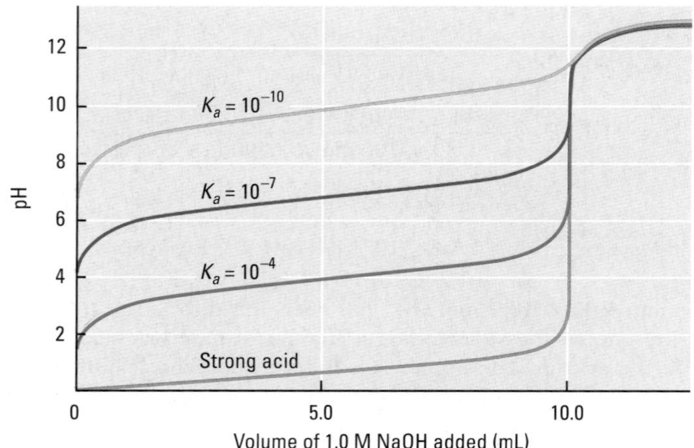

Figure 17.7 The effect of acid strength on the shape of the titration curve. Each curve is for titration of 10.0 mL of a 1.0 M acid with 1.0 M NaOH.

or a weak acid of equal concentration. Comparing Figures 17.5 and 17.6 we see in particular that

- Before the titration, the initial pH of the solution is higher for the weak acid.
- Very near the equivalence point, the length of the rapid rise of the curve is shorter for the weak acid.
- The pH at the equivalence point is higher for the weak acid titration.

A shorter rise in pH means that greater care is needed to select an appropriate indicator.

These features are shown in Figure 17.7. Note especially that near the equivalence point, the weaker the acid, the higher the pH and the shorter the rise in pH.

EXERCISE 17.7 Titration of Acetic Acid with NaOH

Use the K_a expression and value for acetic acid to calculate the pH after 30.0 mL of 0.100 M NaOH has been added to 50.0 mL of 0.100 M acetic acid.

CONCEPTUAL EXERCISE 17.8 Shape of the Titration Curve

Explain why the NaOH–acetic acid titration curve in Figure 17.6 has a relatively flat region between ~10.0 and ~40.0 mL of NaOH added.

Titration of a Weak Base with a Strong Acid

The titration of the weak base NH_3 with the strong acid HCl has the titration curve shown in Figure 17.8. The reaction produces ammonium chloride, NH_4Cl. Notice that because NH_3 is a weak base, the starting pH is greater than 7.0, but less than it would be for 0.100 M NaOH (13.00). Also notice that the pH at the equivalence point, 5.28, is less than 7.0 because ammonium chloride is an acidic salt.

The pH of 0.100 *M* NaOH = $-\log\left(\frac{10^{-14}}{10^{-1}}\right) = 13$

$$NH_4^+(aq) + H_2O(\ell) \rightleftharpoons NH_3(aq) + H_3O^+(aq)$$

Beyond the equivalence point the pH continues to drop as excess acid is added. Although methyl red is a suitable indicator for this titration, phenolphthalein or bromthymol blue are not suitable because their color changes occur well before the equivalence point.

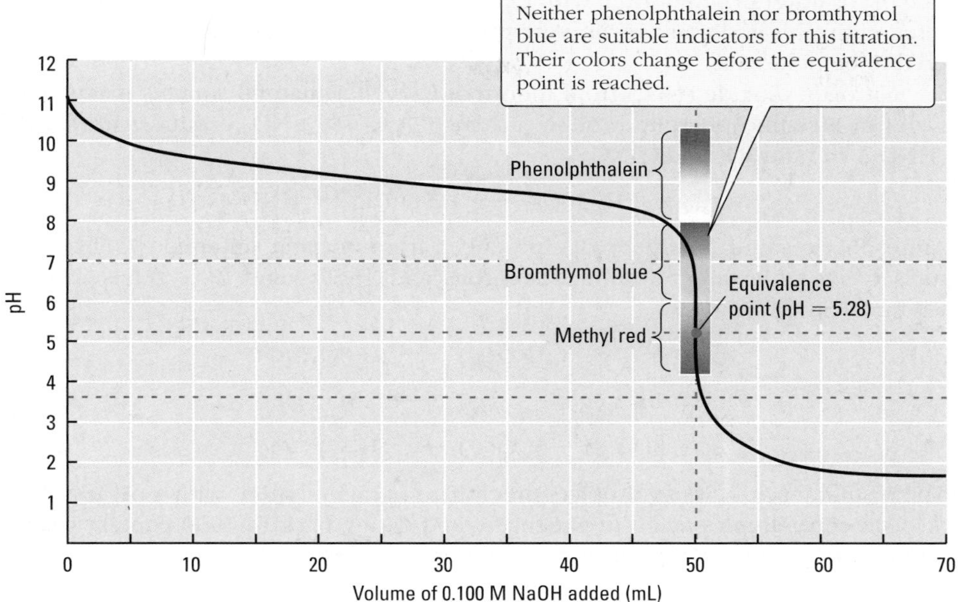

Figure 17.8 The titration of a weak base, NH_3, by a strong acid, HCl. This titration curve is for the titration of 50.0 mL of 0.100 M NH_3 with 0.100 M HCl. It is essentially the inverse of the curve for the titration of a weak acid by a strong base, Figure 17.6.

Titration of a Polyprotic Acid with Base

Polyprotic acids—those with more than one ionizable hydrogen—react stepwise when titrated with bases, one step for each ionizable hydrogen. If the K_a values of the ionizable forms of the acid are sufficiently different, the titration curve has an equivalence point for each of the hydrogens removed from the acid molecule by titration. For example, maleic acid, HOOC—CH=CH—COOH, is a diprotic acid with two ionizable hydrogens, one from each —COOH group. Its titration with NaOH occurs in two steps:

$$\text{HOOC—CH=CH—COOH(aq)} + \text{OH}^-\text{(aq)} \rightleftharpoons \text{HOOC—CH=CH—COO}^-\text{(aq)} + \text{H}_2\text{O}(\ell)$$

$$\text{HOOC—CH=CH—COO}^-\text{(aq)} + \text{OH}^-\text{(aq)} \rightleftharpoons {}^-\text{OOC—CH=CH—COO}^-\text{(aq)} + \text{H}_2\text{O}(\ell)$$

As shown in Figure 17.9, the two equivalence points are at pH = 4.1 and pH = 9.4.

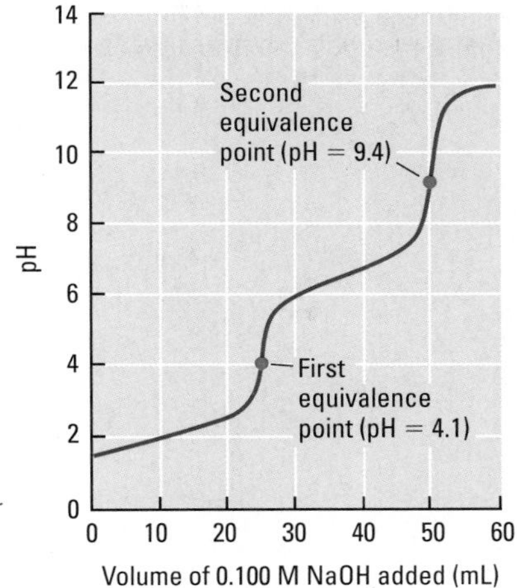

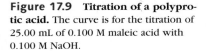

Figure 17.9 Titration of a polyprotic acid. The curve is for the titration of 25.00 mL of 0.100 M maleic acid with 0.100 M NaOH.

17.3 Acid Rain

The term *acid rain* was first used in 1872 by Robert Angus Smith, an English chemist and climatologist. In his book *Air and Rain,* Smith used the term to describe the acidic precipitation that fell on Manchester, England, at the start of the Industrial Revolution. Although neutral water has a pH of 7, water in rain and in some lakes and rivers becomes acidified naturally from dissolved carbon dioxide, a normal component of the atmosphere. The carbon dioxide reacts reversibly with water to form a solution of carbonic acid, a weak acid, which ionizes into hydronium and hydrogen carbonate ions.

$$2\,H_2O(\ell) + CO_2(g) \rightleftharpoons H_2CO_3(aq) + H_2O(\ell) \rightleftharpoons H_3O^+(aq) + HCO_3^-(aq)$$

Although the term "acid rain" is commonly used, the more accurate term is "acid deposition," which takes into account acidic snow, sleet, rain, and fog.

The pH of water in equilibrium with CO_2 from the air is about 5.6, so natural, unpolluted rainwater is slightly acidic. Any precipitation with a pH below 5.6 is considered to be **acid rain.**

Nitrogen dioxide (NO_2) from industrial as well as natural sources reacts with water in the atmosphere to produce acids ***(⇐ p. 469);*** NO_2 produces nitric acid (HNO_3) and nitrous acid (HNO_2).

$$2\,NO_2(g) + H_2O(\ell) \longrightarrow HNO_3(aq) + HNO_2(aq)$$

Atmospheric sulfur dioxide (SO_2), produced from burning sulfur-containing fossil fuels, reacts with water to produce sulfurous acid (H_2SO_3) and, if oxygen is present, sulfuric acid (H_2SO_4).

$$SO_2(g) + H_2O(\ell) \longrightarrow H_2SO_3(aq)$$

$$2\,SO_2(g) + O_2(g) \longrightarrow 2\,SO_3(g)$$

$$SO_3(g) + H_2O(\ell) \longrightarrow H_2SO_4(aq)$$

The resulting acidic water droplets precipitate as rain or snow with a pH less than 5.6. Ice core samples taken in Greenland and dating back to 1900 contain sulfate (SO_4^{2-}) and nitrate (NO_3^-) ions indicating that acid rain has been commonplace, at least from 1900 onward.

Approximately 160 million metric tons of SO_2 and 85 million metric tons of nitrogen oxides are emitted annually into the atmosphere from human activities, largely the burning of fossil fuels. Each year about 43 million metric tons of SO_2 and 59 million metric tons of nitrogen oxides are put into the atmosphere by natural sources.

Acid rain is a problem today due to the large amounts of these acidic oxides being put into the atmosphere by human activities every year (Figure 17.10). When such precipitation falls on areas without naturally occurring bases such as limestone and other carbonate minerals that offset the acidity, serious environmental damage can occur. The average annual pH of precipitation falling on much of the northeastern United States and northeastern Europe is between 4 and 4.5 (Figure 17.11). Rain in those areas has had pH values as low as 1.5. To further complicate matters, acid rain is an international problem because precipitation carried by winds does not observe international borders. Canadian residents are offended by the fact that much of the acid precipitation falling on Canadian cities and forests results from acidic oxides produced in the United States.

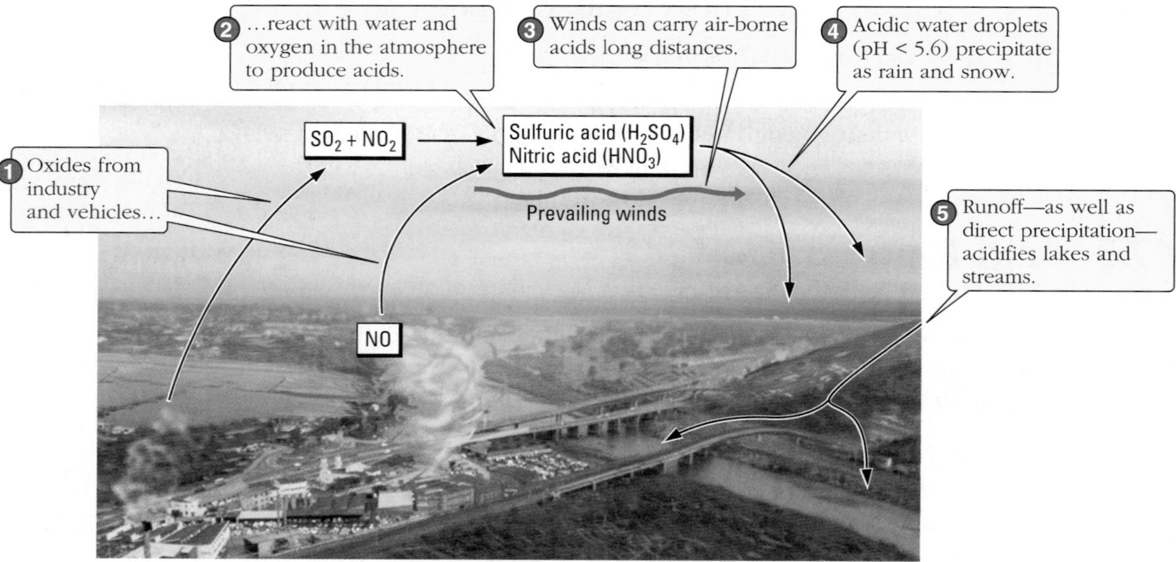

Figure 17.10 **How acid deposition occurs.**

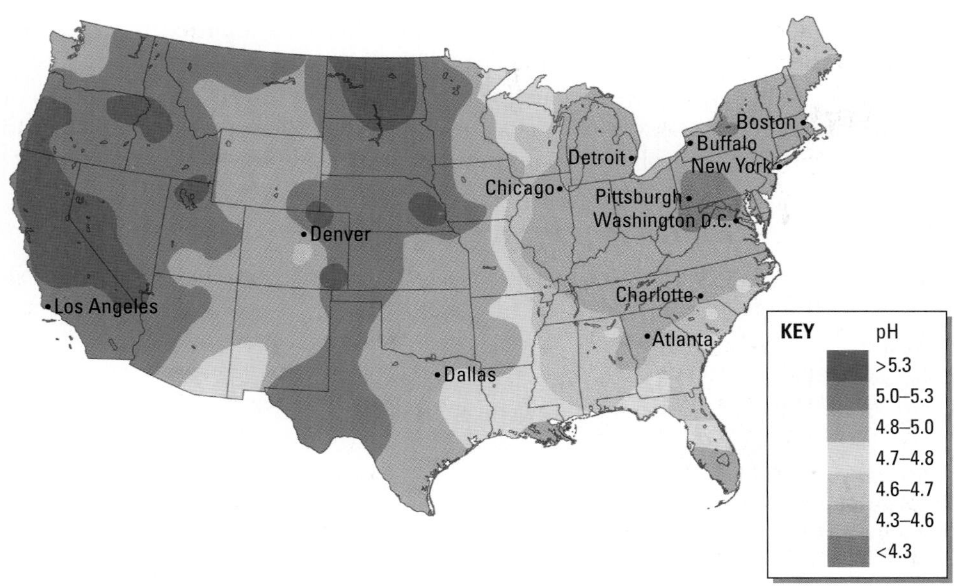

Figure 17.11 **pH of precipitation within the contiguous United States.**

17.4 Solubility Equilibria and the Solubility Product Constant, K_{sp}

Many ionic compounds, such as NH_4Cl and $NaNO_3$, are very soluble in water. Many other ionic compounds, however, are only modestly or slightly water-soluble; they produce *saturated* solutions of 0.001 M or less, far less in some cases. Consider the case of a saturated aqueous solution of silver bromide, AgBr, a light-sensitive ionic compound used in photographic film. When sufficient AgBr is added to water, some of it dissolves to form a saturated solution, and some undissolved AgBr is present. The dissolved AgBr forms aqueous Ag^+ and Br^- ions in solution that are in equilibrium with the undissolved solid AgBr.

ThomsonNOW
Go to the Chemistry Interactive menu for a module on the **precipitation reaction between I^- and Pb^{2+}.**

$$AgBr(s) \rightleftharpoons Ag^+(aq) + Br^-(aq)$$

This balanced chemical equation represents a solubility equilibrium. As with other equilibria (*p. 675*), we can derive an equilibrium constant from the chemical equation. In this case, the equilibrium constant is called the **solubility product constant, K_{sp}**. The magnitude of K_{sp} indicates the extent to which the solid solute dissolves to give ions in solution. To evaluate the equilibrium constant, we first must write a **solubility product constant expression.** For the chemical equation given above, the solubility product constant expression is

The solubility product constant is commonly just called the solubility product.

$$K_{sp} = [Ag^+][Br^-]$$

In general, the balanced chemical equation for dissolving a slightly soluble salt with the general formula A_xB_y is

$$A_xB_y(s) \rightleftharpoons x\,A^{n+}(aq) + y\,B^{m-}(aq)$$

This results in the general K_{sp} expression

$$K_{sp} = [A^{n+}]^x[B^{m-}]^y$$

A solubility product expression has the same general form that other equilibrium constant expressions have, except there is no denominator in the K_{sp} expression, because the reactant is always a pure solid.

Notice that

- The chemical equation related to the solubility product constant expression is written for a solid solute compound as a reactant and its aqueous ions as the products.

ThomsonNOW

Go to the Coached Problems menu for tutorials on:

- **solubility product constant expressions**
- **exploring the solubility product constant**
- **determining the solubility product constant**

- The concentration of the pure solid solute reactant is omitted from the K_{sp} expression because it remains unchanged as the reaction occurs.
- The K_{sp} value equals the product of the *equilibrium* molar concentrations of the cation and the anion, each raised to the power given by the coefficient in the balanced chemical equation representing the solubility equilibrium.

PROBLEM-SOLVING EXAMPLE 17.8 Writing K_{sp} Expressions

Write the K_{sp} expression for each of these slightly soluble salts:
(a) $Fe(OH)_2$ (b) MgC_2O_4 (c) Ag_3PO_4

Answer

(a) $K_{sp} = [Fe^{2+}][OH^-]^2$
(b) $K_{sp} = [Mg^{2+}][C_2O_4^{2-}]$
(c) $K_{sp} = [Ag^+]^3[PO_4^{3-}]$

Strategy and Explanation To write the correct K_{sp} expression we first must write the balanced equation for the dissociation of the solute, and then substitute the products in it appropriately into the K_{sp} expression.

(a) The equilibrium reaction for the solubility of $Fe(OH)_2$ in water is

$$Fe(OH)_2(s) \rightleftharpoons Fe^{2+}(aq) + 2\ OH^-(aq)$$

In this example, the cation:anion relationship is 2:1; for every Fe^{2+} ion there are two OH^- ions produced, so the hydroxide ion concentration is raised to the power of 2 in the K_{sp} expression: $K_{sp} = [Fe^{2+}][OH^-]^2$.

(b) The equilibrium reaction for the solubility of MgC_2O_4 in water is

$$MgC_2O_4(s) \rightleftharpoons Mg^{2+}(aq) + C_2O_4^{2-}(aq)$$

Since the equilibrium chemical equation shows the cations and anions in a one-to-one relationship, the equilibrium constant expression is $K_{sp} = [Mg^{2+}][C_2O_4^{2-}]$.

(c) The equilibrium reaction for the solubility of Ag_3PO_4 in water is

$$Ag_3PO_4(s) \rightleftharpoons 3\ Ag^+(aq) + PO_4^{3-}(aq)$$

Since three Ag^+ ions are produced for every PO_4^{3-} ion, the K_{sp} expression is written $K_{sp} = [Ag^+]^3[PO_4^{3-}]$.

PROBLEM-SOLVING PRACTICE 17.8

Write the K_{sp} expression for each of these slightly soluble salts:
(a) CuBr (b) HgI_2 (c) $SrSO_4$

Table 17.2 K_{sp} Values for Some Slightly Soluble Salts

Compound	K_{sp} at 25 °C
AgBr	3.3×10^{-13}
AuBr	5.0×10^{-17}
$AuBr_3$	4.0×10^{-36}
CuBr	5.3×10^{-9}
Hg_2Br_2*	1.3×10^{-22}
$PbBr_2$	6.3×10^{-6}
AgCl	1.8×10^{-10}
AuCl	2.0×10^{-13}
$AuCl_3$	3.2×10^{-25}
CuCl	1.9×10^{-7}
Hg_2Cl_2*	1.1×10^{-18}
$PbCl_2$	1.7×10^{-5}
AgI	1.5×10^{-13}
AuI	1.6×10^{-13}
AuI_3	1.0×10^{-46}
CuI	5.1×10^{-12}
Hg_2I_2*	4.5×10^{-29}
HgI_2	4.0×10^{-29}
PbI_2	8.7×10^{-9}
Ag_2SO_4	1.7×10^{-5}
$BaSO_4$	1.1×10^{-10}
$PbSO_4$	1.8×10^{-8}
Hg_2SO_4*	6.8×10^{-7}
$SrSO_4$	2.8×10^{-7}

*These compounds contain the diatomic ion Hg_2^{2+}.

Solubility and K_{sp}

The solubility of a sparingly soluble solute and its solubility product constant, K_{sp}, are not the same thing, but they are related. The solubility is the amount of solute per unit volume of solution (mol/L) that dissolves to form a *saturated* solution. On the other hand, the solubility product constant is the equilibrium constant for the chemical equilibrium that exists between a solid ionic solute and its ions in a saturated solution. If the equilibrium concentrations of the ions are known, they can be used to calculate the K_{sp} value for the solute. For example, in a saturated AgCl solution at 10 °C, the molar concentrations of Ag^+ and Cl^- each are experimentally determined to be 6.3×10^{-6} M. This means that K_{sp} at 10 °C is

$$K_{sp} = [Ag^+][Cl^-] = [6.3 \times 10^{-6}][6.3 \times 10^{-6}] = 4.0 \times 10^{-11}$$

The K_{sp} values for selected ionic compounds at 25 °C are listed in Table 17.2. A more extensive listing is in Appendix H. In general, solutes with very low solubility have very small K_{sp} values.

PROBLEM-SOLVING EXAMPLE 17.9 Solubility and K_{sp}

The K_{sp} of $BaSO_4$ is 1.1×10^{-10} at 25 °C. Calculate the solubility of $BaSO_4$, expressing the result in molarity, moles per liter.

Answer 1.0×10^{-5} M

Strategy and Explanation The solubility product constant expression for barium sulfate is derived from the chemical equation

$$BaSO_4(s) \rightleftharpoons Ba^{2+}(aq) + SO_4^{2-}(aq) \qquad K_{sp} = [Ba^{2+}][SO_4^{2-}] = 1.1 \times 10^{-10}$$

The ionization of solid $BaSO_4$ forms Ba^{2+} ions and SO_4^{2-} ions in equal amounts. Therefore, if we let S equal the solubility of $BaSO_4$, then at equilibrium the concentration of Ba^{2+} and SO_4^{2-} ions will each be S.

$$1.1 \times 10^{-10} = (S)(S) = S^2$$

$$S = \sqrt{1.1 \times 10^{-10}} = 1.0 \times 10^{-5}\ \text{M}$$

Consequently, the aqueous solubility of $BaSO_4$ at 25 °C is 1.0×10^{-5} M.

PROBLEM-SOLVING PRACTICE 17.9

The K_{sp} of AgBr at 100 °C is 5×10^{-10}. Calculate the solubility of AgBr at that temperature in moles per liter.

A note of caution is in order here. It might seem perfectly straightforward to calculate the solubility of an ionic compound from its K_{sp}, the calculation just completed in Problem-Solving Example 17.9, or to do the reverse, that is, calculate the K_{sp} from the solubility. Doing so, however, will often lead to incorrect answers. This approach is too simplified and overlooks several complicating factors. One is that ionic solids such as $PbCl_2$ dissociate stepwise, so that $PbCl^+$ ions as well as Pb^{2+} and Cl^- are present in a $PbCl_2$ solution. Also, ion pairs such as $PbCl^+Cl^-$ can exist, reducing the concentrations of unassociated Pb^{2+} and Cl^-. In addition, the solubilities of some solutes, such as metal hydroxides, depend on the acidity or alkalinity of the solution. Also, solutes containing anions such as CO_3^{2-} and PO_4^{3-} that react with water are more soluble than predicted by their K_{sp} values. Solubilities calculated from K_{sp} values, and K_{sp} values calculated from solubilities, best agree with the experimentally measured solubilities of compounds with 1+ and 1− charged ions and ions that do not react with water.

ThomsonNOW
Go to the Chemistry Interactive menu for a module on **precipitation of $PbCl_2$ by addition of Cl^-.**

PROBLEM-SOLVING EXAMPLE 17.10 Solubility and K_{sp}

A saturated aqueous solution of lead(II) iodate, $Pb(IO_3)_2$, contains 3.1×10^{-5} M Pb^{2+} at 25 °C. Assuming that the ions do not react with water, calculate the K_{sp} of lead(II) iodate at that temperature.

Answer $K_{sp} = 1.2 \times 10^{-13}$

Strategy and Explanation We must write the chemical equation for the dissociation of the solute, derive the K_{sp} expression correctly from it, and substitute concentrations of Pb^{2+} and IO_3^- ions into the expression to obtain K_{sp}. In the saturated solution, the chemical equilibrium is

$$Pb(IO_3)_2(s) \rightleftharpoons Pb^{2+}(aq) + 2\ IO_3^-(aq)$$

and the solubility product expression is $K_{sp} = [Pb^{2+}][IO_3^-]^2$.

We next determine the *molar* concentrations of Pb^{2+} and IO_3^-. From the balanced chemical equation we see that when dissociation of the solid lead(II) iodate occurs, one

You can review molarity calculations in Section 5.7.

mole of lead ions is produced for two moles of iodate ions. Since the Pb^{2+} concentration is given as 3.1×10^{-5} M, the iodate ion concentration is 6.2×10^{-5} M, twice that of lead ion. The K_{sp} value can be calculated directly from these equilibrium concentrations and the solubility product equilibrium constant expression.

$$K_{sp} = [Pb^{2+}][IO_3^-]^2 = (3.1 \times 10^{-5})(6.2 \times 10^{-5})^2 = 1.2 \times 10^{-13}$$

✓ **Reasonable Answer Check** Given the fact that the iodate ion concentration is twice that of the lead ion concentration, substituting these values correctly into the proper K_{sp} expression gives the calculated answer, which is reasonable.

PROBLEM-SOLVING PRACTICE 17.10

A saturated solution of silver oxalate, $Ag_2C_2O_4$, contains 6.9×10^{-5} M $C_2O_4^{2-}$ at 25 °C. Calculate the K_{sp} of silver oxalate at that temperature, assuming that the ions do not react with water.

CONCEPTUAL EXERCISE 17.9 Solubility and Le Chatelier's Principle

At 25 °C, 0.014 g calcium carbonate dissolves in 100 mL water. Two equilibria are present in this solution.

(a) $CaCO_3(s) \rightleftharpoons Ca^{2+}(aq) + CO_3^{2-}(aq)$

(b) $CO_3^{2-}(aq) + H_2O(\ell) \rightleftharpoons HCO_3^-(aq) + OH^-(aq)$

Suppose Reaction (b) occurs to an appreciable extent. Use Le Chatelier's principle (***p. 695***) to predict how the extent of Reaction (b) will affect the solubility of $CaCO_3$.

17.5 Factors Affecting Solubility

ThomsonNOW

Go to the Coached Problems menu for tutorials on:

- **estimating the solubility of salts**

The aqueous solubility of ionic compounds is affected by a number of factors, some of which have already been mentioned—temperature (***p. 734***), the formation of ion pairs, and competing equilibria. In this section we will consider four other factors affecting the aqueous solubility of ionic compounds:

- The effect of acids and pH
- The presence of common ions
- The formation of complex ions
- Amphoterism

© Thomson Learning/Charles D. Winters

Reaction of calcium carbonate with acid. A piece of chalk (calcium carbonate) is dissolved by reacting it with hydrochloric acid. Bubbles of CO_2 gas can be seen being formed by the reaction.

pH and Dissolving Slightly Soluble Salts Using Acids

As noted earlier in the discussion of solubility rules, many salts are only slightly soluble in water (***p. 165***). An acid can dissolve an insoluble salt containing a moderately basic ion. As an example, consider calcium carbonate, $CaCO_3$, which is found in minerals such as limestone and marble. $CaCO_3$ is not very soluble in pure water.

$$\text{(a) } CaCO_3(s) \rightleftharpoons Ca^{2+}(aq) + CO_3^{2-}(aq) \qquad K_{sp} = 8.7 \times 10^{-9}$$

Since the solubility of calcium carbonate is so low, the equilibrium concentrations of Ca^{2+} and CO_3^{2-} must also be small. However, if acid is added to the solution, calcium carbonate will dissolve and CO_2 will be released from the solution. Adding acid adds hydronium ions, which react with carbonate and hydrogen carbonate ions.

The total chloride ion concentration at equilibrium is the amount from AgCl (equals S) *plus* what was already there (0.55 M) from the NaCl. Because NaCl is a soluble salt, far more Cl^- comes from NaCl than from AgCl.

Using the equilibrium concentrations from the table gives

$$K_{sp} = 1.8 \times 10^{-10} = [Ag^+][Cl^-] = (S)(S + 0.55)$$

The easiest approach to solve such an equation is to approximate that S is *very* small compared to 0.55; that is, the answer will be approximately the same if we assume that $(S + 0.55) \approx 0.55$. Such an assumption is reasonable because we know that the solubility of AgCl equals only 1.3×10^{-5} mol/L *without* the Cl^- added from NaCl. When NaCl is added, it will further decrease the solubility of AgCl due to the presence of the common ion Cl^-. Therefore,

$$(S)(S + 0.55) \approx (S)(0.55) = K_{sp}$$

or

$$K_{sp} = (S)(0.55) = 1.8 \times 10^{-10}$$

Solving for S, we get

$$S = \frac{1.8 \times 10^{-10}}{0.55} = 3.3 \times 10^{-10}\ \text{M} = [Ag^+]$$

Therefore, the $[Ag^+]$, which is the same as S, is approximately 3.3×10^{-10} mol/L.

Using the molar mass for AgCl, 143.4 g/mol, the solubility of AgCl (g/L) in 0.55 M NaCl is

$$\left(\frac{3.3 \times 10^{-10}\ \text{mol AgCl}}{1\ \text{L}}\right)\left(\frac{143.3\ \text{g AgCl}}{1\ \text{mol AgCl}}\right) = 4.7 \times 10^{-8}\ \text{g AgCl/L}$$

As predicted by Le Chatelier's principle, the solubility of AgCl in the presence of Cl^- added from another source (3.3×10^{-10} M) is clearly less than in pure water (1.3×10^{-5} M).

✓ **Reasonable Answer Check** To check the approximation we made, we substitute the approximate value of S into the exact expression $K_{sp} = (S)(S + 0.55)$. Then, if the product $(S)(S + 0.55)$ is the same as the given value of K_{sp}, the approximation is valid.

$$K_{sp} = (S)(S + 0.55) = (3.3 \times 10^{-10})(3.3 \times 10^{-10} + 0.55) = 1.8 \times 10^{-10}$$

A more accurate solution to this problem can be obtained by solving for S using the quadratic equation described in Appendix A.7 and used in Problem-Solving Example 16.10 (⇐ *p. 797*). When the quadratic equation is used, its answer, to two significant figures, is the same as our approximation.

PROBLEM-SOLVING PRACTICE 17.11

Calculate the solubility of $PbCl_2$ at 25 °C in a solution that is 0.50 M in NaCl.

PROBLEM-SOLVING EXAMPLE 17.12 pH and Common Ion Effect

Manganese(II) hydroxide, $Mn(OH)_2$, is sparingly soluble in water: K_{sp} of $Mn(OH)_2 = 4.6 \times 10^{-14}$ at 25 °C. Calculate the solubility of manganese(II) hydroxide at that temperature in (a) pure water and (b) at a pH of 11.00.

Answer (a) 2.3×10^{-5} M (b) 4.6×10^{-8} M

Strategy and Explanation In part (a) we apply solubility equilibrium concepts to the equilibrium between a solute and its ions in solution in the case where no common ion is present. In part (b) where the initial solution has a pH of 11.00, there will be a significant

ThomsonNOW

Go to the Coached Problems menu for tutorials on:

- **solubility and pH**
- **precipitation reaction systems**
- **determining whether a precipitation reaction will occur**
- **the solubility of metal ions**
- **the common ion effect**

hydroxide ion concentration before any solid $Mn(OH)_2$ has been added to the solution. After the addition, there are two possible sources of hydroxide ion, the common ion: (1) the initial solution and (2) that resulting from the dissolution of solid manganese(II) hydroxide.

(a) The chemical equilibrium is

$$Mn(OH)_2(s) \rightleftharpoons Mn^{2+}(aq) + 2\ OH^-(aq)$$

and the equilibrium constant expression is $K_{sp} = [Mn^{2+}][OH^-]^2 = 4.6 \times 10^{-14}$. Let S equal the solubility of $Mn(OH)_2$. From the solubility equation we see that the concentration of Mn^{2+} is S, that is, S moles per liter of Mn^{2+} for each mole per liter of $Mn(OH)_2$ that dissolves. Hydroxide ion concentration is $2S$ (Mn^{2+} and OH^- are in a 1:2 mole ratio). We can summarize the equilibrium concentrations in a table.

	$Mn(OH)_2(s)$ $\rightleftharpoons$	$Mn^{2+}(aq)$	+	$2\ OH^-(aq)$
Initial concentration (mol/L)		0		0*
Change as reaction occurs (mol/L)		$+S$		$+2S$
Equilibrium concentration (mol/L)		$+S$		$+2S$

*We assume that the contribution of OH^- from the ionization of water is negligible.

We substitute into the K_{sp} expression

$$K_{sp} = [Mn^{2+}][OH^-]^2 = (S)(2S)^2 = 4.6 \times 10^{-14}$$

and solve for S, the solubility of $Mn(OH)_2$.

$$(S)(2S)^2 = 4S^3 = 4.6 \times 10^{-14}$$

$$S = [Mn^{2+}] = \sqrt[3]{4.6 \times 10^{-14}/4} = 2.3 \times 10^{-5}\ M$$

Therefore, the solubility of $Mn(OH)_2$ equals 2.5×10^{-5} M, the Mn^{2+} concentration; the OH^- concentration is $2S$, which is 5.0×10^{-5} M.

(b) Before any $Mn(OH)_2$ is added to the solution, the pH is 11.00. Thus, initially the $[H_3O^+] = 1.0 \times 10^{-11}$ M and the $[OH^-] = \dfrac{1.0 \times 10^{-14}}{1.0 \times 10^{-11}} = 1.0 \times 10^{-3}$ M. When $Mn(OH)_2(s)$ is added, let x equal the concentration of $Mn(OH)_2$ that dissolves, thus forming x mol/L of Mn^{2+} and $2x$ mol/L of OH^-. In the solution there are two possible sources of hydroxide ions: (1) the initial solution in which the hydroxide concentration is 1.0×10^{-3} M and (2) the hydroxide ions arising from the dissociation of dissolved $Mn(OH)_2$. Correspondingly, the OH^- concentration at equilibrium will be $1.0 \times 10^{-3} + 2x$.

	$Mn(OH)_2(s)$ $\rightleftharpoons$	$Mn^{2+}(aq)$	+	$2\ OH^-(aq)$
Initial concentration (mol/L)		0		1.0×10^{-3}
Change as addition of $Mn(OH)_2$ (mol/L)		$+x$		$+2x$
Equilibrium concentration (mol/L)		$+x$		$1.0 \times 10^{-3} + 2x$

To calculate the solubility of $Mn(OH)_2$ we substitute these into the K_{sp} expression.

$$K_{sp} = [Mn^{2+}][OH^-]^2 = (x)(1.0 \times 10^{-3} + 2x)^2 = 4.6 \times 10^{-14}$$

The solubility of $Mn(OH)_2$ in pure water is very slight, as calculated in part (a). In addition, the dissolution will be suppressed by the presence of hydroxide ions (the common ion) in the solution at pH 11.00, so we will assume that $2x << 1.0 \times 10^{-3}$, which simplifies the equation to

$$(x)(1.0 \times 10^{-3})^2 \approx 4.6 \times 10^{-14}$$

$$x \approx \frac{4.6 \times 10^{-14}}{1.0 \times 10^{-6}} = 4.6 \times 10^{-8} \text{ M}$$

Thus, the presence of hydroxide as the common ion decreased the solubility of $Mn(OH)_2$ from that in pure water, 2.5×10^{-5} M, to 4.6×10^{-8} M in a starting solution of pH 11.00.

✓ **Reasonable Answer Check** In part (a) the solubility is described by the expression $(S)(2S)^2 = 4S^3 = 4.6 \times 10^{-14}$. The answer calculated using this expression is reasonable considering how very small the K_{sp} is. We can check the approximation we made in part (b) by substituting the approximate value of x into the actual expression $K_{sp} = (x)(1.0 \times 10^{-3} + 2x)^2$. Doing so we find that the product $(x)(1.0 \times 10^{-3} + 2x)^2$ equals the given K_{sp} value, so the approximation is legitimate.

PROBLEM-SOLVING PRACTICE 17.12

Calculate the solubility of $PbCl_2$ in (a) pure water and (b) 0.20 M NaCl. K_{sp} of $PbCl_2 = 1.7 \times 10^{-5}$.

The addition of aqueous ammonia to aqueous Cu^{2+} ions forms the intense deep-blue/purple $[Cu(NH_3)_4]^{2+}$ complex ion.

Complex Ion Formation

As pointed out in Section 16.9 (***p. 805***), all metal cations are potential Lewis acids because they can accept an electron pair donated by a Lewis base to form a complex ion. The reaction of Cu^{2+} ions with NH_3 is typical.

$$\underset{\text{Lewis acid}}{Cu^{2+}(aq)} + \underset{\text{Lewis base}}{4\,NH_3(aq)} \rightleftharpoons \underset{\text{complex ion}}{[Cu(NH_3)_4]^{2+}(aq)}$$

Many metal salts that are insoluble in water are brought into solution by complex ion formation with Lewis bases such as $S_2O_3^{2-}$, NH_3, OH^-, and CN^-. For example, the solubility of AgBr is very low in water, 1.35×10^{-4} g/L, equivalent to 7.19×10^{-7} M, but AgBr dissolves readily in a sodium thiosulfate ($Na_2S_2O_3$) solution due to the formation of the $[Ag(S_2O_3)_2]^{3-}$ complex ion (Figure 17.12).

$$AgBr(s) + 2\,S_2O_3^{2-}(aq) \rightleftharpoons [Ag(S_2O_3)_2]^{3-}(aq) + Br^-(aq)$$

The dissolving of AgBr in this way can be considered as the sum of two reactions—the solubility equilibrium of aqueous AgBr and the formation of the complex ion.

$$AgBr(s) \rightleftharpoons Ag^+(aq) + Br^-(aq)$$

$$Ag^+(aq) + 2\,S_2O_3^{2-}(aq) \rightleftharpoons [Ag(S_2O_3)_2]^{3-}(aq)$$

$$\textit{Net reaction}: AgBr(s) + 2\,S_2O_3^{2-}(aq) \longrightarrow [Ag(S_2O_3)_2]^{3-}(aq) + Br^-(aq)$$

This reaction is commercially important for removing unreacted AgBr from photographic film, fixing the image (***p. 199***).

The extent to which complex ion formation occurs can be evaluated from the magnitude of the equilibrium constant for the formation of the complex ion, K_f, called the **formation constant.** For example, the formation constant for $[Ag(S_2O_3)_2]^{3-}$ is 2×10^{13}.

$$Ag^+(aq) + 2\,S_2O_3^{2-}(aq) \rightleftharpoons [Ag(S_2O_3)_2]^{3-}(aq)$$

$$K_f = \frac{[[Ag(S_2O_3)_2]^{3-}]}{[Ag^+][S_2O_3^{2-}]^2} = 2 \times 10^{13}$$

ThomsonNOW

Go to the Coached Problems menu for a tutorial on **simultaneous equilibria: dissolution of AgCl via addition of NH_3.**

ThomsonNOW

Go to the Coached Problems menu for tutorials on:

- **solubility of complex ions**

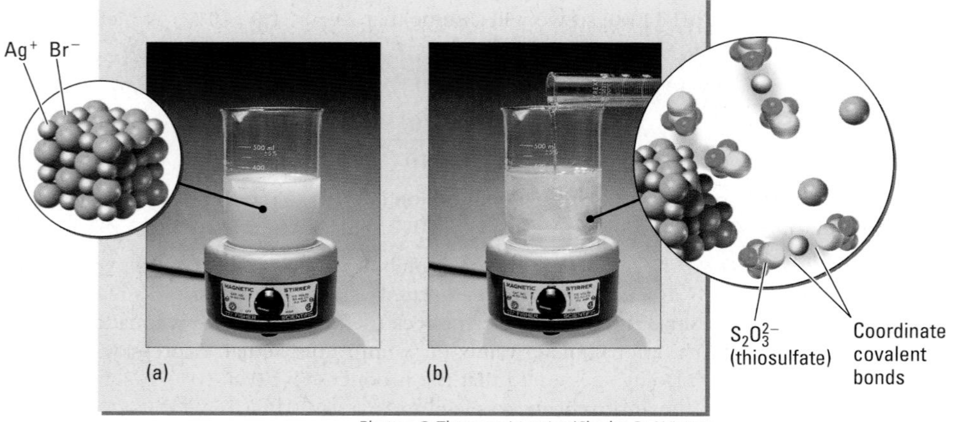

Active Figure 17.12 **Sodium thiosulfate dissolves silver bromide.** (a) Silver bromide (white solid) is insoluble in water. (b) When aqueous sodium thiosulfate is added, the AgBr dissolves. Water molecules have been omitted from the illustration for simplicity's sake. **Go to the Active Figures menu at ThomsonNOW to test your understanding of the concepts in this figure.**

Formation constants for some metal complex ions are given in Table 17.3. The formation and structure of complex ions, which are very important in biochemistry and metallurgy, are considered in more detail in Chapter 22.

PROBLEM-SOLVING EXAMPLE 17.13 Solubility and Complex Ion Formation

The K_{sp} of AgCl is 1.8×10^{-10}. The K_f of $[Ag(CN)_2]^-$ is 5.6×10^{18}. Use these data to show that AgCl will dissolve in aqueous NaCN.

Answer The net equilibrium constant is 1×10^9, which indicates that dissolving AgCl by complex ion formation is highly favored and AgCl will dissolve in NaCN.

Strategy and Explanation To answer this question requires using the solubility product constant for AgCl and the formation constant for $[Ag(CN)_2]^-$ to determine the overall equilibrium constant for the reaction of CN^- with AgCl. The magnitude of an equilibrium constant indicates whether a reaction is product-favored; a $K >> 1$ indicates a very product-favored reaction ***(p. 686).***

The net reaction for dissolving AgCl by $[Ag(CN)_2]^-$ complex ion formation is the sum of the K_{sp} and K_f equations.

$$AgCl(s) \rightleftharpoons Ag^+(aq) + Cl^-(aq) \qquad K_{sp} = 1.8 \times 10^{-10}$$

$$Ag^+(aq) + 2\,S_2O_3^{2-}(aq) \rightleftharpoons Ag(S_2O_3)_2^{3-}(aq) \qquad K_f = 2.0 \times 10^{13}$$

Net reaction: $AgCl(s) + 2\,S_2O_3^{2-}(aq) \rightleftharpoons Ag(S_2O_3)_2^{3-}(aq) + Cl^-(aq)$

Therefore, the equilibrium constant for the net reaction is the product of K_{sp} and K_f: $K_{net} = K_{sp} \times K_f = (1.8 \times 10^{-10})(2.0 \times 10^{13}) = 3.6 \times 10^3$. Because K_{net} is much greater than 1, the net reaction is product-favored, and AgCl is much more soluble in a Na_2SO_4 solution than it is in water.

PROBLEM-SOLVING PRACTICE 17.13

The K_{sp} of AgCl is 1.8×10^{-10}. The K_f of $[Ag(S_2O_3)_2]^{3-}$ is 2×10^{13}. Use these data to show that dissolving AgCl by $[Ag(S_2O_3)_2]^{3-}$ complex ion formation is a product-favored process.

Amphoterism

The majority of metal hydroxides are insoluble in water, but many dissolve in highly acidic or basic solutions. This is because these hydroxides are *amphoteric;* that is,

Table 17.3 Formation Constants for Some Complex Ions in Aqueous Solution

Formation Equilibrium	K_f
$Ag^+ + 2\ Br^- \rightleftharpoons [AgBr_2]^-$	1.3×10^7
$Ag^+ + 2\ Cl^- \rightleftharpoons [AgCl_2]^-$	2.5×10^5
$Ag^+ + 2\ CN^- \rightleftharpoons [Ag(CN)_2]^-$	5.6×10^{18}
$Ag^+ + 2\ S_2O_3^{2-} \rightleftharpoons [Ag(S_2O_3)_2]^{3-}$	2.0×10^{13}
$Ag^+ + 2\ NH_3 \rightleftharpoons [Ag(NH_3)_2]^+$	1.6×10^7
$Al^{3+} + 6\ F^- \rightleftharpoons [AlF_6]^{3-}$	5.0×10^{-3}
$Al^{3+} + 4\ OH^- \rightleftharpoons [Al(OH)_4]^-$	7.7×10^{33}
$Au^+ + 2\ CN^- \rightleftharpoons [Au(CN)_2]^-$	2.0×10^{38}
$Cd^{2+} + 4\ CN^- \rightleftharpoons [Cd(CN)_4]^{2-}$	1.3×10^{17}
$Cd^{2+} + 4\ Cl^- \rightleftharpoons [CdCl_4]^{2-}$	1.0×10^4
$Cd^{2+} + 4\ NH_3 \rightleftharpoons [Cd(NH_3)_4]^{2+}$	1.0×10^7
$Co^{2+} + 6\ NH_3 \rightleftharpoons [Co(NH_3)_6]^{2+}$	7.7×10^4
$Cu^+ + 2\ CN^- \rightleftharpoons [Cu(CN)_2]^-$	1.0×10^{16}
$Cu^+ + 2\ Cl^- \rightleftharpoons [CuCl_2]^-$	1.0×10^5
$Cu^{2+} + 4\ NH_3 \rightleftharpoons [Cu(NH_3)_4]^{2+}$	6.8×10^{12}
$Fe^{2+} + 6\ CN^- \rightleftharpoons [Fe(CN)_6]^{4-}$	7.7×10^{36}
$Hg^{2+} + 4\ Cl^- \rightleftharpoons [HgCl_4]^{2-}$	1.2×10^{15}
$Ni^{2+} + 4\ CN^- \rightleftharpoons [Ni(CN)_4]^{2-}$	1.0×10^{31}
$Ni^{2+} + 6\ NH_3 \rightleftharpoons [Ni(NH_3)_6]^{2+}$	5.6×10^8
$Zn^{2+} + 4\ OH^- \rightleftharpoons [Zn(OH)_4]^{2-}$	2.9×10^{15}
$Zn^{2+} + 4\ NH_3 \rightleftharpoons [Zn(NH_3)_4]^{2+}$	2.9×10^9

they can react with both H_3O^+ ions and OH^- ions (*p. 805*). Aluminum hydroxide, $Al(OH)_3$, is an example of an amphoteric hydroxide (Figure 17.13). When it reacts with acid, aluminum hydroxide dissolves by acting as a base, donating OH^- ions to react with hydronium ions from the acid to form water.

$$Al(OH)_3(s) + 3\ H_3O^+(aq) \longrightarrow Al^{3+}(aq) + 6\ H_2O(\ell)$$

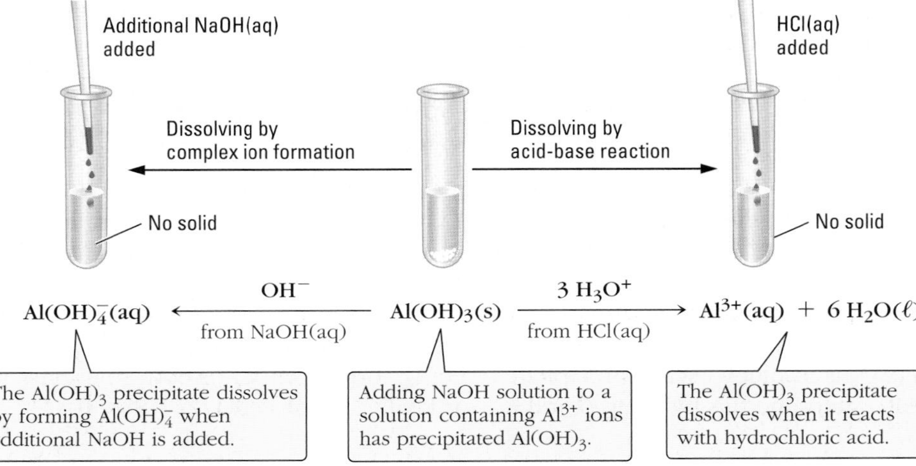

Figure 17.13 The amphoteric nature of $Al(OH)_3$.

© Thomson Learning/Charles D. Winters

Precipitation of lead iodide. Adding a drop of aqueous potassium iodide solution to an aqueous lead(II) nitrate solution precipitates yellow lead(II) iodide. Potassium nitrate, a soluble salt, remains dissolved in the solution.

In highly basic solutions, $Al(OH)_3$ is dissolved through complex ion formation. In this case, Al^{3+} ions act as a Lewis acid by accepting an electron pair from OH^- ions, a Lewis base, to form $[Al(OH)_4]^-$.

$$Al(OH)_3(s) + OH^-(aq) \longrightarrow [Al(OH)_4]^-(aq)$$

17.6 Precipitation: Will It Occur?

Earlier, when writing net ionic equations, we used the solubility rules to predict whether a precipitate will form when ions in two solutions are mixed. Those rules apply to situations where the ions involved are at concentrations of 0.1 M or greater. If the ion concentrations are considerably less than 0.1 M, precipitation may or may not occur. The result depends on the concentrations of the ions in the resulting solution and the K_{sp} value for any precipitate that might form.

For example, AgBr might precipitate when a water-soluble silver salt, such as $AgNO_3$, is added to an aqueous solution of a bromide salt, such as KBr. The net ionic equation for the reaction is

$$Ag^+(aq) + Br^-(aq) \rightleftharpoons AgBr(s)$$

This is the reverse of the equation for K_{sp} of AgBr:

$$AgBr(s) \rightleftharpoons Ag^+(aq) + Br^-(aq)$$

Q is the reaction quotient introduced in Chapter 14 (p. 689).

The bracket notation, [ ], represents molarity at *equilibrium*.

To determine whether a precipitate will form, we compare the magnitude of the ion product, Q, with that of the solubility product constant, K_{sp}. The Q expression has the same form as that for K_{sp}. For Q, however, the *original* concentrations are used, not those at equilibrium as in K_{sp}. For AgBr the two expressions are

$$Q = (\text{conc. } Ag^+)(\text{conc. } Br^-) \qquad K_{sp} = [Ag^+][Br^-]$$

When the value of Q is compared with that of the K_{sp}, three cases are possible (Figure 17.14).

1. $Q < K_{sp}$ ***The solution is unsaturated and no precipitate forms.*** In this case, the solution contains ions at a concentration lower than required for equilibrium with the solid. An equilibrium is not established between a solid solute and its ions because no solid solute is present; more solute can be added to the solution before precipitation occurs. If solid were present, more solid would dissolve.
2. $Q > K_{sp}$ ***The solution contains a higher concentration of ions than it can hold at equilibrium; that is, the solution is supersaturated.*** To reach equilibrium, a precipitate forms, decreasing the concentration of ions until the ion product equals the K_{sp}.
3. $Q = K_{sp}$ ***The solution is saturated with ions and is at equilibrium and at the point of precipitation.***

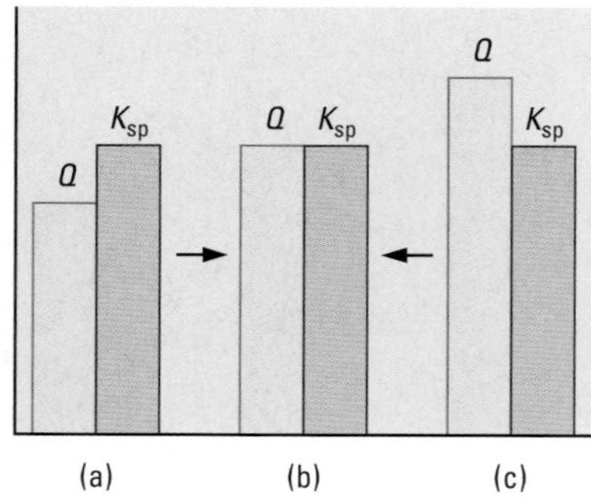

Figure 17.14 Predicting precipitation. (a) When $Q < K_{sp}$, the solution is unsaturated, and no precipitation occurs. (b) $Q = K_{sp}$: the solution is saturated and just at the point of precipitation. (c) $Q > K_{sp}$: the solution is supersaturated, and precipitation occurs until $Q = K_{sp}$.

Consider the case of two solutions, each made by combining $Pb(NO_3)_2$ and Na_2SO_4 solutions. In one case (solution 1) when the solutions are mixed the initial concentrations of Pb^{2+} and SO_4^{2-} are each 1.0×10^{-4} M. In the other case (solution 2) these concentrations are each 2.0×10^{-4}, twice that of the first solution. In each of the two solutions, the products are $NaNO_3$ and $PbSO_4$. The solubility rules indicate that $NaNO_3$ is soluble and remains in solution as Na^+ and NO_3^- ions, whereas $PbSO_4$ is insoluble. Will a precipitate of $PbSO_4$ form in either or both solutions? We can determine this by using Q and K_{sp} for $PbSO_4$; K_{sp} of $PbSO_4 = 1.8 \times 10^{-8}$.

$$Q \text{ of solution 1} = (\text{conc. } Pb^{2+})(\text{conc. } SO_4^{2-})$$
$$= (1.0 \times 10^{-4}\text{ M})(1.0 \times 10^{-4}\text{ M}) = 1.0 \times 10^{-8} < K_{sp} = 1.8 \times 10^{-8}$$

$$Q \text{ of solution 2} = (\text{conc. } Pb^{2+})(\text{conc. } SO_4^{2-})$$
$$= (2.0 \times 10^{-4}\text{ M})(2.0 \times 10^{-4}\text{ M}) = 4.0 \times 10^{-8} > K_{sp} = 1.8 \times 10^{-8}$$

Since Q of solution 1 is less than the K_{sp}, no precipitate will form. In contrast, Q of solution 2 exceeds the K_{sp}, and precipitation will occur.

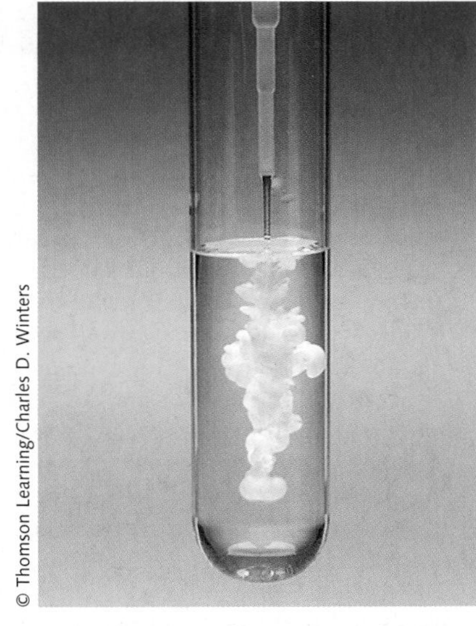
© Thomson Learning/Charles D. Winters

Precipitation of lead sulfate. The addition of sufficiently high concentrations of Pb^{2+} and SO_4^{2-} ions causes $Q > K_{sp}$ and lead sulfate ($PbSO_4$) precipitates.

PROBLEM-SOLVING EXAMPLE 17.14 Q, K_{sp}, and Precipitation

A chemistry student mixes 20.0 mL of 4.5×10^{-3} M $AgNO_3$ with 10.0 mL of 7.5×10^{-2} M $NaBrO_3$. The final volume is 30.0 mL. Will a precipitate of $AgBrO_3$ form? The K_{sp} of $AgBrO_3 = 6.7 \times 10^{-5}$.

Answer Yes, $AgBrO_3$ precipitates.

Strategy and Explanation For a precipitate to form, Q must be greater than K_{sp}. The chemical equilibrium is

$$AgBrO_3(s) \rightleftharpoons Ag^+(aq) + BrO_3^-(aq)$$

The ion product expression is $Q = (\text{conc. } Ag^+)(\text{conc. } BrO_3^-)$; the K_{sp} expression is $K_{sp} = [Ag^+][BrO_3^-]$. To determine whether a precipitate forms, substitute the original concentrations into the Q expression and compare the value with K_{sp}. In calculating the original concentrations, the total volume of 0.0300 L must be taken into account. The number of moles of Ag^+ in 0.0200 L of 4.5×10^{-3} M $AgNO_3$ is

$$(0.0200\text{ L})\left(\frac{4.5 \times 10^{-3}\text{ mol Ag}^+}{1\text{ L}}\right) = 9.0 \times 10^{-5}\text{ mol Ag}^+$$

The Ag^+ concentration in the 0.0300-L mixture is

$$\frac{9.0 \times 10^{-5}\text{ mol Ag}^+}{0.0300\text{ L}} = 3.0 \times 10^{-3}\text{ M Ag}^+$$

Likewise, for the amount of BrO_3^- and its concentration in the 0.0300-L mixture:

$$(0.0100\text{ L})\left(\frac{7.5 \times 10^{-2}\text{ mol BrO}_3^-}{1\text{ L}}\right) = 7.5 \times 10^{-4}\text{ mol BrO}_3^-$$

$$\frac{7.5 \times 10^{-4}\text{ mol BrO}_3^-}{0.0300\text{ L}} = 2.5 \times 10^{-2}\text{ M BrO}_3^-$$

$$Q = (3.0 \times 10^{-3}\text{ M})(2.5 \times 10^{-2}\text{ M}) = 7.5 \times 10^{-5}$$

which is greater than the $K_{sp} = 6.7 \times 10^{-5}$. Precipitation occurs until the ion product equals K_{sp}.

PROBLEM-SOLVING PRACTICE 17.14

(a) Will AgCl precipitate from a solution containing 1.0×10^{-5} M Ag^+ and 1.0×10^{-5} M Cl^-? K_{sp} AgCl $= 1.8 \times 10^{-10}$.

(b) An AgCl precipitate forms from a solution that is 1.0×10^{-5} M Ag^+. What must be the minimum Cl^- concentration in this solution for precipitation to occur?

Kidney Stones—Common Ion Effect and Le Chatelier's Principle

Many ions circulate in our bloodstream, some combinations of which can precipitate to form kidney stones. Such stones can become large enough to be extremely painful and even life-threatening, requiring treatment by drugs, lasers, or surgery.

CHEMISTRY IN THE NEWS

Plant Crystals

Calcium oxalate crystals occur in kidney stones, but can such crystals occur in plants? They can. Calcium oxalate crystals are found in about 75% of flowering plants. The crystals are grown in plants within individual cells in specific tissues and organs, but the biological functions of the crystals are not yet fully understood. In plants, calcium oxalate crystals are found in several crystal forms, the most striking being the druses, which have a carnation-like appearance. The druses and other calcium oxalate crystal forms may be used by plants as a defense against insect predators. The sharp edges of the druses' rigid mineral "petals," and the spear-like points of raphides, another class of the crystals, are pointed enough to cut the mouths of predatory insects.

Calcium oxalate crystals in plants were first observed and identified in the late 1600s by Antonie van Leeuwenhoek, the inventor of the microscope. Since then researchers have discovered that the crystals form within plant cells known as idioblasts. Current researchers conclude that the crystals form as a way for plants to control the concentration of calcium ions absorbed from the soil. Fundamental questions remain such as—How do plants produce the oxalate ions that precipitate along with calcium ions? What roles other than defense do the crystals play? The crystals do play a role in forensic science. By identifying particular crystal types and their location in a plant, along with other cellular characteristics, forensic scientists can identify calcium oxalate crystals when used as adulterants in some herbal medicines.

SOURCE: *Chemical & Engineering News*, February 6, 2006; pp. 26–27.

Photos: Courtesy of Harry T. (Jack) Horner, University Professor, Iowa State University, Ames

(a)

(b)

Two kinds of calcium oxalate crystals. (a) Raphide calcium oxalate crystals isolated from wild flowering coffee plant leaves. (b) Druses calcium oxalate crystals isolated from the leaves of a Yerba Linda plant.

SIU/Visuals Unlimited, Inc.

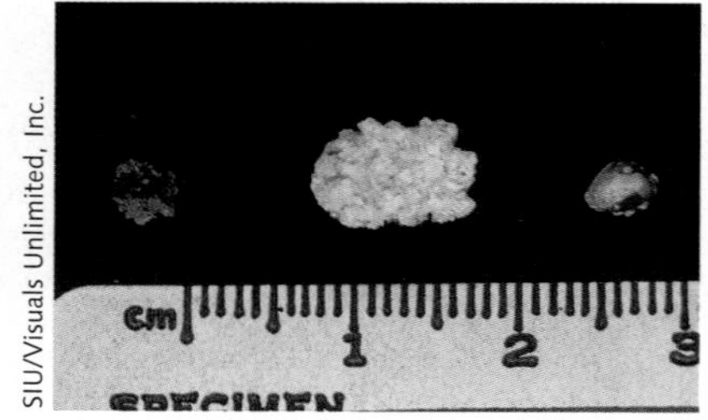

Kidney stones. These kidney stones were surgically removed from a patient.

Leaving chocolate out of a diet seems far more punishment than forgoing spinach.

Kidney stones usually consist of insoluble calcium and magnesium compounds such as calcium oxalate (CaC_2O_4), calcium phosphate $Ca_3(PO_4)_2$, magnesium ammonium phosphate ($MgNH_4PO_4$), or a mixture of these. For calcium oxalate kidney stones, the equilibrium $CaC_2O_4(s) \rightleftharpoons Ca^{2+}(aq) + C_2O_4^{2-}(aq)$ applies. High intake of foods rich in calcium or oxalate can cause a rise in the urinary concentration of either ion (or both) sufficient to make $Q > K_{sp}$ so that the equilibrium shifts to the left. The result is precipitation of calcium oxalate as a kidney stone. Thus, foods rich in Ca^{2+}, such as milk, ice cream, or cheese, or those high in $C_2O_4^{2-}$, such as chocolate, spinach, celery, or black tea, can trigger the onset of a kidney stone through the common ion effect. Such foods are restricted in the diets of individuals prone to developing kidney stones. A high-sugar diet may also create kidney stones because excessive sugar promotes excretion of Ca^{2+} and Mg^{2+}, which increases the concentrations of these ions passing through the kidneys. This can cause kidney stone formation, such as through calcium phosphate precipitation:

$$3\ Ca^{2+}(aq) + 2\ PO_4^{3-}(aq) \longrightarrow Ca_3(PO_4)_2(s)$$

Selective Precipitation of Ions

If their solubilities are sufficiently different, ionic compounds can be precipitated selectively from solution. The more soluble compound remains in solution as the less soluble one starts to precipitate. For example, silver chloride (AgCl) and silver chromate (Ag_2CrO_4) are each only slightly soluble in water. The solubilities of these two solutes differ enough, however, that when they are both in the same solution, one can be precipitated from solution, leaving the other behind in solution.

ThomsonNOW

Go to the Chemistry Interactive menu for modules on:

- **separating the lead, silver, and copper ions in a solution**
- **the conversion of solid $PbCl_2$ to $PbCrO_4$**

PROBLEM-SOLVING EXAMPLE 17.15 Selective Precipitation

Consider a solution containing 0.020 M Cl^- and 0.010 M CrO_4^{2-} ions to which Ag^+ ions are added slowly. Which precipitate forms first—AgCl or Ag_2CrO_4? K_{sp} AgCl = 1.8×10^{-10}; K_{sp} Ag_2CrO_4 = 9×10^{-12}.

Answer AgCl precipitates first.

Strategy and Explanation To answer this question we first find the minimum Ag^+ concentration required to precipitate each compound, which is the molar concentration product of its ions that just barely exceeds the K_{sp}. To precipitate AgCl,

$$K_{sp} \text{ of AgCl} = [Ag^+][Cl^-] = 1.8 \times 10^{-10}$$

$$[Ag^+] = \frac{1.8 \times 10^{-10}}{[Cl^-]} = \frac{1.8 \times 10^{-10}}{2.0 \times 10^{-2}} = 9.0 \times 10^{-9}\text{ M}$$

An Ag^+ concentration of slightly greater than 9.0×10^{-9} M will precipitate some AgCl from the solution. To precipitate Ag_2CrO_4,

$$K_{sp} \text{ of } Ag_2CrO_4 = [Ag^+]^2[CrO_4^-] = 9 \times 10^{-12}$$

$$[Ag^+]^2 = \frac{9 \times 10^{-12}}{[CrO_4^{2-}]} = \frac{9 \times 10^{-12}}{1.0 \times 10^{-2}} = 9 \times 10^{-10}\text{ M}; \qquad [Ag^+] = 3 \times 10^{-5}\text{ M}$$

Silver chromate will precipitate when the Ag^+ concentration slightly exceeds 3×10^{-5} M. Because a *much* smaller concentration of Ag^+ (9.0×10^{-9} M) is required to precipitate AgCl, it will precipitate before Ag_2CrO_4.

✓ **Reasonable Answer Check** The Ag^+ concentration required to precipitate Ag_2CrO_4 is approximately 10,000 times greater than that for AgCl (3×10^{-5} M versus 9×10^{-9} M). Therefore, the answer is reasonable; AgCl will precipitate first. In fact, the difference is so great that essentially all of the AgCl will precipitate before Ag_2CrO_4 precipitation begins.

When the Ag^+ concentration exceeds 3×10^{-5} M, Q = (conc. Ag^+)(conc. CrO_4^{2-}) > K_{sp} Ag_2CrO_4, and Ag_2CrO_4 will precipitate.

PROBLEM-SOLVING PRACTICE 17.15

Hydrochloric acid is slowly added to a solution that is 0.10 M in Pb^{2+} and 0.01 M in Ag^+. Which precipitate is formed first, AgCl or $PbCl_2$?

SUMMARY PROBLEMS

1. (a) Describe how to prepare a pH 3.70 buffer using formic acid (HCOOH) and sodium formate, NaHCOO.
   (b) Calculate the pH of this buffer after the addition of 0.0050 mol HCl.
   (c) How many grams of NaOH could be added to the buffer before its buffer capacity is just exceeded?

In NaHCOO, $HCOO^-$ is the formate ion.

$$H{-}C{\overset{\displaystyle /\!\!/ O}{\underset{\displaystyle \backslash O^-}{}}}$$

2. The K_a of nitrous acid, HNO_2, is 4.5×10^{-4}. In a titration, 50.0 mL of 1.00 M HNO_2 is titrated with 0.750 M NaOH.
   (a) Calculate the pH of the solution:
      (i) Before the titration begins
      (ii) When sufficient NaOH has been added to neutralize half of the nitrous acid originally present

(iii) At the equivalence point
(iv) When 0.05 mL NaOH less than that required to reach the equivalence point has been added
(v) When 0.05 mL NaOH more than that required to reach the equivalence point has been added

(b) Can bromthymol blue be used as the indicator for this titration?
(c) Will methyl red be a satisfactory indicator here?
(d) Use data from part (a) to plot a graph of pH (*y*-axis) versus volume of titrant.

3. A 0.500-L solution contains 0.025 mol Ag^+.
(a) Calculate the minimum mass of NaCl that must be added to precipitate AgCl from the solution.
(b) If excess Cl^- is added to the solution, the AgCl precipitate dissolves due to the formation of $[Ag(Cl)_2]^-$; K_f of $[Ag(Cl)_2]^- = 2.5 \times 10^5$. Calculate the minimum amount of Cl^- that must be added to dissolve the precipitate.

ThomsonNOW™
Sign in to ThomsonNOW at **www.thomsonedu.com** to check your readiness for an exam by taking the Pre-Test and exploring the modules recommended in your Personalized Learning Plan.

IN CLOSING

Having studied this chapter, you should be able to . . .

- Explain how buffers maintain pH, how to calculate their pH, how they are prepared, and the importance of buffer capacity (Section 17.1). ThomsonNOW homework: Study Questions 16, 22, 28
- Use the Henderson-Hasselbalch equation or the K_a expression to calculate the pH of a buffer and the pH change after acid or base has been added to the buffer (Section 17.1). ThomsonNOW homework: Study Questions 26, 30
- Interpret acid-base titration curves and calculate the pH of the solution at various stages of the titration (Section 17.2). ThomsonNOW homework: Study Questions 33, 36, 44, 94
- Explain how acid rain is formed and its effects on the environment (Section 17.3). ThomsonNOW homework: Study Questions 48, 50
- Relate a K_{sp} expression to its chemical equation (Section 17.4). ThomsonNOW homework: Study Question 54
- Use the solubility of a slightly soluble solute to calculate its solubility product (Section 17.4). ThomsonNOW homework: Study Question 56
- Describe the factors affecting the aqueous solubility of ionic compounds (Section 17.5). ThomsonNOW homework: Study Questions 68, 70
- Apply Le Chatelier's principle to the common ion effect (Section 17.5).
- Use the solubility product to calculate the solubility of a sparingly soluble solute in pure water and in the presence of a common ion (Section 17.5). ThomsonNOW homework: Study Question 63
- Describe the effect of complex ion formation on the solubility of a sparingly soluble ionic compound (Section 17.5).
- Relate Q, the ion product, to K_{sp} to determine whether precipitation will occur (Section 17.6).
- Predict which of two sparingly soluble ionic solutes will precipitate first (Section 17.6). ThomsonNOW homework: Study Question 110

KEY TERMS

acid rain *(17.3)*

buffer *(17.1)*

buffer capacity *(17.1)*

buffer solution *(17.1)*

common ion effect *(17.5)*

end point *(17.2)*

formation constant, K_f *(17.5)*

Henderson-Hasselbalch equation *(17.1)*

ion product, Q *(17.6)*

solubility product constant, K_{sp} *(17.4)*

solubility product constant expression *(17.4)*

titrant *(17.2)*

titration curve *(17.2)*

QUESTIONS FOR REVIEW AND THOUGHT

■ denotes questions available in ThomsonNOW and assignable in OWL.

Blue-numbered questions have short answers at the back of this book and fully worked solutions in the *Student Solutions Manual.*

ThomsonNOW™
Assess your understanding of this chapter's topics with sample tests and other resources found by signing in to ThomsonNOW at **www.thomsonedu.com**.

Review Questions

1. What is meant by the term "buffer capacity"?
2. Which would form a buffer?
(a) HCl and CH_3COOH (b) NaH_2PO_4 and Na_2HPO_4
(c) H_2CO_3 and $NaHCO_3$
3. Which would form a buffer?
(a) NaOH and NaCl (b) NaOH and NH_3
(c) Na_3PO_4 and Na_2HPO_4
4. Briefly describe how a buffer solution can control the pH of a solution when strong acid is added and when strong base is added. Use NH_3/NH_4Cl as an example of a buffer and HCl and NaOH as the strong acid and strong base.
5. What is the difference between the end point and the equivalence point in an acid-base titration?
6. What is meant by an indicator range for an acid-base indicator?
7. What are the characteristics of a good acid-base indicator?
8. A strong acid is titrated with a strong base, such as KOH. Describe the changes in the composition of the solution as the titration proceeds: prior to the equivalence point, at the equivalence point, and beyond the equivalence point.
9. Repeat the description for Question 8, but use a weak acid rather than a strong one.
10. Use Le Chatelier's principle to explain why $PbCl_2$ is less soluble in 0.010 M $Pb(NO_3)_2$ than in pure water.
11. Describe what a complex ion is and give an example.
12. What is amphoterism?
13. Distinguish between the ion product and the solubility product constant expression of a sparingly soluble solute.
14. Describe how the solubility of a sparingly soluble metal hydroxide can be changed.

Topical Questions

Buffer Solutions

15. Many natural processes can be studied in the laboratory but only in an environment of controlled pH. Which of these combinations would be the best choice to buffer the pH at approximately 7?
(a) H_3PO_4/NaH_2PO_4
(b) NaH_2PO_4/Na_2HPO_4
(c) Na_2HPO_4/Na_3PO_4
16. ■ Which of these combinations would be the best to buffer the pH at approximately 9?
(a) $CH_3COOH/NaCH_3COO$
(b) HCl/NaCl
(c) NH_3/NH_4Cl
17. Without doing calculations, determine the pH of a buffer made from equimolar amounts of these acid-base pairs.
(a) Nitrous acid and sodium nitrite
(b) Ammonia and ammonium chloride
(c) Formic acid and potassium formate
18. Without doing calculations, determine the pH of a buffer made from equimolar amounts of these acid-base pairs.
(a) Phosphoric acid and sodium dihydrogen phosphate
(b) Sodium monohydrogen phosphate and sodium dihydrogen phosphate
(c) Sodium phosphate and sodium monohydrogen phosphate
19. Select from Table 17.1 a conjugate acid-base pair that would be suitable for preparing a buffer solution whose concentration of hydronium ions is
(a) 4.5×10^{-3} M (b) 5.2×10^{-8} M
(c) 8.3×10^{-6} M (d) 9.7×10^{-11} M
Explain your choices.
20. Select from Table 17.1 a conjugate acid-base pair that would be suitable for preparing a buffer solution with pH equal to
(a) 3.45 (b) 5.48
(c) 8.32 (d) 10.15
Explain your choices.

21. To buffer a solution at a pH of 4.57, what mass of sodium acetate ($NaCH_3COO$) should you add to 500. mL of a 0.150 M solution of acetic acid (CH_3COOH)?

22. ■ How many grams of ammonium chloride (NH_4Cl) would have to be added to 500. mL of 0.10 M NH_3 solution to have a pH of 9.00?

23. A buffer solution can be made from benzoic acid (C_6H_5COOH) and sodium benzoate (NaC_6H_5COO). How many grams of the acid would you have to mix with 14.4 g of the sodium salt to have a liter of a solution with a pH of 3.88?

24. If a buffer solution is prepared from 5.15 g NH_4NO_3 and 0.10 L of 0.15 M NH_3, what is the pH of the solution?

25. You dissolve 0.425 g NaOH in 2.00 L of a solution that originally had $[H_2PO_4^-] = [HPO_4^{2-}] = 0.132$ M. Calculate the resulting pH.

26. ■ A buffer solution is prepared by adding 0.125 mol ammonium chloride to 500. mL of 0.500 M aqueous ammonia. What is the pH of the buffer? If 0.0100 mol HCl gas is bubbled into 500. mL of the buffer, what is the new pH of the solution?

27. If added to 1 L of 0.20 M acetic acid (CH_3COOH), which of these would form a buffer?
(a) 0.10 mol $NaCH_3COO$ (b) 0.10 mol NaOH
(c) 0.10 mol HCl (d) 0.30 mol NaOH
Explain your answers.

28. ■ If added to 1 L of 0.20 M NaOH, which of these would form a buffer?
(a) 0.10 mol acetic acid (b) 0.30 mol acetic acid
(c) 0.20 mol HCl (d) 0.10 mol $NaCH_3COO$
Explain your answers.

29. Calculate the pH change when 10.0 mL of 0.10 M NaOH is added to 90.0 mL pure water, and compare the pH change with that when the same amount of NaOH solution is added to 90.0 mL of a buffer consisting of 1.0 M NH_3 and 1.0 M NH_4Cl. Assume that the volumes are additive. K_b of $NH_3 = 1.8 \times 10^{-5}$.

30. ■ Calculate the pH change when 1.0 mL of 1.0 M NaOH is added to 0.100 L of a solution of
(a) 0.10 M acetic acid and 0.10 M sodium acetate.
(b) 0.010 M acetic acid and 0.010 M sodium acetate.
(c) 0.0010 M acetic acid and 0.0010 M sodium acetate.

31. Calculate the pH change when 1.0 mL of 1.0 M HCl is added to 0.100 L of a solution of
(a) 0.10 M acetic acid and 0.10 M sodium acetate.
(b) 0.010 M acetic acid and 0.010 M sodium acetate.
(c) 0.0010 M acetic acid and 0.0010 M sodium acetate.

32. A buffer consists of 0.20 M propanoic acid ($K_a = 1.4 \times 10^{-5}$) and 0.30 M sodium propanoate.
(a) Calculate the pH of this buffer.
(b) Calculate the pH after the addition of 1.0 mL of 0.10 M HCl to 0.010 L of the buffer.
(c) Calculate the pH after the addition of 3.0 mL of 1.0 M HCl to 0.010 L of the buffer.

Titrations and Titration Curves

33. ■ The titration curves for two acids with the same base are

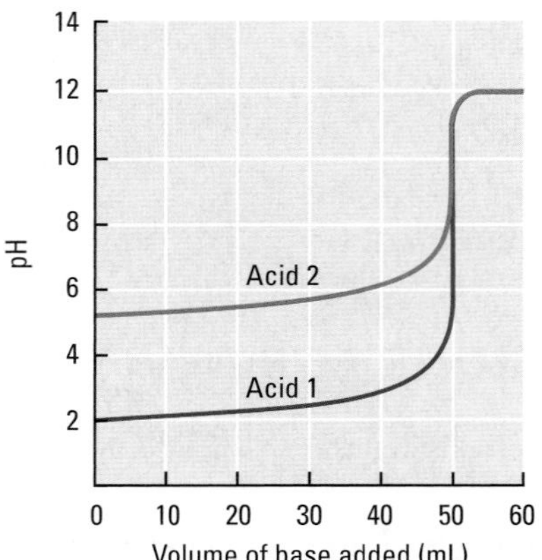

(a) Which is the curve for the weak acid? Explain your choice.
(b) Give the approximate pH at the equivalence point for the titration of each acid.
(c) Explain why the pH at the equivalence point differs for each acid.
(d) Explain why the starting pH values of the two acids differ.
(e) Which indicator could be used for the titration of Acid 1 and for the titration of Acid 2? Explain your choices.

34. Explain why it is that the weaker the acid being titrated, the more alkaline the pH is at the equivalence point.

35. Sketch the titration curve for the titration of 20.0 mL of a 0.100 M solution of a strong acid by a 0.100 M weak base; that is, the base is the titrant. In particular, note the pH of the solution:
(a) Prior to the titration
(b) When half the required volume of titrant has been added
(c) At the equivalence point
(d) 10. mL beyond the equivalence point.

36. ■ Consider all acid-base indicators discussed in this chapter. Which of these indicators would be suitable for the titration of each of these?
(a) NaOH with $HClO_4$ (b) Acetic acid with KOH
(c) NH_3 solution with HBr (d) KOH with HNO_3
Explain your choices.

37. Which of the acid-base indicators discussed in this chapter would be suitable for the titration of
(a) HNO_3 with KOH (b) KOH with acetic acid
(c) HCl with NH_3 (d) KOH with HNO_2
Explain your answers.

38. ■ It required 22.6 mL of 0.0140 M $Ba(OH)_2$ solution to titrate a 25.0-mL sample of HCl to the equivalence point. Calculate the molarity of the HCl solution.

39. It took 12.4 mL of 0.205 M H_2SO_4 solution to titrate 20.0 mL of a sodium hydroxide solution to the equivalence point. Calculate the molarity of the original NaOH solution.

40. Vitamin C is a monoprotic acid. To analyze a vitamin C capsule weighing 0.505 g by titration took 24.4 mL of 0.110 M NaOH. Calculate the percentage of vitamin C, $C_6H_8O_6$, in the capsule. Assume that vitamin C is the only substance in the capsule that reacts with the titrant.

41. An acid-base titration was used to find the percentage of $NaHCO_3$ in 0.310 g of a powdered commercial product used to relieve upset stomachs. The titration required 14.3 mL of 0.101 M HCl to titrate the powder to the equivalence point. Assume that the $NaHCO_3$ in the powder is the only substance that reacted with the titrant. Calculate the percentage of $NaHCO_3$ in the powder.

42. What volume of 0.150 M HCl is required to titrate to the equivalence point each of these samples?
(a) 25.0 mL of 0.175 M KOH
(b) 15.0 mL of 6.00 M NH_3
(c) 15.0 mL of propylamine, $CH_3CH_2CH_2NH_2$, which has a density of 0.712 g/mL
(d) 40.0 mL of 0.0050 M $Ba(OH)_2$

43. What volume of 0.225 M NaOH is required to titrate to the equivalence point each of these samples?
(a) 20.0 mL of 0.315 M HBr
(b) 30.0 mL of 0.250 M $HClO_4$
(c) 6.00 g of concentrated acetic acid, CH_3COOH, which is 99.7% pure

44. ■ A 30.00-mL solution of 0.100 M benzoic acid, a monoprotic acid, is titrated with 0.100 M NaOH. The K_a of benzoic acid is 6.3×10^{-5}. Determine the pH after each of these volumes of titrant has been added:
(a) 10.00 mL (b) 30.00 mL (c) 40.00 mL

45. The titration of 50.00 mL of 0.150 NaOH with 0.150 M HCl is carried out in a chemistry laboratory. Calculate the pH of the solution after these volumes of the titrant have been added:
(a) 0.00 mL (b) 25.00 mL (c) 49.9 mL
(d) 50.00 mL (e) 50.1 mL (f) 75.00 mL
Use the results of your calculations to plot a titration curve for this titration. On the curve indicate the position of the equivalence point.

46. The titration of 50.00 mL of 0.150 HCl with 0.150 M NaOH is carried out in a chemistry laboratory. Calculate the pH of the solution after these volumes of the titrant have been added:
(a) 0.00 mL (b) 25.00 mL (c) 49.9 mL
(d) 50.00 mL (e) 50.1 mL (f) 75.00 mL
Use the results of your calculations to plot a titration curve for this titration. On the curve indicate the position of the equivalence point.

Acid Rain

47. Explain why rain with a pH of 6.7 is not classified as acid rain.

48. ■ Identify two oxides that are key producers of acid rain. Write chemical equations that illustrate how these oxides form acid rain.

49. Acid rain with a pH of 1.5 has been measured. Calculate the hydrogen ion concentration of this rain.

50. ■ Write a chemical equation that shows how limestone neutralizes acid rain.

Solubility Product

51. Write the K_{sp} expressions for these slightly soluble salts.
(a) $BaCrO_4$ (b) $Mn(OH)_2$ (c) Ag_2SO_4

52. The solubility of $PbBr_2$ is 3.74×10^{-3} g per 100.0 mL at 25 °C. Calculate the K_{sp} of $PbBr_2$, assuming that the solute dissociates completely into Pb^{2+} and Br^- ions and that these ions do not react with water.

53. Write a balanced chemical equation for the equilibrium occurring when each of these solutes is added to water, then write the K_{sp} expression for each solute.
(a) Lead(II) carbonate (b) Nickel(II) hydroxide
(c) Strontium phosphate (d) Mercury(I) sulfate

54. ■ Write a balanced chemical equation for the equilibrium occurring when each of these solutes is added to water, then write the K_{sp} expression.
(a) Iron(II) carbonate (b) Silver sulfate
(c) Calcium phosphate (d) Mn(II) hydroxide

55. A saturated solution of silver arsenate (Ag_3AsO_4) contains 8.5×10^{-6} g Ag_3AsO_4 per mL. Calculate the K_{sp} of silver arsenate. Assume that there are no other reactions but the K_{sp} reaction.

56. ■ At 20. °C, 2.03 g $CaSO_4$ dissolves per liter of water. From these data calculate the K_{sp} of calcium sulfate at 20. °C. Assume that there are no other reactions but the K_{sp} reaction.

57. The water solubility of strontium fluoride (SrF_2) is 0.011 g/100. mL. Calculate its solubility product constant. Assume that there are no other reactions but the K_{sp} reaction.

58. The solubility of silver chromate (Ag_2CrO_4) in water is 2.7×10^{-3} g/100. mL. Calculate the K_{sp} of silver chromate. Assume that there are no other reactions but the K_{sp} reaction.

59. Calculate the K_{sp} of HgI_2 given that its solubility in water is 4.0×10^{-29} M. Assume that there are no other reactions but the K_{sp} reaction.

60. The solubility of $PbCl_2$ in water is 1.62×10^{-2} M. Calculate the K_{sp} of $PbCl_2$. Assume that there are no other reactions but the K_{sp} reaction.

61. In a saturated CaF_2 solution at 25 °C, the calcium concentration is analyzed to be 9.1 mg/L. Use this value to calculate the K_{sp} of CaF_2 assuming that the solute dissociates completely into Ca^{2+} and F^- ions and that neither ion reacts with water.

Common Ion Effect

62. What is the molarity of Zn^{2+} ion in a saturated solution of $ZnCO_3$ that contains 0.25 M Na_2CO_3?

63. ■ Calculate the solubility of $ZnCO_3$ in
(a) Water (b) 0.050 M $Zn(NO_3)_2$
(c) 0.050 M K_2CO_3 K_{sp} of $ZnCO_3 = 3 \times 10^{-8}$

64. Calculate the solubility (mol/L) of $SrSO_4$ ($K_{sp} = 3.1 \times 10^{-7}$) in 0.010 M Na_2SO_4.

65. Iron(II) hydroxide, $Fe(OH)_2$, has a solubility in water of 6.0×10^{-1} mg/L at a given temperature.
   (a) Calculate the K_{sp} of iron(II) hydroxide.
   (b) Calculate the hydroxide concentration needed to precipitate Fe^{2+} ions such that no more than 1.0 μg Fe^{2+} per liter remains in the solution.

66. The solubility of $Mg(OH)_2$ in water is approximately 9 mg/L at a given temperature.
   (a) Calculate the K_{sp} of magnesium hydroxide.
   (b) Calculate the hydroxide concentration needed to precipitate Mg^{2+} ions such that no more than 5.0 μg Mg^{2+} per liter remains in the solution.

67. What is the Cl^- concentration (in mol/L) in a solution that is 0.05 M in $AgNO_3$ and contains some undissolved AgCl?

Factors Affecting the Solubility of Sparingly Soluble Solutes

68. ■ What is the maximum concentration of Zn^{2+} in a solution of pH 10.00?

69. Determine the maximum concentration of Mn^{2+} in two solutions with pH as follows:
   (a) 7.81 (b) 11.15

70. ■ Hydrochloric acid is added to dissolve 5.00 g $Mg(OH)_2$ in a liter of water. To what value must the pH be adjusted to do so?

71. When a few drops of 1×10^{-5} M $AgNO_3$ are added to 0.01 M NaCl, a white precipitate immediately forms. When a few drops of 1×10^{-5} M $AgNO_3$ are added to 5 M NaCl, no precipitate forms. Explain these observations.

Complex Ion Formation

72. For these complex ions, write the chemical equation for the formation of the complex ion, and write its formation constant expression.
   (a) $[Ag(CN)_2]^-$ (b) $[Cd(NH_3)_4]^{2+}$

73. For these complex ions, write the chemical equation for the formation of the complex ion and write its formation constant expression.
   (a) $[Co(Cl)_6]^{3-}$ (b) $[Zn(OH)_4]^{2-}$

74. ■ Calculate how many moles of $Na_2S_2O_3$ must be added to dissolve 0.020 mol AgBr in 1.0 L water.

75. Gaseous ammonia is added to a 0.063 M solution of $AgNO_3$ until the aqueous ammonia concentration rises to 0.18 M. Calculate the concentrations of $[Ag(NH_3)_2]^+$ and Ag^+ in the solution.

76. Write chemical equations to illustrate the amphoteric behavior of
   (a) $Zn(OH)_2$ (b) $Sb(OH)_3$

77. Write chemical equations to illustrate the amphoteric behavior of
   (a) $Cr(OH)_3$ (b) $Sn(OH)_2$

General Questions

78. A buffer solution was prepared by adding 4.95 g sodium acetate, $NaCH_3COO$, to 250. mL of 0.150 M acetic acid, CH_3COOH.
   (a) What ions and molecules are present in the solution? List them in order of decreasing concentration.
   (b) What is the resulting pH of the buffer?
   (c) What is the pH of 100. mL of the buffer solution if you add 80. mg NaOH? (Assume negligible change in volume.)
   (d) Write a net ionic equation for the reaction that occurs to change the pH.

79. How many grams of NH_4Cl must be added to 400 mL of a 0.93 M solution of NH_3 to prepare a pH = 9.00 buffer?

80. Calculate the relative concentrations of *o*-ethylbenzoic acid ($pK_a = 3.79$) and potassium *o*-ethylbenzoate that are needed to prepare a pH = 4.0 buffer.

81. Calculate the relative concentrations of aniline ($pK_b = 9.42$) and anilinium chloride that are required to prepare a buffer of pH 5.00.

82. A solution contains 7.50 g KNO_2 per liter. How much HNO_2 must be added to prepare a buffer of pH 4.00? (Assume there is no volume change.)

83. Which of these buffers has the greater resistance to change in pH?
   (a) Conjugate acid concentration = 0.100 M = conjugate base concentration
   (b) Conjugate acid concentration = 0.300 M = conjugate base concentration

   Explain your answer.

84. (a) What is the pH of a 0.15 M acetic acid solution?
   (b) If you add 83 g sodium acetate to 1.50 L of the 0.15 M acetic acid solution, what is the new pH of the solution?

85. (a) Calculate the pH of a 0.050 M solution of HF.
   (b) What is the pH of the solution if you add 1.58 g NaF to 250. mL of the 0.050 M solution?

86. When 40.00 mL of a weak monoprotic acid solution is titrated with 0.100 M NaOH, the equivalence point is reached when 35.00 mL base have been added. After 20.00 mL NaOH solution has been added, the titration mixture has a pH of 5.75. Calculate the ionization constant of the acid.

87. Each of the solutions given in the table has equal volume and the same concentration, 0.1 M.

Acid	pH
HCl	1.1
Formic	2.3
Acetic	2.9
HCN	5.1

   Which solution requires the greatest volume of 0.1 M NaOH to titrate to the equivalence point? Explain your answer.

88. What is the effect on the equilibrium if more solid AgCl is added to a solution saturated with Ag^+ and Cl^- ions?

89. At 20.°C, 2.03 g $CaSO_4$ dissolve per liter of water. From these data calculate the K_{sp} of calcium sulfate at 20.°C.

90. The solubility of silver chromate, Ag_2CrO_4, in water is 2.7×10^{-3} g per 100. mL. Estimate the K_{sp} of silver chromate at that temperature.

Applying Concepts

91. The average normal concentration of Ca^{2+} in urine is 5.33 g/L.
    (a) What concentration of oxalate is needed to precipitate calcium oxalate to initiate formation of a kidney stone? K_{sp} of calcium oxalate = 2.3×10^{-9}.
    (b) What minimum phosphate concentration would it take to precipitate a calcium phosphate kidney stone? K_{sp} of calcium phosphate = 1×10^{-25}.
92. Explain why even though an aqueous acetic acid solution contains acetic acid and acetate ions, it cannot be a buffer.
93. Vinegar must contain at least 4% acetic acid (0.67 M). A 5.00-mL sample of commercial vinegar required 33.5 mL of 0.100 M NaOH to reach the equivalence point. Does the vinegar meet the legal limit of 4% acetic acid?
94. ■ An unknown acid is titrated with base, and the pH is 3.64 at the point where exactly half of the acid in the original sample has been neutralized. What is the value of the ionization constant of the acid?
95. When asked to prepare a carbonate buffer with a pH = 10, a lab technician wrote this equation to determine the ratio of weak acid to conjugate base needed:

$$10 = 10.32 + \log\frac{[HCO_3^-]}{[H_2CO_3]}$$

What is wrong with this setup? If the technician prepared a solution containing equimolar concentrations of HCO_3^- and CO_3^{2-}, what would be the pH of the resulting buffer?

96. When you hold your breath, carbon dioxide gas is trapped in your body. Does this increase or decrease your blood pH? Does it lead to acidosis or alkalosis? Explain your answers.
97. Calcium fluoride, CaF_2, is used to fluoridate a municipal water supply. The water is extremely hard with a Ca^{2+} concentration of 0.070 M. Calculate the fluoride concentration in this solution. K_{sp} of calcium fluoride = 3.9×10^{-11}.
98. Apatite, $Ca_5(PO_4)_3OH$, is the mineral in teeth.

$$Ca_5(PO_4)_3OH(s) \rightleftharpoons 5\,Ca^{2+}(aq) + 3\,PO_4^{3-}(aq) + OH^-(aq)$$

    (a) On a chemical basis explain why drinking milk strengthens young children's teeth.
    (b) Sour milk contains lactic acid. Not removing sour milk from the teeth of young children can lead to tooth decay. Use chemical principles to explain why.
99. Calculate the maximum concentration of Mg^{2+} (molarity) that can exist in a solution of pH 12.00.
100. Choose the words that make this statement true: During a recent television medical drama, a person went into cardiac arrest and stopped breathing. A doctor quickly injected sodium hydrogen carbonate solution into the heart. This would indicate that cardiac arrest leads to (acidosis or alkalosis) and that the sodium hydrogen carbonate helps to (increase or decrease) the pH. Explain your choices clearly.

More Challenging Questions

101. An experiment found that 0.0050 mol $Ca(OH)_2$ dissolved to form 0.100 L of a saturated aqueous solution.
    (a) Calculate the pH of the solution.
    (b) Calculate the K_{sp} of $Ca(OH)_2$.
102. You are given four different aqueous solutions and told that they each contain NaOH, Na_2CO_3, $NaHCO_3$, or a mixture of these solutes. You do some experiments and gather these data about the samples.

Sample A: Phenolphthalein is colorless in the solution.
Sample B: The sample was titrated with HCl until the pink color of phenolphthalein disappeared, then methyl orange was added to the solution. The solution became a pink color. Methyl orange changes color from pH 3.01 (red) to pH 4.4 (orange).
Sample C: Equal volumes of the sample were titrated with standardized acid. Using phenolphthalein as an indicator required 15.26 mL of standardized acid to change the phenolphthalein color; it required 17.90 mL for a color change using methyl orange as the indicator.
Sample D: Two equal volumes of the sample were titrated with standardized HCl. Using phenolphthalein as the indicator, it took 15.00 mL of acid to reach the equivalence point; using methyl orange as the indicator required 30.00 mL HCl to achieve neutralization.

Identify the solute in each of the solutions.

103. An aqueous solution contains these ions:

Ion	Concentration (M)
Br^-	0.0010
Cl^-	0.0010
CrO_4^{2-}	0.0010

A solution of silver nitrate is added dropwise to the aqueous solution. Which silver salt will precipitate first? Last?

104. Pyridine (C_5H_5N) $K_b = 1.5 \times 10^{-9}$, is a base like ammonia. A 25.0-mL sample of 0.085 M pyridine is titrated with 0.102 M HCl. The equivalence point occurs at 21.1 mL. Calculate the pH when 5.5 mL acid has been added.
105. You have 1.00 L of 0.10 M formic acid (HCOOH) whose $K_a = 1.8 \times 10^{-4}$. You want to bubble into the formic acid solution sufficient HCl gas to decrease the pH of the formic acid solution by 1.0 pH unit. Calculate the volume of HCl (liters) that must be used at STP to bring about the desired change in pH. Assume no volume change has occurred in the solution due to the addition of HCl gas.
106. One liter (1.0 L) of distilled water has an initial pH of 5.6 because it is in equilibrium with carbon dioxide in the atmosphere. One drop of concentrated hydrochloric acid, 12 M HCl, is added to the distilled water. Calculate the pH of the resulting solution. 20 drops = 1.0 mL.
107. A 0.0010 M aqueous solution of anisic acid ($C_8H_8O_3$) is prepared and 100 mL of it is left in an uncovered beaker. After some time, 50. mL water has evaporated from the solution in the beaker (assume no solute evaporated). The K_a of anisic acid is 3.38×10^{-5}.
    (a) Calculate the pH of the initial solution and the pH after evaporation has occurred.
    (b) Calculate the degree of dissociation of anisic acid in each of the solutions.

108. You want to prepare a pH 4.50 buffer using sodium acetate and glacial acetic acid. You have on hand 300 mL of 0.100 M sodium acetate. How many grams of glacial acetic acid (d = 1.05 g/mL) should you use to prepare the buffer?

109. The titration curves for the amino acids glutamic acid and lysine are given below.

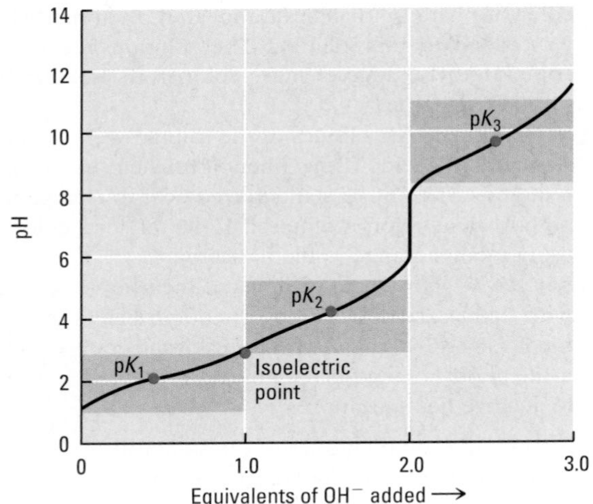

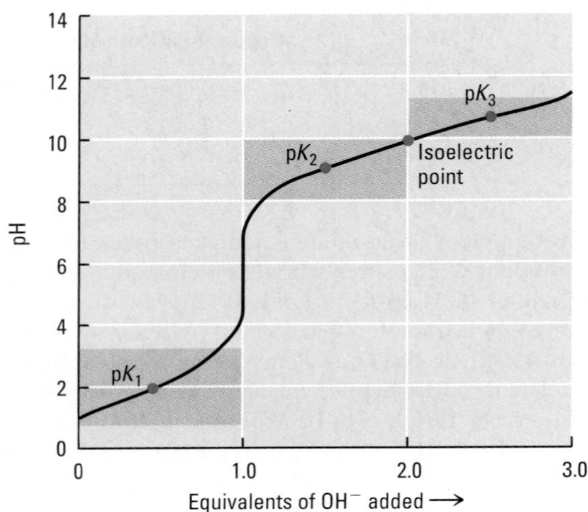

(a) Using the structural formulas given in Table 12.8, write the structural formulas for the forms of each amino acid corresponding to points on the titration curve when 0, 1.0, 2.0, and 3.0 equivalents of OH^- have been added. An equivalent in this case is the amount of base added that neutralizes an amount of H_3O^+ ions equivalent to the amount of amino acid present.

(b) The isoelectric point is the pH at which an amino acid has no net charge. Write the structural formulas of glutamic acid and lysine at their isoelectric points.

(c) What relationship exists between the conjugate acid and conjugate base forms of these amino acids at each of the pK points noted on the titration curve?

110. ■ A 1.00-L solution contains 0.010 M F^- and 0.010 M SO_4^{2-}. Solid barium nitrate is slowly added to the solution.
(a) Calculate the [Ba^{2+}] when $BaSO_4$ begins to precipitate.
(b) Calculate the [Ba^{2+}] when BaF_2 starts to precipitate.
Assume no volume change occurs. K_{sp} values: $BaSO_4$ = 1.1×10^{-11}; BaF_2 = 1.7×10^{-6}.

111. An experiment requires the addition of 0.075 mol gaseous NH_3 to 1.0 L of 0.025 M $Mg(NO_3)_2$. Ammonium chloride, NH_4Cl, is added prior to the addition of the NH_3 to prevent precipitation of $Mg(OH)_2$. Calculate the minimum number of grams of ammonium chloride that must be added. K_{sp} of $Mg(OH)_2$ = 2.1×10^{-5}.

Conceptual Challenge Problems

CP17.A (Section 17.2) Suppose you were asked on a laboratory test to outline a procedure to prepare a buffered solution of pH 8.0 using hydrocyanic acid, HCN. You realize that a pH of 8.0 is basic, and you find that the K_a of hydrocyanic acid is 4.0×10^{-10}. What is your response?

CP17.B (Sections 17.4 and 17.5) Barium sulfate is swallowed to enhance X-ray studies of the gastrointestinal tract.
(a) Calculate the solubility of Ba^{2+} in mol/L in pure water. K_{sp} of $BaSO_4$ = 1.1×10^{-10}.
(b) In the stomach, the HCl concentration is 0.10 M. Calculate the solubility of Ba^{2+} in mol/L in this solution. K_a of HSO_4^- = 1.2×10^{-2}.
(c) The two calculations in parts (a) and (b) were done using data at 25 °C. Repeat part (b) at 37 °C, body temperature. At that temperature, K_{sp} of $BaSO_4$ = 1.5×10^{-10}; K_a of HSO_4^- = 7.1×10^{-3}.

18

Thermodynamics: Directionality of Chemical Reactions

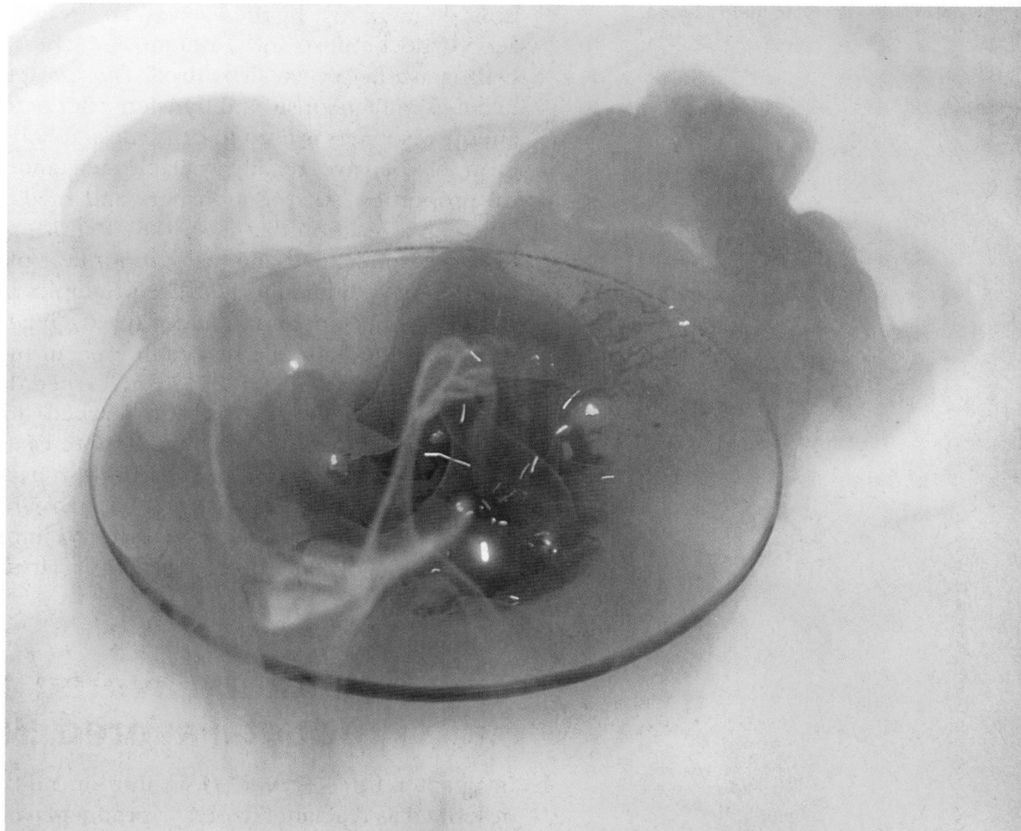

When liquid bromine is poured onto small pieces of aluminum, nothing appears to happen at first. Eventually the mixture becomes hot and a violent exothermic reaction occurs. Sparks fly, and you see flames. The reaction begins slowly as soon as the bromine contacts the aluminum, and its rate increases as the temperature rises. The reaction continues until either the bromine or the aluminum has been completely consumed. Chemical thermodynamics provides methods for predicting whether a reaction will be product-favored, like this one.

ThomsonNOW
Throughout the text, this icon indicates an opportunity to test yourself on key concepts and to explore interactive modules by signing in to ThomsonNOW at **www.thomsonedu.com**.

Because air can slowly oxidize hydrocarbons to alkenes, which can polymerize and form insoluble gums, gasoline is usually drained from lawn mowers or cars that are going to be stored for many months, or else an antioxidant (such as N-nitrosodiethylamine) is added.

The product-favored reaction of bromine with aluminum.

The term "product-favored" designates reactions that many scientists refer to as "spontaneous"; many people use the two terms interchangeably. We prefer "product-favored" because some reactions do begin spontaneously, but produce only tiny quantities of products when equilibrium is reached. "Product-favored" describes clearly a situation in which products predominate over reactants. Also, the nonscientific usage of "spontaneous" implies a rapid change; if the rate of a product-favored reaction is very slow, the reaction does not appear spontaneous at all.

Many chemical reactions behave as the reaction of bromine and aluminum does (see photos in margin). They begin when the reactants come into contact and continue until at least one reactant, (the limiting reactant, (⇐ ***p. 140***), is completely used up. Other reactions, such as the rusting of iron at room temperature, happen much more slowly, but reactants are still converted completely to products. After many years and enough flaking of hydrated iron(III) oxide (rust) from its surface, a piece of iron exposed to air will rust away. Still other reactions are even slower at room temperature. Gasoline reacts so slowly with air at room temperature that it can be stored safely for long periods, although it may go bad after a very long time. However, if its temperature is raised by a spark or flame, gasoline vapor burns rapidly and is essentially completely converted to CO_2 and H_2O.

By contrast with the reaction of bromine and aluminum, many chemical reactions do not occur by themselves. For example, table salt, NaCl, does not of its own accord decompose into sodium and chlorine. Neither does water change into hydrogen and oxygen all by itself. These reactions take place only if another process occurs simultaneously and transfers energy to them. (A significant portion of world energy resources is used to cause desirable reactions to occur—reactions that transform inexpensive, readily available substances into new substances with more useful properties, such as polymers and medicines.) It is important to differentiate between a reaction that is so slow that it *appears* not to occur, such as air oxidation of gasoline, and one that *cannot* take place of its own accord, such as decomposition of sodium chloride. The principles of chemical kinetics ***(⇐ Chapter 13, p. 607)*** can be applied to find ways to speed up a slow reaction, but they are of no use in dealing with one that cannot occur by itself.

In Chapter 6 ***(⇐ p. 235)*** you learned that thermal energy is transferred when most reactions occur. You also learned how to predict whether a reaction is exothermic or endothermic and how to calculate what quantity of energy transfer takes place as a reaction occurs. In this chapter you will learn how thermodynamics helps us to predict what will happen when potential reactants are mixed. Will most or all of the reactants be converted to products, as in the case of bromine and aluminum? Will some be converted? Or virtually none?

18.1 Reactant-Favored and Product-Favored Processes

In Chapter 14 ***(⇐ p. 673)*** we introduced the idea that a chemical process can be described as reactant-favored or product-favored. When products predominate over reactants, we designate the reaction as a *product-favored process.* Examples include the reaction of bromine with aluminum, rusting of iron, and combustion of gasoline. If a process is product-favored, most or all of the reactants will eventually be converted to products without continuous outside intervention, although "eventually" may mean a very, very long time.

Other reactions have virtually no tendency to occur by themselves. Examples include the reactions for which we wrote "N.R." for "no reaction" in Chapter 5 ***(⇐ p. 163)***. For example, nitrogen and oxygen have coexisted in the earth's atmosphere for at least a billion years without appreciable concentrations of nitrogen oxides such as N_2O, NO, or NO_2 building up. Similarly, deposits of salt, NaCl(s), have existed on earth for millions of years without forming the elements Na(s) and Cl_2(g). If, when equilibrium has been reached, reactants predominate over products, we categorize a chemical reaction as a *reactant-favored process.*

A reactant-favored process is always *exactly the opposite* of a product-favored process. For example, the equation

$$2\ Na(s) + Cl_2(g) \longrightarrow 2\ NaCl(s)$$

describes a product-favored reaction, because sodium metal and chlorine gas react readily to produce salt. However, if we had written the same equation in the reverse direction

$$2\ NaCl(s) \longrightarrow 2\ Na(s) + Cl_2(g)$$

the system would be designated as reactant-favored. This equation represents decomposition of sodium chloride to form sodium and chlorine, a reaction that does not occur of its own accord. The designations "product-favored" and "reactant-favored" indicate the direction in which a chemical reaction will take place—either forward or backward based on a given equation.

Unless there is some continuous outside intervention, a reactant-favored process does not produce large quantities of products. What do we mean by continuous outside intervention? Usually it is some flow of energy. For example, if enough energy is provided to a sample of air to keep it at a very high temperature, small but significant quantities of NO can be formed from the N_2 and O_2. Such high temperatures are found in lightning bolts and in combustion chambers of electric power generating plants and automobile engines. A power plant or a large number of automobiles can produce enough NO and other nitrogen oxides to cause significant air pollution problems. Salt can be decomposed to its elements by continuously heating it to keep it molten and passing electricity through it to separate the ions, carry out oxidation and reduction, and form the elements.

$$2\ NaCl(\ell) \xrightarrow{\text{electricity}} 2\ Na(\ell) + Cl_2(g)$$

In each case, a reactant-favored process can be forced to produce products if sufficient electrical energy is continuously supplied. This is in contrast to the situation for a product-favored process such as combustion of gasoline, which requires only a brief spark to initiate the reaction. Once started, gasoline combustion continues of its own accord without a continuous supply of energy from outside.

Although earth's atmosphere is 78% N_2 and 21% O_2, the concentration of N_2O, the most abundant oxide of nitrogen in the atmosphere, is more than two million times smaller than the concentration of N_2.

The air in the immediate vicinity of a lightning bolt can be heated enough to cause a small fraction of the nitrogen and oxygen to combine to form NO, but this reaction takes place only while the lightning is present. A similar reaction can occur in the engine of an automobile, but again only a small fraction of the air is converted to nitrogen oxides, and only while the temperature remains high.

EXERCISE 18.1 Reactant-Favored and Product-Favored Processes

Write a chemical equation for each process described below, and classify each as reactant-favored or product-favored.

(a) A puddle of water evaporates on a summer day.
(b) Silicon dioxide (sand) decomposes to the elements silicon and oxygen.
(c) Paper, which is mainly cellulose $(C_6H_{10}O_5)_n$, burns at a temperature of 451 °F.
(d) A pinch of sugar dissolves in water at room temperature.

18.2 Chemical Reactions and Dispersal of Energy

The fundamental rule that governs whether a process is product-favored is that ***energy will spread out (disperse) unless it is hindered from doing so.*** A simple example is the one-way transfer of energy from a hotter sample to a colder sample (***⇐ p. 218***). As a hot frying pan on a stove cools to room temperature, thermal energy that was concentrated in the pan spreads out over the atoms, molecules, and ions in the stove, the pan, and the surrounding air. When thermal equilibrium is reached, the room has become slightly warmer—energy has been dispersed and the process is product-favored.

Most exothermic reactions are product-favored at room temperature for a similar reason. When an exothermic reaction takes place, energy is transferred to the surroundings. (See, for example, the reaction of bromine with aluminum in the photo at the beginning of this chapter.) Chemical potential energy that has been

A chemical reaction system is usually defined as the collection of atoms that make up the reactants. These same atoms also make up the products, but there they are bonded in a different way. Everything else is designated as the surroundings (***⇐ p. 220***).

stored in bonds between relatively few atoms, ions, and molecules (the reactants) spreads over many more atoms, ions, and molecules as the surroundings (as well as the products) are heated. Therefore it is usually true that after an exothermic reaction, energy is more dispersed than it was before.

Probability and Dispersal of Energy

Dispersal of energy occurs because the probability is much higher that energy will be spread over many particles than that it will be concentrated in a few. To better understand energy dispersal and probability, consider the hypothetical case of a very small sample of matter consisting of two atoms, A and B. Suppose that this sample contains two units of energy, each designated by *. The energy can be distributed over the two atoms in three ways: Atom A could have both units of energy, atom A and atom B could each have one unit, or atom B could have both units. Designate these three situations as

ThomsonNOW™
Go to the Chemistry Interactive menu for a module on **energy dispersal.**

A** A*B* B**

Now suppose that atoms A and B come into contact with two other atoms, C and D, that have no energy. There are ten possibilities for distributing the two units of energy over four atoms.

A** A*B* A*C* A*D* B** B*C* B*D* C** C*D* D**

Only three of these cases (A**, A*B*, and B**) have all the energy in atoms A and B, which was the initial situation. When all four atoms are in contact, there are seven chances out of ten that some energy will have transferred from A and B to C and D. Thus, there is a probability of $7/10 = 0.70$ that the energy will become spread out over more than just the two atoms A and B.

The low probability that a lot of energy will be associated with only a few particles makes a substance with a lot of chemical potential energy valuable. Humans call substances such as coal, oil, and natural gas "energy resources" and sometimes fight wars over them because of their concentrated energy.

CONCEPTUAL EXERCISE 18.2 Probability of Energy Dispersal

Suppose that you have three units of energy to distribute over two atoms, A and B. Designate each possible arrangement. Now suppose that atoms A and B come into contact with three more atoms, C, D, and E. From the possible arrangements of energy over the five atoms, calculate the probability that all the energy will remain confined to atoms A and B.

The probability that energy will become dispersed becomes overwhelming when large numbers of atoms or molecules are involved. For example, suppose that atoms A and B had been brought into contact with a mole of other atoms. There would still be only three arrangements in which all the energy was associated with atoms A and B, but there would be many, many more arrangements (more than 10^{47}) in which all the energy was associated with other atoms. In such a case it is essentially certain that energy will be transferred. ***If energy can be dispersed over a very much larger number of particles, it will be.***

For the same amount of substance or number of particles, higher thermal energy corresponds to higher temperature (← ***p. 220***). Therefore, a substance at a higher temperature has greater energy per particle on average. Dispersal of energy over a larger number of particles corresponds to transfer of energy from a substance at a higher temperature to another substance at a lower temperature.

Dispersal of Energy Accompanies Dispersal of Matter

Energy becomes more dispersed when a system consisting of atoms or molecules expands to occupy a larger volume. This kind of energy dispersal is associated with a characteristic property of gases: A gas expands until it fills a container. Recall that molecules in the gas phase are essentially independent of one another and that only weak forces attract the molecules together (← ***p. 436***). Suppose a sample of bromine gas at room temperature is confined within one flask that is connected through a tube with a barrier to a second flask of equal size from which all gas molecules have been removed (Figure 18.1). What happens if the barrier is removed? The confined bromine expands to fill the vacuum in the second flask.

The conclusions drawn about dispersal of bromine molecules on this page apply in general to particles of a gas, whether they are atoms or molecules.

That dispersal of atoms and molecules also involves dispersal of energy seems obvious, because energy accompanies the particles as they disperse. However, dispersal of energy refers to spreading of energy over a greater number of different *energy levels* **(*quantum levels,* ⇐ *p. 280*)** of the atoms and molecules. When the volume of a gas-phase system increases, the energy levels associated with the motion of each atom or molecule get closer together; that is, there are more levels within the same range of energies. Therefore, at a given temperature, more different energy levels are accessible in the larger volume than in the smaller volume.

When matter spreads out, the number of ways of arranging the energy associated with that matter increases. The energy is more dispersed, and therefore the spreading out of matter is product-favored. This basic idea applies to mixing of different gases and dissolving of one liquid in another, as well as to expansion of a gas. We have already mentioned it in connection with dissolving of solids **(⇐ *p. 734*),** and it is also useful in understanding colligative properties **(⇐ *p. 744*).**

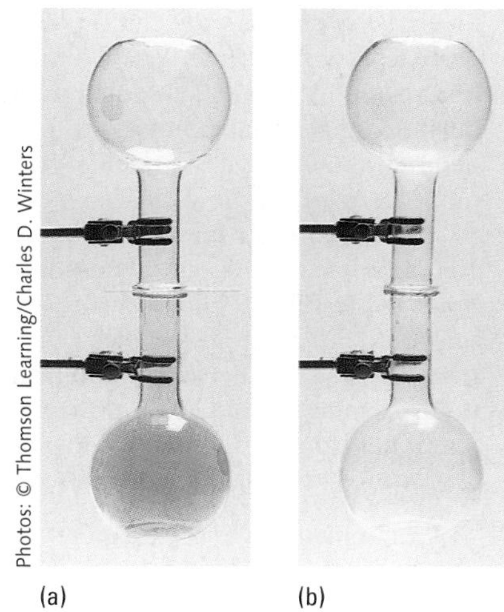

Figure 18.1 Expansion of a gas. A gas will expand to fill any container. (a) Bromine vapor is confined in the lower flask. There is a vacuum in the upper flask. (b) When the barrier between the flasks is removed, the bromine molecules rapidly rush into flask B, and eventually bromine is evenly distributed throughout both flasks.

Expansion of a gas is product-favored, so the opposite process—compression of a gas—should be reactant-favored. If we wanted to reverse the expansion of a gas by concentrating all the particles into a smaller volume, a continuous outside influence such as a pump would be required—the pump could do work on the gas to force it into a less probable arrangement in which energy was less dispersed. The work done by the pump would be stored in the gas and could later be used for some other purpose.

To summarize, any physical or chemical process in which energy is concentrated in a few energy levels in the initial state and energy is dispersed over many energy levels in the final state is product-favored. Two important situations where this is true are

- an exothermic reaction, which disperses potential energy of chemical bonds to thermal energy of a much larger number of atoms or molecules, and
- a process where matter spreads out, which disperses energy as well as matter.

Atoms or molecules of a material that is a solid at room temperature (like the glass flask) do not disperse, because there are strong attractive forces between them. Their tendency to disperse becomes more obvious if the temperature is raised so that they can vaporize.

If both of these situations apply to a reaction, then it will definitely be product-favored, because the final distribution of energy will be more probable. On the other hand, a process that spreads out neither energy nor matter will be reactant-favored—the initial substances will remain no matter how long we wait. If one of these situations applies but not the other, then quantitative information is needed to decide which effect is greater. The remainder of this chapter develops that quantitative information.

18.3 Measuring Dispersal of Energy: Entropy

The nanoscale dispersal of energy in a sample of matter is measured by a thermodynamic quantity called *entropy*, symbolized by S **(⇐ *p. 704*).** Entropy changes can be measured with a calorimeter **(⇐ *p. 242*),** the same instrument used to measure the enthalpy change when a reaction occurs. For a process that takes place at constant temperature and pressure, the entropy change can be calculated by dividing the thermal energy transferred, q_{rev}, by the absolute temperature, T,

The symbol Δ was defined in Section 6.2 **(⇐ *p. 220*).** The quantity of energy transferred by heating, q, was defined in Section 6.3 **(⇐ *p. 223*).**

$$\Delta S = S_{final} - S_{initial} = \frac{q_{rev}}{T} \qquad [18.1]$$

CHEMISTRY YOU CAN DO

Energy Distributions

To envision how energy can be distributed over nanoscale particles, try this game. You will need an even number of friends (about 20) and an equal number of tokens (dollar bills, pennies, or fake money) to play the game. The friends need not be chemistry students, but chemistry students are more likely to still be your friends when you finish!

Arrange your friends in two concentric circles. Each circle must have the same number of people. People in the inner circle should face outward and stand almost shoulder to shoulder. People in the outer circle should pair up with someone in the inner circle and stand directly in front of their partners, facing inward. Everyone should start with one token (dollar bill, penny, or fake money).

A turn in the game goes this way.

1. Each pair plays "rock-paper-scissors" once. On the count of three, both members of a pair simultaneously display "rock" (a closed fist), "paper" (an open palm), or "scissors" (forefinger and middle finger in a V-shape). Rock breaks scissors (rock wins over scissors), scissors cuts paper, and paper wraps rock. If both people display the same thing, there is a tie.
2. If there is a winner and the loser has a token, one token is transferred from loser to winner. (Never exchange more than a single token.) If there is a tie, or if the loser has no token, there is no transfer.
3. When you say "next," each person in the inner circle moves one place to the left, to a new partner in the outer circle, and a new turn begins.

After every fifth turn, ask all players who have zero tokens to raise their hands. Record the number of raised hands. Also ask about one token, two tokens, and so on up to the highest number of tokens anyone has. Record how many players have each number of tokens. Continue playing the game and recording the results until you have recorded at least five sets of data (25 turns).

Think about the meaning of the data you have collected and answer these questions:

1. How does this game simulate what happens when lots of molecules exchange energy with each other? What role does probability play in the distribution of tokens? Of energy?

Courtesy of Robert M. Hanson/St. Olaf College. © *Journal of Chemical Education.* Vol. 83, 2006, p. 581.

2. What do you think is the probability that a particular individual would end the game with more than a single token?
3. Why isn't the most probable distribution of tokens the distribution where everyone has a single token?
4. What do you think would happen if the game were repeated, but a single individual had all of the tokens at the beginning? Try it and find out whether your prediction was correct.
5. How are the two versions of the game (the original and the one in Question 4) related to the rule given in Chapter 14 that the same equilibrium state is reached (at a given temperature) whether you start with reactants or products?
6. Suppose you were the person who had all the tokens at the beginning. What do you think your chances are of coming away from the game with half as many tokens as you started with?

SOURCE: Hanson, R. M. and Michalek, B. *Journal of Chemical Education*, Vol. 83, 2006; pp. 581–588.

The absolute temperature scale is also called the *Kelvin temperature scale or the thermodynamic temperature scale.* It was defined in Section 10.4 *(← p. 441).* The kelvin unit should be used in all thermodynamic calculations involving temperature.

The subscript "rev" has been added to q to indicate that the equation applies only to processes that can be reversed by a very small change in conditions. An example of such a process is melting of ice at 0 °C and normal atmospheric pressure. If the temperature is just a tiny bit below 0 °C, so that energy is transferred from the water to its surroundings, the water will freeze. If the temperature is a tiny bit above 0 °C, the ice will melt. *Any process for which a very small change in conditions can reverse its direction* is called a **reversible process.**

Another case where Equation 18.1 can be used is when energy is transferred to or from a thermal reservoir at constant temperature and pressure. A thermal reservoir is any large sample of matter, such as everything in your dorm room or the surroundings of a chemical reaction. For example, when a cup of hot coffee cools to room temperature, the quantity of energy transferred is small enough that the temperature of the thermal reservoir (your room) hardly changes. This small quantity of energy could be transferred into or out of your room reversibly. Therefore, to calculate the entropy change of your room you could measure room temperature (in kelvins), calculate how much energy transferred to the room as the coffee cooled, and divide energy transfer by temperature. You could not do the same calculation to determine the entropy change of the coffee, because its temperature changed during the process.

Oesper Collection in the History of Chemistry, University of Cincinnati

Ludwig Boltzmann
1844–1906

An Austrian physicist, Ludwig Boltzmann gave us the useful interpretation of entropy as probability. Engraved on his tombstone in Vienna is the equation

$$S = k \log W$$

It relates entropy, S, and thermodynamic probability, W—the number of different nanoscale arrangements of energy that correspond to a given macroscale system. The proportionality constant, k, is called Boltzmann's constant, and log stands for the natural logarithm (in modern symbolism, ln).

Entropy changes are usually reported in units of joules per kelvin (J/K), whereas enthalpy changes are usually given in kilojoules (kJ). This means that you need to be careful about the units to avoid being wrong by a factor of 1000.

PROBLEM-SOLVING EXAMPLE 18.1 Calculating Entropy Change

The melting point of glacial acetic acid (pure acetic acid) is 16.6 °C at 1 bar. Calculate $\Delta S°$ for the process

$$CH_3COOH(s) \longrightarrow CH_3COOH(\ell)$$

given that the molar enthalpy of fusion of acetic acid is 11.53 kJ/mol. Express your answer in units of joules per kelvin.

Answer 39.79 J/K

Strategy and Explanation Melting acetic acid at its melting point of 16.6 °C is reversible, so use the equation $\Delta S = q_{rev}/T$. The pressure is 1 bar, the standard-state pressure ***(p. 248)***, so $\Delta S° = \Delta S$. Remember that for a constant-pressure process, thermal energy transfer, q_{rev}, is the same as the enthalpy change, $\Delta H°$ ***(p. 229)***. Therefore, the entropy change can be calculated from the enthalpy of fusion and the temperature.

$$\Delta S° \text{ (melting acetic acid)} = \frac{q_{rev}}{T} = \frac{\Delta H° \text{ (melting acetic acid)}}{T} = \frac{\Delta H_{fusion}}{T}$$

$$= \frac{11.53 \text{ kJ}}{(16.6 + 273.15) \text{ K}} = 3.979 \times 10^{-2} \text{ kJ/K} = 39.79 \text{ J/K}$$

✓ **Reasonable Answer Check** The result is positive, meaning that entropy increased when solid acetic acid was converted to liquid. Since larger entropy corresponds to greater spreading out of energy, and since molecules move greater distances and in more different ways in a liquid than in a solid ***(p. 20)***, this is reasonable.

PROBLEM-SOLVING PRACTICE 18.1

A chemical reaction transfers 30.8 kJ to a thermal reservoir that has a temperature of 45.3 °C before and after the energy transfer. Calculate the entropy change for the thermal reservoir.

CONCEPTUAL EXERCISE 18.3 The Importance of Absolute Temperature

Consider what would happen if the Celsius temperature scale were used when calculating entropy change by means of $\Delta S = q_{rev}/T$. Suppose, for example, that energy were transferred reversibly to $H_2O(s)$ at a temperature 10° below its melting point and we wanted to calculate the entropy change. Would the value calculated from the Celsius temperature agree with the fact that transfer of thermal energy to a sample always increases its entropy?

Absolute Entropy Values

In Chapter 6 we mentioned that there is no way to measure the total energy content of a sample of matter. Therefore, to summarize a large number of calorimetric measurements of enthalpy changes, we tabulated standard molar enthalpies of formation in Table 6.2 (*p. 250*). The standard enthalpy of formation is the difference between the enthalpy of a substance in its standard state and the enthalpies of the elements that make up that substance, all in their standard states. The elements have enthalpy values, but we do not know what they are. Therefore, the standard enthalpies of formation of elements are by definition set to zero.

Though it is impossible to cool anything all the way to absolute zero, it is possible to get very close. Temperatures of a few nanokelvins can be achieved in a Bose-Einstein condensate—the coldest thing known to science. (See **http://www.nobel.se/physics/laureates/2001/index.html**)

For entropy the situation is simpler, because it is possible to define conditions for which it is logical to assume the entropy of a substance has its lowest possible value—namely, zero. Measuring ΔS for a change from those zero-entropy conditions tells us the absolute entropy value for the substance under new conditions. This follows from the definition $\Delta S = S_{final} - S_{initial}$, because if $S_{initial} = 0$, then $\Delta S = S_{final}$, the absolute entropy of the substance. Because decreasing temperature corresponds to decreasing molecular motion, the minimum possible temperature can reasonably be expected to correspond to minimum motion and thus minimum dispersal of energy. Thus it is logical to assume that a perfect crystal of a substance at 0 K has an entropy value of zero. [In a perfect crystal every nanoscale particle is in exactly the right position in the crystal lattice (*p. 512*), and there are no empty spaces or discontinuities.] To calculate the entropy change, start as close as possible to absolute zero, successively introduce small quantities of energy, and calculate ΔS from the equation $\Delta S = q_{rev}/T$ for each increment of energy at each temperature. Then sum these entropy changes to give the total (or absolute) entropy of a substance at any desired temperature.

The idea that a perfect crystal of any substance at 0 K has minimum entropy is called the **third law of thermodynamics.** Even if absolute zero cannot be achieved, there are ways of estimating how much energy has been dispersed in a substance near 0 K. Thus, accurate entropy values can be obtained for many substances.

The results of such measurements for several substances at 298.15 K are given in Table 18.1. These are standard molar entropy values, and so they apply to 1 mol of each substance at the standard pressure of 1 bar and the specified temperature of 25 °C. The units are joules per kelvin per mole (J K^{-1} mol^{-1}). Because there is a real zero on the entropy scale, the values in Table 18.1 are not measured relative to elements in their most stable form under standard-state conditions. Therefore, absolute entropies can be determined for elements as well as compounds. ***The standard molar entropy of a substance at temperature* T *is a sum of the quantities of energy that must be dispersed in that substance at successive temperatures up to* T, *that is, it is* ΔS *from 0 K to* T.**

The process of introducing small quantities of energy, calculating an entropy increase, and then summing these small entropy increases is actually done by measuring the heat capacity of a substance as a function of temperature and then using integral calculus to calculate the integral of the function q_{rev}/T between the limits of 0 K and the desired temperature.

Table 18.1 Some Standard Molar Entropy Values at 298.15 K*

Compound or Element	Entropy, $S°$ (J K^{-1} mol^{-1})	Compound or Element	Entropy, $S°$ (J K^{-1} mol^{-1})	Compound or Element	Entropy, $S°$ (J K^{-1} mol^{-1})
C(graphite)	5.740	$Br_2(\ell)$	152.231	Ca(s)	41.42
C(g)	158.096	I_2(s)	116.135	NaF(s)	51.5
CH_4(g)	186.264	Ar(g)	154.7	MgO(s)	26.94
CH_3CH_3(g)	229.60	H_2(g)	130.684	NaCl(s)	72.13
$CH_3CH_2CH_3$(g)	269.9	N_2(g)	191.61	KOH(s)	78.9
$CH_3OH(\ell)$	126.8	O_2(g)	205.138	$MgCO_3$(s)	65.7
CO(g)	197.674	NH_3(g)	192.45	NH_4NO_3(s)	151.08
CO_2(g)	213.74	HCl(g)	186.908	NaCl(aq)	115.5
F_2(g)	202.78	H_2O(g)	188.825	NH_4NO_3(aq)	259.8
Cl_2(g)	223.066	$H_2O(\ell)$	69.91	KOH(aq)	91.6

*Data from Wagman, D. D., Evans, W. H., Parker, V. B., Schumm, R. H., Halow, I., Bailey, S. M., Churney, K. L., and Nuttall, R. "The NBS Tables of Chemical Thermodynamic Properties." *Journal of Physical and Chemical Reference Data*, Vol. 11, Suppl. 2, 1982.

Qualitative Guidelines for Entropy

Some useful guidelines can be drawn from the data given in Table 18.1.

- ***Entropies of gases are usually much larger than those of liquids, which in turn are usually larger than those of solids.*** In a solid the particles can vibrate only around their lattice positions. When a solid melts, its particles can move around more freely, and molar entropy increases. When a liquid vaporizes, the position restrictions due to forces between the particles nearly disappear, and another large entropy increase occurs (Figure 18.2). For example, the entropies (in J K^{-1} mol^{-1}) of the halogens $I_2(s)$, $Br_2(\ell)$, and $Cl_2(g)$ are 116.1, 152.2, and 223.1, respectively. Similarly, the entropies of C(s, graphite) and C(g) are 5.7 and 158.1.
- ***Entropies of more complex molecules are larger than those of simpler molecules,*** especially in a series of closely related compounds. In a more complicated molecule there are more ways for the atoms to move about in three-dimensional space, and hence there is greater entropy. For example, the entropies (in J K^{-1} mol^{-1}) of the gases methane (CH_4), ethane (CH_3CH_3), and propane ($CH_3CH_2CH_3$) are 186.26, 229.6, and 269.9, respectively. For atoms or molecules of similar molar mass, compare the gases Ar, CO_2, and $CH_3CH_2CH_3$, which have entropies of 154.7, 213.74, and 269.9, respectively (Figure 18.3).

For a more detailed look at estimating entropy changes, see Craig, N. C. *Journal of Chemical Education*, Vol. 80, 2003; pp. 1432–1436.

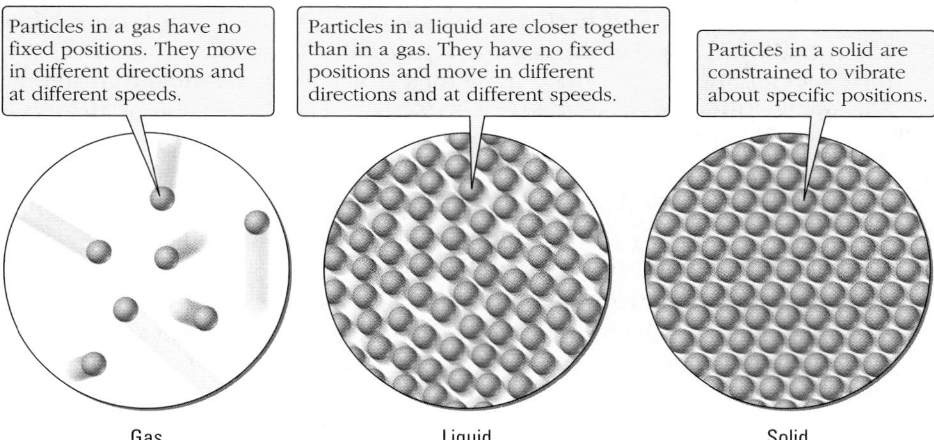

Particles in a gas have no fixed positions. They move in different directions and at different speeds.
Particles in a liquid are closer together than in a gas. They have no fixed positions and move in different directions and at different speeds.
Particles in a solid are constrained to vibrate about specific positions.
Gas
Liquid
Solid

Figure 18.2 Entropies of solid, liquid, and gas phases. The entropy of each phase is related to the freedom of motion of the particles.

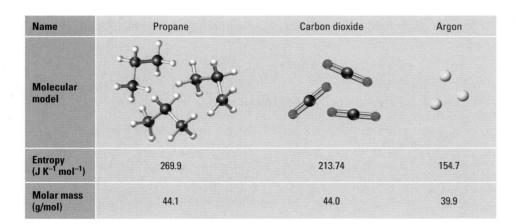

Name	Propane	Carbon dioxide	Argon
Molecular model			
Entropy (J K^{-1} mol^{-1})	269.9	213.74	154.7
Molar mass (g/mol)	44.1	44.0	39.9

Figure 18.3 Entropy and molecular structure. Three groups of particles of similar molar mass are shown. In propane, $CH_3CH_2CH_3$, many different conformations (different ways to arrange the atoms relative to one another within a molecule) and many different bond stretching and bending vibrations are possible *(← p. 392)*. In CO_2 there is a single conformation and many fewer vibrations. The individual Ar atoms can move about, but there are no conformations or vibrations possible.

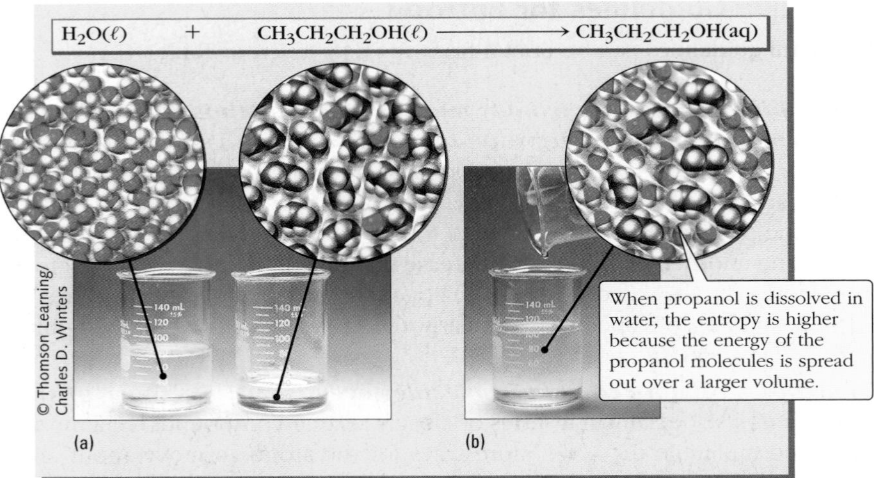

$H_2O(\ell)$ + $CH_3CH_2CH_2OH(\ell)$ → $CH_3CH_2CH_2OH(aq)$
© Thomson Learning/Charles D. Winters
(a)
(b)
When propanol is dissolved in water, the entropy is higher because the energy of the propanol molecules is spread out over a larger volume.

Active Figure 18.4 **Entropy and dissolving.** There is usually an increase in entropy when a solid or liquid dissolves in a liquid solvent, because, as solute particles become dispersed among solvent particles, the energy of motion of the particles is spread out over a larger volume. **Go to the Active Figures menu at ThomsonNOW to test your understanding of the concepts in this figure.**

When ionic solids that consist of very small ions (such as Li^+) or of ions that carry two or more units of charge (such as Mg^{2+} or Al^{3+}) dissolve in water, there is often a decrease in entropy. This happens because very small ions and highly charged ions strongly attract water dipoles. The water molecules surrounding such ions are held in a relatively rigid structure and are no longer as free to move and rotate as they were in pure water. Although the ions become dispersed among the water molecules, the reduced dispersion of energy among the water molecules results in an overall decrease in entropy.

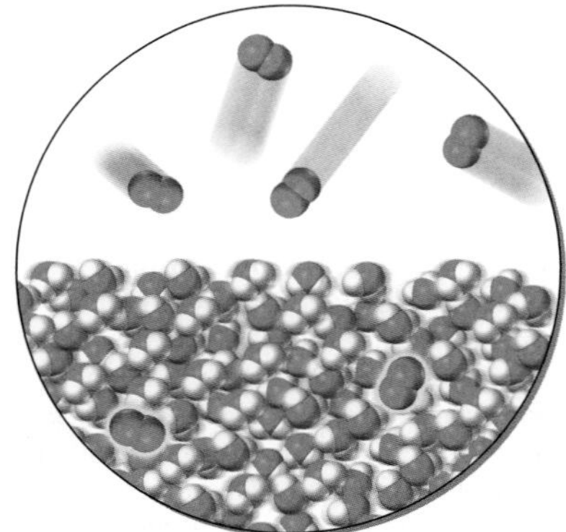

Figure 18.5 **Entropy of solution of a gas.** The very large entropy of the gas exceeds that of the solution. Even though particles are dispersed among each other in the liquid solution, the gas particles are much more widely spread out and have much higher entropy.

- ***Entropies of ionic solids that have similar formulas are larger when the attractions among the ions are weaker.*** The weaker such forces are, the easier it is for ions to vibrate about their lattice positions and the greater the entropy is. The entropy of NaF(s) is 51.5 J K^{-1} mol^{-1}, and that of MgO(s) is 26.94 J K^{-1} mol^{-1}; Na^+ and F^-, with unit positive and negative charges, attract each other less than Mg^{2+} and O^{2-}, each of which has two units of charge ***(← p. 93);*** therefore NaF(s) has higher entropy. NaF(s) and NaCl(s) have entropies of 51.5 and 72.13 J K^{-1} mol^{-1}. Chloride ions, Cl^-, are larger than fluoride ions, F^-, and attractions are smaller when the ions are farther apart.
- ***Entropy usually increases when a pure liquid or solid dissolves in a solvent.*** Energy usually becomes more dispersed when different kinds of molecules mix together and occupy a larger volume (Figure 18.4). An example is $NH_4NO_3(s)$ and $NH_4NO_3(aq)$ with standard molar entropies of 151.08 J K^{-1} mol^{-1} and 259.8 J K^{-1} mol^{-1}, respectively. Some ionic compounds dissolving in water are exceptions to this generalization because the ions are strongly hydrated.
- ***Entropy decreases when a gas dissolves in a liquid.*** Although gas molecules are dispersed among solvent molecules in solution, the very large entropy of the gas phase is lost when the widely separated gas particles become crowded together with solvent particles in the liquid solution (Figure 18.5).

PROBLEM-SOLVING EXAMPLE 18.2 Relative Entropy Values

For each pair of substances below, predict which has greater entropy and give a reason for your choice. (Assume 1-mol samples at 25 °C and 1 bar.)

(a) $H_2C{=}CH_2(g)$ or $CH_3CH_2CH_2CH_3(g)$ (b) $CO_2(aq)$ or $CO_2(g)$
(c) LiF(s) or RbBr(s) (d) $N_2(\ell)$ or $N_2(s)$

Answer

(a) $CH_3CH_2CH_2CH_3(g)$ (b) $CO_2(g)$
(c) RbBr(s) (d) $N_2(\ell)$

Strategy and Explanation Use the rules given on p. 875 and 876.

(a) Larger, more complex molecules have greater entropy than similar smaller ones, so the entropy of $CH_3CH_2CH_2CH_3(g)$ is greater.

(b) The molecules of a gas are free to move, rotate, and vibrate, so when a gas dissolves in a liquid, the entropy decreases; therefore $CO_2(g)$ has greater entropy.

(c) These are ionic solids, both with 1+ and 1− ions. The attractive forces are greater the closer the ions are to each other, and the ions are smaller in LiF ***(see table of ionic radii, p. 314),*** so RbBr has greater entropy.

(d) Entropy increases from solid to liquid to gas for the same substance, so $N_2(\ell)$ has greater entropy.

PROBLEM-SOLVING PRACTICE **18.2**

In each case, predict which of the two substances has greater entropy, assuming 1-mol samples at 25 °C and 1 bar. Check your prediction by looking up each substance's absolute entropy in Table 18.1.

(a) C(g) or C(s, graphite) (b) Ca(s) or Ar(g) (c) KOH(s) or KOH(aq)

Predicting Entropy Changes

The general guidelines about entropy can be used to predict whether an increase or decrease in entropy will occur when reactants are converted to products. For both of the processes

$$H_2O(s) \longrightarrow H_2O(\ell) \qquad \text{and} \qquad H_2O(\ell) \longrightarrow H_2O(g)$$

an entropy increase is expected. Water molecules in the solid phase are more restricted and their energy is more localized than in the liquid, and water molecules in the liquid cannot move, rotate, or vibrate as freely as they can in the gas. This is confirmed by the data in Table 18.1, where $S°\ (H_2O(g)) > S°\ (H_2O(\ell))$.

For the decomposition of iron(III) oxide to its elements,

$$2\ Fe_2O_3(s) \longrightarrow 4\ Fe(s) + 3\ O_2(g)$$

an increase in entropy is also predicted, because 3 mol gaseous oxygen is present in the products and the reactant is a solid. This is confirmed by the experimental $\Delta S°$, 551.7 J/K. Because gases have much higher entropy than solids or liquids, gaseous substances are most important in determining entropy changes.

Predicting entropy changes for chemical processes is usually easier than predicting enthalpy changes. For gas phase reactions, the enthalpy-change guideline is that having more bonds or stronger bonds (or both) in the products gives a negative $\Delta H°$ (***← p. 240);*** however, a table of bond enthalpies is usually needed to tell which bonds are stronger.

An example where a decrease in entropy can be predicted is

$$2\ CO(g) + O_2(g) \longrightarrow 2\ CO_2(g)$$

Here there is 3 mol gaseous substance (2 mol CO and 1 mol O_2) at the beginning but only 2 mol gaseous substance at the end of the reaction. Two moles of gas almost always contains less entropy than three moles of gas, so $\Delta S°$ is negative (experimentally, $\Delta S° = -173.0$ J/K). Another example in which entropy decreases is the process

$$Ag^+(aq) + Cl^-(aq) \longrightarrow AgCl(s)$$

Here the reactant ions are free to move about among water molecules in aqueous solution, but those same ions are held in a crystal lattice in the solid, a situation with greater constraint and thus less spreading out of energy.

Predicting an entropy decrease when ions precipitate from aqueous solution does not always work, especially when the ions have more than single positive or negative charges, such as Mg^{2+} or SO_4^{2-}. The higher the charge on an ion the more tightly water-molecule dipoles are attracted to it. When water molecules are tightly attracted around an ion their motion is restricted and entropy decreases.

CONCEPTUAL EXERCISE **18.4 Predicting Entropy Changes**

For each process, tell whether entropy increases or decreases, and explain how you arrived at your prediction.

(a) $2\ CO_2(g) \longrightarrow 2\ CO(g) + O_2(g)$ (b) $NaCl(s) \longrightarrow NaCl(aq)$

(c) $MgCO_3(s) \xrightarrow{\text{heat}} MgO(s) + CO_2(g)$

18.4 Calculating Entropy Changes

The standard molar entropy values given in Table 18.1 can be used to calculate entropy changes for physical and chemical processes. Assume that each reactant and each product is at the standard pressure of 1 bar and at the temperature given (298.15 K). The number of moles of each substance is specified by its stoichiometric coefficient in the equation for the process. Multiply the entropy of each product substance by the number of moles of that product and add the entropies of all products. Calculate the total entropy of the reactants in the same way and subtract it from the total entropy of the products. This is summarized in the equation

Notice that the equation for calculating $\Delta S°$ has the same form as that for calculating $\Delta H°$ for a reaction (⟵ *p. 251*).

$$\Delta S° = \Sigma\{(\text{moles product}) \times S°(\text{product})\} - \Sigma\{(\text{moles reactant}) \times S°(\text{reactant})\}$$

Note that this calculation gives the entropy change for the chemical reaction *system* only. It tells whether the energy of the atoms that make up the system is more dispersed or less dispersed after the reaction. It does not account for any entropy change in the surroundings.

PROBLEM-SOLVING EXAMPLE 18.3 Calculating an Entropy Change from Tabulated Values

The reaction

$$CO(g) + 2\,H_2(g) \longrightarrow CH_3OH(\ell)$$

is being evaluated as a possible way to manufacture liquid methanol, $CH_3OH(\ell)$, for use in motor fuel. Calculate $\Delta S°$ for the reaction.

Answer $\Delta S° = -332.3$ J/K

Strategy and Explanation Use information in Table 18.1, subtracting the entropies of the reactants from the entropy of the product. Because all substances, including elements, have nonzero absolute entropy values, elements as well as compounds must be included.

$$\begin{aligned}\Delta S° &= \Sigma\{(\text{moles of product}) \times S°(\text{product})\} - \Sigma\{(\text{moles of reactant}) \times S°(\text{reactant})\}\\ &= (1\text{ mol}) \times S°[CH_3OH(\ell)] - \{(1\text{ mol}) \times S°[CO(g)] + (2\text{ mol}) \times S°[H_2(g)]\}\\ &= (1\text{ mol}) \times (126.8\text{ J K}^{-1}\text{ mol}^{-1})\\ &\quad - \{(1\text{ mol}) \times (197.7\text{ J K}^{-1}\text{ mol}^{-1}) + (2\text{ mol}) \times (130.7\text{ J K}^{-1}\text{ mol}^{-1})\}\\ &= -332.3\text{ J/K}\end{aligned}$$

✓ **Reasonable Answer Check** There is a decrease in entropy, which is reasonable because 3 mol gaseous reactants is converted to 1 mol liquid phase product. The product molecule is more complicated, but the fact that it is a liquid makes its entropy much smaller than that of 1 mol gaseous carbon monoxide and 2 mol gaseous hydrogen.

PROBLEM-SOLVING PRACTICE 18.3

Use absolute entropies from Table 18.1 to calculate the entropy change for each of these processes, thereby verifying the predictions made in Conceptual Exercise 18.4.

(a) $2\,CO_2(g) \longrightarrow 2\,CO(g) + O_2(g)$ (b) $NaCl(s) \longrightarrow NaCl(aq)$

(c) $MgCO_3(s) \xrightarrow{\text{heat}} MgO(s) + CO_2(g)$

18.5 Entropy and the Second Law of Thermodynamics

A great deal of experience with many chemical reactions and other processes in which energy is transformed and transferred is consistent with the conclusion that *whenever a product-favored chemical or physical process occurs, energy becomes more dispersed or disordered.* This is summarized in the **second law of thermodynamics,** which states that the ***total entropy of the universe (a system plus its surroundings) is continually increasing.*** Evaluating whether the total entropy increases during a proposed chemical reaction allows us to predict whether reactants will form appreciable quantities of products.

Predicting whether a reaction is product-favored can be done in three steps:

1. Calculate how much entropy changes as a result of transfer of energy between system and surroundings ($\Delta S_{surroundings}$).
2. Calculate how much entropy changes as a result of dispersal of energy within the system (ΔS_{system}).
3. Add these two results to get $\Delta S_{universe} = \Delta S_{system} + \Delta S_{surroundings}$.

Let us apply these steps to the reaction

$$CO(g) + 2\,H_2(g) \longrightarrow CH_3OH(\ell)$$

for which we calculated the entropy change in Problem-Solving Example 18.3. If the reaction is product-favored, it would be a good way to produce methanol for use as automotive fuel. The reactants can be obtained from plentiful resources: coal and water. We base our prediction upon having as reactants 1 mol CO(g) and 2 mol H_2(g) and as product 1 mol liquid methanol, with all substances at 1 bar and 298.15 K (25 °C). Then data from Table 6.2 ***(⇐ p. 250)*** and Table 18.1 (or Appendix J) apply. If the entropy of the universe is predicted to be higher after the product has been produced, then the reaction is product-favored under these conditions and might be useful. If not, perhaps some other conditions could be used, or perhaps we should consider some other reaction altogether.

ThomsonNOW
Log in to ThomsonNOW at **www.thomsonedu.com** and go to the Coached Problems menu for a tutorial and a module on:

- **calculating ΔS for the universe**
- **the second law of thermodynamics**
- **dissolution of NH_4NO_3**

Steps 1 and 2 can be carried out by assuming that reactants under standard conditions of 1 bar and a specified temperature are converted to products under the same standard conditions.

We will consider the effect of changing temperature later in this chapter. The effects of changing other conditions can often be predicted qualitatively by using Le Chatelier's principle ***(⇐ p. 695).***

Step 1: Calculate the reaction's dispersal or concentration of energy to or from the surroundings by calculating $\Delta H°$ and assuming that this quantity of thermal energy is transferred reversibly. The entropy change for the surroundings can be calculated as

$$\Delta S°_{surroundings} = \frac{q_{rev}}{T} = \frac{\Delta H_{surroundings}}{T} = \frac{-\Delta H°_{system}}{T} = \frac{-\Delta H°}{T}$$

The minus sign in this equation comes from the fact that the direction of energy transfer for the surroundings is opposite from the direction of energy transfer for the system. For an exothermic reaction (negative $\Delta H°$) there will be an increase in entropy of the surroundings, a fact that we have already mentioned. For the proposed methanol-producing reaction, $\Delta H°$ (calculated from data in Appendix J) is -128.1 kJ, so the entropy change is

$$\Delta S°_{surroundings} = \frac{-(-128.1\ \text{kJ})}{298\ \text{K}} \times \frac{1000\ \text{J}}{\text{kJ}} = 430.\ \text{J/K}$$

Step 2: Calculate the entropy change for dispersal of energy within the system (that is, for the atoms involved in the methanol-producing reaction). This entropy change can be evaluated from the absolute entropies of the products and reactants as described in the previous section and has already been calculated in Problem-Solving Example 18.3 to be

$$\Delta S°_{system} = \Delta S° = -332.3\ \text{J/K}$$

Step 3: Calculate the total entropy change for the system and the surroundings, $\Delta S°_{universe}$. Because the universe includes both system and surroundings, $\Delta S°_{universe}$ is the sum of the entropy change for the system and the entropy change for the surroundings. (We assume that nothing

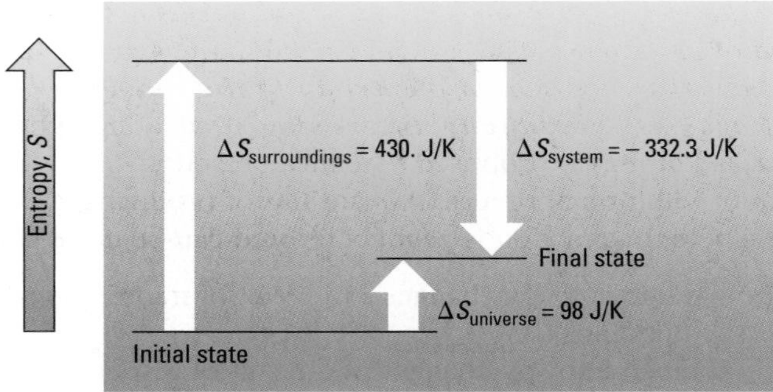

Entropy diagram for combination of carbon monoxide and hydrogen to form methanol.

else but our reaction happens, so there are no other entropy changes.) This total entropy change is

$$\Delta S^\circ_{universe} = \Delta S^\circ_{surroundings} + \Delta S^\circ_{system} = \frac{-\Delta H^\circ_{system}}{T} + \Delta S^\circ_{system} \quad [18.2]$$

$$= (430. - 332.3)\ \text{J/K} = 98\ \text{J/K}$$

Combination of carbon monoxide and hydrogen to form methanol increases the entropy of the universe. Because the process is product-favored, it could be useful for manufacturing methanol.

CONCEPTUAL EXERCISE 18.5 Effect of Temperature on Entropy Change

The reaction of carbon monoxide with hydrogen to form methanol is quite slow at room temperature. As a general rule, reactions go faster at higher temperatures. Suppose that you tried to speed up this reaction by increasing the temperature.

(a) Assuming that ΔH° does not change very much as the temperature changes, what effect would increasing the temperature have on $\Delta S^\circ_{surroundings}$?

(b) Assuming that ΔS° for a reaction system does not change much as the temperature changes, what effect would increasing the temperature have on $\Delta S^\circ_{universe}$?

PROBLEM-SOLVING EXAMPLE 18.4 Calculating $\Delta S^\circ_{universe}$

When gasoline burns one reaction is combustion of octane in air.

$$2\ C_8H_{18}(g) + 25\ O_2(g) \longrightarrow 16\ CO_2(g) + 18\ H_2O(\ell)$$

Calculate $\Delta S^\circ_{universe}$ for this reaction, thereby confirming that the reaction is product-favored. (Assume that reactants and products are at 298.15 K and 1 bar; use data from Appendix J.)

Answer 35,591 J/K

Strategy and Explanation A balanced chemical equation for the reaction is given. Calculate ΔH° and ΔS° by subtracting the sum of reactant values from the sum of product values.

$$\Delta H^\circ = \Sigma\{(\text{moles of product}) \times \Delta H^\circ_f\ (\text{product})\} - \Sigma\{(\text{moles of reactant}) \times \Delta H^\circ_f\ (\text{reactant})\}$$

$$= \{16 \times (-393.509\ \text{kJ}) + 18 \times (-285.830\ \text{kJ})\} - \{2 \times (-208.447\ \text{kJ}) + 25 \times 0\ \text{kJ}\}$$

$$= -11{,}024.2\ \text{kJ}$$

$$\Delta S^\circ = \Sigma\{(\text{moles of product}) \times S^\circ\ (\text{product})\} - \Sigma\{(\text{moles of reactant}) \times S^\circ\ (\text{reactant})\}$$
$$= \{16 \times (213.74\ \text{J/K}) + 18 \times (69.91\ \text{J/K}\} - \{2 \times (466.835\ \text{J/K}) + 25 \times (205.138\ \text{J/K})\}$$
$$= -1383.9\ \text{J/K}$$

Now use Equation 18.2 to calculate $\Delta S^\circ_{\text{universe}}$.

$$\Delta S^\circ_{\text{universe}} = \Delta S_{\text{surroundings}} + \Delta S_{\text{system}} = \frac{-\Delta H^\circ}{T} + \Delta S^\circ$$
$$= \frac{-(-11{,}024.2\ \text{kJ})}{298.15\ \text{K}} + (-1383.9\ \text{J/K})$$
$$= 36.975\ \text{kJ/K} - 1383.9\ \text{J/K}$$
$$= 36{,}975\ \text{J/K} - 1383.9\ \text{J/K} = 35{,}591\ \text{J/K}$$

Notice that for $-\Delta H^\circ/T$ the units were kilojoules per kelvin (kJ/K), while for ΔS° the units were joules per kelvin (J/K). Therefore it was necessary to convert kilojoules to joules by multiplying the first term by 1000 J/kJ.

Because $\Delta S^\circ_{\text{universe}}$ is positive, the reaction is product-favored.

✓ **Reasonable Answer Check** A positive result is reasonable, because you know that gasoline burns in air, which means that the reaction is product-favored. Even though the entropy change for the system is unfavorable, the reaction is highly product-favored because it is strongly exothermic.

PROBLEM-SOLVING PRACTICE 18.4

Use data from Appendix J to determine whether the synthesis of ammonia from nitrogen and hydrogen is product-favored at 298.15 K and 1 bar.

CONCEPTUAL EXERCISE 18.6 Variation of ΔH° and ΔS° with Temperature

Suppose that the combustion of gaseous octane is carried out at 150 °C.

(a) How would this affect the chemical equation for the reaction?

(b) What effect would the change in the chemical equation have on ΔH° and ΔS° for the reaction?

(c) When is it definitely *not* safe to assume that ΔH° and ΔS° for a reaction will be almost the same over a broad range of temperatures?

Predictions of the sort we have just made by calculating $\Delta S^\circ_{\text{universe}}$ can also be made qualitatively, without calculating, if we know whether a reaction is exothermic and if we can predict whether energy is dispersed within the system when the reaction takes place. ***A reaction is certain to be product-favored if it is exothermic and the entropy of the product atoms, molecules, and ions is greater than the entropy of the reactants ($\Delta S_{system} > 0$). Also, a reaction is certainly not product-favored if it is endothermic and there is a decrease in entropy for the system.*** There are two other possible cases, as indicated in Table 18.2, but they are more difficult to predict without quantitative information.

As examples, consider the reactions of carbonates with acids ***(⇐ p. 178)***. These reactions are product-favored because they are exothermic and produce gases. Reaction of limestone with hydrochloric acid is typical.

$$CaCO_3(s) + 2\ HCl(aq) \longrightarrow CaCl_2(aq) + H_2O(\ell) + CO_2(g) \qquad \text{(exothermic)}$$

© Thomson Learning/Charles D. Winters

A product-favored reaction. Carbonates react rapidly with acid to produce carbon dioxide and water.

Similarly, combustion reactions of hydrocarbons such as butane, $CH_3CH_2CH_2CH_3$, are product-favored because they are exothermic and produce a larger number of gas phase product molecules than there were gas phase reactant molecules.

$$2\ CH_3CH_2CH_2CH_3(g) + 13\ O_2(g) \longrightarrow 8\ CO_2(g) + 10\ H_2O(g) \qquad \text{(exothermic)}$$

Table 18.2 Predicting Whether a Reaction Is Product-Favored (at constant T and P)

Sign of ΔH_{system}	Sign of ΔS_{system}	Product-Favored?
Negative (exothermic)	Positive	Yes
Negative (exothermic)	Negative	Yes at low T; no at high T
Positive (endothermic)	Positive	No at low T; yes at high T
Positive (endothermic)	Negative	No

But what about a reaction such as the production of ethylene, $CH_2{=}CH_2$, from ethane, CH_3CH_3? Although entropy is predicted to increase (one gas phase molecule forms two), the reaction is very endothermic. (That the reaction is endothermic might be predicted on the basis of a decrease from seven bonds in the reactant molecule to six bonds—five single, one double—in the products.)

$$CH_3CH_3(g) \longrightarrow H_2(g) + CH_2{=}CH_2(g) \qquad \Delta H° = +137 \text{ kJ}; \;\; \Delta S° = +121 \text{ J/K}$$

Enthalpy change predicts that this process is reactant-favored, while entropy change predicts the opposite. Which is more important? It depends on the temperature.

Calculating $\Delta S°_{surroundings}$ $(= -\Delta H°/T)$ requires dividing the enthalpy change by the temperature. Because $\Delta H°$ stays pretty much the same at different temperatures, the higher the temperature, the smaller the absolute value of $\Delta S°_{surroundings}$ $(= -\Delta H°/T)$. At room temperature $\Delta S°_{surroundings}$ is usually bigger in absolute value than $\Delta S°_{system}$, so exothermic reactions are expected to be product-favored and endothermic reactions (like this one) are expected to be reactant-favored. The ethylene-producing reaction is reactant-favored at 25 °C, because $\Delta S°_{universe} = -339$ J/K. To make a successful industrial process, chemical engineers have designed plants that carry out this reaction at about 1000 °C. At this higher temperature, $\Delta S°_{surroundings}$ is smaller in magnitude than $\Delta S°_{system}$. Thus $\Delta S°_{universe} = 13$ J/K, and products are predicted to predominate over reactants.

At 25 °C:

$$\Delta S°_{surroundings} = \frac{-137 \text{ kJ}}{298 \text{ K}} = -460 \text{ J/K}$$

$$\Delta S°_{system} = 121 \text{ J/K}$$

$$\Delta S°_{universe} = (-460 + 121) \text{ J/K} = -339 \text{ J/K}$$

At 1000 °C:

$$\Delta S°_{surroundings} = \frac{-137 \text{ kJ}}{(1000 + 273) \text{ K}} = -108 \text{ J/K}$$

$$\Delta S°_{system} = 121 \text{ J/K}$$

$$\Delta S°_{universe} = 13 \text{ J/K}$$

In Chapter 14 (*p. 704*) we stated that if a reaction is exothermic or involves an increase in entropy of the system, this favors the products. The entropy effect becomes more important the higher the temperature.

EXERCISE 18.7 Predicting the Direction of a Reaction

Using data from Appendix J, complete the table and then classify each reaction into one of the four types in Table 18.2. Predict whether each reaction is product-favored or reactant-favored at room temperature.

Reaction	$\Delta H°$, 298 K (kJ)	$\Delta S°$, 298 K (J/K)
(a) $C_2H_4(g) + 3\,O_2(g) \longrightarrow 2\,H_2O(\ell) + 2\,CO_2(g)$	______	______
(b) $2\,Fe_2O_3(s) + 3\,C(\text{graphite}) \longrightarrow 4\,Fe(s) + 3\,CO_2(g)$	______	______
(c) $C(\text{graphite}) + O_2(g) \longrightarrow CO_2(g)$	______	______
(d) $2\,Ag(s) + 3\,N_2(g) \longrightarrow 2\,AgN_3(s)$	______	______

EXERCISE 18.8 Product- or Reactant-Favored?

(a) Is the combination reaction of hydrogen gas and chlorine gas to give hydrogen chloride gas (at 1 bar) predicted to be product-favored or reactant-favored at 298 K?

(b) What is the value for $\Delta S°_{universe}$ for the reaction in part (a)?

18.6 Gibbs Free Energy

Calculations of the sort done in the previous section would be simpler if we did not have to separately evaluate the entropy change of the surroundings from T and a table of ΔH_f° values and the entropy change of the system from a table of S° values. To simplify such calculations, a new thermodynamic function was defined by J. Willard Gibbs (1838–1903). It is now called **Gibbs free energy** and it is given the symbol G. Gibbs defined his free energy so that $\Delta G_{system} = -T\Delta S_{universe}$. Because of the minus sign, if the entropy of the universe increases, the Gibbs free energy of the system must decrease. That is, ***a decrease in Gibbs free energy of a system is characteristic of a process that is product-favored at constant temperature and pressure.***

ThomsonNOW™
Go to the Coached Problems menu for a tutorial on **Gibbs free energy.**

In the previous section we showed that the total entropy change accompanying a chemical reaction carried out at constant temperature and pressure is

$$\Delta S_{universe} = \Delta S_{surroundings} + \Delta S_{system} = \frac{-\Delta H_{system}}{T} + \Delta S_{system}$$

Combining this algebraically with Gibbs's definition of free energy, we have

$$\Delta G_{system} = -T\Delta S_{universe} = -T\left[\frac{-\Delta H_{system}}{T} + \Delta S_{system}\right] = \Delta H_{system} - T\Delta S_{system}$$

or, under standard-state conditions.

$$\Delta G^\circ = \Delta H^\circ - T\Delta S^\circ \qquad [18.3]$$

This equation summarizes the ideas about chemical equilibrium that were developed in Chapter 14 (⇐ ***p. 706***). A negative value of ΔG° indicates that a reaction is product-favored, and the equation says that two conditions will make ΔG° more negative: (1) if the reaction is exothermic, ΔH° will be negative, thereby favoring the products; and (2) if the products have greater entropy than the reactants, then ΔS° will be positive and the $-T\Delta S^\circ$ term will be negative, which favors the products. Because ΔS° is multiplied by T, the entropy of the system is more important at higher temperatures.

For ΔH°, ΔS°, and ΔG° you can assume that the values apply to the system—that is, $\Delta G^\circ = \Delta G^\circ_{system}$—unless a subscript is attached to indicate that a value is for the surroundings.

CONCEPTUAL EXERCISE 18.9 Predicting Whether a Process Is Product-Favored

Make a table similar to Table 18.2, but add a new column for the sign of ΔG°. Based on the value of ΔG°, predict whether the reaction is product-favored. If there is insufficient information, indicate whether the products would be favored more at high temperatures or at low temperatures. Check your results against Table 18.2.

The Gibbs free energy change provides a way of predicting whether a reaction will be product-favored that depends only on the system—that is, the chemical substances undergoing reaction. Therefore, we can tabulate values of the standard Gibbs free energy of formation, ΔG_f°, for a variety of substances, and from them calculate

$$\Delta G^\circ = \Sigma\{(\text{moles of product}) \times \Delta G_f^\circ(\text{product})\} - \Sigma\{(\text{moles of reactant}) \times \Delta G_f^\circ(\text{reactant})\} \qquad [18.4]$$

for a great many reactions. The calculation is similar to using ΔH_f° values from Table 6.2 or Appendix J to calculate ΔH° for a reaction (⇐ ***p. 250***). As was the case for ΔH_f° values, there are no ΔG_f° values for elements in their standard states, because forming an element from itself constitutes no change at all. Appendix J contains a table that includes ΔG_f° values for many compounds.

It is important to realize that ΔG° varies significantly as the temperature changes (because of the $-T\Delta S^\circ$ term). Therefore, values of ΔG° calculated from Equation 18.4 apply only to the temperature specified in the table of ΔG_f° values. Appendix J specifies a temperature of 25 °C.

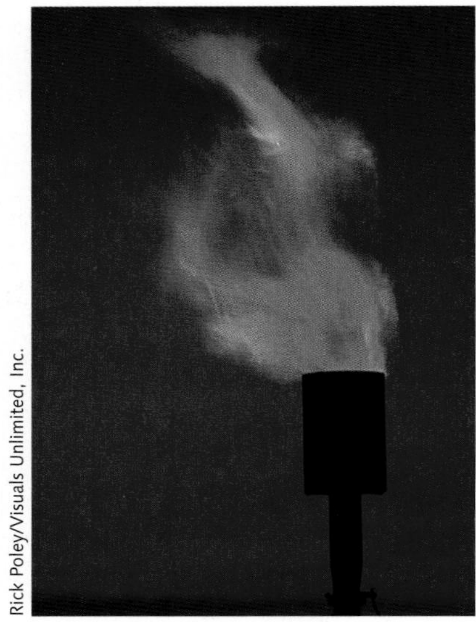

Combustion of natural gas (methane).

PROBLEM-SOLVING EXAMPLE 18.5 Using Standard Gibbs Free Energies of Formation

Calculate the standard Gibbs free energy change for the combustion of octane using values of ΔG_f° from Appendix J. Assume that the initial and final states have the same temperature and pressure so that Equation 18.4 applies.

Answer $-10{,}611$ kJ

Strategy and Explanation Write a balanced equation for the combustion reaction and look up ΔG_f° values in Appendix J.

$$2\,C_8H_{18}(g) + 25\,O_2(g) \longrightarrow 18\,H_2O(\ell) + 16\,CO_2(g)$$

ΔG_f°(kJ/mol): 16.72 0 −237.1 −394.4

(Notice that elements in their standard states have $\Delta G_f^\circ = 0$, just as they have $\Delta H_f^\circ = 0$.)

Now calculate

$$\begin{aligned}
\Delta G^\circ &= \Sigma\{(\text{moles of product}) \times \Delta G_f^\circ\ (\text{product})\}\\
&\quad - \Sigma\{(\text{moles of reactant}) \times \Delta G_f^\circ\ (\text{reactant})\}\\
&= \{(18\text{ mol} \times \Delta G_f^\circ\ [H_2O(g)]\} + \{16\text{ mol} \times \Delta G_f^\circ[CO_2(g)]\}\\
&\quad - \{2\text{ mol} \times \Delta G_f^\circ[C_8H_{18}(g)]\} - \{25\text{ mol} \times \Delta G_f^\circ[O_2(g)]\}\\
&= \{18\text{ mol} \times (-237.1\text{ kJ/mol})\} + \{16\text{ mol} \times (-394.4\text{ kJ/mol})\}\\
&\quad - \{2\text{ mol} \times (16.72\text{ kJ/mol})\} - \{25\text{ mol} \times (0\text{ kJ/mol})\}\\
&= -10{,}611\text{ kJ}
\end{aligned}$$

The Gibbs free energy change, ΔG°, is a large negative number, indicating that the reaction is product-favored under standard-state conditions.

✓ **Reasonable Answer Check** A large, negative Gibbs free energy change is reasonable for a combustion reaction, which, once initiated, occurs of its own accord.

PROBLEM-SOLVING PRACTICE 18.5

In the text we concluded that the reaction to produce methanol from CO and H_2 is product-favored.

$$CO(g) + 2\,H_2(g) \longrightarrow CH_3OH(\ell)$$

(a) Verify this result by calculating ΔG° from ΔH° and ΔS° for the system. Use values of ΔH_f° and S° from Appendix J.

(b) Compare your result in part (a) with the calculated value of ΔG° obtained from ΔG_f° values from Appendix J.

(c) Is the sign of ΔG° positive or negative? Is the reaction product-favored? At all temperatures?

ThomsonNOW™

Go to the Coached Problems menu for tutorials on:

- **putting ΔH and ΔS together**
- **Gibbs free energy: temperatures, ΔS and ΔH**

The Effect of Temperature on Reaction Direction

Many reactions are product-favored at some temperatures and reactant-favored at other temperatures. Thus, it might be possible to make such a reaction produce products by increasing or decreasing the temperature. There is a simple, approximate way to estimate the temperature at which a reactant-favored process becomes product-favored.

In Exercises 18.5 and 18.6 (p. 880 and p. 881) we developed the idea that ΔH° and ΔS° have nearly constant values over a broad range of temperatures, provided that each of the substances involved in a chemical reaction remains in the same state of matter (solid, liquid, or gas). Because ΔH° and ΔS° are nearly constant, the T on the right-hand side of the equation $\Delta G^\circ = \Delta H^\circ - T\Delta S^\circ$ implies that ΔG° must vary with temperature. It also implies that if we know ΔH° and ΔS° at one temperature, we can estimate $\Delta G°$ over a range of temperatures. For example, suppose we want to know whether the reaction

$$2\ HgO(s) \longrightarrow 2\ Hg(\ell) + O_2(g)$$

will produce products at a temperature of 350. °C and a pressure of 1 bar. To find out, calculate $\Delta H°$ and $\Delta S°$ at 298 K and then estimate $\Delta G°$, assuming that $\Delta H°$ and $\Delta S°$ have the same values at 350. °C (623 K) that they do at 298 K. The boiling point of mercury is 356 °C, and mercury(II) oxide does not melt until well above 500. °C, so the substances are all in the same states at 350. °C that they were at 25 °C.

The effect of temperature on reaction direction can be seen in Equation 18.2 (p. 880). Because $\Delta S_{surroundings} = -\Delta H_{system}/T$ and ΔH is nearly constant over a broad range of temperatures, the larger T is the smaller (and therefore less important) the entropy change of the surroundings becomes relative to the entropy change of the system. At high temperatures ΔS_{system} governs the directionality of the reaction.

$$\begin{aligned}
\Delta S°(298.15\text{ K}) &= \{2\text{ mol} \times S°[Hg(\ell)] + (1\text{ mol}) \times S°[O_2(g)]\} \\
&\quad - \{2\text{ mol} \times S°[HgO(s)]\} \\
&= 2 \times 76.02\text{ J/K} + 205.138\text{ J/K} - 2 \times 70.29\text{ J/K} = 216.60\text{ J/K} \\
\Delta H°(298.15\text{ K}) &= \{2\text{ mol} \times \Delta H_f°[Hg(\ell)] + 1\text{ mol} \times \Delta H_f°[O_2(g)]\} \\
&\quad - \{2\text{ mol} \times \Delta H_f°[HgO(s)]\} \\
&= 0\text{ kJ} + 0\text{ kJ} - 2 \times (-90.83)\text{ kJ} = 181.66\text{ kJ} \\
\Delta G°(623\text{ K}) &= \Delta H°(298.15\text{ K}) - T \times \Delta S°(298.15\text{ K}) \\
&= 181.66\text{ kJ} - 623\text{ K} \times 216.60\text{ J/K} \\
&= 181.66\text{ kJ} - 134{,}942\text{ J} = 181.66\text{ kJ} - 134.9\text{ kJ} = 47\text{ kJ}
\end{aligned}$$

Note that 139,242 J has been converted to 134.9 kJ in the last step.

Because $\Delta G°$ is positive, the reaction is not product-favored at 623 K. But at 623 K $\Delta G°$ has a smaller positive value than at 298 K, where it has the value $\Delta G° = -(2\text{ mol}) \times \Delta G_f°(HgO[s]) = +117.1$ kJ. That is, the reaction is more product-favored at 623 K than it is at room temperature. Because the calculated values of $\Delta H°$ and $\Delta S°$ are both positive, the reaction is expected to be product-favored at high temperatures but not at low temperatures. Heating to an even higher temperature than 623 K does decompose mercury(II) oxide (see Figure 18.6).

Joseph Priestley's discovery of oxygen in 1774 involved heating mercury(II) oxide to decompose it.

CONCEPTUAL EXERCISE 18.10 High-Temperature Decomposition

Suppose that a sample of HgO(s) is heated above the boiling point of mercury (356 °C).

(a) Could you use the same method to estimate the Gibbs free energy change that was used in the preceding paragraph? Why or why not? (*Hint:* Write the equation for the process that would occur if the temperature were 400 °C.)

(b) At 400 °C, would you expect the reaction to be more or less product-favored than it was at 350 °C? Give two reasons for your choice.

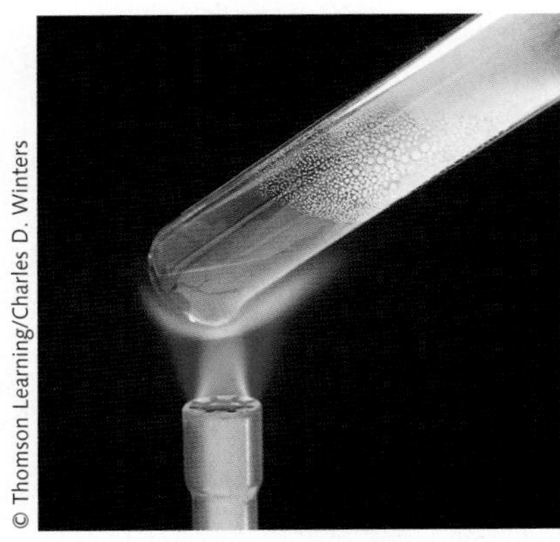

Figure 18.6 Decomposition of HgO(s). When heated, red mercury(II) oxide decomposes to liquid mercury metal and oxygen gas. Shiny droplets of mercury can be seen where they have candensed in the cooler part of the test tube.

For reactions such as decomposition of mercury(II) oxide that are reactant-favored at low temperatures and product-favored at high temperatures, it is possible to calculate the minimum temperature to which the system must be heated to make it product-favored. Below that temperature $\Delta G°$ is positive, and above that temperature $\Delta G°$ is negative. Therefore, $\Delta G°$ must equal zero at the desired temperature. Because $\Delta G° = \Delta H° - T\Delta S°$ (Equation 18.3), we can set $\Delta H° - T\Delta S° = 0$. Solving for T gives

$$T(\text{at which } \Delta G° \text{changes sign}) = \frac{\Delta H°}{\Delta S°} \qquad [18.5]$$

Equation 18.5 also applies to reactions that are product-favored at low temperatures and reactant-favored at high temperatures. For such reactions Equation 18.5 gives the temperature *below* which the reaction is product-favored. Heating the system above this temperature will probably result in insufficient quantities of products being produced. As an example, the contributions of $\Delta H°$ and $-T\Delta S°$ to $\Delta G°$ for the decomposition of silver(I) oxide are shown graphically in Figure 18.7.

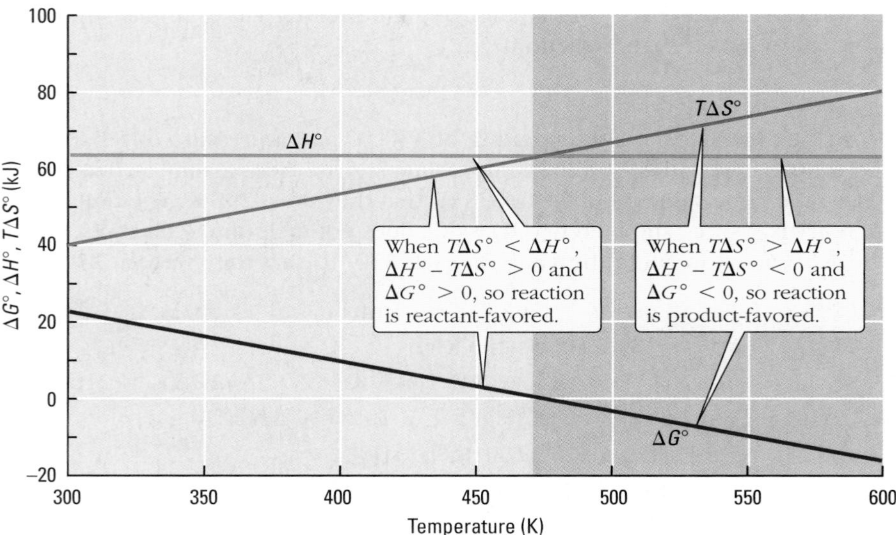

Figure 18.7 Effect of temperature on reaction spontaneity. The two terms that contribute to the Gibbs free energy change, $\Delta H°$ and $T\Delta S°$, are plotted as a function of temperature for the reaction of silver ore, $Ag_2O(s)$, to form silver metal and oxygen gas.

PROBLEM-SOLVING EXAMPLE 18.6 Effect of Temperature on Gibbs Free Energy Change

Calculate the temperature to which silver ore, $Ag_2O(s)$, must be heated to decompose the ore to oxygen gas and solid silver metal.

Answer 468 K (195 °C)

Strategy and Explanation Begin by writing the equation for the reaction, and use values from Appendix J to calculate $\Delta H°$ and $\Delta S°$ at 25 °C. Then use Equation 18.5 to calculate the desired temperature.

$$2\,Ag_2O(s) \longrightarrow 4\,Ag(s) + O_2(g)$$

$$\Delta H° = (4\text{ mol}) \times (0\text{ kJ/mol}) + (1\text{ mol}) \times (0\text{ kJ/mol}) - (2\text{ mol}) \times (-31.05\text{ kJ/mol}) = 62.10\text{ kJ}$$

$$\Delta S° = (4\text{ mol}) \times (42.55\text{ J K}^{-1}\text{ mol}^{-1}) + (1\text{ mol}) \times (205.138\text{ J K}^{-1}\text{ mol}^{-1}) - (2\text{ mol}) \times (121.3\text{ J K}^{-1}\text{ mol}^{-1}) = 132.7\text{ J/K}$$

Because both $\Delta H°$ and $\Delta S°$ are positive, the reaction will become more product-favored as temperature increases.

$$T = \frac{62.1\text{ kJ}}{132.7\text{ J/K}} \times \frac{1000\text{ J}}{1\text{ kJ}} = \frac{62.1\text{ kJ}}{0.1327\text{ kJ/K}} = 467.9\text{ K} = 468\text{ K}$$

Note that $\Delta H°$ has units of kJ, whereas $\Delta S°$ has units of J/K; this necessitates a unit conversion so that the energy units can cancel.

✓ **Reasonable Answer Check** The reaction produces a gas and a solid from a solid, so the entropy change should be positive. Silver is one of the few metals that can be found as the element in nature, so its oxide probably does not require an extremely high temperature to decompose; 468 K (195 °C) is a reasonable answer. Check that each reactant and each product is in the same physical state at 195 °C as it was at room temperature. The *CRC Handbook* gives the melting point of silver as 962 °C and reports that silver(I) oxide decomposes before it melts, so the approximation of nearly constant $\Delta S°$ and $\Delta H°$ is valid.

PROBLEM-SOLVING PRACTICE 18.6

For the reaction

$$2\,CO(g) + O_2(g) \longrightarrow 2\,CO_2(g)$$

(a) Predict the temperature at which the reaction changes from being product-favored to being reactant-favored.

(b) If you wanted this reaction to produce $CO_2(g)$, what temperature conditions would you choose?

18.7 Gibbs Free Energy Changes and Equilibrium Constants

The difference between the standard Gibbs free energies of products and reactants determines whether a reaction is product-favored, but so far we have considered only pure reactants and pure products. It is also important to consider what happens to the Gibbs free energy *during a reaction* (when some, but not all, of the reactants have been converted to products), because that determines the position of equilibrium.

Variation of Gibbs Free Energy During a Reaction

One way to remove carbon dioxide from air is to pass the air over solid sodium hydroxide.

$$NaOH(s) + CO_2(g) \longrightarrow NaHCO_3(s) \qquad \Delta G° \text{ (at 25 °C)} = -77.2 \text{ kJ} \qquad [18.6]$$

Suppose that this reaction is carried out so that the pressure remains at 1 bar and the temperature remains at 25 °C throughout the change from NaOH and CO_2 to $NaHCO_3$. Then all three substances will remain in their standard states throughout the reaction. The concentration of CO_2, which is proportional to its pressure, remains constant, and the concentrations of the two solids also remain constant, as you showed in Exercise 14.1 (⇐ *p. 671*).

Recall that the standard-state pressure is 1 bar, which is nearly the same as 1 atm.

When the reaction is halfway over, half of the reactants have been converted to products, so 0.5 mol NaOH(s) and 0.5 mol CO_2(g) have been converted to 0.5 mol $NaHCO_3$(s). The equation for what has happened so far is

$$0.5\ NaOH(s) + 0.5\ CO_2(g) \longrightarrow 0.5\ NaHCO_3(s)$$
$$\Delta G°\text{(at 25 °C)} = (0.5)(-77.2 \text{ kJ}) = -38.6 \text{ kJ}$$

Recall that a thermochemical expression refers to the exact number of moles of reactants and products indicated by the coefficients.

At this point we say that the extent of reaction is 0.5. The **extent of reaction,** which we will represent by x, *is the fraction of the reactants that has been converted to products.* For the general equation

$$a\,\text{A} + b\,\text{B} \longrightarrow c\,\text{C} + d\,\text{D}$$

if z mol reactant A has been consumed, then the extent of reaction is z/a. If y mol B has been consumed, the extent of reaction is y/b. If w mol D has formed, the extent of reaction is w/d. For a reaction such as the combination of NaOH with CO_2, in which all substances remain in their standard states throughout the chemical change, an extent of reaction of x corresponds to a Gibbs free energy change of x times the Gibbs free energy change for the complete reaction. This is shown graphically in Figure 18.8.

If we start with a mol A, then the maximum quantity of A that can be consumed is a mol A. Therefore $z \leq a$ and $x = z/a$ is always a fraction. Its maximum value is 1.

CONCEPTUAL EXERCISE 18.11 Gibbs Free Energy and Extent of Reaction

For Reaction 18.6,

(a) Calculate the Gibbs free energy change when the extent of reaction is 0.10, 0.40, and 0.80.

(b) Verify the statement in the text that an extent of reaction of x corresponds to a Gibbs free energy change of x times $\Delta G°$ for the complete reaction.

(c) Show that the slope of the line in Figure 18.8 is $\Delta G°$.

Reactions That Reach Equilibrium

When a reaction occurs in which two or more substances form a solution, the situation is different. The reaction will reach an equilibrium state in which both reactants

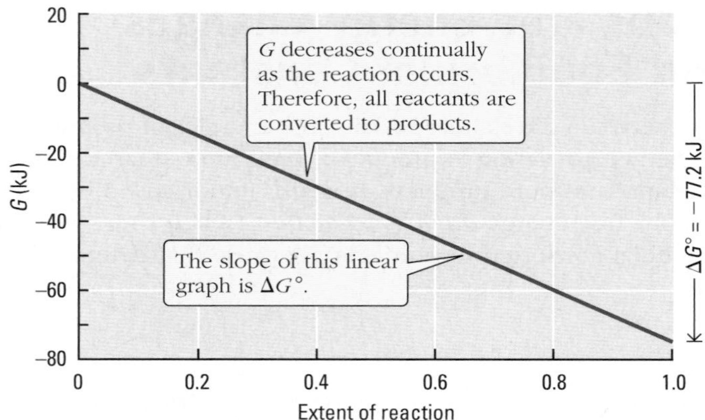

Figure 18.8 Gibbs free energy as a function of extent of reaction. For the reaction of NaOH(s) with CO_2(g), the Gibbs free energy decreases linearly as the reaction proceeds. A similar linear graph is obtained whenever all reactants and products are in their standard states throughout a reaction. The slope of such a graph is equal to $\Delta G°$. The Gibbs free energy of the reactants has been arbitrarily set to zero so that it is easier to see how big the differences in Gibbs free energy are.

and products are present. Consider an equilibrium reaction we discussed in Chapter 14 (***p. 675***), the isomerization of *cis*-2-butene at constant pressure.

$$cis\text{-2-butene(g)} \rightleftharpoons \text{trans-2-butene(g)} \qquad \Delta G° \text{ (at 500 K)} = -2.08 \text{ kJ}$$

If we start with 1.0 mol *cis*-2-butene at 500 K and 1 bar, then as soon as some *trans*-2-butene forms it will mix with (dissolve in) the *cis*-2-butene. When one gas dissolves in the other, each becomes less concentrated. Because the total pressure remains at 1 bar in the constant-pressure process, the partial pressure of each gas must be less than 1 bar, which means that neither gas is in its standard state, and the method you used in Exercise 18.11 to calculate ΔG as a function of extent of reaction no longer applies. However, we can tell something about how G varies as the reaction proceeds. Mixing two ideal gases involves no change in enthalpy, but it does involve an increase in entropy because each gas expands to fill the entire container, thereby causing increased dispersal of energy. (This is in addition to any difference in entropy between products and reactants.) Since $\Delta S_{dilution}$ is positive and $\Delta H_{dilution}$ is zero, then $\Delta G_{dilution}$ must be negative. Therefore, in Figure 18.9, there is *not* a straight line from reactants to products. Because there is a negative component ($\Delta G_{dilution}$) in addition to x times $\Delta G°$, there is a curved line below where the straight line would have been.

Recall that one of the assumptions on which the ideal gas law is based is that there are no forces of attraction among the molecules. Therefore there is no enthalpy change when the gases mix.

The graph in Figure 18.9 is steeper at both ends than a straight line from $G_{reactants}$ to $G_{products}$ and passes through a minimum at $x = 0.62$. This value of x means that the fraction of *trans* molecules is 0.62 and the fraction of *cis* molecules is $1 - 0.62 = 0.38$. Since both *cis*- and *trans*-2-butene occupy the same container, their concentrations and partial pressures are proportional to the number of molecules. Therefore, we can substitute these fractions into the equilibrium constant expression.

If there is the same number of gas phase molecules in the reactants as in the products, $K_c = K_P$.

$$K_P = K_c = \frac{P_{trans}}{P_{cis}} = \frac{[trans]}{[cis]} = \frac{0.62}{0.38} = 1.6$$

This is the same value that was reported in Chapter 14 (***p. 676***). That is, the equilibrium concentrations correspond to the minimum in the graph of G versus extent of reaction. This correspondence makes sense, because any set of concentrations that differs from $x = 0.62$ has a higher value of G and therefore could change

When $x = 0.62$, the system is at equilibrium and $Q = K$. When x differs from 0.62, $Q \neq K$ and the system must react to reach equilibrium.

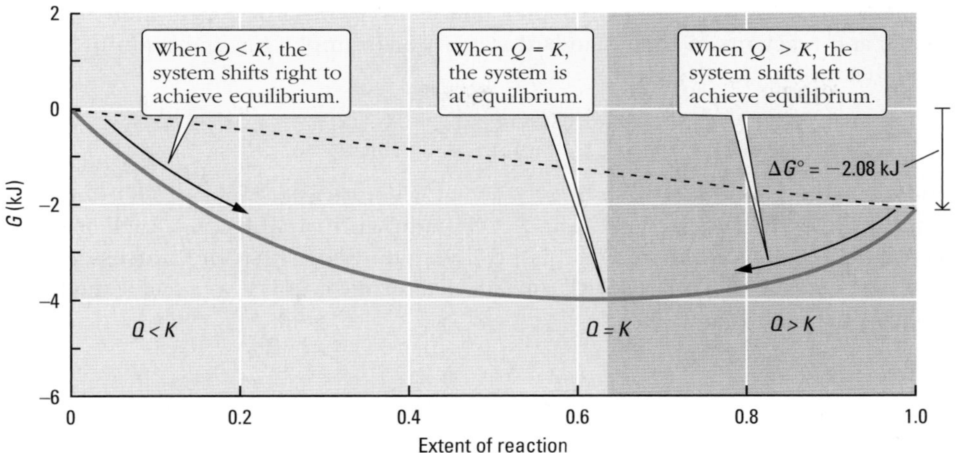

Figure 18.9 Gibbs free energy versus extent of reaction when there is mixing. For the isomerization of *cis*-2-butene, dilution of *cis*-2-butene with *trans*-2-butene increases entropy and decreases Gibbs free energy. This causes the graph of Gibbs free energy versus extent of reaction to curve below a straight line from reactants to products. As in Figure 18.8, the total Gibbs free energy of the reactants has been set arbitrarily to zero.

to the equilibrium concentrations with a decrease in Gibbs free energy (which corresponds to an increase in entropy of the universe).

At the far left of Figure 18.9 the graph drops faster than a straight line from $G_{\text{reactants}}$ to G_{products}. The slope at any point on the curve differs from the slope of the straight line ($\Delta G°$) by a factor that depends on the reaction quotient, Q (*p. 689*).

$$\text{Slope} = \Delta G° + RT \ln Q \qquad [18.7]$$

At equilibrium the slope is zero, and $Q = K°$, where the superscript indicates the **standard equilibrium constant.** The expression for $K°$ is similar to the equilibrium constant as defined in Chapter 14 (*p. 677*) except that each concentration is divided by the standard-state concentration of 1 mol/L and each pressure is divided by the standard-state pressure of 1 bar. That is, even if the equilibrium constant has units (of concentration or pressure), $K°$ is unitless. Substituting into Equation 18.7 gives

$$0 = \Delta G° + RT \ln K°$$

which rearranges to

$$\Delta G° = -RT \ln K° \qquad [18.8]$$

If the reaction occurs in solution, $K°$ has the same form (and the same value) as the concentration equilibrium constant. For gases, $K°$ relates pressures, not concentrations, and has the same value as K_p.

Regardless of the choice of standard state, Equation 18.8 indicates that the Gibbs free energy change for a reaction is the negative of a constant times the temperature times the natural logarithm of the equilibrium constant. If $K°$ is larger than 1, then ln $K°$ is positive and $\Delta G°$ will be negative because of the minus sign. Both of these conditions, a negative $\Delta G°$ and $K° > 1$, indicate that the reaction is *product-favored* under standard-state conditions. Conversely, if $K° < 1$, then ln $K°$ is negative and $\Delta G°$ must be positive, indicating a *reactant-favored* system.

Relation between $\Delta G°$ and $K°$ at 25 °C

$\Delta G°$ (kJ/mol)	$K°$
200	9×10^{-36}
100	3×10^{-18}
10	2×10^{-2}
1	7×10^{-1}
0	1
−1	1.5
−10	6×10^{1}
−100	3×10^{17}
−200	1×10^{35}

$K°$	$\Delta G°$ ($-RT \ln K°$)	Product-Favored?
<1	Positive	No
>1	Negative	Yes
$=1$	0	Neither

$\Delta G°$ is the difference in Gibbs free energy between products in their standard states and reactants in their standard states. For example, for ionization of formic acid in water,

$$\underset{1\ \text{mol/L}}{HCOOH(aq)} \rightleftharpoons \underset{1\ \text{mol/L}}{HCOO^-(aq)} + \underset{1\ \text{mol/L}}{H^+(aq)} \qquad \Delta G° = 21.3\ \text{kJ}$$

the Gibbs free energy change of 21.3 kJ is for converting 1 mol HCOOH(aq) at a concentration of 1 mol/L into 1 mol $HCOO^-$(aq) and 1 mol H^+(aq), each at a concentration of 1 mol/L. Since $\Delta G°$ is positive, we predict that the process will be reactant-favored. This agrees with the fact that formic acid is a weak acid and therefore is only slightly ionized.

PROBLEM-SOLVING EXAMPLE 18.7 Gibbs Free Energy and Equilibrium Constant

In the preceding paragraph you learned that $\Delta G° = 21.3$ kJ at 25 °C for the reaction

$$HCOOH(aq) \rightleftharpoons HCOO^-(aq) + H^+(aq)$$

Use this information to calculate the equilibrium constant for ionization of formic acid in aqueous solution at 25 °C.

Answer $K_c = K° = 1.8 \times 10^{-4}$

Strategy and Explanation The relation between Gibbs free energy and $K°$ was given in Equation 18.8 as

$$\Delta G° = -RT \ln K°$$

To obtain $K°$ from $\Delta G°$, we first divide both sides of the equation by $-RT$:

$$-\Delta G°/RT = \ln K°$$

Next, we make use of the properties of logarithms (which are discussed in Appendix A.6). Since ln represents a logarithm to the base e, we can remove the logarithm function by using each side of the equation as an exponent of e.

$$e^{-\Delta G°/RT} = e^{\ln K°} = K°$$

Now we can substitute the known values into the equation.

$$K° = e^{-\Delta G°/RT} = e^{-(21.3\ \text{kJ/mol})(1000\ \text{J/kJ})/(8.314\ \text{J K}^{-1}\ \text{mol}^{-1})(298\ \text{K})} = 1.8 \times 10^{-4}$$

Thus, the positive value of $\Delta G°$ results in a value of $K°$ less than 1 and, indeed, indicates a reactant-favored system. Because this reaction occurs in aqueous solution, the standard states of reactants and products involve concentrations. Therefore $K° = K_c$.

✓ **Reasonable Answer Check** Formic acid is a weak acid and is not expected to have a very large equilibrium constant. The value calculated appears to be reasonable.

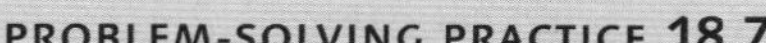

Make certain to check units in calculations like this one. Standard Gibbs free energy changes involve kilojoules, and the gas constant R involves joules, so a unit conversion is needed.

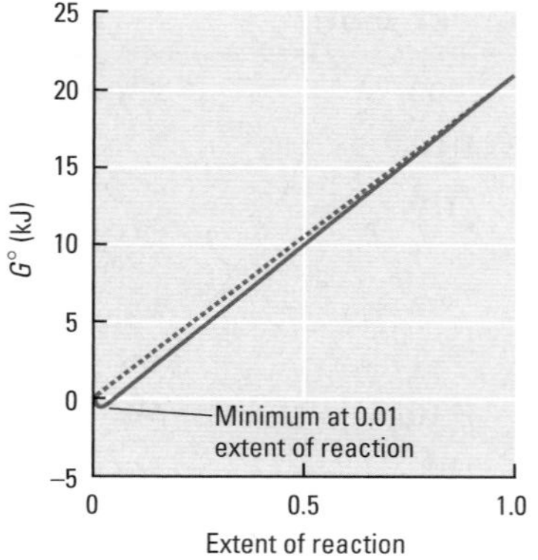

Graph of G as a function of extent of reaction for the reaction in Problem-Solving Example 18.7. $\Delta G° = 21.3$ kJ, a positive value. The minimum in the solid line is very close to zero extent of reaction; that is, the system is reactant-favored. The extent of reaction at equilibrium is 0.01. (The dotted line is a straight line from reactants to products. The solid curve dips below the dotted line at very small extent of reaction.)

PROBLEM-SOLVING PRACTICE 18.7

For each of the following reactions, evaluate $K°$ at 298 K from the standard free energy change. If necessary, obtain data from Appendix J to calculate $\Delta G°$. Check your results against the K_c and K_P values in Table 14.1 (⇐ p. 686). For which of these reactions is $K_c = K°$?

(a) $CaCO_3(s) \rightleftharpoons Ca^{2+}(aq) + CO_3^{2-}(aq)$

(b) $H_2CO_3(aq) \rightleftharpoons HCO_3^-(aq) + H^+(aq)$

(c) $2\ NO_2(g) \rightleftharpoons N_2O_4(g)$

Remember also that $\Delta G°$ can be calculated from the equation

$$\Delta G° = \Delta H° - T\Delta S°$$

If we know or can estimate changes in enthalpy and entropy for a reaction, then we can calculate or estimate the Gibbs free energy change and hence the equilibrium constant. And because $\Delta H°$ and $\Delta S°$ have nearly constant values over a wide range of temperatures, we can estimate equilibrium constants at a variety of temperatures, not just at 25 °C. Problem-Solving Example 18.8 shows how to do this.

Assuming constant values of $\Delta H°$ and $\Delta S°$ over a broad range of temperatures does not work well for reactions involving ions in aqueous solution, such as (a) and (b) in Problem-Solving Practice 18.7, because the extent of hydration of the ions varies with temperature.

PROBLEM-SOLVING EXAMPLE 18.8 Estimating $K°$ at Different Temperatures

Use data from Appendix J to obtain values of $\Delta H°$ and $\Delta S°$ for the reaction

$$N_2(g) + O_2(g) \rightleftharpoons 2\,NO(g)$$

From these data estimate the value of $\Delta G°$ and hence the value of $K°$ at (a) 298 K, (b) 1000. K, and (c) 2300. K.

This reaction is of great importance because it can take place to a significant extent in high-temperature combustion processes. If the temperature is high enough, nitrogen and oxygen form the air pollutant NO.

Answer

(a) $\Delta G° = 173.1$ kJ; $K° = 4.5 \times 10^{-31}$
(b) $\Delta G° = 155.7$ kJ; $K° = 7.3 \times 10^{-9}$
(c) $\Delta G° = 123.5$ kJ; $K° = 1.57 \times 10^{-3}$

Strategy and Explanation At each temperature, use the Gibbs equation to calculate $\Delta G° = \Delta H° - T\Delta S°$. Then calculate $K°$ as was done in Problem-Solving Example 18.7. Part (c) is done below to illustrate the calculations:

$$\Delta G° = \Delta H° - T\Delta S° = 180{,}500\text{ J} - (2300.\text{ K})(24.772\text{ J/K})$$
$$= 1.235 \times 10^5\text{ J} = 123.5\text{ kJ}$$
$$K° = e^{-\Delta G°/RT} = e^{-(1.235\times 10^5\text{ J})/(8.314\text{ J/K})(2300.\text{ K})} = 1.57 \times 10^{-3}$$

✓ **Reasonable Answer Check** The reaction is endothermic, so Le Chatelier's principle predicts that the equilibrium will shift toward products as the temperature increases. This is reflected by the increasing values of $K°$ as the temperature rises.

PROBLEM-SOLVING PRACTICE 18.8

For the ammonia synthesis reaction,

$$N_2(g) + 3\,H_2(g) \rightleftharpoons 2\,NH_3(g)$$

estimate the equilibrium constant at (a) 298. K, (b) 450. K, and (c) 800. K.

Gibbs Free Energy Changes under Nonstandard-State Conditions

In previous sections we calculated $\Delta G°$ for reactions in which reactants in their standard states were converted to products in their standard states. However, substances usually are not at a pressure of 1 bar or at a concentration of 1 mol/L. How do we calculate ΔG if the reactants and products are not at standard concentration or standard pressure? A simple adjustment can be made to $\Delta G°$ to account for the difference between actual pressures or concentrations and standard-state pressures or concentrations. The equation from which ΔG (for nonstandard-state conditions) can be calculated from $\Delta G°$ (for standard-state conditions) is

$$\Delta G = \Delta G° + RT \ln Q \qquad [18.9]$$

For the standard state (concentration of 1 mol/L or pressure of 1 bar), $Q = 1$. Substituting this into Equation 18.9, we get $\ln Q = \ln(1) = 0$ and $\Delta G = \Delta G°$, the correct value for the standard state.

According to Equation 18.9, the bigger Q becomes, the more positive the correction factor $RT \ln Q$ becomes, and the more positive ΔG becomes. This makes sense, because the larger Q is, the more the concentrations (or pressures) of products exceed those of reactants. According to Le Chatelier's principle, increasing the concentrations of products (or decreasing the concentrations of reactants) causes the reaction to shift in the reverse direction. A shift toward more reactants is expected, because the more positive ΔG is, the more reactant-favored a process is.

To determine the change in Gibbs free energy when reactants at nonstandard-state concentrations or pressures are converted to products at nonstandard-state concentrations or pressures, we first write an appropriate chemical equation, then calculate $\Delta G°$ and Q, and finally use Equation 18.9 to correct the value of $\Delta G°$ for the nonstandard-state conditions, giving ΔG.

PROBLEM-SOLVING EXAMPLE 18.9 Gibbs Free Energy Change for Nonstandard-State Conditions

For the ammonia synthesis reaction at 25 °C, calculate the change in Gibbs free energy if 1 mol $N_2(g)$ at 0.23 bar and 3 mol $H_2(g)$ at 0.42 bar are converted to 2 mol $NH_3(g)$ at 1.45 bar.

Answer $\Delta G = -20.96$ kJ/mol

Strategy and Explanation First write a balanced equation, then calculate $\Delta G°$ and Q. Finally, use Equation 18.9 to calculate ΔG.

$$N_2(g) + 3\,H_2(g) \rightleftharpoons 2\,NH_3(g)$$

$$\Delta G° = 2(\Delta G_f° \,(NH_3(g))) - \{(\Delta G_f° \,(N_2(g))) + 3(\Delta G_f° \,(H_2(g)))\}$$
$$= 2(-16.45)\text{ kJ/mol} - (0 + 0) = -32.90\text{ kJ/mol}$$

$$Q = \frac{P^2_{NH_3}}{P_{N_2}P^3_{H_2}} = \frac{(1.45)^2}{(0.23)(0.42)^3} = 123.4$$

$$\Delta G = \Delta G° + RT\ln Q = -32.90\text{ kJ/mol} + (8.314\text{ J mol}^{-1}\text{ K}^{-1})(298.15\text{ K}) \times \ln(123.4)$$
$$= -32.90\text{ kJ/mol} + 11936\text{ J/mol}$$
$$= -32.90\text{ kJ/mol} + 11.936\text{ kJ/mol} = -20.96\text{ kJ/mol}$$

✓ **Reasonable Answer Check** The nonstandard-state conditions involve a concentration of ammonia (the product) well above standard pressure and concentrations of nitrogen and hydrogen (the reactants) well below standard pressure. According to Le Chatelier's principle, if an equilibrium is disturbed by increasing concentrations of products or decreasing concentrations of reactants (both of which apply here), then the equilibrium will shift toward the left—that is, in a reactant-favored direction. The value of ΔG (−20.96 kJ/mol) is negative, but it is less negative than the value of $\Delta G°$ (−32.90 kJ/mol). A less negative ΔG corresponds to a less product-favored (that is, more reactant-favored) process, which corresponds with the prediction from Le Chatelier's principle.

PROBLEM-SOLVING PRACTICE 18.9

Calculate ΔG at 25 °C for a reaction in which $Ca^{2+}(aq)$ combines with $CO_3^{2-}(aq)$ to form a precipitate of $CaCO_3(s)$ if the concentrations of $Ca^{2+}(aq)$ and $CO_3^{2-}(aq)$ are 0.023 M and 0.13 M, respectively.

18.8 Gibbs Free Energy, Maximum Work, and Energy Resources

An important interpretation of the Gibbs free energy is that **ΔG *represents the maximum useful work that can be done by a product-favored system on its surroundings under conditions of constant temperature and pressure.* ΔG *also represents the minimum work that must be done to cause a reactant-favored process to occur.*** Consider the product-favored reaction of hydrogen with oxygen to form liquid water under standard conditions.

$$2\,H_2(g) + O_2(g) \rightleftharpoons 2\,H_2O(\ell) \qquad \Delta G° = -474.258\text{ kJ}$$

This thermochemical expression tells us that for every 2 mol $H_2O(\ell)$ produced, as much as 474.258 kJ of useful work could be done. The negative sign of $\Delta G°$ tells us that the work is done on the surroundings. (Because the system has less Gibbs free energy after the reaction than before it, the surroundings will have more energy.) Even if the reactants and the products are not at standard pressure or concentration, ΔG still equals $-w_{max}$, the maximum work the system can do on its surroundings.

$$\Delta G = -w_{max} \text{ (work done on the surroundings)} \qquad [18.10]$$

It is important to remember that w_{max} is the maximum work the system can do on the *surroundings*. Therefore the sign of w_{max} is opposite to that of ΔG.

Now consider the decomposition of water to form hydrogen and oxygen, which is the reverse of the previous reaction.

$$2\,H_2O(\ell) \rightleftharpoons 2\,H_2(g) + O_2(g) \qquad \Delta G° = 474.258 \text{ kJ}$$

The positive value of $\Delta G°$ indicates that this process is reactant-favored. Because the Gibbs free energy of the products is greater than the Gibbs free energy of the reactant, at least 474.258 kJ must be supplied for every 2 mol $H_2O(\ell)$ that decomposes. This 474.258 kJ is the minimum work that must be done to change 2 mol liquid water into hydrogen gas and oxygen gas. One way to supply this work is to use a direct electric current to carry out electrolysis of the water. In general, a continuous supply of energy is required for a reactant-favored process, such as decomposition of liquid water, to continue.

Because transformations of energy from one form to another are not 100% efficient, we seldom observe anything close to the maximum quantity of useful work given by the value of $\Delta G°$.

PROBLEM-SOLVING EXAMPLE 18.10 Gibbs Free Energy Change and Maximum Work

Use data from Appendix J to predict whether each reaction is product-favored or reactant-favored at 25 °C and 1 bar. For each product-favored reaction, calculate the maximum useful work the reaction could do. For each reactant-favored process, calculate the minimum work needed to cause it to occur.

(a) $2\,Al_2O_3(s) \longrightarrow 4\,Al(s) + 3\,O_2(g)$
(b) $Cl_2(g) + Mg(s) \longrightarrow MgCl_2(s)$

Answer

(a) Reactant-favored; at least 3164.6 kJ must be supplied
(b) Product-favored; can do up to 591.79 kJ of useful work

Strategy and Explanation Use data from Appendix J to calculate $\Delta G°$ for each reaction. If $\Delta G°$ is negative, the process is product-favored, and the value of $\Delta G°$ gives the maximum work that can be done. If $\Delta G°$ is positive, the process is reactant-favored, and the value tells the minimum work that has to be done to force the reaction to occur.

(a) $\Delta G° = 0 + 0 - 2(-1582.3)$ kJ $= 3164.6$ kJ; at least 3164.6 kJ is required.
(b) $\Delta G° = -591.79$ kJ $- 0 - 0 = -591.79$ kJ; up to 591.79 kJ useful work can be done.

✓ **Reasonable Answer Check** Reaction (a) is decomposition of an oxide to a metal and oxygen. Because metals are good reducing agents and oxygen is a strong oxidizing agent, the reverse of this reaction is likely to be product-favored, which would make Reaction (a) reactant-favored. This result agrees with the calculation. Reaction (b) is combination of an alkaline earth element with a halogen, which should form a stable ionic compound. Therefore Reaction (b) should be product-favored, which agrees with the calculation. In both cases the value of $\Delta G°$ is large, which also would be expected based on the arguments just given.

Charging a dead battery. A dead battery can be charged by using electricity from a power plant to cause a reactant-favored process to occur. After the battery has been charged, the reverse of that reactant-favored process (a product-favored process) can generate electricity to start the car. (See discussion on the next page.)

PROBLEM-SOLVING PRACTICE 18.10

Predict whether each reaction is reactant-favored or product-favored at 298 K and 1 bar, and calculate the minimum work that would have to be done to force it to occur, or the maximum work that could be done by the reaction.

(a) $2\,CO_2(g) \longrightarrow 2\,CO(g) + O_2(g)$
(b) $4\,Fe(s) + 3\,O_2(g) \longrightarrow 2\,Fe_2O_3(s)$

Bob Webster, Montana Power Company

Coal-fired electric power plant.

Coupling Reactant-Favored Processes with Product-Favored Processes

A dead car battery will not charge itself. The process that takes place when a battery is charged is reactant-favored. But a battery can be charged if it is connected to a charger that is, in turn, powered by electricity generated in a power plant that burns coal. Coal, which is mainly carbon, burns in air according to the equation

$$C(s) + O_2(g) \longrightarrow CO_2(g) \qquad \Delta G^\circ = -394.4 \text{ kJ}$$

If enough coal is burned, the large negative Gibbs free energy change for its combustion more than offsets the positive Gibbs free energy change of the battery-charging process. An overall decrease in Gibbs free energy occurs, even though the battery-charging part we are interested in has an increase. Once a battery has been charged, the charging reaction's reverse (which is product-favored) can supply electricity to start a car's engine or play its radio. Some of the Gibbs free energy lost when the coal was burned has been stored in the car's battery for use later.

Charging a battery is an example of coupling a product-favored reaction with a reactant-favored process to cause the latter to take place. Both processes occur at the same time and in a way that allows the Gibbs free energy released by the product-favored reaction to be used by the reactant-favored reaction. Other examples include obtaining aluminum or iron from their ores; synthesizing large, complicated molecules from simple reactants to make medicines, plastics, and other useful materials; and maintaining a comfortable temperature in a house on a day when the outside temperature is above 100 °F. All of these processes involve decreasing entropy in the region of our interest, but all can be made to occur provided that there is a larger increase in entropy at a power plant or somewhere else.

All plants and animals, and the earth's ecosystem as a whole, depend on coupling of chemical reactions for their very existence. We will consider this topic in more detail in the next section.

The Gibbs free energy change indicates a chemical reaction's capacity to drive a different reactant-favored system to produce products. The word "free" in the name indicates not "zero cost," but rather "available." ***Gibbs free energy is available to do useful tasks that would not happen on their own.*** Another way of saying this is that Gibbs free energy is a measure of the *quality* of the energy contained in a chemical reaction system. If it contains a lot of Gibbs free energy, a chemical system can do a lot of useful work for us; the energy is of high quality—potentially useful to humankind. When the system's reactants are transformed into products, that available free energy can do useful work, but only if the reaction is coupled to some other, reactant-favored process we want to carry out. If systems are not coupled, then the free energy released by a reaction will be wasted as thermal energy.

© Thomson Learning/Charles Steele

Thermite reaction. Exercise 18.12 deals with the reaction of aluminum with iron(III) oxide. The reaction is very product-favored and releases a large quantity of Gibbs free energy.

CONCEPTUAL EXERCISE 18.12 Coupling Reactions

One way to produce iron metal is to reduce iron(III) oxide with aluminum. This is called the thermite reaction. You can think of the reaction as occurring in two steps. The first is the loss of oxygen from iron(III) oxide,

(i) $Fe_2O_3(s) \longrightarrow 2\,Fe(s) + \frac{3}{2}\,O_2(g)$

The second is the combination of aluminum with the oxygen,

(ii) $2\,Al(s) + \frac{3}{2}\,O_2(g) \longrightarrow Al_2O_3(s)$

(a) Calculate the enthalpy, entropy, and Gibbs free energy changes for each step. Decide whether each step is product- or reactant-favored. Comment on the signs of ΔH°, ΔS°, and ΔG° for each step.

(b) What is the overall net reaction that occurs when aluminum is combined with iron(III) oxide? What are the enthalpy, entropy, and Gibbs free energy changes for the overall reaction? Is it product- or reactant-favored? Comment on the signs of ΔH°, ΔS°, and ΔG° for the overall reaction.

(c) Discuss briefly how coupling Reaction (i) with Reaction (ii) affects our ability to obtain iron metal from iron(III) oxide by reacting it with aluminum.

(d) Suggest a reaction other than oxidation of aluminum that might be used to reduce iron(III) oxide to iron. Test your selection by calculating the Gibbs free energy change for the coupled system.

18.9 Gibbs Free Energy and Biological Systems

Have you ever thought about how unlikely it is that a human being can exist? Your body contains about 100 trillion (10^{14}) cells, all working together to make you what you are. Each of those cells contains trillions of molecules, and many of the molecules contain tens of thousands of atoms. Those molecules and cells are arranged in structures such as organs, bones, and skin that provide for all the functions of your body and that determine its overall shape and size. When necessary, you can synthesize molecules on very short notice. For example, it does not take long to generate the surge of adrenalin your body makes when you are scared. Your body is a very highly organized system, which means that its entropy must be very low. This in turn means that, thermodynamically speaking, you are very, very improbable.

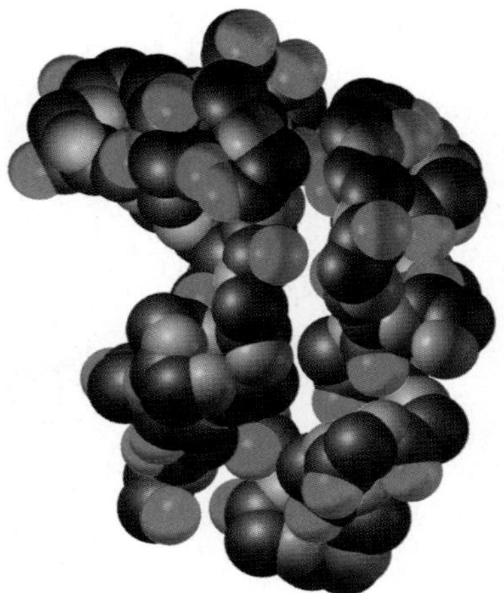

insulin

Insulin, a protein. Like all protein molecules, the insulin molecule is highly ordered. It contains 51 amino acids connected in exactly the correct order and folded into exactly the molecular shape needed for its function in the metabolism of glucose. Hydrogen atoms are not shown for simplicity.

Human Metabolism and Gibbs Free Energy

How can it be, then, that you exist? How can all the molecules from which you are made be synthesized and organized into the organs and other tissues of your body? The answer lies in the coupling of reactions described in the preceding section. Since you are a very-low-entropy system, you must be very high in Gibbs free energy. Your body obtains that Gibbs free energy by oxidizing the food you eat with oxygen you inhale. In the processes of metabolism, foods are oxidized, and the Gibbs free energy released by their oxidation causes other reactions to occur that store Gibbs free energy in specific molecules within your body. Later those molecules can release the Gibbs free energy to cause muscles to contract, nerve signals to be sent, important molecules to be synthesized, and other processes to occur. **Metabolism** refers to all of the chemical changes that occur as food nutrients are processed by an organism to release Gibbs free energy and form the complex chemical constituents of living cells. **Nutrients** are the chemical raw materials needed for survival of an organism.

As an example of metabolism, consider the single nutrient glucose, also known as dextrose or blood sugar (***p. 596***). Glucose can be oxidized to carbon dioxide and water according to the equation

glucose

$$C_6H_{12}O_6(aq) + 6\,O_2(g) \longrightarrow 6\,CO_2(g) + 6\,H_2O(\ell) \qquad \Delta G^{\circ\prime} = -2870 \text{ kJ}$$

Thus, a large quantity of Gibbs free energy can be released when glucose is oxidized. This reaction, which is strongly product-favored, is an example of an **exergonic** reaction—*one that releases Gibbs free energy.* The same quantity of Gibbs free energy is available whether glucose is burned in air or reacts in your body. However, burning glucose would release all of the Gibbs free energy as thermal energy. This process would not be appropriate in your body because it would raise the temperature rapidly, which in turn would kill cells. Instead, your body makes use of a large number of reactions that allow the Gibbs free energy to be released in small steps and stored in small quantities that can be used later.

The prime symbol (′) on $\Delta G^{\circ\prime}$ indicates that the value of the Gibbs free energy change is for pH = 7, the same concentration of $H^+(aq)$ ions as in a typical cell. When aqueous solutions are involved, ΔG° values in tables such as Appendix J refer to H_3O^+ concentrations of 1 mol/L, but such a high concentration of acid would destroy most cells. Consequently, biochemists have calculated a set of $\Delta G^{\circ\prime}$ values that apply to solutions at pH = 7. These values are usually reported for a temperature of 37 °C, human body temperature.

By far the most important way in which Gibbs free energy is stored in your body is through formation of adenosine triphosphate (ATP) from adenosine diphosphate (ADP). The structures of ADP and ATP are shown in Figure 18.10; note that they are closely related.

$$ADP^{3-}(aq) + H_2PO_4^-(aq) \longrightarrow ATP^{4-}(aq) + H_2O(\ell) \qquad \Delta G^{\circ\prime} = 30.5 \text{ kJ}$$

This is an example of an **endergonic** reaction—*one that consumes Gibbs free energy* and is therefore reactant-favored. In a typical bacterial cell, this reaction takes place 38 times for each molecule of glucose that is oxidized. In human cells, it takes place 32 times for each molecule of glucose that is oxidized. That is, in a human cell the Gibbs free energy released by the exergonic glucose oxidation is

The words "exergonic" and "endergonic" have nearly the same prefixes as "exothermic" and "endothermic." In both cases *ex* means "out" and *end* means "into." *Thermic* indicates that thermal energy is released or taken up. *Ergonic* indicates that Gibbs free energy is released or used up.

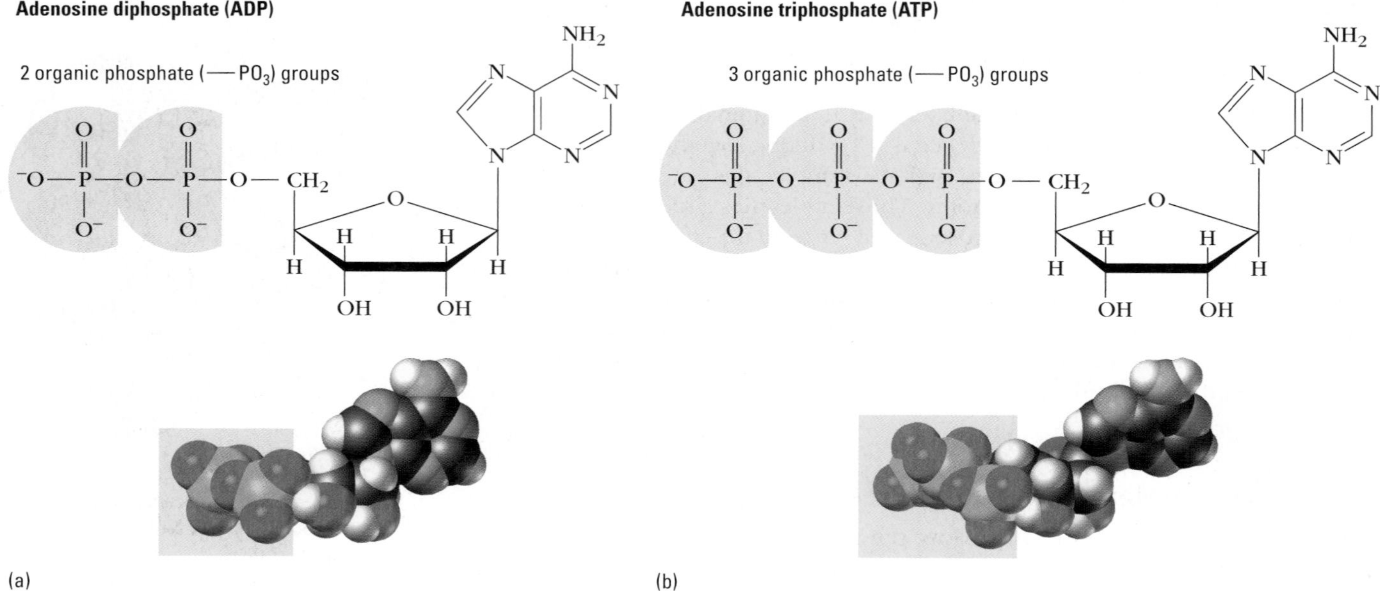

Figure 18.10 Biochemical storage of Gibbs free energy. Structures of (a) adenosine diphosphate (ADP) and (b) adenosine triphosphate (ATP). Notice that ATP has one more organic phosphate (—PO_3) group at the left end of the molecule, but otherwise the structures are identical. Notice also that ADP has three negatively charged oxygen atoms and ATP has four.

Because ATP is high in Gibbs free energy, it is said to be a high-energy molecule (or ion). Sometimes the bonds in ATP are called high-energy bonds, but this description is a misnomer. Actually, the bonds have low bond energies and can break easily to form ADP and release Gibbs free energy.

used to force the endergonic, reactant-favored process of forming ATP from ADP to occur 32 times. The overall process is

$$C_6H_{12}O_6(aq) + 6\,O_2(g) + 32\,ADP^{3-}(aq) + 32\,H_2PO_4^-(aq) \longrightarrow$$
$$6\,CO_2(g) + 32\,ATP^{4-}(aq) + 38\,H_2O(\ell) \qquad \Delta G^{\circ\prime} = -1894\text{ kJ}$$

Since it is exergonic, this reaction must be product-favored, and therefore appreciable quantities of products can be obtained.

EXERCISE 18.13 Coupled Metabolic Reactions

Add the Gibbs free energy change for oxidation of glucose to the appropriate Gibbs free energy change for 32 conversions of ADP to ATP, thereby verifying that the Gibbs free energy change given above for the overall reaction is correct. What happens to the 1894 kJ of Gibbs free energy released by the overall reaction?

The metabolic process by which the Gibbs free energy contained in nutrients is stored in ATP is far more complicated than the overall equation given above makes it seem. Metabolism can be divided into three stages that were first clearly identified by Hans Krebs. The first stage, digestion, breaks apart large molecules, such as carbohydrates (polysaccharides), fats, or proteins, into smaller molecules, such as glucose, glycerol and fatty acids, or amino acids. These smaller molecules are more easily transferred into the blood by the digestive system. In the second stage, the smaller molecules are changed into a few simple units that play a central role in metabolism. The most important of these is the acetyl group in acetyl coenzyme A (acetyl CoA). The structure of acetyl CoA is shown in Figure 18.11. The third stage consists of oxi-

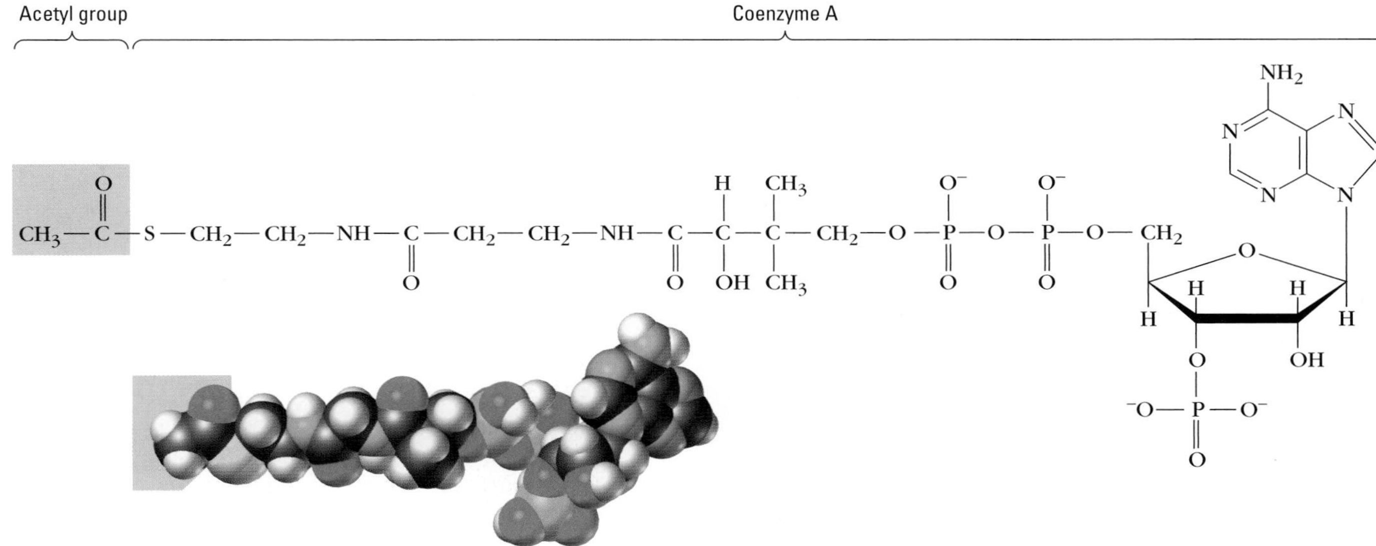

Figure 18.11 Structure of acetyl coenzyme A. Structural formula and space-filling model of acetyl coenzyme A. The acetyl group at the far left end of the molecule can be formed from glucose that came originally from starch.

dation of the acetyl group from acetyl CoA to form carbon dioxide and water. This takes place in an eight-step cycle of reactions called the citric acid cycle, which also transforms ADP into ATP, a process called oxidative phosphorylation. The overall three-stage process is diagrammed in Figure 18.12.

The citric acid cycle is also known as the Krebs cycle or the tricarboxylic acid (TCA) cycle.

Because conversion of ADP to ATP is endergonic, ATP contains stored Gibbs free energy. In your body, ATP generated from glucose or other nutrients is a convenient and readily available Gibbs free energy resource, just as electricity generated from coal or natural gas is a convenient and readily available Gibbs free energy resource in modern society. ATP can release Gibbs free energy in packets of 30.5 kJ for each ATP converted to ADP. This size is convenient for driving many biochemical processes in your body. For example, as part of the metabolism of glucose, it is necessary to attach a phosphate group to the glucose molecule.

Conversion of ATP to ADP is the source of the energy that causes muscles to contract. This answers the question, "Where does the energy come from to make my muscles work?" that was posed in Chapter 1 (*p. 2*).

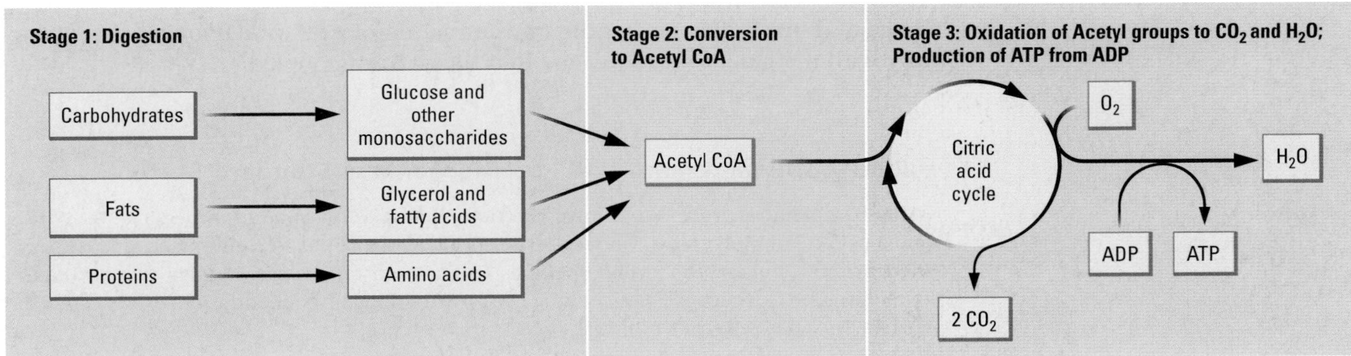

Figure 18.12 Gibbs free energy and nutrients. Extraction of Gibbs free energy from nutrients is a three-stage process. In stage 1 (digestion), large molecules are broken down into smaller ones. In stage 2, smaller molecules are converted to acetyl groups attached to coenzyme A. In stage 3 (the citric acid cycle), the acetyl groups are oxidized to carbon dioxide and water.

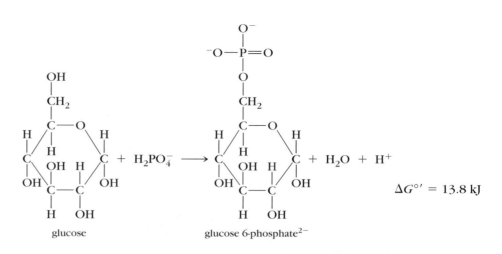

The "6" in glucose 6-phosphate indicates that the phosphate group has been added to the oxygen atom attached to carbon number 6 of glucose.

$$\text{glucose} + H_2PO_4^- \longrightarrow \text{glucose 6-phosphate}^{2-} + H_2O + H^+ \qquad \Delta G°' = 13.8 \text{ kJ}$$

glucose — glucose 6-phosphate^{2-}

This reaction is endergonic by 13.8 kJ and, therefore, is not product-favored. It will not occur unless forced to do so.

The endergonic reaction can be caused to occur by coupling it to the transformation of ATP to ADP.

$$\text{ATP(aq)}^{4-} + H_2O(\ell) \longrightarrow \text{ADP}^{3-}\text{(aq)} + H_2PO_4^-\text{(aq)} \qquad \Delta G°' = -30.5 \text{ kJ}$$

$H_2PO_4^-$ produced by this reaction can react with glucose to form glucose 6-phosphate, coupling the two reactions directly. Also, water produced in the first reaction is used up in the second one. The overall process is

$$\text{Glucose} + \text{ATP}^{4-} \longrightarrow \text{glucose 6-phosphate}^{2-} + \text{ADP}^{3-} + H_3O^+$$
$$\Delta G°' = (-30.5 + 13.8) \text{ kJ} = -16.7 \text{ kJ}$$

The negative value of $\Delta G°'$ indicates that the overall process is exergonic and product-favored. Thus the ATP $\longrightarrow$ ADP transformation can cause glucose to undergo a condensation reaction with dihydrogen phosphate. The 16.7 kJ of Gibbs free energy released appears as thermal energy transferred from the system to its surroundings.

In biochemistry the convention is to write these equations in a shorthand notation that indicates that they are coupled. The process just described is represented as:

Glucose ⟶ Glucose 6-phosphate
(ATP ↷ ADP)

The curved line indicates that the transformation of ATP to ADP occurs simultaneously with the glucose reaction and that the two are coupled.

PROBLEM-SOLVING EXAMPLE 18.11 Biochemical Standard State

Many biochemical processes involve reactions that take place at a temperature of 37 °C and a pH of 7 in body fluids. Under these conditions the Gibbs free energy change is specified as $\Delta G°'$, where the prime specifies that all substances are at their standard-state concentrations except for H_3O^+, which is at a biological concentration of 10^{-7} mol/L (pH = 7). What is $\Delta G°$ (1 mol/L H_3O^+) for this reaction?

$$\text{Glucose} + \text{ATP}^{4-} \longrightarrow \text{glucose 6-phosphate}^{2-} + \text{ADP}^{3-} + H_3O^+$$
$$\Delta G°' = -16.7 \text{ kJ/mol}$$

Answer $\Delta G° = 24.8$ kJ/mol

Strategy and Explanation The $\Delta G°'$ value differs from $\Delta G°$ because one of the concentrations (that of H_3O^+) has a nonstandard value of 10^{-7} mol/L. That is, $\Delta G°'$ is ΔG for conditions such that every concentration is 1 mol/L except for the concentration of H_3O^+. Therefore, set $\Delta G = \Delta G^{\circ\prime}$, calculate Q, and use Equation 18.9 (p. 891) to calculate ΔG°.

$$\Delta G = \Delta G^{\circ\prime} = -16.7 \text{ kJ/mol} = \Delta G^\circ + RT \ln Q$$

$$\Delta G^\circ = -16.7 \text{ kJ/mol} - RT \ln Q$$

$$Q = \frac{(\text{conc. glucose 6-phosphate}^{2-})(\text{conc. ADP}^{3-})(\text{conc. H}_3\text{O}^+)}{(\text{conc. glucose})(\text{conc. ATP}^{4-})}$$

$$= \frac{(1)(1)(1 \times 10^{-7})}{(1)(1)} = 1 \times 10^{-7}$$

$$\Delta G^\circ = -16.7 \text{ kJ/mol} - RT \ln(1 \times 10^{-7})$$

$$= -16.7 \text{ kJ/mol} - (8.314 \text{ J mol}^{-1} \text{ K}^{-1})\{(273 + 37) \text{ K}\}(-16.12)$$

$$= -16.7 \text{ kJ/mol} + 41541 \text{ J/mol} = -16.7 \text{ kJ/mol} + 41.5 \text{ kJ/mol} = 24.8 \text{ kJ/mol}$$

✓ **Reasonable Answer Check** Using Le Chatelier's principle, we predict less shift toward products for a system in which the concentration of H_3O^+, a product, is 1 mol/L than there would be for a system in which the concentration of H_3O^+ is 1×10^{-7} mol/L. Less shift toward products means a more positive ΔG, and the value of ΔG° is indeed more positive than $\Delta G^{\circ\prime}$.

PROBLEM-SOLVING PRACTICE 18.11

Will ΔG° be larger than, smaller than, or the same size as $\Delta G^{\circ\prime}$ for this reaction?

$$C_6H_{12}O_6(aq) + 6\,O_2(g) \longrightarrow 6\,CO_2(g) + 6\,H_2O(\ell) \qquad \Delta G^{\circ\prime} = -2870 \text{ kJ}$$

Explain why you chose the response you did.

Photosynthesis and Gibbs Free Energy

You may be wondering where the nutrients you take into your body get the Gibbs free energy they so obviously have. The answer is from solar energy via photosynthesis. **Photosynthesis** is a series of reactions in a green plant that combines carbon dioxide with water to form carbohydrates and oxygen. The carbohydrates and other constituents you consume in vegetables are derived from photosynthesis. If you eat meat, the animal from which it came probably ate vegetables and grain and therefore derived its nutrients from plant photosynthesis. The overall reaction in photosynthesis is just the opposite of oxidation of glucose.

$$6\,CO_2(g) + 6\,H_2O(\ell) \longrightarrow C_6H_{12}O_6(aq) + 6\,O_2(g) \qquad \Delta G^{\circ\prime} = 2870 \text{ kJ}$$

Photosynthesis is endergonic and can occur only because of an influx of energy in the form of sunlight. That is, the energy in the sunlight causes this reactant-favored process to form appreciable quantities of products, and the sunlight's energy is stored as Gibbs free energy in the glucose and oxygen that are formed. This process is diagrammed in Figure 18.13.

Organisms that can carry out photosynthesis are called *phototrophs* (literally, "light-feeders") because they can use sunlight to supply needed energy. Phototrophs include all green plants, all algae, and some groups of bacteria. The phototrophs capture light by means of photosynthetic pigment systems and store the light energy in chemical bonds in molecules such as glucose. Nearly all other organisms belong to the class of *chemotrophs* (literally, "chemical-feeders"), which must depend on the chemical bonds created by the phototrophs for their energy. All animals, fungi, and most bacteria are chemotrophs. A world composed only of chemotrophs would not last long because without the phototrophs, food supplies would disappear almost immediately. Without sunlight and its ability to drive a reactant-favored system to form products (carbohydrate and oxygen), organisms such as humans and indeed almost the entire biosphere of planet earth could not exist.

A few organisms, the chemautotrophs, which are found near deep-ocean volcanic vents, do not depend on phototrophs for their energy supply.

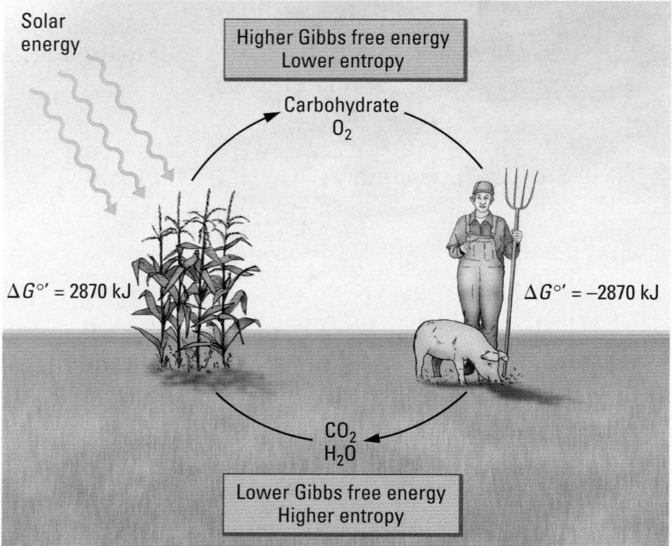

Figure 18.13 Solar energy storage by photosynthesis. The energy in the foods we eat is derived from solar energy via photosynthesis. Organisms that can photosynthesize combine carbon dioxide with water to form carbohydrates and oxygen, which have a much higher Gibbs free energy. That Gibbs free energy is released when the carbohydrates are oxidized in metabolic processes.

Both phototrophs and chemotrophs make use of the Gibbs free energy stored during photosynthesis by using oxidation of glucose to drive a large number of conversions of ADP to ATP and then using the ATP to couple to desired endergonic reactions and force them to occur. Thus, ATP is the minute-to-minute energy currency of living cells. The Gibbs free energy released in these reactions contributes to synthesis of molecules needed by the cell, causes some desirable process such as muscle contraction, or is dissipated as thermal energy. If more Gibbs free energy is taken in than the organism needs, then the excess Gibbs free energy can be stored long-term through the synthesis of fats, which have approximately twice as much Gibbs free energy as an equal mass of carbohydrate.

It is significant that when ATP reacts and causes other reactions to occur, the product ADP is very similar to the reactant ATP. ADP can easily be recycled to form ATP. A reasonable estimate of the quantity of ATP converted to ADP during one day in the life of an average human is 117 mol. Since the molar mass of the sodium salt of ATP is 551 g/mol, we can calculate that

$$117 \text{ mol} \times \frac{551 \text{ g}}{\text{mol}} = 64{,}500 \text{ g ATP}$$

is converted to ADP every day. This is 64.5 kg, which is close to the 70-kg body weight of an average person. Obviously ATP is not a long-term storage molecule for Gibbs free energy. Instead it is recycled from ADP as needed and used almost immediately for some necessary process. The typical 70-kg human body contains only 50 g of ATP and ADP. If we actually had to take in 64.5 kg of ATP per day to provide Gibbs free energy, it would be a very expensive habit. The price of ATP from a laboratory supplier is currently about $10 per gram, which would put the cost of supplying each of us with our daily energy currency at more than half a million dollars!

PROBLEM-SOLVING EXAMPLE 18.12 Coupling of Biological Reactions

ATP undergoes hydrolysis with release of Gibbs free energy according to the equation

(i) adenosine triphosphate + $H_2O(\ell)$ $\longrightarrow$
adenosine diphosphate + dihydrogen phosphate $\Delta G°'$ (i) = −30.5 kJ

Other organophosphates undergo similar hydrolysis reactions. For creatine phosphate and glycerol 3-phosphate the hydrolysis reactions are

(ii) creatine phosphate + $H_2O(\ell)$ $\longrightarrow$ creatine + dihydrogen phosphate
$\Delta G°'$ (ii) = −43.1 kJ

(iii) glycerol 3-phosphate + $H_2O(\ell)$ $\longrightarrow$ glycerol + dihydrogen phosphate
$\Delta G°'$ (iii) = −9.7 kJ

For each reaction below, predict whether the reaction is product-favored and, if it is, calculate the maximum work that could be done if the reaction took place as written.

(a) adenosine triphosphate + creatine $\longrightarrow$ creatine phosphate + adenosine diphosphate
(b) glycerol + adenosine triphosphate $\longrightarrow$
glycerol 3-phosphate + adenosine diphosphate

Answer

(a) Not product-favored
(b) Product-favored; $\Delta G°'$ = −20.8 kJ, so up to 20.8 kJ of work could be done on the surroundings

Strategy and Explanation Use the same procedure as for Hess's law calculations (*p. 246*) to write overall reactions that couple two of the reactions for which $\Delta G°'$ values are known. Then sum the $\Delta G°'$ values to obtain $\Delta G°'$ for the desired reaction. If $\Delta G°'$ for the desired reaction is negative, the process is product-favored and the magnitude of $\Delta G°'$ gives the maximum work. Use part (a) as an example. (The calculation for part (b) follows a similar procedure.) Because ATP is a reactant in the desired equation, use Reaction (i) as written.

(i) adenosine triphosphate + $H_2O(\ell)$ $\longrightarrow$
adenosine diphosphate + dihydrogen phosphate $\Delta G°'$(i) = −30.5 kJ

Because creatine phosphate is a product in the desired reaction, reverse Reaction (ii) and change the sign of $\Delta G°'$.

reverse of (ii) creatine + dihydrogen phosphate $\longrightarrow$
creatine phosphate + $H_2O(\ell)$ $\Delta G°' = -\Delta G°'$ (ii) = +43.1 kJ

The overall reaction is

adenosine triphosphate + creatine $\longrightarrow$ adenosine diphosphate + creatine phosphate
$\Delta G°' = \Delta G°'$ (i) − $\Delta G°'$ (ii) = −30.5 kJ + 43.1 kJ = 12.6 kJ

Therefore, the process (a) is reactant-favored, and at least 12.6 kJ would have to be supplied to force it to occur.

PROBLEM-SOLVING PRACTICE 18.12

ATP, creatine phosphate, and glycerol 3-phosphate could be thought of as phosphate donors (just as Brønsted-Lowry acids can be thought of as proton donors).

(a) Which of the three substances is the strongest phosphate donor?
(b) Which is the weakest?
(c) Explain your choices.

EXERCISE 18.14 Recycling of ATP

From the figures given previously for the daily quantity of ATP converted to ADP by an average human and the quantity of ATP and ADP actually present in the body, calculate the number of times each ADP molecule must be recycled to ATP each day.

Diamond: a material resource. Energy resources are like other natural resources in that they contain high-quality, concentrated energy. An analogy is a material resource such as a diamond, which is pure carbon with the atoms bonded so that each is surrounded tetrahedrally by four others. A diamond is valuable because it consists of a single crystal with atoms arranged in a specific way. If you ground a diamond into dust and spread the carbon it was made of over the area of a city block, the carbon would be nearly worthless, because it would require tremendous expense to collect the carbon and convert it back to diamond. Similarly, an energy resource is valuable not for the energy it contains, but because that energy is concentrated and available to do useful work.

18.10 Conservation of Gibbs Free Energy

When a ton of coal is burned its energy has not been used up. The law of conservation of energy (⇐ ***p. 217***) summarizes many experiments whose results verify that energy cannot be destroyed. When coal is burned in a power plant, its chemical energy is changed to an equal quantity of energy in other forms. These are electrical energy, which can be very useful, and thermal energy in the gases going up the smokestack and in the immediate surroundings of the plant, which is much less useful. However, an *energy resource* has been used up: the coal's ability to store energy and release it to do useful work. When coal burns in air, some of the Gibbs free energy that was in the coal and the oxygen that combined with it has been used up. This fact is indicated by the negative value of $\Delta G°$ for the combustion of coal. The same is true of any other product-favored reaction.

What we commonly refer to as **energy conservation** *is actually conservation of useful energy: Gibbs free energy.* Energy conservation does not mean conserving energy—nature takes care of conserving energy automatically. But nature does not automatically conserve Gibbs free energy. Substances with high Gibbs free energies are energy resources, and it is their *useful* energy that we must take pains to conserve. Once a product-favored reaction with a negative $\Delta G°$ has taken place, it cannot be reversed, thereby restoring the Gibbs free energy of its reactants, without coupling the reverse reaction with some other product-favored reaction. That is, once we have used an energy resource, it cannot be restored, except by using some other energy resource. Analysis of chemical systems in terms of Gibbs free energy can lead to important insights into how energy resources can be conserved effectively.

By comparing Gibbs free energy changes calculated using the equations in this chapter with the actual loss of Gibbs free energy in industrial processes, environmentalists and industrialists can suggest ways to minimize loss of Gibbs free energy. For example, there is a very large quantity of Gibbs free energy stored in aluminum metal and oxygen gas compared with aluminum ore, Al_2O_3. This can be seen from the thermochemical expression

$$2\ Al_2O_3(s) \longrightarrow 4\ Al(s) + 3\ O_2(g) \qquad \Delta G° = 3164.6\ kJ$$

which shows that the Gibbs free energy of 4 mol Al(s) and 3 mol O_2(g) is 3164.6 kJ higher than the Gibbs free energy of 2 mol Al_2O_3(s). If 4 mol Al(s) is oxidized to aluminum oxide, 3164.6 kJ of Gibbs free energy is lost—energy that was expended to manufacture the aluminum is wasted if the aluminum is oxidized. It is not surprising, then, that major programs for recycling aluminum operate throughout the United States. A similar statement can be made about almost every metal: Once reduced from their ores, metals are storehouses of Gibbs free energy that should be maintained in their reduced forms to avoid repeating the expenditure of Gibbs free energy needed to separate them from chemical combination with oxygen.

Energy Conservation and Coupled Reactions

The previous section mentioned that in a typical human cell, oxidation of 1 mol glucose to carbon dioxide and water can cause 32 conversions of adenosine diphosphate to adenosine triphosphate (p. 896). In Exercise 18.13, you calculated the overall change in Gibbs free energy when 1 mol glucose is metabolized and 32 mol ADP is transformed into 32 mol ATP.

$$C_6H_{12}O_6(aq) + 6\ O_2(g) \longrightarrow 6\ CO_2(g) + 6\ H_2O(\ell) \qquad \Delta G°' = -2870\ kJ$$

$$32\ ADP^{3-}(aq) + 32\ H_2PO_4^-(aq) \longrightarrow 32\ ATP^{4-}(aq) + 32\ H_2O(\ell) \qquad \Delta G°' = 976.0\ kJ$$

$$C_6H_{12}O_6(aq) + 6\ O_2(g) + 32\ ADP^{3-}(aq) + 32\ H_2PO_4^-(aq) \longrightarrow 6\ CO_2(g) + 32\ ATP^{4-}(aq) + 38\ H_2O(\ell) \qquad \Delta G°' = -1894\ kJ$$

Some of the Gibbs free energy released when the glucose is oxidized does useful work by causing synthesis of ATP, the energy-storage medium of living cells. However, nearly two thirds of the Gibbs free energy does no useful work and ends up as thermal energy. That is, about two thirds of the original Gibbs free energy is lost, and one third is stored in ATP. The energy-storage process has an efficiency of about 33%, because only about 33% of the Gibbs free energy change actually does useful work.

The more efficient a biochemical process (or an industrial process) is, the less Gibbs free energy is lost and the more energy is used productively. Efficiency is defined as the useful work done per 100 units of energy input. Current energy-conversion systems have a broad range of efficiencies. For example, the generators in a large electrical generating plant convert about 99% of the mechanical energy input to electric energy output; based on the energy of the coal combustion reaction, however, the overall efficiency of such a plant is only about 40%. An incandescent electric light bulb converts only about 5% of the electrical energy it receives into light energy, but a fluorescent light converts four times as much—about 20%. The higher the energy efficiency of the devices we use, the less Gibbs free energy is lost.

Like ATP in your body, many compounds can store Gibbs free energy. An example is ethylene. About 50 billion pounds of this gas is produced in the United States every year from the dehydrogenation of ethane in chemical plants like the one shown in Figure 18.14.

Courtesy of Phillips Petroleum Co.

Figure 18.14 Ethylene production. A chemical plant in Houston, Texas, that produces ethylene. Much of the ethylene is transformed into polyethylene, a plastic used in many consumer items *(p. 576).*

Ethylene manufacturing is only 25% energy-efficient.

$$C_2H_6(g) \longrightarrow H_2(g) + C_2H_4(g) \qquad \Delta G° = 100.97 \text{ kJ}$$

When a mole of hydrogen and a mole of ethylene are produced from a mole of ethane, at least 100.97 kJ of Gibbs free energy must be supplied from an external source. This Gibbs free energy is then stored in the hydrogen and ethylene. Ethylene production is the largest single consumer of Gibbs free energy in the chemical industry, so there has been great interest in improving the process to save energy and money. Since 1960 the Gibbs free energy requirement per pound of ethylene produced has declined by 60%. Even so, the energy used to make 1 mol ethylene from 1 mol ethane (about 400 kJ) is four times the minimum required (100.97 kJ).

ESTIMATION

Gibbs Free Energy and Automobile Travel

Given that $\Delta G° = -5295.74$ kJ per mole of octane burned, estimate the quantity of Gibbs free energy consumed when a typical car makes a 1000-mile round trip on interstate highways.

Assume that the typical car averages 20 miles per gallon and that combustion of gasoline can be approximated by combustion of octane. Because the trip is a round trip, the car ends up exactly where it started out, which means that it has done no useful thermodynamic work. Therefore all of the Gibbs free energy released by combustion of the fuel is lost. The combustion reaction is

$$C_8H_{18}(\ell) + \tfrac{25}{2}\, O_2(g) \longrightarrow 8\, CO_2(g) + 9\, H_2O(\ell) \qquad \Delta G° = -5295.74 \text{ kJ}$$

Fuel economy of 20 miles per gallon means that five gallons of fuel will be used in 100 miles and 50 gallons in 1000 miles. One gallon is four quarts and a quart is about a liter, so the volume of octane is about $4 \times 50 = 200$ L. The density of gasoline must be less than that of water, because gasoline floats on water. Assume that it is about 80% as big. Then the density is 0.8 g/mL or 800 g/L. The 200 L of fuel weighs about $200 \times 800 = 160{,}000$ g.

The molar mass of octane, C_8H_{18}, is about $8 \times 12 + 18 = 114$ g/mol. To make the arithmetic easier, round this value to 100 g/mol. Then 160,000 g of octane corresponds to 1600 mol of octane. The Gibbs free energy released by combustion is about 5000 kJ for every mole of octane burned. Since $1600 \times 5000 = 8{,}000{,}000$ kJ, about 8 million kJ of useful energy is consumed for every 1000 miles a car is driven. Most of us drive ten times that far every year, and there are a lot of cars in the United States, so the energy resources consumed by automobile travel are huge.

ThomsonNOW

Sign in to ThomsonNOW at **www.thomsonedu.com** to work an interactive module based on this material.

CHEMISTRY IN THE NEWS

Biofuels

With gasoline prices increasing, and with the serious problems anticipated from global warming *(⇐ p. 473)*, there is increasing interest in switching to fuels such as ethanol or biodiesel. Such fuels are said to require less fossil fuel input and to produce smaller quantities of greenhouse gases. They might also serve as sources for the many chemicals currently manufactured from petroleum *(⇐ p. 550)*. But are the claims valid? Analyzing the net energy (Gibbs free energy) available from biofuels and the quantities of greenhouse gases emitted in processing the fuels allows us to find out.

To find the net energy, it is necessary to factor in all of the energy inputs, not just the obvious ones, that are required to manufacture the fuel and deliver the fuel to its intended use. For example, suppose that E85 ethanol *(⇐ p. 552)* is made from corn. At first glance the principal energy input is sunlight, which causes the corn to grow. Also, some energy is required for the fermentation process that converts the cornstarch to ethanol and some for transporting the fuel to a gas station. Because corn plants take in carbon dioxide as they grow, the carbon dioxide produced when the ethanol burns will be recycled into the next crop of corn and will not contribute to global warming. So biofuels are great, right? Not necessarily!

This analysis left out the fact that the farmland on which the corn grew was tilled and fertilized. So tractors had to burn gasoline or diesel fuel to do the tilling and distribute the fertilizer, ammonia fertilizer had to be made from hydrogen that came from natural gas *(⇐ p. 554)*, and other fertilizers such as phosphates and lime had to be manufactured. All of these processes consume Gibbs free energy. If the land was irrigated, water had to be pumped and coal had to be burned to produce the electricity to drive the pumps. These and many other processes are needed to create biofuels. All such processes consume Gibbs free energy (they have to, for anything to happen) and many of them create greenhouse gases. When corn is fermented to ethanol, by-products are produced (not all of the corn changes to ethanol), but this may be good. The by-products can be used as animal feed, for example, replacing animal feed that would otherwise have consumed Gibbs free energy during its production.

As you can imagine, there has been much scientific study of this issue. Some studies have indicated that burning corn-derived ethanol in a car actually consumes more Gibbs free energy than burning gasoline. That is, the biofuel was worse for the environment than the fossil fuel. Other studies found that ethanol as currently produced is better than gasoline, but no study found that it was environmentally benign. In January 2006, *Science* magazine published a study that examined and compared six previous studies and attempted to produce a definitive result. A summary of the results of this study is shown in the figure.

The figure shows energy inputs for gasoline and three different ways of generating ethanol: Ethanol Today—from corn as is done today; CO_2-intensive—from Nebraska corn shipped to a lignite-fueled processing plant in North Dakota; and Cellulosic—from cellulose in plants such as switchgrass, not just from starch in corn kernels. The horizontal green bars represent the quantity of Gibbs free energy consumed in each process relative to one unit of Gibbs free energy delivered as fuel. Numbers at the right side of the figure (such as 94 for gasoline) indicate kg CO_2-equivalent per MJ (1 MJ = 10^6 J) of fuel and therefore measure greenhouse-gas (GHG) emissions.

It is clear from the figure that current methods for generating ethanol have a large potential for displacing petroleum (0.05 for ethanol versus 1.1

This is largely due to inefficiencies in energy transfer from external sources to the reaction system.

It is important to recognize that completely eliminating consumption of Gibbs free energy is impossible. Whenever anything happens, whether a chemical reaction or a physical process, the final state must have less Gibbs free energy than was available initially. This is the same as saying that the entropy of the universe must have increased during the change. This statement is true of any system in which the initial substances are changed into something new—any product-favored system. Thus, losses of Gibbs free energy are inevitable. The aim of energy conservation is to minimize—not eliminate—them. This can be done by maximizing the efficiency of coupling exergonic reactions to endergonic processes we want to cause to occur. The ideas of thermodynamics help us figure out how to accomplish that goal and are the most powerful tool we have for conserving energy while maintaining a high standard of living.

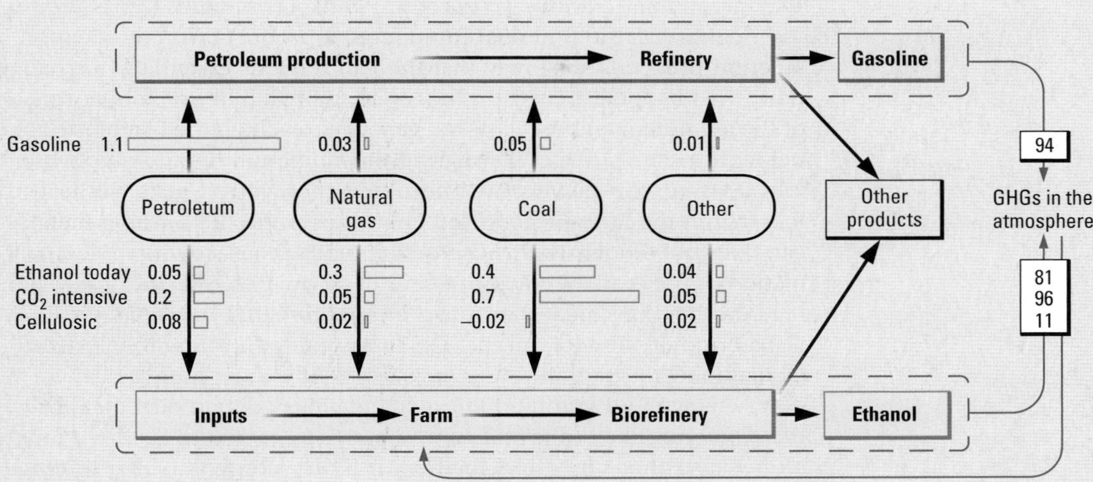

for gasoline), but that producing the ethanol consumes large quantities of natural gas (0.3) and coal (0.4). To produce 1 MJ from ethanol fuel requires $0.05 + 0.3 + 0.4 + 0.04 = 0.79$ MJ of energy input, so only 0.21 (about 20%) of the ethanol energy has actually displaced use of other fuels. That is, producing ethanol as we do today will reduce the need for petroleum, but increase the need for natural gas and coal. The GHG emissions are 81 versus 94 for gasoline, a reduction, but only by 14%—not the complete avoidance of GHG emissions that an incomplete analysis would have suggested.

Perhaps the most significant result of this study is the tremendous reduction in fossil fuel use and in GHG emissions that would accrue if an economically feasible process were developed by which any cellulosic material (wood, brush, grass) could be converted to ethanol. In that case the energy input is only $0.08 + 0.02 - 0.02 + 0.02 = 0.10$ MJ per MJ of fuel produced. (The negative value for coal results because some electricity would be produced and this would reduce burning of coal.)

Even more important, a thermodynamic analysis of the entire system required to generate fuel provides important results that apply to societal decisions. Science is crucial to good public policy.

SOURCES: *New York Times*, November 15, 2005; p. D3; *New York Times*, February 28, 2006; p. C1; Farrell, A. E., Plevin, R. J., Turner, B. T., Jones, A. D., O'Hare, M., and Kammen, D. M. *Science*, Vol. 311, 2006; pp. 506–508. Reprinted with permission AAAS.

18.11 Thermodynamic and Kinetic Stability

Chemists often say that substances are "stable," but what exactly does that statement mean? Usually it means that the substance in question does not decompose or react with other substances that normally come in contact with it. Most chemists, for example, would say that the aluminum can that holds the soda you drink is stable. It will be around for quite a long time. The fact that aluminum cans do not decompose rapidly is one of the reasons you are encouraged to recycle them instead of throwing them away. Some aluminum cans have emerged almost unchanged from landfills after 40 or 50 years.

Strictly speaking, there are two kinds of stability. We discussed one of them earlier in this chapter. A substance is *thermodynamically stable* if it does not undergo product-favored reactions. Such reactions disperse energy and increase the entropy of their surroundings. Although we just said it was stable, the aluminum in a soda can is actually *thermodynamically unstable* compared with its oxide, because its reaction with oxygen in air has a negative Gibbs free energy change.

$$4\,Al(s) + 3\,O_2(g) \longrightarrow 2\,Al_2O_3(s) \qquad \Delta G^\circ = -3164.6\text{ kJ}$$

The aluminum exhibits a different kind of stability—it is ***kinetically stable.*** Although it has the potential to undergo a product-favored oxidation reaction, this reaction proceeds so slowly that the can remains essentially unchanged for a long time. This happens because a thin coating of aluminum oxide forms on the surface of the aluminum and prevents oxygen from reaching the rest of the aluminum atoms below the can's surface. If we grind the aluminum into a fine powder and throw it into a flame, the powder will burn and the evolved heat will lead to an entropy increase in the little piece of the universe around the burning metal. An aluminum can does not oxidize away because the oxidation is slow (kinetics), not because formation of the oxide would not occur of its own accord (thermodynamics).

Another substance that is *thermodynamically unstable* but ***kinetically stable*** is diamond. If you look up the data in Appendix J, you will find that the conversion of diamond to graphite has a negative Gibbs free energy change. But diamonds don't change into graphite. Engagement rings contain diamonds precisely because the diamond (like the love it represents) is expected to last for a long time. It does so because there is a very high activation energy barrier ***(⇐ p. 629)*** for the change from the diamond structure to the graphite structure. When a chemist says something is stable, it usually means that it is kinetically stable—only an activation energy barrier prevents it from reacting fast enough for us to see a change.

PROBLEM-SOLVING EXAMPLE 18.13 Thermodynamic and Kinetic Stability

Whenever air is heated to a very high temperature, the reaction between nitrogen and oxygen to form nitrogen monoxide occurs. It is an important source of nitrogen-containing air pollutants that can be formed in the cylinders of an automobile engine.

(a) Write a balanced equation with minimum whole-number coefficients for the equilibrium reaction of N_2 with O_2 to form NO.

(b) Is this reaction product-favored at room temperature? That is, is NO thermodynamically stable compared to N_2 and O_2?

(c) Estimate the temperature at which the standard equilibrium constant for this reaction equals 1.

(d) If NO is formed at high temperature in an automobile engine, why does it not all change back to N_2 and O_2 when the mixture of gases enters the exhaust system and its temperature falls?

(e) How might the concentration of NO in automobile exhaust be reduced?

Answer

(a) $N_2(g) + O_2(g) \rightleftharpoons 2\,NO(g)$ (b) No (c) 7301 K

(d) The reverse reaction is too slow. (e) Use a suitable catalyst.

Strategy and Explanation

(a) See answer.

(b) Calculate ΔG° at 25 °C using data from Appendix J.
$\Delta G^\circ = 2\{\Delta G_f^\circ[NO(g)]\} = 2(86.55\text{ kJ}) = 173.10\text{ kJ}$. Since $\Delta G^\circ > 0$, the reaction is not product-favored.

(c) If $K^\circ = 1$, then $\Delta G^\circ = -RT\ln K^\circ = -RT\ln(1) = 0$.
Because $\Delta G^\circ = \Delta H^\circ - T\Delta S^\circ = 0$, $\Delta H^\circ = T\Delta S^\circ$ and $T = \Delta H^\circ/\Delta S^\circ$.
Using data from Appendix J gives $\Delta H^\circ = 2\{\Delta H_f^\circ[NO(g)]\} = 2(90.25\text{ kJ}) = 180.50\text{ kJ}$, and

$$\begin{aligned}\Delta S^\circ &= 2\{S^\circ[NO(g)]\} - \{S^\circ[N_2(g)] + S^\circ[O_2(g)]\}\\ &= 2(210.76\text{ J/K}) - 191.66\text{ J/K} - 205.138\text{ J/K} = 24.722\text{ J/K}\end{aligned}$$

Therefore $T = 180.50\text{ kJ}/24.722\text{ J/K} = 180{,}500\text{ J}/24.722\text{ J/K} = 7301\text{ K}$.

(d) When the mixture of gases, which contains some NO as well as N_2 and O_2, leaves the cylinder of the automobile engine and enters the exhaust system, it cools very rapidly to a temperature below 500 K. The reverse reaction should occur, according to thermodynamics, but it does not. The activation energy for the decomposition of NO is quite high, because NO contains a double bond, and it is very difficult to separate the two atoms (which must be done to form N_2 and O_2). Therefore, the reaction rate is greatly affected by temperature. At low temperatures the reverse reaction is very slow, so significant concentrations of NO exist in automobile exhaust, even though N_2 and O_2 are thermodynamically stable compared to NO.

(e) With a suitable catalyst, decomposition of NO to its elements can take place at appreciable rates even at relatively low temperatures. Catalytic converters are installed in the exhaust systems of cars partly to reduce the concentration of NO. Because N_2 and O_2 are more stable thermodynamically, it is reasonable to use a catalyst to speed up their formation.

✓ **Reasonable Answer Check** It is reasonable that $\Delta G°$ for the reaction of N_2 with O_2 is positive, because N_2 and O_2 are the principal components of the atmosphere where their partial pressures are close to standard pressure, and they do not react with each other. It is reasonable that $\Delta S°$ for the reaction is small and positive. The total number of gas phase molecules does not change, but the product molecules have two different atoms, and both reactant molecules have two atoms that are the same, making the product molecules slightly more probable. It is reasonable that the reaction is endothermic, because the reactant molecules have a triple bond and a double bond, and the product molecules have two double bonds. The bonds broken are therefore expected to be stronger than the bonds formed.

PROBLEM-SOLVING PRACTICE 18.13

All of these substances are stable with respect to decomposition to their elements at 25 °C. Which are kinetically stable and which are thermodynamically stable?

(a) $MgO(s)$
(b) $N_2H_4(\ell)$
(c) $C_2H_6(g)$
(d) $N_2O(g)$

Finally, think about whether you yourself are stable (thermodynamically or kinetically). From a thermodynamic standpoint, most of the substances you are made of are unstable with respect to oxidation to carbon dioxide, water, and other substances. That is, based on Gibbs free energy changes, most of the substances that you are made of should undergo product-favored reactions that would completely destroy them. Your protein, fat, carbohydrate, and even DNA should spontaneously change into much smaller, simpler molecules. Fortunately for you, the reactions by which this change would happen are very slow at room temperature and body temperature. Only when enzymes catalyze those reactions do they occur with reasonable speed. It is the combination of thermodynamic instability and kinetic stability that allows those enzymes to control the reactions in your body or in any living organism. Were it not for the kinetic stability of a wide variety of substances, everything would be quickly converted to a small number of very thermodynamically stable substances. Life and the environment as we know them would then be impossible.

The roles of thermodynamics and kinetics in determining chemical reactivity can be summarized by saying that ***thermodynamics tells whether a reaction can produce predominantly products under standard conditions and, if it does, how much useful work can be accomplished by coupling the reaction to another process.*** If a reaction involves dilution of substances in the gas phase or in solution, ***thermodynamics tells the value of the standard equilibrium constant and allows quantitative prediction of how much product is formed.*** Thermodynamics also can be used to predict what will happen under nonstandard conditions. ***Chemical kinetics tells how fast a given reaction goes***

and indicates how we can control the rate of reaction. Together, thermodynamics and kinetics provide the intellectual foundation on which modern chemical industries are based and the principles upon which fundamental understanding of physiology and medicine depends.

SUMMARY PROBLEM

In a blast furnace for making iron from iron ore, large quantities of coke (which is mainly carbon) are dumped into the top of the furnace along with iron ore (which can be assumed to be Fe_2O_3) and limestone (which is used to help remove impurities from the iron). The overall process is

$$2\ Fe_2O_3(s) + 3\ C(s) \longrightarrow 4\ Fe(s) + 3\ CO_2(g)$$

This reaction can be thought of as a combination of several individual steps.

$$2\ Fe_2O_3(s) \longrightarrow 4\ FeO(s) + O_2(g)$$
$$2\ FeO(s) \longrightarrow 2\ Fe(s) + O_2(g)$$
$$2\ C(s) + O_2(g) \longrightarrow 2\ CO(g)$$
$$2\ CO(g) + O_2(g) \longrightarrow 2\ CO_2(g)$$

(a) Calculate the enthalpy change for each step, assuming a temperature of 25 °C. Which steps are exothermic and which are endothermic?

(b) Based on the equations, predict which of the individual steps would involve an increase and which a decrease in the entropy of the system.

(c) Based on your results in parts (a) and (b), what can you say about whether each step is reactant-favored or product-favored at room temperature? At a much higher temperature ($>$1000 K)?

(d) Calculate the entropy change and the Gibbs free energy change for each reaction step, assuming a temperature of 25 °C. (Obtain data from Appendix J.)

(e) Keeping in mind the equation $\Delta G° = \Delta H° - T\Delta S°$ and the fact that the enthalpy change and entropy change for a reaction do not vary much with temperature, what would be the slope of a graph of $\Delta G°$ versus T for each of the reactions? For which of the reactions does $\Delta G°$ become more negative as the temperature increases? For which does it become more positive? Does this agree with what you predicted in part (c)?

(f) For which of these reactions might the assumption of nearly constant $\Delta H°$ and $\Delta S°$ not be valid as the temperature increases from 25 °C? For each reaction you choose, explain why the assumption might not be correct.

(g) Use your results from previous parts of this problem to estimate the Gibbs free energy change for each of these reactions at a temperature of 1500 K.

(h) Which of the two iron oxides is more easily reduced at 1500 K? Which of the reactions involving carbon compounds is more product-favored at 1500 K? What chemical reactions do you think are taking place in the hottest part of the blast furnace?

(i) In portions of the furnace where the temperature is about 800 K, would you predict that the same reactions would be occurring as in the higher-temperature part of the furnace? Why or why not?

(j) Show that the individual steps can be combined to give the overall reaction. From the enthalpy, entropy, and Gibbs free energy changes already calculated, calculate these changes for the overall reaction.

(k) In a typical blast furnace every kilogram of iron produced requires 2.5 kg iron ore, 1 kg coke, and nearly 6 kg air (to provide oxygen for oxidation of the coke to heat the furnace and to combine with carbon in the coke, forming CO(g)). How much Gibbs free energy would be destroyed if the coke were simply burned to form carbon dioxide? Given the quantity of iron produced in a typical furnace, how much Gibbs free energy is stored by coupling the oxidation of coke to the reduction of iron oxides? What percentage of the Gibbs free energy available from combustion of coke is wasted per kilogram of iron produced?

IN CLOSING

Having studied this chapter, you should be able to . . .

- Understand and be able to use the terms "product-favored" and "reactant-favored" (Section 18.1).
- Explain why there is a higher probability that energy will be dispersed than that it will be concentrated in a small number of nanoscale particles (Section 18.2). ThomsonNOW homework: Study Question 19
- Calculate the entropy change for a process occurring at constant temperature (Section 18.3). ThomsonNOW homework: Study Question 35
- Use qualitative rules to predict the sign of the entropy change for a process (Section 18.3). ThomsonNOW homework: Study Questions 25, 31
- Calculate the entropy change for a chemical reaction, given a table of standard molar entropy values for elements and compounds (Section 18.4). ThomsonNOW homework: Study Question 41
- Use entropy and enthalpy changes to predict whether a reaction is product-favored (Section 18.5). ThomsonNOW homework: Study Questions 45, 49
- Describe the connection between enthalpy and entropy changes for a reaction and the Gibbs free energy change; use this relation to estimate quantitatively how temperature affects whether a reaction is product-favored (Section 18.6). ThomsonNOW homework: Study Questions 47, 55, 127, 137
- Calculate the Gibbs free energy change for a reaction from values given in a table of standard molar Gibbs free energies of formation (Section 18.6). ThomsonNOW homework: Study Questions 57, 63, 67, 75
- Relate the Gibbs free energy change and the standard equilibrium constant for the same reaction and be able to calculate one from the other (Section 18.7). ThomsonNOW homework: Study Questions 77, 85, 109
- Describe how a reactant-favored system can be coupled to a product-favored system so that a desired reaction can be carried out (Section 18.8). ThomsonNOW homework: Study Question 89
- Explain how biological systems make use of coupled reactions to maintain the high degree of order found in all living organisms; give examples of coupled reactions that are important in biochemistry (Section 18.9). ThomsonNOW homework: Study Question 97
- Explain the relationship between Gibbs free energy and energy conservation (Sections 18.8 and 18.10).
- Distinguish between thermodynamic stability and kinetic stability and describe the effect of each on whether a reaction is useful in producing products (Section 18.11). ThomsonNOW homework: Study Question 101

ThomsonNOW

Sign in to ThomsonNOW at **www.thomsonedu.com** to check your readiness for an exam by taking the Pre-Test and exploring the modules recommended in your Personalized Learning Plan.

KEY TERMS

endergonic *(18.9)*

energy conservation *(18.10)*

exergonic *(18.9)*

extent of reaction *(18.7)*

Gibbs free energy *(18.6)*

metabolism *(18.9)*

nutrients *(18.9)*

photosynthesis *(18.9)*

reversible process *(18.3)*

second law of thermodynamics *(18.5)*

standard equilibrium constant *(18.7)*

third law of thermodynamics *(18.3)*

QUESTIONS FOR REVIEW AND THOUGHT

■ denotes questions available in ThomsonNOW and assignable in OWL.

Blue-numbered questions have short answers at the back of this book and fully worked solutions in the *Student Solutions Manual.*

ThomsonNOW™
Assess your understanding of this chapter's topics with sample tests and other resources found by signing in to ThomsonNOW at **www.thomsonedu.com**.

Review Questions

1. Define the terms "product-favored system" and "reactant-favored system." Give one example of each.
2. What are the two ways that a final chemical state of a system can be more probable than its initial state?
3. Define the term "entropy," and give an example of a sample of matter that has zero entropy. What are the units of entropy? How do they differ from the units of enthalpy?
4. State five useful qualitative rules for predicting entropy changes when chemical or physical changes occur.
5. State the second law of thermodynamics.
6. In terms of values of $\Delta H°$ and $\Delta S°$, under what conditions can you be sure that a reaction is product-favored? When can you be sure that it is not product-favored?
7. Define the Gibbs free energy change of a chemical reaction in terms of its enthalpy and entropy changes. Why is the Gibbs free energy change especially useful in predicting whether a reaction is product-favored?
8. Why are materials whose reactions release large quantities of Gibbs free energy useful to society? Give two examples of such materials.
9. How are materials whose reactions release large quantities of Gibbs free energy important to you? Give two examples of such materials.
10. Define the terms "endergonic" and "exergonic."
11. What is the citric acid cycle, and why is it important to organisms?
12. Define these important biochemistry terms: metabolism, nutrients, ATP, ADP, oxidative phosphorylation, coupled reactions, phototrophs, chemotrophs, photosynthesis.
13. Describe two ways to cause reactant-favored reactions to form products.
14. Describe the process by which sunlight is employed to convert high-entropy, low-Gibbs-free-energy substances into low-entropy, high-Gibbs-free-energy substances.

Topical Questions

Reactant-Favored and Product-Favored Processes

15. For each process, write a chemical equation and classify the process as reactant-favored or product-favored.
    (a) Water decomposes to its elements, hydrogen and oxygen.
    (b) Gasoline spilled on the ground evaporates (use octane, C_8H_{18}, to represent gasoline).
    (c) Sugar dissolves in water at room temperature.
16. For each process, write a chemical equation and classify the process as reactant-favored or product-favored.
    (a) Carbon dioxide gas decomposes to its elements, carbon and oxygen.
    (b) The steel (mostly iron) body of an automobile rusts.
    (c) Gasoline reacts with oxygen to form carbon dioxide and water (use octane, C_8H_{18}, to represent gasoline).

Chemical Reactions and Dispersal of Energy

17. Suppose you flip a coin.
    (a) What is the probability that the coin will come up heads?
    (b) What is the probability that it will come up tails?
    (c) If you flip the coin 100 times, what is the most likely number of heads and tails you will see?
18. Suppose you make a tetrahedron and put numbers 1, 2, 3, and 4 on each of the four sides. You toss the tetrahedron in the air and observe it after it comes to rest.
    (a) What is the probability that the tetrahedron will come to rest with the numbers 2, 3, and 4 visible?
    (b) What is the probability that the tetrahedron will come to rest with the numbers 1, 2, and 3 visible?
    (c) If you toss the tetrahedron 100 times, what is the most likely number of times you will see a 1 after it comes to rest?
19. ■ Consider two equal-sized flasks connected as in shown in the figure.

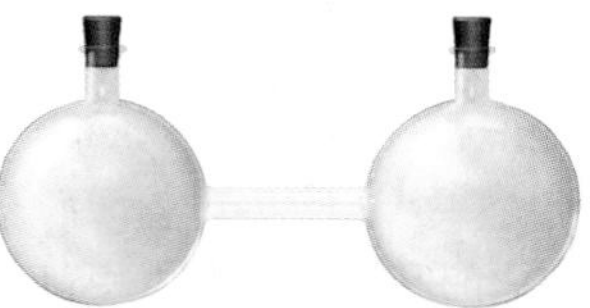

    (a) Suppose you put one molecule inside. What is the probability that the molecule will be in flask A? What is the probability that it will be in flask B?
    (b) If you put 100 molecules into the two-flask system, what is the most likely arrangement of molecules? Which arrangement has the highest entropy?

20. Suppose you have four identical molecules labeled 1, 2, 3, and 4. Draw 16 simple two-flask diagrams as in the figure for Question 19, and draw all possible arrangements of the four molecules in the two flasks. How many of these arrangements have two molecules in each flask? How many have no molecules in one flask? From these results, what is the most probable arrangement of molecules? Which arrangement has the highest entropy?

Measuring Dispersal of Energy: Entropy

21. For each process, tell whether the entropy change of the system is positive or negative.
    (a) Water vapor (the system) deposits as ice crystals on a cold windowpane.
    (b) A can of carbonated beverage loses its fizz. (Consider the beverage but not the can as the system. What happens to the entropy of the dissolved gas?)
    (c) A glassblower heats glass (the system) to its softening temperature.
22. For each process, tell whether the entropy change of the system is positive or negative.
    (a) Water boils.
    (b) A teaspoon of sugar dissolves in a cup of coffee. (The system consists of both sugar and coffee.)
    (c) Calcium carbonate precipitates out of water in a cave to form stalactites and stalagmites. (Consider only the calcium carbonate to be the system.)
23. For each situation described in Question 15, tell whether the entropy of the system increases or decreases.
24. For each situation described in Question 16, tell whether the entropy of the system increases or decreases.
25. ■ For each pair of items, tell which has the higher entropy, and explain why.
    (a) Item 1, a sample of solid CO_2 at −78 °C, or item 2, CO_2 vapor at 0 °C
    (b) Item 1, solid sugar, or item 2, the same sugar dissolved in a cup of tea
    (c) Item 1, a 100-mL sample of pure water and a 100-mL sample of pure alcohol, or item 2, the same samples of water and alcohol after they had been poured together and stirred
26. For each pair of items, tell which has the higher entropy, and explain why.
    (a) Item 1, a sample of pure silicon (to be used in a computer chip), or item 2, a piece of silicon having the same mass but containing a trace of some other element, such as B or P
    (b) Item 1, an ice cube at 0 °C, or item 2, the same mass of liquid water at 0 °C
    (c) Item 1, a sample of pure I_2 solid at room temperature, or item 2, the same mass of iodine vapor at room temperature
27. ■ Comparing the formulas or states for each pair of substances, tell which you would expect to have the higher entropy per mole at the same temperature, and explain why.
    (a) NaCl(s) or CaO(s)
    (b) Cl_2(g) or P_4(g)
    (c) NH_4NO_3(s) or NH_4NO_3(aq)
28. Comparing the formulas or states for each pair of substances, tell which you would expect to have the higher entropy per mole at the same temperature, and explain why.
    (a) CH_3NH_2(g) or $(CH_3)_2NH$(g)
    (b) Au(s) or Hg(ℓ)
    (c) Kr(g) or C_6H_{14}(g)
29. From each pair of substances listed below, select the one having the larger standard molar entropy at 25 °C. Give reasons for your choice.
    (a) Ga(s) or Ga(ℓ)
    (b) AsH_3(g) or Kr(g)
    (c) NaF(s) or MgO(s)
30. From each pair of substances listed below, select the one having the larger standard molar entropy at 25 °C. Give reasons for your choice.
    (a) H_2O(g) or H_2S(g)
    (b) CH_3OH(ℓ) or C_2H_5OH(ℓ)
    (c) Butane or cyclobutane
31. ■ Without doing a calculation, predict whether the entropy change will be positive or negative when each reaction occurs in the direction it is written.
    (a) $C_2H_4(g) + H_2(g) \longrightarrow C_2H_6(g)$
    (b) $CH_3OH(\ell) + \frac{3}{2}O_2(g) \longrightarrow CO_2(g) + 2\ H_2O(g)$
    (c) $N_2(g) + 3\ H_2(g) \longrightarrow 2\ NH_3(g)$
    (d) $CaCO_3(s) \longrightarrow CaO(s) + CO_2(g)$
32. Without doing a calculation, predict whether the entropy change will be positive or negative when each reaction occurs in the direction it is written.
    (a) $CH_3OH(\ell) \longrightarrow CO(g) + 2\ H_2(g)$
    (b) $Br_2(\ell) + H_2(g) \longrightarrow 2\ HBr(g)$
    (c) $C_3H_8(g) \longrightarrow C_2H_4(g) + CH_4(g)$
    (d) $Ag^+(aq) + I^-(aq) \longrightarrow AgI(s)$
33. Without consulting a table of standard molar entropies, predict whether $\Delta S°_{system}$ will be positive or negative for each of these reactions.
    (a) $2\ CO(g) + O_2(g) \longrightarrow 2\ CO_2(g)$
    (b) $2\ H_2(g) + O_2(g) \longrightarrow 2\ H_2O(\ell)$
    (c) $2\ O_3(g) \longrightarrow 3\ O_2(g)$
34. Without consulting a table of standard molar entropies, predict whether $\Delta S°_{system}$ will be positive or negative for each of these reactions.
    (a) $2\ NH_3(g) \longrightarrow N_2(g) + 3\ H_2(g)$
    (b) $2\ Na(s) + Cl_2(g) \longrightarrow 2\ NaCl(s)$
    (c) $H_2(g) + I_2(s) \longrightarrow 2\ HI(g)$

Calculating Entropy Changes

35. ■ Calculate the entropy change, $\Delta S°$, for the vaporization of ethanol, C_2H_5OH, at the boiling point of 78.3 °C. The heat of vaporization of the alcohol is 39.3 kJ/mol.

$$C_2H_5OH(\ell) \longrightarrow C_2H_5OH(g) \qquad \Delta S° = ?$$

36. Diethyl ether, $(C_2H_5)_2O$, was once used as an anesthetic. What is the entropy change, $\Delta S°$, for the vaporization of ether if its heat of vaporization is 26.0 kJ/mol at the boiling point of 35.0 °C?
37. Calculate $\Delta S°$ for each substance when the quantity of thermal energy indicated is transferred reversibly to the system at the temperature specified. Assume that you have enough

of each substance so that its temperature remains constant as the thermal energy is transferred.
(a) $H_2(g)$, 0.775 kJ, 295 K (b) KCl(s), 500. kJ, 500. K
(c) $N_2(g)$, 2.45 kJ, 1000. K

38. Calculate $\Delta S°$ for each of these substances when the quantity of thermal energy indicated is transferred reversibly to the system at the temperature specified. Assume that you have enough of each substance so that its temperature remains constant as the thermal energy is transferred.
(a) NaCl(s), 5.00 kJ, 500. K
(b) $N_2O(g)$, 0.30 kJ, 300. K

39. The standard molar entropy of methanol vapor, $CH_3OH(g)$, is 239.8 J K^{-1} mol^{-1}.
(a) Calculate the entropy change for the vaporization of 1 mol methanol (use data from Table 18.1 or Appendix J).
(b) Calculate the enthalpy of vaporization of methanol, assuming that $\Delta S°$ doesn't depend on temperature and taking the boiling point of methanol to be 64.6 °C.

40. The standard molar entropy of iodine vapor, $I_2(g)$, is 260.7 J K^{-1} mol^{-1} and the standard molar enthalpy of formation is 62.4 kJ/mol.
(a) Calculate the entropy change for vaporization of 1 mol of solid iodine (use data from Table 18.1 or Appendix J).
(b) Calculate the enthalpy change for sublimation of iodine.
(c) Assuming that $\Delta S°$ does not change with temperature, estimate the temperature at which iodine would sublime (change directly from solid to gas).

41. ■ Check your predictions in Question 31 by calculating the entropy change for each reaction. Standard molar entropies not in Table 18.1 can be found in Appendix J.

42. Check your predictions in Question 32 by calculating the entropy change for each reaction. Standard molar entropies not in Table 18.1 can be found in Appendix J.

43. Check your predictions in Question 33 by calculating the entropy change for each reaction. Standard molar entropies not in Table 18.1 can be found in Appendix J.

44. Check your predictions in Question 34 by calculating the entropy change for each reaction. Standard molar entropies not in Table 18.1 can be found in Appendix J.

Entropy and the Second Law of Thermodynamics

45. ■ Calculate $\Delta S°_{system}$ at 25 °C for the reaction

$$C_2H_4(g) + H_2O(g) \longrightarrow C_2H_5OH(\ell)$$

Can you tell from the result of this calculation whether this reaction is product-favored? If you cannot tell, what additional information do you need? Obtain that information and determine whether the reaction is product-favored.

46. Calculate $\Delta S°_{system}$ at 25 °C for the reaction

$$C_6H_6(\ell) + 4\ H_2(g) \longrightarrow C_6H_{14}(\ell)$$

Can you tell from the result of this calculation whether this reaction is product-favored? If you cannot tell, what additional information do you need? Obtain that information and determine whether the reaction is product-favored.

47. Is this reaction predicted to favor the products at low temperatures, at high temperatures, or both? Explain your answer briefly.

$$Mg(s) + \tfrac{1}{2}O_2(g) \longrightarrow MgO(s) \qquad \Delta H° = -601.70\ \text{kJ}$$

48. Is this reaction predicted to favor the products at low temperatures, at high temperatures, or both? Explain your answer briefly.

$$MgCO_3(s) \longrightarrow MgO(s) + CO_2(g) \qquad \Delta H° = 116.48\ \text{kJ}$$

49. ■ Explain briefly why the exothermic combustion of propane is product-favored.

$$C_3H_8(g) + 5\ O_2(g) \longrightarrow 3\ CO_2(g) + 4\ H_2O(g)$$

50. Explain briefly why the exothermic reaction of a metal carbonate with an acid is product-favored.

$$CuCO_3(s) + H_2SO_4(aq) \longrightarrow CuSO_4(aq) + CO_2(g) + H_2O(\ell)$$

51. Sodium reacts violently with water according to the equation

$$Na(s) + H_2O(\ell) \longrightarrow NaOH(aq) + \tfrac{1}{2}H_2(g)$$

(a) Predict the signs of $\Delta H°$ and $\Delta S°$ for the reaction.
(b) Verify your predictions with calculations.

52. Once ignited, magnesium reacts vigorously with oxygen in air according to the equation

$$2\ Mg(s) + O_2(g) \longrightarrow 2\ MgO(s)$$

(a) Predict the signs of $\Delta H°$ and $\Delta S°$ for the reaction.
(b) Verify your predictions with calculations.

53. Hydrogen burns in air with considerable heat transfer to the surroundings. Consider the decomposition of water to gaseous hydrogen and oxygen. Without doing any calculations, and basing your prediction on the enthalpy change and the entropy change, is this reaction product-favored at 25 °C? Explain your answer briefly.

54. Hydrogen gas combines with chlorine gas in an exothermic reaction to form HCl(g). Consider the decomposition of gaseous hydrogen chloride to hydrogen and chlorine. Without doing any calculations, and basing your prediction on the enthalpy change and the entropy change, is this reaction product-favored at 25 °C? Explain your answer briefly.

55. ■ For each reaction, calculate $\Delta H°$ and $\Delta S°$ and predict whether the reaction is always product-favored, product-favored only at low temperatures, product-favored only at high temperatures, or never product-favored.
(a) $Fe_2O_3(s) + 2\ Al(s) \longrightarrow 2\ Fe(s) + Al_2O_3(s)$
(b) $N_2(g) + 2\ O_2(g) \longrightarrow 2\ NO_2(g)$

56. For each reaction, calculate $\Delta H°$ and $\Delta S°$ and predict whether the reaction is always product-favored, product-favored only at low temperatures, product-favored only at high temperatures, or never product-favored.
(a) $C_6H_{12}O_6(s) + 6\ O_2(g) \longrightarrow 6\ CO_2(g) + 6\ H_2O(\ell)$
(b) $MgO(s) + C(\text{graphite}) \longrightarrow Mg(s) + CO(g)$

Gibbs Free Energy

57. Determine whether the combustion of ethane, C_2H_6, is product-favored at 25 °C.

$$C_2H_6(g) + \tfrac{7}{2}O_2(g) \longrightarrow 2\ CO_2(g) + 3\ H_2O(\ell)$$

(a) Calculate $\Delta S_{universe}$. Required values of $\Delta H_f°$ and S° are in Appendix J.
(b) Verify your result by calculating the value of $\Delta G°$ for the reaction.
(c) Do your calculated answers in parts (a) and (b) agree with your preconceived idea of this reaction?

58. The reaction of magnesium with water can be used as a means for heating food.

$$Mg(s) + 2\,H_2O(\ell) \longrightarrow Mg(OH)_2(s) + H_2(g)$$

Determine whether this reaction is product-favored at 25 °C.
(a) Calculate $\Delta S_{universe}$. See Appendix J for the needed data.
(b) Verify your result by calculating $\Delta G°$ for the reaction.

59. Add a column for the sign of the Gibbs free energy to Table 18.2 (p. 882). For the first and last lines in the table, tell whether ΔG is positive or negative.

60. Based on your table from Question 59, when ΔH_{system} and ΔS_{system} are both negative, is ΔG is positive or negative or does the sign depend on temperature? If the sign depends on temperature, does the reaction become product-favored at high or low temperatures?

61. Use a mathematical equation to show how the statement leads to the conclusion cited: If a reaction is exothermic (negative ΔH) and if the entropy of the system increases (positive ΔS), then ΔG must be negative, and the reaction will be product-favored.

62. Use a mathematical equation to show how the statement leads to the conclusion cited: If ΔH and ΔS have the same sign, then the magnitude of T determines whether ΔG will be negative and whether the reaction will be product-favored.

63. ■ Predict whether the reaction below is product-favored or reactant-favored by calculating $\Delta G°$ from the entropy and enthalpy changes for the reaction at 25 °C.

$$H_2(g) + CO_2(g) \longrightarrow H_2O(g) + CO(g) \qquad \Delta H° = 41.17\text{ kJ} \qquad \Delta S° = 42.08\text{ J/K}$$

64. Predict whether this reaction is product-favored at 25 °C by calculating the change in standard Gibbs free energy from the entropy and enthalpy changes.

$$H_2(g) + I_2(g) \rightleftharpoons 2\,HI(g) \qquad \Delta H° = 52.96\text{ kJ} \qquad \Delta S° = 166.4\text{ J/K}$$

65. If this reaction were product-favored, it would be a good way to make pure silicon, crucial in the semiconductor industry, from sand (SiO_2).

$$SiO_2(s) + C(s) \longrightarrow Si(s) + CO_2(g)$$

Calculate $\Delta G°$ from data in Appendix J and decide whether the reaction can be used to produce silicon at 25 °C.

66. From data in Appendix J, calculate $\Delta G°$ for the reactions of sand with hydrogen fluoride and hydrogen chloride. Explain why hydrogen fluoride attacks glass, whereas hydrogen chloride does not.

$$SiO_2(s) + 4\,HF(g) \longrightarrow SiF_4(g) + 2\,H_2O(g)$$
$$SiO_2(s) + 4\,HCl(g) \longrightarrow SiCl_4(g) + 2\,H_2O(g)$$

67. ■ Use data from Appendix J to calculate $\Delta G°$ for each reaction at 25 °C. Which are product-favored?
(a) $C_2H_2(g) + H_2(g) \longrightarrow C_2H_4(g)$
(b) $2\,SO_3(g) \longrightarrow 2\,SO_2(g) + O_2(g)$
(c) $4\,NH_3(g) + 5\,O_2(g) \longrightarrow 4\,NO(g) + 6\,H_2O(g)$

68. Evaluate $\Delta H°$ for each reaction in Question 67. Use your results to calculate standard molar entropies at 25.00 °C for
(a) $C_2H_2(g)$ (b) $SO_3(g)$
(c) $NO(g)$

69. ■ If a system falls within the second or third category in Table 18.2 (p. 882), then there must be a temperature at which it shifts from being reactant-favored to being product-favored. For each reaction, obtain data from Appendix J and calculate what that temperature is.
(a) $CO(g) + 2\,H_2(g) \rightleftharpoons CH_3OH(\ell)$
(b) $2\,Fe_2O_3(s) + 3\,C(\text{graphite}) \rightleftharpoons 4\,Fe(s) + 3\,CO_2(g)$

70. If a system falls within the second or third category in Table 18.2 (p. 882) then there must be a temperature at which it shifts from being reactant-favored to being product-favored. For each reaction, obtain data from Appendix J and calculate what that temperature is.
(a) $2\,H_2O(g) \rightleftharpoons 2\,H_2(g) + O_2(g)$
(b) $N_2(g) + 3\,H_2(g) \rightleftharpoons 2\,NH_3(g)$

71. Estimate $\Delta G°$ at 2000. K for each reaction in Question 69.

72. Estimate $\Delta G°$ at 2000. K for each reaction in Question 70.

73. Many metal carbonates can be decomposed to the metal oxide and carbon dioxide by heating.

$$CaCO_3(s) \longrightarrow CaO(s) + CO_2(g)$$

(a) What are the enthalpy, entropy, and Gibbs free energy changes for this reaction at 25.00 °C?
(b) Is it product-favored or reactant-favored?
(c) Based on the signs of $\Delta H°$ and $\Delta S°$, predict whether the reaction is product-favored at all temperatures.
(d) Predict the lowest temperature at which appreciable quantities of products can be obtained.

74. Some metal oxides, such as lead(II) oxide, can be decomposed to the metal and oxygen simply by heating.

$$PbO(s) \longrightarrow Pb(s) + \tfrac{1}{2}\,O_2(g)$$

(a) Is the decomposition of lead(II) oxide product-favored at 25 °C? Explain.
(b) If not, can it become so if the temperature is raised?
(c) As the temperature increases, at what temperature does the reaction first become product-favored?

75. ■ Use the thermochemical expression

$$CaC_2(s) + 2\,H_2O(\ell) \longrightarrow C_2H_2(g) + Ca(OH)_2(aq) \qquad \Delta G° = -119.282\text{ kJ}$$

and data from Appendix J to calculate $\Delta G_f°$ for $Ca(OH)_2(aq)$ at 25 °C.

76. Use the thermochemical expression

$$PCl_3(g) + Cl_2(g) \longrightarrow PCl_5(g) \qquad \Delta G° = -37.2\text{ kJ}$$

and data from Appendix J to calculate $\Delta G_f°$ for $PCl_5(g)$.

Gibbs Free Energy Changes and Equilibrium Constants

77. ■ Use data from Appendix J to obtain the equilibrium constant K_p for each reaction at 298.15 K.
(a) $2\,HCl(g) \rightleftharpoons H_2(g) + Cl_2(g)$
(b) $N_2(g) + O_2(g) \rightleftharpoons 2\,NO(g)$

78. Use data from Appendix J to obtain the equilibrium constant K_p for each of these reactions at 298 K.
(a) $CH_4(g) + 2\,O_2(g) \rightleftharpoons CO_2(g) + 2\,H_2O(g)$
(b) $2\,NO_2(g) \rightleftharpoons N_2O_4(g)$

79. Ethylene reacts with hydrogen to produce ethane.

$$H_2C{=}CH_2(g) + H_2(g) \rightleftharpoons H_3C{-}CH_3(g)$$

(a) Using the data in Appendix J, calculate $\Delta G°$ for the reaction at 25 °C. Is the reaction predicted to be product-favored under standard conditions?
(b) Calculate K_p from $\Delta G°$. Comment on the connection between the sign of $\Delta G°$ and the magnitude of K_p.

80. Use the data in Appendix J to calculate $\Delta G°$ and K_p at 25 °C for the reaction

$$2\ HBr(g) + Cl_2(g) \rightleftharpoons 2\ HCl(g) + Br_2(\ell)$$

Comment on the connection between the sign of $\Delta G°$ and the magnitude of K_p.

81. For each chemical reaction, calculate the standard equilibrium constant at 298 K and at 1000. K from the thermodynamic data in Appendix J. Indicate whether each reaction is product-favored or reactant-favored at each temperature.
(a) The conversion of nitric oxide to nitrogen dioxide in the atmosphere

$$2\ NO(g) + O_2(g) \rightleftharpoons 2\ NO_2(g)$$

(b) The reaction of an alkali metal with a halogen to produce an alkali metal halide salt

$$2\ Na(s) + Cl_2(g) \rightleftharpoons 2\ NaCl(s)$$

82. For each chemical reaction, calculate the standard equilibrium constant at 298 K and at 1000. K from the thermodynamic data in Appendix J. Indicate whether each reaction is product-favored or reactant-favored at each temperature.
(a) The oxidation of carbon monoxide to carbon dioxide

$$2\ CO(g) + O_2(g) \rightleftharpoons 2\ CO_2(g)$$

(b) The first step in the production of electronic-grade silicon from sand

$$SiO_2(s) + 2\ C(s) \rightleftharpoons Si(s) + 2\ CO(g)$$

83. For each reaction, estimate $K°$ at the temperature indicated.
(a) $2\ H_2(g) + O_2(g) \rightleftharpoons 2\ H_2O(g)$ at 800. K
(b) $2\ SO_2(g) + O_2(g) \rightleftharpoons 2\ SO_3(g)$ at 500. K
(c) $2\ HF(g) \rightleftharpoons H_2(g) + F_2(g)$ at 2000. K

84. For each reaction, estimate $K°$ at the temperature indicated.
(a) $H_2(g) + I_2(g) \rightleftharpoons 2\ HI(g)$ at 500. K
(b) $N_2(g) + 3\ H_2(g) \rightleftharpoons 2\ NH_3(g)$ at 400. K
(c) $CO(g) + 3\ H_2(g) \rightleftharpoons CH_4(g) + H_2O(g)$ at 800. K

85. ■ For each reaction, an equilibrium constant at 298 K is given. Calculate $\Delta G°$ for each reaction.
(a) $Br_2(\ell) + H_2(g) \rightleftharpoons 2\ HBr(g)$ $\quad K_p = 4.4 \times 10^{18}$
(b) $H_2O(\ell) \rightleftharpoons H_2O(g)$ $\quad K_p = 3.17 \times 10^{-2}$
(c) $N_2(g) + 3\ H_2(g) \rightleftharpoons 2\ NH_3(g)$ $\quad K_c = 3.5 \times 10^8$

86. For each reaction, an equilibrium constant at 298 K is given. Calculate $\Delta G°$ for each reaction.
(a) $\frac{1}{8}\ S_8(s) + O_2(g) \rightleftharpoons SO_2(g)$ $\quad K_P = 4.2 \times 10^{52}$
(b) $2\ H_2(g) + O_2(g) \rightleftharpoons 2\ H_2O(g)$ $\quad K_c = 3.3 \times 10^{81}$
(c) $CH_4(g) + H_2O(g) \rightleftharpoons CO(g) + 3\ H_2(g)$ $\quad K_c = 9.4 \times 10^{-1}$

Gibbs Free Energy, Maximum Work, and Energy Resources

87. ■ Which of these reactions are capable of being harnessed to do useful work at 298 K and 1 bar? Which require that work be done to make them occur?
(a) $2\ C_6H_6(\ell) + 15\ O_2(g) \longrightarrow 12\ CO_2(g) + 6\ H_2O(g)$
(b) $2\ NF_3(g) \longrightarrow N_2(g) + 3\ F_2(g)$
(c) $TiO_2(s) \longrightarrow Ti(s) + O_2(g)$

88. Which of these reactions are capable of being harnessed to do useful work at 298 K and 1 bar? Which require that work be done to make them occur?
(a) $Al_2O_3(s) \longrightarrow 2\ Al(s) + \frac{3}{2}\ O_2(g)$
(b) $2\ CO(g) + O_2(g) \longrightarrow 2\ CO_2(g)$
(c) $C_2H_6(g) \longrightarrow C_2H_4(g) + H_2(g)$

89. ■ For each of the reactions in Question 87 that requires work to be done, calculate the minimum mass of graphite that would have to be oxidized to $CO_2(g)$ to provide the necessary work.

90. For each of the reactions in Question 88 that requires work to be done, calculate the minimum mass of hydrogen gas that would have to be burned to form water vapor to provide the necessary work.

91. Titanium is obtained from its ore, $TiO_2(s)$, by heating the ore in the presence of chlorine gas and coke (carbon) to produce gaseous titanium(IV) chloride and carbon monoxide.
(a) Write a balanced equation for this process.
(b) Calculate $\Delta H°$, $\Delta S°$, and $\Delta G°$ for the reaction.
(c) Is this reaction product-favored or reactant-favored at 25 °C?
(d) Does the reaction become more product-favored or more reactant-favored as the temperature increases?

92. To obtain a metal from its ore, the decomposition of the metal oxide to form the metal and oxygen is often coupled with oxidation of coke (carbon) to carbon monoxide. For each metal oxide listed, write a balanced equation for the decomposition of the oxide and for the overall reaction when the decomposition is coupled to oxidation of coke to carbon monoxide. Calculate the overall value of $\Delta G°$ for each coupled reaction at 25 °C. Which of the metals could be obtained from these ores at 25 °C by this method?
(a) $CuO(s)$ (b) $Ag_2O(s)$
(c) $HgO(s)$ (d) $MgO(s)$
(e) $PbO(s)$

93. From which of the metal oxides in Question 92 could the metal be obtained by coupling reduction of the oxide with oxidation of coke to carbon monoxide at 800 °C?

94. From which of the metal oxides in Question 92 could the metal be obtained by coupling reduction of the oxide with oxidation of coke to carbon monoxide at 1500 °C?

Gibbs Free Energy in Biological Systems

95. The molecular structure of one form of glucose, $C_6H_{12}O_6$, looks like this:

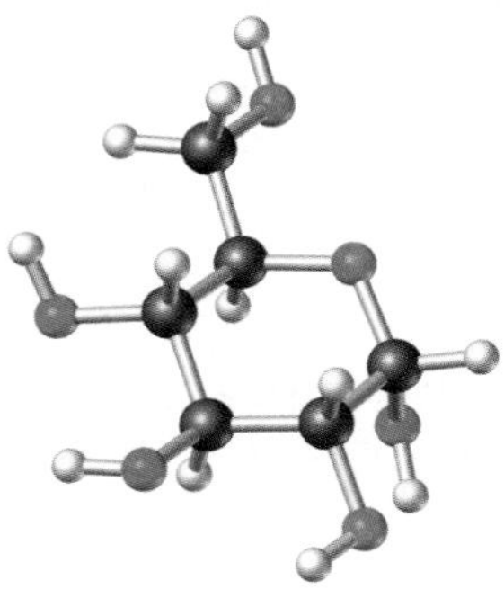

glucose

Glucose can be oxidized to carbon dioxide and water according to the equation

$$C_6H_{12}O_6(s) + 6\ O_2(g) \longrightarrow 6\ CO_2(g) + 6\ H_2O(\ell)$$

(a) Using the method described in Section 8.6 (p. 347) for estimating enthalpy changes from bond energies, estimate $\Delta H°$ for the oxidation of this form of glucose. Make a list of all bonds broken and all bonds formed in this process.
(b) Compare your result with the experimental value of −2816 kJ for combustion of a mole of glucose. Why might there be a difference between this value and the one you calculated in part (a)?

96. Another step in the metabolism of glucose, which occurs after the formation of glucose 6-phosphate, is the conversion of fructose 6-phosphate to fructose 1,6-bisphosphate ("bis" means *two*):

$$\text{fructose 6-phosphate(aq)} + H_2PO_4^-(aq) \longrightarrow \text{fructose 1,6-bisphosphate (aq)} + H_3O^+(aq)$$

(a) This reaction has a Gibbs free energy change of +16.7 kJ per mole of fructose 6-phosphate. Is it endergonic or exergonic?
(b) Write the equation for the formation of 1 mol ADP from ATP, for which $\Delta G° = -30.5$ kJ.
(c) Couple these two reactions to get an exergonic process; write its overall chemical equation, and calculate the Gibbs free energy change.

97. ■ In muscle cells under the condition of vigorous exercise, glucose is converted to lactic acid ("lactate"), $CH_3CHOHCOOH$, by the chemical reaction

$$C_6H_{12}O_6 \longrightarrow 2\ CH_3CHOHCOOH \qquad \Delta G°' = -197\text{ kJ}$$

(a) If all of the Gibbs free energy from this reaction were used to convert ADP to ATP, how many moles of ATP could be produced per mole of glucose?
(b) The actual reaction involves the production of 3 mol ATP per mole of glucose. What is the $\Delta G°$ for this reaction?
(c) Is the overall reaction in part (b) reactant-favored or product-favored?

98. The biological oxidation of ethanol, C_2H_5OH, is also a source of Gibbs free energy.
(a) Does the oxidation of 1 g ethanol give more or less energy than the oxidation of 1 g glucose? (*Hint:* Write the balanced equation for the production of carbon dioxide and water from ethanol and oxygen, and use Appendix J.)
(b) Comment on potential problems of replacing glucose with ethanol in your diet.

Conservation of Gibbs Free Energy

99. What are the resources human society uses to supply Gibbs free energy? (*Hint:* Consider information you learned in Chapter 6.)

100. For one day, keep a log of all the activities you undertake that consume Gibbs free energy. Distinguish between Gibbs free energy provided by nutrient metabolism and that provided by other energy resources.

Thermodynamic and Kinetic Stability

101. ■ Billions of pounds of acetic acid are made each year, much of it by the reaction of methanol with carbon monoxide.

$$CH_3OH(\ell) + CO(g) \longrightarrow CH_3COOH(\ell)$$

(a) By calculating the standard Gibbs free energy change, $\Delta G°$, for this reaction, show that it is product-favored.
(b) Determine the standard Gibbs free energy change, $\Delta G°$, for the reaction of acetic acid with oxygen to form gaseous carbon dioxide and liquid water.
(c) Based on this result, is acetic acid thermodynamically stable compared with $CO_2(g)$ and $H_2O(\ell)$?
(d) Is acetic acid kinetically stable compared with $CO_2(g)$ and $H_2O(\ell)$?

102. Determine the standard Gibbs free energy change, $\Delta G°$, for the reactions of liquid methanol, of CO(g), and of ethyne, $C_2H_2(g)$, with oxygen gas to form gaseous carbon dioxide and (if hydrogen is present) liquid water. Use your calculations to decide which of these substances are kinetically stable and which are thermodynamically stable: $CH_3OH(\ell)$, $CO(g)$, $C_2H_2(g)$, $CO_2(g)$, $H_2O(\ell)$.

103. There are millions of organic compounds known, and new ones are being discovered or made at a rate of more than 100,000 compounds per year. Organic compounds burn readily in air at high temperatures to form carbon dioxide and water. Several classes of organic compounds are listed, with a simple example of each. Write a balanced chemical equation for the combustion in O_2 of each of these compounds, and then use the data in Appendix J to show that each reaction is product-favored at room temperature.

Class of Organics	Simple Example
Aliphatic hydrocarbons	Methane, CH_4
Aromatic hydrocarbons	Benzene, C_6H_6
Alcohols	Methanol, CH_3OH

From these results, it is reasonable to hypothesize that *all* organic compounds are thermodynamically unstable in an oxygen atmosphere (i.e., their room-temperature reaction with $O_2(g)$ to form $CO_2(g)$ and $H_2O(\ell)$ is product-favored). If this hypothesis is true, how can organic compounds exist on earth?

104. Actually, the carbon in $CO_2(g)$ is thermodynamically unstable with respect to the carbon in calcium carbonate (limestone). Verify this by determining the standard Gibbs free energy change for the reaction of lime, CaO(s), with $CO_2(g)$ to make $CaCO_3(s)$.

General Questions

105. ■ This problem will help you understand the dependence of the U.S. economy on energy. Referring to the figure, calculate the energy (in joules) used by the agriculture, mining, and construction industries

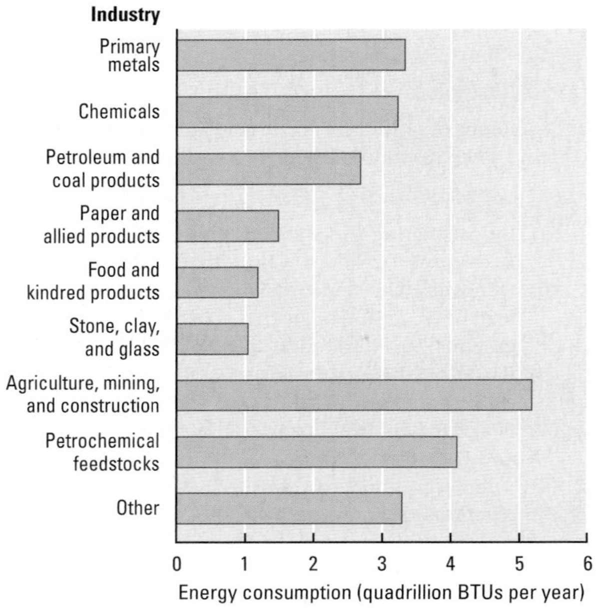

Energy consumption.

(a) in one year.
(b) in one day.
(c) in one second.
(d) Remembering that 1 watt is the expenditure of 1 joule every second, calculate the average power needs of these industries in watts.
(e) Assuming a U.S. population of 300 million people, calculate the power needed by the agriculture, mining, and construction industries *per person in the United States.*

106. Suppose you signed a contract to provide to the agriculture, mining, and construction industries the energy they use each year (see Question 105) by eating glucose and giving them the resulting energy from its oxidation in your body.
(a) How much glucose would you have to eat each day to meet your contract? Assume that it is someone else's job to figure out how to get the energy stored in your ATP to the industries!
(b) An Olympic sprinter uses energy at the rate of 700 to 900 watts in a sprint. Compare this figure with the one you calculated in part (a), and draw conclusions about the feasibility of fulfilling your contract.

Reaction	Chemical Equation	K_c	$\Delta H°$ (kJ)
1	$CH_3OH(g) + H_2(g) \rightleftharpoons CH_4(g) + H_2O(g)$	3.6×10^{20}	−115.4
2	$Mg(OH)_2(s) \rightleftharpoons MgO(s) + H_2O(g)$	1.24×10^{-5}	81.1
3	$2\ CH_4(g) \rightleftharpoons C_2H_6(g) + H_2(g)$	9.5×10^{-13}	64.9
4	$2\ H_2(g) + CO(g) \rightleftharpoons CH_3OH(g)$	3.76	−90.7
5	$H_2(g) + Br_2(g) \rightleftharpoons 2\ HBr(g)$	1.9×10^{24}	−103.7

107. The table above provides data at 25 °C for five reactions. For which (if any) of the reactions 1 through 5 is
(a) K_p greater than K_c?
(b) the reaction product-favored?
(c) there only a single concentration in the K_c expression?
(d) there an increase in the concentrations of products when the temperature increases?
(e) there a change in the sign of $\Delta G°$ if water is liquid instead of gas?

108. The table above provides data at 25 °C for five reactions. For which (if any) of the reactions 1 through 5 is
(a) K_p less than K_c?
(b) there a decrease in the concentrations of products when the pressure increases?
(c) the value of $\Delta S°$ positive?
(d) the sign of $\Delta G°$ dependent on temperature?

109. ■ Consider the gas phase decomposition of sulfur trioxide to sulfur dioxide and oxygen.
(a) Calculate $\Delta G°$ for the reaction at 25 °C.
(b) Is the reaction product-favored under standard conditions at 25 °C?
(c) If the reaction is not product-favored at 25 °C, is there a temperature at which it will become so?
(d) Estimate K_p for the reaction at 1500. °C.
(e) Estimate K_c for the reaction at 1500. °C.

110. The Haber process for the synthesis of ammonia involves the reaction

$$N_2(g) + 3\ H_2(g) \rightleftharpoons 2\ NH_3(g)$$

Using data from Appendix J, estimate the amount (in moles) of $NH_3(g)$ that would be produced from 1 mol $N_2(g)$ and 3 mol $H_2(g)$ once equilibrium is reached at 450 °C and a total pressure of 1000. atm.

111. Mercury is a poison, and its vapor is readily absorbed through the lungs. Therefore it is important that the partial pressure of mercury be kept as low as possible in any area where people could be exposed to it (such as a dentist's office). The relevant equilibrium reaction is

$$Hg(\ell) \rightleftharpoons Hg(g)$$

For Hg(g), $\Delta H_f^\circ = 61.4$ kJ/mol, $S^\circ = 175.0 \text{ J K}^{-1} \text{ mol}^{-1}$, and $\Delta G_f^\circ = 31.8$ kJ/mol. Use data from Appendix J and these values to evaluate the vapor pressure of mercury at different temperatures. (Remember that concentrations of pure liquids and solids do not appear in the equilibrium constant expression, and for gases K° involves pressures in bars.)
(a) Calculate ΔG° for vaporization of mercury at 25 °C.
(b) Write the equilibrium constant expression for vaporization of mercury.
(c) Calculate K° for this reaction at 25 °C.
(d) What is the vapor pressure of mercury at 25 °C?
(e) Estimate the temperature at which the vapor pressure of mercury reaches 10 mm Hg.

Applying Concepts

112. A friend says that the boiling point of water is twice that of cyclopentane, which boils at 50 °C. Write a brief statement about the validity of this observation.
113. Using the second law of thermodynamics, explain why it is very difficult to unscramble an egg. Who was Humpty-Dumpty? Why did his moment of glory illustrate the second law of thermodynamics?
114. Appendix J lists standard molar entropies S°, not standard entropies of formation ΔS_f°. Why is this possible for entropy but not for internal energy, enthalpy, or Gibbs free energy?
115. When calculating ΔS° from S° values, it is necessary to look up all substances, including elements in their standard state, such as O_2(g), H_2(g), and N_2(g). When calculating ΔH° from ΔH_f° values, however, elements in their standard state can be ignored. Why is the situation different for S° values?
116. In the *Chemistry You Can Do* experiment in Chapter 6 (**p. 239**) you considered the heat generated when iron rusts to form iron oxide. Look at the enthalpies of formation of other metal oxides in Table 6.3 or Appendix J and comment on your observations. Are oxidations of metals generally endothermic or exothermic? Are they usually reactant-favored or product-favored?
117. Explain why the entropy of the system increases when solid NaCl dissolves in water.
118. Explain how the entropy of the universe increases when an aluminum metal can is made from aluminum ore. The first step is to extract the ore, which is primarily a form of Al_2O_3, from the ground. After it is purified by freeing it from oxides of silicon and iron, aluminum oxide is changed to the metal by an input of electrical energy.

$$2\,Al_2O_3(s) \xrightarrow{\text{electrical energy}} 4\,Al(s) + 3\,O_2(g)$$

119. Suppose that at a certain temperature T a chemical reaction is found to have a standard equilibrium constant K° of 1.0. Indicate whether each statement is true or false and explain why.
(a) The enthalpy change for the reaction, ΔH°, is zero.
(b) The entropy change for the reaction, ΔS°, is zero.
(c) The Gibbs free energy change for the reaction, ΔG°, is zero.
(d) ΔH° and ΔS° have the same sign.
(e) $\Delta H^\circ/T = \Delta S^\circ$ at the temperature T.
120. When you eat a candy bar, how does your body store the Gibbs free energy that is released during oxidation of the sugars (glucose and other carbohydrates) in the candy bar? What was the original source of the Gibbs free energy needed to synthesize the sugars before they went into the candy bar?
121. Explain how biological systems make use of coupled reactions to maintain the high degree of order found in all living organisms.
122. How can kinetically stable substances exist at all, if they are not thermodynamically stable?
123. Criticize this statement: Provided it occurs at an appreciable rate, any chemical reaction for which $\Delta G < 0$ will proceed until all reactants have been converted to products.
124. Reword the statement in Question 123 so that it is always true.

More Challenging Questions

125. Calculate the entropy change for formation of exactly 1 mol of each of these gaseous hydrocarbons under standard conditions from carbon (graphite) and hydrogen. What trend do you see in these values? Does ΔS° increase or decrease on adding H atoms?
(a) acetylene, C_2H_2(g)
(b) ethylene, C_2H_4(g)
(c) ethane, C_2H_6(g)
126. Calcium hydroxide, $Ca(OH)_2$(s), can be dehydrated to form lime, CaO, by heating. Without doing any calculations, and basing your prediction on the enthalpy change and the entropy change, is this reaction product-favored at 25 °C? Explain your answer briefly.
127. ■ Octane is the product of adding hydrogen to 1-octene.

$$\underset{\text{1-octene}}{C_8H_{16}(g)} + H_2(g) \longrightarrow \underset{\text{octane}}{C_8H_{18}(g)}$$

The enthalpies of formation are

$$\Delta H_f^\circ[C_8H_{18}(g)] = -82.93 \text{ kJ/mol}$$
$$\Delta H_f^\circ[C_8H_{18}(g)] = -208.45 \text{ kJ/mol}$$

Predict whether this reaction is product-favored or reactant-favored at 25 °C and explain your reasoning.
128. This is a group project: Estimate or look up, to the nearest order of magnitude,
(a) the number of kilograms of CH_3OH made each year
(b) the number of kilograms of CO in the entire atmosphere
(c) the number of kilograms of CH_3COOH made each year
(d) the number of kilograms of H_2O on earth
(e) the number of kilograms of CO_2 in the atmosphere
What do these facts tell you about the difference between kinetic stability and thermodynamic stability?
129. From data in Appendix J, estimate
(a) the boiling point of bromine.
(b) the boiling point of tin(IV) chloride.
130. From data in Appendix J, estimate
(a) the boiling point of titanium(IV) chloride.
(b) the boiling point of carbon disulfide, CS_2, which is a liquid at 25 °C and 1 bar.

131. Nitric oxide and chlorine combine at 25 °C to produce nitrosyl chloride, NOCl.

$$2\ NO(g) + Cl_2(g) \longrightarrow 2\ NOCl(g)$$

(a) Calculate the equilibrium constant K_p for the reaction.
(b) Is the reaction product-favored or reactant-favored?
(c) Calculate the equilibrium constant K_c for the reaction.

132. Hydrogen for use in the Haber-Bosch process for ammonia synthesis is generated from natural gas by the reaction

$$CH_4(g) + H_2O(g) \rightleftharpoons CO(g) + 3\ H_2(g)$$

(a) Calculate $\Delta G°$ for this reaction at 25 °C.
(b) Calculate K_p for the reaction at 25 °C.
(c) Is the reaction product-favored under standard conditions? If not, at what temperature will it become so?
(d) Estimate K_c for the reaction at 1000. K.

133. It would be very useful if we could use the inexpensive carbon in coal to make more complex organic molecules such as gaseous or liquid fuels. The formation of methane from coal and water is reactant-favored and thus cannot occur unless there is some energy transfer from outside. This problem examines the feasibility of other reactions using coal and water.
(a) Write three balanced equations for the reactions of coal (carbon) and steam to make ethane gas, $C_2H_6(g)$, propane gas, $C_3H_8(g)$, and liquid methanol, $CH_3OH(\ell)$, with carbon dioxide as a by-product.
(b) Using the data in Appendix J, calculate $\Delta H°$, $\Delta S°$, and $\Delta G°$ for each reaction, and then comment on whether any of them would be a feasible way to make the stated products.

134. You are exploring the marketing possibilities of a scheme by which every family in the United States produces enough water for its own needs by the combustion of hydrogen and oxygen. Would the release of Gibbs free energy from the combination of hydrogen and oxygen be sufficient to supply the family's energy needs? Do not try to collect the actual data you would use, but define the problem well enough so that someone else could collect the necessary data and do the calculations that would be needed.

135. Quite often a graph of ln $K°$ versus $1/T$ is a straight line. Use Equation 18.8 (p. 889) to show how $\Delta H°$ and $\Delta S°$ can be determined from such a graph. Does the fact that such a graph is straight tell you anything about the dependence of $\Delta H°$ and $\Delta S°$ on temperature?

136. Assuming that $\Delta H°$ and $\Delta S°$ do not vary with temperature, use Equation 18.8 (p. 889) to derive a formula relating $K_1°$ at temperature T_1 to $K_2°$ at temperature T_2.

137. ■ Without consulting tables of $\Delta H_f°$, $S°$, or $\Delta G_f°$ values, predict which of these reactions is
(i) always product-favored.
(ii) product-favored at low temperatures, but not product-favored at high temperatures.
(iii) not product-favored at low temperatures, but product-favored at high temperatures.
(iv) never product-favored.

(a) $2\ NO_2(g) \longrightarrow N_2O_4(g)$
(b) $C_5H_{12}(g) + 8\ O_2(g) \longrightarrow 5\ CO_2(g) + 6\ H_2O(g)$
(c) $P_4(g) + 10\ F_2(g) \longrightarrow 4\ PF_5(g)$

[*Hint:* Use the qualitative rules regarding bond enthalpies in Section 6.7 (**p. 241**) to predict the sign of $\Delta H°$.]

138. Using the reactions

$$2\ H_2(g) + O_2(g) \longrightarrow 2\ H_2O(\ell)$$
$$2\ H_2(g) + O_2(g) \longrightarrow 2\ H_2O(g)$$

as an example, explain why it may be incorrect to assume for reactions involving solids or liquids that $\Delta S°$ and $\Delta H°$ do not change appreciably with increasing temperature.

Conceptual Challenge Problems

CP18.A (Section 18.2) Suppose that you are invited to play a game as either the "player" or the "house." A pair of dice is used to determine the winner. Each die is a cube having a different number, one through six, showing on each face. The player rolls two dice and sums the numbers showing on the top side of each die to determine the number rolled. Obviously, the number rolled has a minimum value of 2 (both dice showing a 1) and a maximum of 12 (both dice showing a 6). The player begins the game with his or her initial roll of the dice. If the player rolls a 7 or an 11, he or she wins on the first roll and the house loses. If the player does not roll a 7 or an 11 on the initial roll, then whatever number was rolled is called the point, and the player must roll again. For the player to win, he or she must roll the point again before either a 7 or an 11 is rolled. Should the player roll a 7 or an 11 before rolling the point a second time, the house wins. Which would you choose to be, player or house? Explain clearly in terms of the probabilities of rolling the dice why you chose the role you did.

CP18.B (Section 18.2) Suppose a button is placed in the middle of a football field and a penny is flipped to decide which direction to move the button, up or down the field. Each time the penny comes up heads, the button is moved 10 cm toward your opponent's goal line; and each time it comes up tails, the button is moved 10 cm toward your goal line. Your friend concludes that after many flips of the penny the button is likely to remain within 10 cm of the middle of the field, because numerous flips of the penny will produce heads just as often as tails. You doubt this because you know that perfume molecules and particles diffuse away from their original source, even though, like the button, they are just as likely to be hit from one direction as from any other by the moving molecules around them. How would you explain the error of your friend's conclusion about the movement of the button?

CP18.C (Section 18.3) When thermal energy is transferred to a substance at its standard melting point or boiling point, the substance melts or vaporizes, but its temperature does not change while it is doing so. It is clear then that temperature cannot be a measure of "how much energy is in a sample of matter" or the "intensity of energy in a sample of matter." In "Qualitative Guidelines for Entropy" (p. 874) we noted that atoms and molecules are not stationary, but rather are in constant motion. When heated, their motion increases. If this is true, what can you infer that temperature measures about a sample of matter?

CP18.D (Section 18.10) Suppose that you are a member of an environmental group and have been assigned to evaluate various ways of delivering milk to consumers with respect to Gibbs free energy conservation. Think of all the ways that milk could be delivered, the kinds of containers that could be used, and the ways they could be transported. Consider whether the containers could be reused (refilled) or recycled. Define the problem in terms of the kinds of information you would need to collect, how you would analyze the information, and which criteria you would use to decide which systems are more efficient in use of Gibbs free energy. Do not try to collect the actual data you would use, but define the problem well enough so that someone could collect the necessary data based on your statement of the problem.

CP18.E (Section 18.11) Consider planet earth as a thermodynamic system. Is earth thermodynamically or kinetically stable? Discuss your choice, providing as many arguments as you can to support it.

19 Electrochemistry and Its Applications

Personal electronic devices, such as this iPod, depend on advanced types of batteries to supply their electrical power needs (lithium-ion batteries power iPods). Dependable, long-lasting batteries are important for many of the modern conveniences we use every day such as cellular telephones, calculators, flashlights, computers, CD and MP3 players, and cordless tools. In this chapter we discuss electrochemical reactions and their extraordinary range of applications.

Many of our modern devices are powered by batteries—portable storage devices for electrochemical energy that is produced by product-favored redox reactions. Making these reactions do useful work is the goal of significant chemical research. What chemistry goes on inside a battery? How does it produce electricity? Is this chemistry similar to or different from the other kinds of reactions you have studied?

Oxidation-reduction (redox) reactions are an important class of chemical reactions ***(⇐ p. 179).*** Redox reactions involve the transfer of electrons from one atom, molecule, or ion to another. **Electrochemistry** is the study of the relationship between electron flow and redox reactions, that is, the relationship between electricity and chemical changes. Applications of electrochemistry are numerous and important. In electrochemical cells (commonly called batteries), electrons from a product-favored redox reaction are released and transferred as an electrical current through an external circuit. We rely on batteries to power many useful devices, including CD players, cellular telephones, calculators, flashlights, laptop computers, heart pacemakers, and golf carts.

The voltage produced by an electrochemical reaction depends on the oxidizing agents and reducing agents used as reactants. A knowledge of the strengths of oxidizing and reducing agents helps in the design of better batteries. Product-favored electrochemical reactions are not always beneficial, however. Corrosion of iron, for example, is a product-favored redox reaction. Damage to materials as a result of corrosion is very costly, so preventing corrosion is an important goal.

By contrast, electroplating and electrolysis are applications of reactant-favored redox reactions. In an electrolysis cell, an external energy source creates an electrical current that forces a reactant-favored process to occur. Electrolysis is important in the manufacture of many products, such as aluminum metal and the chlorine used to disinfect water supplies.

ThomsonNOW™

Throughout the text, this icon indicates an opportunity to test yourself on key concepts and to explore interactive modules by signing in to ThomsonNOW at **www.thomsonedu.com**.

19.1 Redox Reactions

Redox reactions form a large class of chemical reactions in which the reactants can be atoms, ions, or molecules. How do you know when a reaction involves oxidation-reduction?

- ***By identifying the presence of strong oxidizing or reducing agents as reactants (⇐ p. 182; Table 5.3).***
- ***By recognizing a change in oxidation number (⇐ p. 184).*** This means you have to determine the oxidation number of each element as it appears in a reactant or a product.
- ***By recognizing the presence of an uncombined element as a reactant or product.*** Producing a free element or incorporating one into a compound almost always results in a change in oxidation number.

An uncombined element is always assigned an oxidation number of 0.

To briefly review the definitions of oxidation and reduction, consider the displacement reaction between magnesium, a relatively reactive metal ***(⇐ p. 188; Table 5.5),*** and hydrochloric acid. The oxidation numbers of the species are shown above their symbols.

You may want to review the definitions of oxidation and reduction in Section 5.3 and the rules for assigning oxidation numbers in Section 5.4.

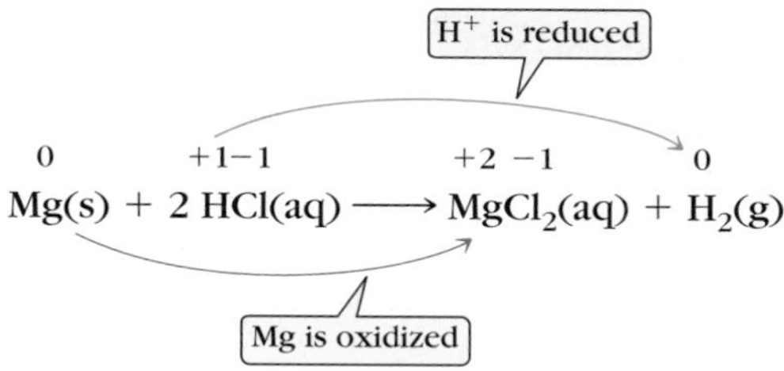

Oxidation: loss of electron(s) and increase in oxidation number

Reduction: gain of electron(s) and decrease in oxidation number

The presence of the uncombined elements Mg and H_2 indicates a redox reaction, as do the changes in oxidation number. Mg(s) is oxidized, indicated by an *increase* in its oxidation number (from 0 to +2). Hydrogen ions from HCl are reduced to H_2, as shown by a *decrease* in oxidation number (from +1 to 0). Magnesium metal is the reducing agent because it causes the H^+ ions in HCl to be reduced (gain electrons). In the process of giving electrons to HCl, metallic Mg is oxidized to Mg^{2+} ions. Hydrochloric acid is the oxidizing agent because it causes magnesium metal to be oxidized (lose electrons). Note that in this and all redox reactions, ***the oxidizing agent is reduced, and the reducing agent is oxidized.*** Oxidation and reduction *always* occur together, with one reactant acting as the oxidizing agent and another acting as the reducing agent. Oxidizing and reducing agents are *always* reactants, never products.

Redox reactions such as the one between Mg and HCl involve complete gain and loss of electrons by the reacting species—the type of reaction that is utilized in electrochemistry. A flow of electrons through the reaction system to make a complete electrical circuit is necessary for the reaction to proceed.

© Thomson Learning/Charles D. Winters

The thermite reaction in progress.

PROBLEM-SOLVING EXAMPLE 19.1 Identifying Oxidizing and Reducing Agents in Redox Reactions

In the thermite reaction, iron(III) oxide and aluminum metal react to give iron metal and aluminum oxide:

$$Fe_2O_3(s) + 2\,Al(s) \longrightarrow 2\,Fe(s) + Al_2O_3(s)$$

Is this a redox reaction? Identify the oxidation numbers of all the atoms that change oxidation number. What gets reduced? What gets oxidized? What is the oxidizing agent? What is the reducing agent?

Answer Yes, this is a redox reaction. Oxidation number changes: Fe, +3 to 0; Al, 0 to +3. The iron in the iron(III) oxide is reduced to metallic iron. The aluminum metal is oxidized to Al^{3+} ions. The oxidizing agent is the iron(III) oxide and the reducing agent is the aluminum metal.

Strategy and Explanation

First, determine the oxidation number of each element on the reactant side of the equation. Oxygen in compounds is normally −2 (***p. 182***). Since the sum of the oxidation numbers of all the atoms in a formula must equal the charge on the formula, each iron in Fe_2O_3 is +3 and each oxygen is −2. The oxidation number for metallic Al is 0, as it is for all uncombined elements.

On the product side, the Al in Al_2O_3 has an oxidation number of +3. Iron is now in its uncombined form, so its oxidation number is now 0.

Thus, the elements that change oxidation number are

$$\overset{+3}{Fe}_2O_3(s) + 2\,\overset{0}{Al}(s) \longrightarrow 2\,\overset{0}{Fe}(s) + \overset{+3}{Al}_2O_3(s)$$

The oxidation state of the iron atoms decreased, while the oxidation state of the aluminum atoms increased. Thus, each iron in Fe_2O_3 has been reduced (+3 to 0) and aluminum has been oxidized (0 to +3). Consequently, Fe_2O_3 is the oxidizing agent and Al metal is the reducing agent. Oxygen is neither reduced nor oxidized in this reaction; its oxidation number remains −2.

PROBLEM-SOLVING PRACTICE 19.1

Give the oxidation number for each atom and identify the oxidizing and reducing agents in these balanced chemical equations.

(a) $2\,Fe(s) + 3\,Cl_2(g) \longrightarrow 2\,FeCl_3(s)$
(b) $2\,H_2(g) + O_2(g) \longrightarrow 2\,H_2O(\ell)$
(c) $Cu(s) + 2\,NO_3^-(aq) + 4\,H_3O^+(aq) \longrightarrow Cu^{2+}(aq) + 2\,NO_2(g) + 6\,H_2O(\ell)$
(d) $C(s) + O_2(g) \longrightarrow CO_2(g)$
(e) $6\,Fe^{2+}(aq) + Cr_2O_7^{2-}(aq) + 14\,H_3O^+(aq) \longrightarrow 6\,Fe^{3+}(aq) + 2\,Cr^{3+}(aq) + 21\,H_2O(\ell)$

19.2 Using Half-Reactions to Understand Redox Reactions

Consider the redox reaction between zinc metal and copper(II) ions shown in Figure 19.1. The net ionic equation is

$$Zn(s) + Cu^{2+}(aq) \longrightarrow Zn^{2+}(aq) + Cu(s)$$

Zinc metal is oxidized to Zn^{2+} ions, and Cu^{2+} ions are reduced to copper metal.

To see more clearly how electrons are transferred, this overall reaction can be thought of as the result of two simultaneous **half-reactions:** one half-reaction for the oxidation of Zn and one half-reaction for the reduction of Cu^{2+} ions. The oxidation half-reaction

$$Zn(s) \longrightarrow Zn^{2+}(aq) + 2\,e^-$$

shows that each zinc atom loses two electrons when it is oxidized to a Zn^{2+} ion. These two electrons are accepted by a Cu^{2+} ion in the reduction half-reaction,

$$Cu^{2+}(aq) + 2\,e^- \longrightarrow Cu(s)$$

Note how the sum of the charges on the left side of the reaction equals the sum of the charges on the right side, even for half-reactions.

As Cu^{2+} ions are converted to Cu(s) in this half-reaction, the blue color of the solution becomes less intense and metallic copper forms on the zinc surface.

The net reaction is the sum of the oxidation and reduction half-reactions.

$$\begin{array}{rll} Zn(s) & \longrightarrow Zn^{2+}(aq) + 2\,e^- & \text{(oxidation half-reaction)} \\ Cu^{2+}(aq) + 2\,e^- & \longrightarrow Cu(s) & \text{(reduction half-reaction)} \\ \hline Zn(s) + Cu^{2+}(aq) & \longrightarrow Zn^{2+}(aq) + Cu(s) & \text{(net reaction)} \end{array}$$

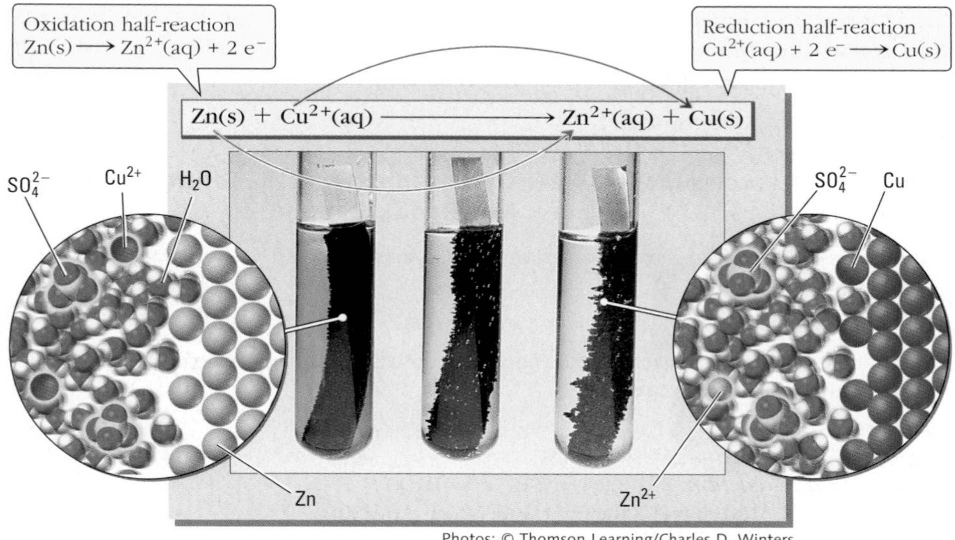

Photos: © Thomson Learning/Charles D. Winters

Active Figure 19.1 **An oxidation-reduction reaction.** A strip of zinc is placed in a solution of copper(II) sulfate (*left*). The zinc reacts with the copper(II) ions to produce copper metal (the brown-colored deposit on the zinc strip) and zinc ions in solution.

$$Zn(s) + Cu^{2+}(aq) \longrightarrow Zn^{2+}(aq) + Cu(s)$$

As copper metal accumulates on the zinc strip, the blue color due to the aqueous copper ions gradually fades (*middle and right*) as Cu^{2+} ions are reduced to metallic copper. The zinc ions in aqueous solution are colorless. **Go to the Active Figures menu at ThomsonNOW to test your understanding of the concepts in this figure.**

Figure 19.2 Copper metal screen in a solution of $AgNO_3$. The blue color intensifies as more copper is oxidized to aqueous Cu^{2+} ion. Ag^+ ions are reduced to silver metal.

ThomsonNOW
Go to the Chemistry Interactive menu for a module on **silver coating copper in a solution of $AgNO_3$.**

Notice that no electrons appear in the equation for the net reaction because the number of electrons produced by the oxidation half-reaction must equal the number of electrons gained by the reduction half-reaction. This must always be true in a net reaction. Otherwise, electrons would be created or destroyed, violating the laws of conservation of mass and conservation of electrical charge.

Consider another example, as shown in Figure 19.2. A piece of metallic copper screen is immersed in a solution of silver nitrate. As the reaction proceeds, the solution gradually turns blue, and fine, silvery, hair-like crystals form on the copper screen. Knowing that Cu^{2+} ions in aqueous solution appear blue, we can conclude that the copper metal is being oxidized to Cu^{2+}. Reduction must also be taking place, so it is reasonable to conclude that the hair-like crystals of silver result from the reduction of Ag^+ ions to metallic silver. The two half-reactions are

$$\mathrm{Cu(s)} \longrightarrow \mathrm{Cu^{2+}(aq)} + 2\,\mathrm{e^-} \qquad \text{(oxidation half-reaction)}$$

$$\mathrm{Ag^+(aq)} + \mathrm{e^-} \longrightarrow \mathrm{Ag(s)} \qquad \text{(reduction half-reaction)}$$

In this case, two electrons are produced in the oxidation half-reaction, but only one is needed for the reduction half-reaction. *One* atom of copper provides enough electrons (two) to reduce *two* Ag^+ ions, so the reduction half-reaction must occur twice every time the oxidation half-reaction occurs once. To indicate this relationship, we multiply the reduction half-reaction by 2.

$$2\,\mathrm{Ag^+(aq)} + 2\,\mathrm{e^-} \longrightarrow 2\,\mathrm{Ag(s)} \qquad \text{(reduction half-reaction)} \times 2$$

Adding this reduction half-reaction to the oxidation half-reaction gives the net equation

$$\begin{array}{rcl} 2\,\mathrm{Ag^+(aq)} + 2\,\mathrm{e^-} & \longrightarrow & 2\,\mathrm{Ag(s)} \\ \mathrm{Cu(s)} & \longrightarrow & \mathrm{Cu^{2+}(aq)} + 2\,\mathrm{e^-} \\ \hline \mathrm{Cu(s)} + 2\,\mathrm{Ag^+(aq)} & \longrightarrow & \mathrm{Cu^{2+}(aq)} + 2\,\mathrm{Ag(s)} \end{array}$$

The method shown here is a general one. A net equation can always be generated by writing oxidation and reduction half-reactions, using coefficients to adjust the half-reaction equations so that the number of electrons released by the oxidation equals the number gained by the reduction, and then adding the two half-reactions to give the equation for the net reaction.

PROBLEM-SOLVING EXAMPLE 19.2 Determining Half-Reactions from Net Redox Reactions

Aluminum will undergo a redox reaction with an acid such as HCl to produce $Al^{3+}(aq)$ and $H_2(g)$.

(unbalanced equation) $\mathrm{Al(s)} + \mathrm{H^+(aq)} \longrightarrow \mathrm{Al^{3+}(aq)} + \mathrm{H_2(g)}$

Write the oxidation half-reaction and the reduction half-reaction equations, and combine them to give the net redox reaction.

Answer

Oxidation half-reaction: $\mathrm{Al(s)} \longrightarrow \mathrm{Al^{3+}(aq)} + 3\,\mathrm{e^-}$
Reduction half-reaction: $2\,\mathrm{H^+(aq)} + 2\,\mathrm{e^-} \longrightarrow \mathrm{H_2(g)}$
Net reaction: $2\,\mathrm{Al(s)} + 6\,\mathrm{H^+(aq)} \longrightarrow 2\,\mathrm{Al^{3+}(aq)} + 3\,\mathrm{H_2(g)}$

Strategy and Explanation To identify the half-reactions, we must first identify the species whose oxidation number increases (it is oxidized) and the species whose oxidation number decreases (it is reduced). Aluminum is oxidized, and its oxidation number increases from 0 to +3.

$$\mathrm{Al(s)} \longrightarrow \mathrm{Al^{3+}(aq)} + 3\,\mathrm{e^-}$$

This half-reaction must have three electrons on the right side to balance the 3+ charge of the aluminum ion and yield equal charges on the right side and the left side of the half-reaction.

Hydrogen is reduced, and its oxidation number decreases from +1 to 0.

$$2\ H^+(aq) + 2\ e^- \longrightarrow H_2(g)$$

The half-reaction must have two electrons on the left side to balance the charge of the two hydrogen ions.

Notice that these two half-reactions contain different numbers of electrons. The two half-reactions are multiplied by 2 and 3, respectively, so that six e^- appear in each half-reaction.

$$2\ [Al(s) \longrightarrow Al^{3+}(aq) + 3\ e^-] \text{ gives } 2\ Al(s) \longrightarrow 2\ Al^{3+}(aq) + 6\ e^-$$

$$3\ [2\ H^+(aq) + 2\ e^- \longrightarrow H_2(g)] \text{ gives } 6\ H^+(aq) + 6\ e^- \longrightarrow 3\ H_2(g)$$

$$\text{Net reaction: } 2\ Al(s) + 6\ H^+(aq) \longrightarrow 2\ Al^{3+}(aq) + 3\ H_2(g)$$

The net reaction is the sum of these two half-reactions multiplied by the proper whole numbers to make the number of electrons produced by the oxidation half-reaction equal to the number of electrons gained in the reduction half-reaction.

PROBLEM-SOLVING PRACTICE 19.2

Write oxidation and reduction half-reactions for these net redox equations. Show that their sum is the net reaction.

(a) $Cd(s) + Cu^{2+}(aq) \longrightarrow Cu(s) + Cd^{2+}(aq)$

(b) $Zn(s) + 2\ H^+(aq) \longrightarrow Zn^{2+}(aq) + H_2(g)$

(c) $2\ Al(s) + 3\ Zn^{2+}(aq) \longrightarrow 2\ Al^{3+}(aq) + 3\ Zn(s)$

Balancing Redox Equations Using Half-Reactions

All of the equations in Problem-Solving Practice 19.2 are balanced. While these particular redox equations could be easily balanced by inspection, this is not always the case. Equations for redox reactions often involve water, hydronium ions, or hydroxide ions as reactants or products. It is difficult to tell by observing the unbalanced equation how many H_2O, H_3O^+, and OH^- are involved, or whether they will be reactants, or products, or even present at all. Fortunately, there are systematic ways to figure this out as shown as follows.

ThomsonNOW™
Go to the Chemistry Interactive menu for a module on the **redox reaction between MnO_4^- and Fe^{2+}.**

Balancing Redox Equations in Acidic Solutions

Consider the reaction of permanganate ion with oxalic acid in an acidic solution. The products are manganese(II) ions and carbon dioxide, so the *unbalanced* equation is

Oxalic acid, HOOC—COOH, is the simplest organic acid containing two carboxylic acid groups.

$$\text{(unbalanced equation)} \quad \underset{\text{oxalic acid}}{H_2C_2O_4(aq)} + \underset{\text{permanganate ion}}{MnO_4^-(aq)} \longrightarrow Mn^{2+}(aq) + CO_2(g)$$

If you try to balance this equation by trial and error, you will almost certainly have a hard time balancing hydrogen and oxygen. You have probably already noticed that no hydrogen-containing species appears on the product side of the unbalanced equation. Because the reaction takes place in an aqueous acidic solution, water and hydronium ions can be involved. Generating the balanced equation for a reaction like this one is best done by following a systematic approach, a series of steps. In each step you must use what you know about the oxidation and reduction half-reactions, as well as conservation of matter and conservation of electrical charge. Problem-Solving Example 19.3 illustrates the steps that produce a balanced equation for a redox reaction occurring in acidic solution.

In acidic solution, balance O by adding H_2O, and balance H by adding H^+.

PROBLEM-SOLVING EXAMPLE 19.3 Balancing Redox Equations for Reactions in Acidic Solutions

Balance the equation for the oxidation of oxalic acid in an acidic permanganate solution. The products of this reaction are CO_2 and Mn^{2+} ions.

Answer $5\ H_2C_2O_4(aq) + 6\ H_3O^+(aq) + 2\ MnO_4^-(aq) \longrightarrow 10\ CO_2(g) + 2\ Mn^{2+}(aq) + 14\ H_2O(\ell)$

Strategy and Explanation It is best to follow a systematic approach listed in the following series of steps to balance the equation for this reaction (and all similar redox reactions).

Step 1: *Recognize whether the reaction is an oxidation-reduction process. If it is, then determine what is reduced and what is oxidized.* This is a redox reaction because the oxidation number of Mn changes from +7 in MnO_4^- to +2 in Mn^{2+}, so the Mn in MnO_4^- is reduced. The oxidation number of each C changes from +3 in $H_2C_2O_4$ to +4 in CO_2, so the C in $H_2C_2O_4$ is oxidized. The oxidation numbers of H (+1) and O (−2) are unchanged.

Step 2: *Break the overall unbalanced equation into half-reactions.*

$$H_2C_2O_4(aq) \longrightarrow CO_2(g) \qquad \text{(oxidation half-reaction)}$$

$$MnO_4^-(aq) \longrightarrow Mn^{2+}(aq) \qquad \text{(reduction half-reaction)}$$

Step 3: *Balance the atoms in each half-reaction.* First balance all atoms except for O and H, then balance O by adding H_2O and balance H by adding H^+. (Hydroxide ion, OH^-, cannot be used here because the reaction occurs in an acidic solution and the OH^- concentration is very low.)

Oxalic acid half-reaction: First, balance the carbon atoms in the half-reaction.

$$H_2C_2O_4(aq) \longrightarrow 2\ CO_2(g)$$

This step balances the O atoms as well (no H_2O needed here), so only H atoms remain to be balanced. Because the product side is deficient by two H, we put $2\ H^+$ there.

$$H_2C_2O_4(aq) \longrightarrow 2\ CO_2(g) + 2\ H^+(aq) \qquad \text{(oxalic acid half-reaction)}$$

Strictly speaking, we ought to use H_3O^+ instead of H^+, but this would result in adding water molecules to each side of the equation, which is rather cumbersome. It is simpler to add H^+ now and add the water molecules at the end.

Permanganate half-reaction: The Mn atoms are already balanced, but the oxygen atoms are not balanced until H_2O is added. Adding $4\ H_2O$ on the product side takes care of the needed oxygen atoms.

$$MnO_4^-(aq) \longrightarrow Mn^{2+}(aq) + 4\ H_2O(\ell)$$

Now there are 8 H atoms on the right and none on the left. To balance hydrogen atoms, $8\ H^+$ are placed on the left side of the half-reaction.

$$8\ H^+(aq) + MnO_4^-(aq) \longrightarrow Mn^{2+}(aq) + 4\ H_2O(\ell) \qquad \text{(permanganate half-reaction)}$$

Step 4: *Balance the half-reactions for charge using electrons (e^-).* The oxalic acid half-reaction has a net charge of 0 on the left side and 2+ on the right. The reactants have lost two electrons. To show this fact, $2\ e^-$ must appear on the right side.

$$H_2C_2O_4(aq) \longrightarrow 2\ CO_2(g) + 2\ H^+(aq) + 2\ e^-$$

This confirms that $H_2C_2O_4$ is the reducing agent (it loses electrons and is oxidized). The loss of two electrons is also in keeping with the increase in the oxidation number of each of two C atoms by 1, from +3 to +4. The $2\ e^-$ also balance the charge on the product side of the equation.

The MnO_4^- half-reaction has a charge of 7+ on the left and 2+ on the right. Therefore, to achieve a net 2+ charge on each side, $5\ e^-$ must appear on the left. The gain of electrons shows that MnO_4^- is the oxidizing agent; it is reduced.

$$5\ e^- + 8\ H^+(aq) + MnO_4^-(aq) \longrightarrow Mn^{2+}(aq) + 4\ H_2O(\ell)$$

Step 5: *Multiply the half-reactions by appropriate factors so that the oxidation half-reaction produces as many electrons as the reduction half-reaction accepts.* In this case, one half-reaction involves two electrons, and the other half-reaction involves five electrons. It takes ten electrons to balance each half-reaction. The oxalic acid half-reaction must be multiplied by 5, and the permanganate half-reaction by 2.

$$5\,[H_2C_2O_4(aq) \longrightarrow 2\,CO_2(g) + 2\,H^+(aq) + 2\,e^-]$$

$$2\,[5\,e^- + 8\,H^+(aq) + MnO_4^-(aq) \longrightarrow Mn^{2+}(aq) + 4\,H_2O(\ell)]$$

Step 6: *Add the half-reactions to give the overall reaction and cancel equal amounts of reactants and products that appear on both sides of the arrow.*

$$5\,H_2C_2O_4(aq) \longrightarrow 10\,CO_2(g) + 10\,H^+(aq) + 10\,e^-$$

$$10\,e^- + 16\,H^+(aq) + 2\,MnO_4^-(aq) \longrightarrow 2\,Mn^{2+}(aq) + 8\,H_2O(\ell)$$

$$5\,H_2C_2O_4(aq) + 16\,H^+(aq) + 2\,MnO_4^-(aq) \longrightarrow 10\,CO_2(g) + 10\,H^+(aq) + 2\,Mn^{2+}(aq) + 8\,H_2O(\ell)$$

Since 16 H^+ appear on the left and 10 H^+ appear on the right, 10 H^+ are canceled, leaving 6 H^+ on the left.

$$5\,H_2C_2O_4(aq) + 6\,H^+(aq) + 2\,MnO_4^-(aq) \longrightarrow 10\,CO_2(g) + 2\,Mn^{2+}(aq) + 8\,H_2O(\ell)$$

Step 7: *Check the balanced equation to make sure both atoms and charge are balanced.*

Atom balance: Each side of the equation has 2 Mn, 28 O, 10 C, and 16 H atoms.

Charge balance: Each side has a net charge of 4+.
On the left side, $(6 \times 1+) + (2 \times 1-) = 4+$.

On the right side, $2(2+) = 4+$.

Step 8: *Add enough water molecules to both sides of the equation to convert all H^+ to H_3O^+.* In this case, six water molecules are needed ($6\,H_2O + 6\,H^+ \rightarrow 6\,H_3O^+$). Six water molecules are added to each side of the equation, which increases the total to 14 on the product side.

$$5\,H_2C_2O_4(aq) + 6\,H_3O^+(aq) + 2\,MnO_4^-(aq) \longrightarrow 10\,CO_2(g) + 2\,Mn^{2+}(aq) + 14\,H_2O(\ell)$$

Step 9: *Check the final results to make sure both atoms and charges are balanced.* The equation is balanced. In the final, balanced equation, the net charges on each side of the reaction are the same, and the numbers of atoms of each kind on each side of the reaction are equal.

PROBLEM-SOLVING PRACTICE 19.3

Balance this equation for the reaction of Zn with $Cr_2O_7^{2-}$ in acidic aqueous solution.

$$Zn(s) + Cr_2O_7^{2-}(aq) \longrightarrow Cr^{3+}(aq) + Zn^{2+}(aq)$$

CONCEPTUAL EXERCISE 19.1 Electrons Lost Equal Electrons Gained

Why must the number of electrons lost always equal the number gained in a redox reaction?

Balancing Redox Equations in Basic Solutions

For redox reactions that occur in basic solutions, the final electrochemical reaction must be completed with water and OH^- ions rather than water and H_3O^+ ions that we used for acidic solutions. During the balancing process, the half-reactions can be balanced as if they occurred in acidic solution. Then, at the end of the series of steps, the H^+ ions can be neutralized by adding an equal number of OH^- ions to both sides of the electrochemical equation and canceling, if necessary, water molecules that appear as both reactants and products. Problem-Solving Example 19.4 illustrates how to balance a redox reaction in basic solution.

PROBLEM-SOLVING EXAMPLE 19.4 Balancing Redox Equations for Reactions in Basic Solutions

In a nickel-cadmium (nicad) battery, cadmium metal forms $Cd(OH)_2$ and Ni_2O_3 forms $Ni(OH)_2$ in an alkaline solution. Write the balanced equation for this reaction.

Answer $Cd(s) + Ni_2O_3(s) + 3\,H_2O(\ell) \longrightarrow Cd(OH)_2(s) + 2\,Ni(OH)_2(s)$

Strategy and Explanation It is best to follow the systematic approach listed in the following series of steps to balance the equation for this reaction.

Step 1: *Recognize whether the reaction is an oxidation-reduction process. Then determine what is reduced and what is oxidized.* This is a redox reaction because the oxidation number of Cd changes from 0 in Cd metal to +2 in $Cd(OH)_2$, so Cd metal is oxidized. The oxidation number of each Ni changes from +3 in Ni_2O_3 to +2 in $Ni(OH)_2$, so the Ni is reduced.

Step 2: *Break the overall unbalanced equation into half-reactions.*

$$Cd(s) \longrightarrow Cd(OH)_2(s) \quad \text{(oxidation half-reaction)}$$

$$Ni_2O_3(s) \longrightarrow Ni(OH)_2(s) \quad \text{(reduction half-reaction)}$$

Step 3: *Balance the atoms in each half-reaction.* First, balance all atoms except the O and H atoms; do them last. Balance each half-reaction as if it were in an *acidic* solution to start. Balance O by adding H_2O and balance H by adding H^+. We will revert to using OH^- ions characteristic of basic solutions later in the process.

In the Cd half-reaction, the Cd atoms are balanced. Adding two water molecules on the left balances the two O atoms on the right, but this leaves four H atoms on the left and only two on the right. Adding two H^+ ions on the right balances H atoms in this half-reaction.

$$2\,H_2O(\ell) + Cd(s) \longrightarrow Cd(OH)_2(s) + 2\,H^+(aq)$$

For the Ni_2O_3 half-reaction, a coefficient of 2 is needed for $Ni(OH)_2$ because there are two Ni atoms on the left. To balance the four O atoms and four H atoms now on the right with the three O atoms on the left requires one water molecule and two H^+ ions on the left.

$$2\,H^+(aq) + H_2O(\ell) + Ni_2O_3(s) \longrightarrow 2\,Ni(OH)_2(s)$$

Step 4: *Balance the half-reactions for charge using electrons.* The Cd half-reaction produces $2\,e^-$ as a product.

$$2\,H_2O(\ell) + Cd(s) \longrightarrow Cd(OH)_2(s) + 2\,H^+(aq) + 2\,e^- \quad \text{(balanced)}$$

The Ni_2O_3 half-reaction requires $2\,e^-$ as a reactant.

$$2\,H^+(aq) + H_2O(\ell) + Ni_2O_3(s) + 2\,e^- \longrightarrow 2\,Ni(OH)_2(s) \quad \text{(balanced)}$$

Step 5: *Multiply the half-reactions by appropriate factors so that the reducing agent produces as many electrons as the oxidizing agent accepts.* The Cd half-reaction produces two electrons, and the Ni_2O_3 half-reaction accepts two, so the electrons are balanced.

Step 6: *Because H^+ does not exist at any appreciable concentration in a basic solution, remove H^+ by adding an appropriate amount of OH^- to both sides of the equation. H^+ and OH^- react to form H_2O.* In the Cd half-reaction, add two OH^- ions to each side to get

$$2\,OH^-(aq) + 2\,H_2O(\ell) + Cd(s) \longrightarrow Cd(OH)_2(s) + 2\,H_2O(\ell) + 2\,e^-$$

On the product side, two OH^- ions plus two H^+ ions form two H_2O molecules. For the Ni_2O_3 half-reaction, add two OH^- ions to each side to get

$$3\,H_2O(\ell) + Ni_2O_3(s) + 2\,e^- \longrightarrow 2\,Ni(OH)_2(s) + 2\,OH^-(aq)$$

Step 7: *Add the half-reactions to give the overall reaction, and cancel reactants and products that appear on both sides of the reaction arrow.*

$$\cancel{2\,OH^-(aq)} + \cancel{2\,H_2O(\ell)} + Cd(s) \longrightarrow Cd(OH)_2(s) + \cancel{2\,H_2O(\ell)} + 2\,e^-$$
$$3\,H_2O(\ell) + Ni_2O_3(s) + 2\,e^- \longrightarrow 2\,Ni(OH)_2(s) + \cancel{2\,OH^-(aq)}$$
$$Cd(s) + Ni_2O_3(s) + 3\,H_2O(\ell) \longrightarrow Cd(OH)_2(s) + 2\,Ni(OH)_2(s)$$

Step 8: *Check the final results to make sure both atoms and charge are balanced.* The equation is balanced. In the final equation, there are no net charges on either side of the reaction arrow, and the numbers of atoms of each kind on each side of the reaction arrow are equal.

PROBLEM-SOLVING PRACTICE 19.4

In a basic solution, aluminum metal forms $Al(OH)_4^-$ ion as it reduces NO_3^- ion to NH_3. Write the balanced equation for this reaction using the steps outlined in Problem-Solving Example 19.4.

19.3 Electrochemical Cells

In a redox reaction, electrons are *transferred* from one kind of atom, molecule, or ion to another. It is easy to see by the color changes in the two redox reactions shown in Figures 19.1 and 19.2 that these reactions favor the formation of products—as soon as the reactants are mixed, changes take place. All product-favored reactions release Gibbs free energy ***(⇐ p. 883),*** energy that can do useful work. An electrochemical cell is a way to capture that useful work as electrical work.

In an **electrochemical cell,** an oxidizing agent and a reducing agent pair are arranged in such a way that they can react only if electrons flow through an outside conductor. Such electrochemical cells are also known as **voltaic cells** or **batteries.** Figure 19.3 diagrams how a voltaic cell can be made from the Zn/Cu^{2+} reaction shown previously in Figure 19.1. The two half-reactions occur in separate beakers, each of which is called a **half-cell.** When Zn atoms are oxidized, the electrons that are given up by zinc pass through the wire and a light bulb to the copper metal. There the electrons are available to reduce Cu^{2+} ions from the solution. The metallic zinc and copper strips are called electrodes. An **electrode** conducts electrical current (electrons) into or out of something, in this case, a solution. An electrode is most often a metal plate or wire, but it can also be a piece of graphite or another conductor. The electrode where oxidation occurs is the **anode;** the electrode where reduction takes place is the **cathode.**

The voltaic cell is named after the Italian scientist Alessandro Volta. In about 1800, Volta constructed the first electrochemical cell, a stack of alternating zinc and silver disks separated by pieces of paper soaked in salt water (an electrolyte). Later, Volta showed that any two different metals and an electrolyte could be used to make a battery. Figure 19.4 shows a cell constructed by sticking a strip of zinc and a strip of copper metal into a grapefruit. In this case, the acidic solution of the grapefruit juice is the electrolyte.

In the electrochemical cell diagrammed in Figure 19.3, electrons are released at the anode by the oxidation half-reaction

$$Zn(s) \longrightarrow Zn^{2+}(aq) + 2\,e^- \qquad \text{(anode reaction)}$$

They then flow from the anode through the filament in the bulb, causing it to glow, and eventually travel to the cathode, where they react with copper(II) ions in the reduction half-reaction

$$Cu^{2+}(aq) + 2\,e^- \longrightarrow Cu(s) \qquad \text{(cathode reaction)}$$

Strictly speaking, many devices we call batteries consist of several voltaic cells connected together, but the term "battery" has taken on the same meaning as "voltaic cell."

Electrochemical cells are sometimes referred to as *galvanic cells* in recognition of the work of Luigi Galvani, who discovered that a frog's leg placed in salt water would twitch when it was touched simultaneously by two dissimilar metals.

To identify the anode and cathode, remember that **O**xidation takes place at the **A**node (both words begin with vowels), and **R**eduction takes place at the **C**athode (both words begin with consonants).

On a flashlight battery, the anode is marked "−" because oxidation produces electrons that make the anode negative. Conversely, the cathode is marked "+" because reduction consumes electrons, leaving the metal electrode positive.

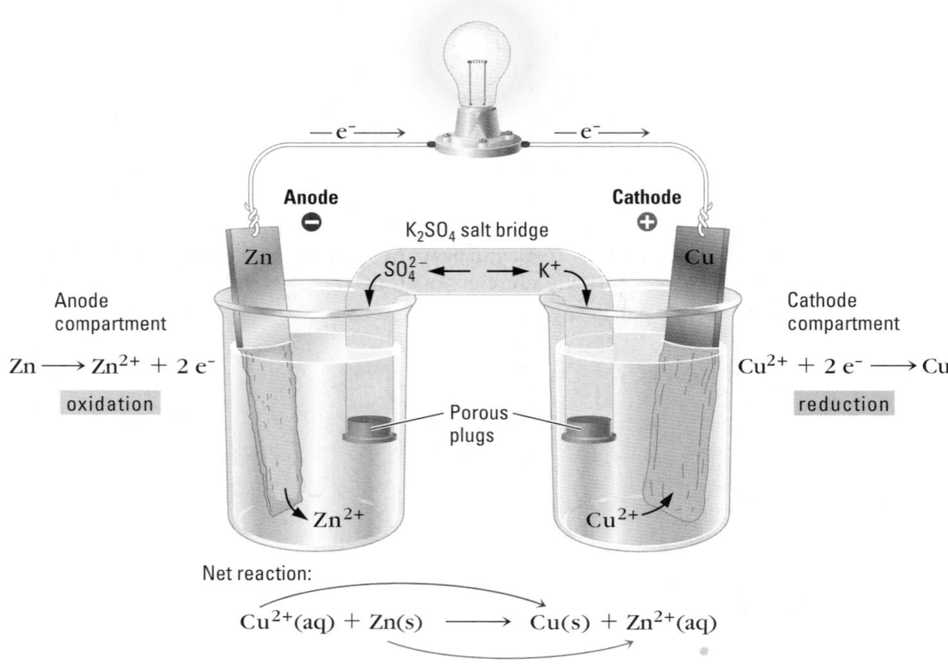

Figure 19.3 A simple electrochemical cell. The cell consists of a zinc electrode in a solution containing Zn^{2+} ions (*left*), a copper electrode in a solution containing Cu^{2+} ions (*right*), and a salt bridge that allows ions to flow into and out of the two solutions. When the two metal electrodes are connected by a conducting circuit, electrons flow through the wires and light bulb from the zinc electrode, where zinc is oxidized, to the copper electrode, where copper ions from the solution are reduced.

Figure 19.4 A grapefruit battery. A voltaic cell can be made by inserting zinc and copper electrodes into a grapefruit. A potential of 0.95 V is obtained. (The water and citric acid of the fruit allow for ion conduction between electrodes.) This cell is more complicated than the one in Figure 19.3. To learn how the grapefruit battery works, see Goodisman, J., *Journal of Chemical Education,* Vol. 78, 2001; p. 516.

In commercial batteries, the salt bridge is often a porous polymer membrane.

If nothing else but electron flow took place, the concentration of Zn^{2+} ions in the anode compartment would increase as zinc metal is oxidized, building up positive charge in the solution. The concentration of Cu^{2+} ions in the cathode compartment would decrease as Cu^{2+} ions are reduced to metallic copper. This makes that solution less positive due to the decrease of positive charge as Cu^{2+} ions are reduced. Excess negative charges from the now unbalanced SO_4^{2-} ions would be present. Because of this negative charge imbalance, the flow of electrons in the wires would very quickly stop. For the cell to work, there must be a way for the positive charge buildup in the anode compartment to be balanced by addition of negative ions or removal of positive ions, and vice versa for the cathode compartment.

The charge buildup can be avoided by using a salt bridge to connect the two compartments. A **salt bridge** is a solution of a salt (K_2SO_4 in Figure 19.3) arranged so that the bulk of that solution cannot flow into the cell solutions, but the ions (K^+ and SO_4^{2-}) can pass freely. As electrons flow through the wire from the zinc electrode to the copper electrode, negative ions (SO_4^{2-}) move through the salt bridge into the anode compartment solution and positive ions (K^+) move in the opposite direction into the cathode compartment solution. In general, anions from the salt bridge flow into the anode cell, and cations from the salt bridge flow into the cathode cell. This flow of ions completes the electrical circuit, allowing current to flow. If the salt bridge is removed from this battery, the flow of electrons will stop.

All voltaic cells and batteries operate in a similar fashion and have these characteristics:

- The oxidation-reduction reaction that occurs must favor the formation of products.
- There must be an external circuit through which electrons flow.
- There must be a salt bridge, porous barrier, or some other means of allowing ions in the salt bridge to flow into the electrode compartments to offset charge buildup.

The components of an electrochemical cell are summarized in Figure 19.5.

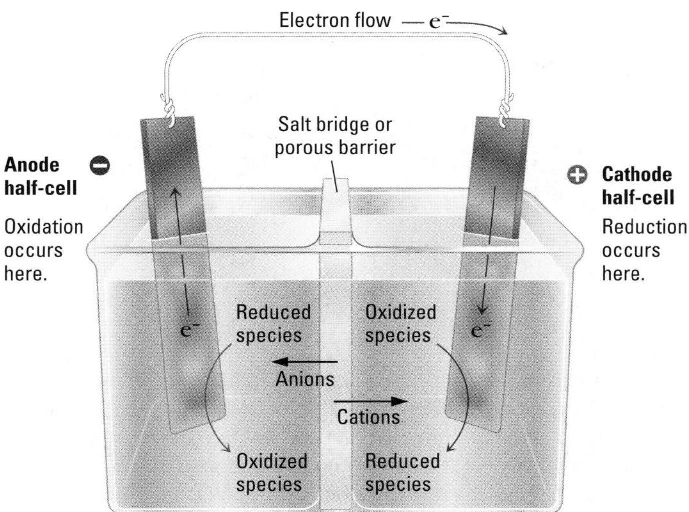

Figure 19.5 Summary of the terminology used in voltaic cells. Oxidation occurs at the anode, and reduction occurs at the cathode. Electrons move from the negative electrode (anode) to the positive electrode (cathode) through the external wire. The electrical circuit is completed in the solution by the movement of ions—anions move from the salt bridge compartment to the anode compartment, and cations move from the salt bridge compartment to the cathode compartment. The half-cells can be separated by either a salt bridge or a porous barrier.

PROBLEM-SOLVING EXAMPLE 19.5 Electrochemical Cells

A simple voltaic cell is assembled with Fe(s) and $Fe(NO_3)_2$(aq) in one compartment and Cu(s) and $Cu(NO_3)_2$(aq) in the other compartment. An external wire connects the two electrodes, and a salt bridge containing $NaNO_3$ connects the two solutions. The overall reaction is

$$Fe(s) + Cu^{2+}(aq) \longrightarrow Cu(s) + Fe^{2+}(aq)$$

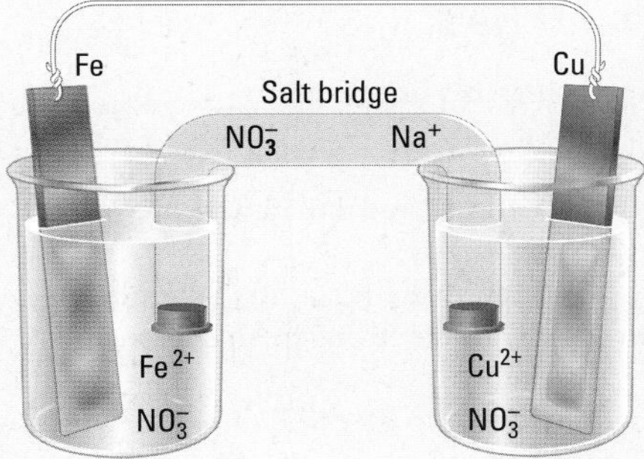

This voltaic cell is shown without an electrical device in the external part of the circuit for simplicity.

What is the reaction at the anode? What is the reaction at the cathode? What is the direction of electron flow in the external wire? What is the direction of ion flow in the salt bridge? Draw a cell diagram, indicating the anode, the cathode, and the directions of electron flow and ion flow.

Answer

Anode reaction: $Fe(s) \longrightarrow Fe^{2+}(aq) + 2\,e^-$

Cathode reaction: $Cu^{2+}(aq) + 2\,e^- \longrightarrow Cu(s)$

The electrons flow through the wire from the anode to the cathode. Nitrate anions move from the salt bridge into the anode compartment. Sodium cations move from the salt bridge into the cathode compartment. The completed cell diagram is shown below.

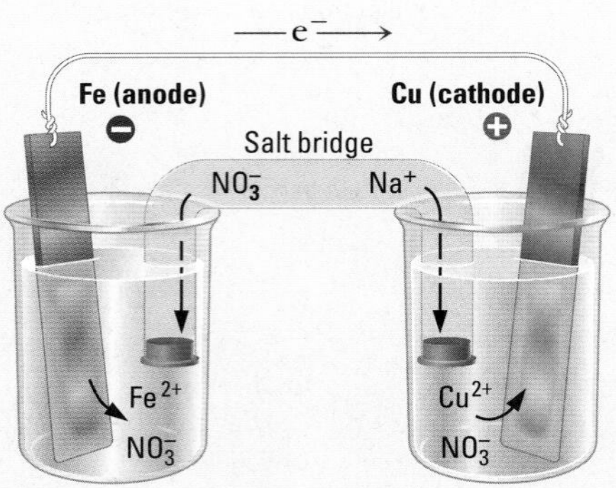

Strategy and Explanation The net reaction shows that iron is being oxidized from Fe(s) to Fe^{2+}(aq) and that Cu^{2+}(aq) is being reduced to Cu(s). We need to decide at which electrodes these reactions occur. Since Fe metal is being oxidized (increase in oxidation number), the electrode in the Fe(s)/Fe^{2+}(aq) compartment is the anode. The electrode in the Cu(s)/Cu^{2+}(aq) compartment must be the cathode because the Cu^{2+} ions are being reduced to Cu metal (decrease in oxidation number).

The half-reactions are

$$\text{Fe(s)} \longrightarrow \text{Fe}^{2+}\text{(aq)} + 2\,e^- \qquad \text{(site of oxidation—anode)}$$
$$\text{Cu}^{2+}\text{(aq)} + 2\,e^- \longrightarrow \text{Cu(s)} \qquad \text{(site of reduction—cathode)}$$

Electrons flow *from* their source (the oxidation of the Fe at the iron electrode which is the anode) through the wire *to* the electrode where they react (the Cu electrode which is the cathode). Because positive Fe^{2+} ions are being produced in the anode compartment, negative NO_3^- ions in the salt bridge move into the anode compartment from the salt bridge to balance the overall charge. Because Cu^{2+} ions are being removed from the cathode compartment, Na^+ ions move into the cathode compartment from the salt bridge to replace the charge of the Cu^{2+} ions reduced to Cu metal.

The external part of the electrical circuit is as follows

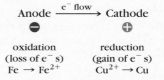

$$\text{Anode} \xrightarrow{e^-\text{ flow}} \text{Cathode}$$

⊖	⊕
oxidation	reduction
(loss of e^- s)	(gain of e^- s)
$Fe \rightarrow Fe^{2+}$	$Cu^{2+} \rightarrow Cu$

PROBLEM-SOLVING PRACTICE 19.5

A voltaic cell is assembled to use this net reaction.

$$\text{Ni(s)} + 2\,\text{Ag}^+\text{(aq)} \longrightarrow \text{Ni}^{2+}\text{(aq)} + 2\,\text{Ag(s)}$$

(a) Write half-reactions for this cell, and indicate which is the oxidation reaction and which is the reduction reaction.
(b) Name the electrodes at which these reactions take place.
(c) What is the direction of flow of electrons in an external wire connected between the electrodes?
(d) If a salt bridge connecting the two electrode compartments contains KNO_3, what is the direction of flow of the K^+ ions and the NO_3^- ions?

There is a shorthand notation for representing an electrochemical cell. For the cell shown in Figure 19.3 with the redox reaction

$$\text{Zn(s)} + \text{Cu}^{2+}\text{(aq)} \longrightarrow \text{Zn}^{2+}\text{(aq)} + \text{Cu(s)}$$

the representation is

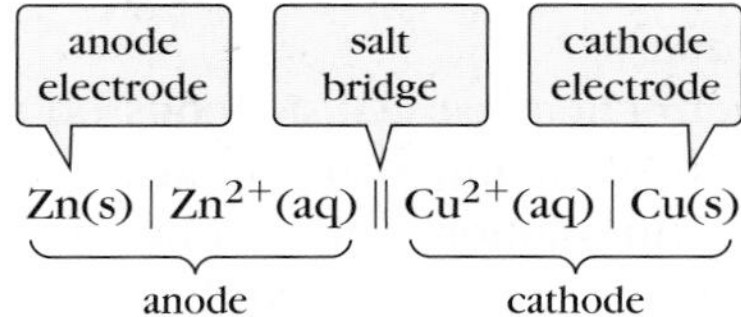

The anode half-cell is represented on the left, and the cathode half-cell is represented on the right. The electrodes are written on the extreme left (anode, Zn) and extreme right (cathode, Cu) of the notation. The single vertical lines denote boundaries between phases, and the double vertical lines denote the salt bridge and the separation between half-cells. Within each half-cell the reactants are written first, followed by the products. The electron flow is from left to right (anode to cathode).

CONCEPTUAL EXERCISE **19.2 Battery Design**

Devise an internal on-off switch for a battery that would not be a part of the flow of electrons.

19.4 Electrochemical Cells and Voltage

Because electrons flow from the anode to the cathode of an electrochemical cell, they can be thought of as being "driven" or "pushed" by an **electromotive force (emf).** The emf is produced by the difference in electrical potential energy between the two electrodes. Just as water flows downhill in response to a difference in gravitational potential energy, so an electron moves from an electrode of higher electrical potential energy to an electrode of lower potential energy. The moving water can do work, as can moving electrons; for example, they could run a motor.

The quantity of electrical work done is proportional to the number of electrons that go from higher to lower potential energy as well as to the size of the potential energy difference.

This is similar to comparing the quantity of work a few drops of water can do when falling 100 m with that possible when a few tons of water fall the same distance.

$$\text{Electrical work} = \text{charge} \times \text{potential energy difference}$$

or

$$\text{Electrical work} = \text{number of electrons} \times \text{potential energy difference}$$

Electrical charge is measured in coulombs. The charge on a single electron is very small (1.6022×10^{-19} C), so it takes 6.24×10^{18} electrons to produce 1 coulomb of charge. A **coulomb (C)** is the quantity of charge that passes a fixed point in an electrical circuit when a current of 1 ampere flows for 1 second. The **ampere (A)** is the unit of electrical current.

$$\frac{1.6022 \times 10^{-19}\text{ C}}{1\text{ e}^-} \times 6.24 \times 10^{18}\text{ e}^- = 1\text{ C}$$

$$1\text{ coulomb} = 1\text{ ampere} \times 1\text{ second}$$

Electrical potential energy difference is measured in volts. The **volt (V)** is defined such that one joule of work is performed when one coulomb of charge moves through a potential difference of one volt:

When a single electron moves through a potential of 1 V, the work done is one electron-volt, abbreviated eV.

$$1\text{ volt} = \frac{1\text{ joule}}{1\text{ coulomb}} \quad \text{or} \quad 1\text{ joule} = 1\text{ volt} \times 1\text{ coulomb}$$

Figure 19.6 Dry cell batteries. The larger batteries are capable of more work since they contain more oxidizing and reducing agents.

The electromotive force of an electrochemical cell, commonly called its **cell voltage,** shows how much work a cell can produce for each coulomb of charge that the chemical reaction produces.

The voltage of an electrochemical cell depends on the temperature and substances that make up the cell, including their pressures if they are gases or their concentrations if they are solutes in solution. The quantity of charge (coulombs) depends on how much of each substance reacts. Look at the 1.5-V batteries shown in Figure 19.6. They have the same voltage because they have electrodes with the same potential difference between them. Yet a larger battery is capable of far more work than a smaller one, because it contains larger quantities of oxidizing and reducing agents. In this section and the next, we consider how cell voltage depends on the materials from which a cell is made. In Section 19.5, we will return to the question of how much electrical work a cell can do.

CONCEPTUAL EXERCISE 19.3 Electrical Charges

Which has the larger charge, a coulomb of charge or Avogadro's number of electrons?

Cell Voltage

A cell's voltage is readily measured by inserting a voltmeter into the circuit. Because the voltage varies with concentrations, **standard conditions** are defined for voltage measurements. These are the same as those used for ΔH° (***⇐ p. 233***): All reactants and products must be present as pure solids, pure liquids, gases at 1 bar pressure, or solutes at 1 M concentration. Voltages measured under these conditions are **standard voltages,** symbolized by $\boldsymbol{E^\circ}$. Unless specified otherwise, all values of E° are for 25 °C (298 K). By definition, *cell voltages for product-favored electrochemical reactions are positive.* For example, the standard cell voltage for the product-favored Zn/Cu^{2+} cell discussed earlier is + 1.10 V at 25 °C.

Since every redox reaction can be thought of as the sum of two half-reactions, it is convenient to assign a voltage to every possible half-reaction. Then the cell voltage for any reaction can be obtained by using the standard voltages of the half-reactions that occur at the cathode and the anode.

$$E^\circ_{\text{cell}} = E^\circ_{\text{cathode}} - E^\circ_{\text{anode}}$$

The two quantities on the right side of the equation, E°_{cathode} and E°_{anode}, are the values for the half-reactions *written as reduction half-reactions.* ***If*** $\boldsymbol{E^\circ_{cell}}$ ***is positive, the reaction is product-favored. If*** $\boldsymbol{E^\circ_{cell}}$ ***is negative, the reaction is reactant-favored.*** However, because only *differences* in potential energy can be measured, it is not possible to measure the voltage for a single half-reaction. Instead, one half-reaction is chosen as the standard, and then all others are compared to it. The half-reaction chosen as the standard is the one that occurs at the **standard hydrogen electrode,** in which hydrogen gas at a pressure of 1 bar is bubbled over a platinum electrode immersed in 1 M aqueous acid at 25 °C (Figure 19.7).

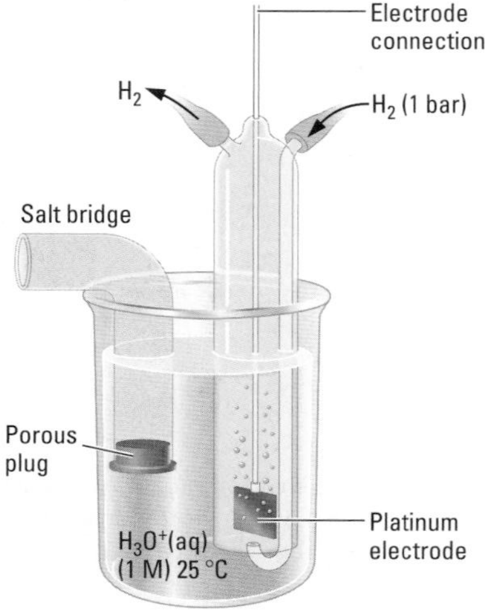

Figure 19.7 The standard hydrogen electrode. Hydrogen gas at 1 bar pressure bubbles over an inert platinum electrode that is immersed in a solution containing exactly 1 M H_3O^+ ions at 25 °C. The potential for this electrode is defined as exactly 0 V.

$$2\,H_3O^+(aq, 1\,M) + 2\,e^- \longrightarrow H_2(g, 1\,bar) + 2\,H_2O(\ell)$$

A voltage of exactly 0 V is *assigned* to this half-cell. In a cell that combines another half-reaction with the standard hydrogen electrode, the overall cell voltage is determined by the difference between the potentials of the two electrodes. Because the potential of the hydrogen electrode is assigned a value of 0, the overall cell voltage equals the voltage of the other electrode.

When the standard hydrogen electrode is paired with a half-cell that contains a better reducing agent than H_2, $H_3O^+(aq)$ is reduced to H_2.

$$H_3O^+ \text{ reduced: } 2\,H_3O^+(aq) + 2\,e^- \longrightarrow H_2(g, 1\,bar) + 2\,H_2O(\ell) \qquad E^\circ = 0.0\,V$$

When the standard hydrogen electrode is paired with a half-cell that contains a better oxidizing agent than H_3O^+, then H_2 is oxidized to H_3O^+.

$$H_2 \text{ oxidized: } H_2(g, 1\ \text{bar}) + 2\,H_2O(\ell) \longrightarrow 2\,H_3O^+(aq, 1\ M) + 2\,e^- \quad E° = 0.0\ V$$

The reaction that occurs at the standard hydrogen electrode is reversible. In either case, the standard hydrogen electrode, by definition, has a potential of 0 V.

Figure 19.8 diagrams a cell in which one compartment contains a standard hydrogen electrode and the other contains a zinc electrode immersed in a 1 M solution of Zn^{2+}. The voltmeter connected between the two electrodes measures the difference in electrical potential energy. For this cell the voltage is +0.76 V. After the cell operates for a time, the zinc electrode decreases in mass as zinc is oxidized to $Zn^{2+}(aq)$. Therefore, the Zn electrode must be the anode; that is, oxidation takes place at this electrode. The hydrogen electrode must be the cathode, where reduction is taking place. The cell reaction is the difference of the half-cell reactions.

$$Zn(s) \longrightarrow Zn^{2+}(aq, 1\ M) + 2\,e^- \qquad E°_{anode} = ?$$

$$2\,H_3O^+(aq, 1\ M) + 2\,e^- \longrightarrow H_2(g, 1\ \text{bar}) + 2\,H_2O(\ell) \qquad E°_{cathode} = 0\ V$$

$$Zn(s) + 2\,H_3O^+(aq, 1\ M) \longrightarrow Zn^{2+}(aq, 1\ M) + H_2(g, 1\ \text{bar}) + 2\,H_2O(\ell)$$

$$E°_{cell} = +0.76\ V\ (\text{cell reaction})$$

The voltmeter tells us that the potential difference between the two electrodes is +0.76 V. Using the equation

$$E°_{cell} = E°_{cathode} - E°_{anode}$$

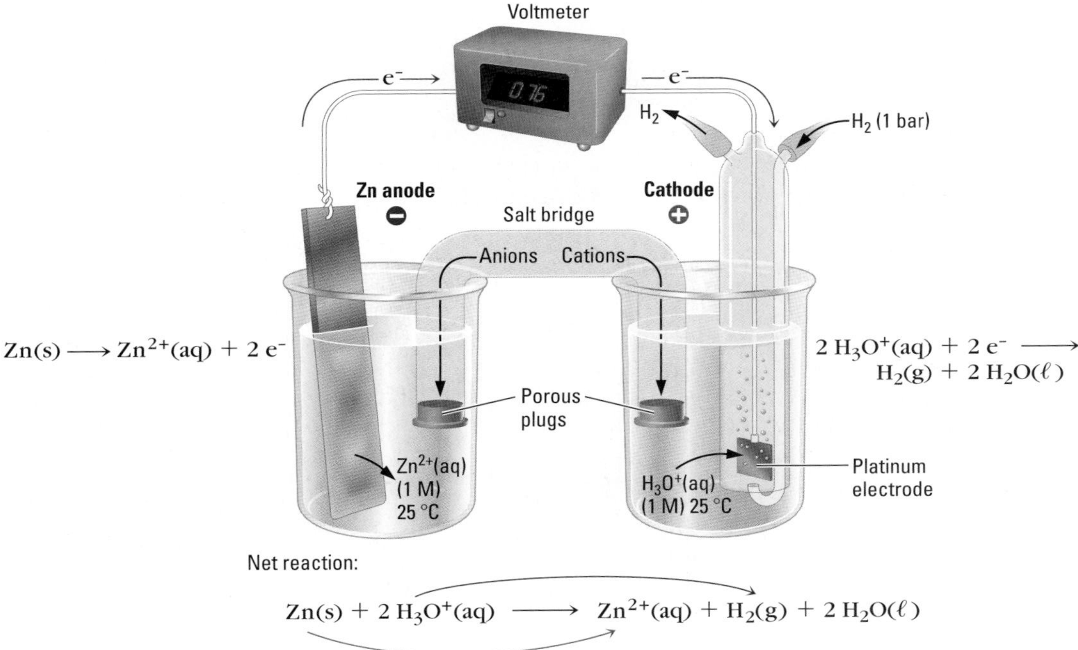

Figure 19.8 An electrochemical cell using a $Zn^{2+}/Zn(s)$ half-cell and a standard hydrogen electrode. In this cell, the zinc electrode is the anode and the standard hydrogen electrode is the cathode. The cell voltage is +0.76 V. Zinc is the reducing agent and is oxidized to Zn^{2+}; H_3O^+ is the oxidizing agent and is reduced to H_2. In the standard hydrogen electrode, reaction occurs only where the three phases—gas, solution, and solid electrode—are in contact. The platinum electrode does not undergo any chemical change, and in the cell pictured here the cathodic half-cell reaction is $2\,H_3O^+(aq) + 2\,e^- \rightarrow H_2(g) + 2\,H_2O(\ell)$. (When the standard hydrogen electrode is the anode, the half-cell reaction is $H_2(g) + 2\,H_2O(\ell) \rightarrow 2\,H_3O^+(aq) + 2\,e^-$.)

we have

$$0.76\ \text{V} = 0\ \text{V} - E^\circ_{\text{anode}}$$
$$E^\circ_{\text{anode}} = E^\circ_{\text{Zn(s)/Zn}^{2+}\text{(aq, 1 M)}} = -0.76\ \text{V}$$

Thus, by using the standard hydrogen electrode, it is possible to assign a standard potential E° value of −0.76 V to the $Zn(s)/Zn^{2+}$(aq, 1 M) electrode.

CONCEPTUAL EXERCISE 19.4 What Is Going On Inside the Electrochemical Cell?

Devise an experiment that would show that Zn is being oxidized in the electrochemical cell shown in Figure 19.8.

The convention of assigning voltages to half-reactions is similar to the convention of tabulating standard enthalpies of formation; in both cases a relatively small table of data can provide information about a large number of different reactions.

The half-cell potentials of many different half-reactions can be measured by comparing them with the standard hydrogen electrode. For example, in a cell consisting of the Cu^{2+}/Cu half-cell connected to a standard hydrogen electrode, the mass of the copper electrode increases and the voltmeter reads +0.34 V (Figure 19.9). This means that the reactions are

$$H_2(\text{g, 1 bar}) + 2\,H_2O(\ell) \longrightarrow 2\,H_3O^+(\text{aq, 1 M}) + 2\,e^- \qquad E^\circ_{\text{anode}} = 0\ \text{V}$$

$$Cu^{2+}(\text{aq, 1 M}) + 2\,e^- \longrightarrow Cu(s) \qquad E^\circ_{\text{cathode}} = ?$$

$$H_2(\text{g, 1 bar}) + Cu^{2+}(\text{aq, 1 M}) + 2\,H_2O(\ell) \longrightarrow 2\,H_3O^+(\text{aq, 1 M}) + Cu(s) \qquad E^\circ_{\text{cell}} = +0.34\ \text{V}$$

$$E^\circ_{\text{cell}} = E^\circ_{\text{cathode}} - E^\circ_{\text{anode}} = E^\circ_{\text{Cu(s)/Cu}^{2+}\text{(aq, 1 M)}} - 0\ \text{V} = +0.34\ \text{V}$$

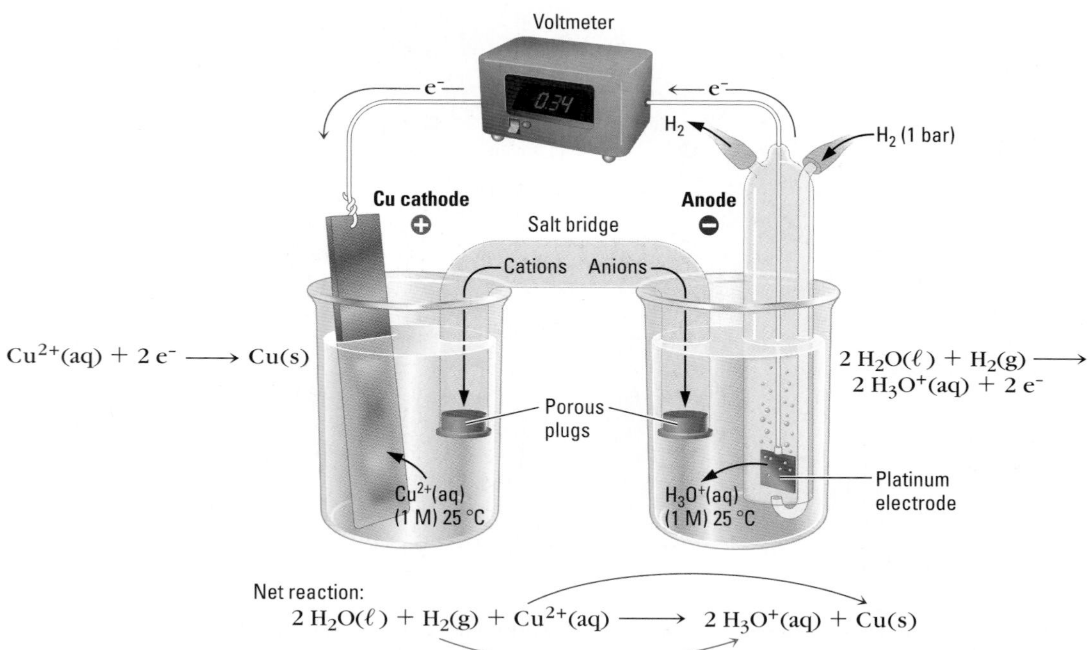

Figure 19.9 An electrochemical cell using the Cu^{2+}/Cu half-cell and the standard hydrogen electrode. A voltage of +0.34 V is produced. In this cell, Cu^{2+} ions are reduced to form Cu metal, and H_2 is oxidized at the standard hydrogen electrode. The reaction at the standard hydrogen electrode is the opposite of that shown in Figure 19.8.

The half-cell potential for the Cu^{2+} (aq, 1 M) + 2 e^- $\longrightarrow$ Cu(s) reduction half-reaction must be +0.34 V. Note that in this cell the standard hydrogen electrode is the anode.

We can now return to the first electrochemical cell we looked at, in which Zn reduces Cu^{2+} ions to Cu. Using the potentials for the half-reactions, we can write

$$\begin{array}{lll} \mathrm{Zn(s)} \longrightarrow \mathrm{Zn^{2+}(aq,\,1\,M)} + 2\,\mathrm{e^-} & & E^\circ_{\text{anode}} = -0.76\text{ V} \\ \mathrm{Cu^{2+}(aq,\,1\,M)} + 2\,\mathrm{e^-} \longrightarrow \mathrm{Cu(s)} & & E^\circ_{\text{cathode}} = +0.34\text{ V} \\ \hline \mathrm{Zn(s)} + \mathrm{Cu^{2+}(aq,\,1\,M)} \longrightarrow \mathrm{Zn^{2+}(aq,\,1\,M)} + \mathrm{Cu(s)} & & E^\circ_{\text{cell}} = ? \end{array}$$

When we combine the standard reduction potentials of the two half-reactions, we have the measured potential for the cell reaction.

$$E^\circ_{\text{cell}} = E^\circ_{\text{cathode}} - E^\circ_{\text{anode}} = (+0.34\text{ V}) - (-0.76\text{ V}) = +1.10\text{ V}$$

The experimentally measured potential for this cell is 1.10 V, confirming that half-cell potentials measured with the standard hydrogen electrode can be subtracted to obtain overall cell potentials.

PROBLEM-SOLVING EXAMPLE 19.6 Determining a Half-Cell Potential

The voltaic cell shown in the drawing below generates a potential of E° = 0.36 V under standard conditions at 25 °C. The net cell reaction is

$$\mathrm{Zn(s)} + \mathrm{Cd^{2+}(aq,\,1\,M)} \longrightarrow \mathrm{Zn^{2+}(aq,\,1\,M)} + \mathrm{Cd(s)}$$

The standard half-cell potential for Zn(s)/Zn^{2+}(aq, 1 M) is −0.76 V.

(a) Determine which electrode is the anode and which is the cathode.
(b) Show the direction of electron flow through the circuit outside the cell, and complete the cell diagram.
(c) Calculate the standard potential for the half-cell Cd^{2+}(aq) + 2 e^- $\longrightarrow$ Cd(s).

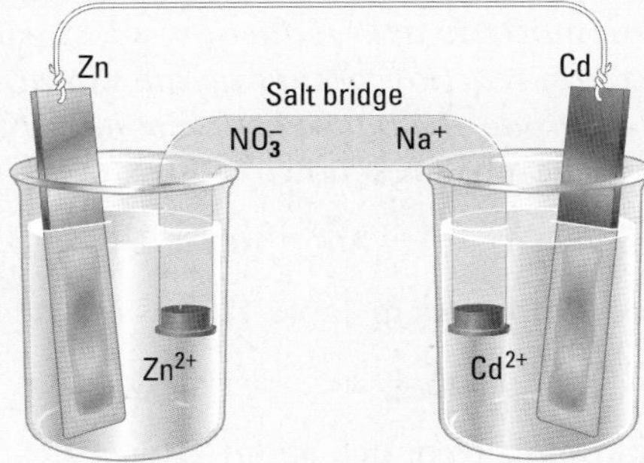

This voltaic cell is shown without an electrical device in the external circuit for simplicity.

Answer

(a) Zinc is the anode, and cadmium is the cathode.
(b) The completed cell diagram is shown below.
(c) The standard cell potential is −0.40 V.

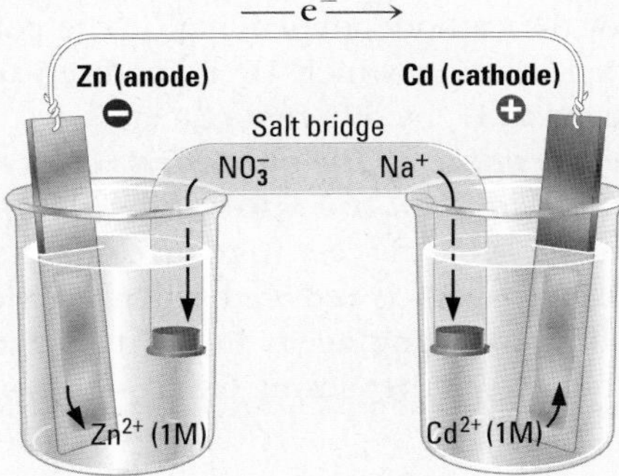

Strategy and Explanation The electrode where oxidation occurs is the anode. Because Zn(s) is oxidized to Zn^{2+}(aq), the Zn electrode is the anode. Cadmium(II) ions are reduced at the Cd electrode, so it is the cathode.

The net cell potential and the potential for the Zn(s)/Zn^{2+}(aq, 1 M) half-cell are known, so the value of $E°$ for Cd^{2+}(aq, 1 M) + 2 e^- ⟶ Cd(s) can be calculated.

$$\begin{array}{lll} Zn(s) \longrightarrow Zn^{2+}(aq, 1\ M) + 2\ e^- & E^\circ_{anode} = -0.76\ V\ (anode) \\ Cd^{2+}(aq, 1\ M) + 2\ e^- \longrightarrow Cd(s) & E^\circ_{cathode} = ?\ V\ (cathode) \\ \hline Zn(s) + Cd^{2+}(aq, 1\ M) \longrightarrow Zn^{2+}(aq, 1\ M) + Cd(s) & E^\circ_{cell} = +0.36\ V \end{array}$$

Using $E^\circ_{cell} = E^\circ_{cathode} - E^\circ_{anode}$, we can solve for $E^\circ_{cathode}$.

$$E^\circ_{cathode} = E^\circ_{cell} + E^\circ_{anode} = 0.36\ V + (-0.76\ V) = -0.40\ V$$

At 25 °C, the value of $E°$ for the Cd^{2+}(aq) + 2 e^- ⟶ Cd(s) half-reaction is −0.40 V.

PROBLEM-SOLVING PRACTICE 19.6

Given that the reaction of aqueous copper(II) ions with iron metal has E°_{cell} = +0.78 V, what is the value of $E°$ for the half-cell Fe(s) → Fe^{2+}(aq) + 2 e^-?

$$Fe(s) + Cu^{2+}(aq, 1\ M) \longrightarrow Fe^{2+}(aq, 1\ M) + Cu(s) \qquad E^\circ_{cell} = +0.78\ V$$

19.5 Using Standard Cell Potentials

The results of a great many measurements of cell potentials such as the ones just described are summarized as **standard reduction potentials** in Table 19.1. A much longer and more complete list of standard reduction potentials appears in Appendix I. The values reported in the tables are called standard reduction potentials because they are the potentials, reported as voltages, that are measured for a cell in which a half-reaction *occurs as a reduction* when paired with the standard hydrogen electrode. *If a half-reaction occurs as an oxidation when paired with the standard hydrogen electrode, the half-reaction voltage has a negative sign.* For example, we saw the oxidation half-reaction

$$Zn(s) \longrightarrow Zn^{2+}(aq) + 2\ e^-$$

in Figure 19.8. This reaction appears in Table 19.1 as the *reduction* half-reaction

$$Zn^{2+}(aq) + 2\ e^- \longrightarrow Zn(s) \qquad E^\circ = -0.76\ V$$

Here are some important points to notice about Table 19.1:

1. ***Each half-reaction is written as a reduction.*** Thus, the species on the left-hand side of each half-reaction is in a higher oxidation state, and the species on the right-hand side is in a lower oxidation state.
2. ***Each half-reaction can occur in either direction.*** A given substance can react at the anode or the cathode, depending on the conditions. For example, we have already seen cases in which H_2 is oxidized to H_3O^+ and others in which H_3O^+ is reduced to H_2 by different reactants.
3. ***The more positive the value of the standard reduction potential, E°, the more easily the substance on the left-hand side of a half-reaction can be reduced.*** When a substance is easy to reduce, it is a strong oxidizing agent. (Recall that an oxidizing agent is reduced when it oxidizes something else.) Thus, F_2(g) is the best oxidizing agent in the table, and Li^+ is the poorest oxidizing agent in the table. Other strong oxidizing agents are at the top left of the table:

$$H_2O_2(aq),\ PbO_2(s),\ Au^{3+}(aq),\ Cl_2(g),\ O_2(g)$$

F_2 is always reduced and Li(s) is always oxidized.

Table 19.1 Standard Reduction Potentials in Aqueous Solution at 25 °C*

Reduction Half-Reaction		$E°$ (V)
$F_2(g) + 2\ e^-$	$\longrightarrow 2\ F^-(aq)$	+2.87
$H_2O_2(aq) + 2\ H_3O^+(aq) + 2\ e^-$	$\longrightarrow 4\ H_2O(\ell)$	+1.77
$PbO_2(s) + SO_4^{2-}(aq) + 4\ H_3O^+(aq) + 2\ e^-$	$\longrightarrow PbSO_4(s) + 6\ H_2O(\ell)$	+1.685
$MnO_4^-(aq) + 8\ H_3O^+(aq) + 5\ e^-$	$\longrightarrow Mn^{2+}(aq) + 12\ H_2O(\ell)$	+1.51
$Au^{3+}(aq) + 3\ e^-$	$\longrightarrow Au(s)$	+1.50
$Cl_2(g) + 2\ e^-$	$\longrightarrow 2\ Cl^-(aq)$	+1.358
$Cr_2O_7^{2-}(aq) + 14\ H_3O^+(aq) + 6\ e^-$	$\longrightarrow 2\ Cr^{3+}(aq) + 21\ H_2O(\ell)$	+1.33
$O_2(g) + 4\ H_3O^+(aq) + 4\ e^-$	$\longrightarrow 6\ H_2O(\ell)$	+1.229
$Br_2(\ell) + 2\ e^-$	$\longrightarrow 2\ Br^-(aq)$	+1.066
$NO_3^-(aq) + 4\ H_3O^+(aq) + 3\ e^-$	$\longrightarrow NO(g) + 6\ H_2O(\ell)$	+0.96
$OCl^-(aq) + H_2O(\ell) + 2\ e^-$	$\longrightarrow Cl^-(aq) + 2\ OH^-(aq)$	+0.89
$Hg^{2+}(aq) + 2\ e^-$	$\longrightarrow Hg(\ell)$	+0.855
$Ag^+(aq) + e^-$	$\longrightarrow Ag(s)$	+0.7994
$Hg_2^{2+}(aq) + 2\ e^-$	$\longrightarrow 2\ Hg(\ell)$	+0.789
$Fe^{3+}(aq) + e^-$	$\longrightarrow Fe^{2+}(aq)$	+0.771
$I_2(s) + 2\ e^-$	$\longrightarrow 2\ I^-(aq)$	+0.535
$O_2(g) + 2\ H_2O(\ell) + 4\ e^-$	$\longrightarrow 4\ OH^-(aq)$	+0.403
$Cu^{2+}(aq) + 2\ e^-$	$\longrightarrow Cu(s)$	+0.337
$Sn^{4+}(aq) + 2\ e^-$	$\longrightarrow Sn^{2+}(aq)$	+0.15
$2\ H_3O^+(aq) + 2\ e^-$	$\longrightarrow H_2(g) + 2\ H_2O(\ell)$	0.00
$Sn^{2+}(aq) + 2\ e^-$	$\longrightarrow Sn(s)$	−0.14
$Ni^{2+}(aq) + 2\ e^-$	$\longrightarrow Ni(s)$	−0.25
$PbSO_4(s) + 2\ e^-$	$\longrightarrow Pb(s) + SO_4^{2-}(aq)$	−0.356
$Cd^{2+}(aq) + 2\ e^-$	$\longrightarrow Cd(s)$	−0.403
$Fe^{2+}(aq) + 2\ e^-$	$\longrightarrow Fe(s)$	−0.44
$Zn^{2+}(aq) + 2\ e^-$	$\longrightarrow Zn(s)$	−0.763
$2\ H_2O(\ell) + 2\ e^-$	$\longrightarrow H_2(g) + 2\ OH^-(aq)$	−0.8277
$Al^{3+}(aq) + 3\ e^-$	$\longrightarrow Al(s)$	−1.66
$Mg^{2+}(aq) + 2\ e^-$	$\longrightarrow Mg(s)$	−2.37
$Na^+(aq) + e^-$	$\longrightarrow Na(s)$	−2.714
$K^+(aq) + e^-$	$\longrightarrow K(s)$	−2.925
$Li^+(aq) + e^-$	$\longrightarrow Li(s)$	−3.045

*In volts (V) versus the standard hydrogen electrode.

4. ***The less positive the value of the standard reduction potential,* E°, *the less likely the reaction will occur as a reduction, and the more likely an oxidation (the reverse reaction) will occur.*** The farther down we go in the table, the better the reducing (electron donating) ability of the atom, ion, or molecule on the right. Thus, Li(s) is the strongest reducing agent in the table, and F^- is the weakest reducing agent in the table. Other strong reducing agents are alkali and alkaline earth metals and hydrogen at the lower right of the table.

ThomsonNOW™
Go to the Coached Problems menu for a module on **voltaic cells under standard conditions.**

5. ***Under standard conditions, any species on the left of a half-reaction will oxidize any species that is below it on the right side of the table.*** For example, we can apply this rule to predict that $Fe^{3+}(aq)$ will oxidize Al(s), $Br_2(\ell)$ will oxidize Mg(s), and $Na^+(aq)$ will oxidize Li(s). The net reaction is found by adding the half-reactions, and the cell voltage can be calculated from the $E^\circ_{cell} = E^\circ_{cathode} - E^\circ_{anode}$ equation, as illustrated below.

$$\begin{array}{lll} Br_2(\ell) + 2\,e^- \longrightarrow 2\,Br^-(aq) & & E^\circ_{cathode} = +1.07\ V \\ Mg(s) \longrightarrow Mg^{2+}(aq) + 2\,e^- & & E^\circ_{anode} = -2.37\ V \\ \hline Br_2(\ell) + Mg(s) \longrightarrow Mg^{2+}(aq) + 2\,Br^-(aq) & & E^\circ_{cell} = +3.45\ V \end{array}$$

$$E^\circ_{cell} = E^\circ_{cathode} - E^\circ_{anode} = 1.07\ V - (-2.37\ V) = +3.44\ V$$

A positive cell potential denotes a product-favored reaction.

6. ***Electrode potentials depend on the nature and concentration of reactants and products, but not on the quantity of each that reacts.*** Changing the stoichiometric coefficients for a half-reaction does *not* change the value of E°. For example, the reduction of Fe^{3+} has an E° of +0.771 V whether the reaction is written as

$$Fe^{3+}(aq, 1\ M) + e^- \longrightarrow Fe^{2+}(aq, 1\ M) \qquad E^\circ = +0.771\ V$$

or

$$2\,Fe^{3+}(aq, 1\ M) + 2\,e^- \longrightarrow 2\,Fe^{2+}(aq, 1\ M) \qquad E^\circ = +0.771\ V$$

In this respect, E° values differ from ΔH° and ΔG° values, which do depend on the coefficients in thermochemical equations.

This fact about half-cell potentials seems unusual at first, but consider that a half-cell voltage is energy per unit charge (1 volt = 1 joule/1 coulomb). When a half-reaction is multiplied by some number, both the energy and the charge are multiplied by that number. Thus the ratio of the energy to the charge (voltage) does not change.

Using the preceding guidelines and the table of standard reduction potentials, we will make some *predictions* about whether reactions will occur and then check our results by calculating E°_{cell}.

CHEMISTRY YOU CAN DO

Remove Tarnish the Easy Way

Silverware tarnishes when exposed to air because the silver reacts with hydrogen sulfide gas in the air to form a thin coating of black silver sulfide, Ag_2S. You can use chemistry to remove the tarnish from silverware and other silver utensils with a solution of baking soda and some aluminum foil. The chemical cleaning of silver is an electrochemical process in which electrons move from aluminum atoms to silver ions in the tarnish, reducing silver ions to silver atoms and oxidizing aluminum atoms to aluminum ions. Look up the position of Ag^+ ion with respect to aluminum metal in Table 19.1. The sodium bicarbonate provides a conductive ionic solution for the flow of electrons and helps to remove the aluminum oxide coating from the surface of the aluminum foil.

Start by putting 1 to 2 L water in a large pan. Add 7 to 8 Tbsp baking soda. Heat the solution, but do not boil it. Place some aluminum foil in the bottom of the pan, and put the tarnished silverware on the aluminum foil. Make sure the silverware is covered with water. Heat the solution almost to boiling. After a few minutes remove the silverware and rinse it in running water.

This method of cleaning silverware is better than using polish, because polish removes the silver sulfide, including the silver it contains. The chemical process described here restores the silver from the tarnish to the surface. If you have aluminum pie pans or aluminum cooking pans, you can use them as both the container and the aluminum source. You may notice that devices for removing silver tarnish are sometimes advertised on television. These devices are actually little more than a piece of aluminum metal and some salt. Would you be willing to pay very much (plus shipping and handling) for such a device after you have done this simple experiment?

1. In the reaction between silver and H_2S, what is being oxidized? What is being reduced?
2. What metal other than aluminum could be used for this reaction?

PROBLEM-SOLVING EXAMPLE 19.7 Predicting Redox Reactions

(a) Will zinc metal react with a 1 M $Ag^+(aq)$ solution? If so, what is $E°$ for the reaction?
(b) Will a 1 M $Fe^{2+}(aq)$ solution react with metallic tin? If so, what is $E°$ for the reaction?

Answer

(a) Yes. $E°_{cell} = +1.56$ V
(b) No. $E°_{cell}$ is negative.

Strategy and Explanation We will answer the questions by referring to Table 19.1 and comparing the positions of the reactants there.

(a) $Ag^+(aq)$ is above metallic zinc in Table 19.1, so it is a better oxidizing agent, and we predict that it can oxidize zinc, causing metallic zinc atoms to form $Zn^{2+}(aq)$ ions. To be certain, we combine the half-cell reactions to give the balanced equation. We subtract the half-cell potentials, which yields a positive $E°_{cell}$, so this reaction is product-favored, as we predicted from Table 19.1.

$$Zn(s) \longrightarrow Zn^{2+}(aq, 1\ M) + 2\ e^- \qquad E°_{anode} = -0.763\ V$$

$$2\ [Ag^+(aq, 1\ M) + e^- \longrightarrow Ag(s)] \qquad E°_{cathode} = +0.80\ V$$

$$Zn(s) + 2\ Ag^+(aq, 1\ M) \longrightarrow Zn^{2+}(aq, 1\ M) + 2\ Ag(s) \qquad E°_{cell} = ?$$

$$E°_{cell} = E°_{cathode} - E°_{anode} = +0.80\ V - (-0.763\ V) = +1.56\ V$$

Note that the half-reaction voltages are not multiplied by the balancing coefficients.

The positive value for $E°_{cell}$ shows that this is a product-favored reaction.

(b) We evaluate the reaction between $Fe^{2+}(aq)$ and metallic Sn the same way. $Fe^{2+}(aq)$ is on the left in Table 19.1 below Sn(s), which is on the right. Therefore, $Fe^{2+}(aq)$ is not a strong enough oxidizing agent to oxidize Sn(s), and we predict that this reaction will not occur. The combined half-reactions are

$$Sn(s) \longrightarrow Sn^{2+}(aq) + 2e^- \qquad E°_{anode} = -0.14\ V$$

$$Fe^{2+}(aq) + 2\ e^- \longrightarrow Fe(s) \qquad E°_{cathode} = -0.44\ V$$

$$Sn(s) + Fe^{2+}(aq) \longrightarrow Sn^{2+}(aq) + Fe(s) \qquad E°_{cell} = ?$$

$$E°_{cell} = E°_{cathode} - E°_{anode} = -0.44\ V - (-0.14\ V) = -0.30\ V$$

The negative value for $E°_{cell}$ shows that this process is reactant-favored, and it will not form appreciable quantities of products under standard conditions. In fact, iron metal will reduce Sn^{2+}, the reverse of the net reaction shown above.

PROBLEM-SOLVING PRACTICE 19.7

Look at Table 19.1 and determine which two half-reactions would produce the largest value of $E°_{cell}$. Write the two half-reactions and the overall cell reaction, and give the $E°$ for the reaction.

CONCEPTUAL EXERCISE 19.5 Using $E°$ Values

Transporting chemicals is of great practical and economic importance. Suppose that you have a large volume of mercury(II) chloride solution, $HgCl_2$, that needs to be transported. A driver brings a tanker truck made of aluminum to the loading dock. Will it be okay to load the truck with your solution? Explain your answer fully.

Standard reduction potentials can be used to explain an annoying experience many of us have had—a pain in a tooth when a filling is touched with a stainless steel fork or a piece of aluminum foil. A common material for dental fillings is a dental amalgam—tin and silver dissolved in mercury to form solid solutions having compositions approximating Ag_2Hg_3, Ag_3Sn, and Sn_xHg (where x ranges from 7 to 9). All of these compounds can undergo electrochemical reactions; for example,

Stainless steel is an alloy of iron.

$$3\ Hg_2^{2+}(aq) + 4\ Ag(s) + 6\ e^- \longrightarrow 2\ Ag_2Hg_3(s) \qquad E° = +0.85\ V$$

$$Sn^{2+}(aq) + 3\ Ag(s) + 2\ e^- \longrightarrow Ag_3Sn(s) \qquad E° = -0.05\ V$$

The $E°$ values in Table 19.1 indicate that both iron and aluminum have much more negative reduction potentials and, therefore, are much better reducing agents than any of the amalgam fillings. Consequently, if a piece of iron or aluminum comes in contact with a dental filling, the saliva and gum tissue will act as a salt bridge, resulting in an electrochemical cell. The iron or aluminum donates electrons, producing a tiny electrical current that activates a nerve and results in pain.

Before we leave this discussion of Table 19.1, consider again the activity series of metals shown in Table 5.5 (⇐ ***p. 188***), which contains many of the same elements as Table 19.1. Looking closely, however, you will notice that the most active metal, lithium, at the top of Table 5.5 is at the very bottom right of Table 19.1. That is because Table 19.1 is arranged by *reduction potential,* and lithium ion has the lowest tendency to be reduced. Table 5.5, on the other hand, lists the metals in order of activity, that is, their tendency to be oxidized. Since oxidation is the opposite of reduction, it is reasonable that lithium is in opposite positions in the two tables.

CONCEPTUAL EXERCISE 19.6 Predicting Redox Reactions Using $E°$ Values

Consider these reduction half-reactions:

Half-Reaction	$E°$ (V)
$Cl_2(g) + 2\,e^- \longrightarrow 2\,Cl^-(aq)$	+1.36
$I_2(s) + 2\,e^- \longrightarrow 2\,I^-(aq)$	+0.535
$Pb^{2+}(aq) + 2\,e^- \longrightarrow Pb(s)$	−0.126
$V^{2+}(aq) + 2\,e^- \longrightarrow V(s)$	−1.18

(a) Which is the weakest oxidizing agent?
(b) Which is the strongest oxidizing agent?
(c) Which is the strongest reducing agent?
(d) Which is the weakest reducing agent?
(e) Will Pb(s) reduce $V^{2+}(aq)$ to V(s)?
(f) Will $I_2(s)$ oxidize $Cl^-(aq)$ to $Cl_2(g)$?
(g) Name the molecules or ions in the above reactions that can be reduced by Pb(s).

CONCEPTUAL EXERCISE 19.7 Predicting $E°$ Values

The two elements on either side of hydrogen in Table 5.5 (lead and antimony) are not listed in Table 19.1. Indicate where they would appear in Table 19.1 and, based on values from the table, estimate the reduction potentials for their positive ions being reduced to the metal atom.

19.6 $E°$ and Gibbs Free Energy

The sign of $E°_{cell}$ indicates whether a redox reaction is product-favored (positive $E°$) or reactant-favored (negative $E°$). Earlier you learned another way to decide whether a reaction is product-favored: The change in standard Gibbs free energy, $\Delta G°$, must be negative (⇐ ***p. 883***). Since both $E°_{cell}$ and $\Delta G°$ tell something about whether a reaction will occur, it should be no surprise that a relationship exists between them.

The "free" in Gibbs free energy indicates that it is energy available to do work. The energy available for electrical work from an electrochemical cell can be calculated by multiplying the quantity of electrical charge transferred times the cell voltage, $E°$. The quantity of charge is given by the number of moles of electrons transferred in the overall reaction, n, multiplied by the number of coulombs per mole of electrons.

Quantity of charge = moles of electrons × coulombs per mole of electrons

The charge on 1 mol of electrons can be calculated from the charge on one electron and Avogadro's number.

$$\text{Charge on 1 mol of e}^- = \left(\frac{1.60218 \times 10^{-19}\ \text{C}}{\text{e}^-}\right)\left(\frac{6.02214 \times 10^{23}\ \text{e}^-}{1\ \text{mol e}^-}\right)$$

$$= 9.6485 \times 10^4\ \text{C/mol e}^-$$

The quantity 9.6485×10^4 C/mol of electrons (commonly rounded to 96,500 C/mol of electrons) is known as the **Faraday constant (*F*)** in honor of Michael Faraday, who first explored the quantitative aspects of electrochemistry.

The electrical work that can be done by a cell is equal to the Faraday constant (F) multiplied by the number of moles of electrons transferred (n) and by the cell voltage ($E°_{cell}$).

$$\text{Electrical work} = nFE°_{cell}$$

Unlike the cell voltage, the electrical work a cell can do *does* depend on the quantity of reactants in the cell reaction. More reactants mean more moles of electrons transferred and hence more work. Equating the electrical work of a cell at standard conditions with $\Delta G°$, we get

$$\Delta G° = -nFE°_{cell}$$

The negative sign on the right side of the equation accounts for the fact that *$\Delta G°$ is always negative for a product-favored process, but $E°_{cell}$ is always positive for a product-favored process.* Thus, these values must have opposite signs.

Using this equation we can calculate $\Delta G°$ for the Cu^{2+}/Zn cell. This value represents the maximum work that the cell can do. The reaction is

$$Cu^{2+}(aq) + Zn(s) \longrightarrow Cu(s) + Zn^{2+}(aq) \qquad E°_{cell} = +1.10\ \text{V}$$

so 2 mol electrons are transferred per mole of copper ions reduced. The Gibbs free energy change when this quantity of reactants is converted is

$$\Delta G° = -\left(\frac{2\ \text{mol e}^-}{\text{transferred}}\right)\left(\frac{9.65 \times 10^4\ \text{C}}{\text{mol e}^-}\right)\left(\frac{1\ \text{J}}{1\ \text{V} \times 1\ \text{C}}\right)\left(\frac{1\ \text{kJ}}{10^3\ \text{J}}\right)(+1.10\ \text{V})$$

$$= -212\ \text{kJ}$$

The positive $E°_{cell}$ and the negative Gibbs free energy change values indicate a very product-favored reaction.

Michael Faraday

1791–1867

As an apprentice to a London bookbinder, Michael Faraday became fascinated by science when he was a boy. At age 22 he was appointed as a laboratory assistant at the Royal Institution and became its director within 12 years. A skilled experimenter in chemistry and physics, he made many important discoveries, the most important of which was electromagnetic induction, the basis of modern electromagnetic technology. Faraday built the first electric motor, generator, and transformer. A popular speaker and educator, he also performed chemical and electrochemical experiments, and he first synthesized benzene.

PROBLEM-SOLVING EXAMPLE 19.8 Determining $E°_{cell}$ and $\Delta G°$

Consider the redox reaction

$$Zn^{2+}(aq) + H_2(g) + 2\ H_2O(\ell) \longrightarrow Zn(s) + 2\ H_3O^+(aq)$$

Use the standard reduction potentials in Table 19.1 to calculate $E°_{cell}$ and $\Delta G°$ and to determine whether the reaction as written favors product formation.

Answer $E°_{cell} = -0.763$ V; $\Delta G° = 147$ kJ. The reaction as written is not product-favored.

ThomsonNOW
Go to the Coached Problems menu for a tutorial on **calculating standard cell potentials.**

Strategy and Explanation The first step is to write the half-reactions for the oxidation and reduction that occur and to use their standard reduction potentials from Table 19.1.

Reduction: $Zn^{2+}(aq) + 2\,e^- \longrightarrow Zn(s)$ $\quad E^\circ_{cathode} = -0.763$ V

Oxidation: $H_2(g) + 2\,H_2O(\ell) \longrightarrow 2\,H_3O^+(aq) + 2\,e^-$ $\quad E^\circ_{anode} = 0$ V

We obtain the E°_{cell} for the reaction from the potentials for the two half-reactions.

$$E^\circ_{cell} = E^\circ_{cathode} - E^\circ_{anode} = -0.763\text{ V} - 0\text{ V} = -0.763\text{ V}$$

Since the value of E°_{cell} is negative, the reaction is not product-favored in the direction written. Zn^{2+} will not oxidize H_2O. The reverse reaction is product-favored; that is, zinc metal reacts with acid.

From the calculated E°_{cell} we can calculate ΔG°.

$$\Delta G^\circ = -nFE^\circ_{cell}$$

$$= -(2\text{ mol e}^-) \times \left(\frac{9.65 \times 10^4\text{ C}}{1\text{ mol e}^-}\right)\left(\frac{1\text{ J}}{1\text{ V} \times 1\text{ C}}\right)(-0.763\text{ V})$$

$$= 1.47 \times 10^5\text{ J} = 147\text{ kJ}$$

The positive value for ΔG° also shows that the reaction as written is not product-favored.

✓ **Reasonable Answer Check** The given oxidation is the half-reaction at the standard hydrogen electrode with a potential of zero, so the overall reaction is governed by the Zn reduction. The negative cell potential and the positive ΔG° are consistent with the reaction not being product-favored as written.

PROBLEM-SOLVING PRACTICE 19.8

Using standard reduction potentials, determine whether this reaction is product-favored as written.

$$Hg^{2+}(aq) + 2\,I^-(aq) \longrightarrow Hg(\ell) + I_2(s)$$

ThomsonNOW
Go to the Coached Problems menu for tutorials on:

- **cell potential and the equilibrium constant**
- **cell potential and Gibbs free energy change**

ΔG°, E°_{cell}, and K°

We have seen that the standard Gibbs free energy change is directly proportional to the E°_{cell} for an electrochemical cell at standard conditions

$$\Delta G^\circ = -nFE^\circ_{cell}$$

Recall from Chapter 18 (p. 889) that the standard Gibbs free energy change is directly proportional to the logarithm of the equilibrium constant of a reaction.

$$\Delta G^\circ = -RT \ln K^\circ$$

Putting these two equations together yields

$$-nFE^\circ_{cell} = -RT \ln K^\circ$$

which, when solved for E°_{cell}, yields

$$E^\circ_{cell} = \frac{RT}{nF} \ln K^\circ$$

Thus by measuring E°_{cell}, the values of ΔG° and K° can be calculated. The relationships linking ΔG°, E°_{cell}, and K° are summarized in Figure 19.10.

The E°_{cell} expression can be simplified by substituting numerical values for R (8.314 J mol^{-1} K^{-1}) and F (96,485 J V^{-1} mol^{-1}) and assuming standard-state temperature (298 K). For n electrons transferred we get

$$E^\circ_{cell} = \frac{RT}{nF} \ln K^\circ = \frac{(8.314\text{ J mol}^{-1}\text{ K}^{-1})(298\text{ K})}{n\,(96{,}485\text{ J V}^{-1}\text{mol}^{-1})} = \frac{0.0257\text{ V}}{n} \ln K^\circ$$

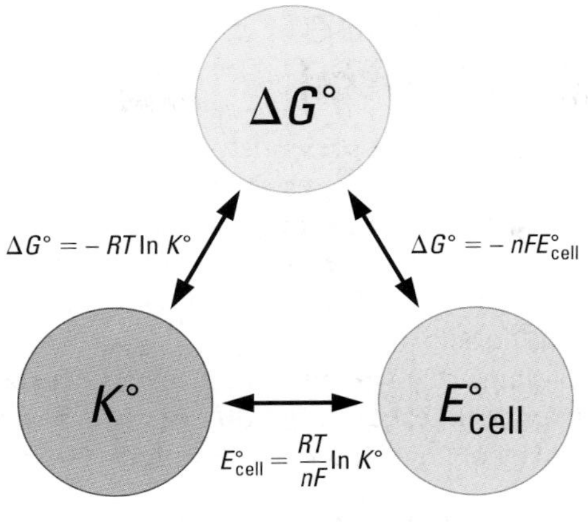

Figure 19.10 The relationships linking $\Delta G°$, $E°_{cell}$, and $K°$. Given any one of the values, the other two can be calculated.

Changing from natural logarithms to base 10 logarithms (multiplying by 2.303) yields

2.303 log(x) = ln(x)

$$E°_{cell} = \frac{0.0592 \text{ V}}{n} \log K° \qquad \text{and} \qquad \log K° = \frac{nE°_{cell}}{0.0592 \text{ V}} \quad \text{(at 298 K)}$$

These equations hold for standard states of all reactants and products.

PROBLEM-SOLVING EXAMPLE 19.9 Equilibrium Constant for a Redox Reaction

Calculate the equilibrium constant K_c for the reaction

$$Fe(s) + Cd^{2+}(aq) \rightleftharpoons Fe^{2+}(aq) + Cd(s)$$

using the standard reduction potentials listed in Table 19.1.

Answer $K_c = K° = 22$

Strategy and Explanation We first need to calculate $E°_{cell}$. To do so we separate the reaction into its two half-reactions.

$$Fe(s) \longrightarrow Fe^{2+}(aq) + 2\ e^- \qquad E°_{anode} = -0.44 \text{ V}$$

$$Cd^{2+}(aq) + 2\ e^- \longrightarrow Cd(s) \qquad E°_{cathode} = -0.40 \text{ V}$$

$$Fe(s) + Cd^{2+}(aq) \longrightarrow Fe^{2+}(aq) + Cd(s) \qquad E°_{cell} = +0.04 \text{ V}$$

Two electrons are transferred.

$$\log K° = \frac{nE°_{cell}}{0.0592 \text{ V}} = \frac{(2)(0.04 \text{ V})}{0.0592 \text{ V}} = 1.35 \qquad \text{and} \qquad K = 10^{1.35} = 22$$

The $K°$ value is larger than 1, which shows that the reaction is product-favored as written. Since the reaction occurs in aqueous solution, $K_c = K° = 22$.

✓ **Reasonable Answer Check** The value for $E°_{cell}$ is positive, which indicates a product-favored reaction, as does $K° > 1$.

PROBLEM-SOLVING PRACTICE 19.9

Using the standard reduction potentials listed in Table 19.1, calculate the equilibrium constant for the reaction

$$I_2(s) + Sn^{2+}(aq) \longrightarrow 2\ I^-(aq) + Sn^{4+}(aq)$$

19.7 Effect of Concentration on Cell Potential

The electrochemical cells discussed previously, all voltaic cells, are based on product-favored chemical reactions. As the reactions proceed in such an electrochemical cell, reactants are consumed and products are generated, so the concentrations of the species change continuously. As the reactant concentrations decrease, the voltage produced by the cell drops. The voltage finally reaches zero when the reactants and products are at equilibrium.

We can relate the voltage of a voltaic cell to the concentrations of the reactants and products of its chemical reaction. To do so, we start with the relationship between Gibbs free energy change and concentration *(⇐ p. 891)*:

$$\Delta G = \Delta G^\circ + RT \ln Q$$

where Q is the reaction quotient ***(⇐ p. 689)***. Q has the same form as the equilibrium constant, but it refers to a reaction mixture at a given instant in time, that is, a reaction mixture not necessarily at equilibrium.

We know that $\Delta G = -nFE_{\text{cell}}$ and $\Delta G^\circ = -nFE^\circ_{\text{cell}}$, so

$$-nFE_{\text{cell}} = -nFE^\circ_{\text{cell}} + RT \ln Q$$

E°_{cell} is the voltage at standard-state conditions; E_{cell} is the voltage at nonstandard conditions.

Rearranging this for E_{cell} gives the relationship we seek:

$$E_{\text{cell}} = E^\circ_{\text{cell}} - \frac{RT}{nF} \ln Q$$

In the Nernst equation, n is the number of moles of electrons transferred in the *balanced equation* of the process.

This is the **Nernst equation.** We can change to base 10 logarithms (multiply by 2.303) to get

$$E_{\text{cell}} = E^\circ_{\text{cell}} - \frac{2.303\,RT}{nF} \log Q$$

ThomsonNOW
Go to the Coached Problems menu for a tutorial on **calculating cell potential under nonstandard conditions.**

We can simplify this expression further by substituting numerical values for R (8.314 J mol^{-1} K^{-1}) and F (96,485 J V^{-1} mol^{-1}) and by assuming 25 °C (298 K) to get

$$E_{\text{cell}} = E^\circ_{\text{cell}} - \frac{(2.303)(8.314\text{ J mol}^{-1}\text{ K}^{-1})(298\text{ K})}{n(96{,}485\text{ J V}^{-1}\text{ mol}^{-1})} \log Q$$

$$E_{\text{cell}} = E^\circ_{\text{cell}} - \frac{0.0592\text{ V}}{n} \log Q \qquad (T = 298\text{ K})$$

If all the concentrations in Q are equal to 1 (which is the standard state), then $Q = 1$ and $\log(1) = 0$, so the Nernst equation reduces to $E_{\text{cell}} = E^\circ_{\text{cell}}$.

The Nernst equation can be used to calculate the voltage produced by an electrochemical cell under nonstandard conditions. It can also be used to calculate the concentration of a reactant or product in an electrochemical reaction from the measured value of the voltage produced.

PROBLEM-SOLVING EXAMPLE 19.10 Using the Nernst Equation

Consider this electrochemical reaction:

$$Zn(s) + Ni^{2+}(aq) \longrightarrow Zn^{2+}(aq) + Ni(s)$$

The standard cell potential $E^\circ_{\text{cell}} = 0.51$ V. Find the cell potential if the Ni^{2+} concentration is 5.0 M and the Zn^{2+} concentration is 0.050 M.

Answer 0.57 V

Strategy and Explanation We use the Nernst equation to solve the problem. Two moles of electrons are transferred from 1 mol Zn to 1 mol Ni^{2+}, giving $n = 2$. At 298 K,

$$E^\circ_{cell} = 0.51\text{ V} - \frac{0.0592\text{ V}}{2}\log\frac{(\text{conc. }Zn^{2+})}{(\text{conc. }Ni^{2+})}$$

$$E^\circ_{cell} = 0.51\text{ V} - \frac{0.0592\text{ V}}{2}\log\left(\frac{0.050}{5.0}\right) = 0.51\text{ V} - \frac{0.0592\text{ V}}{2}\log(10^{-2})$$

$$= 0.51\text{ V} - \frac{0.0592\text{ V}}{2}(-2.00) = 0.57\text{ V}$$

✓ **Reasonable Answer Check** The reactant concentration of Ni^{2+} is 5.0 M, larger than the standard-state value of 1.0 M, and the product concentration of Zn^{2+} is 0.050 M, smaller than the standard-state value. Each of these departures from standard-state conditions tends to make the voltage under these conditions slightly larger than the standard cell potential ($E^\circ_{cell} = 0.51$ V), and it is.

PROBLEM-SOLVING PRACTICE 19.10

What would the cell potential in the chemical system above become if (conc. Zn^{2+}) = 3.0 M and (conc. Ni^{2+}) = 0.010 M?

Concentration Cells

The voltaic cells discussed to this point have different reactions proceeding at the anode and the cathode. However, since the voltage of a cell depends on the concentrations of the reactants, we can construct a voltaic cell that uses the same species in both the anode and the cathode compartments but at different concentrations. A **concentration cell** is a voltaic cell in which the voltage is generated because of a difference in concentrations. As the cell operates, the concentration increases in the dilute cell and decreases in the concentrated cell.

Consider a concentration cell constructed with two identical Cu/Cu^{2+} half-reactions occurring in separated compartments, as shown in Figure 19.11. If the same half-reactions occurred in the two compartments *at standard conditions* of 1 M concentrations, the potentials at the two electrodes would be the same ($E^\circ = +0.337$ V), so the cell potential would be zero. However, in a concentration cell the half-reactions

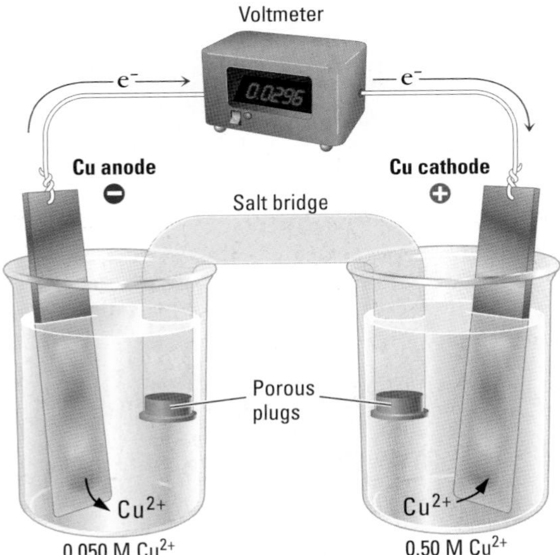

Figure 19.11 **Concentration cell based on Cu/Cu^{2+} half-reactions.** The cell has a positive net cell voltage and operates because the concentrations of Cu^{2+} ion are different in the two half-reaction compartments.

are the same, but the concentrations in the two cells are different. Let's take as an example Cu^{2+} concentrations of 0.050 M in the anode half-cell and 0.50 M in the cathode half-cell. The two half-reactions are

$$\begin{array}{rcll} Cu(s) & \longrightarrow & Cu^{2+}(aq, 0.050\text{ M}) + 2\,e^- & \text{(anode, oxidation)} \\ Cu^{2+}(aq, 0.50\text{ M}) + 2\,e^- & \longrightarrow & Cu(s) & \text{(cathode, reduction)} \\ \hline Cu^{2+}(aq, 0.50\text{ M}) & \longrightarrow & Cu^{2+}(aq, 0.050\text{ M}) & \text{(net reaction)} \end{array}$$

The cell potential is expressed by the Nernst equation:

$$\begin{aligned} E_{cell} &= E^\circ_{cell} - \frac{0.0592\text{ V}}{2}\log\frac{(\text{conc. }Cu^{2+})_{\text{dilute}}}{(\text{conc. }Cu^{2+})_{\text{concentrated}}} \\ &= 0 - 0.0296 \times \log\frac{0.050\text{ M}}{0.50\text{ M}} \\ &= 0 - 0.0296 \times \log(10^{-1}) = 0 - 0.0296 \times (-1.00) = 0.0296\text{ V} \end{aligned}$$

The cell potential is entirely determined by the ratio of concentrations of Cu^{2+} ions in the two half-reaction cells. The value of 0.0296 V is the potential of the cell when the concentrations of Cu^{2+} are at the initial conditions of 0.050 M and 0.50 M. As the cell operates, the concentration of Cu^{2+} in each half-cell changes; it increases in the dilute cell and decreases in the concentrated cell. Eventually, the two concentrations become equal, and the cell potential becomes zero.

Measurement of pH

The H^+ concentration in a solution can be measured using the principles of a concentration cell. Consider a concentration cell based on the H_2/H^+ half-reaction. The cathode compartment contains a standard hydrogen electrode with known H^+ concentration (1.0 M), and the anode compartment contains the same type of electrode in contact with a solution of unknown H^+ concentration. The half-reactions and the overall reaction are

$$\begin{array}{rcll} H_2(g, 1\text{ bar}) & \longrightarrow & 2\,H^+(aq, \text{unknown}) + 2\,e^- & \text{(anode, oxidation)} \\ 2\,H^+(aq, 1\text{ M}) + 2\,e^- & \longrightarrow & H_2(g, 1\text{ bar}) & \text{(cathode, reduction)} \\ \hline 2\,H^+(aq, 1\text{ M}) & \longrightarrow & 2\,H^+(aq, \text{unknown}) & E_{cell} = ? \end{array}$$

The *standard* potential of the cell would be zero: $E^\circ_{cell} = 0$. However, the two half-cells have different hydrogen ion concentrations, so E_{cell} is not zero.

To analyze the cell further we use the Nernst equation with $n = 2$.

$$E_{cell} = E^\circ_{cell} - \frac{0.0592\text{ V}}{2} \times \log\frac{(\text{conc. }H^+)^2_{\text{unknown}}}{(\text{conc. }H^+)^2_{\text{standard}}}$$

$[H^+]_{\text{standard}} = 1$ M and $E^\circ_{cell} = 0$, so

$$E_{cell} = -\frac{0.0592\text{ V}}{2} \times \log\,(\text{conc. }H^+)^2_{\text{unknown}}$$

Since $\log x^2 = 2\log x$,

$$\begin{aligned} E_{cell} &= \frac{-0.0592\text{ V}}{2} \times 2\log\,(\text{conc. }H^+)_{\text{unknown}} \\ &= -0.0592\text{ V} \times \log\,(\text{conc. }H^+)_{\text{unknown}} \end{aligned}$$

Because $-\log\,[H^+] = \text{pH}$, the final expression is

$$E_{cell} = 0.0592\text{ V} \times [-\log\,(\text{conc. }H^+)_{\text{unknown}}] = 0.0592\text{ V} \times \text{pH}_{\text{unknown}}$$

$$\text{pH}_{\text{unknown}} = \frac{E_{cell}}{0.0592\text{ V}}$$

Therefore, by measuring E_{cell}, the $\text{pH}_{\text{unknown}}$ can be determined.

PROBLEM-SOLVING EXAMPLE 19.11 Measuring pH with a Concentration Cell

Consider a concentration cell consisting of two hydrogen electrodes, which can be used to measure pH. One of the cells is a standard hydrogen electrode, and the other cell is in contact with an aqueous solution having an unknown pH. If the unknown concentration of H^+ is less than 1.0 M (which is generally true), then reduction occurs at the standard hydrogen electrode (cathode), and oxidation occurs at the nonstandard hydrogen electrode (anode). If the measured potential of the cell is 0.426 V, what is the pH of the unknown solution at 25 °C?

Answer pH = 7.20

Strategy and Explanation We write the two half-reactions to start.

$$2\,H^+\,(1\text{ M}) + 2\,e^- \longrightarrow H_2(g,\,1\text{ bar}) \qquad \text{(reduction)}$$

$$H_2(g,\,1\text{ bar}) \longrightarrow 2\,H^+\,(\text{unknown M}) + 2\,e^- \qquad \text{(oxidation)}$$

$$2\,H^+(1\text{ M}) \longrightarrow 2\,H^+\,(\text{unknown M}) \qquad E_{cell} = 0.426\text{ V}$$

The Nernst equation at 25 °C is

$$E_{cell} = -\frac{0.0592\text{ V}}{2}\log\frac{(\text{unknown conc. }H^+)^2}{(H^+\ 1\text{ M})^2}$$

$$= -\frac{0.0592\text{ V}}{2} \times \log\,(\text{unknown conc. }H^+)^2 = 0.426\text{ V}$$

$$= -\frac{0.0592\text{ V}}{2} \times 2\log\,(\text{unknown conc. }H^+) = 0.426\text{ V}$$

$$E_{cell} = -0.0592\text{ V} \times \log\,(\text{unknown conc. }H^+) = 0.426\text{ V}$$

By definition pH = $-\log\,[H^+]$, so

$$0.0592\text{ V} \times \text{pH} = 0.426\text{ V} \quad \text{and} \quad \text{pH} = \frac{0.426\text{ V}}{0.0592\text{ V}} = 7.20$$

✓ **Reasonable Answer Check** The general relationship between the variables is pH = $E_{cell}/0.0592$ V, so using approximate values gives $0.43/0.06 \approx 7.2$, which is close to our more exact answer.

PROBLEM-SOLVING PRACTICE 19.11

If the same type of concentration cell were used with a solution of pH of 3.66, what E_{cell} would be measured?

The pH Meter

A concentration cell utilizing two hydrogen electrodes is not the best practical choice for routine pH measurement because it is bulky and difficult to maintain. Commercial pH meters are based on electrochemical principles similar to those described previously, but with more rugged and economical half-cells. A **pH meter** has two electrodes (Figure 19.12). One is a glass electrode using an Ag/AgCl half-cell dipped in an HCl solution of known concentration. At the tip of this indicator electrode is a very thin glass membrane that is sensitive to H^+ ion concentration differences. The other electrode is a reference electrode known as a *saturated calomel electrode.* It consists of a Pt wire dipped in a paste of Hg_2Cl_2 (calomel), liquid Hg, and saturated KCl solution. The glass electrode measures the H^+ ion concentration of the solution relative to its internal hydrogen ion concentration (***p. 781***). The difference in voltage between the two electrodes is then converted electronically to give the pH of the solution.

In commercial instruments, the indicator and reference electrodes are generally combined in a *combination electrode*.

The pH of aqueous solutions is an extremely important indicator of their chemistry. Medical applications of pH measurements abound, as do environmental applications such as measuring the pH of acid rain (***p. 843***).

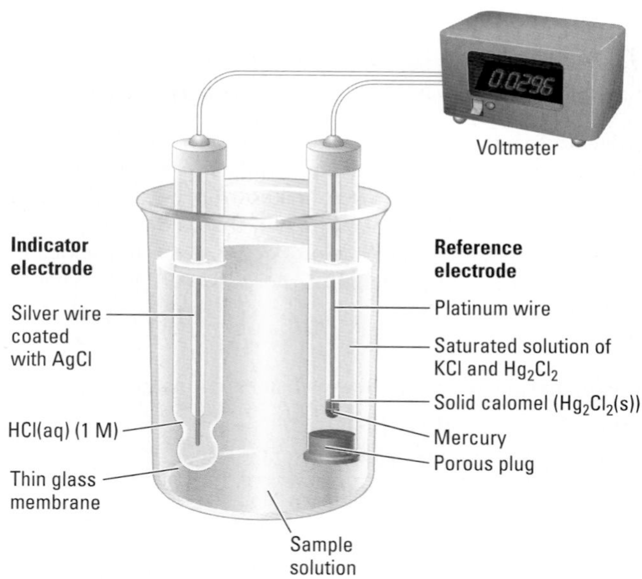

Figure 19.12 **The electrodes and reactions of the pH meter.** The glass electrode is a Ag/AgCl half-cell in a standard HCl solution that is enclosed by a glass membrane. It is sensitive to the external H^+ in the solution relative to the H^+ in the internal standard HCl solution. The saturated calomel electrode is the reference electrode.

19.8 Neuron Cells

Recall (p. 416) that a cell membrane is composed of lipid molecules that separate the aqueous environment inside the cell from the aqueous environment outside the cell.

The central nervous system relies on specialized cells called **neurons** that communicate with each other through chemical signals generated by changes in the intracellular and extracellular ion concentrations (Figure 19.13). The communication between neurons also involves electrical signals generated by millisecond-long alterations in voltage due to changes in ion concentration. These electrical signals depend on the cell membrane of the neuron separating different concentrations of ions inside and outside the cell while it is at rest—that is, while no signals are being sent—as shown in Figure 19.14. These different resting ion concentrations are maintained by a number of processes that move ions across the cell membrane. The main ions of importance are Na^+, K^+, Cl^-, and Ca^{2+}. The intracellular concentrations of these four ions differ markedly from the extracellular concentrations in mammalian cells.

1 mM = 10^{-3} M
1 mV = 10^{-3} V

Ion	Intracellular Concentration (mM)	Extracellular Concentration (mM)	Potentials (mV)
Na^+	18	150	+56
K^+	135	3	−102
Cl^-	7	120	−76
Ca^{2+}	0.0001	1.2	+125

These differences in concentration create potentials across the neuron cell membrane. An associated equilibrium potential for each ion is given by the Nernst equation.

$E°$ has been omitted from the initial Nernst equation because its value is zero as it is for any concentration cell.

$$E_{\text{ion}} = \frac{2.303\,RT}{nF}\log\frac{(\text{conc. ion})_{\text{outside}}}{(\text{conc. ion})_{\text{inside}}}$$

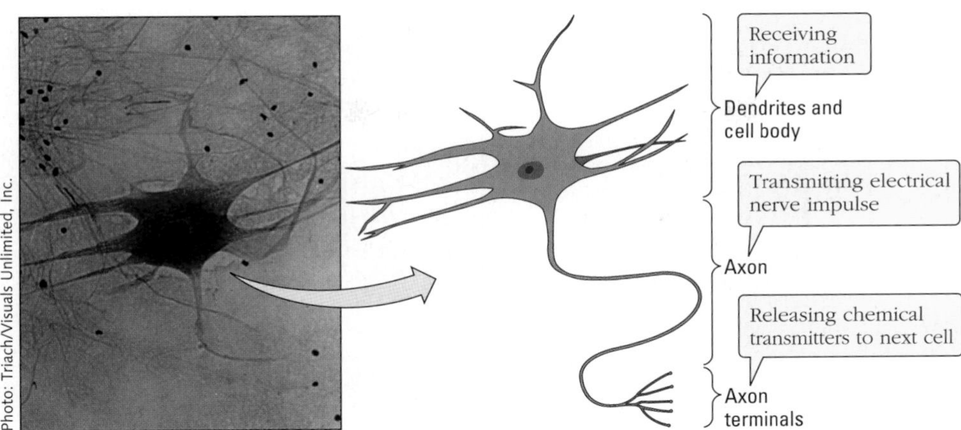

Figure 19.13 **A mammalian neuron cell.**

Substituting numerical values for the constants, R and F, assuming $n = 1$ and a body temperature of 37 °C, and converting to millivolts, we can simplify this expression to

$$E_{\text{ion}} = (61.5 \text{ mV}) \log \frac{(\text{conc. ion})_{\text{outside}}}{(\text{conc. ion})_{\text{inside}}} \qquad \text{(in mV)}$$

This expression computes the potential outside the cell membrane relative to the potential inside the cell membrane for each individual ion. Applying the equation to the specific case of K^+, we have $(\text{conc. } K^+)_{\text{outside}} = 3$ mM and $(\text{conc. } K^+)_{\text{inside}} = 135$ mM, so

$$E_{K^+} = (61.5 \text{ mV}) \log \left(\frac{(\text{conc. } K^+)_{\text{outside}}}{(\text{conc. } K^+)_{\text{inside}}} \right)$$

$$= (61.5 \text{ mV}) \log \left(\frac{3}{135} \right) \text{mV} = 61.5\,(-1.65) \text{ mV} = -102 \text{ mV}$$

The cell membrane has a potential that is 102 mV (0.102 V) more negative on the inside than the outside due to the much higher K^+ concentration inside the cell (Figure 19.15).

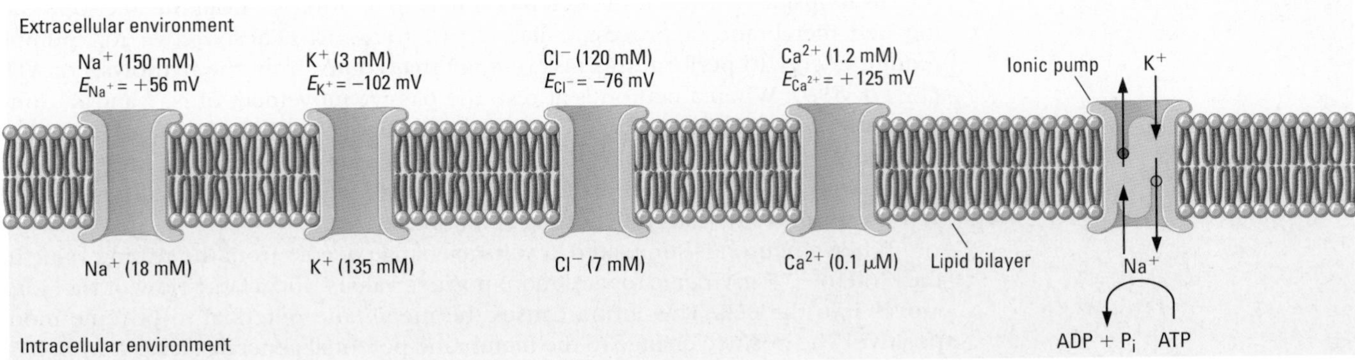

Figure 19.14 **Ion concentrations inside and outside a mammalian neuron cell.** Ion channels for Na^+, K^+, Cl^-, and Ca^{2+} are shown, as is the Na^+-K^+ ion pump. Concentrations are given in millimoles per liter, except for intracellular Ca^{2+}, which is given in micromoles per liter.

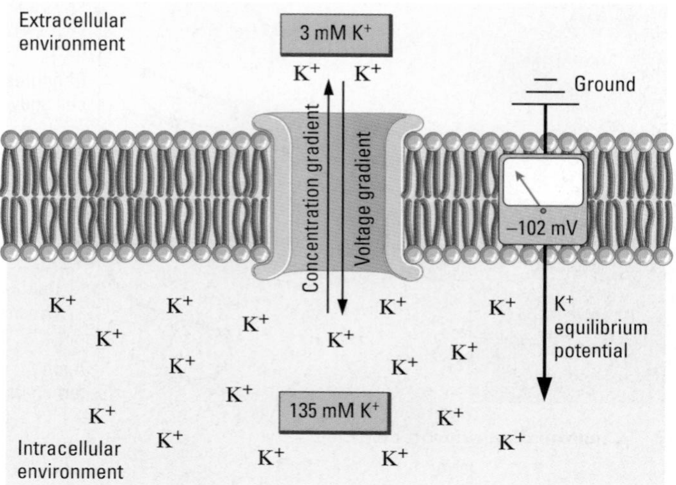

Figure 19.15 **The equilibrium potential for K^+ across a neuron cell membrane.** The K^+ concentration is higher inside the cell than outside it. The Nernst equation explains the −102 mV equilibrium potential for K^+.

Just as for K^+, each important ion shown in Figure 19.14 has a potential that depends on the concentrations of that ion inside and outside the cell membrane. The values for Na^+, K^+, Cl^-, and Ca^{2+} shown in the figure can be used with the Nernst equation to calculate the contribution that each ionic species makes to the final resting potential of the neuron. The *resting membrane potential* for a cell—that is, the potential when no nerve impulse is being transmitted—depends on each of the individual ion potentials. The equilibrium potentials for each of the ions involved are averaged in proportion to the relative permeability of the cell membrane for each ion. The resting membrane potential is different for different types of neuron cells and is in the range of −60 to −75 mV, with the inside of the cell being more negative than the outside.

How do the concentrations of ions inside and outside the cell membrane become different? Ions move across the cell membrane by several mechanisms, but all the ions undergo continual movement. In general, ions tend to move down concentration gradients, that is, from a region of higher to lower concentration. In addition, active *ion pumps* move Na^+ and K^+ *against* their concentration gradients, that is, from regions of lower to higher concentration. Thus, the ion pumps move Na^+ from inside to outside the cell membrane and move K^+ from outside to inside the cell membrane, a process called active transport. These active ion pumps require energy to perform this task, energy that comes from the hydrolysis of ATP (**⟵ *p. 898***). When a neuron is at rest, the passive movement of Na^+ and K^+ ions is exactly counterbalanced by the active movement of Na^+ and K^+ ions via the ion pumps. When a neuron is at rest, no *net* flow of ions occurs, so the ions are all at equilibrium. The concentrations of the ions remain constant, although passive and active transport are both functioning at all times.

When a neuron is stimulated, a voltage change occurs from the resting value in the −60 to −75 mV range toward more positive values, and a large flow of Na^+ ions moves into the cell. This influx causes the membrane potential to become more positive. The positive change in the membrane potential generates an *action potential,* a brief electrical stimulus that results in chemical signaling between the neurons. For a cell to return to its resting potential, the cell membrane potential must return to a more negative value. The membrane potential is changed by a sustained flow of K^+ out of the cell, moving the potential back toward its resting value. The sequence leading to the generation of the action potential occurs during a period of about 1 ms (1×10^{-3} s). Generation of action potentials by this mechanism forms the basis for the transmission of signals along nerve cells in the central nervous system. During this process, the bulk concentrations of K^+ and Na^+ change little, either inside or outside the cell membrane. The number of ions moving across the cell membrane is much less than the total number present. The action potential is an *electrochemical* event related to the change in ion concentrations inside and outside the cell, not to bulk concentrations of the ions.

EXERCISE 19.8 Neuron Equilibrium Potential

What would the membrane potential, E_{Na^+}, across a neuron cell membrane be if Na^+ were the only ion to be considered?

19.9 Common Batteries

Voltaic cells include the convenient, portable sources of energy that we call *batteries.* Some batteries, such as the common flashlight battery, consist of a single cell while others, such as automobile batteries, contain multiple cells. Batteries can be classified as primary or secondary depending on whether the reactions at the anode and cathode can be easily reversed. In a **primary battery** the electrochemical reactions cannot easily be reversed, so when the reactants are used up the battery is "dead" and must be discarded. In contrast, a **secondary battery** (sometimes called a storage battery or a rechargeable battery) uses an electrochemical reaction that can be reversed, so this type of battery can be recharged.

© Richard T. Nowitz/Corbis

Used "dead" batteries.

Primary Batteries

For many years the "dry cell," invented by Georges Leclanché in 1866, was the major source of energy for flashlights and toys. The container of the dry cell is made of zinc, which acts as the anode. The zinc is separated from the other chemicals by a liner of porous paper that functions as the salt bridge (Figure 19.16). In the center of the dry cell is a graphite cathode, which is unreactive, inserted into a moist mixture of ammonium chloride (NH_4Cl), zinc chloride ($ZnCl_2$), and manganese(IV) oxide (MnO_2). As electrons flow from the cell, the zinc is oxidized,

$$Zn(s) \longrightarrow Zn^{2+}(aq) + 2\,e^- \qquad \text{(anode, oxidation)}$$

and the ammonium ions are reduced,

$$2\,NH_4^+(aq) + 2\,e^- \longrightarrow 2\,NH_3(g) + H_2(g) \qquad \text{(cathode, reduction)}$$

The ammonia reacts with zinc ions to form a zinc-ammonia complex ion (Section 17.5 (*p. 853*)), which prevents a buildup of gaseous ammonia.

$$Zn^{2+}(aq) + 2\,NH_3(g) \longrightarrow [Zn(NH_3)_2]^{2+}(aq)$$

The hydrogen produced at the cathode is oxidized by the MnO_2 in the cell, which prevents hydrogen accumulation.

$$H_2(g) + 2\,MnO_2(s) \longrightarrow Mn_2O_3(s) + H_2O(\ell)$$

The overall cell reaction, which produces 1.5 V, is

$$2\,MnO_2(s) + 2\,NH_4^+(aq) + Zn(s) \longrightarrow Mn_2O_3(s) + H_2O(\ell) + [Zn(NH_3)_2]^{2+}(aq)$$

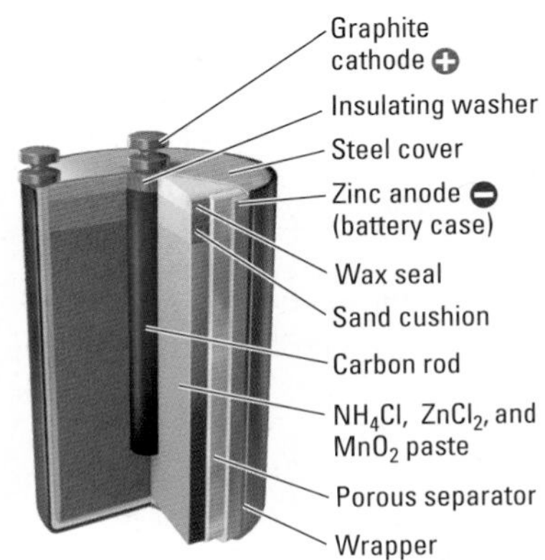

Figure 19.16 Leclanché dry cell. It contains a zinc anode (the battery container), a graphite cathode, and an electrolyte consisting of a moist paste made of NH_4Cl, $ZnCl_2$, and MnO_2.

Courtesy of Wilson Greatbatch Technologies, Inc.

Wilson Greatbatch
1919–

In the early 1960s, Wilson Greatbatch had an idea of how a battery might be used to help an ailing heart keep pumping. His story, told in his own words, is fascinating:

I quit all my jobs, and with two thousand dollars I went out in the barn in the back of my house and built 50 pacemakers in two years. I started making the rounds of all the doctors in Buffalo who were working in this field, and I got consistently negative results. The answer I got was, well, these people all die in a year, you can't do much for them. . . . When I first approached Dr. Shardack with the idea of the pacemaker, he alone thought that it really had a future. He said, "You know—if you can do that—you can save a thousand lives a year."

After the first ten years, we were still only getting one or two years out of pacemakers . . . and the failure mechanism was always the battery. The human body is a very hostile environment. . . . You're trying to run things in a warm salt water environment. . . . So we started looking around for new power sources. And we finally wound up with this lithium battery. It really revolutionized the pacemaker business. The doctors have told me that the introduction of the lithium battery was more significant than the invention of the pacemaker in the first place.

SOURCE: *The World of Chemistry* video, Program 15, The Annenberg/CPB Collection.

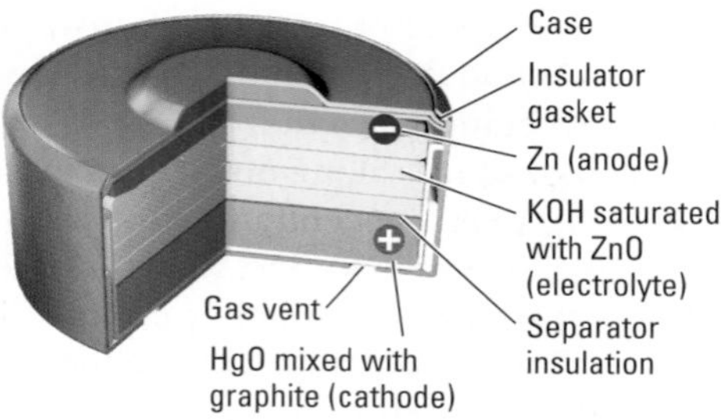

Figure 19.17 Mercury battery. The reducing agent is zinc and the oxidizing agent is mercury(II) oxide.

A major disadvantage of the Leclanché cell is the occurrence of a slow reaction even when current is not being drawn. As a consequence, stored cells run down and tend to have a short shelf life.

Some of the problems of the dry cell are overcome by the newer, more expensive alkaline battery. An ***alkaline battery,*** which produces 1.54 V, also uses the oxidation of zinc as the anode reaction, but under alkaline (pH > 7) conditions.

$$Zn(s) + 2\,OH^-(aq) \longrightarrow ZnO(aq) + H_2O(\ell) + 2\,e^- \qquad \text{(anode, oxidation)}$$

The electrons that pass through the external circuit are consumed by reduction of manganese(IV) oxide at the cathode.

$$MnO_2(s) + H_2O(\ell) + e^- \longrightarrow MnO(OH)(s) + OH^-(aq) \qquad \text{(cathode, reduction)}$$

In the ***mercury battery*** (Figure 19.17), the oxidation of zinc is again the anode reaction. The cathode reaction is the reduction of mercury(II) oxide.

$$HgO(s) + H_2O(\ell) + 2\,e^- \longrightarrow Hg(\ell) + 2\,OH^-(aq)$$

The voltage of this battery is about 1.35 V. Mercury batteries are used in calculators, watches, hearing aids, cameras, and other devices in which small size is an advantage. However, mercury and its compounds are poisonous, so proper disposal of mercury batteries is necessary.

Secondary Batteries

Secondary batteries are rechargeable because, as they discharge, the oxidation products remain at the anode and the reduction products remain at the cathode. As a result, if the direction of electron flow is reversed, the anode and cathode reactions are reversed and the reactants are regenerated. Under favorable conditions, secondary batteries may be discharged and recharged hundreds or even thousands of times. Examples of secondary batteries include automobile batteries, nicad (nickel-cadium) batteries, and lithium ion batteries.

Lead-Acid Storage Batteries

The familiar automobile battery, the lead-acid storage battery, is a secondary battery consisting of six cells, each containing porous metallic lead electrodes and lead(IV) oxide electrodes immersed in aqueous sulfuric acid (Figure 19.18). When this battery produces an electric current, metallic lead is oxidized to lead(II) sulfate at the anode, and lead(IV) oxide is reduced to lead(II) sulfate at the cathode.

Mercury batteries are hermetically sealed to prevent leakage of mercury and should never be heated. Heating increases the pressure of mercury vapor within the battery, ultimately causing the battery to explode.

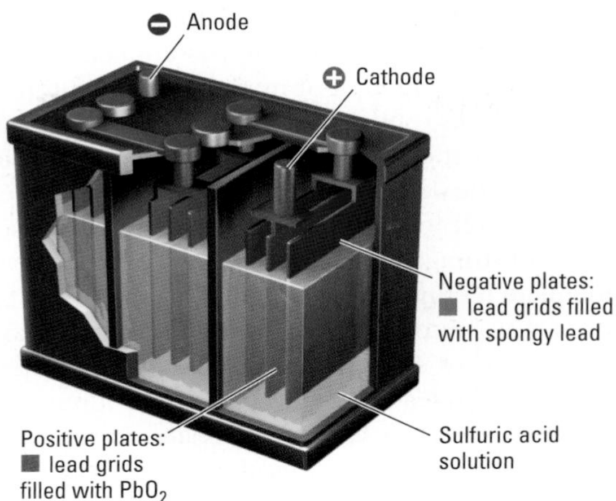

Figure 19.18 Lead-acid storage battery. The anodes are lead grids filled with spongy lead. The cathodes are lead grids filled with lead(IV) oxide, PbO_2. Each cell produces a potential of about 2 V. Six cells connected in series produce the desired overall battery voltage.

$$Pb(s) + HSO_4^-(aq) + H_2O(\ell) \longrightarrow PbSO_4(s) + H_3O^+(aq) + 2\,e^- \qquad E^\circ_{anode} = -0.356 \text{ V}$$

$$PbO_2(s) + 3\,H_3O^+(aq) + HSO_4^-(aq) + 2\,e^- \longrightarrow PbSO_4(s) + 5\,H_2O(\ell) \qquad E^\circ_{cathode} = 1.685 \text{ V}$$

$$Pb(s) + PbO_2(s) + 2\,H_3O^+(aq) + 2\,HSO_4^-(aq) \longrightarrow 2\,PbSO_4(s) + 4\,H_2O(\ell) \qquad E^\circ_{cell} = +2.041 \text{ V}$$

The voltage from the six cells connected in series in a typical automobile battery gives a total voltage of 12 V.

To understand why the lead storage battery is rechargeable, consider that the lead sulfate formed at both electrodes is an insoluble compound that mostly *stays on the electrode surface.* As a result, it remains available for the reverse reaction. To recharge a secondary battery, a source of direct electrical current is supplied so that electrons are forced to flow in the direction opposite from when the battery was discharging. This causes the overall battery reaction to be reversed and regenerates the reactants that originally produced the battery's voltage and current. For the lead-acid storage battery, the overall redox reaction is

Discharging battery produces electricity

$$Pb(s) + PbO_2(s) + 2\,HSO_4^-(aq) + 2\,H_3O^+(aq) \rightleftharpoons 2\,PbSO_4(s) + 4\,H_2O(\ell)$$

Charging battery requires electricity from an external source

The lead-acid battery was first described to the French Academy of Sciences in 1860 by Gaston Planté.

Normal charging of an automobile lead-acid storage battery occurs during driving. In addition to reversing the overall battery reaction, charging reduces a little hydronium ion at the cathode and oxidizes a little water at the anode.

Reduction of hydronium ion: $4\,H_3O^+(aq) + 4\,e^- \longrightarrow 2\,H_2(g) + 4\,H_2O(\ell)$

Oxidation of water: $6\,H_2O(\ell) \longrightarrow O_2(g) + 4\,H_3O^+(aq) + 4\,e^-$

These reactions produce a hydrogen-oxygen mixture inside the battery, which, if accidentally ignited, can explode. Therefore, no sparks or open flames should be brought near a lead-acid storage battery, even the sealed kind.

When a car with a dead battery is jump-started, the last jumper cable connection should be to the car's frame—well away from the battery—to avoid igniting any H_2 in the battery with a spark.

As the battery operates, sulfuric acid is consumed in both the anode and the cathode reactions, thereby decreasing the concentration of the sulfuric acid electrolyte. Before the introduction of modern sealed automotive batteries, the density of this battery acid was routinely measured to indicate the state of charge of the battery. The density of the battery acid decreases as the battery discharges. Consequently, the lower the density, the lower the charge of the battery. In modern sealed batteries, it is difficult to gain access to the acid to measure its density.

The number of electrons that a battery can move from the anode to the cathode is proportional to the amount of reactants involved.

The lead-acid storage battery is relatively inexpensive, reliable, and simple, and it has an adequate lifetime. High weight is its major fault. A typical automobile battery contains about 15 to 20 kg lead, which is required to provide the large number of electrons needed to start an automobile engine, especially on a cold morning. Another problem with lead batteries is that lead mining and manufacturing and disposal of the used batteries can contaminate air and groundwater. Auto batteries should be recycled by companies equipped with the proper safeguards.

ThomsonNOW™
Go to the Chemistry Interactive menu for modules on:

- **the operation of several types of batteries**
- **discharging and recharging a lead storage battery**

Nickel-Cadmium and Nickel-Metal Hydride Batteries

Nickel-cadmium (nicad) secondary batteries are lightweight, can be quite small, and produce a constant voltage until completely discharged, which makes them useful in cordless appliances, video camcorders, portable radios, and other applications (Figure 19.19).

Nicad batteries can be recharged because the reaction products are insoluble hydroxides that remain at the electrode surfaces. The anode reaction during the discharge cycle is the oxidation of cadmium, and the cathode reaction is the reduction of nickel oxyhydroxide, NiO(OH).

$$Cd(s) + 2\,OH^-(aq) \longrightarrow Cd(OH)_2(s) + 2\,e^- \qquad E^\circ_{anode} = -0.809\text{ V}$$

$$2[NiO(OH)(s) + H_2O(\ell) + e^- \longrightarrow Ni(OH)_2(s) + OH^-(aq)] \qquad E^\circ_{cathode} = 0.490\text{ V}$$

$$Cd(s) + 2\,NiO(OH)(s) + 2\,H_2O(\ell) \longrightarrow Cd(OH)_2(s) + 2\,Ni(OH)_2(s) \qquad E^\circ_{cell} = 1.299\text{ V}$$

Like mercury batteries, nicad batteries should be disposed of properly because of the toxicity of cadmium and its compounds.

Another nickel battery that uses the same cathode reaction but a different anode reaction is the *nickel-metal hydride* (NiMH) *battery,* which eliminates the use of cadmium. These batteries are used in portable power tools, cordless shavers, and photoflash units. Here the anode is a metal (M) alloy, often nickel or a rare earth, in a basic electrolyte (KOH). The anode reaction oxidizes hydrogen absorbed in the metal alloy, and water is produced.

$$MH(s) + OH^-(aq) \longrightarrow M(s) + H_2O(\ell) + e^- \qquad \text{(anode reaction)}$$

The overall reaction is

$$MH(s) + NiO(OH)(s) \longrightarrow M(s) + Ni(OH)_2(s) \qquad E_{cell} = 1.4\text{ V}$$

EXERCISE 19.9 Recharging a Nicad Battery

Write the electrode reactions that take place when a nicad battery is recharged; identify the anode and cathode reactions.

Figure 19.19 Nickel-cadmium (nicad) batteries in a battery charger.

Lithium-Ion Batteries

Lithium-ion batteries (Figure 19.20) benefit from the low density and high reducing strength of lithium metal (Table 19.1). The anode in such a battery is made of lithium metal that has been mixed with a conducting carbon polymer. The polymer has tiny spaces in its structure that can hold the lithium atoms as well as the lithium ions formed by the oxidation reaction.

$$\text{Li(s) (in polymer)} \longrightarrow \text{Li}^+\text{(in polymer)} + e^- \qquad \text{(anode reaction)}$$

The cathode also contains lithium ions, but in the lattice of a metal oxide such as CoO_2. This oxide lattice, like the carbon-polymer electrode, has holes in it that accommodate Li^+ ions. The reduction reaction is

$$\text{Li}^+(\text{in CoO}_2) + e^- + \text{CoO}_2 \longrightarrow \text{LiCoO}_2 \qquad \text{(cathode reaction)}$$

The overall reaction in the lithium-ion battery is therefore

$$\text{Li(s)} + \text{CoO}_2\text{(s)} \longrightarrow \text{LiCoO}_2\text{(s)} \qquad E_{\text{cell}} = 3.4 \text{ V}$$

Lithium-ion batteries have a large voltage (3.4 V per cell) and very high energy output for their mass. They can be recharged many hundreds of times. Because of these desirable characteristics, lithium-ion batteries are used in cellular telephones, laptop computers, and digital cameras.

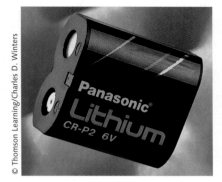

Figure 19.20 Lithium-ion battery. It finds many uses in which a high energy density and low mass are desired.

CONCEPTUAL EXERCISE 19.10 Emergency Batteries

You are stranded on an island and need to communicate your location to receive help. You have a battery-powered radio transmitter, but the lead batteries are discharged. There is a swimming pool nearby and you find a tank of chlorine gas and some plastic tubing that can withstand being oxidized by chlorine. Devise a battery that might be used to power the radio using these items.

CHEMISTRY IN THE NEWS

Hybrid Cars

A number of hybrid cars—including the Toyota Prius, Honda Insight, Honda Civic, Ford Escape, Lexus RX 400H, Mercury Marina, Toyota Highlander—are now for sale in the United States and Japan, and more are to be offered soon. Hybrid cars have two propulsion systems: an electric motor and a gasoline engine. The energy to power such a car comes from gasoline. The electricity comes ultimately from its gasoline engine, which charges the battery that is used to run the electric motor. However, the gasoline engine in a hybrid car is smaller than that in a normal car, and the gasoline engine switches off when the hybrid car is stopped or cruising at low speed. Therefore, fuel efficiency is high and hybrid cars get up to 60 mpg in city driving, twice as much as the gasoline mileage of non-hybrid, conventional cars.

Overall, the hybrid car is much more energy efficient than a conventional car. When the Toyota Prius starts up, the electric motor is used, but when the car is accelerating, and the demand for power is high, both the electric motor and the gasoline engine are used. At speeds of less than about 20 mph, the electric motor alone provides the propulsion, so hybrid cars get their best gasoline mileage in city traffic. The Prius cruises using both propulsion systems, although some of the energy from the engine is used to charge the batteries using the motor-generator. When going downhill, the Prius turns off the gasoline engine. Furthermore, when the brakes are applied, the motor-generator converts some of the kinetic energy of the car into electricity, charging the batteries, and saving energy wasted in a conventional car.

The batteries that power the motor are nickel–metal hydride (NiMH) batteries that are charged by the gasoline engine during normal driving or as the car goes downhill, so the car never needs to be plugged in to be recharged. Eventually, the batteries need to be replaced.

Because a hybrid car does not use the gasoline engine all of the time, it produces much less exhaust, both polluting gases and carbon dioxide, than a conventional car. For example, the Toyota Prius has such low tailpipe emissions that it qualifies for the California Air Resources Board's stringent Super-Ultra-Low-Emission Vehicle class. Due to their environmentally friendly nature, in addition to their excellent gas mileage, demand for hybrid cars is projected to rise as emission standards grow ever stricter.

Toyota Prius, a hybrid car.

SOURCES: Danny Hakim, "Toyota Develops Hybrids with an Eye on the Future," *New York Times*, 8/3/2005; "Motor Trend 2004 Car of the Year Winner: Toyota Prius," *Motor Trend*, January 2004.

19.10 Fuel Cells

A **fuel cell** is an electrochemical cell that converts the chemical energy of fuels directly into electricity. It functions somewhat like a battery, but in contrast to a battery, its reactants are continually supplied from an external reservoir. The best-known fuel cell is the alkaline fuel cell, which is used in the Space Shuttle. Alkaline fuel cells have been used since the 1960s by NASA to power electrical systems on spacecraft, but they are very expensive and unlikely to be commercialized. They require pure hydrogen and oxygen to operate and are susceptible to contamination. Therefore, alternative fuel cells have been developed.

Proton-Exchange Membrane Fuel Cell

The hydrogen-oxygen fuel cell produces electrical power from the oxidation of hydrogen in an electrochemical cell. The two half-reactions are

$$H_2 \longrightarrow 2\,H^+ + 2\,e^- \qquad \text{(anode, oxidation)}$$

$$\tfrac{1}{2}O_2 + 2\,H^+ + 2\,e^- \longrightarrow H_2O \qquad \text{(cathode, reduction)}$$

Hydrogen gas is pumped to the anode, and oxygen gas or air is pumped to the cathode (Figure 19.21). The graphite electrodes are surrounded by catalysts containing platinum. The two electrodes are separated by a semipermeable membrane—a proton-exchange membrane (PEM), which is a thin plastic sheet—that allows the passage of H^+ ions but not electrons. Instead, the electrons must pass through an external circuit, where they can be used to perform work. The electrochemical reactions are aided by platinum catalysts on both sides of the PEM membrane that are in contact with the anode and the cathode. The H^+ ions produced at the anode pass through the PEM to the cathode. At the cathode, hydroxide ions produced by

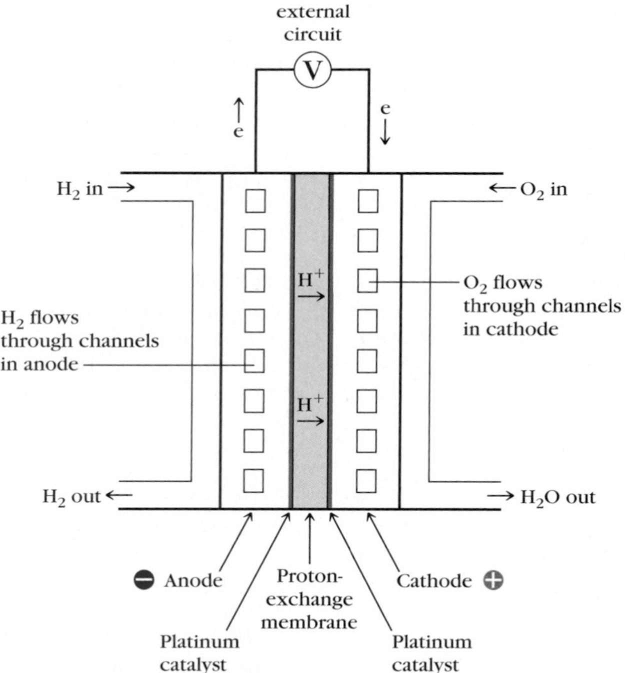

Figure 19.21 A proton-exchange membrane H_2/O_2 fuel cell. H_2 is oxidized in the anode chamber. O_2 is reduced in the cathode chamber.

the reduction of O_2 react with the protons to produce water. The overall reaction of the fuel cell

$$H_2(g) + \frac{1}{2}O_2(g) \longrightarrow H_2O(\ell)$$

produces approximately 0.7 V. To get a higher voltage, fuel cells are stacked and connected in series.

PEM fuel cells can operate at a temperature of 80 °C. Engineering and materials advances have produced PEM fuel cells with power densities such that a device the size of a suitcase can power a car.

The pure hydrogen needed by the PEM fuel cell poses a problem for their more general use. Pure hydrogen is flammable and is difficult to store and distribute. Other hydrogen sources are more attractive. A device called a *reformer* can turn a hydrocarbon (such as natural gas) or alcohol (such as methanol) fuel into hydrogen, which can then be fed to the fuel cell. However, a fuel cell with a reformer has lower overall efficiency and produces products in addition to water.

Hydrogen does not occur in nature as H_2. Therefore, H_2 fuel has to be manufactured. Currently, most H_2 is produced as a by-product of petroleum refining (← *p. 550*) or by treating methane with steam (← *p. 705*).

19.11 Electrolysis—Causing Reactant-Favored Redox Reactions to Occur

ThomsonNOW™
Go to the Chemistry Interactive menu for a module on the **decomposition of water through electrolysis.**

Reactant-favored redox systems can be made to produce products if electrons are forced into the electrochemical system from an external source of electrical current, such as a battery. This process, called **electrolysis,** provides a way to carry out reactant-favored electrochemical reactions that will not take place by themselves. Electrolytic processes are even more important in our economy than the product-favored redox reactions that power batteries. Such electrolytic processes are used in the production and purification of many metals, including copper and aluminum (Sections 22.3 and 21.4), and in electroplating processes that produce a thin coating of metal on many different kinds of items.

Lysis means "splitting," so electrolysis means "splitting with electricity." Electrolysis reactions are chemical reactions caused by the flow of electricity.

Like voltaic cells, electrolysis cells contain electrodes in contact with a conducting medium and an external circuit. As in a voltaic cell, the electrode where reduction takes place in an electrolysis cell is called the cathode, and the electrode where oxidation takes place is called the anode. The electrodes in electrolysis cells are often inert, and their function is to furnish a path for electrons to enter and leave the cell. In contrast to voltaic cells, however, the external circuit connected to an electrolysis cell must contain a direct current *source* of electrons. A battery can serve as a source of electrical current when an electrolysis is carried out on a small scale. The battery forces electrons at a high enough voltage into one of the electrodes (which becomes negative) and removes electrons from the other electrode (which becomes positive). There is often no need for a physical separation of the two electrode reactions, so there is usually no salt bridge. The conducting medium in contact with the electrodes is often the same for both electrodes, and it can be a molten salt or an aqueous solution.

An example of electrolysis is the decomposition of molten sodium chloride. In this process, a pair of electrodes dips into pure sodium chloride that has been heated above its melting temperature (Figure 19.22). In the molten liquid, Na^+ and Cl^- ions are free to move. The Na^+ ions are attracted to the negative electrode, and the Cl^- ions are attracted to the positive electrode. Reduction of Na^+ ions to Na atoms occurs at the cathode (negative electrode). Oxidation of Cl^- ions occurs at the anode (positive electrode).

$$2\,Na^+(\text{in melt}) + 2\,e^- \longrightarrow 2\,Na(\ell) \qquad \text{(cathode, reduction)}$$

$$2\,Cl^-(\text{in melt}) \longrightarrow Cl_2(g) + 2\,e^- \qquad \text{(anode, oxidation)}$$

$$2\,Na^+(\text{in melt}) + 2\,Cl^-(\text{in melt}) \longrightarrow 2\,Na(\ell) + Cl_2(g) \qquad \text{(net cell reaction)}$$

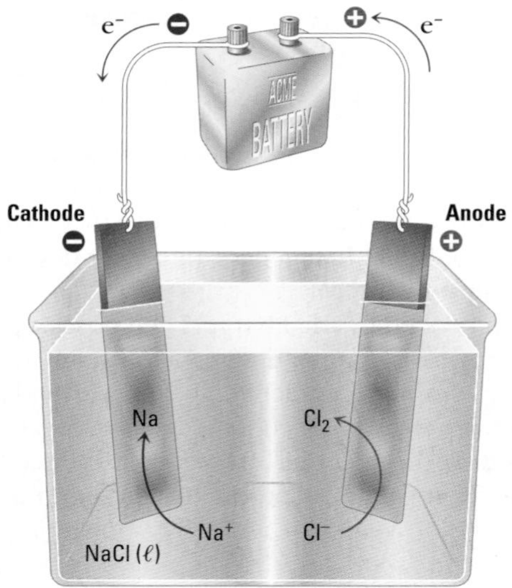

Figure 19.22 **Electrolysis of molten sodium chloride.**

The electrolysis of molten salts is an energy-intensive process because energy is needed to melt the salt as well as to cause the anode and cathode reactions to take place.

What happens if we pass electricity through an *aqueous solution* of a salt, such as potassium iodide, KI, rather than through the molten salt? To predict the outcome of the electrolysis we must first decide what in the solution can be oxidized and reduced. For KI(aq), the solution contains K^+ ions, I^- ions, and H_2O molecules. Potassium is already in its highest common oxidation state in K^+, so it cannot be oxidized. However, both the I^- ion and the H_2O could be oxidized. The possible anode half-reaction oxidations and their potentials are

$$2\,I^-(aq) \longrightarrow I_2(s) + 2\,e^- \qquad E^\circ_{anode} = 0.535\text{ V}$$

$$6\,H_2O(\ell) \longrightarrow O_2(g) + 4\,H_3O^+(aq) + 4\,e^- \qquad E^\circ_{anode} = 1.229\text{ V}$$

Whenever two or more electrochemical reactions are possible at the same electrode, you can use Table 19.1 (p. 939) ***to decide which reaction is more likely to occur under standard-state conditions.*** Considering the two possible anode reactions, item 4 in the discussion of Table 19.1 indicates that the less positive the reduction potential, the more likely a half-reaction is to occur as an oxidation. That is, the farther toward the bottom of Table 19.1 a half-reaction is, the more likely it is to occur as an oxidation. Oxidation of I^- to I_2 has the less positive E°, so this is the more likely anode reaction.

Since I^- is already a reduced form of iodine, there are only two species that can be reduced at the cathode: K^+ ions and water molecules. The possible cathode half-reaction reductions and their potentials are

$$K^+(aq) + e^- \longrightarrow K(s) \qquad E^\circ_{cathode} = -2.925\text{ V}$$

$$2\,H_2O(\ell) + 2\,e^- \longrightarrow H_2(g) + 2\,OH^-(aq) \qquad E^\circ_{cathode} = -0.8277\text{ V}$$

Point 3 in the discussion of Table 19.1 states that the more positive the reduction potential, the more easily a substance on the left-hand side of a half-reaction can be reduced. Since E° for H_2O is more positive (less negative) than E° for K^+, H_2O is more likely to be reduced. Therefore, in aqueous KI solution, the overall reaction and cell potential are

$$2\,I^-(aq) + 2\,H_2O(\ell) \longrightarrow I_2(s) + H_2(g) + 2\,OH^-(aq)$$

$$E^\circ_{cell} = -1.363\text{ V}$$

$$E^\circ_{cell} = E^\circ_{cathode} - E^\circ_{anode}$$
$$= -0.8277\text{ V} - (0.535\text{ V})$$
$$= -1.363\text{ V}$$

The negative value of E°_{cell} indicates that the reaction is not product-favored and that an external energy source is needed for the reaction to occur.

An experiment in which electrical current is passed through aqueous KI (Figure 19.23) shows that this prediction is correct. At the anode (on the right in Figure 19.23a), the I^- ions are oxidized to I_2, which produces a yellow-brown color in the solution. At the cathode, water is reduced to gaseous hydrogen and aqueous hydroxide ions. The formation of excess OH^- ions is shown by the pink color of the phenolphthalein indicator that has been added to the solution.

When electrolysis is carried out by passing electrical current through an aqueous solution, the electrode reactions most likely to take place are those that require the least voltage, that is, the half-reactions that combine to give the least negative overall cell voltage. This means that in aqueous solution the following conditions apply.

1. ***A metal ion or other species can be reduced if it has a reduction potential more positive than −0.8 V, the potential for reduction of water.*** Table 19.1 shows that most metal ions are in this category. If a species has a reduction potential more negative than −0.8 V, then water will preferentially be reduced to $H_2(g)$ and OH^- ions. Metal ions in this latter category include Na^+, K^+, Mg^+, and Al^{3+}. Consequently, producing these metals from their ions requires electrolysis of a molten salt with no water present.

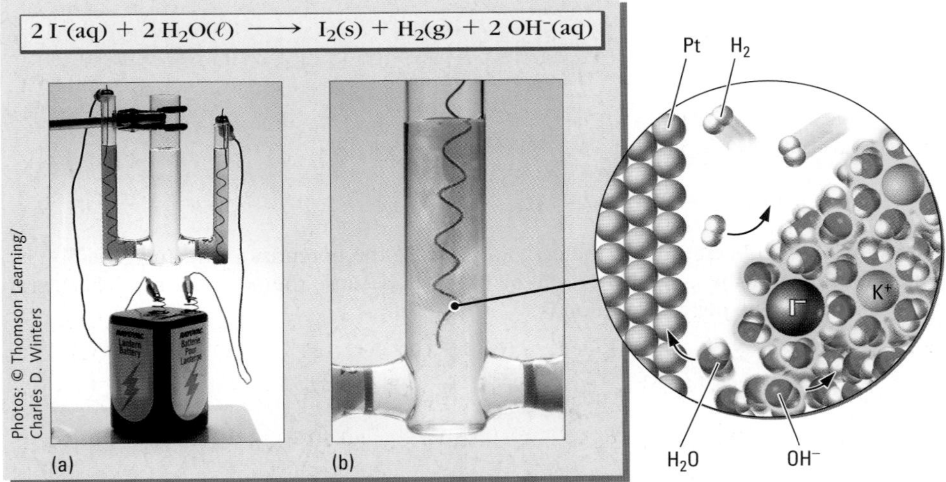

Figure 19.23 The electrolysis of aqueous potassium iodide.
(a) Aqueous KI is found in all three compartments of the cell, and both electrodes are platinum. At the positive electrode, or anode (*right*), the I^- ion is oxidized to iodine, which gives the solution a yellow-brown color.

$$2\,I^-(aq) \longrightarrow I_2(aq) + 2\,e^-$$

At the negative electrode, or cathode (*left*), water is reduced, and the presence of OH^- ion is indicated by the pink color of the acid-base indicator, phenolphthalein.

$$2\,H_2O(\ell) + 2\,e^- \longrightarrow H_2(g) + 2\,OH^-(aq)$$

(b) A close-up of the cathode of a different cell running the same reaction. Bubbles of H_2 and evidence of OH^- generation at the electrode are readily apparent.

2. ***A species can be oxidized in aqueous solution if it has a reduction potential less positive than 1.2 V, the potential for reduction of $O_2(g)$ to water.*** If the reduction potential is less positive than for reduction of water, then oxidation of the species on the right-hand side of a half-reaction is more likely than oxidation of water. Most of the half-equations in Table 19.1 are in this category. If a species has a reduction potential more positive than 1.2 V (that is, if its half-reaction is above the water-oxygen half-reaction in Table 19.1), water will be oxidized preferentially. For example, $F^-(aq)$ cannot be oxidized electrolytically to $F_2(g)$ because water will be oxidized to $O_2(g)$ instead.

The voltage that must be applied to an electrolysis cell is always somewhat greater than the voltage calculated from standard reduction potentials. An *overvoltage* is required, which is an additional voltage needed to overcome limitations in the electron transfer rate at the interface between electrode and solution. Redox reactions that involve the formation of O_2 or H_2 are especially prone to have large overvoltages. Since overvoltages cannot be predicted accurately, the only way to determine with certainty which half-reaction will occur in an electrolysis cell when two possible reactions have similar standard reduction potentials is to perform the experiment.

PROBLEM-SOLVING EXAMPLE 19.12 Electrolysis of Aqueous NaOH

Predict the results of passing a direct electrical current through an aqueous solution of NaOH. Calculate the cell potential.

Answer The net cell reaction is $2\ H_2O(\ell) \rightarrow 2\ H_2(g) + O_2(g)$. Hydrogen is produced at the cathode and oxygen is produced at the anode. The cell potential is -1.23 V.

Strategy and Explanation First, list all the species in the solution: Na^+, OH^-, and H_2O. Next, use Table 19.1 to decide which species can be oxidized and which can be reduced, and note the potential of each possible half-reaction.

Reductions:

$$Na^+(aq) + e^- \longrightarrow Na(s) \qquad E^\circ_{cathode} = -2.71\ V$$

$$2\ H_2O(\ell) + 2\ e^- \longrightarrow H_2(g) + 2\ OH^-(aq) \qquad E^\circ_{cathode} = -0.83\ V$$

Oxidations:

$$4\ OH^-(aq) \longrightarrow O_2(g) + 2\ H_2O(\ell) + 4\ e^- \qquad E^\circ_{anode} = +0.40\ V$$

$$6\ H_2O(\ell) \longrightarrow O_2(g) + 4\ H_3O^+(aq) + 4\ e^- \qquad E^\circ_{anode} = +1.229\ V$$

Water will be reduced to H_2 at the cathode because the potential for this half-reaction is more positive. At the anode, OH^- will be oxidized because the potential is smaller than that for water. The net cell reaction is

$$2\ H_2O(\ell) \longrightarrow 2\ H_2(g) + O_2(g)$$

and the cell potential under standard conditions is

$$E^\circ_{cell} = E^\circ_{cathode} - E^\circ_{anode} = (-0.83\ V) - (+0.40\ V) = -1.23\ V$$

PROBLEM-SOLVING PRACTICE 19.12

Predict the results of passing a direct electrical current through (a) molten NaBr, (b) aqueous NaBr, and (c) aqueous $SnCl_2$.

The electrolysis of aqueous NaCl (brine) is an extremely important industrial reaction in the chlor-alkali process. This process is the major commercial source of chlorine gas and sodium hydroxide; it is described in detail in Section 21.4.

CONCEPTUAL EXERCISE 19.11 Making F_2 Electrolytically

In 1886, Moissan was the first to prepare F_2 by the electrolysis of F^- ions. He electrolyzed KF dissolved in pure HF. No water was present, so only F^- ions were available at the anode. What was produced at the cathode? Write the half-equations for the oxidation and the reduction reactions, and then write the net cell reaction.

19.12 Counting Electrons

When an electric current is passed through an aqueous solution of the soluble salt $AgNO_3$, metallic silver is produced at the cathode. One mole of electrons is required for every mole of Ag^+ reduced.

$$Ag^+(aq) + e^- \longrightarrow Ag(s)$$

If a copper(II) salt in aqueous solution were reduced, 2 mol electrons would be required to produce 1 mol metallic copper from 1 mol copper(II) ions.

$$Cu^{2+}(aq) + 2\,e^- \longrightarrow Cu(s)$$

Each of these balanced half-reactions is like any other balanced chemical equation. That is, each illustrates the fact that both matter and charge are conserved in chemical reactions. Thus, if you could measure the number of moles of electrons flowing through an electrolysis cell, you would know the number of moles of silver or copper produced. Conversely, if you knew the amount of silver or copper produced, you could calculate the number of moles of electrons that had passed through the circuit.

The number of moles of electrons transferred during a redox reaction is usually determined experimentally by measuring the current flowing in the external electrical circuit during a given time. The product of the current (measured in amperes, A) and the time interval (in seconds, s) equals the electric charge (in coulombs, C) that flowed through the circuit.

$$\text{Charge} = \text{current} \times \text{time}$$

$$1 \text{ coulomb} = 1 \text{ ampere} \times 1 \text{ second}$$

The Faraday constant (96,500 C/mol of electrons; p. 943) can then be used to find the number of moles of electrons from a known number of coulombs of charge. This information is of practical significance in chemical analysis and synthesis.

Figure 19.24 shows the relationship between quantity of charge used and the quantities of substances that are oxidized or reduced during electrolysis.

Large electric currents, like those needed to run a hair dryer or refrigerator, are measured in amperes (C/s). Smaller currents, in the milliampere (mA) range, are more commonly used in laboratory electrolysis experiments. 1 mA = 10^{-3}A.

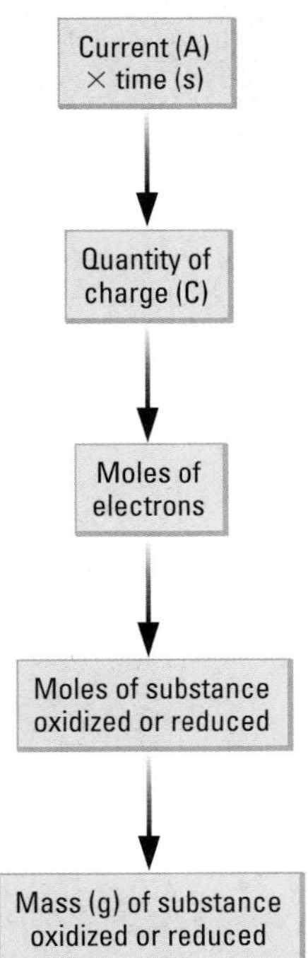

Figure 19.24 Calculation steps for electrolysis. These steps relate the quantity of electrical charge used in electrolysis to the amounts of substances oxidized or reduced.

PROBLEM-SOLVING EXAMPLE 19.13 Using the Faraday Constant

How many grams of copper will be deposited at the cathode of an electrolysis cell if an electric current of 15 mA is applied for 1.0 h through an aqueous solution containing excess Cu^{2+} ions?

Answer 0.018 g C

Strategy and Explanation We use the strategy presented in Figure 19.24. First, we write and balance the relevant half-reaction that occurs at the cathode.

$$Cu^{2+}(aq) + 2\,e^- \rightarrow Cu(s)$$

Then, we calculate the quantity of charge transferred.

$$\text{Charge} = 15 \times 10^{-3}\,\text{A} \times 3600\text{ s} = 15 \times 10^{-3}\,\text{C/s} \times 3600.\text{ s} = 54\text{ C}$$

ThomsonNOW™
Go to the Coached Problems menu for a tutorial on **counting electrons.**

Finally, we determine the mass of copper deposited.

$$(54\text{ C})\left(\frac{1\text{ mol e}^-}{9.65 \times 10^4\text{ C}}\right)\left(\frac{1\text{ mol Cu}}{2\text{ mol e}^-}\right)\left(\frac{63.5\text{ g Cu}}{1\text{ mol Cu}}\right) = 0.018\text{ g Cu}$$

PROBLEM-SOLVING PRACTICE 19.13

In the commercial production of sodium metal by electrolysis, the cell operates at 7.0 V and a current of 25×10^3 A. What mass of metallic sodium can be produced in 1 h?

CONCEPTUAL EXERCISE 19.12 How Many Faradays?

Which would require more Faradays of electricity?

(a) Making 1 mol Al from Al^{3+}
(b) Making 2 mol Na from Na^+
(c) Making 2 mol Cu from Cu^{2+}

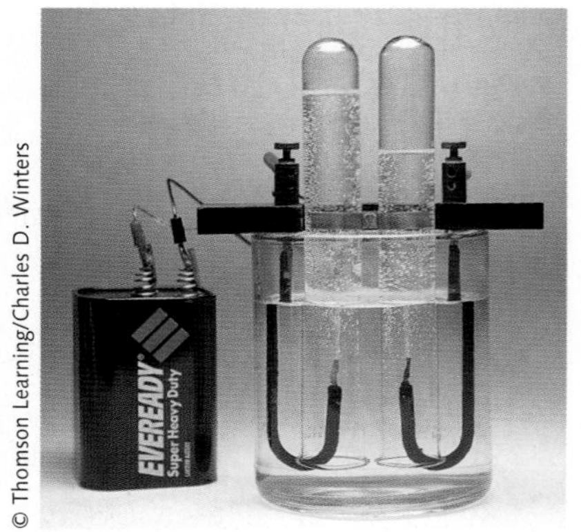

© Thomson Learning/Charles D. Winters

Electrolysis of water. A very dilute solution of sulfuric acid is electrolyzed to produce H_2 at the cathode *(left)* and O_2 at the anode *(right)*.

Electrolytic Production of Hydrogen

Hydrogen can be produced by the electrolysis of water to which a drop or two of sulfuric acid has been added to make the solution conductive. The overall electrochemical reaction is

$$2\,H_2O(\ell) \longrightarrow 2\,H_2(g) + O_2(g) \qquad E^\circ_{cell} = -1.24\text{ V}$$

Oxygen is produced at the anode and hydrogen at the cathode. The minimum voltage required for this reaction is 1.24 V (J/C), but in practice overvoltage requires a higher voltage of about 2 V.

Let's consider how much electrical energy would be required to produce 1.00 kg of gaseous H_2 (about 11,200 L at STP) and at what cost. First we calculate the required charge in coulombs by using the Faraday constant; then we use the fact that 1 joule = 1 volt × 1 coulomb.

The reduction half-reaction shows that 2 mol electrons produces 1 mol (2.02 g) $H_2(g)$.

$$2\,H_3O^+(aq) + 2\,e^- \longrightarrow H_2(g) + 2\,H_2O(\ell)$$

The amount (number of moles) of electrons required to produce 1.00 kg H_2 is calculated as follows:

$$1.00\text{ kg H}_2 \times \left(\frac{1 \times 10^3\text{ g}}{\text{kg}}\right)\left(\frac{1\text{ mol H}_2}{2.016\text{ g H}_2}\right)\left(\frac{2\text{ mol e}^-}{1\text{ mol H}_2}\right) = 9.92 \times 10^2\text{ mol e}^-$$

Now we can calculate the charge using the Faraday constant.

$$(9.92 \times 10^2\text{ mol e}^-) \times \left(\frac{9.65 \times 10^4\text{ C}}{1\text{ mol e}^-}\right) = 9.57 \times 10^7\text{ C}$$

The energy (in joules) can be calculated from the charge and the cell voltage.

$$\text{Energy} = \text{charge} \times \text{voltage} = (9.57 \times 10^7\text{ C})(1.24\text{ J/C}) = 1.19 \times 10^8\text{ J}$$

The kilowatt-hour (kWh) is a unit of energy: 1 kWh = 3.60×10^6 J.

We convert joules to kilowatt-hours (kWh), which is the unit we see when we pay the electric bill.

$$1.19 \times 10^8\text{ J} \times \frac{1\text{ kWh}}{3.60 \times 10^6\text{ J}} = 33.1\text{ kWh}$$

At a rate of 10 cents per kilowatt-hour, the production of 1.00 kg hydrogen would cost \$3.31.

Hydrogen holds great promise as a fuel in our economy because it is a gas and can be transported through pipelines, it burns without producing pollutants, and it could be used in fuel cells to generate electricity on demand. Both water and sulfuric acid are in plentiful supply. The major problem with producing hydrogen in quantities large enough to meet the nation's energy demands is finding a cheap enough source of electricity. Another challenging problem is the development of economical, reliable, and safe methods for hydrogen storage.

EXERCISE 19.13 Calculations Based on Electrolysis

In the production of aluminum metal, Al^{3+} is reduced to Al metal using currents of about 50,000 A and a low voltage of about 4.0 V. How much energy (in kilowatt-hours) is required to produce 2000. metric tons of aluminum metal?

CONCEPTUAL EXERCISE 19.14 How Many Joules?

Think of a battery you just purchased at the store as an energy source that can deliver some number of joules. Name the two pieces of information you need to calculate the number of joules this battery can deliver. Which one is obviously available as you read the label on the battery? Devise a means of determining the other information needed.

Electroplating

If a metal or other electrical conductor serves as the cathode in an electrolysis cell, the metal can be plated with another metal to decorate it or protect it against corrosion. To plate an object with copper, for example, we have only to make the object's surface conducting and use the object as the cathode in an electrolysis cell containing a solution of a soluble copper salt as a source of Cu^{2+} ions. The object will become coated with metallic copper, and the coating will thicken as the electrolysis continues and electrons reduce more Cu^{2+} ions to Cu atoms. If the plated object is a metal, it will conduct electricity by itself. If the object is a nonmetal, its surface can be lightly dusted with graphite powder to make it conducting.

Don Smetzer

Oscar is gold-plated. The Oscar award and most gold jewelry are made by plating a thin coating of gold onto a base metal.

Precious metals such as gold are often plated onto cheaper metals such as copper to make jewelry. If the current and duration of the plating reaction are known, it is possible to calculate the mass of gold that will be reduced onto the cathode surface. For example, suppose the object to be plated is immersed in an aqueous solution of $AuCl_3$ and is made a cathode by connecting it to the negative pole of a battery. The circuit is completed by immersing an inert electrode connected to the positive battery pole in the solution, and gold is reduced at the cathode for 60. min at a current of 0.25 A. The reduction half-reaction is

$$Au^{3+}(aq) + 3\,e^- \longrightarrow Au(s)$$

The mass of gold that is reduced is calculated by

$$(0.25\ C/s)(60.\ min)\left(\frac{60\ s}{1\ min}\right)\left(\frac{1\ mol\ e^-}{9.65 \times 10^4\ C}\right)\left(\frac{1\ mol\ Au}{3\ mol\ e^-}\right)$$
$$\left(\frac{197.\ g\ Au}{1\ mol\ Au}\right) = 0.61\ g\ Au$$

Assuming that gold is selling for \$600 per ounce, that's about \$12.90 worth of gold.

ESTIMATION

The Cost of Aluminum in a Beverage Can

How much does it cost to generate the mass of aluminum (14 g) in one beverage can? The aluminum in these cans is produced by reducing Al^{3+} to Al(s). The reaction is run commercially at 50,000 A at a voltage of 4 V (4 J/C), 1 kWh of electricity costs about 10 cents, and 1 kWh = 3.60×10^6 J.

The charge needed to generate 14 g Al is

$$(14\text{ g Al})\left(\frac{1\text{ mol Al}}{26.98\text{ g Al}}\right)\left(\frac{3\text{ mol e}^-}{1\text{ mol Al}}\right)\left(\frac{96{,}500\text{ C}}{1\text{ mol e}^-}\right) = 1.5 \times 10^5\text{ C}$$

The quantity of energy used is

$$(1.5 \times 10^5\text{ C})(4\text{ J/C})\left(\frac{1\text{ kWh}}{3.60 \times 10^6\text{ J}}\right) = 0.17\text{ kWh}$$

The cost of 0.17 kWh at 10 cents per kWh is 1.7 cents. Thus, the mass of Al in one beverage can could be generated by electrolysis for less than 2 cents.

It is estimated that recycling of aluminum for cans requires overall less than 1% of the cost of extracting aluminum from its ore, which is an energy-intensive and costly process.

ThomsonNOW™
Sign in to ThomsonNOW at **www.thomsonedu.com** to work an interactive module based on this material.

EXERCISE 19.15 Electroplating Silver

Calculate the mass of silver that could be plated from solution with a current of 0.50 A for 20. min. The cathode reaction is $Ag^+(aq) + e^- \rightarrow Ag(s)$.

19.13 Corrosion—Product-Favored Redox Reactions

Corrosion is the oxidation of a metal that is exposed to the environment. Visible corrosion on the steel supports of a bridge, for example, indicates possible structural failure. Corrosion reactions are invariably product-favored, which means that $E°$ for the reaction is positive and $\Delta G°$ is negative. Corrosion of iron, for example, takes place quite readily and is difficult to prevent. It produces the red-brown substance we call rust, which is hydrated iron(III) oxide ($Fe_2O_3 \cdot x\,H_2O$, where x varies from 2 to 4). The rust that forms when iron corrodes does not adhere to the surface of the metal, so it can easily flake off and expose fresh metal surface to corrosion (Figure 19.25). The corrosion of aluminum, a metal that is even more reactive than iron, is also very product-favored. The aluminum oxide that forms as a result of corrosion adheres tightly as a thin coating on the surface of the metal, creating a protective coating that prevents further corrosion.

© George B. Diebold/Corbis

Figure 19.25 **Rusting.** The formation of rust destroys the structural integrity of objects made of iron and steel. Given time, this chain will completely rust away.

For corrosion of a metal (M) to occur, the metal must have an anodic area where the oxidation can occur. The general reaction is

Anode reaction: $$M(s) \longrightarrow M^{n+} + n\,e^-$$

There must also be a cathodic area where electrons are consumed. Frequently, the cathode reactions are reductions of oxygen or water.

Cathode reactions:
$$O_2(g) + 2\,H_2O(\ell) + 4\,e^- \longrightarrow 4\,OH^-(aq)$$
$$2\,H_2O(\ell) + 2\,e^- \longrightarrow 2\,OH^-(aq) + H_2(g)$$

Anodic areas may occur at cracks in the oxide coating that protects the surfaces of many metals or around impurities. Cathodic areas may occur at the metal oxide coating, at less reactive metallic impurity sites, or around other metal compounds trapped at the surface, such as sulfides or carbides.

The other requirements for corrosion are an electrical connection between the anode and the cathode and an electrolyte in contact with both anode and cathode. Both requirements are easily fulfilled—the metal itself is the conductor, and ions dissolved in moisture from the environment provide the electrolyte.

Corrosion is so commonplace that about 25% of the annual steel production in the United States is used to replace material lost to corrosion.

In the corrosion of iron, the anodic reaction is the oxidation of metallic iron (Figure 19.26). If both water and O_2 gas are present, the cathode reaction is the reduction of oxygen, giving the net reaction

$$2\,[\text{Fe(s)} \longrightarrow \text{Fe}^{2+}\text{(aq)} + 2\,e^-] \qquad \text{(anode reaction)}$$

$$\text{O}_2\text{(g)} + 2\,\text{H}_2\text{O}(\ell) + 4\,e^- \longrightarrow 4\,\text{OH}^-\text{(aq)} \qquad \text{(cathode reaction)}$$

$$2\,\text{Fe(s)} + \text{O}_2\text{(g)} + 2\,\text{H}_2\text{O}(\ell) \longrightarrow \underset{\text{iron(II) hydroxide}}{2\,\text{Fe(OH)}_2\text{(s)}}$$

In the presence of an ample supply of oxygen and water, as in open air or flowing water, the iron(II) hydroxide is oxidized to the red-brown iron(III) oxide monohydrate (Figure 19.25).

$$4\,\text{Fe(OH)}_2\text{(s)} + \text{O}_2\text{(g)} \longrightarrow \underset{\text{red-brown}}{2\,\text{Fe}_2\text{O}_3 \cdot 2\,\text{H}_2\text{O(s)}} + \text{H}_2\text{O}(\ell)$$

This hydrated iron oxide is the familiar rust you see on iron and steel objects and the substance that colors the water red in some mountain streams and home water pipes. Rust is easily removed from the metal surface by mechanical shaking, rubbing, or even the action of rain or freeze-thaw cycles, thus exposing more iron at the surface and allowing the objects to eventually deteriorate completely.

ThomsonNOW™

Go to the Chemistry Interactive menu for modules on:

- **the redox reactions involved in rusting of iron**

Other substances in air and water can hasten corrosion. Metal salts, such as the chlorides of sodium and calcium from sea air or from salt spread on roadways in the winter, function as salt bridges between anodic and cathodic regions, thus speeding up corrosion reactions.

CONCEPTUAL EXERCISE

19.16 Do All Metals Corrode?

Do all metals corrode as readily as iron and aluminum? Name three metals that you would expect to corrode about as readily as iron and aluminum, and name three metals that do not corrode readily. Name a use for each of the three noncorroding metals. Explain why metals fall into these two groups.

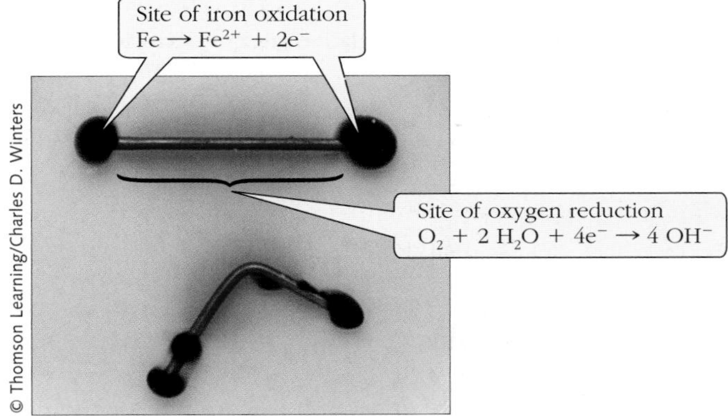
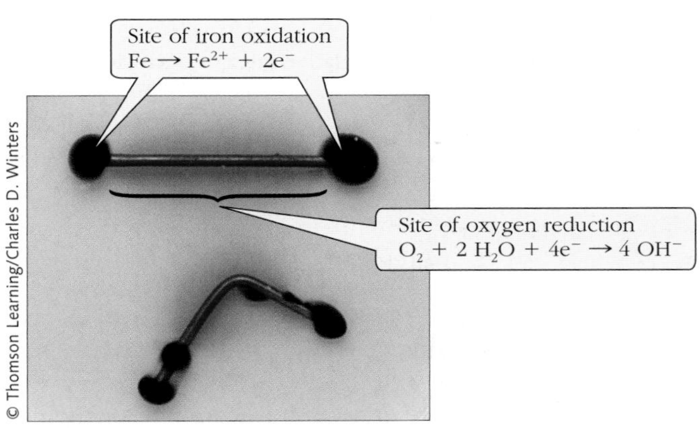

Figure 19.26 Corroding iron nails. Two nails were placed in an agar gel, which also contained the indicator phenolphthalein and $[Fe(CN)_6]^{3-}$. The nails began to corrode and produced Fe^{2+} ions at the tip and where the nail is bent. (These points of stress corrode more quickly.) These points are the anode, as indicated by the formation of the blue-colored compound called Prussian blue, $Fe_3[Fe(CN)_6]_2$. The remainder of the nail is the cathode, since oxygen is reduced in water to give OH^-. The presence of OH^- ions causes the phenolphthalein to turn pink.

Water drop

$Zn^{2+}(aq)$

O_2

Zn (anode) ⊖

Fe (cathode) ⊕

$Zn(s) \rightarrow Zn^{2+}(aq) + 2\ e^-$ $\quad 4\ e^- + O_2(g) + 4\ H_3O^+(aq) \rightleftharpoons 6\ H_2O(\ell)$

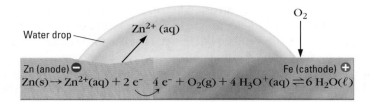

Figure 19.27 Cathodic protection of an iron-containing object. The iron is coated with a film of zinc, a metal more easily oxidized than iron. The zinc acts as the anode and forces iron to become the cathode, thereby preventing the corrosion of the iron.

Corrosion Protection

How can metal corrosion be prevented? The general approaches are to (1) inhibit the anodic process, (2) inhibit the cathodic process, or (3) do both. The most common method is **anodic inhibition,** which directly limits or prevents the oxidation half-reaction by painting the metal surface, coating it with grease or oil, or allowing a thin film of metal oxide to form. More recently developed methods of anodic protection are illustrated by the following reaction, which occurs when the surface is treated with a solution of sodium chromate.

$$2\ Fe(s) + 2\ Na_2CrO_4(aq) + 2\ H_2O(\ell) \longrightarrow Fe_2O_3(s) + Cr_2O_3(s) + 4\ NaOH(aq)$$

The surface iron is oxidized by the chromate salt to give iron(III) and chromium(III) oxides. These form a coating that is impervious to O_2 and water, and further atmospheric oxidation is inhibited.

Cathodic protection is accomplished by forcing the metal to become the cathode instead of the anode. Usually, this goal is achieved by attaching another, more readily oxidized metal to the metal being protected. The best example involves galvanized iron, iron that has been coated with a thin film of zinc (Figure 19.27). $E°$ for zinc is considerably more negative than $E°$ for iron (Zn is lower in Table 19.1 than Fe), so zinc is more easily oxidized. Therefore, the zinc metal film is oxidized before any of the iron and the zinc coating forms a *sacrificial anode.* In addition, when the zinc is corroded, $Zn(OH)_2$ forms an insoluble film on the surface (K_{sp} of $Zn(OH)_2 = 4.5 \times 10^{-17}$) that further slows corrosion.

Galvanized objects. A thin coating of zinc helps prevent the oxidation of iron.

CONCEPTUAL EXERCISE 19.17 Corrosion Rates

Rank these environments in terms of their relative rates of corrosion of iron. Place the fastest first. Explain your answers.

(a) Moist clay
(b) Sand by the seashore
(c) The surface of the moon
(d) Desert sand in Arizona

SUMMARY PROBLEM

Many kinds of secondary batteries are known. If it were not for the density of lead, the lead-acid storage battery would find far greater application. Most electric vehicles currently use lead-acid storage batteries as their source of power, but automotive engineers continue to look longingly at other batteries because of their higher energy-to-mass ratios. When you review Table 19.1 and consider the chemical properties of all of the other oxidizing and reducing agents shown there, you might be tempted to create a hybrid battery that would combine some of the desirable features of, say, a PbO_2 cathode and some other kind of anode rather than the lead anode found in the lead-acid storage battery. In that way, at least the high reduction potential of the half-reaction involving PbO_2 might still be used.

(a) What would be the $E°$ value of a cell made using the PbO_2 reduction reaction and magnesium metal as the reducing agent? Write the two half-reactions and the net cell reaction. Would this cell be the basis for a secondary battery? Explain your answer.

(b) What would be the $E°$ value of a cell made using the PbO_2 reduction reaction and nickel metal as the reducing agent? Write the two half-reactions and the net cell reaction. Would this cell have a voltage greater than or less than that of a single cell of a lead-acid storage battery? Would this cell be the basis of a secondary battery? Explain. What could you do to the chemistry in the anode compartment to make it a secondary battery?

(c) If your Ni/PbO_2 hybrid battery were a success and it was manufactured for use in electric automobiles, how many amperes could it produce, assuming that 500.0 g Ni reacted in exactly 30 min? How much PbO_2 would be reduced during this same period of time?

(d) Of course, batteries must be recharged. How much time would be required to recharge your Ni/PbO_2 battery to its original state (the 500.0 g Ni being converted back to its original form) if a current of 25.5 A is passed through the battery?

(e) Just as you are getting ready to cash in on the success of your new battery, someone announces that it has some serious environmental problems. What could these be? Explain.

IN CLOSING

Having studied this chapter, you should be able to . . .

- Identify the oxidizing and reducing agents in a redox reaction (Section 19.1). ThomsonNOW homework: Study Question 6
- Write equations for oxidation and reduction half-reactions, and use them to balance the net equation (Section 19.2). ThomsonNOW homework: Study Questions 10, 14
- Identify and describe the functions of the parts of an electrochemical cell; describe the direction of electron flow outside the cell and the direction of the ion flow inside the cell (Section 19.3). ThomsonNOW homework: Study Question 22
- Describe how standard reduction potentials are defined and use them to predict whether a reaction will be product-favored as written (Sections 19.4 and 19.5). ThomsonNOW homework: Study Questions 28, 32, 34
- Calculate $\Delta G°$ from the value of $E°$ for a redox reaction (Section 19.6). ThomsonNOW homework: Study Questions 38, 42, 46
- Explain how product-favored electrochemical reactions can be used to do useful work, and list the requirements for using such reactions in rechargeable batteries (Section 19.6).
- Explain how the Nernst equation relates concentrations of redox reactants to E_{cell} (Section 19.7). ThomsonNOW homework: Study Questions 48, 50
- Use the Nernst equation to calculate the potentials of cells that are not at standard conditions (Section 19.7). ThomsonNOW homework: Study Question 52

ThomsonNOW

Sign in to ThomsonNOW at **www.thomsonedu.com** to check your readiness for an exam by taking the Pre-Test and exploring the modules recommended in your Personalized Learning Plan.

- Explain the source of the equilibrium potential across the membrane of a neuron cell (Section 19.8).
- Describe the chemistry of the dry cell, the mercury battery, and the lead-acid storage battery (Section 19.9).
- Describe how a fuel cell works, and indicate how it differs from a battery (Section 19.10). ThomsonNOW homework: Study Question 58
- Use standard reduction potentials to predict the products of electrolysis of an aqueous salt solution (Section 19.11). ThomsonNOW homework: Study Question 64
- Calculate the quantity of product formed at an electrode during an electrolysis reaction, given the current passing through the cell and the time during which the current flows (Section 19.12). ThomsonNOW homework: Study Questions 65, 73, 77
- Explain how electroplating works (Section 19.12).
- Describe what corrosion is and how it can be prevented by cathodic protection (Section 19.13).

KEY TERMS

ampere (A) *(19.4)*

anode *(19.3)*

anodic inhibition *(19.13)*

battery *(19.3)*

cathode *(19.3)*

cathodic protection *(19.13)*

cell voltage *(19.4)*

concentration cell *(19.7)*

corrosion *(19.13)*

coulomb (C) *(19.4)*

electrochemical cell *(19.3)*

electrochemistry *(Introduction)*

electrode *(19.3)*

electrolysis *(19.11)*

electromotive force (emf) *(19.4)*

Faraday constant (*F*) *(19.6)*

fuel cell *(19.10)*

half-cell *(19.3)*

half-reaction *(19.2)*

Nernst equation *(19.7)*

neurons *(19.8)*

pH meter *(19.7)*

primary battery *(19.9)*

salt bridge *(19.3)*

secondary battery *(19.9)*

standard conditions *(19.4)*

standard hydrogen electrode *(19.4)*

standard reduction potentials *(19.5)*

standard voltages ($E°$) *(19.4)*

volt (V) *(19.4)*

voltaic cell *(19.3)*

QUESTIONS FOR REVIEW AND THOUGHT

■ denotes questions available in ThomsonNOW and assignable in OWL.

Blue-numbered questions have short answers at the back of this book and fully worked solutions in the *Student Solutions Manual.*

ThomsonNOW
Assess your understanding of this chapter's topics with sample tests and other resources found by signing in to ThomsonNOW at **www.thomsonedu.com**.

Review Questions

1. Describe the principal parts of an electrochemical cell by drawing a hypothetical cell, indicating the cathode, the anode, the direction of electron flow outside the cell, and the direction of ion flow within the cell.
2. Explain how product-favored electrochemical reactions can be used to do useful work.
3. Explain how reactant-favored electrochemical reactions can be induced to make products.
4. Explain how electroplating works.
5. Tell whether each of these statements is true or false. If false, rewrite it to make it a correct statement.
   (a) Oxidation always occurs at the anode of an electrochemical cell.
   (b) The anode of a battery is the site of reduction and is negative.
   (c) Standard conditions for electrochemical cells are a concentration of 1.0 M for dissolved species and a pressure of 1 bar for gases.
   (d) The potential of a cell does not change with temperature.
   (e) All product-favored oxidation-reduction reactions have a standard cell voltage $E°_{cell}$, with a negative sign.

Topical Questions

Redox Reactions

6. ■ In each of these reactions assign oxidation numbers to all species, and tell which substance is oxidized and which is reduced. Tell which is the oxidizing agent and which is the reducing agent.
   (a) $2\ Al(s) + 3\ Cl_2(g) \longrightarrow 2\ AlCl_3(s)$
   (b) $8\ H_3O^+(aq) + MnO_4^-(aq) + 5\ Fe^{2+}(aq) \longrightarrow 5\ Fe^{3+}(aq) + Mn^{2+}(aq) + 12\ H_2O(\ell)$
   (c) $FeS(s) + 3\ NO_3^-(aq) + 4\ H_3O^+(aq) \longrightarrow 3\ NO(g) + SO_4^{2-}(aq) + Fe^{3+}(aq) + 6\ H_2O(\ell)$
7. In each of these reactions assign oxidation numbers to all species, and tell which substance is oxidized and which is reduced. Tell which is the oxidizing agent and which is the reducing agent.
   (a) $Fe(s) + Br_2(\ell) \longrightarrow FeBr_2(s)$
   (b) $8\ HI(aq) + H_2SO_4(aq) \longrightarrow H_2S(aq) + 4\ I_2(s) + 4\ H_2O(\ell)$
   (c) $H_2O_2(aq) + 2\ Fe^{2+}(aq) + 2\ H_3O^+(aq) \longrightarrow 2\ Fe^{3+}(aq) + 4\ H_2O(\ell)$
8. Choose four elements: a metal that is a representative element, a transition metal, a nonmetal, and a metalloid. Using the index to this text, find a chemical reaction in which each element occurs as a reactant. Assign oxidation numbers to all elements on the reactant and product sides, and identify the oxidizing agent and the reducing agent.
9. Answer Question 8 again, but this time find a chemical reaction in which each element is produced.

Using Half-Reactions to Understand Redox Reactions

10. ■ Write half-reactions for these changes:
    (a) Oxidation of zinc to Zn^{2+} ions
    (b) Reduction of H_3O^+ ions to hydrogen gas
    (c) Reduction of Sn^{4+} ions to Sn^{2+} ions
    (d) Reduction of chlorine to Cl^- ions
    (e) Oxidation of sulfur dioxide to sulfate ions in acidic solution
11. Write half-reactions for these changes:
    (a) Reduction of MnO_4^- ion to Mn^{2+} ion in acid solution
    (b) Reduction of $Cr_2O_7^{2-}$ ion to Cr^{3+} ion in acid solution
    (c) Oxidation of hydrogen gas to H_3O^+ ions
    (d) Reduction of hydrogen peroxide to water in acidic solution
    (e) Oxidation of nitric oxide to nitrogen monoxide in acidic solution
12. For each reaction in Question 6, write balanced half-reactions.
13. For each reaction in Question 7, write balanced half-reactions.
14. ■ Balance this redox reaction in a basic solution:

$$Zn(s) + NO_3^-(aq) \longrightarrow Zn(OH)_4^{2-}(aq) + NH_3(aq)$$

15. Balance this redox reaction in a basic solution:

$$NO_2^-(aq) + Al(s) \longrightarrow NH_3(aq) + Al(OH)_4^-(aq)$$

16. Balance these redox reactions, and identify the oxidizing agent and the reducing agent.
    (a) $CO(g) + O_3(g) \longrightarrow CO_2(g)$
    (b) $H_2(g) + Cl_2(g) \longrightarrow HCl(g)$
    (c) $H_2O_2(aq) + Ti^{2+}(aq) \longrightarrow H_2O(\ell) + Ti^{4+}(aq)$ in acidic solution
    (d) $Cl^-(aq) + MnO_4^-(aq) \longrightarrow Cl_2(g) + MnO_2(s)$ in acidic solution
    (e) $FeS_2(s) + O_2(g) \longrightarrow Fe_2O_3(s) + SO_2(g)$
    (f) $O_3(g) + NO(g) \longrightarrow O_2(g) + NO_2(g)$
    (g) $Zn(Hg)$ (amalgam) $+ HgO(s) \longrightarrow ZnO(s) + Hg(\ell)$ in basic solution (This is the reaction in a mercury battery.)
17. Balance these redox reactions, and identify the oxidizing agent and the reducing agent.
    (a) $FeO(s) + O_3(g) \longrightarrow Fe_2O_3(s)$
    (b) $P_4(s) + Br_2(\ell) \longrightarrow PBr_5(\ell)$
    (c) $H_2O_2(aq) + Co^{2+}(aq) \longrightarrow H_2O(\ell) + Co^{3+}(aq)$ in acidic solution
    (d) $Cl^-(aq) + Cr_2O_7^{2-}(aq) \longrightarrow Cl_2(g) + Cr^{3+}(aq)$ in acidic solution
    (e) $CuFeS_2(s) + O_2(g) \longrightarrow Cu_2S(s) + FeO(s) + SO_2(g)$
    (f) $H_2CO(g) + O_2(g) \longrightarrow CO_2(g) + H_2O(\ell)$
    (g) $C_3H_8(g) + O_2(g) \longrightarrow CO_2(g) + H_2O(\ell)$ in acidic solution (This is the reaction in a propane fuel cell.)

Electrochemical Cells

18. For the redox reaction $Cu^{2+}(aq) + Zn(s) \longrightarrow Cu(s) + Zn^{2+}(aq)$, why can't you generate electric current by placing a piece of copper metal and a piece of zinc metal in a solution containing $CuCl_2(aq)$ and $ZnCl_2(aq)$?
19. Explain the function of a salt bridge in an electrochemical cell.
20. Are standard half-cell reactions always written as oxidation reactions or reduction reactions? Explain.
21. Tell whether this statement is true or false. If false, rewrite it to make it a correct statement: The value of an electrode potential changes when the half-reaction is multiplied by a factor. That is, $E°$ for $Li^+ + e^- \longrightarrow Li$ is different from that for $2\ Li^+ + 2\ e^- \longrightarrow 2\ Li$.
22. ■ A voltaic cell is assembled with $Pb(s)$ and $Pb(NO_3)_2(aq)$ in one compartment and $Zn(s)$ and $ZnCl_2(aq)$ in the other. An external wire connects the two electrodes, and a salt bridge containing KNO_3 connects the two solutions.
    (a) In the product-favored reaction, zinc metal is oxidized to Zn^{2+}. Write a balanced net ionic equation for this reaction.
    (b) Which half-reaction occurs at each electrode? Which is the anode and which is the cathode?
    (c) Draw a diagram of the cell, indicating the direction of electron flow outside the cell and of ion flow within the cell.
23. A voltaic cell is assembled with $Sn(s)$ and $Sn(NO_3)_2(aq)$ in one compartment and $Ag(s)$ and $AgNO_3(aq)$ in the other. An external wire connects the two electrodes, and a salt bridge containing KNO_3 connects the two solutions.
    (a) In the product-favored reaction, Ag^+ is reduced to silver metal. Write a balanced net ionic equation for this reaction.
    (b) Which half-reaction occurs at each electrode? Which is the anode and which is the cathode?
    (c) Draw a diagram of the cell, indicating the direction of electron flow outside the cell and of ion flow within the cell.

■ In ThomsonNOW and OWL   Blue-numbered questions answered at end of this book

Electrochemical Cells and Voltage

24. ■ You light a 25-W light bulb with the current from a 12-V lead-acid storage battery. After 1.0 h of operation, how much energy has the light bulb utilized? How many coulombs have passed through the bulb? Assume 100% efficiency. (A watt is the transfer of 1 J of energy in 1 s.)
25. Draw a diagram of a standard hydrogen electrode and describe how it works.
26. Copper can reduce silver ion to metallic silver, a reaction that could, in principle, be used in a battery.

$$Cu(s) + 2\ Ag^+(aq) \longrightarrow Cu^{2+}(aq) + 2\ Ag(s)$$

(a) Write equations for the half-reactions involved.
(b) Which half-reaction is an oxidation and which is a reduction? Which half-reaction occurs in the anode compartment and which takes place in the cathode compartment?

27. Chlorine gas can oxidize zinc metal in a reaction that has been suggested as the basis of a battery. Write the half-reactions involved. Label which is the oxidation half-reaction and which is the reduction half-reaction.

Using Standard Cell Potentials

28. ■ What is the strongest oxidizing agent in Table 19.1? What is the strongest reducing agent? What is the weakest oxidizing agent? What is the weakest reducing agent?
29. Using the reduction potentials in Table 19.1, place these elements in order of increasing ability to function as reducing agents:

(a) Cl_2 (b) Fe
(c) Ag (d) Na
(e) H_2

30. Using the reduction potentials in Table 19.1, place these elements in order of increasing ability to function as oxidizing agents:

(a) O_2 (b) H_2O_2
(c) $PbSO_4$ (d) H_2O

31. One of the most energetic redox reactions is that between F_2 gas and lithium metal.

(a) Write the half-reactions involved. Label which is the oxidation half-reaction and which is the reduction half-reaction.
(b) According to data from Table 19.1, what is $E°_{cell}$ for this reaction?

32. ■ Calculate the value of $E°_{cell}$ for each of these reactions. Decide whether each is product-favored.

(a) $I_2(s) + Mg(s) \longrightarrow Mg^{2+}(aq) + 2\ I^-(aq)$
(b) $Ag(s) + Fe^{3+}(aq) \longrightarrow Ag^+(aq) + Fe^{2+}(aq)$
(c) $Sn^{2+}(aq) + 2\ Ag^+(aq) \longrightarrow Sn^{4+}(aq) + 2\ Ag(s)$
(d) $2\ Zn(s) + O_2(g) + 2\ H_2O(\ell) \longrightarrow 2\ Zn^{2+}(aq) + 4\ OH^-(aq)$

33. Consider these half-reactions:

Half-Reaction	$E°$ (V)
$Au^{3+}(aq) + 3\ e^- \longrightarrow Au(s)$	1.50
$Pt^{2+}(aq) + 2\ e^- \longrightarrow Pt(s)$	1.2
$Co^{2+}(aq) + 2\ e^- \longrightarrow Co(s)$	−0.28
$Mn^{2+}(aq) + 2\ e^- \longrightarrow Mn(s)$	−1.18

(a) Which is the weakest oxidizing agent?
(b) Which is the strongest oxidizing agent?
(c) Which is the strongest reducing agent?
(d) Which is the weakest reducing agent?
(e) Will Co(s) reduce $Pt^{2+}(aq)$ to Pt(s)?
(f) Will Pt(s) reduce $Co^{2+}(aq)$ to Co(s)?
(g) Which ions can be reduced by Co(s)?

34. ■ Consider these half-reactions:

Half-Reaction	$E°$ (V)
$Ce^{4+}(aq) + e^- \longrightarrow Ce^{3+}(aq)$	1.61
$Ag^+(aq) + e^- \longrightarrow Ag(s)$	0.80
$Hg_2^{2+}(aq) + 2\ e^- \longrightarrow 2\ Hg(\ell)$	0.79
$Sn^{2+}(aq) + 2\ e^- \longrightarrow Sn(s)$	−0.14
$Ni^{2+}(aq) + 2\ e^- \longrightarrow Ni(s)$	−0.25
$Al^{3+}(aq) + 3\ e^- \longrightarrow Al(s)$	−1.66

(a) Which is the weakest oxidizing agent?
(b) Which is the strongest oxidizing agent?
(c) Which is the strongest reducing agent?
(d) Which is the weakest reducing agent?
(e) Will Sn(s) reduce $Ag^+(aq)$ to Ag(s)?
(f) Will $Hg(\ell)$ reduce $Sn^{2+}(aq)$ to Sn(s)?
(g) Name the ions that can be reduced by Sn(s).
(h) Which metals can be oxidized by $Ag^+(aq)$?

35. In principle, a battery could be made from aluminum metal and chlorine gas.

(a) Write a balanced equation for the reaction that would occur in a battery using $Al^{3+}(aq)/Al(s)$ and $Cl_2(g)/Cl^-(aq)$ half-reactions.
(b) Tell which half-reaction occurs at the anode and which at the cathode. What are the polarities of these electrodes?
(c) Calculate the standard potential, $E°_{cell}$, for the battery.

$E°$ and Gibbs Free Energy

36. Choose the correct answers: In a product-favored chemical reaction, the standard cell potential, $E°_{cell}$, is (greater/less) than zero, and the Gibbs free energy change, $\Delta G°$, is (greater/less) than zero.
37. For each of the reactions in Question 32, compute the Gibbs free energy change, $\Delta G°$.
38. ■ Hydrazine, N_2H_4, can be used as the reducing agent in a fuel cell.

$$N_2H_4(aq) + O_2(aq) \longrightarrow H_2(g) + 2\ H_2O(\ell)$$

(a) If $\Delta G°$ for the reaction is −598 kJ, calculate the value of $E°$ expected for the reaction.
(b) Suppose the equation is written with all coefficients doubled. Determine $\Delta G°$ and $E°$ for this new reaction.

39. The standard cell potential for the oxidation of Mg by Br_2 is 3.45 V.

$$Br_2(\ell) + Mg(s) \longrightarrow Mg^{2+}(aq) + 2\ Br^-(aq)$$

(a) Calculate $\Delta G°$ for this reaction.
(b) Suppose the equation is written with all coefficients doubled. Determine $\Delta G°$ and $E°$ for this new reaction.

40. The standard cell potential, $E°$, for the reaction of Zn(s) and Cl_2(g) is 2.12 V. Write the chemical equation for the reaction of 1 mol zinc. What is the standard Gibbs free energy change, $\Delta G°$, for this reaction?
41. What is the equilibrium constant K_c and $\Delta G°$ for the reaction between Cd(s) and Cu^{2+}(aq)?
42. ■ What is the equilibrium constant K_c and $\Delta G°$ for the reaction between I_2(s) and Br^-(aq)?
43. What is the equilibrium constant K_c and $\Delta G°$ for the reaction between Ag(s) and Zn^{2+}(aq)?
44. What is the equilibrium constant K_c and $\Delta G°$ for the reaction between Cl_2(g) and Br^-(aq)?
45. Consider a voltaic cell with the following reaction. As the cell reaction proceeds, what happens to the values of E_{cell}, ΔG, and K_c? Explain your answers.

$$Cu^{2+}(aq, 1\ M) + Zn(s) \longrightarrow Cu(s) + Zn^{2+}(aq, 1\ M) \quad E°_{cell} = 1.10\ V$$

46. ■ Estimate the equilibrium constant K_c for this reaction.

$$Ni(s) + Co^{2+}(aq) \rightleftharpoons Ni^{2+}(aq) + Co(s) \quad E°_{cell} = +0.046\ V$$

Effect of Concentration on Cell Potential

47. Consider the voltaic cell

$$Zn(s) + Cd^{2+}(aq) \longrightarrow Zn^{2+}(aq) + Cd(s)$$

operating at 298 K.
(a) What is the $E°_{cell}$ for this cell?
(b) If E_{cell} = 0.390 and (conc. Cd^{2+}) = 2.00 M, what is (conc. Zn^{2+})?
(c) If (conc. Cd^{2+}) = 0.068 M and (conc. Zn^{2+}) = 1.00 M, what is E_{cell}?

48. ■ Consider the voltaic cell

$$2\ Ag^+(aq) + Cd(s) \longrightarrow 2\ Ag(s) + Cd^{2+}(aq)$$

operating at 298 K.
(a) What is the $E°_{cell}$ for this cell?
(b) If (conc. Cd^{2+}) = 2.0 M and (conc. Ag^+) = 0.25 M, what is E_{cell}?
(c) If E_{cell} = 1.25 V and (conc. Cd^{2+}) = 0.100 M, what is (conc. Ag^+)?

49. Consider the reaction

$$H_2(g) + Sn^{4+}(aq) \longrightarrow 2\ H^+(aq) + Sn^{2+}(aq)$$

operating at 298 K.
(a) What is the $E°_{cell}$ for this cell?
(b) What is the E_{cell} for P_{H_2} = 1.0 bar, (conc. Sn^{2+}) = 6.0×10^{-4} M, (conc. Sn^{4+}) = 5.0×10^{-4} M, and pH = 3.60?

50. ■ What is the cell potential of a concentration cell that contains two hydrogen electrodes if the cathode contacts a solution with pH = 7.8 and the anode contacts a solution with (conc. H^+) = 0.05 M?
51. What is the potential of an electrode made from zinc metal immersed in a solution where (conc. Zn^{2+}) = 0.010 M?
52. ■ For a voltaic cell with the reaction

$$Pb(s) + Sn^{2+}(aq) \longrightarrow Pb^{2+}(aq) + Sn(s)$$

at what ratio of concentrations of lead and tin ions will E_{cell} = 0?

Common Batteries

53. What are the advantages and disadvantages of lead-acid storage batteries?
54. ■ Nicad batteries are rechargeable and are commonly used in cordless appliances. Although such batteries actually function under basic conditions, imagine an electrochemical cell using this setup.

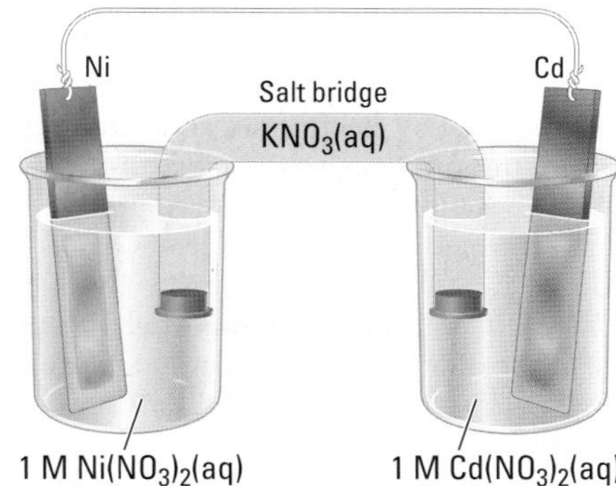

(a) Write a balanced net ionic equation depicting the reaction occurring in the cell.
(b) What is oxidized? What is reduced? What is the reducing agent and what is the oxidizing agent?
(c) Which is the anode and which is the cathode?
(d) What is $E°$ for the cell?
(e) What is the direction of electron flow in the external wire?
(f) If the salt bridge contains KNO_3, toward which compartment will the NO_3^- ions migrate?

55. Consider the nicad cell in Question 54.
(a) If the concentration of Cd^{2+} is reduced to 0.010 M, and (conc. Ni^{2+}) = 1.0 M, will the cell emf be smaller or larger than when the concentration of Cd^{2+}(aq) was 1.0 M? Explain your answer in terms of Le Chatelier's principle.
(b) Begin with 1.0 L of each of the solutions, both initially 1.0 M in dissolved species. Each electrode weighs 50.0 g at the start. If 0.050 A is drawn from the battery, how long can it last?

Fuel Cells

56. How does a fuel cell differ from a battery?
57. Describe the principal parts of an H_2/O_2 fuel cell. What is the reaction at the cathode? At the anode? What is the product of the fuel cell reaction?
58. ■ Hydrazine, N_2H_4, has been proposed as the fuel in a fuel cell in which oxygen is the oxidizing agent. The reactions are

$$N_2H_4(aq) + 4\ OH^-(aq) \longrightarrow N_2(g) + 4\ H_2O(\ell) + 4\ e^-$$
$$O_2(g) + 2\ H_2O(\ell) + 4\ e^- \longrightarrow 4\ OH^-(aq)$$

(a) Which reaction occurs at the anode and which at the cathode?
(b) What is the net cell reaction?
(c) If the cell is to produce 0.50 A of current for 50.0 h, what mass in grams of hydrazine must be present?
(d) What mass in grams of O_2 must be available to react with the mass of N_2H_4 determined in part (c)?

Electrolysis: Reactant-Favored Redox Reactions

59. Consider the electrolysis of water in the presence of very dilute H_2SO_4. What species is produced at the anode? At the cathode? What are the relative amounts of the species produced at the two electrodes?

60. Write chemical equations for the electrolysis of molten salts of three different alkali halides to produce the corresponding halogens and alkali metals.
61. From Table 19.1 write down all of the aqueous metal ions that can be reduced by electrolysis to the corresponding metal.
62. From Table 19.1 write down all of the aqueous species that can be oxidized by electrolysis, and determine the products.
63. What are the products of the electrolysis of a 1 M aqueous solution of NaBr? What species are present in the solution? What is formed at the cathode? What is formed at the anode?
64. ■ For each of these solutions, tell what reactions take place at the anode and at the cathode during electrolysis.
(a) $NiBr_2(aq)$ (b) NaI(aq)
(c) $CdCl_2(aq)$ (d) $CuI_2(aq)$
(e) $MgF_2(aq)$ (f) $HNO_3(aq)$

Counting Electrons

65. ■ A current of 0.015 A is passed through a solution of $AgNO_3$ for 155 min. What mass of silver is deposited at the cathode?
66. Current is passed through a solution containing $Ag^+(aq)$. How much silver was in the solution if all the silver was removed as Ag metal by electrolysis for 14.5 min at a current of 1.0 mA?
67. A current of 2.50 A is passed through a solution of $Cu(NO_3)_2$ for 2.00 h. What mass of copper is deposited at the cathode?
68. A current of 0.0125 A is passed through a solution of $CuCl_2$ for 2.00 h. What mass of copper is deposited at the cathode and what volume of Cl_2 gas (in mL at STP) is produced at the anode?
69. The major reduction half-reaction occurring in the cell in which molten Al_2O_3 and molten aluminum salts are electrolyzed is $Al^{3+}(aq) + 3\ e^- \rightarrow Al(s)$. If the cell operates at 5.0 V and 1.0×10^5 A, what mass (in grams) of aluminum metal can be produced in 8.0 h?
70. The vanadium(II) ion can be produced by electrolysis of a vanadium(III) salt in solution. How long must you carry out an electrolysis if you wish to convert completely 0.125 L of 0.0150 M $V^{3+}(aq)$ to $V^{2+}(aq)$ using a current of 0.268 A?
71. The reactions occurring in a lead-acid storage battery are given in Section 19.9. A typical battery might be rated at 50. ampere-hours (A-h). This means that it has the capacity to deliver 50. A for 1.0 h or 1.0 A for 50. h. If it does deliver 1.0 A for 50. h, what mass of lead would be consumed?
72. An effective battery can be built using the reaction between Al metal and O_2 from the air. If the Al anode of this battery consists of a 3-oz piece of aluminum (84 g), for how many hours can the battery produce 1.0 A of electricity?
73. ■ A dry cell is used to supply a current of 250. mA for 20 min. What mass of Zn is consumed?
74. If the same current as in Question 73 were supplied by a mercury battery, what mass of Hg would be produced at the cathode?
75. Assume that the anode reaction for the lithium battery is

$$Li(s) \longrightarrow Li^+(aq) + e^-$$

and the anode reaction for the lead-acid storage battery is

$$Pb(s) + HSO_4^-(aq) + H_2O(\ell) \longrightarrow PbSO_4(s) + 2\ e^- + H_3O^+(aq)$$

Compare the masses of metals consumed when each of these batteries supplies a current of 1.0 A for 10. min.
76. A hydrogen-oxygen fuel cell operates on the simple reaction

$$2\ H_2(g) + O_2(g) \longrightarrow 2\ H_2O(\ell)$$

If the cell is designed to produce 1.5 A of current, how long can it operate if there is an excess of oxygen and only sufficient hydrogen to fill a 1.0-L tank at 200. bar pressure at 25 °C?
77. ■ How long would it take to electroplate a metal surface with 0.500 g nickel metal from a solution of Ni^{2+} with a current of 4.00 A?
78. How much current is required to electroplate a metal surface with 0.400 g chromium metal from a solution of Cr^{3+} in 1.00 h?

Corrosion: Product-Favored Redox Reactions

79. Explain how rust is formed from iron materials by corrosion.
80. Why does iron corrode faster in salt water than in fresh water?
81. What common metal does not corrode readily under normal conditions?
82. Why does coating a steel object with chromium stop corrosion of the iron?
83. Explain how galvanizing iron stops corrosion of the underlying iron.

General Questions

84. A 12-V automobile battery consists of six cells of the type described in Section 19.9. The cells are connected in series so that the same current flows through all of them. Calculate the theoretical minimum electrical potential difference needed to recharge an automobile battery. (Assume standard-state concentrations.) How does this compare with the maximum voltage that could be delivered by the battery? Assuming that the lead plates in an automobile battery each weigh 2.50 kg and that sufficient PbO_2 is available, what is the maximum possible work that could be obtained from the battery?
85. Three electrolytic cells are connected in series, so that the same current flows through all of them for 20. min. In cell A, 0.0234 g Ag plates out from a solution of $AgNO_3(aq)$; cell B contains $Cu(NO_3)_2(aq)$; cell C contains $Al(NO_3)_3(aq)$. What mass of Cu will plate out in cell B? What mass of Al will plate out in cell C?
86. Fluorinated organic compounds are important commercially; they are used as herbicides, flame retardants, and fire-extinguishing agents, among other things. A reaction such as

$$CH_3SO_2F + 3\ HF \longrightarrow CF_3SO_2F + 3\ H_2$$

is actually carried out electrochemically in liquid HF as the solvent.
(a) Draw the structural formula for CH_3SO_2F. (S is the "central" atom with the O atoms, F atom, and CH_3 group bonded to it.) What is the geometry around the S atom? What are the O—S—O and O—S—F bond angles?
(b) If you electrolyze 150. g CH_3SO_2F, how many grams of HF are required and how many grams of each product can be isolated?
(c) Is H_2 produced at the anode or the cathode of the electrolysis cell?

(d) A typical electrolysis cell operates at 8.0 V and 250 A. How many kilowatt-hours of energy does one such cell consume in 24 h?

Applying Concepts

87. Four metals A, B, C, and D exhibit these properties:
(a) Only A and C react with 1.0 M HCl to give H_2 gas.
(b) When C is added to solutions of ions of the other metals, metallic A, B, and D are formed.
(c) Metal D reduces B^{n+} ions to give metallic B and D^{n+} ions.
On the basis of this information, arrange the four metals in order of increasing ability to act as reducing agents.

88. The table below lists the cell potentials for the ten possible electrochemical cells assembled from the elements A, B, C, D, and E and their respective ions in solutions. Using the data in the table, establish a standard reduction potential table similar to Table 19.1. Assign a reduction potential of 0.00 V to the element that falls in the middle of the series.

	A(s) in A^{n+}(aq)	B(s) in B^{n+}(aq)
E(s) in E^{n+}(aq)	+0.21 V	+0.68 V
D(s) in D^{n+}(aq)	+0.35 V	+1.24 V
C(s) in C^{n+}(aq)	+0.58 V	+0.31 V
B(s) in B^{n+}(aq)	+0.89 V	—

	C(s) in C^{n+}(aq)	D(s) in D^{n+}(aq)
E(s) in E^{n+}(aq)	+0.37 V	+0.56 V
D(s) in D^{n+}(aq)	+0.93 V	—
C(s) in C^{n+}(aq)	—	—
B(s) in B^{n+}(aq)	—	—

89. When this electrochemical cell runs for several hours, the green solution gets lighter and the yellow solution gets darker.

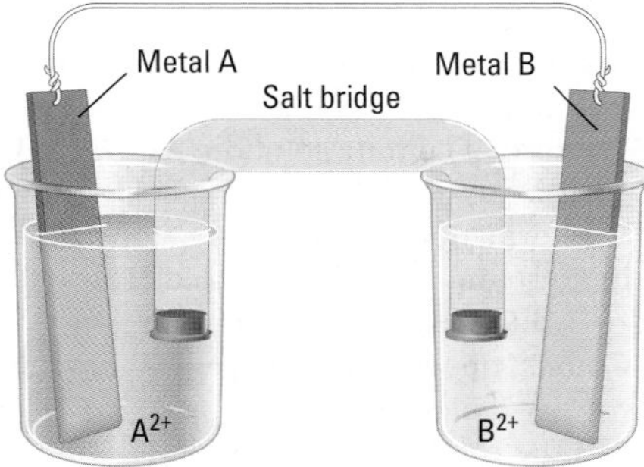

(a) What is oxidized, and what is reduced?
(b) What is the oxidizing agent, and what is the reducing agent?
(c) What is the anode, and what is the cathode?
(d) Write equations for the half-reactions.
(e) Which metal gains mass?
(f) What is the direction of the electron transfer through the external wire?
(g) If the salt bridge contains KNO_3(aq), into which solution will the K^+ ions migrate?

90. An electrolytic cell is set up with Cd(s) in $Cd(NO_3)_2$(aq) and Zn(s) in $Zn(NO_3)_2$(aq). Initially both electrodes weigh 5.00 g. After running the cell for several hours the electrode in the left compartment weighs 4.75 g.
(a) Which electrode is in the left compartment?
(b) Does the mass of the electrode in the right compartment increase, decrease, or stay the same? If the mass changes, what is the new mass?
(c) Does the mass of the solution in the right compartment increase, decrease, or stay the same?
(d) Does the volume of the electrode in the right compartment increase, decrease, or stay the same? If the volume changes, what is the new volume? (The density of Cd is 8.65 g/cm^3.)

91. Using data from Appendix I, show why
(a) Co^{3+} is not stable in aqueous solution.
(b) Fe^{2+} is not stable in air.

92. When H_2O_2 is mixed with Fe^{2+}, which redox reaction will occur—the oxidation of Fe^{2+} to Fe^{3+} or the reduction of Fe^{2+} to Fe? What are the $E°_{cell}$ values for the electrochemical cells corresponding to the two reactions?

93. Calculate the potential of a cell consisting of two hydrogen electrodes, one immersed in a solution with (conc. H^+) = 1.0×10^{-8} M and the other with (conc. H^+) = 0.025 M.

94. The permanganate ion MnO_4^- can be reduced to the manganese(II) ion Mn^{2+} in aqueous acidic solution, and the reduction potential for this half-cell reaction is 1.52 V. If this half-cell is combined with a Zn^{2+}/Zn half-cell to form a galvanic cell at standard conditions,
(a) Write the chemical equation for the half-reaction occurring at the anode.
(b) Write the chemical equation for the half-reaction occurring at the cathode.
(c) Write the overall balanced equation for the reaction.
(d) Calculate the cell voltage.

More Challenging Questions

95. Fluorine, F_2, is made by the electrolysis of anhydrous HF.

$$2\ HF(\ell) \longrightarrow H_2(g) + F_2(g)$$

Typical electrolysis cells operate at 4000 to 6000 A and 8 to 12 V. A large-scale plant can produce about 9.0 metric tons of F_2 gas per day.
(a) What mass in grams of HF is consumed?
(b) Using the conversion factor of 3.60×10^6 J/kWh, how much energy in kilowatt-hours is consumed by a cell operating at 6.0×10^3 A at 12 V for 24 h?

96. What reaction would take place if a 1.0 M solution of $Cr_2O_7^{2-}$ was added to a 1.0 M solution of HBr?

97. If Cl_2 and Br_2 are added to an aqueous solution that contains Cl^- and Br^-, what product-favored reaction will occur?

98. This reaction occurs in a cell with H_2(g) pressure of 1.0 atm and (conc. Cl^-) = 1.0 M at 25 °C; the measured E_{cell} = 0.34 V. What is the pH of the solution?

$$2\ H_2O(\ell) + 2\ H_2(g) + 2\ AgCl(s) \longrightarrow 2\ H_3O^+(aq) + 2\ Cl^-(aq) + 2\ Ag(s)$$

99. An electric current of 2.00 A was passed through a platinum salt solution for 3.00 hours, and 10.9 g of metallic platinum was formed at the cathode. What is the charge on the platinum ions in the solution?

100. $E_{cell} = 0.010$ V for a galvanic cell with this reaction at 25 °C.

$$Sn(s) + Pb^{2+}(aq) \longrightarrow Sn^{2+}(aq) + Pb(s)$$

(a) What is the equilibrium constant K_c for the reaction?
(b) If a solution with (conc. Pb^{2+}) = 1.1 M had excess tin metal added to it, what would the equilibrium concentration of Sn^{2+} and Pb^{2+} be?

101. You wish to electroplate a copper surface having an area of 1200 mm^2 with a 1.0-μm-thick coating of silver from a solution of $Ag(CN)_2^-$ ions. If you use a current of 150.0 mA, what electrolysis time should you use? The density of metallic silver is 10.5 g/cm^3.

102. Will 1.0 M nitric acid, HNO_3, oxidize metallic gold to form a 1 M Au^{3+} solution? Explain why or why not.

103. A student wanted to measure the copper(II) concentration in an aqueous solution. For the cathode half-cell she used a silver electrode with a 1.00 M solution of $AgNO_3$. For the anode half-cell she used a copper electrode dipped into the aqueous sample. If the cell gave $E_{cell} = 0.62$ V at 25 °C, what was the copper(II) concentration of the solution?

104. In a mercury battery, the anode reaction is

$$Zn(s) + 2\,OH^-(aq) \longrightarrow ZnO(aq) + H_2O(\ell) + 2\,e^-$$

and the cathode reaction is

$$HgO(s) + H_2O(\ell) + 2\,e^- \longrightarrow Hg(\ell) + 2\,OH^-(aq)$$

The cell potential is 1.35 V. How many hours can such a battery provide power at a rate of 4.0×10^{-4} watt (1 watt = 1 J s^{-1}) if 1.25 g HgO is available?

Conceptual Challenge Problems

CP19.A (Section 19.6) Most automobiles run on internal combustion engines, in which the energy used to run the vehicle is obtained from the combustion of gasoline. The main component of gasoline is octane (C_8H_{18}). An automobile manufacturer has recently announced a chemical method for generating hydrogen gas from gasoline and proposes to develop a car in which an H_2/O_2 fuel cell powers an electric propulsion motor, thus eliminating the internal combustion engine with its problems (for example, the generation of unwanted by-products that pollute the air). The hydrogen for the fuel cell would be directly generated from gasoline on board the vehicle. There are two steps in this hydrogen generation process:

(i) Partial oxidation of octane by oxygen to carbon monoxide and hydrogen
(ii) Combination of carbon monoxide with additional gaseous water to form carbon dioxide and more hydrogen (the water-gas shift reaction)

(a) Write the chemical equation for the complete combustion of 1 mol octane.
(b) Write balanced chemical equations for the two-step hydrogen generation process. How many moles of H_2 are produced per mole of octane? (Remember that water is a reactant in the two-step process.)
(c) By combining these equations, show that the net *overall* reaction is the same as in the combustion of octane.
(d) Assuming that the entire Gibbs free energy change of the H_2/O_2 fuel cell reaction is available for use by the electric propulsion motor, calculate the energy produced by a fuel cell when it consumes all of the hydrogen produced from 1 mol of octane. Compare this energy with the Gibbs free energy change for the combustion of 1 mol of octane. (*Note:* The Gibbs free energy of formation, ΔG_f°, for $C_8H_{18}(\ell)$ is 6.14 kJ/mol.)

CP19.B (Section 19.4) People obtain energy by oxidizing food. Glucose is a typical foodstuff. This carbohydrate is oxidized to water and carbon dioxide.

$$C_6H_{12}O_6(aq) + 6\,O_2(g) \longrightarrow 6\,CO_2(g) + 6\,H_2O(\ell)$$

The heat of combustion of glucose is 2.80×10^3 kJ/mol, which means that as glucose is oxidized, its electrons lose 2.80×10^3 kJ/mol as they give up potential energy in a complicated series of chemical steps.

(a) Assume that a person requires 2400 food Calories per day and that this energy is obtained from the oxidation of glucose. How much O_2 must a person breathe each day to react with this much glucose?
(b) Each mole of O_2 requires 4 mol electrons, regardless of whether the O atoms become part of CO_2 or H_2O. What would be the average electric current (C/s) in a human body using the above amount of energy described in part (a) per day?
(c) Use the answer from part (b) and calculate the electrical potential this current flows through in a day to produce the 2400 food Calories. (1 Calorie = 4.18 kJ)

CP19.C (Section 19.11) A piece of chromium metal is attached to a battery and dipped into 50 mL of 0.3 M KOH solution in a 250-mL beaker. A stainless steel electrode is connected to the other electrode of the battery and immersed in the same solution. A steady current of 0.50 A is maintained for exactly 2 hours. Several samples of a gas formed at the stainless steel electrode during the electrolysis are captured, and all are found to ignite in air. After the electrolysis, the chromium electrode is weighed and found to have decreased in weight by 0.321 g. The mass of the stainless steel electrode does not change.

After electrolysis, the KOH solution is neutralized with nitric acid to a pH of slightly less than 7, then is heated and reacted with 0.151 M lead(II) nitrate solution. As the lead(II) nitrate solution is added, a yellow precipitate quickly forms from the hot solution. The formation of precipitate stops after 40.4 mL of the lead(II) nitrate solution has been added. The yellow solid is then filtered, dried, and weighed. Its mass is 1.97 g.

(a) How much electrical charge passes through the cell?
(b) How many moles of Cr react?
(c) What is the oxidation state of the Cr after reacting?
(d) Assuming that the yellow compound that precipitates from the solution during the titration contains both Pb and Cr, what do you conclude to be the ratio of the numbers of atoms of Pb and Cr?
(e) If the yellow compound contains an element other than Pb and Cr, what is it and how much is in the compound? What is the formula for the yellow compound?

20 Nuclear Chemistry

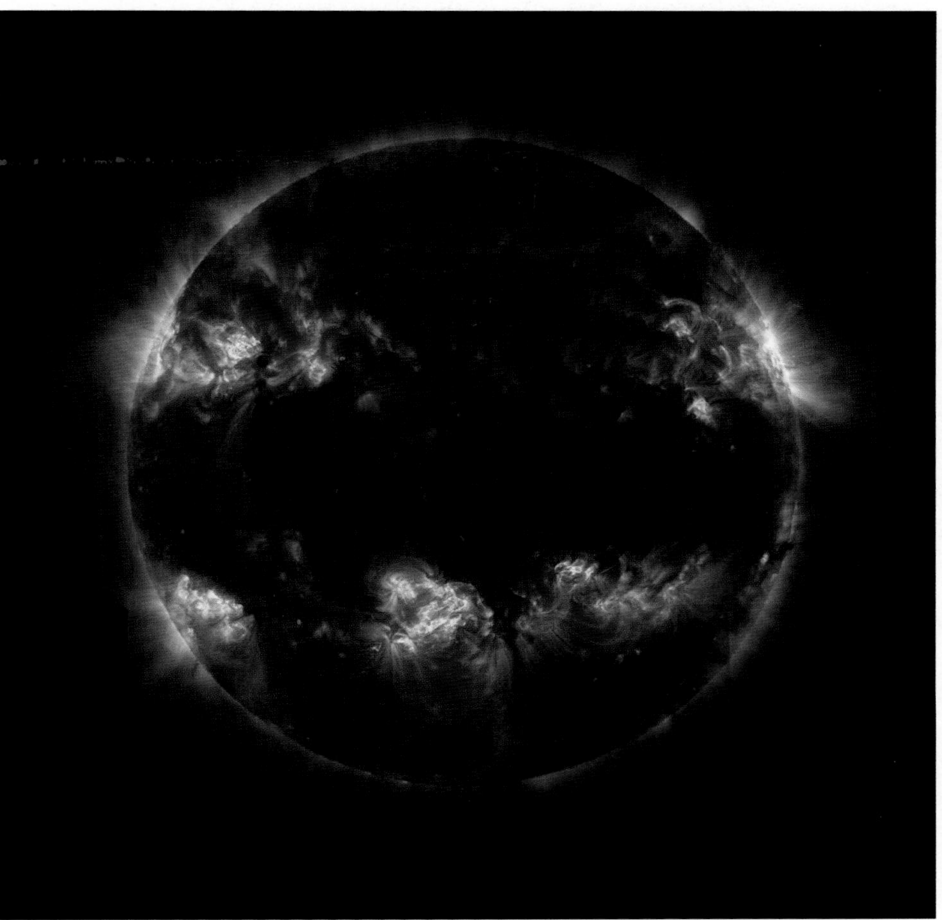

Nuclear fusion reactions are the source of energy in the sun, whose surface is shown here. Such reactions occurring in the sun are the ultimate source of almost all of the energy available on earth. During fusion, light nuclei such as $^{1}_{1}H$ combine to form heavier nuclei such as $^{4}_{2}He$, and tremendous energy is released. Temperatures of 10^6 to 10^7 K are required for fusion to occur.

TRACE Project, Stanford-Lockheed Institute for Space Research, NASA

ThomsonNOW
Throughout the text, this icon indicates an opportunity to test yourself on key concepts and to explore interactive modules by signing in to ThomsonNOW at **www.thomsonedu.com**.

Nuclear chemistry, a subject that bridges chemistry and physics, has a significant impact on our society. No matter what your reason for taking a college course in chemistry—to prepare for a career in one of the sciences or simply to gain knowledge as a concerned citizen—you should know about nuclear chemistry. Radioactive isotopes are widely used in medicine. PET (positron emission tomography) scans depend on radioactivity. Your room may be protected by a smoke detector that uses a radioactive element as part of its sensor, and research in all fields of science employs radioactive elements and their compounds. The national security of the United States since World War II has depended on nuclear weapons, and more than 30 nations use nuclear reactors to produce electricity. This chapter considers several aspects of nuclear chemistry: changes in atomic nuclei and their effects, fission and fusion of nuclei and the energy that can be derived from such changes, units used to measure radioactivity, and beneficial applications of radioactive isotopes.

Einstein's original letter can be seen at **http://www.atomicmuseum.com/tour/atomicage.cfm**.

On August 2, 1939, with the world hovering on the brink of World War II, Albert Einstein sent a letter to President Franklin D. Roosevelt. In this letter, which profoundly changed the course of history, Einstein called attention to work being done on the physics of the atomic nucleus. He noted that he and others believed this work suggested the possibility that "uranium may be turned into a new and important source of energy . . . and [that it was] conceivable . . . that extremely powerful bombs of a new type may thus be constructed. . . ." Einstein's letter was the impetus for the Manhattan Project, which led to the detonation of the first atomic bomb at 5:30 AM on July 16, 1945, in the desert of New Mexico.

The Alsos Digital Library for Nuclear Issues has a Web site at **http://alsos.wlu.edu** with a section on the Manhattan Project.

Since World War II, more powerful nuclear weapons have been developed and stockpiled by a number of nations. With the end of the Cold War, fears of a nuclear holocaust receded and nuclear disarmament treaties were signed between the United States and the former Soviet Union for removing the plutonium-239 and other nuclear fuel from existing nuclear warheads. Unfortunately, those fears have been replaced by the concern that other nations have developed or acquired nuclear weapons. For many years, the respected journal *The Bulletin of the Atomic Scientists* has used the symbol of a clock with its hands near the fateful midnight hour (representing nuclear annihilation) to illustrate the danger faced by the world from atomic weapons. Even with the end of the Cold War, the hands have moved back only a little from midnight.

The Bulletin of the Atomic Scientists can be found on-line at **http://www.thebulletin.org/**.

20.1 The Nature of Radioactivity

Many minerals, called phosphors, glow for some time after being stimulated by exposure to sunlight or ultraviolet light. In 1896, French physicist Antoine Henri Becquerel was studying this phenomenon, called *phosphorescence,* when he made an important and totally unexpected observation that led him to the discovery of radioactivity. While waiting for a sunny day, Becquerel stored a photographic plate wrapped in black paper along with a uranium salt (a material known to phosphoresce) in a dark drawer. To his amazement, the image of the uranium salt appeared on the plate that had been in the drawer, unexposed to sunlight. Becquerel realized that radiation from the uranium salt had penetrated the black paper and exposed the photographic plate even though the uranium salt had not been stimulated by light.

Becquerel performed many more related experiments and found that pure uranium metal produced the same emissions as uranium salts did, but even more strongly. This result would be expected if the radiation were the property of the metal and not dependent on its form of chemical combination. But no pure metal was known to phosphoresce, which mystified Becquerel. What was the source of this radiation? It turned out that the radiation had nothing to do with phosphorescence. Failing to find the reasons for the emissions, Becquerel gave the project to his graduate student Marie Curie. She and her husband Pierre, a physicist, studied the phenomenon intensively and termed it *radioactivity.*

One of Marie Curie's first findings was to confirm Becquerel's observation that uranium metal itself was radioactive and that the degree to which a uranium-containing sample was radioactive depended on the percentage of uranium present. When she tested pitchblende, a common ore containing uranium and other metals (such as lead, bismuth, and copper), she was surprised to find that it was even more radioactive than pure uranium. Only one explanation was possible: pitchblende contained an element (or elements) more radioactive than uranium. Eventually, the Curies discovered the element they named *polonium* after Marie's homeland of Poland. They also discovered radium, another highly radioactive element.

For more on experiments done by Becquerel and the Curies, see Walton, H. F., *Journal of Chemical Education,* Vol. 69, 1992; p. 10.

In England at about the same time, Sir J. J. Thomson and his graduate student Ernest Rutherford were studying the radiation from uranium and thorium (⇦ ***p. 45***). Rutherford found that "There are present at least two distinct types of radiation—one that is readily absorbed, which will be termed for convenience alpha (α) radiation, and the other of a more penetrative character, which will be termed beta (β) radiation." **Alpha radiation,** he discovered, was composed of particles that, when passed through an electric field, were attracted to the negative side of the field (⇦ ***p. 43***). Indeed, his later studies showed these **alpha (α) particles** to be helium nuclei, ${}^{4}_{2}He^{2+}$, which were ejected at high speeds from a radioactive element (Table 20.1). Alpha particles have limited penetrating power and can be absorbed by skin, clothing, or several sheets of ordinary paper.

Encouraged by the Curies, Becquerel returned to the study of radiation. He found that the radiation from uranium was affected by magnetic fields and consisted of two kinds of particles, which we now know to be alpha and beta particles.

In the same experiment with electric fields, Rutherford found that **beta radiation** must be composed of negatively charged particles, since the beam of beta radiation was attracted to the electrically positive plate of an electric field. Later work by Becquerel showed that these particles have an electric charge and mass equal to those of an electron. Thus, **beta (β) particles** are electrons ejected at high speeds from radioactive nuclei. They are more penetrating than alpha particles (Table 20.1), and a $\frac{1}{8}$-inch-thick piece of aluminum is necessary to stop them. Beta particles can penetrate 1 to 2 cm into living bone or tissue.

A third type of radiation was later discovered by P. Villard, a Frenchman, who named it **gamma (γ) radiation**, using the third letter in the Greek alphabet in keeping with Rutherford's scheme. Unlike alpha and beta particles, which are particulate in nature, gamma rays are a form of electromagnetic radiation and are not affected by an electric field. Gamma radiation is the most penetrating, and it can pass completely through the human body. Thick layers of lead or concrete are required to stop a beam of gamma rays.

Table 20.1 Characteristics of α, β, and γ Emissions

Name	Symbol	Charge	Mass (g/particle)	Penetrating Power*
Alpha	${}^{4}_{2}He^{2+}$, ${}^{4}_{2}\alpha$, ${}^{4}_{2}He$	2+	6.65×10^{-24}	0.03 mm
Beta	${}^{0}_{-1}e$, ${}^{0}_{-1}\beta$	1−	9.11×10^{-28}	2 mm
Gamma	${}^{0}_{0}\gamma$, γ	0	0	10 cm

*Distance at which half the radiation has been stopped by water.

20.2 Nuclear Reactions

Equations for Nuclear Reactions

Ernest Rutherford found that radium not only emits alpha particles but also produces the radioactive gas radon in the process. Such observations led Rutherford and Frederick Soddy, in 1902, to propose the revolutionary theory that *radioactivity is the result of a natural change of a radioactive isotope of one element into an isotope of a different element.* In such changes, called **nuclear reactions** or *transmutations,* an unstable nucleus (the *parent nucleus*) spontaneously emits radiation and is converted (decays) into a more stable nucleus of a different element (the *daughter product*). Thus, a nuclear reaction results in a change in atomic number and, in some cases, a change in mass number as well. For example, the reaction of radium studied by Rutherford can be written as

$$^{226}_{88}\text{Ra} \longrightarrow {}^{4}_{2}\text{He} + {}^{222}_{86}\text{Rn}$$

When a radioactive atom decays, the emission of a charged particle leaves behind a charged atom. Thus, when radium-226 decays, it gives a helium-4 cation (${}^{4}_{2}\text{He}^{2+}$) and a radon-222 anion (${}^{222}_{86}\text{Rn}^{2-}$). By convention, the ion charges are not shown in balanced nuclear equations.

In this representation, the subscripts are the atomic numbers and the superscripts are the mass numbers ***(⟸ p. 54).***

In a chemical change, the atoms in molecules and ions are rearranged, but atoms are neither created nor destroyed; the number of atoms remains the same. Similarly, in nuclear reactions the total number of nuclear particles, or **nucleons** (protons and neutrons), remains the same. The essence of nuclear reactions, however, is that one nucleon can change into a different nucleon along with the release of energy. A proton can change to a neutron or a neutron can change to a proton, but the total number of nucleons remains the same. Therefore, ***the sum of the mass numbers of reacting nuclei must equal the sum of the mass numbers of the nuclei produced.*** Furthermore, because electrical charge cannot be created or destroyed, ***the sum of the atomic numbers of the products must equal the sum of the atomic numbers of the reactants.*** These principles can be verified for the preceding nuclear equation.

Recall that the atomic number is the number of protons in an atom's nucleus (that is, the total positive charge on the nucleus), and the mass number is the sum of protons and neutrons in a nucleus.

	${}^{226}_{88}\text{Ra}$ radium-226	⟶	${}^{4}_{2}\text{He}$ alpha particle	+	${}^{222}_{86}\text{Rn}$ radon-222
Mass number:	226	⟶	4	+	222
Atomic number:	88	⟶	2	+	86

Alpha and Beta Particle Emission

One way a radioactive isotope can decay is to eject an alpha particle from the nucleus. This is illustrated by the radium-226 reaction above and by the conversion of uranium-234 to thorium-230 by alpha emission.

	${}^{234}_{92}\text{U}$ uranium-234 (parent nucleus)	⟶	${}^{4}_{2}\text{He}$ alpha particle	+	${}^{230}_{90}\text{Th}$ thorium-230 (daughter product)
Mass number:	234	⟶	4	+	230
Atomic number:	92	⟶	2	+	90

In alpha emission, *the atomic number of the parent nucleus decreases by two units and the mass number decreases by four units for each alpha particle emitted.*

Note that in beta decay the mass number is uncharged.

Emission of a beta particle is another way for a radioactive isotope to decay. For example, loss of a beta particle by uranium-239 (parent nucleus) to form neptunium-239 (daughter product) is represented by

	${}^{239}_{92}\text{U}$ uranium-239	⟶	${}^{0}_{-1}\text{e}$ beta particle	+	${}^{239}_{93}\text{Np}$ neptunium-239
Mass number:	239	⟶	0	+	239
Atomic number:	92	⟶	−1	+	93

How does a nucleus, composed only of protons and neutrons, increase its number of protons by ejecting an electron during beta emission? It is generally accepted that a series of reactions is involved, but the net process is

$$\underset{\text{neutron}}{{}^{1}_{0}\text{n}} \longrightarrow \underset{\text{electron}}{{}^{0}_{-1}\text{e}} + \underset{\text{proton}}{{}^{1}_{1}\text{p}}$$

where we use the symbol p for a proton and n for a neutron. In this process, a neutron is converted to a proton and a beta particle is released. Therefore, *the ejection of a beta particle always means that a different element is formed because a neutron has been converted into a proton. The new element (daughter product) has an atomic number one unit greater than that of the decaying (parent) nucleus.* The mass number does not change, however, because no proton or neutron has been emitted.

If a neutron changes to a proton, conservation of charge requires that a negative particle (a beta particle) be created.

In many cases, the emission of an alpha or beta particle results in the formation of a product nucleus that is also unstable and therefore radioactive. The new radioactive product may undergo a number of successive transformations until a stable, nonradioactive nucleus is finally produced. Such a series of reactions is called a **radioactive series.** One such series begins with uranium-238 and ends with lead-206, as illustrated in Figure 20.1. The first step in the series is

A nucleus formed as a result of alpha or beta emission is often in an excited state and therefore emits a gamma ray.

$${}^{238}_{92}\text{U} \longrightarrow {}^{4}_{2}\text{He} + {}^{234}_{90}\text{Th}$$

The final step, the conversion of polonium-210 to lead-206, is

$${}^{210}_{84}\text{Po} \longrightarrow {}^{4}_{2}\text{He} + {}^{206}_{82}\text{Pb}$$

PROBLEM-SOLVING EXAMPLE 20.1 Radioactive Series

An intermediate species in the uranium-238 decay series shown in Figure 20.1 is polonium-218. It emits an alpha particle, followed by emission of a beta particle, followed by the emission of a beta particle. Write the nuclear equations for these three reactions.

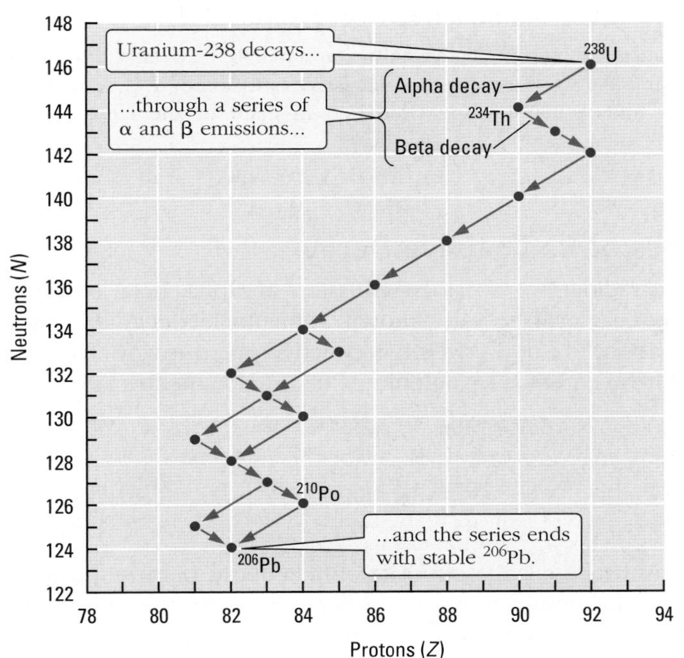

Figure 20.1 The ^{238}U decay series.

Answer $^{218}_{84}\text{Po} \longrightarrow {}^{4}_{2}\text{He} + {}^{214}_{82}\text{Pb}$

$^{214}_{82}\text{Pb} \longrightarrow {}^{0}_{-1}\text{e} + {}^{214}_{83}\text{Bi}$

$^{214}_{83}\text{Bi} \longrightarrow {}^{0}_{-1}\text{e} + {}^{214}_{84}\text{Po}$

Strategy and Explanation The starting point of these linked reactions is polonium-218. When it emits an alpha particle, the atomic number decreases by two and the mass number decreases by four to produce lead-214.

$$\underset{\text{polonium-218}}{^{218}_{84}\text{Po}} \longrightarrow {}^{4}_{2}\text{He} + \underset{\text{lead-214}}{^{214}_{82}\text{Pb}}$$

The second of these linked reactions begins with lead-214. When it emits a beta particle, the atomic number increases by one and the mass number remains constant to produce bismuth-214.

$$\underset{\text{lead-214}}{^{214}_{82}\text{Pb}} \longrightarrow {}^{0}_{-1}\text{e} + \underset{\text{bismuth-214}}{^{214}_{83}\text{Bi}}$$

In the third reaction, bismuth-214 emits a beta particle, so the atomic number increases by one and the mass number remains constant to produce polonium-214.

$$\underset{\text{bismuth-214}}{^{214}_{83}\text{Bi}} \longrightarrow {}^{0}_{-1}\text{e} + \underset{\text{polonium-214}}{^{214}_{84}\text{Po}}$$

✓ **Reasonable Answer Check** An alpha emission decreases the atomic number by two and decreases the mass number by four. Each beta emission increases the atomic number by one and leaves the mass number unchanged. Therefore, one alpha and two beta emissions would leave the atomic number unchanged and decrease the mass number by four, which is what our systematic analysis found.

PROBLEM-SOLVING PRACTICE 20.1

(a) Write an equation showing the emission of an alpha particle by an isotope of neptunium, $^{237}_{93}\text{Np}$, to produce an isotope of protactinium.

(b) Write an equation showing the emission of a beta particle by sulfur-35, $^{35}_{16}\text{S}$, to produce an isotope of chlorine.

EXERCISE 20.1 Radioactive Decay Series

The actinium decay series begins with uranium-235, $^{235}_{92}\text{U}$, and ends with lead-207, $^{207}_{82}\text{Pb}$. The first five steps involve the successive emission of α, β, α, α, and β particles. Identify the radioactive isotope produced in each of the steps, beginning with uranium-235.

Other Types of Radioactive Decay

The positron was discovered by Carl Anderson in 1932. It is sometimes called an "antielectron," one of a group of particles that have become known as "antimatter." Contact between an electron and a positron leads to mutual annihilation of both particles with production of two high-energy photons (gamma rays). This process is the basis of positron emission tomography (PET) scanning to detect tumors (Section 20.9).

In addition to radioactive decay by emission of alpha, beta, or gamma radiation, other nuclear decay processes are known. Some nuclei decay, for example, by emission of a **positron,** $^{0}_{+1}\text{e}$ or β^{+}, which is effectively a positively charged electron. For example, positron emission by polonium-207 leads to the formation of bismuth-207.

	$^{207}_{84}\text{Po}$ polonium-207	$\longrightarrow$	$^{0}_{+1}\text{e}$ positron	+	$^{207}_{83}\text{Bi}$ bismuth-207
Mass number:	207	$\longrightarrow$	0	+	207
Atomic number:	84	$\longrightarrow$	+1	+	83

Notice that this process is the opposite of beta decay, because positron decay leads to a *decrease* in the atomic number. Like beta decay, positron decay does not change the mass number because no proton or neutron is ejected.

Another nuclear process is **electron capture,** in which the atomic number is reduced by one but the mass number remains unchanged. In this process an inner-shell electron (for example, a 1*s* electron) is captured by the nucleus.

	$^{7}_{4}Be$ beryllium-7	+	$^{0}_{-1}e$ electron	$\longrightarrow$	$^{7}_{3}Li$ lithium-7
Mass number:	7	+	0	$\longrightarrow$	7
Atomic number:	4	+	−1	$\longrightarrow$	3

In the old nomenclature of atomic physics, the innermost shell (n = 1 principal quantum number) was called the K-shell, so the electron capture mechanism is sometimes called *K-capture.*

In summary, radioactive nuclei can decay in four ways, as summarized in the figure at right.

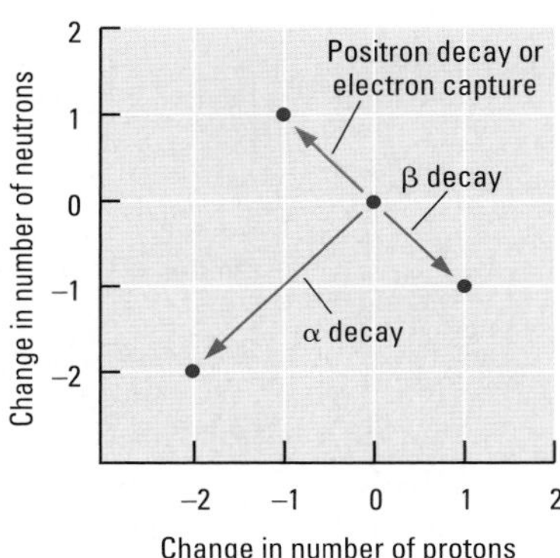

Effects of four radioactive decay processes. The chart shows the changes in the number of protons and neutrons during alpha decay, beta decay, positron emission, and electron capture.

PROBLEM-SOLVING EXAMPLE 20.2 Nuclear Equations

Complete these nuclear equations by filling in the missing symbol, mass number, and atomic number of the product species.

(a) $^{18}_{9}F \longrightarrow ^{18}_{8}O + \_\_\_\_$ (b) $^{26}_{13}Al + ^{0}_{-1}e \longrightarrow \_\_\_\_$

(c) $^{208}_{79}Au \longrightarrow ^{208}_{80}Hg + \_\_\_\_$ (d) $^{218}_{84}Po \longrightarrow ^{4}_{2}He + \_\_\_\_$

Answer

(a) $^{0}_{+1}e$ (b) $^{26}_{12}Mg$ (c) $^{0}_{-1}e$ (d) $^{214}_{82}Pb$

Strategy and Explanation

In each case we deduce the missing species by comparing the atomic numbers and mass numbers before and after the reaction.

(a) The missing particle has a mass number of zero and a charge of +1, so it must be a positron, $^{0}_{+1}e$. When the positron is included in the equation, the atomic mass is 18 on each side, and the atomic numbers sum to 9 on each side.

(b) The missing nucleus must have a mass number of 26 + 0 = 26 and an atomic number of 13 − 1 = 12, so it is $^{26}_{12}Mg$.

(c) The missing particle has a mass number of zero and a charge of −1, so it must be a beta particle, $^{0}_{-1}e$.

(d) The missing nucleus has a mass number of 218 − 4 = 214 and an atomic number of 84 − 2 = 82, so it is $^{214}_{82}Pb$.

PROBLEM-SOLVING PRACTICE 20.2

Complete these nuclear equations by filling in the missing symbol, mass number, and atomic number of the product species.

(a) $^{11}_{6}C \longrightarrow ^{11}_{5}B + ?$ (b) $^{35}_{16}S \longrightarrow ^{35}_{17}Cl + ?$

(c) $^{30}_{15}P \longrightarrow ^{0}_{+1}e + ?$ (d) $^{22}_{11}Na \longrightarrow ^{0}_{-1}e + ?$

EXERCISE 20.2 Nuclear Reactions

Aluminum-26 can undergo either positron emission or electron capture. Write the balanced nuclear equation for each case.

20.3 Stability of Atomic Nuclei

The naturally occurring isotopes of elements from hydrogen to bismuth are shown in Figure 20.2, where the radioactive isotopes are represented by orange circles and

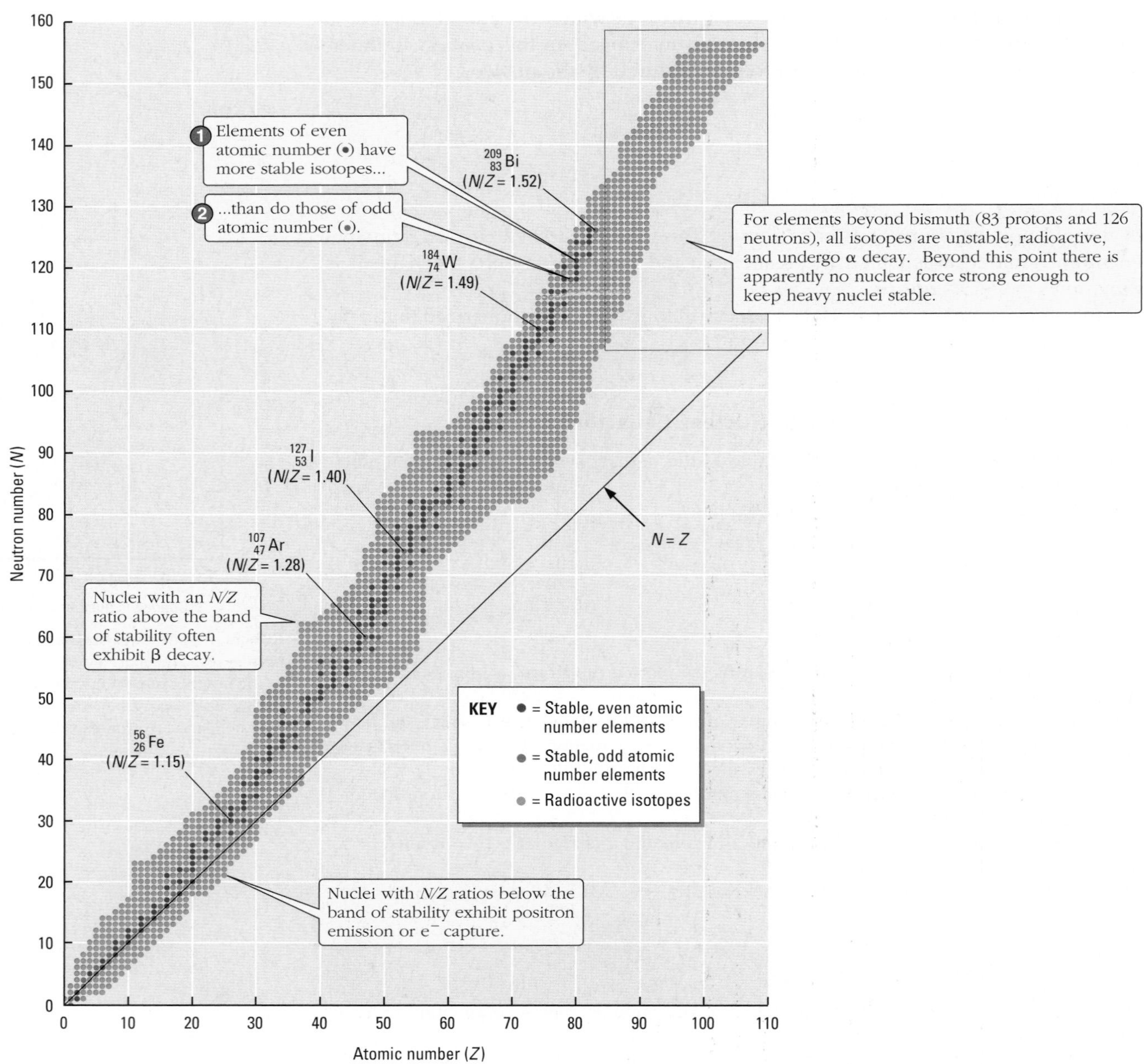

Figure 20.2 A plot of neutrons versus protons for the nuclei from hydrogen ($Z = 1$) through bismuth ($Z = 83$). A narrow band of stability is apparent. The N/Z values for some example stable nuclei are shown.

ThomsonNOW™
Go to the Coached Problems menu for a module **exploring nuclear stability.**

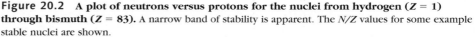

the stable (nonradioactive) isotopes are represented by purple and green circles. It is surprising that so few stable isotopes exist. Why are there not hundreds more? To investigate this question, we will systematically examine the elements, starting with hydrogen.

In its simplest and most abundant form, hydrogen has only one nuclear particle, a single proton. In addition, the element has two other well-known isotopes: nonradioactive deuterium, with one proton and one neutron, ${}^{2}_{1}H = D$, and radioactive tritium, with one proton and two neutrons, ${}^{3}_{1}H = T$. Helium, the next element, has two protons and two neutrons in its most stable isotope. At the end of the actinide series is element 103, lawrencium, one isotope of which has 154 neutrons and a mass number of 257. From hydrogen to lawrencium, except for ${}^{1}_{1}H$ and ${}^{3}_{2}He$, *the mass numbers of stable isotopes are always at least twice as large as the atomic number.* In other words, except for ^1_1H and ^3_2He, every isotope of every element has a nucleus containing *at least* one neutron for every proton. Apparently the tremendous *repulsive* forces between the positively charged protons in the nucleus are moderated by the presence of neutrons, which have no electrical charge. Figure 20.2 illustrates a number of principles:

1. For light elements up to Ca ($Z = 20$), the stable isotopes usually have equal numbers of protons and neutrons, or perhaps one more neutron than protons. Examples include ^7_3Li, $^{12}_6C$, $^{16}_8O$, and $^{32}_{16}S$.
2. Beyond calcium the neutron/proton ratio becomes increasingly greater than 1. The band of stable isotopes deviates more and more from the line $N = Z$ (number of neutrons = number of protons). It is evident that more neutrons are needed for nuclear stability in the heavier elements. For example, whereas one stable isotope of Fe has 26 protons and 30 neutrons ($N/Z = 1.15$), one stable isotope of platinum has 78 protons and 117 neutrons ($N/Z = 1.50$).
3. For elements beyond bismuth-209 (83 protons and 126 neutrons), all nuclei are unstable and radioactive. Furthermore, the rate of disintegration becomes greater the heavier the nucleus. For example, half of a sample of $^{238}_{92}U$ disintegrates in 4.5 billion years, whereas half of a sample of $^{256}_{103}Lr$ decays in only 28 seconds.
4. A careful analysis of Figure 20.2 reveals additional interesting features. First, elements with an even atomic number have a greater number of stable isotopes than do those with an odd atomic number. Second, stable isotopes usually have an even number of neutrons. For elements with an odd atomic number, the most stable isotope has an even number of neutrons. In fact, of the nearly 300 stable isotopes represented in Figure 20.2, roughly 200 have an even number of neutrons *and* an even number of protons. Only about 120 have an odd number of either protons or neutrons. Only four isotopes (^2_1H, ^6_3Li, $^{10}_5B$, and $^{14}_7N$) have odd numbers of *both* protons and neutrons.

The Band of Stability and Type of Radioactive Decay

The narrow band of stable isotopes in Figure 20.2 (the purple and green circles) is sometimes called the *peninsula of stability* in a "sea of instability." Any nucleus (the orange circles) not on this peninsula will decay in such a way that the nucleus can come ashore on the peninsula. The chart can help us predict what type of decay will be observed.

The nuclei of all elements beyond Bi ($Z = 83$) are unstable—that is, radioactive—and most decay by *alpha particle emission.* For example, americium, the radioactive element used in smoke alarms, decays in this manner.

$$^{243}_{95}Am \longrightarrow {}^4_2He + {}^{239}_{93}Np$$

Beta emission occurs in isotopes that have too many neutrons to be stable—that is, isotopes above the peninsula of stability in Figure 20.2. When beta decay converts a neutron to a proton and an electron (beta particle), which is then ejected, the atomic number increases by one, and the mass number remains constant.

$$^{60}_{27}Co \longrightarrow {}^{0}_{-1}e + {}^{60}_{28}Ni$$

Conversely, lighter nuclei—below the peninsula of stability—that have too few neutrons attain stability by *positron emission* or by *electron capture,* because these processes convert a proton to a neutron in one step.

$$^{13}_{7}N \longrightarrow {}^{0}_{+1}e + {}^{13}_{6}C$$

$$^{41}_{20}Ca + {}^{0}_{-1}e \longrightarrow {}^{41}_{19}K$$

Decay by these two routes is observed for elements with atomic numbers ranging from 4 to greater than 100; as Z increases, electron capture becomes more likely than positron emission.

PROBLEM-SOLVING EXAMPLE 20.3 Nuclear Stability

For each of these unstable isotopes, write a nuclear equation for its probable mode of decay.

(a) Silicon-32, $^{32}_{14}Si$ (b) Titanium-43, $^{43}_{22}Ti$

(c) Plutonium-239, $^{239}_{94}Pu$ (d) Manganese-56, $^{56}_{25}Mn$

Answer

(a) $^{32}_{14}Si \longrightarrow ^{0}_{-1}e + ^{32}_{15}P$ (b) $^{43}_{22}Ti \longrightarrow ^{0}_{+1}e + ^{43}_{21}Sc$ or $^{43}_{22}Ti + ^{0}_{-1}e \longrightarrow ^{43}_{21}Sc$

(c) $^{239}_{94}Pu \longrightarrow ^{4}_{2}He + ^{235}_{92}U$ (d) $^{56}_{25}Mn \longrightarrow ^{0}_{-1}e + ^{56}_{26}Fe$

Strategy and Explanation Note the ratio of protons to neutrons. If there are excess neutrons, beta emission is probable. If there are excess protons, either electron capture or positron emission is probable. If the atomic number is greater than 83, then alpha emission is probable.

(a) Silicon-32 has excess neutrons, so beta decay is expected.

(b) Titanium-43 has excess protons, so either positron emission or electron capture is probable.

(c) Plutonium-239 has an atomic number greater than 83, so alpha decay is probable.

(d) Manganese-56 has excess neutrons, so beta decay is expected.

PROBLEM-SOLVING PRACTICE 20.3

For each of these unstable isotopes, write a nuclear equation for its probable mode of decay.

(a) $^{42}_{19}K$ (b) $^{234}_{92}U$ (c) $^{20}_{9}F$

ThomsonNOW™
Go to the Coached Problems menu for a tutorial on **calculating binding energy** and a module on **binding energy.**

Binding Energy

As demonstrated by Ernest Rutherford's alpha particle scattering experiment *(← p. 45)*, the nucleus of the atom is extremely small. Yet the nucleus can contain up to 83 protons before becoming unstable, suggesting that there must be a very strong short-range binding force that can overcome the electrostatic repulsive force of a number of protons packed into such a tiny volume. A measure of the force holding the nucleus together is the nuclear **binding energy.** This energy (E_b) is defined as the negative of the energy change (ΔE) that would occur if a nucleus were formed directly from its component protons and neutrons. For example, if a mole of protons and a mole of neutrons directly formed a mole of deuterium nuclei, the energy change would be more than 200 million kJ, the equivalent of exploding 73 tons of TNT.

The nuclear binding energy is similar to the bond energy for a chemical bond *(← p. 349)* in that the binding energy is the energy that must be supplied to separate all of the particles (protons and neutrons) that make up the atomic nucleus and the bond energy is the energy that must be supplied to separate one mole of two bonded atoms. In both cases the energy change is positive, because work must be done to separate the particles.

$$^{1}_{1}H + ^{1}_{0}n \longrightarrow ^{2}_{1}H \qquad \Delta E = -2.15 \times 10^{8}\ kJ$$

$$\text{Binding energy} = E_b = -\Delta E = 2.15 \times 10^{8} kJ$$

This nuclear synthesis reaction is highly exothermic (so E_b is very positive), an indication of the strong attractive forces holding the nucleus together. The deuterium nucleus is more stable than an isolated proton and an isolated neutron.

To understand the enormous energy released during the formation of atomic nuclei, we turn to an experimental observation and a theory. The experimental observation is that the mass of a nucleus is always slightly less than the sum of the masses of its constituent protons and neutrons.

$$\underset{1.007825\ g/mol}{^{1}_{1}H} + \underset{1.008665\ g/mol}{^{1}_{0}n} \longrightarrow \underset{2.01410\ g/mol}{^{2}_{1}H}$$

Change in mass = Δm = mass of product − sum of masses of reactants

$$\Delta m = 2.01410 \text{ g/mol} - 1.008665 \text{ g/mol} - 1.007825 \text{ g/mol}$$
$$\Delta m = 2.01410 \text{ g/mol} - 2.016490 \text{ g/mol}$$
$$\Delta m = -0.00239 \text{ g/mol} = -2.39 \times 10^{-6} \text{ kg/mol}$$

The theory is that the missing mass, Δm, is released as energy, which we describe as the binding energy.

The relationship between mass and energy is contained in Albert Einstein's 1905 theory of special relativity, which holds that mass and energy are simply different manifestations of the same quantity. Einstein stated that the energy of a body is equivalent to its mass times the square of the speed of light, $E = mc^2$. To calculate the energy change in a process in which the mass has changed, the equation becomes

$$\Delta E = (\Delta m)c^2$$

We can calculate ΔE in joules if the change in mass is given in kilograms and the velocity of light is given in meters per second (because $1 \text{ J} = 1 \text{ kg m}^2/\text{s}^2$). For the formation of 1 mol deuterium nuclei from 1 mol protons and 1 mol neutrons, we have

$$\Delta E = (-2.39 \times 10^{-6} \text{ kg})(3.00 \times 10^{8} \text{ m/s})^2$$
$$= -2.15 \times 10^{11} \text{ J} = -2.15 \times 10^{8} \text{ kJ}$$

This is the value of ΔE given at the beginning of this section for the change in energy when a mole of protons and a mole of neutrons form a mole of deuterium nuclei.

Consider another example, the formation of a helium-4 nucleus from two protons and two neutrons.

$$2\,{}^{1}_{1}\text{H} + 2\,{}^{1}_{0}\text{n} \longrightarrow {}^{4}_{2}\text{He} \qquad E_b = +2.73 \times 10^{9} \text{ kJ/mol } {}^{4}_{2}\text{He nuclei}$$

This binding energy, E_b, is very large, even larger than that for deuterium. To compare nuclear stabilities more directly, nuclear scientists generally calculate the **binding energy per nucleon.** Each ${}^{4}_{2}$He nucleus contains four nucleons—two protons and two neutrons. Therefore, 1 mol ${}^{4}_{2}$He atoms contains 4 mol nucleons.

$$E_b \text{ per mol nucleons} = \frac{2.73 \times 10^{9} \text{ kJ}}{\text{mol } {}^{4}_{2}\text{He nuclei}} \times \frac{1 \text{ mol } {}^{4}_{2}\text{He nuclei}}{4 \text{ mol nucleons}}$$
$$= 6.83 \times 10^{8} \text{ kJ/mol nucleons}$$

The greater the binding energy per nucleon, the greater the stability of the nucleus. The binding energies per nucleon are known for a great number of nuclei and are plotted as a function of mass number in Figure 20.3. It is very interesting and important that the point of maximum stability occurs in the vicinity of iron-56, ${}^{56}_{26}$Fe. This means that *all nuclei are thermodynamically unstable with respect to iron-56.* That is, very heavy nuclei can split, or *fission,* to form smaller, more stable nuclei with atomic numbers nearer to that of iron, while simultaneously releasing enormous quantities of energy (Section 20.6). In contrast, two very light nuclei can come together and undergo *nuclear fusion* exothermically to form heavier nuclei (Section 20.7). Because of its high nuclear stability, *iron is the most abundant of the heavier elements in the universe.*

EXERCISE 20.3 Binding Energy

Calculate the binding energy, in kJ/mol, for the formation of lithium-6.

$$3\,{}^{1}_{1}\text{H} + 3\,{}^{1}_{0}\text{n} \longrightarrow {}^{6}_{3}\text{Li}$$

The necessary masses are ${}^{1}_{1}$H = 1.00783 g/mol, ${}^{1}_{0}$n = 1.00867 g/mol, and ${}^{6}_{3}$Li = 6.015125 g/mol. Is the binding energy greater than or less than that for helium-4? Compare the binding energy per nucleon of lithium-6 and helium-4. Which nucleus is more stable?

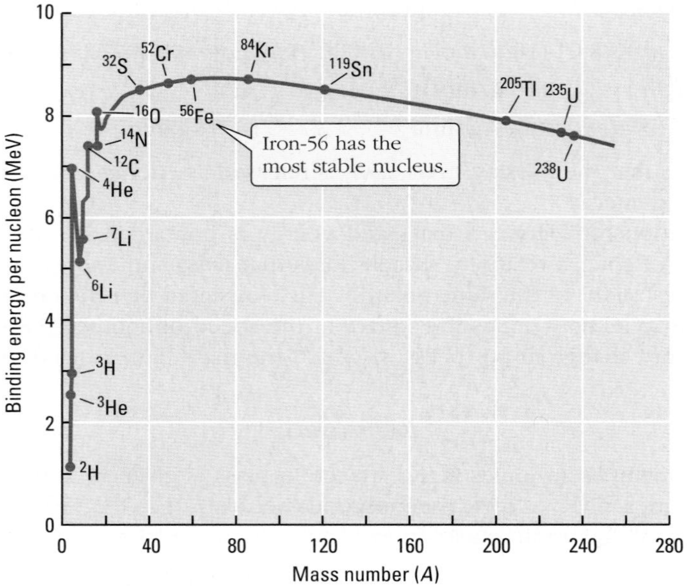

Figure 20.3 Binding energy per nucleon. The values plotted were derived by calculating the binding energy per nucleon in million electron volts (MeV) for the most abundant isotope of each element from hydrogen to uranium (1 MeV = 1.602×10^{-13} J). The nuclei at the top of the curve are most stable.

CONCEPTUAL EXERCISE 20.4 Binding Energy

By interpreting the shape of the curve in Figure 20.3, determine which is more exothermic per gram—fission or fusion. Explain your answer.

20.4 Rates of Disintegration Reactions

Cobalt-60 is radioactive and is used as a source of β particles and γ rays to treat malignancies in the human body. One-half of a sample of cobalt-60 will change via beta decay into nickel-60 in a little more than five years (Table 20.2). On the other

Table 20.2 Half-Lives of Some Common Radioactive Isotopes

Isotope	Decay Process	Half-Life
$^{238}_{92}U$	$^{238}_{92}U \longrightarrow {}^{234}_{90}Th + {}^{4}_{2}He$	4.15×10^{9} yr
$^{3}_{1}H$ (tritium)	$^{3}_{1}H \longrightarrow {}^{3}_{2}He + {}^{0}_{-1}e$	12.3 yr
$^{14}_{6}C$ (carbon-14)	$^{14}_{6}C \longrightarrow {}^{14}_{7}N + {}^{0}_{-1}e$	5730 yr
$^{131}_{53}I$	$^{131}_{53}I \longrightarrow {}^{131}_{54}Xe + {}^{0}_{-1}e$	8.04 d
$^{123}_{53}I$	$^{123}_{53}I + {}^{0}_{-1}e \longrightarrow {}^{123}_{52}Te$	13.2 h
$^{57}_{24}Cr$	$^{57}_{24}Cr \longrightarrow {}^{57}_{25}Mn + {}^{0}_{-1}e$	21 s
$^{28}_{15}P$	$^{28}_{15}P \longrightarrow {}^{28}_{14}Si + {}^{0}_{+1}e$	0.270 s
$^{90}_{38}Sr$	$^{90}_{38}Sr \longrightarrow {}^{90}_{39}Y + {}^{0}_{-1}e$	28.8 yr
$^{60}_{27}Co$	$^{60}_{27}Co \longrightarrow {}^{60}_{28}Ni + {}^{0}_{-1}e$	5.26 yr

hand, copper-64, which is used in the form of copper acetate to detect brain tumors, decays much more rapidly; half of the radioactive copper decays in slightly less than 13 hours. These two radioactive isotopes are clearly different in their rates of decay.

Half-Life

The relative instability of a radioactive isotope is expressed as its **half-life,** the time required for one half of a given quantity of the isotope to undergo radioactive decay. In terms of reaction kinetics ***(⇐ p. 619),*** radioactive decay is a first-order reaction. Therefore, the rate of decay is given by the first-order rate law equation

$$\ln[A]_t = -kt + \ln[A]_0$$

where $[A]_0$ is the initial concentration of isotope A, $[A]_t$ is the concentration of A after time t has passed, and k is the first-order rate constant. Because radioactive decay is first-order, the half-life ($t_{1/2}$) of an isotope is the same no matter what the initial concentration. It is given by

$$t_{1/2} = \frac{\ln 2}{k} = \frac{0.693}{k}$$

The relationship $t_{1/2} = \frac{0.693}{k}$ was introduced in the context of kinetics of reactions ***(⇐ p. 624).***

As illustrated by Table 20.2, isotopes have widely varying half-lives. Some take years, even millennia, for half of the sample to decay (^{238}U, ^{14}C), whereas others decay to half the original number of atoms in fractions of seconds (^{28}P). The unit of half-life is whatever time unit is most appropriate—anything from years to seconds.

As an example of the concept of half-life, consider the decay of plutonium-239, an alpha-emitting isotope formed in nuclear reactors.

$$^{239}_{94}Pu \longrightarrow ^{4}_{2}He + ^{235}_{92}U$$

The half-life of plutonium-239 is 24,400 years. Thus, half of the quantity of $^{239}_{94}Pu$ present at any given time will disintegrate every 24,400 years. For example, if we begin with 1.00 g $^{239}_{94}Pu$, 0.500 g of the isotope will remain after 24,400 years. After 48,800 years (two half-lives), only half of the 0.500 g, or 0.250 g, will remain. After 73,200 years (three half-lives), only half of the 0.250 g will still be present, or 0.125 g. The amounts of $^{239}_{94}Pu$ present at various times are illustrated in Figure 20.4. All radioactive isotopes follow this type of decay curve.

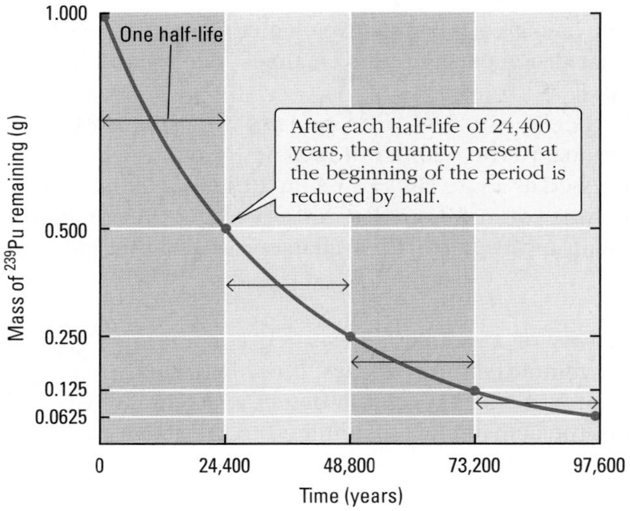

Figure 20.4 **The decay of 1.00 g plutonium-239.**

PROBLEM-SOLVING EXAMPLE 20.4 Half-Life

Iodine-131, used to treat hyperthyroidism, has a half-life of 8.04 days.

$$^{131}_{53}\text{I} \longrightarrow ^{131}_{54}\text{Xe} + ^{0}_{-1}\text{e} \qquad t_{1/2} = 8.04 \text{ days}$$

If you have a sample containing 10.0 μg of iodine-131, what mass of the isotope will remain after 32.2 days?

Answer 0.0625 μg

Strategy and Explanation First, we find the number of half-lives in the given 32.2-day time period. Since the half-life is 8.04 days, the number of half-lives is

$$32.2 \text{ days} \times \frac{1 \text{ half-life}}{8.04 \text{ days}} = 4.00 \text{ half-lives}$$

This means that the initial quantity of 10.0 μg is reduced by half four times.

$$10.0\ \mu\text{g} \times 1/2 \times 1/2 \times 1/2 \times 1/2 = 10.0\ \mu\text{g} \times 1/16 = 0.0625\ \mu\text{g}$$

After 32.2 days, only one sixteenth of the original ^{131}I remains.

✓ **Reasonable Answer Check** After the passage of four half-lives, the remaining ^{131}I should be a small fraction of the starting amount, and it is.

PROBLEM-SOLVING PRACTICE 20.4

Strontium-90 is a radioisotope ($t_{1/2}$ = 29 years) produced in atomic bomb explosions. Its long half-life and tendency to concentrate in bone marrow by replacing calcium make it particularly dangerous to people and animals.

(a) The isotope decays with loss of a β particle. Write a balanced equation showing the other product of decay.

(b) A sample of the isotope emits 2000 β particles per minute. How many half-lives and how many years are necessary to reduce the emission to 125 β particles per minute?

EXERCISE 20.5 Half-Lives

The radioactivity of formerly highly radioactive isotopes is essentially negligible after ten half-lives. What percentage of the original radioisotope remains after this amount of time (ten half-lives)?

Rate of Radioactive Decay

To determine the half-life of a radioactive element, its *rate of decay,* that is, the number of atoms that disintegrate in a given time—per second, per hour, per year—must be measured.

Radioactive decay is a first-order process (*p. 619*), with a rate that is directly proportional to the number of radioactive atoms present (N). This proportionality is expressed as a rate law (Equation 20.1) in which A is the **activity** of the sample—the number of disintegrations observed per unit time—and k is the first-order rate constant or *decay constant* characteristic of that radioisotope.

$$A = kN \qquad [20.1]$$

Suppose the activity of a sample is measured at some time t_0 and then measured again after a few minutes, hours, or days. If the initial activity is A_0 at t_0, then a second measurement at a later time t will detect a smaller activity A. Using Equation 20.1, the ratio of the activity A at some time t to the activity at the beginning of the experiment (A_0) must be equal to the ratio of the number of radioactive atoms N that are present at time t to the number present at the beginning of the experiment (N_0).

$$\frac{A}{A_0} = \frac{kN}{kN_0} \quad \text{or} \quad \frac{A}{A_0} = \frac{N}{N_0}$$

Thus, either A/A_0 or N/N_0 expresses the fraction of radioactive atoms remaining in a sample after some time has elapsed.

The activity of a sample can be measured with a device such as a Geiger counter (Figure 20.5). It detects radioactive emissions as they ionize a gas to form free electrons and cations that can be attracted to a pair of electrodes. In the Geiger counter, a metal tube is filled with low-pressure argon gas. The inside of the tube acts as the cathode. A thin wire running through the center of the tube acts as the anode. When radioactive emissions enter the tube through the thin window at the end, they collide with argon atoms; these collisions produce free electrons and argon cations. As the free electrons accelerate toward the anode, they collide with other argon atoms to generate more free electrons. The free electrons all go to the anode, and they constitute a pulse of current. This current pulse is counted, and the rate of pulses per unit time is the output of the Geiger counter.

The **curie (Ci)** is commonly used as a unit of activity. One curie represents a decay rate of 3.7×10^{10} disintegrations per second (s^{-1}), which is the decay rate of 1 g radium. One millicurie (mCi) $= 10^{-3}$ Ci $= 3.7 \times 10^7\ s^{-1}$. Another unit of radioactivity is the **becquerel (Bq)**; 1 becquerel is equal to one nuclear disintegration per second (1 Bq $= 1\ s^{-1}$).

The *curie* was named for Pierre Curie by his wife, Marie; the *becquerel* honors Henri Becquerel.

The unit for curie and becquerel is s^{-1} because each is a number (of disintegrations) per second.

The change in activity of a radioactive sample over a period of time, or the fraction of radioactive atoms still present in a sample after some time has elapsed, can be calculated using the integrated rate equation for a first-order reaction

$$\ln A = -kt + \ln A_0$$

which can be rearranged to

$$\ln \frac{A}{A_0} = -kt \qquad [20.2]$$

Equation 20.2 can be derived from Equation 20.1 using calculus.

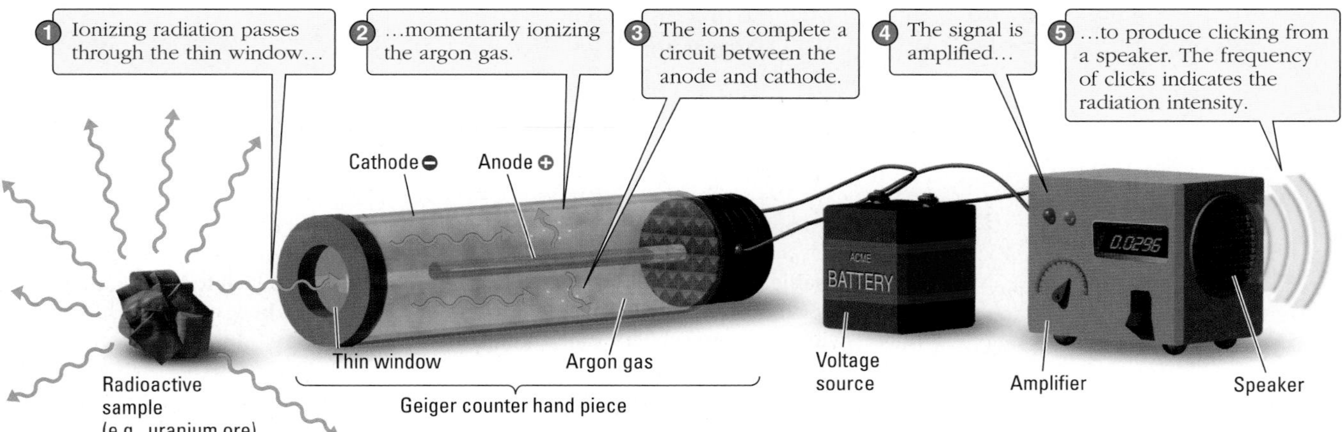

1 Ionizing radiation passes through the thin window...
2 ...momentarily ionizing the argon gas.
3 The ions complete a circuit between the anode and cathode.
4 The signal is amplified...
5 ...to produce clicking from a speaker. The frequency of clicks indicates the radiation intensity.
Cathode ⊖
Anode ⊕
ACME
BATTERY
0.0296
Radioactive sample (e.g., uranium ore)
Thin window
Argon gas
Geiger counter hand piece
Voltage source
Amplifier
Speaker

Active Figure 20.5 **A Geiger counter. Go to the Active Figures menu at ThomsonNOW to test your understanding of the concepts in this figure.**

where A/A_0 is the ratio of activities at time t. Equation 20.2 can also be stated in terms of the fraction of radioactive atoms present in the sample after some time, t, has passed.

$$\ln \frac{N}{N_0} = -kt \quad [20.3]$$

In words, Equation 20.3 says

$$\text{Natural logarithm}\left(\frac{\text{number of radioactive atoms at time } t}{\text{number of radioactive atoms at start of experiment}}\right)$$
$$= \text{natural logarithm(fraction of radioactive atoms remaining at time } t)$$
$$= -(\text{decay constant})(\text{time})$$

As radioactive atoms decay, N becomes a smaller and smaller fraction of N_0.

Notice the negative sign in Equation 20.3. The ratio N/N_0 is less than 1 because N is always less than N_0. This means that the logarithm of N/N_0 is negative, and the other side of the equation has a compensating negative sign because k and t are always positive.

As we have seen, the half-life of an isotope is inversely proportional to the first-order rate constant k:

$$t_{1/2} = \frac{0.693}{k}$$

Thus, the half-life can be found by calculating k from Equation 20.3 using N and N_0 from laboratory measurements over the time period t, as illustrated in Problem-Solving Example 20.5.

PROBLEM-SOLVING EXAMPLE 20.5 Half-Life

A sample of ^{24}Na initially undergoes 3.50×10^4 disintegrations per second (s^{-1}). After 24 h, its disintegration rate has fallen to $1.16 \times 10^4\ s^{-1}$. What is the half-life of ^{24}Na?

Answer 15.0 h

Strategy and Explanation We use Equation 20.2 relating activity (disintegration rate) at time zero and time t with the decay constant k. The experiment provided us with A, A_0, and the time.

$$\ln\left(\frac{1.16 \times 10^4\ s^{-1}}{3.50 \times 10^4\ s^{-1}}\right) = \ln(0.331) = -k(24\ h)$$

$$k = -\frac{\ln(0.331)}{24\ h} = -\left(\frac{-1.104}{24\ h}\right) = 0.0460\ h^{-1}$$

From k we can determine $t_{1/2}$.

$$t_{1/2} = \frac{0.693}{k} = \frac{0.693}{0.0460\ h^{-1}} = 15.0\ h$$

✓ **Reasonable Answer Check** The activity (disintegration rate) fell to between one half and one quarter of its initial value in 24 h, so the half-life must be less than 24 h, and this agrees with our more accurate calculation.

PROBLEM-SOLVING PRACTICE 20.5

The decay of iridium-192, a radioisotope used in cancer radiation therapy, has a rate constant of $9.3 \times 10^{-3}\ d^{-1}$.

(a) What is the half-life of ^{192}Ir?

(b) What fraction of an ^{192}Ir sample remains after 100 days?

PROBLEM-SOLVING EXAMPLE 20.6 Time and Radioactivity

A 1.00-mg sample of ^{131}I ($t_{1/2}$ = 8.04 days) has an initial disintegration rate of $4.7 \times 10^{12}\ s^{-1}$ (disintegrations per second). How long will it take for the disintegration rate of the sample to fall to $2.9 \times 10^{11}\ s^{-1}$?

Answer 776 h

Strategy and Explanation We use the half-life of ^{131}I to find the decay constant, *k*. We convert the known half-life from days to hours.

$$t_{1/2} = 8.04 \text{ days} \times \frac{24 \text{ h}}{1 \text{ day}} = 193 \text{ h}$$

Then we calculate the decay constant, *k*.

$$k = \frac{0.693}{t_{1/2}} = \frac{0.693}{193 \text{ h}} = 3.59 \times 10^{-3}\ \text{h}^{-1}$$

We then use this value of *k* in the equation relating disintegration rate to time. The initial disintegration rate is $A_0 = 4.7 \times 10^{12}\ s^{-1}$, and the disintegration rate after the elapsed time is $A = 2.9 \times 10^{11}\ s^{-1}$. We can use Equation 20.2 to calculate the elapsed time *t*. Both disintegration rates are given in disintegrations per second, but they appear as a ratio in Equation 20.2, so we can use them as provided. If we converted them both to disintegrations per hour, we would get the same numerical result for the ratio.

$$\ln\left(\frac{2.9 \times 10^{11}\ \text{s}^{-1}}{4.7 \times 10^{12}\ \text{s}^{-1}}\right) = -kt = -(3.59 \times 10^{-3}\ \text{h}^{-1})t$$

$$t = \frac{-2.785}{3.59 \times 10^{-3}\ \text{h}^{-1}} = 776 \text{ h}$$

✓ **Reasonable Answer Check** The disintegration rate has fallen by approximately a factor of sixteen ($(4.7 \times 10^{12})/(2.9 \times 10^{11}) = 16.2$), so the elapsed time must be approximately four half-lives of ^{131}I, and it is.

PROBLEM-SOLVING PRACTICE 20.6

In 1921 the women of America honored Marie Curie by giving her a gift of 1.00 g of pure radium, which is now in Paris at the Curie Institute of France. The principal isotope, ^{226}Ra, has a half-life of 1.60×10^3 years. How many grams of radium-226 remain?

Carbon-14 Dating

In 1946 Willard Libby developed a technique for measuring the age of archaeological objects using radioactive carbon-14. Carbon is an important building block of all living systems, and all organisms contain the three isotopes of carbon: ^{12}C, ^{13}C, and ^{14}C. The first two isotopes are stable (nonradioactive) and have been present for billions of years. Carbon-14, however, is radioactive and decays to nitrogen-14 by beta emission.

$$^{14}_{6}\text{C} \longrightarrow {}^{0}_{-1}\text{e} + {}^{14}_{7}\text{N}$$

The half-life of ^{14}C is known by experiment to be 5.73×10^3 years. The number of carbon-14 atoms (*N*) in a carbon-containing sample can be measured from the activity of the sample (*A*). If the number of carbon-14 atoms originally in the sample (N_0) can be determined, or if the initial activity (A_0) can be determined, the age of the sample can be found from Equation 20.2 or 20.3.

This method of age determination clearly depends on knowing how much ^{14}C was originally in the sample. The answer to this question comes from work by physicist Serge Korff, who discovered in 1929 that ^{14}C is continually generated in the upper atmosphere. High-energy cosmic rays collide with gas molecules in the upper atmosphere and cause them to eject neutrons. These free neutrons collide with nitrogen atoms to produce carbon-14.

Willard Libby won the 1960 Nobel Prize in chemistry for his discovery of radiocarbon dating.

© Bettmann/Corbis

Willard Libby and his apparatus for carbon-14 dating.

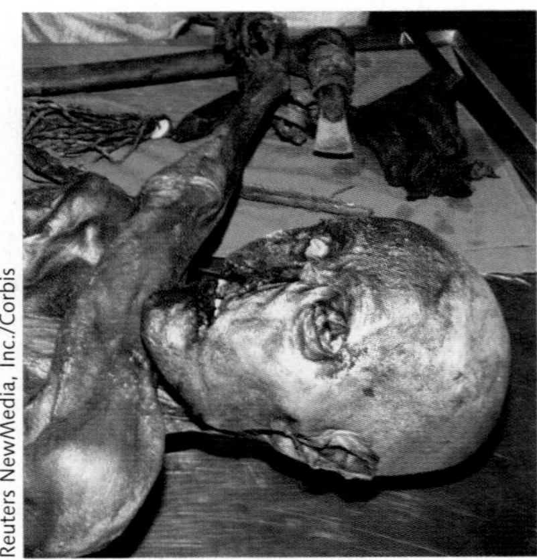

Reuters NewMedia, Inc./Corbis

The Ice Man. This human mummy was found in 1991 in glacial ice high in the Alps. Carbon-14 dating determined that he lived about 5300 years ago. The mummy is exhibited at the South Tyrol Archaeological Museum in Bolzano, Italy.

$$^{14}_{7}N + ^{1}_{0}n \longrightarrow ^{14}_{6}C + ^{1}_{1}H$$

Throughout the *entire* atmosphere, only about 7.5 kg ^{14}C is produced per year. However, this relatively small quantity of radioactive carbon is incorporated into CO_2 and other carbon compounds and then distributed worldwide as part of the carbon cycle. The continual formation of ^{14}C, transfer of the isotope within the oceans, atmosphere, and biosphere, and decay of living matter keep the supply of ^{14}C constant.

Plants absorb carbon dioxide from the atmosphere and convert it into food via photosynthesis (***p. 899***). In this way, the ^{14}C becomes incorporated into living tissue, where radioactive ^{14}C atoms and nonradioactive ^{12}C atoms in CO_2 chemically react in the same way. The beta decay activity of carbon-14 in *living* plants and in the air is constant at 15.3 disintegrations per minute per gram ($min^{-1}\ g^{-1}$) of carbon. When a plant dies, however, carbon-14 disintegration continues *without the ^{14}C being replaced.* Consequently, the ^{14}C activity of the dead plant material decreases with the passage of time. The smaller the activity of carbon-14 in the plant, the longer the period of time between the death of the plant and the present. Assuming that ^{14}C activity in living plants was about the same hundreds or thousands of years ago as it is now, measurement of the ^{14}C beta activity of an artifact can be used to date an article containing carbon. The slight fluctuations of the ^{14}C activity in living plants for the past several thousand years have been measured by studying growth rings of long-lived trees, and the carbon-14 dates of objects can be corrected accordingly.

The time scale accessible to carbon-14 dating is determined by the half-life of ^{14}C. Therefore, this method for dating objects can be extended back approximately 50,000 years. This span of time is almost nine half-lives, during which the number of disintegrations per minute per gram of carbon would fall by a factor of about $(\frac{1}{2})^9 = 1.95 \times 10^{-3}$ from about 15.3 $min^{-1}\ g^{-1}$ to about 0.030 $min^{-1}\ g^{-1}$, which is a disintegration rate so low that it is difficult to measure accurately.

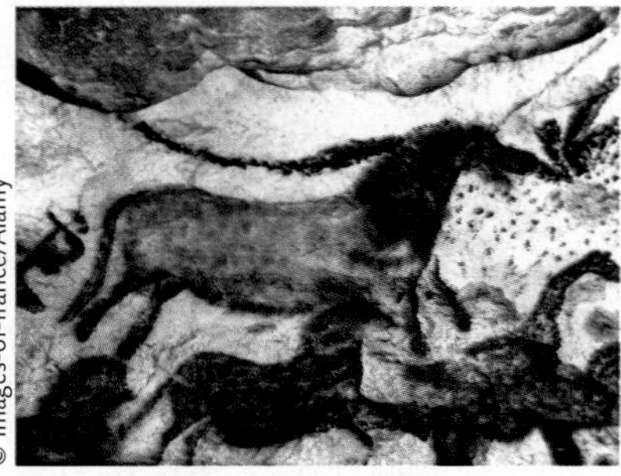

© images-of-france/Alamy

Prehistoric cave paintings from Lascaux, France.

PROBLEM-SOLVING EXAMPLE 20.7 Carbon-14 Dating

Charcoal fragments found in a prehistoric cave in Lascaux, France, had a measured disintegration rate of 2.4 $min^{-1}\ g^{-1}$ carbon. Calculate the approximate age of the charcoal.

Answer 15,300 years old

Strategy and Explanation We will use Equation 20.2 to solve the problem

$$\ln\left(\frac{A}{A_0}\right) = -kt$$

where A is proportional to the known activity of the charcoal (2.4 $min^{-1}\ g^{-1}$) and A_0 is proportional to the activity of the carbon-14 in living material (15.3 $min^{-1}\ g^{-1}$). We first need to calculate k, the rate constant, using the half-life of carbon-14, 5.73×10^3 yr.

$$k = \frac{0.693}{t_{1/2}} = \frac{0.693}{5.73 \times 10^3\ \text{yr}} = 1.21 \times 10^{-4}\ \text{yr}^{-1}$$

Now we are ready to calculate the time, t.

$$\ln\left(\frac{2.4\ \text{min}^{-1}\ \text{g}^{-1}}{15.3\ \text{min}^{-1}\ \text{g}^{-1}}\right) = -kt$$

$$\ln(0.15686) = -(1.21 \times 10^{-4}\ \text{yr}^{-1})t$$

$$t = \frac{1.8524}{1.21 \times 10^{-4}\ \text{yr}^{-1}} = 1.53 \times 10^4\ \text{yr}$$

Thus, the charcoal is approximately 15,300 yrs old.

✓ **Reasonable Answer Check** The disintegration rate has fallen a factor of six from the rate for living material, so more than two but less than three half-lives have elapsed. This agrees with our calculated result.

PROBLEM-SOLVING PRACTICE 20.7

Tritium, ^{3}H ($t_{1/2}$ = 12.3 yr), is produced in the atmosphere and incorporated in living plants in much the same way as ^{14}C. Estimate the age of a sealed sample of Scotch whiskey that has a tritium content 0.60 times that of the water in the area where the whiskey was produced.

EXERCISE 20.6 Radiochemical Dating

The radioactive decay of uranium-238 to lead-206 provides a method of radiochemically dating ancient rocks by using the ratio of ^{206}Pb atoms to ^{238}U atoms in a sample. Using this method, a moon rock was found to have a $^{206}Pb/^{238}U$ ratio of 100/109, that is, 100 ^{206}Pb atoms for every 109 ^{238}U atoms. No other lead isotopes were present in the rock, indicating that all of the ^{206}Pb was produced by ^{238}U decay. Estimate the age of the moon rock. The half-life of ^{238}U is 4.51×10^9 years.

CONCEPTUAL EXERCISE 20.7 Radiochemical Dating

Ethanol, C_2H_5OH, is produced by the fermentation of grains or by the reaction of water with ethylene, which is made from petroleum. The alcohol content of wines can be increased fraudulently beyond the usual 12% available from fermentation by adding ethanol produced from ethylene. How can carbon dating techniques be used to differentiate the ethanol sources in these wines?

20.5 Artificial Transmutations

In the course of his experiments, Rutherford found in 1919 that alpha particles ionize atomic hydrogen, knocking off an electron from each atom. Using atomic nitrogen instead, he found that bombardment with alpha particles *produced protons.* He correctly concluded that the alpha particles had knocked a proton out of the nitrogen nucleus and that a nucleus of another element had been produced. In other words, nitrogen had undergone a *transmutation* to oxygen.

$$^{4}_{2}He + ^{14}_{7}N \longrightarrow ^{17}_{8}O + ^{1}_{1}H$$

Rutherford had proposed that protons and neutrons are the fundamental building blocks of nuclei. Although his search for the neutron was not successful, it was later found by James Chadwick in 1932 as a product of the alpha particle bombardment of beryllium.

$$^{9}_{4}Be + ^{4}_{2}He \longrightarrow ^{12}_{6}C + ^{1}_{0}n$$

Changing one element into another by alpha particle bombardment has its limitations. Before a positively charged bombarding particle (such as the alpha particle) can be captured by a positively charged nucleus, the bombarding particle must have sufficient kinetic energy to overcome the repulsive forces developed as the particle approaches the nucleus. But the neutron is electrically neutral, so Enrico Fermi (in 1934) reasoned that a nucleus would not oppose its entry. By this approach, nearly every element has since been transmuted, and a number of *transuranium elements* (elements beyond uranium) have been prepared. For example, plutonium-239 forms americium-241 by neutron bombardment.

Lawrence Berkeley Laboratory

Glenn Seaborg
1912–1999

A pioneer in developing radioisotopes for medical use (Section 20.9), Glenn Seaborg was the first to produce iodine-131, used subsequently to treat his mother's abnormal thyroid condition. As a result of Seaborg's further research, it became possible to predict accurately the properties of many of the as-yet-undiscovered transuranium elements. In a remarkable 21-year span (1940–1961), Seaborg and his colleagues synthesized ten new transuranium elements (plutonium to lawrencium). He received the Nobel Prize in 1951 for his creation of new elements. In the 1990s Seaborg was honored by having element 106 named for him.

$$^{239}_{94}Pu + ^{1}_{0}n \longrightarrow ^{240}_{94}Pu$$

$$^{240}_{94}Pu + ^{1}_{0}n \longrightarrow ^{241}_{94}Pu$$

$$^{241}_{94}Pu \longrightarrow ^{241}_{95}Am + ^{0}_{-1}e$$

Of the elements currently known, all with atomic numbers of 92 or less exist in nature except Tc and Pm. The transuranium elements, those with atomic numbers greater than 92, are all synthetic. Up to element 101, mendelevium, all the elements can be made by bombarding the nucleus of a lighter element with small particles such as $^{4}_{2}He$ or $^{1}_{0}n$. Beyond element 101, special techniques using heavier particles are required and are still being developed. For example, Roentgenium, element 111, first synthesized in 1994, was made by bombarding bismuth-209 with nickel-64 nuclei.

Roentgenium, Rg, element 111, honors Wilhelm Röntgen for his discovery of X-radiation.

$$^{64}_{28}Ni + ^{209}_{83}Bi \longrightarrow ^{272}_{111}Rg + ^{1}_{0}n$$

The International Union of Pure and Applied Chemistry (IUPAC) has named element 110 darmstadtium (Ds) and element 111 roentgenium (Rg). There have been reported syntheses of elements 112, 113, 114, and 115, but these elements have not yet been named by the IUPAC.

Lawrence Berkeley Laboratory

Darleane C. Hoffman

1926–

After achieving recognition for her investigation of the chemical properties of the heaviest elements at Los Alamos National Laboratory, the University of California, Berkeley, and the Lawrence Berkeley Laboratory, Darleane C. Hoffman was awarded the National Medal of Science in 1997 and the American Chemical Society's highest honor, the Priestley Medal, in 2000. Hoffman first became interested in nuclear chemistry as a student at Iowa State University and searched for new heavy elements and isotopes in nuclear test debris while working at Los Alamos National Laboratory. Later she moved to Berkeley to continue her research and became involved in the search for new, superheavy elements.

EXERCISE 20.8 Nuclear Transmutations

Balance these equations for nuclear reactions, indicating the symbol, the mass number, and the atomic number of the remaining product.

(a) $^{13}_{6}C + ^{1}_{0}n \longrightarrow ^{4}_{2}He + ?$ (b) $^{14}_{7}N + ^{4}_{2}He \longrightarrow ^{1}_{0}n + ?$

(c) $^{253}_{99}Es + ^{4}_{2}He \longrightarrow ^{1}_{0}n + ?$

CONCEPTUAL EXERCISE 20.9 Element Synthesis

In 1996 a team of scientists reported the production of element 112. Bombardment of lead-208 produced a nucleus that emitted a neutron to yield element 112 with mass number = 277. Write a balanced nuclear equation for this process, and identify the bombarding nucleus.

20.6 Nuclear Fission

In 1938, the nuclear chemists Otto Hahn and Fritz Strassman were confounded when they isolated barium from a sample of uranium that had been bombarded with neutrons. Further work by Lise Meitner, Otto Frisch, Niels Bohr, and Leo Szilard confirmed that the bombarded uranium-235 nucleus had formed barium by the capture of a neutron and had undergone **nuclear fission,** a nuclear reaction in which the bombarded nucleus splits into two lighter nuclei (Figure 20.6).

$$^{235}_{92}U + ^{1}_{0}n \longrightarrow ^{236}_{92}U \longrightarrow ^{141}_{56}Ba + ^{92}_{36}Kr + 3\,^{1}_{0}n \qquad \Delta E = -2 \times 10^{10}\ kJ$$

Element 109 (Mt) is named in honor of Lise Meitner.

This nuclear equation shows that bombardment with a single neutron produces three neutrons among the products. The fact that the fission reaction produces more neutrons than are required to begin the process is important. Each of the product neutrons is capable of inducing another fission reaction, so three neutrons would induce three fissions, which would release nine neutrons to induce nine more fissions, from which 27 neutrons are obtained, and so on. Since the neutron-induced fission of uranium-235 is extremely rapid, this sequence of reactions can lead to an explosive chain reaction, as illustrated in Figure 20.7.

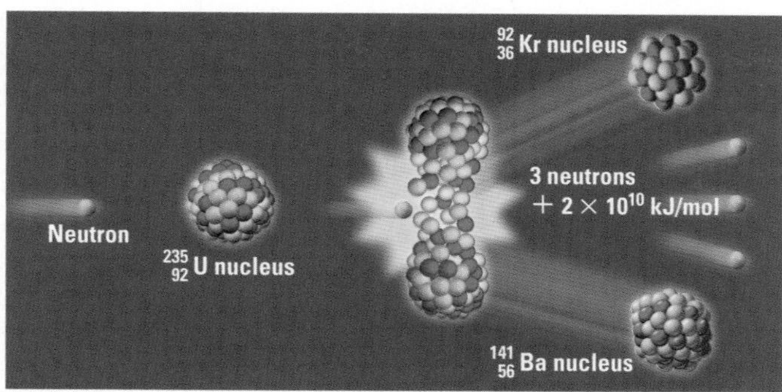

Figure 20.6 **The fission of a $^{235}_{92}$U nucleus from its bombardment with a neutron.**

Nuclear fission of uranium-235 produces a variety of products. Thirty-four elements have been detected among the fission products, including those shown in Figure 20.7. If the quantity of uranium-235 is small, then most of the neutrons escape without hitting a nucleus and so few neutrons are captured by ^{235}U nuclei

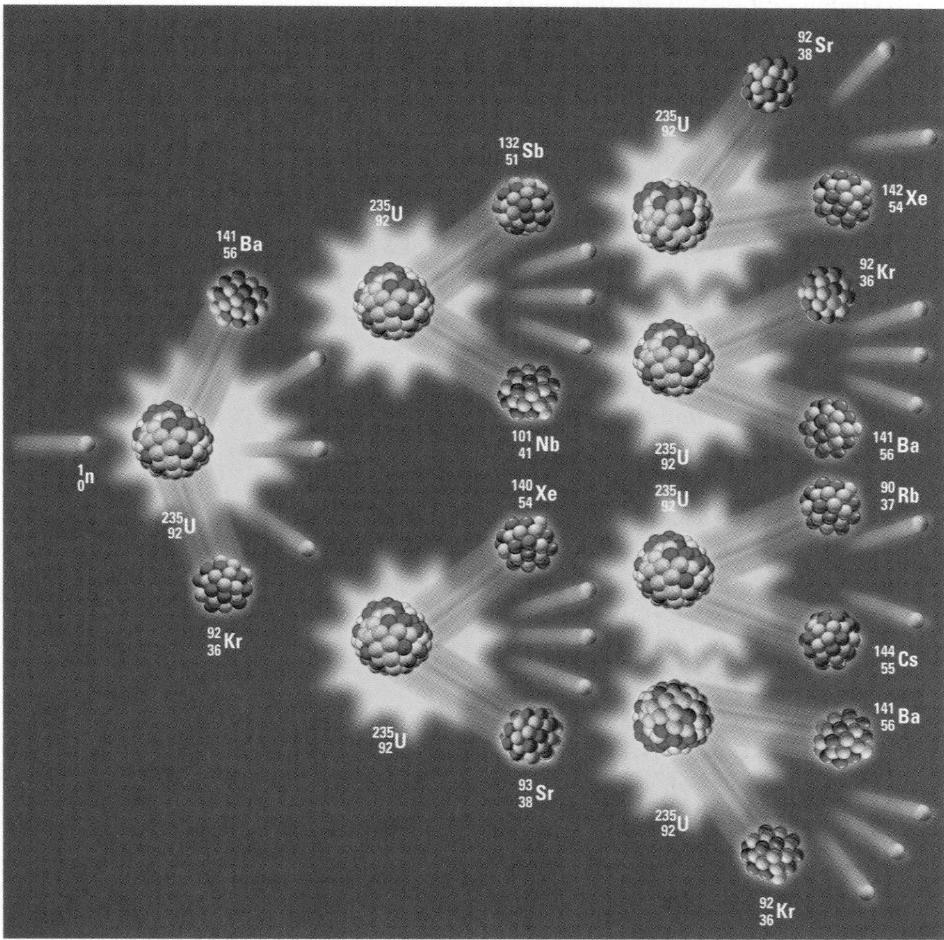

Figure 20.7 **A self-propagating nuclear chain reaction initiated by capture of a neutron.** The nuclear fission of uranium-235 produces a variety of products. Thirty-four elements have been detected among the fission products, including those shown here. Each fission event produces two lighter nuclei plus two or three neutrons.

that the chain reaction cannot be sustained. In an atomic bomb, two small masses of uranium-235, neither capable of sustaining a chain reaction, are brought together rapidly to form one mass capable of supporting a chain reaction. An atomic explosion results. The minimum mass of fissionable material required for a self-sustaining chain reaction is termed the **critical mass.**

Nuclear Reactors

Nuclear fission reactions are the energy source in nuclear power plants. The nuclear reactor fuel is fissionable material, for example, uranium-235. The rate of fission in a nuclear reactor is controlled by limiting the number and energy of neutrons available so that energy derived from fission can be used safely as a heat source for a nuclear power plant (Figure 20.8). In a **nuclear reactor,** the rate of fission is controlled by inserting cadmium rods or other neutron absorbers into the reactor. The rods absorb the neutrons that would otherwise propagate fission reactions. The rate of the fission reaction can be increased or decreased by withdrawing or inserting the control rods, respectively. The materials that control the numbers of neutrons (by absorbing them) or control their energy (by absorbing some of their energy) are known as *moderators.*

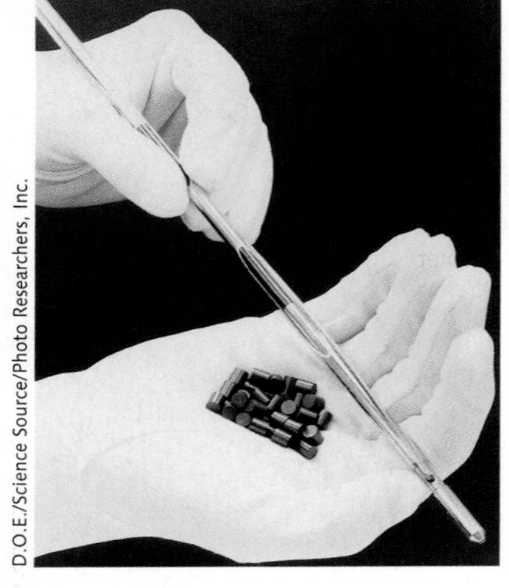

Uranium oxide pellets used in nuclear fuel rods.

D.O.E./Science Source/Photo Researchers, Inc.

The uranium fuel in the reactor core is in the form of uranium dioxide (UO_2) pellets, which are about the size of the eraser on a pencil. The pellets are placed end-to-end in metal alloy tubes, which are then grouped into stainless steel–clad bundles. As pointed out earlier, once a fission reaction is started, it can be sustained as a chain reaction. However, a source of neutrons is needed to initiate the chain reaction. One means of generating these initial neutrons is a nuclear reaction source, such as beryllium-9, and a heavy, alpha-emitting element, such as plutonium or americium. The heavy element emits alpha particles; when they strike a beryllium-9 nucleus, the two nuclei combine to form a carbon-12 nucleus and a neutron is emitted.

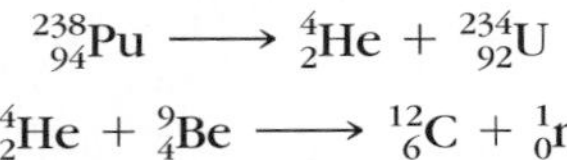

$$^{238}_{94}Pu \longrightarrow {}^{4}_{2}He + {}^{234}_{92}U$$

$$^{4}_{2}He + {}^{9}_{4}Be \longrightarrow {}^{12}_{6}C + {}^{1}_{0}n$$

These neutrons then initiate the nuclear fission of uranium-235 in the reactor core.

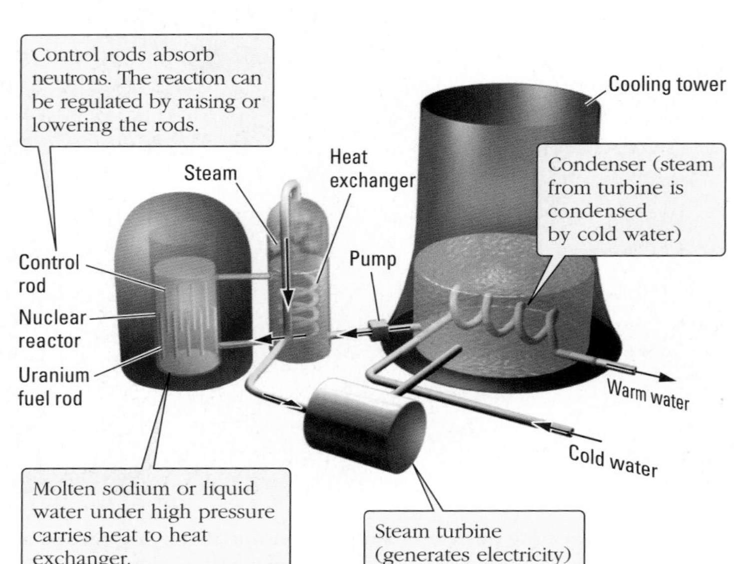

Figure 20.8 **Schematic diagram of a nuclear power plant.**

The tremendous heat generated by the fission reaction heats the primary coolant, a substance with a very high heat capacity, usually water. The primary coolant is at a pressure of more than 150 atm, so it does not boil, even though the temperature is higher than its normal boiling point (⇐ *p. 494*). The hot primary coolant is pumped in a closed loop from the reaction vessel to the steam generators, and back again to the reaction vessel. Heat transfer to water that runs the steam generators lowers the temperature of the primary coolant, which returns to the reactor to be heated again. This closed loop links the nuclear reactor and the rest of the power plant.

Conventional (non-nuclear) power plants burn fossil fuel to generate the heat to produce steam to drive the turbine.

Cooling towers are also used by fossil fuel–burning power plants.

Uranium-238 can fission, but only when bombarded by fast neutrons, unlike those in nuclear reactors. Thus, we consider uranium-238 to be nonfissionable in the context of nuclear reactors.

The water that turns the steam generators is sometimes referred to as the secondary coolant. As the water in the steam generators is vaporized, the steam strikes the large turbine blades, causing the turbine to spin. The turbine shaft is connected to a large metal rod in the generator, which is surrounded by a magnetic field. The rapid spinning of the turbine shaft in a magnetic field produces electricity.

Phil Degginger/Stone/Getty Images

A nuclear power plant with four prominent cooling towers and two nuclear reactor buildings (the domed buildings).

After striking the turbine blades, the steam must be condensed so that the heating/cooling cycle can be repeated to create additional electricity. To do so, cooling water is pumped from a neighboring river or lake to the secondary coolant loop. Enormous amounts of outside cooling water are needed to condense the vast quantity of steam produced by such power plants. For example, the nuclear power reactor at the Entergy Arkansas Unit 1 uses 750,000 gal of cooling water per minute. Having picked up heat from the secondary coolant, the cooling water must then be cooled itself before being returned to its source. Such cooling is done in many nuclear power plants by passing the water through large concave evaporative cooling towers, which are often mistaken for the nuclear reactors themselves.

Not all nuclei can be made to fission on colliding with a neutron, but ^{235}U and ^{239}Pu are both fissionable isotopes. Natural uranium contains an average of only 0.72% of the fissionable ^{235}U isotope. More than 99% of the natural element is nonfissionable uranium-238. Since the percentage of natural ^{235}U is too small to sustain a chain reaction, uranium for nuclear power fuel must be enriched to about 3% uranium-235. To accomplish this goal, some of the ^{238}U isotope in a sample is effectively discarded, thereby raising the concentration of ^{235}U. If sufficient fissionable uranium-235 is present (a critical mass), it can capture enough neutrons to sustain the fission chain reaction. Approximately one third of the fuel rods in a nuclear reactor are replaced annually because fission by-products absorb neutrons, reducing the efficiency of the fission reactions.

Nuclear fuel rods that have become depleted of U-235 are known as *spent* fuel.

A sample is considered to be of weapons-grade quality only if its ^{235}U content is greater than 90%. Even in reactors using weapons-grade quality, the ^{235}U is still too dispersed to produce uncontrolled fission.

Because the mass of uranium-235 in the fuel rods of a nuclear power plant is lower than the critical mass needed for an atomic bomb, the reactor core *cannot* undergo an uncontrolled chain reaction to convert the reactor into an atomic bomb.

Nuclear fission fuels have extremely large energy density (⇐ *p. 255*). For example, fission of 1.0 kg (2.2 lb) uranium-235 releases 9.0×10^{13} J, the equivalent of exploding 33,000 tons (33 kilotons) TNT. Each UO_2 fuel pellet used in a nuclear reactor has the energy equivalent to burning 136 gallons oil, 2.5 tons wood, or 1 ton coal.

EXERCISE 20.10 Energy of Nuclear Fission

Burning 1.0 kg high-grade coal produces 2.8×10^4 kJ, whereas fission of 1.0 mol uranium-235 generates 2.1×10^{10} kJ. How many metric tons of coal (1 metric ton = 10^3 kg) are needed to produce the same quantity of energy as that released by the fission of 1.0 kg uranium-235? (Assume that the processes have equal efficiency.)

There is, of course, substantial controversy surrounding the use of nuclear power plants, and not just in the United States. Their proponents argue that the health of our economy and our standard of living depend on inexpensive, reliable,

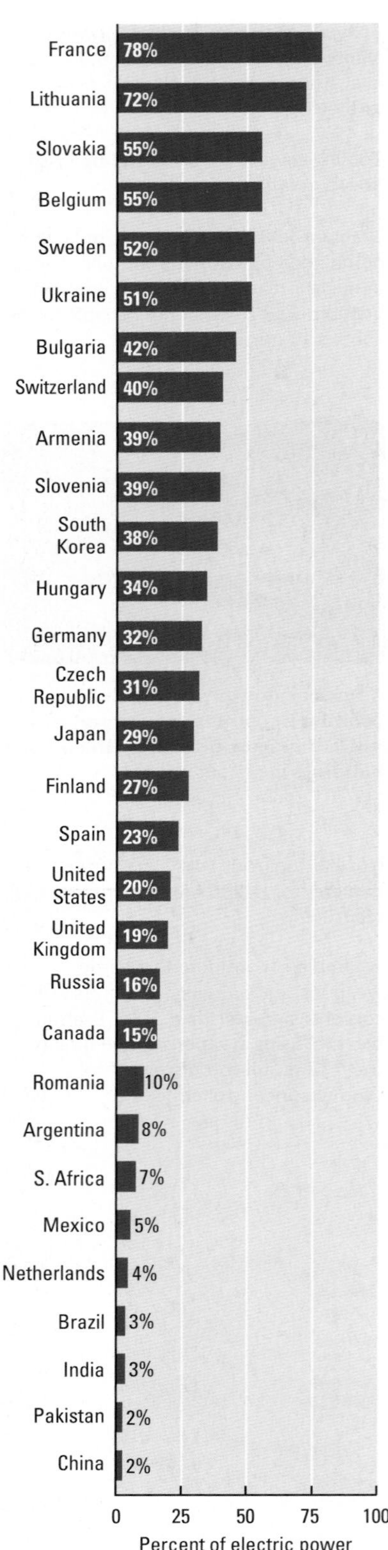

Figure 20.9 The approximate fraction of electricity generated by nuclear power in various countries. About 20% of the electricity in the United States is produced by nuclear power.

and safe sources of energy. Just within the past few years the demand for electric power has at times exceeded the supply in the United States, and many believe nuclear power plants should be built to meet that demand. Nuclear power plants can be the source of "clean" energy, in that they do not pollute the atmosphere with ash, smoke, or oxides of sulfur, nitrogen, or carbon as coal-fired plants do. In addition, nuclear plants help to ensure that our supplies of fossil fuels will not be depleted as quickly in the near future, and they reduce our dependency on buying such fuels from other countries. Currently, 104 nuclear plants operate in the United States. Approximately 440 nuclear power plants worldwide in 31 countries produce about 17% of the world's electricity. The nuclear plants in the United States supply about 20% of our nation's electric energy (Figure 20.9). Around the world, 16 countries use nuclear power to generate at least 25% of their electricity. France, for example, uses 59 nuclear plants to generate more than three out of every four kilowatts of electricity (78%) in that country. In contrast, 72% of the electricity in Lithuania is produced by just one nuclear plant, indicating the differences in demand for electricity between the two countries.

Since the 1979 accident at the Three Mile Island nuclear power plant near Harrisburg, Pennsylvania, *no* construction of new nuclear power plants has begun in the United States. One problem associated with nuclear power plants is the highly radioactive fission products in the spent fuel. In the United States tens of thousands of tons of spent fuel waste are being stored, and the amount is growing steadily. Although some of the products are put to various uses (Section 20.9), many are unsuitable as a fuel or for other purposes. Because these products are often highly radioactive and some have long half-lives (plutonium-239, $t_{1/2}$ = 24,400 yr), proper disposal of this high-level nuclear waste poses an enormous problem. One approach is that high-level radioactive wastes can be encased in a glassy material having a volume of about 2 m^3 per reactor per year. In 1996 the Department of Energy Savannah River site near Augusta, Georgia, began encapsulating radioactive waste in glass, a process called vitrification, in which a mixture of glass particles and radioactive waste is heated to 1200 °C. The molten mixture is poured into stainless steel canisters, cooled, and stored. Eventually, such high-level nuclear wastes may be stored underground in geological formations, such as salt deposits, that are known to be stable for hundreds of millions of years. A site at Yucca Mountain, Nevada, is the designated national repository for high-level nuclear waste, such as that from spent nuclear reactor cores.

EXERCISE 20.11 Radioactive Decay of Fission Products

Unlike the 1979 incident at Three Mile Island, the accident at the Chernobyl nuclear plant in the former Soviet Union in 1986 released significant quantities of radioisotopes into the atmosphere. One of those radioisotopes was strontium-90 ($t_{1/2}$ = 29.1 yr). What fraction of strontium-90 released at that time remains?

EXERCISE 20.12 Nuclear Waste

Cesium-137 ($t_{1/2}$ = 30.2 yr) is produced by ^{235}U fission. If ^{137}Cs is part of nuclear waste stored deep underground, how long will it take for the initial ^{137}Cs activity when it was first buried to drop (a) by 60%? and (b) by 90%?

In recent years, a new type of nuclear reactor design, termed a pebble bed reactor, has emerged. These are high-temperature, gas-cooled reactors. The fuel elements, "pebbles" the size of a tennis ball, contain approximately 4% fissionable uranium encased in silicon carbide and pyrolytic graphite, which act as the neutron

CHEMISTRY IN THE NEWS

Building a Repository for High-Level Nuclear Waste

High-level radioactive waste disposal is a problem that has existed ever since the first nuclear power plants came online and since nuclear weapons were developed. What to do with these wastes? In 1957, the National Academy of Sciences first proposed burying such wastes in geologic formations deep underground for long-term storage. Ultimately, a site at Yucca Mountain, Nevada, 100 miles northwest of Las Vegas, was designated as the national long-term geologic repository, to be designed to hold high-level nuclear wastes safely for 10,000 years.

The plan is to bury the nuclear waste in chambers about 1000 feet below the surface and at least 1000 feet above the water table. The wastes will be encased in ceramic or glass and stored in metal canisters for deep burial. The site would have to accommodate burial of beta-emitting wastes such as Sr-90 and Cs-137 for 300 to 500 years, which is about ten half-lives for these radioactive isotopes. The repository would also have to safely store radioisotopes with much longer half-lives, such as Pu-239 ($t_{1/2}$ = 24,400 years), but with lower radiation intensity. The repository is designed to store 70,000 tons of spent nuclear fuel and 8000 tons of high-level military nuclear waste. At the rate of 20 shipments per day, it would take at least 20 to 25 years just to transport the accumulated waste to Yucca Mountain. Once buried, the nuclear waste could ostensibly be recovered and reprocessed to extract fissionable material from it, such as plutonium-239. There is currently a moratorium in the United States on the reprocessing of waste nuclear fuel.

To fulfill the requirements of the 1982 Nuclear Waste Policy Act, the Department of Energy (DOE) expected to begin receiving nuclear waste at Yucca Mountain in 1998. However, due to political maneuvering and technological delays, no nuclear waste is stored as yet at the site. In January 2002 the Department of Energy Secretary recommended to President George W. Bush that Yucca Mountain be established as a nuclear waste depository, and in February 2002 the DOE completed an Environmental Impact Study. President Bush recommended the site to Congress in February 2002, the U.S. House of Representatives approved the site in May 2002, and the U.S. Senate approved the site in July 2002. The DOE is now preparing an application to the Nuclear Regulatory Commission (NRC) to construct the repository.

The repository has become highly politicized with suits and countersuits filed, and the process is embroiled in controversy. In the summer of 2005 the EPA proposed new radiation exposure limits for the Yucca Mountain repository that are meant to protect public health for up to 1 million years. This new proposal was in response to a federal court ruling that held the prior radiation standard to be inadequate.

Photo Courtesy of the U.S. Department of Energy

The portal of the Yucca Mountain repository as it appears at night.

Political wrangling and technical problems associated with the Yucca Mountain project continue. As a result, the ultimate fate of the project cannot be predicted with certainty at this time.

SOURCES: Hess, G. "Yucca Radiation Limits Proposed." *Chemical & Engineering News*, August 15, 2005; p. 8. Janofsky, M. "Rules for Atomic Waste Would Run Million Years." *New York Times*, August 10, 2005; p. A9. The U.S. Department of Energy maintains a Web site with information about the Yucca Mountain project: **http://www.ocrwm.doe.gov/ymp/index.shtml**. See also the U.S. Environmental Protection Agency Web site: **http://www.epa.gov/radiation/yucca**.

moderators. The core of such a reactor contains hundreds of thousands of these pebbles. They are cooled by flowing inert gas (helium, nitrogen, carbon dioxide), which is heated and then used to drive turbines to generate power. The gas temperature is in the range of 700–900 °C, which provides up to 50% thermal efficiency for power production. Pebbles are withdrawn from the pebble bed, tested to see whether they are spent, and replaced with new ones as needed. In this design, the fuel, its containment, and the moderator are all together in the pebble. This design has been advanced as being inherently safe. Even if the gas flow stops, the reactor core cools rather than increasing its temperature. One criticism of the design is that it produces more radioactive waste than traditional designs. A current pebble bed reactor research project is China's HTR-10 reactor at the Institute of Nuclear and New Energy Technology (INET) at Tsinghua University near Beijing. This research reactor has been running since 2000. INET is planning to start construction of a much larger version in 2007. China's energy plans for the next decade include a major move into the use of nuclear power.

U.S. Department of Energy

These stainless steel canisters (2 feet in diameter and 10 feet tall) hold high-level radioactive waste that has been vitrified into a glassy solid.

20.7 Nuclear Fusion

Tremendous amounts of energy are generated when very light nuclei combine to form heavier nuclei in a reaction called **nuclear fusion.** One of the best examples is the fusion of hydrogen nuclei (protons) to give helium nuclei.

$$4\ {}^{1}_{1}\text{H} \longrightarrow {}^{4}_{2}\text{He} + 2\ {}^{0}_{+1}\text{e} \qquad \Delta E = -2.5 \times 10^{9}\ \text{kJ}$$

The activation energy (p. 629) for nuclear fusion is very large because there is very strong repulsion between two positive nuclei when they are brought very close together—Coulomb's law (p. 93). Thus, very high temperatures are needed for fusion to occur.

The helium nucleus produced by this reaction is more stable (has higher binding energy per nucleon) than the reactant hydrogen nuclei, as shown in Figure 20.3 (p. 988). This fusion reaction is the source of the energy from our sun and other stars, and it is the beginning of the synthesis of the elements in the universe (Section 21.1). Temperatures of 10^6 to 10^7 K, found in the core and radioactive zone of the sun, are required to bring the positively charged ${}^{1}_{1}\text{H}$ nuclei together with enough kinetic energy to overcome nuclear repulsions and react.

Deuterium can also be fused to give helium-3,

$${}^{2}_{1}\text{H} + {}^{2}_{1}\text{H} \longrightarrow {}^{3}_{2}\text{He} + {}^{1}_{0}\text{n} \qquad \Delta E = -3.2 \times 10^{8}\ \text{kJ}$$

which can undergo further fusion with a proton to give helium-4.

$${}^{1}_{1}\text{H} + {}^{3}_{2}\text{He} \longrightarrow {}^{4}_{2}\text{He} + {}^{0}_{+1}\text{e} \qquad \Delta E = -1.9 \times 10^{9}\ \text{kJ}$$

Each of these reactions releases an enormous quantity of energy, so it has been the dream of many nuclear physicists to harness them to provide energy for the people of the world.

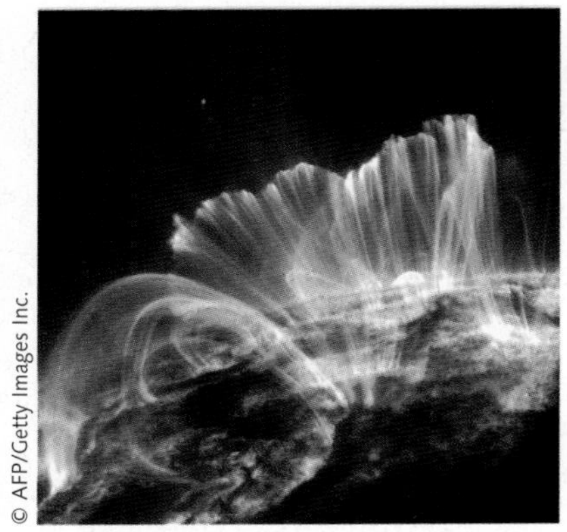

© AFP/Getty Images Inc.

Nuclear fusion reactions power the sun, whose surface is shown here.

Hydrogen bombs are based on fusion. At the very high temperatures that allow fusion reactions to occur rapidly, atoms do not exist as such. Instead, there is a **plasma,** which consists of unbound nuclei and electrons. To achieve the high temperatures required for the fusion reaction of the hydrogen bomb, a fission bomb (atomic bomb) is first set off. In one type of hydrogen bomb, lithium-6 deuteride (LiD, a solid salt) is placed around a ${}^{235}\text{U}$ or ${}^{239}\text{Pu}$ fission bomb, and the fission reaction is set off to initiate the process. Lithium-6 nuclei absorb neutrons produced by the fission and split into tritium and helium.

$${}^{1}_{0}\text{n} + {}^{6}_{3}\text{Li} \longrightarrow {}^{3}_{1}\text{H} + {}^{4}_{2}\text{He}$$

The temperature reached by the fission of uranium or plutonium is high enough to bring about the fusion of tritium and deuterium, accompanied by the release of 1.7×10^{9} kJ per mole of ${}^{3}\text{H}$. A 20-megaton hydrogen bomb usually contains about 300 lb LiD, as well as a considerable mass of plutonium and uranium.

Development of nuclear fusion as a commercial energy source is appealing because hydrogen isotopes are available (from water), and fusion products are usually nonradioactive or have short half-lives, which eliminates the problems associated with the disposal of high-level radioactive fission reactor products. However, controlling a nuclear fusion reaction for peaceful commercial uses has proven to be extraordinarily difficult. Three critical requirements must be met to achieve controlled fusion. First, the temperature must be high enough for fusion to occur sufficiently rapidly. The fusion of deuterium and tritium, for example, requires a temperature of at least 100 million K. Second, the plasma must be confined long enough to generate a net output of energy. Third, the energy must be recovered in some usable form.

Containment is one of the biggest problems in developing controlled nuclear fusion.

Magnetic "bottles" (enclosures in space bounded by magnetic fields) have confined the plasma so that controlled fusion has been achieved. But the energy generated by the fusion has been less than that required to produce the magnetic bottle and control the fusion reaction. Using more energy to produce less energy is not a commercially appealing investment. Thus, commercial fusion reactors are not likely in the near future without a dramatic breakthrough in fusion technology.

EXERCISE 20.13 Nuclear Fusion

Complete the equations for these nuclear fusion reactions.

(a) $^{7}_{3}\text{Li} + ^{1}_{1}\text{H} \longrightarrow ^{1}_{0}\text{n} + \_\_\_\_$

(b) $^{2}_{1}\text{H} + \_\_\_\_ \longrightarrow ^{4}_{2}\text{He} + ^{1}_{1}\text{H}$

20.8 Nuclear Radiation: Effects and Units

The use of nuclear energy and radiation is a double-edged sword that carries both risks and benefits. It can be used to harm (nuclear armaments) or to cure (radioisotopes in medicine).

All technologies carry risks as well as benefits. In the 1800s, railroads were new and the poet William Wordsworth wrote of their risks and benefits in terms of "Weighing the mischief with the promised gain. . . ."

Alpha, beta, and gamma radiation disrupt normal cell processes in living organisms by interacting with key biomolecules, breaking their covalent bonds, and producing energetic free radicals and ions that can lead to further disruptive reactions. The potential for serious radiation damage to humans is well known. The biological effects of the atomic bombs exploded at Hiroshima and Nagasaki, Japan, at the close of World War II in 1945 have been well documented. However, controlled exposure to nuclear radiation can be beneficial in destroying malignant tissue, as in radiation therapy for treating some cancers.

Radiation Units

The SI unit of radioactivity is the becquerel; 1 Bq = 1 s^{-1}. Another common unit of radioactivity is the curie; 1 Ci = 3.70×10^{10} s^{-1}. However, to measure the effects of radiation on tissue, units are needed for radiation dose that take into account the energy absorbed by tissue when radiation passes into it.

To quantify radiation and its effects, particularly on humans, several units have been developed. The SI unit of absorbed radiation dose is the **gray (Gy),** which is equal to the absorption of 1 J per kilogram of material; 1 Gy = 1 J/kg. The **rad** (*r*adiation *a*bsorbed *d*ose) also measures the quantity of radiation energy *absorbed;* 1 rad represents a dose of 1.00×10^{-2} J absorbed per kilogram of material. Thus, 1 Gy = 100 rad. Another unit is the **roentgen (R),** which corresponds to the deposition of 93.3×10^{-7} J per gram of tissue.

The roentgen is named to honor Wilhelm Röntgen, the German physicist who discovered X-rays.

The biological effects of radiation per rad or gray differ with the kind of radiation, which can be quantified more generally using the **rem** (*r*oentgen *e*quivalent in *m*an).

$$\text{Effective dose in rems} = \text{quality factor} \times \text{dose in rads}$$

The quality factor depends on the type of radiation and other factors. It is arbitrarily set as 1 for beta and gamma radiation. It is between 10 and 20 for alpha particles, depending on total dose, dose rate, and type of tissue. Since one rem is a large quantity of radiation, the millirem (mrem) is commonly used (1 mrem = 10^{-3} rem). The SI unit of effective dose is the **sievert (Sv),** which is defined similarly to the rem, except that the absorbed dose is in grays, not rads. Consequently, 1 Sv = 100 rem.

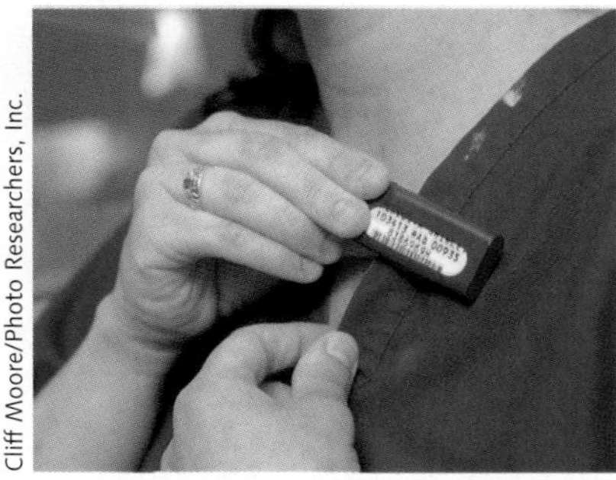
Cliff Moore/Photo Researchers, Inc.

Individuals who work where there is potential danger from exposure to excessive nuclear radiation wear film badges to monitor their radiation dose.

Background Radiation

Humans are constantly exposed to natural and artificial **background radiation,** estimated to be collectively about 360 mrem per year (Figure 20.10), well below 500 mrem, the federal government's background radiation standard for the general public. Note that *most* background radiation, about 300 mrem per year (82%), comes from *natural* background radiation sources: cosmic radiation and radioactive elements and minerals found naturally in the earth, air, and materials around and within us. The remaining 18% comes from artificial sources.

There are no observable physiologic effects from a single dose of radiation less than 25 rem (25×10^3 mrem).

Cosmic radiation, emitted by the sun and other stars, continually bombards the earth and accounts for about 8% of natural background radiation. The remainder comes from radioactive isotopes such as ^{40}K. Potassium is present to the extent of about 0.3 g per kilogram of soil and is essential to all living organisms. We all acquire some radioactive potassium from the foods we eat. For example, a hamburger contains 960 mg of ^{40}K, giving off 29 disintegrations per second (s^{-1}); a hot dog contains 200 mg of ^{40}K and gives off 6 s^{-1}; a serving of french fries has 650 mg ^{40}K and gives off 20 s^{-1}. Other radioactive elements found in some abundance on the earth include thorium-232 and uranium-238. Approximately 8% of the natural background radiation arises from Th-232 and U-238 in rocks and soil. Thorium, for example, is found to the extent of 12 g per 1000 kg soil. Most natural background radiation comes from radon, a by-product of radium decay, as discussed in the next subsection.

On average, roughly 15% of our annual exposure comes from medical procedures such as diagnostic X-rays and the use of radioactive compounds to trace the body's functions. Consumer products account for 3% of our total annual exposure. Contrary to popular belief, less than 1% comes from sources such as the radioactive fission products from testing nuclear explosives in the atmosphere, nuclear power plants and their wastes, nuclear weapons manufacture, and nuclear fuel processing.

Burning fossil fuels (coal and oil) releases into the atmosphere considerable quantities of naturally occurring radioactive isotopes originally in the fossil fuel. Thus, fossil fuel plants add significantly to the background radiation. Far more thorium and uranium are released annually into the atmosphere from fossil fuel–burning plants than from nuclear power plants. The emission from nuclear power plants is essentially zero.

A newly published report by the National Research Council concludes that there is no safe level of radiation for humans, although the risks of low-dose radiation are small. The researchers studied doses of radiation of 0.1 Sv (10,000 mrem) or less, which is much greater than the 360 mrem per year natural background level discussed above. Rules for nuclear workers and others who are systematically exposed to elevated radiation levels may be affected by this type of research finding.

Radon

As shown in Figure 20.10, radon accounts for 55% of natural background radiation. Radon is a chemically inert gas in the same periodic table group as helium, neon, argon, and krypton. Radon-222 is produced in the decay series of uranium-238 (p. 981). Other isotopes of radon are products of other decay series. Although chemically inert, radon is problematic because it is a radioactive gas.

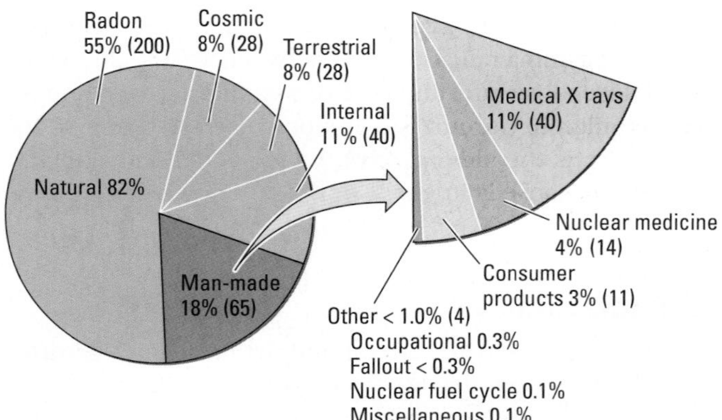

Figure 20.10 Sources of average background radiation exposure in the United States. The sources are expressed as percentages of the total, as well as in millirems per year, the values in parentheses. As seen from the figure, background radiation from natural sources far exceeds that from artificial sources.

CHEMISTRY YOU CAN DO

Counting Millirems: Your Radiation Exposure

In 1987, the Committee on Biological Effects of Ionizing Radiation of the National Academy of Sciences issued a report that contained a survey for an individual to evaluate his or her exposure to ionizing radiation. The table below is adapted from this report and updated. By adding up your exposure, you can compare your annual dose to the U.S. annual average of 360 mrem.

Common Sources of Radiation		**Your Annual Dose (mrem)**
	Location: Cosmic radiation at sea level	27
	For your elevation (in feet), add this number of mrem	______
	Elevation / *mrem*: 1000 / 2; 4000 / 15; 7000 / 40	
Where you live	*Elevation* / *mrem*: 2000 / 5; 5000 / 21; 8000 / 53	
	Elevation / *mrem*: 3000 / 9; 6000 / 29; 9000 / 70	
	Ground: U.S. average	26
	Radon: U.S. average	200
	House construction: For stone, concrete, or masonry building, add 7; for wood, add 30	______
What you eat, drink, and breathe	**Radioisotopes** in the body from	
	Food, air, water: U.S. average........	40
	Weapons test fallout	4
	X-ray and radiopharmaceutical diagnosis	
	Number of chest X-rays _____ × 10........	______
How you live	Number of lower gastrointestinal tract X-rays _____ × 500	______
	Number of radiopharmaceutical examinations _____ × 300........	______
	(Average dose to total U.S. population = 53 mrem)	______
	Jet plane travel: For each 2500 miles add 1 mrem........	______
	TV viewing: Number of hours per day _____ × 0.15	______
	At site boundary: Average number of hours per day _____ × 0.2........	______
How close you live to a nuclear plant	**One mile away:** Average number of hours per day _____ × 0.02........	______
	Five miles away: Average number of hours per day _____ × 0.002........	______
	Over five miles away: None........	______
	Note: Maximum allowable dose determined by "as low as reasonably achievable" (ALARA) criteria established by the U.S. Nuclear Regulatory Commission. Experience shows that your actual dose is substantially less than these limits.	
	Your total annual dose in mrem:	______

Compare your annual dose with the U.S. annual average of 360 mrem.

SOURCE: Based on the BEIR Report III. National Academy of Sciences, Committee on Biological Effects of Ionizing Radiation. *The Effects on Populations of Exposure to Low Levels of Ionizing Radiation.* Washington, DC: National Academy of Sciences, 1987.

It should be kept in mind that radon occurs naturally in our environment. Because it comes from natural uranium deposits, the quantity of radon depends on the nature of the rocks and soil in a given locality. Furthermore, since the gas is chemically inert, it is not trapped by chemical processes in the soil or water. Thus, it is free to seep up from the ground and into underground mines or into homes

An online radiation calculator is available at **http://www.epa.gov/radiation/students/calculate.html**.

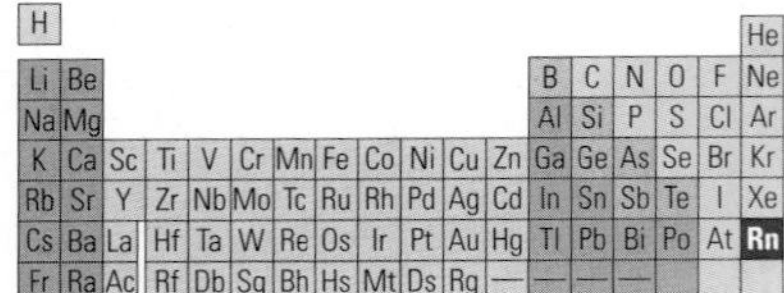

through pores in concrete block walls, cracks in basement floors or walls, and around pipes. Radon-222 decays to give polonium-218, a radioactive, heavy metal element that is not a gas and is not chemically inert.

If radon is inhaled, decay to polonium will occur deep in the lungs, and ^{218}Po will be generated there.

$$^{222}_{86}Rn \longrightarrow {}^{4}_{2}He + {}^{218}_{84}Po \qquad t_{1/2} = 3.82 \text{ days}$$

$$^{218}_{84}Po \longrightarrow {}^{4}_{2}He + {}^{214}_{82}Pb \qquad t_{1/2} = 3.10 \text{ minutes}$$

This answers the question posed in Chapter 1 (p. 3), "I've heard that most homes in the United States contain small quantities of a radioactive gas. How can I find out whether my home is safe?"

Polonium-218 can lodge in lung tissues, where it undergoes alpha decay to give lead-214, itself a radioactive isotope. The range of an alpha particle is quite small, perhaps 0.7 mm (about the thickness of a sheet of paper). However, this is approximately the thickness of the epithelial cells of the lungs, so the radiation can damage these tissues and induce lung cancer.

1 picocurie, pCi = 10^{-12} Ci.

Most homes in the United States are believed to have some level of radon gas. There is currently a great deal of controversy over the level of radon that is considered safe. Estimates indicate that only about 6% of U.S. homes have radon levels above 4 picocuries per liter (pCi/L) of air, the action level standard set by the U.S. Environmental Protection Agency. Some believe that 1.5 pCi/L is more likely the average level and that only about 2% of the homes will contain more than 8 pCi/L. To test for the presence of radon, you can purchase testing kits of various kinds. If your home shows higher levels of radon gas than 4 pCi/L, you should probably have it tested further and perhaps take corrective actions such as sealing cracks around the foundation and in the basement. But keep in mind the relative risks involved. A 1.5 pCi/L level of radon leads to a lung cancer risk about the same as the risk of your dying in an accident in your home.

A commercially available kit to test for radon gas in a residence.

EXERCISE 20.14 Radon Levels

Calculate how long it will take for the activity of a radon-222 sample ($t_{1/2}$ = 3.82 days) initially at 8 pCi to drop

(a) To 4 pCi, the EPA action level.

(b) To 1.5 pCi, approximately the U.S. average.

20.9 Applications of Radioactivity

Food Irradiation, Cold Pasteurization

In addition to preserving foods, gamma irradiation is used to sterilize bandages, contact lens solutions, and many cosmetics.

In some parts of the world, spoilage of stored food may claim up to 50% of the food crop. In Western society, this figure is lowered considerably by refrigeration, canning, and chemical additives. Still, there are problems with food spoilage, and food preservation costs are a substantial fraction of the final cost of food. Food and grains can be preserved by gamma irradiation, also known as cold pasteurization. Contrary to some popular opinion, such irradiation does *not* make foods radioactive, any more than a dental X-ray makes you radioactive.

Food irradiation with gamma rays from ^{60}Co or ^{137}Cs sources is allowed in 40 countries and is endorsed by the World Health Organization and the American Medical Association. Astronauts' food has been preserved by gamma irradiation. The United States and several other countries require that foods preserved by irradiation be labeled with the international symbol for irradiated food, the radura.

The radura, the international symbol for irradiated food.

Bacteria, molds, and yeasts are killed or their growth is retarded by irradiation. As a result, the shelf life of irradiated foods during refrigeration is prolonged in much the same way that heat pasteurization protects milk. In recent years, outbreaks of foodborne illnesses caused by new types of harmful bacteria or inappropriate food handling have heightened interest in the benefits of irradiation as a safety measure, especially for use with meat.

ESTIMATION

Radioactivity of Common Foods

It may be surprising to you, but common foods, such as baked potatoes and bananas, are radioactive. But how radioactive are they?

Potassium ions are essential as a major mineral ***(Section 3.11*** ***p. 109)******.*** They play an important role in conducting electric impulses that lead to muscle contraction and nerve transmission ***(*** ***p. 950)******.*** Many fruits and vegetables are good sources of dietary potassium.

Potassium has three naturally occurring isotopes, ^{39}K, ^{40}K, and ^{41}K, of which ^{40}K is radioactive, emitting betas and gammas with $t_{1/2} = 1.26 \times 10^9$ yr. Its natural abundance is 0.0117%. A typical banana or baked potato contains approximately 500 mg K^+, and therefore about (0.5 g)(0.000117) = 6×10^{-5} g $^{40}K^+$.

We can calculate the rate constant of ^{40}K from its half-life as $k = \frac{0.693}{t_{1/2}} = 5.5 \times 10^{-10}\ yr^{-1}$. We convert the rate constant from years to seconds:

$$k = \frac{5.5 \times 10^{-10}\ yr^{-1}}{3.15 \times 10^7\ s/yr} = 1.7 \times 10^{-17} s^{-1}.$$

We will use the relationship $A = k\,N$ to calculate the disintegration rate, so we need to convert the mass of $^{40}K^+$ to number of $^{40}K^+$ ions:

$$N = \frac{6 \times 10^{-5}\ g}{40\ g/mol} \times 6.022 \times 10^{23}\ ions/mol = 9 \times 10^{17}\ {}^{40}K^+\ ions$$

Now we can calculate the disintegration rate of the baked potato or banana:

$$A = kN = -(1.7 \times 10^{-17} s^{-1})(9 \times 10^{17}\ K^+\ ions) = 15\ s^{-1} = 15\ Bq$$

So a typical baked potato or banana undergoes an average of 15 disintegrations per second due to its $^{40}K^+$ content. This is a small amount of radioactivity. A typical household smoke detector has a disintegration rate (leading to alpha particle emission) about 2000 times as large as that of a baked potato or banana.

SOURCES: Ball, D. W. *Journal of Chemical Education,* Vol. 81, 2004; p. 1440. **http://www.whfoods.com**.

ThomsonNOW
Sign in to ThomsonNOW at **www.thomsonedu.com** to work an interactive module based on this material.

The U.S. Food and Drug Administration (FDA) has approved irradiation of meat, poultry, and a variety of fresh fruits and vegetables and spices. Except for spices, however, irradiated foods are not as yet widely available.

The FDA permits irradiation up to 300 kilorads for the pasteurization of meat and poultry. Radiation levels in the 1- to 5-megarad range (1 megarad = 10^6 rads) sterilize, killing every living organism. Foods irradiated at these levels will keep indefinitely when sealed in plastic or aluminum foil packages. However, FDA approval is unlikely for irradiation sterilization of foods in the near future because of potential problems caused by as-yet-undiscovered, but possible, "unique radiolytic products." For example, irradiation sterilization might produce a substance that is capable of causing genetic damage. To prove or disprove the presence of these substances, animal feeding studies using foods sterilized by irradiation are now being conducted in the United States.

Strawberries preserved by gamma irradiation.

Radioactive Tracers

The chemical behavior of a radioisotope is essentially identical to that of the nonradioactive isotopes of the same element. Compounds containing radioactive atoms are formed and undergo chemical reactions in exactly the same way as compounds containing the same nonradioactive atoms. Therefore, chemists can use radioactive isotopes as **tracers,** radioisotopes used to track the pathway of an element in a chemical process, in both nonbiological and biological chemical reactions. To use a tracer, a chemist prepares a reactant compound in which one of the elements consists of both radioactive and stable (nonradioactive) isotopes, and introduces it into the reaction (or feeds it to an organism). After the reaction, the chemist measures the radioactivity of the products (or determines which parts of the organism contain the radioisotope) by using a Geiger counter or other radiation detector. Several radioisotopes commonly used as tracers are listed in Table 20.3.

Table 20.3 Radioisotopes Used as Tracers

Isotope	Half-Life	Use
^{14}C	5730 years	CO_2 for photosynthesis research
^{3}H	12.33 years	Tag hydrocarbons
^{35}S	87.2 days	Tag pesticides, measure air flow
^{32}P	14.3 days	Measure phosphorus uptake by plants

Melvin Calvin used ^{14}C to monitor the uptake and release of $^{14}CO_2$ to determine the basic biochemical pathways of photosynthesis. This groundbreaking work earned him the 1961 Nobel Prize in chemistry.

For example, plants take up phosphorus-containing compounds from the soil through their roots. The use of the radioactive phosphorus isotope ^{32}P, a beta emitter, provides a way to detect the uptake of phosphorus by a plant, as well as to measure the speed of uptake under various conditions. Plant biologists can grow hybrid strains of plants that quickly absorb phosphorus, an essential nutrient. They can test the new plants by measuring their uptake of the radioactive ^{32}P tracer. This type of research leads to faster-maturing crops, better yields per acre, and more food or fiber at less expense.

Medical Imaging

Radioactive isotopes are used in **nuclear medicine** in two different ways: diagnosis and therapy. In the diagnosis of internal disorders such as tumors, physicians need information on the locations of abnormal tissue. This identification is done by imaging, a technique in which the radioisotope, either alone or combined with some other substance, accumulates at the site of the disorder. There, the radioisotope decays, emitting its characteristic radiation, which is then detected. Modern medical diagnostic instruments determine not only where the radioisotope is located in the patient's body, but also construct an image of the volume within the body where the radioisotope is concentrated.

Four of the most common diagnostic radioisotopes are given in Table 20.4. Most are created in a particle accelerator in which heavy, charged nuclear particles are made to react with other target atoms. Each of these radioisotopes produces gamma radiation, which in low doses is less harmful to tissue than internal ionizing radiations such as beta or alpha particles because gamma rays pass through the tissue without being absorbed. When combined with special carrier compounds, these radioisotopes can be made to accumulate in specific areas of the body. For

Table 20.4 Diagnostic Radioisotopes

Radioisotope	Name	Half-Life (hours)	Site for Diagnosis
^{99m}Tc*	Technetium-99m	6.0	As $^{99m}TcO_4^-$ to the thyroid
^{201}Tl	Thallium-201	72.9	To the heart
^{123}I	Iodine-123	13.2	To the thyroid
^{67}Ga	Gallium-67	78.2	To various tumors and abscesses

* The technetium-99m isotope is the radioisotope most commonly used for diagnostic purposes. The *m* stands for "metastable."

example, the pyrophosphate ion, $P_4O_7^{4-}$, can bond to the technetium-99m radioisotope. Together they accumulate in the skeletal structure where abnormal bone metabolism is occurring (Figure 20.11). The technetium-99m radioisotope is metastable, as denoted by the letter *m*; this term means that the nucleus loses energy by disintegrating to a more stable version of the same isotope,

$$^{99m}\text{Tc} \longrightarrow {}^{99}\text{Tc} + \gamma$$

and the gamma rays are detected. Such investigations often pinpoint bone tumors.

The metastable ^{99m}Tc is in a nuclear excited energy state. This is analogous to a H atom in which an electron is in an excited state (*p. 282*).

EXERCISE 20.15 Rate of Radioactive Decay

Gallium citrate containing radioactive gallium-67 is used medically as a tumor-seeking agent. It has a half-life of 78.2 hours. How much time is needed for a gallium citrate sample to reach 10% of its original activity?

EXERCISE 20.16 Half-Life

Chromium-51 is a radioisotope ($t_{1/2}$ = 27.7 days) used to evaluate the lifetime of red blood cells; the radioisotope iron-59 ($t_{1/2}$ = 44.5 days) is used to assess bone marrow function. A hospital laboratory has 80 mg iron-59 and 100 mg chromium-51. After 90 days, which radioisotope is present in greater mass?

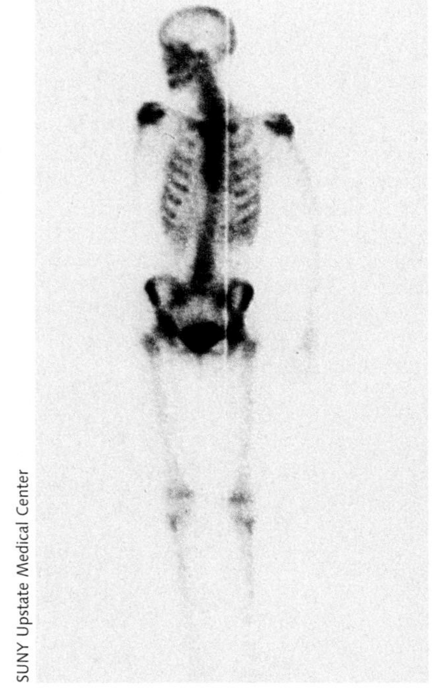

SUNY Upstate Medical Center

Figure 20.11 A whole-body scan. Phosphate with technetium-99m was injected into the blood and then absorbed by the bones and kidneys. This picture was taken three hours after injection.

Paradoxically, high-energy radiation, which can kill healthy cells, is used therapeutically to kill malignant, cancerous cells—those exhibiting rapid, uncontrolled growth. Because they divide more rapidly than normal cells, malignant cells are more susceptible to radiation damage. Thus, malignant cells are more likely to be killed than normal cells. For external radiation therapy, a narrow beam of high-energy gamma radiation from a cobalt-60 or cesium-137 source is focused on the cancerous cells. Internal radiation therapy uses gamma-emitting salts of radioisotopes such as ^{192}Ir ($t_{1/2}$ = 73.8 days). The radioactive salts are encapsulated in platinum or gold "seeds" or needles and surgically implanted into the body. Because the thyroid gland uses iodine, thyroid cancer can be treated internally by oral administration of a sodium iodide solution containing a relatively high concentration of radioactive iodine-131.

Positron emission tomography (PET) is a form of nuclear imaging that uses positron emitters, such as carbon-11, fluorine-18, nitrogen-13, or oxygen-15. These radioisotopes are neutron-deficient, have short half-lives, and therefore must be prepared in a cyclotron immediately before use. When they decay, a proton is converted into a neutron, a positron, and a neutrino; the neutrino is generally not shown in the equation.

The neutrino, first observed experimentally in 1956, is a subatomic particle with zero electrical charge and a mass less than that of an electron.

$$^{1}_{1}\text{p} \longrightarrow {}^{1}_{0}\text{n} + {}^{0}_{+1}\text{e}$$

Since matter is essentially transparent to neutrinos, they escape undetected. However, the positron, $^{0}_{+1}$e, travels on average less than a few millimeters before it encounters an electron, $^{0}_{-1}$e, and undergoes antimatter-matter annihilation.

$$^{0}_{+1}\text{e} + {}^{0}_{-1}\text{e} \longrightarrow 2\,\gamma$$

The annihilation event produces two gamma rays that move in opposite directions and are detected by detectors located 180° apart in the PET scanner. Several million annihilation gamma rays can be detected within a circular field around the subject over approximately 10 min. Computer signal-averaging techniques applied to these data generate an image of the region of tissue containing the radioisotope (Figure 20.12).

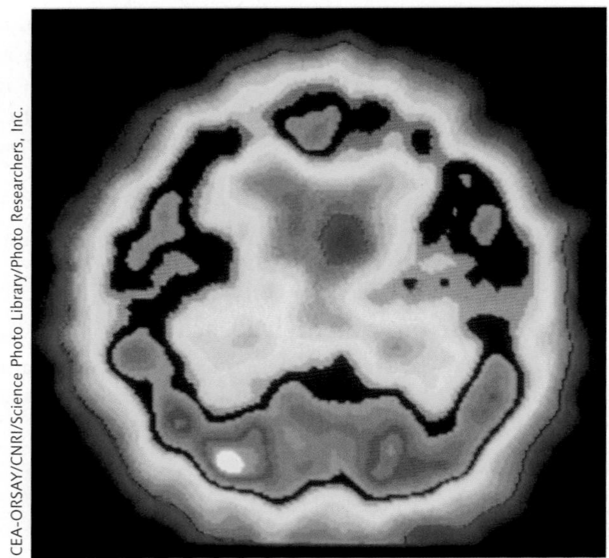

CEA-ORSAY/CNRI/Science Photo Library/Photo Researchers, Inc.

Figure 20.12 PET (positron emission tomography) scan of an axial section through a normal human brain. PET scans are obtained by injecting a tracer labeled with a short-lived radioisotope into the bloodstream. The isotope concentrates in brain tissue and emits positrons. The positrons react with electrons to create gamma rays that are recorded by a circular detector when the scan is performed. Here, radioactive methionine (an amino acid) has been used to show the level of activity of protein synthesis in the brain.

SUMMARY PROBLEM

(a) One of the species in the uranium-238 decay series (Figure 20.1) is radon-222, an alpha emitter. Write the nuclear equation for alpha emission by ^{222}Rn.

(b) Uranium-238 can be converted to plutonium-239 through a series of nuclear reactions involving absorption of a neutron followed by two beta emissions. Write these three nuclear reactions connecting ^{238}U and ^{239}Pu.

(c) Uranium-235 is the main fissionable nucleus used in nuclear reactors. When it fissions, it can produce ^{132}Sb and ^{101}Nb as products. Write the nuclear reaction for this fission reaction.

(d) Hydrogen bombs that use fusion reactions were developed following World War II. One reaction used in a hydrogen bomb was

$$^{2}_{1}H + ^{3}_{1}H \longrightarrow ^{4}_{2}He + ^{1}_{0}n$$

Calculate the energy released, in kilojoules per gram of reactants, for this fusion reaction. The necessary nuclear masses are $^{2}_{1}H = 2.01355$ g/mol; $^{3}_{1}H = 3.01550$ g/mol; $^{4}_{2}He = 4.00150$ g/mol; $^{1}_{0}n = 1.00867$ g/mol.

(e) The goal of recent nuclear arms treaties has been to dismantle the stockpiles of nuclear weapons built up by the United States and the former Soviet Union since World War II, including those containing plutonium-239 ($t_{1/2} = 2.44 \times 10^4$ years). How long will it take for the activity of plutonium-239 in a nuclear warhead to decrease (i) to 75% of its initial activity? (ii) To 10% of its initial activity?

(f) Deep underground burial has been proposed for long-term storage of the ^{239}Pu waste removed from nuclear weapons. Based on the answers to Summary Problem (e), comment on factors that need to be considered for the storage and burial of such nuclear waste.

(g) Iodine-131 emits both beta and gamma rays and is used in treating thyroid cancer. Its half-life is 13.2 h. (i) Write the nuclear reaction for beta decay of ^{131}I. (ii) If a ^{131}I sample had an activity of 5.0×10^{11} Bq (5.0×10^{11} s^{-1}), what would its activity be after 48 h?

IN CLOSING

ThomsonNOW
Sign in to ThomsonNOW at **www.thomsonedu.com** to check your readiness for an exam by taking the Pre-Test and exploring the modules recommended in your Personalized Learning Plan.

Having studied this chapter, you should be able to . . .

- Characterize the three major types of radiation observed in radioactive decay: alpha (α), beta (β), and gamma (γ) (Section 20.1). ThomsonNOW homework: Study Question 11
- Write a balanced equation for a nuclear reaction or transmutation (Section 20.2). ThomsonNOW homework: Study Questions 13, 17
- Decide whether a particular radioactive isotope will decay by α, β, or positron emission or by electron capture (Sections 20.2 and 20.3). ThomsonNOW homework: Study Question 20
- Calculate the binding energy for a particular isotope and understand what this energy means in terms of nuclear stability (Section 20.3). ThomsonNOW homework: Study Question 22
- Use Equation 20.3, $\ln\frac{N}{N_0} = -kt$, which relates (through the decay constant k) the time period over which a sample is observed (t) to the number of radioactive atoms present at the beginning (N_0) and end (N) of the time period (Section 20.4). ThomsonNOW homework: Study Questions 27, 31
- Calculate the half-life of a radioactive isotope ($t_{1/2}$) from the activity of a sample at two times, or use the half-life to find the time required for an isotope to decay to a particular activity (Section 20.4). ThomsonNOW homework: Study Questions 35, 77
- Describe nuclear chain reactions, nuclear fission, and nuclear fusion (Sections 20.6 and 20.7).
- Describe the basic functioning of a nuclear power reactor (Section 20.6).
- Describe some sources of background radiation and the units used to measure radiation (Section 20.8).
- Give examples of some uses of radioisotopes (Section 20.9).

KEY TERMS

activity *(20.4)*
alpha (α) particles *(20.1)*
alpha radiation *(20.1)*
background radiation *(20.8)*
becquerel (Bq) *(20.4)*
beta (β) particles *(20.1)*
beta radiation *(20.1)*
binding energy *(20.3)*
binding energy per nucleon *(20.3)*
critical mass *(20.6)*
curie (Ci) *(20.4)*
electron capture *(20.2)*
gamma (γ) radiation *(20.1)*
gray (Gy) *(20.8)*
half-life *(20.4)*
nuclear fission *(20.6)*
nuclear fusion *(20.7)*
nuclear medicine *(20.9)*
nuclear reactions *(20.2)*
nuclear reactor *(20.6)*
nucleons *(20.2)*
plasma *(20.7)*
positron *(20.2)*
rad *(20.8)*
radioactive series *(20.2)*
rem *(20.8)*
roentgen (R) *(20.8)*
sievert (Sv) *(20.8)*
tracers *(20.9)*

QUESTIONS FOR REVIEW AND THOUGHT

■ denotes questions available in ThomsonNOW and assignable in OWL.

Blue-numbered questions have short answers at the back of this book and fully worked solutions in the *Student Solutions Manual.*

ThomsonNOW™
Assess your understanding of this chapter's topics with sample tests and other resources found by signing in to ThomsonNOW at **www.thomsonedu.com**.

Review Questions

1. Complete the tables.

	Symbol	Mass	Charge
α particle	______	______	______
β particle	______	______	______
γ radiation	______	______	______

	Ionizing Power	Penetrating Power
α particle	______	______
β particle	______	______
γ radiation	______	______

2. Compare nuclear and chemical reactions in terms of changes in reactants, type of products formed, and conservation of matter and energy.
3. What is meant by the "band of stability"?
4. What is the binding energy of a nucleus?
5. If the mass number of an isotope is much greater than twice the atomic number, what type of radioactive decay might you expect?
6. If the number of neutrons in an isotope is much less than the number of protons, what type of radioactive decay might you expect?
7. Define critical mass and chain reaction.
8. What is the difference between nuclear fission and nuclear fusion? Illustrate your answer with an example of each.
9. Use the World Wide Web to locate the nuclear reactor power plant nearest to your college residence. Do you consider it to pose a threat to your health and safety? If so, why? If not, why not?
10. Name at least two uses of radioactive isotopes (outside of their use in power reactors and weapons).

Topical Questions

Nuclear Reactions

11. ■ By what processes do these transformations occur?
    (a) Thorium-230 to radium-226
    (b) Cesium-137 to barium-137
    (c) Potassium-38 to argon-38
    (d) Zirconium-97 to niobium-97
12. By what processes do these transformations occur?
    (a) Uranium-238 to thorium-234
    (b) Iodine-131 to xenon-131
    (c) Nitrogen-13 to carbon-13
    (d) Bismuth-214 to polonium-214
13. ■ Fill in the mass number, atomic number, and symbol for the missing particle in each nuclear equation.
    (a) ${}^{242}_{94}\text{Pu} \longrightarrow {}^{4}_{2}\text{He} + \_\_\_\_$
    (b) $\_\_\_\_ \longrightarrow {}^{32}_{16}\text{S} + {}^{0}_{-1}\text{e}$
    (c) ${}^{252}_{98}\text{Cf} + \_\_\_\_ \longrightarrow 3\,{}^{1}_{0}\text{n} + {}^{259}_{103}\text{Lr}$
    (d) ${}^{55}_{26}\text{Fe} + \_\_\_\_ \longrightarrow {}^{55}_{25}\text{Mn}$
    (e) ${}^{15}_{8}\text{O} \longrightarrow \_\_\_\_ + {}^{0}_{+1}\text{e}$
14. Fill in the mass number, atomic number, and symbol for the missing particle in each nuclear equation.
    (a) $\_\_\_\_ \longrightarrow {}^{22}_{10}\text{Ne} + {}^{0}_{+1}\text{e}$
    (b) ${}^{122}_{53}\text{I} \longrightarrow {}^{122}_{54}\text{Xe} + \_\_\_\_$
    (c) ${}^{210}_{84}\text{Po} \longrightarrow \_\_\_\_ + {}^{4}_{2}\text{He}$
    (d) ${}^{195}_{79}\text{Au} + \_\_\_\_ \longrightarrow {}^{195}_{78}\text{Pt}$
    (e) ${}^{241}_{94}\text{Pu} + {}^{16}_{8}\text{O} \longrightarrow 5\,{}^{1}_{0}\text{n} + \_\_\_\_$
15. Write a balanced nuclear equation for each word statement.
    (a) Magnesium-28 undergoes β emission.
    (b) When uranium-238 is bombarded with carbon-12, four neutrons are emitted and a new element forms.
    (c) Hydrogen-2 and helium-3 react to form helium-4 and another particle.
    (d) Argon-38 forms by positron emission.
    (e) Platinum-175 forms osmium-171 by spontaneous radioactive decay.
16. Write a balanced nuclear equation for each word statement.
    (a) Einsteinium-253 combines with an alpha particle to form a neutron and a new element.
    (b) Nitrogen-13 undergoes positron emission.
    (c) Iridium-178 captures an electron to form a stable nucleus.
    (d) A proton and boron-11 fuse, forming three identical particles.
    (e) Nobelium-252 and six neutrons form when carbon-12 collides with a transuranium isotope.
17. ■ One radioactive series that begins with uranium-235 and ends with lead-207 undergoes this sequence of emission reactions: α, β, α, β, α, α, α, α, β, β, α. Identify the radioisotope produced in each of the *first five steps.*
18. One radioactive series that begins with uranium-235 and ends with lead-207 undergoes this sequence of emission reactions: α, β, α, β, α, α, α, α, β, β, α. Identify the radioisotope produced in each of the *last six steps.*
19. Radon-222 is unstable, and its presence in homes may constitute a health hazard. It decays by this sequence of emissions: α, α, β, β, α, β, β, α. Write out the sequence of nuclear reactions leading to the final product nucleus, which is stable.

Nuclear Stability

20. ■ Write a nuclear equation for the type of decay each of these unstable isotopes is most likely to undergo.
(a) Neon-19 (b) Thorium-230
(c) Bromine-82 (d) Polonium-212

21. Write a nuclear equation for the type of decay each of these unstable isotopes is most likely to undergo.
(a) Silver-114 (b) Sodium-21
(c) Radium-226 (d) Iron-59

22. ■ Boron has two stable isotopes, ^{10}B (abundance = 19.78%) and ^{11}B (abundance = 80.22%). Calculate the binding energies per nucleon of these two nuclei and compare their stabilities.

$$5\,^1_1H + 5\,^1_0n \longrightarrow {}^{10}_5B$$
$$5\,^1_1H + 6\,^1_0n \longrightarrow {}^{11}_5B$$

The required masses (in g/mol) are ^1_1H = 1.00783; ^1_0n = 1.00867; $^{10}_5B$ = 10.01294; and $^{11}_5B$ = 11.00931.

23. Calculate the binding energy in kJ per mole of P for the formation of $^{30}_{15}P$ and $^{31}_{15}P$.

$$15\,^1_1H + 15\,^1_0n \longrightarrow {}^{30}_{15}P$$
$$15\,^1_1H + 16\,^1_0n \longrightarrow {}^{31}_{15}P$$

Which is the more stable isotope? The required masses (in g/mol) are ^1_1H = 1.00783; ^1_0n = 1.00867; $^{30}_{15}P$ = 29.97832; and $^{31}_{15}P$ = 30.97376.

24. The most abundant isotope of uranium is U-238, which has an isotopic mass of 238.0508 g/mol. What is its nuclear binding energy in kJ/mol and binding energy per nucleon?

25. What is the nuclear binding energy in kJ/mol and binding energy per nucleon of chlorine-35, which has an isotopic mass of 34.9689 g/mol?

26. What is the nuclear binding energy and binding energy per nucleon in kJ/mol of iodine-127, which has an isotopic mass of 126.9004 g/mol?

Rates of Disintegration Reactions

27. ■ Sodium-24 is a diagnostic radioisotope used to measure blood circulation time. How much of a 20-mg sample remains after 1 day and 6 hours if sodium-24 has $t_{1/2}$ = 15 hours?

28. Iron-59 in the form of iron(II) citrate is used in iron metabolism studies. Its half-life is 45.6 days. If you start with 0.56 mg iron-59, how much would remain after 1 year?

29. Iodine-131 is used in the form of sodium iodide to treat cancer of the thyroid.
(a) The isotope decays by ejecting a β particle. Write a balanced equation to show this process.
(b) The isotope has a half-life of 8.05 days. If you begin with 25.0 mg of radioactive $Na^{131}I$, what mass remains after 32.2 days?

30. Phosphorus-32 is used in the form of $Na_2H^{32}PO_4$ in the treatment of chronic myeloid leukemia, among other things.
(a) The isotope decays by emitting a β particle. Write a balanced equation to show this process.
(b) The half-life of ^{32}P is 14.3 days. If you begin with 9.6 mg of radioactive $Na_2H^{32}PO_4$, what mass remains after 28.6 days?

31. ■ What is the half-life of a radioisotope if it decays to 12.5% of its radioactivity in 12 years?

32. After 2 hours, tantalum-172 has $\frac{1}{16}$ of its initial radioactivity. How long is its half-life?

33. Radioisotopes of iodine are widely used in medicine. For example, iodine-131 ($t_{1/2}$ = 8.05 days) is used to treat thyroid cancer. If you ingest a sample of $Na^{131}I$, how much time is required for the isotope to decrease to 5.0% of its original activity?

34. The noble gas radon has been the focus of much attention because it may be found in homes. Radon-222 emits α particles and has a half-life of 3.82 days.
(a) Write a balanced equation to show this process.
(b) How long does it take for a sample of radon to decrease to 10.0% of its original activity?

35. ■ A sample of wood from a Thracian chariot found in an excavation in Bulgaria has a ^{14}C activity of 11.2 disintegrations per minute per gram. Estimate the age of the chariot and the year it was made. ($t_{1/2}$ for ^{14}C is 5.73×10^3 years, and the activity of ^{14}C in living material is 15.3 disintegrations per minute per gram.)

36. A piece of charred bone found in the ruins of a Native American village has a $^{14}C/^{12}C$ ratio of 0.72 times that found in living organisms. Calculate the age of the bone fragment. (See Question 35 for required data on carbon-14.)

37. How long will it take for a sample of plutonium-239 with a half-life of 2.4×10^4 years to decay to 1% of its original activity?

38. A 1.00-g sample of wood from an archaeological site gave 4100 disintegrations of ^{14}C in a 10-hour measurement. In the same time, a 1.00-g modern sample gave 9200 disintegrations. What is the age of the wood?

Artificial Transmutations

39. There are two isotopes of americium, both with half-lives sufficiently long to allow the handling of large quantities. Americium-241 has a half-life of 248 years as an α emitter, and it is used in gauging the thickness of materials and in smoke detectors. This isotope is formed from ^{239}Pu by absorption of two neutrons followed by emission of a β particle. Write a balanced equation for this process.

40. Americium-240 is made by bombarding plutonium-239 atoms with α particles. In addition to ^{240}Am, the products are a proton and two neutrons. Write a balanced equation for this process.

41. ■ To synthesize the heavier transuranium elements, one must bombard a lighter nucleus with a relatively large particle. If you know that the products of a bombardment reaction are californium-246 and four neutrons, with what particle would you bombard uranium-238 atoms?

42. The officially named element with the highest atomic number is $^{272}_{111}Rg$. To try to make heavier elements, attempts have been made to force calcium-40 and curium-248 to merge. What would be the atomic number of the element formed?

Nuclear Fission and Fusion

43. Name the fundamental parts of a nuclear fission reactor and describe their functions.

44. Explain why it is easier for a nucleus to capture a neutron than to force a nucleus to capture a proton.

45. ■ What is the missing product in each of these fission equations?

(a) $^{235}_{92}U + ^{1}_{0}n \longrightarrow$ ____ $+ ^{93}_{38}Sr + 3\ ^{1}_{0}n$

(b) $^{235}_{92}U + ^{1}_{0}n \longrightarrow$ ____ $+ ^{132}_{51}Sb + 3\ ^{1}_{0}n$

(c) $^{235}_{92}U + ^{1}_{0}n \longrightarrow$ ____ $+ ^{141}_{56}Ba + 3\ ^{1}_{0}n$

46. Explain why no commercial fusion reactors are in operation today.

47. The average energy output of a barrel of oil is 5.9×10^6 kJ/barrel. Fission of 1 mol ^{235}U releases 2.1×10^{10} kJ of energy. Calculate the number of barrels of oil needed to produce the same energy as 1.0 lb ^{235}U.

48. A concern in the nuclear power industry is that, if nuclear power becomes more widely used, there may be serious shortages in worldwide supplies of fissionable uranium. One solution is to build breeder reactors that manufacture more fuel than they consume. One such cycle works as follows:

(i) A ^{238}U nucleus collides with a neutron to produce ^{239}U.

(ii) ^{239}U decays by β emission ($t_{1/2}$ = 24 minutes) to give an isotope of neptunium.

(iii) The neptunium isotope decays by β emission to give a plutonium isotope.

(iv) The plutonium isotope is fissionable. On its collision with a neutron, fission occurs and gives energy, at least two neutrons, and other nuclei as products.

Write an equation for each of these steps, and explain how this process can be used to breed more fuel than the reactor originally contained and still produce energy.

Effects of Nuclear Radiation

49. Two common units of radiation used in newspaper and news magazine articles are the rad and rem. What does each measure? Which would you use in an article describing the damage an atomic bomb would inflict on a human population? What relationship does the gray have with these units?

50. Which electrical power plant—fossil fuel or nuclear—exposes a community to more nuclear radiation? Explain why.

51. Explain how our own bodies are sources of nuclear radiation.

52. What is the source of radiation exposure during jet plane travel?

Uses of Radioisotopes

53. Why are foods irradiated with gamma rays instead of alpha or beta particles?

54. X-rays and PET scans are two medical imaging techniques. How are they similar and how are they different?

55. To measure the volume of the blood system of an animal, the following experiment was done. A 1.0-mL sample of an aqueous solution containing tritium with an activity of 2.0×10^6 disintegrations per second (s^{-1}) was injected into the bloodstream. After time was allowed for complete circulatory mixing, a 1.0-mL blood sample was withdrawn and found to have an activity of $1.5 \times 10^4\ s^{-1}$. What was the volume of the circulatory system? (The half-life of tritium is 12.3 years, so this experiment assumes that only a negligible quantity of tritium has decayed during the experiment.)

56. Radioactive isotopes are often used as "tracers" to follow an atom through a chemical reaction, and the following is an example. Acetic acid reacts with methanol, CH_3OH, by eliminating a molecule of H_2O to form methyl acetate, CH_3COOCH_3. Explain how you would use the radioactive isotope ^{18}O to show whether the oxygen atom in the water product comes from the —OH of the acid or the —OH of the alcohol.

$$\underset{\text{acetic acid}}{CH_3COOH} + \underset{\text{methanol}}{CH_3OH} \longrightarrow \underset{\text{methyl acetate}}{CH_3COOCH_3} + H_2O$$

General Questions

57. Complete these nuclear equations.

(a) $^{214}Bi \longrightarrow$ ____ $+ ^{214}Po$

(b) $4\ ^{1}_{1}H \longrightarrow$ ____ + 2 positrons

(c) ^{249}Es + neutron ⟶ 2 neutrons + ____ $+ ^{161}Gd$

(d) $^{220}Rn \longrightarrow$ ____ + alpha particle

(e) ^{68}Ge + electron ⟶ ____

58. Complete these nuclear equations.

(a) ____ + neutron ⟶ 2 neutrons + $^{137}Tc + ^{97}Zr$

(b) $^{45}Ti \longrightarrow$ ____ + positron

(c) ____ ⟶ beta particle + ^{59}Co

(d) ^{24}Mg + neutron ⟶ ____ + proton

(e) ^{131}Cs + ____ ⟶ ^{131}Xe

59. Radioactive nitrogen-13 has a half-life of 10 minutes. After an hour, how much of this isotope remains in a sample that originally contained 96 mg?

60. The half-life of molybdenum-99 is 67.0 hours. How much of a 1.000-mg sample of ^{99}Mo is left after 335 hours? How many half-lives did it undergo?

61. The oldest known fossil cells were found in South Africa. The fossil has been dated by the reaction

$$^{87}Rb \longrightarrow ^{87}Sr + ^{0}_{-1}e \qquad t_{1/2} = 4.9 \times 10^{10}\ \text{years}$$

If the ratio of the present quantity of ^{87}Rb to the original quantity is 0.951, calculate the age of the fossil cells.

62. Cobalt-60 is a therapeutic radioisotope used in treating certain cancers. If a sample of cobalt-60 initially disintegrates at a rate of $4.3 \times 10^6\ s^{-1}$ and after 21.2 years the rate has dropped to $2.6 \times 10^5\ s^{-1}$, what is the half-life of cobalt-60?

63. Balance these equations used for the synthesis of transuranium elements.

(a) $^{238}_{92}U + ^{14}_{7}N \longrightarrow$ ____ $+ 5\ ^{1}_{0}n$

(b) $^{238}_{92}U +$ ____ $\longrightarrow ^{249}_{100}Fm + 5\ ^{1}_{0}n$

(c) $^{253}_{99}Es +$ ____ $\longrightarrow ^{256}_{101}Md + ^{1}_{0}n$

(d) $^{246}_{96}Cm +$ ____ $\longrightarrow ^{254}_{102}No + 4\ ^{1}_{0}n$

(e) $^{252}_{98}Cf +$ ____ $\longrightarrow ^{257}_{103}Lr + 5\ ^{1}_{0}n$

64. On December 2, 1942, the first man-made self-sustaining nuclear fission chain reactor was operated by Enrico Fermi and others under the University of Chicago stadium. In June 1972, natural fission reactors, which operated billions of years ago, were discovered in Oklo, Gabon. At present, natural uranium contains 0.72% ^{235}U. How many years ago did natural uranium contain 3.0% ^{235}U, sufficient to sustain a natural reactor? ($t_{1/2}$ for $^{235}U = 7.04 \times 10^8$ years.)

Applying Concepts

65. If a radioisotope is used for diagnosis (e.g., detecting cancer), it should decay by gamma radiation. However, if its use is therapeutic (e.g., treating cancer), it should decay by alpha or beta radiation. Explain why in terms of ionizing and penetrating power.
66. During the Three Mile Island incident, people in central Pennsylvania were concerned that strontium-90 (a beta emitter) released from the reactor could become a health threat (it did not). Where would this isotope collect in the body? If so, what types of problems could it cause?
67. ■ Classify the isotopes ^{17}Ne, ^{20}Ne, and ^{23}Ne as stable or unstable. What type of decay would you expect the unstable isotope(s) to have?
68. The following demonstration was carried out to illustrate the concept of a nuclear chain reaction. Explain the connections between the demonstration and the reaction.

    Eighty mousetraps are arranged side by side in eight rows of ten traps each. Each trap is set with two rubber stoppers for bait. A small plastic mouse is tossed into the middle of the traps, setting off one trap, which in turn sets off two traps and so on until all the traps are sprung.
69. Most students have no trouble understanding that 1.5 g of a 24-g sample of a radioisotope would remain after 8 h if it had $t_{1/2} = 2$ h. What they don't always understand is where the other 22.5 g went. How would you explain this disappearance to another student?
70. Nuclear chemistry is a topic that raises many debatable issues. Briefly discuss your views on the following.
    (a) Twice a year the general public is allowed to visit the Trinity Site in Alamogordo, New Mexico, where the first atomic bomb was tested. If you had the opportunity to do so, would you visit the site? Explain your answer.
    (b) Now that the Cold War has ended, should the United States stockpile nuclear weapons? Explain your answer.
    (c) The practice of irradiating food for sterilization is controversial in the United States. What are some of the possible benefits or deficits of this practice? Explain your answer.

More Challenging Questions

71. All radioactive decays are first-order. Why is this so?
72. An average 70.0-kg adult contains about 170 g potassium. Potassium-40, with a relative abundance of 0.0118%, undergoes beta decay with a half-life of 1.28×10^9 years. What is the total activity due to beta decay of ^{40}K for this average person?
73. If the earth receives 7.0×10^{14} kJ/s of energy from the sun, what mass of solar material is lost per hour to supply this amount of energy?
74. A sample of the alpha emitter ^{222}Ra had an initial activity A_0 of 7.00×10^4 Bq. After 10.0 days its activity A had fallen to 1.15×10^4 Bq. Calculate the decay constant and half-life of radon-222.
75. When a bottle of wine was analyzed for its tritium (^3H) content, it was found to contain 1.45% of the tritium originally present when the wine was produced. How old is the bottle of wine? ($t_{1/2}$ of $^3H = 12.3$ years.)
76. A chemist is setting up an experiment using ^{47}Ca, which has a half-life of 4.5 days. He needs 10.0 μg of the calcium. How many μg of $^{47}CaCO_3$ must he order if the delivery time is 50 hours?
77. ■ To determine the age of the charcoal found at an archaeological site, this sequence of experiments was done: The charcoal sample was burned in oxygen and the CO_2 obtained was captured by bubbling it through lime water, $Ca(OH)_2$, to form a precipitate of $CaCO_3$. This precipitate was filtered, dried, and weighed. A sample of 1.14 g $CaCO_3$ produced 2.17×10^{-2} Bq from carbon-14. Modern carbon produces 15.3 disintegrations min^{-1} g^{-1} of carbon. What is the age of the charcoal? The half-life of carbon-14 is 5730 years.

Conceptual Challenge Problems

CP20.A (Section 20.4) The half-life for the alpha decay of uranium-238 to thorium-234 is 4.5×10^9 years, which happens to be the estimated age of the earth.
(a) How many atoms were decaying per second in a 1.0-g sample of uranium-238 that existed 1.0×10^6 years ago?
(b) How would you find the number of atoms now decaying per second in this sample?

CP20.B (Section 20.4) If the earth is 4.5×10^9 years old and the amount of radioactivity in a sample becomes smaller with time, how is it possible for there to be any radioactive elements left on earth that have half-lives less than a few million years?

CP20.C (Section 20.4) Using experiments based on a sample of living wood, a nuclear chemist estimates that the uncertainty of her measurements of the carbon-14 radioactivity in the sample is 1.0%. The half-life of carbon-14 is 5730 years.
(a) How long must a sample of wood be separated from a living tree before the chemist's radioactivity measurements on the sample provide evidence for the time when it died?
(b) Suppose that the chemist's uncertainty in the radioactivity of carbon-14 continues to be 1.0% of the radioactivity of living wood. How long must a sample of wood be dead before the chemist's measurements support the claim that the time since the wood was separated from the tree is not changing?

CP20.D (Section 20.8) You have read that alpha radiation is the least penetrating type of radiation, followed by beta radiation. Gamma radiation penetrates matter well, and thick samples of matter are required to contain gamma radiation. Knowing these facts, what can you correctly deduce about the harmful effects of these three types of radiation on living tissue?

CP-20.E (Section 20.8) Death will likely occur within weeks for a 150-lb person who receives 500,000 mrem of radiation over a short time, an exposure that is 1000 times the federal government's standard for 1 year (500 mrem/yr). A student realizes that 500,000 mrem is 500 rem, and that 500 rem has the effect of depositing 317 J of energy on the body of the 150-lb person. The student is puzzled. How can the deposition of only 317 J of energy from nuclear radiation—much less energy than that deposited by cooling a cup of coffee 1 °C within a person's body—have such a disastrous effect on the person?

21 The Chemistry of the Main Group Elements

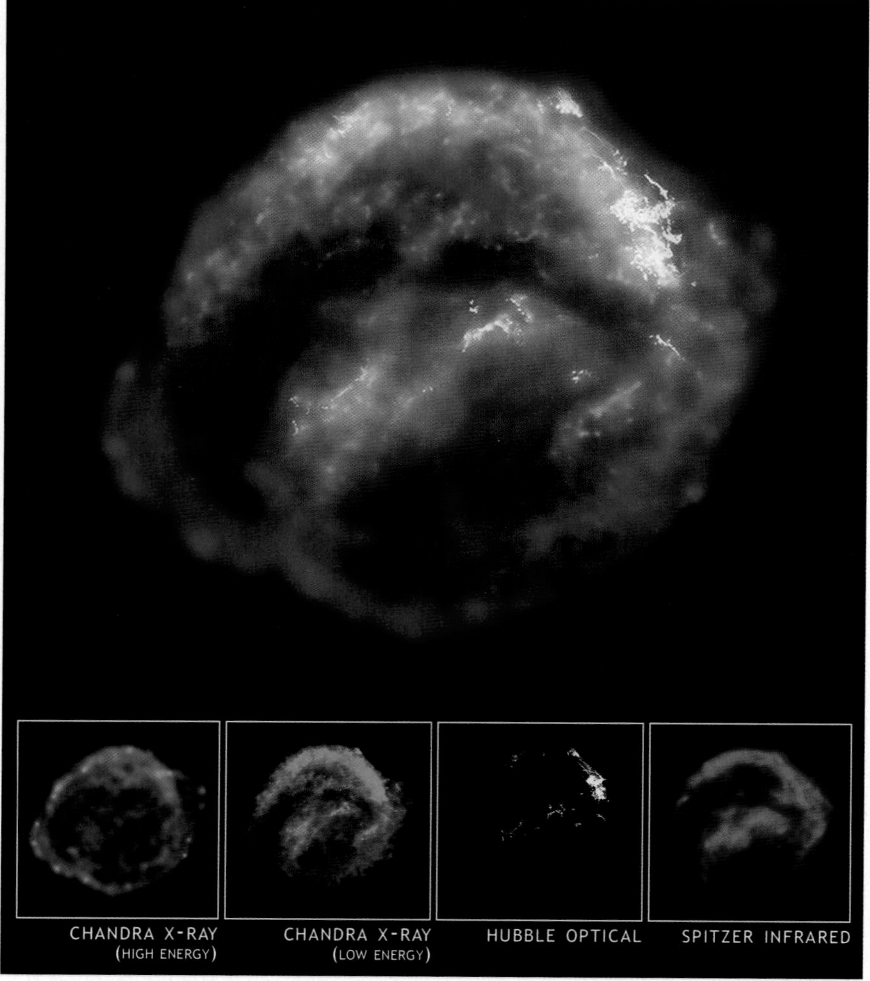

Kepler's supernova remnant. The bubble-shaped cloud of gas and dust envelopes Kepler's supernova remnant, the fast-moving material from an exploded star (supernova) in which chemical elements form. The fast-moving, iron-rich shell of the supernova is surrounded by an expanding shock wave that gathers interstellar gas and dust. Each color in the image represents a different region of the electromagnetic spectrum, from X-rays to infrared light. Those invisible to the naked eye have been color-coded. The formation of the chemical elements is discussed in this chapter.

NASA/ESA/JHU/R. Sankrit & W. Blair

Elements have been discussed throughout this book. Their names, symbols, physical properties, and chemical reactivity have been noted, demonstrating their enormous diversity. Some elements (such as sodium and fluorine) react violently with many other substances; others (the noble gases) are so quiescent that they enter into very few or no chemical combinations. In spite of this diversity, elements in each group of the periodic table have predictable chemical similarities based on their number of valence electrons.

A fundamental question that has not yet been discussed concerns the origin of the elements: How did they form? How, for example, did magnesium atoms acquire a different number of protons from atoms of calcium or helium? This chapter will answer those questions. It will also describe how ten selected main group elements—seven nonmetals and three metals—are extracted from natural sources, the chemical principles associated with the extraction processes, and the properties of those elements. The chapter concludes with an overview of the main group elements, those in Groups 1A through 8A, describing their properties and uses, and relating these to the periodic table. The transition elements, all metals, are discussed in Chapter 22.

ThomsonNOW
Throughout the text, this icon indicates an opportunity to test yourself on key concepts and to explore interactive modules by signing in to ThomsonNOW at **www.thomsonedu.com**.

21.1 Formation of the Elements

Cosmologists—scientists who study the formation of the universe—use spectral evidence as well as knowledge of nuclear reactions to develop theories about the origin of the universe. The cosmologists theorize that about 15 billion years ago all the matter in the universe was contained in a pinpoint-sized region that exploded at inconceivably high temperatures (estimated to be about 10^{32} K) in what is called the "Big Bang." This explosion produced a universe expanding so quickly that in one second it cooled to 10^9 K, a temperature at which the fundamental subatomic particles formed—neutrons, protons, and electrons. Within two hours, the temperature had dropped to 10^7 K, temperatures suitable for the formation of light nuclei—^{2}H, ^{3}He, and ^{4}He.

Expansion is a cooling process; contraction is a heating process.

Nuclear Burning

The elements from hydrogen to iron are formed inside stars by **nuclear burning,** a sequence of nuclear *fusion* reactions not to be confused with chemical combustion. The fusion of protons (hydrogen-1 nuclei) to form helium-4 nuclei is called **hydrogen burning.**

The term "nuclear burning" arises from the fact that nuclear fusion reactions are highly exothermic and that we think of the sun and other stars as burning in the sky.

$$4\,{}^{1}_{1}\mathrm{H} \longrightarrow {}^{4}_{2}\mathrm{He} + 2\,{}^{0}_{+1}\mathrm{e} + 2\,\gamma$$

After billions of years of hydrogen burning, the star contracts and the core becomes sufficiently dense and hot for **helium burning,** the fusion of helium-4 nuclei, to occur.

$${}^{4}_{2}\mathrm{He} + {}^{4}_{2}\mathrm{He} \longrightarrow {}^{8}_{4}\mathrm{Be}$$

Beryllium-8 is unstable, but by fusing with another helium-4 nucleus, beryllium-8 is converted to stable carbon-12.

$${}^{8}_{4}\mathrm{Be} + {}^{4}_{2}\mathrm{He} \longrightarrow {}^{12}_{6}\mathrm{C}$$

The low natural abundance of beryllium is evidence of the instability of beryllium-8 (← ***p. 985***).

When helium burning stops, the star contracts further with the result that the temperature becomes high enough for heavier nuclei to form by fusion. Starting with carbon-12, three successive fusions with helium-4 nuclei form oxygen-16, then neon-20, and then magnesium-24. The process continues up to the formation of calcium-40. Carbon and oxygen burning also occurs:

$$^{12}_{6}C + ^{12}_{6}C \longrightarrow ^{23}_{11}Na + ^{1}_{1}H$$

$$^{12}_{6}C + ^{16}_{8}O \longrightarrow ^{28}_{14}Si$$

Starting with silicon-28, fusion reactions build up heavier nuclei all the way to iron-56 and nickel-58, which are very stable nuclei with the highest binding energies per nucleon (← ***p. 987***). Elements heavier than iron cannot be formed by such nuclear fusion reactions.

Nuclear fusion reactions have very high activation energy barriers (← ***p. 629***) because two positively charged nuclei repel each other strongly. Thus, high temperatures and high kinetic energies of colliding particles are necessary to overcome the high activation energy barrier. The greater the nuclear charge, the higher the energy barrier and the higher the temperature required to overcome the barrier.

EXERCISE 21.1 Fusing Nuclei

Write balanced nuclear equations representing the formation of oxygen-16, neon-20, and magnesium-24 starting from carbon-12 and helium-4 nuclei.

Formation of Heavier Elements

The amount of time a star spends in the various stages of elemental synthesis in relation to the central temperature of the star is given in Figure 21.1. Following helium burning, elements heavier than iron form by neutron capture in massive stars in which the core collapses rapidly. Stable nuclei such as those of iron decompose into neutrons and protons, and the protons are converted into additional neutrons by capturing electrons. The result is the formation of a neutron star whose outer layers explode away as a supernova (chapter-opening photo).

Heavier elements form during supernova generation by one of two processes. In the *s* (slow) *process,* the slow capture of neutrons takes place over many years. Because this capture shifts the neutron/proton ratio, eventually the nuclei will become beta emitters. As noted in Section 20.3 (← ***p. 983***), beta emission causes an increase in the atomic number, so a new element is formed from the parent nucleus. Such is the case for $^{98}_{42}Mo$, for example, which is converted to technetium-99 by this two-step process.

$$^{98}_{42}Mo + ^{1}_{0}n \longrightarrow ^{99}_{42}Mo$$

$$^{99}_{42}Mo \longrightarrow ^{99}_{43}Tc + ^{0}_{-1}e$$

Isotopes with masses up to 209 can form by the *s process.*

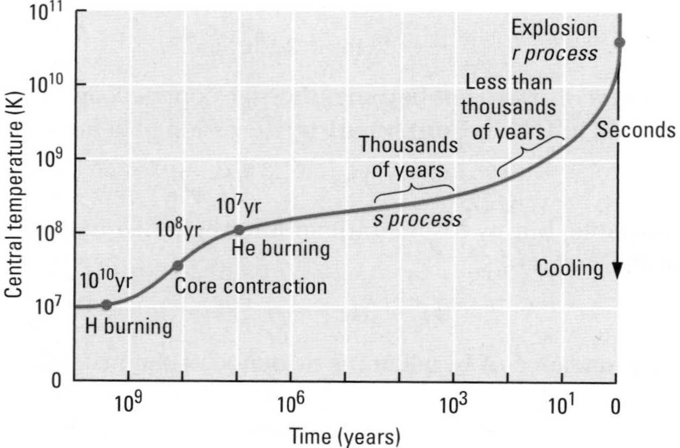

Figure 21.1 The timescale for various stages of elemental syntheses in stars.

Some elements, including the radioactive actinides, are produced by the very rapid *r process,* which occurs during the explosive stage of a star. Because of the speed at which a series of neutrons can be captured one by one in the *r* process, new elements can be produced from nuclei with very short half-lives, too short to react by the *s* process. In the *r* process, a nucleus may capture many neutrons in an extremely short time (0.01 to 10 s) in a series of reactions that produce a nucleus much heavier than the original one. Suppose, for example, that $^{130}_{48}Cd$ is produced in the *r* process. This isotope is highly unstable because it has far too many neutrons; cadmium-116 is the most stable known isotope of cadmium. The cadmium-130 can undergo a rapid series of beta decays, increasing in atomic number until it reaches tellurium-130, the most abundant isotope of tellurium.

$$^{130}_{48}\text{Cd} \xrightarrow{\text{4 beta decays}} {}^{130}_{52}\text{Te}$$

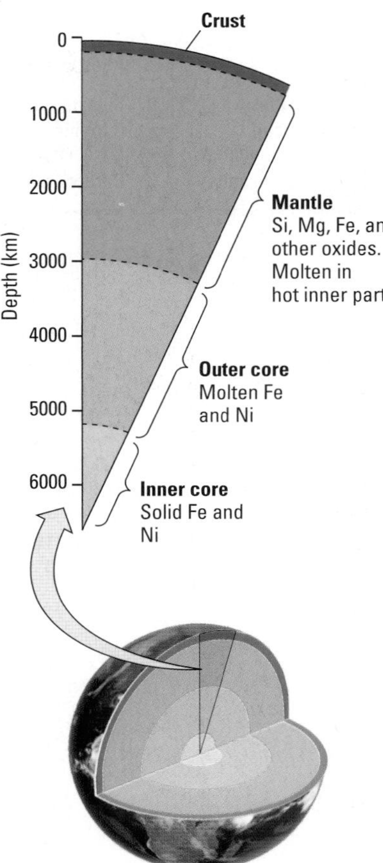

Figure 21.2 A cross section of the earth. Note the thinness of the crust compared with that of the mantle and the core.

EXERCISE 21.2 Cadmium-130 Decay

As noted above, cadmium-130 decays to tellurium-130 through four consecutive beta decays. Write balanced equations for these reactions, starting with cadmium-130 and ending with tellurium-130.

21.2 Terrestrial Elements

In this and the next four sections we describe how nitrogen, oxygen, sulfur, sodium, chlorine, magnesium, aluminum, phosphorus, bromine, and iodine are obtained from their natural sources; we also consider the properties and uses of these elements.

We obtain large quantities of nitrogen and oxygen from the atmosphere. Ocean water is treated to extract commercial quantities of magnesium, bromine, and sodium chloride. And the earth's crust is an indispensable source of most of the other elements. Figure 21.2 illustrates the relation between the earth's crust, mantle, and core. The crust, which extends only from the surface to a depth of about 35 km, is but a tiny fraction of the entire depth of the earth. If you think of the earth as an apple, the crust is akin to the thickness of the skin of the apple.

The average composition of the earth's crust is given in Figure 21.3. All the elements shown in the pie chart are in compounds; because of their reactivity, these elements do not exist in their "free" (elemental) form in the earth's crust. Note the preponderance of oxygen and silicon; they are the major components of silicate minerals, clays, and sand. Aluminum is the most abundant metal ion, followed by ions of iron and several alkali and alkaline earth metals—calcium, sodium, potassium, and magnesium.

Most of the iron on earth is in the core and the mantle, not the crust.

Most elements in the crust of the earth are in chemically combined forms as complex ionic solids known as minerals. A **mineral** is commonly defined as a naturally occurring inorganic compound with a characteristic composition and crystal structure. The major chemical form in which each element occurs on the earth's surface as a source of the element is shown in Figure 21.4. In particular, note the predominance of oxygenated compounds, either binary ones such as MgO and TiO_2, or more complex ones such as carbonates, phosphates, and silicates. The preponderance of such compounds is testimony to the abundance of oxygen and silicon in the earth's surface. The lanthanides and naturally occurring actinides also form oxide minerals. Many transition metals and heavier elements of Groups 3A (13) to 6A (16) are found as sulfides, such as ZnS and Sb_2S_3.

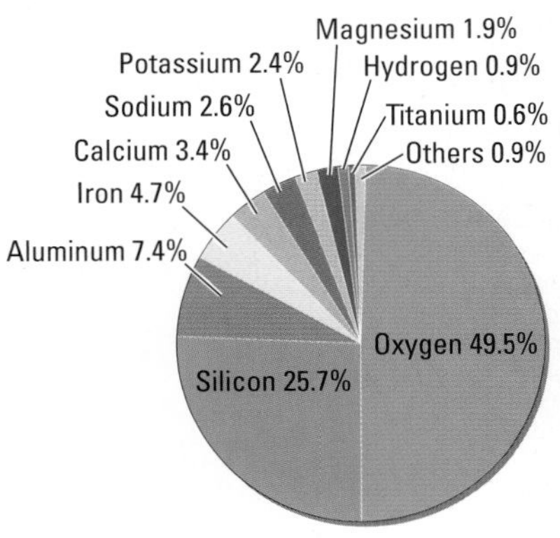

Figure 21.3 Elemental composition of the earth's crust.

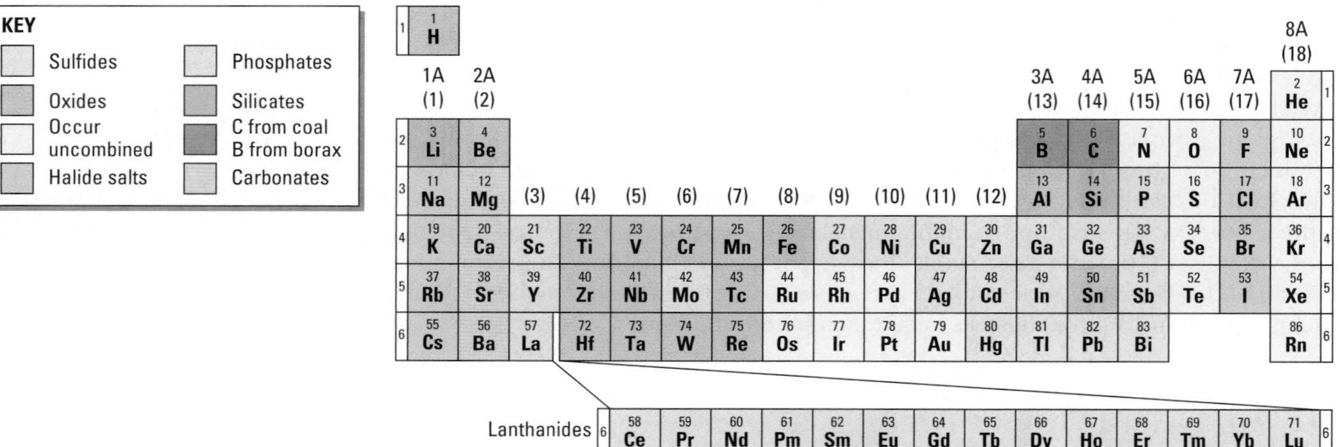

Figure 21.4 **Main types of minerals in the earth's crust used as sources of elements.**

Silica and Silicates

Silica is pure SiO_2. Its most common form is α-quartz, which is a major component of many rocks such as granite and sandstone. Alpha-quartz also occurs as a pure rock crystal (Figure 21.5) and in several less pure forms. Silicon and oxygen, the two most abundant elements in the earth's crust, are combined in the crust as silicate minerals. Such minerals all contain SiO_4 groups in which four oxygen atoms are arranged tetrahedrally around a central Si atom and each oxygen is bonded to another Si atom. The SiO_4 groups are the fundamental building block for all silicate minerals (Figure 21.6). In silicate minerals these tetrahedra typically share one or more oxygens to form chains, sheets, rings, and three-dimensional networks; quartz has a three-dimensional structure, for example, as shown on the next page.

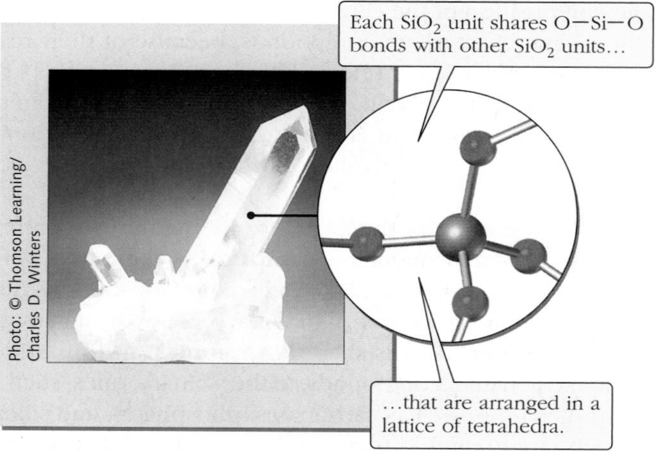

Each SiO_2 unit shares O—Si—O bonds with other SiO_2 units...
...that are arranged in a lattice of tetrahedra.
Photo: © Thomson Learning/ Charles D. Winters

Figure 21.5 **A pure quartz crystal (pure SiO_2).** The formula is derived from the fact that each oxygen atom of a SiO_4 tetrahedron is shared by two silicon atoms. Correspondingly, only half of the oxygens "belong" to a given Si, making the formula SiO_2, not SiO_4. Quartz crystals are used as oscillators in watches, radios, VCRs, and computers.

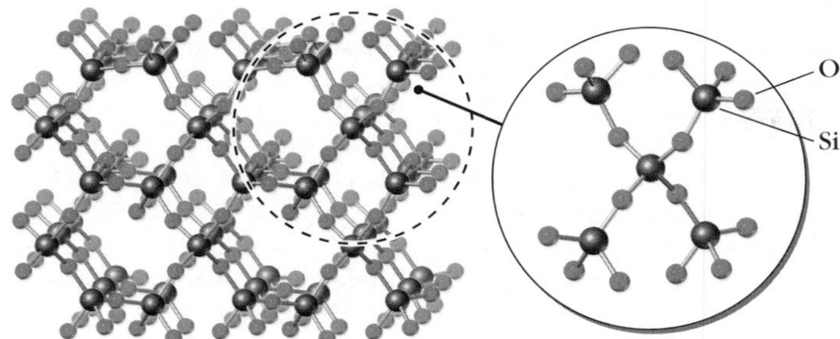

Three-dimensional structure of quartz.

An individual SiO_4 group in which none of the oxygens is shared with another silicon has a charge of 4−. An example is found in the mineral olivine, Mg_2SiO_4, which contains two Mg^{2+} ions for each SiO_4^{4-} ion to balance the charge. The simple silicate minerals olivine, willemite (Zn_2SiO_4), and the gemstone garnet contain discrete SiO_4^{4-} units that do not share oxygens.

In olivine, one out of every ten Mg^{2+} ions is replaced by Fe^{2+}. The Fe^{2+} ions give olivine its characteristic olive-green color.

Condensed silicates contain SiO_4 tetrahedra in which oxygen atoms are shared. *Pyroxenes* contain extended chains of linked SiO_4 tetrahedra, each sharing two oxygen atoms. The repeating unit in the pyroxene polymer is SiO_3, so pyroxene appears to contain the metasilicate ion, SiO_3^{2-}; a typical formula is Na_2SiO_3 (Figure 21.6). Pyroxenes are abundant in the ocean's floor and the earth's mantle. If two pyroxene chains are laid side by side, they can link by sharing oxygen atoms in adjoining chains to form a type of silicate called an *amphibole,* with the typical repeating unit $Si_4O_{11}^{6-}$ (Figure 21.6).

An excellent example of an amphibole is crocidolite, one form of *asbestos.* What is called "asbestos" is not a single substance; rather, this term applies broadly to a family of naturally occurring hydrated silicates that crystallize as fibers. Asbestos minerals are generally subdivided into two forms: serpentine and amphibole fibers. Approximately 5 million tons of the serpentine form of asbestos, chrysotile, are mined each year, chiefly in Canada and the former Soviet Union. Chrysotile is the only form widely used commercially in the United States. Another form, the amphibole crocidolite, is mined in small quantities, mainly in South Africa. The two asbestos minerals differ greatly in composition, color, shape, solubility, and persistence in human tissue. This last property is important in determining their toxicity. Crocidolite is blue, is relatively insoluble, and persists in tissue. Its long,

Garnet is a gemstone containing SiO_4^{4-} tetrahedra.

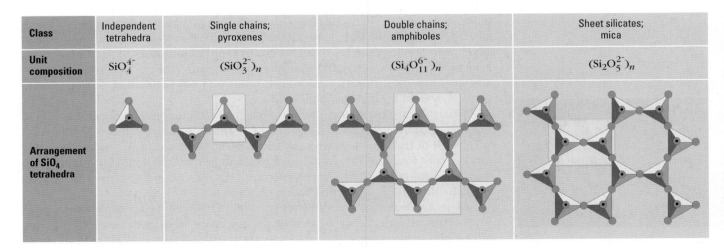

Class	Independent tetrahedra	Single chains; pyroxenes	Double chains; amphiboles	Sheet silicates; mica
Unit composition	SiO_4^{4-}	$(SiO_3^{2-})_n$	$(Si_4O_{11}^{6-})_n$	$(Si_2O_5^{2-})_n$
Arrangement of SiO_4 tetrahedra				

Figure 21.6 Silicate structures. These structures are all based on the tetrahedral SiO_4 unit. The repeating unit of each structure is shown with a tan background.

CHEMISTRY IN THE NEWS

Glass Sea Sponge: Delicate but Strong

The glass sea sponge has a delicately beautiful glassy exoskeleton, yet one that is strong enough to withstand the pressure and strain of the deep ocean. Researchers at Lucent Technologies' Bell Laboratories led by Joanna Aizenberg recently determined how the exoskeleton of the glass sea sponge is so strong in spite of being made of glass, a brittle material. Aizenberg says, "Nature has found a way to perfect inherently fragile materials by employing standard engineering principles from the nano- to the macroscale."

The sponge uses a layered structure incorporating silica, SiO_2, in different arrangements. At the nanoscale level, the skeleton is made up of silica nanospheres around a protein filament. At the next structural level, the nanospheres form needlelike spikes from dozens of concentric silica layers, alternating the silica layers with ones of an organic material. The silica spikes are then joined into larger parallel bundles, a technique used to construct common ceramics. The bundles are arranged horizontally and vertically into a gridwork of glassy threads reinforced by diagonal silica spikes. Such arrangements are used architecturally in structures such as the Eiffel Tower. Aizenberg is enthusiastic about her research team's discovery: "This creature's skeleton is a textbook lesson in mechanical engineering, offering valuable knowledge that could lead to new concepts in materials science and engineering design."

SOURCE: *Chemical & Engineering News,* July 11, 2005; p. 11.

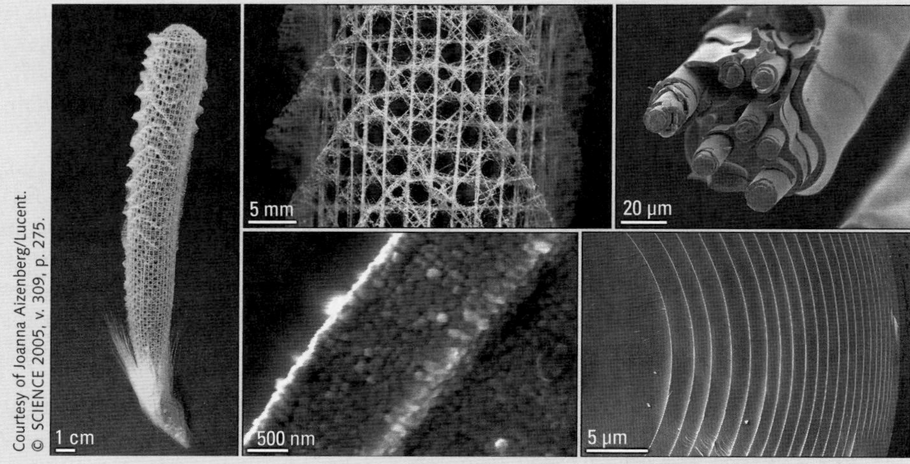

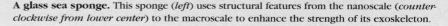

A glass sea sponge. This sponge (*left*) uses structural features from the nanoscale (*counterclockwise from lower center*) to the macroscale to enhance the strength of its exoskeleton.

Because of their double-stranded chain structure, asbestos minerals are fibrous and can even be woven into a cloth-like material.

thin, straight fibers can penetrate narrow lung passages. In contrast, chrysotile is white, and it tends to be soluble and disappear in tissue. Chrysotile fibers are curly, so they ball up like yarn and are more easily rejected by the body.

Long-term occupational exposure to certain asbestos minerals can lead to lung cancer. Although some disagreement persists in the medical and scientific communities, evidence strongly suggests that amphiboles such as crocidolite are much more potent cancer-causing agents than the serpentines such as chrysotile. Most asbestos in public buildings is the chrysotile type, so initiatives to remove asbestos insulation may be misguided overreaction in many cases. Nevertheless, most asbestos-containing materials have been removed from the market, and strict standards now exist for their handling and use.

When silicate chains continue to link in two dimensions, extended sheets of SiO_4 tetrahedral units result (Figure 21.6), characterized by the repeating unit $Si_2O_5^{2-}$. All of the atoms within each sheet are strongly covalent bonded, but each sheet is only weakly bonded to those above and below it. Various clay minerals and mica have this sheet-like silicate structure. Mica, for example, is used to prepare "metallic"-looking paint on new automobiles.

Sheets of mica can be peeled away from each other because the sheets are weakly bonded to each other.

Clays are essential components of soils that come from the weathering of igneous rocks. Since early in human history, clays have been used for pottery, bricks, tiles, and writing materials. Clays are actually *aluminosilicates.* In an aluminosilicate, some of the tetrahedral groups are AlO_4 instead of SiO_4, and some O atoms are shared between an Al atom and a Si atom. An example is feldspar,

$KAlSi_3O_8$, a component of many rocks. When Al^{3+} ions are replaced by other 3+ metal ions, the clay may become colored. For example, a red clay contains Fe^{3+} ions in place of some Al^{3+} ions.

Artists who work with clay first wet it and then mold the clay into a shape. Water molecules strongly interact with the oxygen atoms as well as the metal ions near the surface of clay particles, and the silicate layers slide over one another, making the clay pliable. After the clay has been formed into the desired shape, the object is heated in an oven to remove the water. Bonds form between exposed oxygen atoms and aluminum or silicon atoms on the surfaces of adjacent particles, which causes the clay to harden. Too much water in the wet clay can make it unstable. This instability occurs not only in clay for pottery, but also on a much larger scale in nature. During very heavy rains, entire hillsides of clay can shift and slide downhill, causing massive destruction of property (Figure 21.7).

CONCEPTUAL EXERCISE **21.3 Linking Tetrahedra**

Explain how the silicate unit in pyroxenes has the general formula SiO_3^{2-}, not SiO_4^{4-}.

Methods for Obtaining Pure Elements

Relatively few elements in nature are available directly in their uncombined form: the noble gases, mercury, gold, silver, copper, and sulfur. Nitrogen and oxygen occur as diatomic molecules. All other elements needed for practical applications must be extracted from their compounds. The types of extraction methods used are listed in Table 21.1. In minerals, metals exist as cations that must be reduced to their elemental form. Therefore, chemical and electrochemical oxidation-reduction reactions are needed in the production of metals.

Figure 21.7 A mudslide caused by shifting clay. During heavy rains, clays become saturated with water, causing the aluminosilicate layers to shift, sliding over each other.

Table 21.1 Methods for Extraction of Elements from Their Ores

Extraction Method	Examples of Elements Extracted by This Method
Carbon reduction of oxide	Si, Fe, Sn
Oxidation with Cl_2	Br, I
Reaction of sulfide with O_2	Cu, Hg
Conversion of sulfide to oxide, then reduction with C	Zn, Pb
Halide reduction with sodium or other highly reactive metal	K, Ti, Cr, Cs, U
Halide or oxide reduction with H_2	B, Ni, Mo, W
Electrolysis of solution or molten salt	H, Li, F, Na, Ca, Al, Cl

In the subsequent sections of this chapter we will describe how some elements are extracted from their naturally occurring forms by physical methods (Section 21.3: nitrogen, oxygen, and sulfur), by electrochemical redox reactions (Section 21.4: sodium, chlorine, magnesium, and aluminum), and by chemical redox reactions (Section 21.5: phosphorus, bromine, and iodine).

An **ore** is a mineral that contains a sufficiently high concentration of an element to make its extraction profitable. Because not all elements are used to the same extent, and the quantity used can vary with market demands, a metal is extracted from its ore in response to such demands. Current known reserves of some common elements such as aluminum and iron are sufficient to last hundreds of years at the current rate of use, whereas the known reserves of other widely used elements, such as copper, tin, and lead, are rather slim (Table 21.2). Notice from Table 21.2 that the United States does not have major reserves of several critical metals, for example, the chromium and manganese needed for making steel and other alloys. Thus, we must import and stockpile such metals.

Table 21.2 Known Reserves of Selected Elements

Element	Reserves (10^9 kg)	Lifetime (yr)	Locations of Major Reserves
Al	20,000	220	Australia, Brazil, Guinea
Fe	66,000	120	Australia, Canada, CIS*
Mn	800	100	CIS,* Gabon, S. Africa
Cr	400	100	CIS,* S. Africa, Zimbabwe
Cu	300	36	Chile, CIS,* USA, Zaïre
Zn	150	21	Australia, Canada, USA
Pb	71	20	Australia, Canada, CIS,* USA
Ni	47	55	Canada, CIS,* Cuba, New Caledonia
Sn	5	28	Brazil, China, Indonesia, Malaysia
U	2.8	58	Australia, CIS,* S. Africa, USA

*No individual breakdown is available for nations constituting the Commonwealth of Independent States (formerly the USSR).

21.3 Some Main Group Elements Extracted by Physical Methods: Nitrogen, Oxygen, and Sulfur

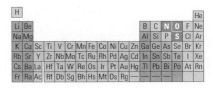

Some elements occur in nature in their elemental form, that is, not combined with any other element. These metals are aloof—gold, silver, mercury, and copper; so is sulfur, a nonmetal. Clearly, nitrogen and oxygen, the principal components of the atmosphere, are such elements, as are the noble gases in the atmosphere. Large quantities of nitrogen, oxygen, and, to a lesser extent, argon are extracted from the atmosphere by the liquefaction of air.

Elements from the Atmosphere

The composition of the atmosphere given in Table 21.3 shows that nitrogen is by far the most abundant component, its concentration being nearly four times that of oxygen, the next most abundant component. The gases of the atmosphere can be separated from one another by liquefying and fractionally distilling air in a process similar to the separation of petroleum fractions ***(⇐ p. 547)***, except at much lower temperatures and higher pressures.

At low temperatures and high pressures, gases no longer behave ideally and the attractive forces between molecules cause the gases in air to condense to liquids. Because of their different boiling points, the liquid components can then be separated from one another by distillation. Before pure oxygen and nitrogen can be obtained from air, water vapor and carbon dioxide are removed. The dry air is then compressed to more than 100 times normal atmospheric pressure, cooled to room temperature, and allowed to expand into a chamber. This expansion produces a cooling effect (the *Joule-Thomson effect*) because energy is required to overcome intermolecular forces as the molecules move farther apart. The expanding gas absorbs kinetic energy from the motion of its own molecules, which cools the gas. If this expansion is repeated and controlled properly, the expanding air cools to the point of liquefaction (Figure 21.8).

The temperature of the liquid air is usually well below the normal boiling points of nitrogen (−195.8 °C), oxygen (−183 °C), and argon (−189 °C). The very cold liquid air is again allowed to vaporize partially. Since N_2 is more volatile and has a lower boiling point than O_2 or Ar, the N_2 evaporates first and the remaining liquid becomes more concentrated in O_2 and Ar. This process, known as the *Linde process*, produces high-purity nitrogen (>99.5%) and oxygen (99.5%). Further processing produces pure Ar.

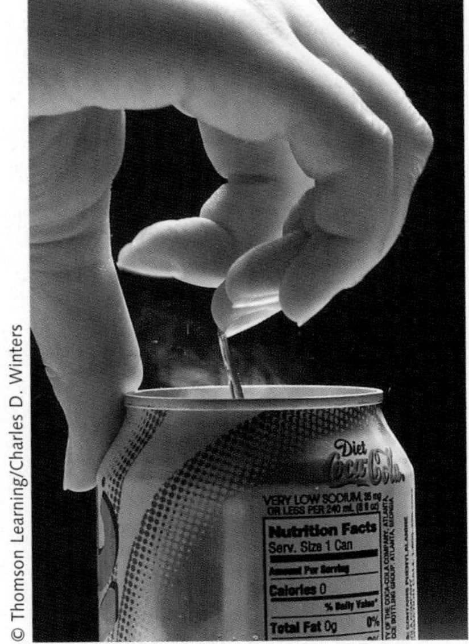

The Joule-Thomson effect. When the tab is opened on a can of carbonated beverage, the gases in the liquid are expelled rapidly enough to cool the water vapor to a liquid in the vicinity of the mouth of the bottle. The water vapor cools to form a tiny visible "cloud."

Table 21.3 Composition of Clean, Dry Air at Sea Level

Component	Percent by Volume	Component	Percent by Volume
N_2	78.09	He, Ne, Kr, Xe	0.002
O_2	20.948	CH_4	0.00015*
Ar	0.93	H_2	0.00005
CO_2	0.03*	All others combined	<0.00004

*Variable.

Ottmar Bierwagen/Spectrum Stock

Pure sulfur. Huge blocks of recently mined sulfur await shipment.

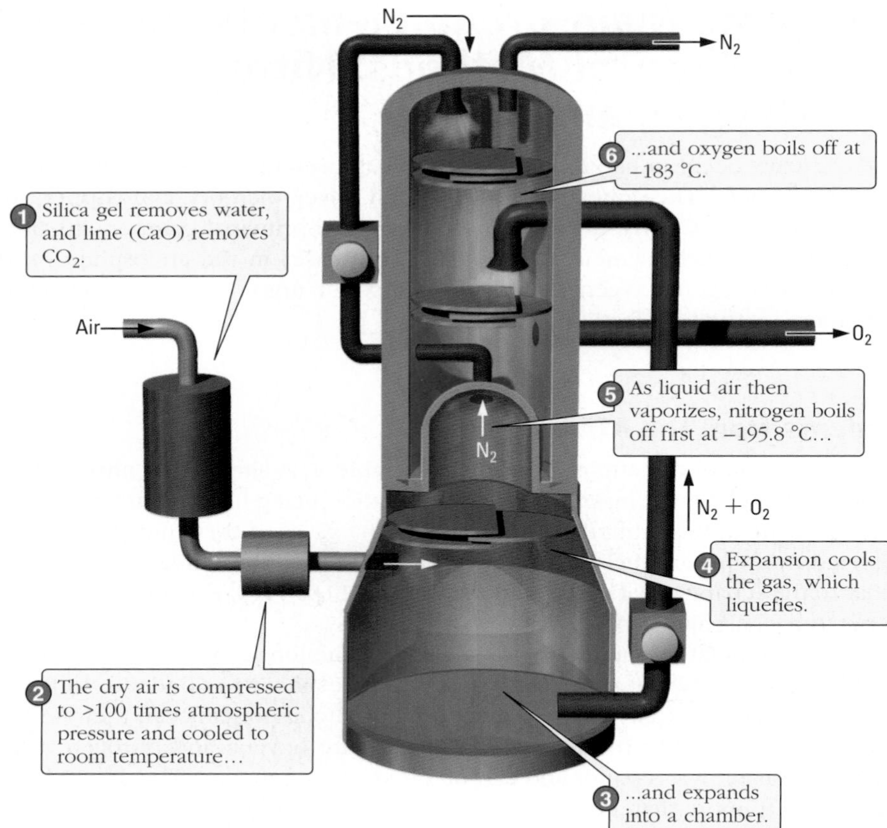

N_2
N_2
6 ...and oxygen boils off at −183 °C.
1 Silica gel removes water, and lime (CaO) removes CO_2.
Air
O_2
5 As liquid air then vaporizes, nitrogen boils off first at −195.8 °C...
N_2
$N_2 + O_2$
4 Expansion cools the gas, which liquefies.
2 The dry air is compressed to >100 times atmospheric pressure and cooled to room temperature...
3 ...and expands into a chamber.

Figure 21.8 Fractional distillation of air. Air can be liquefied by using low temperatures and high pressure. The components of the liquefied air are then separated by taking advantage of their distinctly different boiling points.

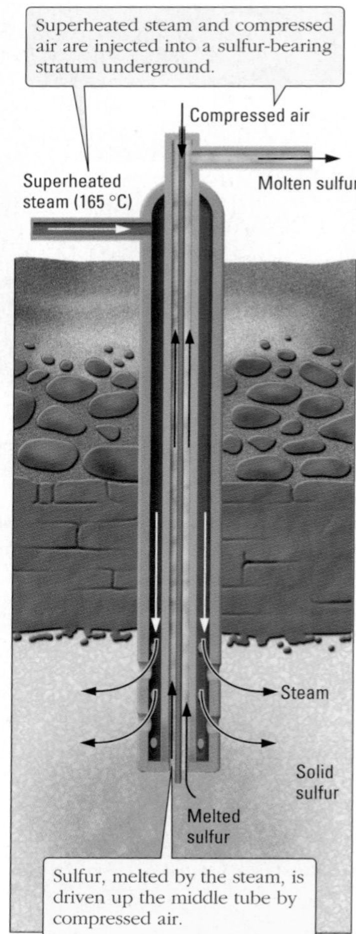

Superheated steam and compressed air are injected into a sulfur-bearing stratum underground.
Compressed air
Superheated steam (165 °C)
Molten sulfur
Steam
Solid sulfur
Melted sulfur
Sulfur, melted by the steam, is driven up the middle tube by compressed air.

Figure 21.9 The Frasch process for mining sulfur. The molten sulfur froths up out of the middle pipe. Most sulfur is used to manufacture sulfuric acid.

EXERCISE 21.4 Liquefied Gases

A cryogenic flask contains 5.0 L of liquid oxygen, which has a density of 1.4 g/mL. What volume will this oxygen occupy at STP if it is allowed to boil?

Sulfur

Sulfur, the element known biblically as brimstone, is a bright yellow solid. Very pure sulfur has been obtained from large deposits in salt domes along the coast of the Gulf of Mexico in the United States and Mexico, and in underground deposits in Poland. Such deposits of sulfur are believed to have been formed by bacterial reduction of sulfur in the naturally occurring mineral gypsum, which is hydrated calcium sulfate, $CaSO_4 \cdot 5\ H_2O$. Millions of tons of sulfur have been recovered from such deposits by the **Frasch process,** developed in the 1890s by Herman Frasch, a petroleum engineer (Figure 21.9). Most sulfur is now produced by extracting it from petroleum and natural gas. Removing the sulfur avoids the formation of sulfur dioxide, an atmospheric pollutant produced when petroleum burns. High-sulfur natural gas from Alberta, Canada, an especially large source of recovered sulfur, has now displaced the Frasch process as the chief source of sulfur.

21.4 Some Main Group Elements Extracted by Electrolysis: Sodium, Chlorine, Magnesium, and Aluminum

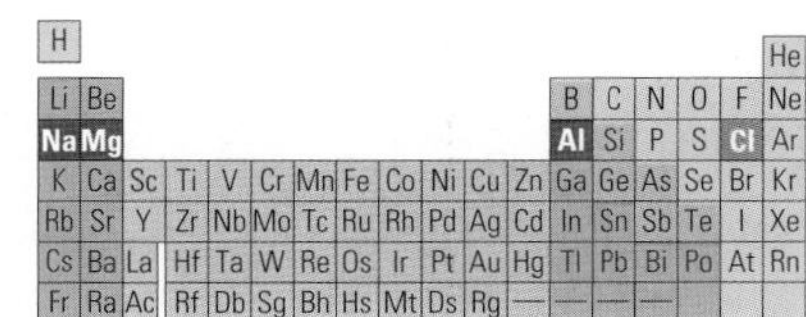

Chapter 19 described how electrolysis is used to force reactant-favored chemical reactions to occur (⇐ *p. 959*). Electrolysis is applied commercially on a vast scale to extract the elements sodium, magnesium, aluminum, and chlorine from their natural sources. These reactive elements exist naturally only in ionic form. Consequently, the metals must be obtained by reduction of their ions from their compounds, and chlorine must be oxidized from Cl^- to Cl_2.

Sodium

Sodium metal was discovered by Humphrey Davy in 1807 by electrolyzing molten NaOH. The half-reactions are

$$4\,OH^-(\text{in the melt}) \longrightarrow O_2(g) + 2\,H_2O(g) + 4\,e^- \quad \text{(anode, oxidation)}$$

$$4\,Na^+(\text{in the melt}) + 4\,e^- \longrightarrow 4\,Na(\text{in the melt}) \quad \text{(cathode, reduction)}$$

$$4\,Na^+(\text{in the melt}) + 4\,OH^-(\text{in the melt}) \longrightarrow 4\,Na(\text{in the melt}) + O_2(g) + 2\,H_2O(g) \quad \text{(net cell reaction)}$$

By the early 1900s, commercial uses for sodium metal had increased so that a large-scale production method was needed. In 1921 the Downs process was developed to meet this demand. In a Downs cell, molten NaCl is electrolyzed at 7 to 8 V and 25,000 to 40,000 A (Figure 21.10). The cell is filled with a 1:3 mixture of NaCl and $CaCl_2$. Pure NaCl is not used because of its high melting point (800 °C). Mixing the two salts lowers the melting point of the mixture to approximately 600 °C.

In the Downs cell, sodium is produced at a cathode made of copper or iron that surrounds a cylindrical graphite anode. Directly over the cathode is an inverted

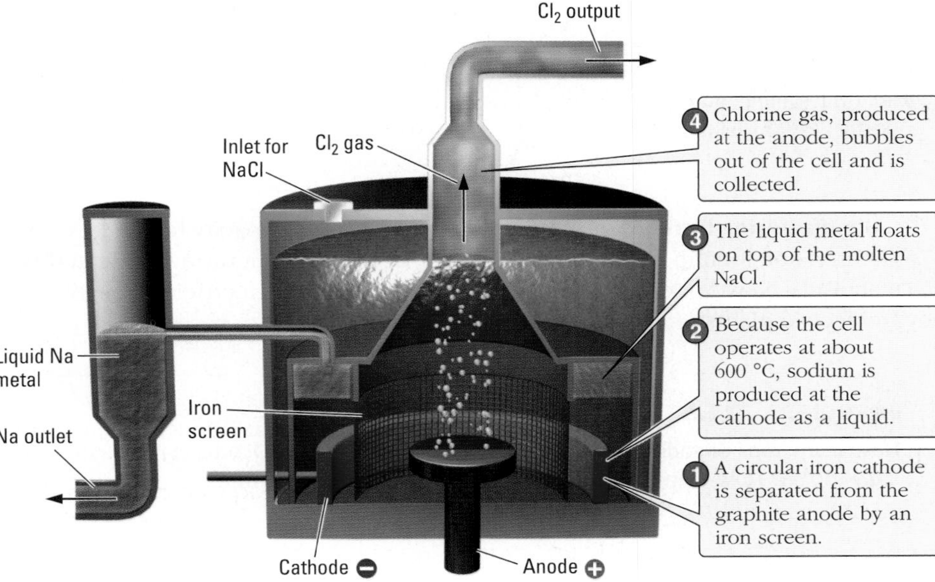
Cl2 output
Inlet for NaCl
Cl2 gas
Liquid Na metal
Na outlet
Iron screen
Cathode
Anode
4 Chlorine gas, produced at the anode, bubbles out of the cell and is collected.
3 The liquid metal floats on top of the molten NaCl.
2 Because the cell operates at about 600 °C, sodium is produced at the cathode as a liquid.
1 A circular iron cathode is separated from the graphite anode by an iron screen.

Figure 21.10 **The Downs cell for the electrolysis of molten NaCl.**

trough through which the molten sodium flows (sodium melts at 97.8 °C); liquid sodium is less dense than the molten mixture and therefore floats on top of it. Gaseous chlorine, the other product of the electrolysis, passes through an inverted cone of nickel metal extending through the molten salt mixture and is collected, cooled, and liquefied.

$$\text{Cl}^-(\text{in the melt}) \longrightarrow \tfrac{1}{2}\,\text{Cl}_2(\text{g}) + \text{e}^- \qquad \text{(anode, oxidation)}$$

$$\text{Na}^+(\text{in the melt}) + \text{e}^- \longrightarrow \text{Na}(\text{in the melt}) \qquad \text{(cathode, reduction)}$$

$$\text{Na}^+(\text{in the melt}) + \text{Cl}^-(\text{in the melt}) \longrightarrow \text{Na}(\text{in the melt}) + \tfrac{1}{2}\,\text{Cl}_2(\text{g}) \qquad \text{(net cell reaction)}$$

PROBLEM-SOLVING EXAMPLE 21.1 Titanium Production

Assume that the annual production of sodium metal in the United States is 76,000 tons. If half of this amount were used to produce titanium from $TiCl_4$, how many tons of titanium could be produced?

$$\text{TiCl}_4(\ell) + 4\,\text{Na}(\ell) \longrightarrow \text{Ti}(\ell) + 4\,\text{NaCl}(\ell)$$

Answer 2.0×10^4 tons

Strategy and Explanation Half of the sodium produced would be 38,000 tons. From the balanced equation, we see that 4 mol sodium are needed to produce 1 mol titanium. Using this mole ratio we can calculate the number of moles of titanium, from which we can then find the mass of titanium.

$$3.8 \times 10^4 \text{ ton Na}\left(\frac{2000 \text{ lb Na}}{1 \text{ ton Na}}\right)\left(\frac{454 \text{ g Na}}{1 \text{ lb Na}}\right)\left(\frac{1 \text{ mol Na}}{23.0 \text{ g Na}}\right) = 1.5 \times 10^9 \text{ mol Na}$$

$$1.5 \times 10^9 \text{ mol Na}\left(\frac{1 \text{ mol Ti}}{4 \text{ mol Na}}\right) = 3.8 \times 10^8 \text{ mol Ti}$$

and

$$3.8 \times 10^8 \text{ mol Ti}\left(\frac{47.9 \text{ g Ti}}{1 \text{ mol Ti}}\right)\left(\frac{1 \text{ lb Ti}}{454 \text{ g Ti}}\right)\left(\frac{1 \text{ ton Ti}}{2000 \text{ lb Ti}}\right) = 2.0 \times 10^4 \text{ ton Ti}$$

✓ **Reasonable Answer Check** It takes about 100 g sodium to produce about 50 g titanium, or approximately 100 tons of sodium per 50 tons of titanium, or about half the mass of titanium per mass of sodium. Therefore, 38,000 tons of sodium would produce about 19,000 tons of Ti, which is close to the calculated answer.

PROBLEM-SOLVING PRACTICE 21.1

Under the same conditions as in Problem-Solving Example 21.1, how many tons of sodium chloride are produced?

Nada Pecnik/Visuals Unlimited

A hydroelectric plant on the Niagara River near Niagara Falls, New York.

Manufacturing facilities in the United States have the capacity to produce about 76,000 tons of sodium metal per year. Much of the manufacturing is located near Niagara Falls, New York, because of the relatively low-cost electricity available from hydroelectric plants.

EXERCISE 21.5 The Downs Cell

How many tons of sodium can be produced in one day by a Downs cell operating at 2.0×10^4 A? How many tons of Cl_2 are produced in this same time?

Chlorine and Sodium Hydroxide

Chlorine is produced by the electrolysis of aqueous sodium chloride in the **chlor-alkali process;** the alkali produced by this process is sodium hydroxide. More than 24 billion pounds of sodium hydroxide are produced annually in the United States, along with a similar quantity of chlorine. These large amounts testify to the usefulness of these two products. The oxidizing and bleaching ability of chlorine is utilized in many industrial and everyday applications, and this element is a raw material in the manufacture of chlorine-containing chemicals. Sodium hydroxide is the base of choice in many industrial chemistry applications because it is inexpensive. It is also used widely to produce soaps, detergents, and other compounds.

The chlor-alkali process electrolyzes brine (saturated aqueous NaCl), as illustrated in Figure 21.11. Chloride ions are oxidized at the anode, and water is reduced at the cathode. The anode and cathode compartments are separated by a special polymeric membrane that allows only cations to pass through it. The brine solution is added to the anode compartment, and sodium ions pass through the membrane into the cathode compartment. The half-reactions are

$$2\,Cl^-(aq) \longrightarrow Cl_2(g) + 2\,e^- \qquad \text{(anode, oxidation)}$$

$$2\,H_2O(\ell) + 2\,e^- \longrightarrow 2\,OH^-(aq) + H_2(g) \qquad \text{(cathode, reduction)}$$

$$2\,Cl^-(aq) + 2\,H_2O(\ell) \longrightarrow Cl_2(g) + 2\,OH^-(aq) + H_2(g) \qquad \text{(net cell reaction)}$$

The reduction potential of Na^+ is more negative than that of water, so water, not sodium ions, is reduced.

The anode is specially treated titanium, and the cathode is stainless steel or nickel. The membrane is not permeable to water and acts as a salt bridge. Thus, as chloride ions are oxidized in the anode compartment, sodium ions must migrate from there to the cathode compartment to maintain charge balance. The resulting NaOH solution in the cathode compartment is 21% to 30% NaOH by weight.

The membrane cell was developed to replace the mercury cell that had been used previously in the chlor-alkali process. A major problem with mercury cells is the environmental damage caused by loss of mercury into the environment during normal operation of the cells. In the past, when mercury cells were cleaned, mercury was routinely allowed to run into neighboring bodies of water.

EXERCISE 21.6 NaOH Production

A chlor-alkali membrane cell operates at 2.0×10^4 A for 100. hours. How many tons of NaOH are produced in this time?

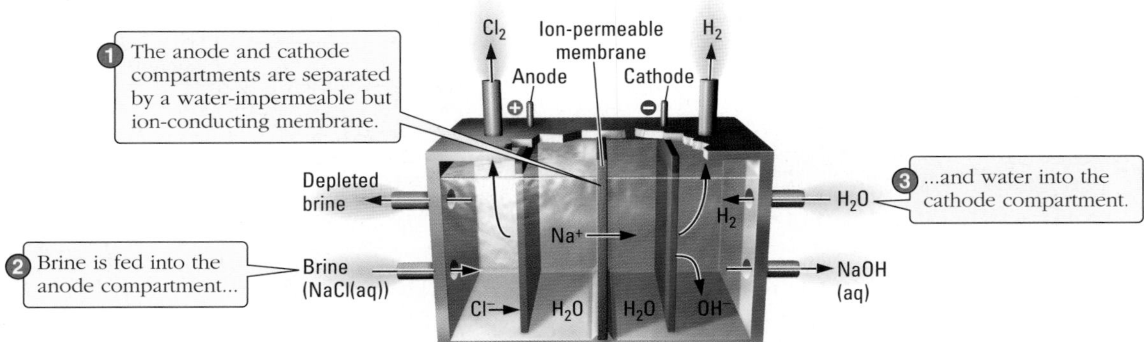

Figure 21.11 **A membrane cell used in the chlor-alkali process.**

Magnesium from Seawater

With a concentration of 1.35 mg Mg^{2+} per liter, the oceans provide a nearly limitless supply of magnesium, containing approximately 6 thousand tons of it per cubic mile. As with other reactive metals, the conversion of the metal ion to the metal is not a product-favored reaction, so electrolysis is required to extract magnesium metal.

The *Dow process* is used to reduce Mg^{2+} ions in seawater into magnesium metal (Figure 21.12). It begins with the precipitation of Mg^{2+} as its insoluble hydroxide ($K_{sp} = 1.5 \times 10^{-11}$). Hydroxide ions come from an inexpensive base, $Ca(OH)_2$, produced by roasting calcium carbonate in seashells to form calcium oxide, which then reacts with water to produce calcium hydroxide. The calcium hydroxide is added to seawater to precipitate magnesium hydroxide.

$$\underset{\text{seashells}}{CaCO_3(s)} \xrightarrow{\text{heat}} \underset{\text{lime}}{CaO(s)} + CO_2(g)$$

$$CaO(s) + H_2O(\ell) \longrightarrow Ca(OH)_2(aq)$$

$$Mg^{2+}(aq) + Ca(OH)_2(aq) \longrightarrow Mg(OH)_2(s) + Ca^{2+}(aq)$$

The magnesium hydroxide is filtered and neutralized by hydrochloric acid, another inexpensive chemical, to produce magnesium chloride.

$$Mg(OH)_2(s) + 2\,HCl(aq) \longrightarrow MgCl_2(aq) + 2\,H_2O(\ell)$$

The dried, anydrous magnesium chloride is then melted and electrolyzed in a steel pot, which serves as the cathode (Figure 21.13). The electrode reactions are

$$2\,Cl^-(\text{in the melt}) \longrightarrow Cl_2(g) + 2\,e^- \quad \text{(anode, oxidation)}$$

$$Mg^{2+}(\text{in the melt}) + 2\,e^- \longrightarrow Mg(\text{in the melt}) \quad \text{(cathode, reduction)}$$

$$Mg^{2+}(\text{in the melt}) + 2\,Cl^-(\text{in the melt}) \longrightarrow Mg(\text{in the melt}) + Cl_2(g) \quad \text{(net cell reaction)}$$

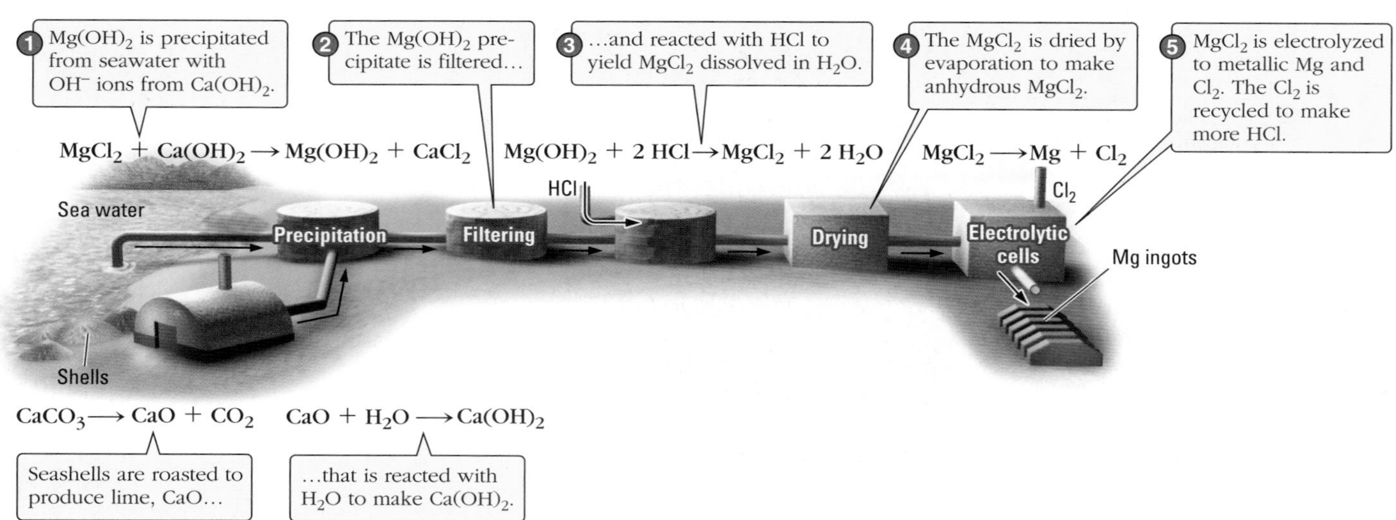

1 Mg(OH)2 is precipitated from seawater with OH− ions from Ca(OH)2.
2 The Mg(OH)2 precipitate is filtered...
3 ...and reacted with HCl to yield MgCl2 dissolved in H2O.
4 The MgCl2 is dried by evaporation to make anhydrous MgCl2.
5 MgCl2 is electrolyzed to metallic Mg and Cl2. The Cl2 is recycled to make more HCl.
MgCl2 + Ca(OH)2 → Mg(OH)2 + CaCl2
Mg(OH)2 + 2 HCl → MgCl2 + 2 H2O
MgCl2 → Mg + Cl2
Sea water
Precipitation
Filtering
HCl
Drying
Electrolytic cells
Cl2
Mg ingots
Shells
CaCO3 → CaO + CO2
CaO + H2O → Ca(OH)2
Seashells are roasted to produce lime, CaO...
...that is reacted with H2O to make Ca(OH)2.

Figure 21.12 **The steps for extracting magnesium metal from seawater.**

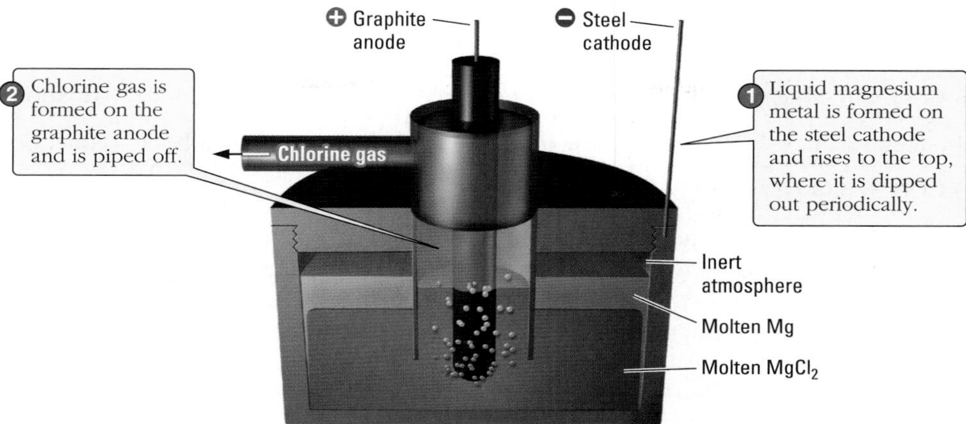

Figure 21.13 **Electrolysis of molten magnesium chloride.**

The molten magnesium is less dense than the molten $MgCl_2$ and floats at the surface, where the metal can be removed. Chlorine produced at the anode is converted to HCl by mixing Cl_2 with methane from natural gas and burning the mixture.

$$4\ Cl_2(g) + 2\ CH_4(g) + O_2(g) \longrightarrow 2\ CO(g) + 8\ HCl(g)$$

The HCl is recycled to neutralize $Mg(OH)_2$, which forms $MgCl_2$.

Aluminum Production

Aluminum is the most abundant metal in the earth's surface (7.4%), but it is present there as Al^{3+} ions, from which the metal must be obtained by reduction. Aluminum was first isolated in metallic form in 1825 by an expensive and potentially dangerous method—using metallic sodium or potassium to reduce Al^{3+} ions in aluminum chloride, $AlCl_3$. Thus, metallic aluminum was very expensive and considered to be a precious metal, like gold or platinum. An early use was in jewelry, including the Danish crown. In the 1855 Exposition in Paris, some of the first aluminum metal pieces produced were displayed along with the French crown jewels. In 1884, a 2.8-kg aluminum cap, produced by sodium reduction, topped the Washington Monument as ornamentation and the tip of a lightning rod system. At that time, the aluminum cap cost considerably more than the same mass of silver.

Napoleon II saw the advantages of using aluminum for military purposes because of its low density, and he commissioned studies on improving its production. Near the town of Les Baux, France, was a ready source of the aluminum-containing ore bauxite (Al_2O_3 combined with oxides of Si, Fe, and other elements); but how could aluminum be extracted from it readily? In 1886 Paul Héroult, a Frenchman, conceived an electrochemical process that is still used today. In a curious coincidence, Charles Martin Hall, an American, independently came up with the identical process two months earlier. Hence, the commercial method is known as the *Hall-Héroult process.* Just five years after the process was first used to produce aluminum commercially, the price of the metal plummeted from $12 per pound, a substantial sum at that time, to 70 cents per pound. What was once a precious metal soon became commonplace.

Remarkably, these two men, linked through their common discovery, also shared the same birth year (1863) and died the same year (1914).

In the Hall-Héroult process, metallic aluminum is obtained by electrolysis of Al_2O_3 dissolved in molten cryolite, Na_3AlF_6. The cryolite allows the electrolysis to be carried out at a lower temperature (1000 °C) than would be required for molten Al_2O_3 (m.p. 2030 °C). The aluminum oxide–cryolite mixture is electrolyzed in a cell using carbon anodes and a carbon cell lining that serves as the cathode (Figure 21.14). The half-reactions for extracting aluminum are

$$3\ C(s) + 3\ O_2(g) \longrightarrow 3\ CO_2(g) + 12\ e^- \quad \text{(anode, oxidation)}$$

$$4\ Al^{3+}(\text{in the melt}) + 12\ e^- \longrightarrow 4\ Al(\text{in the melt}) \quad \text{(cathode, reduction)}$$

$$4\ Al^{3+}(\text{in the melt}) + 3\ C(s) + 3\ O_2(g) \longrightarrow 4\ Al(\text{in the melt}) + 3\ CO_2(g) \quad \text{(net cell reaction)}$$

As the cell operates, molten aluminum deposits on the cathode and sinks to the bottom of the cell, from which it is removed periodically. Such cells operate at a very low voltage of 4.0 to 5.5 V, but at a very high current of 50,000 to 150,000 A.

Aluminum production uses extremely large quantities of electricity, so aluminum production plants are located near hydroelectric power sources, such as those in the Pacific Northwest, because electricity from hydroelectric plants is generally less expensive than that from fossil fuel power plants. Production of each kilogram of aluminum requires about 13 to 16 kWh (4.68×10^4 to 5.76×10^4 kJ) of electric energy, excluding that required to heat the molten mixture. Because of the high energy cost to extract aluminum metal from its ore, there is much interest in recycling aluminum beverage containers and other aluminum objects. It takes far less energy to process recycled aluminum than to produce the metal from bauxite. Putting this into perspective, you could run your television set for three hours on the energy saved by recycling just one aluminum can!

Charles Martin Hall
1863–1914

While a student at Oberlin College (OH), Charles Martin Hall became intrigued with trying to separate aluminum from its ores cheaply. When just 22 years old, using batteries and a blacksmith's forge, Hall succeeded in reducing Al_2O_3 dissolved in cryolite to metallic aluminum. To take advantage of his discovery, he formed the Aluminum Corporation of America (ALCOA), an enterprise that made Hall a multimillionaire.

PROBLEM-SOLVING EXAMPLE 21.2 Aluminum Production

If electricity costs \$0.080 per kilowatt-hour (kWh), how much does the electricity cost to produce 1.00 ton aluminum in a Hall-Héroult cell operating at 5.00 V? 1 kWh = 3.60×10^6 J; 1 V = 1 J/C.

Answer \$1100

Strategy and Explanation First, calculate the moles of aluminum produced.

$$1.00 \text{ ton Al}\left(\frac{2000 \text{ lb Al}}{1 \text{ ton Al}}\right)\left(\frac{454 \text{ g Al}}{1 \text{ lb Al}}\right)\left(\frac{1 \text{ mol Al}}{26.98 \text{ g Al}}\right) = 3.37 \times 10^4 \text{ mol Al}$$

Because the reduction reaction is $Al^{3+} + 3\ e^- \rightarrow Al$, 1 mol aluminum requires 3 mol electrons. Therefore,

$$\text{Total charge} = 3.37 \times 10^4 \text{ mol Al} \times \frac{3 \text{ mol e}^-}{1 \text{ mol Al}} \times \frac{9.65 \times 10^4 \text{ C}}{1 \text{ mol e}^-} = 9.74 \times 10^9 \text{ C}$$

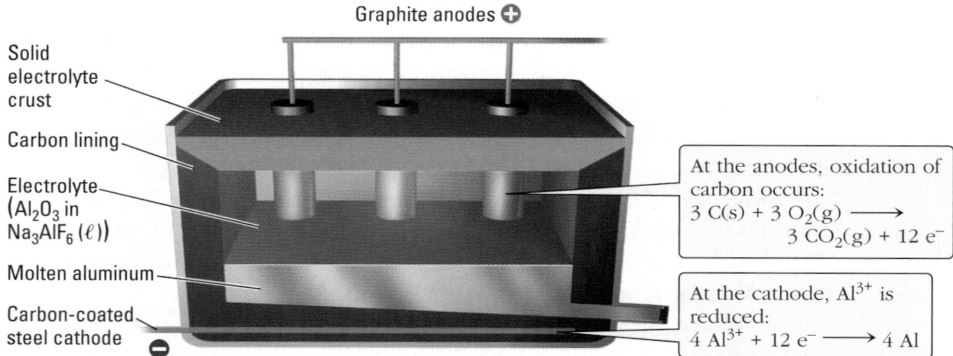

Figure 21.14 A Hall-Héroult process electrolytic cell. Molten aluminum is drawn off from the bottom of the cell into molds.

The number of kilowatt-hours is

$$9.74 \times 10^9 \text{ C} \times \frac{5.00 \text{ J}}{1 \text{ C}} \times \frac{1 \text{ kWh}}{3.60 \times 10^6 \text{ J}} = 1.35 \times 10^4 \text{ kWh}$$

$$\text{Cost} = 1.35 \times 10^4 \text{ kWh} \times \frac{\$0.080}{1 \text{ kWh}} = \$1.1 \times 10^3, \text{ or } \$1100.$$

PROBLEM-SOLVING PRACTICE 21.2

How long would it take a Hall-Héroult cell operating at 1.00×10^5 A to produce 1.00 ton aluminum metal?

21.5 Some Main Group Elements Extracted by Chemical Oxidation-Reduction: Phosphorus, Bromine, and Iodine

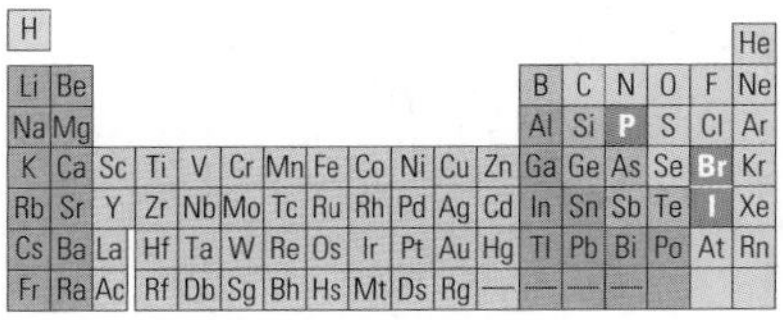

Phosphorus, bromine, and iodine are all produced by chemical redox reactions.

Phosphorus

Elemental phosphorus is extracted from phosphate-bearing rock by heating the rock with sand (SiO_2) and coke in an electric furnace (Figure 21.15). At 1400 to 1500 °C, gaseous phosphorus is formed and evaporates from the mixture, leaving behind insoluble calcium silicate.

$$2\ Ca_3(PO_4)_2(\ell) + 10\ C(s) + 6\ SiO_2(\ell) \longrightarrow P_4(g) + 10\ CO(g) + 6\ CaSiO_3(\ell)$$

The mixture of phosphorus vapor and carbon monoxide gas is passed through water, where the phosphorus condenses and the CO bubbles out.

CONCEPTUAL EXERCISE 21.7 Phosphorus Extraction

The extraction of phosphorus from phosphate rock involves oxidation and reduction. Identify which element is oxidized and which is reduced.

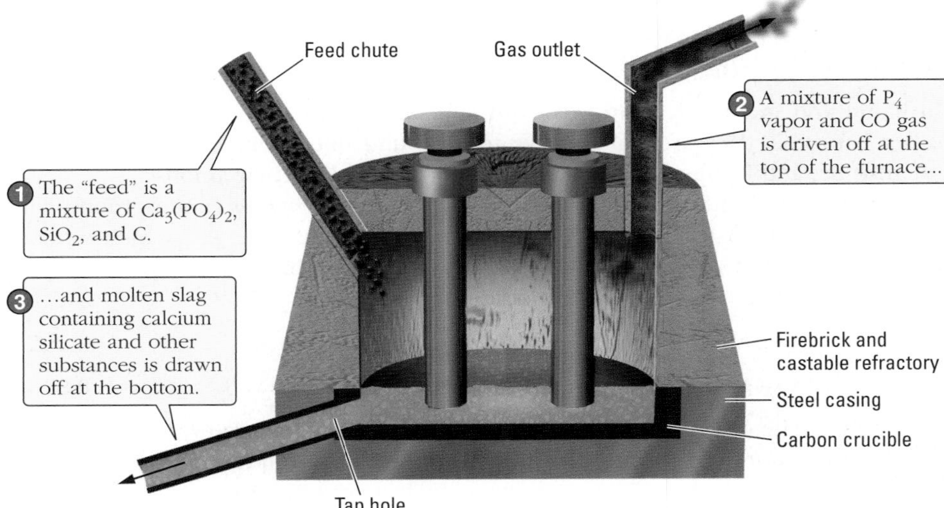

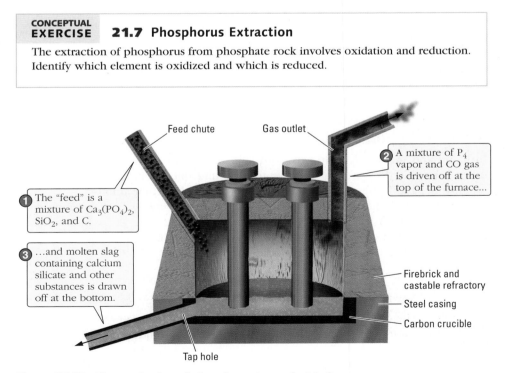

Figure 21.15 **The production of phosphorus in an electric furnace.**

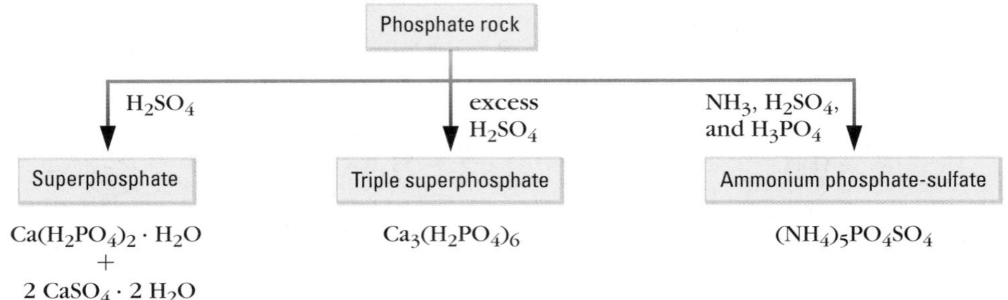

Figure 21.16 **Fertilizers produced from phosphate rock.**

About 90% of the elemental phosphorus produced is oxidized subsequently in air to P_4O_{10}, which reacts with water to produce phosphoric acid, H_3PO_4.

$$P_4(s) + 5\ O_2(g) \longrightarrow P_4O_{10}(s)$$

$$P_4O_{10}(s) + 6\ H_2O(\ell) \longrightarrow 4\ H_3PO_4(aq)$$

Some phosphoric acid is used in soft drinks, baking powder, and detergents.

The principal use of phosphate rock is to make fertilizers directly rather than to produce the element (Figure 21.16). Phosphate rock is reacted with sulfuric acid and converted into a soluble fertilizer. The mixture of hydrated calcium dihydrogen phosphate and calcium sulfate is called "superphosphate."

$$\underset{\text{phosphate rock}}{Ca_3(PO_4)_2(s)} + 2\ H_2SO_4(aq) + 5\ H_2O(\ell) \longrightarrow \underbrace{Ca(H_2PO_4)_2 \cdot H_2O(s) + 2\ CaSO_4 \cdot 2\ H_2O(s)}_{\text{superphosphate}}$$

Fertilizer is also made from phosphoric acid by neutralizing the acid with ammonia to form ammonium hydrogen phosphate, $(NH_4)_2HPO_4$.

EXERCISE 21.8 Phosphorus in Phosphate Rock

Calculate the mass percent of phosphorus present in another form of phosphate rock, hydroxyapatite, $Ca_5(PO_4)_3OH$.

Bromine and iodine.

Bromine and Iodine

Bromine and iodine are halogens with similar but different properties. Like the other halogens, they are too reactive to be found uncombined in nature. Consequently, Br_2 and I_2 must be extracted by the oxidation of their anions.

Bromine and iodine are extracted from seawater or *brines* (underground natural salt water deposits) by treating the solution with chlorine gas, which oxidizes Br^- to Br_2 and I^- to I_2. This is a case of a more reactive halogen (chlorine) displacing a less reactive one (bromide or iodide) from solution.

$$Cl_2(g) + 2\ Br^-(aq) \longrightarrow Br_2(\ell) + 2\ Cl^-(aq) \qquad E° = +0.292\ V$$

$$Cl_2(g) + 2\ I^-(aq) \longrightarrow I_2(aq) + 2\ Cl^-(aq) \qquad E° = +0.823\ V$$

The large positive voltages indicate both as very product-favored reactions.

Another source of iodine is iodate ions, IO_3^-, in Chilean ore deposits. Iodate is converted to I_2 in a two-step process using hydrogen sulfite ions.

Photos: © Thomson Learning/Charles D. Winters

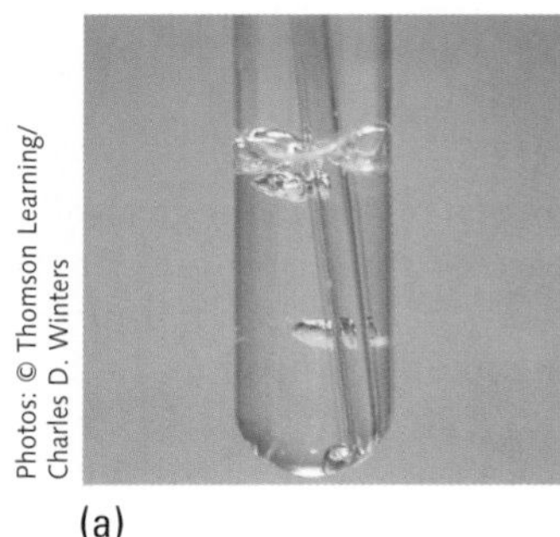
(a)

(b)

(c)

Displacement of Br^- and I^- by Cl_2. (a) Chlorine gas is bubbled into a colorless NaBr or NaI solution. (b) I^- is oxidized by Cl_2 to give I_2. (c) Br^- is oxidized by Cl_2 to give Br_2. Carbon tetrachloride, a dense liquid, is added and extracts the Br_2 and I_2 from the upper aqueous layer into the bottom CCl_4 layer, concentrating the Br_2 *(orange)* and I_2 *(violet)*.

Post Street Archives, Michigan

Herbert H. Dow
1866–1930

The first to produce bromine by the electrolysis of brine (1891) was Herbert H. Dow. In the 1920s, the demand for bromine rose sharply in order to make ethylene bromide, which was starting to be used in the higher octane gasoline required by high-performance automobile engines. Dow realized that ethylene bromide demand would be so great that brine sources could not supply enough bromine. He told the head of General Motors that to meet the demand, ". . . we'll have to go to sea and extract bromine from ocean water."* Herbert Dow died four years before achieving this goal, which was accomplished by his son Willard.
*Brandt, E. N. *Chemical Heritage,* Vol. 18, Number 3, Fall 2000; p. 39.

Step 1: $2\,IO_3^-(aq) + 6\,HSO_3^-(aq) \longrightarrow 2\,I^-(aq) + 6\,HSO_4^-(aq)$

Step 2: $5\,I^-(aq) + IO_3^-(aq) + 3\,H_3O^+ + 3\,HSO_4^-(aq) \longrightarrow 3\,I_2(aq) + 3\,SO_4^{2-}(aq) + 3\,H_2O(\ell)$

PROBLEM-SOLVING EXAMPLE 21.3 Oxidation-Reduction Reactions

Identify the oxidizing and reducing agents in Step 1 of the extraction of iodine from IO_3^--bearing ores.

Answer Oxidizing agent: IO_3^-; reducing agent: HSO_3^-

Strategy and Explanation Recall from Chapter 19 that reduction involves a decrease in oxidation number due to a gain of electrons. The reduction requires a reducing agent, which in this case is hydrogen sulfite ion, HSO_3^-, which donates the electrons. The +4 oxidation number (state) of sulfur in HSO_3^- is increased to +6 in HSO_4^-. This is oxidation, an increase in oxidation number, indicating a loss of electrons. Thus, the reducing agent (HSO_3^-) is oxidized (+4 sulfur to +6 sulfur), while simultaneously iodine in the oxidizing agent, IO_3^-, is reduced (from +5 to −1).

PROBLEM-SOLVING PRACTICE 21.3

Identify the oxidizing and reducing agents in Step 2 of the extraction of I_2 from Chilean ores.

CONCEPTUAL EXERCISE 21.9 Bromine Conversion

Use the terms *oxidation, reduction, oxidizing agent,* and *reducing agent* to explain the extraction of bromine from brines.

EXERCISE 21.10 Iodine Reaction

Calculate $E°$ for the reaction of $I_2(s)$ with $Br^-(aq)$. What does the value of $E°$ indicate about using $I_2(s)$ to oxidize $Br^-(aq)$ to $Br_2(\ell)$?

© Thomson Learning/Charles D. Winters

Chunks of metallic potassium in oil. A layer of oil is used as protective covering to prevent the potassium from reacting with oxygen in the air.

21.6 A Periodic Perspective: The Main Group Elements

In this section we will discuss some of the properties and uses of the main group elements, Groups 1A–8A (1, 2, 13–18), from the perspective of their positions in the periodic table. In addition to providing a general overview of each group, one or more elements in a group will be highlighted for more detailed description.

Group 1A(1): The Alkali Metals

The alkali metals make up the leftmost group of the periodic table. Their densities, melting points, boiling points, atomic radii, and ionic radii are illustrated below. The densities are low because the alkali metals have relatively large atomic radii compared to their molar masses. Weak metallic bonding is responsible for their softness and relatively low melting points.

The chemical behavior of the alkali metals is dominated by loss of the ns^1 outer electron leading solely to the formation of M^+ ions. Therefore, most alkali metal compounds are ionic, except for organometallic compounds containing an alkali metal-to-carbon bond. The Group 1A elements react with air, water, and most nonmetals. The reactions of the heavier alkali metals are particularly vigorous,

1A
(1)

Element	Ionic radius, pm	Atomic radius, pm	Density (g/mL)	MP (°C)	BP (°C)
3 **Li** Lithium	Li^+ 90	Li 152	0.53	181	1347
11 **Na** Sodium	Na^+ 116	Na 186	0.96	98	881
19 **K** Potassium	K^+ 152	K 227	0.85	63	766
37 **Rb** Rubidium	Rb^+ 166	Rb 248	1.53	39	688
55 **Cs** Cesium	Cs^+ 181	Cs 265	1.900	28	705
87 **Fr** Francium	Fr^+ 194	Fr (~270)			

Photos: © Thomson Learning/Charles D. Winters and Jim Marshall, Univ. of N. Texas

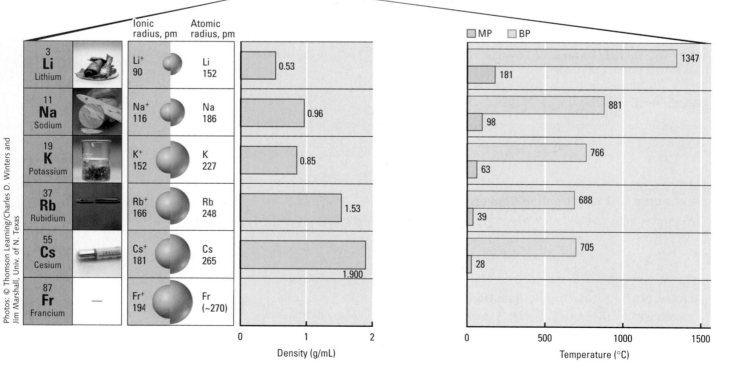

Group 1A(1) elements, the alkali metals: Li, Na, K, Rb, Cs, Fr.

Table 21.4 Reactions of Alkali Metals

Group 1A Metal	Combining Substance	Reaction
Li	Oxygen	$4\ Li(s) + O_2(g) \longrightarrow 2\ Li_2O(s)$
Na		$2\ Na(s) + O_2(g) \longrightarrow Na_2O_2(s)$
K, Rb, Cs		$M(s) + O_2(g) \longrightarrow MO_2(s)$
All	Halogens	$2\ M(s) + X_2 \longrightarrow 2\ MX(s)$; X = F, Cl, Br, I
All	Sulfur	$2\ M(s) + S(s) \longrightarrow M_2S(s)$
Li	Nitrogen	$6\ Li(s) + N_2(g) \longrightarrow 2\ Li_3N(s)$
All	Water	$2\ M(s) + 2\ H_2O(\ell) \longrightarrow 2\ M^+(aq) + 2\ OH^-(aq) + H_2(g)$

even explosive. Reactions of alkali metals with oxygen, sulfur, the halogens, and water are summarized in Table 21.4. The uses of some alkali metal compounds are given in Table 21.5.

The product of the reaction of an alkali metal with oxygen is dependent on the alkali metal, as seen from Table 21.4. Lithium is the only Group 1A metal that reacts directly with oxygen to form in good yield the normal oxide (M_2O) containing M^+ metal ions and O^{2-} oxide ions. In contrast, sodium reacts directly with oxygen to form predominantly sodium peroxide (Na_2O_2), which contains Na^+ and O_2^{2-} (peroxide) ions. The remaining alkali metals produce the metal superoxide (MO_2) in which M^+ and superoxide O_2^- ions are present. The peroxide and superoxide ions contain two covalently bonded oxygen atoms, with superoxide ions having one fewer electron than peroxide ions. Potassium superoxide is a quick source of oxygen used in emergency breathing apparatus, such as for firefighters and miners in rescue circumstances where the concentration of oxygen is low. Water vapor in the breath reacts with superoxide ions to produce oxygen and potassium hydroxide; the latter removes exhaled carbon dioxide.

The molecular orbital theory (← *p. 365*) can be used to describe bonding in O_2^{2-} and O_2^- ions.

$$4\ KO_2(s) + 2\ H_2O(\ell) \longrightarrow 4\ KOH(s) + 3\ O_2(g)$$

$$2\ KOH(s) + CO_2(g) \longrightarrow K_2CO_3(s) + H_2O(\ell)$$

Table 21.5 Uses of Alkali Metals and Some of Their Compounds

Element or Compound	Uses
Lithium	Lithium batteries for computers, cell phones
Lithium carbonate (Li_2CO_3)	Treatment of bipolar disorder
Sodium	Nuclear reactor coolant, manufacture of Ti
Sodium chloride (NaCl)	Production of sodium metal, chlorine, NaOH
Sodium hydroxide (NaOH)	Soaps and detergents, pulp and paper industry, bleach preparation, widely used industrial base
Sodium carbonate (Na_2CO_3)	Glass manufacturing, water softening, detergents, reduction of SO_2 stack gas emission
Sodium hydrogen carbonate ($NaHCO_3$)	Baking powder, baking soda, fire extinguishers, pharmaceuticals
Potassium nitrate (KNO_3)	Gunpowder, fireworks; strong oxidizing agent
Potassium superoxide (KO_2)	Oxygen source in emergency breathing apparatus
Rubidium and cesium	Photoelectric cells

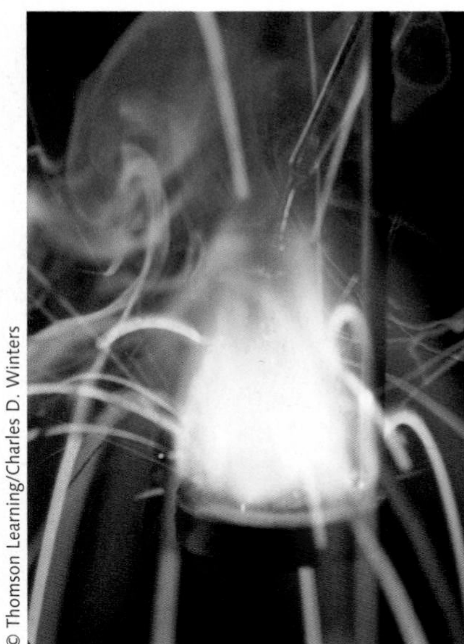

Figure 21.17 Reaction of potassium with water. When water is dripped on to potassium metal, a violent reaction occurs.

EXERCISE 21.11 Lewis Structures

Write the Lewis structures for oxide, peroxide, and superoxide ions. Write the molecular orbital diagrams for peroxide ion and superoxide ion.

Halogens react directly with alkali metals to produce stable binary halide salts whose lattice energies are substantial (⇐ *p. 321*). Examples include NaCl, KBr, and CsI. Likewise, the Group 1A metals all react directly with sulfur to form ionic sulfides with the general formula M_2S. Lithium is the only Group 1A metal that reacts directly with nitrogen gas, forming an ionic nitride, Li_3N.

Water reacts vigorously with the alkali metals, especially the heavier ones (Figure 21.17), to produce hydrogen gas plus a solution of the metal hydroxide. The reaction of sodium with water is exemplary:

$$Na(s) + 2\,H_2O(\ell) \longrightarrow 2\,Na^+(aq) + 2\,OH^-(aq) + H_2(g)$$

This reaction is highly exothermic: $\Delta H° = -367.6$ kJ.

Because sodium metal is a strong reducing agent, it is used to obtain metals from metal halides. In particular, titanium, an element essential in aircraft production, can be prepared from its chloride by reduction with sodium.

$$TiCl_4(s) + 4\,Na(s) \longrightarrow Ti(s) + 4\,NaCl(s)$$

A major use for sodium metal was in the production of tetraethyllead $[Pb(C_2H_5)_4]$, once used as an octane enhancer in leaded gasoline. Although leaded gasoline is still used in some countries, tetraethyllead is banned as a gasoline additive in the United States. Consequently, sodium production has declined.

Liquid sodium has high thermal conductivity and an anomalously high heat capacity. Metallic sodium has a low melting point and can be liquefied easily. These properties make liquid sodium an excellent heat-exchange liquid in nuclear reactors (⇐ *p. 999*).

Group 2A(2): The Alkaline Earth Metals

Like their Group 1A neighbors, the alkaline earth metals (Group 2A) are silvery white, ductile, and malleable metals that are a bit harder than the alkali metals. The densities, melting points, boiling points, atomic radii, and ionic radii of the alkaline earth metals are shown on page 1039. The Group 2A elements, like the adjacent Group 1A elements, show regular changes in properties down the group. Chemical reactivity increases down the group.

The alkaline earth metals are characterized chemically by the loss of the ns^2 outer electrons to yield 2+ ions. Beryllium is the exception, forming no predominantly ionic compounds due to the high charge density of Be^{2+} ions, which are found only as hydrated ions such as $[Be(H_2O)_4]^{2+}$. Anhydrous beryllium compounds are covalent and many are polymeric solids, such as $BeCl_2$, which has a Cl-bridging chain structure. Beryllium and its compounds are poisonous.

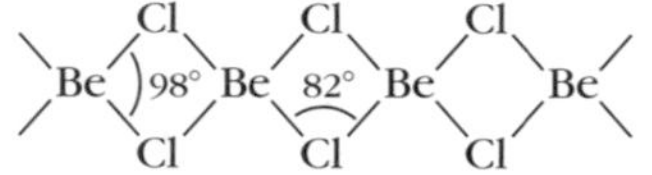

The polymeric structure of beryllium chloride.

The toxicity of Be compounds is considered to be due to beryllium displacing Mg^{2+} ions from Mg-based enzymes.

Some reactions of the alkaline earth metals are summarized in Table 21.6. Like the alkali metals, the lighter alkaline earth metals (Mg and Ca) react with oxygen to form oxides, while the heavier metals (Sr and Ba) form peroxides. Beryllium oxide forms directly only at temperatures higher than 600 °C. The oxides (except BeO) react with water to form the corresponding hydroxides. The vigor of the reaction of the metals with water increases down the group: beryllium does not react; magnesium does so only with steam above 100 °C; calcium and strontium react slowly with liquid water at room temperature, but barium reacts rapidly. All of the Group 2A elements react directly with nitrogen to form nitrides, which react with water to form

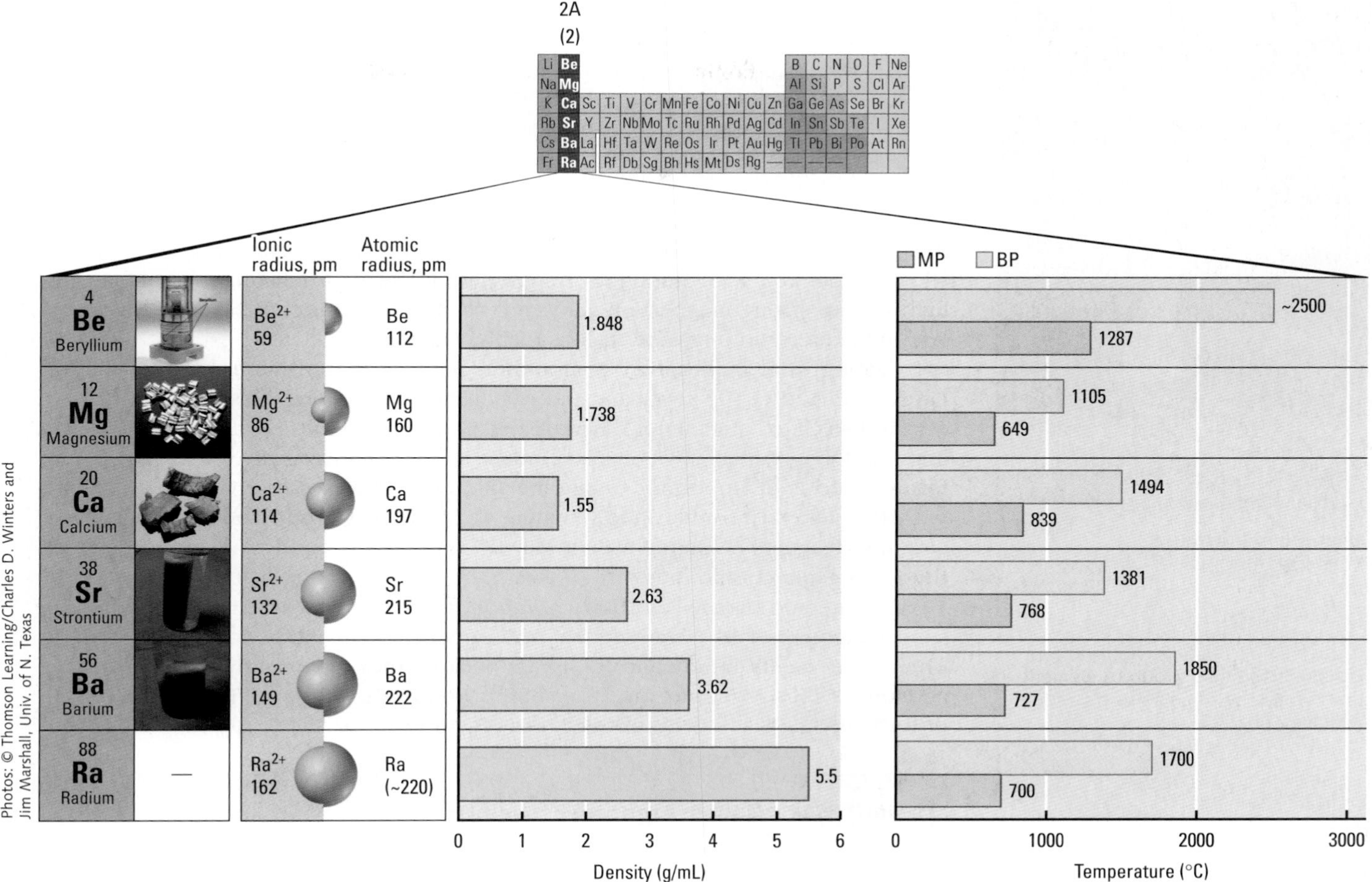

Group 2A(2) elements, the alkaline earths.

aqueous hydroxides and release ammonia. Direct halogenation of the metals forms ionic halide salts of the general formula MX_2. The alkaline earth metals, except beryllium, react directly with hydrogen to form hydrides (MH_2) and with carbon to form carbides (MC_2).

The carbide ion, C_2^{2-}, has the Lewis structure $:C{\equiv}C:^{2-}$.

Table 21.6 Some Reactions of Alkaline Earth Elements

Group 2A Metal	Combining Substance	Reaction
Be, Mg, Ca	Oxygen	$2\,M(s) + O_2(g) \longrightarrow 2\,MO(s)$
Sr, Ba	Oxygen	$M(s) + O_2(g) \longrightarrow MO_2(s)$
All	Halogens	$M(s) + X_2 \longrightarrow MX_2(s)$; X = F, Cl, Br, I
All (high temp.)	Nitrogen	$3\,M(s) + N_2(g) \longrightarrow M_3N_2(s)$; M = Be, Mg, Ca, Sr, Ba, Ra
Ca, Sr, Ba	Water	$M(s) + 2\,H_2O(\ell) \longrightarrow M(OH)_2(aq) + H_2(g)$
Mg, Ca, Sr, Ba	Hydrogen	$M(s) + H_2(g) \longrightarrow MH_2(s)$
Mg, Ca, Sr, Ba	Carbon	$M(s) + 2\,C(s) \longrightarrow MC_2(s)$

Magnesium burning.

EXERCISE 21.12 Group 2A Compounds

Write the formulas for these compounds.

(a) Calcium oxide (b) Barium peroxide
(c) Strontium nitride (d) Calcium carbide

Magnesium metal has limited use in flashbulbs, fireworks, and flares because the metal burns with a brilliant white light. Its most important use is in making alloys, principally with aluminum. Magnesium is the least dense structural material; lightweight, strong magnesium alloys are used to make aircraft wheels, truck bodies, and ladders, among other things (Table 21.7).

Calcium and magnesium compounds are used extensively, as noted below. Limestone ($CaCO_3$) is abundant and, when heated, decomposes to produce lime (CaO), also called quicklime. When treated with water, lime is converted into slaked lime, $Ca(OH)_2$, which is widely used industrially as an inexpensive base. Magnesium forms a series of important organometallic compounds called Grignard reagents with the general formula RMgX, where R is an alkyl group and X is a halogen. Grignard reagents contain covalent Mg—C and Mg—X bonds and are used to synthesize organic compounds.

Lithium and magnesium have a *diagonal relationship*, one in which a pair of similar elements falls on a diagonal line between adjacent periodic table groups. The two elements are chemically similar because they have nearly the same atomic and ionic radii (Li 152 pm, Mg 160 pm; Li^+ 90 pm, Mg^{2+} 86 pm). For example, each of these metals reacts with nitrogen directly to form a nitride.

Lithium is the only Group IA element that forms a nitride by direct reaction with N_2.

EXERCISE 21.13 Lighting Things Up

When magnesium metal burns in air, magnesium nitride and magnesium oxide are produced.

(a) Write the formula for magnesium nitride.
(b) Write a balanced chemical equation for the formation of magnesium nitride from the elements.

Table 21.7 Uses of Alkaline Earth Elements and Some of Their Compounds

Element or Compound	Uses
Beryllium metal	X-ray tube windows
Magnesium metal	Alloys for building materials
Magnesia (MgO)	Firebrick manufacturing, thermal insulator
Lime (CaO)	Steel and paper manufacturing, water treatment
Calcium carbonate ($CaCO_3$)	Limestone and marble for building materials, toothpaste abrasive, antacid
Barium metal	Spark plugs (alloy)
Barium sulfate ($BaSO_4$)	X-ray imaging

Group 3A(13): Boron, Aluminum, Gallium, Indium, Thallium

All elements of Group 3A have an ns^2np^1 outer electron configuration leading to +3 as the stable oxidation state except thallium, for which the +1 oxidation state is the more stable. *Where multiple oxidation states are possible for elements in a group,*

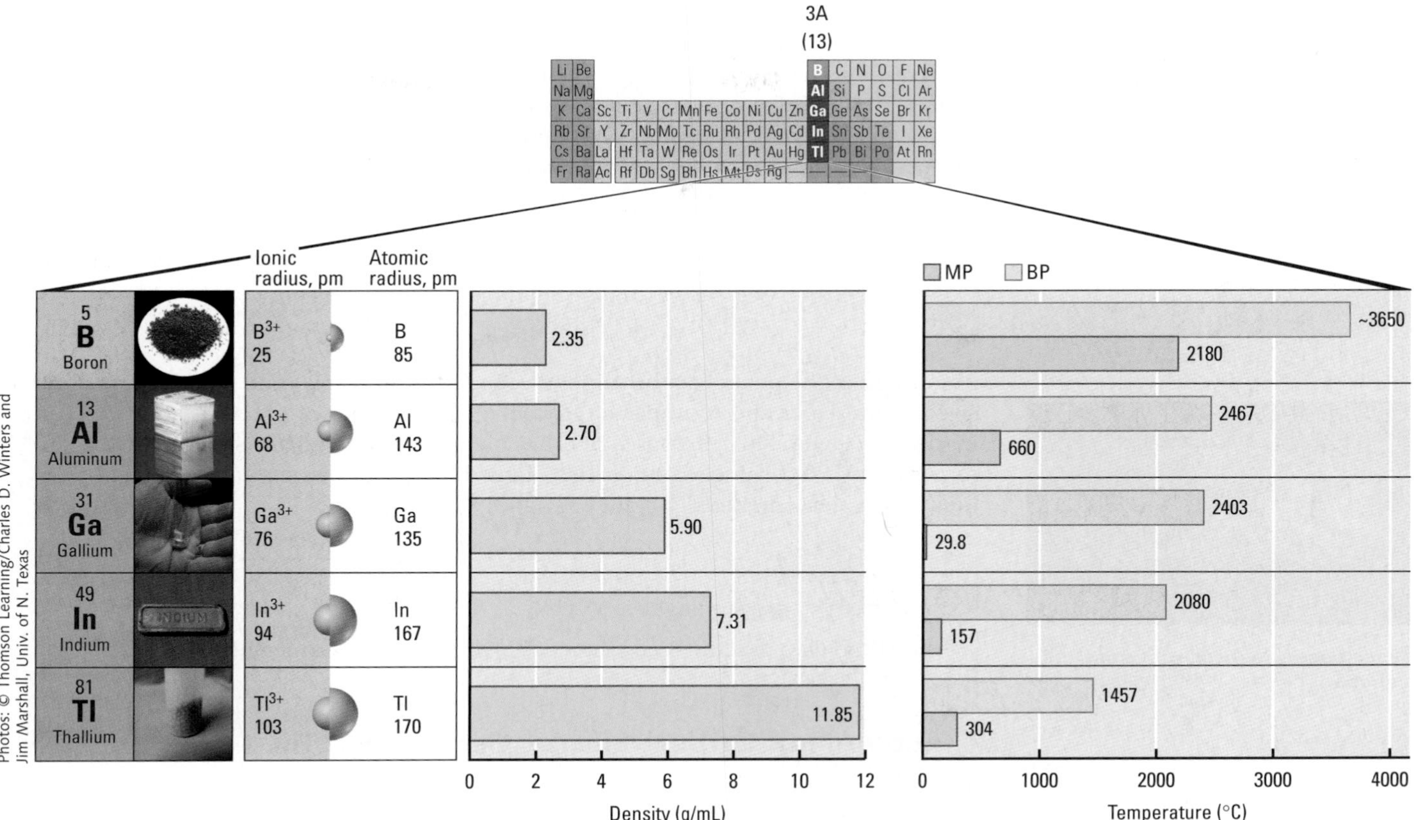

Group 3A(13) elements.

the lower oxidation state is usually the more favored in the heavier elements of the group.

Some of the physical and atomic properties of the elements in Group 3A are given above. The elements of this group exhibit a wider range of properties than those in Group 1A or 2A. Boron, the first member of the group, is a metalloid, an anomaly in a group where all other elements are silvery white metals. Aluminum is more representative of the group and for that reason is discussed in more detail here. The extraction of aluminum metal from bauxite ore was described in Section 21.4. Aluminum is the most abundant metal ion (Al^{3+}) in the earth's crust, exceeded in elemental abundance only by oxygen and silicon. As with Li and Mg, a diagonal relationship exists between Be and Al and between B and Si due to similarities in electronegativity and effective nuclear charge ***(⟵ p. 313)*** within each pair of elements. Table 21.8 summarizes some reactions of the Group 3A elements.

Table 21.8 Some Reactions of Group 3A Elements

Group 3A Element	Combining Substance	Reaction
Al, Ga, In	Oxygen	$4\ M(s) + 3\ O_2(g) \longrightarrow 2\ M_2O_3(s)$
B, Al, Ga, In	Halogens	$2\ M(s) + 3\ X_2 \longrightarrow 2\ MX_3$; X = F, Cl, Br, I
Ga, In (high temperature)	Water	$2\ M(s) + 6\ H_2O(g) \longrightarrow 2\ M^{3+}(aq) + 6\ OH^-(aq) + 3\ H_2(g)$

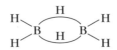

The B—H—B bridge bonding in diborane, B_2H_6.

Boron and hydrogen form an extensive series of covalently bonded hydrides called *boranes* with the general formulas B_nH_{n+4} or B_nH_{n+6} such as B_2H_6 and B_5H_{11}, respectively. These compounds contain boron atoms bridged by hydrogen atoms in what is described as a three-center-two-electron bond.

Aluminum is an economically important, useful metal because of its low density (2.70 g/cm^3) and high strength when alloyed. It can be fashioned into wire, food wrapping sheets, stepladders, aircraft and automotive parts, and many other useful items. Aluminum metal resists corrosion because a transparent, chemically inactive film of aluminum oxide clings avidly to the metal's surface and protects the metal beneath it from further oxidation.

$$4\,Al(s) + 3\,O_2(g) \longrightarrow 2\,Al_2O_3(s)$$

Aluminum oxide occurs as the mineral corundum, which is used widely as an abrasive in sandpaper and toothpaste (Table 21.9). A number of precious gems are primarily Al_2O_3 with small amounts of other metal ions strategically substituted for aluminum ions. Red rubies contain Cr^{3+} ions, blue sapphires contain Fe^{2+} and Fe^{3+} ions, and green emeralds contain Cr^{3+} and V^{3+} ions.

Gemstones: sapphire, ruby, and emerald.

CONCEPTUAL EXERCISE

21.14 Al_4^{4-}

The Al_4^{4-} ion has recently been synthesized. Write the Lewis structure of this ion.

Group 4A(14): Carbon, Silicon, Germanium, Tin, Lead

The relative uniformity of the Group 1A alkali metals is absent from the Group 4A elements, which display the full range of element types from nonmetals to metals. Carbon, the first member, is a nonmetal; silicon and germanium are both metalloids; and tin and lead are metals. Some physical and atomic properties of the Group 4A elements are shown on page 1043.

The Group 4A elements all have an ns^2np^2 outer electron configuration. Promotion of the ns^2 electrons into empty np orbitals allows for hybridization and, through electron sharing, the formation of four bonds as in compounds such as CH_4, $SiBr_4$, $SnCl_4$, and $Pb(C_2H_5)_4$. The bonding in Group 4A compounds shifts from predominantly covalent in earlier members of the group to more ionic with tin and lead. The lower oxidation state (+2) is more important for tin and lead than the +4 state. Sn(II) and Pb(II) compounds are white, crystalline solids, whereas the Sn(IV) and Pb(IV) analogs are volatile liquids consisting of covalently bonded molecules. The np^2 electrons are used to form the lower oxidation state compounds. In such cases, the ns^2 electrons are not involved in bonding, a phenomenon sometimes

$SnCl_2$ and $PbCl_2$ are crystalline solids; $SnCl_4$ and $PbCl_4$ are volatile liquids.

Table 21.9 Some Group 3A Compounds and Their Uses

Element or Compound	Uses
Boron oxide	Borosilicate glass
Boric acid (H_3BO_3)	Eyewash, astringent
Aluminum metal	Foil wrap, alloys, structural material
Aluminum oxide	Pigments, fireworks, refractory bricks, toothpaste
Aluminum sulfate	Water purification
Gallium arsenide (GaAs)	Semiconductor
$Tl_2Ba_2Ca_2Cu_3O_{10}$	High-temperature superconductor

4A (14)

Element	Ionic radius, pm	Atomic radius, pm	Density (g/mL)	MP (°C)	BP (°C)
6 C Carbon		C 77	2.27	4100	Sublimes, 4100
14 Si Silicon		Si 118	2.34	1420	~3280
32 Ge Germanium		Ge 122	5.32	945	2850
50 Sn Tin	Sn^{2+} 118	Sn 140	7.26	232	2623
82 Pb Lead	Pb^{2+} 112	Pb 146	11.34	327	1751

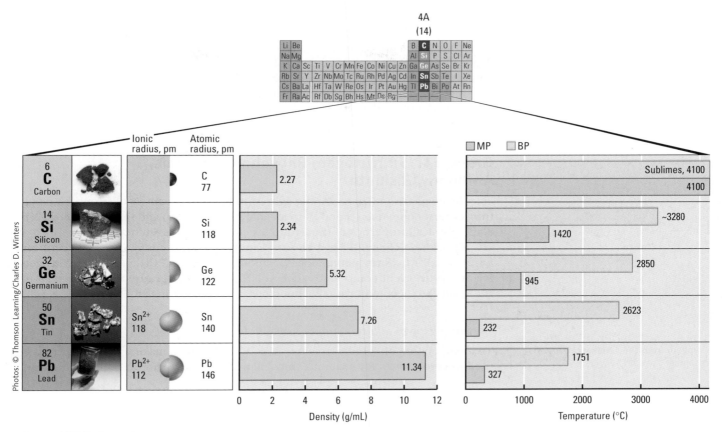

Group 4A(14) elements.

referred to as the *"inert pair effect."* Sn(II) compounds are reducing agents, being converted to Sn(IV) by oxidizing agents. In contrast, Sn(IV) compounds are oxidizing agents that are reduced to Sn(II) by reducing agents.

Carbon compounds, and to a lesser extent silicon compounds, exhibit **catenation,** in which bonds between atoms of the same element form chains or rings. Hydrocarbons containing carbon-to-carbon bonds exemplify this phenomenon (⇐ ***p. 84***).

The chemistry of carbon and its compounds has been described in Chapters 3 and 12. Silicon chemistry was discussed in Section 21.2. Tin and lead have been known since ancient times; lead lined the aqueducts to ancient Rome. In contrast, germanium was not discovered until 1886 after Mendeleev predicted its expected properties in 1871, calling the proposed element *ekasilicon.* Like diamond and silicon, germanium has a covalent network structure and was used in the first transistors due to its semiconductor properties. Because ultrapure silicon is cheaper and more rugged, it has replaced germanium for this application.

Tin is used to make pewter, an alloy of 85% tin and the remainder a combination of copper, zinc and antimony, or lead. Bronze, an alloy of tin (20%) and copper (80%), revolutionized tool making and weaponry because bronze can be fabricated into a sharp, hard edge. Its use for this purpose ushered in the Bronze Age. Tin was available because cassiterite (SnO_2), its ore, can be reduced easily using a charcoal fire.

$$SnO_2(s) + 2\,C(s) \longrightarrow Sn(\ell) + 2\,CO(g)$$

The molten tin was recovered readily as it flowed from the fire.

Pewter and bronze articles.

Lead was used to produce, tetraethyllead, $(C_2H_5)_4Pb$, as a gasoline additive to enhance octane rating (⇐ ***p. 551***), but this use has been phased out in the United States due to the release of toxic lead compounds into the atmosphere and the fact that lead destroys the catalytic effect of automobile catalytic converters. Lead reacts with oxygen and carbon dioxide to form an oxide or carbonate coating on the surface that protects the metal from further reaction. Lead reacts with and dissolves slowly in water. Because lead compounds are toxic, they must not be allowed in water used for human consumption.

Group 5A(15): Nitrogen, Phosphorus, Arsenic, Antimony, Bismuth

Like the elements in Group 4A, the lightest to heaviest members of Group 5A range from typical nonmetals (N and P) to metalloids (As and Sb) and then to a metal (Bi). Some physical properties of the Group 5A elements are given below.

All of these elements have an ns^2np^3 outer electron configuration. Except for nitrogen, all can form pentavalent compounds such as PCl_5 and $BiCl_5$. Nitrogen forms only trivalent compounds such as NH_3 and NCl_3, because the nitrogen atom is too small to accommodate five bonding pairs of electrons around it.

Uses of Nitrogen and Its Compounds

Cryogen, from the Greek word *kryos* meaning "icy cold."

The temperature of liquid nitrogen, −196 °C, is very low, making it a **cryogen.** It is used in cryosurgery, for example, to cool an area of skin prior to removal of a wart

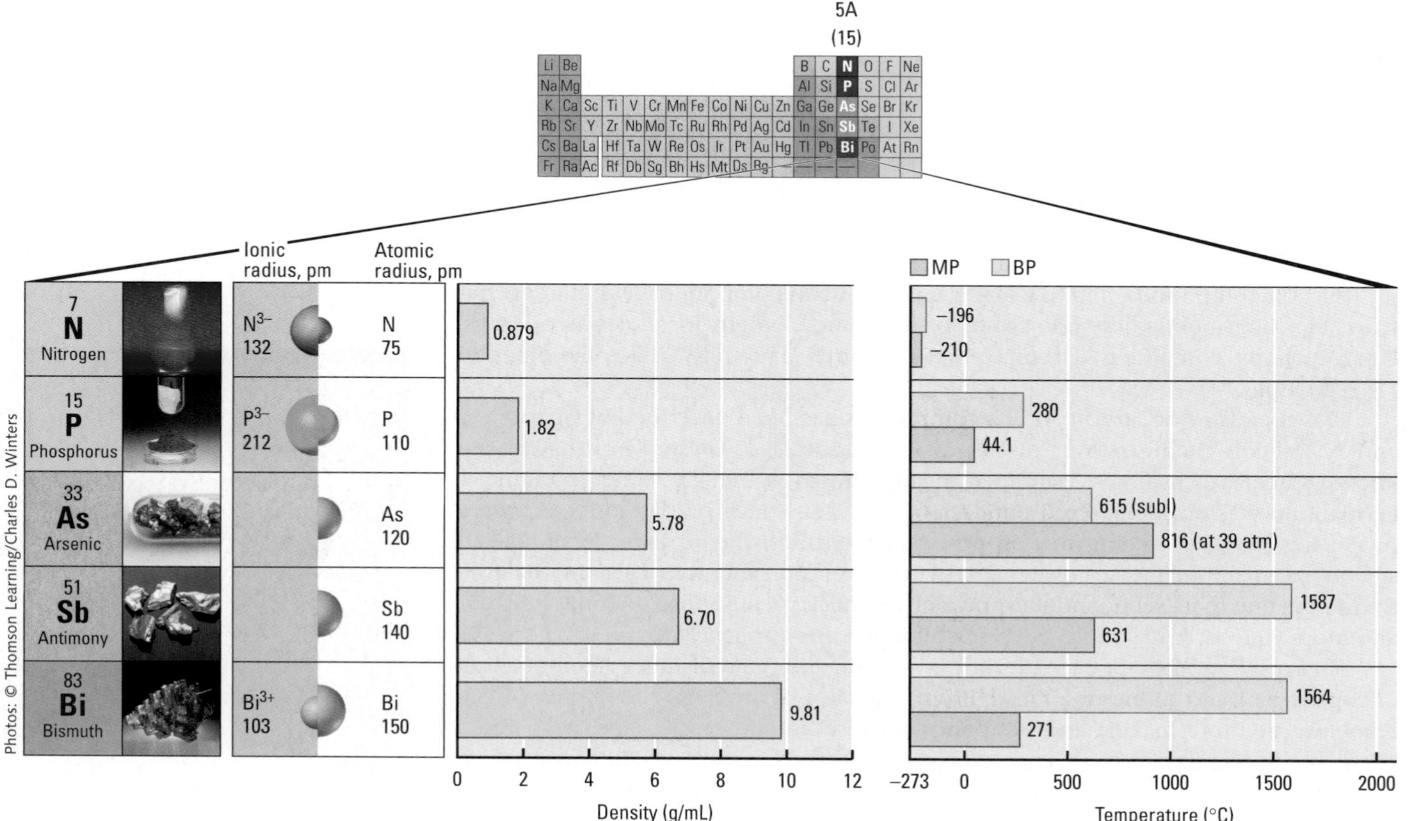

Group 5A(15) elements.

or other unwanted or pathogenic tissue. Because of its low boiling point and inertness at low temperatures, liquid nitrogen has found wide use in frozen-food preparation and preservation during transit. Since nitrogen is so chemically unreactive at room temperature, it is used as an inert atmosphere for applications such as welding.

Nitrogen, phosphorus, and potassium are primary nutrients for plants. Although bathed in an atmosphere containing abundant nitrogen, most plants are unable to use the air directly as a supply of this vital element due to the energy required to break the N≡N triple bond, one of the strongest known. **Nitrogen fixation** is the process of changing atmospheric nitrogen into compounds that can be dissolved in water, absorbed through the plant roots, and assimilated by the plant (Figure 21.18). Nitrogen fixation is part of the **nitrogen cycle,** a natural cycle of chemical pathways involving nitrogen. In nitrogen fixation, nitrogen-fixing bacteria convert N_2 into NH_3. Ammonia is converted by other bacteria into nitrate, NO_3^-, which is used by plants. When an organism dies, bacteria reverse the process by converting nitrate to N_2 and organic nitrogen compounds to ammonia.

Most plants thrive on soils rich in nitrates, but many plants that grow in swamps, where there is a lack of oxidized materials, can use reduced forms of nitrogen such as the ammonium ion. The nitrate ion is the most highly oxidized form of combined nitrogen, and the ammonium ion is the most reduced form of nitrogen.

Liquid nitrogen.

CONCEPTUAL EXERCISE **21.15 Chemically Combined Nitrogen**

Show that the nitrate ion is the most highly oxidized form of combined nitrogen and that the ammonium ion is the most reduced form of combined nitrogen.

Nitrogen is fixed by natural processes on a massive scale in two ways. In the first method, nitrogen is oxidized under highly energetic conditions in the dis-

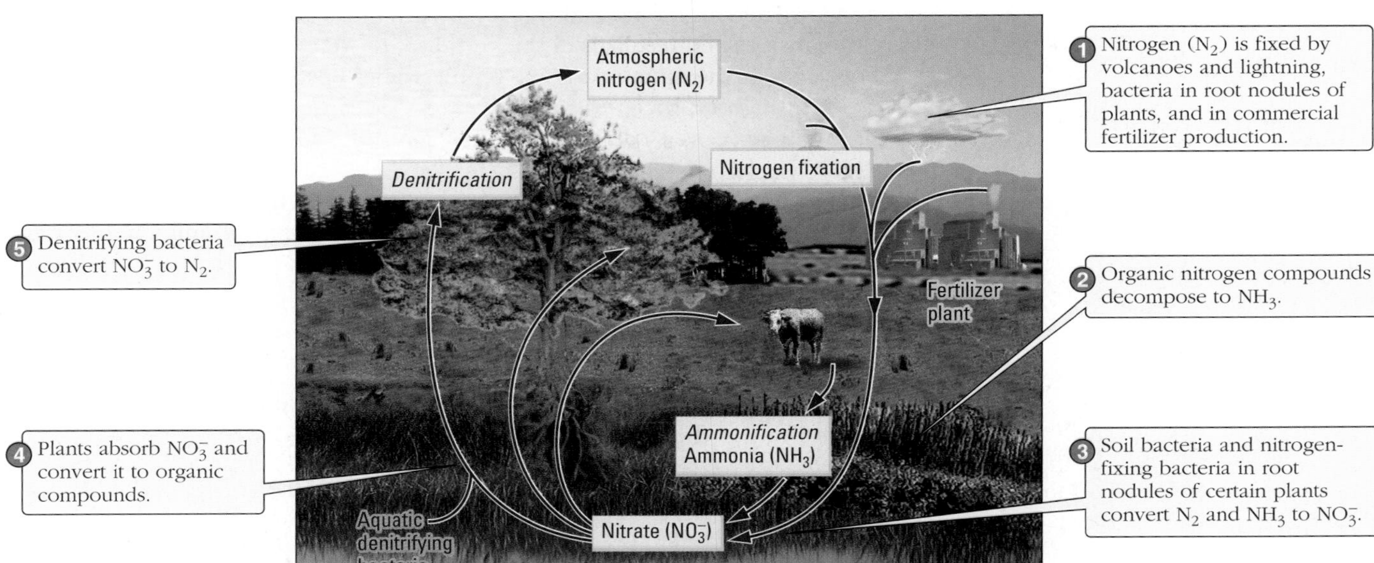

Figure 21.18 Natural nitrogen chemical pathways: the nitrogen cycle. Nitrogen-fixing bacteria convert N_2 into NH_3, which is converted by other bacteria into nitrate, NO_3^-, which is used by plants. When an organism dies, bacteria reverse the process, converting nitrate to N_2 and nitrogen compounds to ammonia.

charge of lightning or, to a lesser extent, in a fire. The initial atmospheric reaction is that of nitrogen with oxygen to form nitrogen monoxide (NO), a colorless, reactive gas.

$$N_2(g) + O_2(g) \rightleftharpoons 2\ NO(g) \qquad K_c = 1.7 \times 10^{-3} \text{ (at 2300 K)}$$

Once formed, nitrogen monoxide is easily oxidized in air to nitrogen dioxide (NO_2), which dissolves in water to form nitrous acid (HNO_2) and nitric acid (HNO_3).

$$H_2O(\ell) + 2\ NO_2(g) \longrightarrow HNO_2(aq) + HNO_3(aq)$$

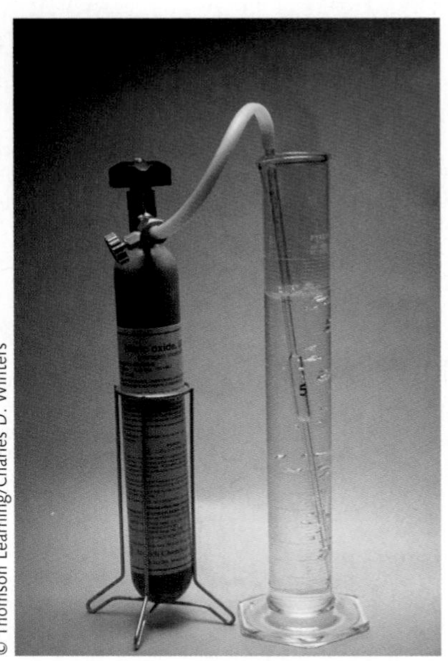

© Thomson Learning/Charles D. Winters

Conversion of colorless NO to red-brown NO_2. Colorless NO is bubbled through water. When NO emerges from the water into the air above the water, oxygen in the air reacts with NO to form reddish-brown nitrogen dioxide, NO_2.

These acids are readily soluble in rain, clouds, or ground moisture, and thus increase nitrogen concentration in soil. They also contribute to the formation of acid rain (⊏ *p. 843*).

In the second natural method of nitrogen fixation, bacteria that live on the roots of plants called *legumes,* such as clover, beans, and peas, convert atmospheric nitrogen into ammonia. This complex series of reactions depends on enzyme catalysis. Under ideal conditions, legume fixation can add more than 100 lb of nitrogen per acre of soil in one growing season.

The main industrial use of nitrogen at present is in the Haber-Bosch process (⊏ *p. 706*), which synthetically fixes nitrogen by combining it with hydrogen to form ammonia.

$$N_2(g) + 3\ H_2(g) \rightleftharpoons 2\ NH_3(g)$$

Millions of tons of ammonia are produced annually by this method. Pure gaseous ammonia is condensed and the liquid anhydrous ammonia is applied directly to fields as a fertilizer. Ammonia is also reacted with nitric acid to produce ammonium nitrate, the major solid fertilizer in the world.

About 15% of the ammonia made by the Haber-Bosch process is converted to nitric acid through a process developed by a German chemist, Wilhelm Ostwald. This three-step process is carried out at pressures of 1 to 10 atm.

Step 1: The ammonia is burned in air over a platinum-rhodium catalyst at about 1000 °C, achieving a greater than 95% conversion of ammonia to nitric oxide, NO.

$$4\ NH_3(g) + 5\ O_2(g) \xrightarrow[\text{catalyst}]{1000\ ^\circ C} 4\ NO(g) + 6\ H_2O(g)$$

Step 2: More air is added to the gaseous mixture, which lowers the temperature and causes the reaction

$$2\ NO(g) + O_2(g) \rightleftharpoons 2\ NO_2(g)$$

Step 3: The nitrogen dioxide produced in the second step is passed through water to produce nitric acid.

$$3\ NO_2(g) + H_2O(\ell) \longrightarrow 2\ HNO_3(aq) + NO(g)$$

The resulting aqueous solution is about 60% nitric acid by mass. The anhydrous acid is produced by adding sulfuric acid and boiling the mixture to distill nearly pure nitric acid from it.

PROBLEM-SOLVING EXAMPLE 21.4 Another Nitrogen-Hydrogen Compound

Compounds of nitrogen and hydrogen other than ammonia exist. For example, *trans*-tetrazene, N_4H_4, has four nitrogen atoms in a chain with the terminal nitrogens each bonded to two hydrogen atoms. Write the Lewis structure of *trans*-tetrazene.

Answer

```
                H
               /
              N̈
             / \
     N̈ == N̈     H
 H       /
  \     /
    N̈
   /
 H
```

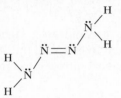

Strategy and Explanation The prefix *trans* indicates that the —NH_2 groups must be across the plane of a double bond from each other, an N═N double bond between the interior nitrogen atoms. The correct Lewis structure is shown above.

PROBLEM-SOLVING PRACTICE 21.4

Dinitrogen pentaoxide is the acid anhydride of nitric acid.

(a) Write the Lewis structure of dinitrogen pentaoxide.
(b) Write the balanced chemical equation for the formation of nitric acid from the reaction of dinitrogen pentaoxide with water.

PROBLEM-SOLVING EXAMPLE 21.5 Producing Nitric Acid

Consider the second step in the Ostwald process.

$$2\,NO(g) + O_2(g) \rightleftharpoons 2\,NO_2(g) \qquad \Delta H^\circ = -113.0\text{ kJ}$$

What would happen to the yield of NO_2 at equilibrium if

(a) The pressure were increased by decreasing the volume?
(b) The temperature were increased?
(c) A catalyst were used?

Answer

(a) The yield should increase.
(b) The yield should decrease.
(c) No effect on yield.

Strategy and Explanation Apply Le Chatelier's principle (← *p. 695*) in each case.

(a) The yield would increase because the pressure change favors the formation of fewer moles of gas (3 mol gaseous reactants, 2 mol gaseous products).
(b) An increase in temperature favors the endothermic reverse reaction, decreasing the yield.
(c) A catalyst will speed up both the forward and reverse reactions equally, so it has no effect on yield.

PROBLEM-SOLVING PRACTICE 21.5

Use Le Chatelier's principle to explain why in the manufacturing of nitric acid

(a) Lowering the temperature after Step 1 favors NO_2 formation.
(b) Coupling Step 2 with Step 3 of the Ostwald process favors the formation of nitric acid.

Richard Olivier/Corbis

Expanding air bags.

A Lifesaving Use of N_2: Automobile Air Bags

Air bags save the lives of thousands of motorists annually in the United States. This is done through the application of a simple decomposition reaction of sodium azide, an ionic compound containing sodium ions and azide ions, N_3^-, which decomposes rapidly to liberate N_2 gas. An uninflated air bag has a small cylinder containing a carefully formulated mixture of the solids sodium azide (NaN_3), potassium nitrate (KNO_3), and silicon dioxide (SiO_2). When a car decelerates rapidly, as in a collision, a sensor sends an electrical signal to the mixture, rapidly igniting and decomposing the sodium azide and releasing nitrogen gas, which inflates the air bag.

In addition to being used in automobile air bags, azides are used as explosives.

$$2\ NaN_3(s) \longrightarrow 2\ Na(s) + 3\ N_2(g)$$

Residual sodium metal must be removed because it could react vigorously with water (page 1036). Removal in this case is achieved by reacting the sodium with potassium nitrate in a reaction that produces additional nitrogen gas to inflate the air bag.

$$10\ Na(s) + 2\ KNO_3(s) \longrightarrow K_2O(s) + 5\ Na_2O(s) + N_2(g)$$

The heat released by these reactions melts the solid products and the silicon dioxide (sand), fusing them into an unreactive glass.

$$K_2O(s) + Na_2O(s) + SiO_2(s) \xrightarrow{\text{heat}} \text{glass}$$

EXERCISE 21.16 Expanding Air Bags

Calculate the volume of N_2 released in an air bag at STP when 150. g sodium azide decompose.

The Discovery of Phosphorus, 1775, William Pether Engraving after Joseph Wright painting. Fisher Collections. Chemical Heritage Foundation. Photo by Will Brown.

The discovery of phosphorus by Herman Brand by extraction from urine. In a very limited oxygen supply, white phosphorus glows with a greenish light, the source of the term "phosphorescence."

Phosphorus and Its Compounds

Phosphorus has two main allotropes, *white* phosphorus and *red* phosphorus, which have very different properties. White phosphorus is highly reactive, igniting spontaneously in air at room temperature. For this reason, white phosphorus is stored under water. The waxy, nonpolar, solid white phosphorus is soft and easily cut, reflecting the fact that it consists of P_4 tetrahedra held together by weak noncovalent intermolecular forces ***(← p. 408).*** Because it is nonpolar, white phosphorus is not soluble in water, but dissolves readily in nonpolar liquids such as carbon disulfide (CS_2) or hexane (C_6H_{14}). Red phosphorus, unlike white phosphorus, does not ignite in air at room temperature and is much less toxic. A third, less common allotrope called black phosphorus is produced by heating white phosphorus at high pressure.

Although white phosphorus is toxic, phosphorus is an essential dietary mineral because of the many ways it is used by the body. Phosphorus is part of phosphate groups that link alternately with deoxyribose units to form the backbone of the DNA double helix ***(← p. 416).*** Phosphate anhydride linkages, which have this structure,

$$-O-\overset{\overset{\displaystyle O}{\|}}{\underset{\underset{\displaystyle O^-}{|}}{P}}-O-\overset{\overset{\displaystyle O}{\|}}{\underset{\underset{\displaystyle O^-}{|}}{P}}-O-$$

are responsible for how cellular energy is stored in ATP ***(← p. 895).***

Tooth enamel and bone contain the mineral hydroxyapatite, $Ca_5(PO_4)_3OH$. Water fluoridation reduces tooth decay because fluoride ions from the fluoridated water substitute for OH^- ions in tooth enamel to form fluoroapatite, $Ca_5(PO_4)_3F$, which is more resistant to decay than hydroxyapatite.

Michael Davidson/ Photo Researchers, Inc.

ATP in cells (electron microscope photo).

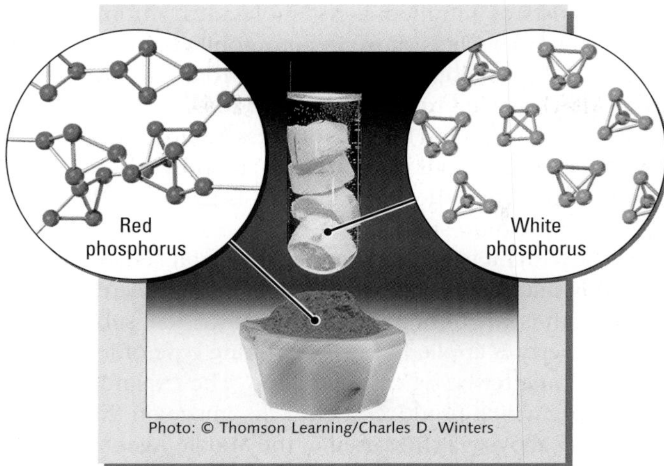

Phosphorus allotropes. Because white phosphorus *(top)* reacts with air at room temperature, it must be stored under water. Red phosphorus *(bottom)* does not react with air at room temperature.

PROBLEM-SOLVING EXAMPLE 21.6 Phosphorus Pentachloride

Although the empirical formula for phosphorus pentachloride is PCl_5, in the solid state the compound actually consists of PCl_4^+ and PCl_6^- ions. Write the Lewis structures of these ions and predict their shapes.

Answer $[PCl_4]^+$ (Lewis structure: P bonded to four :Cl: atoms, each Cl with three lone pairs) Tetrahedral; $[PCl_6]^-$ (Lewis structure: P bonded to six :Cl: atoms, each Cl with three lone pairs) Octahedral

Strategy and Explanation Count the number of valence electrons and allow for the charge of the ion in each case. The PCl_4^+ ion has 32 valence electrons; PCl_6^- has 48 valence electrons. As a central atom, phosphorus can exceed an octet of electrons. The Lewis structures are

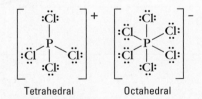

$[PCl_4]^+$ (P bonded to four :Cl: atoms, each Cl with three lone pairs) Tetrahedral; $[PCl_6]^-$ (P bonded to six :Cl: atoms, each Cl with three lone pairs) Octahedral

PCl_4^- is tetrahedral; PCl_6^- is octahedral.

PROBLEM-SOLVING PRACTICE 21.6

Pure phosphoric acid, H_3PO_4, occurs only as a solid. When it melts, phosphoric acid units gradually lose water. Write the Lewis structure for the other product of this dehydration.

Arsenic, Antimony, and Bismuth

Arsenic and antimony each have two allotropes, a metallic form and an amorphous form. Amorphous arsenic is yellow and, like white phosphorus, is soluble in carbon disulfide, where it exists as tetrahedral As_4 molecules. Antimony and arsenic are used to harden lead alloys such as those in automobile battery grids and in bullets. Arsenic compounds are poisonous, a fact often put to use in mystery novels. A 0.1 g or greater dose of As_2O_3 is fatal to humans. All of the trihalides of arsenic and antimony are known, such as AsF_3 and SbI_3, a colorless liquid and a red crystalline solid, respectively. By contrast, only four pentahalides are known—AsF_5, $AsCl_5$, SbF_5, and $SbCl_5$. Failure to produce the other pentahalides is likely due to the strong oxidizing nature of the +5 oxidation state.

Bismuth is a yellowish metal that has limited uses, chief among them being in low-melting alloys used in automatic sprinkler systems and in electrical fuses. One physical property of bismuth is worth noting: it is one of the few substances that expands on freezing. This property is applied to make printing type of low-melting alloys that expand on solidification after being cast in a mold. The expanded metal gives a very sharp edge to the type, which produces a clear ink image on the printed page. Printing type made of these alloys was first used in the Middle Ages shortly after the initial Gutenberg printing press was developed (1440). Bismuth is commonly found in the lower +3 oxidation state, whereas the lighter elements (P, As, and Sb) are often found in the higher +5 oxidation state. This is another example of the inert pair effect, as noted previously for the heavier Group 3A and 4A elements.

Group 6A(16): Oxygen, Sulfur, Selenium, Tellurium, Polonium

The term *chalcogen* is derived from the Greek word *khalos* for copper because copper ores contain compounds of these elements.

Group 6A is the oxygen family of elements; S, Se, and Te are also known as the chalcogens. Some physical properties of the Group 6A elements are given below. Like the preceding two groups, Group 6A is diverse, consisting of true nonmetals (O, S, and Se), a metalloid (Te), and a metal (Po, a rare, radioactive element).

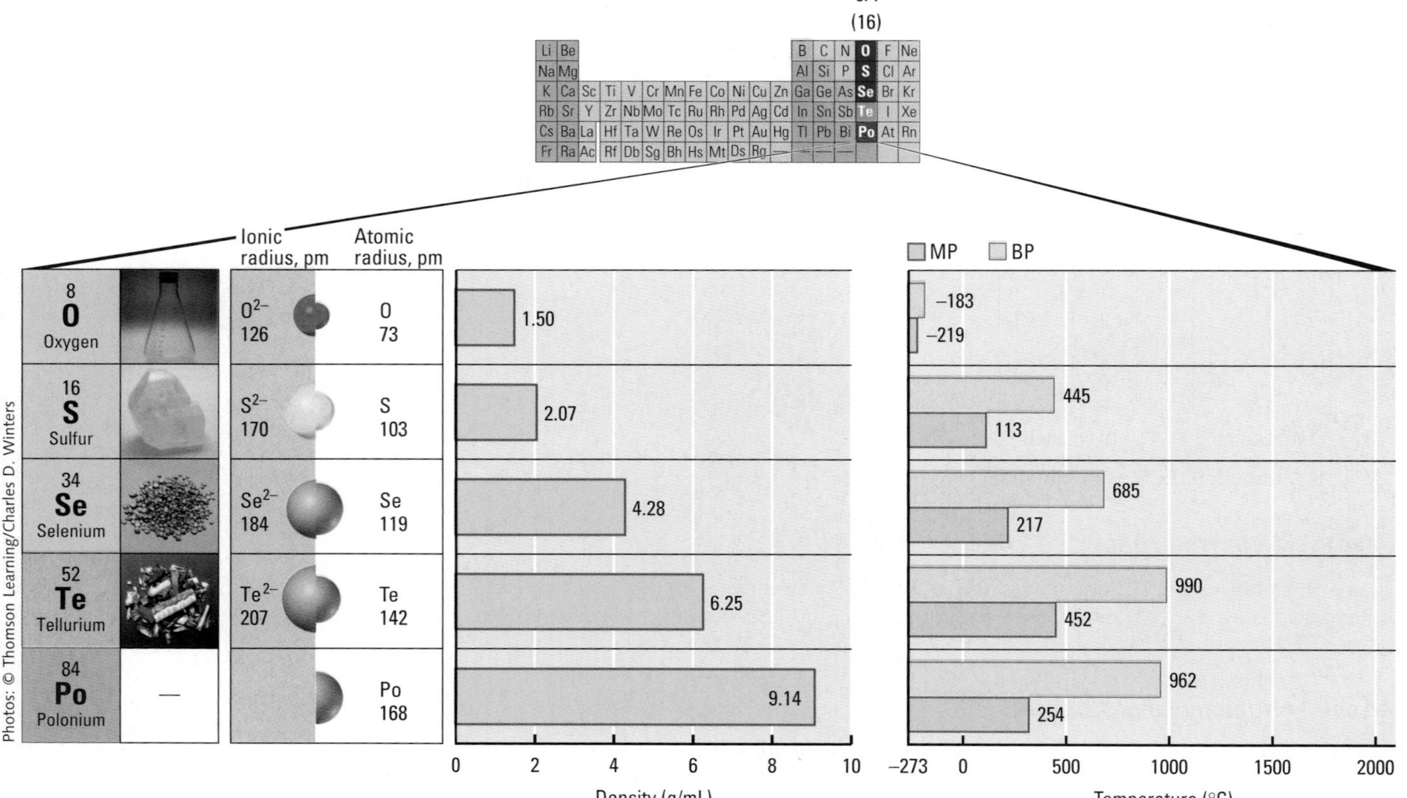

Group 6A(16) elements.

Properties and Uses of Oxygen

Most of the oxygen produced by fractional distillation of liquid air is used as an oxidizing agent and in steel making (Section 22.2), although some is used in rocket propulsion (to oxidize hydrogen) and in controlled oxidation reactions of other types. Liquid oxygen (LOX) can be shipped and stored at its boiling temperature of −183 °C under atmospheric pressure. Cryogens, such as liquid nitrogen and liquid oxygen, present special hazards since contact with them produces instantaneous frostbite and brittleness in structural materials such as plastics, rubber gaskets, and some metals, causing materials to fracture easily at these low temperatures. The high oxygen concentration present in liquid oxygen can, in spite of the low temperature, accelerate oxidation reactions to the point of explosion. For this reason, contact between liquid oxygen and substances that will ignite and burn in air must be prevented.

Special cryogenic containers holding liquid oxygen incorporate huge vacuum-walled bottles much like those used to carry hot soup or hot coffee. These special containers can be seen outside hospitals and industrial complexes, on highways and railroads, and even aboard ocean-going vessels. In hospitals as well as in homes, supplemental oxygen is used to help patients who have difficulty breathing.

© Thomson Learning/Charles D. Winters

Cryogenic containers of liquid oxygen.

Most atmospheric oxygen comes from photosynthesis, in which green plants convert water and carbon dioxide into glucose and oxygen. The concentration of oxygen in the atmosphere has upper and lower limits that are essential for our safety. If the concentration exceeds 25%, the rates of oxidation reactions would increase significantly, potentially endangering us by the increased rates of oxygen-requiring metabolic processes. With too little atmospheric oxygen, less than 17%, we would suffocate.

Properties and Uses of Sulfur

Sulfur exists in two common allotropic forms—rhombic (mp 115 °C) and monoclinic (mp 119 °C), both consisting of S_8 rings in the solid. When sulfur is heated above 150 °C, the S_8 rings break open, forming chains that become entangled, thereby increasing the viscosity of the molten sulfur. Upon continued heating, the color of sulfur changes from yellow to dark red because of unpaired electrons at the ends of the chains. If heated to 210 °C and poured into cold water, the sulfur forms an uncrystallized polymer called "plastic sulfur," which reverts back to the common crystalline forms at 25 °C (Figure 21.19).

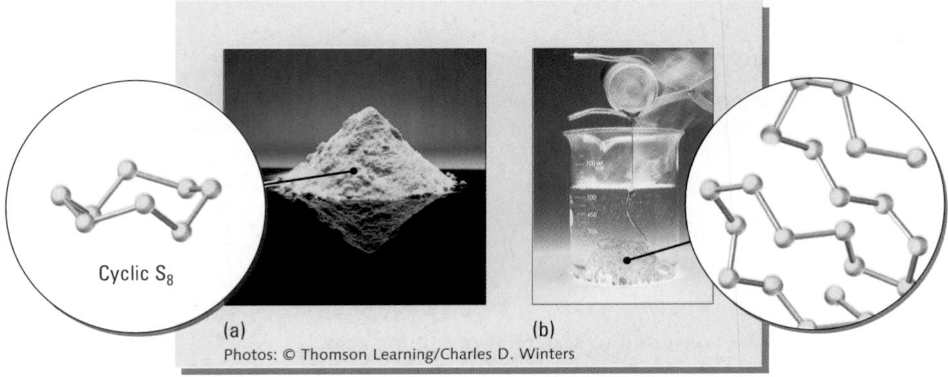

Photos: © Thomson Learning/Charles D. Winters

Active Figure 21.19 **Sulfur allotropes.** (a) At room temperature, sulfur is a bright yellow solid. At the nanoscale it consists of rings of eight sulfur atoms. (b) When melted, the rings break open to form long chains. **Go to the Active Figures menu at ThomsonNOW to test your understanding of the concepts in this figure.**

Sulfur is a critical element in the body, necessary for the formation of methionine, an essential amino acid (← ***p. 591***). Sulfur atoms form —S—S— disulfide linkages among chains of amino acids; the disulfide linkages help to create the essential molecular shapes of proteins and enzymes (← ***p. 594***). Sulfur is also used to cross-link polymer chains in the vulcanization of rubber (← ***p. 581***). The sulfur helps to align the polymer chains, which makes the rubber more elastic and prevents it from becoming sticky in warm weather.

Sulfuric Acid Production

Most sulfur is used to produce sulfuric acid, the workhorse industrial chemical used in steel production, in automobile batteries, in the petroleum industry, and in the manufacture of fertilizers, plastics, drugs, dyes, and many other products. Since sulfuric acid costs less to make than any other acid, it is the first to be considered when an acid is needed in an industrial process.

Sulfur is converted to sulfuric acid in four steps, collectively called the *contact process.* In the first step, sulfur is burned in air to give mostly sulfur dioxide.

$$S_8(s) + 8\,O_2(g) \longrightarrow 8\,SO_2(g)$$

The SO_2 is then converted to SO_3 over a heated catalyst, such as platinum metal or vanadium(V) oxide.

$$2\,SO_2(g) + O_2(g) \xrightarrow{\text{catalyst}} 2\,SO_3(g)$$

The next step converts the sulfur trioxide to sulfuric acid by the addition of water. The best way to do this is to pass the SO_3 into H_2SO_4 to form pyrosulfuric acid, $H_2S_2O_7$, and then to dilute the $H_2S_2O_7$ with water. The net reaction is 1 mol H_2SO_4 for every 1 mol SO_3.

$$SO_3(g) + H_2SO_4(\ell) \longrightarrow H_2S_2O_7(\ell)$$

$$H_2S_2O_7(\ell) + H_2O(\ell) \longrightarrow 2\,H_2SO_4(aq)$$

$$\textit{Net reaction}\text{: } SO_3(g) + H_2O(\ell) \longrightarrow H_2SO_4(aq)$$

Sulfur dioxide for the contact process can also be obtained as a by-product from copper or lead smelting. Unless this sulfur dioxide is recovered, it pollutes the atmosphere.

PROBLEM-SOLVING EXAMPLE 21.7 Sulfur and Sulfuric Acid

In a recent year, 1.3×10^{10} kg sulfur were produced in the United States. If all of this had been converted to sulfuric acid, how many kilograms of sulfuric acid would it have produced?

Answer 4.0×10^{10} kg H_2SO_4

Strategy and Explanation The equations for the formation of sulfuric acid provide the mole-to-mole relationships for the conversion of S_8 to H_2SO_4. The net reaction is the formation of one mole of H_2SO_4 per mole of SO_3. Each mole of SO_3 requires 1 mol SO_2, and each mole of S_8 forms 8 mol SO_2.

$$1.3 \times 10^{10}\text{ kg } S_8\left(\frac{10^3\text{ g } S_8}{1\text{ kg } S_8}\right)\left(\frac{1\text{ mol } S_8}{256.5\text{ g } S_8}\right)\left(\frac{8\text{ mol } SO_2}{1\text{ mol } S_8}\right)\left(\frac{1\text{ mol } SO_3}{1\text{ mol } SO_2}\right) = 4.05 \times 10^{11}\text{ mol } SO_3$$

This amount can be converted to kilograms of sulfuric acid.

$$4.05 \times 10^{11}\text{ mol } SO_3\left(\frac{1\text{ mol } H_2SO_4}{1\text{ mol } SO_3}\right)\left(\frac{98.08\text{ g } H_2SO_4}{1\text{ mol } H_2SO_4}\right)\left(\frac{1\text{ kg } H_2SO_4}{10^3\text{ g } H_2SO_4}\right) = 4.0 \times 10^{10}\text{ kg } H_2SO_4$$

✓ **Reasonable Answer Check** Sulfuric acid is approximately 33% sulfur by mass ([32 g S/98 g H_2SO_4] × 100%). Therefore, 33 kg sulfur forms about 100 kg sulfuric acid, and 1×10^{10} kg S forms about 3×10^{10} kg sulfuric acid, which is close to the calculated value of 4.0×10^{10} kg H_2SO_4.

PROBLEM-SOLVING PRACTICE 21.7

Calculate the number of kilograms of SO_3 produced from 1.3×10^{10} kg S_8.

Group 7A(17): The Halogens

The halogens exhibit a trend in physical states not found in any other group: The lightest halogen (fluorine) is a gas and the heaviest (iodine) is a solid. Some physical properties of the Group 7A elements are given below. At room temperature, fluorine and chlorine, the first two elements, are diatomic pale yellow and yellow-green gases, respectively; bromine, the next element, is a reddish-brown liquid; and then iodine, a violet-black solid. These physical states reflect the increasing strengths of noncovalent intermolecular forces with increasing numbers of electrons in the diatomic molecules. Astatine is an intensely radioactive element with isotopes of very short half-lives, accounting for the scarcity of astatine in nature.

All halogens have an ns^2np^5 outer electron configuration that leads to a gain of one electron to form a 1− ion or a sharing of one electron in a pair to complete an octet. Fluorine is extremely reactive, evidence of its exceptionally high electronegativity, the highest of any element and of the very weak F—F bond in F_2 (bond enthalpy 158 kJ/mol). The halogens are oxidizing agents in many reactions including

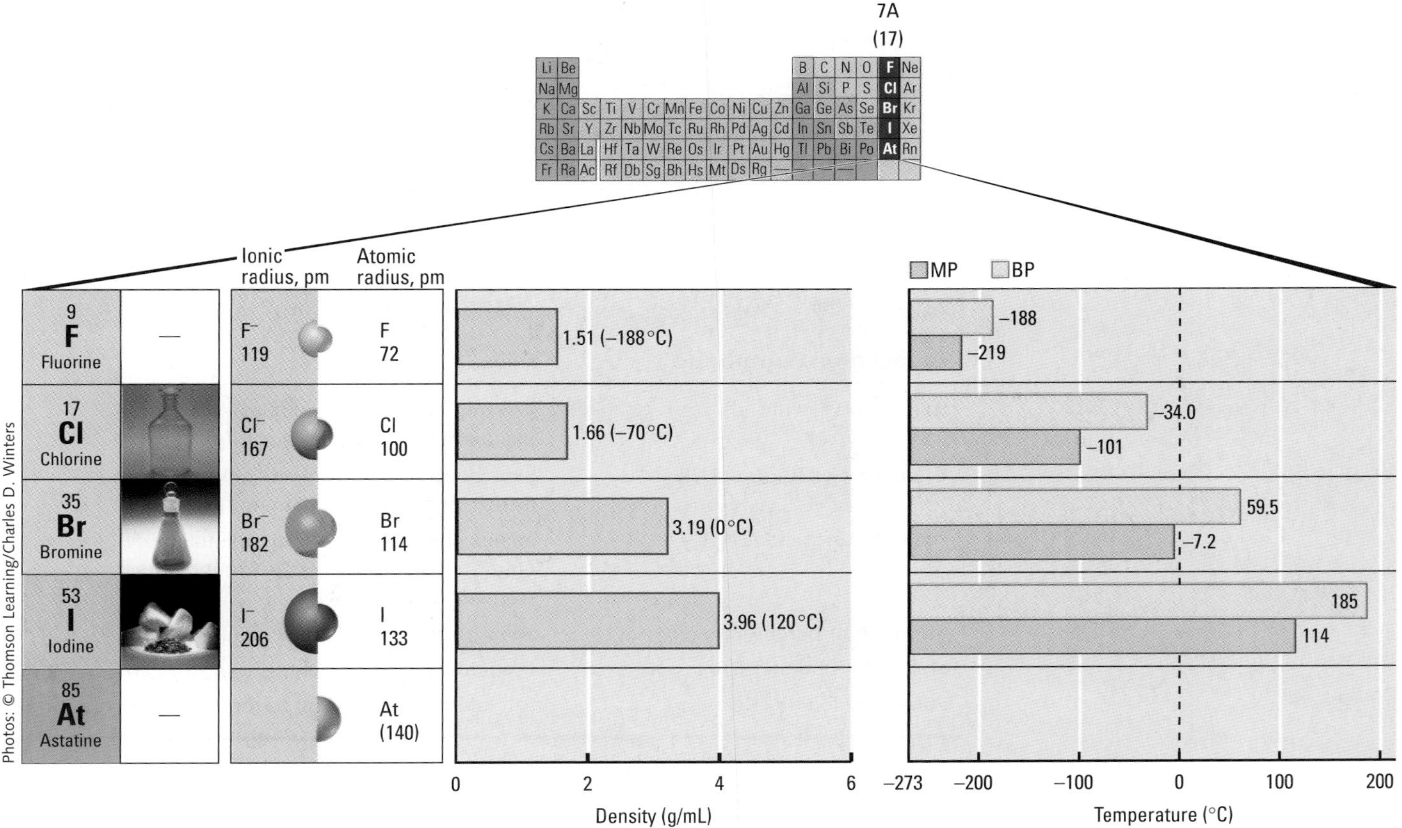

Group 7A(17) elements, the halogens.

Sublimation of iodine. Solid iodine is a gray, metallic-looking solid. When heated, it converts directly by sublimation into gaseous iodine, a purple vapor.

the oxidation of a heavier halide ion by a lighter halogen. For example, chlorine is a strong oxidizing agent that oxidizes bromide ions to bromine.

$$Cl_2(g) + 2\,Br^-(aq) \longrightarrow 2\,Cl^-(aq) + Br_2(aq)$$

This reaction is used to extract bromine from seawater, which contains 30 ppm bromide ions (Section 21.5). Some uses of the halogens and their compounds are given in Table 21.10.

Chlorine, a toxic gas with an irritating odor, is the most important halogen used in industry. Chlorine is used to purify water ***(⇐ p. 761),*** to bleach paper and textiles, to manufacture herbicides, insecticides, and other chlorinated organic compounds, to produce polyvinyl chloride, and to extract titanium metal from its ores.

Bromine is the only nonmetal that is a liquid at room temperature. It is used to prepare methyl bromide, CH_3Br, an efficient fire extinguisher and pesticide. Bromine is also used to make light-sensitive silver bromide for photographic films.

Iodine is the only common halogen that is a solid at room temperature. It is a dark, metallic-looking solid that sublimes to a violet-colored vapor. Iodine was discovered by burning dried seaweed, which contains a relatively high concentration of iodide ions. Iodine is an essential dietary mineral for humans because the iodide ions are necessary for the production of thyroxine, a growth-controlling hormone produced by the thyroid gland.

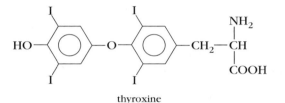

HO–C₆H₂I₂–O–C₆H₂I₂–CH₂–CH(NH₂)–COOH

thyroxine

Insufficient dietary iodine causes enlargement of the thyroid gland, a condition known as *goiter.* Potassium iodide (0.01%) is added to table salt (iodized salt) to prevent goiter.

Because it is so rare in nature, astatine or its compounds are not available in easily manipulated amounts. Elegant tracer experiments have established the existence of the astatide ion, At^-, in keeping with Group 7A behavior.

Table 21.10 Uses of Halogens and Some of Their Compounds

Halogen or Compound	Uses
Hydrogen fluoride	Frosting light bulbs and television tubes, production of uranium hexafluoride
Uranium hexafluoride (UF_6)	Separation of fissionable U-235 from U-238 during processing of uranium ores
Hydrochloric acid (HCl)	Magnesium manufacturing, manufacture of vinyl chloride and chlorinated solvents; human stomach acid (0.1 M)
Ammonium perchlorate (NH_4ClO_4)	Solid propellant for Space Shuttle
Methyl bromide (CH_3Br)	Pesticide and fire extinguisher
Potassium iodide (KI)	Salt additive to prevent goiter, a thyroid condition

PROBLEM-SOLVING EXAMPLE 21.8 Reactive Fluorine

In a Teflon vessel water reacts with hypofluorous acid (HOF) to produce hydrogen fluoride, hydrogen peroxide, and oxygen.
(a) Write a balanced chemical equation for this reaction.
(b) Why is a glass reaction vessel not used?

Answer

(a) $4\ HOF(aq) + 2\ H_2O(\ell) \longrightarrow 4\ HF(aq) + H_2O_2(aq) + O_2(g)$
(b) The HF produced would etch the glass vessel and break it.

Strategy and Explanation

(a) Balance the equation by writing the correct formulas and coefficients for the reactants and products. Hypofluorous acid is analogous to hypochlorous acid, HOCl.

$$4\ HOF(aq) + 2\ H_2O(\ell) \longrightarrow 4\ HF(aq) + H_2O_2(aq) + O_2(g)$$

(b) Teflon is used because it is nonreactive. Glass would react with HF produced by the reaction.

PROBLEM-SOLVING PRACTICE 21.8

Classify each of the following equations as being either a redox reaction or a non-redox reaction. If a redox reaction occurs, also identify the oxidizing agent and the reducing agent.
(a) $2\ NaF(s) + H_2SO_4(aq) \longrightarrow 2\ HF(g) + Na_2SO_4(aq)$
(b) $S_8(s) + 24\ F_2(g) \longrightarrow 8\ SF_6(g)$

Group 8A(18): The Noble Gases

Some physical properties of the Group 8A elements are given on p. 1056. Once called the "rare gases," these elements are not rare. Though relatively low in terrestrial abundance, helium is the second most abundant element in the universe. Argon makes up nearly 1% of the atmosphere.

Evidence of helium was first obtained when a new yellow line was observed spectroscopically during an eclipse of the sun in 1868. Evidence of terrestrial helium first came from observation of the same line in the spectrum of gases released during the eruption of Mount Vesuvius in 1881. Subsequently, the remaining noble gases except radon were isolated from liquid air and characterized by J. Rayleigh, W. Ramsay, and M. Travers between 1895 and 1898. Radon was first isolated and identified in 1902 by Rutherford and Soddy. These findings led to the addition of a new group (8A) to the periodic table.

The stubbornness of Group 8A elements against forming compounds is to be expected given their complete octets of ns^2np^6 outer electrons (only ns^2 for helium). In fact, it was a long-standing accepted chemical truism that these elements, then called the "inert elements," formed no compounds. That view was overthrown dramatically in 1962 with the synthesis of a series of xenon compounds (⇐ *p. 317*). Since then, compounds of xenon containing oxygen and fluorine have been synthesized, as well as fluorides of krypton and radon. Spectroscopic evidence exists for the argon compound HArF.

Helium is used as an inert atmosphere in welding, as a nitrogen substitute in gas mixtures for underwater diving, as a liquid cryogen, and to fill meteorological balloons. Argon is used as a protective atmosphere for arc-welding aluminum, in airport runway lights, and as the gas in incandescent light bulbs, where it decreases the evaporation of the tungsten metal filament.

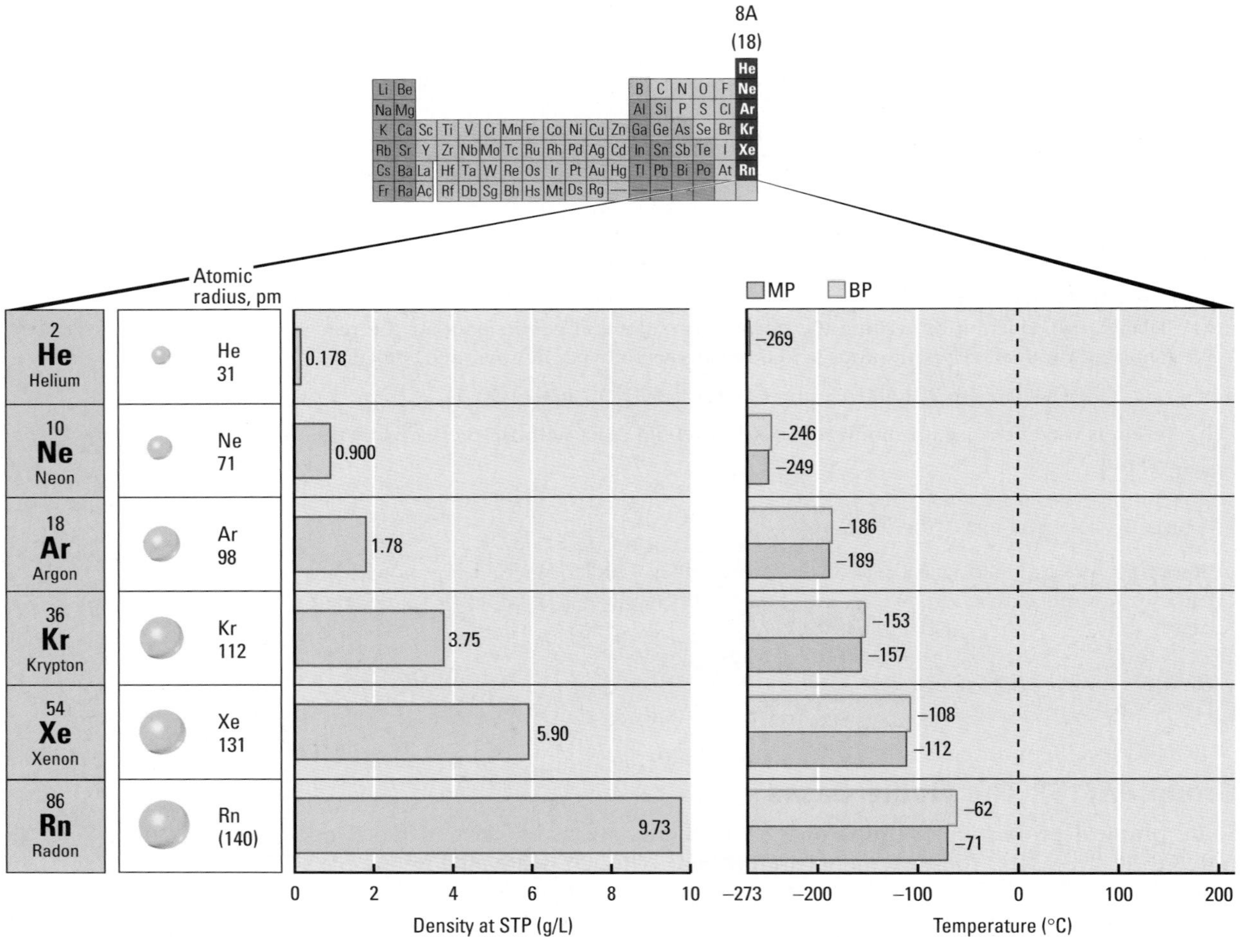

Group 8A(18) elements, the noble gases.

SUMMARY PROBLEM

(a) In 2004, the United States used about 7.6×10^9 barrels of crude oil, enough for 26 barrels per person (1 barrel = 42 gallons; density of crude oil = 0.83 g/mL). Assume that the crude oil is 3% sulfur by mass and that all of the sulfur was removed from the crude oil before it was used. How many liters of SO_2 at 25 °C and 1 atm from just your share of crude oil in 2004 would have been prevented from entering the atmosphere?

(b) Consider the conversion of $SO_2(g)$ to $SO_3(g)$.

$$SO_2(g) + \frac{1}{2}O_2(g) \longrightarrow SO_3(g)$$

(i) Use data from Appendix J to calculate $\Delta H°$ for this reaction.

(ii) The reaction reaches equilibrium. Identify the effect on the concentration of sulfur trioxide produced if these conditions are applied to the equilibrium.

1. The pressure is increased.

2. The temperature is decreased.

3. A catalyst is used.
4. Sulfur dioxide is added.
5. Sulfur trioxide is removed as it forms.

(iii) At 1000. K, there initially were 0.250 mol SO_2, 0.125 mol O_2, and no SO_3 in a 10.0-L reaction chamber. At equilibrium there are 0.136 mol SO_3 and 0.00570 mol O_2. Calculate the equilibrium constant.

(c) Using data from Appendix J, calculate $\Delta G°$ for the conversion of sulfur dioxide gas to sulfur trioxide gas. Then calculate the value for K_p at 800. °C and at 1000. °C.

IN CLOSING

Having studied this chapter, you should be able to . . .

- Give a general explanation of how elements form in stars (Section 21.1).
- Know the principal elements in the earth's crust (Section 21.2).
- Describe the general structure of silicates (Section 21.2).
- Identify the general methods by which elements are extracted from the earth's crust (Section 21.2). ThomsonNOW homework: Study Question 28
- Identify the major components of the atmosphere (Section 21.3).
- Apply chemical principles to the processes for extracting and purifying elements, and the reactions they undergo (Sections 21.3–21.5). ThomsonNOW homework: Study Questions 24, 51
- Describe the Frasch process for obtaining sulfur (Section 21.3).
- Describe how electrolysis is used to obtain sodium, chlorine, magnesium, and aluminum (Section 21.4). ThomsonNOW homework: Study Questions 20, 22
- Explain how chemical redox reactions are used to extract bromine, iodine, and phosphorus from compounds (Section 21.5).
- Explain how elements are obtained by the liquefaction of air (Section 21.6).
- Explain how sulfuric acid is produced (Section 21.6).
- Interpret the periodic trends among the main group elements (Section 21.6).

ThomsonNOW
Sign in to ThomsonNOW at **www.thomsonedu.com** to check your readiness for an exam by taking the Pre-Test and exploring the modules recommended in your Personalized Learning Plan.

KEY TERMS

catenation *(21.6)*
chlor-alkali process *(21.4)*
cryogen *(21.6)*
Frasch process *(21.3)*
helium burning *(21.1)*
hydrogen burning *(21.1)*
mineral *(21.2)*
nitrogen cycle *(21.6)*
nitrogen fixation *(21.6)*
nuclear burning *(21.1)*
ore *(21.2)*

QUESTIONS FOR REVIEW AND THOUGHT

■ denotes questions available in ThomsonNOW and assignable in OWL.

Blue-numbered questions have short answers at the back of this book and fully worked solutions in the *Student Solutions Manual.*

ThomsonNOW
Assess your understanding of this chapter's topics with sample tests and other resources found by signing in to ThomsonNOW at **www.thomsonedu.com**.

Review Questions

1. What is meant by hydrogen burning and helium burning in relation to the formation of elements?
2. Identify the most abundant nonmetallic element in the earth's crust. Identify the most abundant metallic element in the earth's crust.
3. Describe the difference between an ore and a mineral.
4. Give a simple explanation for the abundance of clay minerals in the earth's crust.
5. Differentiate among pyroxenes, amphiboles, and silica.
6. Explain how the silicate unit in amphiboles has the general formula $Si_4O_{11}^{6-}$.
7. Explain how the silicate unit in mica and other sheet silicates has the general formula $Si_2O_5^{2-}$.
8. Identify two major differences between white phosphorus and red phosphorus.
9. Identify
   (a) Two elements obtained from the atmosphere.
   (b) Two elements obtained from the sea.
   (c) Two elements obtained from the earth's crust.
10. Write balanced equations for the recovery of magnesium from seawater. Begin with the precipitation of magnesium hydroxide by addition of calcium hydroxide to seawater.
11. Why are nitrogen and oxygen important industrial chemicals?
12. Describe how nature fixes nitrogen. Why is nitrogen fixation necessary?
13. Briefly explain why different products are obtained from the electrolysis of molten NaCl and the electrolysis of aqueous NaCl.
14. Identify two uses of phosphate rock.
15. Describe the structural changes that occur in sulfur as it is heated from room temperature to 210 °C.
16. Identify the substance or substances produced by each of these commercial processes.
   (a) Hall-Héroult (b) Contact
   (c) Ostwald (d) Dow
17. Identify the substance or substances produced by each of these commercial processes.
   (a) Frasch (b) Chlor-alkali
   (c) Downs cell
18. Why is phosphate rock not applied directly as a phosphorus fertilizer?

Topical Questions

Electrolytic Methods

19. (a) Write the balanced chemical equation for the electrolysis of aqueous NaCl.
   (b) In 2002, 8.98×10^9 kg NaOH and 1.14×10^{10} kg chlorine were produced in the United States. Does the ratio of these masses agree with the ratio of masses from the balanced chemical equation? If not, what does that suggest about the ways that NaOH and Cl_2 are produced?
20. ■ To produce magnesium metal, 1000. kg of molten $MgCl_2$ are electrolyzed.
   (a) At which electrode is magnesium produced?
   (b) What is produced at the other electrode?
   (c) How many moles of electrons are used in the process?
   (d) An industrial process uses 8.4 kWh per pound of Mg. How much energy is required per mole of magnesium?
21. A Downs cell operates at 7.0 V and 4.0×10^4 A.
   (a) How much Na(s) and Cl_2(g) can be produced in 24 hours by such a cell?
   (b) Assuming 100% efficiency, what is the energy consumption (kWh) of this cell?
22. ■ How much energy (kWh) is required to prepare a ton of sodium in a typical Downs cell operating at 25,000 A and 7.0 V?
23. What mass of aluminum can be produced when 6.0×10^4 A is passed through a series of 100 Hall-Héroult electrolytic cells operating at 85% efficiency for 24 hours?
24. What mass of aluminum can be produced from the electrolysis of molten $AlCl_3$ in an electrolytic cell operating at 100. A for 2.00 hr?

General Questions

25. Complete this table.

Formula	Name	Oxidation State of Nitrogen
______	Nitrogen	______
NH_3	______	______
______	Nitrous acid	______
______	Nitrogen dioxide	______
NH_4^+	______	______
______	Ammonium nitrate	______

26. Complete this table.

Formula	Name	Oxidation State of Phosphorus
______	Phosphorus	______
$(NH_4)_2HPO_4$	______	______
______	Phosphoric acid	______
______	Tetraphosphorus decaoxide	______
$Ca_3(PO_4)_2$	______	______
______	Calcium dihydrogen phosphate	______

27. Molten NaCl is electrolyzed in a Downs cell operating at 1.00×10^4 A for 24 hr.
   (a) How much sodium is produced?
   (b) What volume of Cl_2 in liters is collected from the outlet tube at 20. °C and 15 atm?
28. ■ Bauxite, the principal source of aluminum oxide, contains 55% Al_2O_3. How much bauxite is required to produce the 5.0×10^6 tons of Al produced annually by electrolysis?

29. Write a plausible Lewis structure for azide ion, N_3^-.
30. Write a plausible Lewis structure for P_4O_{10}.
31. Write the chemical equation for the
    (a) Combustion of white phosphorus.
    (b) Reaction of the combustion product with water.
32. Calculate the temperature at which the conversion of white phosphorus to red phosphorus occurs. $\Delta H° = -17.6$ kJ/mol; $\Delta S° = -18.3$ J/K at 25 °C.
33. There are two common oxides of sulfur. Name these oxides, and write chemical equations for the reaction of each with water. Identify the products.
34. What raw materials are used to produce sulfuric acid? Write chemical equations to represent the steps in the contact process to produce sulfuric acid.
35. Write Lewis structures for all the resonance forms of sulfuric acid.
36. Write Lewis structures for all the resonance forms of nitric acid.
37. Iodine trichloride, ICl_3, is an interhalogen compound.
    (a) Write the Lewis structure of ICl_3.
    (b) Does the central atom have more than an octet of valence electrons?
    (c) Using VSEPR theory, predict the molecular shape of ICl_3.
38. At some temperature, a gaseous mixture in a 1.00-L vessel originally contained 1.00 mol SO_2 and 5.00 mol O_2. When equilibrium was reached, 77.8% of the SO_2 had been converted to SO_3. Calculate the equilibrium constant (K_c) for this reaction at this temperature.
39. White phosphorus is soluble in carbon disulfide. At 10. °C, 800. g P_4 dissolves in 100. g CS_2. The density of carbon disulfide at 10. °C is 1.26 g/mL.
    (a) Calculate the molarity of phosphorus in the solution.
    (b) Calculate the molality of phosphorus in the solution.
40. ■ Drugstores sell 3% aqueous hydrogen peroxide that is used as an antiseptic. Hydrogen peroxide, H_2O_2, decomposes to water and oxygen. Calculate the volume of oxygen produced if 250. mL of 3% hydrogen peroxide decomposes fully at 750. mm Hg and 22 °C.
41. Calculate the volume of concentrated (98%) sulfuric acid that is needed to produce two tons of phosphoric acid from the reaction of sulfuric acid with sufficient phosphate-bearing rock. Density of conc. sulfuric acid = 1.84 g/mL.

$$Ca_3(PO_4)_2(s) + 3\,H_2SO_4(\ell) \longrightarrow 2\,H_3PO_4(\ell) + 3\,CaSO_4(s)$$

Applying Concepts

42. ■ The K_{sp} of $Ca(OH)_2$ is 7.9×10^{-6}; that of $Mg(OH)_2$ is 1.5×10^{-11}. Calculate the equilibrium constant for the reaction

$$Ca(OH)_2(s) + Mg^{2+}(aq) \longrightarrow Ca^{2+}(aq) + Mg(OH)_2(s)$$

Use it to explain why this reaction can be used commercially to extract magnesium from seawater.

43. Commercial concentrated nitric acid contains 69.5 mass percent HNO_3 and has a density of 1.42 g/mL.
    (a) Calculate the molarity of this solution.
    (b) What volume of the concentrated acid must be used to prepare 10.0 L of 6.00 M HNO_3?
44. The compound nitrosyl azide, N_4O, is a covalent compound with an NNNNO atomic arrangement. Write a plausible Lewis structure for this compound.
45. Hydrazoic acid, HN_3, is very explosive in its pure state but can be studied in aqueous solution. The acid is prepared by the reaction of hydrazine with nitrous acid.

$$N_2H_4(\ell) + HNO_2(aq) \longrightarrow HN_3(aq) + 2\,H_2O(\ell)$$

    (a) Determine the oxidation states of nitrogen in the compounds in this reaction.
    (b) What is the oxidizing agent in this reaction?
    (c) The K_a of hydrazoic acid is 2.4×10^{-5} at 25 °C. Calculate the pH of a 0.010 M solution of HN_3.
46. (a) Write two plausible resonance structures for hydrazoic acid, HN_3.
    (b) Use bond energy data (Table 8.2) to calculate the $\Delta H_f°$ for each resonance form; $\Delta H_f° = 218.0$ kJ/mol for H(g) and 472.7 kJ/mol for N(g). (See Appendix J.)
47. Given the reaction

$$Cl_2(g) + H_2O(\ell) \rightleftharpoons H^+(aq) + Cl^-(aq) + HOCl(aq)$$

    (a) Identify the oxidizing agent and the reducing agent.
    (b) Write the equilibrium constant expression for the reaction.
    (c) Calculate the concentration of HOCl in equilibrium with $Cl_2(g)$ at 1.0 atm. $K_c = 2.7 \times 10^{-5}$.
48. Dinitrogen trioxide, N_2O_3, is a blue liquid formed by the reaction of NO_2 and NO.
    (a) Write a balanced chemical equation for the formation of N_2O_3.
    (b) Write the Lewis structure of N_2O_3 and any plausible resonance forms.
    (c) Predict the O—N—O and the N—N—O bond angles.
49. (a) Write the resonance forms of SO_3.
    (b) Predict the molecular shape of SO_3 and the O—S—O bond angle.
50. Iodine can be produced by the oxidation of iodide ion with permanganate ion.

$$MnO_4^- + 2\,I^-(aq) + 8\,H^+(aq) \longrightarrow I_2(s) + Mn^{2+}(aq) + 4\,H_2O(\ell)$$

Excess HI is added to 0.200 g MnO_4^-. Assuming 100% yield, how many grams of iodine are produced?

51. ■ In a Downs cell, molten NaCl is electrolyzed to sodium metal and chlorine gas.

$$2\,NaCl(\ell) \longrightarrow 2\,Na(\ell) + Cl_2(g)$$

$\Delta H°$ and $\Delta S°$ for the reaction are +820 kJ and +180 J/K, respectively.
    (a) Calculate $\Delta G°$ at 600. °C, the electrolysis temperature.
    (b) Calculate the voltage required for the electrolysis.

52. A 425-gal tank of water contains 175 g NaI. Calculate the number of liters of chlorine gas at 758 mm Hg and 25 °C required to convert all the iodide to iodine.

53. ■ The phase diagram for sulfur is given below. The solid forms of sulfur are rhombic and monoclinic.

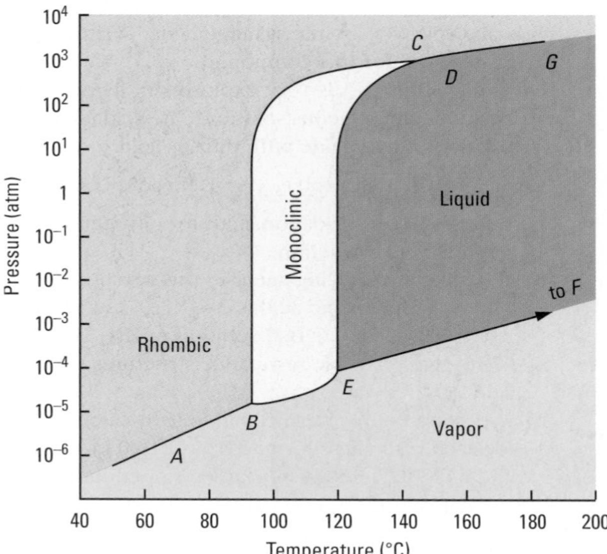

Phase diagram of sulfur.

(a) How many triple points does sulfur have? Indicate the approximate temperature and pressure at each.

(b) Which physical states are present at equilibrium under these conditions?
(i) 10^{-2} atm and 80 °C
(ii) 10^{-1} atm and 140 °C
(iii) 10^{-3} atm and 110 °C
(iv) 10^{-4} atm and 160 °C

54. Refer to the phase diagram in Question 53. Explain how you would sublime monoclinic sulfur.

55. In the laboratory, small amounts of bromine can be produced by the reaction of hydrobromic acid with MnO_2. The unbalanced equation for the reaction is

$$MnO_2(s) + H_3O^+(aq) + Br^-(aq) \longrightarrow Mn^{2+}(aq) + Br_2(\ell) + H_2O(\ell)$$

(a) Balance the equation.
(b) If the reaction is in 100% yield, how many moles of bromide ions must react to produce 6.50 g bromine?
(c) Calculate how many grams of MnO_2 are required in part (b).

56. Lapis lazuli is an aluminum silicate whose brilliant blue color is due to the presence of S_3^- ions. Write a plausible Lewis structure for this ion.

Lapis lazuli jewelry.

57. Bromine is prepared by bubbling 0.240 mol gaseous chlorine into 0.500 L of a solution that is 0.500 M in bromide ion.
(a) Determine the limiting reactant.
(b) Calculate the theoretical number of moles of bromine that could be produced.
(c) If 0.124 mol bromine is produced, calculate the percent yield.

58. Magnesium can be extracted from dolomite, a mineral that contains 13.2% Mg, 21.7% Ca, 13.0% C, and 52.1% O. Determine the simplest formula for this ionic compound.

59. A natural brine found in Arkansas has a bromide ion concentration of 5.00×10^{-3} M. If 210. g Cl_2 were added to 1.00×10^3 L of the brine,
(a) Which would be the limiting reactant?
(b) What would be the theoretical yield of Br_2 (density = 3.12 g/mL)?

60. Chlorine gas was first prepared by Karl Scheele in 1774 by the reaction of sodium chloride, manganese(IV) oxide, and sulfuric acid. In addition to chlorine, the reaction produces water, sodium sulfate, and manganese(II) sulfate. Write the balanced equation for this reaction.

61. Mercury(II) azide, $Hg(N_3)_2$, is an unstable compound used as a detonator in blasting caps. Calculate the volume (L) of nitrogen produced at 1 atm and 25 °C when 2.50 g mercury azide decomposes to liquid mercury and nitrogen.

More Challenging Questions

62. Use the Clausius-Clapeyron equation to calculate the heat of sublimation (J/mol) of monoclinic sulfur. See Question 53.

63. ■ At 20. °C the vapor pressure of white phosphorus is 0.0254 mm Hg; at 40. °C it is 0.133 mm Hg. Use the Clausius-Clapeyron equation to estimate the heat of sublimation (J/mol) of white phosphorus.

64. The density of sulfur vapor at 700. °C and 1.00 atm is 0.8012 g/L. What is the molecular formula of sulfur in the vapor?

65. Assume that the radius of the earth is 6400 km, the crust is 50. km thick, the density of the crust is 3.5 g/cm^3, and 25.7% of the crust is silicon by mass. What is the total mass of silicon in the crust of the earth?

66. A 5.00-g sample of white phosphorus is burned in excess oxygen and the product is dissolved in sufficient water to form 250. mL of solution.
 (a) Write the balanced chemical equation for the burning of phosphorus in excess oxygen.
 (b) Calculate the pH of the resulting solution.
 (c) An excess of aqueous calcium nitrate is added to the solution causing a white precipitate to form. Write a balanced chemical equation for this reaction and calculate the mass of precipitate formed.
 (d) An excess of zinc is added to the remaining solution. The reaction generates a colorless gas. Identify the gas and calculate its volume at STP.

67. Bromine, Br_2, reacts vigorously with hydrogen, H_2, to form hydrogen bromide. At STP, 100. mL hydrogen gas reacts with a stoichiometric amount of bromine. The resulting hydrogen bromide is dissolved in sufficient water to form 250. mL of solution. Calculate the pH of the solution.

68. ■ A solid-fuel rocket booster is used to lift a rocket from its launching pad. The solid fuel contains a mixture of ammonium perchlorate, NH_4ClO_4, and powdered aluminum metal that reacts in the presence of an Fe_2O_3 catalyst.

$$3\,NH_4ClO_4(s) + 3\,Al(s) \longrightarrow Al_2O_3(s) + AlCl_3(s) + 6\,H_2O(g) + 3\,NO(g)$$

 (a) Which element is oxidized? Which is reduced?
 (b) Use the data in Appendix J and calculate the enthalpy change for the reaction. ΔH_f° $NH_4ClO_4(s) = -295$ kJ/mol.

69. At 1 atm and approximately 1800 °C, 50% of P_4 is dissociated into 2 P_2. If equilibrium is established under these conditions, calculate the equilibrium constant.

70. The reaction for the production of white phosphorus is exothermic; $\Delta H^\circ = -3060$ kJ/mol P_4.

$$2\,Ca_3(PO_4)_2(s) + 6\,SiO_2(s) + 10\,C(s) \longrightarrow 6\,CaSiO_3(s) + 10\,CO(g) + P_4(g)$$

Calculate the ΔH_f° of $CaSiO_3(s)$; ΔH_f° ($Ca_3(PO_4)_2(s)$) = -4138 kJ. The enthalpy of sublimation of phosphorus = 13.06 kJ/mol.

71. A typical electric furnace using 500. V and a certain amperage produces four tons of phosphorus per hour using the reaction

$$2\,Ca_3(PO_4)_2(s) + 6\,SiO_2(s) + 10\,C(s) \longrightarrow 6\,CaSiO_3(s) + 10\,CO(g) + P_4(g)$$

Calculate the amperage needed to produce this much phosphorus.

72. ■ The Gibbs free energy of formation, ΔG_f°, of HI is +1.70 kJ/mol at 25 °C. Calculate the equilibrium constant for the reaction $HI(g) \rightleftharpoons \frac{1}{2}H_2(g) + \frac{1}{2}I_2(g)$.

73. Predict whether the ionization of the alkaline earth elements is easier or harder than that of the alkali metals. Explain your answer.

74. Use a Born-Haber cycle to calculate the lattice energy of MgF_2 using these thermodynamic data.

$$\begin{aligned}\Delta H_{sublimation}\ Mg(s) &= +146\ kJ/mol;\\ B.E.\ F_2(g) &= 158\ kJ/mol;\\ I.E._1\ Mg(g) &= +738\ kJ/mol;\\ I.E._2\ Mg^+(g) &= +1451\ kJ/mol;\\ E.A.\ F(g) &= -328\ kJ/mol;\\ \Delta H_f^\circ\ MgF_2(s) &= -1124\ kJ/mol.\end{aligned}$$

Compare this lattice energy with that of SrF_2, -2496 kJ/mol. Explain the difference in the values in structural terms.

75. Aluminum metal reacts rapidly and completely with hydrochloric acid, but only incompletely with nitric acid. Explain.

76. ■ Elemental analysis of a borane indicates this composition: 84.2% B and 15.7% H. The compound has a molar mass of 76.7 g/mol. Determine the molecular formula of the borane.

77. The reaction of iodine with excess liquid chlorine produces the interhalogen compound I_2Cl_6.
 (a) Write the Lewis structure of this molecule.
 (b) Use VSEPR theory to predict the structure of the molecule.

78. Hydroxyapatite is the important compound in tooth enamel. It dissociates according to the equation

$$Ca_5(PO_4)_3OH(s) \rightleftharpoons 5\,Ca^{2+}(aq) + 3\,PO_4^{3+}(aq) + OH^-(aq)$$

Children drink milk to obtain calcium, but the fermentation of the milk produces lactic acid, which remains on the teeth.
 (a) Use the equation above to explain how drinking milk helps babies to produce "strong" teeth.
 (b) Explain why the lactic acid inhibits formation of strong teeth.

22 Chemistry of Selected Transition Elements and Coordination Compounds

Absorption and transmittance of light in the visible region of the spectrum by *d*-to-*d* electron transitions in transition metal ions cause the vivid colors of stained glass windows, such as this one. Transition metal oxides are added to colorless molten glass, which then solidifies to produce colored glass. The oxides and their colors are: Cu_2O (red); Cr_2O_3 (green); Co_2O_3 (blue); and MnO_2 (violet). Transition metals, their complex ions, and coordination compounds are discussed in this chapter.

It is hard to overstate how important metals have been to the development of civilizations. Transitions from the Stone Age (400,000 to 7000 BC) to the Bronze Age (4000 to 3500 BC) to the Iron Age (1800s and beyond) and to the Computer Age (late 20th century) have been marked by humans' ability to extract metals from ores and to process the metals into tools and objects useful in industry, warfare, and homes. In this chapter we consider the transition metals, the *d*-block of elements of the periodic table. Some of these elements and their compounds are of major economic importance. The precious metals gold, silver, and platinum are transition elements used in coinage and jewelry. Others, such as iron and its alloy, steel, are valuable for their structural uses.

We will first consider the transition metals in overview and then look more closely at a few of them, including iron, the most economically important. The chapter closes with coverage of coordination compounds, in which ions or molecules are bonded to transition metal ions or atoms. Such compounds run the gamut from being responsible for the vivid colors of famous oil paintings to their role in significant biomolecules such as hemoglobin, vitamin B_{12}, and critical enzymes.

ThomsonNOW™
Throughout the text, this icon indicates an opportunity to test yourself on key concepts and to explore interactive modules by signing in to ThomsonNOW at **www.thomsonedu.com**.

22.1 Properties of the Transition (*d*-Block) Elements

The four series of *d*-block elements called the ***transition elements*** are in the center of the periodic table in Periods 4 through 7. As the name indicates, these elements lie between the very active metals of the *s* block and the less reactive metals of the *p*-block elements. In this section we discuss properties shared by all the transition elements.

The transition elements in Period 7 beginning with rutherfordium, element 104, are all radioactive. They have been made synthetically in *very* small amounts, just several atoms in some cases, and therefore, little is currently known about their properties.

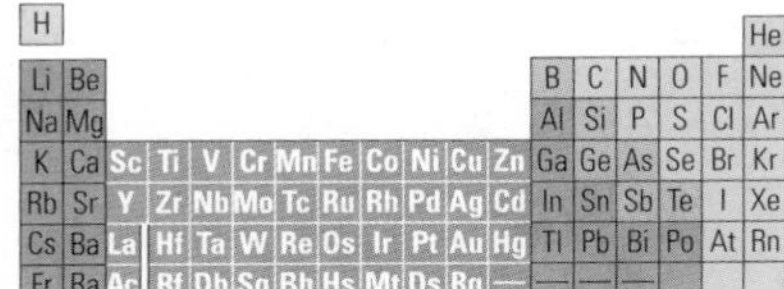

Compared to the *s*-block and *p*-block elements, there is a slow, steady transition in properties from one transition metal to the next. These generalities apply to the transition elements:

- All are metals that conduct electricity well, but to varying degrees.
- Most are ductile (able to be drawn into a wire) and malleable (able to be hammered into thin sheets).
- Except for gold and copper, they are silvery-white or bluish.
- They usually have higher melting and boiling points than the main group elements; tungsten has the highest melting point of any metal (3410 °C). Mercury is an exception, being the only liquid metal at room temperature.

Many transition metal ions form colored aqueous solutions. Concentrated aqueous solutions of nitrate salts containing *(left to right):* Fe^{3+}, Co^{2+}, Ni^{2+}, and Cu^{2+}.

Table 22.1 Outermost Electron Configurations of *d*-Block Elements

Deviations (marked in color) occur when a nonstandard configuration is more stable.

Configuration	$(n-1)d\,ns^2$	$(n-1)d^2\,ns^2$	$(n-1)d^3\,ns^2$	$(n-1)d^4\,ns^2$	$(n-1)d^5\,ns^2$	$(n-1)d^6\,ns^2$	$(n-1)d^7\,ns^2$	$(n-1)d^8\,ns^2$	$(n-1)d^9\,ns^2$	$(n-1)d^{10}\,ns^2$
First series:	$_{21}$**Sc** $3d^1 4s^2$	$_{22}$**Ti** $3d^2 4s^2$	$_{23}$**V** $3d^3 4s^2$	$_{24}$**Cr** $3d^5 4s^1$	$_{25}$**Mn** $3d^5 4s^2$	$_{26}$**Fe** $3d^6 4s^2$	$_{27}$**Co** $3d^7 4s^2$	$_{28}$**Ni** $3d^8 4s^2$	$_{29}$**Cu** $3d^{10} 4s^1$	$_{30}$**Zn** $3d^{10} 4s^2$
Second series:	$_{39}$**Y** $4d^1 5s^2$	$_{40}$**Zr** $4d^2 5s^2$	$_{41}$**Nb** $4d^4 5s^1$	$_{42}$**Mo** $4d^5 5s^1$	$_{43}$**Tc** $4d^5 5s^2$	$_{44}$**Ru** $4d^7 5s^1$	$_{45}$**Rh** $4d^8 5s^1$	$_{46}$**Pd** $4d^{10}$	$_{47}$**Ag** $4d^{10} 5s^1$	$_{48}$**Cd** $4d^{10} 5s^2$
Third series:	$_{57}$**La** $5d^1 6s^2$	$_{72}$**Hf** $4f^{14}5d^2 6s^2$	$_{73}$**Ta** $4f^{14}5d^3 6s^2$	$_{74}$**W** $4f^{14}5d^4 6s^2$	$_{75}$**Re** $4f^{14}5d^5 6s^2$	$_{76}$**Os** $4f^{14}5d^6 6s^2$	$_{77}$**Ir** $4f^{14}5d^7 6s^2$	$_{78}$**Pt** $4f^{14}5d^9 6s^1$	$_{79}$**Au** $4f^{14}5d^{10} 6s^1$	$_{80}$**Hg** $4f^{14}5d^{10} 6s^2$

4*f*-elements intervene

- They generally have high densities; osmium (22.61 g/mL) and iridium (22.65 g/mL) are the most dense metals, even more dense than gold (19.3 g/mL).
- Many form brightly colored compounds.
- Some are paramagnetic; a few are ferromagnetic (⇐ *p. 307*).
- They form complex ions (Section 22.6).
- Most have multiple oxidation states; the scandium (+3) and zinc (+2) groups are exceptions.

The concept of oxidation states (numbers) is described in Sections 5.4 and 19.1.

The transition elements at the beginning of each series show considerable differences in their chemical behavior from those at the end of each series. Such differences in chemical behavior can be attributed to the number and distribution of *d* orbital electrons (Table 22.1). At the left side of the series, members of the scandium group of elements (Sc, Y, La) are reactive metals like the alkaline earth metals, their predecessors in each period. On the other hand, the zinc family members (Zn, Cd, Hg) are not like other transition elements in that the zinc family members have filled $(n-1)\,d$ and ns sublevels. In fact, the elements of the zinc family are sometimes not classified as transition elements.

Electron Configurations

All transition elements have the electron configuration

$$[\text{noble gas}](n-1)\,d^x ns^y$$

where n is the period number (4 through 7), x is the number of d electrons (1 through 10), and y is the number of s electrons (1 or 2, except in palladium), as summarized in Table 22.1. The number of d electrons increases from left to right across a transition metal series. Elements in the zinc group (Zn, Cd, Hg) at the end of the series have filled $(n-1)\,d^{10}$ sublevels. In the preceding group the elements Cu, Ag, and Au also have filled $(n-1)\,d^{10}$ sublevels, along with half-filled ns^1 sublevels, as described below.

The progressive filling of the d orbitals from Sc to Zn in the first series is not uniform, as seen in Table 22.2. The first three elements of the series—Sc, Ti, and

Table 22.2 Orbital Occupancy of the First Transition Series Elements

Element	Partial Orbital Diagram 3*d*					4*s*	4*p*			Unpaired Electrons
Sc	↑					↑↓				1
Ti	↑	↑				↑↓				2
V	↑	↑	↑			↑↓				3
Cr	↑	↑	↑	↑	↑	↑				6
Mn	↑	↑	↑	↑	↑	↑↓				5
Fe	↑↓	↑	↑	↑	↑	↑↓				4
Co	↑↓	↑↓	↑	↑	↑	↑↓				3
Ni	↑↓	↑↓	↑↓	↑	↑	↑↓				2
Cu	↑↓	↑↓	↑↓	↑↓	↑↓	↑				1
Zn	↑↓	↑↓	↑↓	↑↓	↑↓	↑↓				0

V—have [Ar]$3d^14s^2$, [Ar]$3d^24s^2$, and [Ar]$3d^34s^2$ electron configurations, respectively. The filling sequence changes at chromium, which has a ground state electron configuration of [Ar]$3d^54s^1$ with *two* half-filled sublevels, which is a lower energy state than [Ar]$3d^44s^2$. The sequence reverts back to that expected from manganese through nickel, with the pairing of 3*d* electrons. It changes again with copper, which has a filled $3d^{10}$ sublevel and a half-filled $4s^1$ sublevel (Table 22.2) that is more stable than the predicted [Ar]$3d^94s^2$ configuration. The first series ends with zinc, which has filled 3*d and* 4*s* sublevels, [Ar]$3d^{10}4s^2$.

Transition metal atoms lose electrons to form transition metal cations. ***When transition metal atoms lose electrons to form cations, the ns electrons are lost before the (n − 1)d electrons.*** Thus, both 4*s* electrons are lost when Fe^{2+} and Fe^{3+} form. These ions differ in their number of 3*d* electrons, not 4*s* electrons, that are lost; each ion has lost both 4*s* electrons.

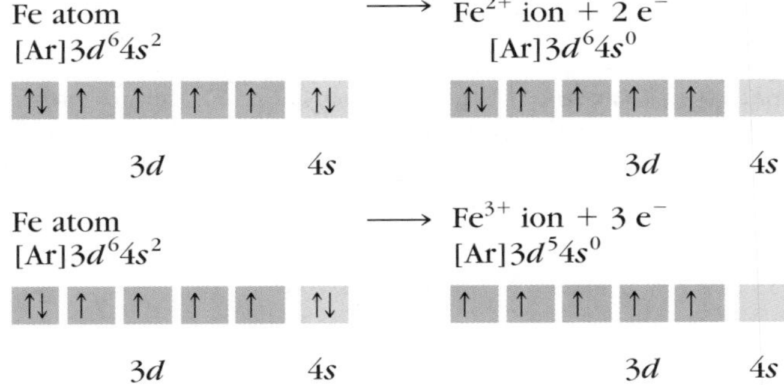

Magnetic susceptibility measurements confirm the electron configurations of the first row and other transition metal ions (Figure 22.1). The *magnetic moment* is a value calculated from the measured paramagnetism of a sample and is indicative of the number of unpaired electrons in the sample (⇐ ***p. 307***). As seen from Figure 22.2, the greater the number of unpaired electrons, the greater the magnetic moment of the substance. Notice from Figure 22.2 the confirming experimental evidence that the 4*s* electrons are removed first to give Fe^{2+} and Fe^{3+} four and five

If a substance is diamagnetic (has no unpaired electrons), its apparent mass is slightly reduced when the magnetic field is "on."

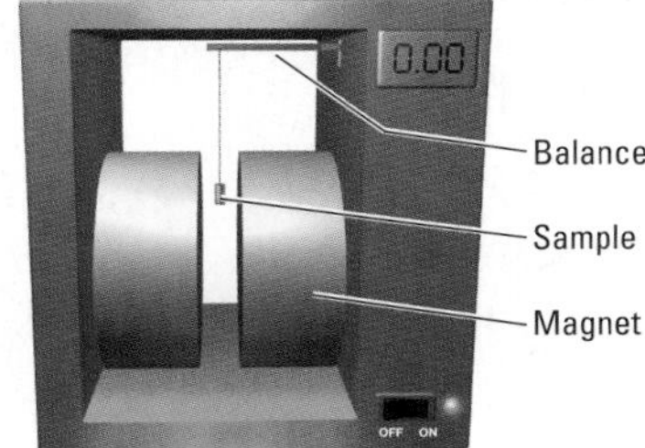

(a)

If a substance is paramagnetic (has unpaired electrons), its apparent mass increases when the field is "on" because the balance arm feels an additional force as the sample is attracted by the magnetic field.

(b)

Figure 22.1 Measurement of magnetic behavior.

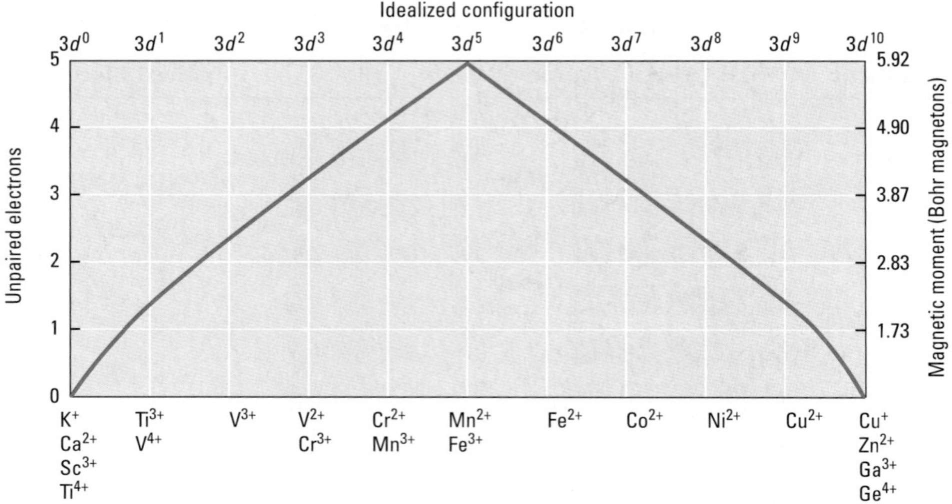

Figure 22.2 **The number of unpaired electrons in the first-row transition metal ions and their magnetic moments.**

unpaired electrons, respectively. If the 3*d* electrons were removed first, these ions would have four (Fe^{2+}) and three (Fe^{3+}) unpaired electrons.

PROBLEM-SOLVING EXAMPLE 22.1 Transition Metal Ion Electron Configurations

Use orbital box diagrams to explain the number of unpaired electrons shown in Figure 22.2 for (a) Ti^{3+}, (b) Cr^{2+}, and (c) Cu^{2+}.

Answer

(a) $[Ar]3d^14s^0$ (b) $[Ar]3d^44s^0$

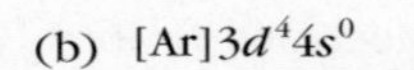

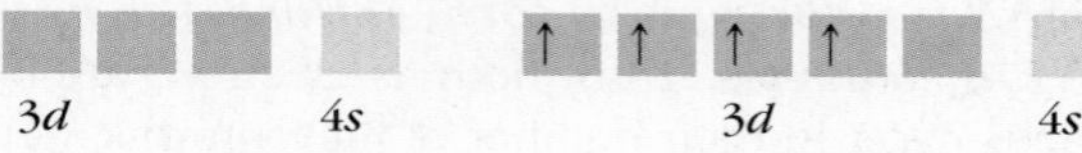

(c) $[Ar]3d^94s^0$

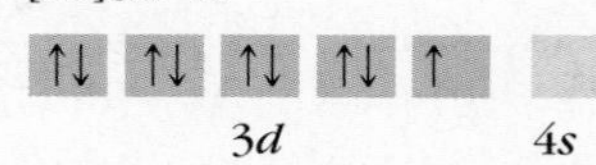

Strategy and Explanation When an ion is formed from a Period 4 transition metal, 4*s* electrons are removed before 3*d* electrons, giving rise to the number of unpaired electrons in the ion.

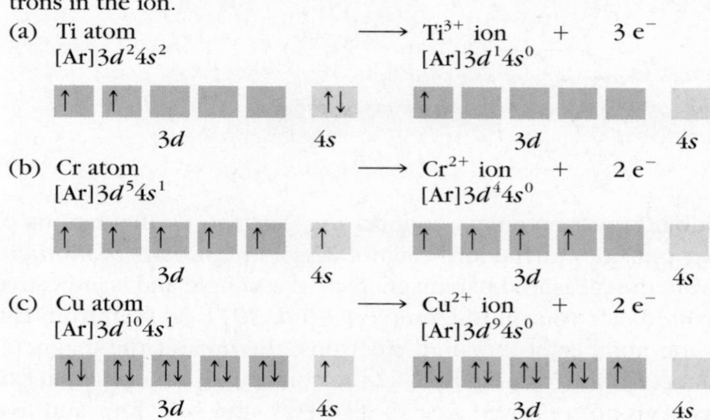

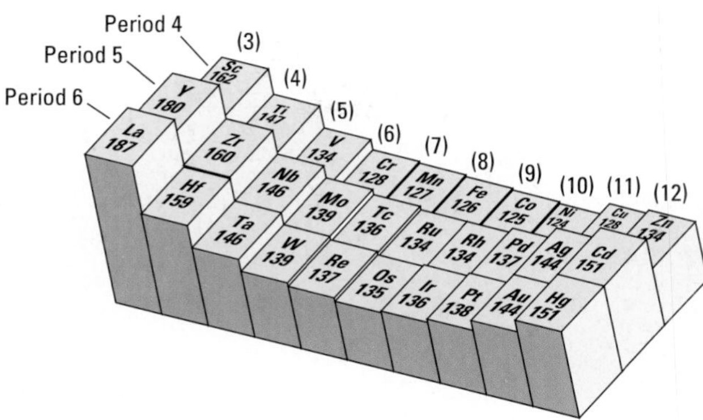

Figure 22.3 **Radii of transition metals (in picometers).**

PROBLEM-SOLVING PRACTICE 22.1

Use orbital box diagrams to explain the number of unpaired electrons shown in Figure 22.2 for (a) Mn^{2+} (b) Cr^{3+}.

Trends in Atomic Radii

Transition metals have less variation in atomic radii across a period than do main group elements. Across a row of transition elements, the atomic radii decrease steadily and then increase slightly (Figure 22.3). The decrease occurs because as each *d* electron is added, there is also an increase of one unit of nuclear charge (a proton). Repulsions among the electrons are not sufficient to counteract the added attraction of the nucleus to the electrons, so the atomic radius decreases. Toward the end of each transition metal series, the radii increase slightly due to several factors, including electron-electron repulsion as electrons are paired in *d* orbitals.

The radii of the second transition series elements (Period 5) are, as expected, greater than those of the first-row transition metals. What is unexpected is what occurs with the atomic radii in going from the second row to the third row (Period 6). Instead of the third-row radii being larger, they are nearly the same as those of the second row. This is a consequence of the lanthanide series of elements (La, element 57, to Lu, element 71) intervening between barium, the Period 6 alkaline earth element, and hafnium, the next transition element of Period 6. In the lanthanide series atoms, the effective nuclear charge builds up, causing a decrease in their atomic radii because all the additional electrons go into the 4*f* orbitals, which do not effectively screen valence electrons from the increasing nuclear charge. The increased nuclear charge pulls the valence electrons closer to the nucleus, decreasing the atomic radii. This decrease in size, known as the **lanthanide contraction,** just offsets the expected size increase going from the second to the third row of transition elements. Consequently, atoms of the second- and third-row transition elements are of similar size, which causes elements in the same group to have similar chemical properties. The transition elements of the second and third rows occur together in ores, and because of their chemical similarities, they are very difficult to separate from each other.

Oxidation States

Except for the scandium group (+3) and zinc group (+2) elements, all transition metals have multiple oxidation states. The oxidation states of the first transition

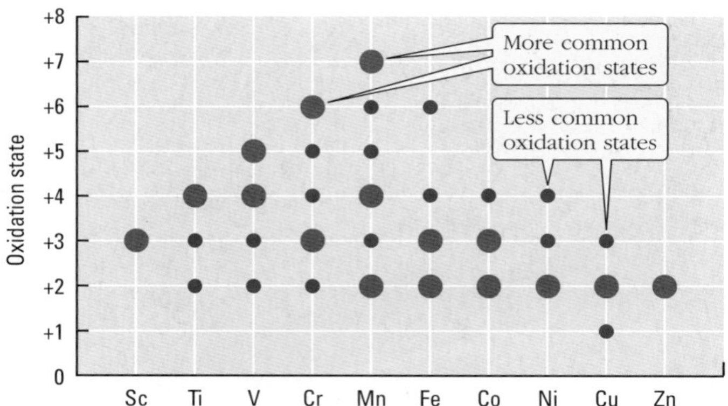

Figure 22.4 **Oxidation states of first transition series elements.**

series elements are listed in Figure 22.4. Manganese, for example, has three common oxidation states: +2 in Mn^{2+}, +4 in MnO_2, and +7 in MnO_4^-. Iron has two common oxidation states: +2 in FeO and +3 in Fe_2O_3. Less common oxidation states are also noted in Figure 22.4.

Transition metals that form 2+ ions do so, in general, by losing two *ns* electrons before losing any $(n - 1)$ *d* electrons. Higher oxidation states involve losing $(n - 1)$ *d* electrons as well. The maximum oxidation state for the first five elements in the first series—Sc through Mn—equals the total of the $(n - 1)$ *d* plus *ns* electrons. Thus, the maximum oxidation state of chromium is +6, which is found in CrO_4^{2-} and $Cr_2O_7^{2-}$, and that of manganese is +7, which is found in MnO_4^-. These high oxidation states make $Cr_2O_7^{2-}$ (in acidic solution) and MnO_4^- strong oxidizing agents. In general, compounds in which the transition metal has a low oxidation state tend to be ionic, whereas compounds with transition metals in high oxidation states are relatively covalent. Thus, $MnCl_2$ (mp 650 °C) is an ionic solid containing Mn^{2+} and Cl^- ions. On the other hand, MnO_4^- is a polyatomic ion containing covalent Mn—O bonds.

Ions of transition elements with partially filled *d* orbitals can accept or donate electrons, a property that makes them effective components of catalysts. In iron compounds, for example, iron can be present as Fe^{2+} (reduced form) or Fe^{3+} (oxidized form). Each iron ion acts as an electron shuttle, losing or gaining electrons between the oxidized and reduced forms when catalyzing electron transfer reactions, such as the production of ammonia by the Haber-Bosch process ***(← p. 706)***.

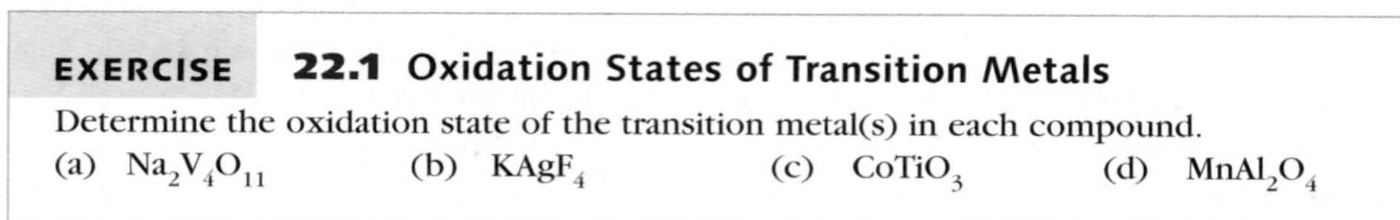
EXERCISE 22.1 Oxidation States of Transition Metals

Determine the oxidation state of the transition metal(s) in each compound.

(a) $Na_2V_4O_{11}$ (b) $KAgF_4$ (c) $CoTiO_3$ (d) $MnAl_2O_4$

EXERCISE 22.2 Oxidation State

Determine the oxidation state of copper in $K_2Pb[Cu(NO_2)_6]$.

22.2 Iron and Steel: The Use of Pyrometallurgy

Iron is the most abundant transition metal and the second most abundant metallic element in the earth's crust (4.7%). Pure iron is a silvery-white, rather soft metal. Its great commercial importance comes with the addition of very small amounts of carbon or other alloying elements to it to form steel.

In air Fe^{2+}(aq) is oxidized to Fe^{3+}(aq).

$$4\ Fe^{2+}(aq) + O_2(g) + 4\ H^+(g) \longrightarrow 4\ Fe^{3+}(aq) + 2\ H_2O(\ell)$$

Aqueous Fe^{3+} reacts with water to form a hydrated oxide known as *rust.*

$$2\ Fe^{3+}(aq) + 4\ H_2O(\ell) \longrightarrow Fe_2O_3 \cdot H_2O(s) + 6\ H^+(aq)$$

Iron reacts with weakly oxidizing acids such as HCl and acetic acid to form the pale green $[Fe(H_2O)_6]^{2+}$ ion.

$$Fe(s) + 2\ H^+(aq) \longrightarrow Fe^{2+}(aq) + H_2(g) \qquad E° = +0.44\ V$$

Fe^{2+}(aq) is the hydrated ion, $[Fe(H_2O)_6]^{2+}$.

When reacted with strongly oxidizing acids such as dilute nitric acid, the metal is oxidized directly to Fe^{3+}.

$$Fe(s) + 4\ H^+(aq) + NO_3^-(aq) \longrightarrow Fe^{3+}(aq) + NO(g) + 2\ H_2O \quad E° = +1.00\ V$$

The principal iron ores are hematite, Fe_2O_3, and magnetite, Fe_3O_4, which are found in large deposits in Minnesota, Commonwealth of Independent States, France, England, and Australia. Iron production involves steps to concentrate and purify the ores. Iron ions in the oxide ores are reduced to the metal by using carbon in the form of coke as the reducing agent at high temperatures in a blast furnace (Figure 22.5). **Pyrometallurgy** is the extraction of a metal from its ore using chemical reactions carried out at high temperatures.

Iron also occurs in nature as the pyrite, FeS_2. This mineral is not used as an ore because in steel making it is difficult to remove all the sulfur, which makes the steel brittle.

To extract iron, a mixture of iron ore, coke, and limestone ($CaCO_3$) is fed into the top of the blast furnace, and a blast of heated air or oxygen is forced up from the bottom of the furnace. The coke reacts exothermically with the heated air, producing a high temperature that speeds up the iron-forming reactions, which makes the process economical. The iron ore is reduced to metallic iron by the reactions

$$2\ C(s) + O_2(g) \longrightarrow 2\ CO(g) \qquad \text{exothermic}$$

$$Fe_2O_3(s) + 3\ CO(g) \longrightarrow 2\ Fe(s) + 3\ CO_2(g) \qquad \text{exothermic}$$

Limestone is added to remove silica-containing impurities in the ore.

$$CaCO_3(s) \longrightarrow CaO(s) + CO_2(g) \qquad \text{endothermic}$$

$$CaO(s) + SiO_2(s) \longrightarrow CaSiO_3(\ell) \qquad \text{endothermic}$$

Iron ore, coke, and limestone are continuously added at the top.
Flue gas
In the reducing zone, CO is oxidized and Fe_2O_3 reduced. $Fe_2O_3 + 3\ CO \longrightarrow 2\ Fe + 3\ CO_2$
230 °C
525 °C
945 °C
1510 °C
Hot gases are used to preheat air.
Heated air
Slag
Molten iron is drawn off the bottom.

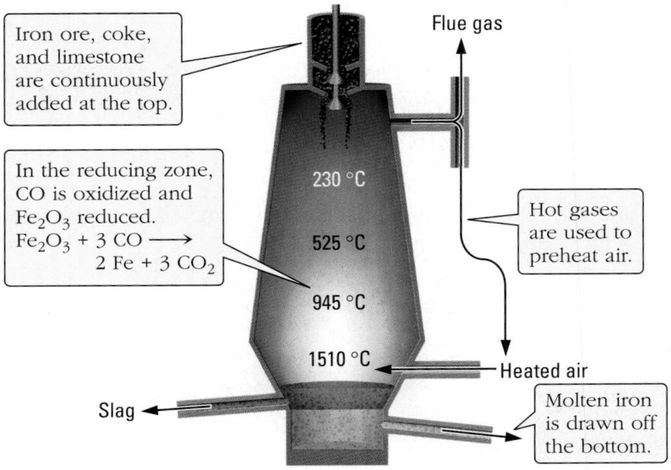

Figure 22.5 Diagram of a blast furnace. Iron ore is reduced to iron in a blast furnace.

Calcium silicate and other metal silicates form *slag*, which is a liquid at the temperature of the blast furnace. The result is the formation of two layers at the bottom of the furnace. The lower, more dense liquid is molten iron that contains a substantial concentration of dissolved carbon and smaller concentrations of other impurities. The upper liquid layer is the slag. Periodically, the blast furnace is tapped from the bottom to draw off the molten iron. The liquid slag is drawn off at a port higher in the furnace (Figure 22.5).

The iron produced by a blast furnace is *pig iron*, a material that is brittle due to impurities of as much as 4.5% carbon, 1.7% manganese, 0.3% phosphorus, 0.04% sulfur, and 1% silicon. The principal embrittling material is cementite, Fe_3C, an iron carbide formed at the temperatures of the blast furnace.

$$3\ Fe(s) + C(s) \longrightarrow Fe_3C(s)$$

Molten pig iron that is poured into molds of a desired shape is called *cast iron*. It can be used directly to make molded automobile engine blocks, brake drums, transmission housings, and the like. Cast iron is too brittle for most structural uses. To convert cast iron or pig iron into **steel,** a much stronger material, the phosphorus, sulfur, and silicon impurities must be removed and the carbon content reduced to about 1.3%.

Steel

Many iron alloys are known collectively as *steels,* each with its own particular structural properties. One of the most common is carbon steel, an iron alloy containing 0.5 to 1.3% carbon. To convert pig iron to steel, the excess carbon is oxidized away using oxygen. Thus, whereas extracting iron from an ore is a reduction process, steel making is an oxidation process. One of several techniques used to make steel is the *basic oxygen process* (Figure 22.6), in which pure oxygen is blown through a ceramic tube that is pushed below the surface of molten, impure iron. At the high temperatures of the melt, the dissolved carbon reacts rapidly with the oxygen to form gaseous carbon monoxide and carbon dioxide, which are vented. The scale of the basic oxygen process operation is impressive. About 200 tons of molten pig iron, 100 tons of scrap iron, and 20 tons of limestone are loaded into the furnace at a time. The steel is produced within an hour using such a process.

The composition of steel is varied by adding silicon, chromium, manganese, molybdenum, nickel, or other metals to give the steel specific physical, chemical, and mechanical properties. Table 22.3 lists the composition and uses of some common steel alloys. Magnetic alloys can be permanent magnets, such as those in audio

ESTIMATION

Steeling Automobiles

Steel is the most recycled consumer product, given that about 95% of steel from automobiles is recycled. The Steel Recycling Institute estimates that 14 million tons of steel scrap from automobiles was recycled in 2000. Estimate how much iron ore (tons) containing approximately 2% Fe_3O_4 did not have to be mined that year due to the use of recycled steel. For this estimation problem, assume that the steel is 100% iron.

Approximately 14×10^6 tons of iron (steel) is recycled. Each 10^3 tons iron ore contains 20 tons Fe_3O_4 in which there are 168 tons Fe per 232 tons Fe_3O_4. About 1×10^9 tons of ore did not have to be mined because of using recycled steel, saving a substantial amount of material and the energy needed to process the ore.

$$14 \times 10^6 \text{ tons iron} \times \frac{232 \text{ tons } Fe_3O_4}{168 \text{ tons iron}} \times \frac{10^3 \text{ tons iron ore}}{20 \text{ tons } Fe_3O_4} = 1 \times 10^9 \text{ tons ore}$$

ThomsonNOW™
Sign in to ThomsonNOW at **www.thomsonedu.com** to work an interactive module based on this material.

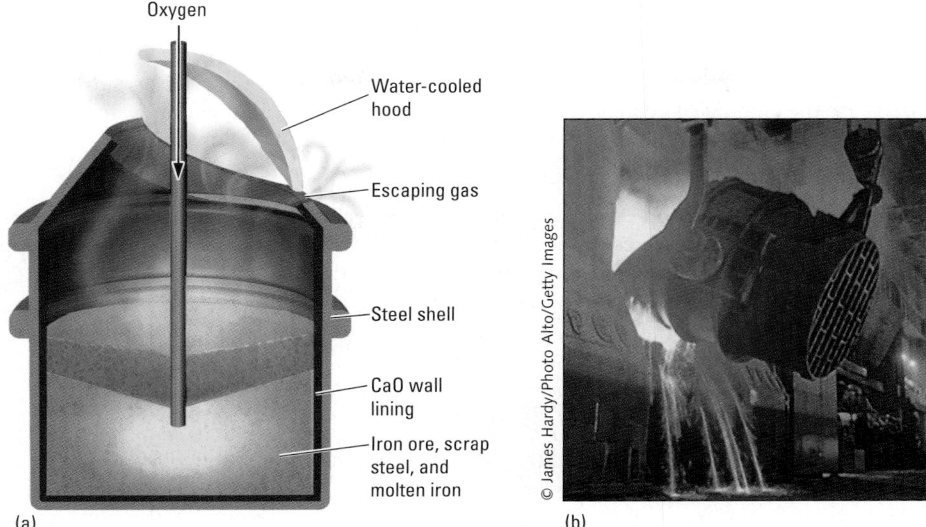

Figure 22.6 The basic oxygen process for making steel. (a) Most of the steel manufactured today is produced by the basic oxygen process. (b) Molten steel being poured from a basic oxygen furnace.

speakers and computer hard drives, or temporary magnets such as those in electric motors, generators, and transformers. Alnico is the general name for a series of popular permanent magnets containing Al, Ni, Co, Fe, and sometimes Cu and Ti. Alnico V, for example, contains 51% Fe, as well as four other elements—14% Ni, 24% Co, 8% Al, and 3% Cu.

Blacksmiths and other steel fabricators have long known that the properties of steel are also affected by the processing temperature, cooling rates, and hammering, rolling, and extrusion. If hot steel is cooled quickly by immersing it in water or oil, the carbon in the steel will remain as cementite, Fe_3C, resulting in hard, but brittle, steel. By rapid cooling, followed by controlled reheating, a process called *tempering,* the cementite-to-graphite ratio is adjusted and the properties of the resultant steel varied further.

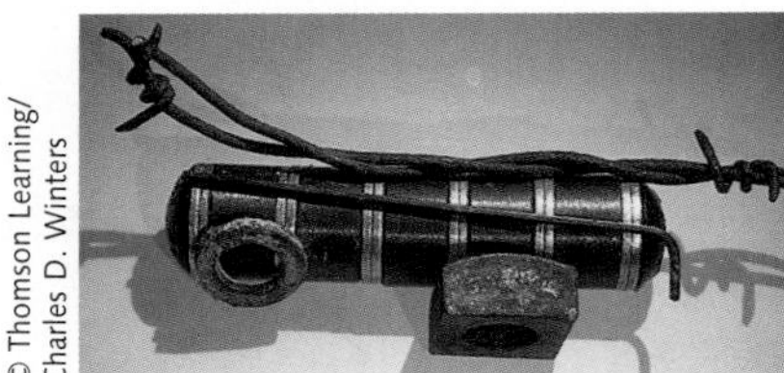

An Alnico magnet picking up iron or steel objects.

The artist Michelangelo wrote about using fire to transform substances:

"It is with fire that blacksmiths iron subdue
Unto fair form, the image of their thought . . ."

Sonnet 59

Table 22.3 Some Steels and Their Uses

Name	Composition	Properties	Uses
Carbon steel	98.7% Fe, 1.3% C	Hard	Sheet steel, tools
Manganese steel	10–18% Mn, 90–82% Fe, 0.5% C	Hard, resistant to wear	Railroad rails, safes, armor plate
Stainless steel	14–18% Cr, 7–9% Ni, 79–73% Fe, 0.2% C	Resistant to corrosion	Cutlery, instruments
Nickel steel	2–4% Ni, 98–96% Fe, 0.5% C	Hard, elastic, resistant to corrosion	Drive shafts, gears, cables
Invar steel	36% Ni, 64% Fe, 0.5% C	Low coefficient of expansion	Meter scales, measuring tapes
Silicon steel	1–5% Si, 99–95% Fe, 0.5% C	Hard, strong, highly magnetic	Magnets
Duriron	12–15% Si, 88–85% Fe, 0.85% C	Resistant to corrosion by acids	Pipes
High-speed steel	14–20% W, 86–80% Fe, 0.5% C	Retains temper when hot	High-speed cutting tools

PROBLEM-SOLVING EXAMPLE 22.2 Cementite

Cementite, Fe_3C, is produced as an impurity during iron production in a blast furnace. To decrease the amount of cementite produced, the blast furnace temperature is reduced. Is the formation of cementite exothermic or endothermic? Explain your answer.

Answer Endothermic

Strategy and Explanation We can apply Le Chatelier's principle by noting that lowering the temperature increases the amount of iron and carbon not converted to cementite. This must mean that raising the temperature increases the amount of cementite formed, indicating cementite formation is an endothermic reaction.

$$3\ Fe(s) + C(s) \rightleftharpoons Fe_3C \quad \text{endothermic}$$

PROBLEM-SOLVING PRACTICE 22.2

In a blast furnace at approximately 1000 °C, much of the carbon is converted into cementite, Fe_3C, during the production of iron, which is used subsequently to produce steel. If the temperature of the steel is decreased rapidly, cementite is trapped in the iron to produce a steel with a high cementite concentration causing the steel to be brittle. Explain why this is the case. If the steel is cooled slowly, not rapidly as before, what would you expect the nature of what is formed to be?

PROBLEM-SOLVING EXAMPLE 22.3 Iron Production

(a) How much carbon monoxide is needed to form 1.00×10^3 kg iron from hematite, Fe_2O_3, and from magnetite, Fe_3O_4?

(b) How much carbon in the form of coke must be used to prepare the total amount of CO needed for the two reductions in part (a)?

Answer

(a) 7.52×10^5 g CO for hematite and 6.68×10^5 g CO for magnetite

(b) 6.09×10^5 g C

Strategy and Explanation Write the balanced chemical equation and use its stoichiometric factors to relate the mass of iron to the mass of carbon monoxide needed. The equation for the reduction of hematite to iron is

$$Fe_2O_3(s) + 3\ CO(g) \longrightarrow 2\ Fe(s) + 3\ CO_2(g)$$

(a) Applying the information from the balanced chemical equation, the mass of CO required to form 1.00×10^3 kg Fe from hematite is

$$1.00 \times 10^3 \text{ kg Fe}\left(\frac{10^3 \text{ g Fe}}{1 \text{ kg Fe}}\right)\left(\frac{1 \text{ mol Fe}}{55.85 \text{ g Fe}}\right)\left(\frac{3 \text{ mol CO}}{2 \text{ mol Fe}}\right)\left(\frac{28.00 \text{ g CO}}{1 \text{ mol CO}}\right) = 7.52 \times 10^5 \text{ g CO}$$

The CO required to reduce magnetite can be calculated the same way based on the equation

$$Fe_3O_4(s) + 4\ CO(g) \longrightarrow 3\ Fe(s) + 4\ CO_2(g)$$

$$1.00 \times 10^3 \text{ kg Fe}\left(\frac{10^3 \text{ g Fe}}{1 \text{ kg Fe}}\right)\left(\frac{1 \text{ mol Fe}}{55.85 \text{ g Fe}}\right)\left(\frac{4 \text{ mol CO}}{3 \text{ mol Fe}}\right)\left(\frac{28.00 \text{ g CO}}{1 \text{ mol CO}}\right) = 6.68 \times 10^5 \text{ g CO}$$

(b) The carbon monoxide is prepared by the reaction

$$2\ C(s) + O_2(g) \longrightarrow 2\ CO(g)$$

From part (a), the total mass of CO required is $(7.52 \times 10^5 \text{ g}) + (6.68 \times 10^5 \text{ g}) = 1.42 \times 10^6$ g. The required amount of carbon is

$$1.42 \times 10^6 \text{ g CO}\left(\frac{1 \text{ mol CO}}{28.00 \text{ g CO}}\right)\left(\frac{2 \text{ mol C}}{2 \text{ mol CO}}\right)\left(\frac{12.01 \text{ g C}}{1 \text{ mol C}}\right) = 6.09 \times 10^5 \text{ g C}$$

✓ **Reasonable Answer Check**

(a) Hematite: 10^3 kg iron is equivalent to about

$$10^6 \text{ g Fe} \times \frac{1 \text{ mol Fe}}{56 \text{ g Fe}} = 1.8 \times 10^4 \text{ mol Fe}$$

which requires 1.5 times that number of moles of CO, about 2.7×10^4 mol CO. This is equivalent to approximately 8×10^5 g CO.

Magnetite: Using the same approach, we estimate that magnetite requires 7×10^5 mol CO. The estimated values for CO needed for each ore are close to the calculated values, so the answers are reasonable.

(b) The sum of the estimated masses of CO is about 1.5×10^6 g CO. This is close to 6×10^5 g C, the actual calculated answer, which is reasonable.

PROBLEM-SOLVING PRACTICE 22.3

Calculate the total mass of iron ore needed to produce 1.00×10^3 kg iron by the process described in Problem-Solving Example 22.3.

EXERCISE 22.3 High-Speed Steel

Ultrahigh-speed steel is used in some saw blades. Such a saw blade, weighing 500. g, contains 0.6% C, 4.0% Cr, 18% W, 1.0% Mo, 1.5% V, and 6.0% Co along with iron.

(a) Calculate the mass of W and of Co in the saw blade.

(b) Which alloying metal is present in the greatest mole percent, that is,

$$\frac{\text{moles alloying metal}}{\text{moles all alloying metals}} \times 100\%?$$

22.3 Copper: A Coinage Metal

Copper is sometimes found in metallic form in nature, and evidence suggests that such naturally occurring copper was known and used during the Stone Age. As early as 10,000 years ago, the metal was hammered into useful items such as coins, jewelry, tools, and weapons. Nearly five millennia later, the Bronze Age was ushered in when humans learned how to alloy copper with tin to make bronze.

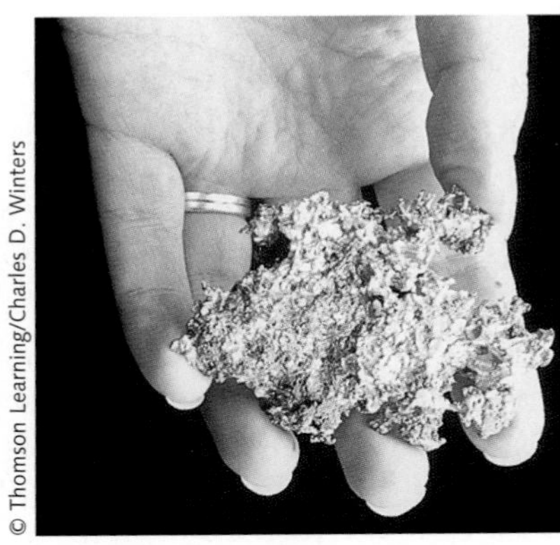

© Thomson Learning/Charles D. Winters

Native copper.

The Metallurgy of Copper

Native copper, that which is found uncombined chemically in nature, is not available in sufficient supply to meet the demands for the metal, and so chemical methods have been developed to extract copper from its ores. The principal ores are chalcocite, Cu_2S, and chalcopyrite, $CuFeS_2$, which occur in conjunction with two iron sulfides, FeS_2 and FeS.

Modern methods to extract copper from its ores begin with crushing the ore and separating it from rocks. The ore is then heated in air to temperatures high enough to drive off the sulfur as sulfur dioxide, a process called *roasting*.

$$3\ FeS_2(s) + 8\ O_2(g) \xrightarrow{\text{heat}} Fe_3O_4(s) + 6\ SO_2(g)$$

$$2\ CuFeS_2(s) + O_2(s) \xrightarrow{\text{heat}} Cu_2S(s) + 2\ FeS(s) + SO_2(g)$$

Copper is separated from the iron by melting the Cu_2S and Fe_3O_4 mixture and then combining it with oxygen and SiO_2 to form a liquid iron silicate slag and molten Cu_2S. The slag is less dense than the molten copper(I) sulfide and floats on it, where it can be drawn off periodically.

James Cowlin

An open-pit copper mine near Bagdad, Arizona.

Copper Hill, Tennessee, a wasteland as a result of SO_2 released by copper smelting.

A copper electrorefining facility showing the refined copper plated on cathodes.

The conversion of copper(I) sulfide to metallic copper takes place in a process similar to the basic oxygen process for steel making. After the iron silicate slag is removed, air is blown through the molten Cu_2S, converting it to Cu_2O, which reacts with the remaining Cu_2S to form copper metal.

$$2\ Cu_2S(\ell) + 3\ O_2(g) \longrightarrow 2\ Cu_2O(\ell) + 2\ SO_2(g)$$

$$2\ Cu_2O(\ell) + Cu_2S(\ell) \longrightarrow 6\ Cu(\ell) + SO_2(g)$$

The resulting copper, called *blister copper,* is about 96% to 99.5% copper, which can be further purified by electrolysis.

The electrorefining of copper is carried out in large electrolytic cells in which anodes of blister copper bars alternate with very thin sheets of pure copper, which are the cathodes (Figure 22.7). A mixture of copper(II) sulfate and dilute sulfuric acid is the electrolyte. Copper metal in the impure mixture is oxidized at the anode to Cu^{2+}, and Cu^{2+} in the electrolyte is reduced to pure metallic copper at the cathode.

Anode:	$Cu(s, \text{impure blister copper}) \longrightarrow Cu^{2+}(aq) + 2\ e^-$
Cathode:	$Cu^{2+}(aq) + 2\ e^- \longrightarrow Cu(s, \text{pure})$
Overall Reaction:	$Cu(s, \text{impure blister copper}) \longrightarrow Cu(s, \text{pure})$

By controlling the voltage, only copper and those impurities (zinc, lead, iron) in the blister copper anode that have a low enough reduction potential (more easily oxidized than Cu) are oxidized and dissolved in the electrolyte. Any less easily oxidized metal impurities, such as metallic gold and silver, are essentially unaffected and drop from the anode as it is consumed during electrolysis, forming an *anode sludge.* The voltage is regulated so that only copper, the least electropositive of the metals in the electrolyte, is plated onto the pure copper cathode. Electrorefined copper is greater than 99.9% pure, a purity required for copper used in electrical applications. Generally, enough gold, silver, and platinum are recovered from the anode sludge to pay for the cost of copper electrorefining.

PROBLEM-SOLVING EXAMPLE 22.4 Copper Electrorefining

Copper is electrorefined by removing impurities such as lead, silver, gold, and zinc. Based on these standard reduction potentials, explain how copper can be separated by electrolysis from these impurities.

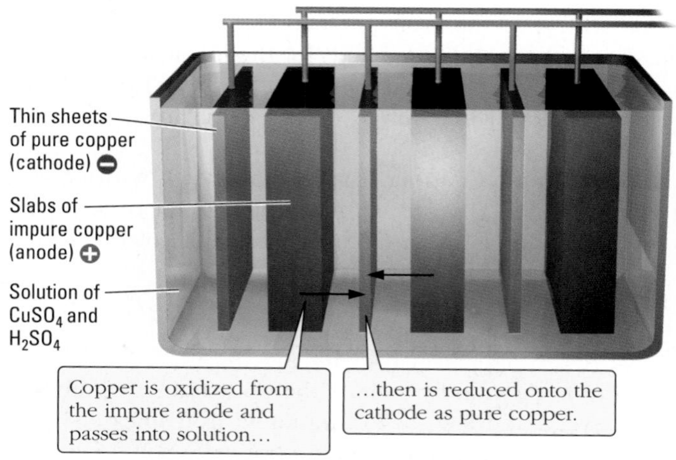

Figure 22.7 **Electrolytic cell for refining copper.**

	$E°$(V)
$Ag^+(aq) + e^- \longrightarrow Ag(s)$	+0.799
$Au^+(aq) + e^- \longrightarrow Au(s)$	+1.68
$Cu^{2+}(aq) + 2\,e^- \longrightarrow Cu(s)$	+0.337
$Pb^{2+}(aq) + 2\,e^- \longrightarrow Pb(s)$	−0.126
$Zn^{2+}(aq) + 2\,e^- \longrightarrow Zn(s)$	−0.763

Answer At the potential that just oxidizes blister copper from the anode (−0.337 V relative to a standard hydrogen electrode), zinc and lead are also oxidized. Silver and gold are not oxidized and form the anode sludge. This potential is not sufficient to reduce Zn^{2+} and Pb^{2+}, so only copper forms at the cathode.

Strategy and Explanation Consider what reactions could occur at the impure blister copper anode and the pure copper cathode. Adjust the cell potential so that only copper and metals that are more easily oxidized than copper will dissolve. Because the other metals that dissolve are easier to oxidize than copper, their ions will be harder to reduce and therefore copper(II) ions will be reduced at the cathode.

Ag^+ and Au^+ have more positive standard reduction potentials than Cu^{2+}, so Ag^+ and Au^+ are easier to reduce than Cu^{2+}. For the reverse (oxidation) reactions, Cu will be easier to oxidize than Ag or Au. A potential that just oxidizes Cu will not oxidize Ag or Au, which will form the anode sludge.

Pb^{2+} and Zn^{2+} are more difficult to reduce than Cu^{2+}. Therefore, Pb and Zn are easier to oxidize than Cu, and a potential that oxidizes Cu at the anode will also oxidize Zn and Pb. Thus, Cu^{2+}, Zn^{2+}, and Pb^{2+} will be available in solution for reduction at the cathode. Because Pb^{2+} and Zn^{2+} are more difficult to reduce than Cu^{2+}, only Cu^{2+} will react at the cathode, forming pure Cu.

PROBLEM-SOLVING PRACTICE 22.4

Explain how zinc and lead could be separated from each other without plating out copper in the electrolytic cell in Problem-Solving Example 22.4.

The mass of pure copper deposited on a cathode by electrorefining can be calculated. For example, a copper electrorefining cell operates at 200. amperes (A) for 24 hours a day for a year. How many kilograms of pure copper are produced? This mass can be calculated by determining the number of coulombs of charge used and recognizing that two moles of electrons are needed for each mole of Cu^{2+} reduced to copper metal. From Section 19.12 we know that

$$1 \text{ ampere} = 1 \text{ coulomb/second (C/s) and } 1 \text{ mol e}^- = 9.65 \times 10^4 \text{ C}$$

$$365 \text{ days} \times 200.\text{ A} \times \left(\frac{1\text{ C/s}}{1\text{ A}}\right)\left(\frac{24\text{ h}}{1\text{ day}}\right)\left(\frac{3600\text{ s}}{1\text{ h}}\right) = 6.31 \times 10^9\text{ C}$$

$$6.31 \times 10^9\text{ C}\left(\frac{1\text{ mol e}^-}{9.65 \times 10^4\text{ C}}\right)\left(\frac{1\text{ mol Cu}}{2\text{ mol e}^-}\right) = 3.27 \times 10^4\text{ mol Cu}$$

$$3.27 \times 10^4\text{ mol Cu}\left(\frac{63.55\text{ g Cu}}{1\text{ mol Cu}}\right)\left(\frac{1\text{ kg Cu}}{10^3\text{ g Cu}}\right) = 2.08 \times 10^3\text{ kg Cu}$$

EXERCISE 22.4 Refining Copper

What mass of pure copper is deposited during electrorefining in a cell operating at 250. A for 12.0 h?

An ancient bronze work of art.

Bronze and Brass

The Bronze Age began about 3800 BC with the discovery that bronze formed when tin combined with copper. This discovery was likely accidental, brought on by the fact that copper ores and tin ores are often found together. In a stroke of good fortune, during the reduction of copper ore by the charcoal of a wood fire some tin ore was present, and tin ions were also reduced to tin metal. The heated mixture of the resulting copper and tin metals formed bronze. The advantage of bronze is that it is sufficiently hard to keep a cutting edge, something copper cannot do. Bronze usually contains 7 to 10% tin, but bronzes with as much as 20% tin are known. Because of its properties, all early civilizations that produced bronze used it to create weapons as well as exquisite works of art.

On prolonged exposure to moist air, copper or bronze forms a green outer coating (a patina) of copper(II) hydroxycarbonate, $Cu_2(OH)_2CO_3$, often seen on statuary, such as the Statue of Liberty, and on copper-clad roofs.

$$2\,Cu(s) + O_2(g) + CO_2(g) + H_2O(g) \longrightarrow Cu_2(OH)_2CO_3(s)$$

Pre-1982 (left) and post-1982 (right) pennies. A copper-clad penny (*right*) with some of the copper cladding removed to expose the silvery-appearing zinc core.

Brass is an alloy made of varying proportions of copper and zinc (20 to 45% Zn), which becomes harder as the percentage of zinc increases. Because brass is easy to forge, cast, and stamp, it is widely used for pipes, valves, and fittings.

Copper is used in all U.S. coins. By 1982, the price of copper had risen to where it was costing the U.S. Treasury Department more than 1 cent to make a penny. Since then, to conserve copper and reduce costs, the penny has been made of 97.5% zinc and 2.5% copper, with the zinc core sandwiched between two thin layers of copper. Silver-colored coins actually contain no silver. The U.S. nickel is made of a copper (75%) and nickel (25%) alloy of uniform composition throughout the coin. The other silver-colored coins are a "sandwich" made of a pure copper core covered with a thin layer of a copper and nickel alloy (91.67% copper). The Sacagawea dollar coin, which looks like gold, contains no gold.

Patina on the Statue of Liberty. The copper hydroxycarbonate coating on the statue is responsible for the green color.

Metallic copper is not attacked by most acids, although it does react with nitric acid. The nitrate ion, NO_3^-, acts as the oxidizing agent. In dilute nitric acid, NO_3^- is reduced to NO; concentrated nitric acid yields NO_2.

$$3\,Cu(s) + 2\,NO_3^-(aq) + 8\,H^+(aq) \longrightarrow 3\,Cu^{2+}(aq) + 2\,NO(g) + 4\,H_2O(\ell)$$

$$Cu(s) + 2\,NO_3^-(aq) + 4\,H^+(aq) \longrightarrow Cu^{2+}(aq) + 2\,NO_2(g) + 2\,H_2O(\ell)$$

Copper(II) sulfate pentahydrate, $CuSO_4 \cdot 5\,H_2O$, is the most widely used copper compound. Commonly called blue vitriol, this compound is used to kill algae and fungi. In the solid it contains the hydrated copper(II) complex ion, $[Cu(H_2O)_4]^{2+}$. The fifth water molecule is bound to the sulfate ion through hydrogen bonding. The water of hydration can be removed by gentle heating or by putting the hydrate into

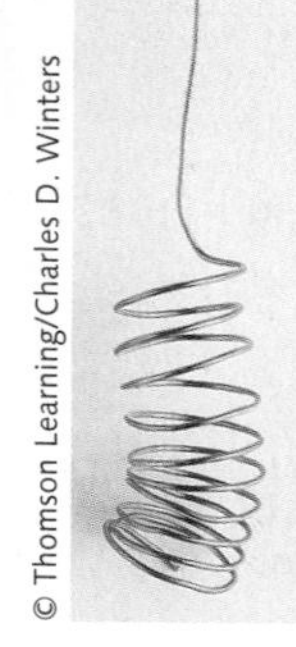

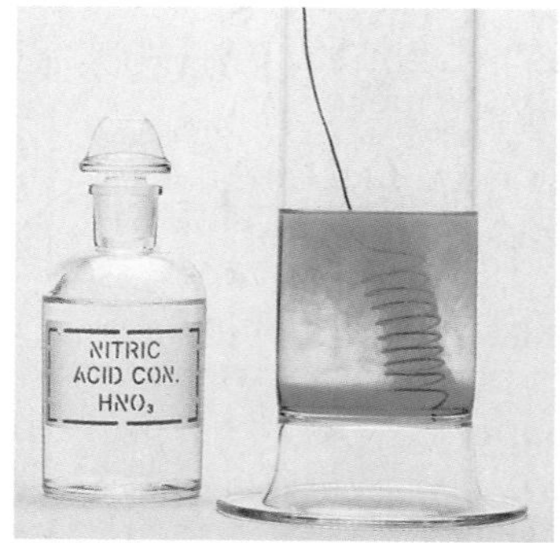

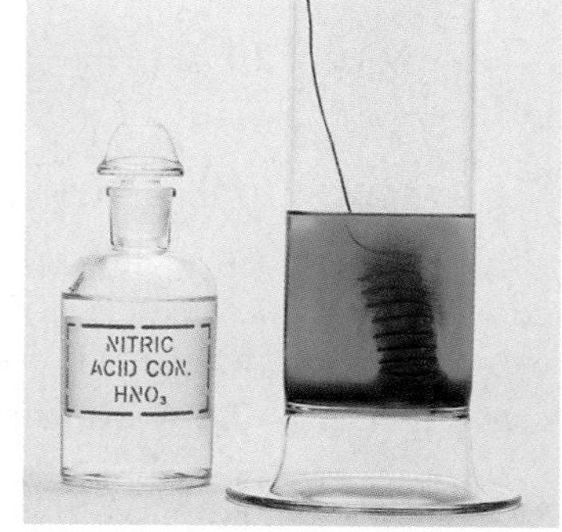

The reaction of metallic copper with concentrated nitric acid produces brown fumes of NO_2. The solution takes on the green color of the other reaction product, aqueous copper(II) nitrate.

a desiccator. Aqueous solutions containing copper(II) ions are blue due to the presence of the $[Cu(H_2O)_6]^{2+}$ ion.

Copper also exists in a +1 oxidation state, Cu^+, although aqueous copper chemistry involves principally Cu^{2+} because Cu^+ is unstable in aqueous solutions and disproportionates (undergoes both oxidation and reduction) into Cu and Cu^{2+}.

$$Cu^+(aq) + e^- \longrightarrow Cu(s) \qquad E° = +0.52 \text{ V}$$
$$Cu^+(aq) \longrightarrow Cu^{2+}(aq) + e^- \qquad E° = +0.15 \text{ V}$$
$$2\, Cu^+(aq) \longrightarrow Cu(s) + Cu^{2+}(aq) \qquad E° = +0.37 \text{ V}$$

In basic solution, however, Cu^{2+} is reduced to copper(I) oxide, Cu_2O, a reaction that serves as a traditional test for glucose in urine. When heated, a basic solution containing blue aqueous Cu^{2+} and a reducing sugar such as glucose react to form a brick-red precipitate of Cu_2O. The reducing sugar is represented here with its aldehyde functional group, RCHO.

$$\underset{\text{blue}}{2\, Cu^{2+}(aq)} + \underset{\text{reducing sugar}}{RCHO(aq)} + 5\, OH^-(aq) \longrightarrow \underset{\text{brick red}}{Cu_2O(s)} + RCOO^-(aq) + 3\, H_2O(\ell)$$

At elevated temperatures copper reacts with oxygen. Below 1000 °C it forms black copper(II) oxide, CuO. Above 1000 °C copper reacts with oxygen to form red copper(I) oxide, Cu_2O, which is found in the mineral cuprite.

(a)

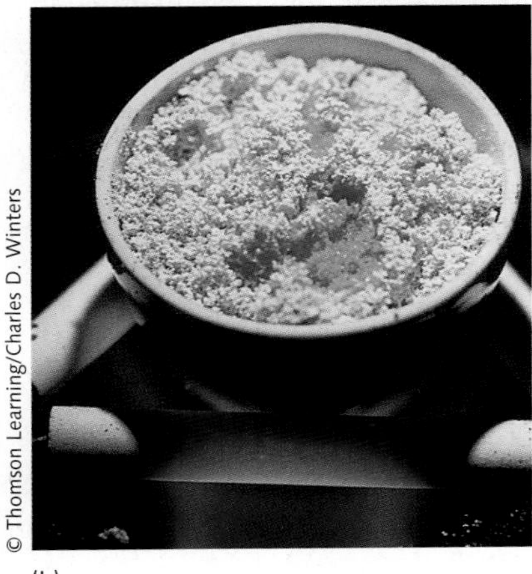

(b)

Hydrated copper(II) sulfate (a) and partially anhydrous copper(II) sulfate (b). Hydrated copper(II) sulfate is converted to anhydrous copper(II) sulfate by heating, which drives off the waters of hydration and causes the color to change.

In a disproportionation reaction, the same substance is oxidized as well as reduced.

22.4 Silver and Gold: The Other Coinage Metals

Silver and gold occur in elemental (uncombined) form in nature, although such sources have dwindled. In the past, gold prospectors, such as the forty-niners in the 1849 California gold rush, panned for gold. They simply swirled gold-bearing rock and gravel from streams in a pan. Because of its high density (19.3 g/cm^3), gold settles out from the sand (about 2.5 g/cm^3) and other rocky impurities. At present, the gold content in such deposits is much too low for panning to be effective.

Early civilizations highly prized silver and gold for their luster, corrosion resistance, and workability in making jewelry, art objects, and coins. Silver is the best metallic conductor of heat and electricity. Long synonymous with wealth and power, gold was thought to be a part of the sun, thus its Latin name *aurum,* meaning "bright dawn," from which its chemical symbol is derived. Gold is the most malleable metal, so malleable that it can be rolled thinly enough to be transparent. Very thin gold leaf is used to cover domes of churches and capitol buildings.

The lack of reactivity of gold and silver is explained by the standard reduction potentials of their ions, which are well above that of hydrogen (0.00 V).

$$Au^+(aq) + e^- \longrightarrow Au(s) \qquad E° = +1.68 \text{ V}$$
$$Ag^+(aq) + e^- \longrightarrow Ag(s) \qquad E° = +0.799 \text{ V}$$

Samples of red copper(I) oxide, Cu_2O, and black copper(II) oxide, CuO.

Both metal ions are fairly strong oxidizing agents. Consequently, neither metal reacts with weak oxidizing acids such as hydrochloric acid, nor do they react readily with oxygen at normal temperatures. When heated in air, gold does not react with oxygen, even at high temperatures. Silver, however, slowly forms silver(I) oxide, Ag_2O, which is unstable and decomposes back to the elements when heated strongly. Silver does react with oxidizing acids such as nitric acid (HNO_3). It takes aqua regia, a 3:1 mixture of concentrated HCl and HNO_3, to dissolve gold.

$$3\, Ag(s) + 4\, H^+(aq) + NO_3^-(aq) \longrightarrow 3\, Ag^+(aq) + NO(g) + 2\, H_2O(\ell)$$
$$Au(s) + 6\, H^+(aq) + 3\, NO_3^-(aq) + 4\, Cl^-(aq) \longrightarrow AuCl_4^-(aq) + 3\, NO_2(g) + 3\, H_2O(\ell)$$

The ability of aqua regia (Latin, "royal water") to dissolve gold has been known since the 1300s.

Gold can be hammered into ultrathin sheets for decorative purposes.

The oxidation of metallic gold to Au^{3+} is highly unfavorable, but the reaction in aqua regia occurs because as Au^{3+} ions form, they are tied up in the complex ion, $AuCl_4^-$, in a highly product-favored reaction.

$$Au^{3+}(aq) + 4\ Cl^-(aq) \longrightarrow AuCl_4^-(aq) \qquad K = 1 \times 10^{38}$$

Complex ions such as $AuCl_4^-$ were discussed in Section 17.5 and are discussed in detail in Section 22.6.

EXERCISE 22.5 Putting Silver and Gold into Solution

(a) Identify the oxidizing and reducing agents in the reaction of nitric acid with silver.
(b) Do the same for the reaction of gold with aqua regia.

Gold and silver are both relatively soft metals whose hardness is increased by alloying with other metals. Sterling silver, for example, is 92.5% silver and 7.5% copper. The proportion of gold in its alloys is expressed in carats. Pure gold (24 carats) is too soft to be used in jewelry. It is alloyed with copper or other metals to make the 18-carat and 14-carat gold for jewelry, which are 75% (18/24 × 100) and 58% (14/24 × 100) gold, respectively.

As noted earlier (p. 1074), silver and gold are by-products of the electro-refining of copper. They are also obtained from ores. Silver is obtained from its principal ore argentite, Ag_2S, by cyanide extraction. The ore is ground and put into a 0.5% solution of NaCN through which air is blown.

$$Ag_2S(s) + 4\ CN^-(aq) \longrightarrow 2\ Ag(CN)_2^-(aq) + S^{2-}(aq)$$

The formation of the $Ag(CN)_2^-$ complex ion brings the insoluble silver sulfide into solution.

$$\begin{aligned} Ag_2S(s) &\rightleftharpoons 2\ Ag^+(aq) + S^{2-}(aq) \\ 2\ Ag^+(aq) + 4\ CN^-(aq) &\rightleftharpoons 2\ Ag(CN)_2^-(aq) \\ \hline Ag_2S(s) + 4\ CN^-(aq) &\longrightarrow 2\ Ag(CN)_2^-(aq) + S^{2-}(aq) \end{aligned}$$

Powdered zinc, a good reducing agent, is added to the solution to convert Ag^+ to metallic silver.

$$Zn(s) + 2\ Ag(CN)_2^-(aq) \longrightarrow 2\ Ag(s) + Zn(CN)_4^{2-}(aq)$$

The silver can be further purified by electrolytic refining.

Like silver, gold is obtained from ores by cyanide extraction followed by reduction with zinc.

$$4\ Au(s) + 8\ CN^-(aq) + O_2(g) + 2\ H_2O(\ell) \longrightarrow 4\ Au(CN)_2^-(aq) + 4\ OH^-(aq)$$

$$Zn(s) + 2\ Au(CN)_2^-(aq) \longrightarrow 2\ Au(s) + Zn(CN)_4^{2-}(aq)$$

The cyanide extraction of silver and gold depends on the formation of the $Ag(CN)_2^-$ and $Au(CN)_2^-$ complex ions. Because of the toxicity of CN^-, waste cyanide solutions must be disposed of properly. In places where such has not been the case and the solutions were simply dumped near the processing site, serious environmental damage has occurred.

22.5 Chromium

Chromium is characteristic of the middle transition elements that exhibit multiple oxidation states (p. 1068). In compounds, chromium has oxidation states from +2 to +6 because of its five 3*d* electrons and one 4*s* electron, although +2 and +3 are

CHEMISTRY IN THE NEWS

Gold Nanoparticles in Drug Delivery

Nanoparticles are approximately the same size as biological materials; thus, they can be used for drug delivery systems. Jennifer L. West, a professor of chemical and biomedical engineering at Rice University, is researching the fabrication and use of nanoshells for drug delivery and cancer treatment. The nanoshells are very tiny spheres with a silica core covered by a layer of gold nanoparticles. Such nanoshells have tunable optical wavelengths, the tunability dependent on the thickness of the gold nanoparticle layer. West and her research colleagues have adjusted the thickness for the wavelength to fall in the near-infrared region, a region where human blood and tissue are relatively transparent.

One such nanoshell system is used for insulin delivery. In this case, the tunable nanoshells are embedded into polymers that respond to temperature changes. The polymers form a capsule-like container in which drugs such as insulin can be held. Irradiating the polymer with the proper wavelength causes the embedded nanoshells to heat up, which in turn activates the polymer causing it to collapse and release the drug. West's research group is also developing nanoshells into which antibodies and peptides have been added, designed to target cancer cells. In such cases, the nanoshells bind to the tumor and, when the proper wavelength of light is used, the nanoshells heat up, destroying the tumor. So far the heating process has been limited to 100 μm from the tumor surface.

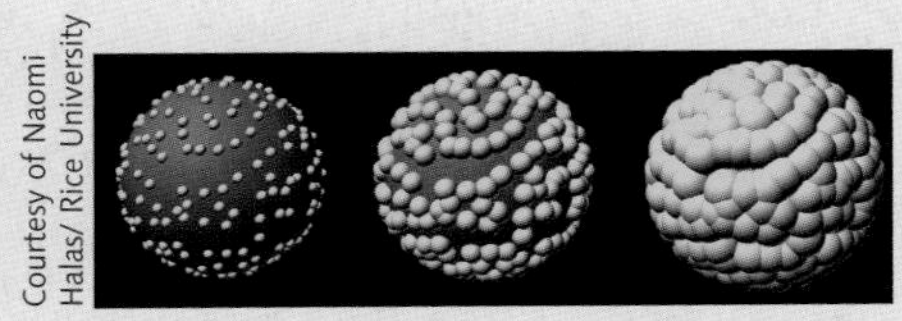

Courtesy of Naomi Halas/ Rice University

Gold-plated nanoshells used for drug delivery systems. The gold is layered around a silicone core.

SOURCE: *Chemical & Engineering News*, May 16, 2005; p. 30.

the most common oxidation states. The +6 oxidation state is found in CrO_4^{2-} and $Cr_2O_7^{2-}$ ions (Figure 22.8). The +4 and +5 oxidation states are uncommon.

Chromium is a hard, brittle metal that is extremely corrosion resistant due to a chromium oxide surface layer that passivates and protects the metal beneath from further oxidation. In this regard, chromium resembles aluminum (***p. 1040***).

Chromium is used in stainless steel and is plated on truck bumpers to give them a bright surface. Its oxides are used in magnetic recording tapes (CrO_2), in abrasives, and as a glass pigment (Cr_2O_3).

The chief chromium ore is chromite, $FeCr_2O_4$, which is treated with lime (CaO), oxygen, and sodium carbonate to form sodium chromate. Water is added to remove the soluble sodium chromate produced.

$$4\ FeCr_2O_4(s) + 8\ CaO(s) + 8\ Na_2CO_3(s) + 7\ O_2(g) \longrightarrow 8\ Na_2CrO_4(s) + 2\ Fe_2O_3(s) + 8\ CaCO_3(s)$$

Sodium chromate is used widely in chrome plating and as an intermediate in the formation of many chromium compounds. One of these compounds is chromium(III) oxide, from which chromium metal is obtained by reduction with aluminum metal.

$$2\ CrO_4^{2-}(aq) + 3\ SO_2(g) + 3\ H_2O(\ell) \longrightarrow Cr_2O_3(s) + 3\ SO_4^{2-} + 2\ H_3O^+(aq)$$

$$Cr_2O_3(s) + 2\ Al(s) \longrightarrow 2\ Cr(s) + Al_2O_3(s)$$

EXERCISE 22.6 Production of Chromium Metal

Use data from Appendix J to calculate the enthalpy change and the Gibbs free energy change for the reduction of chromium(III) oxide by aluminum.

Figure 22.8 Chromium ions in solution. The two flasks on the left contain solutions of chromium ions with oxidation state +3: $Cr(NO_3)_3$ (violet) and $CrCl_3$ (green). The flasks on the right show the colors of the two chromium anions with oxidation state +6: yellow chromate ion (CrO_4^{2-}) and orange dichromate ion ($Cr_2O_7^{2-}$).

In aqueous solution, chromate ions, CrO_4^{2-}, and dichromate ions, $Cr_2O_7^{2-}$, exist in a highly pH-dependent equilibrium.

$$\underset{\text{yellow}}{H^+(aq) + CrO_4^{2-}(aq)} \rightleftharpoons HCrO_4^-(aq)$$

$$2\,HCrO_4^-(aq) \rightleftharpoons \underset{\text{orange}}{Cr_2O_7^{2-}(aq)} + H_2O(\ell)$$

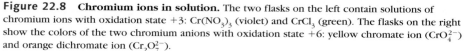

$HCrO_4^-$ $Cr_2O_7^{2-}$ H_2O

Notice from the second equilibrium that dichromate is formed by a condensation reaction in which two $HCrO_4^-$ units are joined by splitting out water.

The net equilibrium is

$$2\,H_3O^+(aq) + 2\,CrO_4^{2-}(aq) \rightleftharpoons Cr_2O_7^{2-}(aq) + 3\,H_2O(\ell) \qquad K_c = 4 \times 10^{14}$$

From the very large value of K, you should recognize that in acid, chromate is converted to dichromate; chromate is stable in basic or neutral solution (Figure 22.8).

CONCEPTUAL EXERCISE 22.7 Applying Le Chatelier's Principle

Use Le Chatelier's principle to explain how the addition of acid or base shifts the equilibrium to favor chromate or dichromate.

In highly acidic solution the dichromate ion is a powerful oxidizing agent.

$$Cr_2O_7^{2-}(aq) + 14\ H_3O^+(aq) + 6\ e^- \longrightarrow 2\ Cr^{3+}(aq) + 21\ H_2O(\ell) \qquad E° = +1.33\ V$$

It is sufficiently strong to oxidize alcohols to aldehydes or ketones (*p. 566*) and aldehydes to carboxylic acids, for example, acetaldehyde to acetic acid.

$$Cr_2O_7^{2-}(aq) + 3\ CH_3CHO(aq) + 8\ H_3O^+(aq) \longrightarrow 2\ Cr^{3+}(aq) + 3\ CH_3COOH(aq) + 12\ H_2O(\ell)$$

The oxidizing strength of dichromate decreases as the pH increases, as shown in Problem-Solving Example 22.5.

PROBLEM-SOLVING EXAMPLE 22.5 Dichromate and pH

Use the Nernst equation (*p. 946*) to calculate the E_{cell} for the reduction of dichromate ion by iodide ion at a pH of 4.0 and 25 °C with all concentrations other than H_3O^+ equal to 1.0 M. $E°_{cell}$ is 0.795 V.

Answer $E_{cell} = +0.242$ V

Strategy and Explanation We can use the balanced equation for the reaction and the Nernst equation to calculate the cell potential at the nonstandard conditions. The balanced equation (*p. 925*) for the reaction is

$$Cr_2O_7^{2-}(aq) + 6\ I^-(aq) + 14\ H_3O^+(aq) \longrightarrow 2\ Cr^{3+}(aq) + 3\ I_2(aq) + 21\ H_2O(\ell)$$

Six moles of electrons are transferred per mole of $Cr_2O_7^{2-}$:

$$2\ Cr^{6+} + 6\ e^- \longrightarrow 2\ Cr^{3+}$$

$$6\ I^- \longrightarrow 3\ I_2 + 6\ e^-$$

The Nernst equation is

$$E_{cell} = E°_{cell} - \frac{RT}{nF} \ln Q = 0.795\ V - \frac{0.0257\ V}{n} \ln \frac{[Cr^{3+}]^2[I_2]^3}{[Cr_2O_7^{2-}][I^-]^6[H_3O^+]^{14}}$$

In the equation, 6 mol e$^-$ are transferred so $n = 6$. All concentrations are 1.0 M except $[H_3O^+]$, which is $10^{-pH} = 10^{-4}$. Therefore, the Nernst equation becomes

$$E_{cell} = 0.795\ V - \frac{0.0257\ V}{6} \ln \frac{1}{(1.0 \times 10^{-4})^{14}}$$

$$= 0.795\ V - \frac{0.0257\ V}{6} \ln (1.0 \times 10^{56})$$

$$= 0.795\ V - 0.553\ V = 0.242\ V$$

Changing the $[H_3O^+]$ from 1.0 M (the standard-state concentration) to 1.0×10^{-4} M (pH = 4.0) decreases the cell potential by 0.553 V. This large change results because of the large coefficient (14) of H_3O^+ in the balanced equation. Dichromate ion is a much weaker oxidizing agent at pH = 4.0 than at standard-state conditions.

PROBLEM-SOLVING PRACTICE 22.5

At what pH does $E_{cell} = 0.00$ V for the reduction of dichromate by iodide ion in acid solution, assuming standard-state concentrations of all species except H_3O^+ ion?

22.6 Coordinate Covalent Bonds: Complex Ions and Coordination Compounds

In Section 8.10, the formation of BF_3NH_3 was described as occurring by the sharing of a lone pair from NH_3 with BF_3. This type of covalent bond, in which one atom contributes *both* electrons for the shared pair, is called a **coordinate covalent bond.** Atoms with lone pairs of electrons, such as nitrogen, phosphorus, and sulfur, can use those lone pairs to form coordinate covalent bonds. For example, the formation of the ammonium ion from ammonia results from formation of a coordinate covalent bond between H^+ and the lone pair of electrons of nitrogen in NH_3.

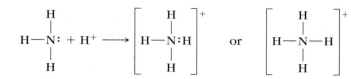

Brackets in the formulas of complex ions do not mean concentration.

$$H-\ddot{N}(H)_2 + H^+ \longrightarrow [H-N{:}H(H)_2]^+ \quad \text{or} \quad [H-N(H)_2-H]^+$$

Once the coordinate covalent bond is formed, it is impossible to distinguish which of the N—H bonds it is; all four bonds are equivalent.

Metals and Coordination Compounds

Much of the chemistry of *d*-block transition metals is related to their ability to form coordinate covalent bonds with molecules or ions that have lone pair electrons. Transition metals with vacant *d* orbitals can accept the lone pairs into those orbitals.

Ligands act as Lewis bases, electron pair donors; transition metal ions are Lewis acids (electron pair acceptors). (Section 16.9).

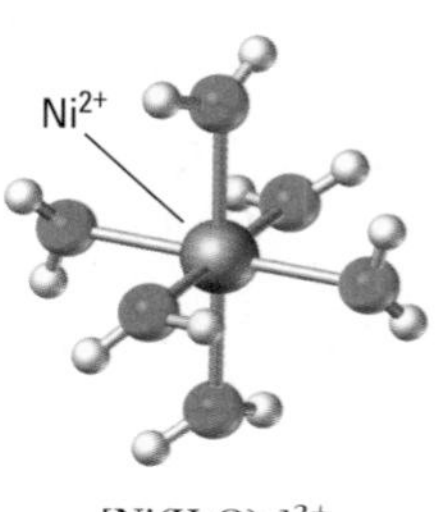

$[Ni(H_2O)_6]^{2+}$

You have seen that metal ions in aqueous solution are surrounded by water molecules; for example, the Ni^{2+} ion in aqueous solution is surrounded by six water molecules. This type of ion, in which several molecules or ions are connected to a central metal ion or atom by coordinate covalent bonds, is known as a *complex ion.* The molecules or ions bonded to the central metal ion are called **ligands,** from the Latin verb *ligare*, "to bind." Each ligand (a water molecule in this example) has one or more atoms with lone pairs that can form coordinate covalent bonds to the metal ion. To write the formula of a complex ion, the ligand formulas are placed in parentheses following the metal ion. The entire complex ion formula is enclosed by brackets, and the ionic charge, if any, is a superscript outside the brackets. For the nickel(II) complex ion with six water ligands this gives $[Ni(H_2O)_6]^{2+}$.

The charge of a complex ion is determined by the charges of the metal ion and the charges of its ligands. In $[Ni(H_2O)_6]^{2+}$ the water ligands have no net charge, so the charge of the complex ion is that of the Ni^{2+} ion. In the complex ion formed by Ni^{2+} with four chloride ions, $[NiCl_4]^{2-}$, the net 2− charge of this complex ion results from the 4− charge of four chloride ions and the 2+ charge of the nickel ion.

Compounds that contain metal ions surrounded by ligands are called **coordination compounds.** Usually, complex ions are combined with oppositely charged ions *(counter ions)* to form neutral compounds. Coordination compounds are generally brightly colored as solids or in solution (Figure 22.8, two left flasks). The complex ion part of a coordination compound's formula is enclosed in brackets; counter ions are outside the brackets, as in the formula $[Ni(H_2O)_6]Cl_2$ of the compound consisting of chloride counter ions with the $[Ni(H_2O)_6]^{2+}$ complex ion. The two Cl^- ions compensate for the 2+ charge of the complex ion. $[Ni(H_2O)_6]Cl_2$ is an ionic compound analogous to $CaCl_2$, which also contains a 2+ cation and two Cl^- ions. In some cases, no compensating ions are needed outside the brackets for a coordination compound. For example, the anticancer drug $[Pt(NH_3)_2Cl_2]$ (cisplatin) is a coordination compound containing NH_3 and Cl^- ligands coordinated to a central Pt^{2+} ion. The two Cl^- ions compensate for the charge of the Pt^{2+} ion, resulting in a neutral coordination compound rather than a complex ion.

When the word *coordinated* is used in chemistry, such as in "the chloride ions in $[NiCl_4]^{2-}$ are coordinated to the nickel ion," it means that coordinate covalent bonds have been formed.

PROBLEM-SOLVING EXAMPLE 22.6 Coordination Compounds

For the coordination compound $K_3[Fe(CN)_6]$, identify

(a) The central metal ion.

(b) The ligands.

(c) The formula and charge of the complex ion and the charge of the central metal ion.

Answer

(a) Iron
(b) Six cyanide ions, CN^-
(c) $[Fe(CN)_6]^{3-}$, Fe^{3+}

Strategy and Explanation We apply the guidelines and concepts described previously. In a formula, a complex ion is enclosed in square brackets; within the brackets, ligands coordinated to the central metal ion are enclosed by parentheses.

(a) The iron ion is the central metal ion, as shown by its placement inside the brackets.
(b) Cyanide ions, CN^-, are coordinated to the central iron ion.
(c) The charge on three potassium ions is $(3 \times 1+) = 3+$. Therefore, the compensating charge of the complex ion must be 3−, arising from the 6− charge of six cyanide ions $(6 \times 1- = 6-)$, combined with the 3+ charge of the central iron(III) ion: $(6-) + (3+) = 3-$.

PROBLEM-SOLVING PRACTICE 22.6

For the coordination compound $[Cu(NH_3)_4]SO_4$, identify

(a) The counter ion.
(b) The central metal ion.
(c) The ligands.
(d) The formula and charge of the complex ion.

CONCEPTUAL EXERCISE 22.8 Coordination Complex Ion

In a complex ion, a central Cr^{3+} ion is bonded to two ammonia molecules, two water molecules, and two hydroxide ions. Give the formula and the net charge of this complex ion.

Naming Complex Ions and Coordination Compounds

Like other compounds, coordination compounds in early times were known by common names, for example, "roseo" salt and Zeise's salt. Since then, a systematic nomenclature has been developed for complex ions and coordination compounds. This nomenclature system indicates the central metal ion and its oxidation state, as well as the number and kinds of ligands. Table 22.4 lists the names and formulas of some common ligands. Although there are extensive rules for such nomenclature, we will consider only some basic aspects of the system by interpreting the names

Table 22.4 Names and Formulas of Some Common Ligands

Neutral Ligand	Ligand Name	Anionic Ligand	Ligand Name
NH_3	Ammine	Br^-	Bromo
CO	Carbonyl	CO_3^{2-}	Carbonato
$H_2NCH_2CH_2NH_2$	Ethylenediamine, en	Cl^-	Chloro
H_2O	Aqua	CN^-	Cyano
		F^-	Fluoro
		OH^-	Hydroxo
		$C_2O_4^{2-}$	Oxalato, ox
		NCS^-	Isothiocyanato
		SCN^-	Thiocyanato

of a neutral coordination compound and two other coordination compounds, one containing a complex cation and the other a complex anion.

Consider the coordination compound $[Co(NH_3)_3(OH)_3]$, which is named triamminetrihydroxocobalt(III). From Table 22.4, we see that the name and formula indicate that three ammonia molecules and three hydroxide ions are bonded to a central cobalt ion. The three hydroxide ions carry a total 3− charge; ammonia molecules have no net charge, and thus cobalt must be Co^{3+} because the compound has no net charge. In naming any coordination compound or complex ion, the ligands are named in alphabetical order—in this case ammine for ammonia precedes hydroxo for hydroxide (for anions *-ide* is changed to *o*). The name and oxidation state (in parentheses) of the metal ion are given last. Greek prefixes *di, tri, tetra,* and so on are used to denote the number of times each of these ligands is used. Such prefixes are ignored when determining the alphabetical order of the ligands.

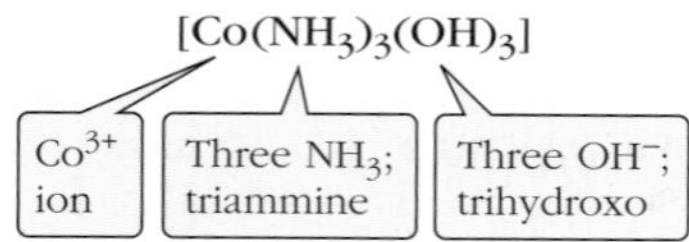

triamminetrihydroxocobalt(III)

Counter ions offset the charge of the complex ion.

Next, consider $[Fe(H_2O)_2(NH_3)_4]Cl_3$, a coordination compound that consists of a complex *cation,* $[Fe(H_2O)_2(NH_3)_4]^{3+}$, and three chloride ions as counter ions. In such cases the complex cation is always named first, followed by the name of the anionic counter ions. The compound's name is tetraamminediaquairon(III) chloride. From Table 22.4, we see that the ligands are ammine (NH_3, four of them) and aqua (H_2O, two of them). For complex cations, the metal ion and its oxidation state follow the names of the ligands.

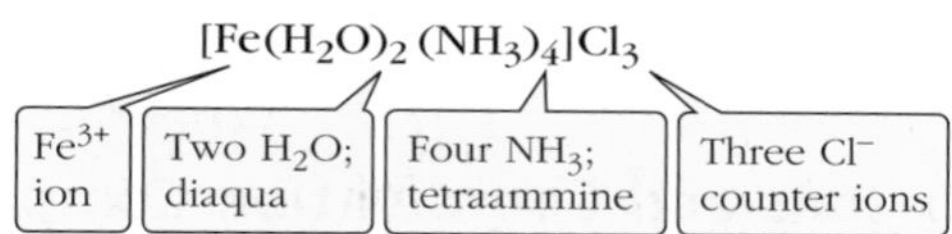

tetraamminediaquairon(III) chloride

The compound $K_2[PtCl_4]$ contains a complex *anion,* $[PtCl_4]^{2-}$, and two K^+ ions as counter ions and is named potassium tetrachloroplatinate(II). As with any ionic compound, the cation is named first, followed by the anion name. For complex anions, the central metal ion's name ends in *-ate* followed by its oxidation state in parentheses.

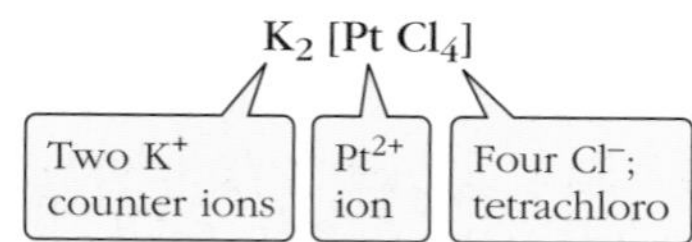

potassium tetrachloroplatinate(II)

PROBLEM-SOLVING EXAMPLE 22.7 Formulas and Names of Coordination Compounds

(a) Write the formula of diamminetriaquahydroxochromium(III) nitrate.
(b) Name $K[Cr(NH_3)_2(C_2O_4)_2]$.

Answer

(a) $[Cr(NH_3)_2(H_2O)_3(OH)](NO_3)_2$
(b) Potassium diamminedioxalatochromate(III)

Strategy and Explanation Use the names and formulas of the ligands in Table 22.4. Compound (a) contains a complex cation, and compound (b) contains a complex anion.

(a) diamminetriaquahydroxochromium(III) nitrate

$[Cr(NH_3)_2(H_2O)_3(OH)](NO_3)_2$

(b) $K[Cr(NH_3)_2(C_2O_4)_2]$

potassium diamminedioxalatochromate(III)

PROBLEM-SOLVING PRACTICE 22.7

(a) Name this coordination compound: $[Ag(NH_3)_2]NO_3$.
(b) Write the formula of pentaaquaisothiocyanatoiron(III) chloride.

CONCEPTUAL EXERCISE 22.9 Coordination Compounds

$CaCl_2$ and $[Ni(H_2O)_6]Cl_2$ have the same formula type, MCl_2. Give the formula of a simple ionic compound (noncoordination) that has a formula analogous to $K_2[NiCl_4]$.

Types of Ligands and Coordination Number

The number of coordinate covalent bonds between the ligands and the central metal ion in a coordination compound is the **coordination number** of the metal ion, usually 2, 4, or 6.

Coordination Number	Examples
2	$[Ag(NH_3)_2]^+$, $[AuCl_2]^-$
4	$[NiCl_4]^{2-}$, $[Pt(NH_3)_4]^{2+}$
6	$[Fe(H_2O)_6]^{2+}$, $[Co(NH_3)_6]^{3+}$

Ligands such as H_2O, NH_3, and Cl^- that form only one coordinate covalent bond to the metal are termed **monodentate** ligands. The word *dentate* derives from the Latin word *dentis,* for tooth, so NH_3 is a "one-toothed" ligand. Common monodentate ligands are shown in Figure 22.9.

Some ligands can form two or more coordinate covalent bonds to the same metal ion because they have two or more atoms with lone pairs separated by several intervening atoms. The general term **polydentate** is used for such ligands. **Bidentate** ligands are those that form *two* coordinate covalent bonds to the central metal ion. A good example is the bidentate ligand 1,2-diaminoethane, $H_2NCH_2CH_2NH_2$, commonly called ethylenediamine and abbreviated *en.* When lone pairs of electrons from both nitrogen atoms in en coordinate to a metal ion, a stable five-membered ring forms (Figure 22.10). Notice that Co^{3+} has a coordination number of 6 in this complex ion.

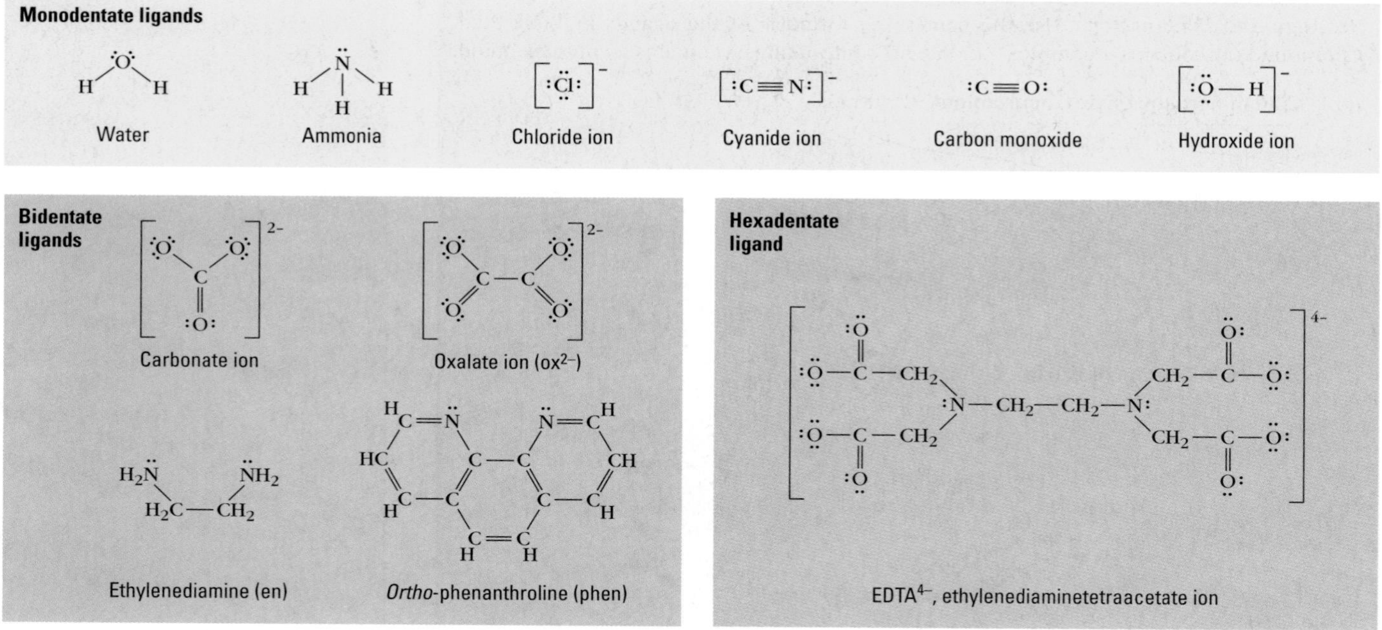

Figure 22.9 Monodentate ligands; bidentate and hexadentate chelating ligands. Ligands with two (bidentate) or more lone pairs to share with a central metal ion are chelating ligands.

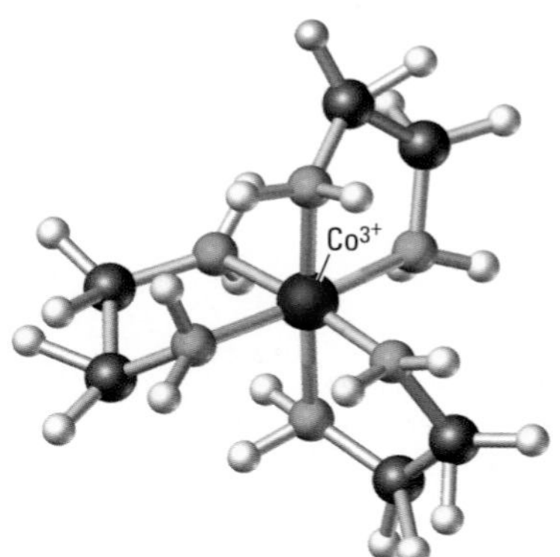

Figure 22.10 The $[Co(en)_3]^{3+}$ complex ion. Cobalt ion (Co^{3+}) forms a coordination complex ion with three ethylenediamine ligands.

Note that Cl_2 in the complex ion's formula represents two chloride ions, not a diatomic chlorine molecule.

The word "chelating," derived from the Greek *chele*, "claw," describes the pincer-like way in which a ligand can grab a metal ion. Some common **chelating ligands,** those that are polydentate ligands and can share two or more electron pairs with the central metal ion, are also shown in Figure 22.9.

PROBLEM-SOLVING EXAMPLE 22.8 Chelating Agents

Two ethylenediamine (en) ligands and two chloride ions form a complex ion with Co^{3+}.

(a) Write the formula of this complex ion.

(b) What is the coordination number of the Co^{3+} ion?

(c) Write the formula of the coordination compound formed by Cl^- counter ions and the Co^{3+} complex ion.

Answer

(a) $[Co(en)_2Cl_2]^+$ (b) 6 (c) $[Co(en)_2Cl_2]Cl$

Strategy and Explanation Recall that ethylenediamine is a bidentate ligand that forms two coordinate covalent bonds per en molecule.

(a) Two en molecules and two chloride ions are bonded to the central cobalt ion, so the formula of the complex ion is $[Co(en)_2Cl_2]^+$. Ethylenediamine is a neutral ligand, each chloride ion is 1−, and cobalt has a 3+ charge. The charge on the complex ion is $2(0) + 2(1-) + (3+) = 1+$.

(b) The coordination number is 6 because there are six coordinate covalent bonds to the central Co^{3+} ion—two from each bidentate ethylenediamine and one from each monodentate chloride ion.

(c) The 1+ charge of the complex ion requires one chloride ion as a counter ion: $[Co(en)_2Cl_2]Cl$.

PROBLEM-SOLVING PRACTICE 22.8

The dimethylglyoximate anion (abbreviated DMG^-),

$$\begin{array}{c} CH_3C{-}CCH_3 \\ \| \quad \| \\ HO{-}\ddot{N} \quad \ddot{N}{-}O^- \end{array}$$

is a bidentate ligand used to test for the presence of nickel. It reacts with Ni^{2+} to form a beautiful red solid in which the Ni^{2+} has a coordination number of 4. DMG^- coordinates to Ni^{2+} by the lone pairs on the nitrogen atoms.

(a) How many DMG^- ions are needed to satisfy a coordination number of 4 on the central Ni^{2+} ion?

(b) What is the net charge after coordination occurs?

(c) How many atoms are in the ring formed by one DMG^- and one Ni^{2+}?

Check your answer to Problem-Solving Practice 22.8 by viewing Figure 22.11 at the Web site.

CONCEPTUAL EXERCISE 22.10 Chelating and Complex Ions

Oxalate ion forms a complex ion with Mn^{2+} by coordinating at the oxygen lone pairs (see Figure 22.9).

(a) How many oxalate ions are needed to satisfy a coordination number of 6 on the central Mn^{2+} ion?

(b) What is the charge on this complex ion?

(c) How many atoms are in the ring formed between one oxalate ion and the central metal ion?

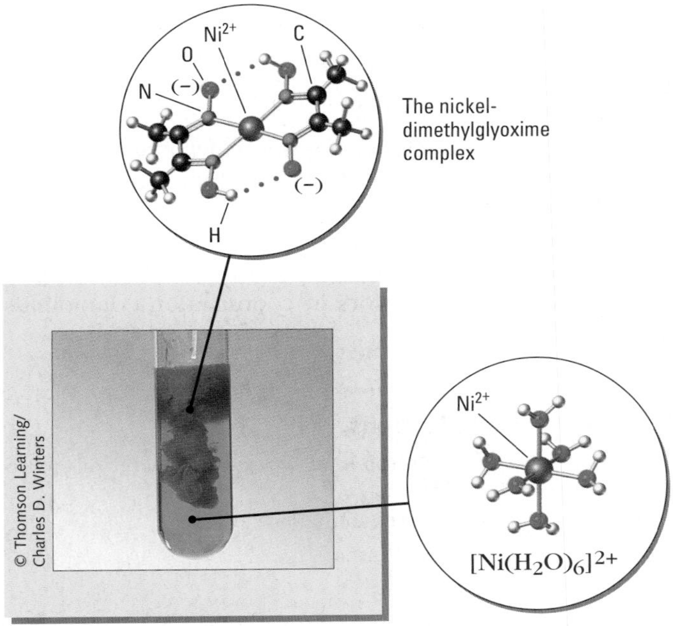

Active Figure 22.11 **The nickel-dimethylglyoxime complex.** Ni^{2+} ions react with the dimethylglyoximate anion (DMG^-) to form a beautiful red solid. **Go to the Active Figures menu at ThomsonNOW to test your understanding of the concepts in this figure.**

CHEMISTRY YOU CAN DO

A Penny for Your Thoughts

You will need the following items to do the experiment:

- Two glasses or plastic cups that will each hold about 50 mL liquid
- About 30 to 40 mL household vinegar
- About 30 to 40 mL household ammonia
- A copper penny

Place the penny in one cup and add 30 to 40 mL vinegar to clean the surface of the penny. Let the penny remain in the vinegar until the surface of the penny is cleaner (reddish-coppery) than it was before (darker copper color). Pour off the vinegar and wash the penny thoroughly in running water.

Next, place the penny in the other cup and add 30 to 40 mL household ammonia. Observe the color of the solution over several hours.

1. What did you observe happening to the penny in the ammonia solution?
2. What did you observe happening to the ammonia solution?
3. Interpret what you observed happening to the solution on the nanoscale level, citing observations to support your conclusions.
4. What is necessary to form a complex ion?
5. Are all of these kinds of reactants present in the solution in this experiment? If so, identify them.
6. How do the terms "ligand," "central metal ion," and "coordination complex" apply to your experiment?
7. Try to write a formula for a complex ion that might form in this experiment.

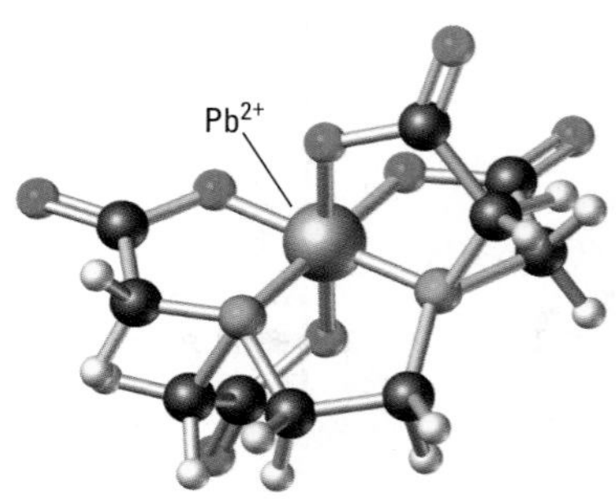

Figure 22.12 A Pb^{2+}-EDTA complex ion. The structure of the chelate formed when the $EDTA^{4-}$ anion forms a complex with Pb^{2+}.

Some household products that contain EDTA. Check the label on your shampoo container. It will likely list disodium EDTA as an ingredient. The EDTA in this case has a 2− charge because two of the four organic acid groups have each lost an H^+, but $EDTA^{2-}$ still coordinates to metal ions in the shampoo.

For metals that display a coordination number of 6, an especially effective ligand is the **hexadentate** ethylenediaminetetraacetate ion (abbreviated EDTA; Figure 22.9) that encapsulates and firmly binds metal ions. It has six lone pair donor atoms (four O atoms and two N atoms) that can coordinate to a single metal ion, so $EDTA^{4-}$ is an excellent chelating ligand. $EDTA^{4-}$ is often added to commercial salad dressing to remove traces of metal ions from solution, because these metal ions could otherwise accelerate the oxidation of oils in the product and make them rancid.

Another use of $EDTA^{4-}$ is in bathroom cleansers, where it removes hard water deposits of insoluble $CaCO_3$ and $MgCO_3$ by chelating Ca^{2+} or Mg^{2+} ions, allowing them to be rinsed away. EDTA is also used in the treatment of lead and mercury poisoning because it has the ability to chelate these metals and aid in their removal from the body (Figure 22.12).

Coordination compounds of *d*-block transition metals are often colored, and the colors of the complexes of a given transition metal ion depend on both the metal ion and the ligand (Figure 22.13). Many transition metal coordination compounds are used as pigments in paints and dyes. For example, Prussian blue, $Fe_4[Fe(CN)_6]_3$, a deep-blue compound known for hundreds of years, is the "bluing agent" in engineering blueprints. The origin of colors in coordination compounds will be discussed in Section 22.7.

CONCEPTUAL EXERCISE **22.11 Complex Ions**

Prussian blue contains two kinds of iron ions. What is the charge of the iron in

(a) The complex ion $[Fe(CN)_6]^{4-}$?

(b) The iron ion not in the complex ion?

Geometry of Coordination Compounds and Complex Ions

The geometry of a complex ion or coordination compound is dictated by the arrangement of the electron donor atoms of the ligands attached to the central metal ion. Although other geometries are possible, we will discuss only the four

Photos: © Thomson Learning/Charles D. Winters

(a)

(b)

Figure 22.13 Color of transition metal compounds. (a) Concentrated aqueous solutions of the nitrate salts containing hydrated transition metal ions of *(left to right)* Fe^{3+}, Co^{2+}, Ni^{2+}, Cu^{2+}, and Zn^{2+}. Aqueous Zn^{2+} compounds are colorless. (b) The colors of the complexes of a given transition metal ion depend on the ligand(s). All of the complexes pictured here contain the Ni^{2+} ion. The green solid is $[Ni(H_2O)_6](NO_3)_2$; the purple solid is $[Ni(NH_3)_6]Cl_2$; the red solid is Ni(dimethylglyoximate)$_2$.

most common ones, those associated with coordination numbers of 2, 4, and 6. To simplify matters, we will consider only monodentate ligands, L, bonded to a central metal ion, M^{n+}.

Coordination Number = 2, ML_2^{n+}

All such complex ions have a *linear* geometry with the two ligands on opposite sides of the central metal ion to give an L—M—L bond angle of 180°, such as that in $[Ag(NH_3)_2]^+$. Other examples are $[CuCl_2]^-$ and $[Au(CN)_2]^-$, the complex ion used to extract gold from ores (p. 1078).

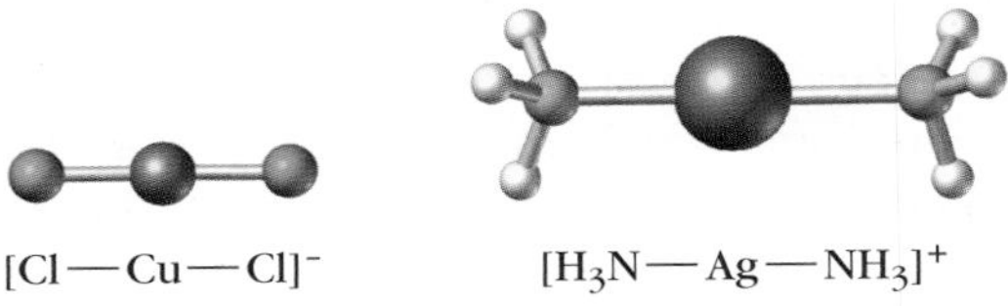

Coordination Number = 4, ML_4^{n+}

Four-coordinate complex ions have either tetrahedral or square planar geometries. In the *tetrahedral* case, the four monodentate ligands are at the corners of a tetrahedron, such as in $[Zn(NH_3)_4]^{2+}$. In *square planar* geometry, the ligands lie in a plane at the corners of a square as in $[Ni(CN)_4]^{2-}$ and $[Pt(NH_3)_4]^{2+}$ ions.

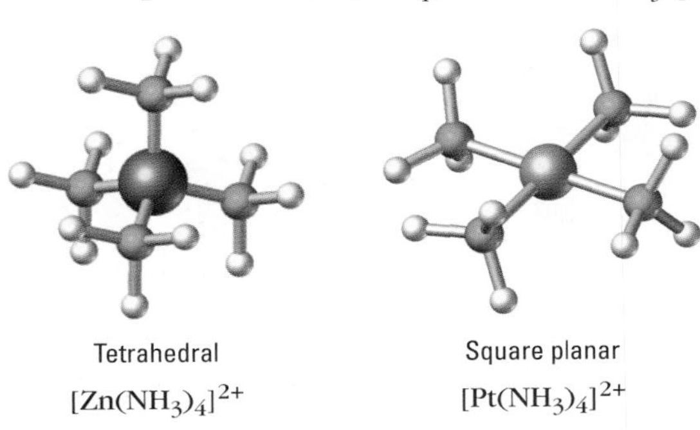

Science Photo Library/Photo Researchers, Inc.

Alfred Werner
1866–1919

In 1893, while still a young professor, Alfred Werner published a revolutionary paper about transition metal compounds. He asserted that transition metal ions could exhibit a secondary valence as well as a primary one, such as in $CoCl_3 \cdot 6\ NH_3$ (now written as $[Co(NH_3)_6]Cl_3$). The primary valence was represented by the ionic bonds between Co^{3+} and the chloride ions; the secondary valence was represented by the coordinate covalent bonds between the metal ion and six NH_3 molecules, what we now called the coordination sphere around the central metal ion. Werner also made the inspired proposal that the ammonia molecules were octahedrally coordinated around the Co^{3+} ion, thereby laying the foundation for understanding the geometry of complex ions. For his groundbreaking work, Werner received the 1913 Nobel Prize in chemistry.

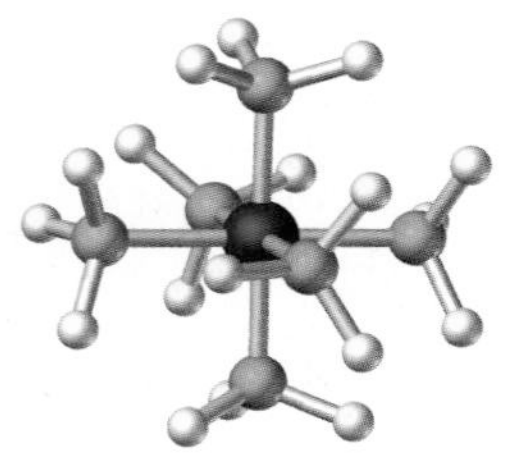

Octahedral $[Co(NH_3)_6]^{3+}$

Coordination Number = 6, ML_6^{n+}

Octahedral geometry is characteristic of this coordination number. The six ligands are at the corners of an octahedron with the central metal ion at its center. Octahedral geometry can be regarded as derived from a square planar geometry by adding two ligands, one above and one below the square plane. Two common octahedral complex ions are $[Co(NH_3)_6]^{3+}$ and $[Fe(CN)_6]^{3-}$, in which the six ligands are equidistant from the central metal ion and all six ligand sites are equivalent.

Isomerism in Coordination Compounds and Complex Ions

Various types of isomerism have been discussed previously with regard to organic compounds. *Constitutional isomerism* occurs with molecules that have the same molecular formula but differ in the way their atoms are connected together, such as occurs with butane and 2-methylpropane **(← p. 88)**. *Stereoisomerism* is a second general category of isomerism in which the isomers have the same bond connections, but the atoms are arranged differently in space. One type of stereoisomerism is *geometric isomerism,* such as that found in *cis-* and *trans*-1,2-dichloroethene **(← p. 345)**. The other type of stereoisomerism is *optical isomerism,* which occurs when mirror images are nonsuperimposable **(← p. 559)**. Constitutional, geometric, and optical isomers also occur with coordination complex ions and coordination compounds.

Linkage Isomerism, a Type of Constitutional Isomerism

Linkage isomerism occurs when a ligand can bond to the central metal using either of two different electron-donating atoms. Thiocyanato $(SCN)^-$ and isothiocyanato $(NCS)^-$ ions are examples of such ligands with coordination to a metal ion by sulfur in the first case and by nitrogen in the second, as illustrated for the Co^{3+} complex ions shown in the margin.

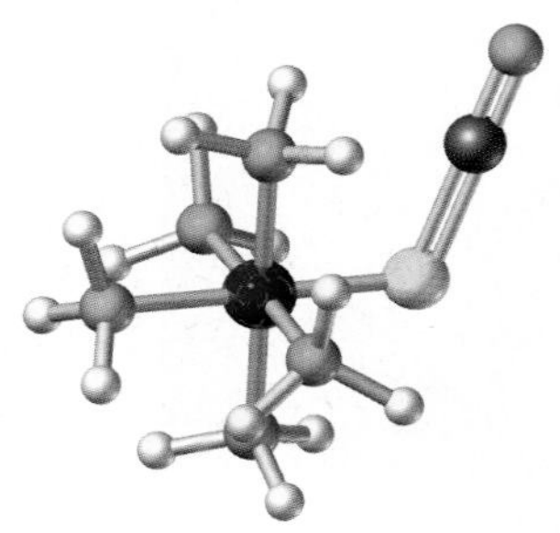

pentaammine-thiocyanatocobalt(III) ion

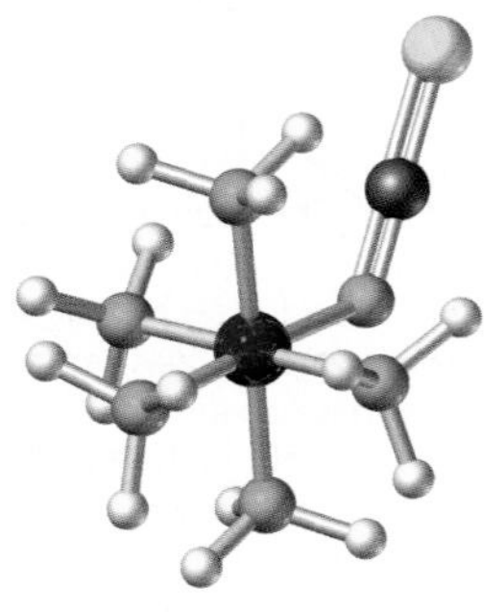

pentaammine-isothiocyanatocobalt(III) ion

Geometric Isomerism

Geometric isomerism does not exist in tetrahedral complex ions because all the corners of a tetrahedron are equivalent. Geometric isomerism, however, does occur with square planar complex ions and compounds of the type Ma_2b_2 or Ma_2bc, where M is the central metal ion and a, b, and c are different ligands. The square planar coordination compound $[Pt(NH_3)_2Cl_2]$, an Ma_2b_2 type, occurs in two geometric forms. The *cis*-$[Pt(NH_3)_2Cl_2]$ isomer has the chloride ligands as close as possible at 90° to one another. In *trans*-$[Pt(NH_3)_2Cl_2]$, the chloride ions are as far apart as possible, directly across the square plane of the molecule at 180° from each other.

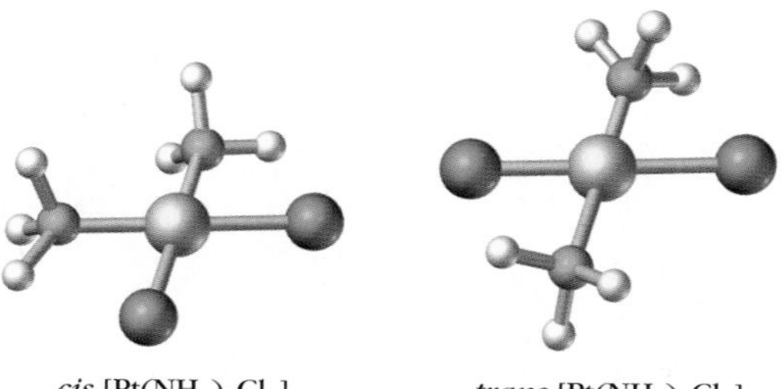

cis-$[Pt(NH_3)_2Cl_2]$ *trans*-$[Pt(NH_3)_2Cl_2]$

These two isomers differ in water solubility, color, melting point, and chemical reactivity. The *cis* isomer is used in cancer chemotherapy, whereas the *trans* form is not effective against cancer.

Cis-trans isomerism is also possible in octahedral complex ions and compounds, as illustrated with $[Co(NH_3)_4Cl_2]^+$. In this complex ion the *cis* isomer has the chloride ions adjacent to each other; the *trans* isomer has them opposite each other. The differences in properties are striking, particularly the color. The *cis* isomer is violet, whereas the *trans* form is green.

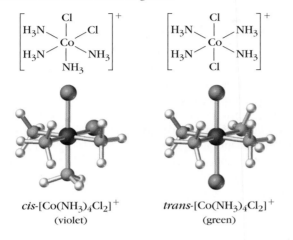

cis-$[Co(NH_3)_4Cl_2]^+$ (violet) *trans*-$[Co(NH_3)_4Cl_2]^+$ (green)

PROBLEM-SOLVING EXAMPLE 22.9 Geometric Isomerism

How many geometric isomers are there for $[Co(en)_2Cl_2]^+$?

Answer Only two geometric isomers are possible, *cis* and *trans*.

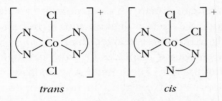

trans *cis*

Strategy and Explanation Start by putting the two Cl^- ions in *trans* positions, that is, one at the "top" of the octahedron and the other at the "bottom." The two ethylenediamine ligands (en), represented here as N⌒N, occupy the other four sites around the cobalt ion. This is the *trans* isomer. The *cis* isomer has the Cl^- ions in adjacent *(cis)* positions.

PROBLEM-SOLVING PRACTICE 22.9

How many geometric isomers are there for the square planar compound $[Pt(NH_3)_2ClBr]$?

EXERCISE 22.12 Geometric Isomerism

How many isomers are possible for $[Co(NH_3)_3Cl_3]$? Write the structural formulas of the isomers.

Optical Isomerism

Optical isomers are mirror images that are not superimposible. Such nonsuperimposable mirror images are known as *enantiomers* (⇐ p. 560). An example of a complex ion that has optical isomerism is $[Cr(en)_2Cl_2]^+$. There are two enantiomers, as shown in Figure 22.14. No matter how they are twisted and turned, the two enantiomers are nonsuperimposable.

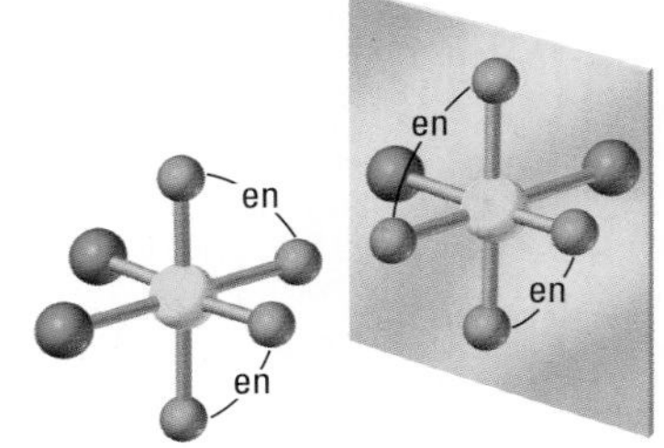

$[Cr(en)_2Cl_2]^+$

Figure 22.14 Optical isomerism in $[Cr(en)_2Cl_2]^+$. The ion on the left cannot be superimposed on its mirror image *(right)*.

Optical isomerism is not possible for square planar complexes based on the geometry around the metal ion; the mirror images are superimposable. Although optical isomers of tetrahedral complex ions with four different ligands are theoretically possible, no such stable complexes are known.

Coordination Compounds and Life

Bioinorganic chemistry, the study that applies chemical principles to inorganic ions and compounds in biological systems, is a rapidly growing field centered mainly around coordination compounds. This is because the very existence of living systems depends on many coordination compounds in which metal ions are chelated to the nitrogen and oxygen atoms in proteins and especially in enzymes. Copper-containing proteins, for example, give the blood of crabs, lobsters, and snails its blue color, as well as transport oxygen.

The blue blood of horseshoe crabs is used to test for bacterial contamination of drugs.

In humans, molecular oxygen (O_2) is transported through the circulatory system by hemoglobin, a very large protein (molecular weight of about 68,000 amu) in red blood cells. Hemoglobin is blue but becomes red when oxygenated. This is why arterial blood is bright red (high O_2 concentration) and blood in veins is bluish (low O_2 concentration).

A hemoglobin molecule carries four O_2 molecules, each of which forms a coordinate covalent bond to one of four Fe^{2+} ions. Each Fe^{2+} ion is at the center of one heme, a nonprotein part of the hemoglobin molecule that consists of four linked nitrogen-containing rings (Figure 22.15). Bound in this way, molecular oxygen is carried to the cells, where it is released as needed by breaking the Fe—O_2 coordinate covalent bond.

Other substances that can donate an electron pair can also bond to the Fe^{2+} in heme. Carbon monoxide is such a ligand and forms an exceptionally strong Fe^{2+}—CO bond, nearly 200 times stronger than the O_2—Fe^{2+} bond. Therefore, when a person breathes in CO, it displaces O_2 from hemoglobin and prevents red blood cells from carrying oxygen. The initial effect is drowsiness. But if CO inhalation continues, cells deprived of oxygen can no longer function and the person suffocates.

It is interesting (and fortunate) that N≡N does not behave chemically like C≡O, even though each contains 14 electrons.

Structures similar to the oxygen-carrying unit in hemoglobin are also found in other biologically important compounds, including such diverse ones as myoglobin and vitamin B-12. Myoglobin, like hemoglobin, contains Fe^{2+} and carries and stores

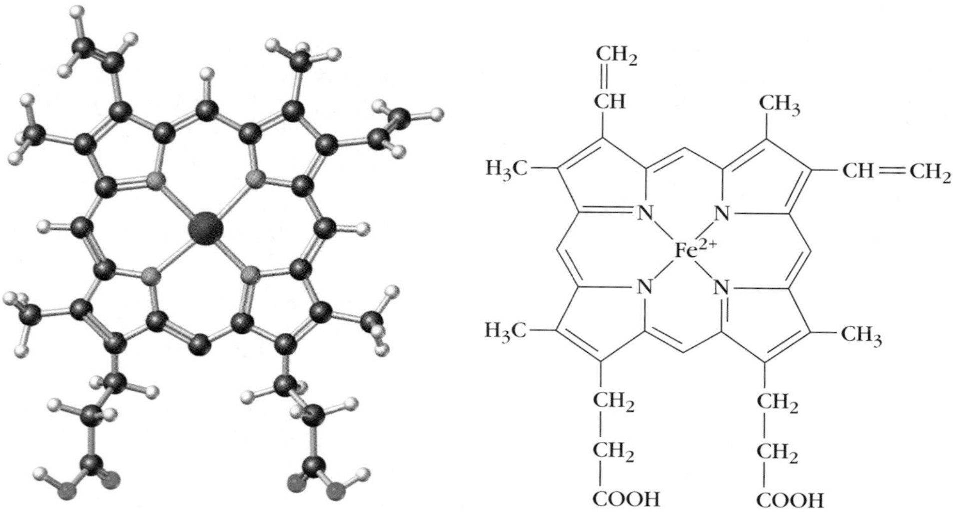

Figure 22.15 Heme, the carrier of Fe^{2+} in hemoglobin. Fe^{2+} is coordinated to four nitrogen atoms in heme. There are four hemes in each hemoglobin molecule.

molecular oxygen, principally in muscles. At the center of a vitamin B-12 molecule is a Co^{3+} ion bonded to the same type of group that Fe^{2+} bonds to in hemoglobin. Vitamin B-12 is the only known dietary use of cobalt, but it makes cobalt an essential mineral *(⇐ p. 111)*.

The dietary necessity of zinc for humans has become established only since the 1980s. Zinc, in the form of Zn^{2+} ions, is essential to the functioning of several hundred enzymes, including those that catalyze the breaking of P—O—P bonds in adenosine triphosphate (ATP), an important energy-releasing compound in cells *(⇐ p. 895)*.

Copper ranks third among biologically important transition metal ions in humans, trailing only iron and zinc. Although we generally excrete any dietary excess of copper, a genetic defect causes Wilson's disease, a condition in which Cu^{2+} accumulates in the liver and brain. Fortunately, Wilson's disease can be treated by administering chelating agents that coordinate excess Cu^{2+} ions, allowing them to be excreted harmlessly.

22.7 Crystal-Field Theory: Color and Magnetism in Coordination Compounds

Bright colors are characteristic of many coordination compounds (Figure 22.13). Any theory about bonding in such compounds needs to address the origin of their colors. One such approach is the crystal-field theory developed by Hans Bethe and J. H. van Vleck. Crystal-field theory explains color as originating from electron transitions between two sets of *d* orbitals in a complex ion, similar to the bright line atomic spectra that originate from electron transitions among atomic orbitals in elements. In many cases of complex ions, the energy difference between the *d* orbitals corresponds with a wavelength within the visible region of the spectrum.

Crystal-Field Theory

The crystal-field theory considers the bonding between ligands and a metal ion to be primarily electrostatic. It assumes that the electron pairs on the ligands create an electrostatic field around the *d* orbitals of the metal ion, such as in the case of the octahedral $[Fe(CN)_6]^{4-}$ complex ion. This electrostatic field alters the relative energies of the various *d* orbitals. Before this interaction occurs, the *d* orbitals in a sublevel of the isolated metal ion, such as the 3*d*, all have the same energy. In the presence of ligands, the *d* orbitals split into two groups of differing energy (Figure 22.16). The higher-energy pair consists of the $d_{x^2-y^2}$ and d_{z^2} orbitals; the lower-energy trio is made up of the d_{xy}, d_{yz}, and d_{xz} orbitals.

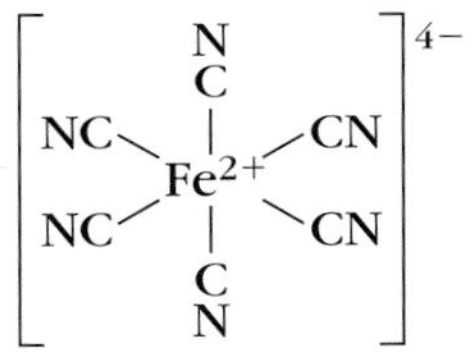

The $[Fe(CN)_6]^{4-}$ complex ion. The six CN^- ions are arranged octahedrally around the central Fe^{2+} ion.

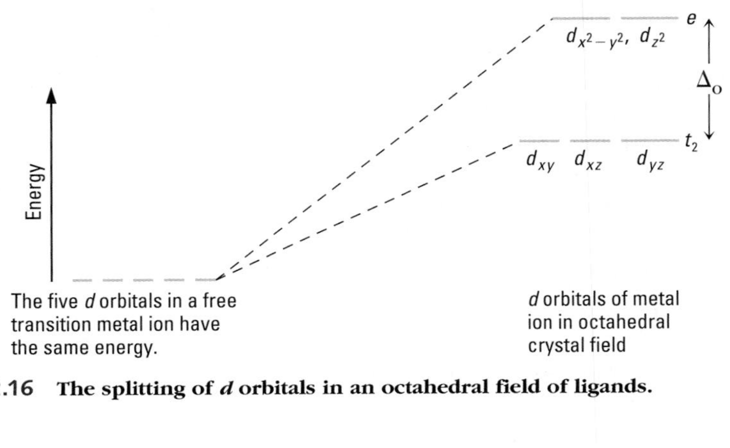

Figure 22.16 **The splitting of *d* orbitals in an octahedral field of ligands.**

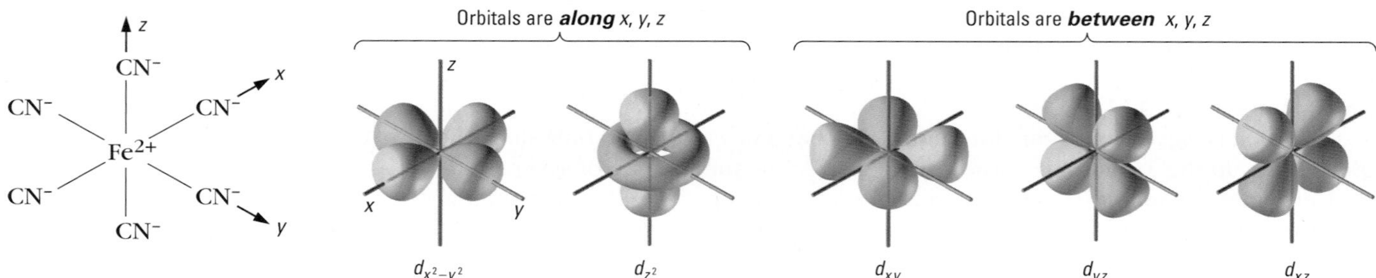

Figure 22.17 The spatial arrangement of *d* orbitals of an Fe^{2+} ion and the octahedral orientation of six CN^- ligands as negative point charges along the *x*-, *y*-, and *z*-axes. The $d_{x^2-y^2}$ and d_{z^2} orbitals of the metal ion are aligned along the axes and oriented toward ligands approaching the six corners of the octahedron. The other three *d* orbitals are oriented between the axes.

Why does this splitting occur? To understand this, we must first consider the orientations of *d* orbitals on the central metal ion and what happens to the energy of the *d* orbitals when they are approached by the six ligands as an assembly of negative charges. From Figure 22.17 note that in a set of five *d* orbitals, two of the *d* orbitals—$d_{x^2-y^2}$ and the d_{z^2}—have their lobes of maximum electron density directly *along* the *x*-, *y*-, and *z*-axes. In contrast, the other three *d* orbitals—d_{xy}, d_{yz}, and d_{xz}—have their lobes of maximum electron density aligned *between* the *x*-, *y*-, and *z*-axes rather than directly along them. Thus, as the CN^- ligands approach along the *x*-, *y*-, and *z*-axes to form an octahedral field of ligands, the electrostatic field created by the electron pairs of the ligands is felt more strongly by electrons in the $d_{x^2-y^2}$ and d_{z^2} orbitals—those directly along the axes. Electrons in orbitals not directly along the axes—the d_{xy}, d_{yz}, and d_{xz} orbitals—are not disturbed as strongly by the electrostatic field. The overall result is the splitting of the 3*d* orbitals in the complex ion into two sets of differing energy; the higher-energy pair of orbitals, labeled *e*, and the lower-energy trio of orbitals, labeled t_2 (Figure 22.16). The energy difference between the sets of *d* orbitals is known as the **crystal-field splitting energy, Δ.** For an octahedral field of ligands, as in $Fe(CN)_6^{4-}$, a subscript o is added, Δ_o.

Electron Configurations and Magnetic Properties of Coordination Complex Ions

The electron configurations and magnetic properties of transition metal ions can be interpreted by using crystal-field theory. Applying Hund's rule ***(⟸ p. 299)***, electrons occupy the vacant orbitals of lowest energy first. If the vacant orbitals all have the same energy, electrons occupy those orbitals unpaired with their spins parallel. Consequently, in a complex ion where the metal ion has a d^1, d^2, or d^3 electron configuration, the electrons will occupy the t_2 lower-energy set of *d* orbitals (Figure 22.18).

We might expect that the trend of unpaired electrons would continue with the filling of the *e* set of higher-energy *d* orbitals for metal ions with d^4 and d^5 electron configurations. This is true in some, but not all cases. Whether these electrons are unpaired or paired depends on the magnitude of the crystal-field splitting energy, Δ_o, the energy gap between the metal ion's two sets of *d* orbitals. That crystal-field splitting energy depends on the metal ion and the ligands. The relative effect of the ligands is given by the **spectrochemical series,** a listing of ligands in the order of their splitting energy. An abbreviated spectrochemical series for octahedral complexes is

The spectrochemical series is determined experimentally.

$$CN^- > NO_2^- > en > NH_3 > NCS^- > H_2O > F^- > Cl^-$$

$$\text{strong field} \longrightarrow \text{decreasing } \Delta_o \longrightarrow \text{weak field}$$

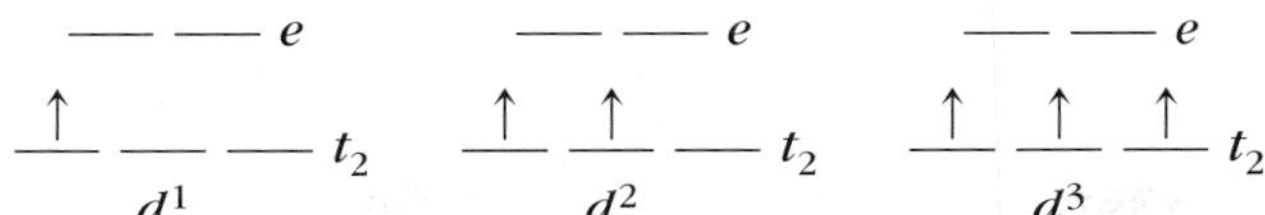

Figure 22.18 Octahedral complex ions with d^1, d^2, and d^3 electron configurations. In each case the electrons are unpaired.

Ligands such as CN^-, NO_2^-, and en that cause a large Δ_o are called *strong-field* ligands; those at the other end of the series, such as Cl^- and F^-, with smaller Δ_o are termed *weak-field* ligands.

Consider two complex ions $[Fe(CN)_6]^{4-}$ and $[Fe(H_2O)_6]^{2+}$, each containing the Fe^{2+} ion with six $3d$ electrons. As seen from the spectrochemical series, cyanide ion, CN^-, is a strong-field ligand with a Δ_o large enough to cause the six d electrons to pair in the three lower-energy t_2 orbitals (Figure 22.19). In $[Fe(H_2O)_6]^{2+}$, water is a weak-field ligand with a Δ_o sufficiently smaller that four of the six d electrons remain unpaired. Using Hund's rule for the aqua complex ion, the first five electrons occupy each of the five d orbitals individually, with pairing occurring when the sixth electron is added into one of the t_2 orbitals. The result is four electrons occupying the t_2 orbitals and the remaining two electrons unpaired in the higher-energy set of d_{z^2} and $d_{x^2-y^2}$ orbitals (Figure 22.19). Thus, $[Fe(H_2O)_6]^{2+}$ is paramagnetic with four unpaired electrons, while $[Fe(CN)_6]^{4-}$ has no unpaired electrons and is diamagnetic (p. 1065).

How paramagnetic a metal ion is can be measured using a special balance as described on p. 1065.

High-spin and Low-spin Complexes

Pairs of complex ions like this are known for many other transition metal complex ions, those in which the metal ion contains between four and seven inner d electrons. The complex ion with the greater number of unpaired electrons is known as the **high-spin complex;** the **low-spin complex** contains the lesser number of unpaired electrons. High-spin complexes are expected with weak-field ligands where the crystal-field splitting energy Δ_o is small. The opposite applies to low-spin complexes in which strong-field ligands cause maximum pairing of electrons in the set of three t_2 orbitals due to large Δ_o.

High-spin: maximum number of unpaired electrons.

Low-spin: minimum number of unpaired electrons. $[Fe(CN)_6]^{4-}$ is low-spin; $[Fe(H_2O)_6]^{2+}$ is high-spin.

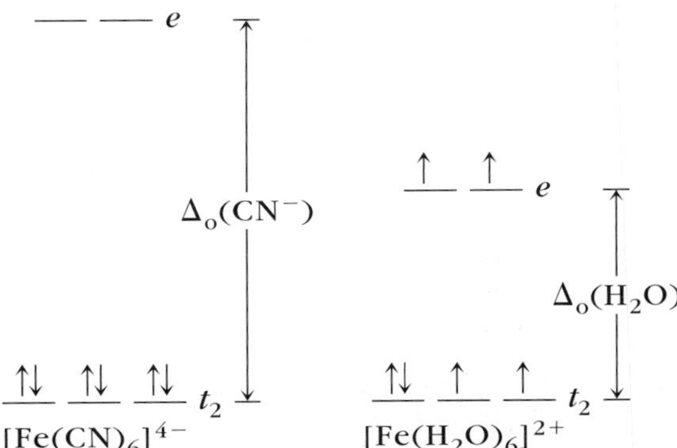

Figure 22.19 Ligands affect the number of unpaired electrons in these two complex ions. Cyanide is a strong-field ligand and the large Δ_o causes maximum electron pairing in Fe^{2+}. Water is a weak-field ligand and the small Δ_o allows the maximum number of unpaired electrons.

PROBLEM-SOLVING EXAMPLE 22.10 High-spin and Low-spin Complex Ions

Experimental data indicate that $[Co(CN)_6]^{3-}$ is diamagnetic and $[CoF_6]^{3-}$ is paramagnetic.

(a) Use the crystal-field model to illustrate the $3d$ electron configuration of the cobalt ion in each complex.

(b) How many unpaired electrons are in the $3d$ orbitals of the paramagnetic complex?

(c) Which is the low-spin complex?

Answer

(a)

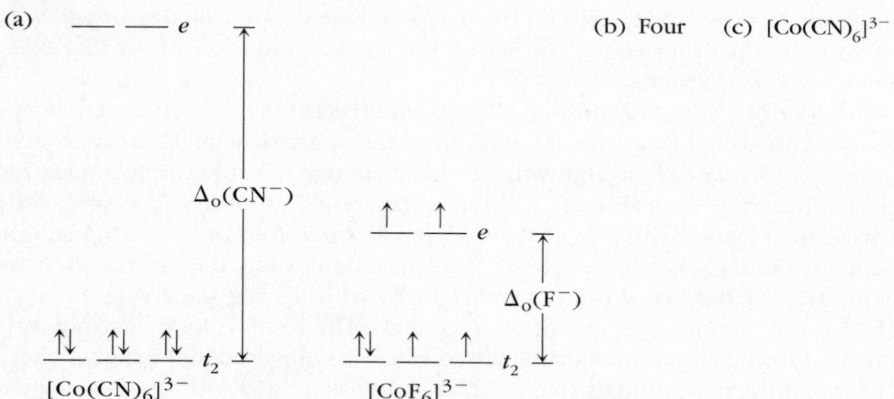

(b) Four (c) $[Co(CN)_6]^{3-}$

Strategy and Explanation We first must determine the number of $3d$ electrons in each case and then use the spectrochemical series to determine the relative ligand field strengths of cyanide and fluoride ions as ligands. Cobalt metal atoms have the $[Ar]4s^23d^7$ electron configuration and lose three electrons to form Co^{3+} ions. Recall that in forming ions, transition metals lose ns electrons before losing $(n - 1)$ d electrons (p. 1065). In this case, two $4s$ electrons are lost plus one $3d$ electron to give Co^{3+} ion an $[Ar]3d^6$ electron configuration.

(a) Cyanide ion is a strong-field ligand that will cause maximum pairing of electrons (low spin). Fluoride ion, a weak-field ligand, is not sufficiently strong to cause maximum d electron pairing; instead, the maximum number of unpaired electrons occurs (high spin).

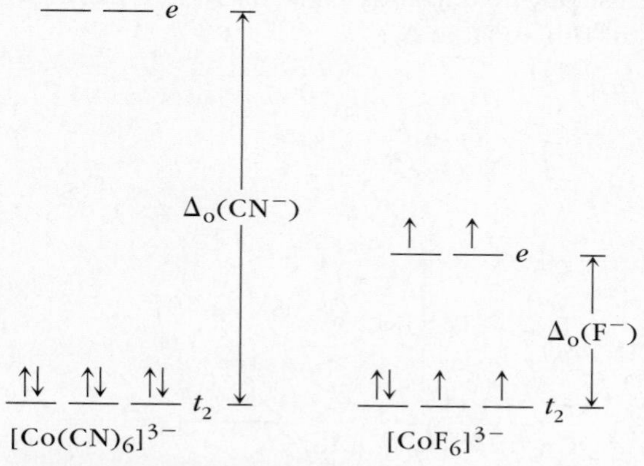

(b) In $[CoF_6]^{3-}$, there are four unpaired $3d$ electrons.

(c) Because $[Co(CN)_6]^{3-}$ is diamagnetic, it contains no unpaired electrons and must be the low-spin complex of these two.

PROBLEM-SOLVING PRACTICE 22.10

A Cr^{2+} ion contains four $3d$ electrons. How many unpaired electrons are in a high-spin octahedral complex of this metal ion? A low-spin complex of this ion? Is either complex paramagnetic? Explain your answer.

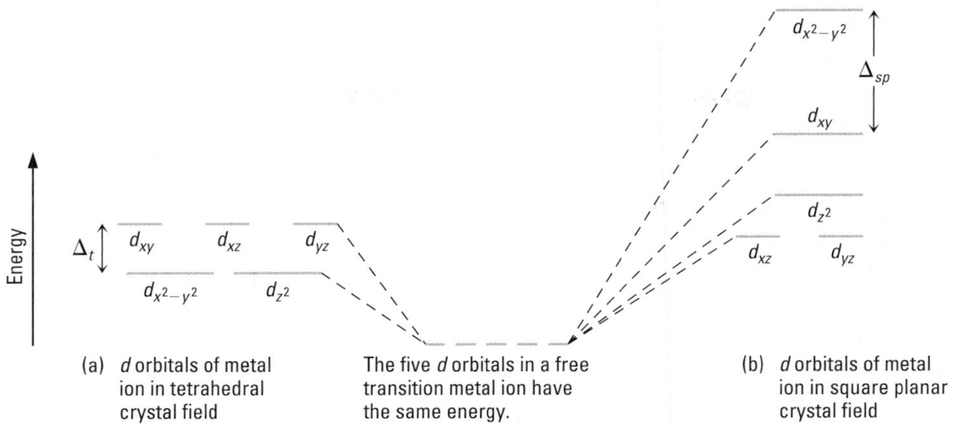

Figure 22.20 Splitting of metal ion *d* orbitals in (a) tetrahedral and (b) square planar complexes.

CONCEPTUAL EXERCISE 22.13 High- and Low-spin Complexes

Explain why Ni^{2+} ions cannot form high- and low-spin complexes.

Tetrahedral and Square Planar Complexes

Up to this point we have focused on the application of crystal-field theory to octahedral complexes. It can be applied to tetrahedral and square planar complexes as well, which have different crystal-field splitting patterns (Figure 22.20). The *d* orbital splitting pattern for tetrahedral complexes is the opposite of the octahedral pattern. In tetrahedral complexes the $d_{x^2-y^2}$ and d_{z^2} orbitals are lower in energy by Δ_t than the set of d_{xz}, d_{xy}, and d_{yz} orbitals.

In square planar complexes, we assume that ligands are in a plane containing the *x*- and *y*-axes. Because the $d_{x^2-y^2}$ orbital points along these axes, it is the highest in energy, followed by the d_{xy} orbital and then the d_{z^2} orbital. The d_{xz} and d_{yz} orbitals are of equal and lowest energy (Figure 22.20). The crystal-field splitting energy for square planar complexes, Δ_{sp}, is the energy difference between the $d_{x^2-y^2}$ and d_{xy} orbitals. A square planar complex can be considered as forming by removing the two ligands along the *z*-axis from an octahedral complex (Figure 22.21).

Remember that the d_{z^2} orbital has a donut of electron density in the *x*-*y* plane (Figure 22.17).

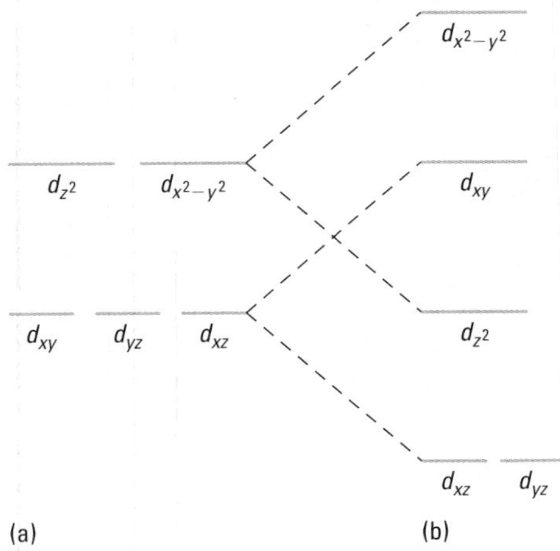

Figure 22.21 The change in relative energy of *d* orbitals in (a) octahedral and (b) square planar complexes. Removing two ligands from the *z*-axis in an octahedral complex and moving the four ligands in the *x*-*y* plane closer to the metal ion yields a square planar complex.

An aqueous solution of Ni^{2+} is green. The $[Ni(H_2O)_6]^{2+}$ complex absorbs red-violet light and transmits green light.

Coordination compounds with metal ions whose *d* orbitals are filled, such as Zn^{2+}, or have no *d* electrons, such as Sc^{3+}, are not colored.

Color in Coordination Complexes

Most transition metal complexes are brightly colored and explaining their color is a major success of crystal-field theory. The colors arise because of transitions of electrons, called ***d*-to-*d* transitions,** from a lower-energy set of *d* orbitals to a higher-energy *d* orbital of the metal ion. In a coordination complex, the energy difference between these sets of *d* orbitals typically corresponds to photon wavelengths within the visible region of the spectrum (400–700 nm), thus giving rise to a color when some wavelengths of visible light are absorbed. The color we see is the color of light that is transmitted; *it is the complementary color of the light absorbed.* On the color wheel in Figure 22.22, complementary colors are directly across from each other. Therefore, as seen from Figure 22.22, a complex appears yellow, for example, because it absorbs blue-violet light and transmits yellow light, the complementary color. A $[Ni(H_2O)_6]^{2+}$ solution is green because the complex absorbs red-violet light and transmits green light.

We can apply crystal-field theory to understand the origin of the vivid purple (red-violet) color of the $[Ti(H_2O)_6]^{3+}$ complex ion. This case is particularly simple because Ti^{3+} has only a single 3*d* electron. The complex ion absorbs light at 510 nm, the green region of the visible spectrum, raising the *d* electron from a lower-energy t_2 orbital to a higher-energy *e* orbital (Figure 22.23).

$$t_2^1 e^0 \xrightarrow{E = \Delta_o = \frac{hc}{\lambda}} t_2^0 e^1$$

In this case, the energy difference is a direct measure of Δ_o. We can calculate Δ_o, the crystal-field splitting energy, for the octahedral $[Ti(H_2O)_6]^{3+}$ complex ion by

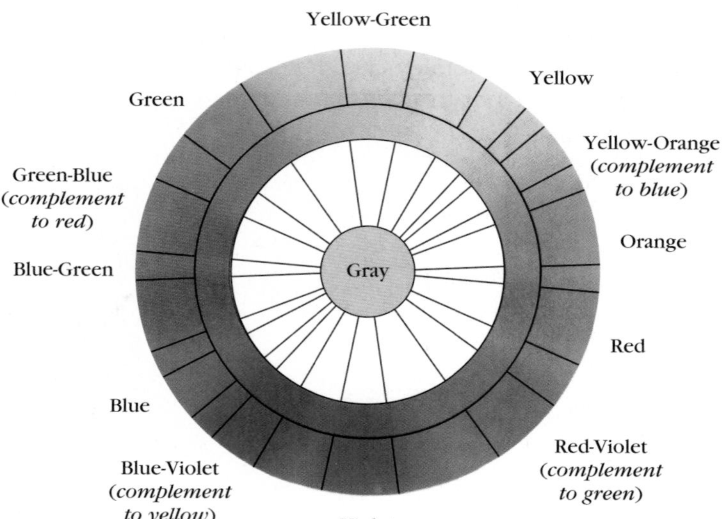

Figure 22.22 The color wheel. The color of a coordination complex is the complementary color to that which is absorbed (directly across from the color on the wheel.) Thus, yellow-orange is the complementary color of blue.

relating it to the wavelength and energy of the absorbed photon. For one photon of 510-nm light,

$$E = \frac{hc}{\lambda} = \frac{(6.626 \times 10^{-34}\ \text{J}\cdot\text{s})(2.998 \times 10^{8}\ \text{m/s})}{510 \times 10^{-9}\ \text{m}} = 3.90 \times 10^{-19}\ \text{J}$$

This translates into 235 kJ/mol, which is the splitting energy, Δ_o, the energy separating the t_2 and e sets of d orbitals.

$$E = 3.90 \times 10^{-19}\ \text{J} \times \frac{1\ \text{kJ}}{10^{3}\ \text{J}} \times \frac{6.022 \times 10^{23}}{1\ \text{mol}} = 2.35 \times 10^{2}\ \text{kJ/mol}$$

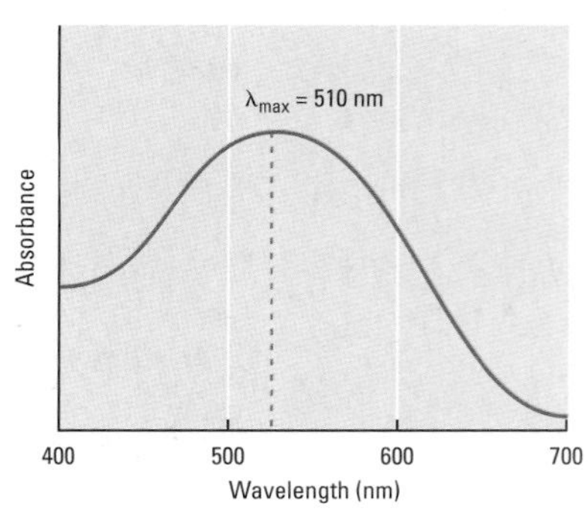

Figure 22.23 The absorption spectrum of $[Ti(H_2O)_6]^{3+}$. The absorption of 510-nm wavelength light causes a *d*-to-*d* transition of the 3*d* electron of $[Ti(H_2O)_6]^{3+}$.

Color and the Spectrochemical Series

The color of a coordination complex depends on the splitting energy, which depends on the metal ion and the field strength of its ligands, as given by the spectrochemical series (p. 1094). Stronger-field ligands cause a larger splitting energy. Consequently, complexes with such ligands absorb light at shorter wavelengths compared to those with weaker-field ligands, as seen vividly in Figure 22.24 and listed in Table 22.5 for a series of Co^{3+} complexes. Note from Table 22.5 that as weaker-field ligands are substituted for NH_3 in the complex, the wavelength of the absorbed radiation becomes longer, indicating a lower splitting energy.

CONCEPTUAL EXERCISE 22.14 Color and Electron Configuration

An aqueous solution of $[Cr(H_2O)_6]^{2+}$ is blue. Predict whether the 3*d* electrons of the central chromium ion are in a low-spin or high-spin configuration.

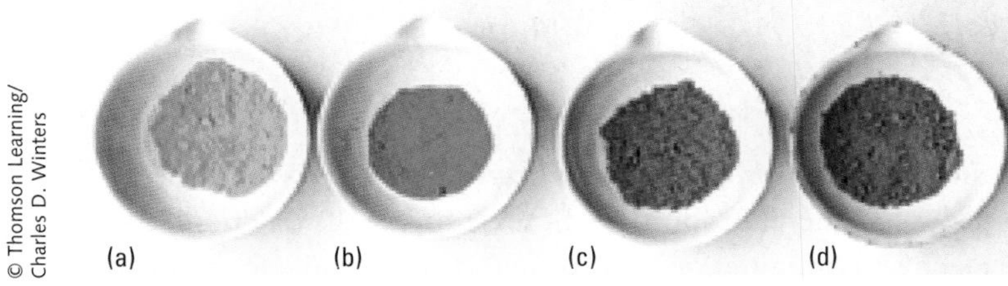

Figure 22.24 The shift in the color of coordination complexes as ligands surrounding a Co^{3+} ion are changed. (See Table 22.5.)

Table 22.5 Colors of Co^{3+} Complexes

Complex	Color Observed	Color Absorbed	Approximate Wavelength Absorbed (nm)
(a) $[Co(NH_3)_6]^{3+}$	Yellow	Blue-violet	430
(b) $[Co(NH_3)_5NCS]^{2+}$	Orange	Blue-green	470
(c) $[Co(NH_3)_5H_2O]^{3+}$	Red	Green-blue	500
(d) $[Co(NH_3)_5Cl]^{2+}$	Reddish purple	Green-blue	522

SUMMARY PROBLEM

I. Solid iron(II) sulfide, FeS, is roasted to form either $Fe_2O_3(s)$ or $Fe_3O_4(s)$.

(a) Write balanced equations for the roasting of FeS(s) to $Fe_2O_3(s)$ and to $Fe_3O_4(s)$.

(b) Use thermochemical data from Appendix J to calculate the enthalpy change for these reactions at 25 °C, given $\Delta H_f^\circ = -101.671$ kJ/mol for FeS(s).

(c) Use thermochemical data from Appendix J to calculate the Gibbs free energy change for these reactions at 25 °C, given $S^\circ = 82.81$ J mol^{-1} K^{-1} for FeS(s). Which reaction is more product-favored at this temperature?

(d) Calculate the Gibbs free energy change for the conversion of FeS to Fe_2O_3 and to Fe_3O_4 at 600 °C. Which reaction is more product-favored at this temperature?

II. Iron(III) forms a deep red coordination compound $[Fe(acac)_3]$ with acetylacetonate ions $(acac)^-$.

$$\left(CH_3-\overset{\overset{:O:}{\|}}{C}-CH=\overset{\overset{:\ddot{O}:}{|}}{C}-CH_3 \right)^-$$

The acetylacetonate anion is a bidentate ligand that furnishes lone pairs from the C=O and C—O oxygens.

(a) Write a structural formula for $[Fe(acac)_3]$.

(b) Give the coordination number of Fe^{3+} in this compound.

(c) Account for the fact that there is no net charge on $[Fe(acac)_3]$.

(d) Why are no counter ions needed for this compound?

ThomsonNOW™
Sign in to ThomsonNOW at **www.thomsonedu.com** to check your readiness for an exam by taking the Pre-Test and exploring the modules recommended in your Personalized Learning Plan.

IN CLOSING

Having studied this chapter, you should be able to . . .

- Recognize the general properties of transition metals (Section 22.1).
- Write electron configurations and orbital box diagrams for transition metals and their ions (Section 22.1). ThomsonNOW homework: Study Question 14
- Explain why most transition metals have multiple oxidation states (Section 22.1). ThomsonNOW homework: Study Question 16
- Explain trends in sizes of transition metal atomic radii (Section 22.1).
- Describe how iron ore is processed into iron and then into steel (Section 22.2).
- Discuss how copper is extracted from its ores and purified (Section 22.3).
- Discuss the chemistry of gold and silver (Section 22.4).
- Discuss the chemistry of chromium (Section 22.5).
- Explain the coordinate covalent bonding of ligands in coordination compounds and complexes (Section 22.6). ThomsonNOW homework: Study Question 28
- Interpret the names and formulas of coordination complex ions and compounds (Section 22.6). ThomsonNOW homework: Study Questions 34, 38, 42
- Discuss isomerism in coordination compounds and complex ions (Section 22.6). ThomsonNOW homework: Study Question 48

- Give examples of coordination compounds and their uses (Section 22.6).
- Describe crystal-field theory and its applications to interpreting colors and magnetic properties of coordination compounds (Section 22.7).

KEY TERMS

bidentate *(22.6)*
chelating ligands *(22.6)*
coordinate covalent bond *(22.6)*
coordination compound *(22.6)*
coordination number *(22.6)*
crystal-field splitting energy, Δ *(22.7)*
***d*-to-*d* transitions** *(22.7)*
hexadentate *(22.6)*
high-spin complex *(22.7)*
lanthanide contraction *(22.1)*
ligands *(22.6)*
low-spin complex *(22.7)*
monodentate *(22.6)*
polydentate *(22.6)*
pyrometallurgy *(22.2)*
spectrochemical series *(22.7)*
steel *(22.2)*

QUESTIONS FOR REVIEW AND THOUGHT

■ denotes questions available in ThomsonNOW and assignable in OWL.

Blue-numbered questions have short answers at the back of this book and fully worked solutions in the *Student Solutions Manual.*

ThomsonNOW
Assess your understanding of this chapter's topics with sample tests and other resources found by signing in to ThomsonNOW at **www.thomsonedu.com**.

Review Questions

1. What is the primary reducing agent in the production of iron from its ores? Write a balanced chemical equation for this reduction process.
2. Why is lime necessary in the blast furnace reduction of iron ore?
3. What is the difference between pig iron and cast iron?
4. Explain the purpose of each of these materials in the blast-furnace conversion of iron ore to iron.
(a) Air (b) Limestone (c) Coke
5. Identify what is produced by each of these processes or operations.
(a) Blast furnace
(b) Basic oxygen process
(c) Roasting
6. Identify a common use for Cr, Cu, Fe, Au, and Ag.
7. Name three transition metals that are found "free" in nature.
8. What is the lanthanide contraction? Why does it occur?
9. In general, how do the atomic radii change across the first transition series (Period 4)?
10. What is the distinguishing chemical feature of a ligand?
11. Distinguish between
(a) A monodentate and a bidentate ligand.
(b) A *cis* and a *trans* isomer.
(c) A coordination compound and a coordination complex ion.
(d) A geometric and an optical isomer.
12. Define these words or phrases and give an example for each.
(a) Coordination compound (b) Complex ion
(c) Ligand (d) Chelate
(e) Bidentate ligand

Topical Questions

Transition Metals

13. Write electron configurations for the 2+ ions of
(a) Iron. (b) Copper. (c) Chromium.
14. ■ Write electron configurations for the common oxidation states of
(a) Silver. (b) Gold.
15. Which Period 4 transition metal ions are isoelectronic with
(a) Zn^{2+} (b) Mn^{2+}
(c) Cr^{3+} (d) Fe^{3+}
16. Which two oxidation states of chromium are paramagnetic?
17. Arrange these in order of decreasing strength as oxidizing agents: Mn^{2+}, MnO_4^-, MnO_2. Explain the trend.
18. ■ Arrange these in order of decreasing strength as oxidizing agents: $Cr_2O_7^{2-}$ (in acid), Cr^{2+}, Cr^{3+}. Explain the trend.
19. Write a balanced equation to represent
(a) The roasting of Cu_2S to copper metal.
(b) The reduction of Fe_2O_3 with aluminum.
20. Write a balanced equation to represent
(a) The reduction of Fe_2O_3 with carbon monoxide in a blast furnace.
(b) The production of hydrogen gas when hydrochloric acid reacts with an iron nail.
21. Balance this redox reaction (in acidic solution).
$$Cu(s) + NO_3^-(aq) \longrightarrow Cu^{2+}(aq) + NO_2(g)$$
22. Balance this redox reaction (in acidic solution).
$$Fe(s) + NO_3^-(aq) \longrightarrow Fe^{3+}(aq) + NO_2(g)$$
23. Determine the oxidation state of the transition metal in each compound.
(a) V_2O_5 (b) $K_2Cr_2O_7$
(c) MnO_2 (d) OsO_4

24. ■ Determine the oxidation state of iron in $KFe[Fe(CN)_6]$. Explain your answer.

Coordination Compounds

25. In a complex ion, a central ruthenium ion, Ru(III), is bonded to six ammonia molecules.
    (a) Give the formula and net charge for this complex ion.
    (b) How many chloride ions are needed to balance the net charge on this complex ion?
    (c) Write the formula for the compound formed by this complex ion and the chloride ions that are not part of it.
26. In a complex ion, a central Cr^{3+} ion is bonded to two ammonia molecules, three water molecules, and a hydroxide ion.
    (a) Give the formula and charge of the complex ion.
    (b) Identify a single counter ion that could be used with the complex ion to form an uncharged compound.
27. Consider the complex ion $[Cr(NH_3)_2(H_2O)_2Br_2]^+$.
    (a) Identify the ligands and their charges (if any).
    (b) What is the charge on the central metal ion?
    (c) What is the formula of the sulfate salt of this cation?
28. ■ Consider the complex ion $[Co(C_2O_4)_2Cl_2]^{3-}$.
    (a) Identify the ligands and their charges (if any).
    (b) What is the charge on the central metal ion?
    (c) What would be the formula and charge of the complex ion if the $C_2O_4^{2-}$ ions were replaced by NH_3 molecules?
29. Determine the charge of the central metal ion in each case.
    (a) $[Zn(H_2O)_3(OH)]^+$
    (b) $[Pt(NH_3)_3Cl_3]^+$
    (c) $[Cr(CN)_6]^{3-}$
30. For coordination compounds $Na_3[IrCl_6]$ and $[Mo(CO)_4]Br_2$, identify in each case
    (a) The ligands.
    (b) The central metal ion and its charge.
    (c) The formula and charge of the complex ion.
    (d) Any ions not part of the complex ion.
31. Give the coordination number of the central metal ion in
    (a) $[Ni(en)_2(NH_3)_2]^{2+}$
    (b) $[Fe(en)(C_2O_4)Cl_2]^-$
32. ■ Give the coordination number of the central metal ion in
    (a) $[Pt(en)_2]^{2+}$
    (b) $[Cu(C_2O_4)_2]^{2-}$
33. Write a structural formula for the coordination compound $[Cr(en)(NH_3)_2I_2]$, and give the coordination number for the central Cr^{2+} ion.
34. Give the formula of each of these coordination compounds formed with Pt^{2+}.
    (a) Two ammonia molecules and two bromide ions
    (b) One ethylenediamine molecule and two nitrite ions, NO_2^-
    (c) One chloride ion, one bromide ion, and two ammonia molecules
35. Give the charge on the central metal ion in each of these.
    (a) $[VCl_6]^{4-}$ (b) $[Sc(H_2O)_3Cl_3]$
    (c) $[Mn(NO)(CN)_5]^{3-}$ (d) $[Ni(en)_2(NH_3)_2]^{2+}$
36. Identify the coordination number of the metal ion in these coordination complexes.
    (a) $[FeCl_4]^-$ (b) $[PtBr_4]^{2-}$
    (c) $[Mn(en)_3]^{2+}$ (d) $[Cr(NH_3)_5H_2O]^{3+}$
37. Using structural formulas, show how the carbonate ion can be either a monodentate or bidentate ligand to a transition metal cation.
38. ■ Classify each ligand as monodentate, bidentate, and so on.
    (a) $(CH_3)_3P$
    (b) 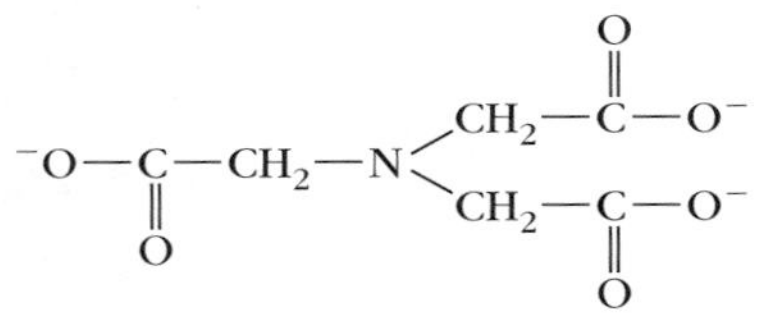 $^-O-\underset{\underset{O}{\|}}{C}-CH_2-N\begin{matrix} CH_2-\overset{\overset{O}{\|}}{C}-O^- \\ CH_2-\underset{\underset{O}{\|}}{C}-O^- \end{matrix}$
    (c) $H_2N-(CH_2)_2-NH-(CH_2)_2-NH_2$
    (d) H_2O
39. Which of these would be expected to be effective chelating agents? Explain your answer.
    (a) CH_3CH_2OH (b) $H_2N-(CH_2)_3-NH_2$
    (c) $HO-\overset{\overset{O}{\|}}{C}-CH_2-\overset{\overset{O}{\|}}{C}-OH$ (d) PH_3
40. Give the formula of a simple ionic (noncoordination) compound analogous to $[Rh(en)_3]Cl_3$.

Naming Complex Ions and Coordination Compounds

41. Write the formula for
    (a) Potassium diaquadioxalatocobaltate(III)
    (b) Diamminetriaquahydroxochromium(II) nitrate
    (c) Ammonium tetrachlorocuprate(II)
    (d) Tetrachloroethylenediaminecobaltate(III)
    (e) Triaquatrifluorocobalt(III)
42. ■ Write the name corresponding to each of these formulas.
    (a) $[MnCl_4]^{2-}$ (b) $K_3[Fe(C_2O_4)_3]$
    (c) $[Pt(NH_3)_2(CN)_2]$ (d) $[Fe(H_2O)_5(OH)]^{2+}$
    (e) $[Mn(en)_2Cl_2]$

Geometry of Coordination Complexes

43. Sketch the geometry of
    (a) *cis*-$[Cu(H_2O)_2Br_4]^{2-}$ (b) *trans*-$[Ni(NH_3)_2(en)_2]^{2+}$
44. Sketch the geometry of
    (a) *cis*-$[Pt(H_2O)_2Cl_2]$ (b) *trans*-$[Cr(H_2O)_4Cl_2]^+$
45. The ligand 1,2-diaminocyclohexane

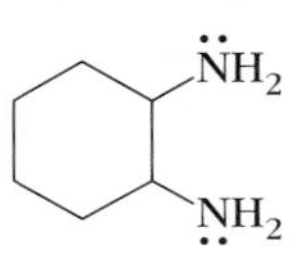

    $\ddot{N}H_2$
    $\ddot{N}H_2$

    is abbreviated "dech." Sketch the geometry of *cis*-$[Pd(H_2O)_2(dech)]^{2+}$.
46. The acetylacetonate ion $(acac)^-$

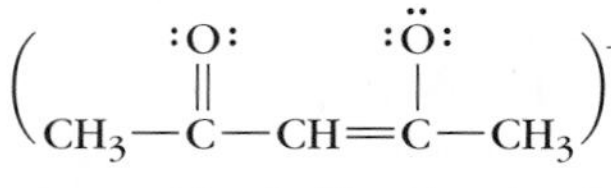

    $\left(CH_3-\overset{\overset{:O:}{\|}}{C}-CH=\overset{\overset{:\ddot{O}:}{|}}{C}-CH_3\right)^-$

    forms a complex with Co^{3+}. Sketch the geometry of $[Co(acac)_3]$.
47. Which of these octahedral coordination complexes can exhibit geometric isomerism?
    (a) $[Pt(H_2O)_2Cl_2Br_2]^{2-}$ (b) $[Pt(H_2O)_2Cl_3Br]^-$

48. ■ Which of these octahedral coordination complexes can exhibit geometric isomerism?
(a) $[Cr(H_2O)_3Cl_3]$ (b) $[Cr(H_2O)_4Cl_2]^+$
49. Draw the possible geometric isomers, if any, of
(a) $[Ni(NH_3)_4Cl_2]$
(b) $[Pt(NH_3)_2(SCN)Br]$ (The S in NCS is bonded to Pt^{2+}.)
(c) $[Co(en)Cl_4]^-$
50. Draw the possible geometric isomers, if any.
(a) $[Co(H_2O)_4Cl_2]^+$ (b) $[Pt(NH_3)Cl_3]^-$
(c) $[Co(H_2O)_3Cl_3]$ (d) $[Co(en)_2(NH_3)Br]^{2+}$

Crystal-Field Theory and Magnetic Properties of Complex Ions

51. Draw the crystal-field splitting diagrams and put in the *d* electrons for these octahedral complexes. In those cases where they are possible, draw diagrams for both low-spin and high-spin cases.
(a) $[Cr(H_2O)_6]^{2+}$ (b) $[Mn(H_2O)_6]^{2+}$
(c) $[FeF_6]^{3-}$ (d) $[Cr(en)_3]^{3+}$
52. Using the spectrochemical series, predict the actual number of unpaired electrons in each complex in Question 51 for which the possibility of low-spin and high-spin forms exist.
53. Fe^{3+} forms octahedral complexes with NCS^- and with NO_2^- ligands. One complex displays a greater paramagnetism than the other.
(a) Write the formula for each of these complex ions.
(b) Use the spectrochemical series to predict whether the complex ions are high-spin or low-spin.
(c) Identify which complex ion is more paramagnetic.
(d) Draw the crystal-field splitting diagram, including *d* electrons, for each complex ion.
54. ■ Explain why Cr^{2+} forms high-spin and low-spin octahedral complexes, but Cr^{3+} does not.
55. How many unpaired electrons are in the high-spin and low-spin octahedral complexes of Cr^{2+}?
56. Use crystal-field theory to explain why some Co^{3+} octahedral complexes are diamagnetic and others are paramagnetic.
57. Use crystal-field theory to explain why some octahedral Co^{2+} complexes are more paramagnetic than others.
58. Use crystal-field theory to explain why Cu^{2+} does not form high-spin and low-spin octahedral complexes.

Crystal-Field Theory and Color in Complex Ions

59. An aqueous solution of $[Ni(NH_3)_6]^{2+}$ is purple. Predict the approximate wavelength and predominant color of light absorbed by the complex.
60. ■ An aqueous solution of $[Rh(C_2O_4)_3]^{3-}$ is yellow. Predict the approximate wavelength and predominant color of light absorbed by the complex.
61. An aqueous solution of $[Rh(C_2O_4)_3]^{3-}$ is yellow; that of $[Rh(en)_3]^{3+}$ is a different color. Oxalate is to the right of en in the spectrochemical series. Predict what the change in color likely will be from $[Rh(C_2O_4)_3]^{3-}$ to $[Rh(en)_3]^{3+}$. In which direction does the absorbed wavelength change?
62. ■ As discussed in Section 22.6, an aqueous solution of $[Ti(H_2O)_6]^{3+}$ is purple (red-violet). Predict how the value of Δ_o would change if all H_2O ligands were replaced by CN^- ligands; by Cl^- ligands. How would the color change in each case?
63. A solution of a complex ion absorbs visible light at a wavelength of 540 nm.
(a) What is the color of the solution?
(b) Calculate the energy of an absorbed photon in joules and in kJ/mol.

General Questions

64. Give the electron configuration of
(a) Ti^{3+} (b) V^{2+}
(c) Ni^{3+} (d) Cu^+
65. Give the electron configuration of
(a) Cr^{2+} (b) Zn^{2+}
(c) Co^{2+} (d) Mn^{4+}
66. Write an orbital box diagram and determine the number of unpaired electrons for each species in Question 64.
67. Write an orbital box diagram and determine the number of unpaired electrons for each species in Question 65.
68. Assuming 100% recovery of the metal, which would yield the greater number of grams of copper?
(a) One kilogram of an ore containing 3.60 mass % azurite, $Cu(OH)_2 \cdot 2\ CuCO_3$
(b) One kilogram of an ore containing 4.95 mass % chalcopyrite, $CuFeS_2$
69. What mass of copper could be electroplated from a $CuSO_4$ solution using an electric current of 2.50 A for 5.00 h? Assume 100% efficiency.
70. Copper metal is obtained directly by roasting covelite, CuS.
(a) Write a balanced equation for this process.
(b) Assume that the roasting is 90.0% efficient. How many tons of SO_2 would be released into the air by roasting 500. tons of covelite?
71. What mass of SO_2 is produced when 1.0 ton of chalcocite, Cu_2S, is roasted to Cu_2O?
72. What is the coordination number of the central metal ion in
(a) $[Pt(NH_3)_2Br_2]$ (b) $[Fe(CN)_6]^{3-}$
(c) $[Ti(H_2O)Cl_5]^{2-}$ (d) $[Mn(C_2O_4)_3]^{4-}$
73. What is the coordination number of the central metal ion in
(a) $[Ni(en)Cl_2]$ (b) $[Mo(CO)_4Br_2]$
(c) $[Cd(CN)_4]^{2-}$ (d) $[Co(CN)_5(OH)]^{3-}$
74. Draw sketches for all octahedral complexes of Co^{3+} using only ethylenediamine and/or Cl^- as ligands.
75. Draw sketches for as many octahedral complexes as you can for the formula $Co(NH_3)_4Cl_2Br$.
76. In your own words explain why
(a) $H_2N{-}(CH_2)_3{-}NH_2$ is a bidentate ligand.
(b) AgCl dissolves in NH_3.
(c) There are no geometric isomers of tetrahedral complexes.
77. Determine whether each statement is true or false. If it is false, correct the statement.
(a) The coordination number of the Fe^{3+} ion in $[Fe(H_2O)_4(C_2O_4)]^+$ is five.
(b) Cu^+ has two unpaired electrons.
(c) The net charge of a coordination complex of Cr^{3+} with two NH_3, one en, and two H_2O is 2+.

78. Determine whether each statement is true or false. If it is false, correct the statement.
   (a) In $[Pt(NH_3)_4Cl_4]$, platinum has a 4+ charge and a coordination number of six.
   (b) In general, Cu^{2+} is more stable than Cu^+ in aqueous solutions.
79. The metal ion in $[Pt(NH_3)_2(C_2O_4)]$ is surrounded by a square planar array of coordinating atoms.
   (a) Give the oxidation number of the central metal ion.
   (b) Draw the structural formula of this coordination compound.
80. Chromium(III) forms three different compounds with water and chloride ions, all of which have the same composition: 19.51% Cr, 39.92% Cl, and 40.57% water. One of the compounds is violet and dissolves in water to give a complex ion with a 3+ charge plus three chloride ions. All three chloride ions precipitate immediately as AgCl when $AgNO_3$ is added to the solution. Draw the structural formula of this complex ion.

Applying Concepts

81. Iron nails are put into Fe^{2+} aqueous solutions to reduce any Fe^{3+} that forms back to Fe^{2+}. Write a balanced chemical equation for this preventative reaction.
82. Use VSEPR theory to predict the shape and bond angles around chromium in
   (a) Chromate ions (b) Dichromate ions
83. The structure of cyclam is given below.

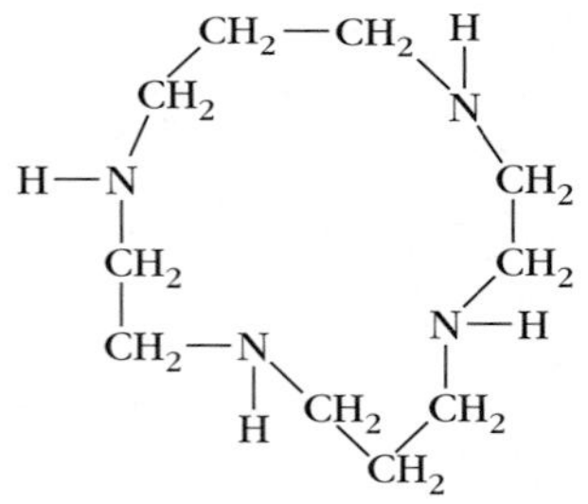

   Cyclam can act as a ligand. How many coordinate covalent bonds can one cyclam molecule form with a central metal ion?
84. The compound 1,10-phenanthroline is a chelating agent used in analytical chemistry. Its isomer 4,7-phenanthroline is not. Use these structural formulas to explain this difference.

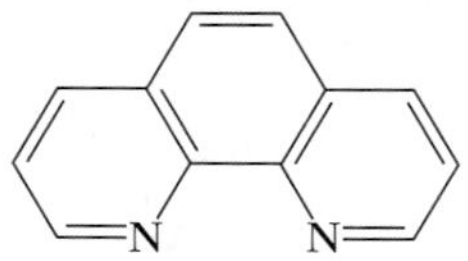

   1,10-phenanthroline 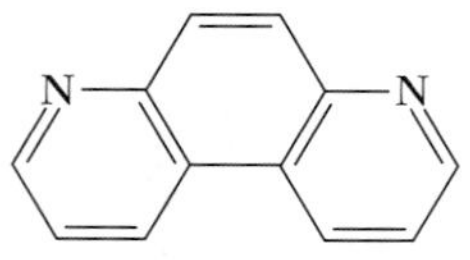 4,7-phenanthroline

85. Two different isomers are known with the formula $[Pt(py)_2Cl_2]$, where py represents pyridine, an uncharged monodentate ligand in which an N atom bonds to the metal ion. There is, however, only one structure known for $[Pt(phen)Cl_2]$, where phen represents *ortho*-phenanthroline, an uncharged bidentate ligand (Figure 22.9). Draw the structural formulas of all three molecules and explain why there are isomers in one case, but not the other.
86. An electrochemical cell is made by immersing a strip of chromium into a 1.0 M solution of Cr^{3+} and a strip of gold into a 1.0 M solution of Au^{3+}. The half-cells are connected by a salt bridge. A wire and light bulb complete the circuit.
   (a) Write the balanced chemical equation for the reaction that is product-favored.
   (b) Calculate the cell potential.
   (c) Draw a sketch of the cell and indicate the anode, cathode, and direction of electron flow.
87. Repeat the directions for Question 86 using a cell constructed of a strip of nickel immersed in a 1.0 M Ni^{2+} solution and a strip of silver dipping into a 1.0 M Ag^+ solution.
88. Calculate $\Delta G°$ for the reduction of Fe_2O_3 with CO gas at 25 °C and at 1000 °C. What application does this have to the conversion of iron ore to iron in a blast furnace?
89. To determine the percent iron in an ore, a 1.500-g sample of the ore containing Fe^{2+} is titrated to the equivalence point with 18.6 mL of 0.05012 M $KMnO_4$. The products of the titration are Fe^{3+} and Mn^{2+}. What is the percent of iron in the ore?
90. Consider the reaction

$$2\,Cu^+(aq) \longrightarrow Cu^{2+}(aq) + Cu(s)$$

   for which $E°_{cell} = +0.37$ V. Use the Nernst equation to calculate
   (a) E when the Cu^{2+} concentration is equal to the Cu^+ concentration $= 1 \times 10^{-4}$ M.
   (b) The concentration of Cu^+ when the Cu^{2+} concentration = 1.0 M and $E = 0.00$ V.
91. Consider the reaction

$$2\,Ag^+(aq) \longrightarrow Ag(s) + Ag^{2+}(aq)$$

   for which $E°_{cell} = -1.18$ V. Use the Nernst equation to calculate
   (a) E when the Ag^+ concentration $= 1 \times 10^{-4}$ M, which is five times the concentration of Ag^{2+}.
   (b) The concentration of Ag^{2+} when the Ag^+ concentration = 1.0 M and $E = 0.00$ V.
92. Use the Nernst equation to calculate E_{cell} for $Cr_2O_7^{2-}$ oxidation of Cl^- in 6.0 M H^+ when the concentration of $Cr_2O_7^{2-}$ = concentration of Cr^{3+} = 0.10 M, and all other concentrations = 1.0 M.
93. If 1.00 mol of each of these compounds is dissolved in separate samples of water sufficient to dissolve the compound, how many moles of ions are present in each solution?
   (a) $[Pt(en)Cl_2]$ (b) $Na[Cr(en)_2(SO_4)_2]$
   (c) $K_3[Au(CN)_4]$ (d) $[Ni(H_2O)_2(NH_3)_4]Cl_2$
94. For each of the compounds in Question 93, state which it would most likely resemble in colligative properties and conductivity: $(NH_2)_2CO$ (urea), KCl, K_2SO_4, or K_3PO_4.
95. In aqueous solution, $[Cr(NH_3)_6]Cl_3$ is yellow, but aqueous $[Cr(NH_3)_5Cl]Cl_2$ is purple. Explain the difference in colors.
96. Early coordination chemists relied on close experimental observation to determine the formulas of coordination compounds. They found, for example, that aqueous $BaCl_2$ did not cause precipitation when added to a solution of a Co^{3+}-containing coordination compound, but precipitation

occurred when aqueous silver nitrate was added to a solution of the coordination compound. The coordination compound was known to contain one Co^{3+} ion, one sulfate ion, one chloride ion, and four ammonia molecules. Write the structural formula of the coordination compound that is consistent with the experimental results.

More Challenging Questions

97. The bidentate oxalate ion, $C_2O_4^{2-}$, forms octahedral complexes with Fe^{3+} and Ru^{3+} ions.
    (a) Write the structural formula for each complex.
    (b) The ruthenium complex is low-spin; the iron complex is high-spin. Write the *d*-orbital splitting diagram for each metal ion.
    (c) Which complex has the higher Δ_o? Explain your answer.
98. Analysis of a coordination compound gives these results: 22.0% Co, 31.4% N, 6.78% H, and 39.8% Cl. One mole of the compound dissociates in water to form 4 mol ions.
    (a) What is the formula of the compound?
    (b) Write the equation for its dissociation in water.
99. A chemist synthesizes two coordination compounds. One compound decomposes at 210 °C, the other at 240 °C. When analyzed, the compounds give the same mass percent data: 52.6% Pt, 7.6% N, 1.63% H, and 38.2% Cl. Both compounds contain a 4+ central metal ion.
    (a) What is the simplest formula of the compounds?
    (b) Draw structural formulas for the complexes present.
100. A coordination compound has the simplest formula $PtN_2H_6Cl_2$ with a molar mass of about 600 g/mol. It contains a complex cation and a complex anion. Draw its structural formula.
101. The glycinate ion (gly) is $H_2NCH_2CO_2^-$. It can act as a ligand coordinating through the nitrogen and one of the oxygens. Using $\overset{\frown}{N\ \ O}$ to represent glycinate ion, draw structural formulas for four stereoisomers of $[Co(gly)_3]$.
102. Five-coordinate coordination complexes are known, including $[CuCl_5]^{3-}$ and $[Ni(CN)_5]^{3-}$. Write the structural formulas and identify a plausible geometry for these complexes.
103. Predict the number of unpaired electrons in a square planar transition metal ion with seven *d* electrons.
104. A coordination compound has the empirical formula $Fe(H_2O)_4(CN)_2$. Its paramagnetism is the equivalent of 2.67 unpaired electrons per Fe ion. Explain how this is possible.
105. Two different compounds are known with the formula $Pd(py)_2Cl_2$, but there is only one compound with the formula $Zn(py)_2Cl_2$. The symbol py is for pyridine, a monodentate ligand. Explain the differences in the Pd and Zn compounds.
106. An octahedral coordination complex ion is formed by the combination of an Fe^{3+} ion and det ligands (det is $H_2NCH_2CH_2NHCH_2CH_2NH_2$). Write a structural formula for the complex ion.

APPENDIX

Problem Solving and Mathematical Operations

In this book we have provided many illustrations of problem solving and many problems for practice. Some are numerical problems that must be solved by mathematical calculations. Others are conceptual problems that must be solved by applying an understanding of the principles of chemistry. Often, it is necessary to use chemical concepts to relate what we know about matter at the nanoscale to the properties of matter at the macroscale. The problems throughout this book are representative of the kinds of problems that chemists and other scientists must regularly solve to pursue their goals, although our problems are often not as difficult as those encountered in the real world.

Problem solving is not a simple skill that can be mastered with a few hours of study or practice. Because there are many different kinds of problems and many different kinds of people who are problem solvers, no hard and fast rules are available that are guaranteed to lead you to solutions. The general guidelines presented in this appendix are, however, helpful in getting you started on any kind of problem and in checking whether your answers are correct. The problem-solving skills you develop in a chemistry course can later be applied to difficult and important problems that may arise in your profession, your personal life, or the society in which you live.

In getting a clear picture of a problem and asking appropriate questions regarding the problem, you need to keep in mind all the principles of chemistry and other subjects that you think may apply. In many real-life problems, not enough information is available for you to arrive at an unambiguous solution; in such cases, try to look up or estimate what is needed and then forge ahead, noting assumptions you have made. Often the hardest part is deciding which principle or idea is most likely to help solve the problem and what information is needed. To some degree this can be a matter of luck or chance. Nevertheless, in the words of Louis Pasteur, "In the field of observation chance only favors those minds which have been prepared." The more practice you have had, and the more principles and facts you can keep in mind, the more likely you are to be able to solve the problems that you face.

A.1 General Problem-Solving Strategies

"The mere formulation of a problem is often far more essential than its solution. To raise new questions, new possibilities, to regard old problems from a new angle, requires creative imagination and marks real advances in science."
—Albert Einstein

1. **Define the problem.** Carefully review the information contained in the problem. What is the problem asking you to find? What key principles are involved? What known information is necessary for solving the problem and what information is there just to place the question in context? Organize the information to see what is necessary and to see the relationships among the known data. Try writing the information in an organized way. If the information is numerical, be sure to include proper units. Can you picture the situation under consideration? Try sketching it and including any relevant dimensions in the sketch.

2. **Develop a plan.** Have you solved a problem of this type before? If you recognize the new problem as similar to ones you know how to solve, you can use the same method that worked before. Try reasoning backward from the units of what is being sought. What data are needed to find an answer in those units?

Can the problem be broken down into smaller pieces, each of which can be solved separately to produce information that can be assembled to solve the entire problem? When a problem can be divided into simpler problems, it often helps to write down a plan that lists the simpler problems and the order in which those problems must be put together to arrive at an overall solution. Many major problems in chemical research have to be solved in this way. In problems in this book we have mostly provided the needed numerical data, but in the laboratory, the first aspect of solving a problem is often devising experiments to gather the data or searching databases to find needed information.

If you are still unsure about what to do, do something anyway. It may not be the right thing to do, but as you work on it, the way to solve the problem may become apparent, or you may see what is wrong with your initial approach, thereby making clearer what a good plan would be.

3. **Execute the plan.** Carefully write down each step of a mathematical problem, being sure to keep track of the units. Do the units cancel to give you the answer in the desired units? Don't skip steps. Don't do any except the simplest steps in your head. Once you've written down the steps for a mathematical problem, check what you've written—is it all correct? Students often say they got a problem wrong because they "made a stupid mistake." Teachers—and textbook authors—make mistakes, too. These errors usually arise because they don't take the time to write down the steps of a problem clearly and correctly. In solving a mathematical problem, remember to apply the principles of dimensional analysis and significant figures. Dimensional analysis is introduced in Sections 1.4 (← ***p. 7)*** and 2.3 (← ***p. 46);*** it is reviewed below. Section 2.4 (← ***p. 52)*** and Appendix A.3 ***(p. A.5)*** introduce significant figures.

4. ✓ **Check the answer to see whether it is reasonable.** As a final check of your solution to any problem, ask yourself whether the answer is reasonable: Are the units of a numerical answer correct? Is a numerical answer of about the right size? Don't just copy a result from your calculator without thinking about whether it makes sense.

Suppose you have been asked to convert 100. yards to a distance in meters. Using dimensional analysis and some well-known factors for converting from the English system to the metric system, you could write

$$100.\ \text{yd} \times \frac{3\ \text{ft}}{1\ \text{yd}} \times \frac{12\ \text{in.}}{1\ \text{ft}} \times \frac{2.54\ \text{cm}}{1\ \text{in.}} \times \frac{1\ \text{m}}{100\ \text{cm}} = 91.4\ \text{m}$$

To check that a distance of 91.4 m is about right, recall that a yard is a little shorter than a meter. Therefore 100 yd should be a little less than 100 m. If you had mistakenly divided instead of multiplied by 3 ft/yd in the first step, your final answer would have been a little more than 10 m. This is equivalent to only about 30 ft, and you probably know a 100-yd football field is longer than that.

A.2 Numbers, Units, and Quantities

Many scientific problems require you to use mathematics to calculate a result or draw a conclusion. Therefore, knowledge of mathematics and its application to problem solving is important. However, one aspect of scientific calculations is often absent from pure mathematical work: Science deals with *measurements* in which an unknown quantity is compared with a standard or unit of measure. For example, using a balance to determine the mass of an object involves comparing the object's mass with standard masses, usually in multiples or fractions of one gram; the result is reported as some number of grams, say 4.357 g. *Both the number and the unit are important.* If the result had been 123.5 g, this would clearly be different, but a result of 4.357 oz (ounces) would also be different, because the unit "ounce" is different from the unit "gram." A *result that describes the quantitative measurement of a property,* such as 4.357 g, is called a *quantity* (or physical quantity), and it consists of a number and a unit. Chemical problem solving requires calculating with quantities. Notice that whether a quantity is large or small depends on the units as well as the number; the two quantities 123.5 g and 4.357 oz, for example, represent the *same* mass.

A quantity is always treated as though the number and the units are multiplied together; that is, 4.357 g can be handled mathematically as $4.357 \times \text{g}$. Using this simple rule, you will see that calculations involving quantities follow the normal rules of algebra and arithmetic: $5\ \text{g} + 7\ \text{g} = 5 \times \text{g} + 7 \times \text{g} = (5 + 7) \times \text{g} = 12\ \text{g}$; or $6\ \text{g} \div 2\ \text{g} = (6\ \cancel{\text{g}})/(2\ \cancel{\text{g}}) = 3$.

(Notice that in the second calculation the unit g appears in both the numerator and the denominator and cancels out, leaving a pure number, 3.) Treating units as algebraic entities has the advantage that *if a calculation is set up correctly, the units will cancel or multiply together so that the final result has appropriate units.* For example, if you measured the size of a sheet of paper and found it to be 8.5 in. by 11 in., the area A of the sheet could be calculated as area = length × width = 11 in. × 8.5 in. = 94 in.^{2}, or 94 square inches. If a calculation is set up incorrectly, the units of the result will be inappropriate. Using units to check whether a calculation has been properly set up is called *dimensional analysis* **(← p. 10)**.

This idea of using algebra on units as well as numbers is useful in all kinds of situations. For example, suppose you are having a party for some friends who like pizza. A large pizza consists of 12 slices and costs \$10.75. You expect to need 36 slices of pizza and want to know how much you will have to spend. A strategy for solving the problem is first to figure out how many pizzas you need and then to figure out the cost in dollars. This solution could be diagrammed as

$$\text{Slices} \xrightarrow[\text{step 1}]{\text{slices per pizza}} \text{pizzas} \xrightarrow[\text{step 2}]{\text{dollars per pizza}} \text{dollars}$$

Step 1: Find the number of pizzas required by dividing the number of slices per pizza into the number of slices, thus converting "units" of slices to "units" of pizzas:

$$\text{Number of pizzas} = 36 \text{ slices}\left(\frac{1 \text{ pizza}}{12 \text{ slices}}\right) = 3 \text{ pizzas}$$

If you had multiplied the number of slices times the number of slices per pizza, the result would have been labeled pizzas × slices^2, which does not make sense. In other words, the labels indicate whether multiplication or division is appropriate.

Strictly speaking, slices and pizzas are not units in the same sense that a gram is a unit. Nevertheless, labeling things this way will often help you keep in mind what a number refers to—pizzas, slices, or dollars in this case.

Step 2: Find the total cost by multiplying the cost per pizza by the number of pizzas needed, thus converting "units" of pizzas to "units" of dollars:

$$\text{Total price} = 3 \text{ pizzas}\left(\frac{\$10.75}{1 \text{ pizza}}\right) = \$32.25$$

Notice that in each step you have multiplied by a factor that allowed the initial units to cancel algebraically, giving the answer in the desired units. A factor such as (1 pizza/12 slices) or (\$10.75/pizza) is referred to as a *proportionality factor* **(← p. 10)**. This name indicates that it comes from a proportion. For instance, in the pizza problem you could set up the proportion

$$\frac{x \text{ pizzas}}{36 \text{ slices}} = \frac{1 \text{ pizza}}{12 \text{ slices}} \quad \text{or} \quad x \text{ pizzas} = 36 \text{ slices}\left(\frac{1 \text{ pizza}}{12 \text{ slices}}\right) = 3 \text{ pizzas}$$

A proportionality factor such as (1 pizza/12 slices) is also called a *conversion factor,* which indicates that it converts one kind of unit or label to another; in this case the label "slices" is converted to the label "pizzas."

Many everyday scientific problems involve proportionality. For example, the bigger the volume of a solid or liquid substance, the bigger its mass. When the volume is zero, the mass is also zero. These facts indicate that mass, m, is directly proportional to volume, V, or, symbolically,

$$m \propto V$$

where the symbol $\propto$ means "is proportional to." Whenever a proportion is expressed this way, it can also be expressed as an equality by using a proportionality constant—for example,

$$m = d \times V$$

In this case the proportionality constant, d, is called the density of the substance. This equation embodies the definition of density as mass per unit volume, since it can be rearranged algebraically to

$$d = \frac{m}{V}$$

As with any algebraic equation involving three variables, it is possible to calculate any one of the three quantities *m*, *V*, or *d*, provided the other two are known. If density is wanted, simply use the definition of mass per unit volume; if mass or volume is to be calculated, the density can be used as a proportionality factor.

Suppose that you are going to buy a ton of gravel and want to know how big a bin you will need to store it. You know the mass of gravel and want to find the volume of the bin; this implies that density will be useful. If the gravel is primarily limestone, you can assume that its density is about the same as for limestone and look it up. Limestone has the chemical formula $CaCO_3$, and its density is 2.7 kg/L. However, these mass units are different from the units for mass of gravel—namely, tons. Therefore you need to recall or look up the mass of 1 ton (exactly 2000 pounds [lb]) and the fact that there are 2.20 lb per kilogram. This provides enough information to calculate the volume needed. Here is a "roadmap" plan for the calculation:

$$\text{Mass of gravel in tons} \xrightarrow[\text{step1}]{\text{change units}} \text{mass of gravel in kilograms} \xrightarrow[\text{step 2}]{\text{density}} \text{volume of bin}$$

Step 1: Figure out how many kilograms of gravel are in a ton.

$$m_{\text{gravel}} = 1 \text{ ton} = 2000 \text{ lb} = 2000 \text{ lb}\left(\frac{1 \text{ kg}}{2.20 \text{ lb}}\right) = 909 \text{ kg}$$

The fact that there are 2.20 pounds per kilogram implies two proportionality factors: (2.20 lb/1 kg) and (1 kg/2.20 lb). The latter was used because it results in appropriate cancellation of units.

Step 2: Use the density to calculate the volume of 909 kg of gravel.

$$V_{\text{gravel}} = \frac{m_{\text{gravel}}}{d_{\text{gravel}}} = \frac{909 \text{ kg}}{2.7 \text{ kg/L}} = 909 \text{ kg}\left(\frac{1 \text{ L}}{2.7 \text{ kg}}\right) = 340 \text{ L}$$

In this step we used the definition of density, solved algebraically for volume, substituted the two known quantities into the equation, and calculated the result. However, it is quicker simply to remember that mass and volume are related by a proportionality factor called density and to use the units of the quantities to decide whether to multiply or divide by that factor. In this case we divided mass by density because the units kilograms canceled, leaving a result in liters, which is a unit of volume.

Also, it is quicker and more accurate to solve a problem like this one by using a single setup. Then all the calculations can be done at once, and no intermediate results need to be written down. The "roadmap" plan given above can serve as a guide to the single-setup solution, which looks like this:

$$V_{\text{gravel}} = 1 \text{ ton}\left(\frac{2000 \text{ lb}}{1 \text{ ton}}\right)\left(\frac{1 \text{ kg}}{2.20 \text{ lb}}\right)\left(\frac{1 \text{ L}}{2.7 \text{ kg}}\right) = 340 \text{ L}$$

To calculate the result, then, you would enter 2000 on your calculator, divide by 2.20, and divide by 2.7. Such a setup makes it easy to see what to multiply and divide by, and the calculation goes more quickly when it can be entered into a calculator all at once.

The liter is not the most convenient volume unit for this problem, however, because it does not relate well to what we want to find out—how big a bin to make. A liter is about the same volume as a quart, but whether you are familiar with liters, quarts, or both, 300 of them is not easy to visualize. Let's convert liters to something we can understand better. A liter is a volume equal to a cube one tenth of a meter (1 dm) on a side; that is, $1\text{L} = 1 \text{ dm}^3$. Consequently,

$$340 \text{ L} = 340 \text{ L}\left(\frac{1 \text{ dm}^3}{1 \text{ L}}\right)\left(\frac{1 \text{ m}}{10 \text{ dm}}\right)^3 = 340 \text{ dm}^3\left(\frac{1 \text{ m}^3}{1000 \text{ dm}^3}\right) = 0.34 \text{ m}^3$$

Thus, the bin would need to have a volume of about one third of a cubic meter; that is, it could be a meter wide, a meter long, and about a third of a meter high and it would hold the ton of gravel.

One more thing should be noted about this example. We don't need to know the volume of the bin very precisely, because being off a bit will make very little difference; it might mean getting a little too much wood to build the bin, or not making the bin quite big enough and having a little gravel spill out, but this isn't a big deal. In other cases, such as calculating the quantity of fuel needed to get a space shuttle into orbit, being off by a few percent could be a life-or-death matter. Because it is important to know how precise data are and to be able to evaluate how important precision is, scientific results usually indicate their precision. The simplest way to do so is by means of significant figures.

A.3 Precision, Accuracy, and Significant Figures

The **precision** of a measurement indicates how well several determinations of the same quantity agree. Some devices can make more precise measurements than others. Consider rulers A and B shown in the margin. Ruler A is marked every centimeter and ruler B is marked every millimeter. Clearly you could measure at least to the nearest millimeter with ruler B, and probably you could estimate to the nearest 0.2 mm or so. However, using ruler A you could probably estimate only to the nearest millimeter or so. We say that ruler B allows more precise measurement than ruler A. If several different people used ruler B to measure the length of an iron bar, their measurements would probably agree more closely than if they used ruler A.

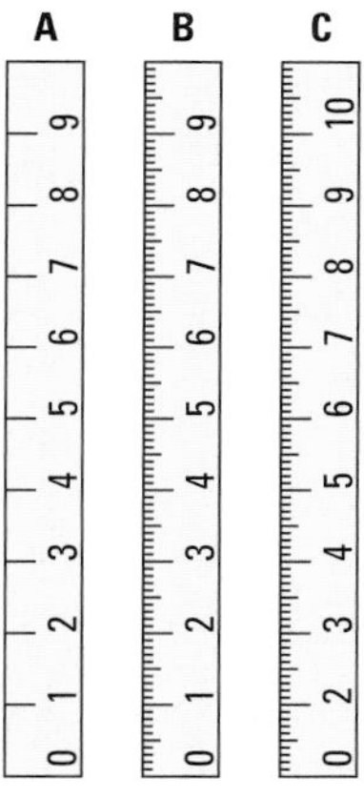

Three rulers.

Precision is also illustrated by the results of throwing darts at a bull's-eye (Figure A.1). In part (a), the darts are scattered all over the board; the dart thrower was apparently not very skillful (or threw the darts from a long distance away from the board), and the precision of their placement on the board is low. This is analogous to the results that would be obtained by several people using ruler A. In part (b), the darts are all clustered together, indicating much better reproducibility on the part of the thrower—that is, greater precision. This is analogous to measurements that might be made using ruler B.

Notice also that in Figure A.1b every dart has come very close to the bull's-eye; we describe this result by saying that the thrower has been quite **accurate**—the average of all throws is very close to the accepted position, the bull's-eye. Figure A.1c illustrates that it is possible to be precise without being accurate—the dart thrower has consistently missed the bull's-eye, but all darts are clustered very precisely around the wrong point on the board. This third case is like an experiment with some flaw (either in its design or in a measuring device) that causes all results to differ from the correct value by the same quantity. An example is measuring length with ruler C, on which the scale is incorrectly labeled. The precision (reproducibility) of measurements with ruler C might be quite good, but the accuracy would be poor because all items longer than 1 cm would be off by a centimeter.

In the laboratory we attempt to set up experiments so that the greatest possible accuracy can be obtained. As a further check on accuracy, results may be compared among different laboratories so that any flaw in experimental design or measurement can be detected. For each individual experiment, several measurements are usually made and their precision determined. In most cases, better precision is taken as an indication of better experimental work, and it is necessary to know the precision to compare results among different experimenters. If two different experimenters both had results like those in Figure A.1a, for example, their average values could differ quite a lot before they would say that their results did not agree within experimental error.

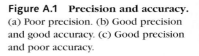

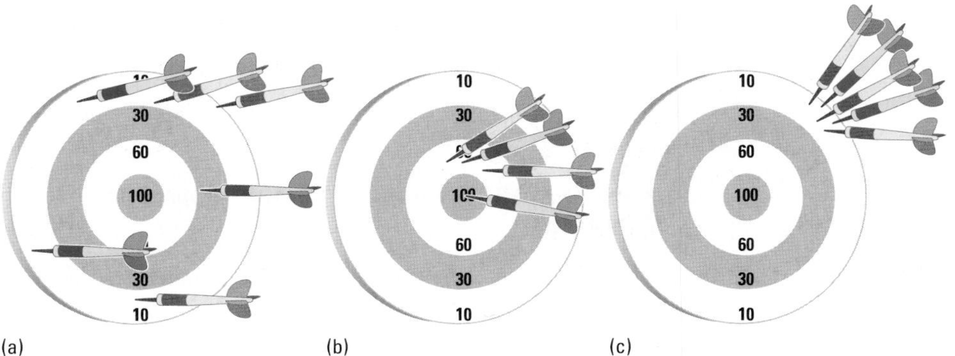

Figure A.1 Precision and accuracy. (a) Poor precision. (b) Good precision and good accuracy. (c) Good precision and poor accuracy.

In most experiments several different kinds of measurements must be made, and some can be done more precisely than others. It is common sense that *a calculated result can be no more precise than the least precise piece of information that went into the calculation.* This is where the rules for significant figures come in. In the last example in the preceding section, the quantity of gravel was described as "a ton." Usually gravel is measured by weighing an empty truck, putting some gravel in the truck, weighing the truck again, and subtracting the weight of the truck from the weight of the truck plus gravel. The quantity of gravel is not adjusted if there is a bit too much or a bit too little, because that would be a lot of trouble. You might end up with as much as 2200 pounds or as little as 1800 pounds, even though you asked for a ton. In terms of significant figures this would be expressed as 2.0×10^3 lb.

The quantity 2.0×10^3 lb is said to have two significant figures; it designates a quantity in which the 2 is taken to be exactly right but the 0 is not known precisely. (In this case, the number could be as large as 2.2 or as small as 1.8, so the 0 obviously is not exactly right.) In general, in a number that represents a scientific measurement, the last digit on the right is taken to be inexact, but all digits farther to the left are assumed to be exact. When you do calculations using such numbers, you must follow some simple rules so that your results will reflect the precision of all the measurements that go into the calculations. Here are the rules:

Rule 1: To determine the number of significant figures in a measurement, read the number from left to right and count all digits, starting with the first digit that is *not* zero.

For a number written in scientific notation, all digits are significant.

The number π is now known to 1,011,196,691 digits. It is doubtful that you will need that much precision in this course—or ever.

Example	Number of Significant Figures
1.23 g	3
0.00123 g	3; the zeros to the left of the 1 simply locate the decimal point. The number of significant figures is more obvious if you write numbers in scientific notation; thus, $0.00123 = 1.23 \times 10^{-3}$.
2.0 g and 0.020 g	2; both have two significant digits. When a number is greater than 1, *all zeros to the right of the decimal point are significant.* For a number less than 1, only zeros to the right of the first significant digit are significant.
100 g	1; in numbers that do not contain a decimal point, "trailing" zeros may or may not be significant. To eliminate possible confusion, the practice followed in this book is to include a decimal point if the zeros are significant. Thus, 100. has three significant digits, while 100 has only one. Alternatively, we write in scientific notation 1.00×10^2 (three significant digits) or 1×10^2 (one significant digit). For a number written in scientific notation, all digits preceding the 10^x term are significant.
100 cm/m	Infinite number of significant figures, because this is a defined quantity. There are *exactly* 100 centimeters in one meter.
$\pi = 3.1415926\ldots$	The value of π is known to a greater number of significant figures than any data you will ever use in a calculation.

Rule 2: When adding or subtracting, the number of decimal places in the answer should be equal to the number of decimal places in the number with the *fewest* places. Suppose you add three numbers:

0.12	2 significant figures	2 decimal places
1.6	2 significant figures	1 decimal place
+ 10.976	5 significant figures	3 decimal places
12.696		

This sum should be reported as 12.7, a number with one decimal place, because 1.6 has only one decimal place.

Rule 3: In multiplication or division, the number of significant figures in the answer should be the same as that in the quantity with the fewest significant figures.

$$\frac{0.1208}{0.0236} = 0.512 \text{ or, in scientific notation, } 5.12 \times 10^{-1}$$

Since 0.0236 has only three significant figures, while 0.01208 has four, the answer is limited to three significant figures.

Rule 4: When a number is rounded (the number of digits is reduced), the last digit retained is increased by 1 only if the following digit is 5 or greater. If the following digit is exactly 5 (or 5 followed by zeros), then increase the last retained digit by 1 if it is *odd* or leave the last digit unchanged if it is *even.* Thus, both 18.35 and 18.45 are rounded to 18.4.

Full Number	Number Rounded to Three Significant Figures
12.696	12.7
16.249	16.2
18.350	18.4
18.351	18.4

One last word regarding significant figures and calculations. In working problems on a pocket calculator, you should do the calculation using all the digits allowed by the calculator and round only at the end of the problem. Rounding in the middle can introduce small errors (called *rounding errors* or *round-off errors*). If your answers do not quite agree with those in the Appendices of this book, rounding errors may be the source of the disagreement.

Now let us consider a problem that is of practical importance and that makes use of all the rules. Suppose you discover that young children are eating chips of paint that flake off a wall in an old house. The paint contains 200. ppm lead (200. milligrams of Pb per kilogram of paint). Suppose that a child eats five such chips. How much lead has the child gotten from the paint?

The ppm unit stands for "parts per million." If a substance is present with a concentration of 1 ppm, there is 1 gram of the substance in 1 million grams of the sample.

As stated, this problem does not include enough information for a solution to be obtained; however, some reasonable assumptions can be made, and they can lead to experiments that could be used to obtain the necessary information. The statement does not say how big the paint chips are. Let's assume that they are 1.0 cm by 1.0 cm so that the area is 1.0 cm^2. Then eating five chips means eating 5.0 cm^2 of paint. (This assumption could be improved by measuring similar chips from the same place.) Since the concentration of lead is reported in units of mass of lead per mass of paint, we need to know the mass of 5.0 cm^2 of paint. This could be determined by measuring the areas of several paint chips and determining the mass of each. Suppose that the results of such measurements were those given in the table.

Mass of Chip (mg)	Area of Chip (cm^2)	Mass per Unit Area (mg/cm^2)
29.6	2.34	12.65
21.9	1.73	12.66
23.6	1.86	12.69

$$\text{Average mass per unit area} = \frac{(12.65 + 12.66 + 12.69)\ \text{mg/cm}^2}{3}$$

$$= 12.67\ \text{mg/cm}^2 = 12.7\ \text{mg/cm}^2$$

The average has been rounded to three significant figures because each experimentally measured mass and area has three significant figures. (Notice that more than three significant figures were kept in the intermediate calculations so as not to lose precision.) Now we can use this information to calculate how much lead the child has consumed.

$$m_{\text{paint}} = 5.0\ \text{cm}^2\ \text{paint}\left(\frac{12.7\ \text{mg paint}}{1\ \text{cm}^2\ \text{paint}}\right)\left(\frac{1\ \text{g}}{1000\ \text{mg}}\right)\left(\frac{1\ \text{kg}}{1000\ \text{g}}\right)$$

$$= 6.35 \times 10^{-5}\ \text{kg paint}$$

$$m_{\text{Pb}} = 6.35 \times 10^{-5}\ \text{kg paint}\left(\frac{200.\ \text{mg Pb}}{1\ \text{kg paint}}\right) = 1.27 \times 10^{-2}\ \text{mg Pb}$$

$$= 1.3 \times 10^{-2}\ \text{mg Pb} = 0.013\ \text{mg Pb}$$

The final result was rounded to two significant figures because there were only two significant figures in the initial 5.0 cm^2 area of the paint chip. This is quite adequate precision, however, for you to determine whether this quantity of lead is likely to harm the child.

The methods of problem solving presented here have been developed over time and represent a good way of keeping track of the precision of results, the units in which those results were obtained, and the correctness of calculations. These methods are not the only way that such goals can be achieved, but they do work well. We recommend that you include units in all calculations and check that they cancel appropriately. It is also important not to overstate the precision of results by keeping too many significant figures. By solving many problems, you should be able to develop your problem-solving skills so that they become second nature and you can use them without thinking about the mechanics. You can then devote all your thought to the logic of a problem solution.

A.4 Electronic Calculators

The directions for calculator use in this section are given for calculators using "algebraic" logic. Such calculators are the most common type used by students in introductory courses. For calculators using RPN logic (such as those made by Hewlett-Packard), the procedure will differ slightly.

The advent of inexpensive electronic calculators has made calculations in introductory chemistry much more straightforward. You are well advised to purchase a calculator that has the capability of performing calculations in scientific notation, has both base 10 and natural logarithms, and is capable of raising any number to any power and of finding any root of any number. In the discussion below, we will point out in general how these functions of your calculator can be used. You should practice using your calculator to carry out arithmetic operations and make certain that you are able to use all of its functions correctly.

Although electronic calculators have greatly simplified calculations, they have also forced us to focus again on significant figures. A calculator easily handles eight or more significant figures, but real laboratory data are rarely known to this precision. Therefore, if you have not already done so, review Appendix A.3 on significant figures, precision, and rounding numbers.

The mathematical skills required to read and study this textbook successfully involve algebra, some geometry, scientific notation, logarithms, and solving quadratic equations. The next three sections review the last three of these topics.

A.5 Exponential or Scientific Notation

In exponential or scientific notation, a number is expressed as a product of two numbers: $N \times 10^n$. The first number, N, is called the digit term and is a number between 1 and 10. The second number, 10^n, the exponential term, is some integer power of 10. For example, 1234 would be written in scientific notation as 1.234×10^3 or 1.234 multiplied by 10 three times.

$$1234 = 1.234 \times 10^1 \times 10^1 \times 10^1 = 1.234 \times 10^3$$

Conversely, a number less than 1, such as 0.01234, would be written as 1.234×10^{-2}. This notation tells us that 1.234 should be divided twice by 10 to obtain 0.01234.

$$0.01234 = \frac{1.234}{10^1 \times 10^1} = 1.234 \times 10^{-1} \times 10^{-1} = 1.234 \times 10^{-2}$$

Some other examples of scientific notation follow:

$10,000 = 1 \times 10^4$	$12,345 = 1.2345 \times 10^4$
$1000 = 1 \times 10^3$	$1234 = 1.234 \times 10^3$
$100 = 1 \times 10^2$	$123 = 1.23 \times 10^2$
$10 = 1 \times 10^1$	$12 = 1.2 \times 10^1$
$1 = 1 \times 10^0$	(any number to the zeroth power = 1)
$1/10 = 1 \times 10^{-1}$	$0.12 = 1.2 \times 10^{-1}$
$1/100 = 1 \times 10^{-2}$	$0.012 = 1.2 \times 10^{-2}$
$1/1000 = 1 \times 10^{-3}$	$0.0012 = 1.2 \times 10^{-3}$
$1/10,000 = 1 \times 10^{-4}$	$0.00012 = 1.2 \times 10^{-4}$

When converting a number to scientific notation, notice that the exponent n is positive if the number is greater than 1 and negative if the number is less than 1. The value of n is the number of places by which the decimal was shifted to obtain the number in scientific notation.

$$1\,2\,3\,4\,5. = 1.2345 \times 10^4$$

Decimal shifted 4 places to the left. Therefore,* n *is positive and equal to 4.

$$0.0\,0\,1\,2 = 1.2 \times 10^{-3}$$

Decimal shifted 3 places to the right. Therefore,* n *is negative and equal to 3.

If you wish to convert a number in scientific notation to the usual form, the procedure above is simply reversed.

$$6\,.\,2\,7\,3 \times 10^2 = 627.3$$

Decimal point shifted 2 places to the right, because* n *is positive and equal to 2.

$$0\,0\,6.273 \times 10^{-3} = 0.006273$$

Decimal point shifted 3 places to the left, because* n *is negative and equal to 3.

To enter a number in scientific notation into a calculator, first enter the number itself. Then press the EE (Enter Exponent) key followed by n, the power of 10. For example, to enter 6.022×10^{23}, you press these keys in succession:

6 . 0 2 2 EE 2 3 .

(Do not enter the number using the multiplication key, the number 10, and the EE key. This will result in a number that is 10 times bigger than you want. For example, pressing these keys

6 . 0 2 2 × 1 0 EE 2 3

enters the number $6.022 \times 10 \times 10^{23} = 6.022 \times 10^{24}$, because EE 2 3 means 10^{23}.)

There are two final points concerning scientific notation. First, if you are used to working on a computer you may be in the habit of writing a number such as 1.23×10^3 as 1.23E3, or 6.45×10^{-5} as 6.45E-5. Second, some electronic calculators allow you to convert numbers readily to scientific notation. If you have such a calculator, you can change a number shown in the usual form to scientific notation by pressing an appropriate key or keys.

Usually you will handle numbers in scientific notation with a calculator. In case you need to work without a calculator, however, the next few sections describe pencil-and-paper calculations as well as calculator methods.

1. Adding and Subtracting

When adding or subtracting numbers in scientific notation without using a calculator, first convert the numbers to the same powers of 10. Then add or subtract the digit terms.

$$(1.234 \times 10^{-3}) + (5.623 \times 10^{-2}) = (0.1234 \times 10^{-2}) + (5.623 \times 10^{-2})$$
$$= 5.746 \times 10^{-2}$$

$$(6.52 \times 10^2) - (1.56 \times 10^3) = (6.52 \times 10^2) - (15.6 \times 10^2)$$
$$= -9.1 \times 10^2$$

In this calculation, the result has only two significant figures, although each of the original numbers had three. Subtracting two numbers that are nearly the same can reduce the number of significant figures appreciably.

2. Multiplying

The digit terms are multiplied in the usual manner, and the exponents are added algebraically. The result is expressed with a digit term with only one nonzero digit to the left of the decimal.

$$(1.23 \times 10^3)(7.60 \times 10^2) = (1.23 \times 7.60)(10^3 \times 10^2) = (1.23)(7.60) \times 10^{3+2}$$
$$= 9.35 \times 10^5$$
$$(6.02 \times 10^{23})(2.32 \times 10^{-2}) = (6.02)(2.32) \times 10^{23-2}$$
$$= 13.966 \times 10^{21} = 1.3966 \times 10^{22}$$
$$= 1.40 \times 10^{22} \quad \text{(rounded to 3 significant figures)}$$

3. Dividing

The digit terms are divided in the usual manner, and the exponents are subtracted algebraically. The quotient is written with one nonzero digit to the left of the decimal in the digit term.

$$\frac{7.60 \times 10^3}{1.23 \times 10^2} = \frac{7.60}{1.23} \times 10^{3-2} = 6.18 \times 10^1$$

$$\frac{6.02 \times 10^{23}}{9.10 \times 10^{-2}} = \frac{6.02}{9.10} \times 10^{(23)-(-2)} = 0.662 \times 10^{25} = 6.62 \times 10^{24}$$

4. Raising Numbers in Scientific Notation to Powers

When raising a number in scientific notation to a power, treat the digit term in the usual manner. The exponent is then multiplied by the number indicating the power.

$$(1.25 \times 10^3)^2 = (1.25)^2 \times (10^3)^2 = (1.25)^2 \times 10^{3\times 2} = 1.5625 \times 10^6 = 1.56 \times 10^6$$
$$(5.6 \times 10^{-10})^3 = (5.6)^3 \times 10^{(-10)\times 3} = 175.6 \times 10^{-30} = 1.8 \times 10^{-28}$$

Electronic calculators usually have two methods of raising a number to a power. To square a number, enter the number and then press the "x^2" key. To raise a number to any power, use the "y^x" key. For example, to raise 1.42×10^2 to the 4th power, that is, to find $(1.42 \times 10^2)^4$,

(a) Enter 1.42×10^2.
(b) Press "y^x."
(c) Enter 4 (this should appear on the display).
(d) Press "=" and $4.0658689. . . \times 10^8$ will appear on the display. (The number of digits depends on the calculator.)

As a final step, express the number in the correct number of significant figures (4.07×10^8 in this case).

5. Taking Roots of Numbers in Scientific Notation

Unless you use an electronic calculator, the number must first be put into a form in which the exponential is exactly divisible by the root. The root of the digit term is found in the usual way, and the exponent is divided by the desired root.

$$\sqrt{3.6 \times 10^7} = \sqrt{36 \times 10^6} = \sqrt{36} \times \sqrt{10^6} = 6.0 \times 10^3$$
$$\sqrt[3]{2.1 \times 10^{-7}} = \sqrt[3]{210 \times 10^{-9}} = \sqrt[3]{210} \times \sqrt[3]{10^{-9}} = 5.9 \times 10^{-3}$$

To take a square root on an electronic calculator, enter the number and then press the "$\sqrt{x}$" key. To find a higher root of a number, such as the fourth root of 5.6×10^{-10},

On some calculators, Steps (a) and (c) may be interchanged.

(a) Enter the number, 5.6×10^{-10} in this case.
(b) Press the "$\sqrt[x]{y}$" key. (On most calculators, the sequence you actually use is to press "2ndF" and then "$\sqrt[x]{y}$." Alternatively, you may have to press "INV" and then "y^x".)
(c) Enter the desired root, 4 in this case.
(d) Press "=". The answer here is 4.8646×10^{-3} or 4.9×10^{-3}.

A general procedure for finding any root is to use the "y^x" key. For a square root, x is 0.5 (or $\frac{1}{2}$), whereas it is 0.33 (or $\frac{1}{3}$) for a cube root, 0.25 (or $\frac{1}{4}$) for a fourth root, and so on.

A.6 Logarithms

There are two types of logarithms used in this text: common logarithms (abbreviated log), whose base is 10, and natural logarithms (abbreviated ln), whose base is e (=2.7182818284).

$$\log x = n \qquad \text{where } x = 10^n$$

$$\ln x = m \qquad \text{where } x = e^m$$

Logarithms to the base 10 are needed when dealing with pH.

Most equations in chemistry and physics were developed in natural or base e logarithms, and this practice is followed in this text. The relation between log and ln is

$$\ln x = 2.303 \log x$$

Aside from the different bases of the two logarithms, they are used in the same manner. What follows is largely a description of the use of common logarithms.

A common logarithm is the power to which you must raise 10 to obtain the number. For example, the log of 100 is 2, since you must raise 10 to the power 2 to obtain 100. Other examples are

$$\log 1000 = \log (10^3) = 3$$

$$\log 10 = \log (10^1) = 1$$

$$\log 1 = \log (10^0) = 0$$

$$\log 1/10 = \log (10^{-1}) = -1$$

$$\log 1/10{,}000 = \log (10^{-4}) = -4$$

To obtain the common logarithm of a number other than a simple power of 10, use an electronic calculator. For example,

$$\log 2.10 = 0.3222, \text{ which means that } 10^{0.3222} = 2.10$$

$$\log 5.16 = 0.7126, \text{ which means that } 10^{0.7126} = 5.16$$

$$\log 3.125 = 0.49485, \text{ which means that } 10^{0.49485} = 3.125$$

To check this result on your calculator, enter the number and then press the "log" key.

To obtain the natural logarithm of the numbers above, use a calculator having this function. Enter each number and press "ln".

$$\ln 2.10 = 0.7419, \text{ which means that } e^{0.7419} = 2.10$$

$$\ln 5.16 = 1.6409, \text{ which means that } e^{1.6409} = 5.16$$

To find the common logarithm of a number greater than 10 or less than 1 with a log table, first express the number in scientific notation. Then find the log of each part of the number and add the logs. For example,

Nomenclature of Logarithms: The number to the left of the decimal in a logarithm is called the *characteristic,* and the number to the right of the decimal is called the *mantissa.*

$$\log 241 = \log (2.41 \times 10^2) = \log 2.4 + \log 10^2$$

$$= 0.382 + 2 = 2.382$$

$$\log 0.00573 = \log (5.73 \times 10^{-3}) = \log 5.73 + \log 10^{-3}$$

$$= 0.758 + (-3) = -2.242$$

Significant Figures and Logarithms

The mantissa (digits to the right of the decimal point in the logarithm) should have as many significant figures as the number whose log was found. (So that you could more clearly see the result obtained with a calculator or a table, this rule was not strictly followed until the last two examples.)

Obtaining Antilogarithms

If you are given the logarithm of a number and need to find the number from it, you need to obtain the "antilogarithm" or "antilog" of the number. There are two common procedures used by electronic calculators to do this:

Procedure A	Procedure B
(a) Enter the log or ln (a number).	(a) Enter the log or ln (a number).
(b) Press 2ndF.	(b) Press INV.
(c) Press 10^x or e^x.	(c) Press log or ln x.

Test one or the other of these procedures with the following examples.

EXAMPLE 1 Find the number whose log is 5.234.

Recall that $\log x = n$, where $x = 10^n$. In this case $n = 5.234$. Enter that number in your calculator and find the value of 10^n, the antilog. In this case,

$$10^{5.234} = 10^{0.234} \times 10^5 = 1.71 \times 10^5$$

Notice that the characteristic (5) sets the decimal point; it is the power of 10 in the exponential form. The mantissa (0.234) gives the value of the number x. Thus, if you use a log table to find x, you need only look up 0.234 in the table and see that it corresponds to 1.71.

EXAMPLE 2 Find the number whose log is −3.456.

$$10^{-3.456} = 10^{0.544} \times 10^{-4} = 3.50 \times 10^{-4}$$

Notice here that −3.456 must be expressed as the sum of −4 and +0.544.

Mathematical Operations Using Logarithms

Because logarithms are exponents, operations involving them follow the same rules as the use of exponents. Thus, multiplying two numbers can be done by adding logarithms.

$$\log xy = \log x + \log y$$

For example, we multiply 563 by 125 by adding their logarithms and finding the antilogarithm of the result.

$$\log 563 = 2.751$$
$$\log 125 = 2.097$$
$$\log (563 \times 125) = 2.751 + 2.097 = 4.848$$
$$563 \times 125 = 10^{4.848} = 10^4 \times 10^{0.848} = 7.05 \times 10^4$$

One number (x) can be divided by another (y) by subtraction of their logarithms.

$$\log \frac{x}{y} = \log x - \log y$$

For example, to divide 125 by 742,

$$\log 125 = 2.097$$
$$\log 742 = 2.870$$
$$\log (125/742) = 2.097 - 2.870 = -0.773$$
$$125/742 = 10^{-0.773} = 10^{0.227} \times 10^{-1} = 1.69 \times 10^{-1}$$

Similarly, powers and roots of numbers can be found using logarithms.

$$\log x^y = y(\log x)$$

$$\log \sqrt[y]{x} = \log x^{1/y} = \frac{1}{y} \log x$$

As an example, find the fourth power of 5.23. First find the log of 5.23 and then multiply it by 4. The result, 2.874, is the log of the answer. Therefore, find the antilog of 2.874.

$$(5.23)^4 = ?$$

$$\log (5.23)^4 = 4 \log 5.23 = 4(0.719) = 2.874$$

$$(5.23)^4 = 10^{2.874} = 748$$

As another example, find the fifth root of 1.89×10^{-9}.

$$\sqrt[5]{1.89 \times 10^{-9}} = (1.89 \times 10^{9})^{1/5} = ?$$

$$\log (1.89 \times 10^{-9})^{1/5} = \frac{1}{5} \log (1.89 \times 10^{-9}) = \frac{1}{5}(-8.724) = -1.745$$

The answer is the antilog of -1.745.

$$(1.89 \times 10^{-9})^{1/5} = 10^{-1.745} = 1.80 \times 10^{-2}$$

A.7 Quadratic Equations

Algebraic equations of the form $ax^2 + bx + c = 0$ are called **quadratic equations.** The coefficients *a, b,* and *c* may be either positive or negative. The two roots of the equation may be found using the *quadratic formula.*

$$x = \frac{-b \pm \sqrt{b^2 - 4ac}}{2a}$$

As an example, solve the equation $5x^2 - 3x - 2 = 0$. Here $a = 5$, $b = -3$, and $c = -2$. Therefore,

$$x = \frac{3 \pm \sqrt{(-3^2) - 4(5)(-2)}}{2(5)}$$

$$= \frac{3 \pm \sqrt{9 - (-40)}}{10} = \frac{3 \pm \sqrt{49}}{10} = \frac{3 \pm 7}{10}$$

$$x = 1 \textit{ and } -0.4$$

How do you know which of the two roots is the correct answer? You have to decide in each case which root has physical significance. However, it is usually true in this course that negative values are not significant.

Many calculators have a built-in quadratic-equation solver. If your calculator offers this capability, it will be convenient to use it, and you should study the manual until you know how to use your calculator to obtain the roots of a quadratic equation.

When you have solved a quadratic expression, you should always check your values by substituting them into the original equation. In the example above, we find that $5(1)^2 - 3(1) - 2 = 0$ and that $5(-0.4)^2 - 3(-0.4) - 2 = 0$.

You will encounter quadratic equations in the chapters on chemical equilibria, particularly in Chapters 14, 16, and 17. Here you may be faced with solving an equation such as

$$1.8 \times 10^{-4} = \frac{x^2}{0.0010 - x}$$

This equation can certainly be solved by using the quadratic formula or your calculator (to give $x = 3.4 \times 10^{-4}$). However, you may find the method of successive approximations to be especially convenient. Here you begin by making a reasonable approximation of x. This approximate value is substituted into the original equation, and this expression is solved to give what is hoped to be a more correct value of x. This process is repeated until the answer converges on a particular value of x—that is, until the value of x derived from two successive approximations is the same.

Step 1: Assume that x is so small that $(0.0010 - x) \approx 0.0010$. This means that

$$x^2 = 1.8 \times 10^{-4}(0.0010)$$

$$x = 4.2 \times 10^{-4} \text{ (to 2 significant figures)}$$

Step 2: Substitute the value of x from Step 1 into the denominator (but not the numerator) of the original equation and again solve for x.

$$x^2 = (1.8 \times 10^{-4})(0.0010 - 0.00042)$$
$$x = 3.2 \times 10^{-4}$$

Step 3: Repeat Step 2 using the value of x found in that step.

$$x = \sqrt{1.8 \times 10^{-4}(0.0010 - 0.00032)} = 3.5 \times 10^{-4}$$

Step 4: Continue by repeating the calculation, using the value of x found in the previous step.

Step 5. $x = \sqrt{1.8 \times 10^{-4}(0.0010 - 0.00034)} = 3.4 \times 10^{-4}$

Here we find that iterations after the fourth step give the same value for x, indicating that we have arrived at a valid answer (and the same one obtained from the quadratic formula).

Some final thoughts on using the method of successive approximations: First, in some cases this method does not work. Successive steps may give answers that are random or that diverge from the correct value. For quadratic equations of the form $K = x^2/(C - x)$, the method of approximations will work only as long as $K < 4C$ (assuming one begins with $x = 0$ as the first guess; that is, $K \approx x^2/C$). This will always be true for weak acids and bases.

Second, values of K in the equation $K = x^2/(C - x)$ are usually known only to two significant figures. Therefore, we are justified in carrying out successive steps until two answers are the same to two significant figures.

Finally, if your calculator does not automatically obtain roots of quadratic equations, we highly recommend this method. If your calculator has a memory function, successive approximations can be carried out easily and very rapidly. Even without a memory function, the method of successive approximations is much quicker than solving a quadratic equation by hand.

A.8 Graphing

Mass (g)	Volume (mL)
2.03	0.76
5.27	1.95
9.57	3.54
11.46	4.25
14.96	5.55
18.02	6.68
21.83	8.10
25.17	9.32
30.08	11.14
32.84	12.17
36.27	13.43
36.77	13.62
39.36	14.58

When analyzing experimental data, chemists and other scientists often graph the data to see whether the data agree with a mathematical equation. If the equation does fit the data (often indicated by a linear graph), then the graph can be used to obtain numerical values (parameters) that can be used in the equation to predict information not specifically included in the data set. For example, suppose that you measured the masses and the volumes of several samples of aluminum. Your data set might look like the table in the margin. When these data are plotted with volume along the horizontal (x) axis and mass on the vertical (y) axis, a graph like this one can be obtained.

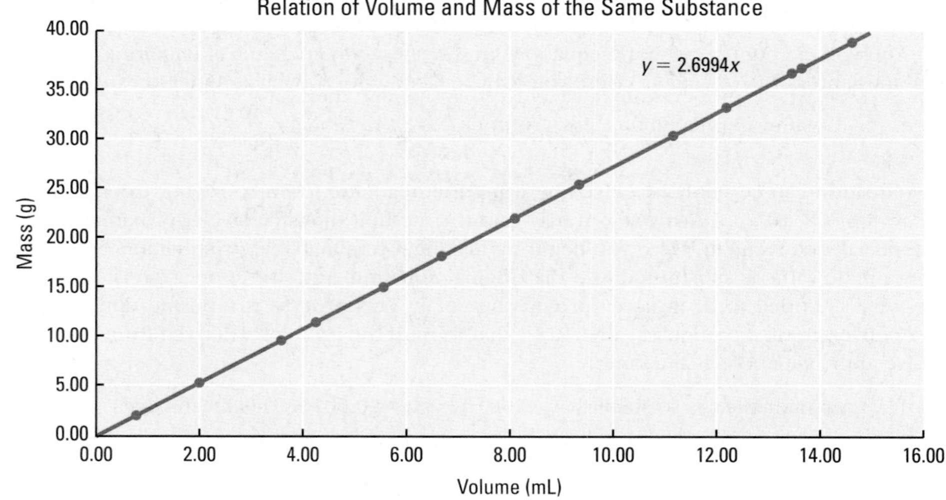

Notice the features of this graph. It has a title that describes its contents, it has a label for each axis that specifies the quantity plotted and the units used, it has equally spaced grid lines in the vertical and horizontal directions, and the numbers associated with those grid lines have the same difference between successive grid lines. (The differences are 2.00 mL on the horizontal axis and 5.00 g on the vertical axis.) The experimental points are indicated by circular markers, and a straight line has been drawn through the points. Because the line passes through all of the points and through the origin (point 0, 0), the data can be represented by the equation $y = 2.70x$, where 2.70 is the slope of the line. (On the graph, the slope is indicated by 2.6994; in the equation, it has been rounded to 2.70 because some of the data have only three significant figures.)

The slope of a graph such as this one can be obtained by choosing two points on the straight line (not two data points). The points should be far apart to obtain the most precise result. The slope is then given by the change in the y-axis variable divided by the change in the x-axis variable from the first point to the second. An example is shown on the graph below. From the calculation of the slope, it is more obvious that the slope has only three significant figures. For this graph, the slope represents mass divided by volume—that is, density. A good way to measure density is to measure the mass and volume of several samples of the same substance and then plot the data. The resulting graph should be linear and should pass very nearly through the origin. The density can be obtained from its slope. When a graph passes through the origin, we say that the y-axis variable is directly proportional to the x-axis variable.

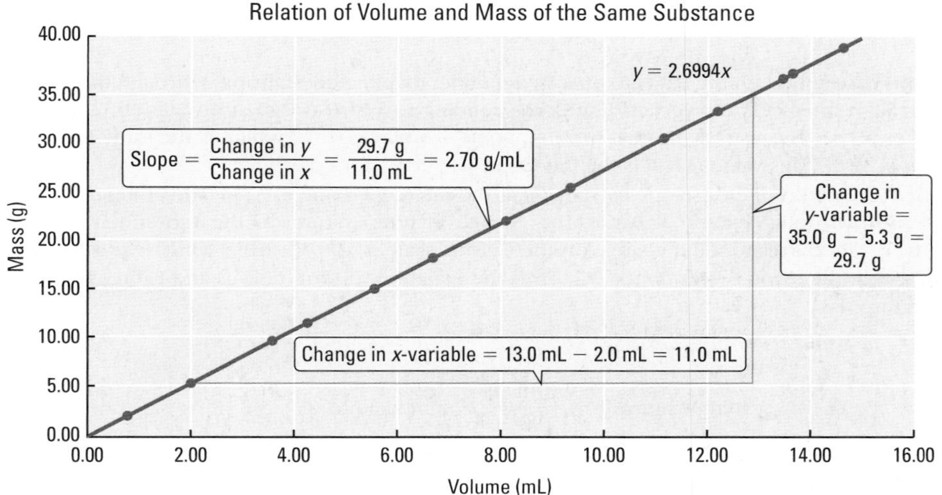

When a graph does not pass through the origin, there is an intercept as well as a slope. The intercept is the value of the y-axis variable when the x-axis variable is zero. A good example is the relation between Celsius and Fahrenheit temperatures. Suppose you have measured a series of temperatures using both a Celsius thermometer and a Fahrenheit thermometer. Your data set might look something like this table.

Temperature (°C)	**Temperature (°F)**
−5.23	22.586
13.54	56.372
32.96	91.328
48.34	119.012
63.59	146.462
74.89	166.802
88.02	190.436
105.34	221.612

When graphed, these data look like this.

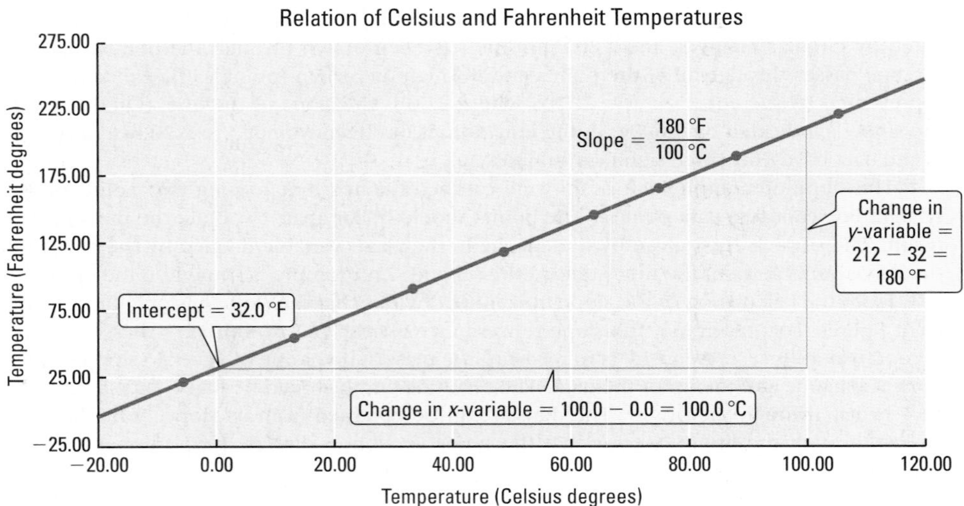

In this case, although the data are on a straight line, that line does not pass through the origin. Instead, it intersects the vertical line corresponding to $x = 0$ (0 °C) at a value of 32 °F. This makes sense, because the normal freezing point of water is 0 °C, which is the same temperature as 32 °F. This value, 32 °F, is the intercept.

If we determine the slope by starting at the intercept value of 32 °F and going to 212° F, the normal boiling point of water, the change in temperature on the Fahrenheit scale is 180 °F. The corresponding change on the Celsius scale is 100 °C, from which we can determine that the slope is 180 °F/100 °C. Thus the equation relating Celsius and Fahrenheit temperatures is

$$y \quad = \quad m \quad x \quad + \quad b$$

$$\text{temperature } ^\circ\text{F} = \frac{180\ ^\circ\text{F}}{100\ ^\circ\text{C}} \times \text{temperature } ^\circ\text{C} + 32\ ^\circ\text{F}$$

QUESTIONS FOR APPENDIX A

Blue-numbered questions have short answers at the end of this book and fully worked solutions in the *Student Solutions Manual.*

General Problem-Solving Strategies

1. List four steps that can be used for guidance in solving problems. Choose a problem that you are interested in solving, and apply the four steps to that problem.
2. You are asked to study a lake in which fish are dying and determine the cause of their deaths. Suggest three things you might do to define this problem.
3. You have calculated the area in square yards of a carpet whose dimensions were originally given to you as 96 in by 72 in. Suggest at least one way to check that the results of your calculation are reasonable. (12 in. = 1 ft; 3 ft = 1 yd.)

Numbers, Units, and Quantities

4. When a calculation is done and the units for the answer do not make sense, what can you conclude about the solution to the problem? What would you do if you were faced with a situation like this?
5. The term "quantity" (or "physical quantity") has a specific meaning in science. What is that meaning, and why is it important?
6. To measure the length of a pencil, you use a tape measure calibrated in inches with marks every sixteenth of an inch. Which of these results would be a suitable record of your observation? Why would the other results be unsuitable? (1/16 in = 0.0625 in)
(a) 8.38 ft (b) 8.38 m (c) 8.38 in
(d) 8.38 (e) 0.698 ft

7. To measure the inseam length of a pair of slacks, you use a tape measure calibrated in centimeters with marks every tenth of a centimeter. Which of these results would be a suitable record of your observation? Why would the other results be unsuitable?
(a) 75.0 cm (b) 75.0 m (c) 75.0
(d) 75.0 in (e) 750 mm
8. What is wrong with each calculation? How would you carry out each calculation correctly? What is the correct result for each calculation?
(a) $4.32\text{ g} + 5.63\text{ g} = 9.95\text{ g}^2$
(b) $5.23\text{ g} \times \frac{4.87\text{ g}}{1.00\text{ mL}} = 25.5\text{ g}$
(c) $3.57\text{ cm}^3 \times \left(\frac{1\text{ m}}{100\text{ cm}}\right) = 3.57 \times 10^{-2}\text{ m}^3$
9. What is wrong with each calculation? How would you carry out each calculation correctly? What is the correct result for each calculation?
(a) $7.86\text{ g} - 5.63\text{ g} = 2.23\text{ g}^2$
(b) $7.37\text{ mL} \times \frac{1.00\text{ mL}}{2.23\text{ g}} = 3.30\text{ mL}$
(c) $9.26\text{ m}^3 \times \left(\frac{100\text{ cm}}{1\text{ m}}\right) = 9.26 \times 10^2\text{ m}^3$

Precision, Accuracy, and Significant Figures

10. These measurements were reported for the length of an eight-foot pole: 95.31 in; 96.44 in; 96.02 in; 95.78 in; 95.94 in (1 ft = 12 in).
(a) Based on these measurements, what would you report as the length of the pole?
(b) How many significant figures should appear in your result?
(c) Assuming that the pole is exactly eight feet long, is the result accurate?
11. These measurements were reported for the mass of a sample in the laboratory: 32.54 g; 32.67 g; 31.98 g; 31.76 g; 32.05 g.
(a) Based on these measurements, what would you report as the mass of the sample?
(b) How many significant figures should appear in your result?
(c) The sample is exactly balanced by a weight known to have a mass of 35.0 g. Is the result accurate?
12. How many significant figures are in each quantity?
(a) 3.274 g (b) 0.0034 L
(c) 43,000 m (d) 6200. ft
13. How many significant figures are in each quantity?
(a) 0.2730 g (b) 8.3 g/mL
(c) 300 m (d) 2030.0 dm^3
14. Round each number to four significant figures.
(a) 43.3250 (b) 43.3165 (c) 43.3237
(d) 43.32499 (e) 43.3150 (f) 43.32501
15. Round each number to three significant figures.
(a) 88.3520 (b) 88.365 (c) 88.45
(d) 88.5500 (e) 88.2490 (f) 88.4501
16. Evaluate each expression and report the result to the appropriate number of significant figures.
(a) $\frac{4.47}{0.3260}$ (b) $\frac{4.03 + 3.325}{29.75}$
(c) $\frac{8.234}{5.673 - 4.987}$
17. Evaluate each expression and report the result to the appropriate number of significant figures.
(a) $\frac{4.47}{0.3260}$ (b) $\frac{4.03 + 3.325}{29.75}$
(c) $\frac{8.234}{5.673 - 4.987}$

Exponential or Scientific Notation

18. Without using a calculator, express each number in scientific notation and with the appropriate number of significant figures.
(a) 76,003 (b) 0.00037 (c) 34,000
19. Without using a calculator, express each number in scientific notation and with the appropriate number of significant figures.
(a) 49,002 (b) 0.0234 (c) 23,400
20. Evaluate each expression using your calculator and report the result to the appropriate number of significant figures with appropriate units.
(a) $\frac{0.7346}{304.2}$
(b) $\frac{(3.45 \times 10^{-3})(1.83 \times 10^{12})}{23.4}$
(c) $3.240 - 4.33 \times 10^{-3}$
(d) $(4.87\text{ cm})^3$
21. Evaluate each expression using your calculator and report the result to the appropriate number of significant figures with appropriate units.
(a) $\frac{893.0}{0.2032}$
(b) $\frac{(5.4 \times 10^3)(8.36 \times 10^{-12})}{5.317 \times 10^{-3}}$
(c) $3.240 \times 10^5 - 8.33 \times 10^3$
(d) $(4.87\text{ cm} + 7.33 \times 10^{-1}\text{ cm})^3$

Logarithms

22. Use your calculator to find the logarithm of each number and report the logarithm to the appropriate number of significant figures.
(a) log(0.7327) (b) ln(34.5)
(c) $\log(6.022 \times 10^{23})$ (d) $\ln(6.022 \times 10^{23})$
(e) $\log\left(\frac{8.34 \times 10^{-5}}{2.38 \times 10^3}\right)$
23. Use your calculator to find the logarithm of each number and report the logarithm to the appropriate number of significant figures.
(a) log(54.3) (b) ln(0.0345)
(c) $\log(4.344 \times 10^{-3})$ (d) $\ln(8.64 \times 10^4)$
(e) $\ln\left(\frac{4.33 \times 10^{24}}{8.32 \times 10^{-2}}\right)$
24. Use your calculator to evaluate each expression and report the result to the appropriate number of significant figures.
(a) antilog(0.7327) (b) antiln(34.5)
(c) $10^{2.043}$ (d) $e^{3.20 \times 10^{-4}}$
(e) exp(4.333/3.275)
25. Use your calculator to evaluate each expression and report the result to the appropriate number of significant figures.
(a) antilog(87.2) (b) antiln(0.0034)
(c) $e^{2.043}$ (d) $10^{(3.20 \times 10^{-4})}$
(e) $\exp(4.3 \times 10^3/8.314)$

Quadratic Equations

26. Find the roots of each quadratic equation.
(a) $3.27x^2 + 4.32x - 2.83 = 0$
(b) $x^2 + 4.32 = 4.57x$
27. Find the roots of each quadratic equation.
(a) $8.33x^2 - 2.32x - 7.53 = 0$
(b) $4.3x^2 - 8.37 = -2.22x$

Graphing

28. Graph these data involving mass and volume, label the graph appropriately, and determine whether m is directly proportional to V.

V (mL)	m (g)
0.347	0.756
1.210	2.638
2.443	5.326
7.234	15.76
11.43	24.90

29. Graph these data for heat transfers during a reaction, label the graph appropriately, and determine whether the heat evolved is directly proportional to the amount of reactant consumed.

Amount of Reactant (mol)	Heat Evolved (J)
94.2	43.2
70.7	32.5
65.7	30.1
34.2	15.7
54.3	24.9

APPENDIX

B Units, Equivalences, and Conversion Factors

B.1 Units of the International System (SI)

The metric system was begun by the French National Assembly in 1790 and has undergone many modifications since its inception. The International System of Units, or *Système International* (SI), which represents an extension of the metric system, was adopted by the 11th General Conference on Weights and Measures in 1960. It is constructed from seven base units, each of which represents a particular physical quantity (Table B.1). More information about the SI is available at **http://physics.nist.gov/cuu/Units/index.html**.

The first five units listed in Table B.1 are particularly useful in chemistry. They are defined as follows:

1. The *meter* is the length of the path traveled by light in a vacuum during a time interval of 1/299,792,458 of a second.
2. The *kilogram* represents the mass of a platinum-iridium block kept at the International Bureau of Weights and Measures in Sevres, France.
3. The *second* is the duration of 9,192,631,770 periods of a certain line in the microwave spectrum of cesium-133.
4. The *kelvin* is 1/273.16 of the temperature interval between absolute zero and the triple point of water (the temperature at which liquid water, ice, and water vapor coexist).
5. The *mole* is the amount of substance that contains as many elementary entities (atoms, molecules, ions, or other particles) as there are atoms in exactly 0.012 kg of carbon-12 (12 g of ^{12}C atoms).

Decimal fractions and multiples of metric and SI units are designated by using the **prefixes** listed in Table B.2. The prefix *kilo-*, for example, means that a unit is multiplied by 10^3.

$$1 \text{ kilogram} = 1 \times 10^3 \text{ grams} = 1000 \text{ grams}$$

The prefix *centi-* means that the unit is multiplied by the factor 10^{-2}.

$$1 \text{ centigram} = 1 \times 10^{-2} \text{ gram} = 0.01 \text{ gram}$$

Table B.1 SI Fundamental Units

Physical Quantity	Name of Unit	Symbol
Length	Meter	m
Mass	Kilogram	kg
Time	Second	s
Temperature	Kelvin	K
Amount of substance	Mole	mol
Electric current	Ampere	A
Luminous intensity	Candela	cd

The prefixes are added to give units of a magnitude appropriate to what is being measured. The distance from New York to London (5.6×10^3 km = 5600 km) is much easier to comprehend measured in kilometers than in meters (5.6×10^6 m = 5,600,000 m). Following Table B.2 is a list of units for measuring very small and very large distances.

Table B.2 Prefixes for Metric and SI Units*

Factor	Prefix	Symbol	Factor	Prefix	Symbol
10^{18}	exa-	E	10^{-1}	*deci-*	d
10^{15}	peta-	P	10^{-2}	*centi-*	c
10^{12}	tera-	T	10^{-3}	*milli-*	m
10^{9}	giga-	G	10^{-6}	*micro-*	μ
10^{6}	mega-	M	10^{-9}	*nano-*	n
10^{3}	*kilo-*	k	10^{-12}	*pico-*	p
10^{2}	hecto-	h	10^{-15}	femto-	f
10^{1}	deka-	da	10^{-18}	atto-	a

*The prefixes most commonly used in chemistry are shown in italics.

attometer (am)	0.000000000000000001 meter
femtometer (fm)	0.000000000000001 meter
picometer (pm)	0.000000000001 meter
nanometer (nm)	0.000000001 meter
micrometer (μm)	0.000001 meter
millimeter (mm)	0.001 meter
centimeter (cm)	0.01 meter
decimeter (dm)	0.1 meter
meter (m)	1 meter
dekameter (dam)	10 meters
hectometer (hm)	100 meters
kilometer (km)	1000 meters
megameter (Mm)	1,000,000 meters
gigameter (Gm)	1,000,000,000 meters
terameter (Tm)	1,000,000,000,000 meters
petameter (Pm)	1,000,000,000,000,000 meters
exameter (Em)	1,000,000,000,000,000,000 meters

In the International System of Units, all physical quantities are represented by appropriate combinations of the base units listed in Table B.1. The result is a derived unit for each kind of measured quantity. The most common derived units are listed in Table B.3. It is easy

Table B.3 Derived SI Units

Physical Quantity	Name of Unit	Symbol	Definition	Expressed in Fundamental Units
Area	Square meter	m^2	—	
Volume	Cubic meter	m^3	—	
Density	Kilogram per cubic meter	kg/m	—	
Force	Newton	N	$\frac{\text{(kilogram)(meter)}}{\text{(second)}^2}$	kg m/s^2
Pressure	Pascal	Pa	$\frac{\text{(newton)}}{\text{(meter)}^2}$	$\text{N/m}^2 = \text{kg m}^{-1}\text{ s}^{-2}$
Energy	Joule	J	$\frac{\text{(kilogram)(meter)}^2}{\text{(second)}^2}$	$\text{kg m}^2\text{ s}^{-2}$
Electric charge	Coulomb	C	(ampere)(second)	A s
Electric potential difference	Volt	V	$\frac{\text{(joule)}}{\text{(ampere)(second)}}$	$\text{J A}^{-1}\text{ s}^{-1} = \text{kg m}^2\text{ s}^{-3}\text{ A}^{-1}$

to see that the derived unit for area is length × length = meter × meter = square meter, m^2, or that the derived unit for volume is length × length × length = meter × meter × meter = cubic meter, m^3. More complex derived units are arrived at by a similar kind of combination of units. Units such as the joule, which measures energy, have been given simple names that represent the combination of fundamental units by which they are defined.

B.2 Conversion of Units for Physical Quantities

The result of a measurement is a physical quantity, which consists of a number and a unit. Algebraically, a physical quantity can be treated as if the number is multiplied by the unit. To convert a physical quantity from one unit of measure to another requires a conversion factor (proportionality factor) based on equivalences between units of measure such as those given in Table B.4. (See Appendix A.2 for more about physical quantities and proportionality factors.) Each equivalence provides two conversion factors that are the reciprocals of each other. For example, the equivalence between a quart and a liter, 1 quart = 0.9463 liter, gives

$$\frac{1\ \text{quart}}{0.9463\ \text{liter}} \quad \text{There is 1 quart per 0.9463 liter.}$$

$$\frac{0.9463\ \text{liter}}{1\ \text{quart}} \quad \text{There is 0.9463 liter per 1 quart.}$$

The method of canceling units described in Appendix A.2 provides the basis for choosing which conversion factor is needed: It is always the one that allows the unit being converted to be canceled and leaves the new unit uncanceled.

To convert 2 quarts to liters:

$$2\ \text{quarts} \times \frac{0.9463\ \text{liter}}{1\ \text{quart}} = 1.893\ \text{liters}$$

To convert 2 liters to quarts:

$$2\ \text{liters} \times \frac{1\ \text{quart}}{0.9463\ \text{liter}} = 2.113\ \text{quarts}$$

Because of the definitions of Celsius degrees and Fahrenheit degrees, conversions between these temperature scales are a bit more complicated. Both units are based on the properties of water. The Celsius unit is defined by assigning 0 °C as the freezing point of pure water and 100 °C as its boiling point, when the pressure is exactly 1 atm. The size of the Fahrenheit degree is equally arbitrary. Fahrenheit defined 0 °F as the freezing point of a solution in which he had dissolved the maximum quantity of ammonium chloride (because this was the lowest temperature he could reproduce reliably), and he intended 100 °F to be the normal human body temperature (but this value turned out to be 98.6 °F). Today, the reference points are set at exactly 32 °F and 212 °F (the freezing and boiling points of pure water, at 1 atm). The number of units between these two Fahrenheit temperatures is 180 °F. Thus, the Celsius degree is almost twice as large as the Fahrenheit degree; it takes only 5 Celsius degrees to cover the same temperature range as 9 Fahrenheit degrees.

To be entirely correct, we must specify that pure water boils at 100 °C and freezes at 0 °C only when the pressure of the surrounding atmosphere is 1 atm.

$$\frac{100\ ^\circ\text{C}}{180\ ^\circ\text{F}} = \frac{5\ ^\circ\text{C}}{9\ ^\circ\text{F}}$$

This relationship is the basis for converting a temperature on one scale to a temperature on the other. If t_C is the numerical value of the temperature in °C and t_F is the numerical value of the temperature in °F, then

$$t_C = \left(\tfrac{5}{9}\right)(t_F - 32)$$

$$t_F = \left(\tfrac{9}{5}\right)t_C + 32$$

Table B.4 Common Units of Measure

Mass and Weight

1 pound = 453.59 grams = 0.45359 kilogram
1 kilogram = 1000 grams = 2.205 pounds
1 gram = 10 decigrams = 100 centigrams = 1000 milligrams
1 gram = 6.022×10^{23} atomic mass units
1 atomic mass unit = 1.6605×10^{-24} grams
1 short ton = 2000 pounds = 907.2 kilograms
1 long ton = 2240 pounds
1 metric tonne = 1000 kilograms = 2205 pounds

Length

1 inch = 2.54 centimeters (exactly)
1 mile = 5280 feet = 1.609 kilometers
1 yard = 36 inches = 0.9144 meter
1 meter = 100 centimeters = 39.37 inches = 3.281 feet = 1.094 yards
1 kilometer = 1000 meters = 1094 yards = 0.6215 miles
1 Angstrom = 1.0×10^{-8} centimeters = 0.10 nanometers = 100 picometers
= 1.0×10^{-10} meters = 3.937×10^{-9} inches

Volume

1 quart = 0.9463 liters
1 liter = 1.0567 quarts
1 liter = 1 cubic decimeter = 10^3 cubic centimeters = 10^{-3} cubic meters
1 milliliter = 1 cubic centimeters = 0.001 liters = 1.056×10^{-3} quarts
1 cubic foot = 28.316 liters = 29.924 quarts = 7.481 gallons

Force and Pressure

1 atmosphere = 760.0 millimeters of mercury = 1.01325×10^5 pascals
= 14.70 pounds per square inch
1 bar = 10^5 pascals = 0.98692 atmospheres
1 torr = 1 millimeter of mercury
1 pascal = 1 kg m^{-1} s^{-2} = 1 N/m^2

Energy

1 joule = 1×10^7 ergs
1 thermochemical calorie = 4.184 joules = 4.184×10^7 ergs
= 4.129×10^{-2} liter-atmospheres
= 2.612×10^{19} electron volts
1 erg = 1×10^{-7} joules = 2.3901×10^{-8} calories
1 electron volt = 1.6022×10^{-19} joules = 1.6022×10^{-12} ergs = 96.85 kJ/mol*
1 liter-atmosphere = 24.217 calories = 101.32 joules = 1.0132×10^9 ergs
1 British thermal unit = 1055.06 joules = 1.05506×10^{10} ergs = 252.2 calories

Temperature

0 K = −273.15 °C

If T_K is the numerical value of the temperature in kelvins, t_C is the numerical value of the temperature in °C, and t_F is the numerical value of the temperature in °F, then

$$T_K = t_C + 273.15$$
$$t_C = \left(\tfrac{5}{9}\right)(t_F - 32)$$
$$t_F = \left(\tfrac{9}{5}\right)t_C + 32$$

*The other units in this line are per particle and must be multiplied by 6.022×10^{23} to be strictly comparable.

For example, to show that your normal body temperature of 98.6 °F corresponds to 37.0 °C, use the first equation.

$$t_C = \left(\tfrac{5}{9}\right)(t_F - 32) = \left(\tfrac{5}{9}\right)(98.6 - 32) = \left(\tfrac{9}{5}\right)(66.6) = 37.0$$

Thus, body temperature in °C = 37.0 °C.

Laboratory work is almost always done using Celsius units, and we rarely need to make conversions to and from Fahrenheit degrees. It is best to try to calibrate your senses to Celsius units; to help you do so, it is useful to know that water freezes at 0 °C, a comfortable room temperature is about 22 °C, your body temperature is 37 °C, and the hottest water you could leave your hand in for some time is about 60 °C.

QUESTIONS FOR APPENDIX B

Blue-numbered questions have short answers at the end of this book and fully worked solutions in the *Student Solutions Manual.*

Units of the International System

1. Which SI unit accompanied by which prefix would be most convenient for describing each quantity?
   (a) Mass of this book
   (b) Volume of a glass of water
   (c) Thickness of this page
2. Which SI unit accompanied by which prefix would be most convenient for describing each quantity?
   (a) Distance from New York to San Francisco
   (b) Mass of a glass of water
   (c) Area of this page
3. Explain the difference between an SI base (fundamental) unit and a derived SI unit.
4. What is the official SI definition of the mole? Describe in your own words each part of the definition and explain why each part of the definition is important.

Conversion of Units for Physical Quantities

5. Express each quantity in SI base (fundamental) units. Use exponential (scientific) notation whenever it is needed.
   (a) 475 pm (b) 56 Gg (c) 4.28 μA
6. Express each quantity in SI base (fundamental) units. Use exponential (scientific) notation whenever it is needed.
   (a) 32.5 ng (b) 56 Mm (c) 439 pm
7. Express each quantity in SI base (fundamental) units. Use exponential (scientific) notation whenever it is needed.
   (a) 8.7 nm^2 (b) 27.3 aJ (c) 27.3 μN
8. Express each quantity in SI base (fundamental) units. Use exponential (scientific) notation whenever it is needed.
   (a) 56.3 cm^3 (b) 5.62 MJ (c) 33.4 kV
9. Express each quantity in the units indicated. Use scientific notation.
   (a) 1.00 kg in pound
   (b) 2.45 ton in kilograms
   (c) 1 L in cubic inches (in^3)
   (d) 1 atm in pascals and in bars
10. Express each quantity in the units indicated. Use scientific notation.
    (a) 24.3 amu in grams
    (b) 87.3 mL in cubic feet (ft^3)
    (c) 24.7 dg in ounces
    (d) 1.02 bar in millimeters of mercury (mm Hg) and in torrs
11. Express each temperature in Fahrenheit degrees.
    (a) 37 °C (b) −23.6 °C (c) −40.0 °C
12. Express each temperature in Celsius degrees.
    (a) 180. °F (b) −40.0 °F (c) 28.3 °F

APPENDIX

C Physical Constants* and Sources of Data

Quantity	Symbol	Traditional Units	SI Units
Acceleration of gravity	g_n	980.665 cm/s^2	9.806 65 m/s^2
Atomic mass unit ($\frac{1}{12}$ the mass of ^{12}C atom)	amu or u	1.660 54 $\times$ 10^{-24} g	1.660 54 $\times$ 10^{-27} kg
Avogadro constant	N_A, L	6.022 142 $\times$ 10^{23} particles/mol	6.022 142 $\times$ 10^{23} particles/mol
Bohr radius	a_o	0.529 177 2108 Å	5.291 772 108 $\times$ 10^{-11} m
Boltzmann constant	k	1.380 650 5 $\times$ 10^{-16} erg/K	1.380 650 5 $\times$ 10^{-23} J/K
Charge-to-mass ratio of electron	e/m	$-$1.758 820 12 $\times$ 10^8 C/g	$-$1.758 820 12 $\times$ 10^{11} C/kg
Elementary charge (electron or proton charge)	e	1.602 176 53 $\times$ 10^{-19} C	1.602 176 53 $\times$ 10^{-19} C
Electron rest mass	m_e	9.109 3826 $\times$ 10^{-28} g	9.109 3826 $\times$ 10^{-31} kg
Faraday constant	F	96 485.3383 C/mol e$^-$	96 485.3383 C/mol e$^-$
		23.06 kcal V^{-1} mol^{-1}	96 485 J V^{-1} mol^{-1}
Gas constant	R	0.082 057 L atm mol^{-1} K^{-1}	8.314 472 dm^3 Pa mol^{-1} K^{-1}
		1.987 cal mol^{-1} K^{-1}	8.314 472 J mol^{-1} K^{-1}
Molar volume (STP)	V_m	22.414 L/mol	22.414 $\times$ 10^{-3} m^3/mol
			22.414 dm^3/mol
Neutron rest mass	m_n	1.674 927 28 $\times$ 10^{-24} g	1.674 927 28 $\times$ 10^{-27} kg
		1.008 664 amu	
Planck's constant	h	6.626 0693 $\times$ 10^{-27} erg s	6.626 0693 $\times$ 10^{-34} J s
Proton rest mass	m_p	1.672 621 71 $\times$ 10^{-24} g	1.672 621 71 $\times$ 10^{-27} kg
		1.007 276 amu	
Rydberg constant	R_∞	3.289 841 960 $\times$ 10^{15} s^{-1}	1.097 373 156 8525 $\times$ 10^7 m^{-1}
		2.179 871 90 $\times$ 10^{-11} erg	2.179 871 90 $\times$ 10^{-18} J
Velocity of light (in a vacuum)	c, c_0	2.997 924 58 $\times$ 10^{10} cm/s	2.997 924 58 $\times$ 10^8 m/s
		186 282 mile/s	

*Data from the National Institute for Standards and Technology reference on constants, units, and uncertainty, **http://physics.nist.gov/cuu/Constants/index.html.**

Online Sources

- Finding Chemical Data in Web and Library Sources. University of Adelaide Library.
  http://www.library.adelaide.edu.au/guide/sci/Chemistry/propindex.html
- SIRCh: Selected Internet Resources for Chemistry
  http://www.indiana.edu/~cheminfo/cis_ca.html
  SIRCh: Physical Property Information
  http://www.indiana.edu/~cheminfo/ca_ppi.html
- Thermodex. University of Texas at Austin.
  http://thermodex.lib.utexas.edu/
- How Many? A Dictionary of Units of Measurement.
  http://www.unc.edu/~rowlett/units/index.html

Print Sources

- Lide, David R., ed. *CRC Handbook of Chemistry and Physics,* 87th edition, Boca Raton, FL: CRC Press, 2006.
- Budavari, Susan; O'Neil, Maryadele J.; Smith, Ann; Heckelman, Patricia E., eds. *The Merck Index: An Encyclopedia of Chemicals, Drugs, and Biologicals,* 13th edition, Rahway, NJ: Merck & Co., 2001.
- Speight, James, ed. *Lange's Handbook of Chemistry,* 16th edition, New York: McGraw-Hill, 2004.
- Perry, Robert H., Green, Don W., eds. *Perry's Chemical Engineer's Handbook,* 7th edition, New York: McGraw-Hill, 1997.
- Zwillinger, Daniel, ed. *CRC Standard Mathematical Tables and Formulae,* 30th edition, Boca Raton, FL: CRC Press, 1991.
- Lewis, Richard J., Jr. *Hawley's Condensed Chemical Dictionary* (with CD-ROM), 14th edition, New York: Wiley, 2002.

APPENDIX

D Ground-State Electron Configurations of Atoms

Z	Element	Configuration	Z	Element	Configuration	Z	Element	Configuration
1	H	$1s^1$	40	Zr	[Kr] $4d^2 5s^2$	78	Pt	[Xe] $4f^{14} 5d^9 6s^1$
2	He	$1s^2$	41	Nb	[Kr] $4d^4 5s^1$	79	Au	[Xe] $4f^{14} 5d^{10} 6s^1$
3	Li	[He] $2s^1$	42	Mo	[Kr] $4d^5 5s^1$	80	Hg	[Xe] $4f^{14} 5d^{10} 6s^2$
4	Be	[He] $2s^2$	43	Tc	[Kr] $4d^5 5s^2$	81	Tl	[Xe] $4f^{14} 5d^{10} 6s^2 6p^1$
5	B	[He] $2s^2 2p^1$	44	Ru	[Kr] $4d^7 5s^1$	82	Pb	[Xe] $4f^{14} 5d^{10} 6s^2 6p^2$
6	C	[He] $2s^2 2p^2$	45	Rh	[Kr] $4d^8 5s^1$	83	Bi	[Xe] $4f^{14} 5d^{10} 6s^2 6p^3$
7	N	[He] $2s^2 2p^3$	46	Pd	[Kr] $4d^{10}$	84	Po	[Xe] $4f^{14} 5d^{10} 6s^2 6p^4$
8	O	[He] $2s^2 2p^4$	47	Ag	[Kr] $4d^{10} 5s^1$	85	At	[Xe] $4f^{14} 5d^{10} 6s^2 6p^5$
9	F	[He] $2s^2 2p^5$	48	Cd	[Kr] $4d^{10} 5s^2$	86	Rn	[Xe] $4f^{14} 5d^{10} 6s^2 6p^6$
10	Ne	[He] $2s^2 2p^6$	49	In	[Kr] $4d^{10} 5s^2 5p^1$	87	Fr	[Rn] $7s^1$
11	Na	[Ne] $3s^1$	50	Sn	[Kr] $4d^{10} 5s^2 5p^2$	88	Ra	[Rn] $7s^2$
12	Mg	[Ne] $3s^2$	51	Sb	[Kr] $4d^{10} 5s^2 5p^3$	89	Ac	[Rn] $6d^1 7s^2$
13	Al	[Ne] $3s^2 3p^1$	52	Te	[Kr] $4d^{10} 5s^2 5p^4$	90	Th	[Rn] $6d^2 7s^2$
14	Si	[Ne] $3s^2 3p^2$	53	I	[Kr] $4d^{10} 5s^2 5p^5$	91	Pa	[Rn] $5f^2 6d^1 7s^2$
15	P	[Ne] $3s^2 3p^3$	54	Xe	[Kr] $4d^{10} 5s^2 5p^6$	92	U	[Rn] $5f^3 6d^1 7s^2$
16	S	[Ne] $3s^2 3p^4$	55	Cs	[Xe] $6s^1$	93	Np	[Rn] $5f^4 6d^1 7s^2$
17	Cl	[Ne] $3s^2 3p^5$	56	Ba	[Xe] $6s^2$	94	Pu	[Rn] $5f^6 7s^2$
18	Ar	[Ne] $3s^2 3p^6$	57	La	[Xe] $5d^1 6s^2$	95	Am	[Rn] $5f^7 7s^2$
19	K	[Ar] $4s^1$	58	Ce	[Xe] $4f^1 5d^1 6s^2$	96	Cm	[Rn] $5f^7 6d^1 7s^2$
20	Ca	[Ar] $4s^2$	59	Pr	[Xe] $4f^3 6s^2$	97	Bk	[Rn] $5f^9 7s^2$
21	Sc	[Ar] $3d^1 4s^2$	60	Nd	[Xe] $4f^4 6s^2$	98	Cf	[Rn] $5f^{10} 7s^2$
22	Ti	[Ar] $3d^2 4s^2$	61	Pm	[Xe] $4f^5 6s^2$	99	Es	[Rn] $5f^{11} 7s^2$
23	V	[Ar] $3d^3 4s^2$	62	Sm	[Xe] $4f^6 6s^2$	100	Fm	[Rn] $5f^{12} 7s^2$
24	Cr	[Ar] $3d^5 4s^1$	63	Eu	[Xe] $4f^7 6s^2$	101	Md	[Rn] $5f^{13} 7s^2$
25	Mn	[Ar] $3d^5 4s^2$	64	Gd	[Xe] $4f^7 5d^1 6s^2$	102	No	[Rn] $5f^{14} 7s^2$
26	Fe	[Ar] $3d^6 4s^2$	65	Tb	[Xe] $4f^9 6s^2$	103	Lr	[Rn] $5f^{14} 6d^1 7s^2$
27	Co	[Ar] $3d^7 4s^2$	66	Dy	[Xe] $4f^{10} 6s^2$	104	Rf	[Rn] $5f^{14} 6d^2 7s^2$
28	Ni	[Ar] $3d^8 4s^2$	67	Ho	[Xe] $4f^{11} 6s^2$	105	Db	[Rn] $5f^{14} 6d^3 7s^2$
29	Cu	[Ar] $3d^{10} 4s^1$	68	Er	[Xe] $4f^{12} 6s^2$	106	Sg	[Rn] $5f^{14} 6d^4 7s^2$
30	Zn	[Ar] $3d^{10} 4s^2$	69	Tm	[Xe] $4f^{13} 6s^2$	107	Bh	[Rn] $5f^{14} 6d^5 7s^2$
31	Ga	[Ar] $3d^{10} 4s^2 4p^1$	70	Yb	[Xe] $4f^{14} 6s^2$	108	Hs	[Rn] $5f^{14} 6d^6 7s^2$
32	Ge	[Ar] $3d^{10} 4s^2 4p^2$	71	Lu	[Xe] $4f^{14} 5d^1 6s^2$	109	Mt	[Rn] $5f^{14} 6d^7 7s^2$
33	As	[Ar] $3d^{10} 4s^2 4p^3$	72	Hf	[Xe] $4f^{14} 5d^2 6s^2$	110	Ds	[Rn] $5f^{14} 6d^8 7s^2$
34	Se	[Ar] $3d^{10} 4s^2 4p^4$	73	Ta	[Xe] $4f^{14} 5d^3 6s^2$	111	Rg	[Rn] $5f^{14} 6d^9 7s^2$
35	Br	[Ar] $3d^{10} 4s^2 4p^5$	74	W	[Xe] $4f^{14} 5d^4 6s^2$	112	—	[Rn] $5f^{14} 6d^{10} 7s^2$
36	Kr	[Ar] $3d^{10} 4s^2 4p^6$	75	Re	[Xe] $4f^{14} 5d^5 6s^2$	113	—	[Rn] $5f^{14} 6d^{10} 7s^2 7p^1$
37	Rb	[Kr] $5s^1$	76	Os	[Xe] $4f^{14} 5d^6 6s^2$	114	—	[Rn] $5f^{14} 6d^{10} 7s^2 7p^2$
38	Sr	[Kr] $5s^2$	77	Ir	[Xe] $4f^{14} 5d^7 6s^2$	115	—	[Rn] $5f^{14} 6d^{10} 7s^2 7p^3$
39	Y	[Kr] $4d^1 5s^2$						

APPENDIX

E Naming Simple Organic Compounds

The systematic nomenclature for organic compounds was proposed by the International Union of Pure and Applied Chemistry (IUPAC). The IUPAC set of rules provides different names for the more than 10 million known organic compounds, and allows names to be assigned to new compounds as they are synthesized. Many organic compounds also have *common* names. Usually the common name came first and is widely known. Many consumer products are labeled with the common name, and when only a few isomers are possible, the common name adequately identifies the product for the consumer. However, as illustrated in Section 3.4 (⟵ ***p. 86***), a system of common names quickly fails when several structural isomers are possible.

E.1 Hydrocarbons

The name of each member of the hydrocarbon classes has two parts. The first part, called the prefix (*meth-, eth-, prop-, but-*, and so on), reflects the number of carbon atoms. When more than four carbons are present, the Greek or Latin number prefixes are used: *pent-, hex-, hept-, oct-, non-*, and *dec-*. The second part of the name, called the suffix, tells the class of hydrocarbon. Alkanes have carbon-carbon single bonds, alkenes have carbon-carbon double bonds, and alkynes have carbon-carbon triple bonds.

Unbranched Alkanes and Alkyl Groups

The names of the first 20 unbranched (straight-chain) alkanes are given in Table E.1.

Alkyl groups are named by dropping *-ane* from the parent alkane and adding *-yl* (see Table 3.5 for examples).

Branched-Chain Alkanes

The rules for naming branched-chain alkanes are as follows:

1. *Find the longest continuous chain of carbon atoms; it determines the parent name for the compound.* For example, the following compound has two methyl groups attached to a *heptane* parent; the longest continuous chain contains seven carbon atoms.

$$\begin{array}{l} CH_3CH_2CH_2\underset{\displaystyle CH_3}{\underset{|}{C}}HCH_2\underset{\displaystyle CH_3}{\underset{|}{C}}HCH_3 \end{array}$$

Table E.1 Names of Unbranched Alkanes

CH_4	Methane	$C_{11}H_{24}$	Undecane
C_2H_6	Ethane	$C_{12}H_{26}$	Dodecane
C_3H_8	Propane	$C_{13}H_{28}$	Tridecane
C_4H_{10}	Butane	$C_{14}H_{30}$	Tetradecane
C_5H_{12}	Pentane	$C_{15}H_{32}$	Pentadecane
C_6H_{14}	Hexane	$C_{16}H_{34}$	Hexadecane
C_7H_{16}	Heptane	$C_{17}H_{36}$	Heptadecane
C_8H_{18}	Octane	$C_{18}H_{38}$	Octadecane
C_9H_{20}	Nonane	$C_{19}H_{40}$	Nonadecane
$C_{10}H_{22}$	Decane	$C_{20}H_{42}$	Eicosane

The longest continuous chain may not be obvious from the way the formula is written, especially for the straight-line format that is commonly used. For example, the longest continuous chain of carbon atoms in the following chain is *eight,* not *four* or *six.*

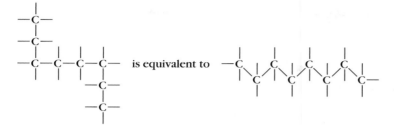

```
    |
   —C—
    |
   —C—
    |    |    |    |
   —C—C—C—C—     is equivalent to   —C         C         C         C
    |    |    |    |                     \  |  / \  |  / \  |  / \  |
              —C—                          C         C         C         C—
               |                           |         |         |         |
              —C—
               |
```

2. *Number the longest chain beginning with the end of the chain nearest the branching. Use these numbers to designate the location of the attached group. When two or more groups are attached to the parent, give each group a number corresponding to its location on the parent chain.* For example, the name of

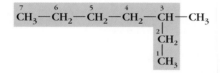

```
  7    6    5    4   3    2  1
CH3CH2CH2CHCH2CHCH3
                |        |
              CH3    CH3
```

is 2,4-dimethylheptane. The name of the compound below is 3-methylheptane, not 5-methylheptane or 2-ethylhexane.

```
  7        6         5         4         3
CH3—CH2—CH2—CH2—CH—CH3
                                    2|
                                    CH2
                                    1|
                                    CH3
```

3. *When two or more substituents are identical, indicate this by the use of the prefixes* di-, tri-, tetra., *and so on. Positional numbers of the substituents should have the smallest possible sum.*

```
                CH3  CH3
  1     2    3|  4    5|   6    7    8
CH3CH2CCH2CHCHCH2CH3
                |            |
              CH3        CH3
```

The correct name of this compound is 3,3,5,6-tetramethyloctane.

4. *If there are two or more different groups, the groups are listed alphabetically.*

```
          CH3
   1    2|  3    4   5    6
CH3CCH2CHCH2CH3
          |         |
        CH3   CH2
                    |
                  CH3
```

The correct name of this compound is 4-ethyl-2,2-dimethylhexane. Note that the prefix *di-* is ignored in determining alphabetical order.

Alkenes

Alkenes are named by using the prefix to indicate the number of carbon atoms and the suffix *-ene* to indicate one or more double bonds. The systematic names for the first two members of the alkene series are *ethene* and *propene.*

$$CH_2{=}CH_2 \qquad CH_3CH{=}CH_2$$

When groups, such as methyl or ethyl, are attached to carbon atoms in an alkene, the longest hydrocarbon chain is numbered from the end that will give the double bond the lowest number, and then numbers are assigned to the attached groups. For example, the name of

$$\underset{5}{CH_3}\overset{\overset{\textstyle CH_3}{|}}{\underset{4}{C}H}\underset{3}{C}H{=}\underset{2}{C}H\underset{1}{C}H_3$$

is 4-methyl-2-pentene. See Section 8.5 for a discussion of *cis-trans* isomers of alkenes.

Alkynes

The naming of alkynes is similar to that of alkenes, with the lowest number possible being used to locate the triple bond. For example, the name of

$$\underset{1}{CH_3}\underset{2}{C}{\equiv}\underset{3}{C}\overset{\overset{\textstyle CH_3}{|}}{\underset{4}{C}H}\underset{5}{C}H_3$$

is 4-methyl-2-pentyne.

Benzene Derivatives

Monosubstituted benzene derivatives are named by using a prefix for the substituent. Some examples are

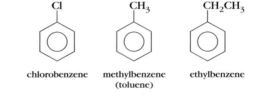

Cl CH$_3$ CH$_2$CH$_3$

chlorobenzene methylbenzene (toluene) ethylbenzene

Three isomers are possible when two groups are substituted for hydrogen atoms on the benzene ring. The relative positions of the substituents are indicated either by the prefixes *ortho-*, *meta-*, and *para-* (abbreviated *o-*, *m-*, and *p-*, respectively) or by numbers. For example,

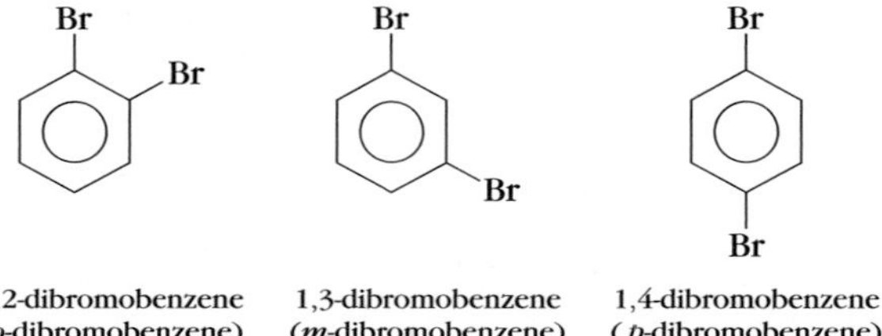

1,2-dibromobenzene (*o*-dibromobenzene) 1,3-dibromobenzene (*m*-dibromobenzene) 1,4-dibromobenzene (*p*-dibromobenzene)

The dimethylbenzenes are called *xylenes.*

If more than two groups are attached to the benzene ring, numbers must be used to identify the positions. The benzene ring is numbered to give the lowest possible numbers to the substituents.

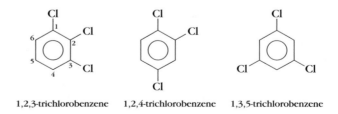

1,2,3-trichlorobenzene 1,2,4-trichlorobenzene 1,3,5-trichlorobenzene

E.2 Functional Groups

An atom or group of atoms that defines the structure of a specific class of organic compounds and determines their properties is called a *functional group* **(⇐ *p. 80*).** The millions of organic compounds include classes of compounds that are obtained by replacing hydrogen atoms of hydrocarbons with functional groups (Sections 3.1, 12.5, 12.6, and 12.7). The important functional groups are shown in Table E.2.

The "R" attached to the functional group represents the hydrocarbon framework with one hydrogen removed for each functional group added. The IUPAC system provides a systematic method for naming all members of a given class. For example, alcohols end in *-ol* (methan*ol*); aldehydes end in *-al* (methan*al*); carboxylic acids end in *-oic* (ethan*oic* acid); and ketones end in *-one* (propan*one*).

Alcohols

Isomers are also possible for molecules containing functional groups. For example, three different alcohols are obtained when a hydrogen atom in pentane is replaced by —OH, depending on which hydrogen atom is replaced. The rules for naming the "R" or hydrocarbon framework are the same as those for hydrocarbon compounds.

$CH_3CH_2CH_2CH_2CH_2OH$ 1-pentanol

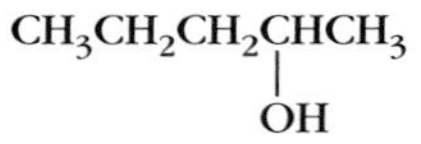

$CH_3CH_2CH_2CH(OH)CH_3$ 2-pentanol

$CH_3CH_2CH(OH)CH_2CH_3$ 3-pentanol

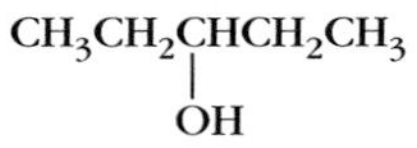

Compounds with one or more functional groups (Table E.2) and alkyl substituents are named so as to give the functional groups the lowest numbers. For example, the correct name of

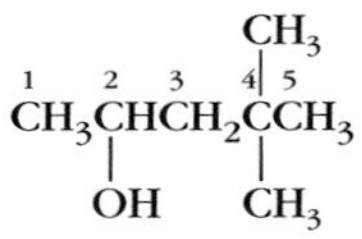

$$\overset{1}{C}H_3\overset{2}{C}H(OH)\overset{3}{C}H_2\overset{4}{C}(CH_3)_2\overset{5}{C}H_3$$

is 4,4-dimethyl-2-pentanol.

Aldehydes and Ketones

The systematic names of the first three aldehydes are methanal, ethanal, and propanal.

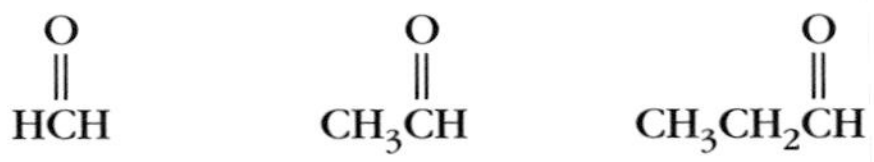

HCH (C=O)	CH_3CH (C=O)	CH_3CH_2CH (C=O)
methanal (formaldehyde)	ethanal (acetaldehyde)	propanol (propionaldehyde)

For ketones, a number is used to designate the position of the carbonyl group, and the chain is numbered in a way that gives the carbonyl carbon the smallest number.

CH_3CCH_3 (C=O)	$CH_3CH_2CCH_3$ (C=O)	$CH_3CCH_2CH{=}CH_2$ (C=O)
2-propanone (acetone)	2-butanone (methyl ethyl ketone)	4-penten-2-one

Carboxylic Acids

The systematic names of carboxylic acids are obtained by dropping the final *e* of the name of the corresponding alkane and adding *-oic acid.* For example, the name of

$$CH_3CH_2CH_2CH_2CH_2COOH$$

is hexanoic acid. Other examples are

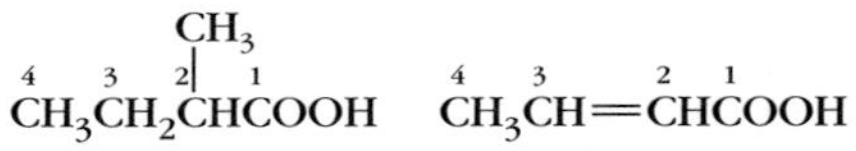

$\overset{4}{C}H_3\overset{3}{C}H_2\overset{2}{C}H(CH_3)\overset{1}{C}OOH$	$\overset{4}{C}H_3\overset{3}{C}H{=}\overset{2}{C}H\overset{1}{C}OOH$
2-methylbutanoic acid	2-butenoic acid

Table E.2 Classes of Organic Compounds Based on Functional Groups*

General formulas of class members	Class name	Typical compound	Compound name	Common use of sample compound
$R{-}X$	Halide	$H{-}CH(Cl){-}Cl$	Dichloromethane (methylene chloride)	Solvent
$R{-}OH$	Alcohol	$H{-}CH_2{-}OH$	Methanol (wood alcohol)	Solvent
$R{-}C({=}O){-}H$	Aldehyde	$H{-}C({=}O){-}H$	Methanal (formaldehyde)	Preservative
$R{-}C({=}O){-}OH$	Carboxylic acid	$H{-}CH_2{-}C({=}O){-}OH$	Ethanoic acid (acetic acid)	Vinegar
$R{-}C({=}O){-}R'$	Ketone	$H{-}CH_2{-}C({=}O){-}CH_2{-}H$	Propanone (acetone)	Solvent
$R{-}O{-}R'$	Ether	$C_2H_5{-}O{-}C_2H_5$	Diethyl ether (ethyl ether)	Anesthetic
$R{-}C({=}O){-}O{-}R'$	Ester	$CH_3{-}C({=}O){-}O{-}C_2H_5$	Ethyl ethanoate (ethyl acetate)	Solvent in fingernail polish
$R{-}N(H){-}H$	Amine	$H{-}CH_2{-}N(H){-}H$	Methylamine	Tanning hides (foul odor)
$R{-}C({=}O){-}N(H){-}R'$	Amide	$CH_3{-}C({=}O){-}N(H){-}H$	Acetamide	Plasticizer

*R stands for an H or a hydrocarbon group such as $-CH_3$ or $-C_2H_5$. R′ could be a different group from R.

Esters

The systematic names of esters are derived from the names of the alcohol and the acid used to prepare the ester. The general formula for esters is

$$R{-}\overset{\overset{\displaystyle O}{\|}}{C}{-}OR'$$

As shown in Section 12.4, the $R{-}\overset{\overset{\displaystyle O}{\|}}{C}$ comes from the acid and the R′O comes from the alcohol. The alcohol part is named first, followed by the name of the acid changed to end in *-ate.* For example,

$$CH_3CH_2\overset{\overset{\displaystyle O}{\|}}{C}{-}OCH_3$$

is named methyl propanoate and

$$CH_3\overset{\overset{\displaystyle O}{\|}}{C}{-}OCH{=}CH_2$$

is named ethenyl ethanoate.

APPENDIX

F Ionization Constants for Weak Acids at 25 °C

Acid	Formula and Ionization Equation	K_a
Acetic	$CH_3COOH + H_2O \rightleftharpoons H_3O^+ + CH_3COO^-$	1.8×10^{-5}
Arsenic	$H_3AsO_4 + H_2O \rightleftharpoons H_3O^+ + H_2AsO_4^-$	$K_1 = 2.5 \times 10^{-4}$
	$H_2AsO_4^- + H_2O \rightleftharpoons H_3O^+ + HAsO_4^{2-}$	$K_2 = 5.6 \times 10^{-8}$
	$HAsO_4^{2-} + H_2O \rightleftharpoons H_3O^+ + AsO_4^{3-}$	$K_3 = 3.0 \times 10^{-13}$
Arsenous	$H_3AsO_3 + H_2O \rightleftharpoons H_3O^+ + H_2AsO_3^-$	$K_1 = 6.0 \times 10^{-10}$
	$H_2AsO_3^- + H_2O \rightleftharpoons H_3O^+ + HAsO_3^{2-}$	$K_2 = 3.0 \times 10^{-14}$
Benzoic	$C_6H_5COOH + H_2O \rightleftharpoons H_3O^+ + C_6H_5COO^-$	6.3×10^{-5}
Boric	$B(OH)_3(H_2O) + H_2O \rightleftharpoons H_3O^+ + B(OH)_4^-$	7.3×10^{-10}
Carbonic	$H_2CO_3 + H_2O \rightleftharpoons H_3O^+ + HCO_3^-$	$K_1 = 4.2 \times 10^{-7}$
	$HCO_3^- + H_2O \rightleftharpoons H_3O^+ + CO_3^{2-}$	$K_2 = 4.8 \times 10^{-11}$
Citric	$H_3C_6H_5O_7 + H_2O \rightleftharpoons H_3O^+ + H_2C_6H_5O_7^-$	$K_1 = 7.4 \times 10^{-3}$
	$H_2C_6H_5O_7^- + H_2O \rightleftharpoons H_3O^+ + HC_6H_5O_7^{2-}$	$K_2 = 1.7 \times 10^{-5}$
	$HC_6H_5O_7^{2-} + H_2O \rightleftharpoons H_3O^+ + C_6H_5O_7^{3-}$	$K_3 = 4.0 \times 10^{-7}$
Cyanic	$HOCN + H_2O \rightleftharpoons H_3O^+ + OCN^-$	3.5×10^{-4}
Formic	$HCOOH + H_2O \rightleftharpoons H_3O^+ + HCOO^-$	1.8×10^{-4}
Hydrazoic	$HN_3 + H_2O \rightleftharpoons H_3O^+ + N_3^-$	1.9×10^{-5}
Hydrocyanic	$HCN + H_2O \rightleftharpoons H_3O^+ + CN^-$	4.0×10^{-10}
Hydrofluoric	$HF + H_2O \rightleftharpoons H_3O^+ + F^-$	7.2×10^{-4}
Hydrogen peroxide	$H_2O_2 + H_2O \rightleftharpoons H_3O^+ + HO_2^-$	2.4×10^{-12}
Hydrosulfuric	$H_2S + H_2O \rightleftharpoons H_3O^+ + HS^-$	$K_1 = 1 \times 10^{-7}$
	$HS^- + H_2O \rightleftharpoons H_3O^+ + S^{2-}$	$K_2 = 1 \times 10^{-19}$
Hypobromous	$HOBr + H_2O \rightleftharpoons H_3O^+ + OBr^-$	2.5×10^{-9}
Hypochlorous	$HOCl + H_2O \rightleftharpoons H_3O^+ + OCl^-$	3.5×10^{-8}
Nitrous	$HNO_2 + H_2O \rightleftharpoons H_3O^+ + NO_2^-H_3O^+$	4.5×10^{-4}
Oxalic	$H_2C_2O_4 + H_2O \rightleftharpoons H_3O^+ + HC_2O_4^-$	$K_1 = 5.9 \times 10^{-2}$
	$HC_2O_4^- + H_2O \rightleftharpoons H_3O^+ + C_2O_4^{2-}$	$K_2 = 6.4 \times 10^{-5}$
Phenol	$HC_6H_5O + H_2O \rightleftharpoons H_3O^+ + C_6H_5O^-$	1.3×10^{-10}
Phosphoric	$H_3PO_4 + H_2O \rightleftharpoons H_3O^+ + H_2PO_4^-$	$K_1 = 7.5 \times 10^{-3}$
	$H_2PO_4^- + H_2O \rightleftharpoons H_3O^+ + HPO_4^{2-}$	$K_2 = 6.2 \times 10^{-8}$
	$HPO_4^{2-} + H_2O \rightleftharpoons H_3O^+ + PO_4^{3-}$	$K_3 = 3.6 \times 10^{-13}$
Phosphorous	$H_3PO_3 + H_2O \rightleftharpoons H_3O^+ + H_2PO_3^-$	$K_1 = 1.6 \times 10^{-2}$
	$H_2PO_3^- + H_2O \rightleftharpoons H_3O^+ + HPO_3^{2-}$	$K_2 = 7.0 \times 10^{-7}$
Selenic	$H_2SeO_4 + H_2O \rightleftharpoons H_3O^+ + HSeO_4^-$	K_1 = very large
	$HSeO_4^- + H_2O \rightleftharpoons H_3O^+ + SeO_4^{2-}$	$K_2 = 1.2 \times 10^{-2}$
Selenous	$H_2SeO_3 + H_2O \rightleftharpoons H_3O^+ + HSeO_3^-$	$K_1 = 2.7 \times 10^{-3}$
	$HSeO_3^- + H_2O \rightleftharpoons H_3O^+ + SeO_3^{2-}$	$K_2 = 2.5 \times 10^{-7}$
Sulfuric	$H_2SO_4 + H_2O \rightleftharpoons H_3O^+ + HSO_4^-$	K_1 = very large
	$HSO_4^- + H_2O \rightleftharpoons H_3O^+ + SO_4^{2-}$	$K_2 = 1.2 \times 10^{-2}$
Sulfurous	$H_2SO_3 + H_2O \rightleftharpoons H_3O^+ + HSO_3^-$	$K_1 = 1.7 \times 10^{-2}$
	$HSO_3^- + H_2O \rightleftharpoons H_3O^+ + SO_3^{2-}$	$K_2 = 6.4 \times 10^{-8}$
Tellurous	$H_2TeO_3 + H_2O \rightleftharpoons H_3O^+ + HTeO_3^-$	$K_1 = 2 \times 10^{-3}$
	$HTeO_3^- + H_2O \rightleftharpoons H_3O^+ + TeO_3^{2-}$	$K_2 = 1 \times 10^{-8}$

APPENDIX

G Ionization Constants for Weak Bases at 25 °C

Base	Formula and Ionization Equation	K_b
Ammonia	$NH_3 + H_2O \rightleftharpoons NH_4^+ + OH^-$	1.8×10^{-5}
Aniline	$C_6H_5NH_2 + H_2O \rightleftharpoons C_6H_5NH_3^+ + OH^-$	4.2×10^{-10}
Dimethylamine	$(CH_3)_2NH + H_2O \rightleftharpoons (CH_3)_2NH_2^+ + OH^-$	7.4×10^{-4}
Ethylenediamine	$(CH_2)_2(NH_2)_2 + H_2O \rightleftharpoons (CH_2)_2(NH_2)_2H^+ + OH^-$	$K_1 = 8.5 \times 10^{-5}$
	$(CH_2)_2(NH_2)_2H^+ + H_2O \rightleftharpoons (CH_2)_2(NH_2)_2H_2^{2+} + OH^-$	$K_2 = 2.7 \times 10^{-8}$
Hydrazine	$N_2H_4 + H_2O \rightleftharpoons N_2H_5^+ + OH^-$	$K_1 = 8.5 \times 10^{-7}$
	$N_2H_5^+ + H_2O \rightleftharpoons N_2H_6^{2+} + OH^-$	$K_2 = 8.9 \times 10^{-16}$
Hydroxylamine	$NH_2OH + H_2O \rightleftharpoons NH_3OH^+ + OH^-$	6.6×10^{-9}
Methylamine	$CH_3NH_2 + H_2O \rightleftharpoons CH_3NH_3^+ + OH^-$	5.0×10^{-4}
Pyridine	$C_5H_5N + H_2O \rightleftharpoons C_5H_5NH^+ + OH^-$	1.5×10^{-9}
Trimethylamine	$(CH_3)_3N + H_2O \rightleftharpoons (CH_3)_3NH^+ + OH^-$	7.4×10^{-5}

APPENDIX H

Solubility Product Constants for Some Inorganic Compounds at 25 °C*

Substance	K_{sp}
Aluminum Compounds	
$AlAsO_4$	1.6×10^{-16}
$Al(OH)_3$	1.9×10^{-33}
$AlPO_4$	1.3×10^{-20}
Barium Compounds	
$Ba_3(AsO_4)_2$	1.1×10^{-13}
$BaCO_3$	8.1×10^{-9}
$BaC_2O_4 \cdot 2\,H_2O$†	1.1×10^{-7}
$BaCrO_4$	2.0×10^{-10}
BaF_2	1.7×10^{-6}
$Ba(OH)_2 \cdot 8\,H_2O$†	5.0×10^{-3}
$Ba_3(PO_4)_2$	1.3×10^{-29}
$BaSeO_4$	2.8×10^{-11}
$BaSO_3$	8.0×10^{-7}
$BaSO_4$	1.1×10^{-10}
Bismuth Compounds	
$BiOCl$	7.0×10^{-9}
$BiO(OH)$	1.0×10^{-12}
$Bi(OH)_3$	3.2×10^{-40}
BiI_3	8.1×10^{-19}
$BiPO_4$	1.3×10^{-23}
Cadmium Compounds	
$Cd_3(AsO_4)_2$	2.2×10^{-32}
$CdCO_3$	2.5×10^{-14}
$Cd(CN)_2$	1.0×10^{-8}
$Cd_2[Fe(CN)_6]$	3.2×10^{-17}
$Cd(OH)_2$	1.2×10^{-14}
Calcium Compounds	
$Ca_3(AsO_4)_2$	6.8×10^{-19}
$CaCO_3$	3.8×10^{-9}
$CaCrO_4$	7.1×10^{-4}
$CaC_2O_4 \cdot H_2O$†	2.3×10^{-9}
CaF_2	3.9×10^{-11}
$Ca(OH)_2$	7.9×10^{-6}
$CaHPO_4$	2.7×10^{-7}
$Ca(H_2PO_4)_2$	1.0×10^{-3}
$Ca_3(PO_4)_2$	1.0×10^{-25}

Substance	K_{sp}
$CaSO_3 \cdot 2\,H_2O$†	1.3×10^{-8}
$CaSO_4 \cdot 2\,H_2O$†	2.4×10^{-5}
Chromium Compounds	
$CrAsO_4$	7.8×10^{-21}
$Cr(OH)_3$	6.7×10^{-31}
$CrPO_4$	2.4×10^{-23}
Cobalt Compounds	
$Co_3(AsO_4)_2$	7.6×10^{-29}
$CoCO_3$	8.0×10^{-13}
$Co(OH)_2$	2.5×10^{-16}
$Co(OH)_3$	4.0×10^{-45}
Copper Compounds	
$CuBr$	5.3×10^{-9}
$CuCl$	1.9×10^{-7}
$CuCN$	3.2×10^{-20}
$Cu_2O(Cu^+ + OH^-)$‡	1.0×10^{-14}
CuI	5.1×10^{-12}
$CuSCN$	1.6×10^{-11}
$Cu_3(AsO_4)_2$	7.6×10^{-36}
$CuCO_3$	2.5×10^{-10}
$Cu_2[Fe(CN)_6]$	1.3×10^{-16}
$Cu(OH)_2$	1.6×10^{-19}
Gold Compounds	
$AuBr$	5.0×10^{-17}
$AuCl$	2.0×10^{-13}
AuI	1.6×10^{-23}
$AuBr_3$	4.0×10^{-36}
$AuCl_3$	3.2×10^{-25}
$Au(OH)_3$	1×10^{-53}
AuI_3	1.0×10^{-46}
Iron Compounds	
$FeCO_3$	3.5×10^{-11}
$Fe(OH)_2$	7.9×10^{-15}
FeS	4.9×10^{-18}
$Fe_4[Fe(CN)_6]_3$	3.0×10^{-41}
$Fe(OH)_3$	6.3×10^{-38}

Substance	K_{sp}
Lead Compounds	
$Pb_3(AsO_4)_2$	4.1×10^{-36}
$PbBr_2$	6.3×10^{-6}
$PbCO_3$	1.5×10^{-13}
$PbCl_2$	1.7×10^{-5}
$PbCrO_4$	1.8×10^{-14}
PbF_2	3.7×10^{-8}
$Pb(OH)_2$	2.8×10^{-16}
PbI_2	8.7×10^{-9}
$Pb_3(PO_4)_2$	3.0×10^{-44}
$PbSeO_4$	1.5×10^{-7}
$PbSO_4$	1.8×10^{-8}
Magnesium Compounds	
$Mg_3(AsO_4)_2$	2.1×10^{-20}
$MgCO_3 \cdot 3\,H_2O$†	4.0×10^{-5}
MgC_2O_4	8.6×10^{-5}
MgF_2	6.4×10^{-9}
$MgNH_4PO_4$	2.5×10^{-12}
Manganese Compounds	
$Mn_3(AsO_4)_2$	1.9×10^{-11}
$MnCO_3$	1.8×10^{-11}
$Mn(OH)_2$	4.6×10^{-14}
$Mn(OH)_3$	$\sim 1 \times 10^{-36}$
Mercury Compounds	
Hg_2Br_2	1.3×10^{-22}
Hg_2CO_3	8.9×10^{-17}
Hg_2Cl_2	1.1×10^{-18}
Hg_2CrO_4	5.0×10^{-9}
Hg_2I_2	4.5×10^{-29}
$Hg_2O \cdot H_2O$ ($Hg_2^{2+} + 2\,OH^-$)†‡	1.6×10^{-23}
Hg_2SO_4	6.8×10^{-7}
$Hg(CN)_2$	3.0×10^{-23}
$Hg(OH)_2$	2.5×10^{-26}
HgI_2	4.0×10^{-29}
Nickel Compounds	
$Ni_3(AsO_4)_2$	1.9×10^{-26}
$NiCO_3$	6.6×10^{-9}
$Ni(CN)_2$	3.0×10^{-23}
$Ni(OH)_2$	2.8×10^{-16}
Silver Compounds	
Ag_3AsO_4	1.1×10^{-20}
$AgBr$	3.3×10^{-13}
Ag_2CO_3	8.1×10^{-12}
$AgCl$	1.8×10^{-10}
Ag_2CrO_4	9.0×10^{-12}
$AgCN$	1.2×10^{-16}
$Ag_4[Fe(CN)_6]$	1.6×10^{-41}
Ag_2O ($Ag^+ + OH^-$)‡	2.0×10^{-8}
AgI	1.5×10^{-16}
Ag_3PO_4	8.9×10^{-17}
Ag_2SO_3	1.5×10^{-14}
Ag_2SO_4	1.7×10^{-5}
$AgSCN$	1.0×10^{-12}
Strontium Compounds	
$Sr_3(AsO_4)_2$	1.3×10^{-18}
$SrCO_3$	9.4×10^{-10}
$SrC_2O_4 \cdot 2\,H_2O$†	5.6×10^{-8}
$SrCrO_4$	3.6×10^{-5}
$Sr(OH)_2 \cdot 8\,H_2O$†	3.2×10^{-4}
$Sr_3(PO_4)_2$	1.0×10^{-31}
$SrSO_3$	4.0×10^{-8}
$SrSO_4$	2.8×10^{-7}
Tin Compounds	
$Sn(OH)_2$	2.0×10^{-26}
SnI_2	1.0×10^{-4}
$Sn(OH)_4$	1×10^{-57}
Zinc Compounds	
$Zn_3(AsO_4)_2$	1.1×10^{-27}
$ZnCO_3$	1.5×10^{-11}
$Zn(CN)_2$	8.0×10^{-12}
$Zn_3[Fe(CN)_6]$	4.1×10^{-16}
$Zn(OH)_2$	4.5×10^{-17}
$Zn_3(PO_4)_2$	9.1×10^{-33}

*No metallic sulfides are listed in this table because sulfide ion is such a strong base that the usual solubility product equilibrium equation does not apply. See Myers, R. J. *Journal of Chemical Education,* Vol. 63, 1986; pp. 687–690.

†Since $[H_2O]$ does not appear in equilibrium constants for equilibria in aqueous solution in general, it does *not* appear in the K_{sp} expressions for hydrated solids.

‡Very small amounts of these oxides dissolve in water to give the ions indicated in parentheses. Solid hydroxides of these metal ions are unstable and decompose to oxides as rapidly as they are formed.

APPENDIX I

Standard Reduction Potentials in Aqueous Solution at 25 °C

Acidic Solution	Standard Reduction Potential, $E°$ (volts)
$F_2(g) + 2e^- \longrightarrow 2\,F^-(aq)$	2.87
$Co^{3+}(aq) + e^- \longrightarrow Co^{2+}(aq)$	1.82
$Pb^{4+}(aq) + 2\,e^- \longrightarrow Pb^{2+}(aq)$	1.8
$H_2O_2(aq) + 2\,H_3O^+(aq) + 2\,e^- \longrightarrow 4\,H_2O(\ell)$	1.77
$NiO_2(s) + 4\,H_3O^+(aq) + 2\,e^- \longrightarrow Ni^{2+}(aq) + 6\,H_2O(\ell)$	1.7
$PbO_2(s) + SO_4^{2-}(aq) + 4\,H_3O^+(aq) + 2\,e^- \longrightarrow PbSO_4(s) + 6\,H_2O(\ell)$	1.685
$Au^+(aq) + e^- \longrightarrow Au(s)$	1.68
$2\,HClO(aq) + 2\,H_3O^+(aq) + 2\,e^- \longrightarrow Cl_2(g) + 4\,H_2O(\ell)$	1.63
$Ce^{4+}(aq) + e^- \longrightarrow Ce^{3+}(aq)$	1.61
$NaBiO_3(s) + 6\,H_3O^+(aq) + 2\,e^- \longrightarrow Bi^{3+}(aq) + Na^+(aq) + 9\,H_2O(\ell)$	~ 1.6
$MnO_4^-(aq) + 8\,H_3O^+(aq) + 5\,e^- \longrightarrow Mn^{2+}(aq) + 12\,H_2O(\ell)$	1.51
$Au^{3+}(aq) + 3\,e^- \longrightarrow Au(s)$	1.50
$2\,ClO_3^-(aq) + 12\,H_3O^+(aq) + 10\,e^- \longrightarrow Cl_2(g) + 18\,H_2O(\ell)$	1.47
$BrO_3^-(aq) + 6\,H_3O^+(aq) + 6\,e^- \longrightarrow Br^-(aq) + 9\,H_2O(\ell)$	1.44
$Cl_2(g) + 2\,e^- \longrightarrow 2\,Cl^-(aq)$	1.358
$Cr_2O_7^{2-}(aq) + 14\,H_3O^+(aq) + 6\,e^- \longrightarrow 2\,Cr^{3+}(aq) + 21\,H_2O(\ell)$	1.33
$N_2H_5^+(aq) + 3\,H_3O^+(aq) + 2\,e^- \longrightarrow 2\,NH_4^+(aq) + 3\,H_2O(\ell)$	1.24
$MnO_2(s) + 4\,H_3O^+(aq) + 2\,e^- \longrightarrow Mn^{2+}(aq) + 6\,H_2O(\ell)$	1.23
$O_2(g) + 4\,H_3O^+(aq) + 4\,e^- \longrightarrow 6\,H_2O(\ell)$	1.229
$Pt^{2+}(aq) + 2\,e^- \longrightarrow Pt(s)$	1.2
$IO_3^-(aq) + 6\,H_3O^+(aq) + 5\,e^- \longrightarrow \frac{1}{2}I_2(aq) + 9\,H_2O(\ell)$	1.195
$ClO_4^-(aq) + 2\,H_3O^+(aq) + 2\,e^- \longrightarrow ClO_3^-(aq) + 3\,H_2O(\ell)$	1.19
$Br_2(\ell) + 2\,e^- \longrightarrow 2\,Br^-(aq)$	1.066
$AuCl_4^-(aq) + 3\,e^- \longrightarrow Au(s) + 4\,Cl^-(aq)$	1.00
$Pd^{2+}(aq) + 2\,e^- \longrightarrow Pd(s)$	0.987
$NO_3^-(aq) + 4\,H_3O^+(aq) + 3\,e^- \longrightarrow NO(g) + 6\,H_2O(\ell)$	0.96
$NO_3^-(aq) + 3\,H_3O^+(aq) + 2\,e^- \longrightarrow HNO_2(aq) + 4\,H_2O(\ell)$	0.94
$2\,Hg^{2+}(aq) + 2\,e^- \longrightarrow Hg_2^{2+}(aq)$	0.920
$Hg^{2+}(aq) + 2\,e^- \longrightarrow Hg(\ell)$	0.855
$Ag^+(aq) + e^- \longrightarrow Ag(s)$	0.7994
$Hg_2^{2+}(aq) + 2\,e^- \longrightarrow 2\,Hg(\ell)$	0.789
$Fe^{3+}(aq) + e^- \longrightarrow Fe^{2+}(aq)$	0.771
$SbCl_6^-(aq) + 2\,e^- \longrightarrow SbCl_4^-(aq) + 2\,Cl^-(aq)$	0.75
$[PtCl_4]^{2-}(aq) + 2\,e^- \longrightarrow Pt(s) + 4\,Cl^-(aq)$	0.73
$O_2(g) + 2\,H_3O^+(aq) + 2\,e^- \longrightarrow H_2O_2(aq) + 2\,H_2O(\ell)$	0.682
$[PtCl_6]^{2-}(aq) + 2\,e^- \longrightarrow [PtCl_4]^{2-}(aq) + 2\,Cl^-(aq)$	0.68
$H_3AsO_4(aq) + 2\,H_3O^+(aq) + 2\,e^- \longrightarrow H_3AsO_3(aq) + 3\,H_2O(\ell)$	0.58
$I_2(s) + 2\,e^- \longrightarrow 2\,I^-(aq)$	0.535

Acidic Solution	Standard Reduction Potential, $E°$ (volts)
$TeO_2(s) + 4\,H_3O^+(aq) + 4\,e^- \longrightarrow Te(s) + 6\,H_2O(\ell)$	0.529
$Cu^+(aq) + e^- \longrightarrow Cu(s)$	0.521
$[RhCl_6]^{3-}(aq) + 3\,e^- \longrightarrow Rh(s) + 6\,Cl^-(aq)$	0.44
$Cu^{2+}(aq) + 2\,e^- \longrightarrow Cu(s)$	0.337
$Hg_2Cl_2(s) + 2\,e^- \longrightarrow 2\,Hg(\ell) + 2\,Cl^-(aq)$	0.27
$AgCl(s) + e^- \longrightarrow Ag(s) + Cl^-(aq)$	0.222
$SO_4^{2-}(aq) + 4\,H_3O^+(aq) + 2\,e^- \longrightarrow SO_2(g) + 6\,H_2O(\ell)$	0.20
$SO_4^{2-}(aq) + 4\,H_3O^+(aq) + 2\,e^- \longrightarrow H_2SO_3(aq) + 5\,H_2O(\ell)$	0.17
$Cu^{2+}(aq) + e^- \longrightarrow Cu^+(aq)$	0.153
$Sn^{4+}(aq) + 2\,e^- \longrightarrow Sn^{2+}(aq)$	0.15
$S(s) + 2\,H_3O^+(aq) + 2\,e^- \longrightarrow H_2S(aq) + 2\,H_2O(\ell)$	0.14
$AgBr(s) + e^- \longrightarrow Ag(s) + Br^-(aq)$	0.0713
$2\,H_3O^+(aq) + 2\,e^- \longrightarrow H_2(g) + 2\,H_2O(\ell)$ (reference electrode)	0.0000
$N_2O(g) + 6\,H_3O^+(aq) + 4\,e^- \longrightarrow 2\,NH_3OH^+(aq) + 5\,H_2O(\ell)$	−0.05
$Pb^{2+}(aq) + 2\,e^- \longrightarrow Pb(s)$	−0.126
$Sn^{2+}(aq) + 2\,e^- \longrightarrow Sn(s)$	−0.14
$AgI(s) + e^- \longrightarrow Ag(s) + I^-(aq)$	−0.15
$[SnF_6]^{2-}(aq) + 4\,e^- \longrightarrow Sn(s) + 6\,F^-(aq)$	−0.25
$Ni^{2+}(aq) + 2\,e^- \longrightarrow Ni(s)$	−0.25
$Co^{2+}(aq) + 2\,e^- \longrightarrow Co(s)$	−0.28
$Tl^+(aq) + e^- \longrightarrow Tl(s)$	−0.34
$PbSO_4(s) + 2\,e^- \longrightarrow Pb(s) + SO_4^{2-}(aq)$	−0.356
$Se(s) + 2\,H_3O^+(aq) + 2\,e^- \longrightarrow H_2Se(aq) + 2\,H_2O(\ell)$	−0.40
$Cd^{2+}(aq) + 2\,e^- \longrightarrow Cd(s)$	−0.403
$Cr^{3+}(aq) + e^- \longrightarrow Cr^{2+}(aq)$	−0.41
$Fe^{2+}(aq) + 2\,e^- \longrightarrow Fe(s)$	−0.44
$2\,CO_2(g) + 2\,H_3O^+(aq) + 2\,e^- \longrightarrow (COOH)_2(aq) + 2\,H_2O(\ell)$	−0.49
$Ga^{3+}(aq) + 3\,e^- \longrightarrow Ga(s)$	−0.53
$HgS(s) + 2\,H_3O^+(aq) + 2\,e^- \longrightarrow Hg(\ell) + H_2S(g) + 2\,H_2O(\ell)$	−0.72
$Cr^{3+}(aq) + 3\,e^- \longrightarrow Cr(s)$	−0.74
$Zn^{2+}(aq) + 2\,e^- \longrightarrow Zn(s)$	−0.763
$Cr^{2+}(aq) + 2\,e^- \longrightarrow Cr(s)$	−0.91
$Mn^{2+}(aq) + 2\,e^- \longrightarrow Mn(s)$	−1.18
$V^{2+}(aq) + 2\,e^- \longrightarrow V(s)$	−1.18
$Zr^{4+}(aq) + 4\,e^- \longrightarrow Zr(s)$	−1.53
$Al^{3+}(aq) + 3\,e^- \longrightarrow Al(s)$	−1.66
$H_2(g) + 2\,e^- \longrightarrow 2\,H^-(aq)$	−2.25
$Mg^{2+}(aq) + 2\,e^- \longrightarrow Mg(s)$	−2.37
$Na^+(aq) + e^- \longrightarrow Na(s)$	−2.714
$Ca^{2+}(aq) + 2\,e^- \longrightarrow Ca(s)$	−2.87
$Sr^{2+}(aq) + 2\,e^- \longrightarrow Sr(s)$	−2.89
$Ba^{2+}(aq) + 2\,e^- \longrightarrow Ba(s)$	−2.90
$Rb^+(aq) + e^- \longrightarrow Rb(s)$	−2.925
$K^+(aq) + e^- \longrightarrow K(s)$	−2.925
$Li^+(aq) + e^- \longrightarrow Li(s)$	−3.045

Basic Solution	Standard Reduction Potential, $E°$ (volts)
$ClO^-(aq) + H_2O(\ell) + 2\,e^- \longrightarrow Cl^-(aq) + 2\,OH^-(aq)$	0.89
$OOH^-(aq) + H_2O(\ell) + 2\,e^- \longrightarrow 3\,OH^-(aq)$	0.88
$2\,NH_2OH(aq) + 2\,e^- \longrightarrow N_2H_4(aq) + 2\,OH^-(aq)$	0.74
$ClO_3^-(aq) + 3\,H_2O(\ell) + 6\,e^- \longrightarrow Cl^-(aq) + 6\,OH^-(aq)$	0.62
$MnO_4^-(aq) + 2\,H_2O(\ell) + 3\,e^- \longrightarrow MnO_2(s) + 4\,OH^-(aq)$	0.588
$MnO_4^-(aq) + e^- \longrightarrow MnO_4^{2-}(aq)$	0.564
$NiO_2(s) + 2\,H_2O(\ell) + 2\,e^- \longrightarrow Ni(OH)_2(s) + 2\,OH^-(aq)$	0.49
$Ag_2CrO_4(s) + 2\,e^- \longrightarrow 2\,Ag(s) + CrO_4^{2-}(aq)$	0.446
$O_2(g) + 2\,H_2O(\ell) + 4\,e^- \longrightarrow 4\,OH^-(aq)$	0.40
$ClO_4^-(aq) + H_2O(\ell) + 2\,e^- \longrightarrow ClO_3^-(aq) + 2\,OH^-(aq)$	0.36
$Ag_2O(s) + H_2O(\ell) + 2\,e^- \longrightarrow 2\,Ag(s) + 2\,OH^-(aq)$	0.34
$2\,NO_2^-(aq) + 3\,H_2O(\ell) + 4\,e^- \longrightarrow N_2O(g) + 6\,OH^-(aq)$	0.15
$N_2H_4(aq) + 2\,H_2O(\ell) + 2\,e^- \longrightarrow 2\,NH_3(aq) + 2\,OH^-(aq)$	0.10
$[Co(NH_3)_6]^{3+}(aq) + e^- \longrightarrow [Co(NH_3)_6]^{2+}(aq)$	0.10
$HgO(s) + H_2O(\ell) + 2\,e^- \longrightarrow Hg(\ell) + 2\,OH^-(aq)$	0.0984
$O_2(g) + H_2O(\ell) + 2\,e^- \longrightarrow OOH^-(aq) + OH^-(aq)$	0.076
$NO_3^-(aq) + H_2O(\ell) + 2\,e^- \longrightarrow NO_2^-(aq) + 2\,OH^-(aq)$	0.01
$MnO_2(s) + 2\,H_2O(\ell) + 2\,e^- \longrightarrow Mn(OH)_2(s) + 2\,OH^-(aq)$	−0.05
$CrO_4^{2-}(aq) + 4\,H_2O(\ell) + 3\,e^- \longrightarrow Cr(OH)_3(s) + 5\,OH^-(aq)$	−0.12
$Cu(OH)_2(s) + 2\,e^- \longrightarrow Cu(s) + 2\,OH^-(aq)$	−0.36
$Fe(OH)_3(s) + e^- \longrightarrow Fe(OH)_2(s) + OH^-(aq)$	−0.56
$2\,H_2O(\ell) + 2\,e^- \longrightarrow H_2(g) + 2\,OH^-(aq)$	−0.8277
$2\,NO_3^-(aq) + 2\,H_2O(\ell) + 2\,e^- \longrightarrow N_2O_4(g) + 4\,OH^-(aq)$	−0.85
$Fe(OH)_2(s) + 2\,e^- \longrightarrow Fe(s) + 2\,OH^-(aq)$	−0.877
$SO_4^{2-}(aq) + H_2O(\ell) + 2\,e^- \longrightarrow SO_3^{2-}(aq) + 2\,OH^-(aq)$	−0.93
$N_2(g) + 4\,H_2O(\ell) + 4\,e^- \longrightarrow N_2H_4(aq) + 4\,OH^-(aq)$	−1.15
$[Zn(OH)_4]^{2-}(aq) + 2\,e^- \longrightarrow Zn(s) + 4\,OH^-(aq)$	−1.22
$Zn(OH)_2(s) + 2\,e^- \longrightarrow Zn(s) + 2\,OH^-(aq)$	−1.245
$[Zn(CN)_4]^{2-}(aq) + 2\,e^- \longrightarrow Zn(s) + 4\,CN^-(aq)$	−1.26
$Cr(OH)_3(s) + 3\,e^- \longrightarrow Cr(s) + 3\,OH^-(aq)$	−1.30
$SiO_3^{2-}(aq) + 3\,H_2O(\ell) + 4\,e^- \longrightarrow Si(s) + 6\,OH^-(aq)$	−1.70

APPENDIX J

Selected Thermodynamic Values*

Species	ΔH_f° (298.15 K) (kJ/mol)	S° (298.15 K) (J K^{-1} mol^{-1})	ΔG_f° (298.15 K) (kJ/mol)
Aluminum			
Al(s)	0	28.275	0
Al^{3+}(aq)	−531	−321.7	−485
$AlCl_3$(s)	−704.2	110.67	−628.8
Al_2O_3(s, corundum)	−1675.7	50.92	−1582.3
Argon			
Ar(g)	0	154.843	0
Ar(aq)	−12.1	59.4	16.4
Barium			
$BaCl_2$(s)	−858.6	123.68	−810.4
BaO(s)	−553.5	70.42	−525.1
$BaSO_4$(s)	−1473.2	132.2	−1362.2
$BaCO_3$(s)	−1216.3	112.1	85.35
Beryllium			
Be(s)	0	9.5	0
$Be(OH)_2$(s)	−902.5	51.9	−815
Bromine			
Br(g)	111.884	175.022	82.396
$Br_2(\ell)$	0	152.231	0
Br_2(g)	30.907	245.463	3.110
Br_2(aq)	−2.59	130.5	3.93
Br^-(aq)	−121.55	82.4	−103.96
BrCl(g)	14.64	240.10	−0.98
BrF_3(g)	−255.6	292.53	−229.43
HBr(g)	−36.40	198.695	−53.45
Calcium			
Ca(s)	0	41.42	0
Ca(g)	178.2	158.884	144.3
Ca^{2+}(g)	1925.9	—	—
Ca^{2+}(aq)	−542.83	−53.1	−553.58
CaC_2(s)	−59.8	69.96	−64.9
$CaCO_3$(s, calcite)	−1206.92	92.9	−1128.79
$CaCl_2$(s)	−795.8	104.6	−748.1
CaF_2(s)	−1219.6	68.87	−1167.3
CaH_2(s)	−186.2	42	−147.2
CaO(s)	−635.09	39.75	−604.03
CaS(s)	−482.4	56.5	−477.4
$Ca(OH)_2$(s)	−986.09	83.39	−898.49
$Ca(OH)_2$(aq)	−1002.82	−74.5	−868.07
$CaSO_4$(s)	−1434.11	106.7	−1321.79

*Taken from Wagman, D. D., Evans, W. H., Parker, V. B., Schumm, R. H., Halow, I., Bailey, S. M., Churney, K. L., and Nuttall, R. The NBS Tables of Chemical Thermodynamic Properties. *Journal of Physical and Chemical Reference Data,* Vol. 11, Suppl. 2, 1982.

Species	ΔH_f° (298.15 K) (kJ/mol)	S° (298.15 K) (J K^{-1} mol^{-1})	ΔG_f° (298.15 K) (kJ/mol)
Carbon			
C(s, graphite)	0	5.74	0
C(s, diamond)	1.895	2.377	2.9
C(g)	716.682	158.096	671.257
$CCl_4(\ell)$	−135.44	216.4	−65.21
$CCl_4(g)$	−102.9	309.85	−60.59
$CHCl_3(\ell)$	−134.47	201.7	−73.66
$CHCl_3(g)$	−103.14	295.71	−70.34
CH_4 (g, methane)	−74.81	186.264	−50.72
C_2H_2 (g, ethyne)	226.73	200.94	209.2
C_2H_4 (g, ethene)	52.26	219.56	68.15
C_2H_6 (g, ethane)	−84.68	229.6	−32.82
C_3H_8 (g, propane)	−103.8	269.9	−23.49
C_4H_{10} (g, butane)	−126.148	310.227	−16.985
$C_6H_6(\ell$, benzene)	49.03	172.8	124.5
$C_6H_{14}(\ell$, hexane)	−198.782	296.018	−4.035
C_8H_{18} (g, octane)	−208.447	466.835	16.718
$C_8H_{18}(\ell$, octane)	−249.952	361.205	6.707
$CH_3OH(\ell$, methanol)	−238.66	126.8	−166.27
CH_3OH(g, methanol)	−200.66	239.81	−161.96
CH_3OH(aq, methanol)	−245.931	133.1	−175.31
$C_2H_5OH(\ell$, ethanol)	−277.69	160.7	−174.78
C_2H_5OH(g, ethanol)	−235.1	282.7	−168.49
C_2H_5OH(aq, ethanol)	−288.3	148.5	−181.64
$C_6H_{12}O_6$ (s, glucose)	−1274.4	235.9	−917.2
$CH_3COO^-(aq)$	−486.01	86.6	−369.31
$CH_3COOH(aq)$	−485.76	178.7	−396.46
$CH_3COOH(\ell)$	−484.5	159.8	−389.9
CO(g)	−110.525	197.674	−137.168
$CO_2(g)$	−393.509	213.74	−394.359
$H_2CO_3(aq)$	−699.65	187.4	−623.08
$HCO_3^-(aq)$	−691.99	91.2	−586.77
$CO_3^{2-}(aq)$	−677.14	−56.9	−527.81
$HCOO^-(aq)$	−425.55	92.0	−351.0
HCOOH(aq)	−425.43	163	−372.3
$HCOOH(\ell)$	−424.72	128.95	−361.35
$CS_2(g)$	117.36	237.84	67.12
$CS_2(\ell)$	89.70	151.34	65.27
$COCl_2(g)$	−218.8	283.53	−204.6
Cesium			
Cs(s)	0	85.23	0
$Cs^+(g)$	457.964	—	—
CsCl(s)	−443.04	101.17	−414.53
Chlorine			
Cl(g)	121.679	165.198	105.68
$Cl^-(g)$	−233.13	—	—
$Cl^-(aq)$	−167.159	56.5	−131.228
$Cl_2(g)$	0	223.066	0

Species	ΔH_f° (298.15 K) (kJ/mol)	S° (298.15 K) (J K^{-1} mol^{-1})	ΔG_f° (298.15 K) (kJ/mol)
Cl_2(aq)	−23.4	121	6.94
HCl(g)	−92.307	186.908	−95.299
HCl(aq)	−167.159	56.5	−131.228
ClO_2(g)	102.5	256.84	120.5
Cl_2O(g)	80.3	266.21	97.9
ClO^-(aq)	−107.1	42.0	−36.8
HClO(aq)	−120.9	142.	−79.9
ClF_3(g)	−163.2	281.61	−123.0
Chromium			
Cr(s)	0	23.77	0
Cr_2O_3(s)	−1139.7	81.2	−1058.1
$CrCl_3$(s)	−556.5	123	−486.1
Copper			
Cu(s)	0	33.15	0
CuO(s)	−157.3	42.63	−129.7
$CuCl_2$(s)	−220.1	108.07	−175.7
$CuSO_4$(s)	−771.36	109.	−661.8
Fluorine			
F_2(g)	0	202.78	0
F(g)	78.99	158.754	61.91
F^-(g)	−255.39	—	—
F^-(aq)	−332.63	−13.8	−278.79
HF(g)	−271.1	173.779	−273.2
HF(aq, un-ionized)	−320.08	88.7	−296.82
HF(aq, ionized)	−332.63	−13.8	−278.79
Hydrogen†			
H_2(g)	0	130.684	0
H_2(aq)	−4.2	57.7	17.6
HD(g)	0.318	143.801	−1.464
D_2(g)	0	144.960	0
H(g)	217.965	114.713	203.247
H^+(g)	1536.202	—	—
H^+(aq)	0	0	0
OH^-(aq)	−229.994	−10.75	−157.244
$H_2O(\ell)$	−285.83	69.91	−237.129
H_2O(g)	−241.818	188.825	−228.572
$H_2O_2(\ell)$	−187.78	109.6	−120.35
H_2O_2(aq)	−191.17	143.9	−134.03
HO_2^-(aq)	−160.33	23.8	−67.3
$HDO(\ell)$	−289.888	79.29	−241.857
$D_2O(\ell)$	−294.600	75.94	−243.439
Iodine			
I_2(s)	0	116.135	0
I_2(g)	62.438	260.69	19.327
I_2(aq)	22.6	137.2	16.40
I(g)	106.838	180.791	70.25

Species	ΔH_f° (298.15 K) (kJ/mol)	S° (298.15 K) (J K^{-1} mol^{-1})	ΔG_f° (298.15 K) (kJ/mol)
I^-(g)	−197	—	—
I^-(aq)	−55.19	111.3	−51.57
I_3^-(aq)	−51.5	239.3	−51.4
HI(g)	26.48	206.594	1.70
HI(aq, ionized)	−55.19	111.3	−51.57
IF(g)	−95.65	236.17	−118.51
ICl(g)	17.78	247.551	−5.46
ICl_3(s)	−89.5	167.4	−22.29
ICl(ℓ)	−23.89	135.1	−13.58
IBr(g)	40.84	258.773	3.69
Iron			
Fe(s)	0	27.78	0
FeO(s, wustite)	−266.27	57.9	−245.12
Fe_2O_3(s, hematite)	−824.2	87.4	−742.2
Fe_3O_4(s, magnetite)	−1118.4	146.4	−1015.4
$FeCl_2$(s)	−341.79	117.95	−302.3
$FeCl_3$(s)	−399.49	142.3	−344
FeS_2 (s, pyrite)	−178.2	52.93	−166.9
$Fe(CO)_5$(ℓ)	−774	338.1	−705.3
Lead			
Pb(s)	0	64.81	0
$PbCl_2$(s)	−359.41	136	−314.1
PbO(s, yellow)	−217.32	68.7	−187.89
PbS(s)	−100.4	91.2	−98.7
Lithium			
Li(s)	0	29.12	0
Li^+(g)	685.783	—	—
LiOH(s)	−484.93	42.8	−438.95
LiOH(aq)	−508.48	2.8	−450.58
LiCl(s)	−408.701	59.33	−384.37
Magnesium			
Mg(s)	0	32.68	0
Mg^{2+}(aq)	−466.85	−138.1	−454.8
$MgCl_2$(g)	−400.4	—	—
$MgCl_2$(s)	−641.32	89.62	−591.79
$MgCl_2$(aq)	−801.15	−25.1	−717.1
MgO(s)	−601.70	26.94	−569.43
$Mg(OH)_2$(s)	−924.54	63.18	−833.51
MgS(s)	−346	50.33	−341.8
$MgSO_4$(s)	−1284.9	91.6	−1170.6
$MgCO_3$(s)	−1095.8	65.7	−1012.1
Mercury			
Hg(ℓ)	0	76.02	0
$HgCl_2$(s)	−224.3	146	−178.6
HgO(s, red)	−90.83	70.29	−58.539
HgS(s, red)	−58.2	82.4	−50.6

Species	ΔH_f° (298.15 K) (kJ/mol)	S° (298.15 K) (J K^{-1} mol^{-1})	ΔG_f° (298.15 K) (kJ/mol)
Nickel			
Ni(s)	0	29.87	0
NiO(s)	−239.7	37.99	−211.7
$NiCl_2$(s)	−305.332	97.65	−259.032
Nitrogen			
N_2(g)	0	191.61	0
N_2(aq)	−10.8	—	—
N(g)	472.704	153.298	455.563
NH_3(g)	−46.11	192.45	−16.45
NH_3(aq)	−80.29	111.3	−26.50
NH_4^+(aq)	−132.51	113.4	−79.31
N_2H_4(ℓ)	50.63	121.21	149.34
NH_4Cl(s)	−314.43	94.6	−202.87
NH_4Cl(aq)	−299.66	169.9	−210.52
NH_4NO_3(s)	−365.56	151.08	−183.87
NH_4NO_3(aq)	−339.87	259.8	−190.56
NO(g)	90.25	210.761	86.55
NO_2(g)	33.18	240.06	51.31
N_2O(g)	82.05	219.85	104.20
N_2O_4(g)	9.16	304.29	97.89
N_2O_4(ℓ)	−19.50	209.2	97.54
NOCl(g)	51.71	261.69	66.08
HNO_3(ℓ)	−174.10	155.60	−80.71
HNO_3(g)	−135.06	266.38	−74.72
HNO_3(aq)	−207.36	146.4	−111.25
NO_3^-(aq)	−205.0	146.4	−108.74
NF_3(g)	−124.7	260.73	−83.2
Oxygen†			
O_2(g)	0	205.138	0
O_2(aq)	−11.7	110.9	16.4
O(g)	249.170	161.055	231.731
O_3(g)	142.7	238.93	163.2
OH^-(aq)	−229.994	−10.75	−157.244
Phosphorus			
P_4(s, white)	0	164.36	0
P_4(s, red)	−70.4	91.2	−48.4
P(g)	314.64	163.193	278.25
PH_3(g)	5.4	310.23	13.4
PCl_3(g)	−287	311.78	−267.8
PCl_3(ℓ)	−319.7	217.1	−272.3
PCl_5(s)	−443.5	—	—
P_4O_{10}(s)	−2984	228.86	−2697.7
H_3PO_4(s)	−1279	110.5	−1119.1
Potassium			
K(s)	0	64.18	0
KF(s)	−567.27	66.57	−537.75

Species	ΔH_f° (298.15 K) (kJ/mol)	S° (298.15 K) (J K^{-1} mol^{-1})	ΔG_f° (298.15 K) (kJ/mol)
KCl(s)	−436.747	82.59	−409.14
KCl(aq)	−419.53	159.0	−414.49
KBr(s)	−393.798	95.90	−380.66
KI(s)	−327.900	106.32	−324.892
$KClO_3$(s)	−397.73	143.1	−296.25
KOH(s)	−424.764	78.9	−379.08
KOH(aq)	−482.37	91.6	−440.5
Silicon			
Si(s)	0	18.83	0
$SiBr_4$(ℓ)	−457.3	277.8	−443.8
SiC(s)	−65.3	16.61	−62.8
$SiCl_4$(g)	−657.01	330.73	−616.98
SiH_4(g)	34.3	204.62	56.9
SiF_4(g)	−1614.94	282.49	−1572.65
SiO_2(s, quartz)	−910.94	41.84	−856.64
Silver			
Ag(s)	0	42.55	0
Ag^+(aq)	105.579	72.68	77.107
Ag_2O(s)	−31.05	121.3	−11.2
AgCl(s)	−127.068	96.2	−109.789
AgI(s)	−61.84	115.5	−66.19
AgN_3(s)	620.60	99.22	591.0
$AgNO_3$(s)	−124.39	140.92	−33.41
$AgNO_3$(aq)	−101.8	219.2	−34.16
Sodium			
Na(s)	0	51.21	0
Na(g)	107.32	153.712	76.761
Na^+(g)	609.358	—	—
Na^+(aq)	−240.12	59.0	−261.905
NaF(s)	−573.647	51.46	−543.494
NaF(aq)	−572.75	45.2	−540.68
NaCl(s)	−411.153	72.13	−384.138
NaCl(g)	−176.65	229.81	−196.66
NaCl(aq)	−407.27	115.5	−393.133
NaBr(s)	−361.062	86.82	−348.983
NaBr(aq)	−361.665	141.4	−365.849
NaI(s)	−287.78	98.53	−286.06
NaI(aq)	−295.31	170.3	−313.47
NaOH(s)	−425.609	64.455	−379.484
NaOH(aq)	−470.114	48.1	−419.15
$NaClO_3$(s)	−365.774	123.4	−262.259
$NaHCO_3$(s)	−950.81	101.7	−851.0
Na_2CO_3(s)	−1130.68	134.98	−1044.44
Na_2SO_4(s)	−1387.08	149.58	−1270.16

Species	ΔH_f° (298.15 K) (kJ/mol)	S° (298.15 K) (J K^{-1} mol^{-1})	ΔG_f° (298.15 K) (kJ/mol)
Sulfur			
S(s, monoclinic)	0.33	—	—
S(s, rhombic)	0	31.80	0
S(g)	278.805	167.821	238.250
S^{2-}(aq)	33.1	−14.6	85.8
S_2Cl_2(g)	−18.4	331.5	−31.8
SF_6(g)	−1209.	291.82	−1105.3
SF_4(g)	−774.9	292.03	−731.3
H_2S(g)	−20.63	205.79	−33.56
H_2S(aq)	−39.7	121	−27.83
HS^-(aq)	−17.6	62.8	12.08
SO_2(g)	−296.830	248.22	−300.194
SO_3(g)	−395.72	256.76	−371.06
$SOCl_2$(g)	−212.5	309.77	−198.3
SO_4^{2-}(aq)	−909.27	20.1	−744.53
H_2SO_4(ℓ)	−813.989	156.904	−690.003
H_2SO_4(aq)	−909.27	20.1	−744.53
HSO_4^-(aq)	−887.34	131.8	−755.91
Tin			
Sn(s, white)	0	51.55	0
Sn(s, gray)	−2.09	44.14	0.13
$SnCl_2$(s)	−325.1	—	—
$SnCl_4$(ℓ)	−511.3	258.6	−440.1
$SnCl_4$(g)	−471.5	365.8	−432.2
SnO_2(s)	−580.7	52.3	−519.6
Titanium			
Ti(s)	0	30.63	0
$TiCl_4$(ℓ)	−804.2	252.34	−737.2
$TiCl_4$(g)	−763.2	354.9	−726.7
TiO_2(s)	−939.7	49.92	−884.5
Uranium			
U(s)	0	50.21	0
UO_2(s)	−1084.9	77.03	−1031.7
UO_3(s)	−1223.8	96.11	−1145.9
UF_4(s)	−1914.2	151.67	−1823.3
UF_6(g)	−2147.4	377.9	−2063.7
UF_6(s)	−2197.0	227.6	−2068.5
Zinc			
Zn(s)	0	41.63	0
$ZnCl_2$(s)	−415.05	111.46	−369.398
ZnO(s)	−348.28	43.64	−318.3
ZnS(s, sphalerite)	−205.98	57.7	−201.29

†Many hydrogen-containing and oxygen-containing compounds are listed only under other elements; for example, HNO_3 appears under nitrogen.

Answers to Problem-Solving Practice Problems

Chapter 1

1.1 (1) *Define the problem:* You are asked to find the volume of the sample, and you know the mass.
(2) *Develop a plan:* Density relates mass and volume and is the appropriate conversion factor, so look up the density in a table. Volume is proportional to mass, so the mass has to be either multiplied by the density or multiplied by the reciprocal of the density. Use the units to decide which.
(3) *Execute the plan:* According to Table 1.1, the density of benzene is 0.880 g/mL. Setting up the calculation so that the unit (grams) cancels gives

$$4.33\ \cancel{g} \times \frac{1\ \text{mL}}{0.880\ \cancel{g}} = 4.92\ \text{mL}$$

Notice that the result is expressed to three significant figures, because both the mass and the density had three significant figures.
(4) *Check your answer:* Because the density is a little less than 1.00 g/mL, the volume in milliliters should be a little larger than the mass in grams. The calculated answer, 4.92 mL, is a little larger than the mass, 4.33 g.

1.2 Substance A must be a mixture since some of it dissolves and some, substance B, does not.
Substance C is the soluble portion of substance A. Since all of substance C dissolves in water there is no way to determine how many components it has. Additionally, it is not possible to determine whether the one or more components themselves are elements or compounds. Therefore it is not possible to say whether C is an element, a compound, or a mixture.
The only thing we know about substance B is that it is insoluble in water. We do not know whether it is one insoluble substance, or more than one insoluble substance. Additionally, we do not know whether the substance or substances of B are elements or compounds. Therefore it is not possible to say whether B is an element, a compound, or a mixture.

1.3 Oxygen is O_2; ozone is O_3. Oxygen is a colorless, odorless gas; ozone is a pale blue gas with a pungent odor.

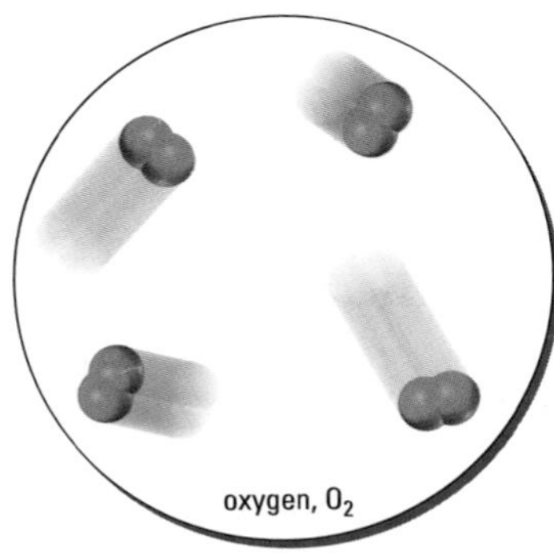
oxygen, O_2

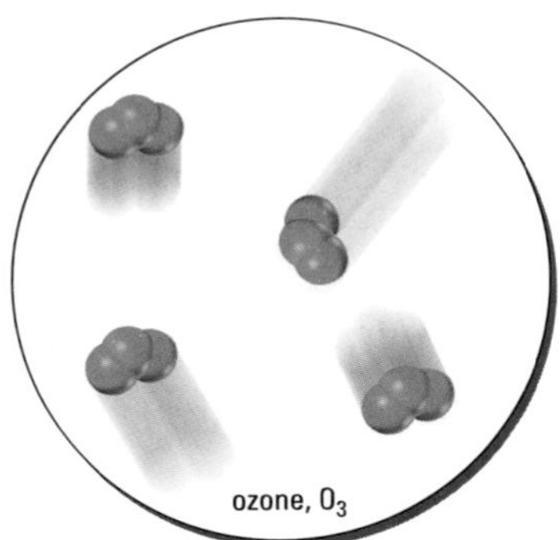
ozone, O_3

Chapter 2

2.1 (a) 10 gal = 40 qt. There are 1.0567 quarts per liter, so

$$40\ \text{qt} \times \frac{1\ \text{L}}{1.0567\ \text{qt}} = 37.9\ \text{L or } 38\ \text{L}$$

(b) $$100\ \text{yds} \times \frac{3\ \text{ft}}{\text{yd}} \times \frac{12\ \text{in}}{\text{ft}} \times \frac{2.54\ \text{cm}}{\text{in}} \times \frac{1\ \text{m}}{100\ \text{cm}} = 91.46\ \text{m}$$

2.2 (a) 1 lb = 453.59 g, so 5 lb = 2268. g
(b) $3\ \text{pt} \times \frac{1\ \text{qt}}{2\ \text{pt}} \times \frac{1\ \text{L}}{1.057\ \text{qt}} \times \frac{1000\ \text{ml}}{1\ \text{L}} = 1420\ \text{ml}$
(c) $\frac{1.420\ \text{L}}{5\ \text{L}} \times 100\% = 28\%$

2.3 Work with the numerator first: 165 mg is 0.165 g. Work with the denominator next: 1 dL = 0.1 L. Therefore, the concentration is

$$\frac{0.165\ \text{g}}{0.1\ \text{L}} = 1.65\ \text{g/L}.$$

2.4 (a) 0.00602 g — 3 sf
(b) 22.871 mg — 5 sf
(c) 344. °C — 3 sf
(d) 100.0 mL — 4 sf
(e) 0.00042 m — 2 sf
(f) 0.002001 L — 4 sf

2.5 (a) 244.2 + 0.1732 = 244.4
(b) $6.19 \times 5.2222 = 32.3$
(c) $\frac{7.2234 - 11.3851}{4.22} = -0.986$

2.6 (a) A phosphorus atom (Z = 15) with 16 neutrons has A = 31.
(b) A neon-22 atom has A = 22 and Z = 10, so the number of electrons must be 10 and the number of neutrons must be $A - Z = 22 - 10 = 12$ neutrons.
(c) The periodic table shows us that the element with 82 protons is lead. The atomic weight of this isotope of lead is 82 + 125 = 207, so the correct symbol is $^{207}_{82}Pb$.

2.7 The magnesium isotope with 12 neutrons has 12 protons, so Z = 12 and the notation is $^{24}_{12}Mg$; the isotope with 13 neutrons has Z = 12 and $^{25}_{12}Mg$; and the isotope with 14 neutrons has Z = 12 and $^{26}_{12}Mg$.

2.8 75 g wire × (fraction Ni) = g Ni, so 75 g Ni × 0.80 = 60 g Ni. For Cr we have 75 g Cr × 0.20 = 15 g Cr. Or, we could have solved for the mass of Cr from 75 g wire − 60 g Ni = 15 g Cr.

2.9 (a) 1 mg Mo = 1×10^{-3} g Mo

$$1 \times 10^{-3}\ \text{g Mo} \times \frac{1\ \text{mol Mo}}{95.94\ \text{g Mo}} = 1.04 \times 10^{-5}\ \text{mol Mo}$$

(b) $$5.00 \times 10^{-3}\ \text{mol Au} \times \frac{196.97\ \text{g Au}}{1\ \text{mol Au}} = 0.985\ \text{g Au}$$

Chapter 3

3.1 (a) $C_{10}H_{11}O_{13}N_5P_3$ (b) $C_{18}H_{27}O_3N$ (c) $C_2H_2O_4$

3.2 (a) Sulfur dioxide
(b) Boron trifluoride
(c) Carbon tetrachloride

3.3 Yes, a similar trend would be expected. The absolute values of the boiling points of the chlorine-containing compounds would be different from their alkane parents, but the differences between successive chlorine-containing compounds should follow a similar trend as for the alkanes themselves.

3.4 (a) A Ca^{4+} charge is unlikely because calcium is in Group 2A, the elements of which lose two electrons to form 2+ ions.
(b) Cr^{2+} is possible because chromium is a transition metal ion that forms 2+ and 3+ ions.
(c) Strontium is a Group 2A metal and forms 2+ ions; thus, a Sr^- ion is highly unlikely.

3.5 (a) CH_4 is formed from two nonmetals and is molecular.
(b) $CaBr_2$ is formed from a metal and a nonmetal, so it is ionic.
(c) $MgCl_2$ is formed from a metal and a nonmetal, so it is ionic.
(d) PCl_3 is formed from two nonmetals and is molecular.
(e) KCl is formed from a metal and a nonmetal and is ionic.

3.6 (a) $In_2(SO_3)_3$ contains two In^{3+} and three SO_3^{2-} ions. There are 14 atoms in this formula unit.
(b) $(NH_4)_3PO_4$ contains three ammonium ions, NH_4^+, and one phosphate ion, PO_4^{3-}, collectively containing 20 atoms.

3.7 (a) One Mg^{2+} ion and two Br^- ions
(b) Two Li^+ ions and one CO_3^{2-} ion
(c) One NH_4^+ ion and one Cl^- ion
(d) Two Fe^{3+} ions and three SO_4^{2-} ions
(e) CuCl and $CuCl_2$

3.8 (a) KNO_2 is potassium nitrite.
(b) $NaHSO_3$ is sodium hydrogen sulfite or sodium bisulfite.
(c) $Mn(OH)_2$ is manganese(II) hydroxide.
(d) $Mn_2(SO_4)_3$ is manganese(III) sulfate.
(e) Ba_3N_2 is barium nitride.
(f) LiH is lithium hydride.

3.9 (a) KH_2PO_4 (b) CuOH (c) NaClO
(d) NH_4ClO_4 (e) $CrCl_3$ (f) $FeSO_3$

3.10 (a) The molar mass of $K_2Cr_2O_7$ is 294.2 g/mol.

$$\frac{12.5\text{ g}}{294.2\text{ g/mol}} = 4.25 \times 10^{-2}\text{ mol}$$

(b) The molar mass of $KMnO_4$ is 158.0 g/mol.

$$\frac{12.5\text{ g}}{158.0\text{ g/mol}} = 7.91 \times 10^{-2}\text{ mol}$$

(c) The molar mass of $(NH_4)_2CO_3$ is 96.1 g/mol.

$$\frac{12.5\text{ g}}{96.1\text{ g/mol}} = 1.30 \times 10^{-1}\text{ mol}$$

3.11 (a) The molar mass of sucrose, $C_{12}H_{22}O_{11}$, is 342.3 g/mol.

$$5.0 \times 10^{-3}\text{ mol sucrose} \times \frac{342.3\text{ g sucrose}}{1\text{ mol sucrose}} = 1.7\text{ g sucrose}$$

(b) $$3.0 \times 10^{-6}\text{ mol ACTH} \times \frac{4600\text{ g ACTH}}{1\text{ mol ACTH}} = 1.4 \times 10^{-2}\text{ g ACTH}$$

$$1.4 \times 10^{-2}\text{ g} \times \frac{10^3\text{ mg}}{1\text{ g}} = 14.\text{ mg ACTH}$$

3.12 (a) The molar mass of cholesterol is 386.7 g/mol.

$$\frac{10.0\text{ g}}{386.7\text{ g/mol}} = 2.59 \times 10^{-2}\text{ mol}$$

The molar mass of $Mn_2(SO_4)_3$ is 398.1 g/mol.

$$\frac{10.0\text{ g}}{398.1\text{ g/mol}} = 2.51 \times 10^{-2}\text{ mol}$$

(b) The molar mass of K_2HPO_4 is 174.2 g/mol.

$$0.25\text{ mol} \times \frac{174.2\text{ g}}{1\text{ mol}} = 44.\text{ g}$$

The molar mass of caffeine is 194.2 g/mol.

$$0.25\text{ mol} \times \frac{194.2\text{ g}}{1\text{ mol}} = 49.\text{ g}$$

3.13 The mass of Si in 1 mol SiO_2 is 28.0855 g. The mass of O in 1 mol SiO_2 is 31.9988 g.

$$\%\text{ Si in SiO}_2 = \frac{28.0855\text{ g}}{60.08\text{ g}} \times 100\% = 46.7\%\text{ Si}$$

$$\%\text{ O in SiO}_2 = \frac{31.9988\text{ g}}{60.08\text{ g}} \times 100\% = 53.3\%\text{ O}$$

3.14 The molar mass of hydrated nickel chloride is

$$58.69\text{ g/mol} + 2(35.45\text{ g/mol}) + 12(1.008\text{ g/mol}) + 6(16.00\text{ g/mol}) = 237.69\text{ g/mol.}$$

The percentages by weight for each element are found from the ratios of the mass of each element in 1 mole of hydrated nickel chloride to the molar mass of hydrated nickel chloride:

$$\frac{58.69\text{ g Ni}}{237.69\text{ g hydrated nickel chloride}} \times 100\% = 24.7\%\text{ Ni}$$

$$\frac{70.90\text{ g Cl}}{237.69\text{ g hydrated nickel chloride}} \times 100\% = 29.8\%\text{ Cl}$$

$$\frac{12.096\text{ g H}}{237.69\text{ g hydrated nickel chloride}} \times 100\% = 5.09\%\text{ H}$$

$$\frac{96.00\text{ g O}}{237.69\text{ g hydrated nickel chloride}} \times 100\% = 40.4\%\text{ O}$$

3.15 A 100-g sample of the phosphorus oxide contains 43.64 g P and 56.36 g O.

$$43.64\text{ g P} \times \frac{1\text{ mol P}}{30.9738\text{ g P}} = 1.41\text{ mol P}$$

$$56.36\text{ g O} \times \frac{1\text{ mol O}}{15.9994\text{ g O}} = 3.52\text{ mol O}$$

The mole ratio is

$$\frac{3.52\text{ mol O}}{1.41\text{ mol P}} = \frac{2.50\text{ mol O}}{1.00\text{ mol P}}$$

There are 2.5 oxygen atoms for every phosphorus atom. Thus, the empirical formula is P_2O_5. The molar mass corresponding to this empirical formula is

$$\left(2\text{ mol P} \times \frac{30.9738\text{ g P}}{1\text{ mol P}}\right) + \left(5\text{ mol O} \times \frac{15.994\text{ g O}}{1\text{ mol O}}\right) = 141.9\text{ g/mol}$$

The known molar mass is 283.89 g/mol. The molar mass is twice as large as the empirical formula mass, so the molecular formula of the oxide is P_4O_{10}.

3.16 Find the number of moles of each element in 100.0 g of vitamin C.

$$40.9\text{ g C} \times \frac{1\text{ mol C}}{12.011\text{ g C}} = 3.405\text{ mol C}$$

$$4.58\text{ g H} \times \frac{1\text{ mol H}}{1.0079\text{ g H}} = 4.544\text{ mol H}$$

$$54.5\text{ g O} \times \frac{1\text{ mol O}}{15.9994\text{ g O}} = 3.406\text{ mol O}$$

Find the mole ratios.

$$\frac{4.544 \text{ mol H}}{3.406 \text{ mol O}} = \frac{1.334 \text{ mol H}}{1.000 \text{ mol O}}$$

The same ratio holds for H to C. Using whole numbers, we have $C_3H_4O_3$ for the empirical formula. The empirical formula weight is (3)(12.011) + (4)(1.0079) + (3)(15.9994) = 88.06 g. The molar mass, however, is 176.13 g/mol, so the molecular formula must be twice the empirical formula: $C_6H_8O_6$.

Chapter 4

4.1 (a) N_2; combination (b) O_2; combination
(c) N_2; decomposition

4.2 (a) Decomposition reaction:

$$2\text{ Al(OH)}_3(s) \longrightarrow \text{Al}_2\text{O}_3(s) + 3\text{ H}_2\text{O}(g)$$

(b) Combination reaction:

$$\text{Na}_2\text{O}(s) + \text{H}_2\text{O}(\ell) \longrightarrow 2\text{ NaOH}(aq)$$

(c) Combination reaction: $S_8(s) + 24\ F_2(g) \rightarrow 8\ SF_6(g)$

(d) Exchange reaction:

$$3\text{ NaOH}(aq) + \text{H}_3\text{PO}_4(aq) \longrightarrow \text{Na}_3\text{PO}_4(aq) + 3\text{ H}_2\text{O}(\ell)$$

(e) Displacement:

$$3\text{ C}(s) + \text{Fe}_2\text{O}_3(s) \longrightarrow 3\text{ CO}(g) + 2\text{ Fe}(\ell)$$

4.3 (a) $2\ Cr(s) + 3\ Cl_2 \rightarrow 2\ CrCl_3(s)$
(b) $As_2O_3(s) + 3\ H_2(g) \rightarrow 2\ As(s) + 3\ H_2O(\ell)$

4.4 (a) $C_2H_5OH + 3\ O_2 \rightarrow 2\ CO_2 + 3\ H_2O$
(b) $C_2H_5OH + 2\ O_2 \rightarrow 2\ CO + 3\ H_2O$

4.5 0.433 mol hematite needs 0.433 × 3 = 1.30 mol CO. Molar mass of CO is 28.01, so 1.30 × 28.01 = 36.4 g CO.

4.6 (a) $$0.300 \text{ mol cassiterite} \times \frac{1 \text{ mol Sn}}{1 \text{ mol cassiterite}} \times \frac{118.7 \text{ g Sn}}{1 \text{ mol Sn}} = 35.6 \text{ g Sn}$$

(b) $$35.6 \text{ g Sn} \times \frac{1 \text{ mol Sn}}{118.7 \text{ g Sn}} \times \frac{2 \text{ mol C}}{1 \text{ mol Sn}} \times \frac{12.01 \text{ g C}}{1 \text{ mol C}} = 7.20 \text{ g C}$$

4.7 (a) $$57. \text{ g C} \times \frac{1 \text{ mol C}}{12.01 \text{ g C}} \times \frac{1 \text{ mol O}_2}{2 \text{ mol C}} \times \frac{32.0 \text{ g O}_2}{1 \text{ mol O}_2} = 76. \text{ g O}_2$$

(b) $$57. \text{ g C} \times \frac{1 \text{ mol C}}{12.01 \text{ g C}} \times \frac{2 \text{ mol CO}}{2 \text{ mol C}} \times \frac{28.0 \text{ g CO}}{1 \text{ mol CO}} = 1.3 \times 10^2 \text{ g CO}$$

4.8 $$6.46 \text{ g MgCl}_2 \times \frac{1 \text{ mol MgCl}_2}{95.2104 \text{ g MgCl}_2} = 6.78 \times 10^{-2} \text{ mol MgCl}_2$$

The same number of moles of Mg as $MgCl_2$ are involved, so

$$6.78 \times 10^{-2} \text{ mol Mg} \times \frac{24.3050 \text{ g Mg}}{1 \text{ mol Mg}} = 1.65 \text{ g Mg}$$

$$\frac{1.65 \text{ g Mg}}{1.72\text{-g sample}} \times 100\% = 95.9\% \text{ Mg in sample}$$

4.9 $$0.75 \text{ mol CO}_2 \times \frac{1 \text{ mol (NH}_2)_2\text{CO}}{1 \text{ mol CO}_2} = 0.75 \text{ mol (NH}_2)_2\text{CO}$$

4.10 (a) $CS_2(\ell) + 3\ O_2(g) \rightarrow CO_2(g) + 2\ SO_2(g)$
(b) Determine the quantity of CO_2 produced by each reactant; the limiting reactant produces the lesser quantity.

$$3.5 \text{ g CS}_2 \times \frac{1 \text{ mol CS}_2}{76.0 \text{ g CS}_2} \times \frac{1 \text{ mol CO}_2}{1 \text{ mol CS}_2} \times \frac{44.01 \text{ g CO}_2}{1 \text{ mol CO}_2} = 2.0 \text{ g CO}_2$$

$$17.5 \text{ g O}_2 \times \frac{1 \text{ mol O}_2}{31.998 \text{ g O}_2} \times \frac{1 \text{ mol CO}_2}{3 \text{ mol O}_2} \times \frac{44.01 \text{ g CO}_2}{1 \text{ mol CO}_2} = 8.02 \text{ g CO}_2$$

Therefore, CS_2 is the limiting reagent.
(c) The yield of SO_2 must be calculated using the limiting reagent, CS_2.

$$3.5 \text{ g CS}_2 \times \frac{1 \text{ mol CS}_2}{76.0 \text{ g CS}_2} \times \frac{2 \text{ mol SO}_2}{1 \text{ mol CS}_2} \times \frac{64.1 \text{ g SO}_2}{1 \text{ mol SO}_2} = 5.9 \text{ g SO}_2$$

4.11 Find the number of moles of each reactant.

$$100. \text{ g SiCl}_4 \times \frac{1 \text{ mol SiCl}_4}{169.90 \text{ g SiCl}_4} = 0.589 \text{ mol SiCl}_4$$

$$100. \text{ g Mg} \times \frac{1 \text{ mol Mg}}{24.3050 \text{ g Mg}} = 4.11 \text{ mol Mg}$$

Find the mass of Si produced, based on the mass available of each reactant.

$$0.589 \text{ mol SiCl}_4 \times \frac{1 \text{ mol Si}}{1 \text{ mol SiCl}_4} \times \frac{28.0855 \text{ g Si}}{1 \text{ mol Si}} = 16.5 \text{ g Si}$$

$$4.11 \text{ mol Mg} \times \frac{1 \text{ mol Si}}{2 \text{ mol Mg}} \times \frac{28.0855 \text{ g Si}}{1 \text{ mol Si}} = 57.7 \text{ g Si}$$

Thus, $SiCl_4$ is the limiting reactant, and the mass of Si produced is 16.5 g.

4.12 To make 1.0 kg CH_3OH with 85% yield will require using enough reactant to produce 1000/0.85, or 1180 g CH_3OH.

$$1180 \text{ g CH}_3\text{OH} \times \frac{1 \text{ mol}}{32.042 \text{ g}} = 36.83 \text{ mol CH}_3\text{OH}$$

$$36.83 \text{ mol CH}_3\text{OH} \times \frac{2 \text{ mol H}_2}{1 \text{ mol CH}_3\text{OH}} = 73.65 \text{ mol H}_2$$

$$73.65 \text{ mol H}_2 \times \frac{2.0158 \text{ g H}_2}{1 \text{ mol H}_2} = 149 \text{ g H}_2$$

4.13 Calculate the mass of Cu_2S you should have produced and compare it with the amount actually produced.

$$2.50 \text{ g Cu} \times \frac{1 \text{ mol Cu}}{63.546 \text{ g Cu}} = 3.93 \times 10^{-2} \text{ mol Cu}$$

$$3.93 \times 10^{-2} \text{ mol Cu} \times \frac{8 \text{ mol Cu}_2\text{S}}{16 \text{ mol Cu}} = 1.97 \times 10^{-2} \text{ mol Cu}_2\text{S}$$

$$1.97 \times 10^{-2} \text{ mol Cu}_2\text{S} \times \frac{159.16 \text{ g Cu}_2\text{S}}{1 \text{ mol Cu}_2\text{S}} = 3.14 \text{ g Cu}_2\text{S}$$

$$\frac{2.53 \text{ g}}{3.14 \text{ g}} \times 100\% = 80.6\% \text{ yield was obtained}$$

Your synthesis met the standard.

4.14 (a) $$491 \text{ mg CO}_2 \times \frac{1 \text{ g CO}_2}{10^3 \text{ mg CO}_2} \times \frac{1 \text{ mol CO}_2}{44.01 \text{ g CO}_2} \times \frac{1 \text{ mol C}}{1 \text{ mol CO}_2} = 1.116 \times 10^{-2} \text{ mol C}$$

$$1.116 \times 10^{-2} \text{ mol C} \times \frac{12.01 \text{ g C}}{1 \text{ mol C}} = 0.1340 \text{ g C} = 134.0 \text{ mg C}$$

$$100 \text{ mg H}_2\text{O} \times \frac{1 \text{ g H}_2\text{O}}{10^3 \text{ mg H}_2\text{O}} \times \frac{1 \text{ mol H}_2\text{O}}{18.02 \text{ g H}_2\text{O}} \times \frac{2 \text{ mol H}}{1 \text{ mol H}_2\text{O}} = 1.110 \times 10^{-2} \text{ mol H}$$

$$1.110 \times 10^{-2} \text{ mol H} \times \frac{1.008 \text{ g H}}{1 \text{ mol H}} = 1.119 \times 10^{-2} \text{ g H} = 11.2 \text{ mg H}$$

The mass of oxygen in the compound

$$= \text{total mass} - (\text{mass C} + \text{mass H})$$

$$= 175 \text{ mg} - (134.0 \text{ mg C} + 11.2 \text{ mg H}) = 29.8 \text{ mg O}$$

The moles of oxygen are

$$29.8 \text{ mg O} \times \frac{1 \text{ g O}}{10^3 \text{ mg O}} \times \frac{1 \text{ mol O}}{16.00 \text{ g O}} = 1.862 \times 10^{-3} \text{ mol O}$$

The empirical formula can be derived from the mole ratios of the elements.

$$\frac{1.116 \times 10^{-2} \text{ mol C}}{1.862 \times 10^{-3} \text{ mol O}} = 5.993 \text{ mol C/mol O}$$

$$\frac{1.110 \times 10^{-2} \text{ mol H}}{1.862 \times 10^{-3} \text{ mol O}} = 5.961 \text{ mol H/mol O}$$

$$\frac{1.862 \times 10^{-3} \text{ mol O}}{1.862 \times 10^{-3} \text{ mol O}} = 1.000 \text{ mol O}$$

The empirical formula of phenol is C_6H_6O.

(b) The molar mass is needed to determine the molecular formula.

4.15 $0.569 \text{ g Sn} \times \frac{1 \text{ mol Sn}}{118.7 \text{ g Sn}} = 4.794 \times 10^{-3} \text{ mol Sn}$

$$2.434 \text{ g I}_2 \times \frac{1 \text{ mol I}_2}{253.81 \text{ g I}_2} \times \frac{2 \text{ mol I}}{1 \text{ mol I}_2} = 1.918 \times 10^{-2} \text{ mol I}$$

$$\frac{1.918 \times 10^{-2} \text{ mol I}}{4.794 \times 10^{-3} \text{ mol Sn}} = \frac{4.001 \text{ mol I}}{1.000 \text{ mol Sn}}$$

Therefore, the empirical formula is SnI_4.

Chapter 5

5.1 (a) NaF is soluble.
(b) $Ca(CH_3COO)_2$ is soluble.
(c) $SrCl_2$ is soluble.
(d) MgO is not soluble.
(e) $PbCl_2$ is not soluble.
(f) HgS is not soluble.

5.2 (a) This exchange reaction forms insoluble nickel hydroxide and aqueous sodium chloride.

$$NiCl_2(aq) + 2\ NaOH(aq) \longrightarrow Ni(OH)_2(s) + 2\ NaCl(aq)$$

(b) This is an exchange reaction that forms aqueous potassium bromide and a precipitate of calcium carbonate.

$$K_2CO_3(aq) + CaBr_2(aq) \longrightarrow CaCO_3(s) + 2\ KBr(aq)$$

5.3 (a) $BaCl_2(aq) + Na_2SO_4(aq) \rightarrow BaSO_4(s) + 2\ NaCl(aq)$

$$Ba^{2+}(aq) + SO_4^{2-}(aq) \longrightarrow BaSO_4(s)$$

(b) $(NH_4)_2S(aq) + FeCl_2(aq) \rightarrow FeS(s) + 2\ NH_4Cl(aq)$

$$Fe^{2+}(aq) + S^{2-}(aq) \rightarrow FeS(s)$$

5.4 Any of the strong acids in Table 5.2 would also be strong electrolytes. Any of the weak acids or bases in Table 5.2 would be weak electrolytes. Any organic compound that yields no ions on dissolution would be a nonelectrolyte.

5.5 $H_3PO_4(aq) + 3\ NaOH(aq) \rightarrow Na_3PO_4(aq) + 3\ H_2O(\ell)$

5.6 (a) Sulfuric acid and magnesium hydroxide
(b) Carbonic acid and strontium hydroxide

5.7

$$2\ HCN(aq) + Ca(OH)_2(aq) \longrightarrow Ca(CN)_2(aq) + 2\ H_2O(\ell)$$

$$2\ HCN(aq) + Ca^{2+}(aq) + 2\ OH^-(aq) \longrightarrow Ca^{2+}(aq) + 2\ CN^-(aq) + 2\ H_2O(\ell)$$

$$HCN(aq) + OH^-(aq) \longrightarrow CN^-(aq) + H_2O(\ell)$$

5.8 The oxidation numbers of Fe and Sb are 0 (Rule 1). The oxidation numbers in Sb_2S_3 are +3 for Sb^{3+} and −2 for S^{2-} (Rules 2 and 4). The oxidation numbers in FeS are +2 for Fe^{2+} and −2 for S^{2-} (Rules 2 and 4).

5.9 In the reaction $PbO(s) + CO(g) \rightarrow Pb(s) + CO_2(g)$, Pb^{2+} is reduced to Pb; Pb^{2+} is the oxidizing agent. C^{2+} is oxidized to C^{4+}; C^{2+} is the reducing agent.

5.10 Reactions (a) and (b) will occur. Aluminum is above copper and chromium in Table 5.5; therefore, aluminum will be oxidized and acts as the reducing agent in reactions (a) and (b). In reaction (a), Cu^{2+} is reduced, and Cu^{2+} is the oxidizing agent. Cr^{3+} is the oxidizing agent in reaction (b) and is reduced to Cr metal. Reactions (c) and (d) do not occur because Pt cannot reduce H^+, and Au cannot reduce Ag^+.

5.11 $36.0 \text{ g Na}_2\text{SO}_4 \times \frac{1 \text{ mol Na}_2\text{SO}_4}{142.0 \text{ g Na}_2\text{SO}_4} = 0.254 \text{ mol Na}_2\text{SO}_4$

$$\text{Molarity} = \frac{0.254 \text{ mol}}{0.750 \text{ L}} = 0.339 \text{ molar}$$

5.12 $V(\text{conc}) = \frac{0.150 \text{ molar} \times 0.050 \text{ L}}{0.500 \text{ molar}} = 0.015 \text{ L} = 15 \text{ mL}$

5.13 (a) 1.00 L of 0.125 M Na_2CO_3 contains 0.125 mol Na_2CO_3.

$$0.125 \text{ mol} \times \frac{105.99 \text{ g}}{1 \text{ mol}} = 13.2 \text{ g Na}_2\text{CO}_3$$

Prepare the solution by adding 13.2 g Na_2CO_3 to a volumetric flask, dissolving it and mixing thoroughly, and adding sufficient water until the solution volume is 1.0 L.

(b) Use water to dilute a specific volume of the 0.125 M solution to 100 mL.

$$V(\text{conc}) = \frac{0.0500 \text{ M} \times 0.100 \text{ L}}{0.125 \text{ M}} = 0.040 \text{ L} = 40. \text{ mL of 0.125 M solution}$$

Therefore, put 40. mL of the more concentrated solution into a container and add water until the solution volume equals 100 mL.

(c) 500 mL of 0.215 M $KMnO_4$ contains 1.70 g $KMnO_4$.

$$0.500 \text{ L} \times \frac{0.0215 \text{ mol KMnO}_4}{1 \text{ L}} = 0.01075 \text{ mol KMnO}_4$$

$$0.01075 \text{ mol KMnO}_4 \times \frac{158.0 \text{ g KMnO}_4}{1 \text{ mol KMnO}_4} = 1.70 \text{ g KMnO}_4$$

Put 1.70 g $KMnO_4$ into a container and add water until the solution volume is 500 mL.

(d) Dilute the more concentrated solution by adding sufficient water to 52.3 mL of 0.0215 M $KMnO_4$ until the solution volume is 250 mL.

$$V(\text{conc}) = \frac{0.00450 \text{ M} \times 0.250 \text{ L}}{0.0215 \text{ M}} = 0.0523 \text{ L} = 52.3 \text{ mL}$$

5.14 $1.2 \times 10^{10} \text{ kg NaOH} \times \frac{1 \text{ mol NaOH}}{0.040 \text{ kg NaOH}} \times \frac{2 \text{ mol NaCl}}{2 \text{ mol NaOH}} \times \frac{58.5 \text{ g NaCl}}{1 \text{ mol NaCl}} \times \frac{1 \text{ L brine}}{360 \text{ g NaCl}} = 4.9 \times 10^{10} \text{ L}$

5.15 The net ionic equation is $AgBr(s) + 2\ S_2O_3^{2-}(aq) \rightarrow Ag(S_2O_3)_2^{3-}(aq) + Br^-(aq)$. Moles $Na_2S_2O_3$ = $(0.0200\ M)(0.125\ L) = 0.0025$ mol $Na_2S_2O_3$.
2 mol $Na_2S_2O_3$ dissolves 1 mol AgBr, so 0.0025 mol $Na_2S_2O_3$ dissolves 0.00125 mol AgBr. The molar mass of AgBr is 187.8 g/mol, so 0.00125 mol AgBr × 187.8 g/mol = 0.235 g AgBr or 235 mg AgBr.

5.16 $H_2SO_4(aq) + 2\ NaOH(aq) \rightarrow Na_2SO_4(aq) + 2\ H_2O(\ell)$

$$\text{Moles NaOH} = (0.0413\ L)(0.100\ M) = 0.00413 \text{ mol NaOH}$$

$$\text{Moles } H_2SO_4 = \tfrac{1}{2} \times 0.00413 = 0.002065 \text{ mol } H_2SO_4$$

$$\text{Molarity} = \frac{0.002065 \text{ mol}}{0.020\ L} = 0.103\ M\ H_2SO_4$$

Chapter 6

6.1 (a) $160 \text{ Cal} \times \dfrac{1000 \text{ cal}}{\text{Cal}} \times \dfrac{4.184\ J}{\text{cal}} = 6.7 \times 10^5\ J$

(b) 75 W = 75 J/s;
$75\ J/s \times 3.0\ h \times 60\ min/h \times 60\ s/min = 8.1 \times 10^5\ J$

(c) $16\ kJ \times \dfrac{1\ kcal}{4.184\ kJ} = 3.8\ kcal$

6.2 $\Delta E = -2400\ J = q + w = -1.89\ kJ + w$
$w = -2400\ J + 1.89\ kJ = -2.4\ kJ + 1.89\ kJ = -0.5\ kJ$

6.3 $q = c \times m \times \Delta T = c \times m \times (T_{final} - T_{initial})$

$$T_{final} = T_{initial} + \frac{q}{c \times m} = 5\ °C + \frac{24{,}100\ J}{(0.902\ J\ g^{-1}\ °C^{-1})(250.\ g)}$$
$$= 5\ °C + 106.8\ °C = 112\ °C$$

6.4 Since no work is done, $\Delta E = q$. Assume that the tea has a mass of 250. g and that its specific heat capacity is the same as that of water, namely 4.184 J g^{-1} $°C^{-1}$.
$\Delta E = q$ = (mass) × (specific heat capacity) × (change in temperature)
$\Delta E = q = (250.\ g) \times (4.184\ J\ g^{-1}\ °C^{-1}) \times [(65 - 37)\ °C]$
$= 2.9 \times 10^4\ J = 2.9 \times 10^1\ kJ$

6.5 $q_{water} = -q_{iron}$
$(4.184\ J/g\ °C)(1000.\ g)(32.8\ °C - 20.0\ °C) = -(0.451\ J/g\ °C)(400.\ g)(32.8\ °C - T_i)$
$T_i = (297 + 32.8)\ °C = 330.\ °C$

6.6 $1.00\ g\ K(s) \times \dfrac{1\ mL}{0.86\ g} = 1.16\ mL$;
$1.00\ g\ K(\ell) \times \dfrac{1\ mL}{0.82\ g} = 1.22\ mL$

The change in volume is (1.22 − 1.16) mL = 0.06 mL.

$$w = 0.06\ mL \times \frac{0.10\ J}{1\ mL} = 6 \times 10^{-3}\ J$$
$$\Delta H = \frac{14.6\ cal}{1\ g} \times 1.00\ g \times \frac{4.184\ J}{1\ cal} = 61.1\ J$$
$$\Delta E = \Delta H + w = 61.1\ J + (6 \times 10^{-3}\ J) = 61.1\ J$$

6.7 (a) $10.0\ g\ I_2 \times \dfrac{1\ mol\ I_2}{253.8\ g\ I_2} \times \dfrac{62.4\ kJ}{1\ mol\ I_2} = 2.46\ kJ$

(b) $3.42\ g\ I_2 \times \dfrac{1\ mol\ I_2}{253.8\ g\ I_2} \times \dfrac{-62.4\ kJ}{1\ mol\ I_2} = -0.841\ kJ = -841\ J$

(c) This process is the reverse of the one in part (a), so $\Delta H°$ is negative. Thus the process is exothermic. The quantity of energy transferred is 841 J.

6.8 The equation as written involves CO_2 as a reactant, but the question asks for CO_2 as a product. Therefore we will have to change the sign of $\Delta H°$. In addition, since the question asks about production of 4 mol CO_2, we must multiply $\Delta H°$ for the endothermic decomposition by 4. So for the production of 4 mol CO_2,

$$\Delta H° = (4) \times (662.8\ kJ) = 2651\ kJ$$

6.9 According to the thermochemical expression, the reaction is endothermic, so 285.8 kJ of energy is transferred into the system per mole of $H_2O(\ell)$ decomposed. Thus,

$$12.6\ g\ H_2O \times \frac{1\ mol\ H_2O}{18.02\ g\ H_2O} \times \frac{285.8\ kJ}{1\ mol\ H_2O} = 200.\ kJ$$

6.10 $\Delta T = (25.43 - 20.64)\ °C = 4.79\ °C$

$$\Delta E_{calorimeter} = \frac{877\ J}{°C} \times 4.79\ °C = 4.200 \times 10^3\ J = 4.200\ kJ$$
$$\Delta E_{water} = 832\ g \times \frac{4.184\ J}{g\ °C} \times 4.79\ °C = 16.67\ kJ$$
$$\Delta E_{reaction} = -(q_{calorimeter} + q_{water}) = -(4.200 + 16.67)\ kJ = -20.87\ kJ$$
$$20.87\ kJ \times \frac{1\ kcal}{4.184\ kJ} \times \frac{1\ Cal}{1\ kcal} = 4.99\ Cal$$

Since metabolizing the Fritos chip corresponds to oxidizing it, the result of 4.99 Cal verifies the statement that one chip provides 5 Cal.

6.11 The total volume of the initial solutions is 200. mL, which corresponds to 200. g of solution. The quantities of reactants are 0.10 mol $H^+(aq)$ and 0.050 mol $OH^-(aq)$, so 0.050 mol H_2O is formed.

$$0.050\ mol\ H_2O \times \frac{-58.6\ kJ}{1\ mol\ H_2O} = -2.93\ kJ$$

Since $\Delta H°$ is negative, energy is transferred to the water, and its temperature will rise.

$$\Delta T = \frac{q}{c \times m} = \frac{2.93 \times 10^3\ J}{(4.184\ J\ g^{-1}\ °C^{-1})(200.\ g)} = 3.5\ °C$$

The final temperature will be (20.4 + 3.5) °C = 23.9 °C.

6.12 The balanced chemical equation is

$$2\ Fe_3O_4(s) \rightarrow 6\ FeO(s) + O_2(g)$$

The equations given in the problem can be arranged so that when added they give this balanced equation:

$2 \times [Fe_3O_4(s) \rightarrow 3\ Fe(s) + 2\ O_2(g)]$ $\quad \Delta H° = 2(1118.4\ kJ)$
$6 \times [Fe(s) + \tfrac{1}{2} O_2(g) \rightarrow FeO(s)]$ $\quad \Delta H° = 6(-272.0\ kJ)$

So

$$\Delta H° = 2236.8\ kJ - 1632.0\ kJ = 604.8\ kJ$$

6.13 (a) $\tfrac{1}{2} N_2(g) + \tfrac{3}{2} H_2(g) \rightarrow NH_3(g)$ $\quad \Delta H_f° = -46.11$ kJ/mol
(b) $C(graphite) + \tfrac{1}{2} O_2(g) \rightarrow CO(g)$ $\Delta H_f° = -110.525$ kJ/mol

6.14 For the reaction given,

$$\Delta H° = \{6\ mol\ CO_2(g)\} \times \Delta H_f°\{CO_2(g)\} + \{5\ mol\ H_2O(g)\} \times \Delta H_f°\{H_2O(g)\} - (2\ mol\ \{C_3H_5(NO_3)_3(\ell)\} \times \Delta H_f°\{C_3H_5(NO_3)_3(\ell)\}$$
$$= \{6(-393.509) + 5(-241.818) - 2(-364)\}\ kJ$$
$$= -2.84 \times 10^3\ kJ$$

For 10.0 g nitroglycerin (nitro),

$$q = 10.0\ g \times \frac{1\ mol\ nitro}{227.09\ g} \times \frac{-2.84 \times 10^3\ kJ}{2\ mol\ nitro} = -62.5\ kJ$$

(The 2 mol nitro in the last factor comes from the coefficient of 2 associated with nitroglycerin in the chemical equation.)

6.15 $SO_2(g) + \frac{1}{2}O_2(g) \rightarrow SO_3(g) \quad \Delta H° = ?$

$$\Delta H° = \Delta H_f°\{SO_3(g)\} - [\Delta H_f°\{SO_2(g)\} + \tfrac{1}{2}\Delta H_f°\{O_2(g)\}]$$

$$= -395.72 \text{ kJ} - (-296.830 + 0) \text{ kJ} = -98.89 \text{ kJ}$$

6.16 (a) $227 \text{ g milk} \times \dfrac{5.0 \text{ g carbohydrate}}{100 \text{ g milk}} \times \dfrac{17 \text{ kJ}}{1 \text{ g carbohydrate}} = 193 \text{ kJ}$

$$227 \text{ g milk} \times \frac{4.0 \text{ g fat}}{100 \text{ g milk}} \times \frac{38 \text{ kJ}}{1 \text{ g fat}} = 345 \text{ kJ}$$

$$227 \text{ g milk} \times \frac{3.3 \text{ g protein}}{100 \text{ g milk}} \times \frac{17 \text{ kJ}}{1 \text{ g protein}} = 127 \text{ kJ}$$

Total caloric intake = (193 + 345 + 127) kJ = 665 kJ

Reduce by 10% to digest, absorb, and metabolize, giving 665 kJ − 66.5 kJ = 598 kJ.

(b) Walking requires $2.5 \times \text{BMR} = 2.5 \times \dfrac{1750 \text{ Cal}}{1 \text{ day}} \times \dfrac{4.184 \text{ kJ}}{1 \text{ Cal}} \times \dfrac{1 \text{ day}}{24 \text{ h}} \times \dfrac{1 \text{ h}}{60 \text{ min}} = 12.7 \text{ kJ/min}$

$$598 \text{ kJ} \times \frac{1 \text{ min}}{12.7 \text{ kJ}} = 47 \text{ min}$$

Chapter 7

7.1 $\nu = \dfrac{c}{\lambda} = \dfrac{2.998 \times 10^8 \text{ m/s}}{4.05 \times 10^{-7} \text{ m}} = 7.40 \times 10^{14} \text{ s}^{-1}$

or 7.40×10^{14} Hz

7.2 (a) One photon of ultraviolet radiation has more energy because ν is larger in the UV spectral region than in the microwave region.

(b) One photon of blue light has more energy because the blue portion of the visible spectrum has a higher frequency than the green portion of the visible spectrum.

7.3 Any from $n_{hi} > 8$ to $n_{lo} = 2$

7.4 (a) $\nu = \dfrac{2.179 \times 10^{-18} \text{ J}}{h}\left(\dfrac{1}{n_i^2} - \dfrac{1}{n_f^2}\right)$

$$= \left(\frac{2.179 \times 10^{-18} \text{ J}}{6.626 \times 10^{-34} \text{ J}\cdot\text{s}}\right)\left(\frac{1}{6^2} - \frac{1}{4^2}\right)$$

$$= (3.289 \times 10^{15} \text{ s}^{-1})\left(\frac{1}{36} - \frac{1}{16}\right)$$

$$= (3.289 \times 10^{15} \text{ s}^{-1})(-3.47 \times 10^{-2})$$

$$= -1.14 \times 10^{14} \text{ s}^{-1}$$

The negative sign indicates that energy is emitted.

$$\lambda = 2.63 \times 10^{-6} \text{ m}$$

(b) Longer than that of the $n = 7$ to $n = 4$ transition.

7.5 $\lambda = \dfrac{h}{mv} = \dfrac{6.626 \times 10^{-34} \text{ J}\cdot\text{s}}{1.67 \times 10^{-27} \text{ kg} \times 2.998 \times 10^7 \text{ m/s}}$

$$= 1.32 \times 10^{-14} \text{ m}$$

7.6 (a) 6d (b) 5 (c) 2, 1, 0, −1, −2

7.7 (a) [Ne]$3s^2 3p^2$ (b) [Ne] 3s [↑↓] 3p [↑] [↑] [ ]

7.8 The electron configurations for :Se: and :Te: are [Ar]$4s^2 3d^{10} 4p^4$ and [Kr]$5s^2 4d^{10} 5p^4$, respectively. Elements in the same main group have similar electron configurations.

7.9 (a) P^{3-} (b) Ca^{2+}

7.10 The ground state Cu atom has a configuration [Ar]$4s^1 3d^{10}$. When it loses one electron, it becomes the Cu^+ ion with configuration [Ar]$3d^{10}$. There is an added stability for the completely filled set of $3d$ orbitals.

7.11 B < Mg < Na < K

7.12 (a) Cs^+ (b) La^{3+}

7.13 F > N > P > Na

Chapter 8

8.1 (a) F—N̈—F with F bonded below N (b) H—N̈—N̈—H with H bonded below each N

(c) $[$:Ö—Cl—Ö: with :Ö: above and :Ö: below Cl$]^-$

8.2 Cl—C(Cl)(Cl)—C(H)(H)—C(H)(H)—H Cl—C(Cl)(H)—C(Cl)(H)—C(H)(H)—H Cl—C(H)(H)—C(Cl)(Cl)—C(H)(H)—H

8.3 (a) $[:N{\equiv}O:]^+$ (b) :C̈l—C(=O:)—C̈l:

8.4 Only (c) can have geometric isomers. Molecules in (a) and (b) each have the same two groups on one of the double-bonded carbons.

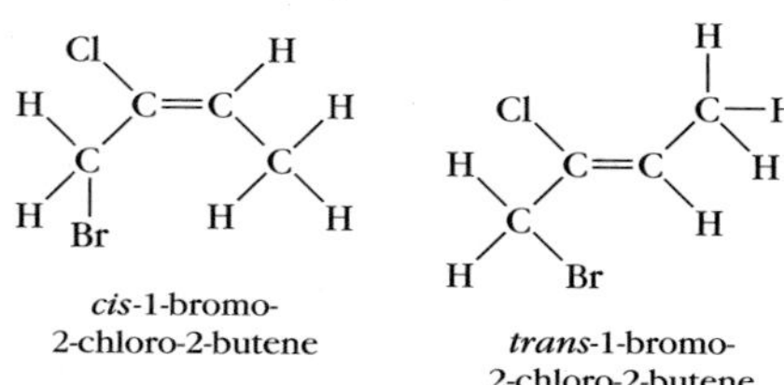

cis-1-bromo-2-chloro-2-butene *trans*-1-bromo-2-chloro-2-butene

8.5 (a) Si is a larger atom than S. (b) Br is a larger atom than Cl.

(c) The greater electron density in the triple bond brings the N≡O atoms closer together than the smaller electron density in the N=O double bond does.

8.6

ΔH = [(4 mol C—H)(416 kJ/mol) + (2 mol O=O)(498 kJ/mol)]
− [(4 mol O—H)(467 kJ/mol) + (2 mol C=O)(803 kJ/mol)]
= (2660 kJ) − (3474 kJ) = −814 kJ

8.7 (a) B—Cl is more polar; $\overset{\delta+}{B}$—$\overset{\delta-}{Cl}$; O—H is more polar; $\overset{\delta-}{O}$—$\overset{\delta+}{H}$.

8.8 The other Lewis structure is :O≡N—N̈:

	O	N	N
Valence electrons	6	5	5
Lone pair electrons	2	0	6
$\frac{1}{2}$ shared electrons	3	4	1
Formal charge	+1	+1	−2

8.9 The N-to-O bond length in NO_2^- is 124 pm. From Table 8.1, N—O is 136 pm; N=O is 115 pm. Thus, the nature of the bond in NO_2^- is between that of a N—O single bond and a N=O double bond.

8.10 (a) :F̈—Be—F̈: (b) :Ö—C̈l—Ö:

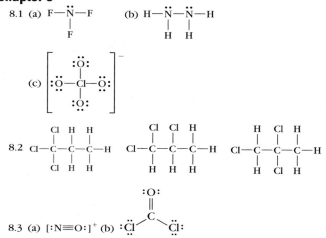

(c) :C̈l—P(:C̈l:)(:C̈l:)(:C̈l:)(:C̈l:) (d) $[H—B—H]^+$

(e)

BeF_2—not an octet around the central Be atom; ClO_2—an odd number of electrons around Cl; PCl_5—more than four electron pairs around the central phosphorus atom; BH_2^+—only two electron pairs around the central B atom; IF_7—iodine has seven shared electron pairs.

8.11 Ten valence electrons. The bond order is $(8 - 2)/2 = 3$. No unpaired electrons. See Table 8.4.

8.12 Bond order $= (8 - 5)/2 = 1.5$. There is one unpaired electron.

Chapter 9

9.1 There are two Be—F bonds in BeF_2 and no lone pairs on beryllium. Therefore, the electron-pair and the molecular geometry are the same, linear, with 180° Be—F bond angles.

9.2

Central Atom (underlined)	Bond Pairs	Lone Pairs	Electron-Pair Geometry	Molecular Shape
$\underline{Br}O_3^-$	3	1	Tetrahedral	Triangular pyramid
$\underline{Se}F_2$	2	2	Tetrahedral	Angular
$\underline{N}O_2^-$	3	1	Triangular planar	Angular

9.3 (a) ClF_2^-: triangular bipyramidal electron-pair geometry and linear molecular geometry

(b) XeO_3: tetrahedral electron-pair geometry and triangular pyramidal molecular geometry

9.4 (a) $BeCl_2$: *sp* hybridization (two bonding electron pairs around the central Be atom), linear geometry

180°

:C̈l— Be— C̈l:

(b) The central N atom has four bonding pairs and no lone pairs in sp^3 hybridized orbitals on N giving tetrahedral electron-pair and molecular geometries.

$[NH_4]^+$ (H—N—H with H above and below; 109.5°)

(c) Each carbon is sp^3 hybridized with no lone pairs, so the electron-pair and molecular geometries are both tetrahedral. The sp^3 hybridized oxygen atom has two bonding pairs, each in single bonds to a carbon atom plus two lone pairs giving it a tetrahedral electron-pair geometry and an angular molecular geometry.

H_3C—Ö—CH_3 (109.5°; <109.5°)

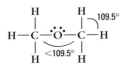

(d) The central boron atom has four single bonds, one to each fluorine atom, and no lone pairs resulting in tetrahedral electron-pair and molecular geometries. The boron is sp^3 hybridized to accommodate the four bonding pairs.

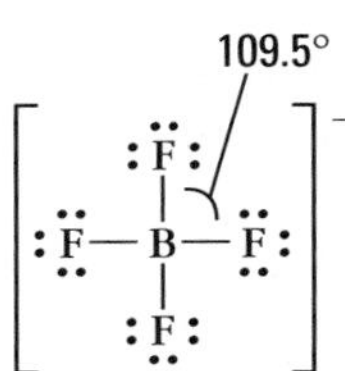

9.5 (a) The central P atom has six bonding pairs and no lone pairs, and is sp^3d^2 hybridized.

(b) The central I atom has three bonding pairs and two lone pairs; these five electron pairs are in sp^3d hybridized orbitals on I.

(c) In ICl_4^-, the central I has four bonding pairs, two lone pairs; these six pairs are in sp^3d^2 orbitals.

9.6 (a) In HCN, the *sp* hybridized carbon atom is sigma bonded to H and to N, as well as having two pi bonds to N. The sigma and two pi bonds form the C≡N triple bond. The nitrogen is *sp* hybridized with a sigma and two pi bonds to carbon; a lone pair is in the nonbonding *sp* hybrid orbital on N.

H—C≡N: (σ, π, π; *sp* orbital)

(b) The double-bonded carbon and nitrogen are both sp^2 hybridized. The sp^2 hybrid orbitals on C form sigma bonds to H and to N; the unhybridized *p* orbital on C forms a pi bond with the unhybridized *p* orbital on N. The sp^2 hybrid orbitals on N form sigma bonds to carbon and to H; the N lone pair is in the nonbonding sp^2 hybrid orbital.

H_2C=N—H (σ, π; sp^2 orbital)

9.7 (a) $BFCl_2$ is a triangular planar molecule with polar B—F and B—Cl bonds. The molecule is polar because the B—F bond is more polar than the B—Cl bonds, resulting in a net dipole.

(b) NH_2Cl is a triangular pyramidal molecule with polar N—H bonds. (N—Cl is a nonpolar bond; N and Cl have the same electronegativity.) It is a polar molecule because the N—H dipoles do not cancel and produce a net dipole.

(c) SCl_2 is an angular polar molecule. The polar S—Cl bond dipoles do not cancel each other because they are not symmetrically arranged due to the two lone pairs on S.

9.8 (a) London forces between Kr atoms must be overcome for krypton to melt.

(b) The C—H covalent bonds in propane must be broken to form C and H atoms; the H atoms covalently bond to form H_2.

9.9 (a) London forces occur between N_2 molecules.

(b) CO_2 is nonpolar, and London forces occur between it and polar water molecules.

(c) London forces occur between the two molecules, but the principal intermolecular forces are the hydrogen bonds between the H on NH_3 with the lone pairs on the OH oxygen, and the hydrogen bonds between the H on the oxygen in CH_3OH and lone pair on nitrogen in NH_3.

Chapter 10

10.1 (a) Pressure in atm =

$$29.5 \text{ in Hg} \times \frac{1 \text{ atm}}{76.0 \text{ cm Hg}} \times \frac{2.54 \text{ cm}}{1 \text{ in}} = 0.986 \text{ atm}$$

(b) Pressure in mm Hg = 29.5 in Hg $\times \dfrac{25.4\text{ mm}}{1\text{ in}}$ = 749 mm Hg

(c) Pressure in bar =

$$29.5\text{ in Hg} \times \frac{1.013\text{ bar}}{760\text{ mm Hg}} \times \frac{25.4\text{ mm}}{1\text{ in}} = 0.999\text{ bar}$$

(d) Pressure in kPa =

$$29.5\text{ in Hg} \times \frac{101.3\text{ kPa}}{760\text{ mm Hg}} \times \frac{25.4\text{ mm}}{1\text{ in}} = 99.9\text{ kPa}$$

10.2 The temperature remains constant, so the average energy of the gas molecules remains constant. If the volume is decreased, then the gas molecules must hit the walls more frequently, and the pressure is increased.

10.3 Volume of NO gas = 1.0 L $O_2 \times \dfrac{2\text{ L NO}}{1\text{ L }O_2}$ = 2.0 L NO

10.4 $V = \dfrac{nRT}{P}$

$$= \frac{(2.64\text{ mol})(0.0821\text{ L atm mol}^{-1}\text{ K}^{-1})(304\text{ K})}{0.640\text{ atm}} = 103\text{ L}$$

10.5 $V_2 = \dfrac{P_1V_1}{P_2} = \dfrac{(1.00\text{ atm})(400.\text{ mL})}{0.750\text{ atm}} = 533\text{ mL}$

10.6 (a) $V_2 = \dfrac{P_1V_1T_2}{P_2T_1}$

$$= \frac{(710\text{ mm Hg})(21\text{ mL})(299.6\text{ K})}{(740\text{ mm Hg})(295.4\text{ K})} = 20.\text{ mL}$$

(b) $V_2 = \dfrac{(21\text{ mL})(299.6\text{ K})}{(295.4\text{ K})} = 21\text{ mL}$

10.7 $\dfrac{10.0\text{ g }NH_4NO_3}{80.043\text{ g/mol}} = 0.1249\text{ mol }NH_4NO_3$

$$0.1249\text{ mol }NH_4NO_3 \times \frac{7\text{ mol product gases}}{2\text{ mol }NH_4NO_3}$$

= 0.437 mol produced

$$V = \frac{(0.437\text{ mol})(0.0821\text{ L atm mol}^{-1}\text{ K}^{-1})(298\text{ K})}{1\text{ atm}} = 10.7\text{ L}$$

10.8 $\dfrac{1.0\text{ g LiOH}}{23.94\text{ g/mol}} = 0.0418\text{ mol LiOH}$

$$0.0418\text{ mol LiOH} \times \frac{1\text{ mol }CO_2}{2\text{ mol LiOH}} = 0.0209\text{ mol }CO_2$$

$$V = \frac{(0.0209\text{ mol})(0.0821\text{ L atm mol}^{-1}\text{ K}^{-1})(295\text{ K})}{1\text{ atm}}$$

$= 0.51\text{ L }CO_2$

10.9 $V = \frac{4}{3}\pi r^3 = \frac{4}{3}\pi(10.\text{ cm})^3 = 4190\text{ cm}^3 = 4.19\text{ L}$

Amount of CO_2 gas, $n = \dfrac{PV}{RT}$

$$= \frac{(2.00\text{ atm})(4.19\text{ L})}{(0.0821\text{ L atm mol}^{-1}\text{ K}^{-1})(293\text{ K})} = 0.348\text{ mol }CO_2$$

Mass of $NaHCO_3$ = 0.348 mol CO_2

$$\times \frac{1\text{ mol }NaHCO_3}{1\text{ mol }CO_2} \times \frac{84.00\text{ g }NaHCO_3}{1\text{ mol }NaHCO_3} = 29\text{ g }NaHCO_3$$

10.10 Amount of gas, $n = \dfrac{PV}{RT}$

$$= \frac{(0.850\text{ atm})(1.00\text{ L})}{(0.0821\text{ L atm mol}^{-1}\text{ K}^{-1})(293\text{ K})} = 0.0353\text{ mol}$$

Molar mass $= \dfrac{1.13\text{ g}}{0.0353\text{ mol}} = 32.0\text{ g/mol}$

The gas is probably oxygen.

10.11 Amount of N_2 = 7.0 g $N_2 \times \dfrac{1\text{ mol }N_2}{28.10\text{ g }N_2}$ = 0.25 mol N_2

Amount of H_2 = 6.0 g $H_2 \times \dfrac{1\text{ mol }H_2}{2.02\text{ g }H_2}$ = 3.0 mol H_2

Total number of moles = 3.0 + 0.25 = 3.25 mol

$$X_{N_2} = \frac{0.25\text{ mol}}{3.25\text{ mol}} = 0.077 \qquad X_{H_2} = \frac{3.0\text{ mol}}{3.25\text{ mol}} = 0.92$$

$$P_{N_2} = \frac{(0.25\text{ mol})(0.0821\text{ L atm mol}^{-1}\text{ K}^{-1})(773\text{ K})}{5.0\text{ L}} = 3.2\text{ atm}$$

$$P_{H_2} = \frac{(3.0\text{ mol})(0.0821\text{ L atm mol}^{-1}\text{ K}^{-1})(773\text{ K})}{5.0\text{ L}} = 38\text{ atm}$$

10.12 For NO:

$$n = \frac{(1.0\text{ atm})(4.0\text{ L})}{(0.0821\text{ L atm mol}^{-1}\text{ K}^{-1})(298\text{ K})} = 0.163\text{ mol NO}$$

For O_2:

$$n = \frac{(0.40\text{ atm})(2.0\text{ L})}{(0.0821\text{ L atm mol}^{-1}\text{ K}^{-1})(298\text{ K})} = 0.0327\text{ mol }O_2$$

All the O_2 is used.

$$0.0327\text{ mol }O_2 \times \frac{2\text{ mol NO}}{1\text{ mol }O_2} = 0.0654\text{ mol NO used}$$

0.163 mol − 0.0654 mol = 0.0976 mol NO remains

$$0.0327\text{ mol }O_2 \times \frac{2\text{ mol }NO_2}{1\text{ mol }O_2} = 0.0654\text{ mol }NO_2\text{ formed}$$

n_{total} = 0.0976 + 0.0654 = 0.163 mol of gas

$$P_{total} = \frac{nRT}{V} = \frac{(0.163\text{ mol})(0.0821\text{ L atm mol}^{-1}\text{ K}^{-1})(298\text{ K})}{6.0\text{ L}}$$

= 0.665 atm

10.13 $P_{HCl} = P_{total} - P_{water}$ = 740 mm Hg − 21 mm Hg = 719 mm Hg

$$n = \frac{PV}{RT} = \frac{(719/760\text{ atm})(0.260\text{ L})}{(0.0821\text{ L atm mol}^{-1}\text{ K}^{-1})(296\text{ K})}$$

$= 0.0101\text{ mol }H_2$

0.0101 mol × 2.0158 g/mol = 0.0204 g = 20.4 mg H_2

10.14 $P = \dfrac{nRT}{V} = \dfrac{(5.00\text{ mol})(0.0821\text{ L atm mol}^{-1}\text{ K}^{-1})(273\text{ K})}{20.0\text{ L}}$

= 5.60 atm

$$P = \left(\frac{nRT}{V - nb}\right) - \left(\frac{n^2a}{V^2}\right)$$

$$P = \left(\frac{(5.00\text{ mol})(0.0821\text{ L atm mol}^{-1}\text{ K}^{-1})(273\text{ K})}{(20.0\text{ L}) - (5.00\text{ mol})(0.0428\text{ L/mol})}\right) - \left(\frac{(5.00\text{ mol})^2(2.25\text{ L}^2\text{ atm/mol}^2)}{(20.0\text{ L})^2}\right) = 5.523\text{ atm}$$

$$\text{Percentage difference in pressure} = \frac{5.60\text{ atm} - 5.52\text{ atm}}{5.60\text{ atm}} \times 100\%$$

= 1.43%

Chapter 11

11.1 $\ln\left(\dfrac{P_2}{143.0\text{ torr}}\right) = \dfrac{3.21 \times 10^4\text{ J/mol}}{8.31\text{ J mol}^{-1}\text{ K}^{-1}}\left[\dfrac{1}{303.15\text{ K}} - \dfrac{1}{333.15\text{ K}}\right]$

$$\ln\left(\frac{P_2}{143.0\text{ torr}}\right) = 1.1473$$

$$\ln P_2 = -1.1473 - \ln(143.0)$$

$$P_2 = 450.\text{ torr}$$

11.2 From Table 11.2, $\Delta H_{vap}(Br_2)$ is 29.54 kJ/mol at its normal boiling point.

So 29.54 kJ/mol × 0.500 mol = 14.77 kJ.

11.3 Heat = $2.5 \times 10^{10}\text{ kg }H_2O \times \dfrac{10^3\text{ g}}{1\text{ kg}} \times \dfrac{1\text{ mol }H_2O}{18.02\text{ g }H_2O}$

$$\times \frac{-44.0\text{ kJ}}{\text{mol}} = -6.10 \times 10^{13}\text{ kJ}$$

This process is exothermic as water vapor condenses, forming rain.

11.4 (21.95)/2 = 10.98 kJ

11.5 The gas phase.

11.6 (a) Solid decane is a molecular solid.

(b) Solid $MgCl_2$ is composed of Mg^{2+} and Cl^- ions and is an ionic solid.

11.7 There are two atoms per bcc unit cell. The diagonal of the bcc unit cell is four times the radius of the atoms in the unit cell, so, solving for the edge,

$$\text{Edge} = \frac{4 \times 144\text{ pm}}{\sqrt{3}} = 333\text{ pm}$$

$$\text{Density} = \frac{\text{mass}}{\text{volume}} = \frac{(2\text{ Au atoms})(196.97\text{ g Au}/6.022 \times 10^{23}\text{ Au atoms})}{[(333\text{ pm})(1\text{ m}/10^{12}\text{ pm})(10^2\text{ cm/m})]^3} = 17.7\text{ g/cm}^3$$

11.8 The edge of the KCl unit cell would be 2 × 152 pm + 2 × 167 pm = 638 pm.
The unit cell of KCl is larger than that of NaCl.

$$\text{Volume of the unit cell} = (638\text{ pm})^3 = 2.60 \times 10^{18}\text{ pm}^3 \times \left(\frac{10^{-10}\text{ cm}}{\text{pm}}\right)^3 = 2.60 \times 10^{-22}\text{ cm}^3$$

$$D = \frac{m}{V} = \frac{(4\text{ formula units KCl}) \times 74.55\text{ g KCl}/6.022 \times 10^{23}\text{ formula units}}{2.60 \times 10^{-22}\text{ cm}^3}$$

$$D = 1.91\text{ g/cm}^3$$

11.9 Energy transfer required

$$= 1.45\text{ g Al} \times \frac{1\text{ mol Al}}{26.98\text{ g Al}} \times \frac{10.7\text{ kJ}}{\text{mol}} = 0.575\text{ kJ}$$

Chapter 12

12.1 The balanced combustion reaction for methanol vapor is

$$CH_3OH(g) + \tfrac{3}{2}O_2(g) \longrightarrow CO_2(g) + 2\,H_2O(g)$$

Using Hess's law, we see that the heat of combustion of methanol vapor is

$$\begin{aligned}\Delta H_{comb} &= [\Delta H_f^\circ CO_2(g)] + 2[\Delta H_f^\circ H_2O(g)] \\ &\quad -1[\Delta H_f^\circ CH_3OH(g)] \\ &= -393.509\text{ kJ/mol} + 2(-241.818\text{ kJ/mol}) \\ &\quad -(-200.66\text{ kJ/mol}) \\ &= -676.49\text{ kJ/mol}\end{aligned}$$

For methanol,

$$-676.49\text{ kJ/mol}\left(\frac{1\text{ mol}}{32.04\text{ g}}\right)\left(\frac{0.791\text{ g}}{1\text{ mL}}\right)\left(\frac{1000\text{ mL}}{\text{L}}\right) = -1.67 \times 10^4\text{ kJ/L}$$

Methanol yields less energy per liter than ethanol.

12.2 (a) $C_8H_{18}(\ell) + \frac{25}{2}O_2(g) \rightarrow 8\,CO_2(g) + 9\,H_2O(g)$

$$\Delta H^\circ = [8(-393.509) + 9(-241.818) - (-249.952)]\text{ kJ} = -5074.48\text{ kJ}$$

$$\frac{5074.48\text{ kJ}}{1\text{ mol }C_8H_{18}} \times \frac{1\text{ mol}}{114.23\text{ g}} = 44.423\text{ kJ/g }C_8H_{18}$$

$$44.423\text{ kJ/g }C_2H_{18} \times 0.699\text{ g/mL} \times \frac{10^3\text{ mL}}{\text{L}} = 3.11 \times 10^4\text{ kJ/L }C_8H_{18}$$

(b) $N_2H_4(\ell) + O_2(g) \rightarrow N_2(g) + 2\,H_2O(g)$

$$\Delta H^\circ = [2(-241.818) - 50.63]\text{ kJ} = -534.26\text{ kJ}$$

$$\frac{534.26\text{ kJ}}{1\text{ mol }N_2H_4} \times \frac{1\text{ mol}}{32.045\text{ g}} = 16.672\text{ kJ/g }N_2H_4$$

$$16.672\text{ kJ/g }N_2H_4 \times 1004\text{ g/L} = 1.67 \times 10^4\text{ kJ/L }N_2H_4$$

(c) $C_6H_{12}O_6(g) + 6\,O_2(g) \rightarrow 6\,CO_2(g) + 6\,H_2O(g)$

$$\Delta H^\circ = [6(-393.509) + 6(-241.818) - (-1274.4)]\text{ kJ} = -2537.56\text{ kJ}$$

$$\frac{2537.56\text{ kJ}}{\text{mol }C_6H_{12}O_6} \times \frac{1\text{ mol}}{180.158\text{ g}} = 14.085\text{ kJ/g}$$

$$14.085\text{ kJ/g} \times 1560\text{ g/L} = 2.20 \times 10^4\text{ kJ/L}$$

12.3 $\text{Energy} = 4.2 \times 10^9\text{ t coal} \times \frac{26.4 \times 10^9\text{ J}}{1\text{ t coal}} = 1.1 \times 10^{20}\text{ J}$

$$\text{ft}^3\text{ natural gas} = 1.1 \times 10^{20}\text{ J} \times \frac{1\text{ ft}^3\text{ natural gas}}{1.055 \times 10^6\text{ J}} = 1.0 \times 10^{14}\text{ ft}^3\text{ natural gas}$$

12.4 The chiral centers are identified by an asterisk in the structural formula. Each of those carbon atoms has four different groups or atoms attached to it.

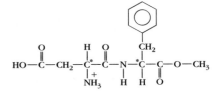

```
                                  (C6H5)
                                    |
      O         H  O               CH2
      ||        |  ||               |
HO—C—CH2—C*—C—N—C*—C—O—CH3
                |         |   |   ||
               NH3+       H   H   O
```

12.5 $CO(g) + 2\,H_2(g) \rightarrow CH_3OH(g)$

$$\begin{aligned}\Delta H^\circ &= [(1\text{ mol C}{\equiv}O)(1073\text{ kJ/mol}) \\ &\quad + (2\text{ mol H—H})(436\text{ kJ/mol})] \\ &\quad - [(3\text{ mol C—H})(416\text{ kJ/mol})] \\ &\quad + (1\text{ mol C—O})(336\text{ kJ/mol}) + (1\text{ mol O—H})(467\text{ kJ/mol}) \\ &= (1945\text{ kJ}) - (2051\text{ kJ}) = -106\text{ kJ}\end{aligned}$$

12.6 (a) The first oxidation product of $CH_3CH_2CH_2OH$ is the aldehyde.

```
        O
        ||
CH3CH2CH
```

The second oxidation product of $CH_3CH_2CH_2OH$ is the acid.

```
       O
       ||
CH3CH2C—OH
```

(b) The oxidation product of this secondary alcohol is the ketone.

```
       O
       ||
CH3—C—CH2CH3
```

12.7 In this case, stearic acid would be on carbon 2 of glycerol where it would be flanked by oleic acids at carbons 1 and 3 of glycerol. See Problem-Solving Example 12.7 (p. 572) for the structural formulas of stearic and oleic acids.

12.8
```
 H   H     H   H   H   H
 |   |     |   |   |   |
—C—C———C—C—C—C—
 |   |     |   |   |   |
 O   H     O   H   H   O—C—CH3
 |         |           ||
 C—CH3     C—CH3       O
 ||        ||
 O         O
```

12.9 (a)
```
  ( H  H  )
  ( |  |  )
 —(-C—C-)—
  ( |  |  )
  ( H  CH3)n
```

(b)
```
  ( H  H        )
  ( |  |        )
 —(-C—C-        )—
  ( |  |        )
  ( H  O—C=O    )
  (      |      )
  (     CH3     )n
```

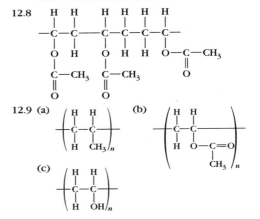

(c)
```
  ( H  H  )
  ( |  |  )
 —(-C—C-)—
  ( |  |  )
  ( H  OH )n
```

12.10 MM vinylchloride, C_2H_3Cl, is 62.5 g/mol.

$$\text{Degree of polymerization} = 1.50 \times 10^5 \text{ g} \times \frac{1 \text{ monomer unit}}{62.5 \text{ g}} = 2.40 \times 10^3 \text{ monomer units}$$

12.11 $HO{-}CH_2{-}C({=}O){-}OH$

12.12 Twelve moles of water

12.13 $H_2N{-}CH(CH_2SH){-}C({=}O){-}NH{-}CH(CH_2C_6H_5){-}C({=}O){-}NH{-}CH(CH_2OH){-}C({=}O){-}NH{-}CH(CH_3){-}C({=}O){-}OH$

Chapter 13

13.1 (a) $\text{Rate} = \dfrac{-\Delta[\text{Cv}^+]}{\Delta t} = 1.27 \times 10^{-6} \text{ mol L}^{-1} \text{ s}^{-1}$

$$\Delta t = \frac{-\Delta[\text{Cv}^+]}{1.27 \times 10^{-6} \text{ mol L}^{-1} \text{ s}^{-1}} = \frac{(4.30 \times 10^{-5} - 3.96 \times 10^{-5}) \text{ mol/L}}{1.27 \times 10^{-6} \text{ mol L}^{-1} \text{ s}^{-1}} = 2.7 \text{ s}$$

(b) No. The rate of reaction depends on the concentration of Cv^+ and, therefore, becomes slower as the reaction progresses. Therefore, the method used in part (a) works only over a small range of concentrations.

13.2 (a) The balanced chemical equation shows that for every mole of O_2 consumed two moles of N_2O_5 are produced. Therefore, the rate of formation of N_2O_5 is twice the rate of disappearance of O_2.
(b) Four moles of NO_2 are consumed for every mole of O_2 consumed. Therefore, if O_2 is consumed at the rate of 0.0037 mol L^{-1} s^{-1} the rate of disappearance of NO_2 is four times this rate.

$$4 \times (0.0037 \text{ mol L}^{-1} \text{ s}^{-1}) = 0.015 \text{ mol L}^{-1} \text{ s}^{-1}$$

13.3 (a) The effect of $[OH^-]$ on the rate of reaction cannot be determined, because the $[OH^-]$ is the same in all three experiments.

(b) $\text{Rate} = k[\text{Cv}^+]$

(c) $k_1 = \dfrac{\text{rate}}{[\text{Cv}^+]} = \dfrac{1.3 \times 10^{-6} \text{ mol L}^{-1} \text{ s}^{-1}}{4.3 \times 10^{-5} \text{ mol/L}} = 3.0 \times 10^{-2} \text{ s}^{-1}$

$k_2 = 3.0 \times 10^{-2} \text{ s}^{-1}$

$k_3 = 3.0 \times 10^{-2} \text{ s}^{-1}$

$k = \dfrac{k_1 + k_2 + k_3}{3} = 3.0 \times 10^{-2} \text{ s}^{-1}$

(d) $\text{Rate} = k[\text{Cv}^+] = (3.0 \times 10^{-2} \text{ s}^{-1})(0.00045 \text{ mol/L}) = 1.4 \times 10^{-5} \text{ mol L}^{-1} \text{ s}^{-1}$

(e) $\text{Rate} = (3.0 \times 10^{-2} \text{ s}^{-1})(0.5 \times 0.00045 \text{ mol/L}) = 6.8 \times 10^{-6} \text{ mol L}^{-1} \text{ s}^{-1}$

13.4 (a) The order of the reaction with respect to each chemical is the exponent associated with the concentration of that chemical. So the reaction is second-order with respect to NO and it is first-order with respect to Cl_2.
(b) Tripling the concentration of NO will make the reaction go $3^2 = 9$ times faster but decreasing the concentration of Cl_2 by a factor of 8 will make the reaction go at 1/8 the initial rate. If these two changes are made simultaneously the relevant factor will be 9/8 = 1.13, so the reaction will occur 13% faster than it did under the initial conditions.

13.5 Make three plots of the data.

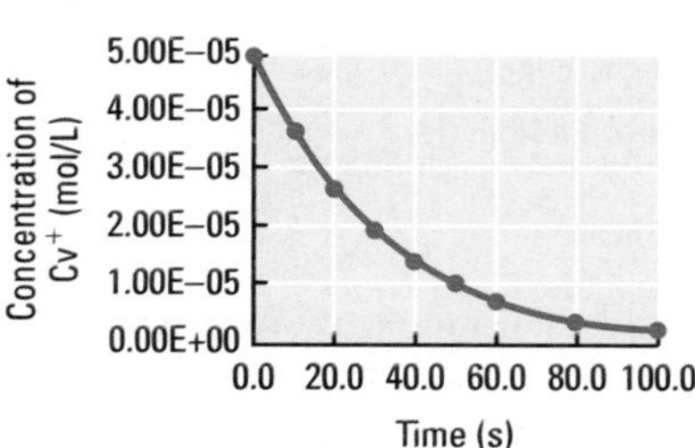

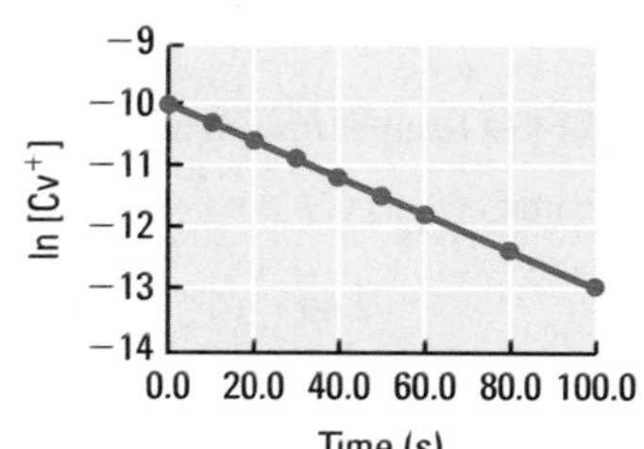

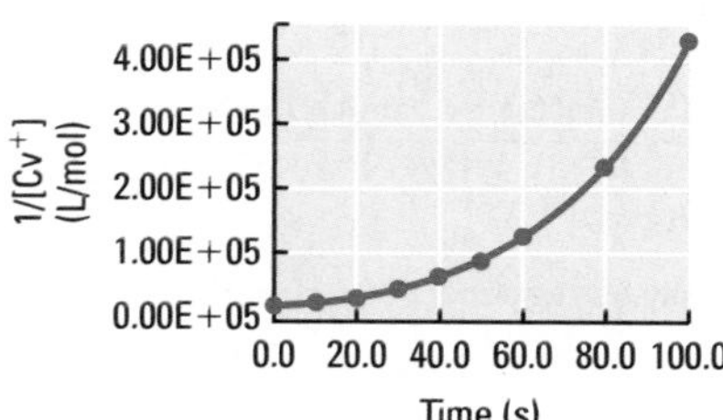

The first-order plot is a straight line and the other plots are curved, so the reaction is first-order. The slope of the first-order plot is -0.0307 s^{-1}, so $k = -\text{slope} = 0.031 \text{ s}^{-1}$.

13.6 Use the integrated first-order rate law from Table 13.2.

$$\ln[\text{A}]_t = -kt + \ln[\text{A}]_0$$

$$\ln\frac{[\text{A}]_t}{[\text{A}]_0} = -kt$$

$$t = -\frac{1}{k}\ln\frac{[\text{A}]_t}{[\text{A}]_0} = -\left(\frac{1}{3.43 \times 10^{-2} \text{ d}^{-1}}\right)\ln\left(\frac{0.1}{1.0}\right)$$

$$= -(29.15 \text{ d})(-2.303) = 67.1 \text{ d}$$

13.7 In Figure 13.3 the $[Cv^+]$ falls from 5.00×10^{-5} M to 2.5×10^{-5} M in 23 s. The $[Cv^+]$ falls from 2.5×10^{-5} M to 1.25×10^{-5} M between 23 s and 46 s. The two times are equal, so $t_{1/2} = 23$ s.

$$k = \frac{0.693}{t_{1/2}} = \frac{0.693}{23 \text{ s}} = 3.0 \times 10^{-2} \text{ s}^{-1}$$

13.8

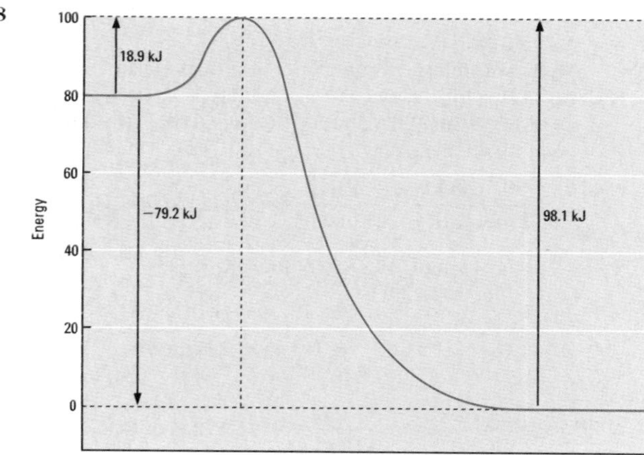

$$Cl_2(g) + 2\,NO(g) \rightarrow 2\,NOCl(g)$$

Since the energy of the products is less than that of reactants, the reaction is exothermic.

13.9 Obtain the value $E_a = 76.3$ kJ/mol from the discussion and analysis of the data in Figure 13.10 and the value $k = 4.18 \times 10^{-5}$ L mol^{-1} s^{-1} at 273 K from Table 13.3.

$$\frac{k_1}{k_2} = e\left[\frac{E_a}{R}\left(\frac{1}{T_2} - \frac{1}{T_1}\right)\right] = e\left[\frac{76{,}300\text{ J/mol}}{8.314\text{ J K}^{-1}\text{mol}^{-1}}\left(\frac{1}{348\text{ K}} - \frac{1}{273\text{ K}}\right)\right]$$

$$= e^{-7.245} = 7.138 \times 10^{-4}$$

$$k_2 = \frac{k_1}{7.138 \times 10^{-4}} = \frac{4.18 \times 10^{-5}\text{ L mol}^{-1}\text{ s}^{-1}}{7.138 \times 10^{-4}}$$

$$= 5.86 \times 10^{-2}\text{ L mol}^{-1}\text{ s}^{-1}$$

13.10 (a) $2\,NH_3(aq) + OCl^-(aq) \rightarrow N_2H_4(aq) + Cl^-(aq) + H_2O(\ell)$

(b) Step 1

(c) NH_2Cl, OH^-, $N_2H_5^+$

(d) Rate = rate of step 1 = $k\,[NH_3][OCl^-]$

13.11 Choose the structure that is most similar to the structure of *p*-aminobenzoic acid. An enzyme might be inhibited by this molecule, which could fit the active site but not be converted to a product similar to folic acid. Or, the molecule might react in an enzyme-catalyzed process, producing a product whose biological function was different from folic acid. The best choice is

$$H_2N-C_6H_4-S(=O)_2-NH_2$$

Chapter 14

14.1 (a) $K_c = [CO_2(g)]$ (b) $K_c = \frac{[H_2]}{[HCl]}$

(c) $K_c = \frac{[CO][H_2]^3}{[CH_4][H_2O]}$ (d) $K_c = \frac{[HCN][OH^-]}{[CN^-]}$

14.2 $K_c = K_{c_1} \times K_{c_2} = (4.2 \times 10^{-7})(4.8 \times 10^{-11})$

$= 2.0 \times 10^{-17}$

14.3 $[CH_3COO^-] = \frac{2.96}{100} \times 0.0200\text{ mol/L} = 5.92 \times 10^{-4}\text{ M}$

$[H_3O^+] = 5.92 \times 10^{-4}$ M

$$[CH_3\,COOH] = \frac{100 - 2.96}{100} \times 0.0200\text{ mol/L}$$

$$= 1.94 \times 10^{-2}\text{ M}$$

$$K_c = \frac{[H_3O^+][CH_3COO^-]}{[CH_3COOH]} = \frac{(5.92 \times 10^{-4})(5.92 \times 10^{-4})}{1.94 \times 10^{-2}}$$

$$= 1.81 \times 10^{-5}$$

The result agrees with the value in Table 14.1.

14.4 (a) $K_c(\text{AgCl}) = 1.8 \times 10^{-10}$; $K_c(\text{AgI}) = 1.5 \times 10^{-16}$. Because $K_c(\text{AgI}) < K_c(\text{AgCl})$, the concentration of silver ions is larger in the beaker of AgCl.

(b) Unless all of the solid AgCl or AgI dissolves (which would mean that there was no equilibrium reaction), the concentrations at equilibrium are independent of the volume.

14.5 $$Q = \frac{(\text{conc. }SO_3)^2}{(\text{conc. }SO_2)^2(\text{conc. }O_2)}$$

$$= \frac{(0.184)^2}{(0.102 \times 2)^2(0.0132)} = 61.6$$

Since $Q < K_c$, the forward reaction should occur.

14.6

	AuI(s) ⇌ $Au^+(aq)$	+ $I^-(aq)$
Initial concentration (mol/L)	0	0
Change as reaction occurs (mol/L)	$+x$	$+x$
Equilibrium concentration (mol/L)	x	x

$$K_c = 1.6 \times 10^{-23} = [Au^+][I^-] = x^2$$

$$x = \sqrt{1.6 \times 10^{-23}} = 4.0 \times 10^{-12} = [Au^+] = [I^-]$$

14.7 The reaction is the reverse of the one in Table 14.1, so

$$K_c = \frac{1}{1.7 \times 10^2} = 5.9 \times 10^{-3}$$

$$Q = \frac{(\text{conc. }NO_2)^2}{(\text{conc. }N_2O_4)} = \frac{\left(\frac{0.500\text{ mol}}{4.00\text{ L}}\right)^2}{\left(\frac{1.00\text{ mol}}{4.00\text{ L}}\right)} = 6.25 \times 10^{-2}$$

Because $Q > K_c$, the reaction should go in the reverse direction. Therefore, let x be the change in concentration of N_2O_4, giving the ICE table:

	N_2O_4 ⇌	$2\,NO_2$
Initial concentration (mol/L)	$\frac{1.00}{4.00} = 0.250$	$\frac{0.500}{4.00} = 0.125$
Change as reaction occurs (mol/L)	x	$-2x$
Equilibrium concentration (mol/L)	$0.250 + x$	$0.125 - 2x$

$$K_c = 5.9 \times 10^{-3} = \frac{(0.125 - 2x)^2}{0.250 + x}$$

$$= \frac{(1.56 \times 10^{-2}) - 0.500x + 4x^2}{0.250 - x}$$

$$(1.48 \times 10^{-3}) + (5.9 \times 10^{-3})x$$
$$= (1.56 \times 10^{-2}) - 0.500x + 4x^2$$

$$4x^2 - 0.5059x + (1.412 \times 10^{-2}) = 0$$

$$x = \frac{-(-0.5059) \pm \sqrt{(-0.5059)^2 - 4 \times 4 \times 1.412 \times 10^{-2}}}{2 \times 4}$$

$$= \frac{0.5059 \pm \sqrt{3.001 \times 10^{-2}}}{8}$$

$$= \frac{0.5059 \pm 0.1732}{8}$$

$$x = 8.49 \times 10^{-2} \text{ or } x = 4.16 \times 10^{-2}$$

If $x = 8.49 \times 10^{-2}$, then $[N_2O_4] = 0.250 + x = 0.335$ and $[NO_2] = 0.125 - 2x = 0.125 - (2 \times 7.86 \times 10^{-2}) = -0.0448$. A negative concentration is impossible, so x must be 4.16×10^{-2}. Then

$$[N_2O_4] = 0.250 + 0.0416 = 0.292$$

$$[NO_2] = 0.125 - (2 \times 0.0416) = 0.0418$$

As predicted by Q, the reverse reaction has occurred, and the concentration of N_2O_4 has increased.

14.8 O_2N-NO_2 and $O_2N\cdot + \cdot NO_2$

Because a bond is broken and because bond breaking is always endothermic (**p. 241**), the reaction must be endothermic. Increasing temperature shifts the equilibrium in the endothermic direction. Figure 14.3 shows that at a higher temperature there is a greater concentration of brown NO_2.

14.9 (a) There are more moles of gas phase reactants than products, so entropy favors the reactants.

(b) Data from Appendix J show that $\Delta H° = -168.66$ kJ. The reaction is exothermic, so the energy effect favors the products.

(c) As T increases the reaction shifts in the endothermic direction, which is toward reactants. The entropy effect also becomes more important at high T, and it favors reactants. There is a greater concentration of SO_2 at low temperature.

Chapter 15

15.1 (a) Ethylene glycol molecules are polar and attracted to each other by dipole-dipole attractions and hydrogen bonding. They will not dissolve in gasoline, a nonpolar substance.
(b) Molecular iodine and carbon tetrachloride are nonpolar; therefore iodine should dissolve readily in carbon tetrachloride.
(c) Motor oil contains a mixture of nonpolar hydrocarbons that will dissolve in carbon tetrachloride, a nonpolar solvent.

15.2 The —OH groups attached to the ring and to the side chain of vitamin C hydrogen bond to water molecules. The oxygen atoms in the ring also form hydrogen bonds to water.

15.3 Use Henry's law, the Henry's law constant, and the fact that air is only 21 mol percent oxygen, a 0.21 mole fraction of oxygen.

$$\text{Pressure of } O_2 = (1.0\text{ atm})\left(\frac{760\text{ mm Hg}}{1\text{ atm}}\right)(0.21) = 160.\text{ mm Hg}$$

$$S_g = k_H P_g = \left(1.66 \times 10^{-6}\,\frac{\text{mol/L}}{\text{mm Hg}}\right)(160.\text{ mm Hg}) = 2.66 \times 10^{-4}\text{ mol/L}$$

$$(2.66 \times 10^{-4}\text{ mol/L})\left(\frac{32.00\text{ g } O_2}{1\text{ mol } O_2}\right) = 0.0085\text{ g/L or } 8.5\text{ mg/L}$$

15.4 Total mass is 750 + 21.5 = 771.5 g.

$$\text{Weight percent glucose} = \frac{21.5\text{ g}}{771.5\text{ g}} \times 100\% = 2.79\%$$

15.5
$$\left(\frac{30\text{ g Se}}{10^9\text{ g } H_2O}\right)\left(\frac{1\text{ g } H_2O}{1\text{ mL } H_2O}\right)\left(\frac{10^6\,\mu\text{g Se}}{1\text{ g Se}}\right) = 3.0 \times 10^{-2}\,\mu\text{g Se/mL } H_2O$$

Se in 100 mL of water

$$= \left(\frac{3.0 \times 10^{-2}\mu\text{g Se}}{1\text{ mL } H_2O}\right)(100\text{ mL } H_2O) = 3.0\,\mu\text{g Se}$$

15.6 (a)
$$(0.250\text{ L})\left(\frac{0.0750\text{ mol NaBr}}{1\text{ L}}\right)\left(\frac{102.9\text{ g NaBr}}{1\text{ mol NaBr}}\right) = 1.93\text{ g NaBr}$$

(b)
$$V_c = \frac{M_d \times V_d}{M_c} = \frac{(0.00150\text{ M})(0.500\text{ L})}{0.0750\text{ M}} = 0.0100\text{ L} = 10.0\text{ mL}$$

15.7 (a)
$$\left(\frac{0.0556\text{ mol } Mg^{2+}}{1\text{ L}}\right)\left(\frac{24.3\text{ g } Mg^{2+}}{1\text{ mol } Mg^{2+}}\right) \times \left(\frac{1.35\text{ g } Mg^{2+}}{1.03 \times 10^{-3}\text{ g solution}}\right) = 0.00131\,\frac{\text{g } Mg^{2+}}{\text{g solution}}$$

(b) 0.00131×10^6 ppm = 1310 ppm

15.8
$$\text{Molarity } H_2O_2 = \frac{\text{moles } H_2O_2}{\text{L solution}} = \frac{30.0\text{ g } H_2O_2}{100.\text{ g solution}} \times \frac{1.11\text{ g solution}}{1\text{ mL solution}} \times \frac{10^3\text{ mL solution}}{1\text{ L solution}} \times \frac{1\text{ mol } H_2O_2}{34.0\text{ g } H_2O_2}$$

$$= \frac{3.33 \times 10^4\text{ mol } H_2O_2}{3.40 \times 10^3\text{ L}} = 9.79\text{ M}$$

15.9
$$\text{Molarity, M} = \left(\frac{20\text{ g NaCl}}{100\text{ g solution}}\right)\left(\frac{1.148\text{ g solution}}{1\text{ mL solution}}\right) \times \left(\frac{1\text{ mol NaCl}}{58.5\text{ g NaCl}}\right)\left(\frac{10^3\text{ mL}}{1\text{ L}}\right) = 3.9\text{ mol NaCl/L}$$

$$\text{Molality, } m = \left(\frac{20\text{ g NaCl}}{0.0080\text{ kg } H_2O}\right)\left(\frac{1\text{ mol NaCl}}{58.5\text{ g NaCl}}\right) = 4.3\text{ mol NaCl/kg } H_2O$$

15.10
$$P_{\text{water}} = (X_{\text{water}})(P^\circ_{\text{water}})$$
$$291.2\text{ mm Hg} = (X_{\text{water}})(355.1\text{ mm Hg})$$
$$X_{\text{water}} = \frac{291.2\text{ mm Hg}}{355.1\text{ mm Hg}} = 0.8201$$
$$X_{\text{urea}} = 1.000 - X_{\text{water}} = 1.000 - 0.8201 = 0.1799$$

15.11
$$\text{Molality of solution} = \left(\frac{1.20\text{ kg ethylene glycol}}{6.50\text{ kg } H_2O}\right)\left(\frac{1000\text{ g}}{\text{kg}}\right)\left(\frac{1\text{ mol ethylene glycol}}{62.068\text{ g ethylene glycol}}\right) = 2.97\text{ mol/kg}$$

Next, calculate the freezing point depression of a 2.97 mol/kg solution.

$$\Delta T_f = (1.86\text{ °C kg mol}^{-1})(2.97\text{ mol/kg}) = 5.52\text{ °C}$$

This solution will freeze at −5.52 °C, so this quantity of ethylene glycol will not protect the 6.5 kg water in the tank if the temperature drops to −25 °C.

15.12 F.P. benzene = 5.50 °C; ΔT_f = 5.50 °C − 5.15 °C = 0.35 °C.

$$\text{Molality} = \frac{\Delta T_f}{K_f} = \frac{0.35\text{ °C}}{5.10\text{ °C kg mol}^{-1}} = 0.0686\text{ mol/kg}$$

$$\frac{0.0686\text{ mol solute}}{1\text{ kg benzene}} \times 0.0500\text{ kg benzene} = 0.00343\text{ mol solute}$$

$$\frac{0.180\text{ g solute}}{0.00343\text{ mol solute}} = 52.5\text{ g/mol}$$

15.13 Hemoglobin is a molecular substance so the factor i is 1.

$$c = \frac{\Pi}{RTi} = \frac{1.8 \times 10^{-3}\text{ atm}}{(0.0821\text{ L atm mol}^{-1}\text{ K}^{-1})(298\text{ K})(1)} = 7.36 \times 10^{-5}\text{ mol/L}$$

$$\text{MM} = \frac{5.0\text{ g}}{7.36 \times 10^{-5}\text{ mol}} = 6.8 \times 10^4\text{ g/mol}$$

Chapter 16

16.1

Acid	Its Conjugate Base	Base	Its Conjugate Acid
$H_2PO_4^-$	HPO_4^{2-}	PO_4^{3-}	HPO_4^{2-}
H_2	H^-	NH_2^-	NH_3
HSO_3^-	SO_3^{2-}	ClO_4^-	$HClO_4$
HF	F^-	Br^-	HBr

16.2 $[OH^-] = 3.0 \times 10^{-8}$ M;

$$[H_3O^+] = \frac{1.0 \times 10^{-14}}{3.0 \times 10^{-8}} = 3.3 \times 10^{-7}\text{ M}$$

Therefore, the solution whose $[H_3O^+]$ is 5.0×10^{-4} M is more acidic.

16.3 The 2.0×10^{-5} M H^+ solution is more acidic.

16.4 In a 0.040 M solution of NaOH, the $[OH^-]$ is 0.040 because the NaOH is 100% dissociated; pH = 12.60.

16.5 (a) $[H_3O^+] = 10^{-7.90} = 1.3 \times 10^{-8}$ M

(b) A pH of 7.90 is basic.

16.6 (a) $HN_3(aq) \rightleftharpoons H^+(aq) + N_3^-(aq)$ $\qquad K_a = \dfrac{[H^+][N_3^-]}{[HN_3]}$

(b) $HCOOH(aq) \rightleftharpoons H^+(aq) + HCOO^-(aq)$

$$K_a = \frac{[H^+][HCOO^-]}{[HCOOH]}$$

(c) $HClO_2(aq) \rightleftharpoons H^+(aq) + ClO_2^-(aq)$

$$K_a = \frac{[H^+][ClO_2^-]}{[HClO_2]}$$

16.7 (a) $CH_3NH_2(aq) + H_2O(\ell) \rightleftharpoons CH_3NH_3^+(aq) + OH^-(aq)$

$$K_b = \frac{[CH_3NH_3^+][OH^-]}{[CH_3NH_2]}$$

(b) $PH_3(aq) + H_2O(\ell) \rightleftharpoons PH_4^+(aq) + OH^-(aq)$

$$K_b = \frac{[PH_4^+][OH^-]}{[PH_3]}$$

(c) $NO_2^-(aq) + H_2O(\ell) \rightleftharpoons HNO_2(aq) + OH^-(aq)$

$$K_b = \frac{[HNO_2][OH^-]}{[NO_2^-]}$$

16.8 Setting up a small table for lactic acid, HLa:

	$HLa + H_2O \rightleftharpoons$	H_3O^+ +	La^-
Initial concentration (mol/L)	0.10	10^{-7}	0
Concentration change due to reaction (mol/L)	$-x$	$+x$	$+x$
Equilibrium concentration (mol/L)	$0.10 - x$	x	x

But $x = 10^{-2.43} = 3.7 \times 10^{-3}$ because $x = [H_3O^+]$. Substituting in the K_a expression,

$$K_a = \frac{[H_3O^+][La^-]}{[HLa]} = \frac{(3.7 \times 10^{-3})^2}{0.10 - (3.7 \times 10^{-3})}$$

$$= \frac{1.4 \times 10^{-5}}{0.1} = 1.4 \times 10^{-4}$$

Lactic acid is a stronger acid than propionic acid, with a K_a of 1.4×10^{-5}.

16.9 (a) Using the same methods as shown in the example,

$$\frac{x^2}{0.015} = 1.9 \times 10^{-5}$$

Solving for x, which is $[H_3O^+]$, we get

$$x = \sqrt{(1.9 \times 10^{-5})(0.015)} = 5.3 \times 10^{-4} = [H_3O^+].$$

So the pH of this solution is $-\log(5.3 \times 10^{-4}) = 3.28$.

(b) $$\% \text{ ionization} = \frac{[H_3O^+]}{[HN_3]_{initial}} \times 100\% = \frac{5.4 \times 10^{-4}}{0.015} \times 100\% = 3.6\%$$

16.10 In such cases use the K_b expression and value to calculate $[OH^-]$ and then pOH from $[OH^-]$. Calculate pH from 14 − pOH.

$C_6H_{11}NH_2(aq) + H_2O(\ell) \rightleftharpoons C_6H_{11}NH_3^+(aq) + OH^-(aq)$

$$K_b = \frac{[C_6H_{11}NH_3^+][OH^-]}{[C_6H_{11}NH_2]} = 4.6 \times 10^{-4}$$

$$K_b = \frac{[C_6H_{11}NH_3^+][OH^-]}{[C_6H_{11}NH_2]} = \frac{x^2}{0.015 - x} = 4.6 \times 10^{-4}$$

$$x^2 = (0.015 - x)(4.6 \times 10^{-4}) = (6.9 \times 10^{-6}) - (4.6 \times 10^{-4}x)$$

Solve the quadratic equation for x.

$$x = \frac{4.8 \times 10^{-3}}{2} = 2.4 \times 10^{-3} = [OH^-]$$

$$pOH = -\log(2.4 \times 10^{-3}) = 2.62; \quad pH = 14.00 - 2.62 = 11.38$$

16.11 Using the same methods as those used in the example, letting $x = [OH^-]$ and $[HCO_3^-]$, and using the value of 2.1×10^{-4} for K_b for CO_3^{2-}, we get

$$\frac{x^2}{1.0} = 2.1 \times 10^{-4} \qquad x = \sqrt{2.1 \times 10^{-4}} = 1.45 \times 10^{-2}$$

pOH = 1.84 and pH = 12.16

16.12 NH_4Cl dissolves by dissociating into NH_4^+ and Cl^- ions. The ammonium ions react with water to produce an acidic solution.

$$NH_4^+(aq) - H_2O(\ell) \rightleftharpoons H_3O^+(aq) + NH_3(aq)$$

$$\text{The } K_a \text{ of } NH_4^+ = \frac{1.0 \times 10^{-14}}{1.8 \times 10^{-5}} = \frac{K_w}{K_b};$$

$$K_a = 5.6 \times 10^{-10} = \frac{[H_3O^+][NH_3]}{[NH_4^+]}$$

NH_4^+	H_2O	H_3O^+	NH_3
Initial	0.10	0	0
Change	$-x$	$+x$	$+x$
Equilibrium	$0.10 - x$	x	x

$$5.6 \times 10^{-10} = \frac{(x)(x)}{(0.10 - x)}$$

Assume $0.10 - x \approx 0.10$ because K_a is so small. Thus, $(5.6 \times 10^{-10})(0.10) = x^2$; $x = [H_3O^+] = 7.48 \times 10^{-6}$ M.

$pH = -\log[H_3O^+] = -\log(7.48 \times 10^{-6}) = 5.16$

$$NH_4Cl(s) \rightleftharpoons NH_4^+(aq) + Cl^-(aq)$$

16.13 The formula weights and moles of acid per gram for the five antacids are as follows:

	Formula Weight	Mol Acid/Gram
$Mg(OH)_2$	58.32	1 mol acid/29.16 g antacid
$CaCO_3$	100.10	1 mol acid/50.05 g antacid
$NaHCO_3$	84.00	1 mol acid/84.00 g antacid
$Al(OH)_3$	78.0034	1 mol acid/26.00 g antacid
$NaAl(OH)_2CO_3$	143.99	1 mol acid/36.00 g antacid

Of these antacids, $Al(OH)_3$ neutralizes the most stomach acid per gram.

Chapter 17

17.1 $$K_a = \frac{[H^+][HCO_3^-]}{[H_2CO_3]} = \frac{H^+(0.025)}{(0.0020)} = [H^+] \times 12.5$$

$$= 4.2 \times 10^{-7}$$

$$[H^+] = \frac{4.2 \times 10^{-7}}{12.5} = 3.4 \times 10^{-8}$$

$$pH = -\log(3.4 \times 10^{-8}) = 7.47$$

17.2 $$7.40 = 7.21 + \log(\text{ratio}) = 7.21 + \log\frac{[HPO_4^{2-}]}{[H_2PO_4^-]}$$

$$\log\frac{[HPO_4^{2-}]}{[H_2PO_4^-]} = 7.40 - 7.21 = 0.19$$

$$\frac{[HPO_4^{2-}]}{[H_2PO_4^-]} = 10^{0.19} = 1.5$$

Therefore, $[HPO_4^{2-}] = 1.5 \times [H_2PO_4^-]$.

17.3 (a) Lactic acid-lactate (b) Acetic acid-acetate
(c) Hypochlorous acid-hypochlorite or $H_2PO_4^-$-HPO_4^{2-}
(d) CO_3^{2-}-HCO_3^-

17.4 The buffer capacity will be exceeded when just over 0.25 mol KOH is added, which will have reacted with the 0.25 mol $H_2PO_4^-$.

$$0.25 \text{ mol } OH^- = 0.25 \text{ mol KOH} \times \frac{56 \text{ g KOH}}{1 \text{ mol KOH}} = 14 \text{ g. Thus,}$$

slightly more than 14 g KOH will exceed the buffer capacity.

17.5 (a) 0.075 mol HCl converts 0.075 mol of lactate to lactic acid (0.075 mol).

$$pH = 3.85 + \log\frac{(0.20 - 0.075)}{(0.15 + 0.075)} = 3.85 + \log\frac{(0.125)}{(0.225)}$$

$$= 3.85 + \log(0.556) = 3.85 + (-0.25) = 3.60$$

(b) 0.025 mol NaOH converts 0.025 mol of lactic acid to 0.025 mol of lactate.

$$pH = 3.85 + \log\frac{(0.20 + 0.025)}{(0.15 - 0.025)} = 3.85 + \log\frac{(0.225)}{(0.125)}$$

$$= 3.85 + \log(1.8) = 3.85 + (0.26) = 4.11$$

17.6 (a) $[H_3O^+] = \frac{(5.00 \times 10^{-3}) - (1.00 \times 10^{-3})}{0.0500 + 0.0100}$

$$= \frac{4.00 \times 10^{-3}}{0.0600} = 6.67 \times 10^{-2}\ M$$

$$pH = 1.176 = 1.18$$

(b) $[H_3O^+] = \frac{(5.00 \times 10^{-3}) - 0.00250}{0.0500 + 0.0250}$

$$= \frac{2.50 \times 10^{-3}}{0.0750} = 3.33 \times 10^{-2}\ M$$

$$pH = -\log(3.33 \times 10^{-2}) = 1.48$$

(c) $[H_3O^+] = \frac{(5.00 \times 10^{-3}) - 0.00450}{0.0500 + 0.0450}$

$$= \frac{5.00 \times 10^{-4}}{0.0950} = 5.26 \times 10^{-3}$$

$$pH = -\log(5.26 \times 10^{-3}) = 2.28$$

(d) $[OH^-] = \frac{0.05 \times 10^{-3}\ mol}{0.0500\ L + 0.0505\ L}$

$$= 5.0 \times 10^{-4}\ mol/L$$

$$pOH = -\log(5.0 \times 10^{-4}) = 3.30$$

$$pH = 14.00 - 3.30 = 10.70$$

17.7 (a) Adding 10.0 mL of 0.100 M NaOH is adding (0.100 mol/L) (0.0100 L) = 0.00100 mol OH^-, which neutralizes 0.00100 mol acetic acid, converting it to 0.00100 mol acetate ion.

$$pH = pH + \log\frac{[\text{acetate}]}{[\text{acetic acid}]}$$

$$= 4.74 + \log\frac{(0.00100/0.0600)}{(0.00400/0.0600)}$$

$$= 4.74 + \log(0.25) = 4.74 + (-0.602) = 4.14$$

(b) $pH = 4.74 + \log\frac{(0.00250/0.0750)}{(0.00250/0.0750)}$

$$= 4.74 + \log(1) = 4.74$$

(c) $pH = 4.74 + \log\frac{(0.00450/0.0950)}{(0.00050/0.0950)}$

$$= 4.74 + \log(9) = 4.74 + 0.95 = 5.70$$

(d) $[OH^-] = \frac{0.10 \times 10^{-3}\ mol}{0.0500\ L + 0.0510\ L}$

$$= 9.9 \times 10^{-4}\ mol/L$$

$$pOH = -\log(9.9 \times 10^{-4})$$

$$= 3.00$$

$$pH = 14.00 - 3.00 = 11.00$$

17.8 (a) $K_{sp} = [Cu^+][Br^-]$ (b) $K_{sp} = [Hg^{2+}][I^-]^2$

(c) $K_{sp} = [Sr^{2+}][SO_4^{2-}]$

17.9 $AgBr(s) \rightleftharpoons Ag^+(aq) + Br^-(aq)$

$K_{sp} = [Ag^+][Br^-] = 5 \times 10^{-10} = x^2$; x = solubility

$x = \sqrt{5 \times 10^{-10}} \cong 2 \times 10^{-5}$

17.10 $Ag_2C_2O_4(s) \rightleftharpoons 2\,Ag^+(aq) + C_2O_4^{2-}(aq)$

$K_{sp} = [Ag^+]^2[C_2O_4^{2-}]$

$K_{sp} = [Ag^+]^2[C_2O_4^{2-}] = (1.4 \times 10^{-4})^2\,(6.9 \times 10^{-5})$

$= 1.4 \times 10^{-12}$

17.11

	$PbCl_2 \rightleftharpoons$	Pb^{2+}	$+\ 2\,Cl^-$
Initially (mol/L)		0	0.5
Change due to dissolving (mol/L)		$+S$	$0.5 + 2S$
At equilibrium (mol/L) because S will be small)		S	0.5 (ignore $2S$)

$$K_{sp} = (S)(0.5)^2 = 1.7 \times 10^{-5}$$

$$S = \frac{1.7 \times 10^{-5}}{(0.5)^2} = 6.8 \times 10^{-5}\ mol/L = Pb^{2+}\ \text{conc}$$

17.12 (a) $PbCl_2\,(s) \rightleftharpoons Pb^{2+}(aq) + 2\,Cl^-(aq)$

$K_{sp} = [Pb^{2+}][Cl^-]^2 = 1.7 \times 10^{-5}$. Let S equal the solubility of lead chloride, which equals $[Pb^{2+}]$

$$K_{sp} = 1.7 \times 10^{-5} = (S)(2S)^2 = 4S^3;\ S = \sqrt[3]{\frac{1.7 \times 10^{-5}}{4}}$$

$$= (4.25 \times 10^{-6})^{1/3} = 1.6 \times 10^{-2}\ mol/L$$

(b) Let solubility of $PbCl_2 = [Pb^{2+}] = S = [Cl^-] = 0.20$ M

$[Pb^{2+}] = S = \frac{1.7 \times 10^{-5}}{(0.20)^2} = 4.3 \times 10^{-4}$ mol/L This is less than that in pure water due to the common ion effect of the presence of chloride ion.

17.13 $AgCl(s) \rightleftharpoons Ag^+(aq) + Cl^-(aq)$ $K_{sp} = 1.8 \times 10^{-10}$

$Ag^+(aq) + 2\,S_2O_3^{2-}(aq) \rightleftharpoons Ag(S_2O_3)_2^{3-}(aq)$ $K_f = 2.0 \times 10^{13}$

Net reaction:

$AgCl(s) + 2\,S_2O_3^{2-}(aq) \rightleftharpoons Ag(S_2O_3)_2^{3-}(aq) + Cl^-(aq)$

Therefore, the equilibrium constant for the net reaction is the product of $K_{sp} \times K_f$: $K_{net} = K_{sp} \times K_f = (1.8 \times 10^{-10})(2.0 \times 10^{13}) = 3.6 \times 10^3$. Because K_{net} is much greater than 1, the net reaction is product-favored, and AgCl is much more soluble in a Na_2SO_4 solution than it is in water.

17.14 (a) $Q = (1.0 \times 10^{-5})(1.0 \times 10^{-5}) = 1.0 \times 10^{-10} < K_{sp}$; no precipitation.

(b) For precipitation to occur, $Q \geq K_{sp}$; Q = conc Ag^+ × conc Cl^-; $K_{sp} = [Ag^+][Cl^-]$; $1.8 \times 10^{-10} = [Ag^+][Cl^-]$;

$$[Cl^-] = \frac{1.8 \times 10^{-10}}{1.0 \times 10^{-5}} = 1.8 \times 10^{-5}\ M$$

the minimum for AgCl precipitation.

17.15 AgCl will precipitate first. $[Cl^-]$ needed to precipitate AgCl:

$$[Cl^-] = \frac{1.8 \times 10^{-10}}{1.0 \times 10^{-2}} = 1.8 \times 10^{-8}\ M$$

$[Cl^-]$ needed to precipitate $PbCl_2$:

$$[Cl^-] = \sqrt{\frac{1.7 \times 10^{-5}}{1.0 \times 10^{-1}}} = 1.3 \times 10^{-2}\ M$$

Chapter 18

18.1 $\Delta S = q_{rev}/T = (30.8 \times 10^3\ J)/(273.15 + 45.3)\ K$

$$= (30.8 \times 10^3\ J)/(318.45\ K)$$

$$\Delta S = 96.7\ J/K$$

18.2 (a) C(g) has higher $S°$, 158.096 J K^{-1} mol^{-1}, versus 5.740 J K^{-1} mol^{-1} for C (graphite).

(b) Ar(g) has higher $S°$, 154.7 J K^{-1} mol^{-1}, versus 41.42 J K^{-1} mol^{-1} for Ca(s).

(c) KOH(aq) has higher $S°$, $91.6\ \text{J K}^{-1}\ \text{mol}^{-1}$, versus $78.9\ \text{J K}^{-1}\ \text{mol}^{-1}$ for KOH(s).

18.3 (a) $\Delta S° = 2\ \text{mol CO(g)} \times S°(\text{CO(g)}) + 1\ \text{mol O}_2\text{(g)} \times S°(\text{O}_2\text{(g)}) - 2\ \text{mol CO}_2\text{(g)} \times S°(\text{CO}_2\text{(g)})$
$= \{2 \times (197.674) + (205.138) - 2 \times (213.74)\}\ \text{J/K}$
$= 173.01\ \text{J/K}$

(b) $\Delta S° = 1\ \text{mol NaCl(aq)} \times S°(\text{NaCl(aq)}) - 1\ \text{mol NaCl(s)} \times S°(\text{NaCl(s)})$
$= (115.5 - 72.13)\ \text{J/K} = 43.4\ \text{J/K}$

(c) $\Delta S° = 1\ \text{mol MgO(s)} \times S°(\text{MgO(s)}) + 1\ \text{mol CO}_2\text{(g)} \times S°(\text{CO}_2\text{(g)}) - 1\ \text{mol MgCO}_3\text{(s)} \times S°(\text{MgCO}_3\text{(s)})$
$= (26.94 + 213.74 - 65.7)\ \text{J/K}$
$= 175.0\ \text{J/K}$

18.4 $\text{N}_2\text{(g)} + 3\ \text{H}_2\text{(g)} \rightarrow 2\ \text{NH}_3\text{(g)}$

$$\Delta H° = 2\ \text{mol NH}_3 \times \Delta H_f°(\text{NH}_3\text{(g)}) = 2(-46.11)\ \text{kJ} = -92.22\ \text{kJ}$$

$$\Delta S° = 2\ \text{mol NH}_3 \times S°(\text{NH}_3\text{(g)}) - 1\ \text{mol N}_2\text{(g)} \times S°(\text{N}_2\text{(g)}) - 3\ \text{mol H}_2 \times S°(\text{H}_2\text{(g)})$$

$$= 2(192.45)\ \text{J/K} - (191.61)\ \text{J/K} - 3(130.684)\ \text{J/K}$$

$$= -198.76\ \text{J/K}$$

$$\Delta S°_{\text{universe}} = \frac{-\Delta H°}{T} + \Delta S° = \frac{92.2\ \text{kJ}}{298.15\ \text{K}} + (-198.76\ \text{J/K})$$

$$= \frac{92{,}200\ \text{J}}{298.15\ \text{K}} - 198.76\ \text{J/K} = 110.5\ \text{J/K}$$

The process is product-favored.

18.5 (a) $\Delta H° = \{(-238.66) - (-110.525)\}\ \text{kJ} = -128.14\ \text{kJ}$
$\Delta S° = \{(126.8) - 197.674 - 2 \times 130.684\}\ \text{J/K}$
$= -332.2\ \text{J/K}$
$\Delta G° = \Delta H° = T\Delta S°$
$= -128.14 \times 10^3\ \text{J} - 298.15\ \text{K} \times (-332.2\ \text{J/K})$
$= -29.09 \times 10^3\ \text{J} = -29.09\ \text{kJ}$

(b) $\Delta G° = [-166.27 - (-137.168)]\ \text{kJ} = -29.10\ \text{kJ}$. The two results agree.

(c) $\Delta G°$ is negative. The reaction is product-favored at 298.15 K. Because $\Delta S°$ is negative, at very high temperatures the reaction will become reactant-favored.

18.6 (a) $T = \Delta H°/\Delta S° = (-565{,}968\ \text{J})/(-173.01\ \text{J/K}) = 3271\ \text{K}$

(b) The reaction is exothermic and therefore is product-favored at temperatures lower than 3271 K.

18.7 (a) $\Delta G° = \{-553.04 - 527.81 - (-1128.79)\}\ \text{kJ}$
$= 47.94\ \text{kJ}$
$K° = e^{-\Delta G°/RT} = e^{-(47.94\ \text{kJ/mol})/(8.314\ \text{J K}^{-1}\ \text{mol}^{-1})(298\ \text{K})}$
$= e^{-(47{,}940\ \text{J/mol})/(8.314\ \text{J K}^{-1}\ \text{mol}^{-1})(298\ \text{K})}$
$= e^{-19.35} = 3.9 \times 10^{-9}$ (close to K_c)

(b) $K° = e^{-14.68} = 4.2 \times 10^{-7}$ (agrees with K_c)

(c) $K° = e^{-(-1.909)} = 6.75$ (agrees with K_P)

For reactions (a) and (b), $K_c = K°$.

18.8 (a) At 298 K,
$\Delta G° = 2 \times (-16.45)\ \text{kJ} = -32.9\ \text{kJ}$
$K° = e^{-(-32{,}900\ \text{J})/(8.314\ \text{J K}^{-1}\ \text{mol}^{-1})(298\ \text{K})} = e^{13.28} = 5.8 \times 10^5$

(b) At 450. K,
$\Delta G° = \Delta H° - T\Delta S° = -92.22\ \text{kJ} - (450.)(-0.19876)\ \text{kJ}$
$= -2.78\ \text{kJ}$
$K° = e^{-(-2780\ \text{J})/(8.314\ \text{J K}^{-1})(450.\ \text{K})} = 2.10$

(c) At 800. K,
$\Delta G° = -92.22\ \text{kJ} - (800.)(-0.19876)\ \text{kJ} = 66.79\ \text{kJ}$
$K° = e^{-(66{,}790\ \text{J})/(8.314\ \text{J K}^{-1}\ \text{mol}^{-1})(800.\ \text{K})} = 4.3 \times 10^{-5}$

18.9 $\Delta G = \Delta G° + RT \ln Q$

$\Delta G°$ was calculated in Problem-Solving Practice 18.7 to be 47.94 kJ for the reverse of this reaction. Therefore,
$\Delta G° = -47.94\ \text{kJ}$

$$\Delta G = -47.94\ \text{kJ/mol} + (8.314\ \text{J K}^{-1}\ \text{mol}^{-1})(298\ \text{K}) \ln\left(\frac{1}{(0.023)(0.13)}\right)$$

$$= -47.94\ \text{kJ/mol} + 1.44 \times 10^4\ \text{J/mol}$$

$$= -47.94\ \text{kJ/mol} + 14.4\ \text{kJ/mol} = -33.5\ \text{kJ/mol}$$

18.10 (a) $\Delta G° = 2(-137.168)\ \text{kJ} - 2(-394.359)\ \text{kJ} = 514.382\ \text{kJ}$. The reaction is reactant-favored, and at least 514.382 kJ of work must be done to make it occur.

(b) $\Delta G° = 2(-742.2)\ \text{kJ} = -1484.4\ \text{kJ}$. The reaction is product-favored and could do up to 1484.4 kJ of useful work.

18.11 $\Delta G°' = \Delta G°$ for this reaction because none of the reactants or products requires a standard state different from 1 bar or 1 mol/L.

18.12 (a) The strongest phosphate donor has the most negative $\Delta G°$ for its reaction with water to produce dihydrogen phosphate. The $\Delta G°'$ values are given in Problem-Solving Example 18.12. Creatine phosphate, at -43.1 kJ, has the most negative value and is the strongest phosphate donor.

(b) Glycerol 3-phosphate has the least negative $\Delta G°'$ at -9.7 kJ and therefore is the weakest phosphate donor.

(c) See parts (a) and (b) for explanation.

18.13 (a) $\Delta G_f°(\text{MgO(s)}) = -569.43\ \text{kJ}$, so formation of MgO(s) is product-favored and MgO(s) is thermodynamically stable.

(b) $\Delta G_f°(\text{N}_2\text{H}_4(\ell)) = 149.34\ \text{kJ}$; kinetically stable.

(c) $\Delta G_f°(\text{C}_2\text{H}_6\text{(g)}) = -32.82\ \text{kJ}$; thermodynamically stable.

(d) $\Delta G_f°(\text{N}_2\text{O(g)}) = 104.20\ \text{kJ}$; kinetically stable.

Chapter 19

19.1 Reducing agents are indicated by "red" and oxidizing agents are indicated by "ox." Oxidation numbers are shown above the symbols for the elements.

(a) $$2\ \underset{\text{red}}{\overset{0}{\text{Fe}}}\text{(s)} + 3\ \underset{\text{ox}}{\overset{0}{\text{Cl}_2}}\text{(g)} \rightarrow 2\ \overset{+3\ -1}{\text{FeCl}_3}\text{(s)}$$

(b) $$2\underset{\text{red}}{\overset{0}{\text{H}_2}}\text{(g)} + \underset{\text{ox}}{\overset{0}{\text{O}_2}}\text{(g)} \rightarrow 2\ \overset{+1\ -2}{\text{H}_2\text{O}}(\ell)$$

(c) $$\underset{\text{red}}{\overset{0}{\text{Cu}}}\text{(S)} + 2\ \underset{\text{ox}}{\overset{+5\ -2}{\text{NO}_3^-}}\text{(aq)} + 4\ \text{H}_3\text{O}^+\text{(aq)} \rightarrow \overset{+2}{\text{Cu}^{2+}}\text{(aq)} + 2\ \overset{+4\ -2}{\text{NO}_2}\text{(g)} + 6\ \overset{+1\ -2}{\text{H}_2\text{O}}(\ell)$$

(d) $$\underset{\text{red}}{\overset{0}{\text{C}}}\text{(s)} + \underset{\text{ox}}{\overset{0}{\text{O}_2}}\text{(g)} \rightarrow \overset{+4\ -2}{\text{CO}_2}\text{(g)}$$

(e) $$6\ \underset{\text{red}}{\overset{+2}{\text{Fe}^{2+}}}\text{(aq)} + \underset{\text{ox}}{\overset{+6\ -2}{\text{Cr}_2\text{O}_7^{2-}}}\text{(aq)} + 14\ \overset{+1\ -2}{\text{H}_3\text{O}^+}\text{(aq)} \rightarrow 6\ \overset{+3}{\text{Fe}^{3+}}\text{(aq)} + 2\ \overset{+3}{\text{Cr}^{3+}}\text{(aq)} + 21\ \overset{+1\ -2}{\text{H}_2\text{O}}(\ell)$$

19.2 (a) Ox: $\text{Cd(s)} \rightarrow \text{Cd}^{2+}\text{(aq)} + 2\ \text{e}^-$
Red: $\text{Cu}^{2+}\text{(aq)} + 2\ \text{e}^- \rightarrow \text{Cu(s)}$
Net: $\text{Cd(s)} + \text{Cu}^{2+}\text{(aq)} \rightarrow \text{Cd}^{2+}\text{(aq)} + \text{Cu(s)}$

(b) Ox: $\text{Zn(s)} \rightarrow \text{Zn}^{2+}\text{(aq)} + 2\ \text{e}^-$
Red: $2\ \text{H}_3\text{O}^+\text{(aq)} + 2\ \text{e}^- \rightarrow \text{H}_2\text{(g)} + 2\ \text{H}_2\text{O}(\ell)$
Net: $\text{Zn(s)} + 2\ \text{H}_3\text{O}^+\text{(aq)} \rightarrow \text{Zn}^{2+}\text{(aq)} + \text{H}_2\text{(g)} + 2\ \text{H}_2\text{O}(\ell)$

(c) Ox: $2\ \text{Al(s)} \rightarrow 2\ \text{Al}^{3+}\text{(aq)} + 6\ \text{e}^-$
Red: $3\ \text{Zn}^{2+}\text{(aq)} + 6\ \text{e}^- \rightarrow 3\ \text{Zn(s)}$
Net: $2\ \text{Al(s)} + 3\ \text{Zn}^{2+}\text{(aq)} \rightarrow 2\ \text{Al}^{3+}\text{(aq)} + 3\ \text{Zn(s)}$

19.3 ***Step 1.*** This is an oxidation-reduction reaction. It is obvious that Zn is oxidized by its change in oxidation state.
Step 2. The half-reactions are

$Zn(s) \longrightarrow Zn^{2+}(aq)$ (This is the oxidation reaction.)

$Cr_2O_7^{2-}(aq) \longrightarrow 2\,Cr^{3+}(aq)$ (This is the reduction reaction.)

Step 3. Balance the atoms in the half-reactions. The atoms are balanced in the Zn half-reaction. We need to add water and H in the $Cr_2O_7^{2-}$ half-reaction. Fourteen H^+ ions are required on the right to combine with the seven O atoms.

$$Cr_2O_7^{2-}(aq) + 14\,H^+(aq) \longrightarrow 2\,Cr^{3+}(aq) + 7\,H_2O(\ell)$$

Step 4. Balance the half-reactions for charge. Write the Zn half-reaction as

$$Zn(s) \longrightarrow Zn^{2+}(aq) + 2\,e^-$$

and write the $Cr_2O_7^{2-}$ half-reaction as

$$Cr_2O_7^{2-}(aq) + 14\,H^+(aq) + 6\,e^- \longrightarrow 2\,Cr^{3+}(aq) + 7\,H_2O(\ell)$$

Step 5. Multiply the half-reactions by factors to make the number of electrons gained equal to the number lost.

$$3\,[Zn(s) \longrightarrow Zn^{2+}(aq) + 2\,e^-]$$

$$1\,[Cr_2O_7^{2-}(aq) + 14\,H^+(aq) + 6\,e^- \longrightarrow 2\,Cr^{3+}(aq) + 7\,H_2O(\ell)]$$

Step 6. Add the two half-reactions, canceling the electrons.

$$3\,Zn(s) \longrightarrow 3\,Zn^{2+}(aq) + 6\,e^-$$

$$Cr_2O_7^{2-}(aq) + 14\,H^+(aq) + 6\,e^- \longrightarrow 2\,Cr^{3+}(aq) + 7\,H_2O(\ell)$$

$$Cr_2O_7^{2-}(aq) + 3\,Zn(s) + 14\,H^+(aq) \longrightarrow 2\,Cr^{3+}(aq) + 3\,Zn^{2+}(aq) + 7\,H_2O(\ell)$$

Step 7. Everything checks.
Step 8. Water was added in Step 3. The balanced equation is

$$Cr_2O_7^{2-}(aq) + 3\,Zn(s) + 14\,H_3O^+(aq) \longrightarrow 2\,Cr^{3+}(aq) + 3\,Zn^{2+}(aq) + 21\,H_2O(\ell)$$

19.4 ***Step 1.*** This is an oxidation-reduction reaction. The wording of the question says Al reduces NO_3^- ion. Al is oxidized.
Step 2. The half-reactions are:

$Al(s) \longrightarrow Al(OH)_4^-(aq)$ (This is the oxidation reaction.)

$NO_3^-(aq) \longrightarrow NH_3(aq)$ (This is the reduction reaction.)

Step 3. Balance the atoms in the half-reactions. For the Al half-reaction, add four H^+ ions on the right and four water molecules on the left.

$$Al(s) + 4\,H_2O(\ell) \longrightarrow Al(OH)_4^- + 4\,H^+(aq)$$

For the NO_3^- half-reaction,

$$NO_3^-(aq) + 9\,H^+(aq) \longrightarrow NH_3(aq) + 3\,H_2O(\ell)$$

Step 4. Balance the half-reactions for charge. Put 3 e^- on the right in the Al half-reaction

$$Al(s) + 4\,H_2O(\ell) \longrightarrow Al(OH)_4^- + 4\,H^+(aq) + 3\,e^-$$

and put 8 e^- on the left side of the NO_3^- half-reaction.

$$NO_3^-(aq) + 9\,H^+(aq) + 8\,e^- \longrightarrow NH_3(aq) + 3\,H_2O(\ell)$$

Step 5. Multiply the half-reactions by factors to make the electrons gained equal to those lost.

$$8[Al(s) + 4\,H_2O(\ell) \longrightarrow Al(OH)_4^- + 4\,H^+(aq) + 3\,e^-]$$

$$3[NO_3^-(aq) + 9\,H^+(aq) + 8\,e^- \longrightarrow NH_3(aq) + 3\,H_2O(\ell)]$$

Step 6. Remove $H^+(aq)$ ions by adding an appropriate amount of OH^-. For the Al half-reaction, add 32 OH^- ions to get

$$8\,Al(s) + 32\,OH^-(aq) + 32\,H_2O(\ell) \longrightarrow 8\,Al(OH)_4^- + 32\,H_2O + 24\,e^-$$

For the NO_3^- half-reactions, add 27 OH^- ions to get

$$3\,NO_3^-(aq) + 27\,H_2O + 24\,e^- \longrightarrow 3\,NH_3(aq) + 9\,H_2O(\ell) + 27\,OH^-(aq)$$

Step 7. Add both half-reactions and cancel the electrons.

$$8\,Al(s) + 32\,OH^-(aq) + 32\,H_2O(\ell) \longrightarrow 8\,Al(OH)_4^- + 32\,H_2O + 24\,e^-$$

$$3\,NO_3^-(aq) + 27\,H_2O + 24\,e^- \longrightarrow 3\,NH_3(aq) + 9\,H_2O(\ell) + 27\,OH^-(aq)$$

$$3\,NO_3^-(aq) + 8\,Al(s) + 59\,H_2O(\ell) + 32\,OH^-(aq) \longrightarrow 8\,Al(OH)_4^-(aq) + 3\,NH_3(aq) + 27\,OH^-(aq) + 41\,H_2O(\ell)$$

Step 8. Make a final check. Since there are OH^- ions and water molecules on both sides of the equation, cancel them out. This gives the final balanced equation.

$$3\,NO_3^-(aq) + 8\,Al(s) + 18\,H_2O(\ell) + 5\,OH^-(aq) \longrightarrow 8\,Al(OH)_4^-(aq) + 3\,NH_3(aq)$$

(This is a fairly complicated equation to balance. If you balanced this one with a minimum of effort, your understanding of balancing redox equations is rather good. If you had to struggle with one or more of the steps, go back and repeat them.)

19.5 (a) $Ni(s) \rightarrow Ni^{2+}(aq) + 2\,e^-$ (This is the oxidation half-reaction.)

$2\,Ag^+(aq) + 2\,e^- \rightarrow 2\,Ag(s)$ (This is the reduction half-reaction.)

(b) The oxidation of Ni takes place at the anode and the reduction of Ag^+ takes place at the cathode.

(c) Electrons would flow through an external circuit from the anode (where Ni is oxidized) to the cathode (where Ag^+ ions are reduced).

(d) Nitrate ions would flow through the salt bridge to the anode compartment. Potassium ions would flow into the cathode compartment.

19.6 Oxidation half-reaction: $Fe(s) \rightarrow Fe^{2+}(aq, 1\,M) + 2\,e^-$ (anode)

Reduction half-reaction: $Cu^{2+}(aq, 1\,M) + 2\,e^- \rightarrow Cu(s)$ (cathode)

$$E^\circ_{cell} = +0.78\text{ V} = E^\circ_{cathode} - E^\circ_{anode}$$

Since $E^\circ_{cathode} = +0.34$ V, E°_{anode} must be -0.44 V.

19.7 $F_2(g) + 2\,e^- \longrightarrow 2\,F^-(aq)$ $\quad E^\circ_{cathode} = +2.87$ V

$2\,Li(s) \longrightarrow 2\,Li^+(aq) + 2\,e^-$ $\quad E^\circ_{anode} = -3.045$ V

$2\,Li(s) + F_2(g) \longrightarrow 2\,Li^+(aq) + 2\,F^-(aq)$ $\quad E^\circ_{cell} = +5.91$ V

$$E^\circ_{cell} = E^\circ_{cathode} - E^\circ_{anode} = +2.87 - (-3.045) = +5.91\text{ V}$$

19.8 The two half-reactions are

$Hg^{2+}(aq) + 2\,e^- \longrightarrow 2\,Hg(\ell)$ $\quad E^\circ_{cathode} = +0.855$ V

$2\,I^-(aq) \longrightarrow I_2(s) + 2\,e^-$ $\quad E^\circ_{anode} = +0.535$ V

$$E^\circ_{cell} = E^\circ_{cathode} - E^\circ_{anode} = +0.855 - 0.535 = +0.320\text{ V}$$

The reaction is product-favored as written.

19.9 We first need to calculate E°_{cell}, and to do this we break the reaction into two half-reactions.

$$\text{Ox: } Sn^{2+}(aq) \longrightarrow Sn^{4+}(aq) + 2\,e^- \qquad E^\circ_{anode} = +0.15 \text{ V}$$

$$\text{Red: } I_2(s) + 2\,e^- \longrightarrow 2\,I^-(aq) \qquad E^\circ_{cathode} = +0.535 \text{ V}$$

$$I_2(s) + Sn^{2+}(aq) \longrightarrow 2\,I^-(aq) + Sn^{4+}(aq) \qquad E^\circ_{cell} = +0.385 \text{ V}$$

$$E^\circ_{cell} = E^\circ_{cathode} - E^\circ_{anode} = +0.535 - 0.15 = +0.385 \text{ V}$$

E°_{cell} is 0.385 V, and 2 mol of electrons are transferred.

$$\log K = \frac{nE^\circ \text{ V}}{0.0592 \text{ V}} = \frac{2 \times 0.385 \text{ V}}{0.0592 \text{ V}}$$

$$= 13.00 \text{ and } K = 1 \times 10^{13}$$

The large value of K indicates that the reaction is strongly product-favored as written.

19.10 $E_{cell} = 0.51 \text{ V} - \left(\frac{0.0592 \text{ V}}{2} \times \log \frac{3}{0.010}\right)$

$$= 0.51 \text{ V} - (0.0296 \text{ V} \times \log(300))$$

$$= 0.51 \text{ V} - 0.073 \text{ V} = +0.44 \text{ V}$$

19.11 If the pH = 3.66, then the E_{cell} = 0.217 V.

19.12 (a) The net cell reaction would be

$$2\,Na^+ + 2\,Br^- \longrightarrow 2\,Na + Br_2$$

Sodium ions would be reduced at the cathode and bromide ions would be oxidized at the anode.

(b) H_2 would be produced at the cathode for the same reasons given in Problem-Solving Example 19.8. That reaction is

$$2\,H_2O(\ell) + 2\,e^- \longrightarrow H_2(g) + 2\,OH^-(aq)$$

At the anode, two reactions are possible: the oxidation of water and the oxidation of Br^- ions.

$$6\,H_2O(\ell) \longrightarrow O_2(g) + 4\,H_3O^+(aq) + 4\,e^- \qquad E^\circ = 1.229 \text{ V}$$

$$2\,Br^-(aq) \longrightarrow Br_2(\ell) + 2\,e^- \qquad E^\circ = 1.08 \text{ V}$$

Bromide ions will be oxidized to Br_2 because that potential is smaller. The net cell reaction is

$$2\,H_2O(\ell) + 2\,Br^-(aq) \longrightarrow Br_2(\ell) + H_2(g) + 2\,OH^-(aq)$$

(c) Sn metal will be formed at the cathode because its reduction potential (−0.14 V) is less negative than the potential for the reduction of water. O_2 will form at the anode because the E° value for the oxidation of water is smaller than the E° value for the oxidation of Cl^-. The net cell reaction is

$$2\,Sn^{2+}(aq) + 6\,H_2O(\ell) \longrightarrow 2\,Sn(s) + O_2(g) + 4\,H_3O^+(aq)$$

19.13 First, calculate the quantity of charge:

$$\text{Charge} = (25 \times 10^3 \text{ A})(1 \text{ h})\left(\frac{60 \text{ s}}{1 \text{ min}}\right)\left(\frac{60 \text{ min}}{1 \text{ h}}\right)$$

$$= 9.0 \times 10^7 \text{ A·s} = 9.0 \times 10^7 \text{ C}$$

Then calculate the mass of Na:

$$\text{Mass of Na} = (9.0 \times 10^7 \text{ C}) \times \left(\frac{1 \text{ mol e}^-}{96{,}500 \text{ C}}\right)\left(\frac{1 \text{ mol Na}}{1 \text{ mol e}^-}\right)\left(\frac{22.99 \text{ g Na}}{1 \text{ mol Na}}\right)$$

$$= 2.1 \times 10^4 \text{ g Na}$$

Chapter 20

20.1 (a) $^{237}_{93}Np \rightarrow\ ^{4}_{2}He +\ ^{233}_{91}Pa$ (b) $^{35}_{16}S \rightarrow\ ^{0}_{-1}e +\ ^{35}_{17}Cl$

20.2 (a) $^{11}_{6}C \rightarrow\ ^{11}_{5}B +\ ^{0}_{1}e$ (b) $^{35}_{16}S \rightarrow\ ^{35}_{17}Cl +\ ^{0}_{-1}e$

(c) $^{30}_{15}P \rightarrow\ ^{0}_{+1}e +\ ^{30}_{14}Si$ (d) $^{22}_{11}Na \rightarrow\ ^{0}_{-1}e +\ ^{22}_{12}Mg$

20.3 (a) $^{42}_{19}K \rightarrow\ ^{0}_{-1}e +\ ^{41}_{20}Ca$ (b) $^{234}_{92}U \rightarrow\ ^{4}_{2}He +\ ^{230}_{90}Th$

(c) $^{20}_{9}F \rightarrow\ ^{0}_{-1}e +\ ^{20}_{10}Ne$

20.4 (a) $^{90}_{38}Sr \rightarrow\ ^{0}_{-1}e +\ ^{90}_{39}Y$

(b) It takes 4 half-lives (4 × 29 y = 116 y) for the activity to decrease to 125 beta particles emitted per minute:

Number of Half-lives	Change of Activity	Total Elapsed Time (y)
1	2000 to 1000	29
2	1000 to 500	58
3	500 to 250	87
4	250 to 125	116

20.5 (a) $t_{1/2} = \frac{0.693}{9.3 \times 10^{-3} \text{ d}^{-1}} = 75 \text{ d}$

(b) ln (fraction remaining) = $-k \times t = -(9.3 \times 10^{-3} \text{ d}^{-1}) \times (100 \text{ d}) = -0.930$

Fraction of iridium-192 remaining = $e^{-0.930}$ = 0.39. Therefore, 39% of the original iridium-192 remains.

20.6 $k = \frac{0.693}{1.60 \times 10^3 \text{ y}} = 4.33 \times 10^{-4} \text{ y}^{-1}$

As of 2004: ln (fraction remaining) = $-k \times t = -(4.33 \times 10^{-4} \text{ y}^{-1}) \times 83 \text{ y} = -3.39 \times 10^{-2}$

Fraction of radium-226 remaining = $e^{-0.0339}$ = 0.965. Therefore, 96.5% of the original radium-226 remains; 0.965 × 1.00 g = 0.965 g.

20.7 $\ln(0.60) = -0.510 = -k \times t$

$$k = \frac{0.693}{t_{1/2}} = \frac{0.693}{12.3 \text{ y}} = 0.0563 \text{ y}^{-1}$$

$$t = \frac{-0.510}{-0.0563 \text{ y}^{-1}} = 9.1 \text{ y}$$

Chapter 21

21.1 $1.5 \times 10^9 \text{ mol Na} \times \frac{4 \text{ mol NaCl}}{4 \text{ mol Na}} \times \frac{58.5 \text{ g NaCl}}{1 \text{ mol NaCl}} \times \frac{1 \text{ lb NaCl}}{454 \text{ g NaCl}} \times \frac{1 \text{ ton NaCl}}{2000 \text{ lb NaCl}} = 9.7 \times 10^4 \text{ ton NaCl}$

21.2 mol of Al $= 1.00 \text{ ton Al} \times \frac{2000 \text{ lb Al}}{1 \text{ ton Al}} \times \frac{454 \text{ g Al}}{1 \text{ ton Al}} \times \frac{1 \text{ mol Al}}{26.98 \text{ g Al}} = 3.37 \times 10^4 \text{ mol Al}$

$Al^{3+} + 3\,e^- \rightarrow Al$; therefore, 3 mol electrons are needed to produce 1 mol Al.

Number of moles of electrons

$$= 3.37 \times 10^4 \text{ mol Al} \times \frac{3 \text{ mol e}^-}{1 \text{ mol Al}}$$

$$= 1.01 \times 10^5 \text{ mol e}^-$$

$$\text{Charge} = 1.01 \times 10^5 \text{ mol e}^- \times \frac{9.65 \times 10^4 \text{ C}}{1 \text{ mol e}^-}$$

$$= 9.75 \times 10^9 \text{ C}$$

$$\text{Time} = 9.75 \times 10^9 \text{ C} \times \frac{1 \text{ s}}{1.00 \times 10^5 \text{ C}} = 9.75 \times 10^4 \text{ s}$$

$$= 1.63 \times 10^3 \text{ min} = 27.0 \text{ h}$$

21.3 I^- is oxidized to I_2; IO_3 is reduced to I_2. IO_3^- is the oxidizing agent and I^- is the reducing agent.

21.4 (a) $:\ddot{O}$ and $\ddot{O}:$ double-bonded to N — $\ddot{O}$ — N, each N also bonded to $:\ddot{O}:$ (Lewis structure O₂N–O–NO₂)

(b) $N_2O_5 + H_2O \rightarrow 2\ HNO_3$

21.5 (a) The formation of NO_2 from NO is exothermic. Thus, lowering the temperature favors the forward reaction (NO_2 formation).

(b) By reacting with water, NO_2 is converted to HNO_3 thereby removing NO_2 from the reaction mixture.

21.6 $:O:$ double-bonded (‖) to P; P bonded to $H—\ddot{O}$ and $\ddot{O}:$

21.7 $S_8(s) + 12\ O_2(g) \longrightarrow 8\ SO_3(g)$

$$1.3 \times 10^{10}\ \text{kg}\ S_8 \times \frac{1\ \text{mol}\ S_8}{0.256\ \text{kg}\ S_8} \times \frac{8\ \text{mol}\ SO_3}{1\ \text{mol}\ S_8}$$

$$\times \frac{0.080\ \text{kg}\ SO_3}{1\ \text{mol}\ SO_3} = 3.3 \times 10^{10}\ \text{kg}\ SO_3$$

21.8 (a) Non-redox reaction; no change in oxidation numbers

(b) Redox reaction; F_2 is the oxidizing agent; S_8 is the reducing agent.

Chapter 22

22.1 (a) Mn $[Ar]\,3d^5 4s^2$

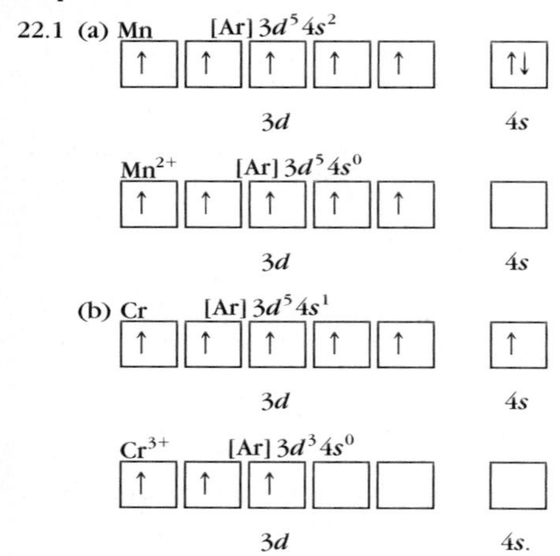

Mn^{2+} $[Ar]\,3d^5 4s^0$

(b) Cr $[Ar]\,3d^5 4s^1$

Cr^{3+} $[Ar]\,3d^3 4s^0$

22.2 Cooling the iron slowly would shift the equilibrium to favor the reverse reaction, the conversion of cementite to iron and carbon (graphite).

22.3 Assume that 50% of the iron comes from each ore.

$$Fe_2O_3\text{: } 5.00 \times 10^2\ \text{kg Fe} \times \frac{159.7\ \text{kg}\ Fe_2O_3}{111.7\ \text{kg Fe}} = 7.15 \times 10^2\ \text{kg}\ Fe_2O_3$$

$$Fe_3O_4\text{: } 5.00 \times 10^2\ \text{kg Fe} \times \frac{231.6\ \text{kg}\ Fe_3O_4}{167.6\ \text{kg Fe}} = 6.91 \times 10^2\ \text{kg}\ Fe_3O_4$$

$$\text{Total} = 7.15 \times 10^2\ \text{kg} + 6.91 \times 10^2\ \text{kg} = 1.41 \times 10^3\ \text{kg}$$

22.4 By controlling the voltage, the zinc could be removed and then the lead.

22.5 $E_{cell} = 0.00\ \text{V} = +0.795\ \text{V} - 0.00985 \log\left(\frac{1}{[H^+]^{14}}\right)$

$$-0.795 = -0.00985 \log \frac{1}{[H^+]^{14}}$$

$$\log \frac{1}{[H^+]^{14}} = \frac{-0.795}{-0.00985} = 80.7$$

$$\frac{1}{[H^+]^{14}} = 10^{80.7} = 5.01 \times 10^{80}$$

$$1 = (5.01 \times 10^{80})[H^+]^{14}$$

$$[H^+]^{14} = \frac{1}{5.01 \times 10^{80}} = 2.00 \times 10^{-81}$$

$$14 \log[H^+] = \log 3.16 \times 10^{-81}$$

$$\log[H^+] = \frac{-80.7}{14} = -5.76$$

$$-\log[H^+] = \text{pH} = 5.76$$

22.6 (a) SO_4^{2-} (b) Cu^{2+} (c) NH_3 (d) $[Cu(NH_3)_4]^{2+}$

22.7 (a) Diamminesilver(I) nitrate

(b) $[Fe(H_2O)_5(NCS)]Cl_2$

22.8 (a) Two (b) Zero (c) Five

22.9 Two

22.10 High-spin: four unpaired electrons; low-spin: two unpaired electrons; both complexes are paramagnetic due to their unpaired electrons.

Answers to Exercises

Chapter 1

1.1 (a) These temperatures can be compared to the boiling point of water, 212 °F or 100 °C. So 110 °C is a higher temperature than 180 °F.
(b) These temperatures can be compared to normal body temperature, 98.6 °F or 37.0 °C. So 36 °C is a lower temperature than 100 °F.
(c) This temperature can be compared to normal body temperature, 37.0 °C. Since body temperature is above the melting point, gallium held in one's hand will melt.

1.2 Reference to the figure on page 10 indicates that kerosene is the top layer, vegetable oil is the middle layer, and water is the bottom layer.
(a) Since the least dense liquid will be the top layer and the densest liquid will be the bottom layer, the densities increase in the order kerosene, vegetable oil, water.
(b) If vegetable oil is added to the tube, the top and bottom layers will remain the same, but the middle layer will become larger.
(c) If kerosene is now added to the tube the top layer will grow, but the middle and bottom layers will remain the same. The order of levels will *not change*. Density does not depend on the quantity of material present. So no matter how much of each liquid is present, the densities increase in the order kerosene, vegetable oil, water.

1.3 (a) Properties: blue (qualitative), melts at 99 °C (quantitative)
Change: melting
(b) Properties: white, cubic (both qualitative)
Change: none
(c) Properties: mass of 0.123 g, melts at 327 °C (both quantitative)
Change: melting
(d) Properties: colorless, vaporizes easily (both qualitative), boils at 78 °C, density of 0.789 g/mL (both quantitative)
Changes: vaporizing, boiling

1.4 Physical change: boiling water
Chemical changes: combustion of propane, cooking the egg

1.5 (a) Homogeneous mixture (solution)
(b) Heterogeneous mixture (contains carbon dioxide gas bubbles in a solution of sugar and other substances in water)
(c) Heterogeneous mixture of dirt and oil
(d) Element; diamond is pure carbon.
(e) Modern quarters (since 1965) are composed of a pure copper core (that can be seen when they are viewed side-on) and an outer layer of 75% Cu, 25% Ni alloy, so they are heterogeneous matter. Pre-1965 quarters are fairly pure silver.
(f) Compound; contains carbon, hydrogen, and oxygen

1.6 (a) Energy from the sun warms the ice and the water molecules vibrate more; eventually they break away from their fixed positions in the solid and liquid water forms. As the temperature of the liquid increases, some of the molecules have enough energy to become widely separated from the other molecules, forming water vapor (gas).
(b) Some of the water molecules in the clothes have enough speed and energy to escape from the liquid state and become water vapor; these molecules are carried away from the clothes by breezes or air currents. Eventually nearly every water molecule in the clothes vaporizes, and the clothes become dry.
(c) Water molecules from the air come into contact with the cold glass, and their speeds are decreased, allowing them to become liquid. As more and more molecules enter the liquid state, droplets form on the glass.
(d) Some water molecules escape from the liquid state, forming water vapor. As more and more molecules escape, the ratio of sugar molecules to water molecules becomes larger and larger, and eventually some sugar molecules start to stick together. As more and more sugar molecules stick to each other, a visible crystal forms. Eventually all of the water molecules escape, leaving sugar crystals behind.

1.7 (a) Tellurium, Te, earth (Latin *tellus* means earth); uranium, U, for Uranus; neptunium, Np, for Neptune; and plutonium, Pu, for Pluto. (Mercury, like the planet Mercury, is named for a Roman god.)
(b) Californium, Cf
(c) Curium, Cm, for Marie Curie; and meitnerium, Mt, for Lise Meitner
(d) Scandium, Sc, for Scandinavia; gallium, Ga, for France (Latin *Gallia* means France); germanium, Ge, for Germany; ruthenium, Ru, for Russia; europium, Eu, for Europe; polonium, Po, for Poland; francium, Fr, for France; americium, Am, for America; californium, Cf, for California
(e) H, He, C, N, O, F, Ne, P, S, Cl, Ar, Se, Br, Kr, I, Xe, At, Rn

1.8 (a) Elements that consist of diatomic molecules are H, N, O, F, Cl, Br, and I; At is radioactive and there is probably less than 50 mg of naturally occurring At on earth, but it does form diatomic molecules; H, N, O, plus group 7A.
(b) Metalloids are B, Si, Ge, As, Sb, and Te; along a zig-zag line from B to Te.

1.9 Tin and lead are two different elements; allotropes are two different forms of the same element, so tin and lead are not allotropes.

Chapter 2

2.1 The movement of the comb though your hair removes some electrons, leaving slight charges on your hair and the comb. The charges must sum to zero; therefore, one must be slightly positive and one must be slightly negative, so they attract each other.

2.2 (a) A nucleus is about one ten-thousandth as large as an atom, so $100 \text{ m} \times (1 \times 10^{-4}) = 1 \times 10^{-2} \text{ m} = 1 \text{ cm}$. (b) Many everyday objects are about 1 cm in size—for example, a grape.

2.3 The statement is wrong because two atoms that are isotopes always have the same number of protons. It is the number of neutrons that varies from one isotope of an element to another.

2.4 Atomic weight of lithium
$= (0.07500)(6.015121 \text{ amu}) + (0.9250)(7.016003 \text{ amu})$
$= 0.451134 \text{ amu} + 6.489802 \text{ amu}$
$= 6.940936 \text{ amu, or } 6.941 \text{ amu}$

2.5 Because the most abundant isotope is magnesium-24 (78.70%), the atomic weight of magnesium is closer to 24 than to 25 or 26, the mass numbers of the other magnesium isotopes, which make up approximately 21% of the remaining mass. The simple arithmetic average is $(24 + 25 + 26)/3 = 25$, which is larger than the atomic weight. In the arithmetic average, the relative abundance of each magnesium isotope is 33%, far less than the actual percent abundance of magnesium-24, and much more than the natural percent abundances of magnesium-25 and magnesium-26.

2.6 There is no reasonable pair of values of the mass numbers for Ga that would have an average value of 69.72.

2.7 Start by calculating the number of moles in 10.00 g of each element.

$$10.00 \text{ g Li} \times \frac{1 \text{ mol Li}}{6.941 \text{ g Li}} = 1.441 \text{ mol Li}$$

$$10.00 \text{ g Ir} \times \frac{1 \text{ mol Ir}}{192.22 \text{ g Ir}} = 0.05202 \text{ mol Ir}$$

Multiply the number of moles of each element by Avogadro's number.

$$1.441 \text{ mol Li} \times 6.022 \times 10^{23} \text{ atoms/mol} = 8.678 \times 10^{23} \text{ atoms Li}$$

$$0.05202 \text{ mol Ir} \times 6.022 \times 10^{23} \text{ atoms/mol} = 3.133 \times 10^{22} \text{ atoms Ir}$$

Find the difference.

$$(8.678 \times 10^{23}) - (0.3133 \times 10^{23})$$
$$= 8.365 \times 10^{23} \text{ more atoms of Li than Ir}$$

2.8 1. (a) 13 metals: potassium (K), calcium (Ca), scandium (Sc), titanium (Ti), vanadium (V), chromium (Cr), manganese (Mn), iron (Fe), cobalt (Co), nickel (Ni), copper (Cu), zinc (Zn), and gallium (Ga)

(b) Three nonmetals: selenium (Se), bromine (Br), and krypton (Kr)

(c) Two metalloids: germanium (Ge) and arsenic (As)

2. (a) Groups 1A (except hydrogen), 2A, 1B, 2B, 3B, 4B, 5B, 6B, 7B, 8B

(b) Groups 7A and 8 (c) None

3. Period 6

2.9 The question should ask: Which of the following men does not have an *element* named after him? The correct answer to the question, after it is properly posed, is Isaac Newton.

Chapter 3

3.1 Propylene glycol structural formula:

```
    H  OH OH
    |  |  |
 H—C—C—C—H
    |  |  |
    H  H  H
```

Condensed formula:

```
    OHOH
     | |
 CH3CHCH2
```

Molecular formula: $C_3H_8O_2$

3.2 (a) CS_2 (b) PCl_3 (c) SBr_2 (d) SeO_2 (e) OF_2 (f) XeO_3

3.3 (a) $C_{16}H_{34}$ and $C_{28}H_{58}$ (b) $C_{14}H_{30}$, 14 carbon atoms and 30 hydrogen atoms

3.4 The structural and condensed formulas for three constitutional isomers of five-carbon alkanes (pentanes) are

```
    H  H  H  H  H
    |  |  |  |  |
 H—C—C—C—C—C—H     CH3CH2CH2CH2CH3
    |  |  |  |  |
    H  H  H  H  H

    H  H  H  H
    |  |  |  |
 H—C—C—C—C—H       CH3CH2CHCH3
    |  |  |  |                 |
    H  H  |  H                CH3
       H—C—H
          |
          H

          H
          |
       H—C—H
    H     |     H             CH3
    |     |     |              |
 H—C—————C—————C—H       CH3—C—CH3
    |     |     |              |
    H     |     H             CH3
       H—C—H
          |
          H
```

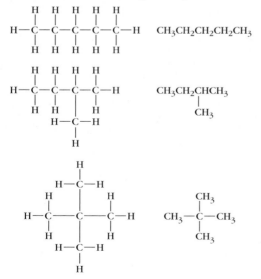

3.5 The compound is a solid at room temperature and is soluble in water, so it is likely to be an ionic compound.

3.6 (a) 174.18 g/mol (b) 386.66 g/mol (c) 398.07 g/mol (d) 194.19 g/mol

3.7 The statement is true. Because both compounds have the same formula, they have the same molar mass. Thus, 100 g of each compound contains the same number of moles.

3.8 Epsom salt is $MgSO_4 \cdot 7H_2O$, which has a molar mass of 246 g/mol.

$$20 \text{ g} \times \frac{1 \text{ mol}}{246 \text{ g}} = 8.1 \times 10^{-2} \text{ mol Epsom salt}$$

3.9 (a) SF_6 molar mass is 146.06 g/mol; 1.000 mol SF_6 contains 32.07 g S and 18.9984 × 6 = 113.99 g F. The mass percents are

$$\frac{32.07 \text{ g S}}{146.06 \text{ g } SF_6} \times 100\% = 21.96\% \text{ S}$$

$$100.0\% - 21.96\% = 78.04\% \text{ F}$$

(b) $C_{12}H_{22}O_{11}$ has a molar mass of 342.3 g/mol; 1.000 mol $C_{12}H_{22}O_{11}$ contains

$$12.011 \times 12 = 144.13 \text{ g C}$$
$$1.0079 \times 22 = 22.174 \text{ g H}$$
$$15.9994 \times 11 = 175.99 \text{ g O}$$

The mass percents of the three elements are

$$\frac{144.13 \text{ g C}}{342.3 \text{ g}} \times 100\% = 42.12\% \text{ C}$$

$$\frac{22.174 \text{ g H}}{342.3 \text{ g}} \times 100\% = 6.478\% \text{ H}$$

$$\frac{175.99 \text{ g O}}{342.3 \text{ g}} \times 100\% = 51.41\% \text{ O}$$

(c) $Al_2(SO_4)_3$ molar mass is 342.15 g/mol; 1.000 mol $Al_2(SO_4)_3$ contains

$$26.9815 \times 2 = 53.96 \text{ g Al}$$
$$32.066 \times 3 = 96.20 \text{ g S}$$
$$15.9994 \times 12 = 192.0 \text{ g O}$$

The mass percents of the three elements are

$$\frac{53.96 \text{ g Al}}{342.15 \text{ g}} \times 100\% = 15.77\% \text{ Al}$$

$$\frac{96.20 \text{ g S}}{342.15 \text{ g}} \times 100\% = 28.12\% \text{ S}$$

$$\frac{192.0 \text{ g O}}{342.15 \text{ g}} \times 100\% = 56.12\% \text{ O}$$

(d) $U(OTeF_5)_6$ molar mass is 1669.6 g/mol; 1.000 mol $U(OTeF_5)_6$ contains 238.0289 g of U and

$$15.9994 \times 6 = 96.00 \text{ g O}$$
$$127.60 \times 6 = 765.6 \text{ g Te}$$
$$18.9984 \times 30 = 570.0 \text{ g F}$$

The mass percents of the four elements are

$$\frac{238.0289 \text{ g U}}{1669.6 \text{ g}} \times 100\% = 14.26\% \text{ U}$$

$$\frac{96.00 \text{ g O}}{1669.6 \text{ g}} \times 100\% = 5.750\% \text{ O}$$

$$\frac{765.6 \text{ g Te}}{1669.6 \text{ g}} \times 100\% = 45.86\% \text{ Te}$$

$$\frac{570.0 \text{ g F}}{1669.6 \text{ g}} \times 100\% = 34.14\% \text{ F}$$

3.10 (a) Carbon, nitrogen, oxygen, phosphorus, hydrogen, selenium, sulfur, Cl, Br, I (b) Calcium and magnesium (c) Chloride, bromide, and iodide (d) Iron, copper, zinc, vanadium (also chromium, manganese, cobalt, nickel, molybdenum, and cadmium)

Chapter 4

4.1 One mol of methane reacts with 2 mol oxygen to produce 1 mol carbon dioxide and 2 mol water.

4.2 (a) The total mass of reactants $\{4\ Fe(s) + 3\ O_2(g)\}$ must equal the total mass of products $\{2\ Fe_2O_3(s)\}$, which is 2.50 g.
(b) The stoichiometric coefficients are 4, 3, and 2.
(c) $1.000 \times 10^4 \text{ O atoms} \times \frac{1\ O_2 \text{ molecule}}{2 \text{ O atoms}} \times \frac{4 \text{ Fe atoms}}{3\ O_2 \text{ molecules}} = 6.667 \times 10^3 \text{ Fe atoms}$

4.3 (a) Not balanced; the number of oxygen atoms do not match.
(b) Not balanced; the number of bromine atoms do not match.
(c) Not balanced; the number of sulfur atoms do not match.

4.4 (a) To predict the product of a combination reaction between two elements, we need to know the ion that will be formed by each element when combined. (b) For calcium, Ca^{2+} ions are formed, and for fluorine, F^- ions are formed. (c) The product is CaF_2.

4.5 (a) Magnesium chloride, $MgCl_2$
(b) Magnesium oxide, MgO, and carbon dioxide, CO_2

4.6 $\frac{2 \text{ mol Al}}{3 \text{ mol } Br_2}$, $\frac{2 \text{ mol Al}}{1 \text{ mol } Al_2Br_6}$, $\frac{3 \text{ mol } Br_2}{1 \text{ mol } Al_2Br_6}$, and their reciprocals

4.7 $0.300 \text{ mol } CH_4 \times \frac{2 \text{ mol } H_2O}{1 \text{ mol } CH_4} = 0.600 \text{ mol } H_2O$

$$0.600 \text{ mol } H_2O \times \frac{18.02 \text{ g } H_2O}{1 \text{ mol } H_2O} = 10.8 \text{ g } H_2O$$

4.8 (a) $300.\text{ g urea} \times \frac{1 \text{ mol urea}}{60.06 \text{ g urea}} \times \frac{2 \text{ mol } NH_3}{1 \text{ mol urea}} \times \frac{17.03 \text{ g } NH_3}{1 \text{ mol } NH_3} = 170.\text{ g urea}$

$$100.\text{ g } H_2O \times \frac{1 \text{ mol } H_2O}{18.02 \text{ g } H_2O} \times \frac{2 \text{ mol } NH_3}{1 \text{ mol } H_2O} \times \frac{17.03 \text{ g } NH_3}{1 \text{ mol } NH_3} = 189.\text{ g } HN_3$$

Therefore, urea is the limiting reactant.
(b) $176.\text{ g } NH_3$

$$300.\text{ g urea} \times \frac{1 \text{ mol urea}}{60.06 \text{ g urea}} \times \frac{1 \text{ mol } H_2O}{1 \text{ mol urea}} \times \frac{44.01 \text{ g } CO_2}{1 \text{ mol } CO_2} = 220.\text{ g } CO_2$$

(c) $300.\text{ g urea} \times \frac{1 \text{ mol urea}}{60.06 \text{ g urea}} \times \frac{1 \text{ mol } H_2O}{1 \text{ mol urea}} \times \frac{18.02 \text{ g } H_2O}{1 \text{ mol } H_2O} = 90.0 \text{ g } H_2O$

$100.\text{ g} - 90.0 \text{ g} = 10.0 \text{ g } H_2O$ remains

4.9 (1) Impure reactants; (2) Inaccurate weighing of reactants and products

4.10 Assuming that the nicotine is pure, weigh a sample of nicotine and burn the sample. Separately collect and weigh the carbon dioxide and water generated, and calculate the moles and grams of carbon and hydrogen collected. By mass difference, determine the mass of nitrogen in the original sample, then calculate the moles of nitrogen. Calculate the mole ratios of carbon, hydrogen, and nitrogen in nicotine to determine its empirical formula.

Chapter 5

5.1 It is possible for an exchange reaction to form two different precipitates—for example, the reaction between barium hydroxide and iron(II) sulfate:

$$Ba(OH)_2(aq) + FeSO_4(aq) \rightarrow BaSO_4(s) + Fe(OH)_2(s)$$

5.2 $H_3PO_4(aq) \rightleftharpoons H_2PO_4^-(aq) + H^+(aq)$
$H_2PO_4^-(aq) \rightleftharpoons HPO_4^{2-}(aq) + H^+(aq)$
$HPO_4^{2-}(aq) \rightleftharpoons PO_4^{3-}(aq) + H^+(aq)$

5.3 (a) Hydrogen ions and perchlorate ions:

$$HClO_4(aq) \rightarrow H^+(aq) + ClO_4^-(aq)$$

(b) $Ca(OH)_2(aq) \rightarrow Ca^{2+}(aq) + 2\ OH^-(aq)$

5.4 (a) $H^+(aq) + Cl^-(aq) + K^+(aq) + OH^-(aq) \rightarrow H_2O(\ell) + K^+(aq) + Cl^-(aq)$
$H^+(aq) + OH^-(aq) \rightarrow H_2O(\ell)$
(b) $2\ H^+(aq) + SO_4^{2-}(aq) + Ba^{2+}(aq) + 2\ OH^-(aq) \rightarrow 2\ H_2O(\ell) + BaSO_4(s)$
$H^+(aq) + OH^-(aq) \rightarrow H_2O(\ell)$
$Ba^{2+}(aq) + SO_4^{2-}(aq) \rightarrow BaSO_4(s)$
(c) $2\ CH_3\ COOH(aq) + Ca^{2+}(aq) + 2\ OH^-(aq) \rightarrow Ca^{2+}(aq) + 2\ CH_3COO^-(aq) + 2\ H_2O(\ell)$
$CH_3\ COOH(aq) + OH^-(aq) \rightarrow H_2O(\ell) + CH_3COO^-(aq)$

5.5 $Al(OH)_3(s) + 3\ H^+(aq) + 3\ Cl^-(aq) \rightarrow 3\ H_2O(\ell) + Al^{3+}(aq) + 3\ Cl^-(aq)$
$H^+(aq) + OH^-(aq) \rightarrow H_2O(\ell)$

5.6 (a) The products are aqueous sodium sulfate, water, and carbon dioxide gas.

$$Na_2CO_3(aq) + H_2SO_4(aq) \rightarrow Na_2SO_4(aq) + H_2O(\ell) + CO_2(g)$$
$$2\ H^+(aq) + CO_3^{2-}(aq) \rightarrow H_2O(\ell) + CO_2(g)$$

(b) The products are aqueous iron(II) chloride and hydrogen sulfide gas.

$$FeS(s) + 2\ HCl(aq) \rightarrow FeCl_2(aq) + H_2S(g)$$
$$2\ H^+(aq) + S^{2-}(aq) \rightarrow H_2S(g)$$

(c) The products are aqueous potassium chloride, water, and sulfur dioxide gas.

$$K_2SO_3(aq) + 2\ HCl(aq) \rightarrow 2\ KCl(aq) + H_2O(\ell) + SO_2(g)$$
$$2\ H^+(aq) + SO_3^{2-}(aq) \rightarrow H_2O(\ell) + SO_2(g)$$

5.7 (a) Gas-forming reaction; the products are aqueous nickel sulfate, water, and carbon dioxide gas.

$$NiCO_3(s) + H_2SO_4(aq) \rightarrow NiSO_4(aq) + H_2O(\ell) + CO_2(g)$$
$$NiCO_3(s) + 2\ H^+(aq) \rightarrow Ni^{2+}(aq) + H_2O(\ell) + CO_2(g)$$

(b) Acid-base reaction; nitric acid reacts with strontium hydroxide, a base, to produce water and strontium nitrate, a salt.

$$2\ HNO_3(aq) + Sr(OH)_2(s) \rightarrow Sr(NO_3)_2(aq) + 2\ H_2O(\ell)$$
$$Sr(OH)_2(s) + 2\ H^+(aq) \rightarrow Sr^{2+}(aq) + 2\ H_2O(\ell)$$

(c) Precipitation reaction; aqueous sodium chloride and insoluble barium oxalate are produced.

$$BaCl_2(aq) + Na_2C_2O_4(aq) \rightarrow BaC_2O_4(s) + 2\ NaCl(aq)$$
$$Ba^{2+}(aq) + C_2O_4^{2-}(aq) \rightarrow BaC_2O_4(s)$$

(d) Precipitation and gas-forming reaction; lead sulfate precipitates and carbon dioxide gas is released.

$$PbCO_3(aq) + H_2SO_4(aq) \rightarrow PbSO_4(s) + H_2O(\ell) + CO_2(g)$$

$$Pb^{2+}(aq) + SO_4^{2-}(aq) \rightarrow PbSO_4(s)$$

$$2\,H^+(aq) + CO_3^{2-}(aq) \rightarrow H_2O(\ell) + CO_2(g)$$

5.8 In the reaction $2\,Ca(s) + O_2(g) \rightarrow 2\,CaO(s)$, Ca loses electrons, is oxidized, and is the reducing agent; O gains electrons, is reduced, and is the oxidizing agent.

5.9 $Cl_2(g) + Ca(s) \rightarrow CaCl_2(s)$. $Cl_2(g)$ is the oxidizing agent.

5.10 (a) This is not a redox reaction. Nitric acid is a strong oxidizing agent, but here it serves as an acid.
(b) In this redox reaction, chromium metal (Cr) is oxidized (loses electrons) to form Cr^{3+} ions in Cr_2O_3; oxygen (O_2) is reduced (gains electrons) to form oxide ions, O^{2-}. Oxygen is the oxidizing agent, and chromium is the reducing agent.
(c) This is an acid-base reaction, but not a redox reaction; there are no strong oxidizing or reducing agents present.
(d) Copper is oxidized and chlorine is reduced in this redox reaction, in which copper is the reducing agent and chlorine is the oxidizing agent. The equations are $Cu \rightarrow Cu^{2+} + 2e^-$ and $Cl_2 + 2e^- \rightarrow 2\,Cl^-$.

5.11 (a) Carbon in oxalate ion, $C_2O_4^{2}$ (oxidation state = 13), is oxidized to oxidation state +4 in CO_2.
(b) Carbon is reduced from +4 in CCl_2F_2 to 0 in C(s).

5.12 (a) $CH_3CH_2OH(\ell) + 3\,O_2(g) \rightarrow 3\,H_2O(\ell) + 2\,CO_2(g)$; redox
(b) $2\,Fe(s) + 6\,HNO_3(aq) \rightarrow 2\,Fe(NO_3)_3(aq) + 3\,H_2(g)$; redox
(c) $AgNO_3(aq) + KBr(aq) \rightarrow AgBr(s) + KNO_3(aq)$; not redox

5.13 Molar mass of cholesterol = 386.7 g/mol

$$240\text{ mg} \times \frac{1\text{ g}}{10^3\text{ mg}} \times \frac{1\text{ mol}}{386.7\text{ g}} = 6.21 \times 10^{-4}\text{ mol cholesterol}$$

$$\frac{6.21 \times 10^{-4}\text{ mol}}{0.100\text{ L}} = 6.2 \times 10^{-3}\text{ M}$$

5.14 (a) $$6.37\text{ g Al(NO}_3)_3 \times \frac{1\text{ mol Al(NO}_3)_3}{213.0\text{ g Al(NO}_3)_3} = 0.0299\text{ mol Al(NO}_3)_3;$$

$$\frac{0.0299\text{ mol}}{0.250\text{ L}} = 0.120\text{ M Al(NO}_3)_3$$

(b) Molarity: $Al^{3+} = 0.120$; $NO_3^- = 3(0.120) = 0.360$

5.15 If the description of solution preparation is always worded in terms of adding enough solvent to make a specific volume of solution, then any possible expansion or contraction has no effect on the molarity of the solution. The denominator of the definition of molarity is liters of *solution.*

5.16 The moles of HCl in the concentrated solution are given by (6.0 mol/L)(0.100 L) = 0.6 mol HCl. The moles of HCl in the dilute solution are given by (1.20 mol/L)(0.500 L) = 0.6 mol HCl.

5.17 The molarity could be increased by evaporating some of the solvent.

5.18 $$0.0193\text{ L} \times \frac{0.200\text{ mol AgNO}_3}{1\text{ L}} \times \frac{1\text{ mol Ag}^+}{1\text{ mol AgNO}_3} \times \frac{1\text{ mol Cl}^-}{1\text{ mol Ag}^+} \times \frac{1\text{ mol NaCl}}{1\text{ mol Cl}^-} = 3.86 \times 10^{-3}\text{ mol NaCl}$$

$$\frac{3.86 \times 10^{-3}\text{ mol NaCl}}{0.0250\text{ L}} = 0.154\text{ M NaCl}$$

Chapter 6

6.1 You transfer some mechanical energy to the ball to accelerate it upward. The ball's potential energy increases the higher it gets, but its kinetic energy decreases by an equal quantity, and eventually it stops rising and begins to fall. As it falls, some of the ball's potential energy changes to kinetic energy, and the ball goes faster and faster until it hits the floor. When the ball hits the floor, some of its kinetic energy is transferred to the atoms, molecules, or ions that make up the floor, causing them to move faster. Eventually all of the ball's kinetic energy is transferred, and the ball stops moving. The nanoscale particles in the floor (and some in the air that the ball fell through) are moving faster on average, and the temperature of the floor (and the air) is slightly higher. The energy has spread out over a much larger number of particles.

6.2
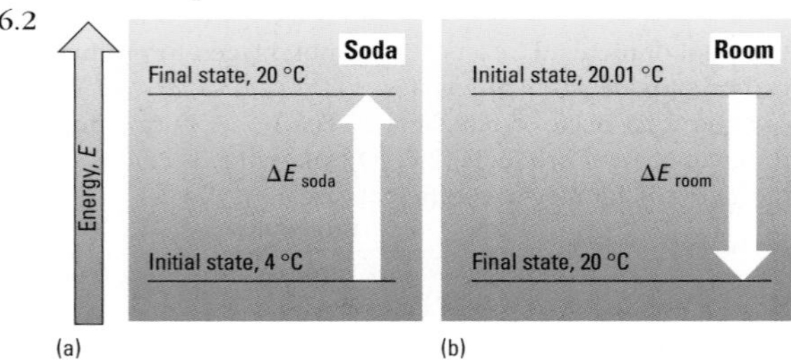

6.3 (a) Heat transfer $= (25.0\ °C - 1.0\ °C)\left(\dfrac{1.5\text{ kJ}}{1.0\ °C}\right) = 36\text{ kJ}$
(b) The system is the can and the liquid it contains.
(c) The surroundings includes the air and other materials in contact with the can, or close to the can.
(d) ΔE is negative because the system transferred energy to the surroundings as it cooled; $\Delta E = -36$ kJ.
(e)
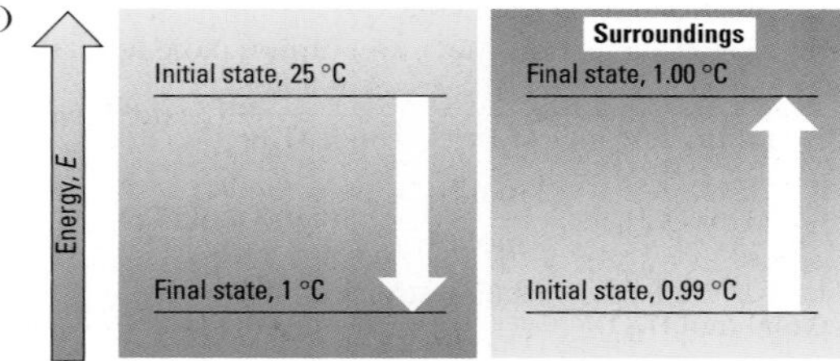

The surroundings are warmed very slightly, say from 0.99 °C to 1.00 °C, by the heat transfer from the can of soda.

6.4 The same quantity of energy is transferred out of each beaker and the mass of each sample is the same. Therefore the sample with the smaller specific heat capacity will cool more. Look up the specific heat capacities in Table 6.1. Because glass has a larger specific heat capacity than carbon, the carbon will cool more and therefore will have the lower temperature.

6.5 The calculation for Al is given as an example.

$$\text{Molar heat capacity} = \frac{0.902\text{ J}}{\text{g °C}} \times \frac{26.98\text{ g}}{1\text{ mol}} = 24.3\text{ J mol}^{-1}\text{ °C}^{-1}$$

Metal	Molar Heat Capacity ($J\ mol^{-1}\ °C^{-1}$)	Metal	Molar Heat Capacity ($J\ mol^{-1}\ °C^{-1}$)
Al	24.3	Cu	24.5
Fe	25.2	Au	25.2

The molar heat capacities of most metals are close to 25 J mol^{-1} $°C^{-1}$. This rule does not work for ethanol or other compounds listed in Table 6.2.

6.6 (a) Since the heat of vaporization is almost seven times larger than the heat of fusion, the temperature stays constant at 100 °C almost seven times longer than it stays constant at 0 °C. It stays constant at 0 °C for slightly less time than it takes to heat the water from 0 °C to 100 °C (see graph). Because the heating is at a constant rate, time is proportional to quantity of energy transferred.

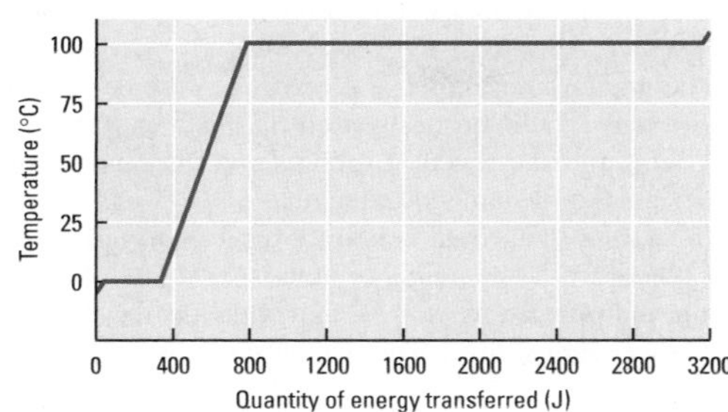

(b) The mass of water is half as great as in part (a), so each process takes half as long. A graph to the same scale as in part (a) begins at 105 °C and reaches −5 °C with half the quantity of energy transferred.

6.7 Heat of fusion: 237 g × 333 J/g = 78.9 kJ
Heating liquid: 237 g × 4.184 J g^{-1} $°C^{-1}$ × 100.0 °C = 99.2 kJ
Heat of vaporization: 237 g × 2260 J/g = 536 kJ
Total heating = (78.9 + 99.2 + 536) kJ = 714 kJ

6.8 The direction of energy transfer is indicated by the sign of the enthalpy change. Transfer to the system corresponds to a positive enthalpy change.

6.9 Because of heats of fusion and heats of vaporization, the enthalpy change is different when a reactant or product is in a different state.

6.10 When 1.0 mol of H_2 reacts (Equation 6.3), $\Delta H = -241.8$ kJ. When half that much H_2 reacts, ΔH is half as great; that is, 0.5 × (−241.8) = −120.9 kJ.

6.11 $\frac{-92.22 \text{ kJ}}{1 \text{ mol } N_2(g)}$ $\frac{-92.22 \text{ kJ}}{3 \text{ mol } H_2(g)}$ $\frac{-92.22 \text{ kJ}}{2 \text{ mol } NH_3(g)}$

$\frac{1 \text{ mol } N_2(g)}{-92.22 \text{ kJ}}$ $\frac{3 \text{ mol } H_2(g)}{-92.22 \text{ kJ}}$ $\frac{2 \text{ mol } NH_3(g)}{-92.22 \text{ kJ}}$

6.12 The reaction used must be exothermic. Because it can be started by opening the package, it probably involves oxygen from the air, and the sealed package prevents the reaction from occurring before it is needed. Many metals can be oxidized easily and exothermically. The reaction of iron with oxygen (as in the *Chemistry You Can Do* experiment) is a good candidate.

6.13 Yes, it would violate the first law of thermodynamics. According to the supposition, we could create energy by starting with 2 mol HCl, breaking all the molecules apart, recombining the atoms to form 1 mol H_2 and 1 mol Cl_2, and then reacting the H_2 and Cl_2 to give 2 HCl.

$$2\ HCl \longrightarrow \text{atoms} \longrightarrow H_2 + Cl_2 \qquad \Delta H° = +185 \text{ kJ}$$
$$H_2 + Cl_2 \longrightarrow 2\ HCl \qquad \Delta H° = -190 \text{ kJ}$$

The net effect of these two processes is that there is still 2 mol HCl, but 5 kJ of energy has been created. This is impossible according to the first law of thermodynamics.

6.14 (a) In the reaction 2 HF → H_2 + F_2 there are two bonds in the two reactant molecules and two bonds in the two product molecules. Since the reaction is endothermic, the bonds in the reactant molecules must be stronger than in the products.
(b) For the reaction 2 H_2O → 2 H_2 + O_2, there are four bonds in the two reactant molecules but only three bonds in the three product molecules. The reaction is endothermic because more bonds are broken than are formed.

6.15 $C_6H_{12}O_6(s) + 6\ O_2(g) \rightarrow 6\ CO_2(g) + 6\ H_2O(\ell)$
Because the volume of any ideal gas is proportional to the amount (moles) of gas, and because there are 6 mol of gaseous reactant and 6 mol of gaseous product, there will be very little change in volume. Almost no work will be done, and $\Delta H \cong \Delta E$.

6.16 In the Problem-Solving Example 6.11 the reaction produced 0.25 mol NaCl and the heat transfer associated with the reaction caused 500 mL of solution to warm by 7 °C.
(a) Here the reaction produces 0.20 mol NaCl by neutralizing 0.20 mol NaOH and the heat transfer warms 400. mL of water. So there is less heat transfer but it will be heating a smaller volume. The quantity of reaction is $\frac{0.20 \text{ mol}}{0.25 \text{ mol}}$ = 0.80 as much, so the heat transfer is 0.80 as much. The quantity of water to be heated is $\frac{400.\text{mL}}{500.\text{mL}}$ = 0.800 as much. Therefore the combined effect on temperature is the same as in Problem-Solving Example 6.11, and the temperature change associated with this process will also be 7 °C.
(b) Here the limiting reactant is 0.10 mol NaOH. (Only half of the H_2SO_4 is used up.) The heat transfer from the reaction warms 200. mL of water. So there is less heat transfer but it will heat a smaller volume. The quantity of reaction is $\frac{0.10 \text{ mol}}{0.25 \text{ mol}}$ = 0.40 as much, so the heat transfer is 0.40 as much. The quantity of water is $\frac{200.}{400.}$ = 0.40 as much. Therefore once again the combined effect on temperature is the same as in Problem-Solving Example 6.11, and the temperature change associated with this process will also be 7 °C.

6.17 $N_2(g) \rightarrow N_2(g)$
(a) The product is the same as the reactant, so there is no change—nothing happens.
(b) Since product and reactant are the same, $\Delta H = 0$.

6.18 (a) It takes 160 kJ to break 1 mol N—N bonds, 4 × 391 kJ to break 4 mol N—H bonds, and 498 kJ to break 1 mol of bonds in O_2.
(b) Forming 1 mol of bonds in N_2 releases 946 kJ, and forming 4 mol O—H bonds releases 4 × 467 kJ.
(c) Therefore, $\Delta H = \{160 + (4 \times 391) + 498\}$ kJ $- \{946 + (4 \times 467)\}$ kJ $= -592$ kJ

$$N_2H_4(g) + O_2(g) \rightarrow N_2(g) + 2\ H_2O(g) \qquad \Delta H° = -592 \text{ kJ}$$

6.19 (a) $CH_4(g) + 2\ O_2(g) \rightarrow CO_2(g) + 2\ H_2O(g)$
$\Delta H° = \{-393.509 + 2(-241.818) - (-74.81)\}$ kJ
$= -802.34$ kJ
$$\frac{802.34 \text{ kJ}}{1 \text{ mol } CH_4} \times \frac{1 \text{ mol}}{16.0426 \text{ g}} = 50.013 \text{ kJ/g } CH_4$$
(b) $C_8H_{18}(\ell) + \frac{25}{2} O_2(g) \rightarrow 8\ CO_2(g) + 9\ H_2O(g)$
$\Delta H° = \{8(-393.509) + 9(-241.818) - (-249.952)\}$ kJ
$= -5074.48$ kJ
$$\frac{5074.48 \text{ kJ}}{1 \text{ mol } C_8H_{18}} \times \frac{1 \text{ mol}}{114.23 \text{ g}} = 44.423 \text{ kJ/g } C_8H_{18}$$
(c) $C_2H_5OH(\ell) + 3\ O_2(g) \rightarrow 2\ CO_2 + 3\ H_2O(g)$
$\Delta H° = \{2(-393.509) + 3(-241.818) - (-277.69)\}$ kJ
$= -1234.782$ kJ
$$\frac{1234.782 \text{ kJ}}{1 \text{ mol } C_2H_5OH(\ell)} \times \frac{1 \text{ mol}}{46.068 \text{ g}} = 26.8 \text{ kJ/g } C_2H_5OH$$
(d) $N_2H_4(\ell) + O_2(g) \rightarrow N_2(g) + 2\ H_2O(g)$
$\Delta H° = \{2(-241.818) - 50.63\}$ kJ $= -534.27$ kJ
$$\frac{534.26 \text{ kJ}}{1 \text{ mol } N_2H_4} \times \frac{1 \text{ mol}}{32.045 \text{ g}} = 16.672 \text{ kJ/g } N_2H_4$$
(e) $H_2(g) + \frac{1}{2} O_2(g) \rightarrow H_2O(g)$ $\qquad \Delta H° = -241.818$ kJ
$$\frac{241.818 \text{ kJ}}{1 \text{ mol } H_2} \times \frac{1 \text{ mol}}{2.0158 \text{ g}} = 119.96 \text{ kJ/g } H_2$$
(f) $C_6H_{12}O_6(s) + 6\ O_2(g) \rightarrow 6\ CO_2(g) + 6\ H_2O(g)$
$\Delta H° = \{6(-393.509) + 6(-241.818) - (-1274.4)\}$ kJ
$= -2537.6$ kJ
$$\frac{2537.6 \text{ kJ}}{1 \text{ mol } C_6H_{12}O_6(s)} \times \frac{1 \text{ mol}}{180.16 \text{ g}} = 14.085 \text{ kJ/g } C_6H_{12}O_6$$
(g) Biomass gives the same result as glucose.
Hydrogen has the greatest fuel value. Octane has the greatest energy density. Its fuel value is more than twice that of the other

liquids and solids, and its density is far greater than for $CH_4(g)$ and $H_2(g)$. Energy density values are

$$CH_4(g) \quad \frac{50.013\ \text{kJ}}{1\ \text{g}} \times \frac{6.9 \times 10^{-4}\ \text{g}}{1\ \text{mL}} = 3.4 \times 10^{-2}\ \text{kJ/mL}$$

$$C_8H_{18}(\ell) \quad \frac{44.423\ \text{kJ}}{1\ \text{g}} \times \frac{0.70\ \text{g}}{1\ \text{mL}} = 31\ \text{kJ/mL}$$

$$C_2H_5OH(\ell) \quad \frac{26.8\ \text{kJ}}{1\ \text{g}} \times \frac{0.80\ \text{g}}{1\ \text{mL}} = 21\ \text{kJ/mL}$$

$$N_2H_4(\ell) \quad \frac{16.672\ \text{kJ}}{1\ \text{g}} \times \frac{1.00\ \text{g}}{1\ \text{mL}} = 16.7\ \text{kJ/mL}$$

$$H_2(g) \quad \frac{119.96\ \text{kJ}}{1\ \text{g}} \times \frac{8.2 \times 10^{-5}\ \text{g}}{1\ \text{mL}} = 9.8 \times 10^{-3}\ \text{kJ/mL}$$

$$\text{Carbohydrate} \quad \frac{14.1\ \text{kJ}}{1\ \text{g}} \times \frac{1.56\ \text{g}}{1\ \text{mL}} = 22.0\ \text{kJ/mL}$$

6.20 According to the thermochemical equation, 2801.6 kJ is released per mole of glucose.

$$\frac{2801.6\ \text{kJ}}{1\ \text{mol}} \times \frac{1\ \text{mol}}{180.16\ \text{g}} \times \frac{1\ \text{kcal}}{4.184\ \text{kJ}} \times \frac{1\ \text{Cal}}{1\ \text{kcal}} = 3.717\ \text{Cal}$$

This rounds to 4 Cal.

6.21 (a) The average BMR is given in the text as 1750 Cal/day for a 70-kg male at rest. So the BMR is

$$\frac{1750\ \text{Cal}}{\text{d}} \times \frac{1000\ \text{cal}}{\text{Cal}} \times \frac{4.184\ \text{J}}{\text{cal}} = 7.32 \times 10^6\ \text{J/day}$$

$$(7.322 \times 10^6\ \text{J/day}) \times (1\ \text{day/24 h}) \times (1\ \text{h/60 min}) \times (1\ \text{min/60 s}) = 84.75\ \text{W}$$

(b) The average BMR for a 70-kg male playing basketball is 7 times the above or 593.2 W. Typical incandescent light bulbs are in the range of 50–250 W.

Chapter 7

7.1 Wavelength and frequency are inversely related. Therefore, low-frequency radiation has long-wavelength radiation.

7.2 Cellular phones use higher frequency radio waves.

7.3 $E = h\nu; \quad \lambda = \dfrac{hc}{E}$

$$E = h\nu = (6.626 \times 10^{-34}\ \text{J}\cdot\text{s})(2.45 \times 10^9\ \text{s}^{-1})$$

$$= 1.62 \times 10^{-24}\ \text{J}$$

$$\lambda = \frac{(6.626 \times 10^{-34}\ \text{J}\cdot\text{s})(2.998 \times 10^8\ \text{m/s})}{1.62 \times 10^{-24}\ \text{J}} = 1.23 \times 10^{-1}\ \text{m}$$

7.4 In a sample of excited hydrogen gas there are many atoms, and each can exist in one of the excited states possible for hydrogen. The observed spectral lines are a result of all the possible transitions of all these hydrogen atoms.

7.5 (a) Emitted (b) Absorbed
(c) Emitted (d) Emitted

7.6 $(2.179 \times 10^{-18}\ \text{J/photon})$

$$\times \left(\frac{1\ \text{kJ}}{10^3\ \text{J}}\right)\left(\frac{6.022 \times 10^{23}\ \text{photons}}{1\ \text{mol}}\right)$$

$$= 1312\ \text{kJ/mol photons}$$

7.7 (a) 5*d* (b) 4*f* (c) 6*p*

7.8 The $n = 3$ level can have only three types of sublevels—*s*, *p*, and *d*. The $n = 2$ level can have only *s* and *p* sublevels, not *d* sublevels ($l = 2$).

7.9 3, 0, 0, $+\frac{1}{2}$; 3, 0, 0 $-\frac{1}{2}$

7.10 (a) 3, 0, 0, $+\frac{1}{2}$ (b) 3, 1, 1, $+\frac{1}{2}$

7.11 Electron a is in the $3p_y$ orbital. Electron b is in the $3p_z$ orbital.

7.12 (a) The maximum number of electrons in the $n = 3$ level is 18 (2 electrons per orbital). The orbitals would be designated 3*s*, $3p_x$, $3p_y$, $3p_z$, $3d_{z^2}$, $3d_{xy}$, $3d_{yz}$, $3d_{xz}$, and $3d_{x^2-y^2}$.
(b) The maximum number of electrons in the $n = 5$ level is 50. The orbitals would be designated 5*s*, $5p_x$, $5p_y$, $5p_z$, $5d_{yz}$, $5d_{xz}$, $5dz^2$, $5d_{xy}$, and the seven 5*f* orbitals and the nine 5*g*, which are not designated by name in the text.

7.13 The first shell that could contain *g* orbitals would be the $n = 5$ shell. There would be nine *g* orbitals.

7.14 For the chlorine atom, $n = 3$, and there are seven electrons in this highest energy level. The configuration is

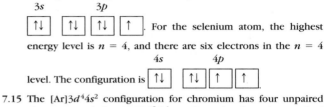

3*s* [↑↓] 3*p* [↑↓][↑↓][↑]. For the selenium atom, the highest energy level is $n = 4$, and there are six electrons in the $n = 4$ level. The configuration is 4*s* [↑↓] 4*p* [↑↓][↑][↑].

7.15 The $[\text{Ar}]3d^44s^2$ configuration for chromium has four unpaired electrons, and the $[\text{Ar}]3d^54s^1$ configuration has six unpaired electrons.

7.16 The $Fe(acac)_2$ contains an Fe^{2+} ion with a 3*d* electron configuration of [↑↓][↑][↑][↑][↑]. This configuration has four unpaired electrons. The compound $Fe(acac)_3$ contains an Fe^{3+}ion, with a 3*d* electron configuration of [↑][↑][↑][↑][↑]. This configuration has five unpaired electrons. The $Fe(acac)_3$, with more unpaired electrons per molecule, would be attracted more strongly into a magnetic field.

Chapter 8

8.1  C_8H_{16}

8.2 N_2 has only 10 valence electrons. The Lewis structure shown has 14 valence electrons.

8.3 None of the structures are correct. (a) is incorrect because sulfur does not have an octet of electrons (it has only six); (b) is incorrect because, although it shows the correct number of valence electrons (26), there is a double bond between F and N rather than a single bond with a lone pair on N; (c) is incorrect because the left carbon has five bonds; (d) is incorrect because COCl should have 17 valence electrons, not 18 as shown.

8.4 (a) C_5H_{10} (b) Two

8.5

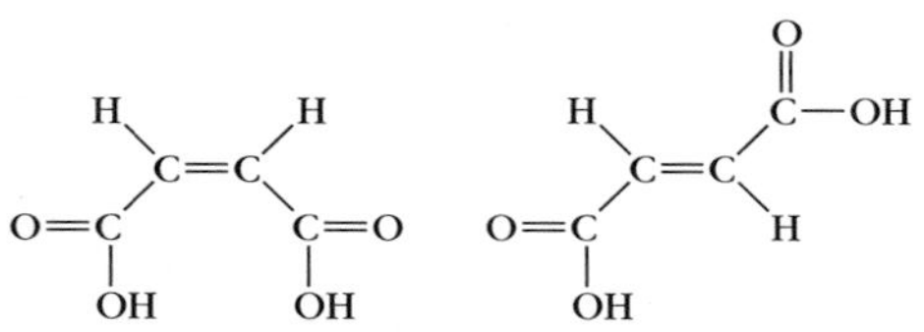

maleic acid
(the *cis* isomer)

fumaric acid
(the *trans* isomer)

8.6 C—N > C=N > C≡N. The order of decreasing bond energy is the reverse order: C≡N > C=N > C—N. See Tables 8.1 and 8.2.

8.7 (a) The electronegativity difference between sodium and chlorine is 2.0, sufficient to cause electron transfer from sodium to chlorine to form Na^+ and Cl^- ions. Molten NaCl conducts an electric current, indicating the presence of ions.

(b) There is an electronegativity difference of 1.2 in BrF, which is sufficient to form a polar covalent bond, but not great enough to cause electron transfer leading to ion formation.

8.8 The Lewis structure of hydrazine is

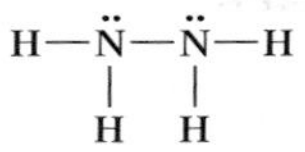

H—N̈—N̈—H
with an H bonded below each N

	H	H	N	N	H	H
Valence electrons	1	1	5	5	1	1
Lone pair electrons	0	0	2	2	0	0
$\frac{1}{2}$ shared electrons	1	1	3	3	1	1
Formal charge	0	0	0	0	0	0

8.9 Atoms cannot be rearranged to derive a resonance structure. There is no N-to-O bond in cyanate ion; therefore, such an arrangement cannot be a resonance structure of cyanate ion.

8.10 CH_3 (benzene ring) ⟷ CH_3 (benzene ring)

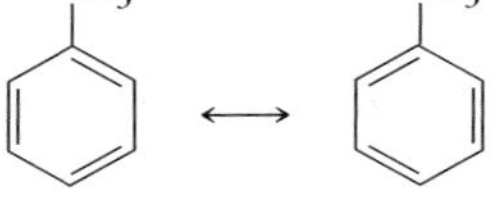

8.11 CH_3 (positions 1, 2, 4 on benzene ring)

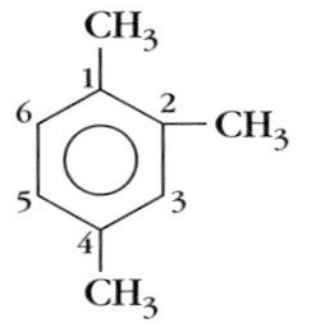

1,2,4-trimethylbenzene

The ring is numbered to give the lowest numbers for the substituents in 1, 2, 3-trimethylbenzene.

8.12 1. (a) 20 carbon atoms and 30 hydrogen atoms
(b) The carbon atom at the top of the six-membered ring
(c) Five C═C double bonds

2. (a) $C_{29}H_{50}O_2$ (b) No C═C double bonds
(c) The H—O bond

Chapter 9

9.1 When the central atom has no lone pairs

9.2 The triangular bipyramidal shape has three of the five pairs situated in equatorial positions 120° apart and the remaining two pairs in axial positions. The square pyramidal shape has four of the atoms bonded to the central atom in a square plane, with the other bonded atom directly above the central atom and equidistant from the other four.

9.3 (a) AX_2E_3 (b) AX_3E_1 (c) AX_2E_3

9.4 Pi bonding is not possible for a carbon atom with sp^3 hybridization because it has no unhybridized $2p$ orbitals. All of its $2p$ orbitals have been hybridized.

9.5 (a) Bromine is more electronegative than iodine, and the H—Br bond is more polar than the H—I bond.

(b) Chlorine is more electronegative than the other two halogens; therefore, the C—Cl bond is more polar than the C—Br and C—I bonds.

9.6 δ^+ δ^- :C≡O:-----:C≡O: (δ^+ δ^-) ----- dipole-dipole forces

δ^- δ^+ :O≡C:-----:O≡C: (δ^- δ^+)

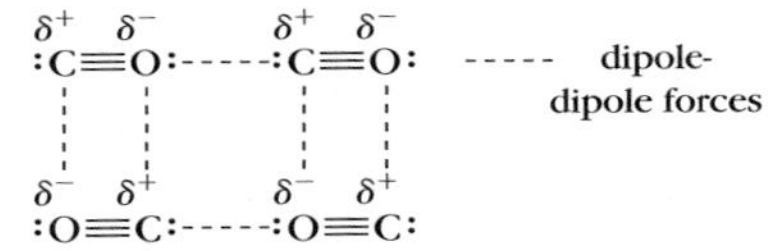

9.7 The F—H · · · F—H hydrogen bond is the strongest because the electronegativity difference between H and F produces a more polar F—H bond than does the lesser electronegativity difference between O and H or N and H in the O—H or N—H bonds.

9.8 H_3C ... T ... G ... hydrogen bond

NH_2 ... C ... A ... hydrogen bond

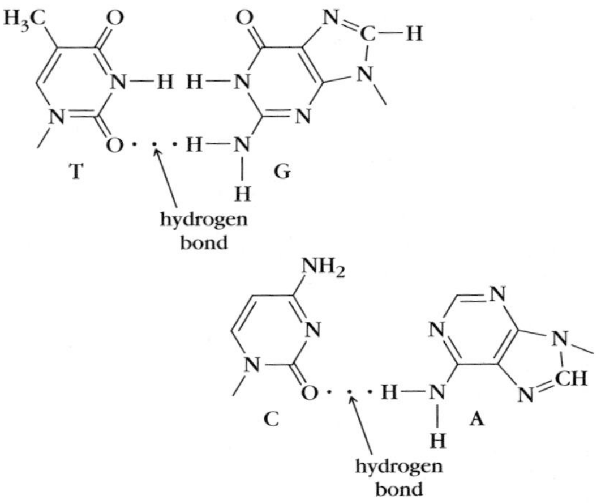

9.9 Replication would be very difficult because covalent bonds would have to be broken. DNA replicates easily because it is hydrogen bonds, not covalent bonds, that occur between the base pairs and hold the two DNA strands together.

Chapter 10

10.1 $(2.7 \times 10^{14}\ \text{molecules}) \times \left(\dfrac{64.06\ \text{g SO}_2}{6.02 \times 10^{23}\ \text{molecules}}\right) = 2.9 \times 10^{-8}\ \text{g SO}_2$

10.2 First, gas molecules are far apart. This allows most light to pass through. Second, molecules are much smaller than the wavelengths of visible light. This means that the waves are not reflected or diffracted by the molecules.

10.3 As more gas molecules are added to a container of fixed volume, there will be more collisions of all of the gas molecules with the container walls. This causes the observed pressure to rise.

10.4 All have the same kinetic energy at the same temperature.

10.5 For a sample of helium, the plot would look like the curve marked He in Figure 10.7. When an equal number of argon molecules, which are heavier, are added to the helium, the distribution of molecular speeds would look like the sum of the curves marked He and O_2 in Figure 10.7, except that the curve for Ar would have its peak a little to the left of the O_2 curve.

10.6 (a) The balloon placed in the freezer will be smaller than the one kept at room temperature because its sample of helium is colder.

(b) Upon warming, the helium balloon that had been in the freezer will be either the same size as the balloon kept at room temperature or perhaps slightly larger because there is a greater chance that He atoms leaked out of the room temperature balloon during the time the other balloon was kept in the freezer. This would be caused by the faster-moving He atoms in the room temperature balloon having more chances to escape from tiny openings in the balloon's walls.

10.7 The gas in the shock absorbers will be more highly compressed. The gas molecules will be closer together. The gas molecules will collide with the walls of the shock absorber more often, and the pressure exerted will be larger.

10.8 Increasing the temperature of a gas causes the gas molecules to move faster, on average. This means that each collision with the container walls involves greater force, because on average, a molecule is moving faster and hits the wall harder. If the container remained the same (constant volume), there would also be more collisions with the container wall because faster-moving molecules would hit the walls more often. Increasing the volume of the container, on the other hand, requires that the faster-moving molecules must travel a greater distance before they strike the container walls. Increasing the volume enough would just balance the greater numbers of harder collisions caused by increased temperature. To maintain a constant volume requires that the pressure increases to match the greater pressure due to more and harder collisions of gas molecules with the walls.

10.9 Two moles of O_2 gas are required for the combustion of one mole of methane gas. If air were pure O_2, the oxygen delivery tube would need to be twice as large as the delivery tube for methane. Since air is only one-fifth O_2, the air delivery tube would need to be 10 times larger than the methane delivery tube to ensure complete combustion.

10.10 Using the ratio of 100 balloons/26.8 g He, calculate the number of balloons 41.8 g of He can fill.

$$\text{Balloons} = (41.8\text{ g He})\left(\frac{100\text{ balloons}}{26.8\text{ g He}}\right) = 155\text{ balloons}$$

This much He will fill more balloons than needed.

10.11 1. Increase the pressure.
2. Decrease the temperature.
3. Remove some of the gas by reaction to form a nongaseous product.

10.12 Density of Cl_2 at 25 °C and 0.750 atm $= \frac{PM}{RT}$

$$= \frac{(0.750\text{ atm})(70.906\text{ g/mol})}{(0.0821\text{ L atm mol}^{-1}\text{ K}^{-1})(298\text{ K})} = 2.17\text{ g/L}$$

Density of SO_2 at 25 °C and 0.750 atm $= \frac{PM}{RT}$

$$= \frac{(0.750\text{ atm})(64.06\text{ g/mol})}{(0.0821\text{ L atm mol}^{-1}\text{ K}^{-1})(298\text{ K})} = 1.96\text{ g/L}$$

Density of Cl_2 at 35 °C and 0.750 atm $= \frac{PM}{RT}$

$$= \frac{(0.750\text{ atm})(70.906\text{ g/mol})}{(0.0821\text{ L atm mol}^{-1}\text{ K}^{-1})(308\text{ K})} = 2.10\text{ g/L}$$

Density of SO_2 at 25 °C and 2.60 atm $= \frac{PM}{RT}$

$$= \frac{(2.60\text{ atm})(64.06\text{ g/mol})}{(0.0821\text{ L atm mol}^{-1}\text{ K}^{-1})(298\text{ K})} = 6.81\text{ g/L}$$

10.13 Density of He $= 1.23 \times 10^{-4}$ g/mL
Density of Li $= 0.53$ g/mL
Since the density of He is so much less than that of Li, the atoms in a sample of He must be much farther apart than the atoms in a sample of Li. This idea is in keeping with the general principle of the kinetic-molecular theory that the particles making up a gas are far from one another.

10.14 A 50-50 mixture of N_2 and O_2 would have less N_2 in it than does air. Since O_2 molecules have greater mass than N_2 molecules, this 50-50 mixture has greater density than air.

10.15 (a) If lowering the temperature causes the volume to decrease, by $PV = nRT$, the pressure can be assumed to be constant. The value of n is unchanged. Since both P and n remain unchanged, the partial pressures of the gases in the mixture remain unchanged.
(b) When the total pressure of a gas mixture increases, the partial pressure of each gas in the mixture increases because the partial pressure of each gas in the mixture is the product of the mole fraction for that gas and the total pressure.

10.16 We can calculate the total number of moles of gas in the flask from the given information.

$$n = \frac{PV}{RT} = \frac{\left(\frac{626}{760}\right)\text{atm }(0.355\text{ L})}{(0.0821\text{ L atm mol}^{-1}\text{ K}^{-1})(308\text{ K})} = 0.01156\text{ mol gas}$$

The number of moles of Ne is

$$0.146\text{ g Ne} \times \frac{1\text{ mol Ne}}{20.18\text{ g/mol}} = 0.007235\text{ mol Ne}$$

We find the number of moles of Ar by subtraction.

$$0.01156\text{ mol gas} - 0.007235\text{ mol Ne} = 0.004325\text{ mol Ar}$$

$$0.004325\text{ mol Ar} \times 39.948\text{ g/mol} = 0.173\text{ g Ar}$$

10.17 The value of n depends directly on the measured pressure, P. Intermolecular attractions in a real gas would cause the measured P to be slightly smaller than for an ideal gas. The lower value of P would cause the calculated number of moles to be somewhat smaller. Using this slightly smaller value of n in the denominator would cause the calculated molar mass to be a little larger than it should be.

10.18 Mass of S burned per hour $= (3.06 \times 10^6\text{ kg})(0.04)$

$$= 1 \times 10^5\text{ kg}$$

$$\text{Mass of }SO_2\text{ per hour} = (1 \times 10^5\text{ kg})\left(\frac{64.06\text{ kg }SO_2}{32.07\text{ kg S}}\right) 2 \times 10^5\text{ kg}$$

$$\text{Mass of }SO_2\text{ per year} = (2 \times 10^5\text{ kg/hr})(8760\text{ hr/yr}) = 2 \times 10^9\text{ kg/yr}$$

10.19 Vol percent SO_2 = (5 parts $SO_2/10^6$ parts air) × 100%

$$= 5 \times 10^{-4}\text{ vol\%}$$

10.20 1 metric ton = 1000 kg = 10^6 g = 1 Mg

$$\text{Mass of }HNO_3 = (400\text{ Mg }N_2)\left(\frac{2\text{ Mg NO}}{\text{Mg }N_2}\right) \times \left(\frac{1\text{ Mg }NO_2}{1\text{ Mg NO}}\right)\left(\frac{1\text{ Mg }HNO_3}{1\text{ Mg }NO_2}\right) = 800\text{ Mg }HNO_3$$

10.21 $\cdot NO_2 \xrightarrow{h\nu} NO\cdot + \cdot O\cdot$

$O_3 \xrightarrow{h\nu} O_2 + \cdot O\cdot$

$\cdot O\cdot + O_2 \longrightarrow O_3$

10.22 Among the many possible molecules would be: NO, which comes from automobile combustion; NO_2, which comes from reactions of NO and O_2 in the atmosphere; O_3, which comes from reactions of NO, NO_2, and O_2 in the atmosphere; and hydrocarbons, and oxidation products of hydrocarbon reactions with O_2 and other reactive species.

10.23 Natural sources: animal respiration, forest fires, decay of cellulose materials, partial digestion of carbohydrates, volcanoes. Human sources: burning fossil fuels, burning agricultural wastes and refined cellulose products such as paper, decay of carbon compounds in landfills.

10.24 (a) 4.7%; (b) 19.0%; (c) 4.2%; 1950–2001 showed the greatest percentage increase in CO_2.

10.25 The fluctuations occur because of the seasons. Photosynthesis, which uses CO_2, is greatest during the spring and summer, accounting for lower CO_2 levels.

10.26 For this calculation, we can use the figure of 450 × 10^9 passenger miles. Using the ratio of 2 × 10^3 kg CO_2/3000 passenger miles, we can calculate the CO_2 released.

$$\text{Quantity of }CO_2 = 450 \times 10^9\text{ passenger mi} \times \frac{2 \times 10^3\text{ kg }CO_2}{3 \times 10^3\text{ passenger mi}} = 3 \times 10^{11}\text{ kg }CO_2$$

If a typical automobile gets 20 mi/gal of fuel and about 1.5 passengers are transported for every mile the automobile travels, then an automobile gets 30 passenger miles per gallon. The number of gallons used is

$$\text{Volume of gasoline} = 450 \times 10^9 \text{ passenger mi} \times \frac{1 \text{ gal gasoline}}{30 \text{ passenger mi}} = 2 \times 10^{10} \text{ gal gasoline}$$

Assume the gasoline produces about the same mass of CO_2 per gallon as does jet fuel, or 2×10^3 kg CO_2/200 gal.

$$\text{Quantity of } CO_2 \text{ from gasoline} = 2 \times 10^{10} \text{ gal gasoline} \times \frac{2 \times 10^3 \text{ kg } CO_2}{200 \text{ gal gasoline}} = 2 \times 10^{11} \text{ kg } CO_2$$

So the numbers are about the same for these two modes of transportation.

10.27 He meant the burning coal converted its carbon into carbon dioxide in the air, thus increasing atmospheric CO_2 concentration.

Chapter 11

11.1 The London forces are greater between bromoform molecules than between chloroform molecules because the bromoform molecules have more electrons. This stronger intermolecular attraction causes the $CHBr_3$ molecules to exhibit a greater surface tension. (The dipole in each molecule contributes less than the London forces to the intermolecular attractions.)

11.2 (a) Water and glycerol would have similar surface tensions because of extensive hydrogen bonding. (b) Octane and decane would have similar surface tensions because both are alkane hydrocarbons.

11.3 (a) 62 °C (b) 0 °C (c) 80 °C

11.4 Bubbles form within a boiling liquid when the vapor pressure of the liquid equals the pressure of the surroundings of the liquid sample. The bubbles are actually filled with vapor of the boiling liquid. One way to prove this would be to trap some of these bubbles and allow them to condense. They would condense to form the liquid that had boiled.

11.5 The evaporating water carries with it thermal energy from the water inside the pot. In addition, a large quantity of thermal energy is required to cause the water to evaporate. Much of this thermal energy comes from the water inside the pot.

11.6 Estimated ΔH°_{vap} for Kr $\approx$ 20 kJ/mol based on HBr, Cl_2, and C_4H_{10}, but based on Xe, the value is probably closer to 10 kJ/mol. Estimated ΔH°_{vap} for $NO_2 \approx$ 20 kJ/mol based on propane and HCl.

11.7 (a) Bromine molecules have more electrons than chlorine molecules. Therefore, bromine molecules are held together by stronger intermolecular attractions.
(b) Ammonia molecules are attracted to one another by hydrogen bonds. This causes ammonia to have a higher boiling point than that of methane, which has no hydrogen bonding.

11.8 Two moles of liquid bromine crystallizing liberates 21.59 kJ of energy. One mole of liquid water crystallizing liberates 6.02 kJ of energy.

11.9 High humidity conditions make the evaporation of water or the sublimation of ice less favorable. Under these conditions, the sublimation of ice required to make the frost-free refrigerator work is less favorable, so the defrost cycle is less effective.

11.10 The impurity molecules are less likely to be converted from the solid phase to the vapor phase. This causes them to be left behind as the molecules that sublime go into the gas phase and then condense at some other place. The molecules that condense are almost all of the same kind, so the sublimed sample is much purer than the original.

11.11 The curve is the vapor pressure curve. Upward: condensation. Downward: vaporization. Left to right: vaporization. Right to left: condensation.

11.12 (a) If liquid CO_2 is slowly released from a cylinder of CO_2, gaseous CO_2 is formed. The temperature remains constant (at room temperature) because there is time for energy to be transferred from the surroundings to separate the CO_2 molecules from their intermolecular attractions. This can be seen from the phase diagram as the phase changes from liquid to vapor as the pressure decreases.
(b) If the pressure is suddenly released, the attractive forces between a large number of CO_2 molecules must be overcome, which requires energy. This energy comes from the surroundings as well as from the CO_2 molecules themselves, causing the temperature of both the surroundings and the CO_2 molecules to decrease. On the phase diagram for CO_2, a decrease in both temperature and pressure moves into a region where only solid CO_2 exists.

11.13 It is predicted that a small concentration of gold will be found in the lead and that a small concentration of lead will be found in the gold. This will occur because of the movement of the metal atoms with time, as predicted by the kinetic molecular theory.

11.14 One Po atom belongs to its unit cell. Two Li atoms belong to its unit cell. Four Ca atoms belong to its unit cell.

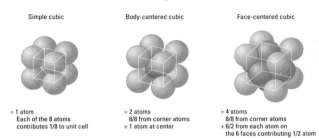

Simple cubic

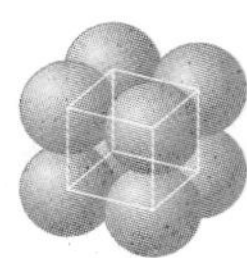

= 1 atom
Each of the 8 atoms contributes 1/8 to unit cell

Body-centered cubic

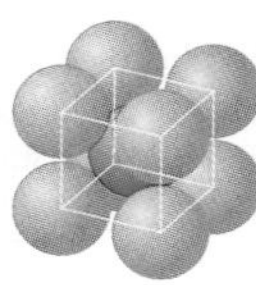

= 2 atoms
8/8 from corner atoms
+ 1 atom at center

Face-centered cubic

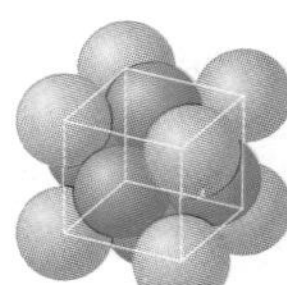

= 4 atoms
8/8 from corner atoms
+ 6/2 from each atom on the 6 faces contributing 1/2 atom

11.15 Each Cs^+ ion at the center of the cube has eight Cl^- ions as its neighbors. One eighth of each Cl^- ion belongs to that Cs^+ ion. So the formula for this salt must be a 1 : 1 ratio of Cs^+ ions to Cl^- ions, or CsCl.

11.16 Cooling a liquid above its freezing point causes the temperature to decrease. When the liquid begins to solidify, energy is released as atoms, molecules, or ions move closer together to form in the solid crystal lattice. This causes the temperature to remain constant until all the molecules in the liquid have positioned themselves in the lattice. Further cooling then causes the temperature to decrease. The shape of this curve is common to all substances that can exist as liquids.

11.17 Increasing strength of metallic bonding is related to increasing numbers of valence electrons. In the transition metals, the presence of *d*-orbital electrons causes stronger metallic bonding. Beyond a half-filled set of *d*-orbitals, however, extra electrons have the effect of decreasing the strength of metallic bonding.

Chapter 12

12.1 (a) $4 H_2$
(b) $C_7H_{16} \rightarrow C_7H_8 + 4 H_2$
(c) Toluene has a much higher octane number and can be used as an octane enhancer.

12.2 $\% \text{ oxygen} = \left(\frac{16.0}{46.0}\right) \times 100\% = 34.8\%$

Ethanol is more highly oxygenated than the hydrocarbons in gasoline.

12.3 (a) $C_2H_6O + 2 O_2 \rightarrow 2 CO + 3 H_2O$

$$\frac{\text{g CO}}{\text{g ethanol}} = \frac{2(28.0 \text{ g})}{46.0 \text{ g}} = 1.22 \text{ g CO/g ethanol}$$

(b) $C_7H_8 + \frac{11}{2} O_2 \rightarrow 7\ CO + 4\ H_2O$

$$\frac{g\ CO}{g\ toluene} = \frac{7(28.0\ g)}{92.0\ g} = 2.13\ g\ CO/g\ toluene$$

(c) Ethanol produces less CO per gram than does toluene.

12.4 $CH_4(g) + \frac{3}{2} O_2(g) \rightarrow CO(g) + 2\ H_2O(g)$
Thus, 28.0 g CO is produced per 16.0 g CH_4;

$$\frac{28.0\ g\ CO}{16.0\ g\ CH_4} = 1.75\ g\ CO/g\ CH_4$$

$C_8H_{18}(\ell) + \frac{17}{2} O_2(g) \rightarrow 8\ CO(g) + 9\ H_2O(g)$
114.0 g octane produces 224.0 g CO;

$$\frac{224\ g\ CO}{114.0\ g\ octane} = 1.96\ g\ CO/g\ C_8H_{18}$$

$\frac{1.96}{1.75} = 1.12$, which indicates 12% more CO is produced per gram of octane than per gram of methane.

12.5 $\text{Thermal energy} = 1.5 \times 10^6\ \text{bbl oil} \times \frac{5.9 \times 10^9\ J}{1\ \text{bbl oil}} = 8.9 \times 10^{15}\ J$

$\text{Electricity delivered} = 1.5 \times 10^6\ \text{bbl} \times \frac{1628\ kWh}{1\ \text{bbl}} \times 0.33 = 8.1 \times 10^8\ kWh$

12.6 Carbon dioxide cannot be burned.

12.7 Ten or so carbon atoms in an alcohol molecule will make it much less water-soluble than alcohols with fewer numbers of carbon atoms.

12.8

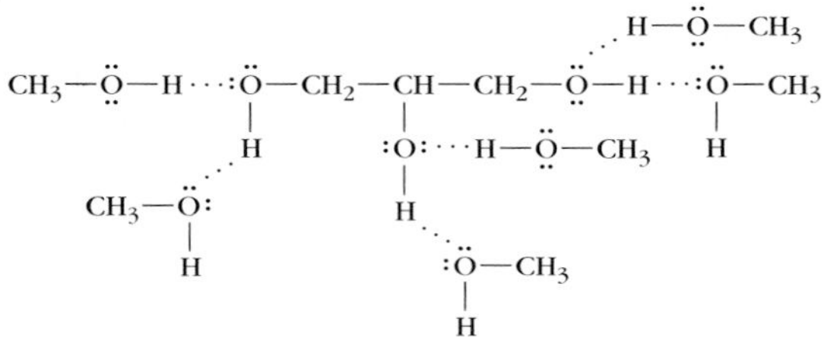

```
                                                  H—Ö—CH3
                                               ..·
CH3—Ö—H···:Ö—CH2—CH—CH2—Ö—H···:Ö—CH3
            |        |              |
            H      :O:···H—Ö—CH3    H
         ..·         |
CH3—Ö:               H
     |                ·..
     H                  :Ö—CH3
                          |
                          H
```

12.9 The acetaldehyde molecule has two fewer hydrogen atoms compared with the ethanol molecule. Loss of hydrogen is oxidation. The acetaldehyde molecule is more oxidized than the ethanol molecule. Comparing the formulas for acetaldehyde and acetic acid, the hydrogen atoms are the same, but the acetic acid molecule has one additional oxygen atom. Gain of oxygen is oxidation. So the acetic acid molecule is more oxidized than the acetaldehyde molecule.

12.10

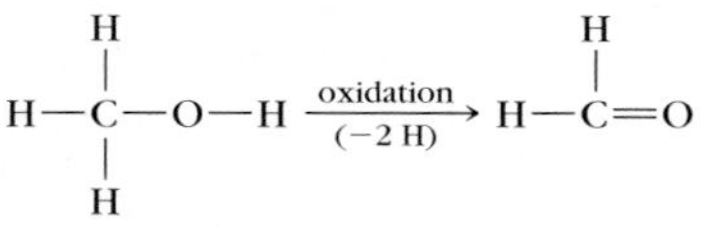

```
   H                            H
   |       oxidation            |
H—C—O—H  ──────────→  H—C=O
   |        (−2 H)
   H
```

12.11 $CH_3CH_2CH_2OH$, 1-propanol

12.12 (a) Estradiol contains two alcohol groups (—OH).
(b) Secondary alcohol
(c) Oxidation (removal of two hydrogens)
(d) Estradiol: Two alcohol groups, aromatic ring, one $—CH_3$ group
Testosterone: One alcohol group; one ketone group; C=C double bond in ring; two $—CH_3$ groups

12.13 Conversion of a ketone on the five-membered ring to a secondary alcohol by reduction (addition of 2 H atoms)

12.14

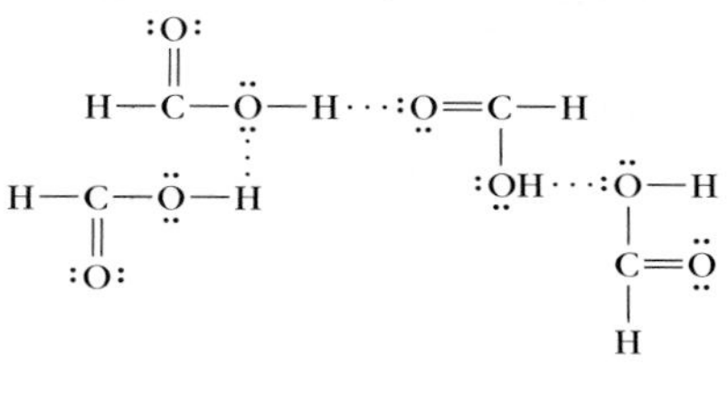

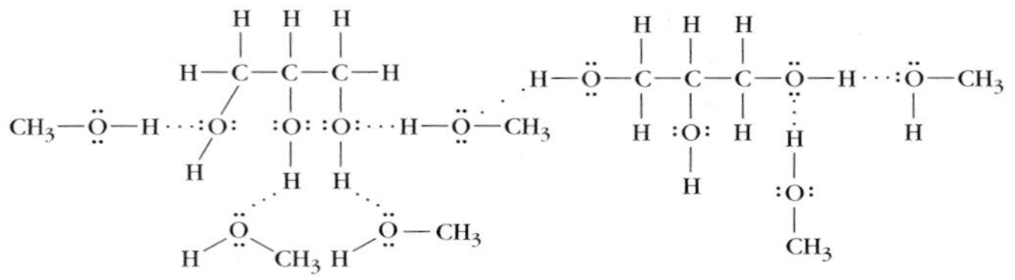

```
      :O:
       ‖
  H—C—Ö—H···:O=C—H
       :            |
H—C—Ö—H       :ÖH···:Ö—H
   ‖                      |
  :O:                    C=Ö
                          |
                          H
```

```
        H  H  H                         H  H  H
        |  |  |                         |  |  |
      H—C——C——C—H              H—Ö—C——C——C—Ö—H···:Ö—CH3
       /   |  |                         |  |  :      |
CH3—Ö—H···:O: :O: :O:···H—Ö—CH3         H :O: H  H     H
        |   |  |                            |     |
        H   H  H                            H    :O:
       ·      ·                                   |
      Ö        Ö—CH3                             CH3
    H   CH3  H
```

12.15 (a)

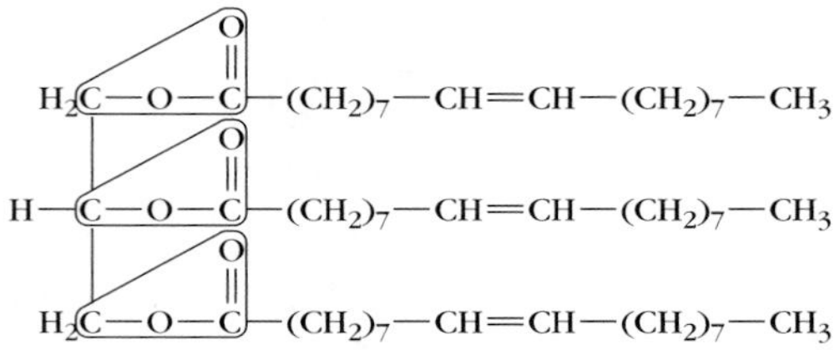

```
          O
          ‖
H2C—O—C—(CH2)7—CH=CH—(CH2)7—CH3
          O
          ‖
H—C—O—C—(CH2)7—CH=CH—(CH2)7—CH3
          O
          ‖
H2C—O—C—(CH2)7—CH=CH—(CH2)7—CH3
```

(b)

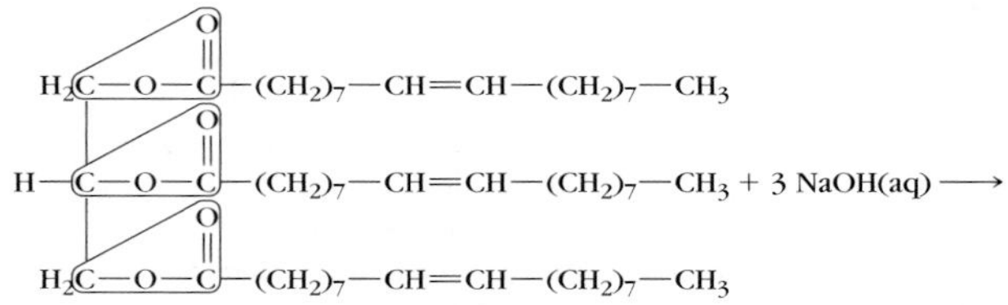

```
          O
          ‖
H2C—O—C—(CH2)7—CH=CH—(CH2)7—CH3
          O
          ‖
H—C—O—C—(CH2)7—CH=CH—(CH2)7—CH3 + 3 NaOH(aq) ——→
          O
          ‖
H2C—O—C—(CH2)7—CH=CH—(CH2)7—CH3
```

(c)

```
CH2OH
 |
HC—OH + 3 Na+[CH3—(CH2)7—CH=CH—(CH2)7—COO]−
 |
CH2OH
```

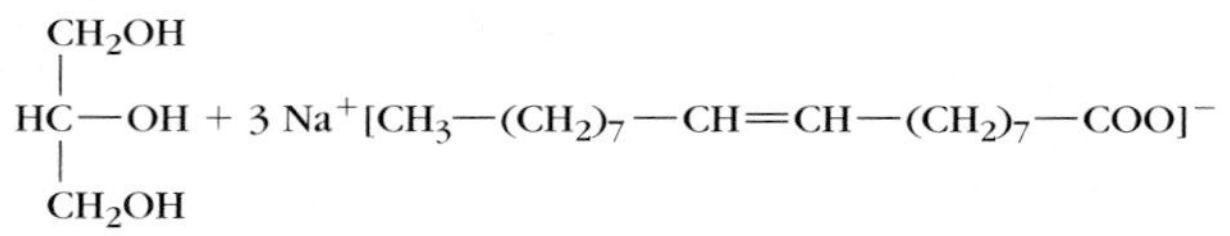

12.16 The ends of the chains are possibly occupied by the OR groups from the initiator molecules.

12.17

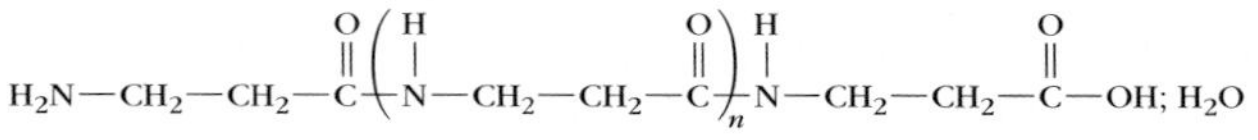

```
                    O  H              O  H                  O
                    ‖  |              ‖  |                  ‖
H2N—CH2—CH2—C—(N—CH2—CH2—C)n—N—CH2—CH2—C—OH; H2O
```

12.18 (1) Amine (2) Carboxylic acid (3) Amide (4) Ester

12.19 Serine and glutamine could hydrogen-bond to one another if they were close in two adjacent protein chains because they have polar groups containing H atoms in their R groups. Glycine and valine would not because they have no additional polar groups in their R groups.

12.20

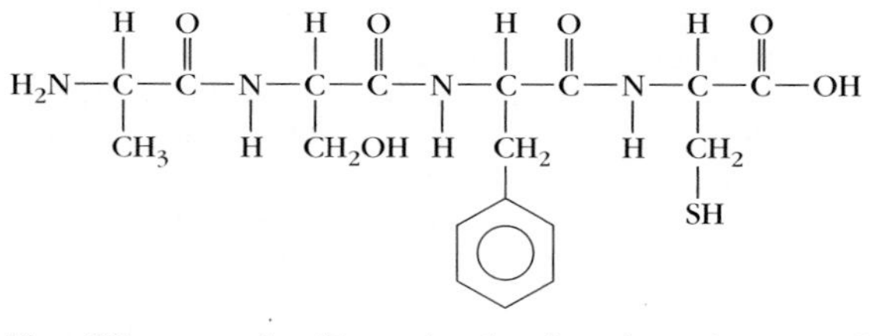

```
       H  O      H  O        H  O      H  O
       |  ‖      |  ‖        |  ‖      |  ‖
H2N—C——C—N—C——C—N—C——C—N—C——C—OH
       |     |   |      |    |     |   |
      CH3    H  CH2OH   H   CH2    H  CH2
                             |         |
                            C6H5       SH
```

12.21 The OH groups in this molecule allow it to be extensively hydrogen-bonded with solvent water molecules.

12.22 Cellulose contains glucose molecules linked together by *trans* 1,4 linkages. Ruminant animals have large colonies of bacteria and protozoa that live in the forestomach and digest cellulose.

12.23 If humans could digest cellulose, then common plants that are easy to grow could become food. There might be less reliance upon cultivation of plants for food. In addition, the entire plant could be used for food rather than just certain parts eaten and the other parts wasted. On the other hand, in times of famine, there might not be enough cellulose to go around. Destroying trees and other plants for food might cause enlargements of desert regions and the disappearance of entire species of plants.

Chapter 13

13.1 (a)

i. $$\text{Rate} = -\frac{\Delta[\text{Cv}^+]}{\Delta t}$$
$$= -\frac{(0.793 \times 10^{-5} - 1.46 \times 10^{-5})\ \text{mol/L}}{(60.0 - 40.0)\ \text{s}}$$
$$= \frac{6.67 \times 10^{-6}\ \text{mol/L}}{20.0\ \text{s}} = 3.3 \times 10^{-7}\ \text{mol L}^{-1}\ \text{s}^{-1}$$

ii. $$\text{Rate} = -\frac{(0.429 \times 10^{-5} - 2.71 \times 10^{-5})\ \text{mol/L}}{(80.0 - 20.0)\ \text{s}}$$
$$= 3.8 \times 10^{-7}\ \text{mol L}^{-1}\ \text{s}^{-1}$$

iii. $$\text{Rate} = -\frac{(0.232 \times 10^{-5} - 5.00 \times 10^{-5})\ \text{mol/L}}{(100.0 - 0.0)\ \text{s}}$$
$$= 4.8 \times 10^{-7}\ \text{mol L}^{-1}\ \text{s}^{-1}$$

(b)

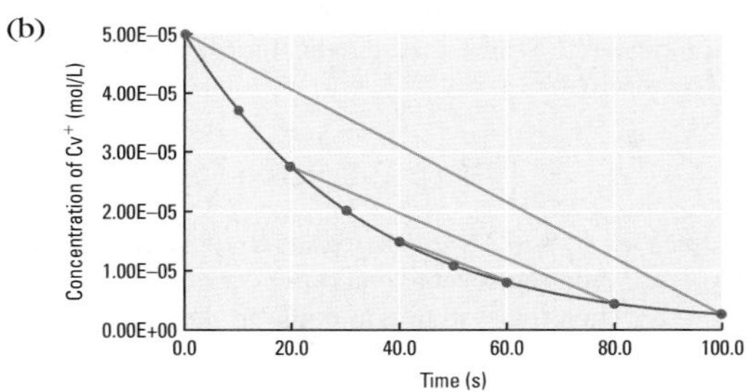

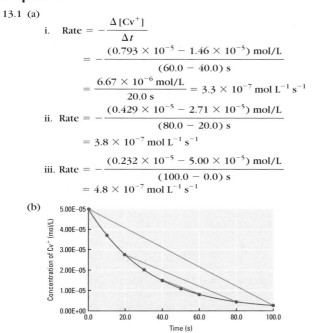

(c) The rate is faster when the concentration of Cv^+ is larger. As the reaction takes place, the rate gets slower. There is a much larger change in concentration from 0 s to 50 s than from 50 s to 100 s. Therefore, the $\Delta\ [\text{Cv}^+]$ is more than three times bigger for the time range from 20 s to 80 s than it is for the time range from 40 s to 60 s, even though Δt is exactly three times larger.

13.2 (a) Rate (1) is twice rate (3); rate (2) is twice rate (4); rate (3) is twice rate (5).

(b) In each case, the $[\text{Cv}^+]$ is twice as great when the rate is twice as great.

(c) Yes, the rate doubles when $[\text{Cv}^+]$ doubles.

13.3

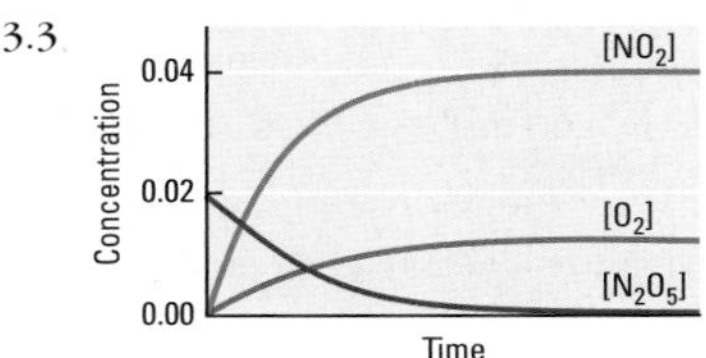

13.4 $\text{Rate}_1 = k[\text{CH}_3\text{COOCH}_3][\text{OH}^-]$

$\text{Rate}_2 = k(2[\text{CH}_3\text{COOCH}_3])(\tfrac{1}{2}[\text{OH}^-]) = k[\text{CH}_3\text{COOCH}_3][\text{OH}^-]$

The rate is unchanged.

13.5

$$\text{Rate} = k[\text{NO}]^x[\text{Cl}_2]^y$$

Taking the log of both sides and applying to experiments 4 and 2 gives

$$\log(\text{Rate}_4) = \log(k) + x \log[\text{NO}]_4 + y \log[\text{Cl}_2]_4$$

$$\log(\text{Rate}_2) = \log(k) + x \log[\text{NO}]_2 + y \log[\text{Cl}_2]_2$$

Recognizing that $[\text{Cl}_2]_4 = [\text{Cl}_2]_2$ and then subtracting the second equation from the first gives

$$\log(\text{Rate}_4) - \log(\text{Rate}_2) = x \log[\text{NO}]_4 - x \log[\text{NO}]_2$$

Using the properties of logarithms gives

$$\log\{\text{Rate}_4/\text{Rate}_2\} = x\{\log[\text{NO}]_4/[\text{NO}]_2\}$$

So

$$x = \log\{\text{Rate}_4/\text{Rate}_2\}/\{\log[\text{NO}]_4/[\text{NO}]_2\} \quad \text{QED}$$

Now we can insert the actual numbers from experiments 4 and 2:

$$x = [\log(6.60 \times 10^{-4}/1.65 \times 10^{-4})]/\log(0.04/0.02)$$

$$x = \log 4/\log 2 = 0.602/0.301 = 2$$

13.6 (a) $\text{CH}_3\text{NC} \rightarrow \text{CH}_3\text{CN}$ unimolecular
(b) $2\ \text{HI} \rightarrow \text{H}_2 + \text{I}_2$ bimolecular
(c) $\text{NO}_2\text{Cl} \rightarrow \text{NO}_2 + \text{Cl}$ unimolecular
(d) $\text{C}_4\text{H}_8 \rightarrow \text{C}_4\text{H}_8$ unimolecular
(e) $\text{NO}_2\text{Cl} + \text{Cl} \rightarrow \text{NO}_2 + \text{Cl}_2$ bimolecular

13.7 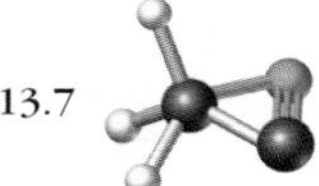

H_3C bonded to both N and C, with N≡C (triangular structure)

The Lewis structure has five pairs of electrons around one C atom, which would not be stable. However, this is a transition state, which is, by definition, unstable.

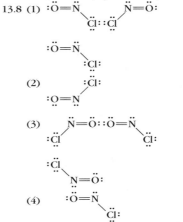

13.8 (1) :Ö=N̈–C̈l::C̈l–N̈=Ö:

(2) :Ö=N̈–C̈l: :C̈l–N̈=Ö: (stacked)

(3) :C̈l–N̈=Ö::Ö=N̈–C̈l:

(4) :C̈l–N̈=Ö: :Ö=N̈–C̈l: (stacked)

(1) and (2) are much more likely to result in a reaction than are (3) and (4).

(1) and (2) are much more likely to result in a reaction than are (3) and (4).

13.9 (a) $k = Ae^{-E_a/RT}$

$$= (6.31 \times 10^8\ \text{L mol}^{-1}\ \text{s}^{-1})\, e^{\frac{-10{,}000\ \text{J/mol}}{(8.314\ \text{J K}^{-1}\ \text{mol}^{-1})(370\ \text{K})}}$$

$$k = 2.4 \times 10^7\ \text{L mol}^{-1}\ \text{S}^{-1}$$

(b) $\text{Rate} = k[\text{NO}][\text{O}_3]$

$$= (2.4 \times 10^7\ \text{L mol}^{-1}\ \text{s}^{-1})(1.0 \times 10^{-3}\ \text{mol/L}) \times (5.0 \times 10^{-4}\ \text{mol/L})$$

$$= 12\ \text{mol L}^{-1}\ \text{s}^{-1}$$

13.10 The reaction does not occur in a single step. If it did, the rate law would be Rate $= k[\text{NO}_2][\text{CO}]$.

13.11 Rate $= k[\text{HOI}][\text{I}^-]$

13.12 (a) $2\ \text{ICl(g)} + \text{H}_2\text{(g)} \rightarrow 2\ \text{HCl(g)} + \text{I}_2\text{(g)}$

(b) Rate $= k_2[\text{HI}][\text{ICl}]$

However, HI is an intermediate. Assume that the concentration of HI reaches a steady state.

Rate of step +1 = rate of step −1 + rate of step 2

Also assume that step 2 is much slower than step +1 and step −1. Then

$$k_1[\text{ICl}][\text{H}_2] = k_{-1}[\text{HI}][\text{HCl}]$$

$$[\text{HI}] = \frac{k_1[\text{ICl}][\text{H}_2]}{k_{-1}[\text{HCl}]}$$

$$\text{Rate} = k_2[\text{ICl}][\text{HI}] = k_2[\text{ICl}] \times \frac{k_1[\text{ICl}][\text{H}_2]}{k_{-1}[\text{HCl}]}$$
$$= \frac{k_1k_2}{k_{-1}}[\text{ICl}]^2[\text{H}_2][\text{HCl}]^{-1}$$

(c) The rate is inversely proportional to the concentration of the product, HCl.
(d) Because [HCl] increases as the reaction proceeds, the rate of reaction will decrease more quickly over time than it would if $[HCl]^{-1}$ were not in the rate law. However, the rate constant will not change.

13.13 (a) $Ce^{4+} + Mn^{2+} \rightarrow Ce^{3+} + Mn^{3+}$
$Ce^{4+} + Mn^{3+} \rightarrow Ce^{3+} + Mn^{4+}$
$Mn^{4+} + Tl^{+} \rightarrow Mn^{2+} + Tl^{3+}$
$2\,Ce^{4+} + Tl^{+} \rightarrow 2\,Ce^{3+} + Tl^{3+}$
(b) Intermediates are Mn^{3+} and Mn^{4+}.
(c) The catalyst is Mn^{2+}.
(d) Rate $= k[Ce^{4+}][Mn^{2+}]$
(e) Rate $= k[Ce^{4+}]^2[Mn^{2+}][Ce^{3+}]^{-1}$

13.14 (a) The concentration of a homogeneous catalyst *must* appear in the rate law.
(b) A catalyst does not appear in the equation for an overall reaction.
(c) A *homogeneous* catalyst must always be in the same phase as the reactants.

Chapter 14

14.1 (a) $(\text{conc. Al}) = \frac{2.70\text{ g}}{\text{mL}} \times \frac{1000\text{ mL}}{1\text{ L}} \times \frac{1\text{ mol}}{26.98\text{ g}} = 100.\text{ mol/L}$

(b) $(\text{conc. benzene}) = \frac{0.880\text{ g}}{\text{mL}} \times \frac{1000\text{ mL}}{1\text{ L}} \times \frac{1\text{ mol}}{78.11\text{ g}} = 11.3\text{ mol/L}$

(c) $(\text{conc. water}) = \frac{0.998\text{ g}}{\text{mL}} \times \frac{1000\text{ mL}}{1\text{ L}} \times \frac{1\text{ mol}}{18.02\text{ g}} = 55.4\text{ mol/L}$

(d) $(\text{conc. Au}) = \frac{19.32\text{ g}}{\text{mL}} \times \frac{1000\text{ mL}}{1\text{ L}} \times \frac{1\text{ mol}}{196.97\text{ g}} = 98.1\text{ mol/L}$

14.2 The mixture is not at equilibrium, but the reaction is so slow that there is no change in concentrations. You could show that the system was not at equilibrium by providing a catalyst or by raising the temperature to speed up the reaction.

14.3 (a) The new mixture is not at equilibrium because the quotient (conc. *trans*)/(conc. *cis*) no longer equals the equilibrium constant. Because (conc. *cis*) was halved, the quotient is twice K_c.
(b) The rate *trans* → *cis* remains the same as before, because (conc. *trans*) did not change. The rate *cis* → *trans* is only half as great, because (conc. *cis*) is half as great as at equilibrium.
(c) At 600 K, K_c is 1.47. Thus, [*trans*] = 1.47[*cis*].
(d) 0.15 mol/L

14.4 (a) If the coefficients of an equation are halved, the numerical value for the new equilibrium constant is the square root of the previous equilibrium constant. So the new equilibrium constant is $(6.25 \times 10^{-58})^{1/2} = 2.50 \times 10^{-29}$.
(b) If a chemical equation is reversed, the value for the new equilibrium constant is the reciprocal of the previous equilibrium constant. So the new equilibrium constant is $1/(6.25 \times 10^{-58}) = 1.60 \times 10^{57}$.

14.5 (a) $K_P = K_c \times (RT)^{\Delta n} = (3.5 \times 10^8\text{ L}^2\text{ mol}^{-2}) \times \{(0.082057\text{ L atm K}^{-1}\text{ mol}^{-1})(298)\}^{-2} = 5.8 \times 10^5\text{ atm}^{-2}$

(b) $K_P = (3.2 \times 10^{81}\text{ L mol}^{-1}) \times \{(0.082057\text{ L atm mol}^{-1}\text{ K}^{-1})(298\text{ K})\}^{-1} = 1.3 \times 10^{80}\text{ atm}^{-1}$
(c) $K_P = K_c = 1.7 \times 10^{-3}$
(d) $K_P = (1.7 \times 10^2\text{ L mol}^{-1}) \times \{(0.082057\text{ L atm mol}^{-1}\text{ K}^{-1})(298\text{ K})\}^{-1} = 6.9\text{ atm}^{-1}$

14.6 (a) Because K_c for the forward reaction is small, K_c' for the reverse reaction is large.
(b) $K_c' = \frac{1}{K_c} = \frac{1}{1.8 \times 10^{-5}} = 5.6 \times 10^4$
(c) Ammonium ions and hydroxide ions should react, using up nearly all of whichever is the limiting reactant.
(d) $NH_4^+(aq) + OH^-(aq) \rightleftharpoons NH_3(aq) + H_2O(\ell)$
$NH_3(aq) \rightleftharpoons NH_3(g)$
You might detect the odor of $NH_3(g)$ above the solution. A piece of moist red litmus paper above the solution would turn blue.

14.7 Q should have the same mathematical form as K_P, so for the general Equation 14.1

$$Q_P = \frac{P_C^c \times P_D^d}{P_A^a \times P_B^b}$$

The rules for Q_P and K_P are analogous to those for Q and K_c:
If $Q_P > K_P$, then the reverse reaction occurs.
If $Q_P = K_P$, then the system is at equilibrium.
If $Q_P < K_P$, the forward reaction occurs.

14.8 $PCl_5(s) \rightleftharpoons PCl_3(g) + Cl_2(g)$
(a) Adding Cl_2 shifts the equilibrium to the left.
(b) Adding PCl_3 to the container shifts the equilibrium to the left.
(c) Because $PCl_5(s)$ does not appear in the K_c expression, adding some will not affect the equilibrium.

14.9 $Q = \frac{(\text{conc. H}_2)(\text{conc. I}_2)}{(\text{conc. HI})^2} = \frac{(0.102 \times 3)(0.0018 \times 3)}{(0.096 \times 3)^2} = \frac{(0.102)(0.0018)(9)}{(0.096)^2(9)} = 0.020 = K_c$

Since $Q = K_c$, the system is at equilibrium under the new conditions. No shift is needed and none occurs.

14.10 From Problem-Solving Practice 14.7, $[NO_2] = 0.0418$ mol/L and $[N_2O_4] = 0.292$ mol/L.
Decreasing the volume from 4.00 to 1.33 L increases the concentrations as

$$(\text{conc. NO}_2) = 0.0418 \times \frac{4.00}{1.33} = 0.1257\text{ mol/L}$$

$$(\text{conc. N}_2\text{O}_4) = 0.292 \times \frac{4.00}{1.33} = 0.8782\text{ mol/L}$$

$$Q = \frac{(\text{conc. NO}_2)^2}{(\text{conc. N}_2\text{O}_4)} = \frac{(0.1257)^2}{0.8782} = 1.80 \times 10^{-2}$$

which is greater than K_c. Therefore, the equilibrium should shift to the left. Let x be the change in concentration of NO_2, (which should be negative).

	N_2O_4	$\rightleftharpoons$	$2\,NO_2$
Initial concentration (mol/L)	0.8782		0.1257
Change as reaction occurs (mol/L)	$-\frac{1}{2}x$		x
New equilibrium concentration (mol/L)	$0.8782 - \frac{1}{2}x$		$0.1257 + x$

$$K_c = 5.9 \times 10^{-3} = \frac{(0.1257 + x)^2}{0.8782 - 0.500x}$$

$$= \frac{(1.580 \times 10^{-2}) + (2.514 \times 10^{-1})x + x^2}{0.8782 - 0.500x}$$

$$5.18 \times 10^{-3} - (2.95 \times 10^{-3})x$$

$$= (1.580 \times 10^{-2}) + (2.514 \times 10^{-1})x + x^2$$

$$x^2 + 0.2544x + (1.062 \times 10^{-2}) = 0$$

$$x = \frac{-0.2544 \pm \sqrt{6.472 \times 10^{-2} - (4 \times 1 \times 1.062 \times 10^{-2})}}{2}$$

$$x = \frac{-0.2544 \pm 0.1491}{2}$$

$$x = -0.0526 \quad \text{or} \quad x = -0.2018$$

The second root is mathematically reasonable, but results in a negative value for the $[NO_2]$. The new equilibrium concentrations are

$$[NO_2] = 0.1257 - 0.0526 = 0.0730 \text{ mol/L}$$

$$[N_2O_4] = 0.8782 - \tfrac{1}{2}(-0.0526) = 0.9045 \text{ mol/L}$$

Compared to the initial equilibrium, the concentrations have changed by

$$NO_2\text{: } \frac{0.0730}{0.0418} = 1.75 \qquad N_2O_4\text{: } \frac{0.945}{0.292} = 3.10$$

The concentration of N_2O_4 did increase by more than a factor of 3. The concentration of NO_2 increased by 1.75, which is less than a factor of 3 increase.

14.11

	How Reaction System Changes	Equilibrium Shifts	Change in K_c?
(a) Add reactant	Some reactants consumed	To right	No
(b) Remove reactant	More reactants formed	To left	No
(c) Add product	More reactants formed	To left	No
(d) Remove product	More products formed	To right	No
(e) Increase P by decreasing V	Total pressure decreases	Toward fewer gas molecules	No
(f) Decrease P by increasing V	Total pressure increases	Toward more gas molecules	No
(g) Increase T	Heat transfer into system	In endothermic direction	Yes
(h) Decrease T	Heat transfer out of system	In exothermic direction	Yes

If a substance is added or removed, the equilibrium is affected only if the substance's concentration appears in the equilibrium constant expression, or if its addition or removal changes concentrations that appear in the equilibrium constant expression. Changing pressure by changing volume affects an equilibrium only for gas phase reactions in which there is a difference in the number of moles of gaseous reactants and products.

14.12 (a) The reaction is exothermic. $\Delta H° = -46.11$ kJ.
(b) The reaction is not favored by entropy.
(c) The reaction produces more products at low temperatures.
(d) If you increase T the reaction will go faster, but a smaller amount of products will be produced.

Chapter 15

15.1 The data in Table 15.2 indicate that the solubility of alcohols decrease as the hydrocarbon chain lengthens. Thus, 1-octanol is less soluble in water than 1-heptanol and 1-decanol should be even less soluble than 1-octanol.

15.2 Methanol is more water soluble than is octanol, but octanol is more soluble in gasoline. The octanol molecule is more hydrocarbon-like and this explains its solubility in gasoline. The methanol molecule is more water-like and this explains its greater solubility in water.

15.3 34 g NH_4Cl would crystallize from solution at 20 °C.

15.4 (a) Unsaturated (b) Supersaturated (c) Saturated

15.5 The solubility of CO_2 decreases with increasing temperature, and the beverage loses its carbonation, causing it to go "flat."

15.6 Putting back water that is too warm would decrease the solubility of oxygen in the lake or river, thereby decreasing the oxygen concentration. This could cause a fish kill if the oxygen concentration dropped sufficiently.

15.7 Hot solvent would cause more of the solute to dissolve because Le Chatelier's principle states that at higher temperature an equilibrium will shift in the endothermic direction.

15.8 (a) Mass fraction of $NaHCO_3$

$$= \frac{0.20}{1000 + 6.5 + 0.20 + 0.1 + 0.10} = 2.0 \times 10^{-4}$$

$$\text{Wt. fraction} = 2.0 \times 10^{-4} \times 100\% = 2.0 \times 10^{-2}$$

(b) KCl and $CaCl_2$ each have the lowest mass fraction, 0.015.

15.9 (a) The 20-ppb sample has the higher lead concentration. (The other sample is 3 ppb lead.)
(b) 0.015 mg/L is equivalent to 0.015 ppm, which is 15 ppb. The 20-ppb sample exceeds the EPA limit; the 3-ppb sample does not.

15.10 $$\frac{280 \text{ mg}}{1 \text{ d}} \times \frac{0.500 \text{ L}}{12 \text{ mg}} = \frac{12 \text{ L}}{\text{d}} = 12 \text{ bottles per day}$$

15.11 $$3.5 \times 10^{20} \text{ gal} \times \frac{3.785 \text{ L}}{\text{gal}} \times \frac{1 \times 10^{-3} \text{ mg Au}}{1 \text{ L}} = 1 \times 10^{18} \text{ mg Au} = 1 \times 10^{15} \text{ g Au}$$

$$1 \times 10^{15} \text{ g Au} \times \frac{1 \text{ lb Au}}{454 \text{ g Au}} = 2 \times 10^{12} \text{ lb Au}$$

15.12 90 proof = 45 mL ethanol/100 mL beverage

$$\frac{45 \text{ mL ethanol}}{100 \text{ mL BVG}} \times \frac{1 \text{ mL BVG}}{0.861 \text{ g BVG}} \times \frac{0.79 \text{ g ethanol}}{1 \text{ mL ethanol}} = \frac{0.414 \text{ g ethanol}}{\text{g BVG}}$$

$$\text{Mass of BVG: } 1 \text{ qt} \times \frac{1 \text{ L}}{1.057 \text{ qt}} \times \frac{861 \text{ g}}{1 \text{ L}} = 814.5 \text{ g BVG.}$$

$$\text{Moles of ethanol: } 814.5 \text{ g BVG} \times \frac{0.414 \text{ g ethanol}}{\text{g BVG}} \times \frac{1 \text{ mol ethanol}}{46.0 \text{ g ethanol}} = 7.32 \text{ mol ethanol}$$

$$\text{Mass of ethanol: } 814.5 \text{ g BVG} \times \frac{0.414 \text{ g ethanol}}{\text{g BVG}} = 336.8 \text{ g ethanol}$$

$$\text{Mass of solvent: } 814.5 \text{ g BVG} - 336.8 \text{ g ethanol} = 477.7 \text{ g solvent} = 0.4777 \text{ kg solvent}$$

$$m_{\text{ethanol}} = \frac{7.32 \text{ mol ethanol}}{0.4777 \text{ kg solvent}} = 15.3 \text{ mol/kg}$$

15.13 (a) Moles of solute and kilograms of solvent
(b) Molar mass of the solute and the density of the solution

15.14 $50.0 \text{ g sucrose} \times \left(\frac{1 \text{ mol sucrose}}{342 \text{ g sucrose}}\right) = 0.146 \text{ mol sucrose}$

$$100.0 \text{ g water} \times \left(\frac{1 \text{ mol water}}{18.0 \text{ g water}}\right) = 5.56 \text{ mol water}$$

$$X_{H_2O} = \frac{5.56}{5.56 + 0.146} = \frac{5.56}{5.706} = 0.974$$

Applying Raoult's law:

$$P_{water} = (X_{water})(P^0_{water}) = (0.974)(71.88 \text{ mm Hg}) = 70.0 \text{ mm Hg}$$

15.15 $\Delta T_b = (2.53\ °C \text{ kg mol}^{-1})(0.10 \text{ mol/kg}) = 0.25\ °C$. The boiling point of the solution is 80.10 °C + 0.25 °C = 80.35 °C.

15.16 First, calculate the required molality of the solution that would have a freezing point of −30 °C.

$$\Delta T_f = -30\ °C = (-1.86\ °C\cdot\text{kg mol}^{-1}) \times m$$

$$m = \frac{-30\ °C}{-1.86\ °C \text{ kg mol}^{-1}} = 16.1 \text{ mol/kg}$$

To protect 4 kg of water from this freezing temperature, you would need 4 × 16.1 mol of ethylene glycol, or 65 mol.

$$65 \text{ mol}\left(\frac{62.1 \text{ g}}{\text{mol}}\right)\left(\frac{1 \text{ mL}}{1.113 \text{ g}}\right)\left(\frac{1 \text{ L}}{1000 \text{ mL}}\right) = 3.6 \text{ L}$$

15.17 $\Delta T_f = K_f \times m \times i$;

$$4.78\ °C = (1.86\ °C \text{ kg mol}^{-1})(2.0 \text{ mol/kg})(i)$$

$$i = \frac{4.78\ °C}{(2.0 \text{ mol/kg})(1.86\ °C \text{ kg mol}^{-1})} = 1.28$$

Degree of dissociation: If completely dissociated, 1 mol $CaCl_2$ should yield 3 mol ions

$$CaCl_2 \longrightarrow Ca^{2+} + 2\ Cl^-$$

and i should be 3. The degree of dissociation in this solution is $\frac{1.28}{3} = 0.427 \approx 43\%$.

15.18 The 0.02 mol/kg solution of ordinary soap would contain more particles since a soap is a salt of a fatty acid while sucrose is a nonelectrolyte.

15.19 $\frac{3 \text{ qt}}{\text{d}} \times \frac{1 \text{ L}}{1.06 \text{ qt}} \times \frac{0.050 \text{ mg Se}}{\text{L}} \times \frac{10^3 \text{ mg}}{1 \text{ mg}} = \frac{142 \text{ mg Se}}{\text{d}}$

15.20 0.025 ppm Pb means that for every liter (1 kg) of water, there is 0.025 mg, or 25 μg of Pb. Using this factor

$$\text{Volume of water} = 100.0\ \mu\text{g Pb}\left(\frac{1 \text{ L}}{25\ \mu\text{g Pb}}\right) = 4.0 \text{ L}$$

Chapter 16

16.1 (a) Acid (b) Base
(c) Acid (d) Base

16.2 More water molecules are available per NH_3 molecule in a very dilute solution of NH_3.

16.3 (a) $H_3O^+(aq) + CN^-(aq)$
(b) $H_3O^+\ (aq) + Br^-\ (aq)$
(c) $CH_3NH_3^+(aq) + OH^-(aq)$
(d) $(CH_3)_2NH_2^+(aq) + OH^-(aq)$

16.4 (a) $HSO_4^-(aq) + H_2O(\ell) \rightarrow H_2SO_4(aq) + OH^-(aq)$
(b) H_2SO_4 is the conjugate acid of HSO_4^-; water is the conjugate acid of OH^-.
(c) Yes; it can be an H^+ donor or acceptor.

16.5 The reverse reaction is favored because HSO_4^- is a stronger acid than CH_3COOH.

16.6 (a) (aq) + $H_2O \rightleftharpoons$ [ ]$^+$(aq) + $OH^-(aq)$
(b) (aq) + HCl(aq) $\longrightarrow$ [ ]$^+$(aq) + $Cl^-(aq)$

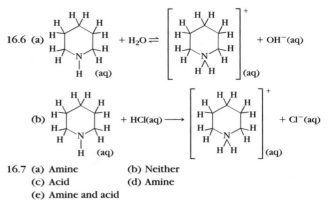

16.7 (a) Amine (b) Neither
(c) Acid (d) Amine
(e) Amine and acid

16.8 The pH values of 0.1 M solutions of these two strong acids would be essentially the same since they both are 100% ionized, resulting in $[H_3O^+]$ values that are the same.

16.9 $[H^+] = 10^{-pH} = 10^{-(-3.6)} = 10^{3.6} = 4 \times 10^3$ M

16.10 Because pH + pOH = 14.0, both solutions have a pOH of 8.5. The $[H_3O^+] = 10^{-pH} = 10^{-5.5} = 3.16 \times 10^{-6}$ M.

16.11 Pyruvic acid is the stronger acid, as indicated by its larger K_a value. Lactic acid's ionization reaction is more reactant-favored (less acid ionizes).

16.12 Being negatively charged, the HSO_4^- ion has a lower tendency to lose a positively charged proton because of the electrostatic attractions of opposite charges.

16.13 (a) Step 1: $HOOC{-}COOH(aq) + H_2O(\ell) \rightleftharpoons H_3O^+(aq) + HOOC{-}COO^-(aq)$
Step 2: $HOOC{-}COO^-(aq) + H_2O(\ell) \rightleftharpoons H_3O^+(aq) + {}^-OOC{-}COO^-(aq)$
(b) Step 1: $C_3H_5(COOH)_3(aq) + H_2O \rightleftharpoons H_3O^+(aq) + C_3H_5(COOH)_2COO^-(aq)$
Step 2: $C_3H_5(COOH)_2COO^-(aq) + H_2O \rightleftharpoons H_3O^+(aq) + C_3H_5(COOH)(COO)_2^{2-}(aq)$
Step 3: $C_3H_5(COOH)(COO)_2^{2-}(aq) + H_2O \rightleftharpoons H_3O^+(aq) + C_3H_5(COO)_3^{3-}(aq)$

16.14 (a) Fluorobenzoic acid (b) Chloroacetic acid
In both cases, the more electronegative halogen atom increases electron withdrawal from the acidic hydrogen, thereby increasing its partial positive charge.

16.15 Oxalic acid is the stronger acid because it has a greater number of oxygens.

16.16 (a) $H{-}C(H)(NH_3^+){-}C(=O){-}O^-$

(b) $H{-}C(H)({}^+NH_3){-}C(=O){-}OH$ at pH 2; $H{-}C(H)(NH_2){-}C(=O){-}O^-$ at pH 10

16.17 $C_6H_5{-}CH_2{-}CH(COO^-){-}NH{-}C(=O){-}CH(CH_3){-}NH{-}C(=O){-}CH({}^+NH_3){-}CH(OH){-}CH_3$

16.18 A reaction had these conditions:

	$Ni(H_2O)_6^{2+}(aq) + H_2O(\ell) \rightleftharpoons$	$Ni(H_2O)_5(OH)^+(aq) +$	$H_3O^+(aq)$
Initial concentration (mol/L)	0.15	0	10^{-7}
Change in concentration on reaction (mol/L)	$-x$	$+x$	$+x$
Concentration at equilibrium (mol/L)	$0.15 - x$	x	x

Substituting these values in the equilibrium constant expression, and simplifying $0.15 - x$ to be 0.15 because the value of K_a is so small

$$K_a = \frac{[Ni(H_2O)_5(OH)^+][H_3O^+]}{[Ni(H_2O)_6^{2+}]}$$

$$= \frac{(x)(x)}{(0.15 - x)} \approx \frac{x^2}{0.15} = 2.5 \times 10^{-11}$$

Solving for x, which is the $[H_3O^+]$

$$x = \sqrt{(0.15)(2.5 \times 10^{-11})} = 1.9 \times 10^{-6}$$

So, the pH of this solution is $\log(1.9 \times 10^{-6}) = 5.72$.

16.19 $K_b = \frac{1.0 \times 10^{-14}}{K_a} = \frac{1.0 \times 10^{-14}}{1.3 \times 10^{-10}} = 7.7 \times 10^{-5}$; carbonate or pentaaquairon(II) ion; by comparing K_b values.

16.20 The pH of soaps is >7 due to the reaction with water of the conjugate base in the soap to form a basic solution.

16.21 $C_6H_5-CH(OH)-CH(CH_3)-NH_2^+-CH_3 \quad Cl^-$

16.22 Ammonium acetate. The pH of a solution of this salt will be 7.

16.23 (a) Lewis base (b) Lewis acid
(c) Lewis acid and base (d) Lewis acid
(e) Lewis acid (f) Lewis base

16.24 Strong bases would cause damage to tissue.

16.25 Baking powder and baking soda are sources of bicarbonate; buttermilk and baking powder are sources of acid.

16.26 Set up a small table for the hydrolysis reaction.

	CO_3^{2-} + H_2O ⇌	HCO_3^- +	OH^-
Initial concentration (mol/L)	5.2	0	10^{-7}
Concentration change due to reaction (mol/L)	$-x$	$+x$	$+x$
Concentration at equilibrium (mol/L)	$5.2 - x$	x	x

Using the K_b expression and substituting the values from the table

$$K_b = \frac{K_w}{K_a(HCO_3^-)} = \frac{[HCO_3^-][OH^-]}{[CO_3^{2-}]} = \frac{x^2}{5.2 - x}$$

$$\approx \frac{x^2}{5.2} = 2.1 \times 10^{-4}$$

$$x = [OH^-] = \sqrt{(5.2)(2.1 \times 10^{-4})} = 3.3 \times 10^{-2}$$

$$pOH = -\log(3.3 \times 10^{-2}) = 1.48$$

$$pH = 14.00 - 1.48 = 12.52$$

Chapter 17

17.1 HCl and NaCl: no; has no significant H^+ acceptor (Cl^- is a very poor base).
KOH and KCl: no; has no H^+ donor.

17.2 $pH = 7.21 + \log\frac{(0.0025)}{(0.00015)} = 7.21 + \log(1.67)$
$= 7.21 + 0.22 = 7.43$

17.3 $pH = pK_a + \log\frac{[acetate]}{[acetic\ acid]}$

$$4.68 = 4.74 + \log\frac{[acetate]}{[acetic\ acid]}$$

$$\log\frac{[acetate]}{[acetic\ acid]} = 4.68 - 4.74 = -0.06;$$

$$\frac{[acetate]}{[acetic\ acid]} = 10^{-0.06} = 0.86$$

Therefore, [acetate] = 0.86 × [acetic acid].

17.4 (a) $pH = 6.38 + \log\frac{[0.25]}{[0.10]} = 6.38 + 0.398 = 6.78$

(b) $pH = 7.21 + \log\frac{[HPO_4^{2-}]}{[H_2PO_4^-]}$

$$= 7.21 + \log\frac{(0.25)}{(0.10)} = 7.21 + 0.398 = 7.61$$

17.5 Since CO_2 reacts to form an acid, H_2CO_3, the phosphate ion that is the stronger base, HPO_4^{2-}, will be used to counteract its presence.

17.6

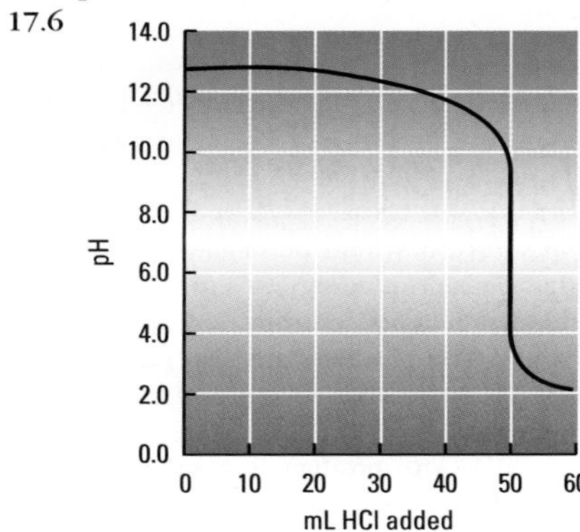

17.7 The addition of 30.0 mL of 0.100 M NaOH neutralizes 30.0 mL of 0.100 M acetic acid, forming 0.0030 mol of acetate ions, which is in 80.0 mL of solution. There is (0.0200 L)(0.100 M) = 0.00200 mol of acetic acid that is unreacted.

$$K_a = 1.8 \times 10^{-5} = \frac{[H^+][C_2H_3O_2^-]}{[HC_2H_3O_2]}$$

$$= \frac{[H^+] \times \left(\frac{0.00300\ mol}{0.0800\ L}\right)}{\left(\frac{0.00200\ mol}{0.0800\ L}\right)}$$

$$1.8 \times 10^{-5} = \frac{[H^+] \times (0.0375)}{(0.025)} = [H^+] \times 1.5$$

$$[H^+] = \frac{1.8 \times 10^{-5}}{1.5} = 1.2 \times 10^{-5}$$

$$pH = -\log(1.2 \times 10^{-5}) = 4.92$$

17.8 As NaOH is added, it reacts with acetic acid to form sodium acetate. After 20.0 mL NaOH has been added, just less than half of the acetic acid has been converted to sodium acetate; when 30.0 mL NaOH has been added, just over half of the acetic acid has been neutralized. Thus, after 20.0 mL and 30.0 mL base have been added, the solution contains approximately equal amounts of acetic acid and acetate ion, its conjugate base, which acts as a buffer.

17.9 Because Reaction (b) occurs to an appreciable extent, CO_3^{2-} is used as it forms by Reaction (a), causing additional $CaCO_3(s)$ to dissolve.

17.10 (a) The excess iodide would create a stress on the equilibrium and shift it to the left; some AgI and some PbI_2 would precipitate from solution.
(b) The added SO_4^{2-} would cause the precipitation of $BaSO_4$.

Chapter 18

18.1 (a) $H_2O(\ell) \rightarrow H_2O(g)$ Product-favored
(b) $SiO_2(s) \rightarrow Si(s) + O_2(g)$ Reactant-favored
(c) $(C_6H_{10}O_5)_n(s) + 6n\ O_2(g) \rightarrow$
$6n\ CO_2(g) + 5n\ H_2O(g)$ Product-favored
(d) $C_{12}H_{22}O_{11}(s) \rightarrow C_{12}H_{22}O_{11}(aq)$ Product-favored

18.2 A*** A**B* A*B** B***
If C, D, and E are added, there are many more arrangements in addition to these:

A*B*C* A*B*D* A*B*E* A*C*D* A*C*E*
A*D*E* B*C*D* B*C*E* B*D*E* C*D*E*
A**C* A**D* A**E* B**C* B**D*
B**E* C**A* C**B* C**D* C**E*
D**A* D**B* D**C* D**E* E**A*
E**B* E**C* E**D* C*** D***
E***

There are 35 possible arrangements, but only 4 of them have the energy confined to atoms A and B. The probability that all energy remains with A and B is thus 4/35 = 0.114, or a little more than 11%.

18.3 Using Celsius temperature and $\Delta S = q_{rev}/T$, if the temperature were -10 °C, the value of ΔS would be negative, in disagreement with the fact that transfer of energy to a sample should increase molecular motion and, hence, entropy.

18.4 (a) The reactant is a gas. The products are also gases, but the number of molecules has increased, so entropy is greater for products. (Entropy increases.)
(b) The reactant is a solid. The product is a solution. Mixing sodium and chloride ions among water molecules results in greater entropy for the product. (Entropy increases.)
(c) The reactant is a solid. The products are a solid and a gas. The much larger entropy of the gas results in greater entropy for the products. (Entropy increases.)

18.5 (a) Because $\Delta S_{surroundings} = -\Delta H/T$ at a given temperature, the larger the value of T, the smaller the value of $\Delta S_{surroundings}$.
(b) If ΔS_{system} does not change much with temperature, then $S_{universe}$ must also get smaller. In this case, because ΔS_{system} is negative, $\Delta S_{universe}$ would become negative at a high enough temperature.

18.6 (a) The reaction would have gaseous water as a product.
(b) Both $\Delta H°$ and $\Delta S°$ would change. $\Delta H° = -10{,}232.197$ kJ and $\Delta S° = 756.57$ J/K.
(c) If any of the reactants or products change to a different phase (s, ℓ, or g) over the range of temperature, $\Delta H°$ and $\Delta S°$ will change significantly at the temperature of the phase transition.

18.7

Reaction	$\Delta H°$, 298 K (kJ)	$\Delta S°$, 298 K (J/K)
(a)	−1410.94	−267.67
(b)	467.87	560.32
(c)	−393.509	2.862
(d)	1241.2	−461.50

Reaction (a) is product-favored at low T (room temperature) and reactant-favored at high T.
Reaction (b) is reactant-favored at low T, product-favored at high T.
Reaction (c) is product-favored at all values of T.
Reaction (d) is reactant-favored at all values of T.

18.8 (a)
$$\Delta S°_{system} = 2 \text{ mol HCl(g)} \times S°(\text{HCl[g]}) - 1 \text{ mol H}_2\text{[g]} \times S°(\text{H}_2\text{[g]}) - 1 \text{ mol Cl}_2\text{[g]} \times S°(\text{Cl}_2\text{[g]})$$
$$= (2 \times 186.908 - 130.684 - 223.066) \text{ J/K}$$
$$= 20.055 \text{ J/K}$$
$$\Delta S°_{surroundings} = -\Delta H°/T = -\Delta H°/T$$
$$= -[2 \text{ mol HCl(g)} \times (-92.307 \text{ kJ/mol})]/298.15 \text{ K}$$
$$= 619.20 \text{ J/K}$$

(b) $\Delta S°_{universe} = (619.20 + 20.066)$ J/K $= 639.27$ J/K (Product-favored)

18.9

Sign of $\Delta H°$	Sign of $\Delta S°$	Sign of $\Delta G°$	Product-favored?
Negative (exothermic)	Positive	Negative	Yes
Negative (exothermic)	Negative	Depends on T	Yes at low T; no at high T
Positive (endothermic)	Positive	Depends on T	No at low T; yes at high T
Positive (endothermic)	Negative	Positive	No

18.10 (a) At 400 °C the equation is

$$2\ \text{HgO(s)} \longrightarrow 2\ \text{Hg(g)} + \text{O}_2\text{(g)}$$

Because Hg(g) is a product, instead of Hg(ℓ), both $\Delta H°$ and $\Delta S°$ will have significantly different values above 356 °C from their values below 356 °C. Therefore, the method of estimating $\Delta G°$ would not work above 356 °C.
(b) At 400 °C the entropy change should be more positive, which would make the reaction more product-favored. Because $\Delta H°$ and $\Delta S°$ are both positive, the reaction is product-favored at high temperatures.

18.11 (a) If the extent of reaction is 0.10, then 0.10 mol of NaOH has reacted with 0.10 mol of CO_2 to produce 0.10 mol of $NaHCO_3$.

$$\Delta G°(0.10 \text{ extent}) = 0.10 \text{ mol} \times \Delta G_f°(\text{NaHCO}_3\text{[s]}) - 0.10 \text{ mol} \times \Delta G_f°(\text{NaOH[s]}) - 0.10 \text{ mol} \times \Delta G_f°(\text{CO}_2\text{[g]})$$
$$= -0.10 \times 851.0 \text{ kJ} + 0.10 \times 379.484 \text{ kJ} + 0.10 \times 394.359 \text{ kJ}$$
$$= -7.72 \text{ kJ}$$

Similarly,

$$\Delta G°(0.40 \text{ extent}) = -30.9 \text{ kJ} \ (= 0.40[-72.2 \text{ kJ}])$$
$$\Delta G°(0.80 \text{ extent}) = -61.8 \text{ kJ} \ (= 0.80[-72.2 \text{ kJ}])$$

(b) In each case, $\Delta G°(x \text{ extent}) = x\Delta G°(\text{full extent})$, which verifies the statement.
(c) Since $\Delta G°(x \text{ extent}) = x\Delta G°(\text{full extent})$,

$$y = xm + b$$

where b = 0 and m = $\Delta G°$(full extent) = slope.

18.12 (a)
$$\Delta S°(\text{i}) = \{2 \times (27.78) + \tfrac{3}{2} \times (205.138) - (87.40)\} \text{ J/K} = 275.86 \text{ kJ}$$
$$\Delta H°(\text{i}) = -\Delta H_f°(\text{Fe}_2\text{O}_3\text{[s]}) = 824.2 \text{ kJ}$$
$$\Delta G°(\text{i}) = -\Delta G_f°(\text{Fe}_2\text{O}_3\text{[s]}) = 742.2 \text{ kJ}$$
$$\Delta S°(\text{ii}) = \{50.92 - 2 \times (28.3) - \tfrac{3}{2} \times (205.138)\} \text{ J/K}$$
$$= -313.4 \text{ kJ}$$
$$\Delta H°(\text{ii}) = \Delta H_f°(\text{Al}_2\text{O}_3\text{[s]}) = -1675.7 \text{ kJ}$$
$$\Delta G°(\text{ii}) = \Delta G_f°(\text{Al}_2\text{O}_3\text{[s]}) = -1582.3 \text{ kJ}$$

Step (i) is reactant-favored. Step (ii) is product-favored.
(b) Net reaction
$Fe_2O_3(s) + 2\ Al(s) \rightarrow 2\ Fe(s) + Al_2O_3(s)$
$\Delta S° = 275.86$ J/K $+ (-313.4$ J/K$) = -37.5$ J/K
$\Delta H° = 824.2$ kJ $+ (-1675.7$ kJ$) = -851.5$ kJ
$\Delta G° = 742.2$ kJ $+ (-1582.3$ kJ$) = -840.1$ kJ

The net reaction has negative $\Delta G°$ and is therefore product-favored. For the *net* reaction, $\Delta S°$, $\Delta H°$, and $\Delta G°$ are all negative. (c) If the two reactions are coupled, it is possible to obtain iron from iron(III) oxide even though that reaction is not product-favored by itself. The large negative $\Delta G°$ for formation of $Al_2O_3(s)$ makes the overall $\Delta G°$ negative for the coupled reactions.

(d) $Mg(s) + \frac{1}{2} O_2(g) \rightarrow MgO(s)$

$$\Delta G° = \Delta G_f°(MgO[s]) = -569.43 \text{ kJ}$$

Coupling the reactions, we have

$$Fe_2O_3(s) \longrightarrow 2\,Fe(s) + \tfrac{3}{2} O_2(g) \qquad \Delta G_1° = 742.2 \text{ kJ}$$

$$3 \times (Mg[s] + \tfrac{1}{2} O_2[g] \longrightarrow MgO[s])$$

$$\Delta G_2° = 3(-569.43) \text{ kJ} = -1708.29 \text{ kJ}$$

$$Fe_2O_3(s) + 3\,Mg(s) \longrightarrow 2\,Fe(s) + 3\,MgO(s)$$

$$\Delta G_3° = -966.1 \text{ kJ}$$

18.13 $\Delta G° = -2870$ kJ + 32 × (30.5 kJ) = −1894 kJ. The 1894 kJ of Gibbs free energy is transformed into thermal energy.

18.14 64,500 g ATP/50 g ATP = 1290 times each ADP must be recycled to ATP on average each day.

Chapter 19

19.1 This is an application of the law of conservation of matter. If the number of electrons gained were different from the number of electrons lost, some electrons must have been created or destroyed.

19.2 Removal of the salt bridge would effectively switch off the flow of electricity from the battery.

19.3 Avogadro's number of electrons is 96,500 coulombs of charge, so it is 96,500 times as large as one coulomb of charge.

19.4 The zinc anode could be weighed before the battery was put into use. After a period of time, the zinc anode could be dried and reweighed. A loss in weight would be interpreted as being caused by the loss of Zn atoms from the surface through oxidation.

19.5 No, because Hg^{2+} ions can oxidize Al metal to Al^{3+} ions. The net cell reaction is

$$2\,Al(s) + 3\,Hg^{2+}(aq) \longrightarrow 2\,Al^{3+}(aq) + 3\,Hg(\ell)$$

$$E_{cell} = +2.51 \text{ V}$$

19.6 For this table,
(a) V^{2+} ion is the weakest oxidizing agent.
(b) Cl_2 is the strongest oxidizing agent.
(c) V is the strongest reducing agent.
(d) Cl^- is the weakest reducing agent.
(e) No, E_{cell} for that reaction would be <0.
(f) No, E_{cell} for that reaction would be <0.
(g) Pb can reduce I_2 and Cl_2.

19.7 In Table 19.1, Sb would be above H_2 and Pb would be below H_2. For Sb, the reduction potential would be between 0.00 and +0.337 V, and for Pb the value would be between 0.00 and −0.14 V.

19.8 For Na^+: $E_{ion} = 61.5 \log\left(\frac{150}{18}\right) = 61.5 \log(8.33)$

$$= 61.5 \times 0.921 = 57 \text{ mV}$$

19.9 During charging, the reactions at each electrode are reversed. At the electrode that is normally the anode, the charging reaction is

$$Cd(OH)_2(s) + 2\,e^- \longrightarrow Cd(s) + 2\,OH^-(aq)$$

This is reduction, so this electrode is now a cathode.
At the electrode that is normally the cathode, the charging reaction is

$$Ni(OH)_2 + OH^-(aq) \longrightarrow NiO(OH)(s) + H_2O(\ell) + e^-$$

This is oxidation, so this electrode is now an anode.

19.10 Remove the lead cathodes and as much sulfuric acid as you can from the discharged battery. Find some steel and construct a battery with Cl_2 gas flowing across a piece of steel. The two half-reactions would be

$$Cl_2(g) + 2\,e^- \longrightarrow 2\,Cl^-(aq) \qquad +1.36 \text{ V}$$

$$Pb(s) + SO_4^{2-}(aq) \longrightarrow PbSO_4(s) + 2\,e^- \qquad +0.356 \text{ V}$$

$$E_{cell} = 1.36 + 0.356 = 1.71 \text{ V}$$

19.11 Potassium metal was produced at the cathode.
Oxidation reaction: $2\,F^-(\text{molten}) \rightarrow F_2(g) + 2\,e^-$
Reduction reaction: $2(K^+[\text{molten}] + e^- \rightarrow K[\ell])$
Net cell reaction:
$2\,K^+(\text{molten}) + 2F^-(\text{molten}) \rightarrow 2\,K(\ell) + F_2(g)$

19.12 Reaction (c) making 2 mol of Cu from Cu^{2+} would require 4 Faradays of electricity. Two F are required for part (b), and 3 F are required for part (a).

19.13 First, calculate how many coulombs of electricity are required to make this much aluminum.

$$(2000.\text{ ton Al})\left(\frac{2000 \text{ lb Al}}{1 \text{ ton Al}}\right)\left(\frac{454.6 \text{ g Al}}{1 \text{ lb Al}}\right)\left(\frac{1 \text{ mol Al}}{26.982 \text{ g Al}}\right)$$

$$\times \left(\frac{3 \text{ mol e}^-}{1 \text{ mol Al}}\right)\left(\frac{96{,}500 \text{ C}}{1 \text{ mol e}^-}\right) = 1.950 \times 10^{13} \text{ C}$$

Next, using the product of charge and voltage, calculate how many joules are required; then convert to kilowatt-hours.

$$\text{Energy} = (1.950 \times 10^{13} \text{ C})(4.0 \text{ V})\left(\frac{1 \text{ J}}{1 \text{ C} - \times 1 \text{ V}}\right)$$

$$\times \left(\frac{1 \text{ kWh}}{3.60 \times 10^6 \text{ J}}\right) = 2.2 \times 10^7 \text{ kWh}$$

19.14 To calculate how much energy is stored in a battery, you need the voltage and the number of coulombs of charge the battery can provide. The voltage is generally given on the battery label. To determine the number of coulombs available, you would have to disassemble the battery and determine the masses of the chemicals at the cathode and anode.

19.15 $(0.50 \text{ A})(20.\text{ min})\left(\frac{60 \text{ s}}{1 \text{ min}}\right)\left(\frac{1 \text{ C}}{1 \text{ A s}}\right)\left(\frac{1 \text{ mol e}^-}{96{,}500 \text{ C}}\right)$

$$\times \left(\frac{1 \text{ mol Ag}}{1 \text{ mol e}^-}\right)\left(\frac{107.9 \text{ g Ag}}{1 \text{ mol Ag}}\right) = 0.67 \text{ g Ag}$$

19.16 No, not all metals corrode as easily. Three metals that would corrode about as readily as Fe and Al are Zn, Mg, and Cd. Three metals that do not corrode as readily as Fe and Al are Cu, Ag, and Au. These three metals are used in making coins and jewelry. Metals fall into these two broad groups because of their relative ease of oxidation compared with the oxidation of H_2. In Table 19.1, you can see this breakdown easily.

19.17 (b) > (a) > (d) > (c)
Sand by the seashore, (b), would contain both moisture and salts, which would aid corrosion. Moist clay, (a), would contain water but less dissolved salts. If an iron object were embedded within the clay, its impervious nature might prevent oxygen from getting to the iron, which would also lower the rate of corrosion. Desert sand in Arizona, (d), would be quite dry, and this low-moisture environment would not lead to a rapid rate of corrosion. On the moon, (c), there would be a lack of moisture and oxygen. This would lead to a very low rate of corrosion.

Chapter 20

20.1 $^{235}_{92}U \rightarrow ^{4}_{2}He + ^{231}_{90}Th$
$^{231}_{90}Th \rightarrow ^{0}_{-1}e + ^{231}_{91}Pa$
$^{231}_{91}Pa \rightarrow ^{4}_{2}He + ^{227}_{89}Ac$
$^{227}_{89}Ac \rightarrow ^{4}_{2}He + ^{223}_{87}Fr$
$^{223}_{87}Fr \rightarrow ^{0}_{-1}e + ^{223}_{88}Ra$

20.2 $^{26}_{13}Al \rightarrow ^{0}_{+1}e + ^{26}_{12}Mg$
$^{26}_{13}Al + ^{0}_{-1}e \rightarrow ^{26}_{12}Mg$

20.3 Mass difference = $\Delta m = -0.03438$ g/mol
$\Delta E = (-3.438 \times 10^{-5}$ kg/mol$)(2.998 \times 10^8$ m/s$)^2$
$= -3.090 \times 10^{12}$ J/mol $= -3.090 \times 10^8$ kJ/mol
E_b per nucleon = 5.150×10^8 kJ/nucleon
E_b for ^{6}Li is smaller than E_b for ^{4}He; therefore, helium-4 is more stable than lithium-6.

20.4 From the graph it can be seen that the binding energy per nucleon increases more sharply for the fusion of lighter elements than it does for heavy elements undergoing fission. Therefore, fusion is more exothermic per gram than fission.

20.5 $(\frac{1}{2})^{10} = 9.8 \times 10^{-4}$; this is equivalent to 0.098% of the radioisotope remaining.

20.6 All the lead came from the decay of ^{238}U; therefore, at the time the rock was dated, $N = 100$ and $N_0 = 109$. The decay constant, k, can be determined:

$$k = \frac{0.693}{4.51 \times 10^9 \text{ y}} = 1.54 \times 10^{-10} \text{ y}^{-1}$$

The age of the rock (t) can be calculated using Equation 20.3:

$$\ln \frac{100}{209} = -(1.54 \times 10^{-10} \text{ y}^{-1}) \times t$$
$$t = 4.80 \times 10^9 \text{y}$$

20.7 Ethylene is derived from petroleum, which was formed millennia ago. The half-life of ^{14}C is 5730 y, and thus much of ethylene's ^{14}C would have decayed and would be much less than that of the ^{14}C alcohol produced by fermentation.

20.8 (a) $^{13}_{6}C + ^{1}_{0}n \rightarrow ^{4}_{2}He + ^{10}_{4}Be$
(b) $^{14}_{7}N + ^{4}_{2}He \rightarrow ^{1}_{0}n + ^{17}_{9}F$
(c) $^{253}_{99}Es + ^{4}_{2}He \rightarrow ^{1}_{0}n + ^{256}_{101}Md$

20.9 $^{208}_{82}Pb + ^{70}_{30}Zn \rightarrow ^{277}_{112}E + ^{1}_{0}n$

20.10 Burning a metric ton of coal produces 2.8×10^7 kJ of energy.

$$\left(\frac{2.8 \times 10^4 \text{ kJ}}{1.0 \text{ kg}}\right)\left(\frac{10^3 \text{ kg}}{\text{metric ton}}\right) = 2.8 \times 10^7 \text{ kJ of energy}$$

The fission of 1.0 kg of ^{235}U produces

$$\frac{2.1 \times 10^{10} \text{ kJ}}{0.235 \text{ kg } ^{235}\text{U}} = 8.93 \times 10^{10} \text{ kJ}$$

It would require burning 3.2×10^3 metric tons of coal to equal the amount of energy from 1.0 kg of ^{235}U:

$$8.93 \times 10^{10} \text{ kJ from } ^{235}\text{U} \times \frac{1 \text{ metric ton coal}}{2.9 \times 10^7 \text{ kJ}}$$
$$= 3.2 \times 10^3 \text{metric tons}$$

20.11 $k = \dfrac{0.693}{29.1 \text{ y}} = 2.38 \times 10^{-2} \text{ y}^{-1}$

ln(fraction) $= -(2.38 \times 10^{-2}$ y$^{-1})$ (18 y, as of 2004) $= -0.4284$
fraction $= e^{-0.4284} = 0.652 = 65.2\%$

20.12 $k = \dfrac{0.693}{30.2 \text{ y}} = 2.29 \times 10^{-2} \text{ y}^{-1}$

(a) 60% drop in activity; 40% activity remaining
$\ln(0.40) = -0.916 = -(2.29 \times 10^{-2} \text{ y}^{-1}) \times t$

$$t = \frac{-0.916}{-2.29 \times 10^{-2} \text{ y}^{-1}} = 40 \text{ y}$$

(b) 90% drop in activity, 10% remains
$\ln(0.10) = -2.30 = -(2.29 \times 10^{-2} \text{ y}^{-1}) \times t$

$$t = \frac{-2.30}{-2.29 \times 10^{-2} \text{ y}^{-1}} = 100 \text{ y}$$

20.13 (a) $^{7}_{3}Li + ^{1}_{1}H \rightarrow ^{1}_{0}n + ^{7}_{4}Be$
(b) $^{2}_{1}H + ^{3}_{2}He \rightarrow ^{4}_{2}He + ^{1}_{1}H$

20.14 $k = \dfrac{0.693}{3.82 \text{ d}} = 0.181 \text{ d}^{-1}$

(a) The drop from 8 to 4 pCi represents one half-life, 3.82 days.
(b) $\ln\left(\dfrac{1.5}{8}\right) = -1.67 = -(0.181 \text{ d}^{-1}) \times t$

$$t = \frac{-1.67}{-0.181 \text{ d}^{-1}} = 9.25 \text{ d}$$

20.15 $k = \dfrac{0.693}{78.2 \text{ h}} = 8.86 \times 10^{-3} \text{ h}^{-1}$

$$\ln(0.10) = -2.30 = -(8.86 \times 10^{-3} \text{ h}^{-1}) \times t$$
$$t = 260 \text{ h}$$

20.16 Iron-59 $k = \dfrac{0.693}{44.5 \text{ d}} = 1.557 \times 10^{-2} \text{ d}^{-1}$

Chromium-51 $k = \dfrac{0.693}{27.7 \text{ d}} = 0.0250 \text{ d}^{-1}$

Fractions remaining:

^{59}Fe: ln(fraction) $= -(1.557 \times 10^{-2} \text{ d}^{-1}) \times 90$ d
fraction $= e^{-1.40} = 0.246$; 80 mg $\times$ 0.246 = 19.7 mg left
^{51}Cr: ln(fraction) $= -(0.0250 \text{ d}^{-1}) \times 90$ d; 10.5 mg left

Alternatively, consider the fact that 90 days is approximately two half-lives of ^{59}Fe. Therefore, approximately $\frac{3}{4}$ of it (about 60 mg) has decayed after 90 days, and about 20 mg remains. In that same time, ^{51}Cr has undergone more than three half-lives, so that less than $\frac{1}{8}$ remains (less than 12.5 mg).

Chapter 21

21.1 $^{12}_{6}C + ^{4}_{2}He \rightarrow ^{16}_{8}O$
$^{16}_{8}O + ^{4}_{2}He \rightarrow ^{20}_{10}Ne$
$^{20}_{10}Ne + ^{4}_{2}He \rightarrow ^{24}_{12}Mg$

21.2
$^{130}_{48}Cd \rightarrow ^{0}_{-1}e + ^{130}_{49}In$
$^{130}_{49}In \rightarrow ^{0}_{-1}e + ^{130}_{50}Sn$
$^{130}_{50}Sn \rightarrow ^{0}_{-1}e + ^{130}_{51}Sb$
$^{130}_{51}Sb \rightarrow ^{0}_{-1}e + ^{130}_{52}Te$

21.3 Two of the four oxygens in an SiO_4 unit are shared with other SiO_4 tetrahedra. Therefore, for each SiO_4 unit,

1 Si + two oxygen not shared + 2 oxygen shared = SiO_{2+1}
$= SiO_3^{2-}$

21.4 $5.0 \text{ L} \times \dfrac{1.4 \text{ g}}{\text{mL}} \times \dfrac{10^3 \text{ mL}}{1 \text{ L}} = 7.0 \times 10^3 \text{ g } O_2$

$$7.0 \times 10^3 \text{ g} \times \frac{1 \text{ mol}}{32.0 \text{ g}} = 2.2 \times 10^2 \text{ mol}; PV = nRT$$

$$V = \frac{(2.2 \times 10^2 \text{ mol})(0.0821 \text{ L}\cdot\text{atm}\cdot\text{mol}^{-1}\cdot\text{K}^{-1})(273 \text{ K})}{1 \text{ atm}}$$

$$= 4.9 \times 10^3 \text{ L}$$

21.5 $2\text{ NaCl}(\ell) \rightarrow 2\text{ Na}(\ell) + Cl_2(g)$
Coulombs (C) = (A)(s)

$$= 2.0 \times 10^4 \text{ A} \times 24 \text{ h} \times \frac{3600 \text{ s}}{\text{h}} = 1.7 \times 10^9 \text{ C}$$

$$1.7 \times 10^9 \text{ C} \times \frac{1 \text{ mol e}^-}{9.65 \times 10^4 \text{ C}} = 1.8 \times 10^4 \text{ mol e}^-$$

$$1.8 \times 10^4 \text{ mol e}^- \times \frac{1 \text{ mol Na}}{1 \text{ mol e}^-} \times \frac{23.0 \text{ g Na}}{1 \text{ mol Na}} \times \frac{1 \text{ lb}}{454 \text{ g}} \times \frac{1 \text{ ton}}{2000 \text{ lb}} = 0.46 \text{ tons Na}$$

$$1.8 \times 10^4 \text{ mol e}^- \times \frac{1 \text{ mol Cl}_2}{2 \text{ mol e}^-} \times \frac{70.9 \text{ g Cl}_2}{1 \text{ mol Cl}_2} \times \frac{1 \text{ lb}}{454 \text{ g}} \times \frac{1 \text{ ton}}{2000 \text{ lb}} = 0.70 \text{ tons Cl}_2$$

21.6 $$C = 2.00 \times 10^4 \text{ A} \times 100. \text{ h} \times \frac{3600 \text{ s}}{\text{h}} = 7.20 \times 10^9 \text{ C}$$

$$7.20 \times 10^9 \text{ C} \times \frac{1 \text{ mol e}^-}{9.65 \times 10^4 \text{ C}} \times \frac{1 \text{ mol NaOH}}{1 \text{ mol e}^-} \times \frac{40.00 \text{ g NaOH}}{1 \text{ mol NaOH}} \times \frac{1 \text{ lb}}{454 \text{ g}} \times \frac{1 \text{ ton}}{2000 \text{ lb}} = 3.29 \text{ tons NaOH}$$

21.7 Phosphorus in $Ca_3(PO_4)_2$ has an oxidation state of +5; it is reduced to zero in P_4. Carbon is oxidized to CO (oxidation state changes from zero to −2).

21.8 $$\frac{93 \text{ g}}{502 \text{ g}} \times 100\% = 19\% \text{ P}$$

21.9 Br^- (reducing agent) is oxidized to Br_2; Cl_2 (oxidizing agent) is reduced to Cl^-.

21.10 $I_2(s) + 2\text{ e}^- \rightarrow 2\text{ I}^-(aq)$ $E° = +0.535$ V
$Br_2(\ell) + 2\text{ e}^- \rightarrow 2\text{ Br}^-(aq)$ $E° = +1.08$ V
$I_2(s)$ is below $Br^-(aq)$ so iodine will not oxidize bromide ion to bromine, as seen by the negative cell voltage. Rather, bromine will oxidize iodide to iodine.

$$I_2(s) + 2\text{ e}^- \rightarrow 2\text{ I}^-(aq) \quad E°_{cathode} = +0.535 \text{ V}$$
$$2\text{ Br}^-(aq) \rightarrow Br_2(\ell) + 2\text{ e}^- + \quad E°_{anode} = -1.08 \text{ V}$$
$$I_2(s) + 2\text{ Br}^-(aq) \rightarrow 2\text{ I}^-(aq) + Br_2(\ell) \quad E°_{cell} = -0.545 \text{ V}$$

21.11 Oxide, O^{2-}, [:Ö:]$^{2+}$; peroxide, O_2^{2-}, [:Ö:Ö:]$^{2-}$; superperoxide, O_2^-, [:Ö:Ö·]$^-$

21.12 (a) CaO (b) BaO_2 (c) Sr_3N_2
(d) CaC_2

21.13 (a) Mg_3N_2 (b) $3\text{ Mg}(s) + N_2(g) \rightarrow Mg_3N_2(s)$

21.14 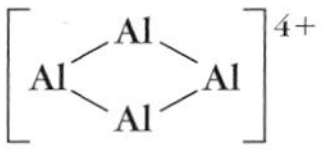

21.15 NO_3^-; oxidation number of oxygen is −2; oxidation number of N is +5.
NH_4^+; oxidation number of hydrogen is +1; oxidation number of nitrogen is −3.

21.16 $$150. \text{ g NaN}_3 \times \frac{1 \text{ mol NaN}_3}{65.0 \text{ g NaN}_3} \times \frac{3 \text{ mol N}_2}{2 \text{ mol NaN}_3} = 3.46 \text{ mol N}_2$$

$$V = \frac{nRT}{P} = \frac{(3.46)(0.0821)(273)}{(1)} = 77.6 \text{ L}$$

Chapter 22

22.1 (a) V^{+5} (b) Ag^{+3} (c) Co^{+2} (d) Mn^{+2}

22.2 Cu^{+2}

22.3 (a) $$500 \text{ g} \times \frac{18 \text{ g W}}{100 \text{ g}} = 90 \text{ g W}$$

$$500 \text{ g} \times \frac{6.0 \text{ g Co}}{100 \text{ g}} = 30 \text{ g Co}$$

(b) Iron is present in the greatest mole percent.

$$500 \text{ g} \times \frac{68.9 \text{ g Fe}}{100 \text{ g}} = 344 \text{ g Fe} = 6.2 \text{ mol}$$

$$\frac{6.2}{8.0} \times 100\% = 77.5\% \text{ Fe}$$

22.4 $Cu^{2+}(aq) + 2\text{ e}^- \rightarrow Cu(s)$

$$12.0 \text{ h}\left(\frac{3600 \text{ s}}{\text{h}}\right)\left(\frac{250 \text{ C}}{\text{s}}\right)\left(\frac{1 \text{ mol e}^-}{9.65 \times 10^4 \text{ C}}\right) \times \left(\frac{1 \text{ mol Cu}}{2 \text{ mol e}^-}\right)\left(\frac{63.55 \text{ g Cu}}{1 \text{ mol Cu}}\right) = 3.56 \times 10^3 \text{ g Cu}$$

22.5 (a) NO_3^- is the oxidizing agent; silver metal is the reducing agent.
(b) NO_3^- is the oxidizing agent; gold is the reducing agent.

22.6 $2\text{ Al}(s) + Cr_2O_3(s) \rightarrow Al_2O_3(s) + 2\text{ Cr}(s)$

$$\Delta H_f° = \Delta H°_{f_{Al_2O_3(s)}} - \Delta H°_{f_{Cr_2O_3(s)}}$$
$$= (-1675.7 \text{ kJ}) - (-1139.7 \text{ kJ})$$
$$= -536.0 \text{ kJ/mol Cr}_2\text{O}_3$$

$$\Delta G° = \Delta G°_{f_{Al_2O_3(s)}} - \Delta G°_{f_{Cr_2O_3(s)}}$$
$$= (-1582.3 \text{ kJ}) - (-1058.1 \text{ kJ})$$
$$-524.2 \text{ kJ/mol Cr}_2\text{O}_3$$

22.7 The addition of acid shifts the equilibrium to the right, converting CrO_4^{2-} to $Cr_2O_7^{2-}$. Added base reacts with H^+ to form water, causing the equilibrium to shift to the left, converting $Cr_2O_7^{2-}$ to CrO_4^{2-}.

22.8 $[Cr(H_2O)_2(NH_3)_2(OH)_2]^+$

22.9 Should be $M_2^+X^{2-}$; such as K_2SO_4

22.10 (a) Three (b) 4− (c) Five

22.11 (a) Fe^2 (b) Fe^{3+}

22.12 Two isomers

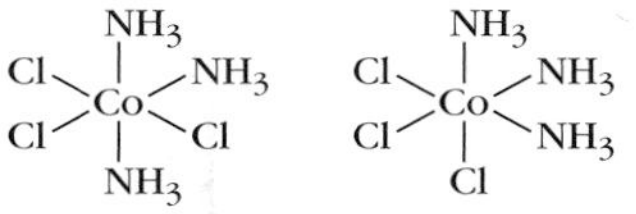

isomer 1 isomer 2

22.13 Ni^{2+} ions have a $3d^8$ configuration, which has three pairs of electrons in the t_2 orbitals and an unpaired electron in each of the two e orbitals. No other electron configuration is possible, so high- and low-spin Ni^{2+} complex ions are not formed.

22.14 Water is a weaker-field ligand (spectrochemical series), as indicated by the blue color of the complex ion due to absorption of longer-wavelength light. Therefore, the *d* orbital electrons would have the maximum number of unpaired electrons and the complex ion would be high-spin.

Answers to Selected Questions for Review and Thought

Chapter 1

11. (a) Quantitative (b) Qualitative (c) Qualitative
(d) Quantitative and qualitative (e) Qualitative
13. (a) Qualitative (b) Quantitative
(c) Quantitative and qualitative (d) Qualitative
15. Sulfur is a pale yellow, powdery solid. Bromine is a dark, red-brown liquid and a red-brown gas that fills the upper part of the flask. Both the melting point and the boiling point of sulfur must be above room temperature. The boiling point, but not the melting point, of bromine must be above room temperature. Both substances are colored. Most of their other properties appear to be different.
17. The liquid will boil because your body temperature of 37 °C is above the boiling point of 20 °C.
19. (a) 20 °C (b) 100 °C
(c) 60 °C (d) 20 °F
21. Copper
23. Aluminum
25. 3.9×10^3 g
27. (a) Physical (b) Chemical
(c) Chemical (d) Physical
29. (a) Chemical (b) Chemical (c) Physical
31. (a) An outside source of energy is forcing a chemical reaction to occur.
(b) A chemical reaction is releasing energy and causing work to be done.
(c) A chemical reaction is releasing energy and causing work to be done.
(d) An outside source of energy is forcing a chemical reaction to occur.
33. Heterogeneous; use a magnet.
35. (a) Homogeneous (b) Heterogeneous
(c) Heterogeneous (d) Heterogeneous
37. (a) A compound decomposed
(b) A compound decomposed
39. (a) Heterogeneous mixture (b) Pure compound
(c) Heterogeneous mixture (d) Homogeneous mixture
41. (a) Heterogeneous mixture (b) Pure compound
(c) Element (d) Homogeneous mixture
43. (a) No (b) Maybe
45. The macroscopic world; a parallelepiped shape; the atom crystal arrangement is a parallelepiped shape.
47. Microscale world
49. Carbon dioxide molecules are crowded in the unopened can. When opened, the molecules quickly escape through the hole.
51. When sucrose is heated, the motion of the atoms increases. Only when that motion is extreme enough, will the bonds in the sucrose break, allowing for the formation of new bonds to produce the "caramelization" products.
53. (a) 3.275×10^4 m (b) 3.42×10^4 nm (c) 1.21×10^{-3} μm
55. Because atoms in the starting materials must all be accounted for in the substances produced, and because the mass of each atom does not change, there would be no change in the mass.
57. (Remember, you are instructed to use your own words to answer this question.) Consider two compounds that both contain the same two elements. In each compound, the proportion of these two elements is a whole-number integer ratio. Because they are different compounds, these ratios must be different. If you pick a sample of each of these compounds such that both samples contain the same number of atoms of the first element, and then you count the number of atoms of the second type, you will find that a small integer relationship exists between the number of atoms of the second type in the first compound and the number of atoms of the second type in the second compound.
59. If two compounds contain the same elements and samples of those two compounds both contain the same mass of one element, then the ratio of the masses of the other elements will be small whole numbers.
61. Many responses are equally valid here. Common examples given here: (a) iron, Fe; gold, Au (b) carbon, C; hydrogen, H
(c) boron, B; silicon, Si (d) nitrogen, N_2; oxygen, O_2.
63.

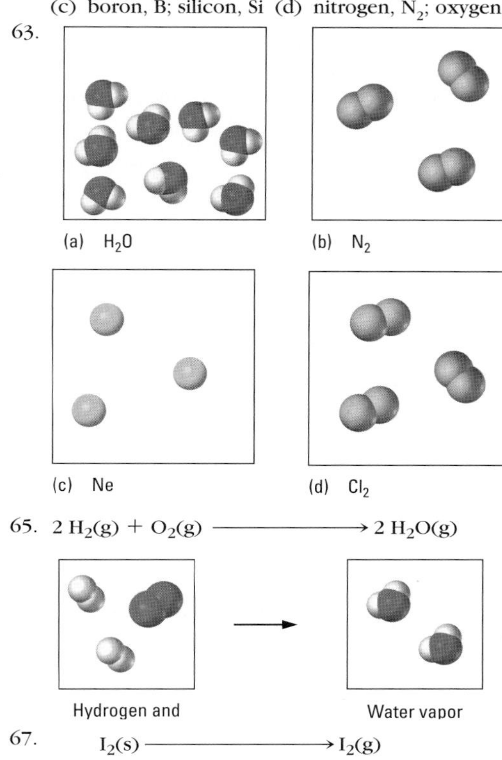

65. $2\ H_2(g) + O_2(g) \longrightarrow 2\ H_2O(g)$

Hydrogen and
Water vapor

67. $I_2(s) \longrightarrow I_2(g)$

Solid iodine
Iodine gas

69. (a) Mass is quantitative and related to a physical property. Colors are qualitative and related to physical properties. Reaction is qualitative and related to a chemical property.
(b) Mass is quantitative and related to a physical property. The fact that a chemical reaction occurs between substances is qualitative information and related to a chemical property.
71. In solid calcium, smaller radius atoms are more closely packed, making a smaller volume. In solid potassium, larger radius atoms are less closely packed, making a larger volume.
73. They are different by how the atoms are organized and bonded together.
75. (a) Bromobenzene (b) Gold (c) Lead
77. (a) 2.7×10^2 mL ice
(b) Bulging, cracking, deformed, or broken
79. Gold
81. (a) Water layer on top of bromobenzene layer.
(b) If it is poured slowly and carefully, ethanol will float on top of the water and slowly dissolve in the water. Both ethanol and water will float on the bromobenzene.
(c) Stirring will speed up dissolving of ethanol and water in each other. After stirring only two layers will remain.

83. Drawing (b)
85. 6.02×10^{-29} m^3
87. (a) Gray and blue (b) Lavender (c) Orange
89. It is difficult to prove that something cannot be broken down (see Section 1.3).
91. (a) Nickel, lead, and magnesium (b) Titanium
93. Obtain four or more lemons. Keep one unaltered to be the "control" case. Perform designated tasks to others, including applying both tasks to the same lemon; then juice all of them, recording results, such as juice volume and ease of task. Repeat with more lemons to achieve better reliability. Hypothesis: Disrupting the "juice sacks" inside the pulp helps to release the juice more easily.

Chapter 2

7. 40,000 cm
9. 614 cm, 242 cm, 20.1 ft
11. 76.2 kg
13. 2.0×10^3 cm^3, 2.0 L
15. 1550 in^2
17. 2.8×10^9 m^3
22. (a) 4 (b) 3 (c) 4
    (d) 4 (e) 3
24. (a) 4.33×10^{-4} (b) 4.47×10^1
    (c) 2.25×10^1 (d) 8.84×10^{-3}
26. (a) 1.9 g/mL (b) 218.4 cm^3
    (c) 0.0217 (d) 5.21×10^{-5}
28. 80.1% silver, 19.9% copper
30. 245 g sulfuric acid
32. 0.9% Na
37. Number of neutrons
39. 27 protons, 27 electrons, and 33 neutrons
41. 78.92 amu/atom
43. (a) 9 (b) 48 (c) 70
45. (a) $^{23}_{11}Na$ (b) $^{39}_{18}Ar$ (c) $^{69}_{31}Ga$
47. (a) 20 e$^-$, 20 p$^+$, 20 n^0 (b) 50 e$^-$, 50 p$^+$, 69 n^0
    (c) 94 e$^-$, 94 p$^+$, 150 n^0
49.

Z	*A*	Number of Neutrons	Element
35	81	46	Br
46	108	62	Pd
77	192	115	Ir
63	151	88	Eu

51. $^{18}_{9}X$, $^{20}_{9}X$ and $^{15}_{9}X$
53. Ions
55. Using a magnetic field
57. 6.941 amu/atom
59. 60.12% ^{69}Ga, 39.87% ^{71}Ga
61. ^{7}Li
63. Pair (2), dozen (12), gross (144), million (1,000,000) (Other answers are possible.)
66. (a) 27 g B (b) 0.48 g O_2
    (c) 6.98×10^{-2} g Fe (d) 2.61×10^3 g H
68. (a) 1.9998 mol Cu (b) 0.499 mol Ca
    (c) 0.6208 mol Al (d) 3.1×10^{-4} mol K
    (e) 2.1×10^{-5} mol Am
70. 2.19 mol Na
72. 9.42×10^{-5} mol Kr
74. 4.131×10^{23} Cr atoms
76. 1.055×10^{-22} g Cu
78. In a group, elements share the same vertical column. In a period, elements share the same horizontal row.
83. Transition elements: iron, copper, chromium
    Halogens: fluorine and chlorine
    Alkali metal: sodium
    (Other answers are possible.)
85. Five; nonmetal: carbon (C), metalloids: silicon (Si) and germanium (Ge), and metals: tin (Sn) and lead (Pb).
87. (a) I (b) In (c) Ir (d) Fe
89. Sixth period
91. (a) Mg (b) Na (c) C (d) S
    (e) I (f) Mg (g) Kr (h) S
    (i) Ge [Other answers are possible for (a), (b), and (i).]
93. (a) Iron or magnesium (b) Hydrogen (c) Silicon
    (d) Iron (e) Chlorine
97. (a) 0.197 nm (b) 197 pm
99. (a) 0.178 nm^3 (b) 1.78×10^{-22} cm^3
102. 89 tons/yr
104. ^{39}K
107. (a) Ti; 22; 47.88 (b) Group 4B; Period 4; zirconium, hafnium, rutherfordium (c) light-weight and strong (d) strong, low-density, highly corrosion resistant, occurs widely, light weight, high-temperature stability (Other answers are possible.)
109. 0.038 mol
111. \$3,800
113. 3.4 mol, 2.0×10^{24} atoms
114. (a) Not possible (b) Possible (c) Not possible
     (d) Not possible (e) Possible (f) Not possible
116. (a) Same (b) Second (c) Same
     (d) Same (e) Same (f) Second
     (g) Same (h) First (i) Second
     (j) First
118. (a) 270 mL (b) No
121. (a) ^{79}Br—^{79}Br, ^{79}Br—^{81}Br, ^{81}Br—^{81}Br
     (b) 78.918 g/mol, 80.196 g/mol
     (c) 79.90 g/mol (d) 51.1% ^{79}Br, 48.9% ^{81}Br
123. (a) K (b) Ar (c) Cu (d) Ge
     (e) H (f) Al (g) O (h) Ca
     (i) Br (j) P
125. (a) Se (b) ^{39}K (c) ^{79}Br (d) ^{20}Ne

Chapter 3

9. (a) BrF_3 (b) XeF_2 (c) P_2F_4 (d) $C_{15}H_{32}$ (e) N_2H_4
11. Butanol, $C_4H_{10}O$, $CH_3CH_2CH_2CH_2OH$

```
    H   H   H   H
    |   |   |   |
H — C — C — C — C — OH
    |   |   |   |
    H   H   H   H
```

Pentanol, $C_5H_{12}O$, $CH_3CH_2CH_2CH_2CH_2OH$

```
    H   H   H   H   H
    |   |   |   |   |
H — C — C — C — C — C — OH
    |   |   |   |   |
    H   H   H   H   H
```

14. (a) C_6H_6 (b) $C_6H_8O_6$
16. (a) 1 calcium atom, 2 carbon atoms, and 4 oxygen atoms
    (b) 8 carbon atoms and 8 hydrogen atoms
    (c) 2 nitrogen atoms, 8 hydrogen atoms, 1 sulfur atom, and 4 oxygen atoms
    (d) 1 platinum atom, 2 nitrogen atoms, 6 hydrogen atoms, and 2 chlorine atoms
    (e) 4 potassium atoms, 1 iron atom, 6 carbon atoms, and 6 nitrogen atoms

18. (a) Same number of atoms of each kind
(b) Different bonding arrangements
20. (a) Li^+ (b) Sr^{2+} (c) Al^{3+} (d) Ca^{2+}
(e) Zn^{2+}
22. Ba: 2+, Br: −1
24. (a) +2 (b) +2 (c) +2 or +3 (d) +3
26. CoO, Co_2O_3
28. (c) and (d) are correct formulas. (a) $AlCl_3$ (b) NaF
30. Mn
32. (a) 1 Pb^{2+} and 2 NO_3^- (b) 1 Ni^{2+} and 1 CO_3^{2-}
(c) 3 NH_4^+ and 1 PO_4^{3-} (d) 2 K^+ and 1 SO_4^{2-}
34. $BaSO_4$, barium ion, 2+, sulfate, 2−; $Mg(NO_3)_2$, magnesium ion, 2+, nitrate, 1−; $NaCH_3CO_2$, sodium ion, 1+, acetate, 1−
36. (a) $Ni(NO_3)_2$ (b) $NaHCO_3$ (c) LiClO
(d) $Mg(ClO_3)_2$ (e) $CaSO_3$
38. (b), (c), and (e) are ionic.
40. Only (e) is ionic, with metal and nonmetal combined. (a)–(d) are composed of only nonmetals.
42. (a) $(NH_4)_2CO_3$ (b) CaI_2 (c) $CuBr_2$ (d) $AlPO_4$
44. (a) Potassium sulfide (b) Nickel(II) sulfate (c) Ammonium phosphate (d) Aluminum hydroxide (e) Cobalt(III) sulfate
46. MgO; MgO has higher ionic charges and smaller ion sizes than NaCl.
48. Conducts electricity in water; check electrical conductivity; examples: NaCl and $C_{12}H_{22}O_{11}$. (Other answers are possible.)
50. Molecular compounds are generally not ionic compounds and, therefore, would not ionize in water.
52. (a) K^+ and OH^- (b) K^+ and SO_4^{2-}
(c) Na^+ and NO_3^- (d) NH_4^+ and Cl^-
54. (a) and (d)
56.

	CH_3OH	Carbon
No. of moles	One	One
No. of molecules or atoms	6.022×10^{23} molecules	6.022×10^{23} atoms
Molar mass	32.0417 g/mol	12.0107 g/mol

	Hydrogen	Oxygen
No. of moles	Four	One
No. of molecules or atoms	2.409×10^{24} atoms	6.022×10^{23} atoms
Molar mass	4.0316 g/mol	15.9994 g/mol

58. (a) 159.688 g/mol (b) 67.806 g/mol (c) 44.0128 g/mol
(d) 197.905 g/mol (e) 176.1238 g/mol
60. (a) 0.0312 mol (b) 0.0101 mol (c) 0.0125 mol
(d) 0.00406 mol (e) 0.00599 mol
62. (a) 179.855 g/mol (b) 36.0 g (c) 0.0259 mol
64. (a) 151.1622 g/mol (b) 0.0352 mol (c) 25.1 g
66. (a) 0.400 mol (b) 0.250 mol (c) 0.628 mol
68. 2.7×10^{23} atoms
70. 1.2×10^{24} molecules
72. (a) 0.250 mol CF_3CH_2F
(b) 6.02×10^{23} F atoms
75. (a) 239.3 g/mol PbS, 86.60% Pb, 13.40% S
(b) 30.0688 g/mol C_2H_6, 79.8881% C, 20.1119% H
(c) 60.0518 g/mol CH_3CO_2H, 40.0011% C, 6.7135% H, 53.2854% O
(d) 80.0432 g/mol NH_4NO_3, 34.9979% C, 5.0368% H, 59.9654% O

77. 58.0% M in MO
79. 245.745 g/mol, 25.858% Cu, 22.7992% N, 5.74197% H, 13.048% S, 32.5528% O
81. One
83. (a) −2 (b) Al_2X_3 (c) Se
86. An empirical formula shows the simplest whole-number ratio of the elements, e.g., CH_3; a molecular formula shows the actual number of atoms of each element, e.g., C_2H_6.
88. $C_4H_4O_4$
90. $C_2H_3OF_3$
92. KNO_3
94. B_5H_7
96. (a) C_3H_4 (b) C_9H_{12}
98. (a) C_5H_7N (b) $C_{10}H_{14}N_2$
100. $C_5H_{14}N_2$
102. $x = 7$
110. (a) $C_7H_5N_3O_6$ (b) $C_3H_7NO_3$
113. (a) (i) Chlorine tribromide (ii) Nitrogen trichloride
(iii) Calcium sulfate (iv) Heptane (v) Xenon tetrafluoride
(vi) Oxygen difluoride (vii) Sodium iodide
(viii) Aluminum sulfide (ix) Phosphorus pentachloride
(x) Potassium phosphate
(b) (iii), (vii), (viii), and (x)
115. (a) 2.39 mol (b) 21.7 cm^3
117. (a) CO_2F_2 (b) CO_2F_2
119. $MnC_9H_7O_3$
121. (a) 1.66×10^{-3} mol (b) 0.346 g
123. 2.14×10^5 g
126.

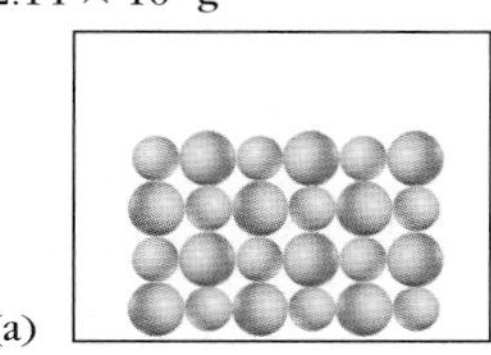
(a)

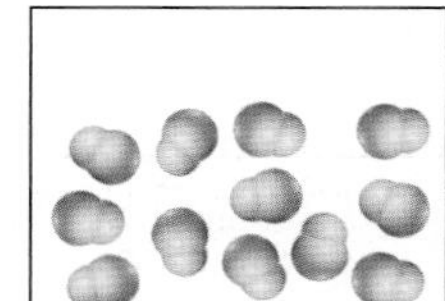
(b)

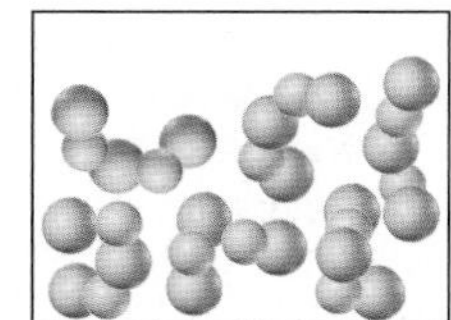
(c)

128. (a) Three (b) Three pairs are identical:

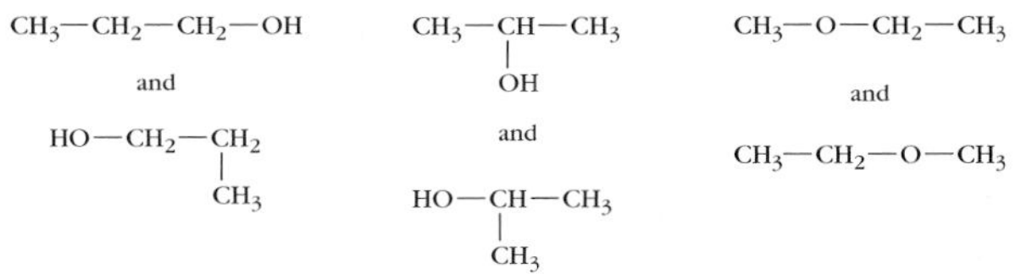

$CH_3—CH_2—CH_2—OH$ and $HO—CH_2—CH_2$ with CH_3 on the second carbon

$CH_3—CH—CH_3$ with OH on the central carbon, and $HO—CH—CH_3$ with CH_3 on the CH

$CH_3—O—CH_2—CH_3$ and $CH_3—CH_2—O—CH_3$

130. Tl_2CO_3, Tl_2SO_4
132. (a) Perbromate, bromate, bromite, hypobromite
(b) Selenate, selenite
135. (a) 0.0130 mol Ni (b) NiF_2
(c) Nickel(II) fluoride
137. CO_2
139. $MgSO_3$, magnesium sulfite
141. 5.0 lb N, 4.4 lb P, 4.2 lb K

Chapter 4

10.

	KOH	HCl
No. molecules	1	1
No. atoms	3	2
No. moles of molecules	1	1
Mass	56.1056 g	36.4609 g
Total mass of reactants	92.5665 g	
Total mass of products		

	KCl	H_2O
No. molecules	1	1
No. atoms	2	3
No. moles of molecules	1	1
Mass	74.5513 g	18.0152 g
Total mass of reactants		
Total mass of products	92.5665 g	

12. (a) 1.00 g (b) 2 for Mg, 1 for O_2, and 2 for MgO
 (c) 50 atoms of Mg
14. $4\ Fe(s) + 3\ O_2(g) \rightarrow 2\ Fe_2O_3(s)$
16. Equation (b)
18.

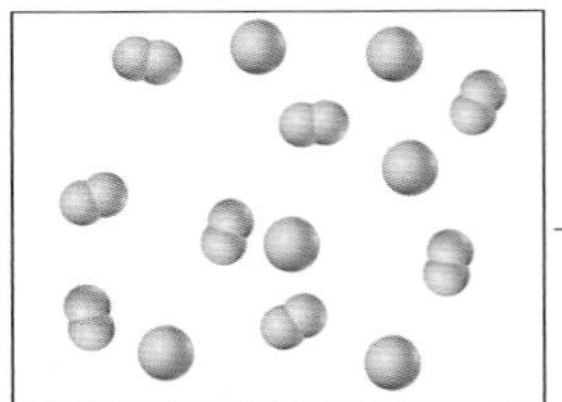

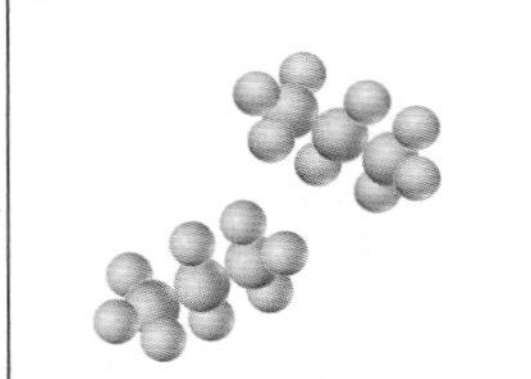

20. (a) Combination (b) Decomposition (c) Exchange
 (d) Displacement
22. (a) Decomposition (b) Displacement (c) Combination
 (d) Exchange
24. (a) $2\ C_4H_{10}(g) + 13\ O_2(g) \rightarrow 8\ CO_2(g) + 10\ H_2O(g)$
 (b) $C_6H_{12}O_6(s) + 6\ O_2(g) \rightarrow 6\ CO_2(g) + 6\ H_2O(g)$
 (c) $2\ C_4H_8O(\ell) + 11\ O_2(g) \rightarrow 8\ CO_2(g) + 8\ H_2O(g)$
26. (a) $2\ Mg(s) + O_2(g) \rightarrow 2\ MgO(s)$, magnesium oxide
 (b) $2\ Ca(s) + O_2(g) \rightarrow 2\ CaO(s)$, calcium oxide
 (c) $4\ In(s) + 3\ O_2(g) \rightarrow 2\ In_2O_3(s)$, indium oxide
28. (a) $2\ K(s) + Cl_2(g) \rightarrow 2\ KCl(s)$, potassium chloride
 (b) $Mg(s) + Br_2(\ell) \rightarrow MgBr_2(s)$, magnesium bromide
 (c) $2\ Al(s) + 3\ F_2(g) \rightarrow 2\ AlF_3(s)$, aluminum fluoride
30. (a) $4\ Al(s) + 3\ O_2(g) \rightarrow 2\ Al_2O_3(s)$
 (b) $N_2(g) + 3\ H_2(g) \rightarrow 2\ NH_3(g)$
 (c) $2\ C_6H_6(\ell) + 15\ O_2(g) \rightarrow 6\ H_2O(\ell) + 12\ CO_2(g)$
32. (a) $UO_2(s) + 4\ HF(\ell) \rightarrow UF_4(s) + 2\ H_2O(\ell)$
 (b) $B_2O_3(s) + 6\ HF(\ell) \rightarrow 2\ BF_3(s) + 3\ H_2O(\ell)$
 (c) $BF_3(g) + 3\ H_2O(\ell) \rightarrow 3\ HF(\ell) + H_3BO_3(s)$
34. (a) $H_2NCl(aq) + 2\ NH_3(g) \rightarrow NH_4Cl(aq) + N_2H_4(aq)$
 (b) $(CH_3)_2N_2H_2(\ell) + 2\ N_2O_4(g)$
 $\rightarrow 3\ N_2(g) + 4\ H_2O(g) + 2\ CO_2(g)$
 (c) $CaC_2(s) + 2\ H_2O(\ell) \rightarrow Ca(OH)_2(s) + C_2H_2(g)$
36. (a) $C_6H_{12}O_6 + 6\ O_2 \rightarrow 6\ CO_2 + 6\ H_2O$
 (b) $C_5H_{12} + 8\ O_2 \rightarrow 5\ CO_2 + 6\ H_2O$
 (c) $2\ C_7H_{14}O_2 + 19\ O_2 \rightarrow 14\ CO_2 + 14\ H_2O$
 (d) $C_2H_4O_2 + 2\ O_2 \rightarrow 2\ CO_2 + 2\ H_2O$
38. 50.0 mol HCl
40. 12.8 g
42. 1.1 mol O_2, 35 g O_2, 1.0×10^2 g NO_2
44. 12.7 g Cl_2, 0.179 mol $FeCl_2$, 22.7 g $FeCl_2$ expected
46.

$(NH_4)_2PtCl_6$	Pt	HCl
12.35 g	5.428 g	5.410 g
0.02782 mol	0.02782 mol	0.1484 mol

48. (a) 0.148 mol H_2O (b) 5.89 g TiO_2, 10.8 g HCl
50. 2.0 mol, 36.0304 g
52. 0.699 g Ga and 0.751 g As
54. (a) $4\ Fe(s) + 3\ O_2(g) \rightarrow 2\ Fe_2O_3(s)$
 (b) 7.98 g (c) 2.40 g
56. (a) $CCl_2F_2 + 2\ Na_2C_2O_4 \rightarrow C + 4\ CO_2 + 2\ NaCl + 2\ NaF$
 (b) 170. g $Na_2C_2O_4$ (b) 112 g CO_2
58. (a) 699 g (b) 526 g
60. $BaCl_2$, 1.12081 g $BaSO_4$
62. (a) Cl_2 is limiting. (b) 5.08 g Al_2Cl_6
 (c) 1.67 g Al unreacted
64. (a) CO (b) 1.3 g H_2 (c) 85.2 g
66. 0 mol CaO, 0.19 mol NH_4Cl, 2.00 mol H_2O, 4.00 mol NH_3,
 2.00 mol $CaCl_2$
68. 1.40 kg Fe
71. 699 g, 93.5%
73. 56.0%
75. 8.8%
77. 5.3 g SCl_2
80. SO_3
82. CH
84. $C_3H_6O_2$
87. 21.6 g N_2
89. Element (b)
91. 12.5 g $Pt(NH_3)_2Cl_2$
93. SiH_4
95. KOH, KOH
97. Two butane molecules react with 13 diatomic oxygen molecules to produce 8 carbon dioxide molecules and 10 water molecules. Two moles of gaseous butane molecules react with 13 moles of gaseous diatomic oxygen molecules to produce 8 moles of gaseous carbon dioxide molecules and 10 moles of liquid water molecules.
99. A_3B
101. Ag^+, Cu^{2+}, and NO_3^-
104. Equation (b)
106. When the metal mass is less than 1.2 g, the metal is the limiting reactant. When the metal mass is greater than 1.2 g, the bromine is the limiting reactant.
108. $H_2(g) + 3\ Fe_2O_3(s) \rightarrow H_2O(\ell) + 2\ Fe_3O_4(s)$
110. 86.3 g
112. (a) CH_4 (b) 200 g (c) 700 g
114. 44.9 amu
116. 0 g $AgNO_3$, 9.82 g Na_2CO_3, 6.79 g Ag_2CO_3, 4.19 g $NaNO_3$
118. 99.7% CH_3OH, 0.3% C_2H_5OH
120. (a) $C_9H_{11}NO_4$ (b) $C_9H_{11}NO_4$

Chapter 5

11. All soluble (a) Fe^{2+} and ClO_4^- (b) Na^+ and SO_4^{2-} (c) K^+ and Br^-
 (d) Na^+ and CO_3^{2-}
13. All soluble (a) K^+ and HPO_4^{2-} (b) Na^+ and ClO^- (c) Mg^{2+} and Cl^-
 (d) Ca^{2+} and OH^- (e) Al^{3+} and Br^-
15. $2\ HNO_3(aq) + Ca(OH)_2(aq) \rightarrow 2\ H_2O(\ell) + Ca(NO_3)_2(aq)$
17. (a) $MnCl_2(aq) + Na_2S(aq) \rightarrow MnS(s) + 2\ NaCl(aq)$
 (b) No precipitate
 (c) No precipitate
 (d) $Hg(NO_3)_2(aq) + Na_2S(aq) \rightarrow HgS(s) + 2\ NaNO_3(aq)$
 (e) $Pb(NO_3)_2(aq) + 2\ HCl(aq) \rightarrow PbCl_2(s) + 2\ HNO_3(aq)$
 (f) $BaCl_2(aq) + H_2SO_4(aq) \rightarrow BaSO_4(s) + 2\ HCl(aq)$

19. (a) CuS insoluble;
$Cu^{2+} + H_2S(aq) \rightarrow CuS(s) + 2\,H^+$;
spectator ion is Cl^-.
(b) $CaCO_3$ insoluble;
$Ca^{2+} + CO_3^{2-} \rightarrow CaCO_3(s)$; spectator ions are K^+ and Cl^-.
(c) AgI insoluble;
$Ag^+ + I^- \rightarrow AgI(s)$; spectator ions are Na^+ and NO_3^-.

21. Complete ionic equation:
$2\,K^+ + CO_3^{2-} + Cu^{2+} + 2\,NO_3^- \rightarrow CuCO_3(s) + 2\,K^+ + 2\,NO_3^-$
Net ionic equation:
$CO_3^{2-} + Cu^{2+} \rightarrow CuCO_3(s)$
Precipitate is copper(II) carbonate.

23. (a) $Zn(s) + 2\,HCl(aq) \rightarrow H_2(g) + ZnCl_2(aq)$
$Zn(s) + 2\,H^+(aq) + 2\,Cl^-(aq) \rightarrow H_2(g) + Zn^{2+}(aq) + 2\,Cl^-(aq)$
$Zn(s) + 2\,H^+(aq) \rightarrow H_2(g) + Zn^{2+}(aq)$
(b) $Mg(OH)_2(s) + 2\,HCl(aq) \rightarrow MgCl_2(aq) + 2\,H_2O(\ell)$
$Mg(OH)_2(s) + 2\,H^+(aq) + 2\,Cl^-(aq) \rightarrow Mg^{2+}(aq) + 2\,Cl^-(aq) + 2\,H_2O(\ell)$
$Mg(OH)_2(s) + 2\,H^+(aq) \rightarrow Mg^{2+}(aq) + 2\,H_2O(\ell)$
(c) $2\,HNO_3(aq) + CaCO_3(s) \rightarrow Ca(NO_3)_2(aq) + H_2O(\ell) + CO_2(g)$
$2\,H^+(aq) + 2\,NO_3^-(aq) + CaCO_3(s) \rightarrow Ca^{2+}(aq) + 2\,NO_3^-(aq) + H_2O(\ell) + CO_2(g)$
$2\,H^+(aq) + CaCO_3(s) \rightarrow Ca^{2+}(aq) + H_2O(\ell) + CO_2(g)$
(d) $4\,HCl(aq) + MnO_2(s) \rightarrow MnCl_2(aq) + Cl_2(g) + 2\,H_2O(\ell)$
$4\,H^+(aq) + 4\,Cl^-(aq) + MnO_2(s) \rightarrow Mn^{2+}(aq) + 2\,Cl^-(aq) + Cl_2(g) + 2\,H_2O(\ell)$
$4\,H^+(aq) + 2\,Cl^-(aq) + MnO_2(s) \rightarrow Mn^{2+}(aq) + Cl_2(g) + 2\,H_2O(\ell)$

25. (a) $Ca(OH)_2(s) + 2\,HNO_3(aq) \rightarrow Ca(NO_3)_2(aq) + 2\,H_2O(\ell)$
$Ca(OH)_2(s) + 2\,H^+(aq) + 2\,NO_3^-(aq) \rightarrow Ca^{2+}(aq) + 2\,NO_3^-(aq) + 2\,H_2O(\ell)$
$Ca(OH)_2(s) + 2\,H^+(aq) \rightarrow Ca^{2+}(aq) + 2\,H_2O(\ell)$
(b) $BaCl_2(aq) + Na_2CO_3(aq) \rightarrow BaCO_3(s) + 2\,NaCl(aq)$
$Ba^{2+}(aq) + 2\,Cl^-(aq) + 2\,Na^+(aq) + CO_3^{2-}(aq) \rightarrow BaCO_3(s) + 2\,Na^+(aq) + 2\,Cl^-(aq)$
$Ba^{2+}(aq) + CO_3^{2-}(aq) \rightarrow BaCO_3(s)$
(c) $2\,Na_3PO_4(aq) + 3\,Ni(NO_3)_2(aq) \rightarrow Ni_3(PO_4)_2(s) + 6\,NaNO_3(aq)$
$6\,Na^+(aq) + 2\,PO_4^{3-}(aq) + 3\,Ni^{2+}(aq) + 6\,NO_3^-(aq) \rightarrow Ni_3(PO_4)_2(s) + 6\,Na^+(aq) + 6\,NO_3^-(aq)$
$2\,PO_4^{3-}(aq) + 3\,Ni^{2+}(aq) \rightarrow Ni_3(PO_4)_2(s)$

27. $Ba(OH)_2(aq) + 2\,HNO_3(aq) \longrightarrow Ba(NO_3)_2(aq) + 2\,H_2O(\ell)$

29. $CdCl_2(aq) + 2\,NaOH(aq) \longrightarrow Cd(OH)_2(s) + 2\,NaCl(aq)$
$Cd^{2+}(aq) + 2\,Cl^-(aq) + 2\,Na^+(aq) + 2\,OH^-(aq) \longrightarrow Cd(OH)_2(s) + 2\,Na^+(aq) + 2\,Cl^-(aq)$
$Cd^{2+}(aq) + 2\,OH^-(aq) \longrightarrow Cd(OH)_2(s)$

31. $Pb(NO_3)_2(aq) + 2\,KCl(aq) \rightarrow PbCl_2(s) + 2\,KNO_3(aq)$
Reactants: lead(II) nitrate and potassium chloride
Products: lead(II) chloride and potassium nitrate

34. (a) Base strong, K^+ and OH^-
(b) Base strong, Mg^{2+} and OH^-
(c) Acid weak, small amounts of H^+ and ClO^-
(d) Acid strong, H^+ and Br^-
(e) Base strong, Li^+ and OH^-
(f) Acid weak, small amounts of H^+, HSO_3^-, and SO_3^{2-}

36. (a) Acid: HNO_2, base: NaOH
complete ionic form:
$HNO_2(aq) + Na^+ + OH^- \longrightarrow H_2O(\ell) + Na^+ + NO_2^-$
net ionic form:
$HNO_2(aq) + OH^- \longrightarrow H_2O(\ell) + NO_2^-$
(b) Acid: H_2SO_4 base: $Ca(OH)_2$
complete ionic and net ionic forms:
$H^+ + HSO_4^- + Ca(OH)_2(s) \longrightarrow 2\,H_2O(\ell) + CaSO_4(s)$
(c) Acid: HI, base: NaOH
complete ionic form:
$H^+ + I^- + Na^+ + OH^- \longrightarrow H_2O(\ell) + Na^+ + I^-$
net ionic form: $H^+ + OH^- \longrightarrow H_2O(\ell)$
(d) Acid: H_3PO_4, base: $Mg(OH)_2$
complete ionic and net ionic forms:
$2\,H_3PO_4(aq) + 3\,Mg(OH)_2(s) \longrightarrow 6\,H_2O(\ell) + Mg_3(PO_4)_2(s)$

38. (a) Precipitation reaction; products are NaCl and MnS;
$MnCl_2(aq) + Na_2S(aq) \rightarrow 2\,NaCl(aq) + MnS(s)$ (b) Precipitation reaction; products are NaCl and $ZnCO_3$;
$Na_2CO_3(aq) + ZnCl_2(aq) \rightarrow 2\,NaCl(aq) + ZnCO_3(s)$
(c) Gas-forming reaction; products are $KClO_4$, H_2O, and CO_2;
$K_2CO_3(aq) + 2\,HClO_4(aq) \rightarrow 2\,KClO_4(aq) + H_2O(\ell) + CO_2(g)$

40. (a) Ox. # O = −2, Ox. # S = +6
(b) Ox. # O = −2, Ox. # H = +1, Ox. # N = +5
(c) Ox. # K = +1, Ox. # O = −2, Ox. # Mn = +7
(d) Ox. # O = −2, Ox. # H = +1
(e) Ox. # Li = +1, Ox. # O = −2, Ox. # H = +1
(f) Ox. # Cl = −1, Ox. # H = +1, Ox. # C = 0

42. (a) S: +6, O: −2 (b) N: +5, O: −2
(c) Mn: +7, O: −2 (d) Cr: +3, O: −2, H: +1
(e) P: +5, O: −2, H: +1 (f) S: +2, O: −2

44. (a) −1 (b) +1 (c) +3 (d) +5
(e) +7

46. (a) −2 (b) 0 (c) +2 (d) +4
(e) +6

48. Only reaction (b) is an oxidation-reduction reaction; oxidation numbers change. Reaction (a) is precipitation and reaction; (c) is acid-base neutralization.

51. Substances (b), (c), and (d)

53. (a) $CO_2(g)$ or $CO(g)$ (b) $PCl_3(g)$ or $PCl_5(g)$
(c) $TiCl_2(s)$ or $TiCl_4(s)$ (d) $Mg_3N_2(s)$
(e) $Fe_2O_3(s)$ (f) $NO_2(g)$

55. F_2 is the best oxidizing agent; I^- anion is the best reducing agent.

57. (b) Br_2 is reduced. NaI is oxidized. Br_2 is oxidizing agent. NaI is reducing agent.
(c) F_2 is reduced. NaCl is oxidized. F_2 is oxidizing agent. NaCl is reducing agent.
(d) Cl_2 is reduced. NaBr is oxidized. Cl_2 is oxidizing agent. NaBr is reducing agent.

59. This is just an example:
$Fe(s) + 2\,HCl(aq) \longrightarrow FeCl_2(aq) + H_2(g)$
(a) Fe is oxidized. (b) HCl is reduced.
(c) The reducing agent is Fe(s).
(d) The oxidizing agent is HCl(aq).
(Other answers are possible.)

61. (a) N.R. (b) N.R. (c) N.R.
(d) $Au^{3+}(aq) + 3\,Ag(s) \longrightarrow 3\,Ag^+(aq) + Au(s)$

63. 0.12 M Ba^{2+}, 0.24 M Cl^-

65. (a) 0.254 M Na_2CO_3 (b) 0.508 M Na^+, 0.254 M CO_3^{2-}

67. 0.494 g $KMnO_4$

69. 5.08×10^3 mL

71. 0.0150 M $CuSO_4$

73. Method (b)

75. 39.4 g $NiSO_4 \cdot 6H_2O$

77. 0.205 g Na_2CO_3

79. 121 mL HNO_3 solution

81. 22.9 mL NaOH solution

83. 0.18 g AgCl, NaCl, 0.0080 M NaCl

85. (a) Step (ii) is wrong. Steps (iii) and (iv) are wrong because Step (ii) is wrong.
(b) 0.00394 g citric acid

87. 1.192 M HCl

89. 96.8% pure

91. Chloride is the spectator ion.
$CaCO_3(s) + 2\,H^+(aq) \longrightarrow CO_2(g) + Ca^{2+}(aq) + H_2O(\ell)$
gas-forming exchange reaction

93. (a) $(NH_4)_2S(aq) + Hg(NO_3)_2(aq) \rightarrow HgS(s) + 2\,NH_4NO_3(aq)$
(b) Reactants: ammonium sulfide, mercury(II) nitrate; products: mercury(II) sulfide, ammonium nitrate
(c) $S^{2-}(aq) + Hg^{2+}(aq) \longrightarrow HgS(s)$
(d) Precipitation reaction

95. (a) Combination reaction; product: $H_2SO_4(aq)$
(b) Combination reaction; product: $SrH_2(s)$
(c) Displacement reaction; products: $MgSO_4(aq)$ and $H_2(g)$
(d) Exchange (precipitation) reaction; products: $Ag_3PO_4(s)$ and $NaNO_3(aq)$
(e) Decomposition and gas-forming reaction; products: CaO(s), $H_2O(\ell)$, and $CO_2(g)$
(f) Oxidation-reduction reaction; products: $Fe^{2+}(aq)$ and $Sn^{4+}(aq)$

97. (a) $NH_3(aq)$, H_2O (b) $CH_3CO_2H(aq)$, H_2O
(c) Na^+, OH^-, H_2O (d) H^+, Br^-, H_2O

99. (c) is a redox reaction. The oxidizing agent is Ti. The reducing agent is Mg.

101. Case 1 (a) Before: clear colorless solution; after: solid at the bottom of beaker with colorless solution above it
(b)

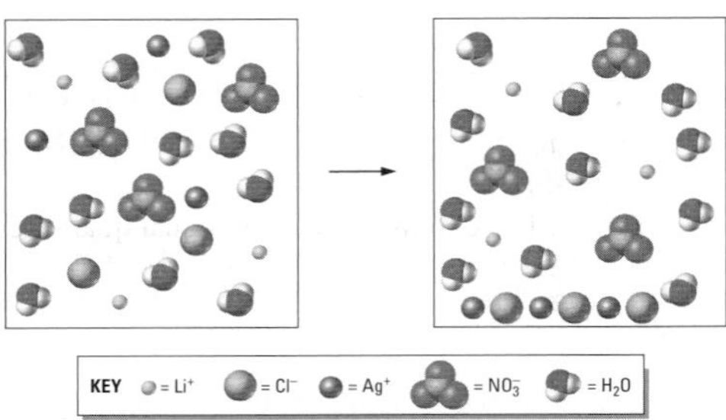

(c) $Li^+ + Cl^- + Ag^+ + NO_3^- \rightarrow Li^+ + AgCl(s) + NO_3^-$
Case 2 (a) Before and after: clear colorless solutions
(b)

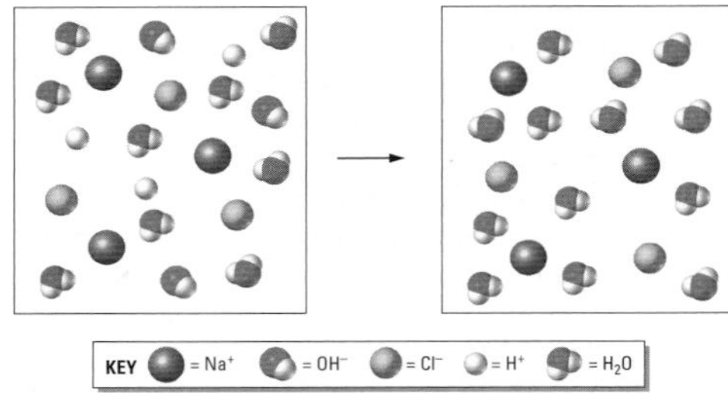

(c) $Na^+ + OH^- + H^+ + Cl^- \rightarrow Na^+ + H_2O(\ell) + Cl^-$

103. (a) Combine $Ba(OH)_2(aq)$ and $H_2SO_4(aq)$
(b) Combine $Na_2SO_4(aq)$ and $Ba(NO_3)_2(aq)$
(c) Combine $BaCO_3(s)$ and $H_2SO_4(aq)$

105. HCl(aq); insoluble lead(II) chloride would precipitate if Pb^{2+} was present, but no precipitate would be seen if only Ba^{2+} was present.

107. The products are the result of something being oxidized and something being reduced. Only the reactants are oxidized and reduced.

109. (d)

111. (a) and (d) are correct.

113. (a) $MgBr_2$, magnesium bromide; $CaBr_2$, calcium bromide; $SrBr_2$, strontium bromide
(b) $Mg(s) + Br_2(\ell) \rightarrow MgBr_2(s)$
$Ca(s) + Br_2(\ell) \rightarrow CaBr_2(s)$
$Sr(s) + Br_2(\ell) \rightarrow SrBr_2(s)$
(c) Oxidation-reduction
(d) The point where increase stops gives: $\frac{\text{g metal}}{\text{g product}}$. The different metals have different molar masses, so the ratios will be different. Use grams to find moles and then set up a ratio.

115. (a) Groups C and D: $Ag^+ + Cl^- \rightarrow AgCl(s)$, Groups A and B: $Ag^+ + Br^- \rightarrow AgBr(s)$
(b) Different silver halide produced in Group C and D than in Groups A and B
(c) Bromide is heavier than chloride.

117. 104 g/mol

119. (a) $CaF_2(s) + H_2SO_4(aq) \rightarrow 2\,HF(g) + CaSO_4(s)$
Reactant: calcium fluoride and sulfuric acid
Products: hydrogen fluoride and calcium sulfate
(b) Precipitation reaction.
(c) Carbon tetrachloride, antimony(V) chloride, hydrogen chloride
(d) CCl_3F

121. 2.26 g, 1.45 g

123. 0.0154 M $CaSO_4$; 0.341 g $CaSO_4$ undissolved

125. 2.6×10^{-3} M

127. 184 mL

129. 6.28% impurity

Chapter 6

11. (a) 399 Cal (b) 5.0×10^6 J/day

13. 1.12×10^4 J

14. 3×10^8 J

15. 3.60×10^6 J, $0.03

17. (a) The chemical potential energy of the atoms in the match, fuse, and fuel are converted to thermal energy (due to the combustion reaction), potential energy (as the rocket's altitude increases), and light energy (colorful sparkles). (b) The chemical potential energy of the atoms in the fuel is converted to thermal energy (due to the combustion reaction), some of which is converted to kinetic energy (for the movement of the vehicle).

19. (a) The system: the plant (stem, leaves, roots, etc.); the surroundings: anything not the plant (air, soil, water, sun, etc.) (b) To study the plant growing, we must isolate it and see how it interacts with its surroundings. (c) Light energy and carbon dioxide are absorbed by the leaves and are converted to other molecules storing the energy as chemical energy that is used to increase the size of the plant. Minerals and water are absorbed through the soil. CO_2 is absorbed from the air. The plant expels oxygen and other waste materials into the surroundings.

21. (a) The system: NH_4 Cl; the surroundings: anything not NH_4Cl, including the water. (b) To study the release of energy during the phase change of this chemical, we must isolate it and see how it interacts with the surroundings. (c) The system's interaction with the surroundings causes heat to be transferred into the system and from the surroundings. There is no material transfer in this process, but there is a change in the specific interaction between the water and system. (d) Endothermic

23. $\Delta E = +32$ J

25.

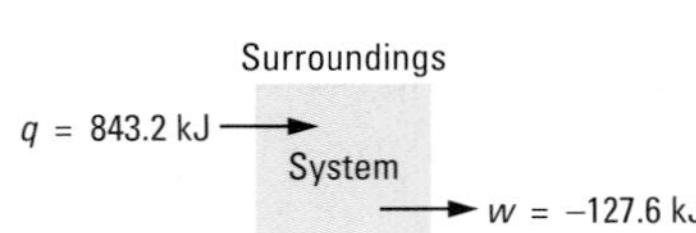

$\Delta E = 715.6$ J

27. Process (a) requires greater transfer of energy than (b).

29. It takes less time to raise the Cu sample to body temperature.
31. 1.0×10^2 kJ
33. More energy (1.48×10^6 J) is absorbed by the water than by the ethylene glycol (9.56×10^5 J).
35. 330. °C
37. (a) 0.45 J g^{-1} °C^{-1}, 25 J mol^{-1} °C^{-1}
39. Gold
41. 160 °C
43. Positive; Negative
45. 4.13×10^5 J
47. 5.00×10^5 J
49. 270 J
51.

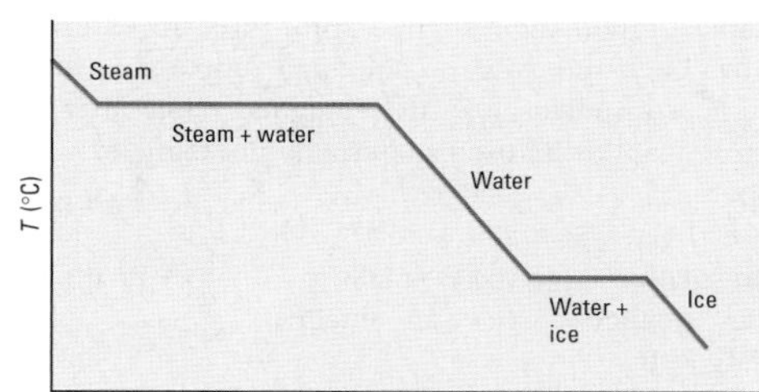

53. (a) X(ℓ) (b) Heat of fusion
(c) Enthalpy of vaporization is positive.
55. Endothermic
57. Endothermic
59. 6.0 kJ of energy is used to convert one mole of ice into liquid water.
61. (a) −210 kJ (b) −33 kJ
63. (a) $\frac{1}{2}$ $C_8H_{18}(\ell) + \frac{25}{4} O_2(g) \rightarrow 4\ CO_2(g) + \frac{9}{2} H_2O(\ell)$
$\Delta H° = -2748.0$ kJ
(b) $100\ C_8H_{18}(\ell) + 1250\ O_2(g) \rightarrow 800\ CO_2(g) + 900\ H_2O(\ell)$
$\Delta H° = -5.4960 \times 10^5$ kJ
(c) $C_8H_{18}(\ell) + \frac{25}{2} O_2(g) \rightarrow 8\ CO_2(g) + 9\ H_2O(\ell)$
$\Delta H° = -5496.0$ kJ
65. $\frac{464.8 \text{ kJ}}{1 \text{ mol CaO}}, \frac{464.8 \text{ kJ}}{3 \text{ mol C}}, \frac{464.8 \text{ kJ}}{1 \text{ mol CaC}_2}, \frac{464.8 \text{ kJ}}{1 \text{ mol CO}}, \frac{1 \text{ mol CaO}}{464.8 \text{ kJ}},$
$\frac{3 \text{ mol C}}{464.8 \text{ kJ}}, \frac{1 \text{ mol CaC}_2}{464.8 \text{ kJ}}, \frac{1 \text{ mol CO}}{464.8 \text{ kJ}}$
67. (a) 2.83×10^5 kJ into system (b) 3.63×10^4 kJ into system (c) 139 kJ out of system
69. $\Delta H = -1450$ kJ/mol
71. 35.5 kJ
73. 6×10^4 kJ released
75. HF
77. For $H_2 + F_2 \longrightarrow 2\ HF$ reaction: (a) 594 kJ (b) −1132 kJ
(c) −538 kJ
For $H_2 + Cl_2 \longrightarrow 2\ HCl$ reaction: (a) 678 kJ (b) −862 kJ
(c) −184 kJ (d) ΔH of HF is most exothermic.
79. 18 °C
81. (a) 1.4×10^4 J (b) −42 kJ
83. 6.6 kJ
85. 394 kJ/mol evolved
87. $\Delta H_f°(SrCO_3) = -1220.$ kJ/mol
89. $\Delta H_f°(PbO) = -217.3$ kJ/mol, 2.6×10^2 kJ evolved
91. $Ag(s) + \frac{1}{2} Cl_2(g) \rightarrow AgCl(s)$ $\Delta H° = -127.1$ kJ
93. (a) $2\ Al(s) + \frac{3}{2} O_2(g) \rightarrow Al_2O_3(s)$ $\Delta H° = -1675.7$ kJ
(b) $Ti(s) + 2\ Cl_2(g) \rightarrow TiCl_4(\ell)$ $\Delta H° = -804.2$ kJ
(c) $N_2(g) + 2\ H_2(g) + \frac{3}{2} O_2(g) \rightarrow NH_4NO_3(s)$
$\Delta H° = -365.56$ kJ
95. $\Delta H° = -98.89$ kJ
97. (a) $\Delta H° = 1372.5$ kJ (b) Endothermic
99. $\Delta H° = -228$ kJ
101. 41.2 kJ evolved
103. 0.78 g propane
105. 44.422 kJ/g octane > 19.927 kJ/g methanol
107. 720 kJ
109. 2.2 hours walking
111. Gold reaches 100 °C first.
113. 75.4 g ice melted
115. $\Delta H_f°(B_2H_6) = 36$ kJ/mol
117. 2.19×10^7 kJ
119. $\Delta H_f°(C_2H_4Cl_2(\ell)) = -165.2$ kJ/mol
121. $\Delta H_f°(CH_3OH) = -200.660$ kJ/mol
123. (a) 36.03 kJ evolved (b) 1.18×10^4 kJ evolved
125. Step 1: −106.32 kJ, Step 2: 275.341 kJ, Step 3: 72.80 kJ
$H_2O\ (g) \longrightarrow H_2(g) + \frac{1}{2} O_2(g)$
$\Delta H° = 241.82$ kJ, endothermic.
127. Melting is endothermic. Freezing is exothermic.
129. Substance A
131. Greater; a larger mass of water will contain a larger quantity of thermal energy at a given temperature.
133. The given reaction produces 2 mol SO_3. Formation enthalpy from Table 6.2 is for the production of 1 mol SO_3.
135. $\Delta E = 310$ J, $w = 0$ J
137. $\Delta H_f°(OF_2) = 18$ kJ/mol
139. CH_4, 50.014 kJ/g; C_2H_6, 47.484 kJ/g; C_3H_8, 46.354 kJ/g; C_4H_{10}, 45.7140 kJ/g; $CH_4 > C_2H_6 > C_3H_8 > C_4H_{10}$
141. (a) 26.6 °C (Above $C_6H_8O_6$ masses of 8.81 g, NaOH is the limiting reactant.)
(b) $C_6H_8O_6$ limits in Experiments 1–3 and NaOH limits in Experiments 4 and 5.
(c) Ascorbic acid has one hydrogen ion; equal quantities of reactant in Exp. 3 at stoichiometric equivalence point.
143. Note: Great variability in answers may result from specific assumptions made.
(a) Approx. 1.0×10^{10} kJ
(b) Approx. 2.1 metric kilotons TNT
(c) 15 kilotons for Hiroshima bomb, 20 kilotons for Nagasaki; max ever = 50,000 kilotons; approx. 14% Hiroshima bomb or 10% Nagasaki bomb
(d) Approx. 20 hours of continuous hurricane damage

Chapter 7

10. Short wavelength and high frequency
12. (a) Radio waves have less energy than infrared.
(b) Microwaves are higher frequency than radio waves.
14. (a) 3.00×10^{-3} m (b) 6.63×10^{-23} J/photon
(c) 39.9 J/mol
16. (d) < (c) < (a) < (b)
18. 6.06×10^{14} Hz
20. 4.4×10^{-19} J/photon
22. Many types of electromagnetic radiation, including visible, ultraviolet, infrared, microwaves
24. 1.1×10^{15} Hz, 7.4×10^{-19} J/photon
26. X-ray (8.42×10^{-17} J/photon) energy is larger than that of orange light (3.18×10^{-19} J/photon).
28. 1.20×10^8 J/mol
30. Photons of this light are too low in energy. Increasing the intensity only increases the number of photons, not their individual energy.
32. No
34. Line emission spectra are mostly dark, with discrete bands of light. Sunlight is a continuous rainbow of colors.
36. The higher-energy state to the lower-energy state; difference
38. (a) Absorbed (b) Emitted (c) Absorbed
(d) Emitted

40. (a), (b), (d)
42. $\Delta E = -6.19 \times 10^{-19}$ J; $\lambda = 321$ nm; ultraviolet
44. $\Delta E = -1.84 \times 10^{-17}$ J; 10.8 nm; ultraviolet
46. 4.576×10^{-19} J absorbed, 434.0 nm
48. 0.05 nm
50. (a) First electron: $n = 1$, $\ell = 0$, $m_\ell = 0$, $m_s = +\frac{1}{2}$; second electron: $n = 1$, $\ell = 0$, $m_\ell = 0$, $m_s = -\frac{1}{2}$; third electron: $n = 2$, $\ell = 0$, $m_\ell = 0$, $m_s = +\frac{1}{2}$; fourth electron: $n = 2$, $\ell = 0$, $m_\ell = 0$, $m_s = -\frac{1}{2}$; fifth electron: $n = 2$, $\ell = 1$, $m_\ell = 1$, $m_s = +\frac{1}{2}$ (The fifth electron could have different m_ℓ and m_s values.)
(b) $n = 3$, $\ell = 0$, $m_\ell = 0$, $m_s = +\frac{1}{2}$ and $n = 3$, $\ell = 0$, $m_\ell = 0$, $m_s = -\frac{1}{2}$
(c) $n = 3$, $\ell = 2$, $m_\ell = 2$, $m_s = +\frac{1}{2}$ (Different m_ℓ and m_s values are also possible.)
52. (a) Cannot occur, m_ℓ too large (b) Can occur
(c) Cannot occur, m_s cannot be 1, here.
(d) Cannot occur, ℓ must be less than n. (e) Can occur
54. (a) $n = 4$, $\ell = 0$, $m_\ell = 0$, $m_s = +\frac{1}{2}$
(b) $n = 3$, $\ell = 1$, $m_\ell = 1$, $m_s = -\frac{1}{2}$
(c) $n = 3$, $\ell = 2$, $m_\ell = 0$, $m_s = +\frac{1}{2}$
56. Four subshells
58. Electrons do not follow simple paths as planets do.
60. Orbits have predetermined paths—position and momentum are both exactly known at all times. Heisenberg's uncertainty principle says that we cannot know both simultaneously.
62. d, p, and s orbitals; nine orbitals total
64. ${}_{13}$Al: $1s^22s^22p^63s^23p^1$; ${}_{16}$S: $1s^22s^22p^63s^23p^4$
66. ${}_{32}$Ge: $1s^22s^22p^63s^23p^63d^{10}4s^24p^2$
68. Oxygen; Group 6A has 6 valence electrons with valence electron configuration of ns^2np^4.
70. (a) $4s$ orbital must be full.
(b) Orbital labels must be 3, not 2; electrons in $3p$ subshell should be in separate orbitals with parallel spin.
(c) $4d$ orbitals must be completely filled before $5p$ orbitals start filling.
72. (a) V

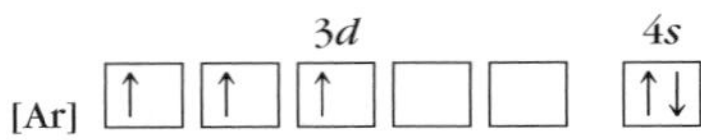

(b) V^{2+}

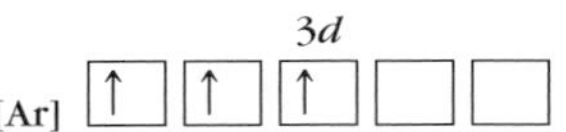

(c) V^{4+}

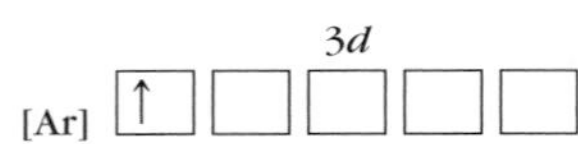

74. 18 elements, all possible orbital electron combinations are already used
76. Mn: $1s^22s^22p^63s^23p^63d^54s^2$; it has 5 unpaired electrons:

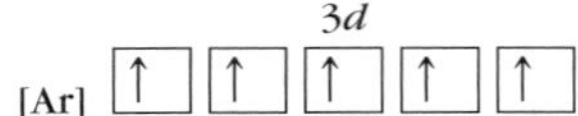

Mn^{2+}: $1s^22s^22p^63s^23p^63d^5$; it has 5 unpaired electrons:

3d

[Ar] ↑ ↑ ↑ ↑ ↑

Mn^{3+}: $1s^22s^22p^63s^23p^63d^4$; it has 4 unpaired electrons:

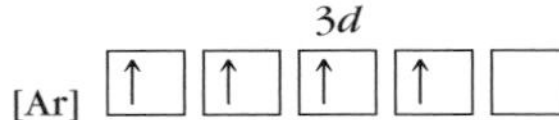

78. [Ne]$3s^23p^4$
80. (a) ${}_{63}$Eu: [Xe]$4f^76s^2$ (b) ${}_{70}$Eu: [Xe]$4f^{14}6s^2$
82. (a) ·Sr· (b) :B̈r· (c) ·Ġa· (d) ·S̈b·
84. (a) [Ar] (b) [Ar]
(c) [Ne]; Ca^{2+} and K^+ are isoelectronic.
86. ${}_{50}$Sn: [Kr]$4d^{10}5s^25p^2$
${}_{50}Sn^{2+}$: [Kr]$4d^{10}5s^2$
${}_{50}Sn^{4+}$: [Kr]$4d^{10}$
88. Ferromagnetism is a property of permanent magnets. It occurs when the spins of unpaired electrons in a cluster of atoms (called a domain) in the solid are all aligned in the same direction. Only metals in the Fe, Co, Ni subgroup (Group 8B) exhibit this property.
90. In both paramagnetic and ferromagnetic substances, atoms have unpaired spins and thus are attracted to magnets. Ferromagnetic substances retain their aligned spins after an external magnetic field has been removed, so they can function as magnets. Paramagnetic substances lose their aligned spins after a time and, therefore, cannot be used as permanent magnets.
92. Number
94. P < Ge < Ca < Sr < Rb
96. (a) Rb smaller (b) O smaller (c) Br smaller
(d) Ba^{2+} smaller (e) Ca^{2+} smaller
98. Al < Mg < P < F
100. (c)
102. Na; it must have one valence electron.
104. (a) Al (b) Al (c) Al < B < C
106. (a) H^- (b) N^{3-} (c) F^-
108. Adding a negative electron to a negatively charged ion requires additional energy to overcome the coulombic charge repulsion.
110. −862 kJ
112. Red < Yellow < Green < Violet
114. Seven pairs
116. (a) He (b) Sc (c) Na
118. (a) F < O < S (b) S
120. (a) Sulfur (b) Radium (c) Nitrogen
(d) Ruthenium (e) Copper
122. (a) Z (b) Z
124. In^{4+}, Fe^{6+}, and Sn^{5+}; very high successive ionization energies
126. (a) False; shorter, not longer (b) True
(c) True (d) False; inversely, not directly
128. (a) Directly related, not inversely related
(b) Inversely proportional to *the square of* the principle quantum number, not inversely proportional to the principle quantum number
(c) Before, not as soon as
(d) Wavelength, not frequency
130. Ultraviolet; 91 nm or shorter wavelength is needed to ionize hydrogen.
132. (a) [Rn]$5f^{14}6d^{10}7s^27p^68s^2$
(b) Magnesium (c) EtO, $EtCl_2$
134. $1s^22s^22p^4$
136. XCl
138. (a) Ground state (b) Could be ground state or excited state
(c) Excited state (d) Impossible
(e) Excited state (f) Excited state
140. [Rn]$5f^{14}5g^{18}6d^{10}6f^{14}7s^27p^67d^{10}8s^28p^2$, Group 4A
141. (a) Increase, decrease
(b) Helium
(c) 5 and 13
(d) He has only two electrons; there are no electrons left.
(e) The first electron is a valence electron, but the second electron is a core electron.
(f) $Mg^{2+}(g) \rightarrow Mg^{3+}(g) + e^-$
144. (a) 4.34×10^{-19} J (b) 6.54×10^{14} Hz
146. 2.18×10^{-18} J
148. To remove an electron from a N atom requires disrupting a half-filled subshell (p^3), which is relatively stable, so the ionization of O requires less energy than the ionization of N.

150. (a) Ir (b) $[Rn]7s^25f^{14}6d^7$
152. (a) Transition elements (b) $[Rn]7s^25f^{14}6d^{10}$

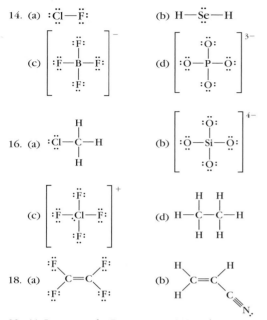

Chapter 8

14. (a) :C̤̈l—F̤̈: (b) H—S̤̈e—H

(c) $[BF_4]^-$: [:F̤̈—B(—F̤̈:)(—F̤̈:)—F̤̈:]$^-$ (d) $[PO_4]^{3-}$: [:Ö—P(—Ö:)(—Ö:)—Ö:]$^{3-}$

16. (a) :C̤̈l—C(H)(H)—H (b) [:Ö—Si(—Ö:)(—Ö:)—Ö:]$^{4-}$

(c) [:F̤̈—C̤l(—F̤̈:)(—F̤̈:)—F̤̈:]$^+$ (d) H—C(H)(H)—C(H)(H)—H

18. (a) (:F̤̈)(:F̤̈)C=C(F̤̈:)(F̤̈:) (b) (H)(H)C=C(H)—C≡N:

22. (a) Incorrect; the F atoms are missing electrons.
(b) Incorrect; the structure has 10 electrons, but needs 12 electrons.
(c) Incorrect. The structure has three too many electrons; Carbon has nine electrons. The single electron should be deleted. Oxygen has ten electrons; delete one pair of electrons.
(d) Incorrect; one hydrogen atom has more than two electrons. One hydrogen atom is completely missing.
(e) Incorrect; the structure has 16 electrons, but needs 18 electrons. The N atom doesn't follow the octet rule. It needs another pair of electrons in the form of a lone pair.
24. Four branched-chain compounds

$CH_3—CH_2—CH(CH_3)—CH_2—CH_3$ (drawn with all H atoms)

$CH_3—CH(CH_3)—CH_2—CH_2—CH_3$ (drawn with all H atoms)

$CH_3—C(CH_3)_2—CH_2—CH_3$ (drawn with all H atoms)

$CH_3—CH(CH_3)—CH(CH_3)—CH_3$ (drawn with all H atoms)

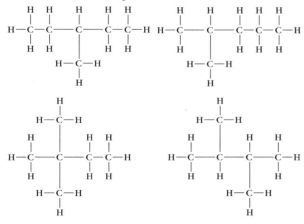

26. (a) Alkyne (b) Alkane (c) Alkene
28. *cis*-2-pentene: H_3C(H)C=C(H)$CH_2—CH_3$ *trans*-2-pentene: H_3C(H)C=C(H)$CH_2—CH_3$

30. (a) No
(b) Yes

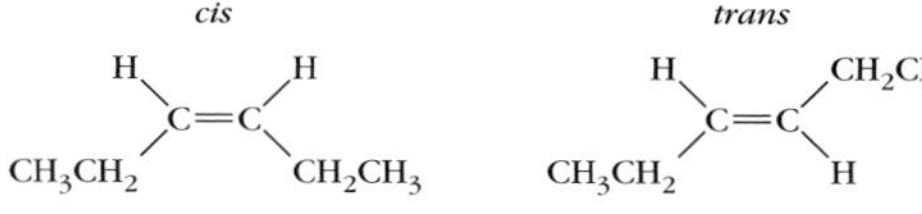

(c) Yes

cis: (H)(CH_3)C=C(H)(CH_3) *trans*: (H)(CH_3)C=C(CH_3)(H)

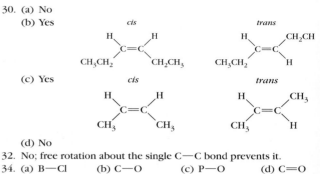

(d) No
32. No; free rotation about the single C—C bond prevents it.
34. (a) B—Cl (b) C—O (c) P—O (d) C=O
36. (a)
38. CO
40. CO_3^{2-} has longer C—O bonds.
42. −92 kJ; the reaction is exothermic.
44. HF has the strongest bond; −538. kJ; −184 kJ; −103 kJ; −11 kJ. The reaction of H_2 with F_2 is most exothermic.
46. (a) N, C, Br, O
(b) S—O is the most polar.
48. (a) All the bonds in urea are somewhat polar.
(b) The most polar bond is C=O; the O end is partially negative.
50. The total formal charge on a molecule is always zero. The total formal charge on an ion is the ionic charge.

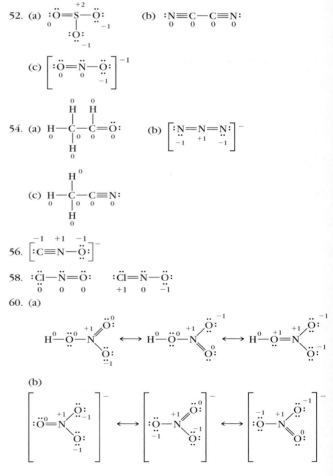

52. (a) :Ö=S(—Ö:)—Ö: (formal charges: O 0, S +2, O −1, O −1) (b) :N≡C—C≡N: (0 0 0 0)

(c) [:Ö=N̈—Ö:]$^{-1}$ (formal charges: 0, 0, −1)

54. (a) H—C(H)(H)—C(H)=Ö: (all formal charges 0) (b) [:N̈=N=N̈:]$^-$ (−1, +1, −1)

(c) H—C(H)(H)—C≡N: (all formal charges 0)

56. [:C≡N—Ö:]$^-$ (−1, +1, −1)

58. :C̤̈l—N̈=Ö: (0 0 0) :C̤̈l=N̈—Ö: (+1 0 −1)

60. (a) H—Ö—N(=Ö:)—Ö: ⟷ H—Ö—N(—Ö:)=Ö: ⟷ H—O=N(—Ö:)—Ö: (formal charges shown for each atom: H 0, O 0 / +1, N +1, terminal O 0 or −1)

(b) [:Ö=N(—Ö:)—Ö:]$^-$ ⟷ [:Ö—N(=Ö:)—Ö:]$^-$ ⟷ [:Ö—N(—Ö:)=Ö:]$^-$ (N +1; double-bonded O 0; single-bonded O −1)

62. $[:O-Br(-O)(-O)-O:]^-$ (formal charges: O −1, O −1, O −1, O −1, Br +3) ⟷ $[O=Br...]^-$ (O 0, O −1, O −1, O −1, Br +2) ⟷

Most plausible: $[O=Br(=O)(-O)-O]^-$ (O 0, O 0, O −1, O −1, Br +1) ⟷ $[O=Br(=O)(=O)-O]^-$ (O 0, O 0, O 0, O −1, Br −1)

65. $[S_2O_3]^{2-}$ resonance structures:
$[O-S(-O)(=O)(=S)]^{2-}$ ⟷ $[O-S(=O)(-O)(=S)]^{2-}$ ⟷ $[O=S(-O)(-O)(=S)]^{2-}$

$[O=S(=O)(-O)(-S)]^{2-}$ ⟷ $[O=S(-O)(=O)(-S)]^{2-}$ ⟷ $[O-S(=O)(=O)(-S)]^{2-}$

67. (a) BrF_5 (b) IF_5 (c) $[:Br-I-Br:]^-$

69. P, Cl, Se, Sn

71. Against; with C=C and C—C, the bond lengths would be different.

73. $C_{14}H_{10}$

75.

	σ_{2s}	σ^*_{2s}	π_{2p}	π_{2p}	σ_{2p}	π^*_{2p}	π^*_{2p}	σ^*_{2p}
NO^+	(↑↓)	(↑↓)	(↑↓)	(↑↓)	(↑↓)	( )	( )	( )

NO^+ has three bonds and no unpaired electrons.

78. Si—F; Si is in the third period and less electronegative than C, S, or O.

80. Yes; "close" means similar electronegativities, therefore covalent bonds. "Far apart" means different electronegativities, therefore ionic bonds.

82. (a) The C=C is shorter. (b) The C=C is stronger.
(c) C≡N (δ^+ δ^-)

84. (a) :Cl—S—Cl:
(b) $[:Cl-Cl-Cl:]^+$
(c) :Cl—O—Cl(—O:)(=O)=O:
(d) :O=S(—Cl:)—Cl:

86. O—O < Cl—O < O—H < O=O < O=C

88. (a) Aromatic (b) Aromatic (c) Neither category
(d) Alkane (e) Alkane (f) Neither category

90. The student forgot to subtract one electron for the positive charge.

94. Atoms are not bonded to the same atoms.

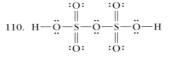

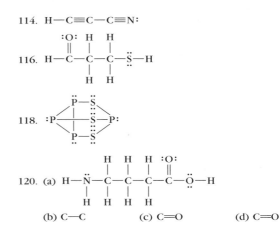

96. (a) H:O:H (with H above O) (b) H:F:H

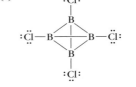

98. Cl: 3.0, S: 2.5, Br: 2.5, Se: 2.4, As: 2.1

100. NF_5; you cannot expand the octet of N.

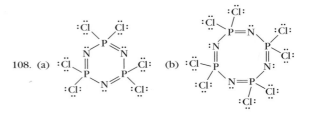

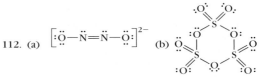

102. :F—N=O:
(a) N—F (b) N=O (c) F—N

104. (a) C—O (b) C≡N (c) C—O

106. (a) B_4Cl_4: tetrahedral B_4 cage, each B bonded to one :Cl:
(b) H—O—N=N—O—H

108. (a) $(PNCl_2)_3$ six-membered P=N ring, each P bearing two :Cl: (b) $(PNCl_2)_4$ eight-membered P=N ring, each P bearing two :Cl:

110. H—O—S(=O)(=O)—O—S(=O)(=O)—O—H

112. (a) $[:O-N=N-O:]^{2-}$ (b) S_3O_9 ring: O—S—O—S—O—S ring, each S bearing two =O

114. H—C≡C—C≡N:

116. H—C(=O)—CH₂—CH₂—S—H

118. P_4S_3 cage: P—S, S—P: bonds (:P, P—S, S—P:, P—S)

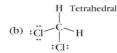

120. (a) H—N(H)—C(H)(H)—C(H)(H)—C(H)(H)—C(=O)—O—H
(b) C—C (c) C=O (d) C=O

Chapter 9

13. (a) H—Be—H
Linear
(b) $CHCl_2$-type: C bonded to H, H, :Cl, :Cl: — Tetrahedral

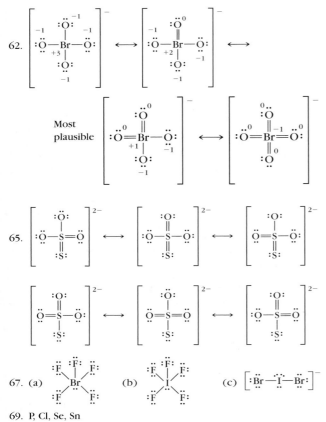

(c) H—B(—H)—H Triangular planar (d) $SeCl_6$ Octahedral

(e) PF_5 Triangular bipyramid

15. (a) Both tetrahedral (b) Tetrahedral, angular (109.5°)
(c) Both triangular planar (d) Both triangular planar
17. (a) Both triangular planar (b) Both triangular planar
(c) Tetrahedral, triangular pyramid
(d) Tetrahedral, triangular pyramidal
Three atoms bonded to the central atom gives triangular shape; however, structures with 26 electrons have the shape of a triangular pyramid and structures with 24 electrons have the shape of a triangular planar.
19. (a) Both octahedral
(b) Triangular bipyramid and see-saw
(c) Both triangular bipyramid
(d) Octahedral and square planar
21. (a) 120° (b) 120°
(c) H—C—H angle is 120°, C—C—N angle is 180°
(d) H—O—N angle = 109.5°, O—N—O angle = 120°
23. (a) 90°, 120°, and 180° (b) 120° and 90° (c) 90° and 120°
25. The O—N—O angle in NO_2^+ is larger than in NO_2.
27. (a) sp^3 (b) sp^3d^2 (c) sp^2
29. Tetrahedral, sp^3 hybridized carbon atom
31. The central S atom in SF_4 has sp^3d hybridization; the central S atom in SF_6 has sp^3d^2 hybridization.
33. PCl_4^+ is tetrahedral; Ge is sp^3 hybridized. PCl_5 is triangular bipyramidal. P atom is sp^3d hybridized. PCl_6^- is isoctahedral. P atom is sp^3d^2 hybridized.
35. There are no *d* orbitals in the $n = 2$ shell and the student did not use all the valence *p* orbitals before including a *d* orbital.
38. The N atom is sp^3 hybridized with 109.5° angles. The first two carbons are sp^3 hybridized with 109.5° angles. The third carbon is sp^2 hybridized with 120° angles. The single-bonded oxygen is sp^3 hybridized with 109.5° angles. The double-bonded O is sp^2 hybridized with 120° angles.
40. (a) The first two carbon atoms are sp^3 hybridized with 109.5° angles. The third and fourth carbon atoms are *sp* hybridized with 180° angles.
(b) C≡C (c) C≡C

42. (a) :O=C=S: (σ and π bonds) (b) H—N(H)—O—H (σ bonds) (c) H—C(H)=C(H)—C(H)=O (d) H—C(H)(H)—C(H)(OH)—C(=O)—O—H

44. (a) Six (b) Three (c) *sp* (d) *sp*
(e) Both sp^2
46. (a) H_2O (b) CO_2 and CCl_4 (c) F
48. (a) Nonpolar
(b) Polar; the H-side of the molecule is more positive.
(c) Polar; the H-side of the molecule is more positive.
(d) Nonpolar
50. (a) The Br—F bond has a larger electronegativity difference.
(b) The H—O bond has a larger electronegativity difference.

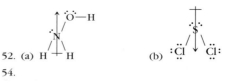

52. (a) NH_2OH (b)

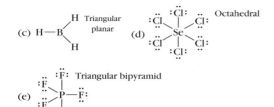

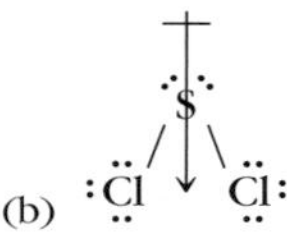

54.

Interaction	Distance	Example
Ion-ion	Longest range	Na^+ interaction with Cl^-
Ion-dipole	Long range	Na^+ ions in H_2O
Dipole-dipole	Medium range	H_2O interaction with H_2O
Dipole-induced dipole	Short range	H_2O interaction with Br_2
Induced dipole-induced dipole	Shortest range	Br_2 interaction with Br_2

56. Wax molecules interact using London forces. Water molecules interact using hydrogen bonding. The water molecules interact much more strongly with other water molecules; hence, beads form as the water tries to avoid contact with the surface of the wax.
58. (c), (d), and (e)
60. Vitamin C is capable of forming hydrogen bonds with water.
62. (a) London forces (b) London forces
(c) Intramolecular (covalent) forces
(d) Dipole-dipole force (e) Hydrogen bonding
65. (a) C—C stretch (b) O—H stretch
67. 427 nm, 483 nm; 427 nm is the more energetic transition.
69.

T A C G G C

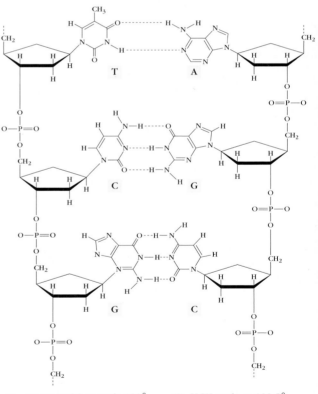

71. (a) (1) NCC = angle 180°, (2) HCH angle = 109.5°, (3) COC angle = 109.5°
(b) C=O
(c) C=O
73. (a) Ö=C=C=C=Ö (b) 180° (c) 180°
75. It has an extended conjugated pi-system, so yes, it should absorb visible light.

77. (a) $[C_4O_4]^{2-}$ (Lewis structure: four-membered C ring, C=C double bond, each C bonded to an O)

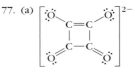

(b) Each C is sp^2 hybridized.

79. HCl has $\delta^+ = 2.69 \times 10^{-20}$ C and HF has $\delta^+ = 6.95 \times 10^{-20}$ C; therefore, F is more electronegative than Cl.

81. KF has $\delta^+ = 1.32 \times 10^{-19}$ C, which is 81.6% of the charge of the electron, so KF is not completely ionic.

83. If the polarity of the bonds exactly cancel each other, a molecule will be nonpolar.

87. (a) Nitrogen (b) Boron (c) Phosphorus
(d) Iodine (There are other right answers.)

89. Five

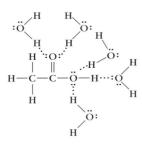

(Structure: acetic acid, H—C—C(=O)—O—H, hydrogen-bonded to five water molecules)

91. Diagram (d) is correct.

93. H:Ö:H (with H above) H:F:H

95. (a)

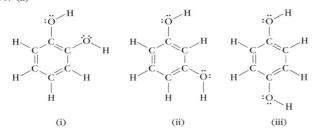

(i) (ii) (iii)

(b) (i) < (ii) < (iii)

97. (a) H—C=C=Ö: (H below the first C)

(b) Triangular planar, linear, H—C—C angle = 120°, C—C—O angle = 180°

(c) sp^2, sp, sp (d) Polar

99. (a) :N≡N—N̈—H N̈=N̈ (ring) N: with H

(b) sp, sp, sp^2 (c) sp^2, sp^2, sp^3
(d) 3, 4 (e) 2, 1
(f) molecule 1 120°, 180°; molecule 2 120°, 120°, 109.5°

101. (a) Angle 1: 120°, angle 2: 120°, angle 3: 109.5° (b) sp^3
(c) For O with two single bonds, sp^3; for O with one double bond, sp^2

Chapter 10

11. Nitrogen serves to moderate the reactiveness of oxygen by diluting it. Oxygen sustains animal life as a reactant in the conversion of food to energy. Oxygen is produced by plants in the process of photosynthesis.

13.

Molecule	ppm	ppb	
N_2	780,840	780,840,000	↑
O_2	209,480	209,480,000	
Ar	9,340	9,340,000	
CO_2	330	330,000	
Ne	18.2	18,200	>1 ppm
H_2	10.	10,000	
He	5.2	5,200	
CH_4	2	2,000	↓
Kr	1	1,000	↑
CO	0.1	100	between
Xe	0.08	80	1 ppm
O_3	0.02	20	and
NH_3	0.01	10	1 ppb
NO_2	0.001	1	↓
SO_2	0.0002	0.2	<1 ppb

15. 1.5×10^8 metric tons SO_2, 2×10^6 metric tons SO_2

16. (a) 0.947 atm (b) 950. mm Hg (c) 542 torr
(d) 98.7 kPa (e) 6.91 atm

18. 14 m

20. With a perfect vacuum at the top of the well, atmospheric pressure can only push water up to about 34 feet. So, the well cannot be deeper than that, not even using a high-quality vacuum pump.

22. I. A gas is composed of molecules whose size is much smaller than the distances between them.
II. Gas molecules move randomly at various speeds and in every possible direction.
III. Except when molecules collide, forces of attraction and repulsion between them are negligible.
IV. When collisions occur, they are elastic.
V. The average kinetic energy of gas molecules is proportional to the absolute temperature.

Postulate I will become false at very high pressures. Postulates III and IV will become false at very high pressures or very low temperatures. Postulate II is most likely to always be correct.

24. Slowest CH_2Cl_2 < Kr < N_2 < CH_4 fastest

26. Ne will arrive first.

29. 4.2×10^{-5} mol CO

30. 25.5 L

32. 62.5 mm Hg

35. 172 mm Hg

37. 26.5 mL

39. −96 °C

41. 4.00 atm

43. 501 mL

45. 0.507 atm

47. Largest number in (d); smallest number in (c)

49. 6.0 L H_2

51. 1.9 L CO_2; about half the loaf is CO_2.

53. 10.4 L O_2, 10.4 L H_2O

55. 21 mm Hg

57. 10.0 L Br_2

59. 1.44 g $Ni(CO)_4$

61. (a) $2\ CH_3OH(\ell) + 3\ O_2(g) \rightarrow 2\ CO_2(g) + 4\ H_2O(g)$
(b) 1.1×10^3 L CO_2

63. 130. g/mol

65. 2.7×10^3 mL

67. P_{He} is 7.000 times greater than P_{N_2}.

69. 3.7×10^{-4} g/L

71. $P_{tot} = 4.51$ atm

73. (a) 154 mm Hg
(b) $X_{N_2} = 0.777$, $X_{O_2} = 0.208$, $X_{Ar} = 0.0093$, $X_{H_2O} = 0.0053$, $X_{CO_2} = 0.0003$
(c) 77.7% N_2, 20.8% O_2, 0.93% Ar, 0.03% CO_2, 0.54% H_2O. This sample is wet. Table 1.3 gives percentages for dry air, so the percentages are slightly different due to the water in this sample.

75. (a) $P_{tot} = 1.98$ atm
(b) $P_{O_2} = 0.438$ atm, $P_{N_2} = 0.182$ atm, $P_{Ar} = 1.36$ atm
77. Membrane irritation: 1×10^{-4} atm; Fatal narcosis: 0.02 atm
79. 0.0041; 3.1 mm Hg; the mean partial pressure includes humid and dry air, summer and winter, worldwide.
82. 18. mL $H_2O(\ell)$, 22.4 L $H_2O(g)$; No, because the vapor pressure of water at 0 °C is 4.6 mm Hg; we cannot achieve 1 atm pressure for this gas at this temperature.
84. Molecular attractions become larger at higher pressures. Molecules hitting the walls hit them with somewhat less force due to the opposing attractions of other molecules.
86. N_2 is more like an ideal gas at high P than CO_2.
88. A free radical is an atom or group of atoms with one or more unpaired electrons; as a result, it is highly reactive. Example: $O_2 \rightarrow \cdot O\cdot + \cdot O\cdot$ (there are other possible examples.)
90. (a) NH_3; Ox. # N = −3 (b) NH_4^+; Ox. # N = −3
(c) N_2O; Ox. # N = +1 (d) N_2; Ox. # N = 0
(e) HNO_3; Ox. # N = +5 (f) HNO_2; Ox # N = +3
(g) NO_2; Ox. # N = +4
92. (a) $\cdot O\cdot + H_2O \rightarrow \cdot OH + \cdot OH$
(b) $\cdot NO_2 + \cdot OH \rightarrow HNO_3$
(c) $RH + \cdot OH \rightarrow \cdot R + H_2O$
94. These reactions can be found in Section 10.11.
(a) $CF_3Cl \xrightarrow{h\nu} \cdot CF_3 + \cdot Cl$ (b) $\cdot Cl + \cdot O\cdot \rightarrow ClO\cdot$
(c) $ClO\cdot + \cdot O\cdot \rightarrow \cdot Cl + O_2$
96. CH_3F has no C—Cl bonds, which in CH_3Cl are readily broken when exposed to UV light to produce ozone-depleting radical halogen, ·Cl.
98. CFCs are not toxic. Refrigerants used before CFCs were very dangerous. One example is NH_3, a strong-smelling, reactive chemical. Use the keywords "CFCs" or "refrigerants" and "toxicity" in any web browser.
101. Primary pollutants (e.g., particle pollutants, including aerosols and particulates, sulfur dioxide, nitrogen oxides, hydrocarbons); secondary pollutants (e.g., ozone); (see Section 10.12 for details).
103. Adsorption means firmly attaching to a surface. Absorption means drawing into the bulk of a solid or liquid.
105. 1.6×10^9 metric tons, 2.3×10^6 hours
107. $O_2 \xrightarrow{h\nu} \cdot O\cdot + \cdot O\cdot$, No, the reaction will not occur if the light is visible light.
109. SO_2, coal and oil, $2\ SO_2 + O_2 \longrightarrow 2\ SO_3$
111. $N_2 + O_2 \xrightarrow{heat} 2\ NO$; reaction takes elemental nitrogen and makes a compound of nitrogen.
113. (a) $P_i = 1.33 \times 10^{-4}$ mm Hg
(b) $X_{SO_2} = 1.75 \times 10^{-7}$
(c) 500. μg SO_2
114. Greenhouse effect = trapping of heat by atmospheric gases. Global warming = increase of the average global temperature. Global warming is caused by an increase in the amount of greenhouse gases in the atmosphere.
116. Examples of CO_2 sources: animal respiration, burning fossil fuels and other plant materials, decomposition of organic matter, etc. Examples of CO_2 removal: photosynthesis in plants, being dissolved in rain water, incorporation into carbonate and bicarbonate compounds in the oceans, etc. Currently, CO_2 production exceeds CO_2 removal.
118. (a) Before: $P_{H_2} = 3.7$ atm; $P_{Cl_2} = 4.9$ atm
(b) Before: $P_{tot} = 8.6$ atm (c) After: $P_{tot} = 8.6$ atm
(d) Cl_2; 0.5 moles remain (e) $P_{HCl} = 7.4$ atm; $P_{Cl_2} = 1.2$ atm
(f) $P = 8.9$ atm
120. Statements (a), (b), (c), and (d) are true.
122.

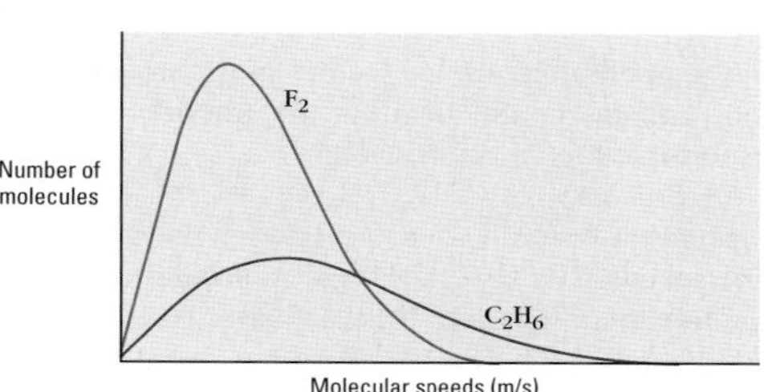

124. For reference, the initial state looks like this:

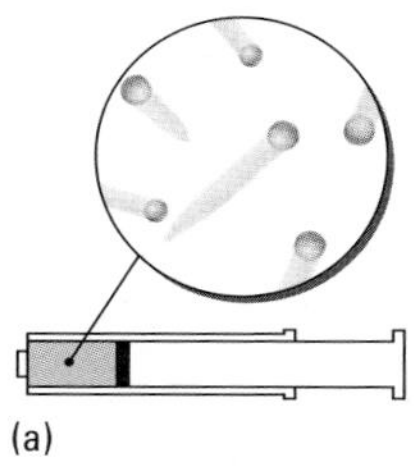

(a)

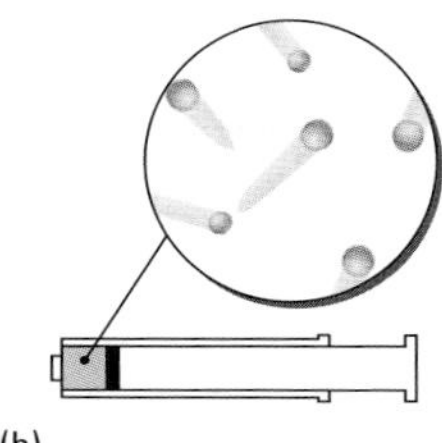

(b)

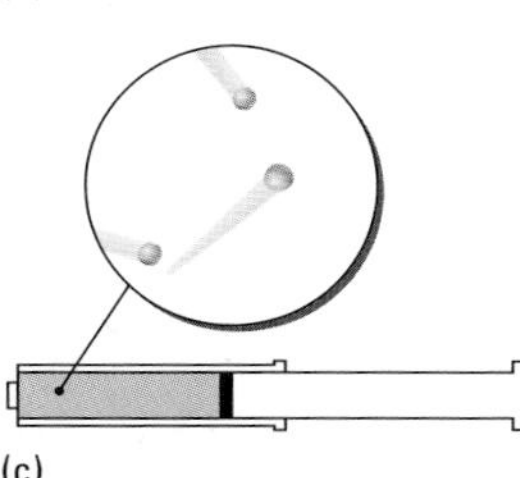

(c)

126. Box (b). The initial-to-final volume ratio is 2:1, so, for every two molecules of gas reactants, there must be one molecule of gas products. 6 reactant molecules must produce 3 product molecules.
128. (a) 64.1 g/mol
(b) Empirical formula = CHF; molecular formula = $C_2H_2F_2$
(c) If the F atoms are on different C atoms

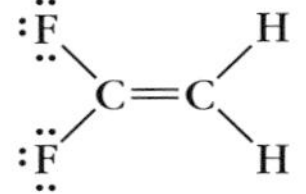
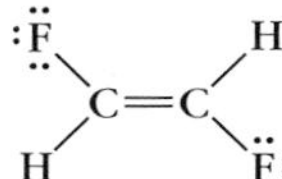
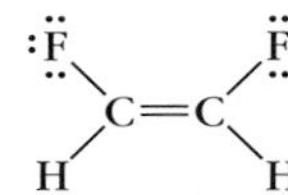

130. $P_{tot} = 0.88$ atm
132. $\dfrac{m_{Ne}}{m_{Ar}} = 0.2$
134. (a) 29.1 mol CO_2; 14.6 mol N_2; 2.43 mol O_2
(b) $1.1 = 10^3$ L
(c) $P_{N_2} = 0.317$ atm; $P_{CO_2} = 0.631$ atm; $P_{O_2} = 0.0527$ atm
136. 458 torr
138. 4.5 mm^3
140. (a) More significant, because of more collisions
(b) More significant, because of more collisions
(c) Less significant, because the molecules will move faster

Chapter 11

13. At higher temperatures, the molecules move around more. The increased random motion disrupts the intermolecular interactions responsible for surface tension.
15. Reduce the pressure
17. The molecules of water in your sweat have a wide distribution of molecular speeds. The fastest of these molecules are more likely to escape the liquid state into the gas phase. The low-speed molecules left behind will have a lower average speed and, therefore, a lower average kinetic energy that, according to kinetic molecular theory, implies that the temperature lowers.
19. 1.5×10^6 kJ
21. 233 kJ
23. 2.00 kJ required for Hg is more than 1.13 kJ required for H_2O sample. Hg has a much greater mass.
25. NH_3 has a relatively large boiling point because the molecules interact using relatively strong hydrogen-bonding intermolecular forces. The increase in the boiling points of the series PH_3, AsH_3, and SbH_3 is related to the increasing London dispersion intermolecular forces experienced, due to the larger central atom in the molecule. (Size: P < As < Sb)
27. Methanol molecules are capable of hydrogen bonding, whereas formaldehyde molecules use dipole-dipole forces to interact. Molecules experiencing stronger intermolecular forces (such as methanol here) will have higher boiling points and lower vapor pressures compared to molecules experiencing weaker intermolecular forces (such as formaldehyde here).
30. 0.21 atm, approximately 57 °C
32. 1600 mm Hg
34. $\Delta H_{vap} = 70.$ kJ/mol
36. The interparticle forces in the solid are very strong.
38. The intermolecular forces between the molecules of H_2O in the solid (hydrogen bonding) are stronger than the intermolecular forces between the molecules of H_2S (dipole-dipole).
40. 27 kJ
42. 51.9 g CCl_2F_2
44. LiF, because the ions are smaller, making the charges more localized and closer together, causing a higher coulombic interaction between the ions.
46. The highest melting point is (a). The extended network of covalent bonds in SiC are the strongest interparticle forces of the listed choices. The lowest melting point is (d). Both I_2 and $CH_3CH_2CH_2CH_3$ interact using the weakest intermolecular forces—London dispersion forces—but large I_2 has many more electrons, so its London forces are relatively stronger.
48. The freezer compartment of a frost-free refrigerator keeps the air so cold and dry that ice inside the freezer compartment sublimes [(s) → (g)]. The hail stones would eventually disappear.
51. (a) Gas phase
    (b) Liquid phase
    (c) Solid phase
54. (a) Molecular (b) Metallic
    (c) Network (d) Ionic
56. (a) Amorphous; it decomposes before melting and does not conduct electricity. (b) Molecular; low melting point and nonconducting (c) Ionic; high melting point and only liquid (molten ions) conduct electricity. (d) Metallic; both solid and liquid conduct electricity.
58. (a) Molecular (b) Ionic
    (c) Metallic or network (d) Amorphous
60. See Figure 11.21 and its description
62. 220 pm
65. Diagonal is 696 pm, side length is 401 pm.
67. No; the ratio of ions in the unit cell must reflect the empirical formula of the compound.
69. (a) 152 pm
    (b) Radius of I^- = 212 pm and radius of Li^+ = 88.0 pm
    (c) It is reasonable that the atom is larger than the cation. The assumption that anions touch anions seems unreasonable; there would be some repulsion and probably a small gap or distortion.
70. Carbon atoms in diamond are sp^3 hybridized and are tetrahedrally bonded to four other carbon atoms. Carbon atoms in pure graphite are sp^2 hybridized and bonded with a triangle planar shape to other carbon atoms. These bonds are partially double bonded so they are shorter than the single bonds in diamond. However, the planar sheets of sp^2 hybridized carbon atoms are only weakly attracted by intermolecular forces to adjacent layers, so these interplanar distances are much longer than the C—C single bonds. The net result is that graphite is less dense than diamond.
72. Diamond is an electrical insulator because all the electrons are in single bonds that are shared between two specific atoms and cannot move around. However, graphite is a good conductor of electricity because its electrons are delocalized in conjugated double bonds that allow the electrons to move easily through the graphite sheets.
74. $\nu = 5.30 \times 10^{17} s^{-1}$; (a) 3.51×10^{-16} J/photon
    (b) 2.11×10^8 J/mol, X ray
76. 361 pm
79. In a conductor, the valence band is only partially filled, whereas, in an insulator, the valence band is completely full, the conduction band is empty, and there is a wide energy gap between the two. In a semiconductor, the gap between the valence band and the conduction band is very small so that electrons are easily excited into the conduction band.
81. Substance (c), Ag, has the greatest electrical conductivity because it is a metal. Substance (d), P_4, has the smallest electrical conductivity because it is a nonmetal. (The other two are metalloids.)
83. A superconductor is a substance that is able to conduct electricity with no resistance. Two examples are $YBa_2Cu_3O_7$ and $Hg_{0.8}Tl_{0.2}Ba_2Ca_2Cu_3O_{8.23}$.
85. $SiO_2(s) + 2\ C(s) \rightarrow Si(s) + 2\ CO(g)$; Si is being reduced and C is being oxidized. $SiCl_4(\ell) + 2\ Mg(s) \rightarrow Si(s) + 2\ MgCl_2(s)$; Si is being reduced and Mg is being oxidized.
87. Doping is the intentional addition of small amounts of specific impurities into very pure silicon. Group III elements are used because they have one less electron per atom than the Group IV silicon. Group V elements are used because they have one more electron per atom.
90. The amorphous solids known as glasses are different from NaCl because they lack symmetry or long-range order, whereas ionic solids such as NaCl are extremely symmetrical. NaCl must be heated to melting temperatures and then cooled very slowly to make a glass.
92. (a) Two examples of oxide ceramics: Al_2O_3 and MgO
    (b) Two examples of nonoxide ceramics: Si_3N_2 and SiC.
    (other answers are possible)
94. 780 kJ, 1.6×10^4 kJ
96. (a) Dipole-dipole forces and London forces
    (b) $CH_4 < NH_3 < SO_2 < H_2O$
97. (a) Approx. 80 mm Hg (b) Approx. 18 °C
    (c) Approx. 640 mm Hg (d) Diethyl ether and ethanol
    (e) Diethyl ether evaporates immediately. Ethanol and water remain liquid.
    (f) Water
99. The butane in the lighter is under great enough pressure so that the vapor pressure of butane at room temperature is less than the pressure inside the light. Hence, it exists as a liquid.
101. 1 and C, 2 and E, 3 and B, 4 and F, 5 and G, 6 and H, 7 and A
103. Each has the same fraction of filled space. The fraction of spaces filled by the closest packed, equal-sized spheres is the same, no matter what the size of the spheres.

104. Vapor-phase water condenses on contact with the skin and the condensation is exothermic, which imparts more energy to the skin.
107. (a) Condensation, freezing (b) Triple point
(c) Melting point curve
109. (a) Approx. 560 mm Hg (b) Benzene
(c) 73 °C
(d) Methyl ethyl ether, approx. 7 °C; carbon disulfide, approx. 47 °C; benzene, approx. 81 °C
111. 22.2 °C
113. 0.533 g/cm^3

Chapter 12

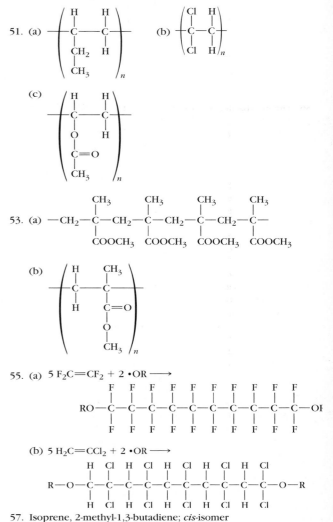

22. (a) 20–200 °C
(b) The octane rating of the straight-run gasoline fraction is 55.
(c) No; because the octane rating is far lower than regular gasoline we buy at the pump (86–94), that means it would cause far more pre-ignition than we expect from the gasoline. It would need to be reformulated to make it an acceptable motor fuel.
24. Gasolines contain molecules in the liquid phase that, at ambient temperatures, can easily overcome their intermolecular forces and escape into the vapor phase. That means all gasolines evaporate easily.
26. (a) $HO{-}C({=}O){-}CH(OH){-}CH(OH){-}C({=}O){-}H$ (b) No chiral centers
(c) $CH_3{-}CH_2{-}CH(NH_2){-}C({=}O){-}OH$
28. (a) No chiral centers (b) No chiral centers
(c) $H{-}CH_2{-}CHBr{-}CH_2{-}CH_2{-}CH_2{-}H$ (drawn as H–C–C–C–C–C–H with H, Br, H, H, H above and H, H, H, H, H below) (d) No chiral centers
30. If four different things are bonded to one atom in the structural formula, then the compound can exist as a pair of enantiomers.
32. (b), (c), (e), (g), (h), and (j)
33. (a) $CH_3{-}CH_2{-}CH_2{-}OH$
(b) $CH_3{-}CH(OH){-}CH_3$ (c) $CH_3{-}C(OH)(CH_3){-}CH_3$
(There are other correct answers)
35. (a) Tertiary (b) Primary (c) Secondary
(d) Secondary (e) Tertiary (f) Secondary
37. (a) $CH_3{-}C({=}O){-}H$
First oxidation product
$CH_3{-}C({=}O){-}OH$
Second oxidation product
(b) $CH_3{-}CH_2{-}CH_2{-}C({=}O){-}H$
First oxidation product
$CH_3{-}CH_2{-}CH_2{-}C({=}O){-}OH$
Second oxidation product
39. (a) $CH_3{-}CH_2{-}CH(CH_3){-}CH_2{-}OH$
(b) $CH_3{-}CH_2{-}CH(OH){-}CH_2{-}CH_3$
(c) $CH_3{-}CH(CH_3){-}CH_2{-}CH_2{-}OH$
41. Wood alcohol (methanol) is made by heating hardwoods such as beech, hickory, maple, or birch. Grain alcohol (ethanol) is made from the fermentation of plant materials such as grains.
43. —OH groups are a common site of hydrogen-bonding intermolecular forces. Their presence would increase the solubility of the biological molecule in water and create specific interactions with other biological molecules.
45. (a) $CH_3COOCH_2CH_3$ (b) $CH_3CH_2COOCH_2CH_2CH_3$
(c) $CH_3CH_2COOCH_3$
47. (a) $CH_3CH_2COOH + CH_3OH$
(b) $HCOOH + CH_3CH_2OH$
(c) $CH_3COOH + CH_3CH_2OH$
49. Examples of thermoplastics are milk jugs (polyethylene), sunglasses and toys (polystyrene), and CD audio discs (polycarbonates). Thermoplastics soften and flow when heated.
51. (a) $\left({-}CH(CH_2CH_3){-}CH_2{-}\right)_n$ (b) $\left({-}CCl_2{-}CH_2{-}\right)_n$
(c) $\left({-}CH(O{-}C({=}O){-}CH_3){-}CH_2{-}\right)_n$
53. (a) $-CH_2{-}C(CH_3)(COOCH_3){-}CH_2{-}C(CH_3)(COOCH_3){-}CH_2{-}C(CH_3)(COOCH_3){-}CH_2{-}C(CH_3)(COOCH_3){-}$
(b) $\left({-}CH_2{-}C(CH_3)(C({=}O){-}O{-}CH_3){-}\right)_n$
55. (a) $5\ F_2C{=}CF_2 + 2\ {\bullet}OR \longrightarrow$
$RO{-}(CF_2)_{10}{-}OR$ (drawn as RO–C–C–C–C–C–C–C–C–C–C–O with F above and below each C)
(b) $5\ H_2C{=}CCl_2 + 2\ {\bullet}OR \longrightarrow$
$R{-}O{-}(CH_2{-}CCl_2)_5{-}O{-}R$ (drawn as R–O–C–C–C–C–C–C–C–C–C–C–O–R with H, Cl above and H, Cl below alternately)
57. Isoprene, 2-methyl-1,3-butadiene; *cis*-isomer

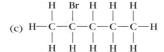

59.

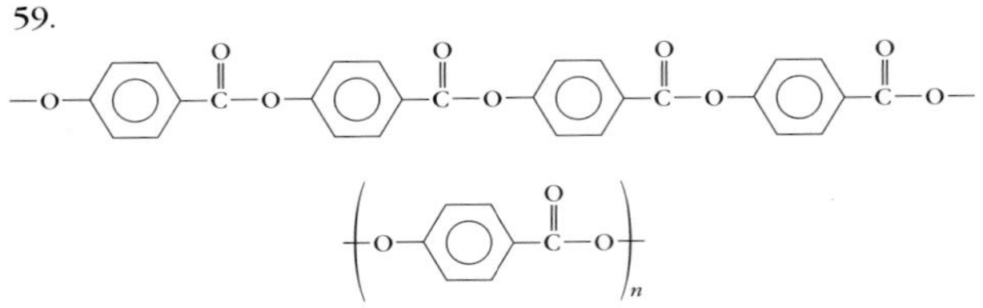

$-O-C_6H_4-C(=O)-O-C_6H_4-C(=O)-O-C_6H_4-C(=O)-O-C_6H_4-C(=O)-O-$

$\left[O-C_6H_4-C(=O)-O \right]_n$

61. Carboxylic acid and alcohol
63. Carboxylic acid and amine; nylon
65. One major difference is that the protein polymer's monomers are not all alike. Different side chains change the properties of the protein.
67. $CH_2{=}CHCN$
69. 16,000 monomer units
71. (a) $\left[CH_2-C(CH_3){=}CH-CH_2-CH_2-C(CH_3){=}CH-CH_2 \right]_n$

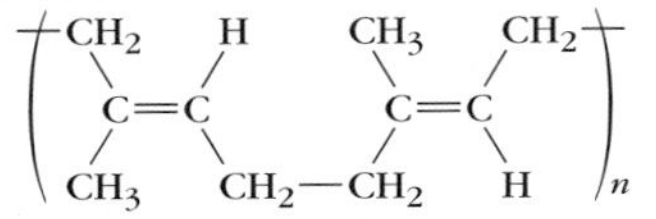

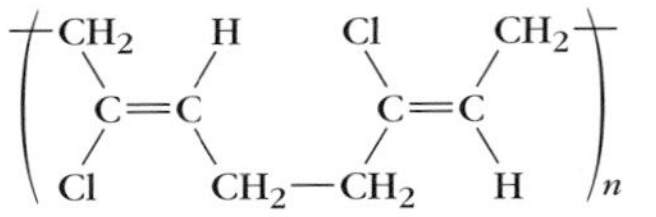

(b) $\left[CH_2-C(Cl){=}CH-CH_2-CH_2-C(Cl){=}CH-CH_2 \right]_n$

(c) $\left[CH_2-CH{=}CH-CH_2-CH_2-CH{=}CH-CH_2 \right]_n$

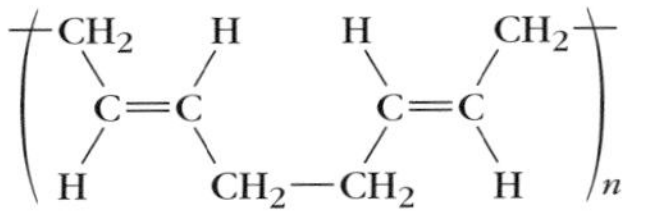

72. Major end uses for recycled PET include fiberfill for ski jackets and sleeping bags, carpet fibers, and tennis balls. HDPE is converted into a fiber used for sportswear, insulating wrap for new buildings, and very durable shipping containers.
74. Proteins, DNA, RNA
76.

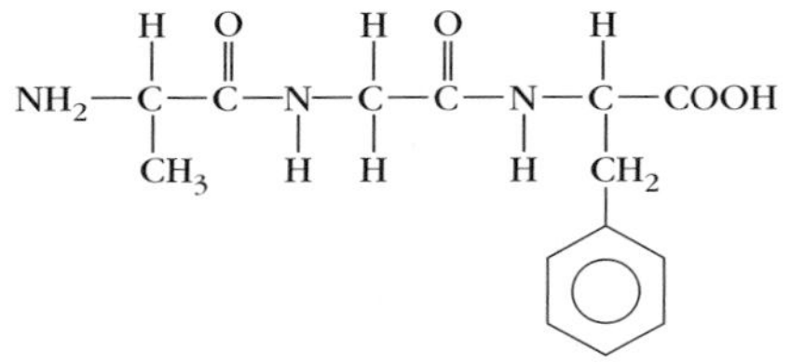

$NH_2-CH(CH_3)-C(=O)-NH-CH_2-C(=O)-NH-CH(CH_2C_6H_5)-COOH$

78. (a) A monosaccharide is a molecule composed of one simple sugar molecule, while disaccharides are molecules composed of two simple sugar molecules.
(b) Disaccharides have only two simple sugar molecules whereas polysaccharides have many.
80. Starch and cellulose
82. (a) Glycogen contains glucose linked together with the glycosidic linkages in "*cis*-positions," and cellulose contains glucose with the glycosidic linkages in "*trans*-positions." Humans do not have the enzyme required to break the *trans*-linkage in cellulose.
(b) Cows have the enzymes for breaking the *trans*-linkage of cellulose.
84. It will show up earlier on the chromatograph. Polar molecules would not be attracted to the stationary phase, so they would exit the chamber more quickly.
86. $CH_3CH_2CH_2CH_2CH_2CH_2CH_2CH_2CH_2CH_2OH$ is a larger molecule than CH_3CH_2OH. The polar alcohol group will interact well with the water; however, the nonpolar end of the molecule will not. The longer nonpolar end of the decanol will not be miscible in water, lowering the solubility compared to smaller, more polar ethanol.
88. Vulcanized rubber has short chains of sulfur atoms that bond together the polymer chains of natural rubber.
90.

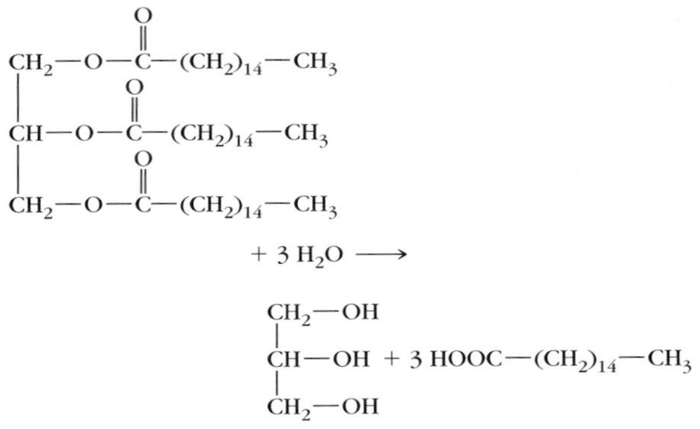

$$\begin{array}{l} CH_2-O-C(=O)-(CH_2)_{14}-CH_3 \\ CH-O-C(=O)-(CH_2)_{14}-CH_3 \\ CH_2-O-C(=O)-(CH_2)_{14}-CH_3 \end{array} + 3\,H_2O \longrightarrow \begin{array}{l} CH_2-OH \\ CH-OH \\ CH_2-OH \end{array} + 3\,HOOC-(CH_2)_{14}-CH_3$$

92. (a) $2\,C_8H_{18}(\ell) + 25\,O_2(g) \rightarrow 16\,CO_2(g) + 18\,H_2O(g)$
(b) 1.4×10^3 L CO_2
94. Glycogen has the glycosidic linkages in "*cis*-positions" whereas cellulose has glycosidic linkages in "*trans*-positions."

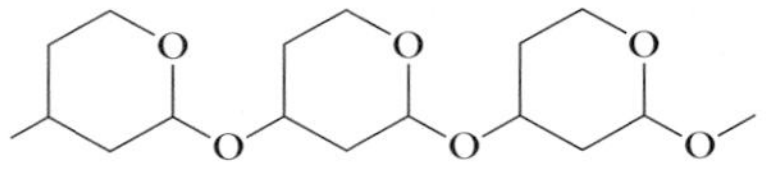

Glycogen

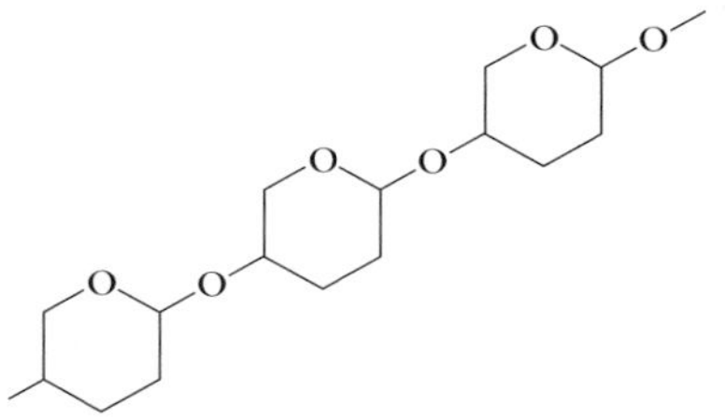

Cellulose

96.

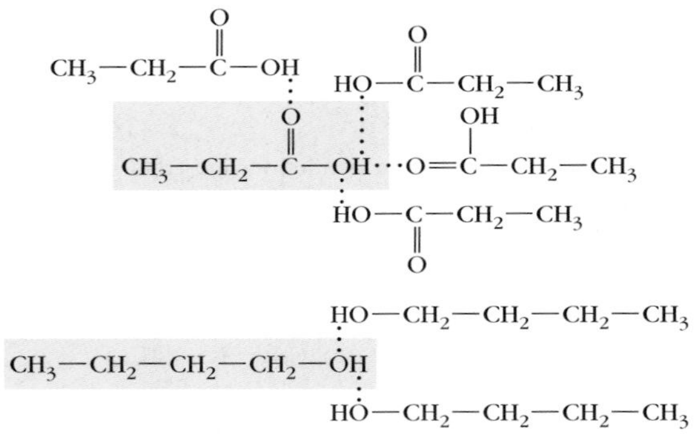

$CH_3-CH_2-C(=O)-OH \cdots$ hydrogen bonded to neighboring propanoic acid molecules: $CH_3-CH_2-C(=O)-OH \cdots O{=}C(OH)-CH_2-CH_3$; $HO-C(=O)-CH_2-CH_3$

$HO-CH_2-CH_2-CH_2-CH_3 \cdots CH_3-CH_2-CH_2-CH_2-OH \cdots HO-CH_2-CH_2-CH_2-CH_3$

Propanoic acid boils at higher *T*, due to more H-bonding.

98. $CH_3-C{\equiv}C-H$
100.

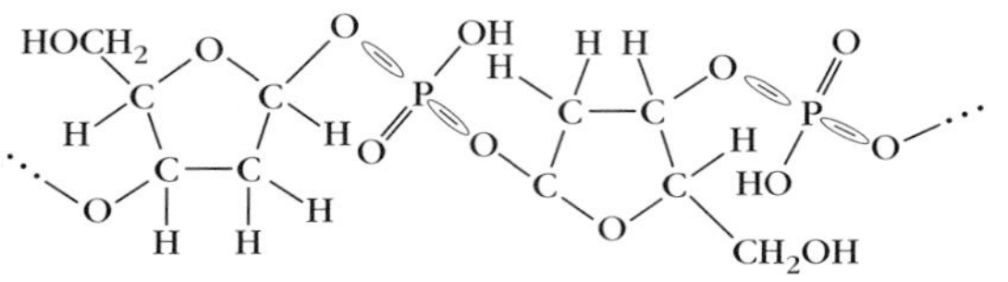

(Sugar–phosphate backbone structure: $HOCH_2$, O, P, OH, HO, CH_2OH, H)

102. Some data that you would need to know are: sources and amounts of CO_2 generated over time to determine additional CO_2; photosynthesis rate of depletion per tree per year; average number of trees per acre; the number of acres of land in Australia that could support trees; and the allowable tree density. Other data besides these may need to be contemplated, so consider it a challenge to think of other things you would need to know.

103. (a) 3×10^8 light bulbs (b) 2×10^3 toasters
105. Less energy; single bonds are easier to stretch.
107. (a) CH_2 (b) Alkene
(c) $H_2C{=}CH_2$ (H and H on each C)
(Other structures also work.)
109.

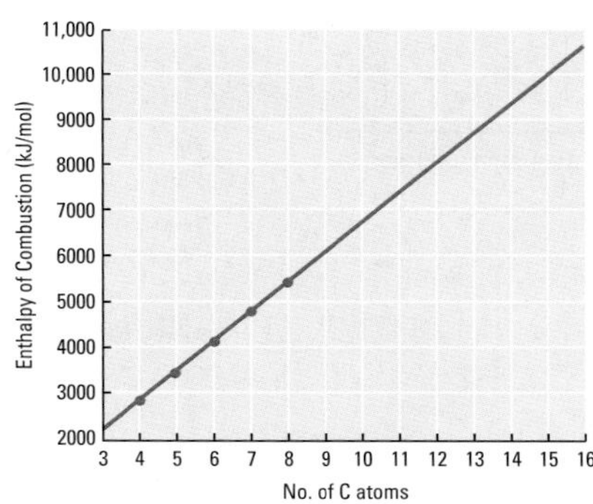

Estimate: C_3, 2200 kJ/mol; C_9, 6100 kJ/mol; C_{16}, 10,700 kJ/mol

111.

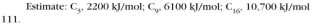

$-\!\!(CH_2-CHCl-CH_2-CHCl)_n\!\!-$ $-\!\!(CH_2-CHCl-CHCl-CH_2)_n\!\!-$

$-\!\!(CH_2-CHCl-CHCl-CH_2-CHCl-CH_2)_n\!\!-$

113. (a) $-O-Si(CH_3)_2-O-Si(CH_3)_2-O-Si(CH_3)_2-O-Si(CH_3)_2-O-$

(b) Condensation polymerization

115.

$-O-CH_2-C(=O)-O-CH(CH_3)-C(=O)-O-CH_2-C(=O)-O-CH(CH_3)-C(=O)-O-CH_2-C(=O)-O-CH(CH_3)-C(=O)-O-CH_2-C(=O)-O-CH(CH_3)-C(=O)-O-$

Chapter 13

9. (d) dissolves fastest; (a) dissolves slowest. Rate of dissolving is larger when the grains of sugar are smaller, because there is more surface contact with the solvent.
11. (a)

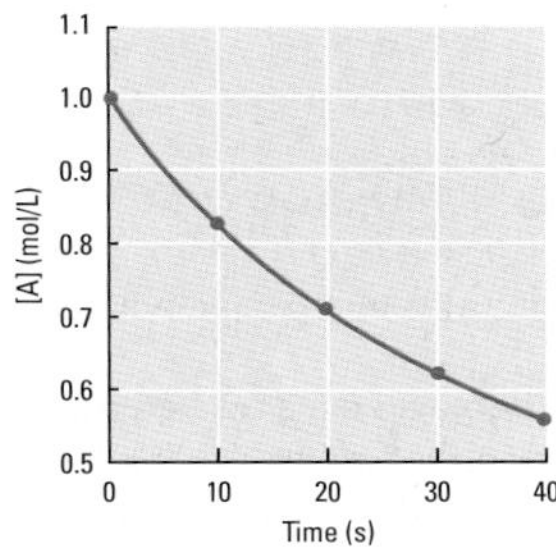

0.0167 M/s, 0.0119 M/s, 0.0089 M/s, 0.0070 M/s

The concentration of the reactant is decreasing.
(b) The rate of change of [B] is twice as fast as the rate of change of [A].
(c) 0.0238 M/s

13. (a) 0.23 mol/L·h (b) 0.20 mol/L·h (c) 0.161 mol/L·h
(d) 0.12 mol/L·h (e) 0.090 mol/L·h (f) 0.066 mol/L·h
15. (a) Calculate the average concentration for each time interval and plot average concentration vs. average rate:

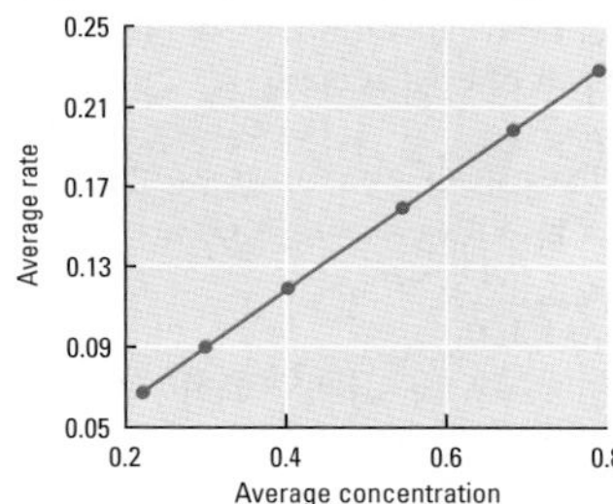

The linear relationship shows: Rate = $k[N_2O_5]$.
(b) $k = 0.29\ h^{-1}$

17.

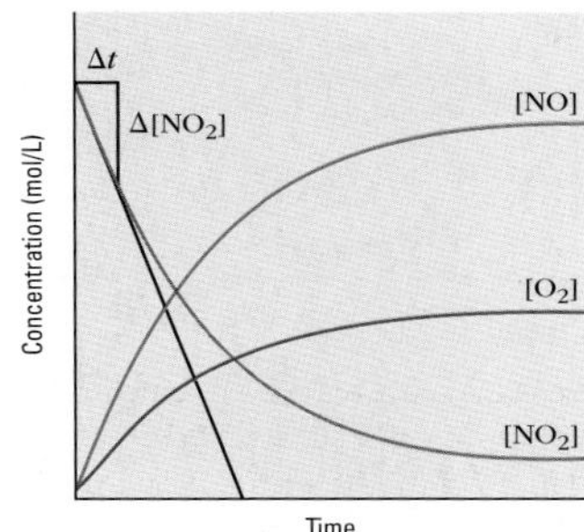

(a) To calculate initial rate, obtain values for Δt and Δ(concentration) near initial time, where the curve can still be approximated by a straight line (the dashed line, above) then divide Δ(concentration) by Δt and change the sign to make it positive, e.g.,

$$\text{Initial rate} = \frac{-\Delta[NO_2]}{\Delta t}$$

(b) The curves on the graph become horizontal after a very long time, so the final rate will be zero.

19. (a) The rate increases by a factor of nine.
(b) The rate will be one fourth as fast.
21. (a) Rate = $k[NO_2]^2$
(b) The rate will be one fourth as fast.
(c) The rate is unchanged.
23. (a) (i) 9.0×10^{-4} M/h, (ii) 1.8×10^{-3} M/h, (iii) 3.6×10^{-3} M/h
(b) If the initial concentration of $Pt(NH_3)_2Cl_2$ is high, the rate of disappearance of $Pt(NH_3)_2Cl_2$ is high. If the initial concentration of $Pt(NH_3)_2Cl_2$ is low, the rate of disappearance of $Pt(NH_3)_2Cl_2$ is low. The rate of disappearance of $Pt(NH_3)_2Cl_2$ is directly proportional to $[Pt(NH_3)_2Cl_2]$.
(c) The rate law shows direct proportionality between rate and $[Pt(NH_3)_2Cl_2]$.
(d) When the initial $[Pt(NH_3)_2Cl_2]$ is high, the rate of appearance of Cl^- is high. When the initial concentration is low, the rate of appearance of Cl^- is low. The rate of appearance of Cl^- is directly proportional to $[Pt(NH_3)_2Cl_2]$.
25. (a) Rate = k[I][II]
(b) $k = 1.04 \dfrac{L}{mol \cdot s}$
27. (a) The order with respect to NO is one and with respect to H_2 is one.
(b) The reaction is second-order.

(c) Rate = k[NO][H_2]
(d) $k = 2.4 \times 10^2$ L mol^{-1} s^{-1}
(e) Initial rate = 1.5×10^{-2} mol L^{-1} s^{-1}

29. (a) First-order in A and third-order in B, fourth-order overall
(b) First-order in A and first-order in B, second-order overall
(c) First-order in A and zero-order in B, first-order overall
(d) Third-order in A and first-order in B, fourth-order overall

31. Equation (a) cannot be right

33. (a) Rate = k[phenyl acetate]
(b) First-order in phenyl acetate
(c) $k = 1.259$ s^{-1} (d) $0.13 \frac{\text{mol}}{\text{L}\cdot\text{s}}$

35. (a) Rate = k[NO_2][CO]
(b) First-order in both NO_2 and CO
(c) $4.2 \times 10^8 \frac{\text{L}}{\text{mol}\cdot\text{h}}$

37. (a) Rate = $k[CH_3COCH_3][H_3O^+]$; first-order in both H_3O^+ and CH_3COCH_3, and zero-order in Br_2
(b) $4 \times 10^{-3} \frac{\text{L}}{\text{mol}\cdot\text{s}}$ (c) $2 \times 10^{-5} \frac{\text{mol}}{\text{L}\cdot\text{s}}$

39. 

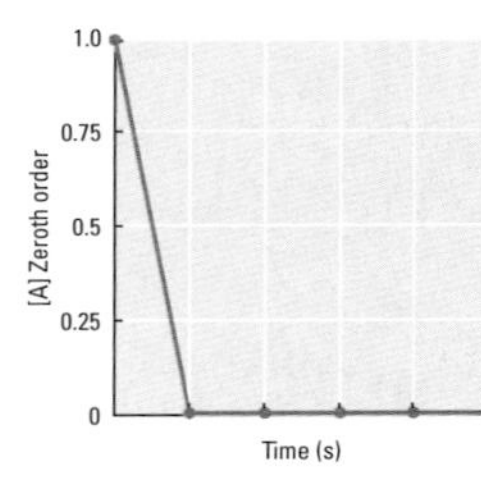

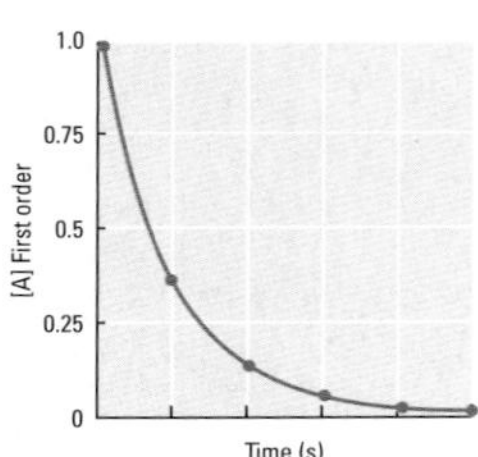

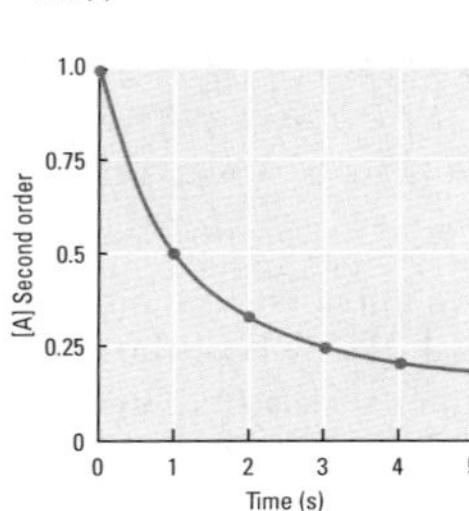

Only the first graph constructed here can be compared directly with one of the graphs in Figure 13.5—in particular, Figure 13.5(c). They both show a downward linear functional dependence of [A] versus time, which is characteristic of zero-order reactions.

41. (a) 0.16 mol/L (b) 90. s (c) 120 s

43. $t = 5.49 \times 10^4$ s

45. (a) Not elementary (b) Bimolecular and elementary
(c) Not elementary (d) Unimolecular and elementary

47. NO + O_3, NO is an asymmetric molecule and Cl is a symmetric atom.

49. E_a = 19 kJ/mol, Ratio = 1.8

51. 10.7 times faster.

53. (a) E_a = 120. kJ/mol, A = 1.22×10^{14} s^{-1}
(b) $k = 1.7 \times 10^{-3}$ s^{-1}

55. (a) E_a = 22.2 kJ/mol, $A = 6.66 \times 10^7 \frac{\text{L}^2}{\text{mol}^2\cdot\text{s}}$
(b) $k = 8.39 \times 10^4 \frac{\text{L}^2}{\text{mol}^2\cdot\text{s}}$

57. (a) 3×10^{-20} (b) 4×10^{-16}
(c) 4×10^{-10} (d) 1.9×10^{-6}

59. (a) $8 \times 10^{-4} \frac{\text{mol}}{\text{L}\cdot\text{s}}$ (b) $3 \times 10^1 \frac{\text{mol}}{\text{L}\cdot\text{s}}$

61. 3×10^2 kJ/mol

63. Exothermic

65. (a) 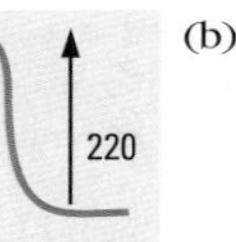

(b) 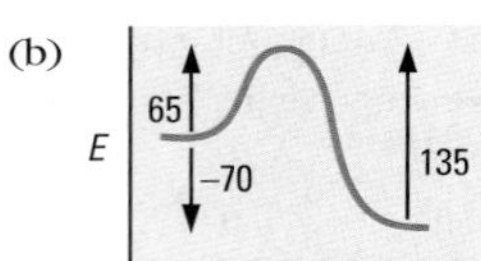

(c) 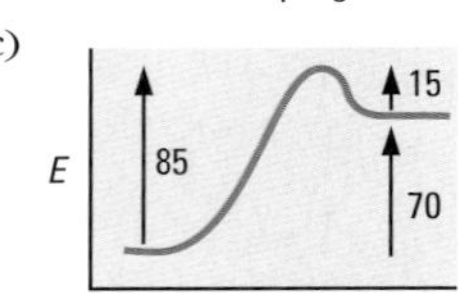

66. (a) 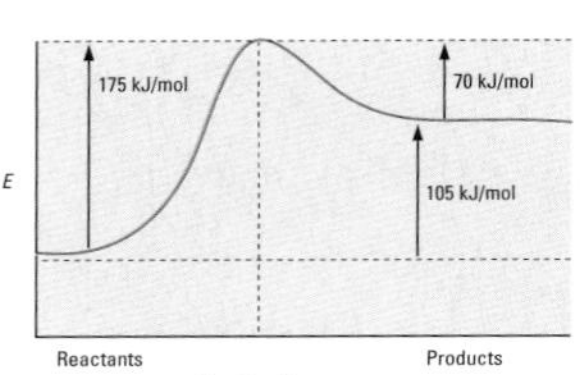

(b) 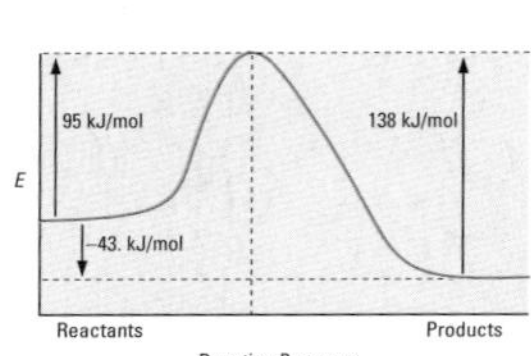

(c) 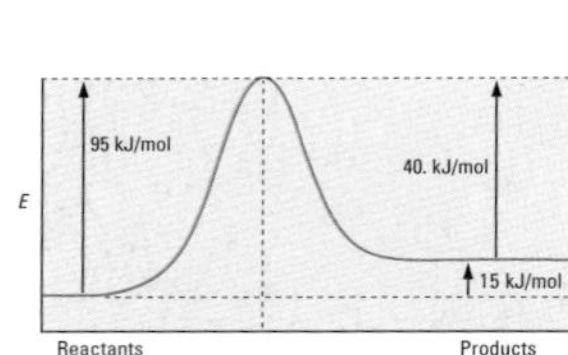

67. (a) Reaction (b) (b) Reaction (c)

69. (a) Reaction (c) (b) Reaction (a)

71. (a) Rate = k[NO][NO_3] (b) Rate = k[O][O_3]
(c) Rate = k[$(CH_3)_3CBr$] (d) Rate = k[HI]^2

73. (a)
$$NO_2 + F_2 \longrightarrow FNO_2 + \cancel{F}$$
$$+ NO_2 + \cancel{F} \longrightarrow FNO_2$$
$$2\,NO_2 + F_2 \longrightarrow 2\,FNO_2$$
(b) The first step is rate-determining.

75. (a) Rate = k'[NO]^2[Cl_2]
(b) $NO + NO \rightleftharpoons N_2O_2$ fast
$N_2O_2 + Cl_2 \rightarrow 2\,NOCl$ slow
(c) $NO + Cl_2 \rightarrow NOCl + Cl$ slow
$NO + Cl \rightarrow NOCl$ fast
There can be other answers for (b) and (c).

77. (a) $CH_3COOH + H_2O \rightleftharpoons CH_3COOH + CH_3OH$
(b) rate = $k'[CH_3COOH][H_3O^+]$
(c) Catalyst: H_3O^+
(d) Intermediates: $H_3C(OH)OCH_3^+$, $H_3C(H_2O)(OH)OCH_3^+$, $H_3C(OH)_2OHCH_3^+$, and H_2O.

79. Only mechanism (a) is compatible with the observed rate law.

81. (a) is true.

83. (a) and (c); (a) homogeneous (c) heterogeneous

85. Enzyme: A protein that catalyzes a biological reaction.
Cofactor: A small molecule that interacts with an enzyme to allow it to catalyze a biological reaction.
Polypeptides: Molecules made by polymerizing several amino acids.
Monomer: The small molecule reactant in the formation of a polymer.

Polysaccharide: An ether-linked chain of monomer sugars.
Lysozyme: An enzyme that speeds the hydrolysis of certain polysaccharides.
Substrate: The reactant of a reaction catalyzed by an enzyme.
Active site: That part of an enzyme where the substrate binds to the enzyme in preparation for conversion into products.
Proteins: Large polypeptide molecules that serve as structural and functional molecules in living organisms.
Inhibition: When a molecule that is not the substrate enters and occupies the active site of an enzyme preventing substrate binding and reaction.

87. 30. times faster
89. The active site on the enzyme (where the substrate binds to form an intermediate complex) is designed to accommodate the substrate's four terminal O atoms. Malonate and oxalate also have four terminal O atoms. When either of these two ions occupy the active site of an enzyme, they prevent the substrate from binding. With a smaller number of active sites free for reaction, the rate of the succinate dehydrogenation reaction decreases.
91. Catalysts make possible the production of vital products. Without them, many necessities and luxuries could not be made efficiently, if at all.
93. (a) To maximize the surface area and increase contact with the heterogeneous catalyst
(b) Catalysis happens only at the surface of the metal; strips or rods with less surface area would be less efficient and more costly.
95.

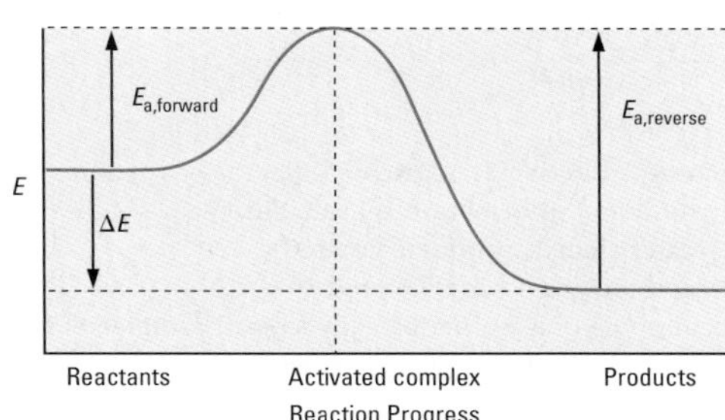

$\Delta E = E_{a,\text{ forward}} - E_{a,\text{ reverse}}$

97. (a) True
(b) False. "The reaction rate decreases as a first-order reaction proceeds at a constant temperature."
(c) True
(d) False. "As a second-order reaction proceeds at a constant temperature, the rate constant does not change."
(Other corrections for (b) and (d) are also possible.)
99. Rate $= k[H_2][NO]^2$
101. (a) NO is second-order, O_2 is first-order.
(b) Rate $= k[NO]^2[O_2]$ (c) $25 \dfrac{L^2}{mol^2 \cdot s}$
(d) $7.8 \times 10^{-4} \dfrac{mol}{L \cdot s}$ (e) $-\dfrac{\Delta[NO]}{\Delta t} = 2.0 \times 10^{-4} \dfrac{mol}{L \cdot s}$,
$+\dfrac{\Delta[NO_2]}{\Delta t} = 2.0 \times 10^{-4} \dfrac{mol}{L \cdot s}$
103. (a) First-order in $HCrO_4^-$, first-order in H_2O_2, and first-order in H_3O^+
(b) Cancel intermediates, H_2CrO_4 and $H_2CrO(O_2)_2$, and add the three reactions.
(c) Second step
105. The catalytic role of the chlorine atom (produced from the light decomposition of chlorofluorocarbons) in the mechanism for the destruction of ozone indicates that even small amounts of CFCs released into the atmosphere pose a serious risk.
107. Curve A represents $[H_2O(g)]$ increase with time, Curve B represents $[O_2(g)]$ increase with time, and Curve C represents $[H_2O_2(g)]$ decrease with time.
109. Snapshot (b)
111. Rate $= k[A]^3[B][C]^2$
113. E_a is very, very small—approximately zero.
115. The Pb must permanently bind with the Pt atoms on the surface of the metal occupying the active sites of the catalyst.
117. (a) Rate $= 2.4 \times 10^{-7}$ mol L^{-1} s^{-1} $+ k[BSC][F^-]$
(b) 0.3 L mol^{-1} s^{-1}
119. (a) 2.8×10^3 s (b) 1.4×10^4 s (c) 2.0×10^4 s
121. (a) $E_a = 3 \times 10^2$ kJ (b) $t = 2 \times 10^3$ s
123. Note: there are other correct answers to this question; the following are examples:
(a) $CH_3CO_2CH_3 + H_3O^+ \longrightarrow CH_3COHOCH_3 + H_2O$ slow
$CH_3COHOCH_3 + H_2O \longrightarrow CH_3COH(OH_2)OCH_3$ fast
$CH_3COH(OH_2)OCH_3 \longrightarrow CH_3C(OH)_2 + CH_3OH$ fast
$CH_3C(OH)_2 + H_2O \longrightarrow CH_3COOH + H_3O^+$ fast
(b) $H_2 + I_2 \longrightarrow 2\ HI$ slow
(c) $H_2 + Pt(s) \longrightarrow PtH_2$ fast
$PtH_2 + I_2 \longrightarrow PtH_2I_2$ slow
$PtH_2I_2 + O_2 \longrightarrow PtI_2O + H_2O$ fast
$PtI_2O + H_2 \longrightarrow Pt(s) + I_2 + H_2O$ fast
(d) $H_2 \longrightarrow 2\ H$ fast
$H + CO \longrightarrow HCO$ slow
$HCO + H \longrightarrow H_2CO$ fast
125. Approximately 26 times faster
127. Estimated $E_a = 402 \times 10^{-21}$ J/molecule; it is the same to one significant figure as the E_a given in Figure 13.7.
131. (i) Define the reaction rate in terms of [A]: Rate $= -\dfrac{\Delta[A]}{\Delta t}$
(ii) Write the rate law using [A], k and t: $-\dfrac{\Delta[A]}{\Delta t} = k[A]$.

Calculus uses "d" instead of "Δ": $-\dfrac{d[A]}{dt} = k[A]$.

(iii) Separate the variables: $\dfrac{d[A]}{[A]} = -kdt$
(iv) Do a separate definite integral on each side of the equation:
$$\int_0^t \frac{d[A]}{[A]} = -k\int_0^t dt$$
$\ln[A]_t - \ln[A]_0 = -k(t_t - t_0)$
(v) With $t = t_t - t_0$:
$\ln[A]_t - \ln[A]_0 = -kt$
$\ln[A]_t = -kt + \ln[A]_0$

The above equation is in the proper form, with $y = \ln[A]_t$, $m = -k$, $x = t$, and $b = \ln[A]_0$.

(vi) $t = t_{1/2}$, when $[A]_t = \dfrac{1}{2}[A]_0$

$\ln\dfrac{1}{2}[A]_0 - \ln[A]_0 = -kt_{1/2}$
$\ln[A]_0 - \ln 2 - \ln[A]_0 = -kt_{1/2}$
$-\ln 2 = -kt_{1/2}$
$t_{1/2} = \dfrac{-\ln 2}{-k} = \dfrac{\ln 2}{k}$

Chapter 14

7. There are many answers to this question; this is an example: Prepare a sample of N_2O_4 in which the N atoms are the heavier isotopes ^{15}N. Introduce the heavy isotope of N_2O_4 into an equilibrium mixture of N_2O_4 and NO_2. Use spectroscopic methods,

such as infrared spectroscopy to observe the distribution of the radioisotope among the reactants and products.

9. (a) 0 °C (b) Dynamic equilibrium; Molecules at the interface between the water and the ice may detach from the ice and enter the liquid phase or may attach to the solid phase leaving the liquid phase.

11.

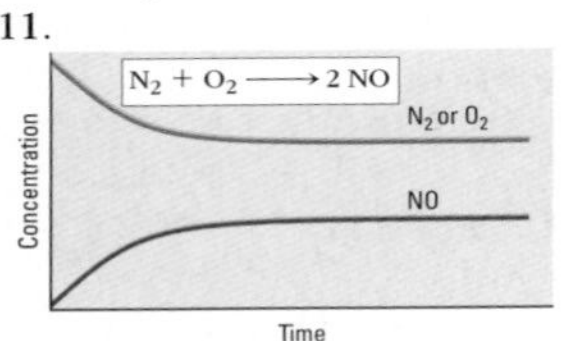

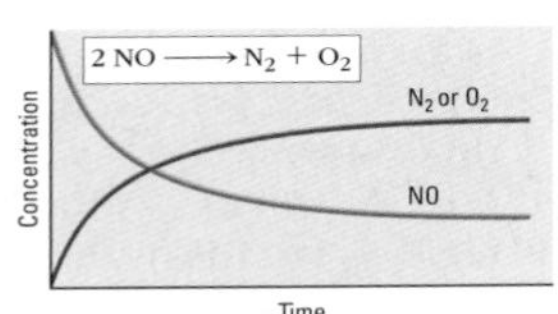

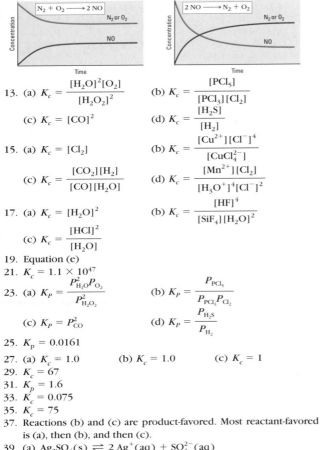

13. (a) $K_c = \dfrac{[H_2O]^2[O_2]}{[H_2O_2]^2}$ (b) $K_c = \dfrac{[PCl_5]}{[PCl_3][Cl_2]}$

(c) $K_c = [CO]^2$ (d) $K_c = \dfrac{[H_2S]}{[H_2]}$

15. (a) $K_c = [Cl_2]$ (b) $K_c = \dfrac{[Cu^{2+}][Cl^-]^4}{[CuCl_4^{2-}]}$

(c) $K_c = \dfrac{[CO_2][H_2]}{[CO][H_2O]}$ (d) $K_c = \dfrac{[Mn^{2+}][Cl_2]}{[H_3O^+]^4[Cl^-]^2}$

17. (a) $K_c = [H_2O]^2$ (b) $K_c = \dfrac{[HF]^4}{[SiF_4][H_2O]^2}$

(c) $K_c = \dfrac{[HCl]^2}{[H_2O]}$

19. Equation (e)

21. $K_c = 1.1 \times 10^{47}$

23. (a) $K_P = \dfrac{P^2_{H_2O}P_{O_2}}{P^2_{H_2O_2}}$ (b) $K_P = \dfrac{P_{PCl_5}}{P_{PCl_3}P_{Cl_2}}$

(c) $K_P = P^2_{CO}$ (d) $K_P = \dfrac{P_{H_2S}}{P_{H_2}}$

25. $K_p = 0.0161$

27. (a) $K_c = 1.0$ (b) $K_c = 1.0$ (c) $K_c = 1$

29. $K_c = 67$

31. $K_p = 1.6$

33. $K_c = 0.075$

35. $K_c = 75$

37. Reactions (b) and (c) are product-favored. Most reactant-favored is (a), then (b), and then (c).

39. (a) $Ag_2SO_4(s) \rightleftharpoons 2\ Ag^+(aq) + SO_4^{2-}(aq)$
$K = [Ag^+]^2[SO_4^{2-}] = 1.7 \times 10^{-5}$
$Ag_2S(s) \rightleftharpoons 2\ Ag^+(aq) + S^{2-}(aq)$
$K = [Ag^+]^2[S^{2-}] = 6 \times 10^{-30}$
(b) $Ag_2SO_4(s)$ (c) $Ag_2S(s)$

41. (a)

	butane $\rightleftharpoons$	2-methylpropane
Conc. initial	0.100 mol/L	0.100 mol/L
Change conc.	$-x$	$+x$
Equilibrium conc.	$0.100 - x$	$0.100 + x$

(b) $K_c = \dfrac{[\text{2-methylpropane}]}{[\text{butane}]} = 2.5 = \dfrac{0.100 + x}{0.100 - x}$ $x = 0.043$

(c) [2-methylpropane] = 0.024 mol/L, [butane] = 0.010 mol/L

43. 3.39 g $C_6H_{12}(g)$

45. $[HI] = 4 \times 10^{-2}$ M; $[H_2] = [I_2] = 5 \times 10^{-3}$ M

47. (a) 1.94 mol HI (b) 1.92 mol HI (c) 1.98 mol HI

49. (a) $[CO] = [H_2O] = 0.33$ mol/L, $[CO_2] = [H_2] = 0.667$ mol/L
(b) $[CO] = [H_2O] = 1.33$ mol/L, $[CO_2] = [H_2] = 0.67$ mol/L

51. 1.15%

53. (a) No (b) Proceed toward products

55. (a) No (b) Proceed toward reactants
(c) $[N_2] = [O_2] = 0.002$ M; $[NO] = 0.061$ M

57. Choice (b); because heat is a reactant, increasing the temperature drives the reaction toward the products.

59.

Change	$[Br_2]$	[HBr]	K_c	K_p
Some H_2 is added to the container.	Decrease	Increase	No change	No change
Temperature of the gases in the container is increased.	Increase	Decrease	Decrease	Decrease
The pressure of HBr is increased.	Increase	Increase	No change	No change

61. (a) No change (b) Left (c) Left

63. (a) Right (b) Left (c) Right

65. (a) Decrease
(b)

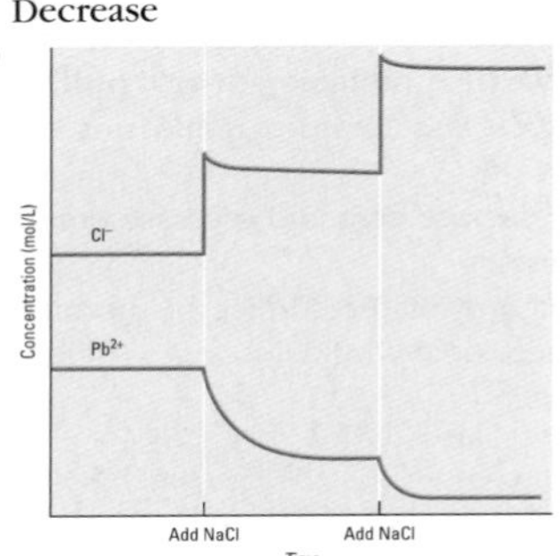

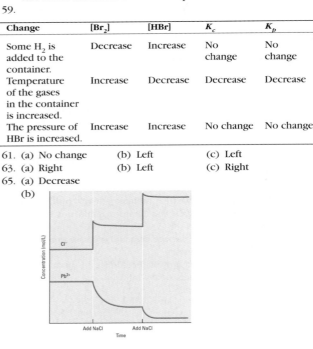

67. (a) Energy effect (b) Entropy effect (c) Neither

69. (a) Insufficient information is available.
(b) Greater than 1, products favored
(c) Less than 1, reactants favored

71. A reaction will only go significantly toward products if it is product-favored. If a reaction is accompanied by an increase in entropy (a favorable entropy effect) and the products are lower in energy (a favorable enthalpy effect), then products are favored. However, if the reaction is favored only by an entropy effect, then it needs high temperatures to assist the endothermic process. If the reaction is favored only by an enthalpy effect, then it needs low temperatures to keep randomness at a minimum.

73. (a) First reaction: $\Delta H = -296.830$ kJ, second reaction: $\Delta H = -197.78$ kJ, third reaction: $\Delta H = -132.44$ kJ
(b) All three are exothermic, none are endothermic.
(c) None of the reactions have entropy increase, the second and third reaction have entropy decrease, and the first reaction entropy is about the same.
(d) Low temperature favors all three reactions.

75. $K_c = [H^+][OH^-]$, $K_c = \dfrac{[CH_3COO^-][H^+]}{[CH_3COOH]}$,

$K_c = \dfrac{[NH_3]^2}{[N_2][H_2]^3}$, $K_c = \dfrac{[CO_2][H_2]}{[CO][H_2O]}$

First reaction
(a) $(1.0 + x)(1.0 + x) = 1.0 \times 10^{-14}$
(b) Quadratic (c) N/A

Second reaction
(a) $\dfrac{(1.0 + x)(1.0 + x)}{(1.0 - x)} = 1.8 \times 10^{-5}$
(b) Quadratic (c) N/A

Third reaction
(a) $\dfrac{(1.0 + 2x)^2}{(1.0 - x)(1.0 - 3x)^3} = 3.5 \times 10^8$

(b) Not quadratic
(c) Use approximation techniques
Fourth reaction
(a) $\frac{(1.0 + x)(1.0 + x)}{(1.00 - x)(1.00 - x)} = 4.00$
(b) This equation is quadratic.
(c) Not applicable

77. (a)

Species	Br_2	Cl_2	F_2	H_2	N_2	O_2
[*E*] (mol/L)	0.28	0.057	1.44	1.76×10^{-5}	4×10^{-14}	4.0×10^{-6}

(b) F_2; (At this temperature, the lowest bond energy is predicted from the reaction that gives the most products.) Compared to Table 8.2, the product production decreases as the bond energy increases: 156 kJ F_2, 193 kJ Br_2, 242 kJ Cl_2, 436 kJ H_2, 498 kJ O_2, 946 kJ N_2. Lewis structures of F_2, Br_2, Cl_2, H_2 have a single bond and more products are produced than O_2 with double bond and N_2 with triple bonds.

79. (a) $K_c = 0.67$ (b) Same result
81. Cases a and b will have decreased [HI] at equilibrium, and Cases c and d will have increased [HI].
83. It is at equilibrium at 600. K. No more experiments are needed.
85. LeChatlier's Principle says the equilibrium will shift to compensate for changes made. The pressure causes the H_2O molecules to be pushed together. Because water is more dense than ice, some of the ice will melt to the liquid state, reducing the pressure.
87. (a) (iii) (b) (i)
89. In the warmer sample, the molecules would be moving faster and more NO_2 molecules would be seen. In the cooler sample, the molecules would be moving somewhat slower and fewer NO_2 molecules would be seen. In both samples, the molecules are moving very fast. The average speed of gas molecules is commonly hundreds of miles per hour. In both samples, one would see a dynamic equilibrium with some N_2O_4 molecules decomposing and some NO_2 molecules reacting with each other, at equal rates.
91. Diagrams (b), (c), and (d)
93. The following is an example of a diagram. (Other answers are possible.)

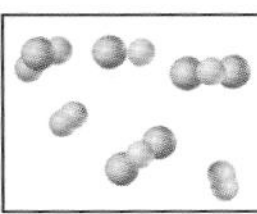

95. Dynamic equilibria with small values of K introduce a small amount of D^+ ions in place of H^+ ions in the place of the acidic hydrogen.
97. Dynamic equilibria representing the decomposition of the dimer $N_2O_4(g)$ produce $NO_2(g)$ and $*NO_2(g)$, which will occasionally recombine into the mixed dimer, $O_2N*{-}NO_2(g)$
$O_2N{-}NO_2(g) \rightleftharpoons 2\,NO_2(g)$
$O_2N*{-}*NO_2(g) \rightleftharpoons 2\,*NO_2(g)$
$O_2N*{-}NO_2(g) \rightleftharpoons *NO_2(g) + NO_2(g)$
99. (a) $K_c = 0.168$ (b) 0.661 atm (c) 0.514
101. (a) $[PCl_3] = [Cl_2] = 0.14$ mol/L, $[PCl_5] = 0.005$ mol/L
(b) $[PCl_3] = 0.29$ mol/L, $[Cl_2] = 0.14$ mol/L, $[PCl_5] = 0.01$ mol/L
103. (a) (i) Right (ii) Right (iii) Left
(b) (i) Left (ii) No shift (iii) Right
(c) (i) Right (ii) Right (iii) No shift
(d) (i) Right (ii) Right (iii) Left
105. $K_c = \frac{A_f e^{(-E_{a,f}/RT)}}{A_r e^{(-E_{a,r}/RT)}} = \frac{A_f}{A_r} e^{[(E_{a,r}-E_{a,f})/RT]}$

When a catalyst is added, the activation energies of both the forward and reverse reactions get smaller:

$$K_{c,\,cat} = \frac{A_{f,cat}}{A_{r,cat}} e^{[(E_{a,r,cat}-E_{a,f,cat})/RT]}$$

The frequency factors did not change and, because the reaction has to conserve energy, the E_a values must each be reduced by the same amount of energy X:

$A_{f,cat} = A_f$ $\quad E_{a,f,cat} = E_{a,f} - X$
$A_{r,cat} = A_r$ $\quad E_{a,r,cat} = E_{a,r} - X$

Notice that

$$E_{a,f,cat} - E_{a,r,cat} = (E_{a,f} - X) - (E_{a,r} - X) = E_{a,f} - X - E_{a,r} + X = E_{a,f} - E_{a,r}$$

Substituting into the $K_{c,cat}$ equation gives

$$K_{c,cat} = \frac{A_f}{A_r} e^{[(E_{a,r}-E_{a,f})/RT]} = K_c$$

Therefore, the equilibrium state, as described quantitatively by the equilibrium constant, does not change with the addition of a catalyst.

107. (a) $H_2(g) + Br_2(g) \rightleftharpoons 2\,HBr(g)$ (b) -103 kJ
(c) Energy effect (d) To the left (e) No effect
(f) The reactants must achieve activation energy. Bromine is a liquid at room temperature.
109. (a) $K_p = 0.03126$; $K_c = 0.0128$
(b) $K_p = 1$
(c) $K_c = 1/(RT_{bp})$

Chapter 15

25. If the solid interacts with the solvent using similar (or stronger) intermolecular forces, it will dissolve readily. If the solute interacts with the solvent using different intermolecular forces than those experienced in the solvent, it will be almost insoluble. For example, consider dissolving an ionic solid in water and oil. The interactions between the ions in the solid and water are very strong, since ions would be attracted to the highly polar water molecule; hence, the solid would have a high solubility. However, the ions in the solid interact with each other much more strongly than the London dispersion forces experienced between the nonpolar hydrocarbons in the oil; hence, the solid would have a low solubility. (Other examples exist.)
27. The dissolving process was endothermic, so the temperature dropped as more solute was added. The solubility of the solid at the lower temperature is lower, so some of the solid did not dissolve. As the solution warmed up, however, the solubility increased again. What remained of the solid dissolved. The solution was saturated at the lower temperature, but is no longer saturated at the current temperature.
29. When an organic acid has a large (non-polar) piece, it interacts primarily using London dispersion intermolecular forces. Since water interacts via hydrogen-bonding intermolecular forces, it would rather interact with itself than with the acid. Hence, the solubility of the large organic acids drops, and some are completely insoluble.
31. The positive H end of the very polar water molecule interacts with the negative ions. The negative O end of the very polar water molecule interacts with the positive ions.

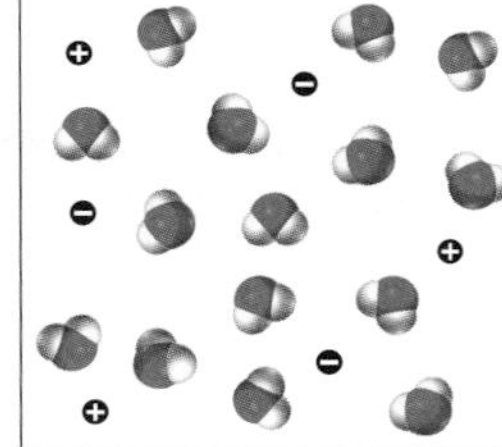

33. 1×10^{-3} M
36. 0.00732% by weight
38. $1 \text{ ppb} = \frac{1 \text{ g part}}{10^9 \text{ g whole}} \times \frac{1 \text{ μg part}}{10^{-6} \text{ g part}} \times \frac{1000 \text{ g whole}}{1 \text{ kg whole}} = \frac{1 \text{ μg part}}{1 \text{ kg whole}}$
40. 90. g ethanol
43. 1.6×10^{-6} g Pb
45. (a) 160. g NH_4Cl (b) 83.9 g KCl (c) 7.46 g Na_2SO_4
46. (a) 0.762 M (b) 0.174 M (c) 0.0126 M (d) 0.167 M
47. (a) 0.180 M (b) 0.683 M (c) 0.0260 M (d) 0.0416 M
49. 96% H_2SO_4
51. 0.1 M NaCl
53. 59 g
55. 1.2×10^{-7} M
59. 100.26 °C
61. (a) < (d) < (b) < (c)
63. $T_f = -1.65$ °C, $T_b = 100.46$ °C
65. $X_{H_2O} = 0.79999$, 712 g sucrose
67. 1.9×10^2 g/mol
69. 3.6×10^2 g/mol; $C_{20}H_{16}Fe_2$
71. 1.8×10^2 g/mol, $C_{14}H_{10}$
73. (a) 2.5 kg (b) 104.2 °C
75. 29 atm
78. 5×10^{-4} g As
80. No
82. The lime-soda process relies on the precipitation of insoluble compounds to remove the "hard water" ions. The ion-exchange process relies on the high charge of the "hard water" ions to attract them to an ion-exchange resin, thereby removing them from the water.
84. Water is sprayed into the air to oxidize organic substances dissolved in it.
86. Fish take in oxygen by extracting it from the water, while plants take in carbon dioxide by extracting it from the water. The concentrations of these gases in calm water drop unless they are replenished. The concentration of dissolved gases in the water is replenished by bubbling air through the water in the aquarium.
88. 4%
90. Water in the cells of the wood leaked out, since the osmotic pressure inside the cells was less than that of the seawater in which the wood was sitting.
92. 28% NH_3
94. 0.982 *m*, 10.2%
96. (a) = (c) < (d) < (b)
98. 1.77 g
100. (a) $Na_2SO_4(aq) + Ba(NO_3)_2(aq) \rightarrow 2\,NaNO_3(aq) + BaSO_4(s)$
(b) 15.0 mL (c) 12.0 mL
102. (a) (b)

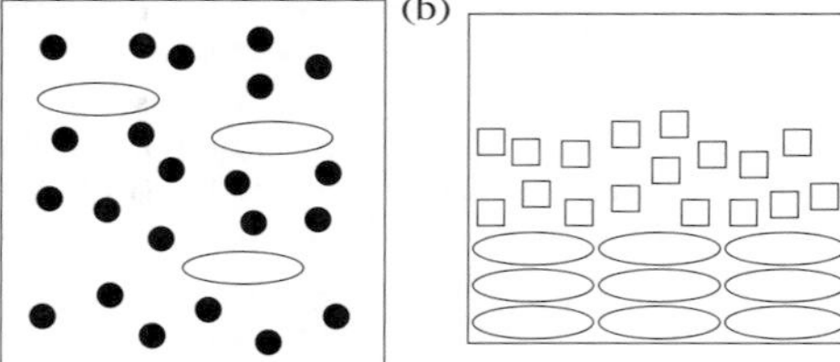

104. (a) Unsaturated (b) Supersaturated (c) Supersaturated (d) Unsaturated

106.

Compound	Mass of Compound	Mass of Water	Mass Fraction of Solute	Weight Percent of Solute	Conc of Solute
Lye	**75.0 g**	125 g	0.375	37.5%	**3.75×10^5 ppm**
Glycerol	33 g	200. g	**0.14**	14%	**1.4×10^5 ppm**
Acetylene	0.0015	**2×10^2 g**	**0.000009**	0.0009%	**9 ppm**

108. Molecules slow down. The reduced motion prevents them from randomly translocating as they had in the liquid state. As a result, the intermolecular forces between one molecule and the next begin to organize them into a crystal form. The presence of a nonvolatile solute disrupts the formation of the crystal. Its size and shape will be different from that of the solute. Intermolecular forces between the solute and solvent are also different from those of solvent molecules with each other. To form the crystalline solid, the solute has to be excluded. If the ice in an iceberg is in regular crystalline form, the water will be pure. Only if the ice is crushed or dirty will other particles be included. So, melting an iceberg will produce relatively pure water.
110. (a) Seawater contains more dissolved solutes than fresh water. The presence of a solute lowers the freezing point. That means a lower temperature is required to freeze the seawater than to freeze fresh water.
(b) Salt added to a mixture of ice and water will lower the freezing point of the water. If the ice cream is mixed at a lower temperature, its temperature will drop faster; hence, it will freeze faster.
112. The empirical and molecular formulas are both $C_{18}H_{24}Cr$.
114. (a) No (b) 108.9°
116. 28 m
118. 0.30 M
120. (a)

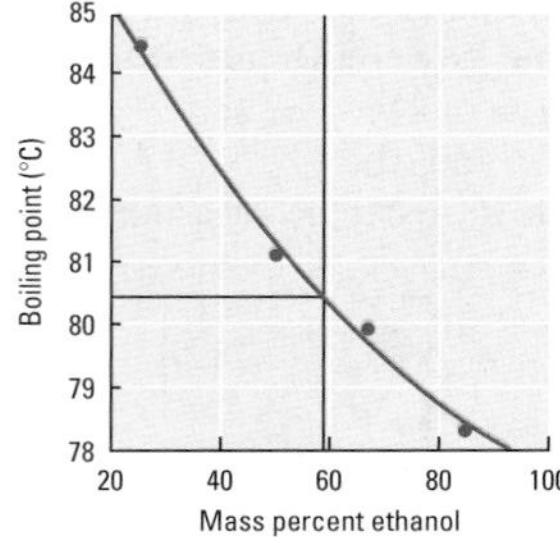

Approx. 80.4 °C
(b) 12.3 M; 44.1 mol/kg
122. (a) Freezes at −3.36°, when the solute concentration is 0.50 *m*.

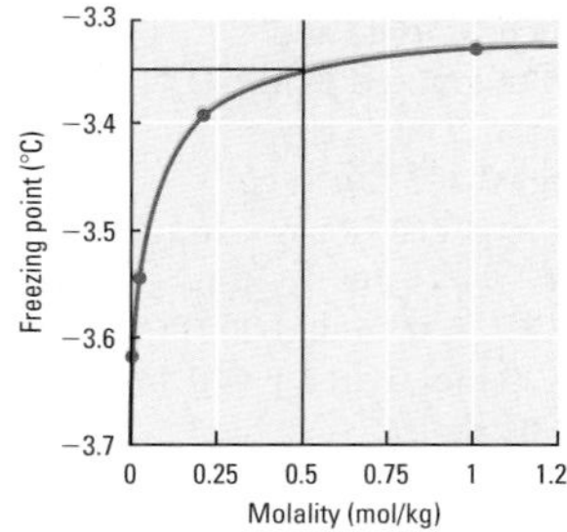

(b) 39,000%, 9,400%, 810%, 80%
124. (a) 6300 ppm, 6300000 ppb
(b) 0.040 M (c) 4.99×10^5 bottles

Chapter 16

12. (a) $HCO_3^- + H_2O \rightleftharpoons CO_3^{2-} + H_3O^+$
(b) $HCl + H_2O \rightleftharpoons Cl^- + H_3O^+$
(c) $CH_3COOH + H_2O \rightleftharpoons CH_3COO^- + H_3O^+$
(d) $HCN + H_2O \rightleftharpoons CN^- + H_3O^+$
14. (a) $HIO + H_2O \rightleftharpoons IO^- + H_3O^+$
(b) $CH_3(CH_2)_4COOH + H_2O \rightleftharpoons CH_3(CH_2)_4COO^- + H_3O^+$
(c) $HOOCCOOH + H_2O \rightleftharpoons HOOCCOO^- + H_3O^+$
$HOOCCOO^- + H_2O \rightleftharpoons {}^-OOCCOO^- + H_3O^+$
(d) $CH_3NH_3^+ + H_2O \rightleftharpoons CH_3NH_2 + H_3O^+$
16. (a) $HSO_4^- + H_2O \rightleftharpoons H_2SO_4 + OH^-$
(b) $CH_3NH_2 + H_2O \rightleftharpoons CH_3NH_3^+ + OH^-$
(c) $I^- + H_2O \rightleftharpoons HI + OH^-$
(d) $H_2PO_4^- + H_2O \rightleftharpoons H_3PO_4 + OH^-$
20. (a) H_2SO_4 (b) HNO_3 (c) $HClO_4$
(d) $HClO_3$ (e) H_2SO_4
22. (a) I^-, iodide, conjugate base
(b) HNO_3, nitric acid, conjugate acid
(c) HCO_3^-, hydrogen carbonate ion, conjugate acid
(d) HCO_3^-, hydrogen carbonate ion, conjugate base
(e) SO_4^{2-}, sulfate ion, conjugate base, H_2SO_4, sulfuric acid, conjugate acid
(f) HSO_3^-, hydrogen sulfite ion, conjugate acid
24. Pairs (b), (c), and (d)
26. (a) Reactant acid = H_2O, reactant base = HS^-, product conjugate acid = H_2S, product conjugate base = OH^-
(b) Reactant acid = NH_4^+, reactant base = S^{2-}, product conjugate acid = HS^-, product conjugate base = NH_3
(c) Reactant acid = HSO_4^-, reactant base = HCO_3^-, product conjugate acid = H_2CO_3, product conjugate base = SO_4^{2-}
(d) Reactant acid = NH_3, reactant base = NH_2^-, product conjugate base = NH_2^-, product conjugate acid = NH_3
28. (a) Reactant acid = CH_3COOH, reactant base = CN^-, product conjugate acid = HCN, product conjugate base = CH_3COO^-
(b) Reactant acid = H_2O, reactant base = O^{2-}, product conjugate acid = OH^-, product conjugate base = OH^-
(c) Reactant acid = H_2O, reactant base = HCO_2^-, product conjugate acid = HCOOH, product conjugate base = OH^-
31. (a) $CO_3^{2-} + H_2O \rightleftharpoons HCO_3^- + OH^-$
$HCO_3^- + H_2O \rightleftharpoons H_2CO_3 + OH^-$
(b) $H_3AsO_4 + H_2O \rightleftharpoons H_2AsO_4^- + H_3O^+$
$H_2AsO_4^- + H_2O \rightleftharpoons HAsO_4^{2-} + H_3O^+$
$HAsO_4^{2-} + H_2O \rightleftharpoons AsO_4^{3-} + H_3O^+$
(c) $NH_2CH_2COO^- + H_2O \rightleftharpoons {}^+NH_3CH_2COO^- + OH^-$
${}^+NH_3CH_2COO^- + H_2O \rightleftharpoons {}^+NH_3CH_2COOH + OH^-$
33. 3×10^{-11} M, basic
35. 3.6×10^{-3} M, acidic
37. pH = 12.40, pOH = 1.60
39. pOH = 12.51
41. 5×10^{-2} M, 2 g HCl
43.

pH	$[H_3O^+]$ (M)	$[OH^-]$ (M)	Acidic or Basic
(a) **6.21**	6.1×10^{-7}	**1.6×10^{-8}**	**Acidic**
(b) **5.34**	**4.5×10^{-6}**	2.2×10^{-9}	**Acidic**
(c) 4.67	**2.1×10^{-5}**	**4.7×10^{-10}**	**Acidic**
(d) **1.60**	2.5×10^{-2}	**4.0×10^{-13}**	**Acidic**
(e) 9.12	**7.6×10^{-10}**	**1.3×10^{-5}**	**Basic**

45. (a) 100 times more acidic (b) 10^4 times more basic
(c) 16 times more basic (d) 2×10^4 times more acidic
47. (a) 5.0×10^{-9} M (b) Basic
49. (a) $F^-(aq) + H_2O(\ell) \rightleftharpoons HF(aq) + OH^-(aq)$ $K = \dfrac{[HF][OH^-]}{[F^-]}$
(b) $NH_3(aq) + H_2O(\ell) \rightleftharpoons NH_4^+(aq) + OH^-(aq)$
$K = \dfrac{[NH_4^+][OH^-]}{[NH_3]}$
(c) $H_2CO_3(aq) + H_2O(\ell) \rightleftharpoons HCO_3^-(aq) + H_3O^+(aq)$
$K = \dfrac{[HCO_3^-][H_3O^+]}{[H_2CO_3]}$
(d) $H_3PO_4(aq) + H_2O(\ell) \rightleftharpoons H_2PO_4^-(aq) + H_3O^+(aq)$
$K = \dfrac{[H_2PO_4^-][H_3O^+]}{[H_3PO_4]}$
(e) $CH_3COO^-(aq) + H_2O(\ell) \rightleftharpoons CH_3COOH(aq) + OH^-(aq)$
$K = \dfrac{[CH_3COOH][OH^-]}{[CH_3COO^-]}$
(f) $S^{2-}(aq) + H_2O(\ell) \rightleftharpoons HS^-(aq) + OH^-(aq)$
$K = \dfrac{[HS^-][OH^-]}{[S^{2-}]}$
51. (a) NH_3 (b) K_2S (c) $NaCH_3COO$ (d) KCN
54. (a) 2.19 (b) 5.13 (c) 8.07 (d) 9.02
(e) 13.30 (f) 1.54 (g) 9.69 (h) 7.00
56. $K_a = 1.6 \times 10^{-5}$
58. $[H_3O^+] = [A^-] = 1.3 \times 10^{-5}$ M, [HA] = 0.040 M
60. $K_a = 1.4 \times 10^{-5}$
62. 8.85
64. $K_b = 1.3 \times 10^{-3}$
66. (a) $C_{10}H_{15}NH_2 + H_2O \rightleftharpoons C_{10}H_{15}NH_3^+ + OH^-$
(b) 10.55
68. 3.28
72. (a) CN^-, product-favored (b) HS^-, reactant-favored
(c) $H_2(g)$, reactant-favored
74. (a) $NH_4^+ + HPO_4^{2-} \rightarrow NH_3 + H_2PO_4^-$, reactant-favored
(b) $CH_3COOH + OH^- \rightarrow CH_3COO^- + H_2O$, product-favored
(c) $HSO_4^- + H_2PO_4^- \rightarrow H_3PO_4 + SO_4^{2-}$, product-favored
(d) $CH_3COOH + F^- \rightarrow CH_3COO^- + HF$, reactant-favored
76. (a) pH < 7 (b) pH > 7 (c) pH = 7
78. (a) pH > 7 (b) pH > 7 (c) pH > 7
80. $[Zn(H_2O)_4]^{2+}$
82. All of them
84. $Na_2CO_3 + 2\ CH_3(CH_2)_{16}COOH \rightarrow$
$2\ CH_3(CH_2)_{16}COONa + H_2O + CO_2$
86. Dishwasher detergent is very basic and should not be used to wash anything by hand, including a car. If it gets into the engine area, it can also dissolve automobile grease and oil, which could prevent the engine from running correctly.
88. Lemon juice is acidic and neutralizes the basic amines. The acid formed from the neutralized base is an ion and not volatile.
90. All three are Lewis bases; CO_2 is a Lewis acid.
92. Cr^{3+} and SO_3 are Lewis acids. CH_3NH_2 is a Lewis base.
94. (a) I_2 is a Lewis acid and I^- is a Lewis base.
(b) BF_3 is a Lewis acid and SO_2 is a Lewis base.
(c) Au^+ is a Lewis acid and CN^- is a Lewis base.
(d) CO_2 is a Lewis acid and H_2O is a Lewis base.
96.

:C̈l—Ï—C̈l: (with :C̈l: bonded below I) T-shaped

It functions as a Lewis acid.

ICl_4^-: four Cl atoms bonded to I Square planar

98. (a) Weak acid (b) Strong base (c) Strong acid
(d) Weak base (e) Strong base (f) Amphiprotic
100. (a) Less than 7 (b) Equal to 7 (c) Greater than 7
102. 2.85
104. (a) Increase (b) Stays the same (c) Stays the same
106. 9.73
108. At 10 °C, pH = 7.27; at 25 °C, pH = 6.998; and at 50 °C, pH = 6.631. At 10 °C, at 25 °C, and at 50 °C the solutions are neutral, since $[H_3O^+] = [OH^-]$.
110. (a) $H_2O > Cl^- = H_3O^+ \gg OH^-$
(b) $H_2O > Na^+ = ClO_4^- \gg H_3O^+ = OH^-$
(c) $H_2O > HNO_2 > H_3O^+ = NO_2^- \gg OH^-$
(d) $H_2O > Na^+ \cong ClO^- > OH^- = HClO \gg H_3O^+$
(e) $H_2O > NH_4^+ \cong Cl^- > H_3O^+ = NH_3 \gg OH^-$
(f) $H_2O > Na^+ = OH^- \gg H_3O^+$
112. Conjugates must differ by just one H^+.
114. Arrhenius Theory: Electron pairs on the solvent water molecules (Lewis base) form a bond with the hydrogen ion (Lewis acid) producing aqueous H_3O^+ ions. Electron pairs on the OH^- ions (Lewis base) form a bond with the hydrogen ion (Lewis acid) in the solvent water molecule, producing aqueous OH^- ions.
Brønsted-Lowry Theory: The H^+ ion from the Brønsted-Lowry acid is bonded to a Brønsted-Lowry base using an electron pair on the base. The electron-pair acceptor, the H1 ion, is the Lewis acid and the electron-pair donor is the Lewis base.
116. (a) 2.01 (b) 23 times
119. Lactic acid sample
120. 0.76 L; No
121. Yes, higher pH
123. Br has a higher electronegativity than H. $ClNH_2$ is weaker, because Cl has a higher electronegativity than Br.
125. Strongest HM > HQ > HZ weakest; $K_{a,HZ} = 1 \times 10^{-5}$, $K_{a,HQ} = 1 \times 10^{-3}$, $K_{a,HM} = 1 \times 10^{-1}$ or larger
127. (a) Weak (b) Weak
(c) Acid/base conjugates (d) 6.26
129. Dilute solution is 9%, saturated solution is 2%.
131. (a) $NH_2CH_2CH_2CH_2CH_2CH_2NH_2$
(b) $NH_3CH_2CH_2CH_2CH_2CH_2NH_3^{2+}$ (c) 9.13
133. The carboxylic acid substituent is an acid (related to pK_a = 2.18). The two NH_2 groups are bases (related to pK_a = 8.95 and 10.53).

Chapter 17

16. (c)
18. (a) pH = 2.1 (b) pH = 7.21 (c) pH = 12.44
20. (a) Lactic acid and lactate ion
(b) Acetic acid and acetate ion
(c) HClO and ClO^-
(d) HCO_3^- and CO_3^{2-}
22. 4.8 g
24. pH = 8.62
26. pH = 9.55, pH = 9.51
28. Sample (b). Only (b) results in a solution containing a conjugate acid/base pair.
30. (a) ΔpH = 0.1 (b) Δ pH = 3.8 (c) Δ pH = 7.25
32. (a) pH = 5.02 (b) pH = 4.99 (c) pH = 4.06
34. At the equivalence point, the solution contains the conjugate base of the weak acid. Weaker acids have stronger conjugate bases. Stronger bases have a higher pH, so the titration of a weaker acid will have a more basic equivalence point.
36. (a) Bromothymol blue
(b) Phenolphthalein
(c) Methyl red
(d) Bromothymol blue. Suitable pH color changes

38. 0.0253 M HCl
40. 93.6%
42. (a) 29.2 mL (b) 600. mL (c) 1.20 L (d) 2.7 mL
44. (a) pH = 3.90 (b) pH = 8.45 (c) pH = 12.15
46. (a) pH = 0.824 (b) pH = 1.30 (c) pH = 3.8
(d) pH = 7.000 (e) pH = 10.2 (f) 12.48

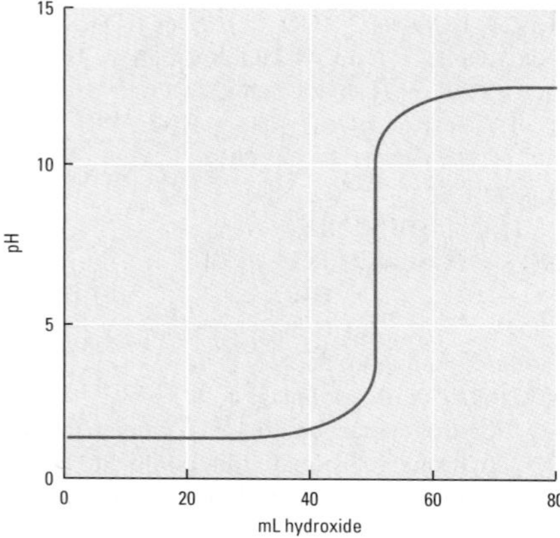

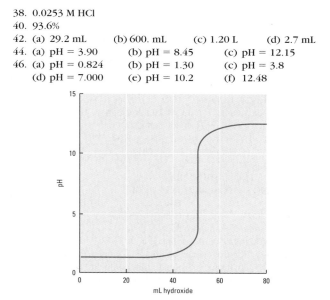

48. NO_2 reaction: $2\,NO_2(g) + H_2O(g) \rightarrow HNO_3(g) + HNO_2(g)$
SO_3 reactions: $2\,SO_2(g) + O_2(g) \rightarrow 2\,SO_3(g)$
$SO_3(g) + H_2O(g) \rightarrow H_2SO_4(g)$
50. $CaCO_3(s) + 2\,H^+(aq) \rightarrow Ca^{2+}(aq) + CO_2(g) + H_2O(\ell)$
54. (a) $FeCO_3(s) \rightleftharpoons Fe^{2+}(aq) + CO_3^{2-}(aq)$ $K_{sp} = [Fe^{2+}][CO_3^{2-}]$
(b) $Ag_2SO_4(s) \rightleftharpoons 2\,Ag^+(aq) + SO_4^{2-}(aq)$ $K_{sp} = [Ag^+]^2[SO_4^{2-}]$
(c) $Ca_3(PO_4)_2(s) \rightleftharpoons 3\,Ca^{2+}(aq) + 2\,PO_4^{3-}(aq)$
$K_{sp} = [Ca^{2+}]^3[PO_4^{3-}]^2$
(d) $Mn(OH)_2(s) \rightleftharpoons Mn^{2+}(aq) + 2\,OH^-(aq)$
$K_{sp} = [Mn^{2+}][OH^-]^2$
55. $K_{sp} = 3.0 \times 10^{-18}$
56. $K_{sp} = 2.22 \times 10^{-4}$
58. $K_{sp} = 2.2 \times 10^{-12}$
60. $K_{sp} = 1.7 \times 10^{-5}$
62. 6.2×10^{-11} M Zn^{2+}
64. 3.1×10^{-5} mol/L
66. (a) $K_{sp} = 1 \times 10^{-11}$ (b) $[OH^-]$ must be 0.008 M or higher.
68. 4.5×10^{-9} M
70. pH = 9.0
72. (a) $Ag^+(aq) + 2\,CN^-(aq) \rightleftharpoons [Ag(CN)_2]^-(aq)$
$$K_f = \frac{[[Ag(CN)_2]^-]}{[Ag^+][CN^-]^2}$$
(b) $Cd^{2+}(aq) + 4\,NH_3(aq) \rightleftharpoons [Cd(NH_3)_4]^{2+}(aq)$
$$K_f = \frac{[[Cd(NH_3)_4]^{2+}]}{[Cd^{2+}][NH_3]^4}$$
74. 0.0078 mol or more
76. (a) $Zn(OH)_2(s) + 2\,H_3O^+(aq) \rightarrow Zn^{2+}(aq) + 4\,H_2O(\ell)$
$Zn(OH)_2(s) + 2\,OH^-(aq) \rightarrow [Zn(OH)_4]^{2-}(aq)$
(b) $Sb(OH)_3(s) + 3\,H_3O^+(aq) \rightarrow Sb^{3+}(aq) + 6\,H_2O(\ell)$
$Sb(OH)_3(s) + OH^-(aq) \rightarrow [Sb(OH)_4]^-(aq)$
78. (a) H_2O, CH_3COO^-, Na^+, CH_3COOH, H_3O^+, OH^-
(b) pH = 4.95
(c) pH = 5.05
(d) $CH_3COOH(aq) + H_2O(\ell) \rightleftharpoons CH_3COO^-(aq) + H_3O^+(aq)$
80. Ratio = 1.6
82. 0.020 mol added to 1 L
84. (a) pH = 2.78 (b) pH = 5.39
86. $K_a = 3.5 \times 10^{-6}$
88. No change
90. $K_{sp} = 2.2 \times 10^{-12}$
92. The tiny amount of base (CH_3COO^-) present is insufficient to prevent the pH from changing dramatically if a strong acid is introduced into the solution.

94. $K_a = 2.3 \times 10^{-4}$
96. Blood pH decreases; acidosis
98. (a) Adding Ca^{2+} drives the reaction more toward reactants, making more apatite.
(b) Acid reacts with OH^-, removing a product and driving the reaction toward the products, causing apatite to decompose.
100. Acidosis; increase
102. Sample A: $NaHCO_3$; Sample B: NaOH; Sample C: Mixture of NaOH and Na_2CO_3 and/or $NaHCO_3$; Sample D: Na_2CO_3
104. 5.61
106. 3.22
108. 3.2 g

Chapter 18

15. (a) $2\ H_2O(\ell) \rightarrow 2\ H_2(g) + O_2(g)$, reactant-favored
(b) $C_8H_{18}(\ell) \rightarrow C_8H_{18}(g)$, product-favored
(c) $C_{12}H_{22}O_{11}(s) \rightarrow C_{12}H_{22}O_{11}(aq)$, product-favored
17. (a) $\frac{1}{2}$ (b) $\frac{1}{2}$ (c) 50 of each
19. (a) Probability of $\frac{1}{2}$ in Flask A; probability of $\frac{1}{2}$ in flask B
(b) 50 in flask A and 50 in flask B
21. (a) Negative (b) Positive (c) Positive
23. (a) Positive (b) Positive (c) Positive
25. (a) Item 2 (b) Item 2 (c) Item 2
27. (a) NaCl(s) (b) $P_4(g)$ (c) $NH_4NO_3(aq)$
29. (a) Ga(ℓ) (b) $AsH_3(g)$ (c) NaF(s)
31. (a) Negative (b) Positive
(c) Negative (d) Positive
33. (a) Negative (b) Negative (c) Positive
35. 112 J K^{-1} mol^{-1}
37. (a) 2.63 J/K (b) 1000 J/K (c) 2.45 J/K
39. (a) +113.0 J/K (b) +38.17 J/K
41. (a) −120.64 J/K (b) 156.9 J/K
(c) −198.76 J/K (d) 160.6 J/K
43. (a) −173.01 J K^{-1} (b) −326.69 J K^{-1} (c) 137.55 J K^{-1}
45. $\Delta S^\circ = -247.7$ J/K; Cannot tell without ΔH° also, since that is needed to calculate ΔG°. $\Delta H^\circ = -88.13$ kJ and $\Delta G^\circ = -14.3$ kJ, so it is product-favored.
47. Product-favored at low temperatures; the exothermicity is sufficient to favor products if the temperature is low enough to overcome the decrease in entropy.
49. Exothermic reactions with an increase in disorder, exhibited by a larger number of gas-phase products than gas-phase reactants, never need help from the surroundings to favor products.
51. (a) Enthalpy change (ΔH) is negative; entropy change (ΔS) is positive.
(b) Using enthalpy of formations and Hess's law: $\Delta H^\circ = -184.28$ kJ; $\Delta S^\circ = -7.7$ J K^{-1}
The enthalpy change is negative, as predicted in (a). But the entropy change is negative, not as predicted in (a). The aqueous solute must have sufficient order to compensate for the high disorder of the gas. The value of −7.7 J K^{-1} is pretty small.
53. Entropy increase is insufficient to drive this highly endothermic reaction to form products without assistance from the surroundings at this temperature.
55. (a) $\Delta S^\circ = -37.52$ J/K, $\Delta H^\circ = -851.5$ kJ, product-favored at low temperature
(b) $\Delta S^\circ = -21.77$ J/K, $\Delta H^\circ = 66.36$ kJ, never product-favored
57. (a) $\Delta S_{univ} = 4.92 \times 10^3$ J/K (b) −1467.3 kJ
(c) Yes; ethane is used as a fuel; hence, we might expect that its combustion reaction is product-favored.
59.

Sign of ΔH°_{system}	Sign of ΔS°_{system}	Product-favored?	Sign of ΔG°_{system}
Negative	Positive	Yes	Negative
Negative	Negative	Yes at low T; no at high T	—
Positive	Positive	No at low T; yes at high T	—
Positive	Negative	No	Positive

61. $\Delta G^\circ = \Delta H^\circ - T\Delta S^\circ$ Here ΔH° is negative, and ΔS° is positive, so $\Delta G^\circ = -|\Delta H^\circ| - |T\Delta S^\circ| = -(|\Delta H^\circ| + |T\Delta S^\circ|) < 0$.
63. $\Delta G^\circ = 28.63$ kJ; reactant-favored
65. $\Delta G^\circ = 462.28$ kJ; uncatalyzed; it would not be a good way to make Si.
67. (a) −141.05 kJ (b) 141.73 kJ (c) −959.43 kJ
Reactions (a) and (c) are product-favored.
69. (a) 385.7 K (b) 835.1 K
71. (a) 49.7 kJ (b) −178.2 kJ (c) −1267.5 kJ
73. (a) $\Delta H^\circ = 178.32$ kJ, $\Delta S^\circ = 160.6$ J/K, $\Delta G^\circ = 130.5$ kJ
(b) Reactant-favored
(c) No; it is only product-favored at high temperatures.
(d) 1110. K
75. $\Delta G^\circ_f(Ca(OH)_2) = 2867.8$ kJ
77. (a) $K = 4 \times 10^{-34}$ (b) $K = 5 \times 10^{-31}$
79. (a) $\Delta G^\circ = -100.97$ kJ, product-favored
(b) $K = 5 \times 10^{17}$. When ΔG° is negative, K is larger than 1.
81. (a) $K^\circ_{298} = 2 \times 10^{12}$, product-favored
$K^\circ_{1000} = 2.0 \times 10^{-2}$, reactant-favored
(b) $K^\circ_{298} = 10^{135}$, product-favored
$K^\circ_{1000} = 3 \times 10^{33}$, product-favored
85. (a) $\Delta G^\circ = -106$ kJ/mol (b) $\Delta G^\circ = 8.55$ kJ/mol
(c) $\Delta G^\circ = -33.8$ kJ/mol
87. (a) can be harnessed; (b) and (c) require work to be done.
89. Reaction 87(b), 5.068 g graphite oxidized; reaction 87(c), 26.94 g graphite oxidized
91. (a) $2\ C(s) + 2\ Cl_2(g) + TiO_2(s) \rightarrow TiCl_4(g) + 2\ CO(g)$
(b) ΔH° −44.6 kJ; $\Delta S^\circ = 242.7$ J K^{-1}; $\Delta G^\circ = -116.5$ kJ
(c) Product-favored
93. CuO, Ag_2O, HgO, and PbO
95. (a) Five O—H bonds, seven C—O bonds, seven C—H bonds, five C—C bonds, and six O═O bonds are broken. Twelve C═O bonds and 12 O—H bonds are formed. $\Delta H^\circ \cong -2873$ kJ.
(b) Interactive forces in condensed phases (solid glucose and liquid water) are neglected in this calculation.
97. (a) 6.46 mol ATP per mol of glucose
(b) $\Delta G^\circ = -106$ kJ (c) Product-favored
99. The combustion of coal, petroleum, and natural gas are the most common sources used to supply free energy. We also use solar and nuclear energy as well as the kinetic energy of wind and water. (There may be other answers.)
101. (a) $\Delta G^\circ = -86.5$ kJ, product-favored
(b) $\Delta G^\circ = -873.1$ kJ (c) No (d) Yes
103. For $CH_4(g)$, $\Delta G^\circ = -817.90$ kJ
For $C_6H_6(\ell)$, $\Delta G^\circ = -3202.0$ kJ
For $CH_3OH(\ell)$, $\Delta G^\circ = -702.34$ kJ
Organic compounds are complex molecular systems that require significant rearrangement of atoms and bonds to undergo combustion. This makes them likely candidates for being kinetically stable.
105. (a) 5.5×10^{18} J/yr
(b) 1.5×10^{16} J/day
(c) 1.7×10^{11} J/s
(d) 1.7×10^{11} W
(e) 6×10^2 W/person

107. (a) Reaction 2 (b) Reactions 1 and 5
(c) Reaction 2 (d) Reactions 2 and 3
(e) None of them

109. (a) $\Delta G° = 141.73$ kJ (b) No (c) Yes
(d) $K_P = 1 \times 10^4$ (e) $K_c = 7 \times 10^1$

111. (a) $\Delta G° = 31.8$ kJ (b) $K_P = P_{Hg(g)}$
(c) $K° = 2.7 \times 10^{-6}$ (d) $P_{Hg(g)} = 2.7 \times 10^{-6}$ atm
(e) $T = 450$ K

113. Scrambled is a very disordered state for an egg. The second law of thermodynamics says that the more disordered state is the more probable state. Putting the delicate tissues and fluids back where they were before the scrambling occurred would take a great deal of energy. Humpty Dumpty is a fictional character who was also an egg. He fell off a wall. A very probable result of that fall is for an egg to become scrambled. The story goes on to tell that all the energy of the king's horses and men was not sufficient to put Humpty together again.

114. Absolute entropies can be determined because the minimum value of $S°$ is zero at $T = 0$ K. It is not possible to define conditions for a specific minimum value for internal energy, enthalpy, or Gibbs free energy of a substance, so relative quantities must be used.

116. Many of the oxides have negative enthalpies of reaction, which means their oxidations are exothermic. These are probably product-favored reactions.

117. NaCl, in an orderly crystal structure, and pure water, with only O—H hydrogen bonding interactions in the liquid state, are far more ordered than the dispersed hydrated sodium and chloride ions interacting with the water molecules.

119. (a) False (b) False (c) True
(d) True (e) True

121. The energy obtained from nutrients is stored as ATP. The source of the energy needed to synthesize the sugars was sunlight used by plants to produce the sugars and other carbohydrates.

123. $\Delta G = 0$ means products are favored; however, the equilibrium state will always have some reactants present, too. To get all the reactants to go away requires the removal of the products from the reactants, so the reaction continues forward.

125. (a) 58.78 J/K (b) −53.29 J/K (c) −173.93 J/K
Adding more hydrogen makes the $\Delta S°$ more negative.

127. Product favored; $\Delta H°$ is negative. Assuming the $S°$ of C_8H_{16} and C_8H_{18} are approximately the same value, $\Delta S°$ is negative, but not large enough to give $\Delta G°$ a different sign; hence, $\Delta G°$ is negative, also.

129. (a) 331.51 K (b) 371. K

131. (a) $K_P = 1.5 \times 10^7$ (b) Product-favored (c) $K_c = 3.7 \times 10^8$

133. (a) $7\ C(s) + 6\ H_2O(g) \rightarrow 2\ C_2H_6(g) + 3\ CO_2(g)$
$5\ C(s) + 4\ H_2O(g) \rightarrow C_3H_8(g) + 2\ CO_2(g)$
$3\ C(s) + 4\ H_2O(g) \rightarrow 2\ CH_3OH(\ell) + CO_2(g)$
(b) For C_2H_6, $\Delta H° = 101.02$ kJ, $\Delta G° = 122.72$ kJ, $\Delta S° = -72.71$ J/K
For C_3H_8, $\Delta H° = 76.5$ kJ, $\Delta G° = 102.08$ kJ, $\Delta S° = -86.6$ J/K
For CH_3OH, $\Delta H° = 96.44$ kJ, $\Delta G° = 187.39$ kJ, $\Delta S° = -305.2$ J/K
None of these are feasible. $\Delta G°$ is positive. In addition, $\Delta H°$ is positive and $\Delta S°$ is negative, suggesting that there is no temperature at which the products would be favored.

135. $\Delta G° = -RT \ln K = \Delta H° - T\Delta S°$

$$\ln K = -\frac{\Delta H°}{RT} + \frac{\Delta S°}{R}$$

If ln K is plotted against $1/T$, the slope would be $-\Delta H°/R$, and the y-intercept would be $\Delta S°/R$. The linear graph shows that $\Delta S°$ and $\Delta H°$ are independent of temperature.

137. (a) (ii) (b) (i) (c) (ii)

Chapter 19

10. (a) $Zn(s) \rightarrow Zn^{2+}(aq) + 2\ e^-$
(b) $2\ H_3O^+(aq) + 2\ e^- \rightarrow 2\ H_2O(\ell) + H_2(g)$
(c) $Sn^{4+}(aq) + 2\ e^- \rightarrow Sn^{2+}(aq)$
(d) $Cl_2(g) + 2\ e^- \rightarrow 2\ Cl^-(aq)$
(e) $6\ H_2O(\ell) + SO_2(g) \rightarrow SO_4^{2-}(aq) + 4\ H_3O^+(aq) + 2\ e^-$

12. (a) $Al(s) \rightarrow Al^{3+}(aq) + 3\ e^-$
$Cl_2(g) + 2\ e^- \rightarrow 2\ Cl^-(aq)$
(b) $Fe^{2+}(aq) \rightarrow Fe^{3+}(aq) + e^-$
$MnO_4^-(aq) + 8\ H_3O^+(aq) + 5\ e^- \rightarrow Mn^{2+}(aq) + 12\ H_2O(\ell)$
(c) $FeS(s) + 12\ H_2O(\ell) \rightarrow Fe^{3+}(aq) + SO_4^{2-}(aq) + 8\ H_3O^+(aq) + 9\ e^-$
$NO_3^-(aq) + 4\ H_3O^+(aq) + 3\ e^- \rightarrow NO(g) + 6\ H_2O(\ell)$

14. $4\ Zn(s) + 7\ OH^-(aq) + NO_3^-(aq) + 6\ H_2O(\ell) \rightarrow 4\ Zn(OH)_4^{2-} + NH_3(aq)$

16. (a) $3\ CO(g) + O_3(g) \rightarrow 3\ CO_2(g)$
O_3 is the oxidizing agent; CO is the reducing agent.
(b) $H_2(g) + Cl_2(g) \rightarrow 2\ HCl(g)$
Cl_2 is the oxidizing agent; H_2 is the reducing agent.
(c) $H_2O_2(aq) + Ti^{2+}(aq) + 2\ H_3O^+(aq) \rightarrow 4\ H_2O(\ell) + Ti^{4+}(aq)$
H_2O_2 is the oxidizing agent; Ti^{2+} is the reducing agent.
(d) $2\ MnO_4^-(aq) + 6\ Cl^-(aq) + 8\ H_3O^+(aq) \rightarrow 2\ MnO_2(s) + 3\ Cl_2(g) + 12\ H_2O(\ell)$
MnO_4^- is the oxidizing agent; Cl^- is the reducing agent.
(e) $4\ FeS_2(s) + 11\ O_2(g) \rightarrow 2\ Fe_2O_3(s) + 8\ SO_2(g)$
O_2 is the oxidizing agent; FeS_2 is the reducing agent.
(f) $O_3(g) + NO(g) \rightarrow O_2(g) + NO_2(g)$
O_3 is the oxidizing agent; NO is the reducing agent.
(g) $Zn(Hg)(amalgam) + HgO(s) \rightarrow ZnO(s) + 2\ Hg(\ell)$
HgO is the oxidizing agent; Zn(Hg) is the reducing agent.

18. The generation of electricity occurs when electrons are transmitted through a wire from the metal to the cation. Here, the transfer of electrons would occur directly from the metal to the cation and the electrons would not flow through any wire.

20. In chemistry, they are conventionally written as reduction reactions.

22. (a) $Zn(s) + Pb^{2+}(aq) \rightarrow Zn^{2+}(aq) + Pb(s)$
(b) Oxidation of zinc occurs at the anode. The reduction of lead occurs at the cathode. The anode is metallic zinc. The cathode is metallic lead.
(c)

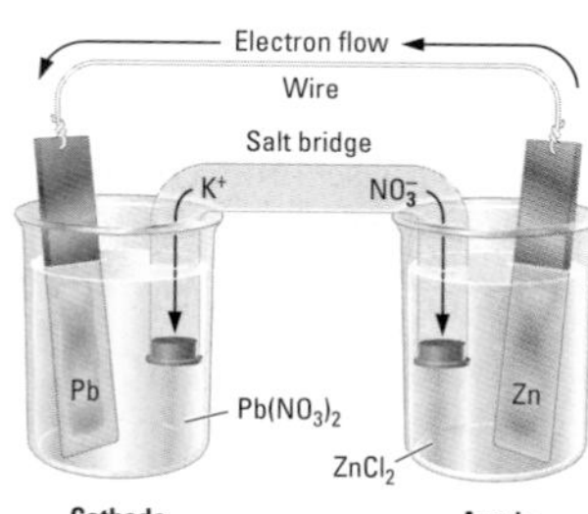

24. 7.5×10^3 C

26. (a) $Cu(s) \rightarrow Cu^{2+}(aq) + 2\ e^-$
$Ag^+(aq) + e^- \rightarrow Ag(s)$
(b) The copper half-reaction is oxidation and it occurs in the anode compartment. The silver half-reaction is reduction and it occurs in the cathode compartment.

28. Li is the strongest reducing agent and Li^+ is the weakest oxidizing agent. F_2 is the strongest oxidizing agent and F^- is the weakest reducing agent.

30. Worst oxidizing agent (d) < (c) < (a) < (b) best oxidizing agent

32. (a) 2.91 V (b) −0.028 V
(c) 0.65 V (d) 1.16 V
Reactions (a), (c), and (d) are product-favored.

34. (a) Al^{3+} (b) Ce^{4+} (c) Al
(d) Ce^{3+} (e) Yes (f) No
(g) Mercury(I) ion, silver ion, and cerium(IV) ion
(h) Hg, Sn, Ni, Al
36. Greater; less
38. (a) 1.55 V (b) −1196 kJ, 1.55 V
40. −409 kJ
42. 1×10^{-18}, 102 kJ
46. $K° = 4 \times 10^1$
48. (a) 1.20 V (b) 1.16 V (c) 3 M
50. −0.378 V
52. (conc. Pb^{2+})/(conc. Sn^{2+}) = 0.3
54. (a) $Ni^{2+}(aq) + Cd(s) \rightarrow Ni(s) + Cd^{2+}(aq)$
(b) Cd is oxidized, Ni^{2+} is reduced, Ni^{2+} is the oxidizing agent, and Cd is the reducing agent.
(c) Metallic Cd is the anode and metallic Ni is the cathode.
(d) 0.15 V
(e) Electrons flow from the Cd electrode to the Ni electrode.
(f) Toward the anode compartment
56. A fuel cell has a continuous supply of reactants and will be useable for as long as the reactants are supplied. A battery contains all the reactants of the reaction. Once the reactants are gone, the battery is no longer useable.
58. (a) The N_2H_4 half-reaction occurs at the anode and the O_2 half-reaction occurs at the cathode.
(b) $N_2H_4(g) + O_2(g) \rightarrow N_2(g) + 2\,H_2O(\ell)$
(c) 7.5 g N_2H_4 (d) 7.5 g O_2
59. $O_2(g)$ produced at anode; $H_2(g)$ produced at cathode; 2 mol H_2 per mol O_2
61. Au^{3+}, Hg^{2+}, Ag^+, Hg_2^{2+}, Fe^{3+}, Cu^{2+}, Sn^{4+}, Sn^{2+}, Ni^{2+}, Cd^{2+}, Fe^{2+}, Zn^{2+}
63. H_2, Br_2, and OH^- are produced. After the reaction is complete, the solution contains Na^+, OH^-, a small amount of dissolved Br_2 (though it has low solubility in water), and a very small amount of H_3O^+. H_2 is formed at the cathode. Br_2 is formed at the anode.
65. 0.16 g Ag
67. 5.93 g Cu
69. 2.7×10^5 g Al
71. 1.9×10^2 g Pb
73. 0.10 g Zn
75. 0.043 g Li; 0.64 g Pb
77. 6.85 min
80. Ions increase the electrolytic capacity of the solution.
82. Chromium is highly resistant to corrosion and protects iron in steel from oxidizing.
85. 0.00689 g Cu, 0.00195 g Al
87. Worst reducing agent B < D < A < C best reducing agent
89. (a) B is oxidized and A^{2+} is reduced.
(b) A^{2+} is the oxidizing agent and B is the reducing agent.
(c) B is the anode and A is the cathode.
(d) $A^{2+} + 2e^- \rightarrow A$
$B \rightarrow B^{2+} + 2e^-$
(e) A gains mass.
(f) Electrons flow from B to A.
(g) K^+ ions will migrate toward the A^{2+} solution.
91. (a) The reaction with water is spontaneous:
$$4\,Co^{3+}(aq) + 6\,H_2O(\ell) \rightarrow 4\,Co^{2+}(aq) + O_2(aq) + 4\,H_3O^+(aq) \quad E°_{cell} = 0.591\text{ V}$$
(b) The reaction with oxygen in air is spontaneous:
$$4\,Fe^{2+}(aq) + O_2(aq) + 4\,H_3O^+(aq) \rightarrow 4\,Fe^{3+}(aq) + 6\,H_2O(\ell) \quad E°_{cell} = 0.458\text{ V}$$
93. $E_{cell} = 0.379$ V
95. (a) 9.5×10^6 g HF (b) 1.7×10^3 kWh
97. $Cl_2(g) + 2\,Br^-(aq) \rightarrow Br_2(\ell) + 2\,Cl^-(aq)$
99. 4+
101. 75 s
103. 5×10^{-6} M Cu^{2+}

Chapter 20

11. (a) α emission (b) β emission (c) Electron capture or positron emission (d) β emission
13. (a) $^{238}_{92}U$ (b) $^{32}_{15}P$ (c) $^{10}_{5}B$ (d) $^{0}_{-1}e$
(e) $^{15}_{7}N$
15. (a) $^{28}_{12}Mg \rightarrow {}^{28}_{13}Al + {}^{0}_{-1}e$
(b) $^{238}_{92}U + {}^{12}_{6}C \rightarrow 4\,{}^{1}_{0}n + {}^{246}_{98}Cf$
(c) $^{2}_{1}H + {}^{3}_{2}He \rightarrow {}^{4}_{2}He + {}^{1}_{1}H$
(d) $^{38}_{19}K \rightarrow {}^{38}_{18}Ar + {}^{0}_{+1}e$
(e) $^{175}_{78}Pt \rightarrow {}^{4}_{2}He + {}^{171}_{76}Os$
17. $^{231}_{90}Th$, $^{231}_{91}Pa$, $^{227}_{89}Ac$, $^{227}_{90}Th$, $^{223}_{88}Ra$
19. $^{222}_{86}Rn \rightarrow {}^{4}_{2}He + {}^{218}_{84}Po$
$^{218}_{84}Po \rightarrow {}^{4}_{2}He + {}^{214}_{82}Pb$
$^{214}_{82}Pb \rightarrow {}^{0}_{-1}e + {}^{214}_{83}Bi$
$^{214}_{83}Bi \rightarrow {}^{0}_{-1}e + {}^{214}_{84}Po$
$^{214}_{84}Po \rightarrow {}^{4}_{2}He + {}^{210}_{82}Pb$
$^{210}_{82}Pb \rightarrow {}^{0}_{-1}e + {}^{210}_{83}Bi$
$^{210}_{83}Bi \rightarrow {}^{0}_{-1}e + {}^{210}_{84}Po$
$^{210}_{84}Po \rightarrow {}^{4}_{2}He + {}^{206}_{82}Pb$
20. (a) $^{19}_{10}Ne \rightarrow {}^{19}_{9}F + {}^{0}_{+1}e$ (b) $^{230}_{90}Th \rightarrow {}^{0}_{-1}e + {}^{230}_{91}Pa$
(c) $^{82}_{35}Br \rightarrow {}^{0}_{-1}e + {}^{82}_{36}Kr$ (d) $^{212}_{84}Po \rightarrow {}^{4}_{2}He + {}^{208}_{82}Pb$
22. The binding energy per nucleon of ^{10}B is 6.252×10^8 kJ/mol nucleon. The binding energy per nucleon of ^{11}B is 6.688×10^8 kJ/mol nucleon. ^{11}B is more stable than ^{10}B because its binding energy is larger.
24. -1.7394×10^{11} kJ/mol ^{238}U; 7.3086×10^8 kJ/mol nucleon
27. 5 mg
29. (a) $^{131}_{53}I \rightarrow {}^{131}_{54}Xe + {}^{0}_{-1}e$ (b) 1.56 mg
31. 4.0 y
33. 34.8 d
35. 2.72×10^3 y; Approx. 710 B.C.
37. 1.6×10^5 y
39. $^{239}_{94}Pu + 2\,{}^{1}_{0}n \rightarrow {}^{0}_{-1}e + {}^{241}_{95}Am$
41. $^{12}_{6}C$
43. Cadmium rods (a neutron absorber to control the rate of the fission reaction), uranium rods (a source of fuel, since uranium is a reactant in the nuclear equation), and water (used for cooling by removing excess heat and used in steam/water cycle for the production of turning torque for the generator)
45. (a) $^{140}_{54}Xe$ (b) $^{101}_{41}Nb$ (c) $^{92}_{36}Kr$
47. 6.9×10^3 barrels
49. A rad is a measure of the amount of radiation absorbed. A rem includes a quality factor that better describes the biological impact of a radiation dose. The unit rem would be more appropriate when talking about the effects of an atomic bomb on humans. The unit gray (Gy) is 100 rad.
51. Since most elements have some proportion of unstable isotopes that decay and we are composed of these elements (e.g., ^{14}C), our bodies emit radiation particles.
53. The gamma ray is a high-energy photon. Its interaction with matter is most likely just to impart large quantities of energy. The alpha and beta particles are charged particles of matter, which could interact, and possibly react, with the matter composing the food.
55. 0.13 L

57. (a) $^{0}_{-1}e$ (b) $^{4}_{2}He$ (c) $^{87}_{35}Br$ (d) $^{216}_{284}Po$
(e) $^{68}_{31}Ga$
59. 2 mg
61. 3.6×10^9 y
63. (a) $^{247}_{299}Es$ (b) $^{16}_{8}O$ (c) $^{4}_{2}He$ (d) $^{12}_{6}C$
(e) $^{10}_{15}B$
65. Alpha and beta radiation decay particles are charged ($^{4}_{2}He^{21}$ and $^{0}_{-1}e^{-}$), so they are better able to interact with and ionize tissues, disrupting the function of the cancer cells. Gamma radiation, like X rays, goes through soft tissue without much of it being absorbed. This is less likely to interfere with the cancerous cells.
67. ^{20}Ne is stable. The ^{17}Ne is likely to decay by positron emission, increasing the ratio of neutrons to protons. The ^{23}Ne is likely to decay by beta emission, decreasing the ratio of neutrons to protons.
69. A nuclear reaction occurred, making products. Therefore, some of the lost mass is found in the decay particles, if the decay is alpha or beta decay, and almost all of the rest is found in the element produced by the reaction.
71. They are spontaneous reactions due to the nature of the nucleus. No other species are involved.
73. 2.8×10^4 kg
75. 75.1 y
77. 3.92×10^3 y

Chapter 21

20. (a) Cathode (b) Chlorine gas (c) 2.027×10^9 C
(d) 1.6×10^3 kJ/mol
22. 7×10^3 kWh
24. 67.1 g Al
26. Answers are in **bold.**

Formula	Name	Oxidation State of Phosphorus
P_4	Phosphorus	**0**
$(NH_4)_2HPO_4$	**Ammonium hydrogen phosphate**	+5
H_3PO_4	Phosphoric acid	**+5**
P_4O_{10}	Tetraphosphorus decaoxide	**+5**
$Ca_3(PO_4)_2$	**Calcium phosphate**	**+5**
$Ca(H_2PO_4)_2$	Calcium dihydrogen phosphate	**+5**

28. 1.7×10^7 tons
30.

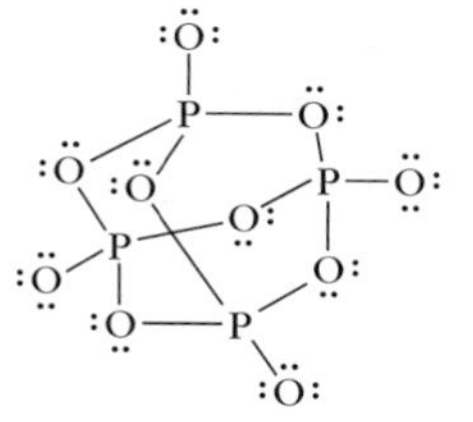

32. 962 K
34. Raw materials: sulfur, water, oxygen, and catalyst (Pt or VO_5)
$S_8(s) + 8\ O_2\ (g) \rightarrow 8\ SO_2\ (g)$
$2\ SO_2\ (g) + O_2\ (g) \rightarrow 2\ SO_3\ (g)$
$SO_3\ (g) + H_2SO_4(\ell) \rightarrow H_2S_2O_7(\ell)$
$H_2S_2O_7\ (\ell) + H_2O(\ell) \rightarrow 2\ H_2SO_4\ (aq)$
36. H—Ö—N(=O)(—O:) ⟷ H—Ö—N(—O:)(=O) ⟷ H—Ö=N(—O:)(—O:)

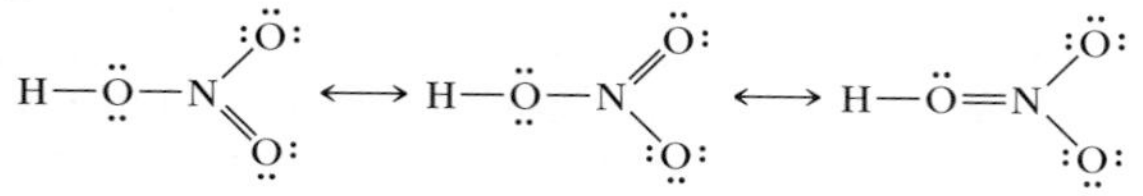

38. $K_c = 2.7$
40. 3 L
42. $K = 5.3 \times 10^5$
Putting seawater in the presence of $Ca(OH)_2$ will cause a precipitate of $Mg(OH)_2$ to form. The solid can be isolated after it settles.
44. N=N=N—N=O (Other structures are possible.)
46. (a) H—N=N=N ⟷ H—N—N≡N:
(b) +140. kJ with single and triple bond; +410. kJ with two double bonds
48. (a) $NO_2(g) + NO(g) \rightarrow N_2O_3(g)$
(b) O=N—N(=O)—O: ⟷ O=N—N(—O:)=O
(c) The O—N—O angle is 120° and the N—N—O angle is slightly less than 120°.
51. (a) 7×10^2 kJ (b) 3 V must be used to overcome the cell potential.
53. (a) One; 120 °C, 1×10^{-4} atm (b) (i) Solid rhombic (ii) Liquid (iii) Solid monoclinic (iv) Vapor
55. (a) $4\ H_3O^+\ (aq) + MnO(s) + 2\ Br^-(aq) \rightarrow Mn^{2+}(aq) + 6\ H_2O(\ell) + Br_2(\ell)$
(b) 0.0813 mol Br^- (c) 3.54 g MnO_2
63. 6.3×10^5 J/mol
66. (a) $P_4(s) + 5\ O_2(g) \rightarrow P_4O_{10}(s)$ (b) pH = 1.18
(c) 25.0 g $Ca_3(PO_4)_2$ (d) 5.42 L
68. (a) Cl is reduced; Al and N are oxidized. (b) −2674 kJ
70. −2708. kJ/mol
72. $K_p = 1.985$
74. 2961 kJ, Mg^{2+} is a much smaller ion with higher charge density than Sr^{2+}, so the $Mg^{2+} + F^-$ ions are closer together and the ionic bond is much stronger.
76. B_6H_{12}

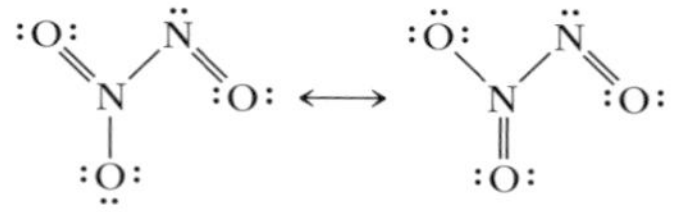

Chapter 22

14. (a) Ag [Kr] $4d^{10}5s^1$; Ag^+ [Kr] $4d^{10}$
(b) Au [Xe] $4f^{14}5d^{10}6s^1$; Au^+ [Xe]$4f^{14}\ 5d^{10}$; Au^{3+} [Xe] $4f^{14}\ 5d^8$
16. Cr^{2+} and Cr^{3+}
18. $Cr_2O_7^{2-}$ (in acid) > Cr^{3+} > Cr^{2+}; the species with a more positive oxidation state of Cr has greater tendency to be reduced. E°_{red}, respectively, is +1.33 V > −0.74 V > −0.91 V, so the more positive the oxidation state, the better the oxidizing agent.
20. (a) $Fe_2O_3(s) + 3\ CO(g) \rightarrow 2\ Fe(s) + 3\ CO_2(g)$
(b) $Fe(s) + 2\ H_3O^+(aq) \rightarrow Fe^{2+}\ (aq) + 2\ H_2O(\ell) + H_2(g)$ or $2\ Fe(s) + 6\ H_3O^+\ (aq) \rightarrow 2\ Fe^{3+}(aq) + 6\ H_2O(\ell) + 3\ H_2(g)$
22. $6\ H_3O^+(aq) + 3\ NO_3^-(aq) + Fe(s) \rightarrow Fe^{3+}(aq) + 3\ NO_2(g) + 9\ H_2O(\ell)$
26. (a) $[Cr(NH_3)_2(H_2O)_3(OH)]^{2+}$; +2
(b) The counter ion would be an anion, such as Cl^-.
28. (a) $C_2O_4^{2-}$ with 2− charge and Cl− with 1− charge
(b) 3+ charge
(c) $[Co(NH_3)_4Cl_2]^+$, with 1+ charge
30. For $Na_3[IrCl_6]$: (a) Six Cl^- (b) Ir with 3+ charge
(c) $[IrCl_6]^{3-}$ with 3− charge (d) Na^+
For $[Mo(CO)_4Br_2]$: (a) Four CO (b) Mo with 2+ charge
(c) $[Mo(CO)_4]^{2+}$ with 2+ charge (d) Br^-
32. (a) Four (b) Four
34. (a) $[Pt(NH_3)_2Br_2]$ (b) $[Pt(en)(NO_2)_2]$
(c) $[Pt(NH_3)_2BrCl]$

36. (a) Four (b) Four
(c) Six (d) Six
38. (a) Monodentate (b) Tetradentate
(c) Tridentate (d) Monodentate
40. For example, $FeCl_3$ (Other answers are possible.)
42. (a) Tetrachloromanganate (II)
(b) Potassium trioxalatoferrate (III)
(c) Diamminedicyanoplatinum (II)
(d) Pentaquahydroxoiron (III)
(e) Diethylenediaminedichloromanganese (II)
44. (a) (b)
46.
48. (a) and (b)
50. (a), (c), and (d)
52. (a) 4 (b) 5 (c) 5 (d) 3
54. Cr^{2+} has four electrons. Cr^{3+} has three electrons. Since there are three low-energy t_2 orbitals, the first three electrons can fill one into each t_2 orbital. The fourth electron (only found in Cr^{2+}) is required to pair up with one of the other electrons (low spin) or to span the Δ_o gap to go into an *e* orbital (high spin).
56. If Δ_o is large, electrons fill the t_2 orbitals and end up all paired. If Δ_o is small, all five orbitals are half-filled before any pairing begins, resulting in four unpaired electrons. So, high-spin Co^{3+} complexes are paramagnetic and low-spin Co^{3+} complexes are diamagnetic.
58. Cu^{2+} has nine *d*-electrons. All t_2 orbitals are filled, so high- and low-spin complexes do not form.
60. violet light; Approximately 400 nm
62. With CN^-: Δ_o increases and the observed color shifts closer to yellow-orange.
With Cl^-: Δ_o decreases and the observed color shifts closer to blue.
65. (a) [Ar] $3d^4$ (b) [Ar] $3d^{10}$ (c) [Ar] $3d^7$ (d) [Ar] $3d^3$

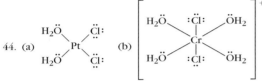

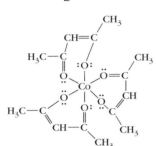

67.
(a)

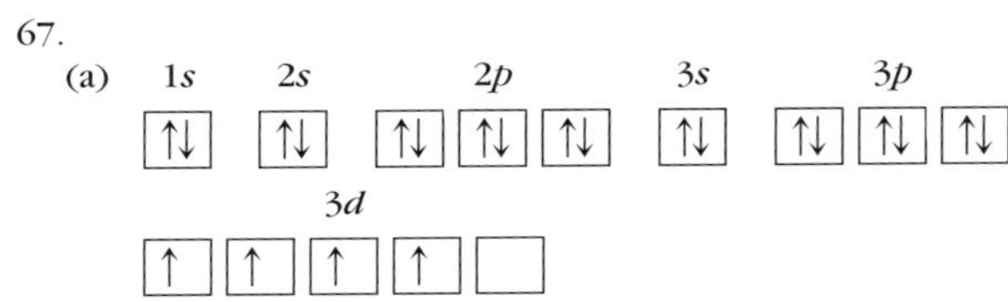

Four unpaired electrons

(b) 1s 2s 2p 3s 3p 3d

Zero unpaired electrons

(c) 1s 2s 2p 3s 3p 3d

Three unpaired electrons

(d) 1s 2s 2p 3s 3p 3d

Three unpaired electrons

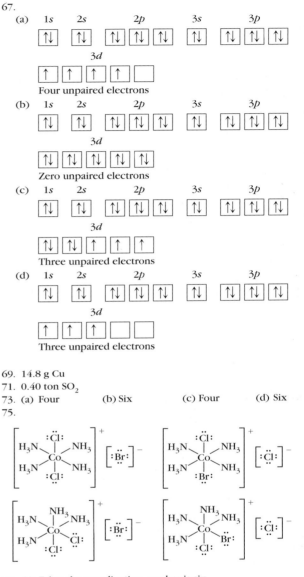

69. 14.8 g Cu
71. 0.40 ton SO_2
73. (a) Four (b) Six (c) Four (d) Six
75.
77. (a) False; the coordination number is six.
(b) False; Cu^+ has no unpaired electrons.
(c) False; the net charge is 3+.
78. (a) False; "In$[Pt(NH_3)_4]Cl_4$ the Pt has a 4+ charge and a coordination number of four."
(b) True
79. (a) +2
(b)
81. $Fe(s) + 2\ Fe^{3+}(aq) \rightarrow 3\ Fe^{2+}(aq)$
83. Four

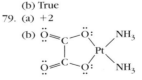

85.

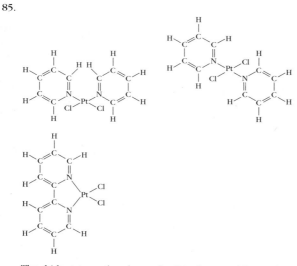

The bidentate ortho-phenanthroline is not able to form the *trans*-isomer.

87. (a) $2\ Ag^+ + Ni \rightarrow Ni^{2+} + 2\ Ag$
(b) +1.05V

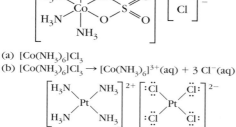

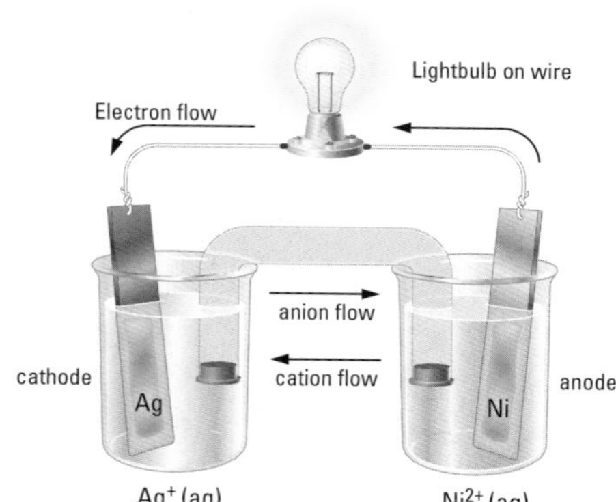

89. 17.4% Fe

91. (a) −1.38 V (b) 1×10^{-20} M Ag^{2+}

93. (a) No ions (b) 2.00 mol ions
(c) 4.00 mol ions (d) 3.00 mol ions

96. $[Co(NH_3)_4(SO_4)]^+\ [Cl]^-$

98. (a) $[Co(NH_3)_6]Cl_3$
(b) $[Co(NH_3)_6]Cl_3 \rightarrow [Co(NH_3)_6]^{3+}(aq) + 3\ Cl^-(aq)$

100. $[Pt(NH_3)_4]^{2+}\ [PtCl_4]^{2-}$

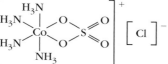

102. $[CuCl_5]^{3-}$ $[Ni(CN)_5]^{3-}$

Triangular bipyramid or square pyramid

104. A mixture of high spin and low spin complexes, in which the high spin species is favored by a 2:1 ratio, would give the observed 2.67 unpaired electrons per iron ion. High spin iron(II) has four unpaired electrons, while low spin iron(II) has zero unpaired electrons. This translates to 8 unpaired electrons for every three iron ions or 8/3 = 2.67 unpaired electrons per iron ion.

106.

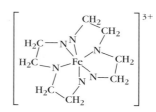

(One other structure is possible, too.)

Appendix A

4. The calculation is incomplete or incorrect. Check for incomplete unit conversions (e.g., cm^3 to L, g to kg, mm Hg to atm, etc.) and check to see that the numerator and denominator of unit factors are placed such that the unwanted units can cancel (e.g., $\frac{g}{mL}$ or $\frac{mL}{g}$).

6. Answer (c) gives the properly reported observation. Answers (a) and (b) do not have the proper units. Answer (d) is incomplete, with no units. Answer (e) shows the conversion of the observed measurement to new units, making (e) the result of a calculation, not the observed measurement.

8. (a) The units should not be squared. To do this correctly, add values with common units (g) and give the answer the same units (g). The result is 9.95 g.
(b) The unit factor is not set up to cancel unwanted units (g); the units reported (g) are not the units resulting from this calculation ($\frac{g^2}{mL}$). To do this correctly, $5.23\ g \times \frac{1.00\ mL}{4.87\ g} = 1.07$ mL.
(c) The unit factor must be cubed to cancel all unwanted units; the units reported (m^3) are not the units resulting from this calculation ($m \cdot cm^2$). To do this correctly, $3.57\ cm^3 \times (\frac{1\ m}{100\ cm})^3 = 3.57 \times 10^{-6}\ m^3$

10. (a) 7.994 ft ± 0.043 ft
(b) Three, because uncertainty is in the hundredths place
(c) The result is accurate (i.e., the true answer, 8.000 ft, is within the range of the uncertainty described by these values, 7.951 ft– 8.037 ft), though not very precise.

12. (a) 4 (b) 2 (c) 2 (d) 4

14. (a) 43.32 (b) 43.32 (c) 43.32
(d) 43.32 (e) 43.32 (f) 43.33

16. (a) 13.7 (b) 0.247 (c) 12.0

18. (a) 7.6003×10^4 (b) 3.7×10^{-4} (c) 3.4000×10^4

20. (a) 2.415×10^{-3} (b) 2.70×10^8 (c) 3.236

22. (a) −0.1351 (b) 3.541 (c) 23.7797
(d) 54.7549 (e) −7.455

24. (a) 5.404 (b) 9×10^{14}
(c) 110. (d) 3.75

26. (a) 0.708, −1.80 (b) 3.66, 1.38

28.

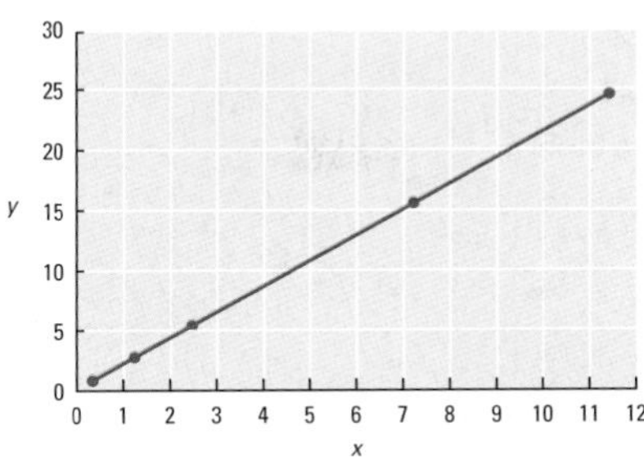

Yes, y is proportional to x.

Appendix B

1. (a) Meter, mega-; Mm (b) Kilogram, no additional prefix; kg (c) Square meter, deci-; dm^2
3. SI base (fundamental) units are set by convention; derived units are based on the fundamental units.
5. (a) 4.75×10^{-10} m (b) 5.6×10^{7} kg (c) 4.28×10^{-6} A
7. (a) 8.7×10^{-18} m^2 (b) 2.73×10^{-17} $kg \cdot m^2 \cdot s^{-2}$ (c) 2.73×10^{-5} $kg \cdot m/s^2$
9. (a) 2.20 lb (b) 2.22×10^{3} kg (c) 60 in^3 (d) 1×10^{5} Pa and 1 bar
11. (a) 99 °F (b) −10.5 °F (c) −40.0 °F

Glossary

absolute temperature scale (Kelvin temperature scale) A temperature scale on which the zero is the lowest possible temperature and the degree is the same size as the Celsius degree.

absolute zero Lowest possible temperature, 0 K.

absorb To draw a substance into the bulk of a liquid or a solid (compare with adsorb).

acid (Arrhenius) A substance that increases the concentration of hydronium ions, H_3O^+, in aqueous solution. (See also **Brønsted-Lowry acid, Lewis acid.**)

acid ionization constant (K_a) The equilibrium constant for the reaction of a weak acid with water to produce hydronium ions and the conjugate base of the weak acid.

acid ionization constant expression Mathematical expression in which the product of the equilibrium concentrations of hydronium ion and conjugate base is divided by the equilibrium concentration of the un-ionized conjugate acid.

acid rain Rain (or other precipitation) with a pH below about 5.6 (the pH of unpolluted rain water).

acidic solution An aqueous solution in which the concentration of hydronium ion exceeds the concentration of hydroxide ion.

actinides The elements after actinium in the seventh period; in actinides the 5*f* subshell is being filled.

activated complex A molecular structure corresponding to the maximum of a plot of energy versus reaction progress; also known as the transition state.

activation energy (E_a) The potential energy difference between reactants and activated complex; the minimum energy that reactant molecules must have to be converted to product molecules.

active site The part of an enzyme molecule that binds the substrate to help it to react.

activity (*A*) A measure of the rate of nuclear decay, given as disintegrations per unit time.

actual yield The quantity of a reaction product obtained experimentally; less than the theoretical yield.

addition polymer A polymer made when monomer molecules join directly with one another, with no other products formed in the reaction.

adsorb To attract and hold a substance on a surface (compare with absorb).

aerosols Small particles (1 nm to about 10,000 nm in diameter) that remain suspended indefinitely in air.

air pollutant A substance that degrades air quality.

alcohol An organic compound containing a hydroxyl group (—OH) covalently bonded to a saturated carbon atom.

aldehyde An organic compound characterized by a carbonyl group in which the carbon atom is bonded to a hydrogen atom; a molecule containing the —CHO functional group.

alkali metals The Group 1A elements in the periodic table (except hydrogen).

alkaline earth metals The elements in Group 2A of the periodic table.

alkane Any of a class of hydrocarbons characterized by the presence of only single carbon-carbon bonds.

alkene Any of a class of hydrocarbons characterized by the presence of a carbon-carbon double bond.

alkyl group A fragment of an alkane structure that results from the removal of a hydrogen atom from the alkane.

alkyne Any of a class of hydrocarbons characterized by the presence of a carbon-carbon triple bond.

allotropes Different forms of the same element that exist in the same physical state under the same conditions of temperature and pressure.

alpha carbon The carbon adjacent to the acid group (—COOH) in an amino acid.

alpha (α) particles Positively charged (2+) particles ejected at high speeds from certain radioactive nuclei; the nuclei of helium atoms.

alpha radiation Radiation composed of alpha particles (helium nuclei).

amide An organic compound characterized by the presence of a carbonyl group in which the carbon atom is bonded to a nitrogen atom ($—CONH_2$, —CONHR, $—CONR_2$); the product of the reaction of an amine with a carboxylic acid.

amide linkage A linkage consisting of a carbonyl group bonded to a nitrogen atom that connects monomers in a polymer.

amine An organic compound containing an $—NH_2$, —NHR, or $—NR_2$ functional group.

amino acids Organic molecules containing a carboxyl group with an R group and an amine group on the alpha carbon; building block monomers of proteins.

amorphous solid A solid whose constituent nanoscale particles have no long-range repeating structure.

amount A measure of the number of elementary entities (such as atoms, ions, molecules) in a sample of matter compared with the number of elementary entities in exactly 0.012 kg pure ^{12}C. Also called molar amount.

ampere The SI unit of electrical current; involves the flow of one coulomb of charge per second.

amphoteric Refers to a substance that can act as either an acid or a base.

anion An ion with a negative electrical charge.

anode The electrode of an electrochemical cell at which oxidation occurs.

anodic inhibition The prevention of oxidation of an active metal by painting it, coating it with grease or oil, or allowing a thin film of metal oxide to form.

antibonding molecular orbital A higher-energy molecular orbital that, if occupied by electrons, does not result in attraction between the atoms.

antioxidant Reducing agent that converts free radicals and other reactive oxygen species into less reactive substances.

aqueous solution A solution in which water is the solvent.

aromatic compound Any of a class of organic compounds characterized by the presence of one or more benzene rings or benzene-like rings.

Arrhenius equation Mathematical relation that gives the temperature dependence of the reaction rate constant; $k = Ae^{-E_a/RT}$.

asymmetric Describes a molecule or object that is not symmetrical.

atom The smallest particle of an element that can be involved in chemical combination with another element.

atom economy The fraction of atoms of starting materials incorporated into the desired final product in a chemical reaction.

atomic mass units (amu) The unit of a scale of relative atomic masses of the elements; 1 amu = 1/12 the mass of a six-proton, six-neutron carbon atom.

atomic number The number of protons in the nucleus of an atom of an element.

atomic radius One-half the distance between the nuclei centers of two like atoms in a molecule.

atomic structure The identity and arrangement of subatomic particles in an atom.

atomic weight The average mass of an atom in a representative sample of atoms of an element.

autoionization The equilibrium reaction in which water molecules react with each other to form hydronium ions and hydroxide ions.

average reaction rate A reaction rate calculated from a change in concentration divided by a change in time.

Avogadro's law The volume of a gas, at a given temperature and pressure, is directly proportional to the amount of gas.

Avogadro's number The number of particles in a mole of any substance (6.022×10^{23}).

axial position(s) Positions above and below the equatorial plane in a triangular bipyramidal structure.

background radiation Radiation from natural and synthetic radioactive sources to which all members of a population are exposed.

balanced chemical equation A chemical equation that shows equal numbers of atoms of each kind in the products and the reactants.

bar A pressure unit equal to 100,000 Pa.

barometer A device for measuring atmospheric pressure.

basal metabolic rate The energy required to maintain an organism that is awake, at rest, and not digesting or metabolizing food.

base (Arrhenius) Substance that increases concentration of hydroxide ions, OH^-, in aqueous solution. (See also **Brønsted–Lowry base, Lewis base.**)

base ionization constant (K_b) The equilibrium constant for the reaction of a weak base with water to produce hydroxide ions and the conjugate acid of the weak base.

base ionization constant expression Mathematical expression in which the product of the equilibrium concentrations of hydroxide ion and conjugate acid is divided by the equilibrium concentration of the conjugate base.

basic solution An aqueous solution in which the concentration of hydroxide ion is greater than the concentration of hydronium ion.

battery (voltaic cell) An electrochemical cell (or group of cells) in which a product-favored oxidation-reduction reaction is used to produce an electric current.

becquerel A unit of radioactivity equal to 1 nuclear disintegration per second.

beta particles Electrons ejected from certain radioactive nuclei.

beta radiation Radiation composed of electrons.

bidentate ligand A ligand that has two atoms with lone pairs that can form coordinate covalent bonds to the same metal ion.

bimolecular reaction An elementary reaction in which two particles must collide for products to be formed.

binary molecular compound A molecular compound whose molecules contain atoms of only two elements.

binding energy The energy required to separate all nucleons in an atomic nucleus.

binding energy per nucleon The energy per nucleon required to separate all nucleons in an atomic nucleus.

biodegradable Capable of being decomposed by biological means, especially by bacterial action.

boiling The process whereby a liquid vaporizes throughout when its vapor pressure equals atmospheric pressure.

boiling point The temperature at which the equilibrium vapor pressure of a liquid equals the external pressure on the liquid.

boiling-point elevation A colligative property; the difference between the normal boiling point of a pure solvent and the higher boiling point of a solution in which a nonvolatile nonelectrolyte solute is dissolved in that solvent.

bond Attractive force between two atoms holding them together, for example, as part of a molecule.

bond angle The angle between the bonds to two atoms that are bonded to the same third atom.

bond enthalpy (bond energy) The change in enthalpy when a mole of chemical bonds of a given type is broken, separating the bonded atoms; the atoms and molecules must be in the gas phase.

bond length The distance between the centers of the nuclei of two bonded atoms.

bonding electrons Electron pairs shared in covalent bonds.

bonding molecular orbital A lower-energy molecular orbital that can be occupied by bonding electrons.

bonding pair A pair of valence electrons that are shared between two atoms.

Born-Haber cycle A stepwise thermochemical cycle in which the constituent elements are converted to ions and combined to form an ionic compound.

boundary surface A surface within which there is a specified probability (often 90%) that an electron will be found.

Boyle's law The volume of a confined ideal gas varies inversely with the applied pressure, at constant temperature and amount of gas.

Brønsted-Lowry acid A hydrogen ion donor.

Brønsted-Lowry acid-base reaction A reaction in which an acid donates a hydrogen ion and a base accepts the hydrogen ion.

Brønsted-Lowry base A hydrogen ion acceptor.

buckyball Buckminsterfullerene; an allotrope of carbon consisting of molecules in which 60 carbon atoms are arranged in a cage-like structure consisting of five-membered rings sharing edges with six-membered rings.

buffer See **buffer solution.**

buffer capacity The quantity of acid or base a buffer can accommodate without a significant pH change (more than one pH unit).

buffer solution A solution that resists changes in pH when limited amounts of acids or bases are added; it contains a weak acid and its conjugate base, or a weak base and its conjugate acid.

caloric value The energy of complete combustion of a stated size sample of a food, usually reported in Calories (kilocalories).

calorie (cal) A unit of energy equal to 4.184 J. Approximately 1 cal is required to raise the temperature of 1 g of liquid water by 1 °C.

Calorie (Cal) A unit of energy equal to 4.184 kJ = 1 kcal. (See also **kilocalorie.**)

calorimeter A device for measuring the quantity of thermal energy transferred during a chemical reaction or some other process.

capillary action The process whereby a liquid rises in a small-diameter tube due to noncovalent interactions between the liquid and the tube's material.

carbohydrates Biochemical compounds with the general formula $C_x(H_2O)_y$, in which x and y are whole numbers.

carbonyl group An organic functional group consisting of carbon bonded to two other atoms and double bonded to oxygen; $>C{=}O$.

carboxylic acid An organic compound characterized by the presence of the carboxyl group (—COOH).

catalyst A substance that increases the rate of a reaction but is not consumed in the overall reaction.

catalytic cracking A petroleum refining process using a catalyst, heat, and pressure to break long-chain hydrocarbons into shorter-chain hydrocarbons, including both alkanes and alkenes suitable for gasoline.

catalytic reforming A petroleum refining process in which straight-chain hydrocarbons are converted to branched-chain hydrocarbons and aromatics for use in gasoline and the manufacture of other organic compounds.

catenation Formation of chains and rings by bonds between atoms of the same element.

cathode The electrode of an electrochemical cell at which reduction occurs.

cathodic protection A process of protecting a metal from corrosion whereby it is made the cathode by connecting it electrically to a more reactive metal.

cation An ion with a positive electrical charge.

cell voltage The electromotive force of an electrochemical cell; the quantity of work a cell can produce per coulomb of charge that the chemical reaction produces.

Celsius temperature scale A scale defined by the freezing (0 °C) and boiling (100 °C) points of pure water, at 1 atm.

cement A solid consisting of microscopic particles containing compounds of calcium, iron, aluminum, silicon, and oxygen in varying proportions and tightly bound to one another.

ceramics Materials fashioned from clay or other natural materials at room temperature and then hardened by heat.

CFCs See **chlorofluorocarbons.**

change of state A physical process in which one state of matter is changed into another (such as melting a solid to form a liquid).

Charles's law The volume of an ideal gas at constant pressure and amount of gas varies directly with its absolute temperature.

chelating ligand A ligand that uses more than one atom to bind to the same metal ion in a complex ion.

chemical change (chemical reaction) A process in which substances (reactants) change into other substances (products) by rearrangement, combination, or separation of atoms.

chemical compound (compound) A pure substance (e.g., sucrose or water) that can be decomposed into two or more different pure substances; homogeneous, constant-composition matter that consists of two or more chemically combined elements.

chemical element (element) A substance (e.g., carbon, hydrogen, or oxygen) that cannot be decomposed into two or more new substances by chemical or physical means.

chemical equilibrium A state in which the concentrations of reactants and products remain constant because the rates of forward and reverse reactions are equal.

chemical formula (formula) A notation combining element symbols and numerical subscripts that shows the relative numbers of each kind of atom in a molecule or formula unit of a substance.

chemical fuel A substance that reacts exothermically with atmospheric oxygen and is available at reasonable cost and in reasonable quantity.

chemical kinetics The study of the speeds of chemical reactions and the nanoscale pathways or rearrangements by which atoms, ions, and molecules are converted from reactants to products.

chemical periodicity, law of Law stating that the properties of the elements are periodic functions of atomic number.

chemical property Describes the kinds of chemical reactions that chemical elements or compounds can undergo.

chemical reaction (chemical change) A process in which substances (reactants) change into other substances (products) by rearrangements, combination, or separation of atoms.

chemistry The study of matter and the changes it can undergo.

chemotrophs (See also **phototrophs.**) Organisms that must depend on phototrophs to create the chemical substances from which they obtain Gibbs free energy.

chiral Describes a molecule or object that is not superimposable on its mirror image.

chlor-alkali process Electrolysis process for producing chlorine and sodium hydroxide from aqueous sodium chloride.

chlorination Addition of chlorine or a chlorine compound to kill bacteria in municipal water supplies; HOCl formed in water is the antibacterial agent.

chlorofluorocarbons (CFCs) Compounds of carbon, fluorine, and chlorine. CFCs have been implicated in stratospheric ozone depletion.

***cis* isomer** The isomer in which two like substituents are on the same side of a carbon-carbon double bond, the same side of a ring of carbon atoms, or the same side of a complex ion.

***cis-trans* isomerism** A form of stereoisomerism in which the isomers have the same molecular formula and the same atom-to-atom bonding sequence, but the atoms differ in the location of pairs of substituents on the same side or on opposite sides of a molecule.

Clausius-Clapeyron equation Equation that gives the relationship between vapor pressure and temperature.

closest packing Arranging atoms so that they are packed into the minimum volume.

coagulation The process in which the protective charge layer on colloidal particles is overcome, causing them to aggregate into a soft, semisolid, or solid mass.

coefficients (stoichiometric coefficients) The multiplying numbers assigned to the formulas in a chemical equation in order to balance the equation.

cofactor An inorganic or organic molecule or ion required by an enzyme to carry out its catalytic function.

colligative properties Properties of solutions that depend only on the concentration of solute particles in the solution, not on the nature of the solute particles.

colloid A state intermediate between a solution and a suspension, in which solute particles are large enough to scatter light, but too small to settle out; found in gas, liquid, and solid states.

combination reaction A reaction in which two reactants combine to give a single product.

combined gas law A form of the ideal gas law that relates the *P, V, T* of a given amount of gas before and after a change: $P_1V_1/T_1 = P_2V_2/T_2$.

combining volumes, law of At constant temperature and pressure, the volumes of reacting gases are always in the ratios of small whole numbers.

combustion analysis A quantitative method to obtain percent composition data for compounds that can burn in oxygen.

combustion reaction A reaction in which an element or compound burns in air or oxygen.

common ion effect Shift in equilibrium position that results from addition of an ion identical to one in the equilibrium.

complementary base pair Bases, each in a different DNA strand, that hydrogen-bond to each other: guanine with cytosine and adenine with thymine.

complex ion An ion with several molecules or ions connected to a central metal ion by coordinate covalent bonds.

composites Materials with components that may be metals, polymers, and ceramics.

compound See **chemical compound.**

compressibility The property of a gas that allows it to be compacted into a smaller volume by application of pressure.

concentration The relative quantities of solute and solvent in a solution.

concentration cell An electrochemical cell in which the voltage is generated because of a difference in concentrations of the same chemical species.

concrete A mixture of cement, sand, and aggregate (crushed stone or pebbles) in varying proportions that reacts with water and carbon dioxide to form a rock-hard solid.

condensation Process whereby a molecule in the gas phase enters the liquid phase.

condensation polymer A polymer made from the condensation reaction of monomer molecules that contain two or more functional groups.

condensation reaction A chemical reaction in which two (or more) molecules combine to form a larger molecule, simultaneously producing a small molecule such as water.

condensed formula A chemical formula of an organic compound indicating how atoms are grouped together in a molecule.

conduction band In a solid, an energy band that contains electrons of higher energy than those in the valence band.

conductor A material that conducts electric current; has an overlapping valence band and conduction band.

conjugate acid-base pair A pair of molecules or ions related to one another by the loss and gain of a single hydrogen ion.

conjugated Refers to a system of alternating single and double bonds in a molecule.

conservation of energy, law of (first law of thermodynamics) Law stating that energy can be neither created nor destroyed—the total energy of the universe is constant.

conservation of mass, law of Law stating that there is no detectable change in mass during an ordinary chemical reaction.

constant composition, law of Law stating that a chemical compound always contains the same elements in the same proportions by mass.

constitutional isomers (structural isomers) Compounds with the same molecular formula that differ in the order in which their atoms are bonded together.

continuous phase The solvent-like dispersing medium in a colloid.

continuous spectrum A spectrum consisting of all possible wavelengths.

conversion factor (proportionality factor) A relationship between two measurement units derived from the proportionality of one quantity to another; e.g., density is the conversion factor between mass and volume.

coordinate covalent bond A chemical bond in which both of the two electrons forming the bond were originally associated with the same one of the two bonded atoms.

coordination compound A compound in which complex ions are combined with oppositely charged ions to form a neutral compound.

coordination number The number of coordinate covalent bonds between ligands and a central metal ion in a complex ion.

copolymer A polymer formed by combining two different types of monomers.

core electrons The electrons in the filled inner shells of an atom.

corrosion Oxidation of a metal exposed to the environment.

coulomb The unit of electrical charge equal to the quantity of charge that passes a fixed point in an electrical circuit when a current of one ampere flows for one second.

Coulomb's law Law that represents the force of attraction between two charged particles; $F = k(q_1q_2/d^2)$.

covalent bond Interatomic attraction resulting from the sharing of electrons between two atoms.

critical mass The minimum quantity of fissionable material needed to support a self-sustaining chain reaction.

critical pressure The vapor pressure of a liquid at its critical temperature.

critical temperature The temperature above which there is no distinction between liquid and vapor phases.

cryogens Liquefied gases that have temperatures below -150 °C.

crystal-field splitting energy Energy difference between sets of *d* orbitals on the central metal ion in a coordination compound.

crystal-field theory Theory that predicts spectra and magnetism of coordination compounds based on electrostatic bonding between ligands and a metal ion.

crystal lattice The ordered, repeating arrangement of ions, molecules, or atoms in a crystalline solid.

crystalline solids Solids with an ordered arrangement of atoms, molecules, or ions that results in planar faces and sharp angles of the crystals.

crystallization The process in which mobile atoms, molecules, or ions in a liquid or solution convert into a crystalline solid.

cubic close packing The three-dimensional structure that results when atoms or ions are closest packed in the *abcabc* arrangement.

cubic unit cell A unit cell with equal-length edges that meet at 90° angles.

curie (Ci) A unit of radioactivity equal to 3.7×10^{10} disintegrations per second.

Dalton's law of partial pressures The total pressure exerted by a mixture of gases is the sum of the partial pressures of the individual gases in the mixture.

decomposition reaction A reaction in which a compound breaks down chemically to form two or more simpler substances.

degree of polymerization Number of repeating units in a polymer chain.

delocalized electrons Electrons, such as in benzene, that are spread over several atoms in a molecule or polyatomic ion.

denaturation Disruption in protein secondary and tertiary structure brought on by high temperature, heavy metals, and other substances.

density The ratio of the mass of an object to its volume.

deoxyribonucleic acid (DNA) A double-stranded polymer of nucleotides that stores genetic information.

deposition The process of a gas converting directly to a solid.

detergent(s) Molecules whose structure contains a long hydrocarbon portion that is hydrophobic and a polar end that is hydrophilic.

dew point Temperature at which the actual partial pressure of water vapor equals the equilibrium vapor pressure.

diamagnetic Describes atoms or ions in which all the electrons are paired in filled shells so their magnetic fields effectively cancel each other.

diatomic molecule A molecule that contains two atoms.

dietary minerals Essential elements that are not carbon, hydrogen, oxygen, or nitrogen.

diffusion Spread of gas molecules of one type through those of another type.

dimensional analysis A method of using units in calculations to check for correctness.

dimer A molecule made from two smaller units.

dipole moment The product of the magnitude of the partial charges ($\delta+$ and $\delta-$) of a molecule times the distance of separation between the charges.

dipole-dipole attraction The noncovalent force of attraction between any two polar molecules or polar regions in the same large molecule.

disaccharides Carbohydrates such as sucrose consisting of two monosaccharide units.

dispersed phase The larger-than-molecule-sized particles that are distributed uniformly throughout a colloid.

displacement reaction A reaction in which one element reacts with a compound to form a new compound and release a different element.

doping Adding a tiny concentration of one substance (a *dopant*) to improve the semiconducting properties of another.

double bond A bond formed by sharing two pairs of electrons between the same two atoms.

dynamic equilibrium A balance between opposing reactions occurring at equal rates.

effective nuclear charge The nuclear positive charge experienced by outer-shell electrons in a many-electron atom.

effusion Escape of gas molecules from a container through a tiny hole into a vacuum.

electrochemical cell A combination of anode, cathode, and other materials arranged so that a product-favored oxidation-reduction reaction can cause a current to flow or an electric current can cause a reactant-favored redox reaction to occur.

electrochemistry The study of the relationship between electron flow and oxidation-reduction reactions.

electrode A device such as a metal plate or wire that conducts electrons into and out of a system.

electrolysis The use of electrical energy to produce a chemical change.

electrolyte A substance that ionizes or dissociates when dissolved in water to form an electrically conducting solution.

electromagnetic radiation Radiation that consists of oscillating electric and magnetic fields that travel through space at the same rate (the speed of light: 186,000 miles/s or 10^8 m/s in a vacuum).

electromotive force (emf) The difference in electrical potential energy between the two electrodes in an electrochemical cell, measured in volts.

electron A negatively charged subatomic particle that occupies most of the volume of an atom.

electron affinity The energy change when a mole of electrons is added to a mole of atoms in the gas phase.

electron capture A radioactive decay process in which one of an atom's inner-shell electrons is captured by the nucleus, which decreases the atomic number by 1.

electron configuration The complete description of the orbitals occupied by all the electrons in an atom or ion.

electron density The probability of finding an electron within a tiny volume in an atom; determined by the square of the wave function.

electron-pair geometry The geometry around a central atom including the spatial positions of bonding and lone electron pairs.

electronegativity A measure of the ability of an atom in a molecule to attract bonding electrons to itself.

electronically excited molecule A molecule whose potential energy is greater than the minimum (ground-state) energy because of a change in its electronic structure.

element (chemical element) A substance (e.g., carbon, hydrogen, and oxygen) that cannot be decomposed into two or more new substances by chemical or physical means.

elementary reaction A nanoscale reaction whose equation indicates exactly which atoms, ions, or molecules collide or change as the reaction occurs.

empirical formula A formula showing the simplest possible ratio of atoms of elements in a compound.

emulsion A colloid consisting of a liquid dispersed in a second liquid; formed by the presence of an *emulsifier* that coats and stabilizes dispersed-phase particles.

enantiomers A pair of molecules consisting of a chiral molecule and its mirror-image isomer.

end point The point at which the indicator changes color during a titration.

endergonic Refers to a reaction that requires input of Gibbs free energy; applies to biochemical reactions that are reactant-favored at body temperature.

endothermic (process) A process in which thermal energy must be transferred into a thermodynamic system in order to maintain constant temperature.

energy The capacity to do work.

energy band In a solid, a large group of orbitals whose energies are closely spaced; in an atomic solid the average energy of a band equals the energy of the corresponding orbital in an individual atom.

energy conservation The conservation of useful energy, that is, of Gibbs free energy.

energy density The quantity of energy released per unit volume of a fuel.

enthalpy change (ΔH) The quantity of thermal energy transferred when a process takes place at constant temperature and pressure.

enthalpy of fusion The enthalpy change when a substance melts; the quantity of energy that must be transferred when a substance melts at constant temperature and pressure.

enthalpy of solution The quantity of thermal energy transferred when a solution is formed at constant T and P.

enthalpy of sublimation The enthalpy change when a solid sublimes; the quantity of energy, at constant pressure, that must be transferred to cause a solid to vaporize.

enthalpy of vaporization The enthalpy change when a substance vaporizes: the quantity of energy that must be transferred when a liquid vaporizes at constant temperature and pressure.

entropy A measure of the number of ways energy can be distributed in a system; a measure of the dispersal of energy in a system.

enzyme A highly efficient biochemical catalyst for one or more reactions in a living system.

enzyme-substrate complex The combination formed by the binding of an enzyme with a substrate through noncovalent forces.

equatorial position Position lying on the equator of an imaginary sphere around a triangular bipyramidal molecular or ionic structure.

equilibrium concentration The concentration of a substance (usually expressed as molarity) in a system that has reached the equilibrium state.

equilibrium constant (K) A quotient of equilibrium concentrations of product and reactant substances that has a constant value for a given reaction at a given temperature.

equilibrium constant expression The mathematical expression associated with an equilibrium constant.

equilibrium vapor pressure Pressure of the vapor of a substance in equilibrium with its liquid or solid in a closed container.

equivalence point The point in a titration at which a stoichiometrically equivalent amount of one substance has been added to another substance.

ester An organic compound formed by the reaction of an alcohol and a carboxylic acid.

evaporation The process of conversion of a liquid to a gas.

exchange reaction A reaction in which cations and anions that were partners in the reactants are interchanged in the products.

excited state The unstable state of an atom or molecule in which at least one electron does not have its lowest possible energy.

exergonic Refers to a reaction that releases Gibbs free energy; applies to biochemical reactions that are product-favored at body temperature.

exothermic Refers to a process in which thermal energy must be transferred out of a thermodynamic system in order to maintain constant temperature.

extent of reaction The fraction of reactants that has been converted to products.

Faraday constant (F) The quantity of electric charge on one mole of electrons, 9.6485×10^4 C/mol.

fat A solid triester of fatty acids with glycerol.

ferromagnetic A substance that contains clusters of atoms with unpaired electrons whose magnetic spins become aligned, causing permanent magnetism.

first law of thermodynamics (law of conservation of energy) Energy can neither be created nor destroyed—the total energy of the universe is constant.

formal charge The charge a bonded atom would have if its electrons were shared equally.

formation constant (K_f) The equilibrium constant for the formation of a complex ion.

formula (chemical formula) A notation combining element symbols and numerical subscripts that shows the relative numbers of each kind of atom in a molecule or formula unit of a substance.

formula unit The simplest cation-anion grouping represented by the formula of an ionic compound; also the collection of atoms represented by any formula.

formula weight The sum of the atomic weights in amu of all the atoms in a compound's formula.

fractional distillation The process of refining petroleum (or another mixture) by distillation to separate it into groups (fractions) of compounds having distinctive boiling point ranges.

Frasch process Process for recovering sulfur from underground deposits by melting the sulfur with superheated water.

free radical An atom, ion, or molecule that contains one or more unpaired electrons; usually highly reactive.

freezing-point lowering A colligative property; the difference between the freezing point of a pure solvent and the freezing point of a solution in which a nonvolatile nonelectrolyte solute is dissolved in the solvent.

frequency The number of complete traveling waves passing a point in a given period of time (cycles per second).

frequency factor The factor (A) in the Arrhenius equation that depends on how often molecules collide when all concentrations are 1 mol/L and on whether the molecules are properly oriented to react when they collide.

fuel cell An electrochemical cell that converts the chemical energy of fuels directly into electricity.

fuel value The quantity of energy released when 1 g of a fuel is burned.

fullerenes Allotropic forms of carbon that consist of many five- and six-membered rings of carbon atoms sharing edges.

functional group An atom or group of atoms that imparts characteristic properties and defines a given class of organic compounds (e.g., the —OH group is present in all alcohols).

galvanized Has a thin coating of zinc metal that forms an oxide coating impervious to oxygen, thereby protecting a less active metal, such as iron, from corrosion.

gamma radiation Radiation composed of highly energetic photons.

gas A phase or state of matter in which a substance has no definite shape and has a volume determined by the volume of its container.

gasohol A blended motor fuel consisting of 90% gasoline and 10% ethanol.

gene The unique sequence of nitrogen bases in DNA that codes for the synthesis of a specific protein; carrier of a genetic trait.

Gibbs free energy A thermodynamic function that decreases for any product-favored system. For a process at constant temperature and pressure, $\Delta G = \Delta H - T\Delta S$.

glass Amorphous, clear solids formed from silicates and other oxides.

global warming Increase in temperature at earth's surface as a result of the greenhouse effect amplified by increasing concentrations of carbon dioxide and other greenhouse gases.

glycogen A highly branched, high-molar-mass polymer of glucose found in animals.

glycosidic linkage The C—O—C bond that connects monosaccharides in disaccharides and polysaccharides; forms between carbons 1 and 4 or 1 and 6 of linked monosaccharides.

gram(s) The basic unit of mass in the metric system; equal to 1×10^{-3} kg.

gray The SI unit of absorbed radiation dose equal to the absorption of 1 joule per kilogram of material.

greenhouse effect Atmospheric warming caused when atmospheric carbon dioxide, water vapor, methane, ozone, and other greenhouse gases absorb infrared radiation reradiated from earth.

ground state The state of an atom or molecule in which all of the electrons are in their lowest possible energy levels.

groups The vertical columns of the periodic table of the elements.

Haber-Bosch process The process developed by Fritz Haber and Carl Bosch for the direct synthesis of ammonia from its elements.

half-cell One half of an electrochemical cell in which only the anode or cathode is located.

half-life, $t_{1/2}$ The time required for the concentration of one reactant to reach half its original value; radioactivity—the time required for the activity of a radioactive sample to reach half of its original value.

half-reaction A reaction that represents either an oxidation or a reduction process.

halide ion A monatomic ion (1−) of a halogen.

halogens The elements in Group 7A of the periodic table.

heat (heating) The energy-transfer process between two samples of matter at different temperatures.

heat capacity The quantity of energy that must be transferred to an object to raise its temperature by 1 °C.

heat of See **enthalpy of.**

heating curve A plot of the temperature of a substance versus the quantity of energy transferred to it by heating.

helium burning The fusion of helium nuclei to form beryllium-8, as it occurs in stars.

Henderson-Hasselbalch equation The equation describing the relationships among the pH of a buffer solution, the pK_a of the acid, and the concentrations of the acid and its conjugate base.

Henry's law A mathematical expression for the relationship of gas pressure and solubility; $S_g = k_H P_g$.

Hess's law If two or more chemical equations can be combined to give another equation, the enthalpy change for that equation will be the sum of the enthalpy changes for the equations that were combined.

heterogeneous catalyst A catalyst that is in a different phase from that of the reaction mixture.

heterogeneous mixture A mixture in which components remain separate and can be observed as individual substances or phases.

heterogeneous reaction A reaction that takes place at an interface between two phases, solid and gas for example.

hexadentate A ligand in which each of six different atoms donates an electron pair to a coordinated central metal ion.

hexagonal close packing The three-dimensional structure that results when layers of atoms in a solid are closest packed in the *ababab* arrangement.

high-spin complex A complex ion that has the maximum possible number of unpaired electrons.

homogeneous catalyst A catalyst that is in the same phase as that of the reaction mixture.

homogeneous mixture A mixture of two or more substances in a single phase that is uniform throughout.

homogeneous reaction A reaction in which the reactants and products are all in the same phase.

Hund's rule Electrons pair only after each orbital in a subshell is occupied by a single electron.

hybrid orbitals Orbitals formed by combining atomic orbitals of appropriate energy and orientation.

hybridized Refers to atomic orbitals of proper energy and orientation that have combined to form hybrid orbitals.

hydrate A solid compound that has a stoichiometric amount of water molecules bonded to metal ions or trapped within its crystal lattice.

hydration The binding of one or more water molecules to an ion or molecule within a solution or within a crystal lattice.

hydrocarbon An organic compound composed only of carbon and hydrogen.

hydrogen bond Noncovalent interaction between a hydrogen atom and a very electronegative atom to produce an unusually strong dipole-dipole force.

hydrogen burning The fusion of hydrogen nuclei (protons) to form helium, as it occurs in stars.

hydrogenation An addition reaction in which hydrogen adds to the double bond of an alkene; the catalyzed reaction of H_2 with a liquid triglyceride to produce saturated fatty acid chains, which convert the triglyceride into a semisolid or solid.

hydrolysis A reaction in which a bond is broken by reaction with a water molecule and the —H and —OH of the water add to the atoms of the broken bond.

hydronium ion H_3O^+; the simplest proton-water complex; responsible for acidity.

hydrophilic "Water-loving," a term describing a polar molecule or part of a molecule that is strongly attracted to water molecules.

hydrophobic "Water-fearing," a term describing a nonpolar molecule or part of a molecule that is not attracted to water molecules.

hydroxide ion OH^- ion; bases increase the concentration of hydroxide ions in solution.

hypertonic Refers to a solution having a higher concentration of nanoscale particles and therefore a higher osmotic pressure than another solution.

hypothesis A tentative explanation for an observation and a basis for experimentation.

hypotonic Refers to a solution having a lower solute concentration of nanoscale particles and therefore a lower osmotic pressure than another solution.

ideal gas A gas that behaves exactly as described by the ideal gas law, and by Boyle's, Charles's, and Avogadro's laws.

ideal gas constant The proportionality constant, R, in the equation $PV = nRT$; $R = 0.0821 \text{ L atm mol}^{-1} \text{ K}^{-1} = 8.314 \text{ J K}^{-1} \text{ mol}^{-1}$.

ideal gas law A law that relates pressure, volume, amount (moles), and temperature for an ideal gas; the relationship expressed by the equation $PV = nRT$.

immiscible Describes two liquids that form two separate phases when mixed because each is only slightly soluble in the other.

induced dipole A temporary dipole created by a momentary uneven distribution of electrons in a molecule or atom.

induced fit The change in the shape of an enzyme, its substrate, or both when they bind.

inhibitor A molecule or ion other than the substrate that causes a decrease in catalytic activity of an enzyme.

initial rate The instantaneous rate of a reaction determined at the very beginning of the reaction.

initiation The breaking of a carbon-carbon double bond in a polymerization reaction to produce a molecule with highly reactive sites that react with other molecules to produce a polymer; first step in a chain reaction.

inorganic compound A chemical compound that is not an organic compound; usually of mineral or nonbiological origin.

insoluble Describes a solute, almost none of which dissolves in a solvent.

instantaneous reaction rate The rate at a particular time after a reaction has begun.

insulator A material that has a large energy gap between fully occupied and empty energy bands, and does not conduct electricity.

intermolecular forces Noncovalent attractions between separate molecules.

internal energy The sum of the individual energies (kinetic and potential) of all of the nanoscale particles (atoms, molecules, or ions) in a sample of matter.

ion An atom or group of atoms that has lost or gained one or more electrons so that it is no longer electrically neutral.

ion product (Q) A value found from an expression with the same mathematical form as the solubility product expression (K_{sp}) but using the actual concentrations rather than equilibrium concentrations of the species involved. (See **reaction quotient.**)

ionic compound A compound that consists of positive and negative ions (cations and anions).

ionic hydrate Ionic compounds that incorporate water molecules in the ionic crystal lattice.

ionic radius Radius of an anion or cation in an ionic compound.

ionization constant for water (K_w) The equilibrium constant that is the mathematical product of the hydronium ion concentration and the concentration of hydroxide ion in any aqueous solution; $K_w = 1 \times 10^{-14}$ at 25 °C.

ionization energy The energy needed to remove a mole of electrons from a mole of atoms in the gas phase.

isoelectronic Refers to atoms and ions that have identical electron configurations.

isomers Compounds that have the same molecular formula but different arrangements of atoms.

isotonic Refers to a solution having the same concentration of nanoscale particles and therefore the same osmotic pressure as another solution.

isotopes Forms of an element composed of atoms with the same atomic number but different mass numbers owing to a difference in the number of neutrons.

joule (J) A unit of energy equal to 1 kg m^2/s^2. The kinetic energy of a 2-kg object traveling at a speed of 1 m/s.

Kelvin temperature scale (See also **absolute temperature scale.**) A temperature scale on which the zero is the lowest possible temperature and the degree is the same size as a Celsius degree.

ketone An organic compound characterized by the presence of a carbonyl group in which the carbon atom is bonded to two other carbon atoms ($R_2C{=}O$).

kilocalorie (kcal or Cal) (See also **calorie.**) A unit of energy equal to 4.184 kJ. Approximately 1 kcal (1 Cal) is required to raise the temperature of 1 kg of liquid water by 1 °C. The food Calorie.

kinetic energy Energy that an object has because of its motion. Equal to $\frac{1}{2}mv^2$, where m is the object's mass and v is its velocity.

kinetic-molecular theory The theory that matter consists of nanoscale particles that are in constant, random motion.

lanthanide contraction The decrease in atomic radii across the fourth-period and the fifth-period *f*-block elements (lanthanides and actinides) due to the lack of electron screening by electrons in the 4*f* and 5*f* orbitals.

lanthanides The elements after lanthanum in the sixth period in which the 4*f* subshell is being filled.

lattice energy Enthalpy of formation of 1 mol of an ionic solid from its separated gaseous ions.

law A statement that summarizes a wide range of experimental results and has not been contradicted by experiments.

law of chemical periodicity Law stating that the properties of the elements are periodic functions of atomic number.

law of combining volumes At constant temperature and pressure, the volumes of reacting gases are always in the ratios of small whole numbers.

law of conservation of energy (first law of thermodynamics) Law stating that energy can be neither created nor destroyed—the total energy of the universe is constant.

law of conservation of mass Law stating that there is no detectable change in mass in an ordinary chemical reaction.

law of constant composition Law stating that a chemical compound always contains the same elements in the same proportions by mass.

law of multiple proportions When two elements A and B can combine in two or more ways, the mass ratio A : B in one compound is a small-whole-number multiple of the mass ratio A : B in the other compound.

Le Chatelier's principle If a system is at equilibrium and the conditions are changed so that it is no longer at equilibrium, the system will react to give a new equilibrium in a way that partially counteracts the change.

Lewis acid A molecule or ion that can accept an electron pair from another atom, molecule, or ion to form a new bond.

Lewis base A molecule or ion that can donate an electron pair to another atom, molecule, or ion to form a new bond.

Lewis structure Structural formula for a molecule that shows all valence electrons as dots or as lines that represent covalent bonds.

Lewis dot symbol An atomic symbol with dots representing valence electrons.

ligands Atoms, molecules, or ions bonded to a central atom, such as the central metal ion in a coordination complex.

limiting reactant The reactant present in limited supply that controls the amount of product formed in a reaction.

line emission spectrum A spectrum produced by excited atoms and consisting of discrete wavelengths of light.

linear Molecular geometry in which there is a angle between bonded atoms.

lipid bilayer The structure of cell membranes that are composed of two aligned layers of phospholipids with their hydrophobic regions within the bilayer.

liquid A phase of matter in which a substance has no definite shape but a definite volume.

London forces Forces resulting from the attraction between positive and negative regions of momentary (induced) dipoles in neighboring molecules.

lone-pair electrons Paired valence electrons unused in bond formation; also called nonbonding pairs.

low-spin complex A complex ion that has the minimum possible number of unpaired electrons.

major minerals Dietary minerals present in humans in quantities greater than 100 mg per kg of body weight.

macromolecule A very large polymer molecule made by chemically joining many small molecules (monomers).

macroscale Refers to samples of matter that can be observed by the unaided human senses; samples of matter large enough to be seen, measured, and handled.

main-group elements Elements in the eight A groups to the left and right of the transition elements in the periodic table; the *s*- and *p*-block elements.

mass A measure of an object's resistance to acceleration.

mass fraction The ratio of the mass of one component to the total mass of a sample.

mass number The number of protons plus neutrons in the nucleus of an atom of an element.

mass percent The mass fraction multiplied by 100%.

mass spectrometer An analytical instrument used to measure atomic and molecular masses directly.

mass spectrum A plot of ion abundance versus the mass of the ions; produced by a mass spectrometer.

materials science The science of the relationships between the structure and the chemical and physical properties of materials.

matter Anything that has mass and occupies space.

melting point The temperature at which the structure of a solid collapses and the solid changes to a liquid.

meniscus A concave or convex surface that forms on a liquid as a result of the balance of noncovalent forces in a narrow container.

metabolism All of the chemical reactions that occur as an organism converts food nutrients into constituents of living cells, to stored Gibbs free energy, or to thermal energy.

metal An element that is malleable, ductile, forms alloys, and conducts an electric current.

metal activity series A ranking of relative reactivity of metals in displacement and other kinds of reactions.

metallic bonding In solid metals, the nondirectional attraction between positive metal ions and the surrounding sea of negatively charged electrons.

metalloid An element that has some typically metallic properties and other properties that are more characteristic of nonmetals.

methyl group A $-CH_3$ group.

metric system A decimalized measurement system.

micelles Colloid-sized particles built up from many surfactant molecules; micelles can transport various materials within them.

microscale Refers to samples of matter so small that they have to be viewed with a microscope.

millimeters of mercury (mm Hg) A unit of pressure related to the height of a column of mercury in a mercury barometer (760 mm Hg = 1 atm = 101.3 kPa).

mineral A naturally occurring inorganic compound with a characteristic composition and crystal structure.

miscible Describes two liquids that will dissolve in each other in any proportion.

model A mechanical or mathematical way to make a theory more concrete, such as a molecular model.

molality (*m*) A concentration term equal to the molar amount of solute per kilogram of solvent.

molar amount See **amount.**

molar enthalpy of fusion The energy transfer required to melt 1 mol of a pure solid.

molar heat capacity The quantity of energy that must be transferred to 1 mol of a substance to increase its temperature by 1 °C.

molar mass The mass in grams of 1 mol of atoms, molecules, or formula units of one kind, numerically equal to the atomic or molecular weight in amu.

molar solubility The solubility of a solute in a solvent, expressed in moles per liter.

molarity Solute concentration expressed as the molar amount of solute per liter of solution.

mole (mol) The amount of substance that contains as many elementary particles as there are atoms in exactly 0.012 kg of carbon-12 isotope.

mole fraction (X) The ratio of number of moles of one component to the total number of moles in a mixture of substances.

mole ratio (stoichiometric factor) A mole-to-mole ratio relating the molar amount of a reactant or product to the molar amount of another reactant or product.

molecular compound A compound composed of atoms of two or more elements chemically combined in molecules.

molecular formula A formula that expresses the number of atoms of each type within one molecule of a substance.

molecular geometry The three-dimensional arrangement of atoms in a molecule.

molecular orbitals Orbitals extending over an entire molecule generated by combining atomic orbitals.

molecular weight The sum of the atomic weights of all the atoms in a substance's formula.

molecule The smallest particle of an element or compound that exists independently, and retains the chemical properties of that element or compound.

momentum The product of the mass (m) times the velocity (v) of an object in motion.

monatomic ion An ion consisting of one atom bearing an electrical charge.

monodentate ligand A ligand that donates one electron pair to a coordinated metal ion.

monomer The small repeating unit from which a polymer is formed.

monoprotic acid An acid that can donate a single hydrogen ion per molecule.

monosaccharides The simplest carbohydrates, composed of one saccharide unit.

monounsaturated fatty acid Refers to fatty acids, such as oleic acid, that contain only one carbon-carbon double bond.

mortar A mixture of cement, sand, and lime that reacts with water and carbon dioxide to form a rock-hard solid.

multiple covalent bonds Double or triple covalent bonds.

multiple proportions, law of When two elements A and B can combine in two or more ways, the mass ratio A : B in one compound is a small-whole-number multiple of the mass ratio A : B in the other compound.

***n*-type semiconductor** A material made by doping a semiconductor with an impurity that leaves extra valence electrons.

nanoscale Refers to samples of matter (e.g., atoms and molecules) whose normal dimensions are in the 1–100 nanometer range.

nanotubes Members of the family of fullerenes in which graphite-like layers of carbon atoms form cylindrical shapes.

Nernst equation The equation relating the potential of an electrochemical cell to the concentrations of the chemical species involved in the oxidation-reduction reactions occurring in the cell.

net ionic equation A chemical equation in which only those molecules or ions undergoing chemical changes in the course of the reaction are represented.

network solid A solid consisting of one huge molecule in which all atoms are connected via a network of covalent bonds.

neurons Specialized cells that are part of animals' nervous systems and that function according to electrochemical principles.

neutral solution A solution containing equal concentrations of H_3O^+ and OH^-; a solution that is neither acidic nor basic.

neutron An electrically neutral subatomic particle found in the nucleus.

newton (N) The SI unit of force; equal to 1 kg times an acceleration of $1\ m/s^2$; $1\ kg\ m/s^2$.

nitrogen cycle The natural cycle of chemical transformations involving nitrogen and its compounds.

nitrogen fixation The conversion of atmospheric nitrogen (N_2) to nitrogen compounds utilizable by plants or industry.

noble gas notation An abbreviated electron configuration of an element in which filled inner shells are represented by the symbol of the preceding noble gas in brackets. For Al, this would be $[Ne]3s^23p^1$.

noble gases Gaseous elements in Group 8A; the least reactive elements.

nonbiodegradable Not capable of being decomposed by microorganisms.

noncovalent interactions All forces of attraction other than covalent, ionic, or metallic bonding.

nonelectrolyte A substance that dissolves in water to form a solution that does not conduct electricity.

nonmetal Element that does not have the chemical and physical properties of a metal.

nonpolar covalent bond A bond in which the electron pair is shared equally by the bonded atoms.

nonpolar molecule A molecule that is not polar either because it has no polar bonds or because its polar bonds are oriented symmetrically so that they cancel each other.

NO_x Oxides of nitrogen.

normal boiling point The temperature at which the vapor pressure of a liquid equals 1 atm.

nuclear burning The nuclear fusion reactions by which elements are formed in stars.

nuclear decay Spontaneous emission of radioactivity by an unstable nucleus that is converted into a more stable nucleus.

nuclear fission The highly exothermic process by which very heavy fissionable nuclei split to form lighter nuclei.

nuclear fusion The highly exothermic process by which very light nuclei combine to form heavier nuclei.

nuclear magnetic resonance The process in which the nuclear spins of atoms align in a magnetic field and absorb radio frequency photons to become excited. These excited atoms then return to a lower energy state when they emit the absorbed radio frequency photons.

nuclear medicine The use of radioisotopes in medical diagnosis and therapy.

nuclear reaction A process in which one or more atomic nuclei change into one or more different nuclei.

nuclear (atomic) reactor A container in which a controlled nuclear reaction takes place.

nucleon A nuclear particle, either a neutron or a proton.

nucleotide Repeating unit in DNA, composed of one sugar unit, one phosphate group, and one cyclic nitrogen base.

nucleus (atomic) The tiny central core of an atom; contains protons and neutrons. (There are no neutrons in hydrogen-1.)

nutrients The chemical raw materials, eaten as food, that are needed for survival of an organism.

octahedral Molecular geometry of six groups around a central atom in which all groups are at angles of 90° to other groups.

octane number A measure of the ability of a gasoline to burn smoothly in an internal-combustion engine.

octet rule In forming bonds, many main group elements gain, lose, or share electrons to achieve a stable electron configuration characterized by eight valence electrons.

optical fiber A fiber made of glass constructed so that light can pass through it with little loss of intensity; used for transmission of information.

orbital A region of an atom or molecule within which there is a significant probability that an electron will be found.

orbital shape The shape of an electron density distribution determined by an orbital.

order of reaction The reaction rate dependency on the concentration of a reactant or product, expressed as an exponent of a concentration term in the rate equation.

ores Minerals containing a sufficiently high concentration of an element to make its extraction profitable.

organic compound A compound of carbon with hydrogen, possibly also oxygen, nitrogen, sulfur, phosphorus, or other elements.

osmosis The movement of a solvent (water) through a semipermeable membrane from a region of lower solute concentration to a region of higher solute concentration.

osmotic pressure (Π) The pressure that must be applied to a solution to stop osmosis from a sample of pure solvent.

overall reaction order The sum of the exponents for all concentration terms in the rate equation.

oxidation The loss of electrons by an atom, ion, or molecule, leading to an increase in oxidation number.

oxidation number (oxidation state) The hypothetical charge an atom would have if all bonds to that atom were completely ionic.

oxidation-reduction reaction (redox reaction) A reaction involving the transfer of one or more electrons from one species to another so that oxidation numbers change.

oxides Compounds of oxygen combined with another element.

oxidized The result when an atom, molecule, or ion loses one or more electrons.

oxidizing agent The substance that accepts electron(s) and is reduced in an oxidation-reduction reaction.

oxoacids Acids in which the acidic hydrogen is bonded directly to an oxygen atom.

oxoanion A polyatomic anion that contains oxygen.

oxygenated gasolines Blends of gasoline with oxygen-containing organic compounds such as methanol, ethanol, and *tertiary*-butyl alcohol.

ozone hole Regions of ozone depletion in the stratosphere centered on the earth's poles, most significantly the South Pole.

ozone layer Region of maximum ozone concentration in the stratosphere.

***p*-block elements** Main-group elements in Groups 3A through 8A whose valence electrons consist of outermost *s* and *p* electrons.

***p-n* junction** An interface between *p*-type and *n*-type semiconductors that produces a rectifier that allows current to flow in only one direction.

***p*-type semiconductor** A material made by doping a semiconductor with an impurity that leaves a deficiency of valence electrons.

paramagnetic Refers to atoms, molecules, or ions that are attracted to a magnetic field because they have unpaired electrons in incompletely filled electron subshells.

partial hydrogenation Addition of hydrogen to some of the carbon-carbon double bonds in a triglyceride (a fat or oil).

partial pressure The pressure that one gas in a mixture of gases would exert if it occupied the same volume at the same temperature as the mixture.

particulate Atmospheric solid particles, generally larger than 10,000 nm in diameter.

parts per billion (ppb) One part in one billion (10^9) parts.

parts per million (ppm) One part in one million (10^6) parts.

parts per trillion (ppt) One part in one trillion (10^{12}) parts.

pascal (Pa) The SI unit of pressure; $1 \text{ Pa} = 1 \text{ N/m}^2$.

Pauli exclusion principle An atomic principle that states that, at most, two electrons can be assigned to the same orbital in the same atom or molecule, and these two electrons must have opposite spins.

peptide linkage The amide linkage between two amino acid molecules; found in proteins.

percent abundance The percentage of atoms of a particular isotope in a natural sample of a pure element.

percent composition by mass The percentage of the mass of a compound represented by each of its constituent elements.

percent yield The ratio of actual yield to theoretical yield, multiplied by 100%.

periodic table A table of elements arranged in order of increasing atomic number so that those with similar chemical and physical properties fall in the same vertical groups.

periods The horizontal rows of the periodic table of the elements.

petroleum fractions The mixtures of hundreds of hydrocarbons in the same boiling point range obtained from the fractional distillation of petroleum.

pH The negative logarithm of the hydronium ion concentration ($-\log [H_3O^+]$).

pH meter An instrument for measuring pH of solutions using electrochemical principles.

phase Any of the three states of matter: gas, liquid, solid. Also, one of two or more solid-state structures of the same substance, such as iron in a body-centered cubic or face-centered cubic structure.

phase change A physical process in which one state or phase of matter is changed into another (such as melting a solid to form a liquid).

phase diagram A diagram showing the relationships among the phases of a substance (solid, liquid, and gas), at different temperatures and pressures.

phospholipid Glycerol derivative with two long, nonpolar fatty-acid chains and a polar phosphate group; present in cell membranes.

photochemical reactions Chemical reactions that take place as a result of absorption of photons by reactant molecules.

photochemical smog Smog produced by strong oxidizing agents, such as ozone and oxides of nitrogen, NO_x, that undergo light-initiated reactions with hydrocarbons.

photodissociation Splitting of a molecule into two free radicals by absorption of an ultraviolet photon.

photoelectric effect The emission of electrons by some metals when illuminated by light of certain wavelengths.

photon A massless particle of light whose energy is given by $h\nu$, where ν is the frequency of the light and h is Planck's constant.

photosynthesis A series of reactions in a green plant that combines carbon dioxide with water to form carbohydrate and oxygen.

phototrophs (See also chemotrophs.) Organisms that can carry out photosynthesis and therefore can use sunlight to supply their Gibbs free energy needs.

physical changes Changes in the physical properties of a substance, such as the transformation of a solid to a liquid.

physical properties Properties (e.g., melting point or density) that can be observed and measured without changing the composition of a substance.

pi (π) bond A bond formed by the sideways overlap of parallel *p* orbitals.

Planck's constant The proportionality constant, h, that relates energy of a photon to its frequency. The value of h is $6.626 \times 10^{-34} \text{ J} \cdot \text{s}$.

plasma A state of matter consisting of unbound nuclei and electrons.

plastic A polymeric material that has a soft or liquid state in which it can be molded or otherwise shaped. See also **thermoplastic** and **thermosetting plastic.**

polar covalent bond A covalent bond between atoms with different electronegativities; bonding electrons are shared unequally between the atoms.

polar molecule A molecule that is polar because it has polar bonds arranged so that electron density is concentrated at one end of the molecule.

polarization The induction of a temporary dipole in a molecule or atom by shifting of electron distribution.

polluted water Water that is unsuitable for an intended use, such as drinking, washing, irrigation, or industrial use.

polyamides Polymers in which the monomer units are connected by amide bonds.

polyatomic ion An ion consisting of more than one atom.

polydentate ligand Refers to ligands that can form two or more coordinate covalent bonds to the same metal ion.

polyester A polymer in which the monomer units are connected by ester bonds.

polymer A large molecule composed of many smaller repeating units, usually arranged in a chain-like structure.

polypeptide A polymer composed of 20–50 amino acid residues joined by peptide linkages (amide linkages).

polyprotic acids Acids that can donate more than one hydrogen ion per molecule.

polysaccharides Carbohydrates that consist of many monosaccharide units.

polyunsaturated acid A carboxylic acid containing two or more carbon-carbon double bonds; commonly refers to a fatty acid.

positron A nuclear particle having the same mass as an electron, but a positive charge.

potential energy Energy that an object has because of its position.

precipitate An insoluble product of an exchange reaction in aqueous solution.

pressure The force exerted on an object divided by the area over which the force is exerted.

primary battery A voltaic cell (or battery of cells) in which the oxidation and reduction half-reactions cannot easily be reversed to restore the cell to its original state.

primary pollutants Pollutants that enter the environment directly from their sources.

primary structure of proteins The sequence of amino acids along the polymer chain in a protein molecule.
principal energy level An energy level containing orbitals with the same quantum number (n = 1, 2, 3 . . .).
principal quantum number An integer assigned to each of the allowed main electron energy levels in an atom.
product A substance formed as a result of a chemical reaction.
product-favored system A system in which, when a reaction appears to be over, products predominate over reactants.
proportionality factor (conversion factor) A relationship between two measurement units derived from the proportionality of one quantity to another; e.g., density is the conversion factor between mass and volume.
proton A positively charged subatomic particle found in the nucleus.
pyrometallurgy The extraction of a metal from its ore using chemical reactions carried out at high temperatures.

qualitative In observations, nonnumerical experimental information, such as a description of color or texture.
quantitative Numerical information, such as the mass or volume of a substance, expressed in appropriate units.
quantum The smallest possible unit of a distinct quantity; for example, the smallest possible unit of energy for electromagnetic radiation of a given frequency.
quantum theory The theory that energy comes in very small packets (quanta); this is analogous to matter occurring in very small particles—atoms.

racemic mixture A mixture of equal amounts of enantiomers of a chiral compound.
rad A unit of radioactivity; a measure of the energy of radiation absorbed by a substance, 1.00×10^{-2} J per kilogram.
radial distribution plot A graph showing the probability of finding an electron as a function of distance from the nucleus of an atom.
radioactive series A series of nuclear reactions in which a radioactive isotope undergoes successive nuclear transformations resulting ultimately in a stable, nonradioactive isotope.
radioactivity The spontaneous emission of energy and/or subatomic particles by unstable atomic nuclei; the energy or particles so emitted.
Raoult's law A mathematical expression for the vapor pressure of the solvent in a solution; $P_1 = X_1 P_1^0$.
rate The change in some measurable quantity per unit time.
rate constant (k) A proportionality constant relating reaction rate and concentrations of reactants and other species that affect the rate of a specific reaction.
rate law (rate equation) A mathematical equation that summarizes the relationship between concentrations and reaction rate.
rate-limiting step The slowest step in a reaction mechanism.
reactant A substance that is initially present and undergoes change in a chemical reaction.
reactant-favored system A system in which, when a reaction appears to be over, reactants predominate over products.
reaction intermediate An atom, molecule, or ion produced in one step and used in a later step in a reaction mechanism; does not appear in the equation for the overall reaction.
reaction mechanism A sequence of unimolecular and bimolecular elementary reactions by which an overall reaction may occur.
reaction quotient (Q) A value found from an expression with the same mathematical form as the equilibrium constant expression but with the actual concentrations in a mixture not at equilibrium.
reaction rate The change in concentration of a reactant or product per unit time.
redox reaction (oxidation-reduction reaction) A reaction involving the transfer of one or more electrons from one species to another so that oxidation numbers change.
reduced The result when an atom, molecule, or ion gains one or more electrons.
reducing agent The atom, molecule, or ion that donates electron(s) and is oxidized in an oxidation-reduction reaction.
reduction The gain of electrons by an atom, ion, or molecule, leading to a decrease in its oxidation number.
reformulated gasolines Oxygenated gasolines with lower volatility and containing a lower percentage of aromatic hydrocarbons than regular gasoline.
relative humidity In the atmosphere, the ratio of actual partial pressure to equilibrium vapor pressure of water at the prevailing temperature.
rem A unit of radioactivity; 1 rem has the physiological effect of 1 roentgen of radiation.
replication The copying of DNA during regular cell division.
resonance hybrid The actual structure of a molecule that can be represented by more than one Lewis structure.
resonance structures The possible structures of a molecule for which more than one Lewis structure can be written, differing by the arrangement of electrons but having the same arrangement of atomic nuclei.
reverse osmosis Application of pressure greater than the osmotic pressure to cause solvent to flow through a semipermeable membrane from a concentrated solution to a solution of lower solute concentration.
reversible process A process for which a very small change in conditions will cause a reversal in direction.
roentgen (R) A unit of radioactivity; 1 R corresponds to deposition of 93.3×10^{-7} J per gram of tissue.

***s*-block elements** Main-group elements in Groups 1A and 2A whose valence electrons are *s* electrons.
salt An ionic compound whose cation comes from a base and whose anion comes from an acid.
salt bridge A device for maintaining balance of ion charges in the compartments of an electrochemical cell.
saponification The hydrolysis of a triglyceride (a fat or oil) by reaction with NaOH to give sodium salts that are soaps.
saturated fats Fats (or oils) that contain only carbon-carbon single bonds in their hydrocarbon chains.
saturated hydrocarbon Hydrocarbon in which carbon atoms are bonded to the maximum number of hydrogen atoms.
saturated solution A solution in which the concentration of solute is the concentration that would be in equilibrium with undissolved solute at a given temperature.
scanning tunneling microscope An analytical instrument that produces images of individual atoms or molecules on a surface.
screening effect Reduction of the effective attraction between nucleus and valence electrons as a result of repulsion of the outer valence electrons by electrons in inner shells.
second law of thermodynamics The total entropy of the universe (the system and surroundings) is continually increasing. In any product-favored system, the entropy of the universe is greater after a reaction than it was before.
secondary battery A voltaic cell (or battery of cells) in which the oxidation and reduction half-reactions can be reversed to restore the cell to its original state after discharge.
secondary pollutants Pollutants that are formed by chemical reactions of primary pollutants.
secondary structure of proteins Regular repeating patterns of molecular structure in proteins, such as α-helix, β-sheet.
semiconductor Material with a narrow energy gap between the valence band and the conduction band; a conductor when an electric field or a higher temperature is applied.
semipermeable membrane A thin layer of material through which only certain kinds of molecules can pass.
shell A collection of orbitals with the same value of the principal quantum number, n.
shifting an equilibrium Changing the conditions of an equilibrium system so that the system is no longer at equilibrium and there is a net reaction in either the forward or reverse direction until equilibrium is reestablished.
sievert The SI unit of effective dose of absorbed radiation, 1 Sv = 100 rem.

sigma (σ) bond A bond formed by head-to-head orbital overlap along the bond axis.

simple sugars Monosaccharides and disaccharides.

single covalent bond A bond formed by sharing one pair of electrons between the same two atoms.

smog A mixture of smoke (particulate matter), fog (an aerosol), and other substances that degrade air quality.

solar cell A device that converts solar photons into electricity; based on doped silicon.

solid A state of matter in which a substance has a definite shape and volume.

solubility The maximum amount of solute that will dissolve in a given volume of solvent at a given temperature when pure solute is in equilibrium with the solution.

solubility product constant (K_{sp}) An equilibrium constant that is the product of concentrations of ions in a solution in equilibrium with a solid ionic compound.

solubility product expression Molar concentrations of a cation and anion, each raised to a power equal to its coefficient in the balanced chemical equation for the solubility equilibrium.

solute The material dissolved in a solution.

solution A homogeneous mixture of two or more substances in a single phase.

solvent The medium in which a solute is dissolved to form a solution.

***sp* hybrid orbitals** Orbitals of the same atom formed by the combination of one *s* orbital and one *p* orbital.

sp^2 hybrid orbitals Orbitals of the same atom formed by the combination of one *s* orbital and two *p* orbitals.

sp^3 hybrid orbitals Orbitals of the same atom formed by the combination of one *s* orbital and three *p* orbitals.

sp^3d hybrid orbitals Orbitals of the same atom formed by the combination of one *s* orbital, three *p* orbitals, and one *d* orbital.

sp^3d^2 hybrid orbitals Orbitals of the same atom formed by the combination of one *s* orbital, three *p* orbitals, and two *d* orbitals.

specific heat capacity The quantity of energy that must be transferred to 1 g of a substance to increase its temperature by 1 °C.

spectator ion An ion that is present in a solution in which a reaction takes place, but is not involved in the net process.

spectrochemical series A list of ligands in the order of their crystal-field splitting energy.

spectroscopy Use of electromagnetic radiation to study the nature of matter.

spectrum A plot of the intensity of light (photons per unit time) as a function of the wavelength or frequency of light.

standard atmosphere (atm) A unit of pressure; 1 atm = 101.325 kPa = 1.01325 bar = 760 mm Hg exactly.

standard-state conditions These are 1 bar pressure for all gases, 1 M concentration for all solutes, at a specified temperature.

standard enthalpy change The enthalpy change when a process occurs with reactants and products all in their standard states.

standard equilibrium constant ($K°$) An equilibrium constant in which each concentration (or pressure) is divided by the standard-state concentration (or pressure); if concentrations are expressed in moles per liter (or pressures in bars) then the concentration (or pressure) equilibrium constant equals the standard equilibrium constant $\Delta G° = -RT \ln K°$.

standard hydrogen electrode The electrode against which standard reduction potentials are measured, consisting of a platinum electrode at which 1 M hydronium ion is reduced to hydrogen gas at 1 bar.

standard molar enthalpy of formation The standard enthalpy change for forming 1 mol of a compound from its elements, with all substances in their standard states.

standard molar volume The volume occupied by exactly 1 mol of an ideal gas at standard temperature (0 °C) and pressure (1 atm), equal to 22.414 L.

standard reduction potential ($E°$) The potential of an electrochemical cell when a given electrode is paired with a standard hydrogen electrode under standard conditions.

standard state The most stable form of a substance in the physical state in which it exists at 1 bar and a specified temperature.

standard solution A solution whose concentration is known accurately.

standard temperature and pressure (STP) A temperature of 0 °C and a pressure of 1 atm.

standard voltages Electrochemical cell voltages measured under standard conditions.

state function A property whose value is invariably the same if a system is in the same state.

steel A material made from iron with most P, S, and Si impurities removed, a low carbon content, and possibly other alloying metals.

steric factor A factor in the expression for rate of reaction that reflects the fact that some three-dimensional orientations of colliding molecules are more likely to result in reaction than others.

stoichiometric coefficients The multiplying numbers assigned to the species in a chemical equation in order to balance the equation.

stoichiometric factor (mole ratio) A factor relating number of moles of a reactant or product to number of moles of another reactant or product.

stoichiometry The study of the quantitative relations between amounts of reactants and products in chemical reactions.

stratosphere The region of the atmosphere approximately 12 to 50 km above sea level.

strong acid An acid that ionizes completely in aqueous solution.

strong base A base that ionizes completely in aqueous solution.

strong electrolyte An electrolyte that consists solely of ions in aqueous solution.

structural formulas Formulas written to show how atoms in a molecule or polyatomic ion are connected to each other.

structural isomers (constitutional isomers) Compounds with the same molecular formula that differ in the order in which their atoms are bonded together.

sublimation Conversion of a solid directly to a gas with no formation of liquid.

subshell A group of atomic orbitals with the same *n* and ℓ quantum numbers.

substance Matter of a particular kind; each substance, when pure, has a well-defined composition and a set of characteristic properties that differ from the properties of any other substance.

substrate A molecule or molecules whose reaction is catalyzed by an enzyme.

superconductor A substance that, below some temperature, offers no resistance to the flow of electric current.

supercritical fluid A substance above its critical temperature and pressure; has density characteristic of a liquid, but flow properties of a gas.

supersaturated solution A solution that temporarily contains more solute per unit volume than a saturated solution at a given temperature.

surface tension The energy required to overcome the attractive forces between molecules at the surface of a liquid.

surfactant(s) Compounds consisting of molecules that have both a hydrophobic part and a hydrophilic part.

surroundings Everything that can exchange energy with a thermodynamic system.

system In thermodynamics, that part of the universe that is singled out for observation and analysis. The region of primary concern.

temperature The physical property of matter that determines whether one object can heat another.

tertiary structure of proteins The overall three-dimensional folding of protein molecules.

tetrahedral Molecular geometry of four atoms or groups of atoms around a central atom with bond angles of 109.5°.

theoretical yield The maximum quantity of product theoretically obtainable from a given quantity of reactant in a chemical reaction.

theory A unifying principle that explains a body of facts and the laws based on them.

thermal equilibrium The condition of equal temperatures achieved between two samples of matter that are in contact.

thermochemical expression A balanced chemical equation, including specification of the states of matter of reactants and products, together with the corresponding value of the enthalpy change.

thermodynamics The science of heat, work, and the transformations of each into the other.

thermoplastic A plastic that can be repeatedly softened by heating and hardened by cooling.

thermosetting plastic A polymer that melts upon initial heating and forms cross-links so that it cannot be melted again without decomposition.

third law of thermodynamics A perfect crystal of any substance at 0 K has the lowest possible entropy.

titrant The solution being added from a buret to another solution during a titration.

titration A procedure whereby a substance in a standard solution reacts with a known stoichiometry with a substance whose concentration is to be determined.

titration curve A plot of the progress of a titration as a function of the volume of titrant added.

torr A unit of pressure equivalent to 1 mm Hg.

trace elements (See also **major minerals.**) The dietary minerals that are present in smaller concentrations than the major minerals, sometimes far smaller concentrations.

tracer A radioisotope used to track the pathway of a chemical reaction, industrial process, or medical procedure.

***trans* isomer** The isomer in which two like substituents are on opposite sides of a carbon-carbon double bond, a ring of carbon atoms, or a coordination complex.

transition elements Elements that lie in rows 4 through 7 of the periodic table in which *d* or *f* subshells are being filled; comprising scandium through zinc, yttrium through cadmium, lanthanum through mercury, and actinium and elements of higher atomic number.

transition state A molecular structure corresponding to the maximum of a plot of energy versus reaction progress; also known as the activated complex.

triangular bipyramidal Molecular geometry of five groups around a central atom in which three groups are in equatorial positions and two are in axial positions.

triangular planar Molecular geometry of three groups at the corners of an equilateral triangle around a central atom at the center of the triangle.

triglycerides Esters in which a glycerol molecule is joined with three fatty acid molecules.

triple bond A bond formed by sharing three pairs of electrons between two atoms.

triple point The point on a temperature/pressure phase diagram of a substance where solid, liquid, and gas phases are all in equilibrium.

troposphere The lowest region of the atmosphere, extending from the Earth's surface to an altitude of about 12 km.

Tyndall effect Scattering of visible light by a colloid.

uncertainty principle The statement that it is impossible to determine simultaneously the exact position and the exact momentum of an electron.

unimolecular reaction A reaction in which the rearrangement of the structure of a single molecule produces the product molecule or molecules.

unit cell A small portion of a crystal lattice that can be replicated in each of three directions to generate the entire lattice.

unsaturated fats Fats (or oils) that contain one or more carbon-carbon double bonds in their hydrocarbon chains.

unsaturated hydrocarbon A hydrocarbon containing double or triple carbon-carbon bonds.

unsaturated solution A solution that contains a smaller concentration of solute than the concentration of a saturated solution at a given temperature.

valence band In a solid, an energy band (group of closely spaced orbitals) that contains valence electrons.

valence bond model A theoretical model that describes a covalent bond as resulting from an overlap of one orbital on each of the bonded atoms.

valence electrons Electrons in an atom's highest occupied principal shell and in partially filled subshells of lower principal shells; electrons available to participate in bonding.

valence-shell electron-pair repulsion model (VSEPR) A simple model used to predict the shapes of molecules and polyatomic ions based on repulsions between bonding pairs and lone pairs around a central atom.

van der Waals equation An equation of state for gases that takes into account the volume occupied by molecules and noncovalent attractions between molecules:

$$\left[P + a\left(\frac{n}{V}\right)^2\right][V - bn] = nRT.$$

vaporization The change of a substance from the liquid to the gas phase.

vapor pressure The pressure of the vapor of a substance in equilibrium with its liquid or solid in a sealed container.

viscosity The resistance of a liquid to flow.

volatility The tendency of a liquid to vaporize.

volt (V) Electrical potential energy difference defined so that 1 joule (work) is performed when 1 coulomb (charge) moves through 1 volt (potential difference).

voltaic cell An electrochemical cell in which a product-favored oxidation-reduction reaction is used to produce an electric current.

water of hydration The water molecules trapped within the crystal lattice of an ionic hydrate or coordinated to a metal ion in a crystal lattice or in solution.

wave functions Solutions to the Schrödinger wave equation that describe the behavior of an electron in an atom.

wavelength The distance between adjacent crests (or troughs) in a wave.

weak acid An acid that is only partially ionized in aqueous solution.

weak base A base that is only partially ionized in aqueous solution.

weak electrolyte An electrolyte that is only partially ionized in aqueous solution.

weight percent A mass fraction expressed as a percent by multiplying by 100%; used for elemental composition of a compound, composition of a solute in solution.

work (working) A mechanical process that transfers energy to or from an object.

X-ray crystallography The science of determining nanoscale crystal structure by measuring the diffraction of X rays by a crystal.

zone refining A purification process in which a molten zone is moved through a solid sample, causing the impurities to concentrate in the liquefied portion.

zwitterion A structure containing both a positive charge and a negative charge, commonly due to loss and gain of a hydrogen ion within the same molecule.

Index

Page numbers in **bold face** refer to key term definitions
Page numbers in *italics* refer to images or diagrams
Page numbers followed by a "t" indicate a table.

A

D

E

F

G

H

L

M

N

Q

R

Useful Conversion Factors and Relationships

Length

SI unit: meter (m)

1 kilometer = 1000. meters
= 0.62137 mile
1 meter = 100. centimeters
1 centimeter = 10. millimeters
1 nanometer = 1×10^{-9} meter
1 picometer = 1×10^{-12} meter
1 inch = 2.54 centimeter (exactly)
1 Ångström = 1×10^{-10} meter

Mass

SI unit: kilogram (kg)

1 kilogram = 1000. grams
1 gram = 1000. milligrams
1 pound = 453.59237 grams = 16 ounces
1 ton = 2000. pounds

Volume

SI unit: cubic meter (m^3)

1 liter (L) = 1×10^{-3} m^3
= 1 dm^3 = 1000. cm^3
= 1.056710 quarts
1 gallon = 4 quarts

Energy

SI unit: joule (J)

1 joule = 1 kg m^2/s^2
= 0.23901 calorie
= 1 C × 1 V
1 calorie = 4.184 joules

Pressure

SI unit: pascal (Pa)

1 pascal = 1 N/m^2
= 1 kg m^{-1} s^{-1}
1 atmosphere = 101.325 kilopascals
= 760. mm Hg = 760. torr
= 14.70 lb/in.^{2}
1 bar = 1×10^5 Pa

Temperature

SI unit: kelvin (K)

If T_K is the numeric value of the temperature in kelvins, t_C the numeric value of the temperature in °C, and t_F the numeric value of the temperature in °F, then

$$T_K = t_C + 273.15$$
$$t_C = (\tfrac{5}{9})(t_F - 32)$$
$$t_F = (\tfrac{9}{5})t_C + 32$$

Location of Useful Tables and Figures

Atomic, Ionic, and Molecular Properties

Atomic radii	Figures 7.21, 7.22
Bond energies	Table 8.2
Bond lengths	Table 8.1
Electron affinities	Table 7.9
Electron configurations	Table 7.5, Figure 7.17, Appendix D
Electronegativity	Figure 8.6
Ionic radii	Figure 7.23
Ionization energies	Figures 7.24, 7.25
Molecular geometries	Table 9.1
Properties of molecular and ionic compounds	Table 3.9
Properties of solids	Table 11.5

Thermodynamic Properties

Specific heat capacities	Table 6.1
Standard molar enthalpies of formation	Table 6.2, App. J
Composition and caloric values of some foods	Table 6.3
Standard molar entropies	Table 18.1, App. J
Enthalpy, Gibbs free energy, entropy	Appendix J
Equilibrium constants	Table 14.1 (Also see Appendices F, G, and H)

Acids, Bases, and Salts

Acid-base properties of ions in solution	Table 16.5
Common acids and bases	Table 5.2
Common polyatomic ions	Table 3.7
Solubility guidelines for ionic compounds	Table 5.1
Common monatomic ions	Figure 3.2
Amino acids	Table 12.8
Buffer systems	Table 17.1
Acid and base ionization constants	Table 16.2 (See also Appendices F, G)
Solubility product constants	Appendix H

Other Useful Tables and Figures

Common oxidizing and reducing agents	Table 5.3
Common ligands	Table 22.4, Figure 22.9
Electromagnetic spectrum	Figure 7.1
Energy units	Table 12.2
Composition of dry air	Table 10.1
Standard reduction potentials	Table 19.1 (See also Appendix I)